MOTOR
AUTO ENGINES
AND
ELECTRICAL SYSTEMS

7th Edition

Editor
Louis C. Forier, SAE

Managing Editor
Larry Solnik, SAE

Associate Editors
Michael Kromida, SAE • Dan Irizarry, SAE • Warren Schildknecht, SAE

Editorial Assistants
Bob Noskowicz • Katherine Keen

(D.E.)

MOTOR Auto Engines & Electrical Systems
Library of Congress Catalog Number: 77-88821
ISBN 0-910992-73-8

Published by

MOTOR

1790 Broadway, New York, N.Y. 10019

The Automotive Business Magazine

Printed in the U.S.A. © Copyright 1977 by The Hearst Corporation

Contents

A complete index begins on page vi

How Engines Work 5
Engine Valve System 38
Engine Piston System 117
Crankshaft and Bearings.167
Engine Oiling Systems...190
Emission Control
   Systems 220
The Diesel and How
   It Works 280
The Wankel and
   How It Works........308
The Turbine and
   How It Works........320
Cooling System 327
Fuel System 366

Fuel Injection 480
Electricity 511
Magnetism 518
Electrical System 530
Ignition System 619
Transistorized and
   Electronic
   Ignition Systems 621
D.C. Generators and
   Regulators 642
Alternator Systems 670
Starting Motors 731
Starting Switches 761
The Storage Battery.....770
Dash Gauges 788

Foreword

The authors believe they are fulfilling a need long felt in the field of automotive service instruction. MOTOR'S Auto Repair Manual, which the authors edited for years, is addressed to the mechanic who already possesses a sound knowledge of the automobile; however, it does not provide the groundwork of basic knowledge which a mechanic should have.

The present work has been written for those seeking this basic knowledge and for the student and teacher of auto service work. Its aim is to present these fundamentals in a sequence that makes them easiest for the student to grasp, to keep the text simple and concise and to illustrate it so abundantly that full understanding of the text is almost automatic. And, above all, the goal was to make the instruction intensely practical.

Everywhere generalizations are avoided. Specific pieces of equipment are dealt with in turn, so the student will acquire not generalized knowledge of a type of component but a practical insight that is of immediate and practical use.

In offering this work, which they confidently believe will simplify the training of auto mechanics, the authors readily acknowledge their debt to the many automotive associations, persons and organizations who contributed of their time and knowledge to make this volume possible. Not the least among them were the many operators of general repair shops and many specialized shops where the authors were permitted to observe the latest repair procedures in actual use.

Books Published by MOTOR

Auto Repair Manual

This most-widely used and respected book of its kind covers 2,300 car models of 37 series of American-make cars from 1973-78. The big 1978 edition has more than 1,400 double-size pages featuring 55,000 essential service specifications and over 300 quick-check specification charts, 225,000 service and repair facts, and over 3,000 "how-to-do-it" pictures.

Foreign Car Repair Manual

For the "Foreign Car" owner who is totally involved with his car and is interested in Do-It-Yourself car repairs; the maintenance of peak performance . . . and in saving money. Also included is a section (not available elsewhere) on parts and tool availability that guides the car owner as to where he can buy the parts and special tools needed to keep his car in top running condition.

Truck Repair Manual

Service and repair is surprisingly easy with this new editon that covers 2,800 Truck Models, 1966 through 1977. Over 1,400 pages of step-by-step instructions, 2,000 cutaway pictures, 3,000 service and repair facts, specs and adjustments. Covers all popular makes of trucks. Manual also gives specs for gasoline and diesel engines used in off-highway equipment and farm tractors.

Auto Engines & Electrical Systems

Ideal basic book for car buffs, students, engineers, mechanics. Over 700 pages, 1,300 pictures and diagrams explain the workings of engines, fuel and electrical systems. Special chapters on the Rotary and Turbine engines.

Automobile Trouble Shooter

This new hardback edition is a must for the glove compartment or tool box of every do-it-yourself car enthusiast. A handy guide to finding out what's wrong, it pinpoints over 2,000 causes of car trouble.

How To Use This Book

You've all heard the old saying that you must have the proper tools to do a proper job. The field of auto repair is, perhaps, where that saying was invented. In no other field is the use of tools more important. Tools are the most important possession an automotive technician has—except his knowledge of how to use the tools.

Improper use of tools can cause even the most skilled automotive technician to badly foul up a job. Improper use of tools can cause a job to be done the wrong way, or even not done at all, simply because the user of the tool may have had all the skills he needed except one—how to properly use that tool.

To someone learning applied auto mechanics, or just studying automotive technology theory, a good textbook is a tool, just as a ratchet or a Phillips head screwdriver is a tool.

The study of automotive technology can be a rewarding experience. It can fulfill the need for practical knowledge of techniques needed by those involved with auto mechanics. It can also fulfill the thirst for pure knowledge that is within all of us. If you suddenly understand how an alternator works, that moment when the whole thing becomes crystal clear can be an extremely fulfilling experience. Even if you never work on an alternator, that moment of gained knowledge can give you a sense of personal satisfaction that can remain with you forever.

Which brings us back to proper use of tools. This textbook is the tool through which you can efficiently complete a job (learning automotive technology) or through which you can easily foul it up. It all comes down to how you use this book.

Why not take a few minutes now to familiarize yourself with the structure and content of *Auto Engines and Electrical Systems*. If you flip through it quickly, you'll find a recurring structure that can help guide you quickly and easily to just what you're looking for in the book.

Table of Contents

The first important thing you'll come to in the book is the Table of Contents. You'll find it on page i. The Table of Contents gives you a broad overview of everything in the entire book. Here, you'll find chapter titles, nothing more. However, the chapter titles will give you a pretty good indication of what the chapter is about.

Some of the chapter titles are self-explanatory. For instance, the chapter entitled "The Wankel and How It Works" is about, right, Wankel rotary engines and how they work.

But some of the chapter titles in the Table of Contents are not as obvious. For instance, the chapter entitled "Crankshaft and Bearings" could be about how they're designed, how they work or how to repair them. Or all of these things.

In this case, you'd have to know more about what's in that chapter. And so, you can turn to the chapter synopsis.

Chapter Synopsis

Beginning on page xiii, you'll find a brief, condensed compendium of the information contained in each chapter. Naturally, in a condensation such as this, details are left out. The purpose of these synopses is to give you a good, but not detailed, idea of what is contained in each chapter and also how the information is presented.

This section proceeds chapter by chapter, presenting a skeleton of each. It not only gives you an overview of each chapter but also can help you find a particular section or explanation without having to wade through the entire chapter.

Master Index

There may be times when, in the course of your studies, you need to find a specific fact, or a group of facts about a specific subject.

In this case, even going through a chapter synopsis can be tedious and time consuming. And one thing a student doesn't have enough of is time.

So we've set up a master index for you that is a much more time-efficient way of finding specific subjects or facts. The Master Index is arranged alphabetically by subject and is quite detailed. Using the Master Index, you should be able to quickly find just about any fact or subject covered in this volume.

Review Questions

There is another important part of this book that can be extremely beneficial to the serious student. At the end of each chapter, there are a set of review questions that test your knowledge and comprehension of the subject matter of that particular chapter.

If you can answer the chapter review questions, then you know the subject matter in that chapter. No question about it.

But what if you don't know the answer? Or you don't know whether your answers are correct? Where are the correct answers?

Frankly, they aren't anywhere except in the text of that chapter. We don't believe in giving you the answers for a very specific reason. Some people get hung up on the answers to questions and study only the answers. Obviously, this is not the best way to get a broad-based foundation of knowledge in any subject. You *will* find the information you need to answer the questions in that chapter. You *won't* find a list of answers.

To help you find the answers in the chapter, we do tell you what page to refer to. It may be necessary to re-read and understand that whole page or more. Or you might find the answer in a single sentence on that page. The point is, if you look for only the answer out of the chapter, you're only cheating yourself. Here is a case when you should make sure that you have a thorough understanding of the question and its answer, plus the background subject matter. If you don't, it's your knowledge that will fall short.

Study Plan

There are several ways you could use this book. And no matter how you use it, you will probably get something out of it. However, there is a *best* way to learn the information contained here. And by using this best method, you will be making proper use of this book.

Before beginning the study of any chapter, look over the Table of Contents to find out what page it starts on.

Next, turn to the chapter synopsis section to get an overview of what you'll be studying.

Then turn to the chapter itself. Read it once through quickly. Then go back for a thorough ingestion of everything contained in the chapter. There will probably be sections of information that you already know. And there will be sections that you may have to read several times before you can honestly say you understand it. This is OK. It's part of the learning process.

Lastly, answer the review questions. Check your answers by referring to the pages next to each review question. Re-read the section if necessary until you understand the correct answer.

Refer to the Master Index whenever there is a specific fact or subject about which you have a question. It's a handy key to several thousand subject points.

Motor Auto Repair Manual

We would be remiss if we didn't mention the bible of the auto repair field, the *Motor Auto Repair Manual.*

You're probably quite familiar with this blue book that is considered a necessity by anyone who repairs cars. It packs over 1400 pages, covers over 2300 car models, contains 55,000 essential service and repair facts and over 2800 how-to-do-it diagrams and illustrations. We don't want to number you to death, but we do want you to know how indispensable the *Motor Auto Repair Manual* is. In short, it contains step-by-step instructions on how to repair almost anything on any car.

As someone studying automotive technology, you should think of the *Motor Auto Repair Manual* and *Auto Engines and Electrical Systems* as a set. One complements the other. This book tells you the *why* while the *Auto Repair Manual* tells you the *how.*

If you're a student of automobile mechanics, you'll soon learn that there is really only one way to learn it—by hands-on experience. But first you have to know where to put your hands. *Auto Engines and Electrical Systems* will show you.

INDEX

Alternators,
  construction, ... 671
  changing A.C. to D.C., ... 676
  Chrysler system, ... 677
  Delco-Remy system, ... 690
  Ford system, ... 700
  functions of a diode, ... 676
  how system functions, ... 672
  Leece-Neville system, ... 712
  Motorola system, ... 717
  Prestolite system, ... 723
  service precautions, ... 676
  single phase alternating voltage, ... 674
  three phase alternating voltage, ... 675
  three phase connections, ... 676
Battery,
  adding water, ... 780
  care of new and rental, ... 785
  removing and installing, ... 781
  servicing in the car, ... 784
  capacity, ... 774
  charging, ... 782
  construction of, ... 771
  how it works, ... 771
  troubles, ... 776
  electrical tests, ... 780
  hydrometer test, ... 777
Bearings and crankshaft, ... 167
  bearings, main, ... 167
  bearing bores out-of-round, main, ... 186
  bearing caps, warped main, ... 186
  bearing clearance, methods of checking, ... 175
  bearing crush fit, ... 179
  bearings, doweled main, ... 181
  bearing installation check, ... 180
  bearings, install rod, ... 179
  bearing journals, "miking" main, ... 173
  bearing knocks, ... 172
  bearing oil leak detector, ... 170
  bearing oil seal, rear main, ... 182
  bearings, replace, main, ... 180
  bearings, rod, ... 167
  bearing spread, ... 179
  bearings to use, what size, ... 174
  bearing troubles, ... 172
  clearance, rod size, ... 179
  counterweights, crankshaft, ... 34
  crankcase, warped, ... 186
  crankshaft, checking for wear, ... 173
  crankshaft end play, ... 181
  crankshaft grinder, ... 185
  crankshaft, hand polishing, ... 185
  crankshaft, preparing for high performance, . . 187
  crankshaft removal, ... 184
  crankshaft, straightness check, ... 186
  engine oil passages clean, ... 178
  flywheel ring gear, replace, ... 187
  flywheel service, ... 187
  main bearing caps, warped, ... 186
  rod journals, "miking," 173
  rod side clearance, ... 179
  taper shims for worn bearings, ... 178

Bore and stroke, ... 10
Brake horsepower, ... 10
Cams, ... 24
Combustion, ... 6
Compression ratio, ... 11
Compression stroke, ... 5
Construction, engine, ... 16
Construction, Corvair engine, ... 16
Cooling system,
  air suction test, ... 351
  anti-freeze compounds, ... 340
  anti-freeze data, ... 340
  anti-freeze protection, ... 356
  anti-freeze, testing, ... 356
  belt adjustments, ... 359
  belts, drive, ... 358
  belts, fan blades and, ... 348
  capacity, extra cooling, ... 340
  caps, radiator, ... 332
  cap, radiator pressure, ... 332
  cap, test pressure, ... 352
  cleaning the system, ... 354
  combustion leakage test, ... 351
  compounds, stop-leak, ... 360
  coolant, frozen, ... 349
  cooling system, Corvair, ... 338
  cooling system damage, ... 340
  cooling system, how it works, ... 327
  cooling system, corrosion, ... 341
  cooling system, flushing, ... 355
  cooling troubles, air cooled engine, ... 353
  corrosion damage, ... 342
  cylinder head joint leakage, ... 346
  drain cocks and plugs, ... 332
  distributing tube, water, ... 331
  fan, ... 330
  fan belt, ... 358
  fan blades and belts, ... 348
  flow tester, radiator, ... 359
  flushing the system, ... 355
  foaming, ... 350
  gauge, test temperature, ... 353
  hose, ... 337
  hose inspection, ... 357
  hose leakage, ... 345
  hose, replace, ... 358
  hydrometer, use of, ... 357
  leakage, ... 344
  overcooling, ... 341
  overheating, ... 341
  overheating in traffic, ... 339
  plug, welch, renew, ... 359
  pressure cap, radiator, ... 332
  pressure cap test, ... 352
  radiator, ... 331
  radiator caps, ... 332
  radiator clogging, ... 353
  radiator flow tester, ... 359
  radiator leakage, ... 344
  radiator removal, ... 363
  rust formation, effects of, ... 342
  rust prevention, importance of, ... 342

stop-leak compounds, ... 360
temperature control, water, ... 336
temperature, test, gauge, ... 353
test, air suction, ... 351
test, combustion leakage, ... 351
test in emergencies, ... 348
testing anti-freeze, ... 356
test, pressure cap, ... 352
test, temperature gauge, ... 353
test, thermostat, ... 352
thermostats, ... 334
tube, water distributing, ... 331
water boils, when, ... 338
water distributing tube, ... 331
water jacket leakage, ... 346
water level, ... 343
water pump, ... 329
water pump service, ... 360
water pump leakage, ... 346
water temperature control, ... 336
welch plug, renew, ... 359
Corvair engine construction, ... 16
Crankshaft and Bearings—see Bearings and
   Crankshaft,
Cylinder bore distortion, ... 28
Cylinder head gaskets, ... 20
Cylinder numbering, ... 14
Dampers, vibration, ... 34
Dash gauges—see Gauges, dash,
Definition of terms, ... 10
Detonation, ... 11
Diesel,
   characteristics of diesel combustion, ... 282
   diesel engine compared to gasoline engine, .. 280
   diesel fuel system, ... 282
   distributor type system, ... 295
   fuel filters, ... 304
   fuel supply units, ... 298
   governors, ... 300
   injection systems, ... 286
   Mercedes-Benz 5 cylinder diesel, ... 305
   Oldsmobile 350 V8, ... 306
   recent developments in diesel engineering, .. 305
   unit injection system, ... 296
   wobble late pump system, ... 292
Displacement, piston, ... 10
Distortion, cylinder bore, ... 28
Electrical system,
   12 volt electrical systems, ... 539
   cable, how to determine size of, ... 544
   cable, selecting proper size, ... 551
   cables, high-tension, ... 545
   cable, ignition, ... 545
   cable size requirements, ... 544
   circuit, charging, ... 535
   circuit, horn, ... 537
   circuit, ignition, ... 537
   circuit, instrument, ... 538
   circuit, lighting, ... 538
   circuit, starting, ... 535
   circuit, wipers, ... 537
   diagrams, wiring, ... 531
   electrical system service, ... 540
   locating trouble with voltmeter 540
Electricity,
   electrical conductors, ... 516

electrical measurements, ... 513
types of circuits, ... 512
Electronic ignition,
   advantages of, ... 621
   circuit components, ... 625
   Chrysler Corporation, ... 633
   Delco-Remy type, ...628, 636
   a brief history, ... 621
   Ford Motor Company, ...624, 635
   Ford types, ...624, 635
   General Motors, ...628, 636
   glossary, ... 625
   high energy ignition system, ... 637
   imported car systems, ... 640
   maintenance, ... 631
   Micro-Processed Sensing and Automatic
      Regulation (MISAR), ... 639
   operating principles, ... 628
   operation, ... 625
   Prestolite (American Motors), ... 632
   service recommendations and specifications, . 626
   trouble diagnosis, ... 626
   trouble shooting, ... 629
Emission control systems,
   air pump systems, ... 225
   American Motors Exhaust Gas
      Recirculation (EGR), ... 236
   American Motors Engine "MOD" System, ... 236
   aspirator air system, ... 236
   catalytic converters, ... 269
   choke hot air modulator, ... 224
   Chrysler CAP, ... 250
   Chrysler CAS, ... 253
   Chrysler Exhaust Gas Recirculation (EGR), ... 257
   Chrysler ignition, ... 257
   Chrysler Lean Burn Engine Electronic
      Spark Advance, ... 249
   Chrysler NO_x, ... 255
   Chrysler Orifice Spark Advance
      Control (OSAC), ... 257
   dual area diaphragm, ... 266
   electric assist choke, ... 221
   Ford CTAV system, ... 268
   Ford decel valve, ... 266
   Ford DVB system, ... 267
   Ford ESC system, ... 261
   Ford EGR system, ... 262
   Ford electronic distributor modulator, ... 260
   Ford HCV system, ... 269
   Ford high speed EGR modulator sub-system, . 261
   Ford IMCO system, ... 260
   Ford spark delay valve, ... 262
   Ford TAV system, ... 267
   Ford TRS system, ... 261
   Ford TRS + 1, ... 262
   fuel evaporative emission controls, ... 271
   General Motors CCS, ... 242
   General Motors CEC, ... 245
   General Motors EFE, ... 248
   General Motors EGR, ... 246
   General Motors SCS, ... 246
   non-air pump systems, ... 235
   Pulse Air Injection Reactor (PAIR), ... 235
   temperature operated vacuum by-pass valve, . 257
   thermostatic controlled air cleaner, TAC and
      auto therm air cleaner systems, ... 239

transmission controlled spark (TCS),	243
Engine construction,	16
Engine, feeding the,	7
Engine operation,	5
Engine torque	10
Exhaust stroke,	5
Feeding the engine,	7
Firing order,	15
Flywheel,	35
Fuel injection,	
acceleration system,	486
air induction subsystem,	489
air meter,	483
auxiliary starting assembly,	508
cold enrichment system,	486
compensation units,	507
components,	482
construction of system components,	509
control system,	498
correction factors,	498
electrical circuit,	510
electronic control unit,	498
engine sensors,	491
fuel delivery subsystem,	488
fuel supply,	503
fuel system,	495
fuel meter,	483
hot idle compensator,	487
idle system,	486
induction system,	496
operating principles,	485
power system,	486
pressure sensor with full load enrichment,	497
starting system,	485
system components,	482
system description,	487
temperature sensor (intake air),	498
Fuel system,	
balanced carburetor,	386
carburetor,	373
carburetor accessories,	388
carburetor service,	436
carburetors, Carter,	389
carburetors, Ford,	398
carburetors, Holley,	399
carburetors, Rochester,	404
carburetors, Stromberg,	417
choke system,	385
combustion,	378
compression pressure,	378
detonation,	379
dual exhaust system,	377
excessive fuel consumption,	434
exhaust system,	376
flexible fuel lines,	428
float system,	382
fuel filters,	368
fuel gauges,	373
fuel lines,	373
fuel pipes,	425
fuel pump,	368
fuel pump overhaul,	466
fuel pump tests,	464
fuel tanks,	367
how altitude affects carburetion,	388
idle and low speed system,	383

injection, fuel—see Fuel injection,	
intake manifolds,	373
manifold heat control,	375
manifold heat control valve,	430
muffler and tail pipe,	430
octane rating,	379
part throttle system,	383
performance tips,	469
power system,	384
ram induction manifolds,	374
throttle valve,	382
types of carburetors,	388
vacuum pump operation,	371
vacuum pump test,	466
vacuum pump troubles,	466
vaporization,	380
vaporization by heat,	381
vaporization by spraying,	381
vaporization by vacuum,	381
Venturi action,	385
weight of air,	380
when capacity is low,	465
when engine won't idle,	432
when engine loses power,	433
when engine won't start,	431
when pressure is high,	465
when pressure is low,	464
windshield wiper tubing,	430
Gaskets, cylinder head,	20
Gauges, dash,	
AC gauges, fuel gauges,	788
AC gauges, oil pressure gauge,	790
AC gauges, temperature gauge,	791
ammeters, trouble shooting,	800
Auto-Lite gauges, thermal fuel gauge,	792
Auto-Lite gauges, magnetic fuel gauge,	794
Auto-Lite gauges, oil pressure gauge,	794
Auto-Lite gauges, temperature gauge,	795
electric clocks,	806
electroluminescent lighting, trouble shooting,	805
fibre optics,	806
generator indicator light, light circuit with D.C. generator,	801
generator indicator light, light circuit with alternator,	801
King-Seeley gauges, voltage regulator,	796
King-Seeley gauges, fuel gauge,	796
King-Seeley gauges, oil pressure gauge,	797
King-Seeley gauges, temperature gauge,	797
oil pressure indicator light, trouble shooting,	802
pressure expansion type oil gauge, trouble shooting,	803
speedometer cable,	805
speedometers,	805
Stewart-Warner gauges,	799
temperature indicator light, trouble shooting,	802
testing, constant voltage type, dash gauge,	799
testing, constant voltage type, fuel tank gauge,	799
testing, constant voltage type, oil gauge sending unit,	799
testing, constant voltage type, temperature gauge sending unit,	799
testing, constant voltage type, voltage regulator,	799
testing, variable voltage type, fuel tank gauge method,	800

testing, variable voltage type gauge, grounded
wire method, ... 800
trouble shooting, ... 806
vapor pressure type temperature gauge,
trouble shooting, ... 804
Generators,
brushes, replace, ... 647
construction, generator, ... 642
generator, construction, ... 642
generator, motoring, ... 653
generator not charging, ... 646
generator, polarizing, ... 653
generator, removing, ... 646
generator, replacing with new, ... 652
generator service, ... 646
motoring generator, ... 653
not charging, generator, ... 646
polarizing generator, ... 653
removing generator, ... 646
replacing brushes, ... 647
replacing with new generator, ... 652
service generator, ... 646
Heating the mixture, ... 8
Horsepower, brake, ... 10
Horsepower, taxable, ... 11
Ignition, electronic, ... 621
Ignition system,
advance spark, ... 578
analysis of spark plug condition, ... 588
breaker point service, ... 596
causes of coil failure, ... 584
Chrysler distributor overhaul, ... 598
coil, causes of failure, ... 584
coil, ignition, ... 561
coil polarity, importance of correct, ... 586
coil, service ignition, ... 583
coil testers, types of, ... 587
coil, test ignition, ... 586
condenser, ... 565
condenser service, ... 592
current, paths of flow in ignition circuit, ... 567
Delco-Remy Corvair distributor, ... 615
Delco-Remy distributor overhaul, ... 598
Delco-Remy distributor service, ... 598
distributor, ... 568
distributor inspection and tests, ... 596
distributor, overhaul Prestolite, ... 598
distributor, overhaul Chrysler, ... 598
distributor, overhaul Delco-Remy, ... 598
distributor, overhaul Holley, ... 611
distributor, service Prestolite, ... 598
distributor, service Chrysler, ... 598
distributor, service Delco-Remy, ... 598
distributor, service Holley, ... 611
distributor, tests, ... 609
Holley distributor overhaul, ... 614
Holley distributor service, ... 611
ignition circuit, paths of current flow in, ... 567
ignition coil, ... 561
ignition coil, causes of failure, ... 584
ignition coil service, ... 583
ignition coil tests, ... 586
ignition miss, ... 583
ignition system service, ... 581
ignition system, trouble shooting, ... 581

ignition timing, ... 611
miss, ignition, ... 583
overhauling Prestolite distributors, ... 598
overhauling Chrysler distributors, ... 598
overhauling Delco-Remy distributors, ... 598
overhauling Holley distributors, ... 614
paths of current flow in ignition circuit, ... 567
performance tips, ... 617
plug, condition analysis of spark, ... 588
plug, resistor, ... 591
plugs, service spark, ... 587
plug, spark, ... 565
points, service breaker, ... 596
polarity, importance of correct coil, ... 586
Prestolite distributor overhaul, ... 598
Prestolite distributor service, ... 598
resistor spark plug, ... 591
service, Prestolite distributor, ... 598
service, breaker points, ... 596
service, condensers, ... 592
service, Delco-Remy distributor, ... 598
service, Holley distributor, ... 611
service, ignition coil, ... 583
service, ignition system, ... 581
service, spark plugs, ... 587
spark advance, ... 578
spark plugs, ... 565
spark plug, resistor, ... 591
spark plug condition, analysis of, ... 588
spark plug service, ... 587
system, trouble shooting ignition, ... 581
test and inspect distributor, ... 596
tests, ignition coil, ... 586
timing, ignition, ... 611
trouble shooting ignition system, ... 581
types of coil testers, ... 587
Intake stroke, ... 5
Knock, spark, ... 12
Lifters, valve, ... 23
Locating valves, ... 25
Lubrication system—see Oiling system,
Magnetism,
electrical motors, ... 527
how generator works, ... 522
Main bearings, ... 32
Mixture, heating the, ... 8
Numbering, cylinder, ... 14
Oiling system,
by-pass oil filters, ... 214
crankcase ventilation, ... 216
draining procedure, oil, ... 201
engine, how lubricated, ... 198
engine oil, ... 191
engine oil contaminants, ... 199
engine oil pan, ... 199
engine wear, general, ... 203
filler pipe, oil, ... 199
filter elements, replace, ... 214
filter, oil, installing new, ... 214
filter, oil, service, ... 214
filters, oil, ... 214
filters, oil by-pass, ... 214
filters, oil, full flow, ... 214
filters, sealed container, replace, ... 214
friction, ... 197
full flow oil filters, ... 214

gear oil pumps, ... 209
leakage preventives, oil, ... 199
lubrication, ... 190
multi-viscosity oils, ... 192
oil change intervals, ... 201
oil classifications, ... 196
oil consumption, analysis of high, ... 203
oil consumption, causes of excessive, ... 201
oil contaminants, ... 199
oil draining procedure, ... 201
oil, engine, ... 191
oil filler pipe, ... 199
oil filter, installing new, ... 214
oil filters, ... 214
oil filters, by-pass, ... 214
oil filter service, ... 214
oil filters, full flow, ... 219
oil leakage preventives, ... 199
oil leaks, external, ... 203
oil level, checking, ... 200
oil lines, ... 204
oil pan, engine, ... 199
oil pan, removing, ... 205
oil pan, replacing, ... 205
oil pressure, ... 203
oil pressure, high, ... 204
oil pressure, low, ... 204
oil pressure, no, ... 203
oil pressure relief valves, ... 213
oil pressure warning light, ... 204
oil pumps, gear, ... 209
oil pumps, installing, ... 213
oil pumps, removing, ... 207
oil pumps, rotor, ... 211
oil screen clogging, ... 207
oil screens, ... 206
oils, multi-viscosity, ... 192
oil viscosity, ... 192
relief valves, oil pressure, ... 213
rotor oil pumps, ... 211
PCV system tests, ... 217
screen, oil, clogging, ... 207
screens, oil, ... 206
test PCV systems, ... 217
vacuum diaphragm, check for leaky, ... 203
valves, oil pressure relief, ... 213
ventilation, crankcase, ... 216
Operation, engine, ... 5
Pins, piston, ... 31
Pistons,
block inspection cylinder, ... 123
blocks, repairing cracked, ... 123
bore measuring instruments, ... 142
bore, out-of-round cylinder, ... 141
bores, check cylinder, ... 141
bore, tapered cylinder, ... 141
bore, wavy cylinder, ... 141
bore wear, cylinder, ... 118
bore, welded, ... 152
boring bar construction, ... 151
boring bars, use of, ... 151
boring, preparation for, ... 151
boring procedure, ... 152
bushings, connecting rod, ... 159
cleaning pistons, ... 133
cold engines wear fast, why, ... 119

combustion chamber, oil loss into, ... 121
compressors, ring, ... 132
connecting rod alignment, checking, ... 162
connecting rod bushings, install, ... 159
connecting rod bushings, remove, ... 159
connecting rod bushings, swaging, ... 160
connecting rod bushings, ... 159
cracked blocks, repairing, ... 123
cylinder block inspection, ... 123
cylinder bores, check, ... 141
cylinder bore measuring instruments, ... 142
cylinder bore taper, ... 141
cylinder bore wear, ... 118
cylinder sleeves or liners, ... 152
distorted cylinder bores, ... 119
engine, disassemble for ring job, ... 124
expanders, piston skirt, ... 137
external oil loss, checking for, ... 122
fitting piston rings, ... 129
flexible hone, ... 145
glaze buster or breaker, ... 144
groove depth for ring expanders, ... 135
hand honing piston pin holes, ... 160
high performance, preparing cylinder
block for, ... 152
hone, flexible, ... 145
hone, rigid, ... 147
honing piston pin holes, hand, ... 160
install rings on piston, ... 131
instruments, cylinder bore measuring, ... 142
internal oil loss, checking for, ... 122
oil consumption, excessive, ... 121
oil ring, U-flex, ... 130
oil loss by leakage, ... 121
oil loss, checking for external, ... 122
oil loss, checking for internal, ... 122
oil loss into combustion chamber, ... 121
pin fits, piston, ... 157
pin holes, hand honing piston 160
pin holes, reaming piston, ... 158
pin honing machine, piston, ... 161
pins, check for loose piston, ... 154
pins, reaming fixtures for piston, ... 158
pins, remove piston, ... 155
pin sizes, piston, ... 157
pins too tight, piston, ... 155
piston inspection, ... 133
piston skirts, check for collapsed, ... 136
piston skirt expanders, ... 137
piston slap, ... 120
piston to rod, assemble, ... 164
preparing cylinder block for high
performance, ... 152
piston pins, check for loose, ... 154
piston pin fits, ... 157
piston pin holes, hand honing, ... 160
piston pin holes, reaming, ... 158
piston pin honing machine, ... 161
piston pins, reaming fixtures for, ... 158
piston pins, remove, ... 155
piston pin sizes, ... 157
piston pins too tight, ... 155
piston rings, fitting, ... 129
piston rings, removing, ... 128
piston ring wear, ... 120
pistons and rods, remove, ... 127

pistons, cleaning, ... 133
reamer fixtures for piston pins, ... 158
reamer, use of ridge, ... 125
reaming piston pin holes, ... 158
reboring, ... 150
ridge reamer, use of, ... 125
ridge, remove ring, ... 124
rigid hone, ... 147
ring compressors, ... 132
ring expanders, groove depth for, ... 135
ring groove clearance, ... 134
ring grooves widening, ... 134
ring job, disassemble engine for, ... 124
ring land diameter, ... 133
ring ridge, remove, ... 124
rings on piston, install, ... 131
rings, fitting piston, ... 129
rings, removing piston, ... 128
rings to use, what, ... 127
rings, U-flex oil, ... 130
rod alignment, checking, ... 162
rod bushings, ... 159
rod bushings, install, ... 159
rod bushings, remove, ... 159
rod bushings, swaging, ... 160
rods and pistons, remove, ... 127
rod to piston, assemble, ... 164
seating rings, ... 121
sleeves or liners, cylinder, ... 152
taper, cylinder bore, ... 141
U-flex oil ring, ... 130
widening ring grooves, ... 134
Power stroke, ... 5
Pre-ignition, ... 12
Ratio, compression, ... 11
Reasons for cooling system, ... 13
Regulators,
adjustments, Bosch electrical, ... 658
adjustments, Bosch mechanical, ... 658
adjustments, Delco-Remy electrical, ... 662
adjustments, Delco-Remy mechanical, ... 665
adjustments, Ford electrical, ... 658
adjustments, Ford mechanical, ... 668
adjustments, Prestolite electrical, ... 658
adjustments, Prestolite mechanical, ... 665
Bosch, electrical tests and adjustments, ... 658
Bosch, mechanical adjustments, ... 668
Delco-Remy, electrical tests and adjustments, . 662
Delco-Remy, mechanical adjustments, ... 665
electrical adjustments, Bosch, ... 658
electrical adjustments, Delco-Remy, ... 662
electrical adjustments, Ford, ... 658
electrical adjustments, Prestolite, ... 658
electrical tests and adjustments, ... 658
electrical tests, Bosch, ... 658
electrical tests, Delco-Remy, ... 662
electrical tests, Ford, ... 658
electrical tests, Prestolite, ... 658
Ford, electrical tests and adjustments, ... 658
Ford, mechanical adjustments, ... 668
operation of regulator, ... 654
Prestolite, electrical tests and adjustments, .. 658
Prestolite, mechanical adjustments, ... 665
purpose of regulator, ... 654
regulator, operation of, ... 654
regulator, purpose of, ... 654

regulator, replace, ... 658
tests, Bosch electrical, ... 658
tests, Delco-Remy electrical, ... 662
tests, electrical, ... 658
tests, Ford electrical, ... 658
tests, Prestolite electrical, ... 658
Rings, piston, ... 31
Rotary engine—see Wankel engine
Spark knock, ... 12
Springs, valve, ... 25
Starting motors,
Bendix drives, ... 736
checking circuit with voltmeter, ... 738
Chrysler direct drive starter, ... 742
Chrysler reduction gear starter, ... 744
construction of overrunning clutch, ... 733
construction of starting motor, ... 734
Delco-Remy starters, ... 749
description of starting motor, ... 733
Ford starter with folo-thru drive, ... 754
Ford starter with integral positive
    engagement drive, ... 752
fundamentals of starter operation, ... 731
how engine is disconnected from starter, ... 731
if lights dim, ... 737
if lights go out, ... 737
lights stay bright, no cranking action, ... 738
overrunning clutch drive operation, ... 733
Prestolite starting motors, ... 739
troubles, ... 756
Starting switches,
Carter vacuum switch, ... 763
Carter vacuum switch service, ... 768
magnetic switches, ... 761
solenoid switches, ... 761
starter switch service, ... 766
Stromberg vacuum switch, ... 764
Stromberg vacuum switch service, ... 768
switches, magnetic, ... 761
switch operation, checking, ... 766
switch service, starter, ... 766
switches, solenoid, ... 761
vacuum switch, Carter, ... 763
vacuum switch service, Carter, ... 768
vacuum switch, Stromberg, ... 764
vacuum switch service, Stromberg, ... 768
Stroke, bore and, ... 10
Stroke, compression, ... 5
Stroke, exhaust, ... 5
Stroke, intake, ... 5
Stroke, power, ... 5
Tappets, valve, ... 23
Taxable horsepower, ... 11
Terms, definition of, ... 10
Timing, valve, ... 9
Torque, engine, ... 10
Turbine engine,
cold weather operation, ... 324
fuel consumption, ... 324
how the turbine works, ... 320
introduction, ... 320
reduced air pollution, ... 324
regenerative turbines, ... 322
the burner, ... 325
the fuel system, ... 325
the ignition system, ... 325

Valve action, ... 21
Valves, locating, ... 25
Valve service, ... 49
   adjustments, valve lash, ... 52
   arms, rocker, ... 55
   camshaft and timing gear, bearings,
     camshaft, 111
   camshaft and timing gear, camshaft
     bearings, 111
   camshaft and timing gear, camshaft gear, ... 98
   camshaft and timing gear, camshaft,
     replace, ... 105
   camshaft and timing gear, camshaft
     sprocket, ... 98
   camshaft and timing gear, camshaft
     thrust plate, ... 96
   camshaft and timing gear, chain,
     timing and sprockets, ... 103
   camshaft and timing gear, cover, timing case, . 99
   camshaft and timing gear, crankshaft gear, . . . 99
   camshaft and timing gear, crankshaft
     sprocket, ... 99
   checking compression, ... 49
   cleaning carbon, ... 71
   clearance, valve stem-to-guide, ... 73
   compression, checking, ... 49
   compression gauge, use of, ... 50
   compression, other methods of checking, ... 51
   Corvair cylinder heads, ... 62
   cover, rocker arm, ... 58
   cylinder heads, Corvair, ... 62
   cylinder head, install, ... 60
   cylinder head, Jeep overhead camshaft
     engine, ... 64
   cylinder head, Pontiac Tempest overhead
     camshaft engine, ... 64
   cylinder head, remove, ... 59
   cylinder head, Volkswagen, ... 64
   freeing stuck valve, ... 51
   gear, camshaft, ... 98
   gear, crankshaft, ... 99
   gears, timing, replace, ... 101
   grinding, valve, ... 80
   guide, replace valve, ... 75
   hot valves, causes of troubles, ... 45
   hot valve seats, causes of troubles, ... 44
   hydraulic lash adjuster, ... 90
   hydraulic valve lifters, ... 84
   inserts, valve seat, ... 83
   inspection, valve, ... 72
   lash adjuster, hydraulic, ... 90
   L-head valve cover, ... 58
   L-head valves, install, ... 68
   L-head valves, remove, ... 67
   lifters, hydraulic valve, ... 84

modifying valve gear for high performance, . . . 90
overhead valves, install, ... 70
overhead valves, remove, ... 69
plate, camshaft thrust, ... 96
reaming valve guides, ... 74
rocker arm cover, ... 58
rocker arms, ... 55
sluggish valve closing, causes of troubles, . . . 48
spring, replace valve, ... 78
spring, testing valve, ... 78
sprocket, camshaft, ... 98
sprocket, crankshaft, ... 99
sprockets, timing chain, ... 103
sticking valves, ... 51
timing belt and sprocket, ... 105
timing case cover, ... 99
timing chain and sprockets, replace, ... 103
timing gears, replace, ... 101
valve cover, L-head, ... 58
valve, freeing stuck, ... 51
valve grinding, ... 80
valve guides, reaming, ... 74
valve guides, replace, ... 75
valve inspection, ... 72
valve lash adjustment, ... 52
valve leakage, causes of troubles, ... 42
valve lifters, hydraulic, ... 84
valves, L-head, install, ... 68
valves, L-head, remove, ... 67
valve operating clearance, introduction, ... 41
valve operating conditions, introduction, ... 40
valves, overhead, install, ... 70
valves, overhead, remove, ... 69
valves, replace, ... 66
valve rotators, ... 90
valve seat inserts, ... 83
valve spring, replace, ... 78
valve spring testing, ... 78
valve, stem-to-guide clearance, ... 73
valves, sticking, ... 51
valve system parts, introduction, ... 38
valve temperature, ... 44
valve timing, ... 93
Wankel rotary engine,
   compression phase, ... 314
   development of engine, ... 318
   engine construction, ... 317
   engine performance, ... 317
   engine specifications, ... 319
   exhaust phase, ... 315
   intake phase, ... 313
   major engine components, ... 308
   power phase, ... 314
   rotary combustion process, ... 309

Chapter Synopsis

How Engines Work

This chapter is an overview of the entire engine and its system components before you begin studying each individual system in detail.

Almost all automotive engines used in passenger cars and trucks today operate on the 4-stroke principle. All four of the cycles are described and explained with illustrations for each.

The theory of combustion is a necessary principle to grasp if you want to understand how an engine works. Here, the theory is detailed to include getting air into the engine via the carburetor, heating the mixture, the timing of the valves to let air and fuel in, and exhaust gases out.

More basics are covered—cylinder numbering, firing order, and engine construction.

Then the chapter goes into the component systems. The cooling system is necessary to dissipate the tremendous heat built up during the combustion process. The valvetrain is covered in a section which explains camshaft, valves, tappets, valve springs and valve action.

Next, you'll find a discussion of piston construction, piston rings and piston pins. The crankshaft is the very heart of the engine. The crank components are discussed, including the crank itself, vibration dampers, the flywheel, bearings and more.

There are also separate sections in this chapter on the Corvair air-cooled engine and also quite extensive definitions of terms.

Review questions are at the end of the chapter.

Engine Valve System

Beginning with this chapter the *Auto Engines and Electrical Systems* textbook begins to delve in detail into each engine system.

Here, you start at the beginning. What are the valve system parts? Discussions on valve operating conditions and valve clearances are next followed by an extensive section on causes of valve troubles such as leakage, high temperature, seats, and heat-caused problems.

Valve system service makes up the rest of the chapter. There are complete instructions on how to check compression, how to use a compression gauge, how to free stuck valves and how to adjust valve lash, plus complete rocker arm service information.

The next part of the chapter explains what is involved with, and how to do, a typical valve job. This is effectively detailed and includes the following: removing cylinder head, replacing valves, cleaning carbon deposits, valve inspection, measuring valve-stem-to-valve guide clearance, reaming valve guides, replacing valve guides, replacing valve springs, grinding valves, servicing valve seat inserts, and servicing hydraulic lifters.

You'll also find a special section on modifying valve gear for high performance.

The second large section in this chapter covers timing gear, timing chain, belt and camshaft.

First is a discussion of the theory of valve timing. Then each component that affects valve timing is discussed individually including the timing chain or belt, camshaft thrust plate, cam gears or sprocket and crank gear or sprocket.

The balance of the chapter covers the servicing of these components. There are sections covering the replacement of timing gears, timing chains and sprockets, camshaft replacement and cam bearing replacement. There's a special section on high performance camshafts and related components, too.

Review questions are at the end of the chapter.

Engine Piston System

The piston system involves all those parts which transmit the power of combustion to the flywheel. In this chapter, you'll learn the system components and their servicing.

The first section of the chapter discusses how to determine cylinder bore wear and distorted cylinder bores in a block. Also discussed are other piston-related problems such as piston ring wear, piston slap, improperly seated rings, excessive oil consumption, and oil loss via various forms of leakage.

The second section discusses piston system service. Here you'll find the proper method of checking for external and internal oil loss, how to inspect a cylinder block, how to repair cracked blocks, how to disassemble an engine to perform a ring job, the proper use of the reamer, how to choose proper replacement rings, proper removal and replacement of pistons and rings, and the use of the ring compressor.

In addition to piston rings, you'll find a complete section on pistons themselves—how to inspect, how to clean, how to check piston skirts for collapse.

The third large section in this chapter deals with cylinder bore service. You'll learn how to check cylinder bores, and how to recognize various conditions such as too much or too little taper, out-of-round bores, wavy bores, etc. Also here is a discussion of various tools used to service cylinder bores. There are three methods commonly used—the dial gauge, the inside micrometer, and the telescope gauge. Also covered are major machine tools used in refurbishing cylinder bores such as the glaze buster, flexible hone, rigid hone, and boring bars. You'll also find a section devoted to reboring of cylinder blocks. The section explains when to bore, how much, and the use of the boring bar. There is also information covering the fitting of cylinder sleeves or liners when reboring isn't practical plus how to set up a block for high performance use.

The fourth major section in this chapter covers piston pin and connecting rod service. Here, you'll find a discussion of piston pin sizes and how to check whether piston pins are too loose or too tight.

Moving onto connecting rods, the chapter discusses how to remove and replace rod bushings, how to hone piston holes, how to check rod alignment and assemble pistons and connecting rods.

Review questions are at the end of the chapter.

Crankshaft and Bearings

Bearings are covered first. Main and rod bearings are identified and discussed. Then the chapter turns to the servicing of crankshafts and bearings.

Oil leakage causing low oil pressure is one of the most common problems found in this part of the engine and there's a discussion here on how to detect bearing oil leaks. Other bearing troubles discussed include binding, out-of-round bearings and knocks.

Before checking bearings for wear, the crankshaft journals should be inspected for surface smoothness, nicks, and for out-of-round and taper. There is a complete discussion on how this can be accomplished, including procedures showing the use of micrometers.

There are three methods of checking bearing clearances—the shim method, Plastigage method and the micrometer method. You'll find all three methods discussed in detail here.

Installing bearings requires specific techniques. So this chapter covers complete installation techniques for both rod and main bearings.

A complete crankshaft service requires removal of the crank. Polishing, grinding and other service procedures are discussed here including the important check for straightness.

Also included is a discussion on servicing flywheels and preparing crankshafts for high performance use.

Review questions are at the end of the chapter.

Engine Oiling Systems

The many parts that make up the automobile engine are either tightly fastened together or move against one another. If they move against one another, friction becomes a concern and therefore lubrication to keep wear to a minimum and to provide for smooth operation is required.

This chapter delves into the lubrication of the engine, starting with basics. Characteristics of engine oil, what oils do in an engine, how they work, and their classifications are covered.

How an engine is lubricated and the components that make up the oiling system are covered, unit by unit.

The oiling system service information starts right at the basics too, with how to check the oil level in an engine. Also covered are causes of excessive oil consumption, oil changing intervals, how to change oil, checking for oil leaks, high or low oil pressure and what it means. Then down into the rest of the oiling system: the oil pan, and how to replace it, oil screens, and removing and replacing oil pumps. Oil filters and their workings are explained, too, as are the crankcase ventilation and PCV systems.

Review questions are at the end of the chapter.

Emission Control Systems

Since 1968, emission control systems have been mandated by federal law to be on all new cars sold in this country. This chapter takes you system by system and component by component through the oftentimes complex network of emission controls used on cars since 1968.

The early systems such as the air pump and electric assist choke are covered first. Other early systems include engine modification systems employed by all four of the major U.S. manufacturers.

Then the chapter turns to more modern systems of emission control such as exhaust gas recirculation, catalytic converters, controlled combustion systems, various ignition spark retard devices and also delay vacuum controlled valves that retard distributor advance.

Next, there is a detailed discussion on evaporative emission controls including both the carbon storage type and the crankcase storage type.

Review questions are at the end of the chapter.

Diesel Engine

As gasoline has become more scarce, and as emission controls have gotten more stringent, the diesel engine has been used more extensively than ever as an automotive powerplant.

This chapter is a detailed explanation of how diesel engines work and how they differ from conventional gasoline engines. There is also a discussion of the characteristics of diesel combustion.

As the diesel fuel system is the most complex part of the diesel engine, the fuel system gets the most coverage here. You'll find information that includes fuel supply units, injection, governors, distributor type systems, and the wobble plate pump system.

Also included is a discussion of recent engineering developments in the diesel engine field including Mercedes-Benz's 5-cylinder diesel engine and General Motors' Oldsmobile 350 V8 diesel.

The Wankel and How It Works

The Wankel rotary engine is a complete depar-ture from ordinary internal combustion engine design. Despite the differences of design, the operation of the rotary engine is quite easy to understand.

This chapter takes you step by step through the Wankel rotary.

A brief history of the development of the engine is discussed and the tremendous performance potential of the rotary engine is touched upon.

Major engine components are discussed and compared to the equivalent unit in a piston engine, followed by the combustion process in the rotary engine. And here is where most of the similarities and differences occur between a rotary and piston engine.

Review questions are at the end of the chapter.

The Turbine and How It Works

In its basic concept, the turbine engine is the simplest of all engine designs. Yet, because the turbine hasn't been used in production automobiles, most students of automotive technology feel that they have to be Aeronautical Engineers to understand this complex mystery surrounding the turbine engine.

Not so, as this chapter quickly proves. The chapter leads off with an explanation of exactly how the turbine engine works. It then covers the fuel system, the burner and the ignition system separately.

Also included is a discussion of the possible future uses of turbine engines and their performance characteristics.

Review questions are at the end of the chapter.

Cooling System

The purpose of the cooling system is to prevent temperatures of cylinders, pistons, valves and other engine parts from rising high enough to destroy the parts or the oil which lubricates them.

The cooling system chapter explains how this is accomplished. Each component's function in the system is described. Components covered include the water pump, fan, radiator, drain cocks and plugs, radiator cap, thermostat, and hoses.

There is a discussion of the theory of water physics which explains what happens when water freezes and boils and how anti-freezes alter the chemistry of the water solution, changing the freezing and boiling points.

Cooling system damage is discussed to help in spotting damage from overheating, overcooling, corrosion and other conditions.

Preventive maintenance is as important as repair. So here you'll find a section devoted to that subject. Proper water level, leakage (both internal and external), proper selection and maintenance of thermostats, fan belts and radiator caps are all covered.

The third part of the chapter covers trouble-shooting—how to test for such conditions as frozen coolant, overheating, foaming, air suction, combustion leakage, improperly functioning thermostat, improperly functioning pressure cap, improperly functioning temperature gauge or warning light, and radiator clogging.

The fourth part of the chapter covers cooling system service operations—how to clean a system, how to flush it out, how to test and add anti-freeze, how to inspect and replace hoses, how to inspect and replace belts, how to service the water pump and how to remove and service the radiator itself.

Review questions are at the end of the chapter.

Fuel System

Feeding the engine is the fuel system—one of the simplest, yet most complex, systems on an automobile. The chapter starts with the basics and explains the components of the fuel system and the function of each. Fuel tanks, filters, pumps and lines, air cleaner, carburetor, intake manifold, exhaust system, even fuel gauges are covered here. Also note that fuel injection is covered in the next chapter.

Once you know the components, you should know how an engine uses them to transport fuel which burns and produces power. So the next part of the chapter covers the theory of fuels and how they produce power. Such topics as combustion, pressure detonation and octane rating are covered.

The physics of carburetion is a section that stays with theory but zeroes in on the actual air-fuel mixture as it proceeds through the carburetor. This section discusses vaporization, atmospheric pressure, vacuum and heat.

The actual carburetor is next. And here you'll get a thorough understanding of how this most important component works. The text explains the float system, throttle valve, idle and low speed system, power system, accelerator system, and choke. There's also a discussion of venturi action and how it affects carburetion principles plus an explanation of the concept of the balanced carburetor that handles all types of loads.

Also in the chapter is a detailed look at the repair and servicing of the most common carburetors found in the field—Carter, Holley, Rochester and Stromberg.

Another section is devoted to the fuel system

service. Here you'll find information on how to maintain the air cleaner, carburetor, fuel lines, fuel filter, fuel pump, intake manifold and exhaust system in top operating condition.

However, should trouble develop, you have to learn how to spot it. The troubleshooting section that follows explains how to spot various fuel system troubles including an inoperative choke, water in the fuel system, a clogged fuel tank or pipe, fuel pump failure, vapor lock, and leaks. Other troubles covered include diagnosing an engine that won't idle or start, an engine that loses power, and excessive fuel consumption.

The next major section covers carburetor service. There is overhaul information on most commonly used carburetors. Carter, Holley, Rochester and even Stromberg models are covered.

The last section covers fuel pump service—testing, overhauling, and diagnosing troubles. There is quite an extensive discussion of high performance fuel systems and their setup.

Review questions are at the end of the chapter.

Fuel Injection

Although it's really part of a car's fuel system, a fuel injection unit is so different from conventional carburetion that a separate chapter was necessary to accurately cover the subject.

First, a brief history of fuel injection and its automotive uses is given. Then a separate section for each type of unit details information on components, their function and the overall operation of each system.

Generally, fuel injection units can be divided into two groups—mechanical and electronic. Both types are covered here.

The separate sections include the Rochester unit used on older General Motors cars, the Bendix electronic type and three Bosch systems.

The Bosch systems are the older D-Jetronic, the newer L-Jetronic, both of which are electronically controlled units, and the mechanically controlled K-Jetronic continuous system.

Electricity

If electricity seems more mysterious than some of our more prosaic tools like hammers and pliers, it is mainly because electricity is invisible. There is no way of seeing an electric current flow through a wire.

This chapter pulls the wraps off the mystery. It covers the theory of electricity from a practical viewpoint, looking briefly into atomic theory and electrons, but then moves quickly onto Ohm's law, volts, amps, types of circuits, electrical measurements and electrical conductors.

Review questions are at the end of the chapter.

Magnetism

Most of the electrical units in a car employ not only electricity but also magnetism (produced by electricity) to make them work. The list of electromagnetic units includes the starting motor (and other electric motors), ignition coil, horn, horn relay, solenoids, cutout relay, current regulator, voltage regulator (unless it's a new electronic type), etc.

This chapter explains the theory of magnetism right from the basics—magnetic fields, moving electrons, opposite poles, etc. Then, the next moves on to cover how an electric generator and electric motors work.

Review questions are at the end of the chapter.

Electrical System

Once you know the basics of electricity and magnetism, it's time to apply that knowledge to the workings of an automobile. With this chapter, we move away from basic theory and into the realm of practical knowledge.

Starting right in with how to read a wiring diagram, the chapter takes you step by step through the various circuits that make up an automobile's electrical system. Covered are the starting circuit, charging circuit, ignition circuit, horn circuit, windshield wiper circuit, instrument circuit and lighting circuit.

The chapter then discusses electrical system service. Such topics as electrical wiring, cable gauge, ignition cable, replacing cables, terminal connections, and selecting proper cable are all covered.

Review questions are at the end of the chapter.

Ignition System

The purpose of the ignition system is to supply sparks across the electrodes of the spark plugs to ignite the air-fuel mixture in the cylinders.

This chapter first explains the components of the ignition system and their function. The ignition coil, condenser, spark plugs, and distributor are covered.

The distributor is detailed separately and the principles of centrifugal and vacuum spark advance are discussed.

The service section begins with a troubleshooting guide that helps pinpoint such troubles as a bad condenser, grease or oil in the distributor, a loose connection, incorrect breaker point gap, worn distributor shaft bushings, worn breaker arm pivot bushings, improper breaker spring tension, hard starting, and miss.

Causes of ignition coil failure are discussed as is the importance of correct coil polarity. Instructions for testing coils are included.

In order to help you judge the condition of spark plugs accurately, there is a section covering analysis of plug conditions plus information on gaskets, cleaning, adjusting gaps and proper installation techniques.

Condensers and breaker points are covered separately in the service section. Causes of condenser breakdown are given as is information on how to read breaker points.

Distributor inspection procedure is explained including methods of testing centrifugal and vacuum advance and testing the condenser circuit.

The balance of the chapter consists of detailed overhaul information for Delco-Remy, Ford, Holley, Prestolite, and Chrysler distributors. Each make is covered in a separate section.

Review questions are at the end of the chapter.

Transistorized and Electronic Ignition Systems

Both pure electronic and also transistorized systems are covered in this chapter.

Since transistorized systems were in use earlier, they're covered first. The first section of this chapter explains what a transistor is and how it is applied to an ignition system. Then, specific systems are covered including the Delco-Remy and Ford systems. The operation of each system is described as are the service recommendations and troubleshooting procedures for each.

The pure electronic systems are in the second part of this chapter. A brief history of electronic ignition systems is followed by a discussion of the advantages of an electronic system over a conventional breaker point ignition system.

Then each electronic system's theory of operation is covered. Included in the discussion are the Chrysler, Ford, General Motors and Prestolite systems plus the systems currently being used on imported cars.

You'll also find a discussion of General Motors' Micro-Processed Sensing and Automatic Regulation (MISAR) system.

Review questions are at the end of the chapter.

Generators and Regulators

In the early days of automobiles, there weren't many electrical devices used and the need for power generation was small. As the need for electrical power grew, so did the importance of the generator. Although no longer in use, this book would not be complete if we didn't cover DC generators because there are probably still several million cars on the road using an older style DC generator.

This chapter first explains generator construction and how a generator actually produces power There is also a section on generator service including rebuilding information.

A regulator is designed for one purpose only—to regulate or control the charging rate in the generator-battery circuit.

The second part of the chapter covers regulators from their construction to their operation. Following the theory discussion, there is information on how to replace regulators and also how to test specific models made by Ford, Prestolite, Bosch and Delco-Remy.

Separate review questions for generators and regulators appear at the end of the chapter.

Alternator Systems

As the electrical demands of the modern automobile became too much for the old style DC generator, automobiles were equipped with AC generators, or alternators. Previously, alternators were used only on city delivery trucks, heavy duty vehicles, buses and other vehicles where the demand on the electrical system was extreme.

This chapter first covers the differences between an alternator and generator, then details the construction of the alternator. There is also a complete discussion on how the alternator functions to produce power.

An explanation of how the alternator changes AC current to DC so that the other parts of the electrical system can use the power is also covered here.

The second part of the chapter covers alternator service. The Chrysler, Delco-Remy, Ford, Leece-Neville, Motorola, and Prestolite systems are all described, their system components detailed, then covered completely for service and overhaul. There is a very extensive testing section for each brand of alternator.

Review questions are at the end of the chapter.

Starting Motors

Starting with the fundamentals of starter operation, this chapter covers the subject completely. You'll find details on starter motor construction, Bendix drives, how to check out a starter, and then how to disassemble, service and reassemble the most commonly found starter motors.

Included here are the Chrysler reduction gear type, the Chrysler direct drive type, Delco-Remy type, Ford starter with integral positive engagement drive, Ford type with Folo-Thru drive and all Prestolite starters.

The end of this chapter contains a guide to starter motor problems.

Review questions are at the end of the chapter.

Starting Switches

Magnetic and solenoid switches are designed to perform mechanical jobs electromagnetically such as closing a heavy circuit or shifting the starter drive pinion with the engine flywheel ring gear for cranking.

There are several types in common usage and this chapter covers them all—magnetic and solenoid types. There are separate sections for the Carter vacuum switch and the Stromberg vacuum switch.

The balance of the chapter consists of starter switch service information including checking switch operation.

Review questions are at the end of the chapter.

The Storage Battery

The first question is, what does a battery do? That's where this chapter starts. You'll also find complete information on how a battery works, how a battery is constructed and how a battery's power capacity is determined.

There is a detailed troubleshooting guide to various battery troubles including sulphation, insufficient electrolyte, overheating, overcharging, buckled plates, freezing, corrosion, and vibration.

The theory and the how to of hydrometer testing is explained. Other service information includes the proper way to add water, how to test for electrical capacity, the proper method of removing and installing a battery, and proper charging techniques.

Review questions are at the end of the chapter.

Dash Gauges

All vehicles have at least four gauges or indicator lights to tell the driver how various systems are functioning. This chapter describes the construction and operation of the commonly used automotive gauges—fuel level, oil pressure, coolant temperature, and ammeter.

There is also servicing information for these gauges plus information on dash indicator lights, speedometers, electric clocks and other accessories.

Review questions are at the end of the chapter.

How Engines Work

Review Questions for This Chapter on Page 36

INDEX

Aluminum pistons	29	Engine operation	5	Power stroke	5
Bearings, main	32	Engine torque	10	Pre-ignition	12
Bore and stroke	10	Exhaust stroke	5	Ratio, compression	11
Brake horsepower	10	Feeding the engine	7	Reasons for cooling system	13
Cams	24	Firing order	15	Rings, piston	31
Combustion	6	Flywheel	35	Spark knock	12
Compression ratio	11	Gaskets, cylinder head	20	Springs, valve	25
Compression stroke	5	Heating the mixture	8	Stroke, bore and	10
Construction, engine	16	Horsepower, brake	10	Stroke, compression	5
Construction, Corvair engine	16	Horsepower, taxable	11	Stroke, exhaust	5
Cooling system, reasons for	13	Intake stroke	5	Stroke, intake	5
Corvair engine construction	16	Knock, spark	12	Stroke, power	5
Counterweights, crankshaft	34	Lifters, valve	23	Tappets, valve	23
Crankshaft	32	Locating valves	25	Taxable horsepower	11
Crankshaft counterweights	34	Main bearings	32	Terms, definition of	10
Cylinder bore distortion	28	Mixture, heating the	8	Timing, valve	9
Cylinder head gaskets	20	Numbering, cylinder	14	Torque, engine	10
Cylinder numbering	14	Operation, engine	5	Valve action	20
Dampers, vibration	34	Pins, piston	31	Valves, locating	25
Definition of terms	10	Pistons	28	Valve lifters	23
Detonation	11	Pistons, aluminum	29	Valve springs	25
Displacement, piston	10	Piston displacement	10	Valve system	18
Distortion, cylinder bore	28	Piston pins	31	Valve tappets	23
Engine construction	16	Piston rings	31	Valve timing	9
Engine, feeding the	7	Piston side thrust	30	Vibration dampers	34

ENGINE OPERATION

All engines used in American passenger cars operate on the four-stroke cycle principle. This means the engine requires four piston strokes (two up and two down) to complete the operating cycle. Thus, during one half of its operation the four (stroke) cycle engine functions merely as an air pump. The names of the strokes of the operating cycle are as follows:

1. Intake (pump in) stroke.
2. Compression stroke.
3. Power stroke.
4. Exhaust (pump out) stroke.

Intake Stroke

On the intake stroke, Fig. 1, the piston moves *down*, allowing a mixture of gasoline and air to flow into the cylinder through the intake valve, filling the space vacated by the piston. The fuel-air mixture is produced in the carburetor and then flows through the intake manifold through the open intake valve into the cylinder. *The exhaust valve is closed during the intake stroke.*

Compression Stroke

On the compression stroke, Fig. 2, both the intake and exhaust valves are closed. The piston moves *up* to compress the mixture upwards of 110 lbs. per sq. in. (psi). In order to produce this pressure, the throttle valve must be wide open. The exact pressure depends principally on the compression ratio of the engine. The spark occurs near the end of the compression stroke, igniting the mixture.

Power Stroke

On the power stroke, Fig. 3, both intake and exhaust valves are closed. Combustion pressure may reach about 600 psi on open throttle shortly after the piston starts *down* at the beginning of the stroke.

Exhaust Stroke

Just before the piston reaches the lower end of the power stroke the exhaust valve opens, Fig. 4.

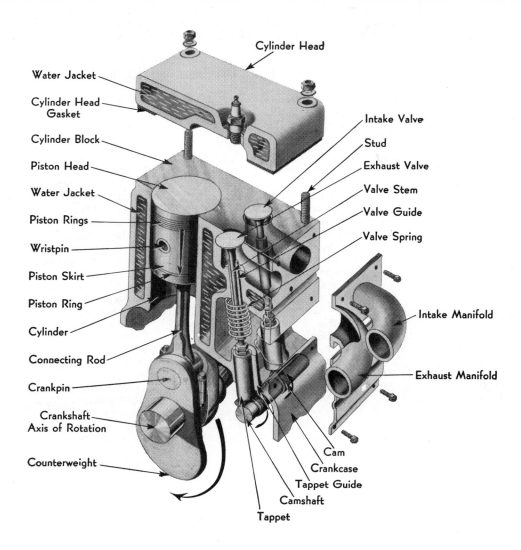

Water Jacket

Cylinder Head Gasket

Cylinder Block

Piston Head

Water Jacket

Piston Rings

Wristpin

Piston Skirt

Piston Ring

Cylinder

Connecting Rod

Crankpin

Crankshaft Axis of Rotation

Counterweight

Cylinder Head

Intake Valve

Stud

Exhaust Valve

Valve Stem

Valve Guide

Valve Spring

Intake Manifold

Exhaust Manifold

Cam

Crankcase

Tappet Guide

Camshaft

Tappet

Fig. 1 Intake stroke. Note that intake valve is open and exhaust valve closed

The pressure in the cylinder at this time causes the exhaust gas to rush into the exhaust manifold at high speed. Most of the gas which remains is pushed out of the cylinder by the upward motion of the piston on its exhaust stroke.

When the piston comes to a momentary stop at the top of the exhaust stroke, the exhaust gases are moving so fast in the exhaust manifold that their inertia helps remove any remaining gas in the combustion chamber. However, a small amount of gas always remains and mixes with the incoming mixture when the intake valve opens. In explanation, it should be realized that in order to evacuate all the gas there would have to be no clearance whatever between the piston and cylinder head.

Obviously, each up and down piston stroke requires one-half revolution of the crankshaft. Therefore, one combustion cycle requires four one-half revolutions or a total of two revolutions.

COMBUSTION

Power in an automobile engine is produced by the combustion of a mixture of gasoline and air within the cylinders. The mixture is set afire by an electric spark which jumps or flows across a gap between the two electrodes of the spark plug, Fig. 5.

From the spark, which initiates combustion, the flame spreads progressively in all directions throughout the combustion chamber, Fig. 6, until all the mixture is consumed. The combustion process requires approximately 3/1000 of a second. It is often called an explosion although it is a very slow one compared with gun powder or dynamite—the latter explodes in 1/50,000 of a second. Pound for pound, gasoline contains more energy than dynamite.

The materials involved in combustion include

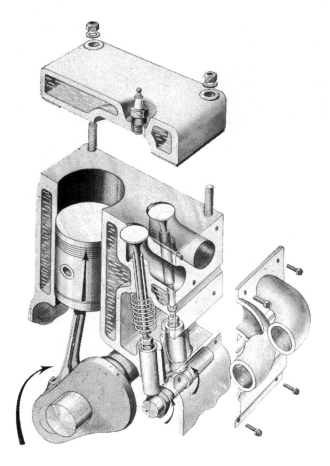

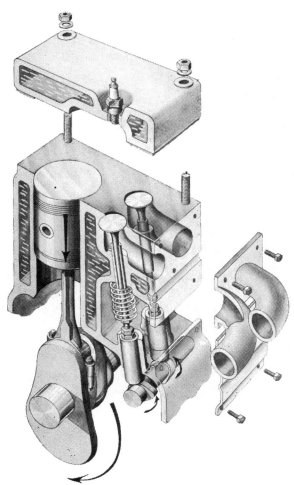

Fig. 2 *Compression stroke.*
Both valves closed

Fig. 3 *Power stroke.*
Both valves closed

the oxygen in the air and hydrogen and carbon which are the two ingredients in gasoline. Combustion takes place in the combustion chamber. Part of this chamber is formed in the cylinder head, the remaining part is the movable piston. Combustion can be referred to as a chemical marriage: The hydrogen in the gasoline is mated with oxygen to form steam. The carbon in the gasoline is also mated with the oxygen. This marriage results mostly in the production of carbon dioxide gas (which forms the bubbles in sparkling water). However, a small amount of the carbon is mated with insufficient oxygen and forms carbon monoxide which is a poisonous gas.

FEEDING THE ENGINE

In describing the operation of an automobile engine it is customary to say that the mixture is sucked into the open intake valve while the piston moves down on its intake stroke.

In order to explain what actually happens, bear in mind that the atmospheric pressure at sea level is 14.7 psi. Therefore, when the piston moves down on its intake stroke, the air pressure pushes the mixture through the carburetor and through the intake manifold into a space vacated by the downward motion of the piston.

This air pressure not only pushes air through the carburetor throat into the cylinder but the same pressure sprays the fuel in the carburetor into the air passing through the carburetor throat, Fig. 7.

The contraction in the throat of the carburetor, as shown in Fig. 7, results in a reduction in pressure which means that the pressure in the throat is always less than one atmosphere (14.7 psi at sea level). On the other hand, the pressure acting on the surface of the fuel in the float bowl is always one atmosphere. It should be obvious that fuel must flow through the jet from the higher pressure in the bowl to the lower pressure in the throat.

7

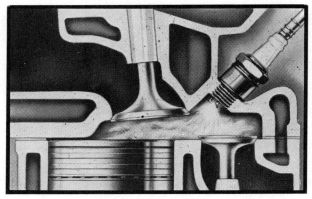

Fig. 5 The mixture is ignited by a spark which flows between the points of the spark plug

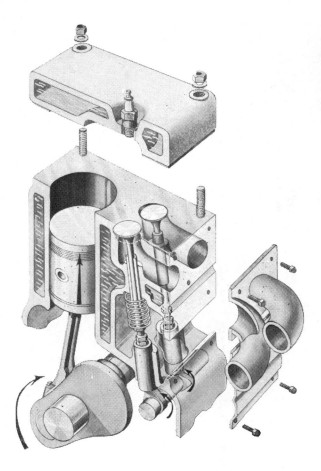

Fig. 4 Exhaust stroke. Exhaust valve is open and intake valve closed

HEATING THE MIXTURE

The evaporation of the fuel in the carburetor and intake manifold absorbs heat from the air, thereby reducing mixture temperature somewhat. Unfortunately, the air does not contain sufficient heat to evaporate the fuel completely. Therefore exhaust heat is applied to the intake manifold, Fig. 8, in order to warm the air enough to cause fairly complete evaporation. Evaporation of the remaining fuel droplets continues as the mixture passes through the intake manifold into the cylinder. Evaporation of most droplets still remaining occurs during the compression stroke. It should be noted, however, that any unevaporated fuel that exists when combustion takes place is wasted because the fuel must be a gas to explode.

Poisonous carbon monoxide gas in the exhaust is due either to not enough air in the mixture or to less than a thorough mixing of the air and fuel.

The manifold heater, Fig. 8, raises the temperature of the mixture. The heater is equipped with a

thermostatically controlled valve which permits maximum exhaust heat to be applied to the mixture when the engine is cold and minimum heat when the exhaust manifold is completely warm. The valve, which is controlled by a thermostatic spring (not shown), is fully closed with a cold engine and fully open when the engine reaches normal operating temperature.

Although the basic principle of manifold heating for both In-Line and V-type engines is the same, differences in design make it necessary to alter the construction of heat control components to suit V-type engines.

A typical example would be one with the heat control located in the right-hand exhaust manifold. When the engine is cold and the thermostat closes the valve, the resulting back pressure in the

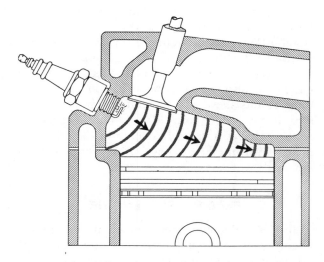

Fig. 6 The progress of combustion is indicated by the expanding circles radiating from the spark plug points

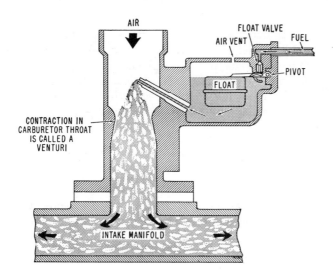

Fig. 7 *Atmospheric air pressure sprays fuel into the carburetor throat*

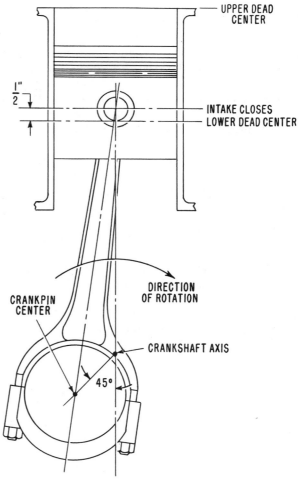

manifold forces the exhaust gas through a crossover passage in the intake manifold to the left-hand exhaust manifold. As the engine warms up and the thermostat releases the valve, the flow of hot gas through the crossover chamber is reduced, permitting the exhaust gas to flow through both exhaust manifolds. This explains why the exhaust comes out of only one muffler tail pipe when the engine is cold on cars with dual exhaust systems.

VALVE TIMING

In order to obtain maximum power, the intake and exhaust valves do not open and close exactly at the dead center positions of the piston. The time of opening and closing varies with different makes and models of engines but the average is about as follows:

Intake valve opens 7 degrees before top dead

Fig. 9 *On the average engine, the intake valve closes 45 degrees after the intake stroke is completed. In other words the valve closes 45 degrees after the piston has started upward on its compression stroke. This means that the piston has moved upward on its compression stroke about ½" when the intake valve closes*

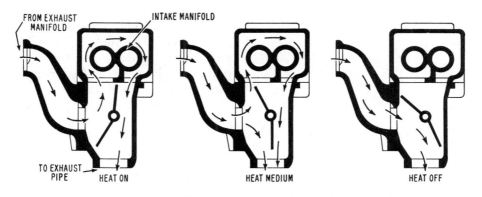

Fig. 8 *Exhaust heat is used to warm intake manifold*

center and closes 45 degrees after bottom dead center on the compression stroke, Fig. 9. Exhaust valve opens 50 degrees before bottom dead center on the power stroke, remains open during the exhaust stroke and closes 9 degrees after top dead center on the intake stroke. Therefore, on the average, the intake valve opens 16 degrees before the exhaust valve closes. This is called valve overlap.

One reason for opening the valves earlier and closing them later is because it takes time to open or close a valve. However, an even more important reason is the effect of gas inertia. At maximum engine power the mixture flowing through the intake manifold may reach a speed of 150 mph or more. Therefore, the mixture has a considerable ramming effect and in consequence a much greater weight of mixture can be crammed into the cylinder when the closing of the intake valve is delayed a reasonable amount, such as 45 degrees. This delay means that the piston has moved about ½ inch upward on its compression stroke before the valve closes.

Similarly, in order to evacuate the burned gases from the cylinder without delay the exhaust valve is opened toward the end of the power stroke, or at about 50 degrees before bottom dead center on the average.

Delaying the closing of the exhaust valve to an average of 9 degrees enables the inertia of the exhaust gases to more completely evacuate the cylinder. Likewise, opening the intake valve an average of 7 degrees before the intake stroke begins, gives the valve a head start in admitting the mixture to the cylinder.

Valve timing in any engine is a compromise between low and high-speed operation. An early intake valve closing is favorable to power at low speeds whereas a later closing of the intake valve assists power production at high engine speed. Likewise late opening of the exhaust valve helps power at low speed whereas earlier opening of the exhaust valve improves power at top speed.

DEFINITION OF TERMS

Bore and Stroke

Bore is the diameter of the cylinder in inches.

Stroke is the distance the piston moves between upper and lower dead center in inches.

Piston Displacement

Piston displacement for one cylinder is the cubic volume through which the piston sweeps in moving the length of one stroke. This volume in cubic inches multiplied by the number of cylinders gives the piston displacement of the whole engine. This total piston displacement indicates the "size" of the engine. Piston displacement may be figured by means of the following simple formula:

Piston displacement equals bore × bore × stroke × number of cylinders × .785.

All other things being equal, the piston displacement of an engine is an index of the power it may be expected to produce.

Engine Torque

Engine torque or turning effort is the rotating force developed at the flywheel. In the case of a typical automobile engine this force might be 200 pounds when measured at a radius of one foot from the center of the flywheel. When the torque radius is one foot it is customary to say that the torque developed by an engine is 200 pounds-feet. This term should not be confused with foot-pounds although there is a relationship between them, namely one pound-foot of torque during one revolution represents 6.28 foot-pounds of mechanical energy.

Brake Horsepower

Brake horsepower is the actual horsepower of the engine delivered at the flywheel. Many years ago it was customary to measure horsepower by a device called a Prony brake (named after the man who invented it)—hence the term brake horsepower. Today, however, horsepower is determined by a dynamometer of which there are several types. Both the Prony brake and the dynamometer measure torque. The horsepower is calculated from the torque.

By definition one horsepower represents the production of 33,000 foot-pounds of mechanical energy per minute. Therefore horsepower equals foot-pounds per minute divided by 33,000. Foot pounds may be defined as the mechanical energy developing by a force of so many pounds acting through a certain distance per minute. Thus a force of, say, 33,000 pounds acting through a distance of ten feet in one minute would represent 330,000 foot-pounds or 10 hp (330,000/33,000).

When an engine is tested on a dynamometer, if it develops 200 pounds-feet of torque it delivers 200 × 6.28 foot-pounds of mechanical energy per revolution or 1256 foot-pounds (200 × 6.28).

If the engine is running 1,000 rpm, the foot-pound production per minute is 1,256,000 foot-pounds (1256 × 1,000) or 38 hp (1,256,000/33,000).

The torque is measured at various other engine speeds from, say, 500 to 4,000 rpm and the horsepower for each speed is calculated.

Taxable Horsepower

In some states, automobiles are taxed according to a taxable horsepower rating based on the following formula:

Taxable horsepower equals bore × bore × number of cylinders × .4.

This ancient formula was fairly true 40 years ago but has no relationship to the horsepower production of modern engines. The taxable horsepower formula is often called the S.A.E. horsepower formula but the Society of Automotive Engineers had nothing to do with it. The formula was originated by The Royal Automobile Club of England.

Compression Ratio

Compression ratio is the ratio between the total volume of the interior of a cylinder with the piston at bottom dead center divided by the volume at top dead center, Fig. 10. For example, in a given engine the total volume with piston at bottom dead center might be 80 cubic inches while the volume at top center might be 10 cubic inches, in which case the compression ratio would be 8 to 1 (80/10). Piston displacement is the difference between these two volumes or 70 cubic inches.

Another way of expressing compression ratio is as follows: Determine the volume of the combustion chamber with piston at top dead center—say this volume is 10 cubic inches for example and that the calculated piston displacement is 70 cubic inches. Therefore the compression ratio is

$$\frac{10 + 70}{10} = \frac{8}{1}$$

Compression ratios on modern cars range from 7.50 to 1 up to 11.0 or more. Power and fuel economy are increased as the compression ratio is raised.

Other things being equal, the higher the compression ratio the higher the *compression pressure* obtained at the end of the compression stroke. Compression pressures on modern cars range roughly from 110 to 200 pounds per square inch on open throttle. In a given make and model of engine, compression pressure is a maximum at

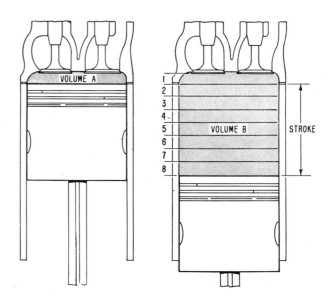

Fig. 10 *The compression ratio of this engine is 8 to 1 since volume B is 8 times volume A*

some certain speed, say, 1,800 rpm, for example. Above and below this speed the compression pressure falls off gradually. Maximum torque is produced when compression pressure is at maximum. Compression pressure is reduced as the throttle is closed, being about 35 pounds at idling speed.

When the spark ignites the mixture, combustion should proceed smoothly in expanding circles, Fig. 6, until all the mixture in the combustion chamber is consumed. Pressure in the combustion chamber rises as combustion proceeds and reaches a maximum about the time all of the mixture is burned. The higher the compression pressure the greater the maximum combustion pressure.

Detonation

The preceding paragraph describes normal combustion. However, under certain conditions, after combustion has proceeded part way through the chamber, the remaining mixture may go off with a bang like so much dynamite. This explosion causes a sharp rise in pressure which shakes the cylinder head and causes the head to vibrate with the result that a metallic knock is heard. The phenomenon just described is called *detonation*. Detonation (pinging) occurs only when the throttle is wide open or nearly so and it is more likely to occur at moderate engine speed (15 to 20 mph) than at high engine speed.

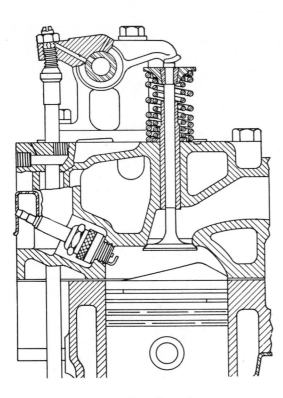

Fig. 11 *Buick engine*

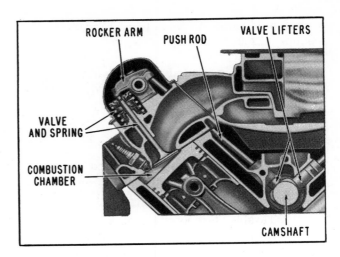

Fig. 12 *Wedge-shaped combustion chamber formed in block with underside of cylinder head being flat. A number of recent engines use this design*

There are certain factors that encourage detonation and others that discourage or completely eliminate it. Factors that encourage detonation include high compression pressure resulting from high compression ratio. A hot spot in the combustion chamber such as an unduly hot exhaust valve, piston head, cylinder head, and so forth. Severe detonation will be heard when the cooling water is boiling in the cylinder head or when there is no water in the head. A heavy coating of carbon on cylinder head or piston head may cause detonation. The carbon prevents adequate cooling of the surfaces and thus detonation is encouraged. Carbon also increases the compression ratio which in turn encourages detonation. If the spark occurs too early as the piston moves upward on its compression stroke, detonation will result. All the factors mentioned in this paragraph including high compression pressure increase the temperature of the mixture and the higher the mixture temperature the greater is the tendency to detonate. Also, the lower the octane number of the fuel the more likely it is that detonation will take place. A lean mixture encourages detonation.

On the other hand, detonation can be satisfactorily suppressed or eliminated by using a sufficiently high octane fuel; by reducing the compression pressure; by timing the spark so that it occurs at the proper instant rather than too soon; by

keeping the combustion chamber surfaces adequately cool, and free of carbon.

Usually detonation does not occur until most of mixture has been burned. Therefore, if this mixture is cooled somewhat its detonation is avoided. Cooling this last portion of the mixture is obtained by suitable combustion chamber design. In the case of the combustion chamber shown in Fig. 6, by the time the flame front has reached the valve area the burned gases have a pressure of 400 to 600 pounds per square inch and that therefore the remaining unburned gas has likewise been compressed to this high figure. Therefore, the remaining mixture is quite hot because mixture temperature increases with pressure. But then the mixture encounters the comparatively cool cylinder head and piston head where its temperature is reduced below the detonating point. The same result can be obtained by bulging the piston head upward, Fig. 11, so that the last portion of the mixture to burn is forced into close contact with the cylinder head and piston head, and in this way the remaining mixture is cooled below the detonation point.

Other examples of combustion chamber design are shown in Figs. 12 to 15.

Spark Knock and Pre-Ignition

Spark knock, as previously indicated, is caused by allowing the spark to occur too early. Pre-ignition is caused by a red hot spot in the combustion chamber. Such a spot might be a piece of carbon which is hot enough to ignite the mixture before the spark occurs, or a red hot spark plug, red hot exhaust valve or gasket edge too hot.

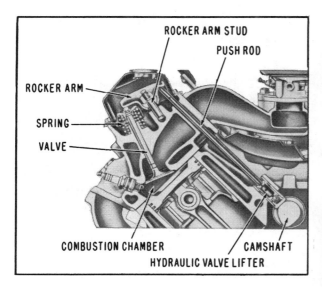

Fig. 13 Combustion chamber of Pontiac Tempest four-cylinder engine

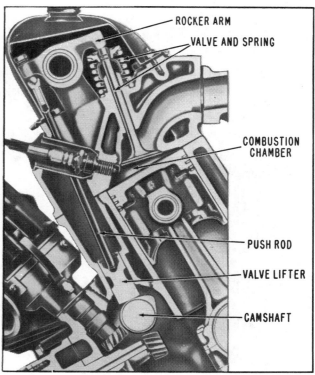

Fig. 14 Combustion chamber of Slant Six Chrysler engine

REASONS FOR COOLING SYSTEM

Cylinder block, cylinder head, piston and rings plus valves must be kept reasonably cool for a variety of reasons. Likewise these parts must be lubricated. The maximum allowable working temperature of the different parts varies, which means that an exhaust valve head may perform satisfactorily at 1200 degrees F or higher whereas numerous other engine parts must not exceed half this temperature.

Cooling, lubrication and combustion are tied together. Except in the Corvair engine, cooling in modern American automobiles is obtained first by circulating water through jackets which surround the cylinder bore and jackets within the cylinder head. The piston, piston rings, valves and other parts are cooled by their contact with the cylinder block.

The water in the cooling system when the engine is warm usually has a temperature ranging from 150 to 195 degrees F. The piston is cooled by its contact with the cylinder bore, Fig. 16, as well as by the contact of the rings with the bore.

The valves are cooled partly by the contact of the valve stem with its guide and partly by the contact of the valve face with the cylinder block whenever the valve is seated. Good contact between valve and seat is most important for satisfactory valve cooling.

Too high temperature of the cooling water may result in knocking or pinging of the engine on wide open throttle.

Excessive piston or ring temperature may result in the destruction of the lubricant. There are two stages: (1) The oil may become gummy and therefore cause piston rings to stick in their grooves with the result that the rings are no longer

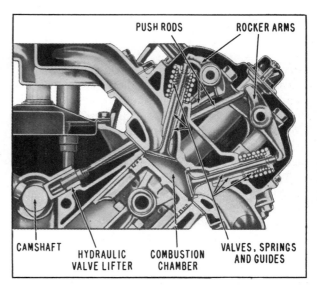

Fig. 15 Chrysler engine with hemispherical combustion chamber

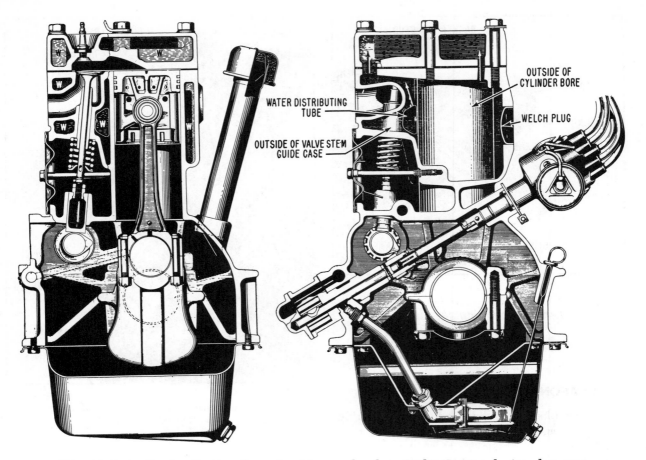

WATER DISTRIBUTING TUBE

OUTSIDE OF VALVE STEM GUIDE CASE

OUTSIDE OF CYLINDER BORE

WELCH PLUG

Fig. 16 Left—Sectional view through piston and valve. Right—Sectional view between cylinder bores. These two pictures should give the reader a good idea as to how both the cylinder block and cylinder head are water-jacketed. The light shading indicates the spaces occupied by water. The left picture shows that the cylinder bore is surrounded by water; the valve stem guide is surrounded by water for much of its length. This picture also shows that the water passages in the cylinder head keep the combustion chamber from becoming too hot. The right picture illustrates how the cylinder bore is completely encased in water. The picture also shows the water holes which permit water to flow from the cylinder block to the cylinder head. The water distributing tube (used on some engines) delivers a stream of water to the underside of each exhaust valve seat

able to seal the piston to its bore or gum may cause valve stems to stick in their guides. (2) The temperature may be high enough to drive the hydrogen out of the oil. In this case, since oil is composed of hydrogen and carbon, a carbon deposit is left wherever the excessive temperature occurs. In other words the oil chars.

The points where carbon (or gum) may be deposited include piston rings and grooves, slots in oil control rings, piston head, cylinder head, exhaust valve face and seat, valve stems and guides. Excessive carbon deposit on cylinder head or piston head may result in detonation or pre-ignition. Carbon deposit on the porcelain insulator of a spark plug will prevent ignition if the high-tension current flows across the carbon film on the insulator instead of jumping between the spark

plug points. Carbon (or dust) deposited on the face of the exhaust valve or its seat will prevent it from seating tightly. The resulting leakage of hot exhaust gases through the opening will eventually burn both the exhaust valve and its seat and increased leakage will result.

The operation of the cooling system on the Corvair is described in the *Cooling System Chapter.* Suffice it to say here that the engine is a horizontally-opposed, air-cooled six-cylinder type. The two aluminum cylinder heads incorporate cooling fins as do the individual cylinders.

CYLINDER NUMBERING

In speaking of the various cylinders in an engine it is customary to assign numbers to them, running from front to rear (Corvair excepted). Therefore the cylinders in a four-cylinder engine are designated 1-2-3-4, in an In-Line six they are 1-2-3-4-5-6 and similarly the cylinders in an In-Line eight are 1-2-3-4-5-6-7-8. In a V8 a similar system may be used by adding the letters R and L for right and left blocks, respectively. Therefore, in a V8 the cylinders in the right block (or bank) would be called 1R, 2R, 3R, 4R. With the exception of the Corvair the terms right and left in an automobile mean right and left as viewed from the driver's seat.

In the Corvair, which has the engine in the rear, the right rear cylinder is #1 and the left rear cylinder is #2. Thus reading in order from the rear, the right bank is numbered 1-3-5 and the left bank is 2-4-6.

However, there is obviously another way of numbering a V engine and that is to call the first cylinder in the left block No. 1 and continue through this block consecutively. The numbers for the right block then run consecutively after the left block. In this case, for example, the cylinders in a V8 would be numbered 1-2-3-4 for the left block and 5-6-7-8 for the right block. Still another way of numbering cylinders in a V engine is to use odd numbers for one bank and even numbers for the other, in which case one bank in a V8 would be numbered 1-3-5-7 and the other bank 2-4-6-8.

The method of numbering V engines is a choice made by the car maker and all his instructions concerning service work on the various cylinders is based on the numbering system which has been chosen. The numbering system involves various service operations including ignition work and valve adjustment, and so forth. With respect to V engines it is quite necessary to know the car maker's numbering system before the service operations can be intelligently performed.

FIRING ORDER

The sequence in which the cylinders fire is called the firing order. There are two possible firing orders for an in-line four cylinder engine, namely 1-2-4-3 and 1-3-4-2. However, the latter, 1-3-4-2, is the one that is used almost without exception. The firing order also indicates the order in which the intake valves open and close; also the exhaust valves.

When an engine is designed, the firing order is determined by the position of the cams on the camshaft, which means that the intake and exhaust cams are positioned, in a four cylinder engine, for example, to give either a firing order of 1-2-4-3 or 1-3-4-2. In the case of a four cylinder

engine, it makes no difference which of these two firing orders is chosen, but in engines with more cylinders, the firing order makes a considerable difference.

The reason for this is that the mixture of gasoline and air in the intake manifold has a certain amount of inertia, and it is undesirable to change its direction of flow more than is necessary. So every effort is made to arrange the intake manifold and the firing order so that there is as little interference with gas flow as possible. Similarly with the exhaust gases. Certain combinations of exhaust gas flow can cause increased back pressure in the system, resulting in reduced power output. Other combinations can produce much increased power. So the firing order and exhaust system are designed to give the best possible result.

When the firing order is determined, the high tension cables from the distributor cap to the spark plugs must be attached according to the order demanded by the camshaft.

Except for the Corvair, Buick and Oldsmobile V-6 engine, all six cylinder passenger car engines have a firing order of 1-5-3-6-2-4, with the numbers running from front to rear. There is only one exception to this rule. The Jaguar engine has the same firing order but the cylinders are numbered from rear to front, with the No. 1 at the rear.

The Corvair firing order is 1-4-5-2-3-6 with numbering system as described above, while the Buick and Olds V-6 uses 1-6-5-4-3-2, with the left bank of cylinders numbered 1-3-5, starting from the front, and the right bank 2-4-6.

The majority of V-8's use a firing order of 1-8-4-3-6-5-7-2, with the left bank of cylinders numbered 1-3-5-7 and the right bank 2-4-6-8, again starting from the front. Exceptions are:

Buick 401 and 425 cu. in. engines of 1963 thru 1966, which use a firing order of 1-2-7-8-4-5-6-3, with the right bank of cylinders numbered 1-3-5-7 and the left bank 2-4-6-8.

Cadillacs from 1963 thru 1967 had a firing order of 1-8-7-2-6-5-4-3, with odd cylinder numbers in the left bank, even numbers in the right.

Cadillacs from 1968 onwards have a firing order of 1-5-6-3-4-2-7-8, with odd numbers in the right bank and even numbers in the left.

Ford and Mercury engines use a firing order of 1-5-4-2-6-3-7-8, with cylinders 1-2-3-4 in the right bank, and 5-6-7-8 in the left bank. The exception is the 351 and 400 cu. in. engine, which has a firing order of 1-3-7-2-6-5-4-8 with the same cylinder numbering system. Pre 1959 engines used 1-5-4-8-6-3-7-2.

Oldsmobile 394 cu. in. engine has a firing order

of 1-8-7-3-6-5-4-2, with odd numbers in the left bank, even numbers in the right.

ENGINE CONSTRUCTION

The foundation of a water-cooled engine is a piece of cast iron which is called the cylinder block. This casting usually includes the upper half of the crankcase. Bolted to the top of the block is a cylinder head which is usually made of cast iron.

The oil pan, which is made of pressed steel, is bolted to the bottom of the cylinder block. The crankcase on a modern car consists of two portions: (1) the upper half which is an integral part of the cylinder block and (2) the lower half which is the oil pan just mentioned.

The block contains cylindrical holes which are correctly called cylinders, although the terms barrels or bores are often used and are almost equally descriptive.

Plungers called pistons move up and down in the cylinders. The motion of the pistons is transmitted to the crankshaft by means of the connecting rods which are pivoted at their upper ends to

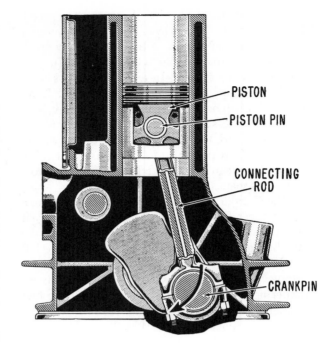

Fig. 17 *Piston and connecting rod assembly installed in engine*

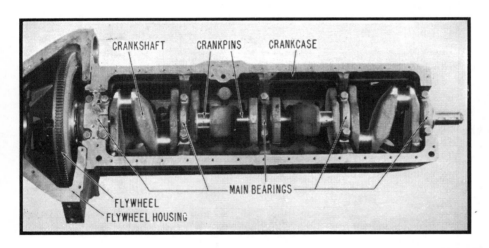

Fig. 18 *Crankshaft and main bearings. Worms-eye view with oil pan off*

the piston pins in the pistons, Fig. 17. The lower ends of the connecting rods are attached to crankpins on the crankshaft.

The crankshaft rotates in main bearings, Fig. 18, which are attached to the upper half of the crankcase.

It should be noted that the up and down motion of the pistons is transformed into rotary motion of the crankshaft by means of the connecting rods.

The rotary motion of the crankshaft is transmitted to the flywheel which in turn rotates clutch, transmission, propeller shaft, rear axle shafts, wheels and tires. The horizontal thrust of the tires against the road pushes the car.

CORVAIR ENGINE CONSTRUCTION

As shown in Fig. 19, this engine is the horizontal opposed air-cooled six-cylinder type. The cast aluminum alloy crankcase is vertically divided into two halves which are held together by bolts at the parting line. Each crankcase half has three pilot openings for individual cast iron cylinders which are positioned to the opening by means of four long studs at each cylinder. These studs pass through holes in the cylinder cooling fin structure and cylinder head and serve to secure the cylinders and head to the crankcase.

The two opposing cylinder heads are made of aluminum. The heads incorporate cooling fins,

Fig. 19 Corvair engine cross section as viewed from the top

built-in intake manifolds, wedge-shaped combustion chambers, and valves for each cylinder. The valves are actuated by push rods through stamped rocker arms.

Steel tubes are used to house the push rods in the open area between the crankcase and cylinder heads adjacent to the cylinders. These tubes serve to protect the exposed push rods as well as drain back oil from the cylinder heads to a relatively shallow oil pan bolted to the bottom of the crankcase.

The camshaft, which actuates the push rods through hydraulic valve lifters, is nested between the two halves of the crankcase below the crankshaft. This camshaft differs from the conventional

in that each of the three exhaust valve cam lobes are twice the width of the intake valve cam lobes and actuate a pair of exhaust valve lifters. The camshaft journals ride directly on the machined base metal of the crankcase.

No separate main bearing caps are required since the four bearings are supported entirely by the crankcase halves.

The engine rear housing mounts to the rear of the crankcase over four long studs. This housing contains the oil pump, crankshaft seal and primary oil passages. It provides the mounting for the distributor and the generator adapter to which the fuel pump, oil fill pipe, generator, oil filter and idler pulley are mounted. A rectangular crankcase

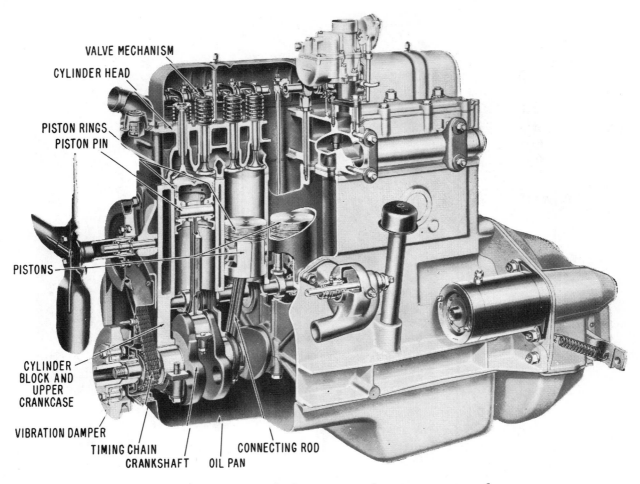

VALVE MECHANISM
CYLINDER HEAD
PISTON RINGS
PISTON PIN
PISTONS
CYLINDER BLOCK AND UPPER CRANKCASE
VIBRATION DAMPER
TIMING CHAIN
CRANKSHAFT
OIL PAN
CONNECTING ROD

Fig. 20 Rambler 6-cylinder engine with major parts named

cover mounts to the top of the crankcase and forms a base for the centrifugal blower.

VALVE SYSTEM

All modern American car engines have valves which are opened by a cam on the camshaft, Fig. 21 and closed by a spring. These valves are called poppet valves because they pop *up* to open and pop *down* to close. The angle between the seat and the valve is usually 45 degrees, although on a few engines it is 30 degrees, Fig. 22.

The intake valve, when it is opened, connects the carburetor with the cylinder; when the exhaust valve is opened it connects the cylinder with the exhaust passages from manifold to muffler. All passenger cars currently produced have one intake valve and one exhaust valve per cylinder.

Valves may be located in the cylinder block, Fig. 21, or in the cylinder head, Fig. 20. If the valves are located in the cylinder block, the engine

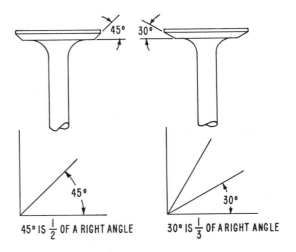

45°

30°

45°

30°

45° IS $\frac{1}{2}$ OF A RIGHT ANGLE

30° IS $\frac{1}{3}$ OF A RIGHT ANGLE

Fig. 22 Valves with 30-degree and 45-degree angles

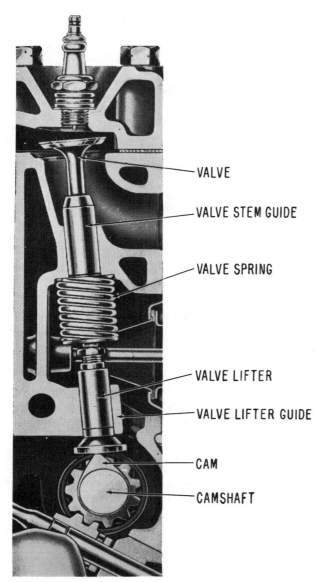

Fig. 21 *Valve mecha-nism. Willys L-head six*

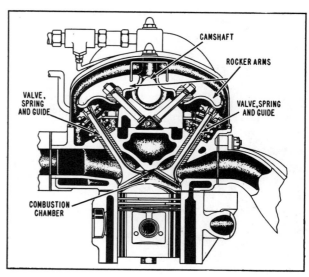

Fig. 23 *Jeep Tornado engine with overhead camshaft and valves*

VALVE

VALVE STEM GUIDE

VALVE SPRING

VALVE LIFTER

VALVE LIFTER GUIDE

CAM

CAMSHAFT

camshaft on top of the cylinder head. The cams act directly on the rocker arms which in turn push the valves open. This design is known as an overhead valve overhead camshaft, or simply overhead camshaft engine.

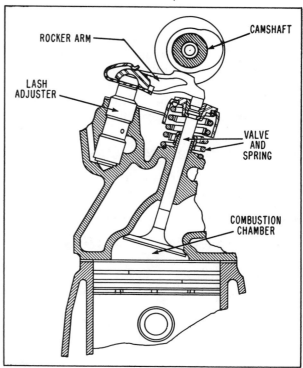

Fig. 24 *Pontiac Tempest engine with overhead camshaft and valves*

is called an L-head design because the combustion space roughly resembles an inverted L when the piston is at lower dead center. The valves are opened by cams acting directly on tappets which push against the valve stems.

If the valves are located in the cylinder head, the engine is called an overhead valve engine.

With two recent exceptions, this type of engine has long push rods at the side of the block which operate rocker arms that push the valves open.

The exceptions are the Jeep overhead valve engine, Fig. 23 and the Pontiac Tempest overhead valve engine, Fig. 24. These engines have the

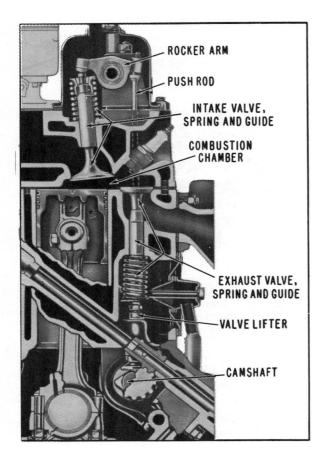

Fig. 25 *Valve mechanism of Jeep F-head engine*

ROCKER ARM
PUSH ROD
INTAKE VALVE, SPRING AND GUIDE
COMBUSTION CHAMBER
EXHAUST VALVE, SPRING AND GUIDE
VALVE LIFTER
CAMSHAFT

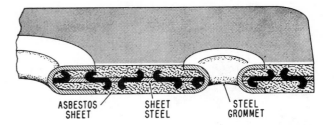

ASBESTOS SHEET   SHEET STEEL   STEEL GROMMET

Fig. 27 *This is a "steel-inserted" asbestos gasket. A sheet of steel is perforated and the prongs are so formed and clinched into the asbestos that they cannot separate from the steel*

Still another type of engine is the so-called F-head, Fig. 25. From a common camshaft in the crankcase, the intake valves are overhead, operated by push rods and rocker arms, and the exhaust valves are in the L of the cylinder block, operated by tappets or valve lifters.

CYLINDER HEAD GASKETS

In spite of the fact that the top surface of the cylinder block and the lower surface of the cylin-

der head may be smooth to the eye, the two surfaces never fit together closely enough to prevent leakage. Therefore a layer of rather soft flexible material, called a gasket, must be inserted between them. There are various designs. The cylinder head gasket, Fig. 26, consists of a layer of asbestos fibre encased in thin sheet copper.

This cylinder head gasket has sufficient resiliency so that it seals the cylinder head to the block while its sheet copper casing is able to withstand the high temperature and pressure of the combustion chamber. Steel-asbestos gaskets of this type are also used.

In another design, a sheet of steel is sandwiched between two layers of asbestos fibre, Fig. 27. There are prongs sticking out of both sides of the steel sheet which hold the asbestos layers securely in place.

In still another design, the asbestos is elim-

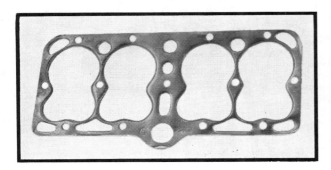

Fig. 26 *Typical copper-asbestos cylinder head gasket*

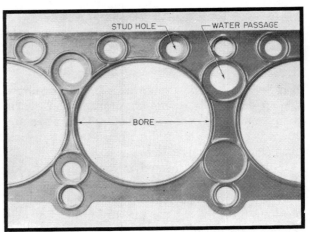

STUD HOLE   WATER PASSAGE
BORE

Fig. 28 *Steel gasket which is crimped at cylinder bores, stud holes and water passages*

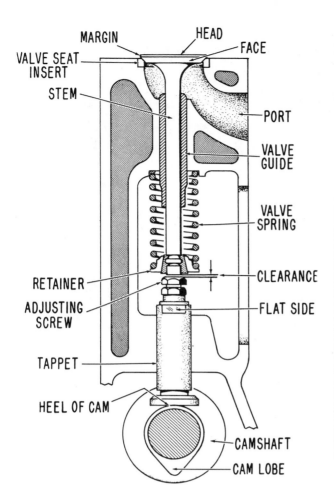

Fig. 29 L-head valve mechanism

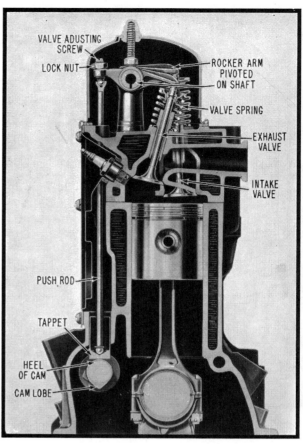

Fig. 30 Overhead valve mechanism

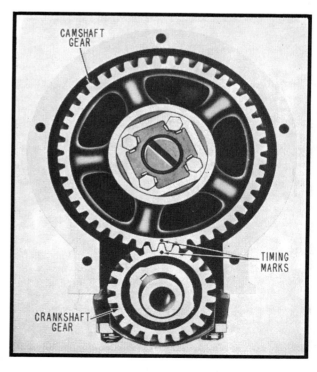

Fig. 31 Camshaft driven from crankshaft by gears

inated. A thin sheet of steel is crimped as shown in Fig. 28, and then coated with a special lacquer cement on both sides. Special gasket cements are often used with the other cylinder head gaskets mentioned.

VALVE ACTION

As explained previously, there are two common types of poppet valve designs in use, namely L-heads and overheads. Fig. 29 shows an L-head with a mushroom tappet and Fig. 30 shows the overhead design with barrel tappets and long push rods. A variation of the overhead design is the overhead camshaft which does not use push rods or valve lifters, Figs. 23 and 24.

The valve is opened by a cam on the camshaft and closed by the valve spring.

The camshaft is rotated by the crankshaft by gears, Fig. 31, a link type chain, Fig. 32, or by a belt, Fig. 33. In either case the camshaft rotates at

Fig. 32 *Camshaft driven by silent chain*

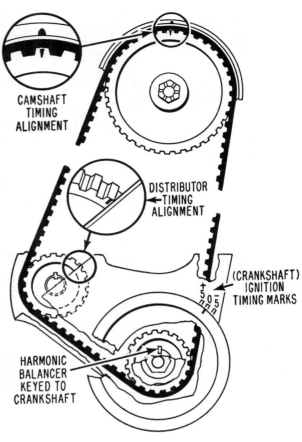

CAMSHAFT TIMING ALIGNMENT

DISTRIBUTOR TIMING ALIGNMENT

(CRANKSHAFT) IGNITION TIMING MARKS

HARMONIC BALANCER KEYED TO CRANKSHAFT

Fig. 33 *Camshaft driven by a timing belt*

When installing timing chain, marks must line up as shown

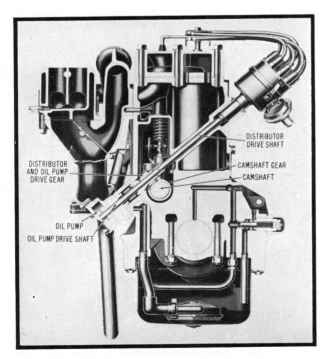

DISTRIBUTOR DRIVE SHAFT

DISTRIBUTOR AND OIL PUMP DRIVE GEAR

CAMSHAFT GEAR

CAMSHAFT

OIL PUMP

OIL PUMP DRIVE SHAFT

Fig. 34 *Distributor and oil pump are driven by a gear which is rotated by the camshaft*

half the speed of the crankshaft. This 2 to 1 speed ratio between crankshaft and camshaft is obviously necessary because the crankshaft produces two piston strokes per crankshaft revolution. It should be remembered that four piston strokes, two down and two up, are required to complete the combustion cycle. Therefore, each valve must open and close once for each two revolutions of the crankshaft. This means that the camshaft must run at half the speed of the crankshaft.

The camshaft drives the oil pump, distributor and fuel pump. In most cases the distributor and oil pump are driven by the same shaft, Fig. 34. The fuel pump is driven by an eccentric (or cam) on the camshaft.

The power of the engine depends on the free flow of gases into the cylinders and out through the exhaust system. The greater the weight of air pumped into the cylinders and the less the exhaust back pressure the greater the power developed. Power increases with weight of air. Free flow of mixture into the cylinder depends on the

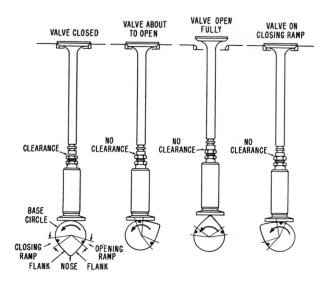

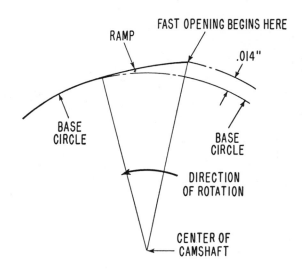

Fig. 35 *Modern cam with ramp. A typical ramp has a gentle slope of only about .0004″ per degree for a distance of about 35 degrees, giving a total ramp lift of .014″. Black dots on the cams indicate points at which the valve starts to open and close. If valve clearance is .010″, the first .010″ of ramp rise moves the tappet into contact with the stem and then the remaining .004″ of ramp rise opens the valve this amount. From this point the opening of the valve is extremely rapid*

Fig. 36 *Diagram of an opening ramp. The steepness of the ramp is exaggerated so that it can be seen clearly. The "Ramp" arrow shows the point at which the valve clearance has been taken up and the valve starts to open*

size of the opening the intake valve provides—the larger its diameter and the higher its lift, the larger the opening. The same is true of the exhaust valve.

However, there is a practical limit to valve diameters and lift. The diameter is limited because the cylinders are crowded together so that the engine won't be too long or heavy. Space and wear considerations also limit the lift. Therefore the lift is usually about ⅓″ and the valve diameter is, for a typical 240 cubic inch engine, 1½ to 1⅝ inches. Sometimes both valves are made the same diameter but on most engines the diameter of the intake is made larger than the exhaust valve because a large intake valve is more important than a large exhaust valve from a power standpoint. Therefore, on the typical engine just mentioned the intake valve is likely to be 1⅝ inches in diameter and the exhaust valve 1½ inches diameter.

Valve seat angle is usually 45 degrees although on some engines it is 30 degrees (or a combination of both angles). The 30-degree seat gives a larger valve opening for a given lift, and therefore

more breathing capacity and more power. On the other hand the 45-degree design gives a better wedging and centering action between valve face and seat and therefore provides a more leak-proof valve closing.

VALVE TAPPETS OR LIFTERS

It is not desirable to allow the cam to act directly on the lower end of the valve stem and therefore a lifter (tappet) is inserted between cam and valve stem. The slim valve stem is incapable of taking the side thrust of the cam without undue wear of stem and stem guide. Also the area of the bottom of the stem is too small to properly resist the wear of the cam which rubs against it. Finally, a means of adjustment is required between cam and valve stem on L-head engines and this adjustment is most conveniently located in the top end of the tappet (unless hydraulic lifters are used).

The adjustment is required to provide slight-clearance between tappet and stem so that the valve can close tightly even when the stem is very hot and therefore fully expanded or lengthened. Clearance for a warm engine may range from approximately .006″ to .022″ or more, depending on the engine design.

On most L-head engines the adjustment consists of a screw and lock nut in the top of the tappet, Fig. 29. In this case there are opposite flat sides on the lifter so that it can be held against

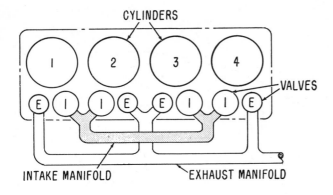

Fig. 37 *Arrangement of intake and exhaust valves on a 4-cylinder engine*

Fig. 38 *Examples of valve springs. Coils are unevenly spaced to suppress spring surge or shimmy. Conical shape of spring at right is designed to eliminate surge entirely*

rotation while the screw is adjusted and the lock nut is tightened. Three wrenches are required for this job.

On some L-head engines, a self-locking screw is used thus eliminating the lock nut. In this case the tappet is held with one wrench until the desired clearance is obtained.

On overhead valve engines, the adjustment is on the rocker arm, Fig. 30. The adjustment is made by turning the screw after loosening the lock nut.

CAMS

The cam which raises the valve tappet is a very carefully designed part, Fig. 35, which is accurately finished. Part of it is circular. This portion is called the base circle or heel. The pointed part of the cam which opens the valve is called the lobe while the tip of the cam is often referred to as the nose. The valve is fully open when the nose is in contact with the center of the tappet.

Modern cams, Fig. 35, are divided into five parts: (1) base circle, (2) opening ramp, (3) opening flank, (4) closing flank, (5) closing ramp. An opening ramp is diagrammed in Fig. 36. The closing ramp is similar.

The speed at which the flank of a cam opens a valve is very rapid. To soften the blow the cam is provided with an opening ramp. Without a ramp the flank of the cam starts to bump the valve open at an almost incredible rate. In the case of an average engine running 4,000 rpm, the flank of the cam opens the valve in approximately 1/100 second. Closing of the valve requires about the same amount of time. Compared with the flank of the cam, the opening ramp has a very gentle slope which raises the lifter only about .014″ while the camshaft rotates about one-tenth of a revolution.

The rise of the ramp is much less with hydraulic valve tappets because valve clearance is zero. This comparatively slow motion eases the lifter gently up against the valve stem and then eases the valve open. After this happens, the flank of the cam starts pushing the lifter at an extremely rapid rate. Obviously the ramp prevents undue wear of the valve mechanism. However, the principal purpose of the ramps is to reduce valve noise.

The main purpose of the closing ramp is to reduce the speed at which the valve hits the seat, thereby reducing valve noise as well as wear of valve face and seat caused by the hammering action. Cocking of the valve stem in its guide causes the valve head to hammer the seat at one point, tending to increase noise and wear. The closing ramp largely nullifies these troubles.

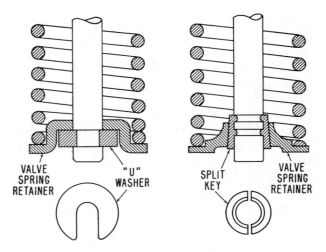

Fig. 39 *The valve spring retainer is locked to the valve stem by a split key at right and a U-shaped washer at left*

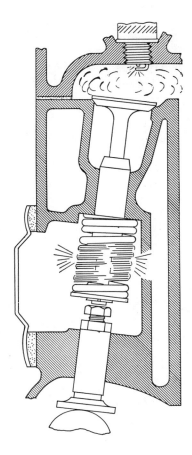

Fig. 40 *Valve spring*
surge or shimmy

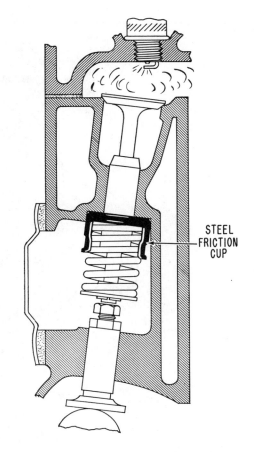

STEEL
FRICTION
CUP

Fig. 41 *Spring damper to*
prevent surge or shimmy

LOCATING VALVES

The arrangement of the intake and exhaust valves varies on different engines. On all engines the valve arrangement can be determined by an inspection of the branches of the intake and exhaust manifolds. It should be obvious that it is necessary to be able to distinguish between intake and exhaust valves in connection with various engine repair operations such as adjusting valve operating clearances.

Fig. 37 illustrates the arrangement of intake and exhaust valves on a typical four-cylinder engine and on both banks of some popular V8's as well.

E-I-I-E-E-I-I-E

Among the arrangements used on six-cylinder engines, there is the one used on Chevrolet, Dodge and Plymouth which is:

E-I-I-E-E-I-I-E-E-I-I-E

The valve arrangement on some Ford Co. and late Rambler six-cylinder engines is:

E-I-I-E-I-E-E-I-E-I-I-E

Late Ford six-cylinder arrangement is:

E-I-E-I-E-I-E-I-E-I-E-I

Early Rambler six-cylinder engines use:

I-E-E-I-I-E-E-I-I-E-E-I

VALVE SPRINGS

Valve springs are made of round steel wire about ⅛ inch in diameter. The wire is wound as shown in Fig. 38. The spring is ground flat on both ends so that it seats squarely. On a typical engine, the free or uncompressed length of the spring is 2-3/32". When the spring is installed and the valve is closed, the length is 1¾" and the spring pressure is 40 pounds for a typical L-head engine. When the valve is open, the spring length is 1-7/16" and the pressure is 83 pounds. The

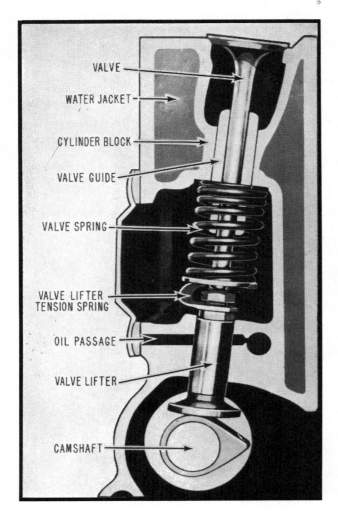

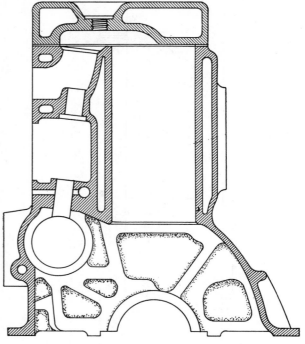

Fig. 43 Unsymmetrical engine causes distortion

Fig. 42   U-shaped flat spring holds lifter against heel of cam

open valve obviously has been lifted 5/16″ (1¾ minus 1-7/16).

On a comparable overhead valve engine, the free length of the spring is 2-3/32″ as before. The installed length of the spring when the valve is closed is 1-11/16″ and the pressure exerted is 53 pounds. When the valve is fully open the spring length is 1-11/32″ and the pressure is 146 pounds. The lift of this valve is 11/32″.

After many miles of use, valve springs may weaken, in which case valves and tappets may not keep in contact with their cams. This means that (1) a valve may bounce open farther than the lift it was designed for and (2) it may not close promptly. In other words, weak springs may completely upset valve action and thereby reduce the power production of the engine. Therefore it is important to check the strength of valve springs whenever a valve grinding job is done or when the valves are removed for any other reason.

In an L-head engine, the upper end of the valve spring, Fig. 29, presses against the cylinder block while the lower end of the spring pushes against the valve spring retainer which is attached to the lower end of the valve spring. In an overhead valve engine, Fig. 30, the lower end of the spring presses against the cylinder head while the upper end presses against the spring retainer attached to the valve stem. On most engines the retainer is held in place by a tapered split key although a U-shaped washer will be found on some of the older cars, Fig. 39.

A valve spring may break for two reasons: it may be overstressed or it may have a tiny crack in its surface. Under use, the crack is gradually enlarged until the spring breaks. This is called fatigue.

Overstressing of a spring is usually caused by an up-and-down vibration in the spring, Fig. 40, which occurs at certain engine speeds. This phenomenon is called spring surge or shimmy. There are various ways of preventing or suppressing spring shimmy. One method is to provide a frictional vibration damper, Fig. 41, which prevents the spring from going into a dance. Another method is to use a spring which is so stiff that it will not vibrate at engine operating speeds, say up to 4,000 rpm. A third method is to use a spring

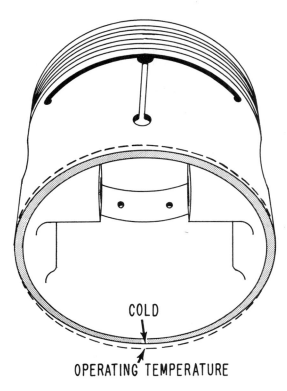

COLD

OPERATING TEMPERATURE

Fig. 44 Aluminum alloy piston with T-slot and cam-ground (oval) skirt

with unevenly spaced coils, Fig. 38. Each coil wants to vibrate at a different engine speed with the result that the spring as a whole is unable to vibrate at any engine speed. A fourth cure is to employ two concentric springs. The vibration rates of the two springs are different at all engine speeds. Therefore if one spring tries to get into a dance at some certain engine speed, its act is discouraged by the other spring.

Suppression or elimination of valve spring surge is most important for two reasons: (1), The vibrating spring is overstressed and therefore may break; (2) spring shimmy interferes with correct valve action in about the same way that a weak spring does. In other words, an upsurge at the right instant may permit the valve to open too far (while the lifter bounces away from the cam) or if an upsurge should occur while the valve is closing, the closing may be delayed with resulting loss of power. The maximum speed of some engines is limited by valve spring surge. In other words, at some certain high speed the springs go into a dance which completely upsets valve action and then power production is sharply reduced.

On some Studebaker engines, a U-shaped flat spring, Fig. 42, is placed between the valve spring retainer and the valve tappet. Its purpose is to hold the tappet against the cam when the valve is just closed and thus avoid the possibility of a slight noise due to the lifter clattering on the heel of the cam.

STRUT

STRUT

Fig. 45 Aluminum alloy piston with invar steel struts. The picture at right shows the piston cut in two vertically. The picture at left shows the piston cut in two horizontally at the center of the piston pin boss. The invar struts are placed in the piston mould and the molten aluminum alloy poured in, thus firmly fastening the struts to the piston skirt

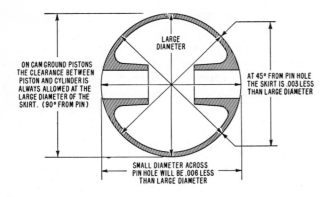

Fig. 46 Piston with oval skirt

ON CAM GROUND PISTONS THE CLEARANCE BETWEEN PISTON AND CYLINDER IS ALWAYS ALLOWED AT THE LARGE DIAMETER OF THE SKIRT. (90° FROM PIN)

LARGE DIAMETER

AT 45° FROM PIN HOLE THE SKIRT IS .003 LESS THAN LARGE DIAMETER

SMALL DIAMETER ACROSS PIN HOLE WILL BE .006 LESS THAN LARGE DIAMETER

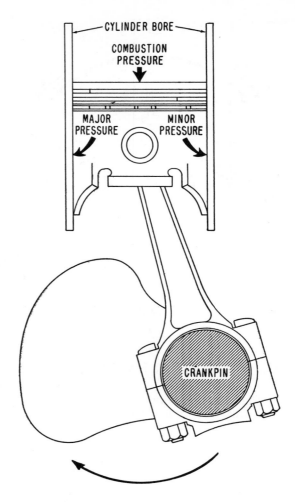

CYLINDER BORE
COMBUSTION PRESSURE
MAJOR PRESSURE
MINOR PRESSURE
CRANKPIN

Fig. 47 How piston presses against bore (viewed from front)

CYLINDER BORE DISTORTION

If someone were able to make an ideal engine it would consist of perfectly round and straight cylinder bores and perfectly round and straight pistons which slide up and down in these bores. Unfortunately this ideal can only be approximately realized.

When the cylinder block is manufactured the bores are round and straight within much less than a thousandth of an inch. But bolting the cylinder head in place distorts the straightness of the bore and may even pull it a little out of round. To keep this distortion to a minimum, car manufacturers specify just how tight the cylinder head bolts should be. Bolts too tight will cause undue distortion. On the other hand, if bolts are too loose, leakage will occur at the gasket. In order to tighten the bolts just the proper amount it is desirable to use a torque wrench. This wrench indicates the pull in pounds that are applied to the wrench. Each car maker issues instructions as to how much the pull should be.

The cylinder bores are machined at room temperature but in an actual engine the temperature is a few hundred degrees higher. Metals expand when they are heated. If an engine had bores which were not attached to the block, no appreciable distortion would occur when they were heated but unfortunately the bores are usually an integral part of the cylinder block which is an unsymmetrical piece of metal which includes water jackets, valve chambers and upper half of crankcase, Fig. 43. These parts expand in different ways with the result that the bores are distorted when heated.

PISTONS

The piston likewise is straight when it is manufactured (at room temperature). Whether it is made round or slightly oval depends on its design. An oval piston when cold may be practically round when hot.

Neither the cylinder bore nor the piston skirt is perfectly stiff which means that both the gas pressures and the inertia pressures acting on the piston slightly distort the wearing surfaces of both bore and piston skirt.

A small amount of clearance between piston and cylinder bore must be allowed for the oil film which must coat both surfaces in order to suppress wear. Some additional clearance between piston and cylinder bore is required when the engine is cold because the piston expands more rapidly than the cylinder bore as the engine warms up because the piston is hotter than the bore when the engine is operating.

Although cylinder bore distortion and piston skirt distortion only amount to a few thousandths

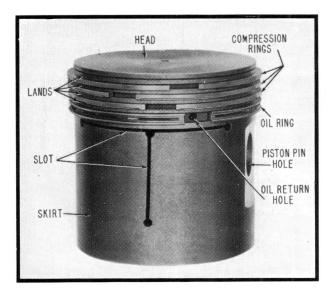

Fig. 48 *Piston with four rings*

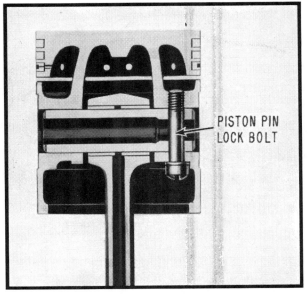

Fig. 49 *Piston pin locked to piston*

of an inch, the result is important when it is remembered that a piston in a typical engine operating at a speed of 4,000 rpm travels at an average speed of half a mile per minute.

Aluminum Pistons

Ordinary aluminum alloy pistons expand at about twice the rate of the cast iron bore. Therefore, special provisions are made in aluminum alloy piston designs to control piston expansion. There is, however, a special silicon-aluminum alloy with a low expansion rate used in some pistons.

All other aluminum alloy pistons have special design features so that the clearance when cold is not enough to permit objectionable slap and yet when the piston becomes hot there is still enough clearance so there is no danger of the piston sticking in the bore.

Numerous types of aluminum alloy pistons have been introduced, all designed to take care of the expansion problem in one way or another. However, it would serve no useful purpose to describe all the various designs here. Instead, the two types in common use will be discussed, namely: The T-slot piston with "cam-ground" skirt, Fig. 44 and the invar strut piston, Fig. 45. Invar is a steel alloy containing about 36 per cent of nickel. The expansion rate of invar is practically zero. Both pistons are slotted horizontally between the head and skirt in order to provide a heat dam to reduce the amount of heat flowing from piston head to skirt and thereby reduce skirt expansion.

Returning to the T-slot design, Fig. 44, it will be seen that the T-slot is formed by a horizontal slot between the piston head and piston skirt which connects with a nearly vertical slot in the piston skirt. The opposite side of the piston has a horizontal slot but no vertical slot. The horizontal slots act as heat dams to prevent undue heat from flowing from piston head to skirt. The nearly vertical slot provides enough flexibility in the skirt to discourage the possibility of the skirt sticking fast in the bore. The skirt is relieved (under cut) in the region of the piston pin hole so that any undue expansion in this region will not cause the piston to seize in the bore. Too much flexibility in the skirt, however, is undesirable and it is for this reason that the slot is not continued all the way down through the lower edge of the skirt. The slot length shown gives the desired flexibility. Too much flexibility would permit rocking of the piston in the bore and would in time slightly round off the outer surfaces of the piston rings and thus render them less effective against leakage of both gas pressure and oil.

If the slot in the skirt were vertical it would gradually make a ridge in the bore. To avoid this the slot is slightly diagonal.

The skirt of the piston is ground slightly elliptical, Fig. 46. In a typical case, the piston diameter is .006″ less when measured across the piston pin holes than when measured across the "large" diameter. The elliptical piston is called a cam-ground piston because the grinding wheel position is controlled by an elliptical cam. Note that with a

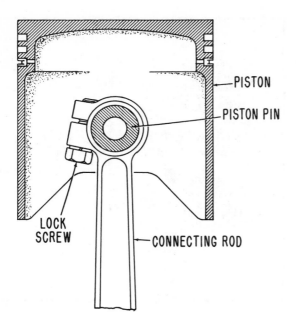

Fig. 50 *Connecting rod is locked to piston*

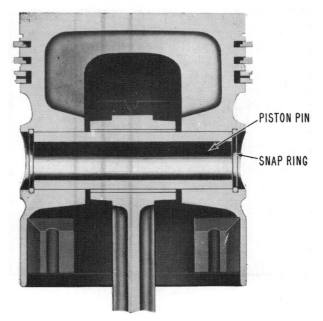

Fig. 51 *Piston pin is locked to neither piston nor connecting rod. In other words the pin is said to float*

cam-ground piston the front and rear surfaces may not contact the bore at all. The terms front and rear mean front and rear with respect to the engine block.

The skirt of the cam-ground T-slotted piston may be fitted to the cylinder bore with close clearance in a cold engine. When the engine becomes warm and the piston expands, the skirt (because of the T-slot) tends to shape itself to the cylinder bore without any danger of the piston sticking in the bore.

Piston Side Thrust

When a piston moves down on its power stroke, the angularity of the connecting rod causes it to be pressed against the cylinder bore surface which is opposite the crankpin, Fig. 47. This side of the piston is often called the major pressure side or major thrust side while the other side is called the minor pressure side. Many car makers recommend that the T-slot be located on the camshaft side, thus enabling the stiffer major pressure side of the piston to resist the side thrust caused by combustion pressure.

However, at high speed the inertia forces acting on the piston are greater than those caused by gas pressure and they act alternately on both sides of the piston as it moves up and down. As the piston rushes down it presses against one side of the bore and as it rushes upward it presses against the other. Therefore both sides of the piston are sub-

jected to the same inertia side thrusts during each revolution of the engine. Gas pressures and inertia pressures should be considered separately.

At low speed, when inertia forces are low, gas pressure is high when the throttle is open. Therefore, some engine manufacturers place the T-slot on the major pressure side because the resilient T-slotted skirt cushions tendency to piston slap.

To sum up, the T-slotted aluminum cam ground piston is able to work successfully in a cast iron cylinder bore for three reasons, namely: (1) Flexible skirt obtained by (nearly) vertical slot. (2) Cam grinding (oval) which enables the piston to fit itself to the bore automatically. (3) Heat dam which prevents most of the heat on the piston head from reaching the skirt.

The skirt is flexible circumferentially. Longitudinally the skirt is rigid enough so that it does not wobble in the bore; also the skirt has sufficient lengthwise rigidity so the outer faces of the rings move straight up and down in the bore. By contrast, a skirt that is flexible lengthwise would cause the outer faces of the rings to wear round and thus reduce their effectiveness.

Coated Pistons

On some engines the piston skirt is coated with some special material in order to prevent scuffing (scratching) of the skirt and bore. In addition some of these coatings are porous and therefore absorb oil which assists in lubricating both the skirt and the bore, thereby reducing wear. The coating may be a thin film of tin or cadmium plating. Tin and cadmium are soft and ductile metals and therefore are unable to scuff the bore.

Chemical coatings are used on many aluminum pistons. The coating is chemically attached to the aluminum. It is soft, brittle and porous. Because it is soft and brittle, scuffing of the bore is avoided. Being porous, it permits the coating to soak up oil which aids lubrication.

These special coatings for both cast iron and aluminum pistons practically prevent scuffing under the following three conditions:

(1) When the engine is first started from cold before the oil is warm enough to reach the cylinder bores and piston skirts.

(2) When the engine is new at which time the coatings are almost a perfect guarantee against scuffing and piston sticking.

(3) When the engine is operated under extremely high temperature conditions caused by continuous fast running, defects in cooling and lubrication system, and excessively high air temperature, the possibility of scuffing or piston sticking will either be tremendously reduced or entirely eliminated by the coating.

When the engine is new, the piston should move up and down exactly parallel to the cylinder bore. The clearance between skirt and bore is small enough so that there is no appreciable rocking of the piston in the bore.

A straight up and down motion of the piston enables the outer surfaces of the piston rings to press effectively against the bore and also permits them to function exactly the way they were designed to do. On the other hand, after excessive wear of skirts and bore, the rocking of the piston will round the outer faces of the piston rings and therefore reduce their ability to prevent gas leakage as well as their ability to control oil consumption.

PISTON RINGS

Piston rings are circular springs, Fig. 48, which are cut in two at one point so that they may be slipped over the piston. They are made somewhat larger than the cylinder bore so that when installed on the piston and inserted in the bore they will press against the bore. With the piston installed in the bore, the gap between the two ends of the ring range upwards of .007″ depending on the engine. As to the actual gap for a given engine, car maker's instructions should be followed. The purpose of the gap is to prevent the two ends of the ring from butting and therefore binding when the ring expands when the engine is hot. Gaps on adjoining rings should be half a circumference apart in order to minimize leakage from one ring gap to another. In Fig. 48, the gaps of all four rings are shown close together for purposes of illustration. Actually, the gaps of any two adjacent rings should be staggered one-half a circumference apart.

At low engine speed, the rings are able to follow the slightly wavy surface of the bore exactly but at high speeds they may leap and bounce, hitting the high spots but skipping the low ones. This action results in extra leakage of gas pressure from the combustion chamber to the crankcase, and also it results in extra oil flowing from the crankcase to the combustion chamber where it is burned and therefore wasted.

Automobile pistons ordinarily are equipped with three or four rings. Only rarely has a piston with five rings been used.

With a three-ring piston, the upper two rings are intended particularly to seal the piston against leakage of gas pressure while the lower ring is designed to prevent too much oil from passing into the combustion chamber. The upper two rings are called compression rings; the lower ring is called an oil control ring or simply an oil ring. If the piston has four rings, the two upper ones are usually compression rings and the two lower for oil control. However, the piston shown in Fig. 48 has three compression rings and one oil ring.

Oil control rings are slotted. Behind the slots are drain holes in the piston skirt so that excess oil scraped from the cylinder bore by the ring is returned to the crankcase. If the slots in the oil rings become filled with carbon, gum or sludge, the oil consumption may be excessive because the extra oil scraped off the cylinder walls cannot drain back through the piston skirt into the crankcase. The same remark holds true concerning clogging of the drain holes in the bottom of the piston ring grooves.

PISTON PINS

There are four ways that the connecting rod may be attached to the piston, as follows: (1) The pin may be locked to one of the piston bosses, Fig. 49, in which case the rod oscillates on the pin. (2) The rod may be locked to the pin which oscillates in the bosses, Fig. 50. (3) The pin may be free to oscillate within its bosses and within the rod, Fig. 51. In this case the pin is prevented from touching

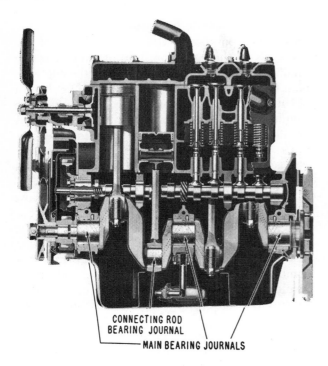

Fig. 52 *Four-cylinder engine with three main bearings*

piston. Due to its pressed fit the necessity of the snap rings shown in Fig. 51 is eliminated.

CRANKSHAFT & MAIN BEARINGS

Crankshafts are supported by two or more plain bearings. The term "plain" simply means that the bearing has a smooth surface of suitable material. (Ball or roller bearings are not used on crankshafts on modern American automobiles). Fig. 52 illustrates a crankshaft for a four-cylinder engine supported in three plain bearings. These bearings are called main bearings in order to distinguish them from the plain bearings in the lower ends of the connecting rods (called connecting-rod bearings or crankpin bearings).

Those portions of the crankshaft which are enclosed by main or rod bearings are called main bearing journals or connecting-rod journals, respectively, Fig. 53.

Nowadays all automobile engines use extremely thin shell main and rod bearings, Fig. 54. These bearings are called precision insert bearings. Their thickness varies but may be as little as .050″. The back of the bearing is made of a strip of steel which is coated with a bearing alloy (such as babbitt) which ranges in thickness from .002″ to .006″ depending on the bearing. The strip is then cut to length and pressed into semi-circular form.

Ideally, a bearing never touches its journal because the two are separated by an oil film. However, in actual practice they do contact each other

(and scuffing or scoring) the cylinder bore by the use of snap rings at its ends. This is called a floating pin. (4) The pin may be a tight press fit in the connecting rod and is free to oscillate in the

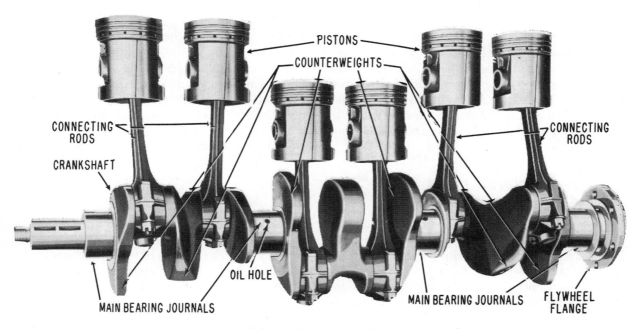

Fig. 53 *Crankshaft showing main bearing journals*

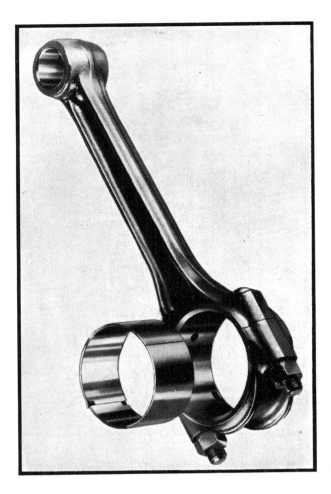

Fig. 54 *Connecting rod with detachable bearing shell*

be scratched (scuffed) and eventually ruined. In extreme cases, the friction of the journal against the bearing surface may melt the bearing alloy which means that the bearing is probably ruined.

A thin shell bearing, consisting of a thin coating of bearing metal on a thin steel back, is generally considered superior because the relatively ductile (plastic) bearing metal has a firm steel foundation immediately beneath it. In short, the extremely thin bearing surface provides adequate softness and ductility combined with a firm foundation.

The thin shell bearing also conducts heat to the cap more readily because the heat path is thin (short). Bearings can be ruined if they become too hot, hence the importance of maximum heat conductivity. Circulation of oil through the bearing helps keep it cool.

Each half of a shell type bearing is usually provided with a locking lip or tang at one end, Fig. 56. The lip fits into a mating notch in the connecting rod and thus aligns the bearing. The same construction is used for main bearings.

For main and rod bearings, there are two alloys in common use in addition to babbitt, namely copper-lead and cadmium. In recent years aluminum faced bearings have been used in some engines. In all cases the bearing metal is bonded to a steel back. The copper-lead alloy contains 60% copper and 25% lead or even higher percentages of one or the other metal. Cadmium bearings usually contain more than 98% of cadmium.

Bronze is an alloy which is principally composed of copper and tin. Bronze bushings are used for piston pins.

under favorable conditions. In a typical automobile engine the clearance between journal and bearing may range from .0005″ to .003″ depending on the engine design. The clearance just mentioned is the total clearance between journal and bearing which means that it is measured when the journal is eccentric with respect to its bearing, Fig. 55. On the other hand, when the engine is assembled and is in running condition, the journal (theoretically) is concentric with its bearing. This means that the clearances just mentioned are halved. The "running" clearance is occupied by oil.

The bearing material is always softer than the steel crankshaft journals. Therefore, if, because of high pressures between journals and bearing or thin oil, the crankshaft journal breaks through the oil film and presses against the bearing surface, the surface can flow slightly in order to accommodate itself to the crankshaft journal. However, if the journal rubs against the bearing surface either too hard and/or too long the bearing surface will

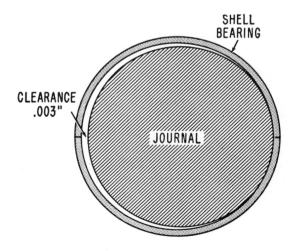

Fig. 55 *Clearance between journal and bearing ranges from .0005″ to .003″ depending on the engine design*

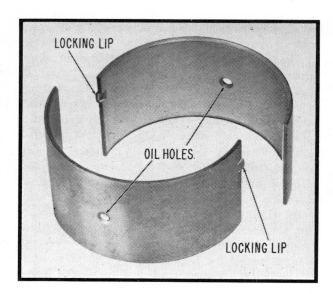

Fig. 56 *Locking lip with thin shell bearing*

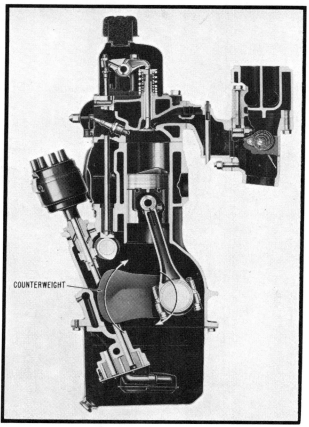

Fig. 57 *Typical crank-shaft counterweight*

Camshaft bearings are made of a variety of alloys such as bronze, bronze-backed babbitt-lined, steel-backed babbitt-lined. These bearings may be made in one piece like a bushing or they may be made of the precision type in which case they are split in two halves.

On most engines in use today there are three main bearings for four cylinders, seven main bearings for in-line sixes, and five main bearings for V8's.

CRANKSHAFT COUNTERWEIGHTS

Most modern crankshafts are equipped with counterweights, Fig. 57. These weights are an integral part of the crankshaft. Their principal purpose is to counterbalance the weight of the crankpins and their cheeks. In other words, the centrifugal force of the latter is counterbalanced by the former.

In further explanation it should be noted that in spite of the fact that a crankshaft is a relatively stiff or rigid piece of steel two or three inches in diameter, nevertheless the weight of the crankpins and their cheeks (or throws) produces such a high centrifugal force at high speeds that a crankshaft without counterweights might be bent sufficiently to cause excessive wear of the main bearings. Therefore the purpose of the counterweights is to counteract this bending. Also the elimination of bending makes a smoother and quieter running engine because a crankshaft which bends in its bearings also bends the crankcase, thus increasing noise and vibration.

VIBRATION DAMPERS

Crankshafts are subject to torsional vibration. This vibration is a very rapid circumferential winding and unwinding of the front end of the crankshaft. Torsional vibration occurs at certain definite speeds which vary with different crankshafts. In a typical case, objectionable torsional vibration might occur at 1200, 1600 and 2400 rpm, the intensity being approximately twice as much at 1600 as at 1200; likewise twice as much at 2400 as at 1600.

Torsional vibration is caused by the steadily recurring impulses of the pistons which cause the front end of the crankshaft to wind and unwind at a very rapid rate. Torsional vibration is suppressed by a vibration damper (also called harmonic balancer). The damper consists of a small flywheel installed on the front end of the crankshaft, usually rotated by the crankshaft through rubber bushings, Fig. 58. As soon as the front end of the crankshaft starts vibrating, the flywheel in its rubber bushings starts vibrating oppositely, thus bucking and discouraging the build up of vibration in the crankshaft. Because of the energetic opposition of this damper flywheel the tor-

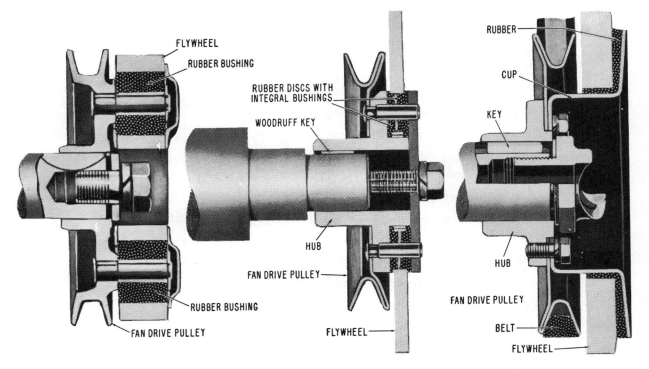

Fig. 58 *Three types of vibration dampers. Left—Ample rubber bushings drive a small, compact flywheel. Center—The vibration damper is a steel disc driven by rubber bushings in two rubber discs. Right—A ring-shaped flywheel is bonded by rubber to a cup-shaped flange attached to the fan pulley. The rubber layer acts in the same manner as the rubber bushings*

sional vibrations in the crankshaft never become powerful enough to be noticed. The damper is usually incorporated in the crankshaft pulley.

Torsional vibration is considered negligible in American four-cylinder engines as well as in some V8's, in which case no vibration damper is required, nor is one always needed in small six-cylinder engines with short crankshafts. But larger in-line six and eight cylinder engines invariably have vibration dampers.

FLYWHEEL

The flywheel is slightly affected by torsional vibration, fluctuating slightly but rapidly back and forth while the crankshaft rotates. This vibration may produce an objectionable screech somewhere between the clutch driven disc and the rear wheels. In order to nullify the effect of this vibration it is customary to cushion the hub of the clutch disc with springs, Fig. 59. The clutch disc drives the hub through these springs. At the same time, friction material clamped between the disc and the hub prevents the hub from picking up the torsional vibration of the flywheel. This type of hub is used on practically all cars with manual shift transmissions.

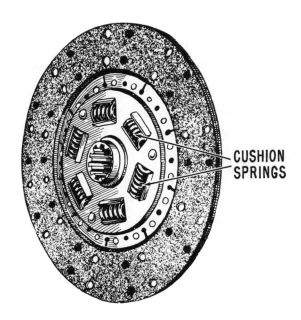

Fig. 59 *Vibration damper in clutch disc*

Review Questions

		Page
1.	Name the four strokes comprising the operating cycle of an automobile engine.	5
2.	When does a four-stroke cycle engine operate as an air pump?	5
3.	Why does a small amount of exhaust gas always remain in the combustion chamber after the exhaust stroke is completed?	6
4.	How is power in an automobile engine produced?	6
5.	How does the speed of the combustion process in an automobile engine compare with the speed of exploding dynamite?	6
6.	What are the basic ingredients of the combustible mixture in an automobile engine?	6
7.	Name the two parts that form the combustion chamber.	7
8.	How is steam formed during the combustion process?	7
9.	The mixture of what two elements produces carbon dioxide?	7
10.	How is carbon monoxide formed?	7
11.	Why is it incorrect to say that the mixture is sucked into the engine?	7
12.	Explain the purpose of the manifold heater.	8
13.	On cars with dual exhaust systems, explain why the exhaust gas comes out of only one muffler tail pipe when the engine is cold?	8
14.	What is valve overlap?	9
15.	What are the advantages of delaying the closing of intake valves?	10
16.	Early intake valve closing is favorable to power at what speeds?	10
17.	Later intake valve closing assists power production at what speeds?	10
18.	What is the piston displacement of a six-cylinder engine having a bore and stroke of 3″ x 4″?	10
19.	What sort of horsepower is brake horsepower? Taxable horsepower?	10
20.	Define compression ratio.	11
21.	What causes detonation? Spark knock? Pre-ignition?	11, 12
22.	How does the cooling system control the temperature of pistons, rings and valves?	13
23.	How does the Corvair cylinder numbering system differ from those of other American engines?	14
24.	Define the meaning of "Firing Order."	15
25.	Explain how piston travel is changed to rotary motion at the crankshaft.	16
26.	Why are no separate main bearing caps required on the Corvair engine?	16
27.	Why is an L-head engine so called?	18
28.	Where are the valves located in an F-head engine?	20
29.	What is the function of a cylinder head gasket?	20
30.	How do barrel type and mushroom type tappets differ?	21

31. What is the advantage of a 30° valve seat; 45° valve seat? 23

32. In a four-cycle engine, why does the camshaft operate at one half the crankshaft speed? 21

33. Why is it desirable to make valves with head diameters as large as possible in an engine? 22

34. Name the five different surfaces on a cam. 23

35. Why is it necessary to suppress valve surge? 26

36. How is cylinder head and block distortion kept to a minimum when cylinder head bolts are tightened? 28

37. Why is it necessary to have a little clearance between piston and cylinder bore? 28

38. Name two common types of aluminum pistons. 29, 30

39. Name three reasons why a cam-ground (oval) piston works successfully. 30

40. What are the advantages of coated pistons? 30

41. Why should piston ring gaps be staggered around the piston? 31

42. Why is it important to have the drain holes in piston ring grooves clean and open? 31

43. Describe four ways a connecting rod may be attached to a piston. 31

44. What is the difference between a main bearing and a main bearing journal? 32

45. Why is it necessary to allow for a running clearance when fitting main or rod bearings? 33

46. How many main bearings would you expect to find in a modern V8 engine? 34

47. What would happen to the main bearings in an engine without counterweights on the crankshaft? 34

48. Describe torsional vibration in a crankshaft. 34

49. Explain how a vibration damper opposes torsional vibration? 34

50. What would result if there were no cushion springs in a clutch driven plate? 35

Engine Valve System

Review Questions for This Chapter on Page 115

INDEX

INTRODUCTION

Valve system parts 38
Valve operating clearance 41
Valve operating conditions 40

CAUSES OF VALVE TROUBLES

Hot valves 45
Hot valve seats 44
Sluggish valve closing 48
Valve leakage 42
Valve temperature 44

CAMSHAFT AND TIMING GEAR SERVICE

Bearings, camshaft 96
Camshaft, replace 105
Camshaft bearings 111
Camshaft gear 98
Camshaft sprocket 98
Camshaft thrust plate 96
Chain, timing, and sprockets .. 103
Cover, timing case 99
Crankshaft gear 99
Crankshaft sprocket 99
Gear, camshaft 98
Gear, crankshaft 99
Gears, timing, replace 101
Plate, camshaft thrust 96
Sprocket, camshaft 98
Sprocket, crankshaft 99

Sprockets, timing chain ... 98, 103
Timing belt and sprocket 105
Timing case cover 99
Timing chain and sprockets ... 103
Timing gears, replace 101
Valve timing 93

VALVE SERVICE

Adjustment, valve lash 52
Arms, rocker 55
Checking compression 49
Cleaning carbon 71
Clearance, valve stem-to-guide 73
Compression, checking 49
Compression gauge, use of 50
Compression, other methods of
   checking 51
Corvair cylinder heads 62
Cover, rocker arm 58
Cylinder heads, Corvair 62
Cylinder head, install 60
Cylinder head, Jeep overhead
   camshaft engine 64
Cylinder head, Pontiac Tempest
   overhead camshaft engine .. 64
Cylinder head, remove 59
Cylinder head, Volkswagen 64
Freeing stuck valve 51
Grinding, valve 80
Guide, replace valve 75
Hydraulic lash adjuster 90
Hydraulic valve lifters 84
Inserts, valve seat 83

Inspection, valve 72
Lash adjuster, hydraulic 90
L-head valve cover 58
L-head valves, install 68
L-head valves, remove 67
Lifters, hydraulic valve 84
Modifying valve gear for high
   performance 90
Overhead valves, install 70
Overhead valves, remove 69
Reaming valve guides 74
Rocker arm cover 58
Rocker arms 55
Spring, replace valve 78
Spring, testing valve 78
Sticking valves 51
Valve cover, L-head 58
Valve, freeing stuck 51
Valve grinding 80
valve guides, reaming 74
Valve guides, replace 75
Valve inspection 72
Valve lash adjustment 52
Valve lifters, hydraulic 84
Valves, L-head, install 68
Valves, L-head, remove 67
Valves, overhead, install 70
Valves, overhead, remove ..:.. 69
Valves, replace 66
Valve rotators 90
Valve seat inserts 83
Valve spring, replace 78
Valve spring testing 78
Valve, stem-to-guide clearance 73
Valves, sticking 51

Introduction

VALVE SYSTEM PARTS

The names of the various parts in the valve system will be found on the numerous illustrations in this chapter. The basic parts, however, are indicated in Figs. 1 to 5. A more detailed description follows:

Valve head is the top portion of the valve, sometimes called the valve crown.

Valve face is the beveled portion of the valve head which contacts the cylinder block when the valve is closed (L-head engines), or which contacts the cylinder head on overhead valve engines.

Margin is the narrow rim on the circumference between the valve face and the top of the valve, Fig. 2.

Valve seat is the beveled surface on the cylinder block or cylinder head which the valve face contacts when the valve is closed. On L-head engines the seat is in the block; on overhead valve engines the seat is in the cylinder head.

Valve stem is the round, rod-like piece attached to the valve head.

Valve stem guide is the part through which the

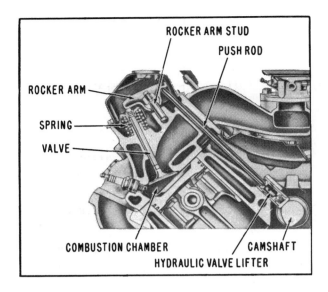

Fig. 1 *Pontiac overhead valve engine*

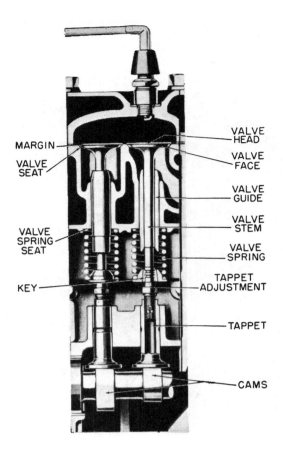

Fig. 2 *Plymouth L-head valve mechanism*

valve stem operates. These guides are usually a separate piece resembling a bushing and pressed in the guide hole. However, in some modern overhead valve engines the guides are machined holes bored directly in the cylinder head.

Valve stem clearance is the clearance between the valve stem and its guide. A certain amount of clearance is necessary between these parts to provide free running (without binding) when the engine is operating. This clearance is also called valve guide clearance or valve stem-to-guide clearance.

Valve operating clearance is the clearance between the tip of the valve stem and the valve lifter tappet. This clearance is necessary so that the valves will fully close when the engine is operating with the parts expanded due to heat. This clearance is also called tappet clearance, valve clearance, or valve lash.

Valve spring is used to pull the valves closed after they have been opened by the cams on the camshaft.

Valve spring retainer, as the name implies, retains the valve spring on the valve.

Valve lock or key is the device which keeps the valve spring and its retainer in place on the valve.

Valve chamber is the space in which the valve springs are located on L-head engines.

Valve cover on which the valve chamber is enclosed is called a valve cover plate, valve chamber cover or just valve cover.

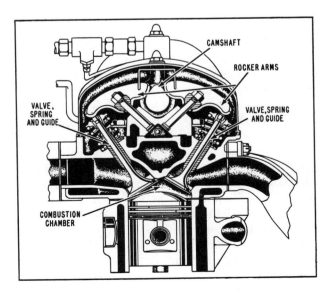

Fig. 3 *Jeep overhead camshaft engine*

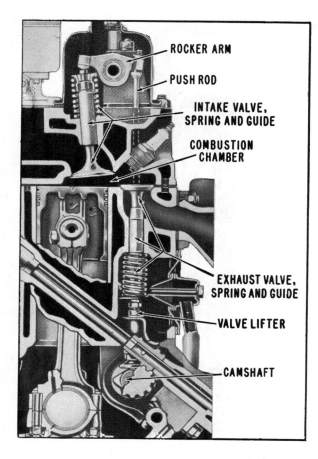

Fig. 4   Jeep F-head engine

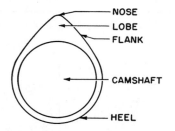

Fig. 5   Names of cam surfaces

Rocker arm cover is the cover which encloses the rocker arms on overhead valve engines.

Push rod cover is the cover which encloses the push rods.

Push rods are long rods which operate between the valve lifters and the rocker arms.

Valve lifters or tappets are the parts which transmit the motion of the cam to the valve stem on L-head engines, or to the push rod on overhead valve engines. These devices are properly called valve lifters because that is exactly what they do. The word tappet, although still used by some car manufacturers, originally meant the adjusting screw which screws into the top of the valve lifter. To be more precise, therefore, the body of the assembly is called the valve lifter while the screw which actually contacts the valve tip or push rod may be called the tappet screw. However, inasmuch as many modern engines have hydraulic valve lifters which require no screw for adjustment purposes the term valve lifter seems more appropriate.

Valve lifter guide is a machined hole bored in the cylinder block in which the valve lifter operates.

Mushroom type valve lifter has a base which is considerably larger than the guide hole in which it operates, Fig. 2.

Barrel type valve lifter is one which has the same diameter throughout its length.

Rocker arm, which is pivoted on a hollow rocker arm shaft, is operated on one side by the push rod to open the valve and on the other side by the valve spring which closes the valve. Some recent engines do not have rocker arm shafts; instead each rocker arm pivots on a stud, Fig. 1.

Cams on camshaft consist of two portions, Fig. 5, the base circle or heel and the lobe which juts out from the heel. It is the lobe which opens the valve. The lobe consists of a rounded nose and two flanks.

VALVE OPERATING CONDITIONS

In order to diagnose valve troubles intelligently and then cure them it is necessary to have a clear idea as to the conditions under which the valves work.

To begin with, it is most important to realize that in a typical engine the valves are open about one-third of the time and closed two-thirds of the time. Each valve opens and closes every two revolutions of the crankshaft.

Therefore, at 60 mph there are about 1500 openings and closings per mile or per minute, which means that each valve in a typical engine is open 1/75 second and closed 2/75 second. This gives an idea how fast the valves move.

But it is of equal importance to realize that these valves are closed 40 minutes in each hour of operation and open 20 minutes, regardless of what the car speed is, or how fast the engine runs. These thoughts should be particularly helpful in understanding the temperature conditions under which the valves operate.

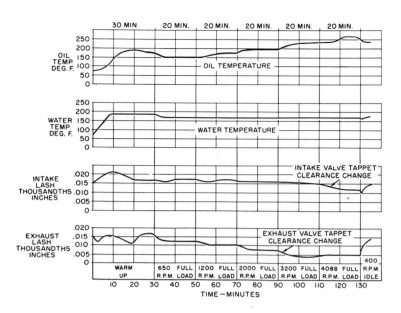

Fig. 6 Chart showing how valve clearance changed during a 130-minute test. The two upper curves show how water and oil temperatures varied during test

Impelled by strong springs, the valves, when they close, strike their seats with violent blows. The force of these blows increases with engine speed because the cams are rotating faster. When the engine is running fast, the collision of a valve with its seat may represent a force of several hundred pounds. The blows should be thought of as hammer blows which are no different in principle from hammer blows striking an anvil. The cam which operates the valve is provided with opening and closing ramps. The closing ramp slows down the final closing of the valve and reduces the force of the hammer blows but even so these blows are still sharp.

VALVE OPERATING CLEARANCE

In order to maintain properly cooled valves it is desirable that the valves be set according to the clearances recommended by the car factory. The more the clearance is reduced below the correct figure the longer the valve remains open and likewise the shorter the time that it rests on its cool seat. In explanation, it should be obvious that the less the clearance the sooner the valve will be opened and the later it will be closed. Thus an exhaust valve is exposed to hot exhaust for a longer period and similarly it sits on its cool seat for a shorter period. Both effects increase the temperature of the valve and may raise the temperature beyond the danger point especially if there are other operating conditions which increase the valve tmperature above normal.

Valve clearance grows less as the engine warms up. Just exactly how the clearance varies in a

typical engine is shown in a 130 minute test which was made by Thompson Products, Inc. Fig. 6 shows how intake and exhaust valve clearance varied during the test. The engine was allowed to idle for 20 minutes and then it was operated at full throttle for 20 minutes at each of the following speeds: 650, 1200, 2000, 3200, 4088 rpm. After 130 minutes the engine was allowed to idle again.

The valve clearance when the engine was cold was .015″ for both intake and exhaust. After the engine had idled for 10 minutes the vertical expansion of the block increased the clearance to .020″. From 20 to 100 minutes the intake clearance was just a trifle more than .015″. Exhaust valve clearance was normal after 10 minutes idling; decreased to .011″ after 20 minutes because of stem expansion; was .017″ at 30 minutes because of block expansion counteracting stem expansion. At this point the block reached its maximum expansion but the exhaust valve temperature continued to increase as engine speed was raised. Therefore the continued expansion of the valve stem reduced the clearance to a minimum of .005″ during the 100 to 130 minute interval.

Oil temperature reached a maximum of more than 250 degrees at 135 minutes. Water temperature was a maximum of 180 degrees during the latter part of the idling period and then dropped to 170 degrees during the remainder of the run.

The test just described emphasizes the importance of correct valve clearances because in this particular engine the exhaust valve clearance is reduced to .005″ when operating at high speed at full throttle. If the valves had been incorrectly adjusted in the first place to .010″ or less they would have leaked at full load.

Causes of Valve Troubles

VALVE LEAKAGE

Valve leakage involves two things: (1) Leaks caused by mechanical defects in the valve system and (2) leaks due to deposits.

To simplify this discussion we shall assume temporarily that all the parts in the valve system are clean, which means that they are not covered with deposits of any description such as carbon, gummy oil, and so forth. With this assumption, if a valve is to be leak-proof, the following mechanical conditions must be present:

Valve face must be circular and concentric with stem. The face, obviously, must be concentric with the stem if the valve is to fit snugly into the valve seat with absolutely leakproof contact. Fig. 7 shows how a concentric valve head fits tight against its seat and also how an eccentric valve head fails to seat perfectly.

Valve head must be square with stem. Obviously the valve will leak if the head is not at right angles to the stem, Fig. 8.

Valve stem must be straight. A valve stem with even a slight kink in it will retard valve closing by binding in the guide, Fig. 9. And the kink may be sufficient to cause the valve to stick open.

Valve guide must be true and clearance correct. If valve stem-to-guide clearance is too little, the speed of the valve closing may be retarded or there may be insufficient clearance for an adequate oil film. If valve clearance is excessive, the guide is no longer able properly to guide the valve onto its seat. The extra clearance permits the valve stem to shuttle crosswise, with the result that the valve head may be somewhat eccentric to the seat when it makes contact, as shown in the left diagram, Fig. 10. Because of the conical face and seat this condition is immediately rectified as shown in the right diagram. In other words, the conical faces force the valve to seat tightly but note that it takes an appreciable instant for the valve to slide sideways into its fully seated position. Thus valve closing is delayed. Prompt valve closing is extremely important for maximum exhaust valve life. Delayed closing increases the temperature.

Eccentric valve seating because of too much guide clearance causes extra wear on both valve face and seat because the brunt of the powerful impact of valve closing is taken by a small portion of the circumference instead of by the whole circumference.

On overhead valve engines, excessive valve stem-to-guide clearance will not only cause the troubles just described but also it should be realized that the wiping action of the rocker arm, acting on the tip of the valve stem, may cock the valve in its guide and cause leakage, Fig. 11.

Fig. 12 shows that if the valve seat is concentric with the valve guide, the valve will close tight, as indicated at the left, whereas if the valve seat is eccentric the valve will be held off its seat and will leak, Fig. 12 right. However, if valve guide clear-

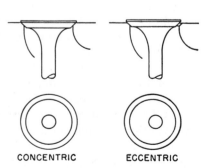

CONCENTRIC   ECCENTRIC

Fig. 7 Right view shows how valve leaks when head is not concentric with valve stem. Concentric valve head at left seats perfectly

WARPED

Fig. 8 Valve leaks when not square with stem

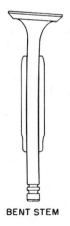

BENT STEM

Fig. 9
Bent valve stem interferes with normal closing because it binds in guide

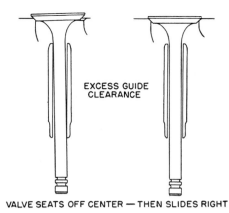

EXCESS GUIDE
CLEARANCE

VALVE SEATS OFF CENTER — THEN SLIDES RIGHT

Fig. 10 If valve guide clearance is too great the valve may seat eccentrically (left) and then it slides over to the concentric position (right)

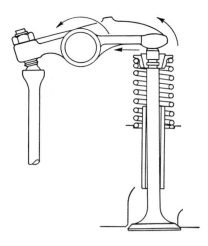

Fig. 11 As valve closes, tip of rocker arm moves to left as shown by arrow, rubbing across stem and thus tilting valve

ance is sufficient to permit it, the valve will slide diagonally to give tight seating. The action is the same as illustrated in Fig. 10.

Valve spring strength must be correct. If valve springs have become weak, valve closing will be delayed, especially at full engine speed, at which time a full strength spring is required to make the valve and valve lifter follow the cam. A weak spring lacks the force to close the valve promptly. Prompt closing is essential for maximum valve life as well as for maximum power. Springs that are too strong will cause undue wear on the valve mechanism because their extra strength causes more powerful battering of the valve against its seat. Likewise cam and valve lifter wear are increased.

Valve springs must be straight. A valve spring which does not stand perpendicular to a flat surface will bulge as shown in Fig. 13 when the valve is open. This places a side thrust on the valve stem and gradually wears the guide as shown in Fig. 14, preventing valve from closing tight. It should also be obvious that if a deformed valve spring is installed on a valve with a worn guide, the valve will start leaking immediately because the spring cocks the valve to one side. If the valve spring is improperly seated, it will cause the same troubles as mentioned above. Fig. 15 shows that the top coil of the spring is not resting in the machined recess at one side.

Valve lifter guide clearance must be correct. On L-head engines, if the clearance between the valve lifter and its guide is excessive, the lifter will be cocked as shown in Fig. 16 each time the valve is opened. At this moment, the lifter thrust will try

to cock the valve stem in the opposite direction as shown. When the valve is closing, all these forces are reversed. Thus a loose valve lifter eventually wears the valve guide into a bell-mouthed shape at top and bottom. The worn valve guide interferes with accurate seating of the valve and thus its hours of leak-proof service are reduced.

Valve guide wear caused by rocker arms. On overhead valve engines, the tip on the valve stem and the tip of the rocker arm which contacts the valve stem should be smooth and should make good contact with each other. When the rocker opens and closes the valve, it creates a side thrust (as shown in Fig. 11), and if the contacting

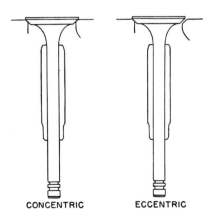

CONCENTRIC   ECCENTRIC

Fig. 12 Valve seat in block or head must be concentric with valve guide. Valve seat is concentric at left and eccentric at right

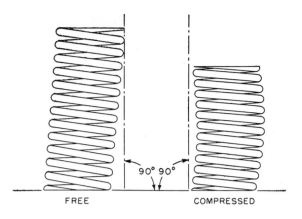

FREE   90° 90°   COMPRESSED

Fig. 13 The distorted valve spring at left buckles when compressed as shown at right

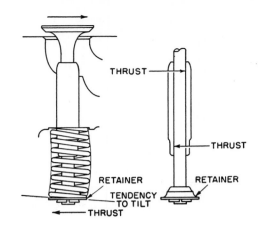

THRUST

THRUST

RETAINER   TENDENCY TO TILT   RETAINER

THRUST

Fig. 14 When spring is compressed it buckles as shown, and lower end tries to tilt the retainer, thrusting lower end of stem to the left and upper end to the right

surfaces are not smooth, the side thrust will be excessive and the valve guide will gradually be worn as illustrated. This wear cocks the valve and prevents it from seating perfectly.

Correct valve operating clearance. In connection with the preceding items, it is assumed that the valve operating clearance is correct. If the clearance is too great, an extra strain is put on the valve parts during valve opening. Likewise, when the valve closes it may batter against the seat with blows of greatly increased intensity. Too much clearance also eliminates the beneficial ramps on the cam.

If the clearance is too little, the valve will be held off its seat for a longer period than necessary. And if there is no clearance at all the valve will leak because it is unable to close tight.

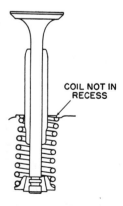

COIL NOT IN RECESS

Fig. 15 Spring is tilted when it is improperly seated in machined recess

VALVE TEMPERATURE

Most valve troubles are directly or indirectly caused by the high temperatures to which they are exposed. This is true even though the valve mechanism is mechanically perfect. High temperatures make it difficult to maintain mechanical perfection for long. And if the valve system is less than mechanically perfect to begin with, the heat hastens the development of valve ailments.

HOT VALVE SEATS

Distortion of either a valve or its seat changes it from the desired circular shape to an oval one, and if the distortion is much more than .001″, leakage must result.

The distortion of the valve seat may be due to improper tightening of the cylinder head. If some cylinder head bolts are too tight and others too loose, the metal surrounding the seats may be pulled out of shape and thus the seat becomes oval instead of circular; or the seat may be tilted so that it is no longer at right angles to the guide. These troubles may occur with either L-head or overhead valve engines if the cylinder head bolts are not tightened carefully with a torque wrench as specified by the car maker.

Metals expand as they are heated. If the seat were a ring which was free from the head or block it would expand equally in all directions and thus remain circular. However, the seat is an integral part of the cylinder block or head and the casting may not permit circular expansion of the seat. Thus the seat may become oval and leakage will result.

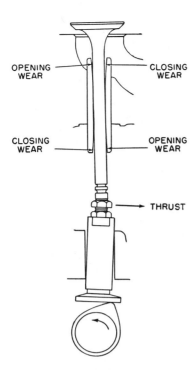

Fig. 16 *Worn valve lifter guide causes valve stem guide wear*

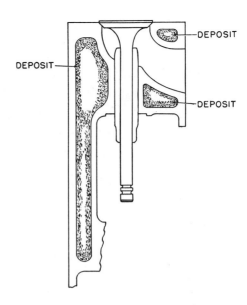

Fig. 17 *Deposits in water passages prevent normal flow of heat from hot valve*

Under normal operating conditions the seat will not depart from its circular shape sufficiently to cause leakage, but serious seat distortion and leakage may occur if the seat temperature becomes considerably higher than its normal maximum operating temperature.

Any defect in the cooling system which reduces its ability to remove heat from the seats may result in overheated seats. These defects include low water, clogged radiator or any other clogged passages, dirty radiator inside or out, loose fan belt. But probably the most serious of all is mineral deposits in the water passages surrounding the valve seats, Fig. 17. These deposits insulate the valve seats from the water and thus prevent proper cooling of the seats with the result that seat temperature rises and thus excessive distortion and leakage may occur. In extreme cases, the temperature may be sufficient to crack the casting at the valve seat, Fig. 18.

An extremely hot exhaust valve seat may distort the adjacent intake valve seat sufficiently to cause leakage, because of flow of heat from the exhaust seat to the intake seat.

Excessively hot exhaust valve seats may be due to a lean mixture which burns so slowly that it is still very hot when the exhaust valve opens.

HOT VALVES

Fig. 19 is a diagram of the head of an exhaust valve at normal operating temperature. The temperature is highest at the center (1350 degrees) and gradually decreases to 1200 degrees at the outside edge. This is to be expected because the face of the valve is in contact with the comparatively cool seat two-thirds of the time. Remember that the entire head of the valve is exposed to the

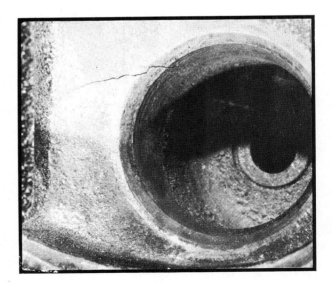

Fig. 18 *Cracked exhaust valve seat due to excessive heat*

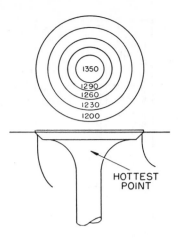

Fig. 19 *Exhaust valve at normal temperature. Note that it is hottest at the center*

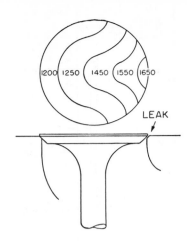

Fig. 20 *How slight valve leakage causes marked increase in temperature*

high temperature of the combustion gases which may exceed 4000 degrees.

The valve head is hottest at its center but is even hotter near the upper end of the valve stem because the hot exhaust gases which flow past this point when the valve is open cause the absorption of additional heat, raising the temperature beyond 1350 degrees.

Now let us assume that something causes this valve to leak very slightly. It could be a distorted seat or valve head or a small piece of dirt between the valve face and the seat. Or perhaps the face of the valve has acquired a smooth, even coating of carbon or iron oxide. The coating does not cause leakage but eventually a piece of the coating breaks off and leakage occurs through this small channel. But, regardless of how the leakage starts, the leakage of 4000 degree combustion gases soon raises the temperature of the valve at the leakage point by several hundred degrees.

Fig. 20 shows the temperature diagram for the head of this slightly leaking valve. At the point of leakage the temperature has risen to 1605 degrees while the opposite side of the valve is a normal 1200 degrees because of its perfect contact with the cool seat. The two pairs of pictures in Fig. 21 show what happens to a normal exhaust valve when it starts to leak.

If the initial valve leakage is caused by a distorted seat, a burned spot would soon appear provided the valve did not rotate. But many valves rotate slightly, in which case the whole face of the valve would eventually be roughened by slight burning.

Note also that when a valve is overheated at one point, as in Fig. 20, the elevated temperature expands the valve head outward so that the hot edge of the valve is eccentric to the valve guide and, therefore, when expansion is sufficient, the valve seats at the hot point but nowhere else. When this happens, the valve is well on its way to ruin.

Deposits which may become attached to the valve face and sometimes also to the valve seat include road dust, carbon from charred oil or fuel, iron oxide formed on the valve face by the hot

NORMAL   BURNED

Fig. 21 *Pictures of valves mentioned in Figs. 19 and 20*

Fig. 22 *These valve stems have been destroyed by sulphurous acid*

CRACK

Fig. 23 *Example of a cracked valve*

exhaust gases lead compounds from Ethyl fluid. And all of these compounds may be firmly attached to the hot valve face by gummy particles of oil. The heat bakes the mixture to a hard, solid film.

Valve face and seat may be corroded by the hot exhaust gases which eat the metal away, often forming small indentations called pits. In some cases, the chemical reaction between the metal and the hot gases may produce small particles of iron carbide on valve face or seat. The iron carbide is chemically attached to the surface. It is extremely hard—harder than a file. These particles cause valve leakage. They are so hard that they are difficult to remove when the valve and seat are ground.

When an exhaust valve develops a leak, for any of the reasons mentioned, its temperature may rise so high that not only is hot corrosion invited but also the temperature may be high enough to weaken the valve. The strength of all metals is reduced as the temperature rises. Special alloy steels are used for exhaust valves because these steels retain their strength at high temperature better than do ordinary carbon steels. But even these steels weaken when the temperature is high enough. A valve steel at 2000 degrees may be only 60 to 80 per cent as strong as it is at room temperature; and it is quite common for a valve to reach 2000 degrees or more if the car is driven hard when a valve is leaking.

The more the valve leaks, the hotter it gets. Here it should be remembered that a leaking valve is exposed to the extreme temperatures of the combustion gases sweeping through the leakage path. All this heat is in addition to that contributed by the exhaust gases which flow past the valve after it opens. No wonder the valve burns and sometimes melts with all this extra heat to contend with.

When leakage causes an exhaust valve to reach a temperature of about 2000 degrees at the center of the head or near the joint between the upper end of the stem and the head, the weakening may cause the valve to warp. The warpage may be caused by any factor which permits the valve to strike the seat at one particular point, thus placing a heavy bending stress on both the valve head and the upper stem.

The point of contact may be due to a distorted seat, dirt at one point of the valve face or seat or to any of the mechanical difficulties illustrated in the preceding pictures.

Any of the factors just mentioned will gradually bend the head of an overheated valve whose metal has been weakened by an elevated temperature. The bending or warping of the head does not occur all at once but gradually—under the repeated impact of hundreds or even thousands of blows. Hot metal "creeps" and thus the head is gradually distorted.

Fig. 24 This valve is not only cracked but the whole face of the valve is burned

An extra hot upper stem will also creep, slowly but surely, because of the continuous downward pressure of the valve spring. Thus the stems of overheated valves are likely to become longer and may even stretch enough to reduce the tappet clearance to zero.

When an exhaust valve leaks and the engine is driven hard, the high temperature gases of combustion may overheat the seat, causing burning or pitting and the expansion of the hot metal in the vicinity of the seat may crack the block. Burning of the seat and cracking are both more likely to occur if the water passages surrounding the seat are coated with mineral deposits which prevent free flow of heat from the hot seat to the cooling water. The same troubles are very likely to occur if the engine is driven hard when the water level in the cooling system is so low that no water reaches the water passages surrounding the valve seats.

SLUGGISH VALVE CLOSING

Next to leakage, sluggish valve closing is the most important cause of valve trouble and it may lead to outright leakage. As previously mentioned, valve leakage and retarded valve closing are the two major valve troubles. A weak or broken spring delays valve closing progressively as engine speed increases.

With prompt or normal valve closing, the valve spring presses on the valve and the valve presses on the tappet. Thus the valve spring forces the tappet against the "closing flank" of the cam. Therefore valve closing closely follows the cam motion.

Extra friction between valve stem and guide may slow down the closing of the valve so that it closes some time after the flank of the cam has moved out of the way.

Retarded intake valve closing causes a reduction in power because some of the mixture in the cylinder is pushed back out into the intake manifold when the piston moves upward on its compression stroke.

Retarded exhaust valve closing also results in loss of power because the valve is open while the intake valve is open on the intake stroke of the piston. Thus the piston draws exhaust gas into the cylinder as well as mixture. The greater the percentage of exhaust gas in the mixture the greater the reduction in power.

Delayed exhaust valve closing increases the valve temperature above normal because the interval that the valve is in contact with its cool seat is appreciably reduced, and because the exposure to hot exhaust gases is correspondingly increased.

Retarded valve closing, as mentioned above, may be due to a weak or broken spring. It may also be due to excessive friction between valve stem and guide; or between the tappet and its guide, or both.

Extra friction between valve stem and guide is due to various causes. In the case of intake valves the most frequent cause is a gradual accumulation of gummy gasoline on the valve stem and guide surfaces. All gasolines contain some gummy substances. In the better grade gasolines the gum content is held to such a small fraction of a percent that the quantity of gum is not sufficient to cause sticking valves. However, even the most carefully refined gasolines are subject to a gradually increasing gum content while in storage. The gum formation is due to a chemical change in some of the ingredients in the fuel. Hence, if it should happen that the gasoline is stored for a long time before it is used, gummy valves may result.

The air in the valve chamber contains small particles of oil. Some of this oil coats the valve stem and valve guide. In the case of exhaust valves, the stems may be at such high temperature that the oil is soon charred to form hard carbon on stem and guide surfaces. Thus the clearance is slowly reduced and the time may arrive when the fit is so tight that valve closing is retarded. Serious carbon formation is unlikely as long as the valve stem guide clearance is normal. But if the clearance is excessive, more oil reaches the clearance space and, of course, valves with loose guides are not likely to seat properly and therefore their stems will be abnormally hot. These two factors encourage the formation of carbon on stem and guide. In fact, some authorities state flatly that a deposit of carbon on valve stem and guide are proof of undue wear.

Note that high oil consumption results in oil particles in the exhaust. These particles may become baked to the exhaust valve stem.

There are other causes of retarded valve closing. Too little guide clearance may slow down valve closing. This is only likely to happen when new valves and guides have been installed or when the old guides have been reamed out and valves with oversize stems installed with too little clearance. Obviously a valve with a bent stem can stick. The bend may be due to stem warpage or to dropping the valve.

When exhaust valves leak badly, the upper ends of the guides are exposed to the high temperature of the gases of combustion. Sometimes they become so overheated that they warp or even melt at the upper end. Either of these conditions may cause valve sticking.

Retarded valve closing may be due to too tight a fit between valve lifters and their guides or to gummy oil on the lifter and guide surfaces.

It should be realized that retarded valve closing may range all the way from a very slight delay up to complete stoppage of the valve before it reaches its seat. The amount of retardation increases as time goes on.

A moderate degree of sluggishness in the closing of either intake or exhaust valves will cause rough idling. With increased retardation of closing some reduction of power may be noted. If an intake valve definitely sticks open, back-firing through the carburetor will occur whenever the valve sticks. If an exhaust valve sticks open more than just a little, the cylinder will misfire. Definite sticking of either an intake valve or exhaust valve will produce a noticeable click caused by the valve lifter banging against the valve stem.

Retarded valve closing may be intermittent. The valve may close normally after one opening and close slowly on the next. The same is true of outright valve sticking. A valve may stick open for one or more revolutions of the camshaft but it may be jarred loose sooner or later by the continual banging of the valve lifter against the stem.

Slow valve closing may be due to a weak or broken valve spring.

Fig. 22 shows a pair of valves whose stems have been eaten away by sulphurous acid. If there is more than a trace of sulphur in the fuel or oil, sulphur dioxide is formed. When this compound is dissolved in water it becomes sulphurous acid. The formation of this acid is encouraged by inadequate crankcase ventilation, humid air and low crankcase oil temperature. Valves which have failed because of excessive temperatures are shown in Figs. 23 and 24.

Valve System Service

CHECKING COMPRESSION

If an inspection shows that the valve operating clearance is correct, that no valve springs are broken, and that no valves are sticking in their guides, a compression test should be made to determine whether the source of trouble is at the valve seats or is a broken cylinder head gasket. The procedure for making this test is herewith given but bear in mind that if the compression is low in two adjacent cylinders, a broken cylinder head gasket is indicated.

In using a compression gauge, the location of the combustion chamber must be taken into consideration. Except for engines in which the combustion chamber is formed in the block, Fig. 25, all other engines have the combustion chamber cast in the cylinder head. When checking compression pressure on engines with cylinder head combustion chambers, it must be realized that the compression pressures in the cylinders of such an engine in perfect condition are not uniform. Car company engineers will tell you that the pressures in the different cylinders of an engine may vary up to 20 pounds. The variation in pressure in some makes of engines is small and large in others. The variation is due principally to lack of uniformity in combustion chamber volumes since it is

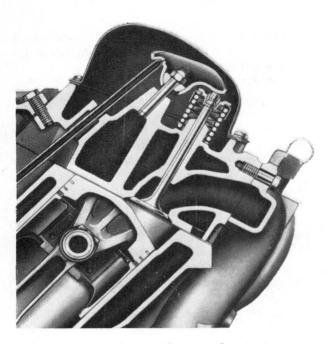

Fig. 25 *Showing how combustion chamber volume is made uniform between cylinders because of the flat cylinder head surface. This example illustrates the wedge-shaped combustion chamber formed in the block on a number of engines introduced in 1958*

impossible to make all the combustion chambers in a cylinder head exactly the same size.

In a given engine with a 10 to 1 compression ratio with all combustion chambers the same volume, the compression pressure might be about 160 pounds in all cylinders. However, if the combustion chamber is ⅓ cubic inch too small the pressure will be 168 pounds and if ⅓ cubic inch too large it will be 152 pounds. This is a variation of 16 pounds.

In the engines where the combustion chamber is formed in the block, this variation is largely eliminated. In the example shown in Fig. 25 the top of the block, instead of being perpendicular to the cylinder bore, forms an angle of 16 degrees from perpendicular, creating a circular wedge with its thin side toward the center of the engine. The underside of the cylinder head is flat, except for slight recesses which provide for valves and spark plugs. The top of the piston is peaked. Half of the top surface is parallel to the cylinder head, and the other half forms an angle of 32 degrees from the lower surface of the head. Inasmuch as this design forms a smooth machined surface, combustion chamber volumes are naturally more uniform and, therefore, not subject to possible vari-

ations as are cast cylinder head combustion chambers.

Just to satisfy the reader's curiosity, below is a table showing the approximate relationship between compression ratio and compression pressure at cranking speeds:

Ratio	Pressure
6.5	110
7.0	120
7.5	130
8.0	140
9.0	150
10.0	160

Various design factors affect the compression pressure. Therefore this table may apply to some engines but not to all. Note also that a carbon deposit will raise the compression pressure at any given ratio by reducing the combustion chamber volume. The greater the deposit the higher the pressure.

However, even if a table such as this could be trusted, there is the question of gauge accuracy. A gauge passes inspection at the factory if it is not more than two pounds high or low when the pressure is 100 pounds. Thus there is a possible error of as much as four pounds to begin with (98 to 102) although of course some of the gauges will be almost perfectly accurate.

But even if the gauge is accurate when made it is not likely to remain so. It is a delicate instrument and the first time it is dropped it may read five pounds too high or too low.

Use of Compression Gauge

A compression test should be made with all the spark plugs out. When this is done it is unnecessary to remove the air cleaner or to hold the throttle open because there are always two or more intake valves open when the engine is being cranked.

When testing the compression of a cylinder, hold the gauge in place until the pointer reaches a maximum reading, even though this may require cranking the engine through ten revolutions or so. The pressure in the curved tube within the gauge builds up slowly because the compressed air must pass through a small orifice at the entrance to the tube.

Note that if a compression test is made after putting oil in the cylinder to seal the rings, the compression reading may be much too high because the oil reduces the volume of the combustion chamber.

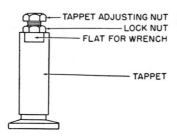

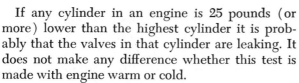

Fig. 26 *Typical valve lifter or tappet for L-head engine. The tappet is prevented from rotating by applying a wrench to the flat surface*

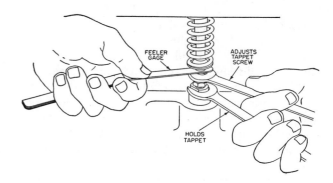

Fig. 27 *Adjusting tappet on L-head engine with two wrenches*

If any cylinder in an engine is 25 pounds (or more) lower than the highest cylinder it is probably that the valves in that cylinder are leaking. It does not make any difference whether this test is made with engine warm or cold.

On engines with cylinder head combustion chambers, it is a mistake to assume that a cylinder which is five to ten pounds lower than the highest has leaky valves because, as previously stated, the variation with tight valves may be much more than ten pounds.

Other Compression Testing Methods

A more positive method of determining cylinder leakage is to employ a leak tester.

Another positive method is to apply air pressure to the cylinders one by one and then listen for leakage. This is an old and tried method although not well known. Remove the porcelain from an old spark plug shell and braze a tire valve in the shell.

Remove all spark plugs. Bring No. 1 piston up to top dead center on the compression stroke. Screw the device just described into the spark plug hole. Apply air pressure. If necessary have an assistant hold the air chuck on the valve throughout the test.

The piston must not move from its top dead center position while making test with air pressure. It won't move if the crankpin is within about five degrees before or after top dead center. The fact that it is not necessary to have the piston exactly on center will speed up the work. When the air is applied, listen for a hiss:

1. At the muffler tail pipe for a leaking exhaust valve.
2. At the air cleaner for a leaking intake valve.
3. At the oil filler opening for leaking rings.
4. Remove the filler cap on the radiator and

look for bubbles which indicate a leaking cylinder head gasket.

Bring No. 2 piston up to top dead center on the compression stroke and repeat the tests—and so on for the other cylinders.

STICKING VALVES

A valve sticking open in the valve guide will cause the engine to misfire and, at times, cause backfiring. The smooth running of the engine will be radically interrupted and accompanied by a loud, sharp, tapping sound. When a sticking valve is suspected, crank the engine and observe whether any valve is stuck open, regardless of the action of the valve lifter or rocker arm. A valve that sticks in its guide may show some movement but this movement will be slow when compared with the rapid movement of a free-acting valve. A stuck valve will have excessive clearance between the valve stem tip and rocker arm tip, or between the valve tip and valve lifter on L-head engines.

FREEING STUCK VALVE

There are several ways that stuck valves can usually be freed but before trying any of them be sure that the engine is warm. If, after the treatments described below, the valve still sticks, remove the cylinder head and the valve and thoroughly clean both the valve stem and its guide.

One way to free a stuck valve is to squirt kerosene or penetrating oil onto the valve stem where it enters the guide. This is readily done on overhead valve engines. On L-head engines, remove the spark plug and use a squirt can with a long nozzle to reach the valve stem. Allow several minutes for the oil to penetrate. Twist the blade of a screwdriver between the coils of the valve spring to separate the coils. This will increase the spring

51

tension and seat the valve. Apply an additional dose of oil and crank the engine several times. As the valve moves, the oil will flow along the stem and guide, flushing dirt and gum from the stem.

Another method of freeing a stuck valve is to pour a pint of special oil made just for this purpose in the crankcase. The oil will often free the valve almost immediately. On the other hand, considerable running of the engine may be necessary.

Still another method of freeing a stuck valve is to start the engine and allow it to run at a fast idle speed, otherwise the engine will stall. After removing the air cleaner, slowly pour kerosene, light engine oil or one of the several preparations made especially for the purpose into the carburetor. While the engine is running the liquid works around the valve stem and guide and should free the valve.

VALVE LASH ADJUSTMENT

Valves must be adjusted after they have been ground or refaced. Also note that after valves have been in use for some thousands of miles they may go out of adjustment. Thus, if the clearance becomes too small, the valve eventually may leak because it is unable to seat fully. If valves are too noisy the clearance should be checked and the valves adjusted to the proper specifications.

On some cars the clearance specifications for intake and exhaust valves are the same while on others more clearance is specified for the exhaust valves than for intake valves.

If the car is often driven at high speed for long periods it may be desirable to increase the exhaust valve clearance by .002″ for L-head engines and .003″ for overhead valve engines.

If intake and exhaust valves have different clearance specifications, it is necessary to determine the location of intake and exhaust valves in each cylinder. Usually this can be done by inspecting the branch arms of the intake and exhaust manifolds where they are bolted to the engine. Make a chalk mark to indicate the location of each exhaust valve.

Cold Valve Adjustment

On some engines it is recommended that the valves be adjusted cold although many car makers specify a warm adjustment.

In most cases the need for adjusting valves is because they have just been ground. Therefore, if the car maker recommends adjusting the valves with cold engine, the work should be done on L-head engines before the head is replaced. The adjustment must be made with valves closed and this is readily done by bringing the piston up on top dead center at the end of the compression stroke.

If a cold adjustment is required on L-head engines when the cylinder head is on, there are two methods of determining whether the valves are closed in a given cylinder. The valves are closed when the valve lifter can be turned freely by hand. This condition exists on the compression stroke and also on the power stroke.

Remove the distributor cap and turn the engine until the rotor points to the distributor segment to which the spark plug wire leading to No. 1 cylinder is connected. When the valves in this cylinder are adjusted, turn the engine slowly in the direction it rotates when running until the rotor points to the next segment. If, for example, the firing order is 153624, the rotor, after pointing to No. 1, successively points to 5, 3, 6, 2, 4.

Warm Valve Adjustment

Before adjusting valve clearance, it is extremely important that the engine be thoroughly warmed up for about 30 minutes to normalize the expansion of all parts. This is very important because during the warm-up period the valve clearances will change considerably. To adjust the valves during or before this warm-up period will produce clearances which will be far from correct after the engine reaches normal operating temperature.

Covering the radiator will not materially hasten this normalizing process because even with the water temperature quickly raised, it does not change the rate at which the oil temperature increases and becomes stabilized, or the engine parts became normalized.

The actual temperature of the oil is not as important as stabilizing the oil temperature. The expansion or contraction of the valve mechanism, cylinder head and block are relative to this oil temperature. These parts stop expanding and valve clearance changes cease to take place only after the oil temperature is stabilized. Then the valves are adjusted with the engine running at normal idling speed.

Some mechanics prefer to adjust the exhaust valves first and then the intake. This is sound practice if the exhaust valves require a different clearance than the intake valves. But if the clearance is the same for both intake and exhaust valves, it is better to start at the front and work toward the rear of the engine.

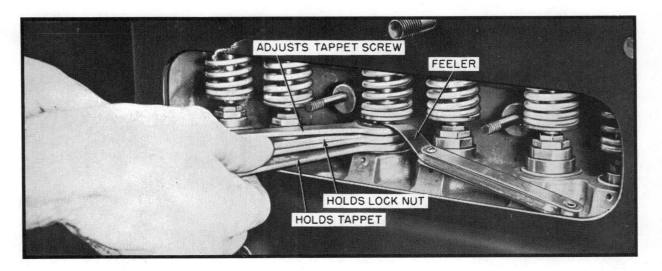

Fig. 28 *Method of adjusting tappet with three wrenches*

Whether it be an L-head or overhead valve engine, if the exhaust valves require a greater clearance than the intake valves, set either the intakes first and then all the exhaust valves. In this way only one thickness feeler gauge is in use at a time and the chances of using the wrong feeler gauge are largely eliminated.

L-HEAD VALVE ADJUSTMENT

The valve lifter (tappet) adjusting screw is threaded into the tappet body, Fig. 26, and is held against rotation by a lock nut. The adjustment may be accomplished with two wrenches after loosening the lock nut slightly, Fig. 27. One hand manipulates the wrenches while the other holds the feeler gauge.

Fig. 28 illustrates how to make the adjustment with three wrenches. The two lower wrenches are held in one hand and the upper wrench which operates the adjusting screw is in the other. After an adjustment is made, all three wrenches are held in one hand while the clearance is checked with the feeler gauge with the other hand. Most mechanics prefer to insert the feeler only when checking the clearance. To leave the feeler in place as shown is likely to damage the feeler blade. When the adjustment is correct a slight drag should be felt on the feeler.

Self-Locking Tappet Screw

In the self-locking tappet screw, Fig. 29, the lower portion of the screw is split and is a sufficiently tight fit in the lifter body so that the screw may be adjusted without any danger of the adjustment changing during its operation.

Studebaker recommends that when replacing a screw that it be checked with a spring scale, Fig. 30, to see that it will hold its adjustment effectively.

If a pull of less than 25 inch-pounds will turn the screw a new one should be tried but it also should be checked. When checking the screw tension the scale should be at right angles to the wrench. To determine the inch-pounds, first measure the distance in inches from the center of the

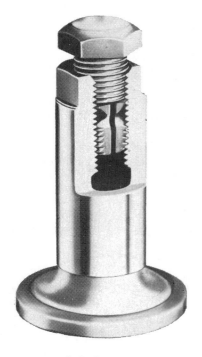

Fig. 29 *Self-locking tappet screw*

Fig. 30 *Checking effectiveness of self-locking tappet screw with spring scale*

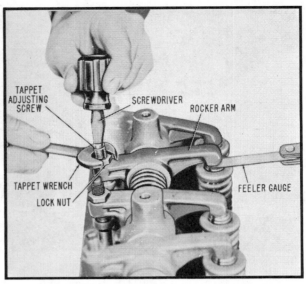

Fig. 31 *Adjusting valve operating clearance on overhead valve engine*

Fig. 32 *Screwdriver with notched blade for adjusting overhead valves*

screw to the point on the wrench where the spring scale is attached.

If a pull of five pounds is required to turn the screw, then the torque is 25 inch-pounds (5 pounds times 5 inches). A reading of five pounds indicates that the screw is satisfactory. Similarly, if a longer wrench were used and the spring scale attached eight inches from the screw, a three-pound reading on the scale would indicate a torque of 24 inch-pounds (3 times 8) which is very close to 25 inch-pounds. If the reading is less than 25 inch-pounds, the screw may work loose during operation, in which case install a new screw.

Ford L-Head Valve Adjustment

On Ford and Mercury engines up to and including 1953 as well as Ford Sixes up to 1947, valve clearance is adjusted by grinding the tips of the valve stems if the clearance is too small. In the rare event that the clearance is too great, it may be reduced by grinding the valve further into its seat.

OVERHEAD VALVE ADJUSTMENT

In some overhead valve engines, the valves are adjusted by turning the screw in the end of the rocker arm after loosening the lock nut just enough to free the screw, Fig. 31. In other engines of this type, where the rocker arms are mounted on individual studs instead of on a shaft, the adjustment is made by turning the self-locking rocker arm stud nut.

Normally, when the manufacturer specifies that valve adjustment should be carried out when the engine is hot, it is done with the engine running. However, with some engines this is not easy as they are liable to squirt oil over the mechanic, or splash oil all over the outside of the engine. So either a guard of some sort must be rigged to protect the mechanic, or the adjustment must be made with the engine stopped. On those engines where it is specified that the clearance should be measured when cold, the adjustment is always made with the engine stopped.

To make the adjustment, insert a feeler gauge of the correct thickness between the valve stem and the tip of the rocker. During the instant that the valve is closed (engine running), if a very slight drag is felt on the feeler the adjustment is correct. To "feel" the drag, pull gently on the

Two Turns, After Slack is Removed

Fig. 33 *Making an initial valve adjustment on 1953–57 Lincoln engine with hydraulic valve lifters. This adjustment places hydraulic lifter plunger in center of its travel*

Fig. 34 *Valve adjustment on Chevrolet V-8 engine*

feeler while turning the screw. When the adjustment is correct, hold the screw against turning and tighten the lock nut. Then check the clearance again in order to be sure that tightening the lock nut did not change the adjustment. To prevent the screwdriver from slipping off the screw it is a good idea to alter the blade of a wide-edged screwdriver as shown in Fig. 32.

Some engines use hex head self locking screws in the rocker arms instead of the slotted head screws and lock nuts.

If adjustment is to be carried out with engine stopped, it pays to remove the spark plugs and turn the engine over by pulling on the fan belt or on the fan blades. Take care, however, not to leave your fingers between the belt and its pulley. The engine can also be "bumped" over with the starter, using a remote control starter switch. In either case, watch the distributor rotor arm when turning the engine. When it points to the segment in the distributor cap that is connected to the cylinder whose valves are to be adjusted, the engine is in the correct position with those valves closed.

Adjustment with Hydraulic Lifters

On some overhead valve engines with hydraulic valve lifters, an initial adjustment is specified after the valves have been ground or whenever the rocker mechanism has been disturbed.

Adjustment is made with the engine stopped, and the valve that is to be adjusted closed. The procedure is to turn the adjusting screw down (clockwise) while moving the push rod up and down until all slack is removed. The correct position can best be found by spinning the push rod between thumb and finger while turning the nut down until it just cannot be spun any more. Be careful not to force the hydraulic plunger down into the lifter body by tightening the valve adjustment too much.

After removing all slack, turn the adjusting screw or nut the additional number of turns specified for the particular engine, which will place the lifter plunger in the center of its travel in the lifter body. When the adjustment is correct, tighten the locknut, if any. Fig. 33 shows the adjustment being made on a Lincoln engine.

ROCKER ARMS

On some engines, rocker arms oscillate on stationary tubular shafts mounted on suitable supports fastened to the cylinder head. On other engines, the rocker arms are mounted on individual studs pressed or screwed into the head, and fastened by a nut which is also used for adjusting the valve clearance, as already explained. This design, Fig. 34, eliminates the conventional shaft and gives the engine designer more freedom to position the valves and coolant passages to the best advantage.

On engines using rocker arm shafts, the upper end of the push rod usually terminates in a cup which fits around a ball on the rocker arm, Fig. 35. The lower end of the push rod rests on the valve lifter, or tappet. The engine oil pump supplies oil under pressure to the hollow rocker arm shafts. Drilled openings in the shaft deliver the oil to the rocker arm bearings. From here oil flows through a channel in the rocker arm which supplies lubricant to the cup and ball at the top of the push rod.

With individually mounted rocker arms the arrangement is somewhat different. Usually these rocker arms are stamped from sheet steel and their upper surface forms a sort of trough (see Fig. 34). With this type of rocker arm, the push rod upper end is ball-shaped, and fits into an upside down cup in the end of the rocker arm. The push rods are hollow, and oil is pumped up from the hydraulic lifters through them into the rocker arms. It then runs down to the rocker arm pivot and to the valve end of the arm.

This type of rocker arm and push rod is also sometimes used in conjunction with rocker arm shafts.

Other variations include adjustable and non-adjustable rocker arms, adjustable and non-adjustable push rods and various combinations of the two.

HIGH PERFORMANCE PUSH RODS AND ROCKER ARMS

Push Rods

Push rods can deflect under load and this must be taken into account when special replacement units are being considered. Most high performance push rods are made of a chrome-moly steel tube, which is both light in weight and rigid, so that it offers good high speed operating characteristics. It should be pointed out that if a push rod should bend or whip in operation, it will affect the valve operation, altering the timing and possibly affecting the ability of the valve to seal the cylinder properly.

Rocker Arms

Rocker arms serve a dual purpose. They receive the motion of the cam at one end and multiply it by the ratio of the rocker arm while transferring it to the valve at the other end. The ratio of the rocker arm is the distance from the center of the push rod contact point to the center of the rocker shaft, divided into the distance from the center of the valve stem to the center of the rocker shaft. (Measurements are taken to the center of the stud hole when the rocker arm is mounted on a stud.) This ratio is usually about 1.5 to 1.

Rocker arms may or may not be adjustable, depending on the design of camshaft and the type of lifter used. Most high performance camshafts use adjustable rocker arms and a solid (non-hydraulic) lifter cam to maintain proper valve lash at high engine speeds.

For high engine speeds, pressed in rocker studs should be replaced by threaded ones.

Some ultra-high performance rocker arms are made of light weight, high strength aluminum and have a needle bearing roller to contact the valve stem. This reduces the friction between rocker arm and valve stem to a negligible quantity, and at the same time reduces the side force on the valve stem resulting from the arc in which the rocker arm operates. This in turn reduces friction between valve stem and guide, and wear on the guide and stem.

When modifying an engine for high performance, it is important to remember that rocker arm geometry, or rocker arm-to-valve stem relationship, must not be changed. Incorrect geometry can result in increased side loads on the valve, rapid wear of the valve guides and stems, and other problems.

ROCKER ARM SERVICE

After removing the rocker arm cover, if the rocker arms are found to be heavily sludged, stuck or damaged, the assembly should be removed for cleaning and opening up of oil passages. If none of these conditions is apparent, run the engine with the rocker arm cover off. While the engine is running, check the rocker arms and shafts to see that oil is being distributed to all the shaft bearings and to the outer end of the arms. Correct lubrication of these parts is indicated when oil drips from the bearing edge of each rocker arm and spreads over the push rod ends.

Oil failing to show at these points is usually caused by dirt in the oil feed pipe, in which case it should be removed and cleaned, preferably by blowing out with compressed air.

Too much oil flowing is a definite indication of worn rocker arms and shafts, which calls for replacement of these parts.

Before removing or disassembling a rocker arm assembly, note carefully the arrangement of the

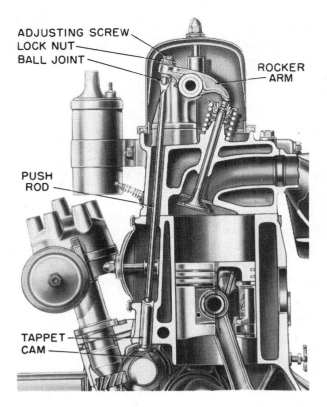

Fig. 35 *Conventional overhead valve mechanism (Ford Six)*

Labels on Fig. 35:
ADJUSTING SCREW
LOCK NUT
BALL JOINT
ROCKER ARM
PUSH ROD
TAPPET
CAM

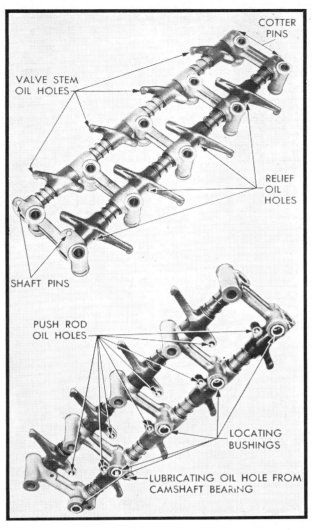

Labels on Fig. 37:
COTTER PINS
VALVE STEM OIL HOLES
RELIEF OIL HOLES
SHAFT PINS
PUSH ROD OIL HOLES
LOCATING BUSHINGS
LUBRICATING OIL HOLE FROM CAMSHAFT BEARING

Fig. 37 *Top and bottom views of Chrysler-built V-8 engine employing double rocker arm assemblies for each cylinder head*

parts and look for any identifying marks which indicate their location. Figs. 36 to 40 illustrate the parts that make up a representative group of rocker arm assemblies.

The rocker arm assembly should always be removed before removing the cylinder head. The first step is to remove the rocker arm cover plate. Then remove the push rod cover (if used). De-

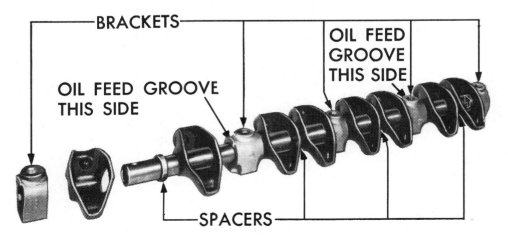

Labels on Fig. 36:
BRACKETS
OIL FEED GROOVE THIS SIDE
OIL FEED GROOVE THIS SIDE
SPACERS

Fig. 36 *Rocker arm assembly on Chrysler engines introduced in 1958*

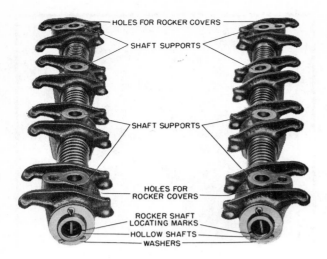

Fig. 38 Cadillac rocker arm assemblies. The shaft springs keep the rocker arms pressed against the shaft supports

tach the oil connections at the rocker arm shafts or supports. Then unscrew the bolts holding the supports and lift off the rocker arm assembly. Remove the push rods before removing the head and replace them after installing the head to avoid accidentally damaging them.

In disassembling the rocker arms from their shafts, it is most important to place the parts along the back of the work bench in exactly the same relationship they assume when on the engine, otherwise there may be difficulty in assembling them properly. This is particularly important because rocker arms are not all alike;

some are at right angles to the rocker shafts and others may be a few degrees more or less than a right angle.

When assembling rocker arms to shafts, squirt oil on the shafts to provide initial lubrication when the engine is started.

ROCKER ARM COVER

Remove the nuts from the top of the rocker arm cover and lift it off. Clean the cover both inside and outside. Also clean the rocker arm mechanism, and the strainer in the oil line to the rockers (if one is present).

When replacing rocker arm covers always use a new gasket and torque the bolts or cap screws to proper specifications to prevent oil leakage.

L-HEAD VALVE COVER

Before removing the covers, wipe away all dirt and scratch a small identifying mark on each cover so that it can readily be returned to its proper place. Remove the nuts or screws with washers and take off the covers. Thoroughly clean the covers with gasoline or other cleaning fluid and wipe dry. As a precaution against subsequent leakage, remove the old gasket, being careful to scrape all remnants from the cover and engine. Remove dirt, oil and sludge from the valve chamber and the valve mechanism by either brushing with gasoline or by the use of cleaning equipment. If the bottom of the valve chamber is open to the crankcase, stuff the openings with rags before cleaning the valve mechanism.

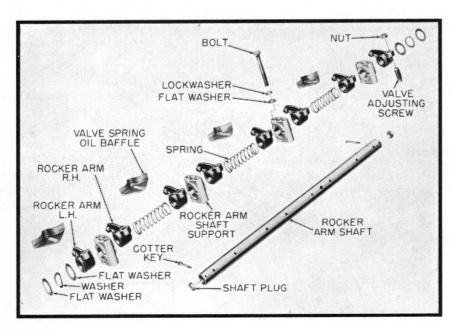

Fig. 39 Layout of rocker arm parts for a Ford V-8 engine

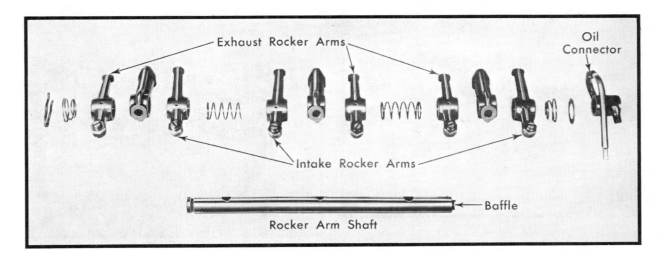

Fig. 40 Rocker arm assembly for three rear cylinders on Chevrolet 6-cylinder engine. Front and rear rocker shafts fit into the oil connector

CYLINDER HEAD, REMOVE

Except Corvair, Jeep, Pontiac Tempest OHC & Volkswagen Engines

A cylinder head is fastened to the cylinder block by studs and nuts or by cap screws. To remove the head, the nuts or cap screws are taken out and the head is lifted off. *Do not remove a cylinder head when the engine is hot because cylinder head warpage may result.*

Before removing a cylinder head, it is necessary to remove any parts that are in the way and to detach any connections between the head and the rest of the car.

Disconnect the starter cable at the battery to prevent accidental shorts.

Disconnect the hose between cylinder head and radiator. Before doing so, drain enough liquid from the cooling system to prevent leakage when the hose connections are loosened. If the liquid contains anti-freeze, drain it into a clean pail so that it can be put back into the system after the repair work is done.

Detach the temperature gauge wire or tube.

Remove high tension cables from spark plugs.

Remove carburetor air cleaner.

If water pump is attached to cylinder head, remove fan belt.

Remove or detach any other item which may interfere with taking off the cylinder head.

The preceding remarks apply generally to both L-head and overhead valve engines. However, it should be noted that there is some additional preliminary work to be done on overhead valve engines because the intake and exhaust manifolds are attached to the head instead of to the block as on the L-head.

On overhead valve engines any tubing or rods which interfere with the removal of the head must be disconnected as follows:

Detach vacuum advance tube at carburetor.

Detach windshield wiper tube at the intake manifold or detach the tube running from the intake manifold to the vacuum side of the fuel pump (if so equipped).

If the coil is attached to the cylinder head, detach the high-tension cable running from coil to distributor at the distributor.

Detach the fuel line at the carburetor.

Detach the throttle rod at the carburetor.

Remove the carburetor and intake manifold as a unit.

Detach the exhaust manifold from the cylinder head or disconnect the exhaust pipe from the exhaust manifold—whichever is easier.

Remove the rocker arm cover.

Remove the rocker arm assemblies.

Remove the push rod cover.

Pull the long push rods up through the head as they are likely to be damaged if the head is removed first.

Remove the cylinder head nuts or cap screws with a socket wrench. If cap screws are used, carefully note their length as they are removed. If there are some long and some short ones, be sure to remember which holes the long ones came from so that when the head is replaced the long ones will be installed in the long holes. If a long bolt is installed in a short hole and turned up tight, the cylinder block may be broken beyond repair, in which case a new one will be required. However, if you forget which holes are the long ones, measure the depth of all holes with any handy rod—a pencil will do. If still in doubt, insert the long

Fig. 41 *Gasket cement is readily applied with a roller*

Fig. 42 *The limber torque wrench handle bends when it is pulled by hand but the pointer remains stationary. Thus the pointer indicates the torque in pounds-feet*

bolts in the holes that are longest and turn them up gently. Then if a bolt refuses to go in all the way (without using force) it is in the wrong hole.

Once the cylinder head nuts or cap screws have been removed, the head is ready to be lifted off. If the head is large, such as a straight eight, two men may be required. If it does not come off readily, jolt it crosswise and lengthwise with a rawhide hammer. If it still sticks, crank the engine with the throttle open and ignition off to loosen the head by the force of compression pressure. Before doing this, however, install spark plugs in the end cylinders to hold the compression, and install a pair of nuts or cap screws. Draw them down finger tight and then back off about two turns. Crank the engine with the starter to see if the head will break loose.

If this does not work, install two eye bolts in the head and apply a chain hoist or a crane just as though you were going to life the engine out of the car.

If the block is equipped with studs, the sticking may be due to corrosion between the studs and head. In this case, squirt some special penetrating oil made for the purpose on the studs so that it will seep down and loosen up the corrosion. A really stubborn case of corrosion may require the removal of the stud; there are a number of stud removers on the market.

Some combination of these methods should loosen any cast iron head and most aluminum heads. However, in the case of some of the older engines equipped with aluminum heads, if the head is badly corroded, no available force will remove it. Therefore, it may be necessary to demolish the head and remove it piece by piece.

CYLINDER HEAD, INSTALL

Except Corvair, Jeep, Pontiac Tempest OHC & Volkswagen Engines

Handle the head carefully. Do not allow it to strike any open valves as this may bend them. Before replacing the head be sure the gasket surfaces of the head and block are perfectly clean, otherwise the gasket may leak when the head is installed. Also be sure that there is no carbon or dirt in the cylinder head bolt holes, or any foreign matter in the bolt holes in the block. If there is, blow them out with compressed air. Dirt in the holes may prevent drawing the bolts up tight.

On some engines, the bolt holes in the block extend into the water space, in which case the end of the bolt should be coated with white lead or sealing compound to prevent leakage.

If there are any burrs around the edges of the bolt holes in the block it is advisable to remove them with a medium file as these burrs may prevent the gasket from fitting tightly. In fact, all burrs on both the head and block should be removed.

It should be noted that the cylinder head gasket may have bolt or stud holes which are slightly enlarged on the under face to take care of burrs. If the holes are slightly larger on one side, this side should contact the block. Some gaskets are marked "Top" or have lettering on the top side. Always install a new gasket as the old one may leak even though it appears to be perfect.

Place the gasket on the cylinder block to be certain that it fits. If all bolt holes and water

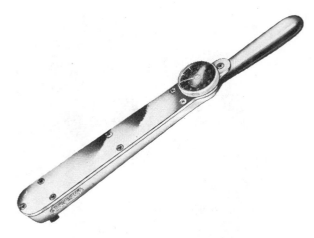

Fig. 43 Torque wrench, with dial which reads in pounds-feet

passage holes do not line up, turn the gasket upside down or end for end and inspect it again. Some gaskets are marked "Front" at the front end; if so, be sure to install it this way.

Apply a coat of gasket cement to both sides of the gasket, using a roller as shown in Fig. 41. Place gasket on a piece of heavy paper. Coat one side. Allow cement to dry tacky. Then turn gasket over and coat other side. If a roller is not available, use a brush. In the case of some Buick models the gasket is coated with a lacquer when it is made, and no additional cement should be used.

On cars that use cylinder head cap screws instead of studs it is desirable (but not absolutely necessary) to install two "pilot" studs in the block to guide the head into place. These studs are made by sawing the heads off of two spare cylinder head cap screws. Screw them lightly into bolt holes at opposite ends of the block. If screwed in too tight it will be difficult to remove them later. After these studs are installed, put the gasket in place and then install the head.

Lubricate the threads on the cap screws and the bosses on the cylinder head. Install the cap screws and turn them until they are nearly tight. Then remove the two pilot studs and install cap screws in these holes. Use a torque wrench, Figs. 42 and 43, to draw the bolts up tight. It is important to tighten the bolts in proper sequence and to tighten them just the right amount in order to prevent distortion of the cylinder block or cylinder head, as well as to insure a tight gasket.

Usually the scale on the torque wrench reads in pounds-feet of torque. But if the wrench reads according to some arbitrary scale, it is customary

for the wrench manufacturer to supply instructions with the tool so that the wrench scale readings can be translated into pounds-feet. The term "pounds-feet" simply means the force in pounds applied to the wrench handle at a radius of one foot. Note that the correct term is pounds-feet and not foot-pounds, although foot-pounds is almost universally used.

All car manufacturers specify the pounds-feet torque which should be used in tightening their cylinder heads and also the sequence in which final tightening should be done.

Be careful to tighten the cylinder head cap screws or stud nuts exactly to the car maker's specifications. Tightening them a little more or less than the specified amount is objectionable because "more" may cause distortion and "less" may permit gasket leakage.

If the cylinder block is equipped with studs, put the gasket on the block, install the cylinder head and lubricate the threads on the studs. Put on the nuts and draw them up until they are nearly tight. Then finish the job with the torque wrench as previously described.

After the head is installed, attach all other parts which have been removed from the engine except the valve covers or the rocker arm cover, whichever is used.

Be sure to fill the radiator.

Start the engine and allow it to run at a fast idle for at least 15 minutes; half an hour is better. After the engine has been run until thoroughly warm, and if it has a cast iron head, tighten all cap screws or stud nuts but do not use more than the specified torque. If the engine has an aluminum head, shut the engine off after it has been warmed up and tighten the cap screws and stud

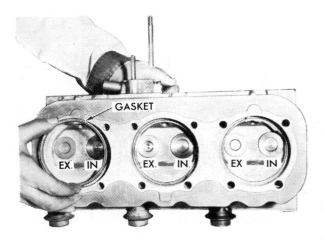

Fig. 44 Installing Corvair cylinder head gaskets

Fig. 45 *Installing Corvair cylinder head*

Fig. 46 *Installing "O" rings, washers and rocker arm studs on Corvair engine*

nuts to the specified torque after engine has cooled.

On overhead valve engines with mechanical valve lifters, always check the valve operating clearance after installing the cylinder head.

CORVAIR CYLINDER HEADS

The procedure for removing a Corvair cylinder head differs a great deal from that of a conventional car, chiefly because the engine is enclosed in metal shields and shrouds and because the engine must be lowered to provide access to the head for its removal. Because of its relative complexity, a step-by-step procedure is herewith given for 1960–64 models. On 1965 models it is not necessary to lower the engine.

Removing Left Bank Head

1. Drain crankcase oil. This is necessary because when the push rod drain tubes are removed, their entrance holes in the cylinder block will be open and the oil may spill out as the engine is being manipulated.
2. Disconnect battery cables, engine ground connection, radio ground strap, left side carburetor cross-shaft support from carburetor, and fuel line at carburetor.
3. Remove carburetor intake hose, carburetor accelerator return spring, and disconnect accelerator rod from carburetor.
4. Remove carburetor and the long stud from the carburetor mounting.

5. Remove wires and spark plugs.
6. Loosen (do not remove yet) all engine side shield retaining screws. Remove screw from engine side shield under carburetor (attached to cylinder head) and then remove engine side shield.
7. Remove oil cooler; this allows engine rear shroud freedom of movement during cylinder head removal.
8. Raise car on hoist and attach a suitable lifting fixture to the power train (engine, transmission and rear axle assembly).
9. Remove both engine side seal retainers and engine rear seal retainers.
10. Remove engine rear center shield and seal assembly.
11. Remove lower engine shroud.
12. Remove exhaust pipe-to-manifold nuts.
13. Open french locks on exhaust manifold and remove holding clamp nuts.
14. Remove exhaust manifold.
15. Remove engine rear mounting cotter pin, nut and washer from below body rear mounting bracket.
16. Remove rocker arm cover while holding a pan below to catch oil draining from cylinder head.
17. Remove rocker arms, push rods and push rod guides.
18. Remove rocker arm studs, washers and "O" rings from head.
19. Remove "O" rings from bottom of push rod drain tubes with a pair of hooked tweezers.

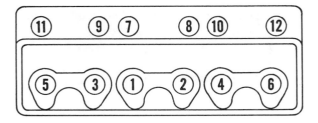

Fig. 47 Corvair cylinder head tightening sequence

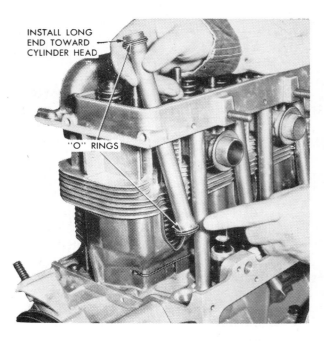

Fig. 48 Installing Corvair push rod oil drain tubes

20. Remove retaining nuts and washers retaining head.
21. Carefully lower engine assembly approximately 3″ to clear cylinder head carburetor flange.
22. Remove cylinder head from crankcase studs.

Removing Right Bank Head

Removal and installation of the right bank head is essentially the same as the left head, except the coil, oil pressure and oil temperature sending units must be disconnected. Also remove the choke heat pipe and choke fresh air pipe from the exhaust manifold.

Replacing Corvair Cylinder Head

The procedure for replacing the head is accomplished by reversing the removal operations. However, in order that the reader may gain some knowledge of the unique construction of this engine, a sequence of operations for replacing the head when the engine is removed from the chassis is hereby outlined:

1. Be sure all cylinder head gaskets are in the combustion chambers, as shown in Fig. 44. Then install head, Fig. 45.

Fig. 49 Installing Corvair push rods

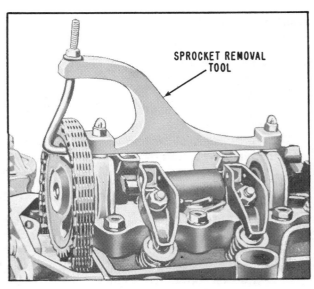

Fig. 50 Camshaft sprocket removal and installation tool for use when removing cylinder head and timing chain on Jeep overhead camshaft engine

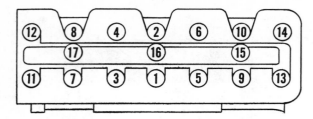

Fig. 51 *Cylinder head tightening sequence on Jeep overhead camshaft engine*

2. Install six flat washers and nuts on long studs, adjacent to intake manifold.
3. Install six new "O" rings in counterbore of cylinder head, Fig. 46. Coat rings with Lubriplate and rocker stud bore with anti-seize compound. Then install rocker arm studs.
4. Tighten nuts and rocker studs in the sequence shown in Fig. 47.
5. Lightly lubricate valve lifters and install in their proper bores.
6. Install push rod oil drain tubes, Fig. 48. Oil the "O" rings and push in place at lifter bore at crankcase and cylinder head.
7. Install push rods with side oil hole up into valve rocker socket, Fig. 49.
8. Install push rod guides over rocker studs and push rods and tighten nuts.
9. Install rocker arms, balls, and nuts. Adjust valve lash as outlined previously after installing distributor.

Replacing Jeep Overhead Camshaft Cyl. Head

1. To remove head with timing chain cover installed, first take off rocker arm cover.
2. Install camshaft sprocket removal and installation tool on rocker arm cover studs, Fig. 50. Fasten with rocker arm cover nuts. Install hook of removal tool in the sprocket and tighten the nut to relieve the tension on the camshaft.
3. Remove fuel pump eccentric from camshaft sprocket.
4. Pull forward on camshaft sprocket to remove it from pilot on camshaft. With sprocket still engaged in timing chain, release tension on tool by loosening nut. Gently allow sprocket to rest on bosses in timing chain cover. *Caution: Do not turn over engine when sprocket is removed from camshaft and is resting on bosses in cover. This will severely damage cover. Do not attempt to remove*

camshaft sprocket from timing chain since this will upset the valve timing.

5. Disconnect lubrication tube from head and block.
6. Remove cylinder head bolts.
7. Lift off head assembly.
8. Reverse the removal procedure to install the head. Position the all-metal head gasket on the block so that the coolant ports and vent ports are properly aligned. Use no sealer or other compound on the gasket. When head is installed, tighten it down in the sequence shown in Fig. 51.

Replacing Pontiac Tempest Overhead Camshaft Cylinder Head

1. Drain cooling system and remove air cleaner.
2. Disconnect accelerator pedal cable at bell-crank on manifold and fuel and vacuum lines at carburetor.
3. Disconnect exhaust pipe at manifold flange, then remove manifold bolts and clamps. Remove manifolds and carburetor as an assembly.
4. Remove rocker arm cover.
5. Remove timing belt upper front cover mounting support bracket and rear lower cover.
6. Disconnect spark plug wires.
7. Remove rocker arms and hydraulic valve lash adjusters. *Store rocker arms and lash adjusters so they can be replaced in exactly the same location.*
8. Remove head bolts and lift off head. Notice that head bolts are of two different lengths. When inserted in proper holes all bolts will project an equal distance from head. Do not use sealer of any kind on threads.
9. Reverse the removal procedure to install the head.

VOLKSWAGEN CYLINDER HEADS
1200, 1300 and 1967–68 1500

To remove Volkswagen cylinder heads, it is first necessary to remove the engine from the car. The procedure is as follows:

1. Disconnect battery ground strap.
2. On early models, turn off fuel tap. On later models pull off fuel hose and clip it.
3. Remove carburetor air cleaner.
4. Remove engine rear cover plate.
5. Disconnect wires from generator, ignition coil and oil pressure switch.
6. Disconnect accelerator cable from carburetor.

7. Loosen and turn distributor to allow vacuum unit to clear rear cover plate when engine is being removed.
8. Raise vehicle at least three feet off the ground.
9. Disconnect both heater control cables and loosen flexible heater pipes from engine.
10. Disconnect fuel hose from engine.
11. Remove nuts from both engine lower mounting bolts.
12. Pull accelerator cable from guide tube.
13. Place a roller jack under engine.
14. While holding the two upper engine mounting bolts, have an assistant remove the nuts.
15. Raise jack to support engine.
16. Withdraw engine far enough to allow clutch release plate to clear main drive shaft. Then lower jack and tilt rear of engine downward until it can be completely removed.

1965–1966 1500 and All 1600 Models

1. Disconnect battery ground strap.
2. Remove air cleaner.
3. Remove cables from generator, coil, oil pressure switch and automatic choke.
4. Detach accelerator cable at connecting link.
5. Remove oil dipstick and take off rubber boot between oil filler and body.
6. Loosen clip on bellows at cooling air intake housing and pull off bellows.
7. Remove rear engine support.
8. Raise and support vehicle on stands.
9. Disconnect flexible pipes between engine and warm air mixing boxes.
10. Disconnect heater flap cables.
11. Take warm air hose from carburetor pre-heater off engine.
12. Pull off fuel hose at front engine cover plate and seal it with a suitable plug.
13. Remove two lower engine mounting nuts.
14. Support engine with roller jack.
15. Hold two upper engine mounting bolts while an assistant removes nuts.
16. Pull engine back slightly until release plate clears main drive shaft.
17. Lower engine, being sure that clutch release plate or main drive shaft are not damaged.

Installation, All Models

When installing the engine, take care to prevent damage to the flywheel gland nut needle bearing and clutch release bearing. To ease entering the main drive shaft into the clutch plate and gland nut needle bearing, engage a transmission gear to keep the mainshaft from turning and rotate the engine by the fan belt as required.

When installing the engine, first install the mounting bolts into the transmission case flange. Then press the engine against the flange, being sure it seats properly around the flange before installing the nuts. Tighten the upper nuts finger tight first, then the lower nuts. Final tightening should be done in the same order.

To remove a cylinder head, with the engine out of the car:

1. Remove hoses between fan housing and heat exchangers.
2. Remove front and rear engine cover plates.
3. Remove muffler.
4. Remove fan belt.
5. Pull off coil-to-distributor cable.
6. Remove fan housing with generator.
7. Remove intake manifold with pre-heating pipe.
8. Remove both heat exchangers.
9. Disassemble and remove cylinder deflector plates.
10. Remove cylinder head cover.
11. Remove rocker shafts.
12. Remove push rods.
13. Remove cylinder heads.

Cylinder Head

Each pair of cylinders has one detachable cylinder head. The head has cooling fins and shrunk-in valve seat inserts and valve guides. No gasket is used between the joining faces of the cylinder and head. Gaskets are used, however, between the flanges of the cylinder and cylinder head to prevent leakage of combustion gases.

NOTE. If a cylinder head is to be removed, a suitable cylinder retainer can be made locally to prevent the cylinders from being pulled off accidentally, and to eliminate the danger of dirt entering the crankshaft, Fig. 52.

When installing heads, the following points should be observed:

1. Check heads for cracks in combustion chamber and exhaust ports. Cracked heads must be replaced.
2. Check spark plug threads in heads. If necessary, install Heli-Coil threaded inserts. (These are a proprietary method of restoring stripped or damaged female threads in a casting.)
3. There is no gasket between upper edge of cylinder and cylinder head.

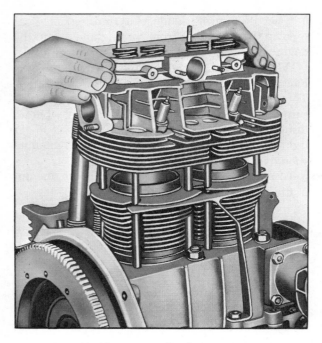

Fig. 52 Showing cylinder retainer in place when removing cylinder head

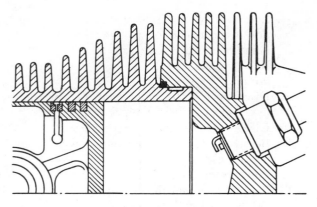

Fig. 53 Showing location of sealing ring (deep black) between cylinder shoulder and cylinder head. Series 1200, 1300 and 1967–68 1500

4. On 1200, 1300 and 1967–69 1500 models, fit new sealing ring between cylinder shoulder and cylinder head, Fig. 53.
5. On all engines, when installing cylinder head, make sure that oil seals at ends of push rod tubes are properly seated, Fig. 54.
6. Push rod tubes can best be centered by using a centering tool of the type shown (made locally), Fig. 55.

7. Turn the tubes so the seam is facing upwards. For perfect sealing, used tubes must be stretched to the correct length before they are installed. This operation must be carried out very carefully to avoid cracking the tubes, Fig. 56.
8. Coat attaching nuts with graphite paste and screw them down until resistance can be felt. Then tighten initially with a torque wrench to 7 ft-lbs in the sequence shown, Fig. 57. Final tightening should be made in the sequence shown in Fig. 58 to the specified torque.

VALVES, REPLACE

If a numbered rack is not available to receive the valves as they are removed, one can be made by drilling enough holes in a board to accom-

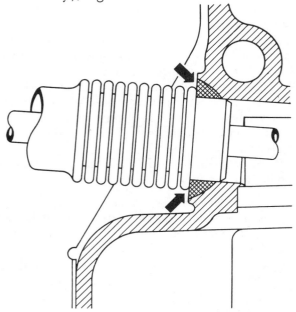

Fig. 54 Oil seals at ends of push rod tubes. All engines

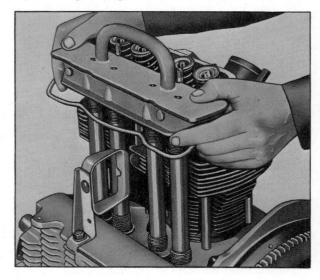

Fig. 55 Centering push rod tubes with tool shown

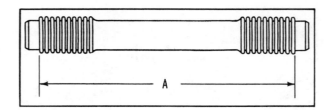

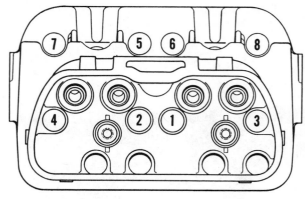

Fig. 56 *When installed, push rod tubes should be stretched, if necessary, to the following lengths (dimension "A"):*

1200	1963–65	7.105–7.144″
1300	1966	7.480–7.520″
1500, 1600	1965–69	7.480–7.520″

Fig. 57 Initial tightening sequence of cylinder head

modate them, Fig. 59. This is not essential when valve faces and seats are to be reconditioned and valve guides replaced. But when valve guides are still serviceable and not removed, each valve should be returned to its own guide because the extent of wear between one guide and another may vary enough to cause trouble later on if valves are switched. Then, too, exhaust valves are often made of greater heat resisting material than are intake valves. In most cases, however, intake valve heads are larger in diameter than exhaust valve heads, in which case switching them would readily be noticed since they would not seat in the head or block properly.

L-Head Valves, Remove

Inasmuch as only 6-cylinder L-head and the Willys 4-cylinder L-head engines are currently in use the following material pertains to them.

Take off the cylinder head and valve covers. Clean the valve chambers. Cover any openings in the floor of the valve chambers with clean cloth or cardboard so that the valve locks will not fall into the crankcase as they are removed.

Fig. 59 Valve rack consisting of a board with numbered holes

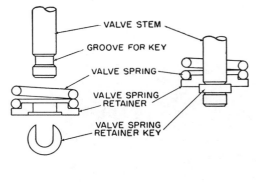

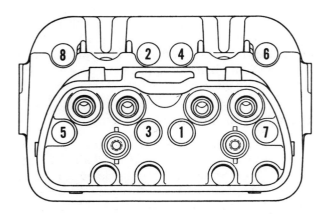

Fig. 58 Final tightening sequence of cylinder head

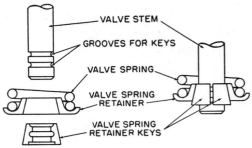

Fig. 60 Valve spring retainer keys

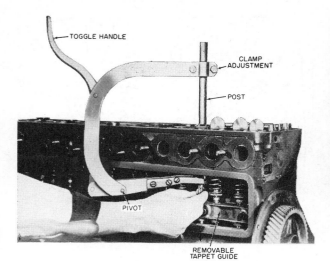

Fig. 61 *C-shaped valve spring compressor. The adjustable post rests on the head of the valve. The end of the lower arm is forked to slip under the valve spring seat. The spring is compressed by the toggle handle which automatically locks in the compressed position. The hand is shown removing the two halves of the retainer lock. Note that this Studebaker engine has removable tappet guides*

Fig. 62 *Valve spring compressor. The lower forked arm rests on the tappet. The upper forked arm is brought into contact with the valve spring retainer. Compressing the handles compresses the valve spring. Then the lock lever is pushed to the right to lock the tool in the position shown. The retainer lock is being removed with a pair of snap ring pliers*

The valve spring retainer is locked to the valve stem by means of a lock or key. Some keys are U-shaped although most are split in two halves. Both types are shown in Fig. 60. The key must be removed in order to remove the spring and valve.

The key is readily removed after the valve spring is compressed.

First remove the valves and springs from those valves which are closed. Then, turn over the engine until another group of valves close and then remove them. Continue to rotate the engine in this manner to close the remaining valves and remove them.

Valves are removed with the aid of a spring compressor, Figs. 61, 62, 63. Compress the spring sufficiently so that the retainer lock can be removed. Release the compressor so that the spring can expand. Remove the valve and then take out the spring, Fig. 64. Remove the rest of the valves and springs in like manner, being sure to place them in the valve rack in the proper sequence, Fig. 59.

In case a valve refuses to come out, it may be removed with the tool shown in Fig. 65. The claw of the tool is slipped under the head of the valve and the sliding hammer is banged up against the nut, breaking the valve loose from its guide.

L-Head Valves, Install

Before installing valves, be sure they are perfectly clean. Coat the valve stems with light engine oil. When installing a spring, do not compress it any more than necessary to insert the retainer key as too much compression may permanently shorten its length and thus weaken the spring action.

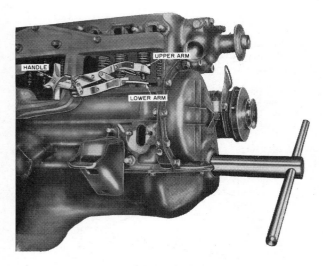

Fig. 63 *Valve spring compressor. Lower forked arm rests on tappet. Upper forked arm compresses valve spring when handle is turned clockwise*

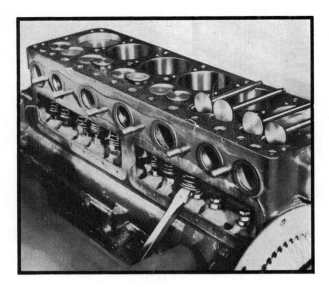

Fig. 64 *After retainer lock is removed, valve spring is pried out with screwdriver*

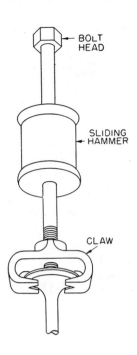

Fig. 65
Tool for removing stuck valves

There are a number of tools for inserting the retainer locks. Fig. 66 shows one design. The retainer lock is placed in the jaw end of the inserter. Then, while the valve spring is compressed, the retainer lock is slipped up over the end of the valve stem. When the lock is in the correct position, the plunger on the handle end of the inserter is pushed inward to eject the lock and at the same time the spring compressor is released so that the spring presses down on the retainer and "seats" the retainer lock.

Fig. 67 illustrates an inserting tool which has permanent magnet jaws to grip the two halves of the retainer lock.

Ford & Mercury V-8s (1949–53)

Fig. 68 shows the valve assembly used on these engines. The valve guide has a forked retainer while the valve spring retainer at the tip of the stem is equipped with a split valve lock. The valve assembly is removed as a unit by means of the forked lever which engages the groove in the lower end of the valve guide, Fig. 69.

Overhead Valves, Remove

There are various types of tools for compressing the valve springs but the type most generally used is the C-shaped design, Fig. 70. The forked jaw is applied to the valve spring seat and then the screw plunger is turned clockwise until it touches the head of the valve. Pushing the operating handle compresses the spring and permits removal of the retainer lock.

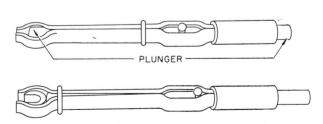

Fig. 66 *Tool for installing valve spring retainer lock*

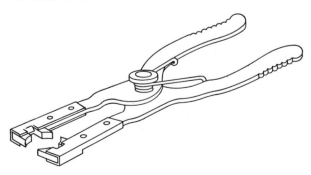

Fig. 67 *Tool with magnet jaws for holding split valve locks*

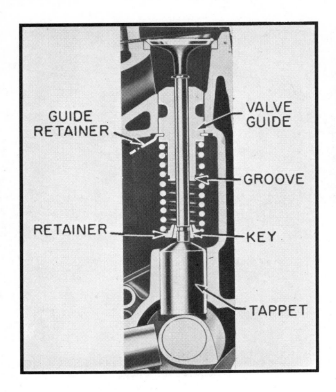

Fig. 68 *One-piece valve guide used on 1949–53 Ford engines*

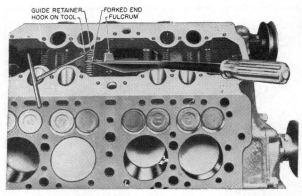

Fig. 69 *To remove valve guide retainer the valve spring compressor is inserted in the middle of the valve spring. Pulling up on the handle relieves the pressure on the retainer so it can be removed*

Another design of a C-shaped tool is shown in Fig. 71. The hollow hex is placed against the valve spring seat. The knurled knob is turned left to release the pointed square shaft. The point is brought into contact with the head of the valve and the knurled knob is locked by turning it to the

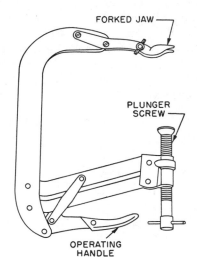

Fig. 70 *C-shaped valve spring compressor for overhead valve engines and most L-head engines*

right or clockwise. Then the handle is moved clockwise to compress the spring.

Special valve tools of different designs are made for modern overhead valve V-8 engines, each designed especially for a particular engine. Fig. 72 illustrates the one for Cadillac. The tool is placed on the bench and the cylinder head is placed on the frame of the tool with the valves touching the valve supports. The arm is aligned with the valve spring as shown and then the spring is compressed by the foot inserted in a stirrup.

Overhead Valves, Install

On most modern overhead valve engines, oil seals are used on the valve stems to prevent the seeping of oil into the combustion chambers. On the Rambler Six shown in Fig. 73 there are three seals. There is a round seal near the upper end of the valve stem and there is a dual seal at the top of the valve guide. There is also a neoprene seal which prevents excessive passage of oil and above it is a felt seal which provides valve stem lubrication during the initial starting period.

A number of modern overhead valve engines have a round rubber oil seal near the upper end of the intake valve stem to prevent undue leakage of oil down the stem and into the engine, Fig. 74. Other overhead valve engines have similar oil seals to provide this protection.

When new seals are installed, they should be checked with a small suction cup which is placed on top of the valve spring retainer. If there is no leakage the cup should stick in place.

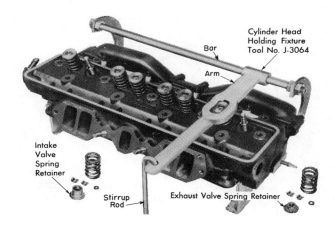

Fig. 72 Special design valve
spring compressor for Cadillac overhead valve engines

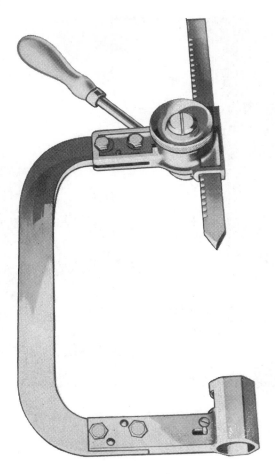

Fig. 71 Another type of
valve spring compressor

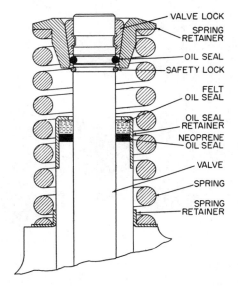

Fig. 73 Oil seals on valve stem prevent
oil leakage into the combustion chamber.
Rambler Six overhead valve engine

CARBON CLEANING

After the valves have been removed, all the
parts should be thoroughly cleaned of carbon and
any other deposits. The work may be done with a
small putty knife and a wire brush with a handle,
obtainable in a hardware store, or with a wire
scraper. However, electrically-driven brushes are
faster and more thorough. Fig. 75 shows the type
of brush used for cleaning the top of the cylinder
block.

To clean the top of a piston, use the brush
shown in Fig. 76 but first move the piston until it
is ½ inch below the top of the block.

The valve stem and valve head should be
cleaned either with a hand brush or with the
rotary brushes shown in Figs. 75, 76, while the
valve is held in a vise; or better yet, clean the
valve with a rotary brush on a bench grinder,
Fig. 77.

After cleaning the valve it is a good plan to
polish the upper half of the stem with a fine emery
or crocus cloth (about 300 grit). The polishing
can be done by hand while the valve is held in a
vise or the valve may be rotated in a valve refacing tool. In either case use a narrow strip of
cloth. Hold the two ends in the fingers to form a
half loop around the valve stem.

Sometimes layers of lead oxide and carbon will
build up on the exhaust valve stem just below the
head, Fig. 78. This deposit is baked into a very
hard substance. The deposit may extend far
enough down the stem to interfere with the closing of the valve.

Valve stem guides should be cleaned by means
of the wire cleaner shown in Fig. 79. This cleaner
can be operated by hand if an electric drill is not

Fig. 74 *Rubber oil seal on intake valve stem on Oldsmobile overhead valve engine*

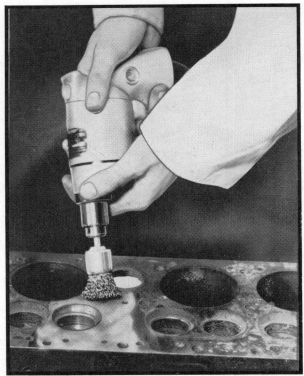

Fig. 75 *Cleaning top of cylinder block with wire brush. The cylinder head is cleaned in the same way*

available. It is moved up and down in the valve guide while it is being rotated.

If the upper end of the valve guide is counterbored be sure that the counter bore is thoroughly cleaned.

After all surfaces have been cleaned of carbon, wipe them with a cloth or use an air nozzle to blow away any remaining particles of carbon. Be careful to blow out all bolt holes in which carbon dust may have settled. This remark applies particularly to cylinder bolt holes. If one of these holes becomes partially filled with carbon it will prevent full tightening of the bolt when it is replaced.

Rice is used to remove carbon from engine combustion chambers without taking off the head in a process developed by Oldsmobile. Called the "Head-On" Carbon Blaster, this device utilizes specially prepared rice, under air pressure, directed through the spark plug openings.

VALVE INSPECTION

After the valves have been cleaned they should be inspected carefully for defects. Excessive heat may cause the valve to bend or warp, Fig. 80. Such a valve should be discarded.

Examine both the upper and lower sides of the head for cracks. If even a small crack is found, scrap the valve.

Inspect the face of each exhaust valve for burning. Even if only a small burned spot is found the valve should be scrapped because the metal adjacent to the spot has probably been damaged and further trouble can be expected if such a valve is

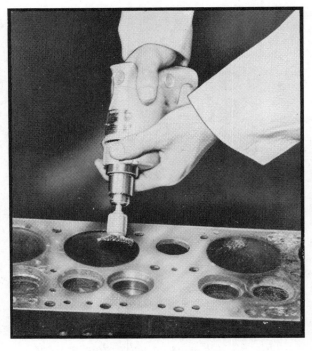

Fig. 76 *Cleaning carbon from top of piston*

Fig. 77 *Cleaning valve with a rotary brush on a bench grinder*

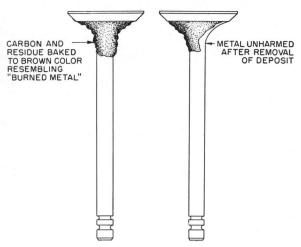

CARBON AND RESIDUE BAKED TO BROWN COLOR RESEMBLING "BURNED METAL"

METAL UNHARMED AFTER REMOVAL OF DEPOSIT

Fig. 78 *Upper end of valve with heavy deposits*

used again. The appearance of a burned valve is readily recognized. It looks like a poker which has remained too long in a hot fire. A burned valve, however, should not be confused with a pitted valve. Pits are caused by slight corrosion resulting from hot exhaust gases sweeping past the face of the valve and gouging out shallow depressions which are then partially filled with carbon. Such a valve is readily refinished so that it is as good as new.

If the valve stem is noticeably corroded the valve should be discarded.

If the tip of the valve stem is scratched it should be smoothed up with a valve refacing tool. If the top face of the tappet adjusting screw, Fig. 81, has a depression worn in it (caused by loose valve stem guide, loose tappet guide or both) a new adjusting screw should be installed or the old screw should be smoothed up in a valve refacing tool.

Both the valve stem and its guide should be checked for wear. Valve stem and guide wear at the top is caused by off-center seating of the valve. Wear of stem and guide at the bottom is caused by too much clearance between the valve tappet and its guide.

Valve Stem-to-Guide Clearance

The exact stem-to-guide clearance varies considerably in different makes of engines. Usually, greater clearance is specified for the exhaust valve than for the intake by the engine manufacturer because the exhaust runs much hotter and therefore expands more.

The valve stem-to-guide clearances recom-

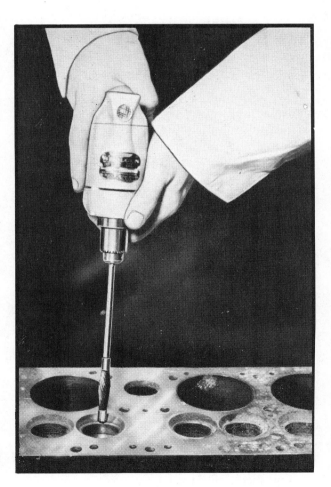

Fig. 79 *Wire cleaner for valve stem guide*

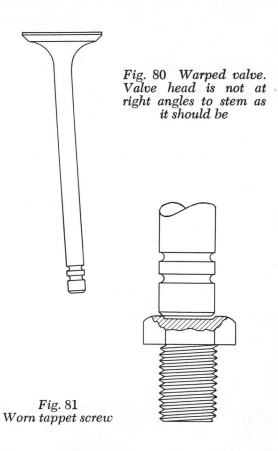

Fig. 80 Warped valve. Valve head is not at right angles to stem as it should be

Fig. 81
Worn tappet screw

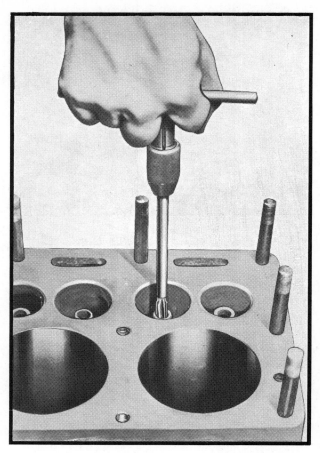

Fig. 82 Reaming a valve guide

mended by a leading valve manufacturer are as follows:

Valve Stem Diameter	Clearance Intake	Exhaust
5/16″	.002″	.003″
11/32″	.0025″	.0035″
3/8″	.003″	.004″
7/16″	.0035″	.0045″
1/2″	.004″	.005″

If the clearance is more than .002″ greater than given above, new valves and guides should be installed. On engines without removable valve guides, valves with overside stems should be installed and the valve guides reamed to provide the specified clearance.

Reaming Valve Guides

After installing new guides, or if valves with oversize stems are to be installed, the new guides or guide holes should be reamed. Lubricate the reamer blades with light oil. Insert the reamer in the valve guide, Fig. 82, and turn it slowly clockwise while pressing down gently. On L-head engines, place a rag underneath the valve guide to prevent metal cuttings falling into the crankcase.

When reaming is completed, remove the reamer by continuing to turn it clockwise while lifting it out of the hole. Never turn a reamer in the opposite direction as it will damage the cutting blades.

Non-adjustable reamers are made for some popular makes of engines, the reamer being the correct diameter for a given standard size or oversize valve stem.

Measuring Valve Stem-to-Guide Clearance

The stem-to-guide clearance may be measured in four ways: With a micrometer; with a dial gauge; with 1/16″ wide shim stock; with Go and No-Go gauges. The surfaces should be perfectly clean before any measurements are made.

After measuring the clearance and it is found to be excessive, install a new valve. If this does not reduce the clearance appreciably, install a new guide. If the engine is one which has no removable valve guides, ream the guide hole to accommodate the next oversize valve.

When using a dial gauge to measure valve-to-guide clearance, install a suitable sleeve over the valve stem to hold the valve at working height in the head or block. Attach a dial indicator having a stem at right angle with the edge of the valve, Fig.

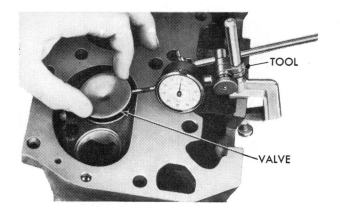

Fig. 83 *Measuring valve stem-to-guide clearance with a dial gauge*

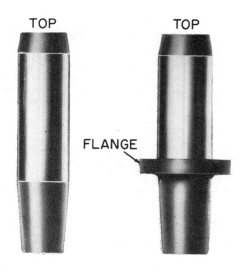

Fig. 84 *Two typical valve stem guides*

83. Move the valve to and from the gauge. If the clearance is more than it should be install a new valve and guide.

The Go and No-Go type plug gauges are quite popular because of the facility with which guides can be checked. However, there are two undesirable guide conditions which are difficult to check with gauges of this type: elliptical or egg-shaped bore wear, and bell-mouthing at the port end of the guide. Careful guide inspection will detect egg-shaped wear, and the careful use of the No-Go gauge will tend to show the degree of bell-mouthing.

Measure the diameter of the valve stem with a micrometer and write down the reading. Then measure the diameter of the valve guide with an inside caliper and write down the reading. Subtract the former reading from the latter in order to determine the clearance. If the clearance is more than it should be a new valve and/or guide should be installed.

Stem-to-guide clearance may be checked with a 1/16″ wide shim stock which is available in various thicknesses. Insert a strip of the shim stock in the guide and then insert the valve stem. By trial and error, select a thickness of shim stock which provides a very slight drag when the valve is moved up and down in the guide. If, for example, a piece of shim stock .003″ thick provides a slight drag, then the stem-to-guide clearance is .003″.

VALVE GUIDE, REPLACE

The method to use in removing and replacing a valve guide depends on its construction. The most common type is shown at left in Fig. 84. It is pressed into the guide hole and held in position by friction. The guide shown in the right view,

Fig. 84, is used on a few of the older engines. Its flange is intended to fit snugly against the roof of the valve chamber on L-head engines or on top of the cylinder head on overhead valve engines. The flanged valve guide must be removed and replaced from below on L-head engines.

On overhead valve engines the straight valve guide can be removed from either the top or bottom of the head, whichever is more convenient.

If it is desirable to remove L-head valve guides by driving them downward, the valve lifters may interfere unless they can be removed. Barrel type lifters are readily lifted out after the valves and springs are removed but mushroom type lifters cannot be taken out without removing the camshaft unless the lifter guides operate in removable guide brackets.

Therefore, if the engine is equipped with mushroom type lifters and the lifter guides are not removable, there are three possibilities: (1) There may be room enough to drive the guide downward and out after removing the lifter adjusting screw. (2) With a suitable puller the guide may be pulled upward and out. (3) If a puller is not available, drive the guide downward almost to the point where it contacts the lifter screw. Cut a sharp nick in the guide with a chisel near the "roof" of the valve chamber and then break the protruding portion of the guide off with a hammer. The remaining portion of the guide in the guide hole can then be driven down and out.

Before removing a guide, measure the distance it sticks out into the valve port with a steel scale and record this figure. Note that the distance may be different for intake and exhaust guides, Fig. 85.

75

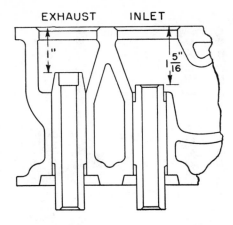

Fig. 85 Valve guide position for intake and exhaust may be different

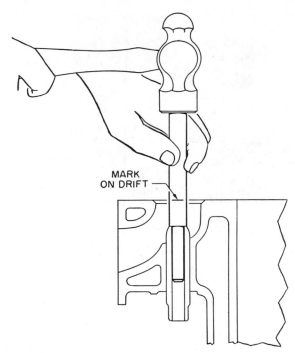

Fig. 86 Installing valve guide with marked drift

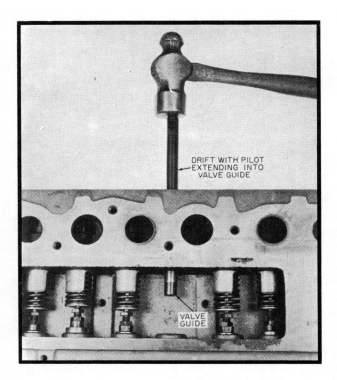

Fig. 87 Removing a valve guide with hammer and drift

Instead of measuring the distance, insert a drift in the guide, Fig. 86, and mark the drift at the top of the block. When the new guide is installed, drive in to this mark.

If the guide has a flange, Fig. 84, it is not necessary to make this measurement because the flange automatically positions the guide.

Fig. 87 shows how a valve guide is driven out with a drift and hammer. The drift has a pilot on

its lower end which should be the same diameter as the valve stem.

If the valve lifters interfere with valve guide removal, the guides may be pulled out from the top. Fig. 88 shows one type of puller; another puller is shown in Fig. 89.

To install a valve guide, first lubricate its outer surface with engine oil. It may be installed with a hammer and drift, Fig. 87. The pilot on the drift should fit the valve guide hole in order to align the guide while it is being hammered into place. Center the guide in the hole and tap the drift gently until the guide is firmly entered in the hole. The force of the hammer blows can then be increased somewhat. Be careful not to drive the guide in too far. If by accident the guide is driven in too far, it may be brought back to its correct position on L-head engines by the use of a puller. On overhead valve engines, the guide can be driven back by turning the head over and using a drift.

To insure that the guide is driven in just the right distance, special tools are made for given engines. Fig. 90 shows such a tool with a large shoulder which butts against the block when the valve guide is driven in far enough. Fig. 91 illustrates a tool for Cadillac engines. When the shoulder on the drift handle contacts the plate, the guide is correctly positioned.

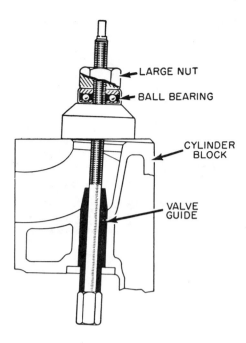

Fig. 88 *This valve guide is pulled out from the top by turning the large nut*

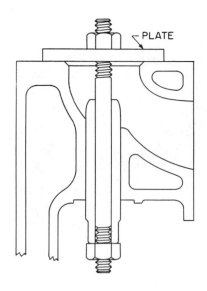

Fig. 89 *Home-made tool for valve guide removal. Turn up both nuts finger tight. Then tighten upper nut to pull out guide*

Check the clearance between the new guides and valves, and ream the guides if the clearance is too little. Light reaming may also be required if the guides have been slightly distorted during installation.

NON REMOVABLE VALVE GUIDES

Most modern overhead valve engines do not have removable valve guides. The valves operate in guide holes bored directly in the material of the cylinder head.

If the clearance becomes excessive, the usual method of rectifying it is to use the next oversize valve and ream the guide bore to fit. Valves with oversize stems are usually available in .003" or .005", .015" and .030" oversizes.

If the wear is not excessive, it is possible to recondition both replaceable and non-replaceable guides by knurling their inside surfaces. This process raises ridges of metal on which the valve stem rides. The ridges also are very effective in retaining lubricant in the guides.

It is also possible in some engines to drill out the non-replaceable guide holes and press in standard type guides, but this is a highly skilled job as the drilling may go through into the water passages in the head. If this happens, it does not always mean scrapping the head, as it is sometimes possible to seal the new guide to the head. But great care and experience are needed, so this work should not be undertaken without careful consideration of the possible consequences.

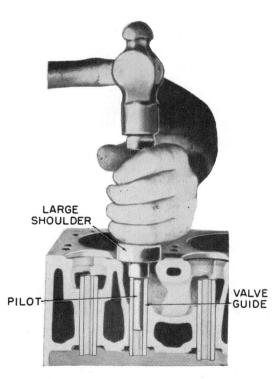

Fig. 90 *Large shoulder on drift prevents driving guide in too far*

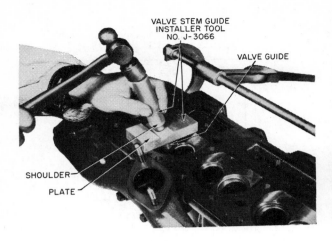

Fig. 91 Valve guide is correctly installed with this Cadillac tool

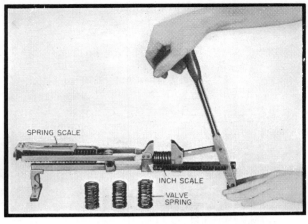

Fig. 92 Valve spring tester

VALVE SPRING TESTING

Valves will not operate properly if their springs are weak. Full strength springs are essential if the valves are to follow their cams while closing. If the springs are weak, the valves close later than they should, the result being loss of power, overheated exhaust valves, and increased valve noise. Even slightly weak springs may cause trouble at maximum engine speed.

It is not good practice to do a valve job without testing the springs, because weak springs may be the real cause of the necessity for the valve job, in which case, another valve job will be required long before it would otherwise be necessary.

For example, if a given spring calls for a pressure of 125 pounds when compressed to 1½″ and is less than 112 pounds, a new spring should be installed.

There are a number of tools on the market for measuring spring pressure. Figs. 92, 93, and 94 illustrate three of them. In Fig. 92, the valve spring is placed between the two jaws, and the handle is pushed forward until the inch scale shows that the spring length has been reduced to 1½″. The spring scale should then register 125 pounds—in the case of the spring just mentioned.

In the tool shown in Fig. 93, the valve spring is installed between the two plates. Then the handle is pressed down until the spring is compressed to the specified length. The pounds pressure exerted by the spring is registered on the dial.

Fig. 94 shows a fixture which permits using a torque wrench to test the valve spring strength. The hand wheel is adjusted so that the distance from it to the top of the screw equals the specified compressed length of the valve spring. Then the valve spring is put in place and the handle pulled

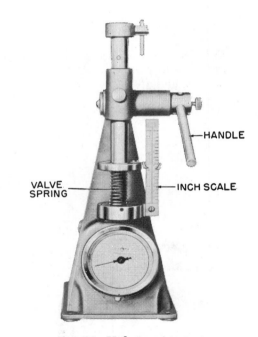

Fig. 93 Valve spring tester

down until its arm contacts the top of the screw. Then the number of pounds pull on the torque wrench is read.

When testing springs, be careful not to compress them beyond the specified length as further compression may permanently weaken them.

VALVE SPRING, REPLACE

A new spring can be installed without removing the cylinder head on all overhead valve engines as well as on some L-head engines.

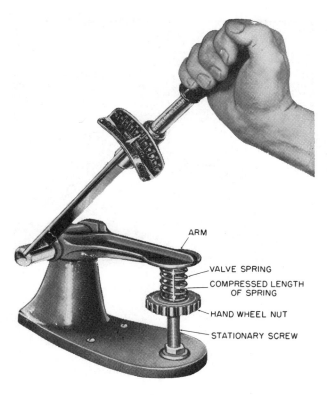

ARM

VALVE SPRING

COMPRESSED LENGTH
OF SPRING

HAND WHEEL NUT

STATIONARY SCREW

Fig. 94   Valve spring tester

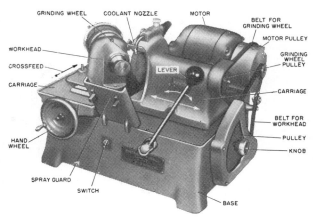

GRINDING WHEEL   COOLANT NOZZLE   MOTOR

BELT FOR
GRINDING WHEEL

MOTOR PULLEY

WORKHEAD

CROSSFEED

CARRIAGE

LEVER

GRINDING
WHEEL
PULLEY

CARRIAGE

BELT FOR
WORKHEAD

PULLEY

KNOB

HAND
WHEEL

SPRAY GUARD

SWITCH

BASE

Fig. 95   Electric valve grinding tool

If the coils on the spring are closer together at one end, be sure to install the spring so that this end is in the same position as on the other springs.

Overhead Valve Springs

On overhead valve engines remove the rocker arm cover. Before attempting to remove the spring the valve should be closed so that minimum compression of the spring is necessary. The piston should also be at or near the top of its stroke to prevent the possibility of the valve falling down into the bore while the broken spring is being removed and a new one installed.

Loosen the locknut on the ball adjusting stud, and screw the stud upward so that the top of the push rod can be moved out of the rocker arm. The rocker arm can now be moved out of the way of the valve spring, either by tilting it upward or pushing it to one side.

Using a suitable valve spring compressor, remove the valve spring lock and spring. Then install the new spring. To hold the valve up while removing or replacing the spring, insert a bent rod through the spark plug hole, pressing its inner end against the head of the valve.

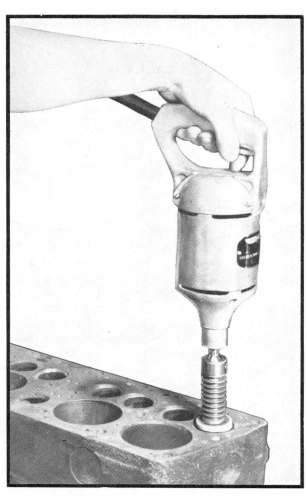

*Fig. 96   Valve seat tool driven by
portable electric drill*

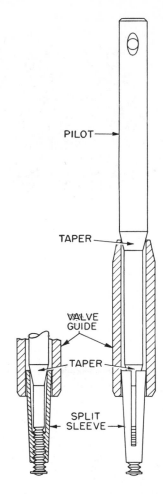

Fig. 97 *Self-centering pilot for valve seat grinder*

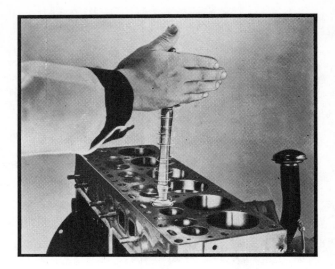

Fig. 98 Hand-grinding a valve into its seat

compressor is used, remove a spark plug and hold the valve down on its seat with a small rod.

After the retainer has been removed, use a small screwdriver to hold the tip of the valve up while the lower end of the spring is started through the clearance space between the valve stem and the lifter. A broken spring, of course, should be easier to remove than a complete spring. Therefore, there is the chance that the broken spring can be removed but the new spring can't be installed without removing the cylinder head.

VALVE GRINDING

It seems almost unnecessary to explain that the purpose of grinding valves is to cure valve leakage. Valve grinding smooths up the face of the valve seat and valve face so that the two refinished surfaces mate so perfectly that leakage is eliminated.

It is customary to use two specially designed electrically-driven grinding tools, one for the valve and the other for the seat.

Electric Valve Grinder

There are two kinds of tools. One refaces the valve, Fig. 95, and the other refinishes the valve seat, Fig. 96. The valve seat refinishing tool is equipped with a grinding wheel that has the same angle as the valve seat.

In the valve refinishing tool, Fig. 95, the face of the valve is parallel to the face of the grinding wheel. This means that the stem of a 45-degree angle valve makes a 45-degree angle with the face of the wheel.

L-Head Valve Springs

On L-head engines whether or not a broken valve spring can be replaced without removing the cylinder head depends on whether enough clearance is available between valve stem and valve lifter to work the old spring out and work the new spring in. The valve is pushed up as far as it will go and if the lifter is adjustable, it is screwed down as far as possible to obtain maximum clearance between stem and lifter. If the lifter has a detachable guide, there is no problem. Just remove the guide and then there is ample room for removing and replacing the spring.

On a given L-head engine it is not easy to determine beforehand whether the spring can be removed without removing the head. But nothing is lost by trying. Before attempting to remove the spring, crank the engine until the valve is closed. Then compress the spring and remove the retainer lock and retainer.

If the valve moves upward when the spring

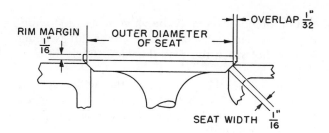

Fig. 99 Important valve and seat dimensions

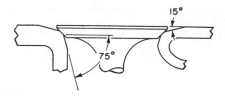

Fig. 102 If seat is too wide after beveling top (Fig. 101) the throat is beveled with a 75-degree stone as indicated

Fig. 100 Grinding the seat alters the dimensions shown in Fig. 99. This diagram illustrates an extreme case

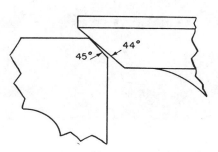

Fig. 103 Diagram showing a 44-degree face resting on a 45-degree seat

Fig. 101 Overlap of valve is restored by a 15-degree bevel

Both the grinding wheel and the valve rotate. In the tool shown, the grinding wheel and its carriage can be moved right or left by swinging the lever. In this way the rim of the wheel is aligned with the face of the valve.

The workhead is mounted on a cross-feed carriage which can be moved toward the grinding wheel or away from it by turning the hand wheel. To grind a valve, the valve face is moved into light contact with the grinding wheel.

To sum up, the grinding wheel carriage can be moved left or right and the workhead can be moved at right angles to the carriage.

Electric Valve Seat Grinder

A grinding wheel operated by an electric drill is used to reface valve seats. A steel cutter may be used except in the case of valve seat inserts which are too hard for the cutter to do a satisfactory job.

One such grinder has a concentric grinding wheel which contacts the seat all around its rim, Fig. 96. This design is often made so that the wheel vibrates or bounces off of the seat about once every revolution. The grinding wheel is lined up on the seat by means of a pilot which is inserted in the valve guide. Only a few seconds are required to reface a cast iron seat, a little more time is necessary for refacing the much harder valve seat inserts. This type tool is always used in conjunction with a pilot inserted through the valve guide so that a concentric refinishing job will be obtained, Fig. 97.

Hand Grinding

To hand-grind a valve into its seat, a thin coating of valve grinding compound is spread on the face of the valve. A common valve grinding tool, Fig. 98, has a rubber suction (vacuum) cup which is pressed against the valve and then it is rotated between the two hands while pressing down just enough to force the valve against its seat. A light spring, just strong enough to lift the

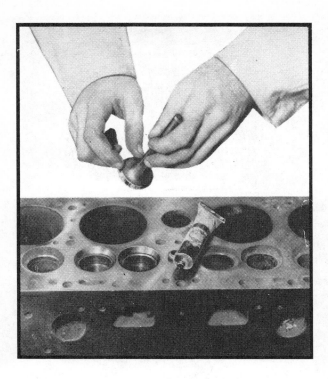

Fig. 104 Using Prussian Blue to test valve tightness

valve, should be placed under the head of the valve. After every few oscillations the spring is allowed to lift the valve and the valve is rotated through a part of a revolution. Then the valve is ground for a few more oscillations, and so on. About every minute the valve is removed and both the valve face and the seat are inspected. When the seat has been ground down so that bright metal shows all around its circumference, and when a similar band of smooth surface appears on the valve face, the valve may be considered finished.

Hand grinding is a slow, time-consuming job and the accuracy obtained is not likely to be as good as with suitable power tools, especially if the valve stem is somewhat loose in its guide.

After each valve is ground, both the valve and its seat should be carefully cleaned of abrasive with a rag moistened with gasoline or kerosene.

Valve Seat Widths

When grinding valve seats it is important to have the seat the correct width. If the seat is too narrow, it may wear unduly because it is not wide enough properly to resist the continual pounding against it. Also, an exhaust valve seat that is too narrow may overheat the valve because of inadequate contact area between valve and seat.

On the other hand, if the valve seat is too wide, the valve may not seat tightly. Other things being equal the wider the seat the more likely it is to leak.

The reason the valve with the wider seat is more likely to leak is that it has less pressure per square inch to hold it down. If the valve spring pull is 100 pounds and the valve face area for the 1/16″ width were .293 square inch, then the pressure on the seat would be 341 pounds per square inch. However, if this seat were enlarged to a width of 3/16″, the pressure would be reduced to 114 pounds per square inch because the area would be three times as great for the same pounds pressure of the spring.

Therefore, for best valve service, the width of the seat should be neither too much nor too little. The recommendations of the various car makers should be carefully followed, but if in doubt as to what width is best, you can't go wrong if you make the seat width 1/16″, Fig. 99. Note that this picture represents a typical valve in a *new* engine.

The face of the valve should extend about 1/32″ beyond the face of the seat, Fig. 99. This overlapping is desirable so that the valve face will have full contact with the seat. On the other hand, too much overlap invites overheating of the rim of the exhaust valve.

After a valve is ground, if its rim margin is less than 1/32″ thick, the valve should be discarded because a thinner rim may become so hot that it will cause pre-ignition.

When a valve seat is ground, both the width of the seat and the outer diameter of the seat are increased. Fig. 100 shows the outer diameter of the seat has increased, the width of the seat is greater, the rim of the valve no longer overlaps the outer circumference of the seat, the valve has moved closer to its tappet.

Usually it is only necessary to remove a few thousandths from the seat to smooth it up. In fact so little may be removed that seat width etc., may not be appreciably affected. On the other hand, if the seat is badly burned, it is necessary to grind off all the burned metal, in which case, the dimensions mentioned may be greatly increased as shown.

The outer diameter of the seat can be reduced by means of a 15-degree grinding wheel, Fig. 101. Enough material is ground off the seat so that its outer diameter is reduced to what it was in a new engine and overlap is now correct. This operation also reduces the width of the seat.

But if the seat is still too wide, its width should be reduced to the correct dimension with a 75-degree grinding wheel, Fig. 102.

Removing material from the top of the seat is called topping while removing it from the throat of the valve port is called throating. The wheel angles for topping usually are either 15 or 20 degrees and for throating 60 to 75 degrees.

When considerable material is removed from the valve seat, the inner rim of the seat may extend down into the rough casting which forms the valve port. If this rim is uneven, the throating wheel should be used to make the rim perfectly circular but in doing so be careful not to make the seat too narrow.

In some cases where exhaust valve burning is difficult to overcome, it has been found very satisfactory to make the angle of the valve face ¼ to 1 degree flatter than the seat face, Fig. 103. In other words, with a 45 degree seat, the angle of the valve face is made somewhere between 44¾ and 44 degrees. (The angle between the valve face and the seat is called (the interference angle.) Thus the valve seats tightly only at the outer edge of the seat as shown. The narrow contact between valve and seat means that the unit pressure is very high between the narrow contacting surfaces, hence the tight seating. Also because of the high contact pressure any particles lodging on face or seat are likely to be quickly pulverized and blown away.

The use of an interference angle naturally reduces the contact area between valve face and seat and therefore it might be assumed that the flow of heat from the hot exhaust valve would be throttled and exhaust temperature would be dangerously increased. Nevertheless, it is apparent that with a narrow contact the pressure with which the valve face pushes against the seat is so high and the contact is so intimate that heat flows from valve to seat very readily whereas with valves and seat having exactly the same angle the full width contact means that the intensity of pressure between them is relatively low and therefore the contact is less intimate. Hence, heat transfer from valve to seat is not as good as might otherwise be expected.

It should also be noted that exhaust valves usually cause no trouble until they begin to leak. Hence, since the narrow contact suppresses leakage the valve job is likely to last longer. And with leakage almost entirely absent, the actual exhaust valve temperature may be less with a narrow contact than with a wider one which permits some leakage right from the start. This leakage may not be enough to be noticed in the operation of the engine but it may be sufficient to elevate the valve temperature to a point which hastens the destruction of the valve job.

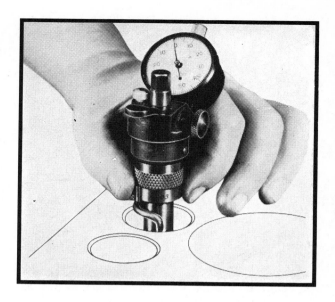

Fig. 105 *Testing accuracy of valve seat with dial gauge*

The interference angle, as it is called, is obtained by adjusting the workhead on the valve grinding tool to somewhere between 44 and 44¾ degrees (in the case of a 45-degree seat). For a 30-degree seat, the valve face angle should be 29 to 29¾ degrees.

After the valve grinding job is finished, it is desirable to determine how tightly the valve fits into its seat. Two simple methods are as follows: (1) Spread a thin coating of Prussian Blue evenly on the valve face, Fig. 104. Put the valve on its seat and rotate it a few times. If the valve is tight, there should be a continuous band of blue on the seat. (2) Make a few pencil marks across the width of the valve face. The marks should be fairly evenly spaced around the circumference. Put the valve on its seat and rotate it a few times. Remove the valve and inspect the pencil marks. If all of them show signs of being rubbed by the seat, the job is satisfactory.

After refinishing valve seats, the accuracy of the job may be checked with a dial gauge, Fig. 105. The gauge is mounted in a fixture which includes a pilot which goes into the valve guide. An arm on the gauge is brought into contact with the valve seat and is moved around the full circumference of the seat. If the pointer on the gauge varies more than .002″ the seat is not sufficiently accurate. For a really good job the variation should be .001″ or less.

VALVE SEAT INSERTS

Some engines have hard steel valve seat inserts for the exhaust valves while others are equipped

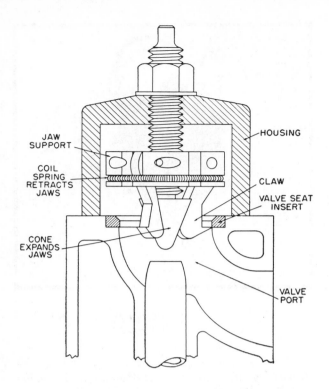

Fig. 106   Valve seat insert puller

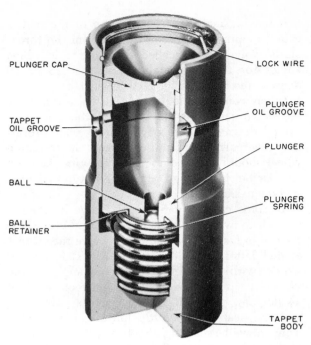

Fig. 107   Unit type hydraulic valve lifter

with these inserts for both intake and exhaust valves.

Seat inserts, especially for exhaust valves, occasionally require replacement. Hard driving with a badly leaking exhaust valve may burn the exhaust seat beyond repair.

The insert is readily removed with the puller shown in Fig. 106. The tool consists of an inverted cup-like housing in which is mounted three claws or jaws which are carried in a jaw support threated on the lower end of the screw. The position of the claws can be moved up or down by turning the nut. The claws are expanded against the sides of the valve port by turning the screw while holding the nut stationary.

To pull out the insert, the device should be adjusted so that the edges of the claws contact the valve port firmly about ⅛″ below the insert. Then pull the claws upward by turning the nut while at the same time turn the screw so that the claws will be certain to grip the insert properly. Then keep on turning the nut until the insert is removed.

If such an insert puller is not available, the insert can be broken and removed in two pieces. Using a drill somewhat smaller than the insert, drill two holes at opposite sides of the insert but be careful not to drill all the way through the seat as damage to the block (or head) will result. Use a sharp cold chisel to cut through the holes and thus split the seat in two halves.

After the seat has been removed, clean the counterbore in the block so that the new seat, when installed, will sit squarely in the recess. Note also, that dirt in the recess will prevent proper conduction of heat from the seat to the block.

The seat is slightly larger in diameter than the recess. Therefore, to install a new seat, pack it in "dry ice" for at least 15 minutes in order to contract it. Line the cold seat up squarely with the recess and use a hammer and block of wood to force it in place.

If a standard size insert fits too loose, the recess should be enlarged by counterboring to fit an oversize seat. Check the diameter of the oversize seat before counterboring. For best results, the insert should be .003″ larger than the diameter of the recess when both are at the same temperature (room temperature) although an insert which is anywhere from .002″ to .004″ larger is considered satisfactory.

Valve seat inserts are so hard that when they need refacing a high-speed grinding stone should be used. They cannot be ground satisfactorily by lapping with valve grinding compound.

HYDRAULIC VALVE LIFTERS

The purpose of the hydraulic valve lifter (or tappet) is to provide perfect adjustment at all times between the valve stem and the lifter on L-head engines and between the rocker arm and the

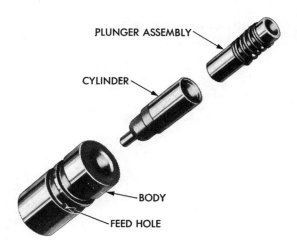

Fig. 108 *Layout of a hydraulic valve lifter used in some Chrysler-built engines. In this type the plunger and cylinder constitute an assembly which is inserted in the body. This design may be called the removable plunger type lifter*

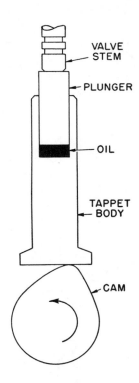

Fig. 109 *This diagram shows how oil is trapped between the plunger and the tappet. The oil is incompressible and, therefore, provides a rigid drive from tappet to plunger*

valve stem on overhead valve engines. The oil-operated hydraulic lifter provides this perfect adjustment automatically. Fig. 107 illustrates the unit lifter while Fig. 108 illustrates a Chrysler lifter in which the plunger and the cylinder constitute an assembly which is inserted in the lifter body.

An ideal hydraulic lifter would be one which keeps in constant contact with the valve stem or push rod regardless of how the valve stem or push rod and cylinder block expand and contract under variations in temperature. With constant contact, the clearance is always zero when the valve is closed.

With this ideal of zero clearance, valve noise is reduced to a minimum. During the valve opening process, the lifter is already in contact with the stem and therefore the noise of contact is eliminated. During the closing process, the valve comes gradually to rest on its seat instead of hitting it with a sharp noisy blow.

This zero clearance obviously cannot be obtained with screw-adjusted valve lifters but is obtained with automatically adjusted hydraulic lifters. This statement, however, should not be construed as unqualified praise for the hydraulic lifter. Both the hydraulic lifter and the screw-adjusted lifter have their advantages and their disadvantages.

The principle of the hydraulic lifter is easy to understand. Fig. 109 is a simplified diagram in which a plunger is slidably mounted in a tappet. At the base of the plunger is a small chamber filled with oil. Since oil (and all other liquids) is practically incompressible, the oil trapped in this chamber provides a positive drive from cam to valve stem. When the valve is being opened by the cam, the drive is just as positive as it would be if a steel spacer were substituted for the oil. It is a mistake to assume that the oil gives a cushioning effect. On the contrary, the oil provides a perfectly rigid drive.

Although the plunger diameter, Fig. 109, has a clearance of a mere fraction of a thousandth of an inch, it is obvious that there must be some slight leakage between the plunger and its bore whenever the valve is open because of the high pressure of the valve spring which pushes down on the plunger.

To compensate for this leakage, oil is supplied under pressure, Fig. 110, to the oil chamber beneath the plunger.

With engine running and the valve momentarily

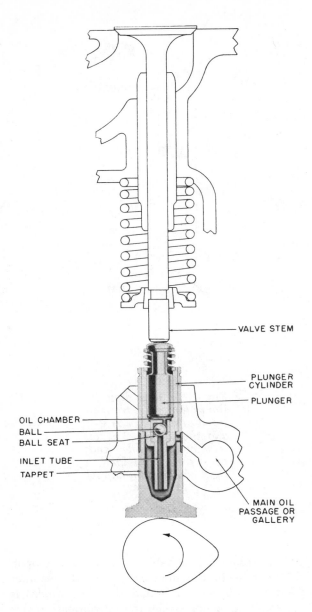

VALVE STEM

PLUNGER CYLINDER

PLUNGER

OIL CHAMBER
BALL
BALL SEAT

INLET TUBE
TAPPET

MAIN OIL
PASSAGE OR
GALLERY

Fig. 110 Oil flows into the oil chamber as shown by the arrow. The gallery is supplied with oil under pressure by the oil pump. Note that with valve closed as shown, an oil groove in the tappet registers with the oil passage leading from the gallery. Oil flows from the gallery through tappet oil groove, through tappet oil hole to check valve to oil chamber. If the oil chamber needs oil, oil pressure unseats the ball and forces oil into it

resting on its seat, its stem is not exerting pressure on the plunger. Therefore the oil pressure in the gallery is more than enough to force the ball check valve off of its seat so that replacement oil may be forced into the oil chamber.

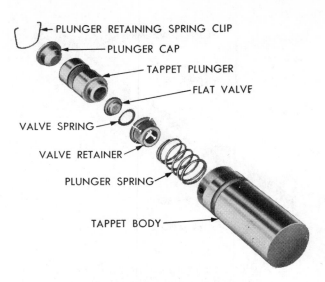

PLUNGER RETAINING SPRING CLIP

PLUNGER CAP

TAPPET PLUNGER

FLAT VALVE

VALVE SPRING

VALVE RETAINER

PLUNGER SPRING

TAPPET BODY

Fig. 111 Unit type hydraulic valve lifter of the flat valve type

However, just as soon as the cam starts to open the valve, the valve stem presses on the plunger which in turn presses on the oil. This oil pressure forces the ball check valve tight into its seat. Then the oil, trapped in the oil chamber, rigidly transmits the upward tappet motion to the plunger which transmits it to the valve stem and in this way the valve is opened. There is a rather light spring between the top of the plunger and the tappet. It pushes upward on the plunger and downward on the tappet so that the plunger is always in contact with the valve stem and of course the tappet is always in contact with the cam.

We are now in a position to explain how the automatic tappet adjustment is obtained regardless of whether the valve parts have been lengthened by a hot engine or shortened by a cold engine.

The oil chamber ought to be called an adjusting chamber. To explain how the chamber works, let us assume that the engine is cold but that it was thoroughly warm when it was last used.

The valve stems have contracted during the cooling off period. During contraction, the plunger springs pushed the plunger upward, thus keeping them in constant contact with their valve stems. The upward movement of the plungers therefore has increased the volume of the adjusting chambers and these chambers are now not completely filled with oil. In other words the valves are out of adjustment on the loose side.

However, when the engine is started it only requires a few seconds for the oil pressure to deliver oil to the adjusting chambers to fill them up and

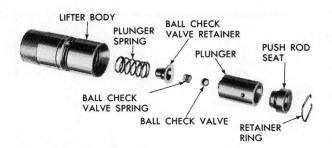

LIFTER BODY
PLUNGER SPRING
BALL CHECK VALVE RETAINER
PLUNGER
PUSH ROD SEAT
BALL CHECK VALVE SPRING
BALL CHECK VALVE
RETAINER RING

Fig. 112 Unit type hydraulic valve lifter of the ball valve type

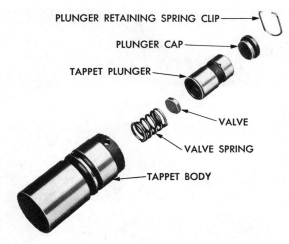

PLUNGER RETAINING SPRING CLIP
PLUNGER CAP
TAPPET PLUNGER
VALVE
VALVE SPRING
TAPPET BODY

Fig. 113 Another unit type lifter with a flat valve. The flat valve is somtimes called valve disc

thus provide the necessary rigid drive from tappet to plunger.

As the engine warms up, the valve stems expand lengthwise, the natural leakage past the plungers permitting them to do so. The leakage is always enough so that there is no danger of a valve being held off of its seat because of rapid valve stem expansion. Actually the leakage is more than adequate.

The correct leakage rate is carefully determined when the hydraulic lifter is manufactured. And it is most important to realize that the leakage rate, although small, is deliberately designed to keep the valve adjustment on the loose side. But this tendency toward looseness is prevented by the engine oil pump which is ready to fill the oil chamber whenever the valve is closed. Therefore the valves are always in perfect adjustment—that is the plungers are always in contact with their valve stems, except for the first few seconds after the engine is started from cold.

The leakage rate is determined by the amount of clearance between the plunger and the plunger bore. The desired clearance is a fraction of a thousandth of an inch and this clearance must be closely approximated on all lifters made for a given model of engine. Therefore after the plungers and bores have been manufactured as accurately as possible they are classified into a series of sizes from a trifle small to a trifle large. Small plungers are assembled to small bores, large to large, and likewise for the sizes in between. In this way satisfactory uniform clearance is obtained in all plungers and bores even though individual plungers and bores may vary in size by a small amount. This "selective" fitting of plungers and bores explains why it is important during repair work that a given plunger always be returned to its own bore.

If the plunger clearance is too great, either because of wear or because of mismating of plunger and bore, the oil may leak away faster

than it is supplied to the oil chamber. The chamber will never be full and therefore the lifter will be noisy.

If the plunger is too tight in its bore because of mismating, the plunger action may be sluggish and noisy lifters may result. Or more likely, the leakage rate will be so slow that the valve is held off its seat when it should be closed.

Most engines have an oil passage drilled the length of the crankcase. This passage is often called a gallery (or galley) and is used to distribute oil from the engine's oil pump to drilled passages leading to the hydraulic valve lifters as well as to the main and camshaft bearings.

Hydraulic valve lifters can be classified as being the unit type and removable plunger type. In the unit type, Fig. 107, the plunger is installed directly in the lifter body. Note that there is an oil groove in both the body and plunger. Oil under pressure is sent from the gallery through the tappet oil groove, tappet oil hole, plunger oil groove, plunger oil hole, past ball check valve to plunger oil chamber.

With the removable plunger type, Fig. 110, the plunger is mounted in a plunger cylinder which is installed in the tappet body. The plunger cylinder assembly is removed as a unit by simply pulling it out of the body.

Other examples of the unit type lifter are shown in Figs. 111, 112, 113.

Hydraulic Lifter Troubles

The most frequent causes of hydraulic valve lifter troubles are dirty oil or failure to change the

87

Fig. 114 Using a hooked wire to remove stuck valve lifter

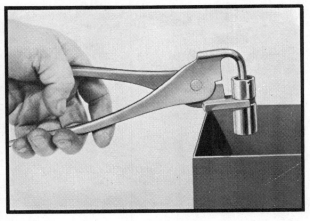

Fig. 115 Special pliers used for checking operation of unit type lifters

oil filter at suggested intervals. When these causes are eliminated, the trouble will generally correct itself without costly disassembly.

The easiest method for locating a noisy valve lifter is by the use of a piece of garden hose about 4 feet long. Place one end of the hose near each valve in progression and listen through the other end. In this manner the sound is localized, making it easy to determine which lifter is at fault.

Another method is to place a finger on the valve spring retainer. If the lifter is not functioning properly, a distinct shock will be felt when the valve returns to its seat.

In most cases where noise exists in one or more lifters, all lifter units should be removed and cleaned. If dirt, varnish or carbon is found to exist

in one unit, it more than likely exists in all the units.

The oil level in the oil pan should never be above the "Full" mark on the dipstick nor below the "Add Oil" mark. Either condition could be responsible for noisy valve lifters.

Oil Level Too High—If the oil level is too high it is possible that the connecting rods can dip into the oil when the engine is running and create foaming. This foam (air mixed with oil) is fed to the hydraulic lifter by the oil pump, and the air gets into the lifters, causing them to go flat and allow the valves to seat noisily.

Oil Level Too Low—Low oil level may allow the oil pump to take in air which, when fed to the lifters, causes them to lose length and allows the valves to seat noisily. Any leaks on the intake side of the oil pump through which air can be drawn will create the same lifter action. When the lifter noise is due to aeration, it may be intermittent or constant, and it will always cause more than one lifter to be noisy. When the oil level and leaks have been corrected, the engine should be run at a fast idle speed for sufficient time to allow all of the air inside of the lifter to be worked out.

Hydraulic Lifter Noises

It should be remembered that worn valve guides or cocked valve springs are sometimes mistaken for noisy lifters. If such is the case, the noise in all probability will be dampened by applying side thrust with your thumb on the valve spring. If the noise is not appreciably reduced, it can then be assumed that the noise is in the hydraulic lifter.

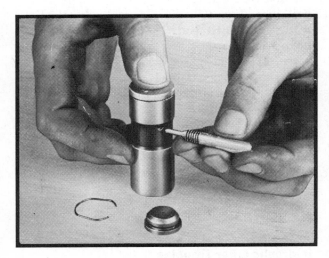

Fig. 116 Using punch to line up oil holes in plunger and body when assembling unit type lifter

Servicing Hydraulic Lifters

Inasmuch as the removable plunger type lifters have not been in use for quite a number of years, the following discussion is concerned only with the unit type lifter.

With this type lifter, Figs. 107 to 113, the units may be lifted out with the fingers. However, if the lifter is stuck in its guide due to varnish or gum substances, use pliers with taped jaws or pull it out with a hooked wire as shown in Fig. 114.

If there is a ring of carbon on the guide just above the top of the lifter it will be necessary to remove the carbon before the lifter can be taken out. Soften up the carbon with a solvent and remove with a stiff bristle brush.

As shown in the illustrations, these units can be completely disassembled. The first step is to remove the lock wire by prying it out of its groove with a small screwdriver or awl. Hold the thumb over the opening to prevent the lock wire from flying away when it comes loose.

If there is a rim of carbon at the upper end of the plunger bore, soak the unit in solvent (lacquer thinner is recommended) and remove the carbon with a stiff bristle brush. The carbon will interfere with the removal of the plunger.

Next turn the unit upside down and strike it against the palm of the hand to force the assembly out. Do this carefully so that the parts, when they do come out, do not scatter. The ball check valve is especially likely to roll out of sight. If pounding the lifter against the palm fails to break the plunger loose, pound it on a block of wood.

Be careful not to mix up plungers and bodies. To avoid this possibility, disassemble and assemble them one at a time. Wash the parts in solvent and then examine them for defects such as scratches, pits and dents. If any are found, discard the complete unit.

Insert the bare plunger in the body. It should move smoothly when pushed gently with a finger. In some cases it will even fall of its own weight if the unit is held vertical. If there is extra resistance, discard the assembly.

To check for leakage, make an air compression test. To do this, turn the plunger upside down and install the ball check valve and the retainer but do not install the spring. While holding the plunger assembly with one hand, slide the body down over the plunger. Still holding the unit upside down, pump the plunger with the little finger or with a screwdriver. If the plunger bounces, the unit is all right. On the other hand if it does not bounce, examine the ball valve or flat valve for damage. If a ball valve is used, examine it for a flat spot and, if one is found, install a new ball of

the same size. However, if the ball or flat valve is in good condition and the compression is poor, the plunger leaks and therefore the complete unit should be discarded.

The tool shown in Fig. 115 is also used to check the compression of unit type lifters; it is used in the same manner described for the removable plunger type.

In testing lifters of this type, the container must be deep enough to completely immerse the lifter assembly. Fill the container with clean kerosene. Remove the cap from the plunger and submerge the lifter. Allow the cylinder to fill with kerosene. Then remove the lifter and replace the cap.

Then, with the tool applied to the lifter as shown in Fig. 115, check the leakdown by compressing the tool handles. If the plunger collapses almost instantly, disassemble the unit, clean it again and retest. If the lifter still does not function satisfactorily, install a new unit, being sure to test the new one before installing it in the engine.

Assembling Lifters

To assemble, turn the plunger upside down and install the check valve, its retainer and plunger spring. Slip the lifter body on the plunger assembly. Turn the assembly over and install the lock wire, being sure it snaps into its groove in the lifter body. A discarded plunger or a special tool for the purpose may be used to press the plunger down in order to insert the lock wire.

If neither a special tool nor a discarded plunger is available, the following method may be used. Assemble the parts as before. Then slide the body down over the plunger. Take a small piece of wire (a paper clip will do) or a small punch, Fig. 116, and insert it in the oil hole in the body. Press the plunger down slowly almost as far as it will go with one finger of the other hand until the wire or punch slips into the oil hole in the plunger. This locks the plunger. Install the plunger cap. Insert the open end of the lock ring in the lifter body. Then force the opposite side of the lock ring into the groove with a small, blunt tool such as a screwdriver.

It is recommended that lifters of this type be filled with engine oil before replacing them in the engine. However, many mechanics rarely take the time to do this. In other words, they are installed dry and then filled by the oil pressure supplied by the idling engine. Apparently no harm is done by neglecting to fill the lifters. However, as there is some risk of scuffing, at least dip the plungers in engine oil before installing them in the engine.

A better method, which is quick and easy, is to

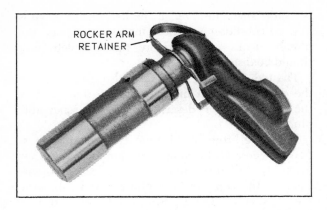

*Fig. 117 Rocker arm retainer
and lash adjuster*

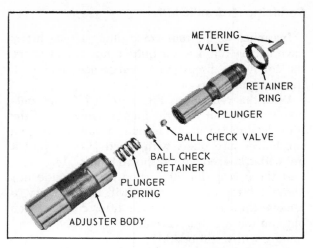

*Fig. 118 Hydraulic lash
adjuster disassembled*

put the dry lifter in a pan of engine oil and pump the plunger until the lifter is reasonably full. The oil level in the pan, of course, should be high enough to cover the oil hole in the side of the lifter body. The plunger should be pumped just a few strokes until the plunger travel becomes a minimum. It is possible to pump the plunger until it is rigid.

HYDRAULIC LASH ADJUSTERS

The Pontiac Tempest overhead camshaft engine uses this device to maintain zero lash automatically. Located in the cylinder head, it serves as the fulcrum of the rocker arm and also locates the rocker arm accurately with the camshaft. A spherical radius on the lash adjuster, which mates with a spherical socket on the rocker arm, absorbs the transverse thrust due to friction between the cam lobe and rocker arm. The lash adjuster is attached to the rocker arm by means of a retainer, Fig. 117. Operation and service of the adjuster is similar to that of a hydraulic valve lifter used in a conventional push rod engine. Fig. 118 shows the adjuster disassembled.

VALVE ROTATORS

In its 1970 models, Oldsmobile adopted valve rotators, Fig. 119. These are devices fitted to the valve stem which force the valve to rotate about three degrees each time it is lifted off its seat. Purpose of the device is to keep the valve seat clean so that the valve can close properly every time the cam permits it to drop on its seat. It prevents the formation of deposits on the valve face and seat, improves valve cooling because of the better contact with the seat, and greatly lengthens the life of the valve. Valve rotators have

been used for many years on truck engines, but this is the first use in passenger cars.

MODIFYING VALVE GEAR FOR HIGH PERFORMANCE

Many of the operations required to modify an engine for high performance are no different from those used in normal engine service. These include checking valve seat face runout, valve stem clearance, reaming valve guides, etc. Other operations that are of particular importance in modifying for high performance include:

Refinishing Valve Seats

Refinishing the valve seat should be closely coordinated with the refacing of the valve so that the finished seat and valve face will be concentric and the specified interference angle will be maintained. This is important so that the valve and seat will have a compression-tight fit. Be sure that the refacer grinding wheels are properly dressed. For high performance applications, some specifications call for an intake valve seat and face angle of 30° and exhaust valve seat and face angle of 45°. When grinding valve seats remove only enough stock to clean up pits and grooves or correct the valve seat runout. On the valve seats of most engines, a 60° angle grinding wheel can be used to remove stock from the bottom of the seat (raise the seats) and a 30° angle wheel can be used to remove stock from the top of the seats (lower the seats).

On stock engines, the finished valve seat should contact the approximate center of the valve face. However, for high performance engines, valve seats should contact the valve face as far out on

Fig. 119 Bottom and top views of valve rotator

BOTTOM

TOP

the edge of the valve as possible. This provides a larger effective valve opening and improves engine breathing. The point of valve seat and face contact can be found by using machinist blueing. Coat the seat with blueing, set the valve in place and rotate it with light pressure. The blueing will transfer to the valve face, showing where the contact occurs.

Valve Seat Width

After the seat has been finished, use a seat width scale or a machinist scale to measure the width. Stock engine seat widths usually run between .050″ and .080″. This allows for more efficient heat transfer during periods when the engine is idling or operating at low speeds. For drag strip operating, Ford, as an example, specifies an intake seat width of .035″ and an exhaust seat width of .050″. It is important to remember that valve seat width is critical. Too wide a seat can collect deposits of carbon; too narrow a seat won't dissipate heat fast enough.

Refacing Valves

If the valve face runout is excessive, or if the face is pitted or grooved, the valves should be refaced. Remove only enough stock to correct runout or to clean up the pits and grooves. If the edge of the valve head is less than 1/32″ thick after grinding, replace the valve as it will run too hot. The interference angle of the valve and seat should not be lapped out. Remove all grooves and score marks from the end of the valve stem and chamfer it as necessary. Do not remove more than .010″ from the end of the stem. The valve stem,

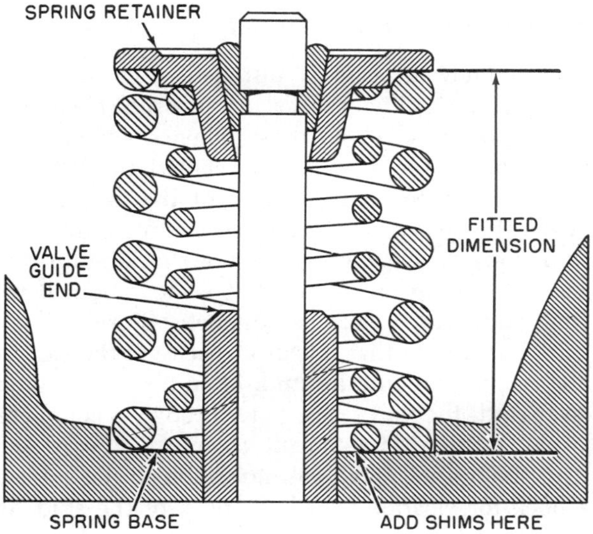

SPRING RETAINER

VALVE GUIDE END

FITTED DIMENSION

SPRING BASE

ADD SHIMS HERE

Fig. 120   Spring and retainer

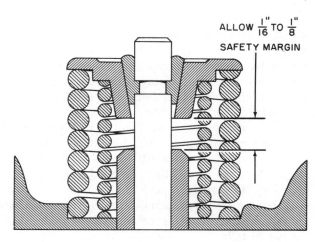

ALLOW $\frac{1}{16}$″ TO $\frac{1}{8}$″

SAFETY MARGIN

Fig. 121   Checking spring retainer-to-guide clearance

from the valve lock groove to the end, must not be shorter than the minimum specified length.

If the valve and/or valve seat has been refaced, it will be necessary to check the clearance between the rocker arm pad and the valve stem with the valve gear installed in the engine.

Installing Larger Valves

Using larger valves may or may not improve engine breathing ability. Larger valves run hotter than smaller ones and they are heavier as well. This could mean burned valves or high speed valve float if the proper precautions are not taken. In addition, valve shrouding by the combustion chamber walls may occur.

In many cases, increasing valve size will necessitate other modifications such as chamfering or boring the cylinders and using deflector or dome type pistons to achieve the best results. If larger valves are to be installed, the following points should be borne in mind:

1. Use a hand-operated reamer to remove large amounts of stock from the valve ports.
2. Measure the face of the new valve and stop reaming when the diameter of the reamed hole equals the smallest diameter of the valve face.
3. It may be necessary to remove some stock from the combustion chamber walls to reduce valve shrouding and improve mixture flow.
4. Check piston-to-valve clearance.

Valve Springs

For best high speed performance, it is important that valve springs be tested and properly installed. A spring testing fixture should be used to measure "full open" pressure and "seated" (fitted dimension) pressure. When checking the inner spring on a dual spring arrangement, remember that the inner spring often rides on a stepped portion of the spring retainer. This must be allowed for by compressing the inner spring more than the outer spring. For example, if the step is 1/16" thick, the inner spring must be compressed 1/16" more than the outer one.

When the spring is at "Full open" pressure, check for coil stacking or coil bind. Coil bind can ruin the valve train. If a .010" to .012" feeler will go between each coil, there is no danger of coil stacking. On some high performance cam kits with inner and outer valve springs, it may be necessary to reduce the diameter of the valve guides to permit installation of the inner springs. Also, the outer spring seats may have to be enlarged to accommodate the larger spring diameter. This can

be done by using a special counterboring tool available from most speed shops. Once the springs are installed, check the fitted dimension. Measure the spring only, Fig. 120, and do not include the thickness of the retainer. If the measurement is less than that specified, the spring seat will have to be counterbored to obtain the correct dimension. If the measurement is greater than that specified, use spacer washers under the spring.

Another important item which should be checked is spring retainer to valve guide clearance, Fig. 121. This is critical when a high lift cam is fitted. Checking should be done with the valve gear assembled and adjusted to specifications. Rotate the engine until the valve is in the full open position. There should be 1/16" to ⅛" clearance between the valve guide end and the spring retainer. It may sometimes be necessary to machine the valve guide slightly to obtain the proper clearance.

Rocker Arm Geometry

Correct rocker arm geometry will allow longer engine operation at high speeds with little valve guide bore wear. With the valve positioned at 40% to 50% of maximum lift, the rocker arm tip radius should coincide with the center line of the valve stem. If the rocker arm tip coincides with the center of the valve stem at less than 40% of total lift, the push rod should be shortened by the amount of the difference between the two center lines. Where the rocker arms are mounted on a rocker shaft, the rocker arm geometry can be off if the adjusting screw is screwed in to the limit of its travel. A longer pushrod should be used to correct for this condition.

Rocker Arm-to-Stud Clearance

If a high lift cam is installed, exceeding approximately .450", rocker arm-to-stud clearance should be checked on those engines which have ball-type rocker arms. There should be .040" to .060" clearance on both sides of the rocker arm when the valve goes through a complete opening and closing cycle. One method of checking for this clearance is to use a piece of solder about 6 in. long and 1/16" (.062") in diameter. Form a hook in the solder so that it will fit between the slot in the rocker arm and the stud. Then turn the engine over until the valve has gone through a complete cycle. If there is sufficient clearance, the rocker arm should not make an impression in the solder. Check for clearance on both sides of the stud. If the clearance is not great enough, be careful only to elongate the slot in the arm and not widen it.

Timing Gear, Chain, Belt & Camshaft Service

VALVE TIMING

Except Pontiac Tempest OHC Engine

When the camshaft is driven by gears it is the custom to mark the gears in one of two ways, Figs. 122 and 123. In Fig. 122 there are three circular marks on the meshing teeth whereas in Fig. 123 there are two marks which should be lined up as shown.

In the case of timing chains and sprockets, there are also two common methods of marking. In Fig. 124 it will be seen that marks on the sprockets should line up with a line drawn between the centers of the crankshaft and the camshaft. In Fig. 125, the correct camshaft setting is determined by the number of links between the two marks on the sprockets.

It is important to note that when meshing a timing chain with its sprockets that the marks should always be on the taut side of the chain. This advice applies to the markings illustrated in Fig. 125 and not to Fig. 124. To determine the taut side, just remember that the rotation of the crankshaft sprocket pulls on the chain and thus rotates the camshaft sprocket. Obviously the side of the chain that does the pulling is the taut side.

The crankshaft gear or sprocket is keyed to the crankshaft and cannot slip unless perchance the key is sheared off—a rare and unlikely possibility. Similarly the camshaft gear or sprocket is firmly

and positively attached to the camshaft. It is either keyed to the camshaft or bolted to a hub on the camshaft.

Chain slippage is quite rare on modern cars because the chains are relatively short—the shorter the chain the less it stretches as it wears. A possible exception would seem to be the chain used in the Jeep overhead camshaft engine, Fig. 126. However, this is prevented by the tensioner device shown. The chain tensioner maintains a constant pressure against the chain to compensate for normal stretch and wear of the chain. Whenever the chain case cover is removed for any reason, the chain should be checked for wear as suggested in Fig. 127. If the space between the arrows indicated is less than the minimum, a new chain should be installed. The procedure for installing the chain is covered further on in this chapter.

Timing marks are located either on the vibration damper, crankshaft pulley or on the front face of the flywheel and usually the marks apply to No. 1 cylinder. If no marks are found on the vibration damper or pulley look for an inspection hole on the front side of the flywheel housing. Remove the hole cover and then turn the flywheel until the marks on the flywheel appear.

Turn the engine over until the distributor rotor

Fig. 122   The valves are correctly timed when the timing marks mesh

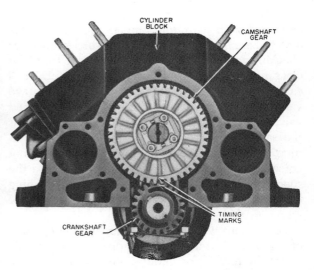

Fig. 123   Valve timing is correct when the timing marks are lined up as shown

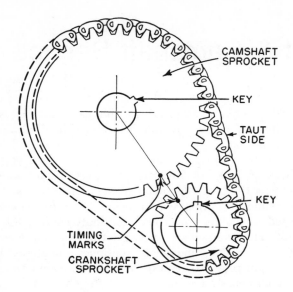

Fig. 124 *The marks on the sprockets should be lined up as shown*

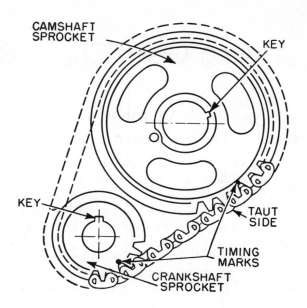

Fig. 125 *Valve timing is correct when there are ten chain teeth between timing marks*

points to the segment connecting with the spark plug cable running to the last cylinder. Then rotate the crankshaft slightly by pulling on the fan blades until No. 1 intake valve is just ready to open.

The reference mark on the flywheel housing may be a pointer or may be merely a straight line cut into the flywheel housing. If there is no pointer or mark, it is assumed that the "zero" point is at the center of the opening. In the case of vibration dampers or pulleys there is a pointer or an indicating mark on the timing case cover to indicate the "zero" point.

To check valve timing, the valve stem-to-tappet clearance must be known. Note that on cars equipped with hydraulic tappets the clearance is always zero. On other cars the valve clearance for checking valve timing is often different from the clearance required for running the engine. In such cases the factories recommend that No. 1 intake valve, for example, be adjusted to the specified valve timing clearance. If the factory recommends the same clearance for valve timing as for running, the clearance should be checked and if not correct it should be adjusted. Be sure to use the specified clearance, as otherwise the valve timing check may be in error by five or ten degrees or more.

On engines with hydraulic tappets, the opening of No. 1 intake valve can be approximately determined by feeling for initial motion in the body of the tappet while the engine is slowly rotated to open the valve. The slightest upward motion of the tappet indicates the valve opening point.

A more accurate method is to mount a dial indicator on the engine. The tip of the gauge should contact the lifter body, *not the plunger*. Then turn the engine over slowly until the lifter body causes the gauge pointer to move ever so slightly.

If a dial gauge is not available, drain the oil from the hydraulic lifter and then eliminate the clearance between the plunger and the valve stem with feeler stock. When doing this be sure that the valve is fully closed.

After adjusting No. 1 intake valve, check the timing as follows: Remove the distributor cap and turn the engine over until the rotor approaches the distributor segment connected to the spark plug wire running to the last cylinder in the block. Then turn the engine slowly by pulling on the fan blades with one hand while the other hand is used to rotate the tappet on an L-head engine or the push rod on an overhead engine. When the tappet or push rod begins to bind, No. 1 intake valve is ready to open. If valve timing is correct the valve opening mark should be in line with the reference mark or pointer. If the timing is correct within a few degrees, it should be considered satisfactory. But if the timing is wrong by 10 or more degrees, remove the cover and replace the gears or chain.

To determine how much the error of one tooth amounts to in degrees, count the number of teeth on the crankshaft gear or sprocket and divide this figure into 360 degrees. For example, an error of one tooth is 15 degrees if there are 24 teeth in the crankshaft gear (360 divided by 24).

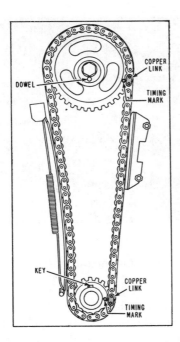

Fig. 126 *Valve timing marks for Jeep overhead camshaft engine*

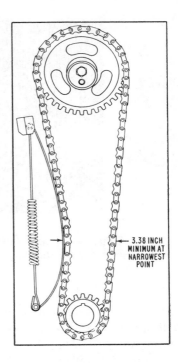

Fig. 127 *Measurement for limit of timing chain wear on Jeep overhead camshaft engine*

Any mistake in interpreting the meshing marks must result in an error of at least one tooth. It can never be less.

Note also that valve timing cannot be "adjusted" less than one crankshaft tooth or 15 degrees in the example mentioned. To make this remark perfectly clear let us state that it is impossible to adjust valve timing 2 or 3 or 5 degrees. (However, it is possible to change the valve timing any desired number of degrees by cutting a new keyway in the crankshaft, camshaft or both.)

Pontiac Tempest OHC Engine

In this engine, the camshaft is driven by a fiber glass reinforced rubber timing belt. Valve timing is correct when all of the marks are lined up as shown in Fig. 128.

The timing belt can be adjusted as follows:

1. Remove 3 screws on front of top cover. Lift up cover to disengage side clips. Remove retaining clips from cover.
2. Using the tool shown in Fig. 129, set the pointer of the fixture on the zero mark. *This calibration must be performed prior to each use of the fixture to insure an accurate belt adjustment.*
3. Remove camshaft sprocket-to-camshaft bolt and install the fixture on the belt with the rollers on the outside (smooth) surface of

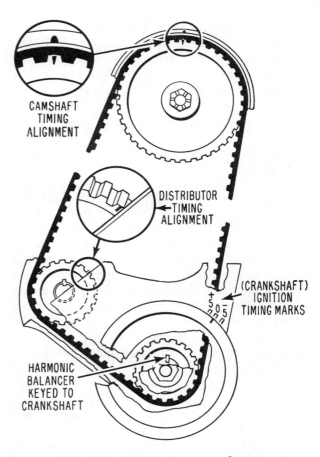

Fig. 128 *Valve timing marks for Pontiac Tempest OHC engine*

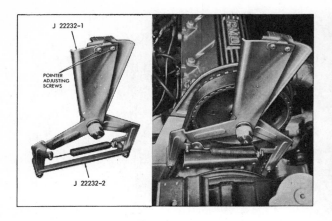

Fig. 129 *Adjusting timing belt with tension fixture shown*

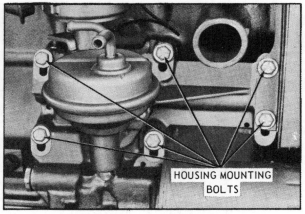

Fig. 130 *Accessory drive housing mounting bolts*

belt. Thread the fixture mounting bolt into camshaft sprocket bolt location finger tight.

4. Squeeze indicator end (upper) of fixture and quickly release so the fixture assumes released or relaxed position.

5. With the tool installed as directed, adjust accessory drive housing, Fig. 130, up or down as required to obtain a tension adjuster indicator centered in the green range with drive housing mounting bolts torqued to 15 ft.-lbs.

6. Remove tension fixture and install sprocket retaining bolt, making sure bolt threads and washer are free of dirt. Install cover.

CAMSHAFT THRUST PLATE

It should be realized that the camshaft must be held in a fore and aft position so that camshaft bearings and bearing journals line up and so that the cams are in line with their tappets. There are the following common ways of doing this.

In Fig. 131, a thrust plate or washer is bolted to the front of the cylinder block. The hole in the plate is smaller than the outer diameter of the hub of the sprocket and also smaller than the front camshaft journal. Therefore the plate prevents the camshaft from moving either to front or rear of its correct position.

On all engines equipped with thrust plates it is desirable to install a new thrust plate whenever a new camshaft gear is installed. There should be a slight amount of clearance at the thrust plate, say .002".

However, on most engines with thrust plates the parts are so designed that when the camshaft gear is properly installed the correct clearance is obtained. For example, in Fig. 131, there is a narrow ring spacer washer between the camshaft

and the hole in the thrust plate. The washer insures correct clearance because it is a trifle thicker than the thrust plate and thus prevents drawing up the gear so tight that it might clamp itself to the thrust plate.

Fig. 132 illustrates a different construction which accomplishes the same purpose. There is a shoulder on the camshaft sprocket hub which butts against the camshaft and thus provides the required thrust plate clearance.

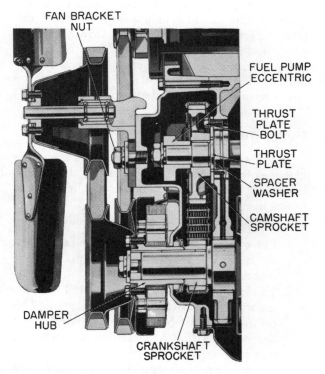

Fig. 131 *Eccentric for driving fuel pump is to left of camshaft sprocket*

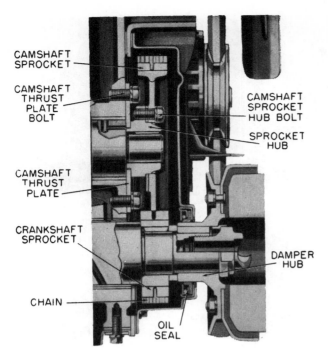

CAMSHAFT SPROCKET

CAMSHAFT THRUST PLATE BOLT

CAMSHAFT THRUST PLATE

CRANKSHAFT SPROCKET

CHAIN

CAMSHAFT SPROCKET HUB BOLT

SPROCKET HUB

DAMPER HUB

OIL SEAL

Fig. 132 Camshaft sprocket is bolted to pressed-on hub

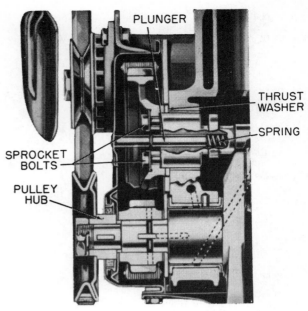

PLUNGER

THRUST WASHER

SPRING

SPROCKET BOLTS

PULLEY HUB

Fig. 133 Cam sprocket is bolted directly to camshaft

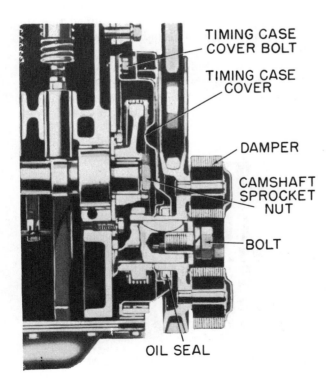

TIMING CASE COVER BOLT

TIMING CASE COVER

DAMPER

CAMSHAFT SPROCKET NUT

BOLT

OIL SEAL

Fig. 134 Camshaft and crankshaft sprockets are retained by a nut and bolt, respectively

In Fig. 133, there is a thrust washer which prevents the cam sprocket from moving rearward while a spring-backed plunger prevents the camshaft from moving forward.

In Fig. 134, the thrust plate clearance is correct when the nut on the end of the camshaft is drawn up tight, the length of the hub being just right to provide the necessary clearance.

The thrust plate or washer should not be confused with the engine front plate. Fig. 135 makes this point clear. The thrust washer, which is between 3 and 4 inches in diameter, lies within a hole in the front plate. A thrust spacer is located between the camshaft and a hole in the thrust plate. This spacer insures correct lengthwise clearance between cam gear hub and thrust washer. The latter, it should be noted, is bolted to the front of the block.

Depending on the engine design the fit between the hub of the timing gear or sprocket and its camshaft or crankshaft may be any one of three, namely: (1) snug fit, .001″ clearance; (2) tight fit, .0004″ clearance; (3) press fit: shaft is .003″ or more *larger* than hub hole.

With No. 1 it should be possible to remove and replace the gear by hand although gentle tapping with a soft hammer may be required.

The same may hold true of No. 2 although a pry bar, Fig. 136, or a puller may be required to remove the gear. Likewise, installation of the gear may require a hollow drift and a hammer.

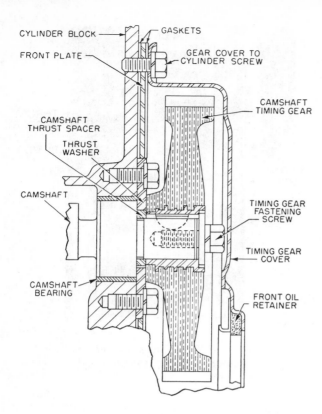

Fig. 135 *Drawing showing relationship between thrust washer and front plate*

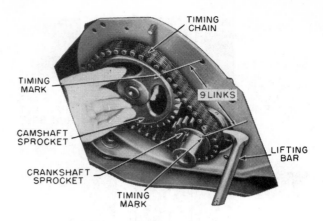

Fig. 136 *Pry bar; also called lifting bar*

No. 3 demands an arbor press. If the clearance is somewhere between No. 2 and 3, a puller or an arbor press may be required depending on how tight is the fit. In all three cases corrosion or rust may make the gear more difficult to remove. Sometimes, both camshaft and crankshaft hubs are press fits, as is the case with Fig. 137, and then again, only the camshaft hub will be a press fit. If an arbor press must be used the shaft or shafts must be removed from the engine.

CAMSHAFT GEAR OR SPROCKET

The camshaft gear or sprocket may be attached to the camshaft in various ways.

In Fig. 131, the sprocket is keyed to the camshaft. In front of the sprocket is the fuel pump eccentric. The eccentric and sprocket are held onto the shaft by a large nut and lockwasher. The hub of the eccentric pushes against the hub of the cam gear to hold it firmly in place.

In Fig. 133, the sprocket is attached directly to the front end of the camshaft by four cap screws with lock washers.

In Fig. 132, the camshaft sprocket is bolted to a hub which is keyed to the camshaft. The hub is

pressed on tightly so that no nut or bolt is required to keep it in place.

The engine shown in Fig. 134 has the camshaft sprocket keyed to the camshaft and retained by a large nut.

If the camshaft gear or sprocket is bolted to a hub on the camshaft, it is obviously readily removed by taking out the bolts, Fig. 132, and the same is true of a sprocket which is bolted directly to the camshaft, Fig. 133.

If the gear or sprocket is keyed directly to the hub and if it is retained by a nut or bolt on the end of the camshaft it may be assumed that the gear is not too tight a fit and therefore can readily be removed by a puller, or pulled off by hand. This construction is shown in Figs. 131 and 134.

If force is necessary to install a gear or sprocket which is keyed and bolted to the camshaft, a

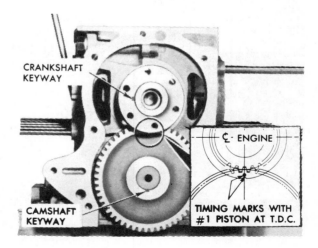

Fig. 137 *Corvair timing gears are pressed on the shafts, requiring no additional fastening device*

Fig. 138 Removing fan pulley. Same type of puller is sometimes used for removing vibration damper

hammer and hollow drift may be used provided the camshaft is braced against moving rearward by inserting a block of hard wood of just the right thickness between the front side of camshaft bearing No. 2 and the adjacent cam. Without this precaution the camshaft will be driven back against the welch plug at the rear. This will either drive it out completely or cause it to leak.

If the gear goes on very easily, it may be feasible to tap the gear gently until it is on far enough so that the bolt (or nut) can be "started." Then the bolt is turned up tight to push the gear home. Also note that a longer bolt with suitable washers may be used in similar fashion to push the gear on.

If there is no retaining nut or bolt, it may be assumed that the gear is a very tight press fit and that an arbor press must be used to remove the old gear as well as to install the new one. This means, of course, that the camshaft must be removed from the engine. The same remark holds true if the hub (of a bolted on camshaft gear) must be removed.

CRANKSHAFT GEAR OR SPROCKET

The crankshaft gear or sprocket is always keyed to the crankshaft and is pressed on. There is usually a nut or bolt on the end of the crankshaft to hold the gear in place. The nut or bolt presses

against the vibration damper or fan pulley which in turn presses against the gear or sprocket. An exception is that of the Corvair, Fig. 137.

TIMING CASE COVER

The amount of preliminary work necessary to gain access to the cover depends on the amount of working space available. If an inspection reveals that there is enough room to remove the cover by taking off the radiator only, then this procedure should be followed.

In determining what has to be done, consideration should also be given as to whether or not the grille must be removed in order to be able to use pulling equipment on the vibration damper, crankshaft pulley or timing gears.

Fig. 138 illustrates a pulley being removed by a screw-type universal puller. The puller is attached to the pulley by the two cap screws which are threaded into two holes in the pulley. The long center screw is then turned clockwise until its end butts against the crankshaft. Further turning of the screw forces the pulley off the shaft.

Another universal puller, called the clamp type, is illustrated in Fig. 139. The two hooked clamps are slipped behind the rim of the damper; turning the center screw pulls the damper off the crankshaft.

A lifter bar, Fig. 136, may be used to pry off the damper or pulley if it is not on too tight.

If the lower edge of the damper or pulley extends below the frame front cross member, it may be necessary to unfasten the front engine mountings and jack up the engine until the cross member no longer interferes. Usually the engine is raised by means of a jack placed under the front end of the oil pan. If such a procedure is necessary, the oil pan should be protected from damage by placing a board or wooden block between the pan and the head of the jack.

With the foregoing remarks digested, the general procedure, therefore, for removing a timing case cover is as follows:

1. Remove radiator, if necessary.
2. Remove vibration damper or pulley.
3. Remove oil pan if it interferes with taking off the cover.
4. Remove water pump if it or any other units interfere with the work.
5. Remove the cap screws which fasten the cover to the engine block. If any of the cap screws are longer or larger than the others, note which holes they come from and be sure to put them back where they belong when installing the cover.

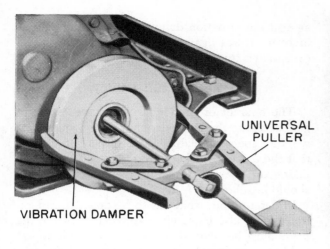

Fig. 139   *Typical puller for vibration damper*

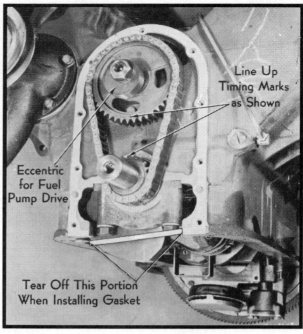

Fig. 140   *The lower portion of the timing cover gasket should be removed*

On some engines, note that the timing case cover is not bolted directly to the cylinder block but instead is bolted to a front end plate which in turn is bolted to the block. Fig. 135 shows the construction used on Studebaker 6-cylinder engines. Note that there is a gasket between the plate and the block as well as between the plate and the timing case cover.

The front end plate rarely if ever needs to be removed unless the gasket between it and the block starts to leak oil, in which case a new gasket should be installed.

Installation

Clean off all remnants of the gasket stuck to the timing case cover and the cylinder block (or front end plate). Use a suitable solvent if necessary to loosen the gasket material and scrape the surfaces clean. If the timing case cover is sheet steel instead of a casting, note whether the flange is bulged around the bolt holes. Such "dents" may occur if the gasket material is soft or if the bolts were previously drawn up too tight. Smooth out the dents with a hammer. Failure to do so may result in oil leakage when the cover is re-installed.

Before installing the cover gasket, coat both sides of the cover with a cement made for the purpose. Then paste the gasket against the front end plate or cylinder block, being careful that all bolt holes in the gasket accurately register with the bolt holes in the block.

Sometimes the gasket consists of more than one piece, but if so the procedure is just the same. Also note that if the gasket is continuous but if not all of it touches the cylinder block, Fig. 140, tear

off the piece which is not in contact with the block.

Before installing the cover, a new oil seal should be inserted in its pocket in the cover, Fig. 141. Depending on the fit between oil seal and pocket as well as on the construction of the oil seal, there

Fig. 141   *This oil seal has a gasket*

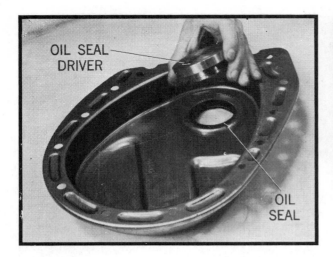

Fig. 142 *Oil seal in Fig.*
141 is installed with a driver

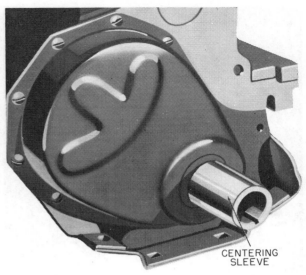

Fig. 143 *Oil seal centering sleeve*
for Chevrolet 6-cylinder engine

are three customary methods of installing the seal: It may be cemented in place; if it cannot be pressed in by hand a suitable driver or drift should be used such as shown in Fig. 142. An appraisal of the old oil seal should indicate which method to use in installing the new one.

The new oil seal, of course, should be identical with the one removed. If it is a felt seal it should be soaked in engine oil for 30 minutes before seating it in the cover.

Instead of using the vibration damper to align the oil seal and cover, a special sleeve made for the purpose may be used. Fig. 143 shows the tool made for Chevrolet 6-cylinder engines.

Fig. 144 shows a special centering tool for Chrysler-built L-head engines. Install the cover and tighten the cap screws only enough to hold cover in place. Thread the tool on the crankshaft and turn it up finger tight. Then tighten the cap screws gradually and at the same time draw up on the nut on the tool gradually. When all screws are tight, remove the tool and install vibration damper.

TIMING GEARS, REPLACE

The camshaft gear on automobile engines is usually made of a highly compressed fibrous composition material and therefore is called a fibre or composition gear. Some cars, however, have aluminum gears. The crankshaft gear is always steel.

The fibre gear provides a quieter drive than a metal gear. The fibre gear wears somewhat faster than the steel crankshaft gear. Therefore, if the gears are noisy it may be sufficient to install only a camshaft gear. In some cases camshaft gears with

teeth of oversize thickness are available. Thus, this so-called oversize gear compensates for the wear of the teeth on the camshaft gear, assuming it is not replaced.

Clearance between teeth of a new pair of gears when installed, Fig. 145, should be somewhere between .002″ and .006″. The same clearance limits are desirable if an oversize camshaft gear is mated with a worn crankshaft gear.

However, it should not be forgotten that a gear with a slight wobble may be noisy. Therefore, if tooth clearance is satisfactory, check the gears for runout before removing them, Fig. 146. If runout

Fig. 144 *Oil seal centering tool*
for Chrysler-built L-head en-
gines indicated by double arrow

Fig. 145 *Checking clearance with feeler gauge*

Fig. 146 *Using a dial gauge to check camshaft gear for runout. The screwdriver is used as a pry bar to rotate the gear*

is more than .004″, check for burrs on the gear hub and camshaft where the gear is mounted. If the runout cannot be brought below .004″, it is desirable to install a new gear. The runout of the crankshaft gear should not be more than .003″.

Removal

If the camshaft gear is not retained by a nut or bolt in the end of the shaft, it is a safe assumption that the gear is a press fit and that the power of an arbor press will be necessary to remove the gear. This means that the camshaft must be removed from the car.

If the camshaft gear is bolted on it is readily removed by taking off the bolts or screws holding it.

If the camshaft gear is retained by a nut on the end of the shaft, either a lifter bar may be used to remove it, Fig. 136, if it is not on too tight. Otherwise use a puller of which there are various types. Fig. 147 shows one type. Another type is shown in Fig. 139.

In Fig. 147, there are two threaded holes in the hub of the gear. The puller screws are inserted in these holes to the full depth of the hub, otherwise the screw threads may be damaged or stripped. Then turn the center screw of the puller to remove the gear.

Fig. 148 shows a puller ready to remove the crankshaft gear. It is the same tool used in Fig. 147 for cam gear removal.

Installation

If a pusher tool is not available the new crankshaft gear can be installed with a soft hammer or hammer and hollow drift, such as a piece of pipe.

After the camshaft and crankshaft gears have

Fig. 147 *Puller for camshaft gear. The puller screws are threaded into holes in the cam gear hub and then the big screw is turned clockwise to pull off gear*

Fig. 148   Puller for crankshaft gear

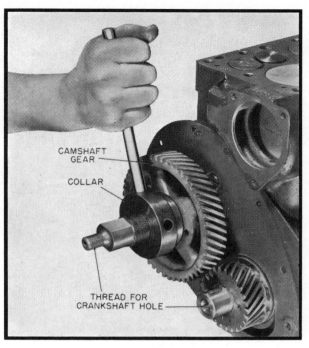

Fig. 149   Pusher tool for installing camshaft gear

been reinstalled on the engine, it is a good plan to check the gears for runout with a dial indicator, Fig. 146.

Camshaft gear run-out of more than .004″ is not desirable. The figure is .003″ for the crankshaft gear. To check either gear for runout place the tip of the dial gauge on contact with the rim of the gear as shown and rotate the gear through one revolution. High and low readings for the camshaft gear should not vary by more than .004″ for the cam gear and .003″ for the crank gear.

Fig. 149 shows a tool for installing the cam gear while Fig. 150 shows the same tool ready to install the crankshaft gear. Looking at the two pictures it will be seen that both ends of the center screw are threaded. The screw end visible in Fig. 150 is threaded into the hole in the camshaft, Fig. 149, while the screw visible in Fig. 149 is threaded into the hole in the crankshaft, Fig. 150. In either case the gear is installed by turning the collar clockwise.

TIMING CHAIN AND SPROCKETS, REPLACE

Except Jeep Overhead Camshaft Engine

When both the chain and the sprockets are new there should be a slight sag or slack as shown by the broken white line, Fig. 151, if the chain is pushed as indicated by the white arrow. Note that this test is made on the slack side of the chain and that the other side of the chain is under tension because the rotation of the crankshaft sprocket pulls on the chain to rotate the camshaft sprocket. The motion of the sprockets and chain is shown by the white arrows. A chain that is too tight will be noisy.

It is desirable to put on a new chain when wear allows the chain to sag more than ½″ when pressed inward on the slack side.

Fig. 150   Pusher tool for installing crankshaft gear

Fig. 151 *The slack in a new chain may be ½ inch*

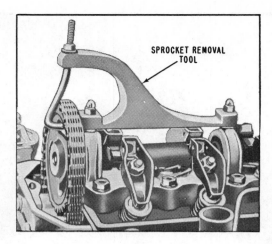

Fig. 152 *Camshaft sprocket removal and installation tool for use when removing cylinder head and timing chain on Jeep overhead camshaft engine*

The camshaft and crankshaft sprockets should also be examined for wear and replaced if visibly worn or scored. Badly worn sprockets will shorten the life of a new chain and may cause the drive to become objectionably noisy.

To remove the chain it is always necessary to remove the camshaft sprocket and sometimes it is necessary to remove both sprockets and the chain as a unit.

The method of removing the sprockets and replacing them is similar to removing and replacing the timing gears as previously explained.

Jeep Overhead Camshaft Engine

Removal
1. Remove timing chain cover.
2. Install tool shown in Fig. 152 on rocker arm cover studs. Install hook of tool in camshaft sprocket and tighten nut to relieve tension on camshaft.
3. Remove fuel pump eccentric from sprocket. Pull forward on sprocket to remove it from camshaft pilot.
4. Release hook of tool and slide sprocket off camshaft.
5. Remove chain tensioner spring to release tension on tensioner blade. Remove blade and spring from lower tensioner mounting stud.

Inspect Chain & Sprockets
Check the chain for excessive wear or stretch.

When the chain is installed with the chain tensioner in place, measure the distance between the chain sides at the narrowest point, Fig. 153. If the distance is less than the required minimum as shown, the chain must be replaced. If sprockets appear to be excessively worn, replace them also.

Timing Chain Tensioner
This device maintains a constant pressure against the chain to compensate for normal stretch and wear of the chain. Check the contact face of the tensioner blade to make sure the rubber facing material is not worn through. Replace if badly worn. Check the tensioner spring to make sure it is not distorted or elongated.

Install Chain & Sprockets
1. Turn over engine until air starts to blow from No. 1 spark plug port to indicate that No. 1 piston is on the compression stroke.
2. Continue to turn over engine until keyways in crankshaft are in the 12 o'clock position, which will indicate that No. 1 piston is at top dead center, Fig. 154.
3. Temporarily install camshaft sprocket and turn camshaft until nose of No. 1 cam and dowel hole on camshaft are pointing downward at the 6 o'clock position.
4. Remove camshaft sprocket.
5. Install key in crankshaft keyway nearest cylinder block.
6. Install tool shown in Fig. 152.
7. Position sprockets in opposite ends of chain so that keyway in crankshaft sprocket is up and dowel of camshaft is down. Position parts so that copper links of the chain are aligned with timing marks on sprockets.

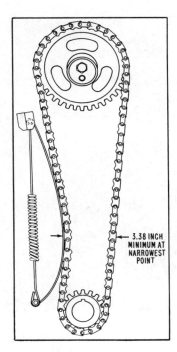

Fig. 153 Measurement for limit of timing chain wear on Jeep overhead camshaft engine

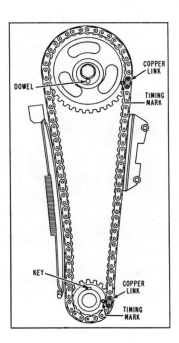

Fig. 154 Valve timing marks on Jeep overhead camshaft engine

8. Lift up assembled chain and sprockets and slide crankshaft sprocket on crankshaft so that the sprocket is fully seated.

9. Engage hook of tool, Fig. 152, on sprocket and tighten nut to tension the chain and pull the mounting hole of the sprocket into alignment with the pilot on end of camshaft.

10. Push sprocket into camshaft pilot so that dowel engages hole in flange on camshaft. This may require slight rotation of crankshaft to secure perfect alignment.

11. Install fuel pump eccentric on camshaft sprocket.

12. Release tension of special tool, Fig. 152, and remove tool from rocker arm cover studs.

TIMING BELT AND CAMSHAFT SPROCKET, REPLACE

Pontiac Tempest Overhead Camshaft Engine

1. Remove timing belt top front cover.
2. Remove sprocket to camshaft retaining bolt. *For ease of reassembly, index timing marks shown in Fig. 128.*
3. Loosen six accessory drive to engine block mounting bolts, Fig. 130.
4. Remove timing belt from camshaft sprocket. *Caution: Do not use tools of any type, other than hands, to pry on the timing belt during belt removal or replacement or during other service operations.*

5. Remove camshaft sprocket.
   If necessary to replace camshaft seal, reinstall sprocket bolt. Thread a tool of the type shown, Fig. 155, into camshaft seal. Tighten center bolt on tool until seal is extracted. Install a suitable seal protector and pilot, Fig. 156, on end of camshaft. Slide seal over tool, then drive it in place, Fig. 157.

CAMSHAFT, REPLACE

Except Jeep & Pontiac Tempest Overhead Camshaft Engine

The removal of the camshaft may be required for a variety of reasons, such as:

1. If new mushroom type valve tappets must be installed, the camshaft must be taken out unless the tappets are mounted in detachable guides.
2. If the camshaft gear is worn excessively and is pressed on to the camshaft so tightly that it cannot be removed by a puller, the camshaft must be removed and the gear pressed off by means of an arbor press.
3. If new camshaft bearings are required, obviously the camshaft must be taken out to replace the bearings.
4. The camshaft may be worn, bent, twisted or broken.

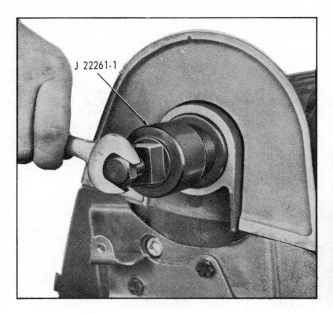

Fig. 155 *Removing camshaft seal*

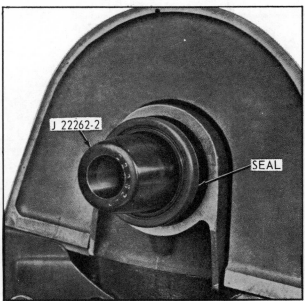

Fig. 156 *Seal and protector tool installed*

It may be quicker or even necessary to remove the engine from the chassis to get the camshaft out. Indeed, it is necessary to remove the engine if camshaft bearings are to be replaced. However, if the camshaft is to be removed with the engine in the car, the radiator, and usually the grille has to be taken off, as well as the fuel pump, distributor and/or oil pump. The valve tappets must either be removed or lifted up and secured out of the way.

On overhead valve engines with barrel-type tappets, lift them out before removing the camshaft.

On engines with mushroom tappets, the tappets can be held up out of the way by various methods without removing the valves. These methods can be used whether the tappets are mushroom or barrel type. Fig. 158 shows a method suggested by Studebaker wherein special U-shaped wedges are made from sheet steel. The right end is slotted at the top so that it slips under the head of the tappet adjusting screw. When the wedge is driven into place it raises the tappet out of the way of the cams. It is not necessary to remove the cylinder head.

If the cylinder head is removed, the valves may be blocked open as shown in Fig. 159. Then the tappets are lifted up and held there with clothes pins.

On almost all automobile engines the oil pump and/or the distributor are driven by a gear on the camshaft. Usually this gear drive interferes with the removal of the camshaft which means that the distributor or the oil pump or both must be removed before the camshaft can be taken out.

On most cars the oil pump and distributor are driven from opposite ends of the same shaft, Figs. 160 and 161. Distributor and oil pump shafts are separate and are connected together by a tongue and slot construction. In Fig. 160 the drive gear is attached to the distributor while in Fig. 161 it is on the oil pump shaft. Therefore in Fig. 160 the distributor must be removed in order to allow the camshaft to be taken out while in Fig. 161 the oil pump must be removed before camshaft removal.

The reason why this drive gear gets in the way is easy to explain. The gear on the camshaft which rotates the drive gear must be no greater in di-

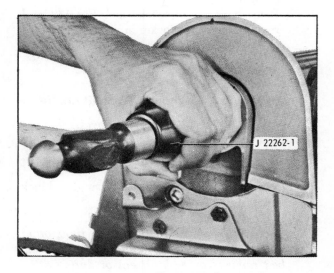

Fig. 157 *Installing camshaft seal*

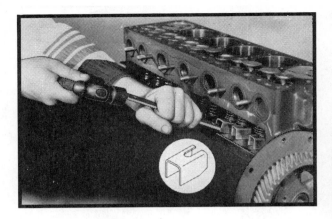

Fig. 158 Tappets are blocked up with wedges

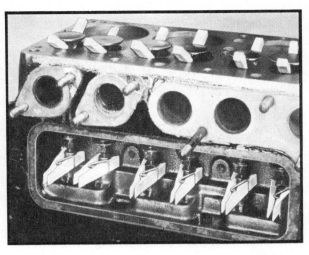

Fig. 159 Clothes pins hold up tappets

ameter than the camshaft bearings, otherwise it will not go through the bearings. Therefore, the teeth of the drive gear which meshes with it obviously must extend inward toward the camshaft ⅛″ or so and therefore will interfere with cams and journals whenever the camshaft is removed or replaced.

Before attempting the removal of the camshaft, some further explanation of oil pump and distributor drives is desirable. In all cases, remove the distributor first. If the drive gear is attached to the distributor shaft, Fig. 160, it is obvious that removal of the oil pump will not be necessary. On

the other hand, if the drive gear is on the oil pump shaft, the oil pump must be removed.

In Fig. 161, the drive gear is fastened to the oil pump shaft and the oil pump is readily accessible because it is mounted on the outside of the crankcase.

After carrying out all the directions in the preceding paragraphs the camshaft should come out

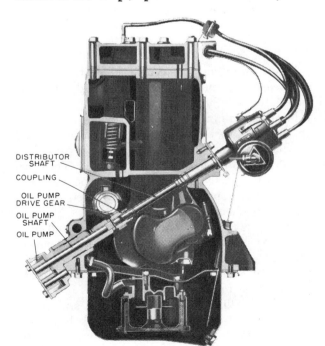

DISTRIBUTOR
SHAFT

COUPLING

OIL PUMP
DRIVE GEAR

OIL PUMP
SHAFT

OIL PUMP

Fig. 160 The camshaft drives a gear attached to distributor shaft

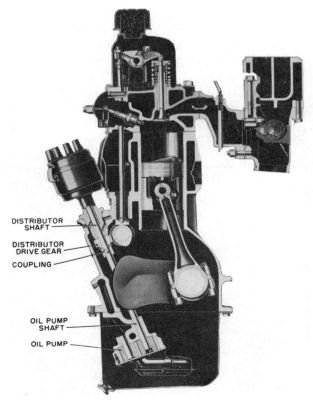

DISTRIBUTOR
SHAFT

DISTRIBUTOR
DRIVE GEAR

COUPLING

OIL PUMP
SHAFT

OIL PUMP

Fig. 161 The camshaft drives a gear attached to the oil pump shaft

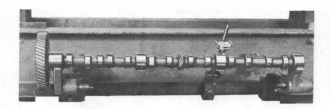

Fig. 162 *This camshaft is mounted on V blocks on the bed of an arbor press to check runout with a dial gauge*

readily. The camshaft should be handled with care as its finely finished surfaces are rather easily damaged. Be careful not to drop the camshaft, as it is easily bent.

Inspection

For best valve action the camshaft should not be out of true more than .002″. If the deviation is more than .002″ the camshaft should be straightened or replaced. The check for this condition is by the use of V blocks mounted on the bed of a wide arbor press, Fig. 162, or on a flat plate. Mount the camshaft end journals on the V blocks and then check the intermediate journals with a dial indicator. Check these journals by rotating the camshaft a full revolution. The high reading of the indicator indicates the high point of the shaft. Chalk mark this point and apply pressure by means of the arbor press to straighten the shaft. Press and re-check until the shaft is less than .002″ out of true. When pressing, avoid damage to the journal by inserting a brass or copper plate or a block of hard wood between the journal and the head of the arbor press.

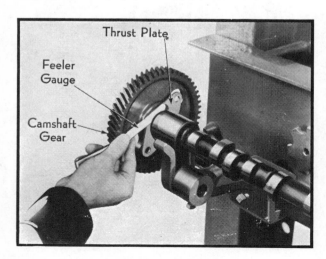

Fig. 164 *Checking clearance at thrust plate on Chevrolet 6-cylinder engine*

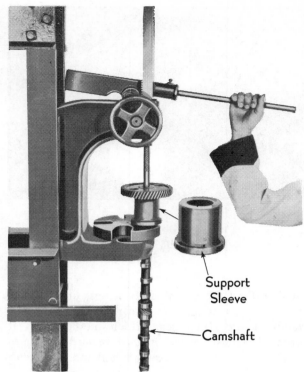

Fig. 163 *Arbor press for removing pressed-on camshaft gear*

Excessive camshaft run-out may cause missing at low speed by preventing full closing of some of the valves because the heel of the cam interferes.

Installation

Before installing the camshaft, coat the cams and bearing journals with engine oil and insert the camshaft carefully.

Mesh the timing gears according to the marks on them or mesh the timing chain according to instructions previously given.

The next job is to time the distributor. Turn the crankshaft until it is in the correct position for breaker point opening on No. 1 cylinder, just the same as for checking spark timing.

Then take the distributor in your hand and turn the rotor until it points to the segment which connects with No. 1 spark plug cable. Turn the distributor shaft slightly until the breaker points are just ready to open.

Install the distributor on the engine and check spark timing to be sure that the work has been properly done. Also check valve timing, if in doubt as to whether the timing gears or chain have been properly meshed.

Before permanently installing the distributor, if

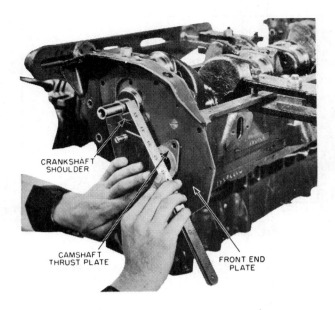

CRANKSHAFT SHOULDER

CAMSHAFT THRUST PLATE

FRONT END PLATE

Fig. 165 Checking alignment of camshaft thrust plate with respect to crankshaft shoulder

Fig. 166 Checking bearing diameter of cam bearing support deck on Jeep overhead camshaft engine

the oil pump has not been removed, turn its drive shaft so that its tongue and groove coupling will mesh with the distributor shaft when the points are just ready to open to fire No. 1 cylinder.

If the oil pump was removed, install it after the distributor has been installed, being careful to mesh the tongue and groove coupling by turning the oil pump shaft until the meshing of the coupling takes place.

If the camshaft gear is a press fit on the camshaft, as in the case of Corvair and Chevrolet 6-cylinder engines, the camshaft must be removed to install a new camshaft gear. Use an arbor press, Fig. 163, to press the gear off. The gear should be mounted on the support sleeve shown. Press the new gear on the shaft by means of the arbor press.

When the camshaft gear is designed to be a tight press fit on the camshaft, a new thrust plate (if used) should be put on the shaft before pressing the gear on with an arbor press and remember that there must be a little clearance between the hub of the gear and the thrust plate. The clearance should be somewhere between a free fit as a minimum and .003" as a maximum. When the gear has been pressed on nearly all the way, check the clearance between thrust plate and gear until it is .003" or less but also make sure that the clearance is enough so that the thrust plate can be rotated freely on the shaft, Fig. 164.

On Chevrolet 6-cylinder engines, before installing the camshaft, check the alignment of the camshaft thrust plate with the shoulder on the crankshaft, Fig. 165. The alignment should also be

checked in case new gaskets are required between the front end plate and the cylinder block in order to cure oil leakage at this point. In either case, proceed as follows:

In assembling the front end plate to the cylinder block, first use two gaskets and hold the plate with three screws. Then place a new camshaft thrust plate over the camshaft hole in the end plate. Using a steel scale as a straight edge, place it across the face of the thrust plate and note whether it is flush with the crankshaft thrust shoulder. If not, either add one gasket or remove one gasket so that the scale will be practically flush with the shoulder.

Jeep Overhead Camshaft Engine

Removal

1. With camshaft sprocket removed, lift rocker arm guide from cylinder head.
2. Check rocker arms to determine which do

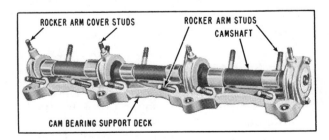

ROCKER ARM COVER STUDS

ROCKER ARM STUDS CAMSHAFT

CAM BEARING SUPPORT DECK

Fig. 167 Camshaft and bearing support deck. Jeep overhead camshaft engine

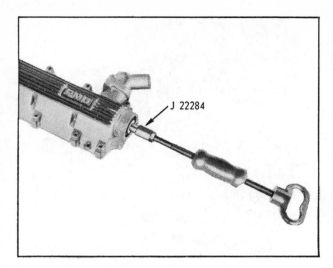

Fig. 168   Removing camshaft

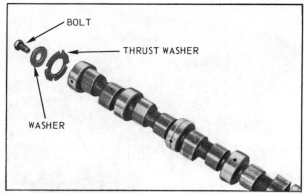

Fig. 169   Camshaft and related parts

not have cam tension against them. *Turn these parallel to camshaft.*

3. Temporarily install camshaft sprocket on camshaft and rotate camshaft to release tension from remaining rocker arms. Turn these parallel to the camshaft also. Continue to do this until all rocker arms are out of engagement with the camshaft.

4. Remove two nuts that attach camshaft retainer to cam bearing support deck and remove retainer.

5. Pull forward on camshaft to remove it from cam bearing support deck.

6. Remove cam bearing support deck from cylinder head.

Inspection

Clean the camshaft thoroughly with a suitable solvent. Make sure all oil passages are free. Check the cams for scoring or wear. The cam face must be perfectly smooth throughout their contact areas. Runout of the camshaft, measured with a dial indicator at the intermediate bearing journals, must not exceed .0005".

Check the camshaft journal diameters with a micrometer. The specified diameters are as follows:

Front	. . . .	1.9965–1.9975"
No. 2	. . . .	1.8715–1.8725"
No. 3	. . . .	1.7465–1.7475"
Rear	. . . .	1.3715–1.3725"

Check the cam bearing support deck for cracks or distortion. Check the bearing surfaces of the cam deck for visible wear or scoring. Using a telescope gauge and micrometer, check the internal diameters of the cam bearing deck bearings

as shown in Fig. 166. The specified diameters are as follows:

Front	. . . .	1.9995–2.0005"
No. 2	. . . .	1.8745–1.8755"
No. 3	. . . .	1.7495–1.7505"
Rear	. . . .	1.3745–1.3755"

Compare each journal diameter with the corresponding bearing diameter. If the bearings are defective or permit more than .04" running clearance, the cam bearing deck and/or the camshaft must be replaced, Fig. 167.

Installation

1. Install cam bearing support deck on cylinder head. Torque nuts to 15–20 ft-lb.

2. Lubricate bearing with engine oil and carefully slide camshaft into bearing support deck from the front, being careful not to damage

Fig. 170   *Thrust washer installation*

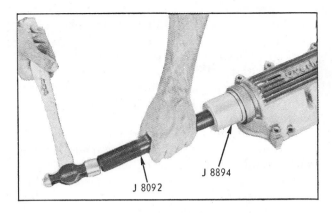

Fig. 171   *Installing camshaft bore plug*

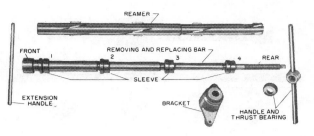

Fig. 173   *Equipment for removing, replacing and reaming Chevrolet 6-cylinder camshaft bearings*

bearings in support deck or journals on camshaft.

3. When camshaft is fully seated, install retainer on front of support deck.

Pontiac Tempest Overhead Camshaft Engine

Removal

1. Remove camshaft sprocket and seal as outlined previously.

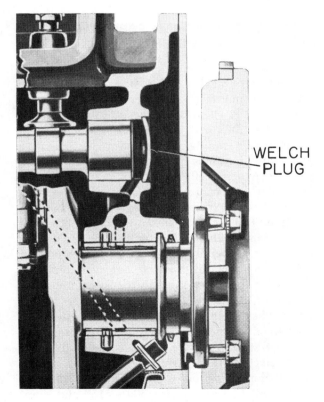

Fig. 172   *Camshaft rear bearing hole is sealed off by a welch plug*

2. Remove rocker arm cover.
3. Using tools of the type shown in Fig. 168 drive camshaft from rocker cover. *Caution: Do not allow camshaft to damage bearing surfaces of rocker cover.*
4. Remove parts shown from rear of camshaft, Fig. 169.
5. Remove water outlet fitting and thermostat from rocker cover.
6. Clean all parts and inspect for wear or damage. Minor nicks or scratches on edge of bearing surface can be corrected with a suitable scraper or file.

Installation

1. Install camshaft into rocker cover.
2. Install thrust washer, Fig. 170.
3. Install retaining bolt and washer and torque to 40 ft-lbs.
4. Using a tool of the type shown in Fig. 171, drive plug in so it is fully seated. *Note: A camshaft bore plug not fully seated could result in excessive camshaft end play which is .003″ to .009″ when read at the sprocket end with a dial indicator.*
5. Replace water outlet fitting and thermostat.

CAMSHAFT BEARINGS, REPLACE

The installation of new camshaft bearings is not a common operation. They have comparatively light loads to carry as compared with crankshaft bearings. Therefore if the camshaft bearings are properly lubricated with clean oil they will usually last for the life of the engine.

If, for some reason, the camshaft bearings do not receive sufficient oil they may burn out, that is, the bearing surface may melt and thus be ruined, in which case new ones will be required.

Also, if examination with a narrow feeler gauge shows that the clearance is .006″ or more, new bearings are called for—if satisfactory valve action and valve life are to be obtained.

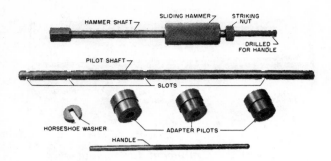

HAMMER SHAFT / SLIDING HAMMER / STRIKING NUT / DRILLED FOR HANDLE / PILOT SHAFT / SLOTS / HORSESHOE WASHER / ADAPTER PILOTS / HANDLE

Fig. 174 Simple equipment for removing and replacing camshaft bearings

If the engine has been removed from the car for an overhaul, it is desirable to install new camshaft bearings if the clearance exceeds .0035″ or .004″.

Various tools are made for removing and replacing camshaft bearings, but usually they consist of piloted drifts to pull out the bearings and to push them in place. However, in one case a jaw-type puller is used for the removal of the rear camshaft bearing.

The rear camshaft bearing rests in a hole in the cylinder block which must be sealed at the rear to prevent oil leakage. A bolted-on plate is sometimes used although on most engines a disk of metal with a slight crown is driven into the hole, Fig. 172, and thus expanded. The disk is called a welch plug or expansion plug. Once the camshaft is removed the plug can be driven out so that the rear bearing may be removed.

If an oil leak should develop at the plug it can be pried out after drilling or punching a hole in its center and then a new one can be driven into place.

Camshaft Bearing Tools

Each car manufacturer makes available to its dealers special tool sets to remove and install camshaft bearings together with instructions on how the set is used. Fig. 173 illustrates the tool set designed especially for Chevrolet 6-cylinder engines, and Fig. 174 shows the set furnished for Chrysler engines. In general they all work in the same manner. However, for the reader's information we will describe the operation of the Chrysler tool set which, incidentally, can be used on a number of non-Chrysler engines.

Before describing the use of this equipment there are two things to remember: (1) The adapter pilots must be just the right size for the camshaft bearings they are to be used on. (2) Regardless of which bearing is being removed or replaced all three pilots should be in place, two of

the pilots being used to align the pilot shaft while the third pilot is used to remove or replace the bearing. Each pilot has two diameters. The larger diameter fits the bearing bore while the smaller diameter fits the bearing. Therefore the pilots can be used for aligning the pilot shaft either with the bearings in place or removed.

Bearings are removed one by one and likewise are replaced one by one. The tool is ready for use when the pilot shaft is screwed into the nut on the hammer shaft. To remove the rear bearing, for example, in a four-bearing engine, install adapter pilots with large diameter to rear in bearings No. 2 and 3, to provide alignment for the tool. Then insert the tool through these bearings and then through No. 4 bearing. Install the remaining adapter pilot with its small diameter pointing forward and then slip the horseshoe washer into the rear slot. The rear bearing is ready to be removed (to front) by pounding the sliding hammer against the striking nut.

Remove No. 3 bearing in the same way after placing adapter pilots in bearings No. 1 and 2. To remove No. 2 bearing place one adapter in this bearing and then place the large diameters of the other two adapters in No. 3 and 4 bearing bores. The procedure for removing No. 1 bearing is similar.

SELECTING A HIGH PERFORMANCE CAMSHAFT

Overcamming an engine is a common problem. If your interests lie in building a "street" only machine, concentrate on moderate increases in net valve lift. Net valve lift is the actual amount the valve lifts off its seat, measured in thousandths of an inch. Depending on engine displacement, a cam with durations from 260°–300° should improve volumetric efficiency at street speeds. The "duration" of a cam is the actual period the valve is off its seat, expressed in degrees of crankshaft rotation. The big problem is one of selecting the proper cam for the engine and the specific driving conditions. Generally speaking, cams with mild or medium amounts of duration and overlap should be used with cars equipped with automatic transmissions. "Overlap" is the period of the engine cycle when both intake and exhaust valves are open together, measured in crankshaft degrees.

Street/strip cars offer a little more flexibility in cam selection than do strictly street machines. Again, depending on engine displacement, look for net lift figures of from .500″–.515″ and durations extending a bit beyond 300°.

For all-out competition, the field is wide open.

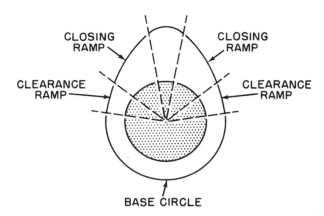

*Fig. 175 Names of solid
cam lobe lift areas*

Net valve lift can reach as high as .600″ with duration exceeding 330°.

Camshaft Installation Tips

When replacing a stock cam with a high performance cam, always replace related components such as lifters, push rods, rocker arms, valve springs and spring retainers, etc. This insures that the proper relationship between camshaft and valve train components is maintained

Never install a cam dry. Always use additional lubricant on wearing surfaces of the cam, tappets and other valve train components. Much of the cam's wear comes within the first few minutes of operation. Lubrication will prevent this when the engine is started for the first time. Most manufacturers specify or recommend use of a particular type of lubricant.

The timing chain and gear should always be replaced to help insure proper cam timing. Extra precautions should be taken when installing and tightening the camshaft sprocket, since high speed vibrations can cause bolts to loosen.

Never use old lifters on a new cam and do not use solid lifters on a hydraulic cam. Hydraulic lifters do not require clearance ramps as do solid lifters, Fig. 175.

Checking Valve-to-Piston Clearance

Always check for ample clearance between the valves and piston head, particularly when installing a cam with increased net lift. If the proper clearance is not maintained, it could result in severe damage to the engine through the valves hitting the pistons. To check the clearance, lay small soft strips of clay across the top of the piston near the valves. With the cam, valves, valve train and cylinder head gasket installed, bolt the head

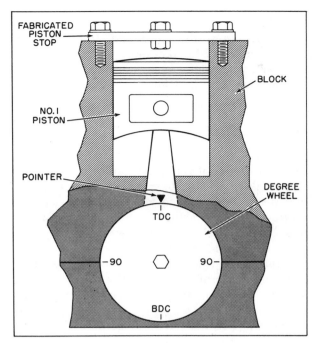

Fig. 176 Locating top dead center (TDC)

into place, and hand turn the engine through two complete revolutions of the crankshaft. This will insure that both valves in each cylinder move through their complete cycle of operations. Remove the head and measure the thickness of the impressions left by the valves in the clay. In many engines, a clearance of .060″–.080″ minimum is sufficient, but Ford calls for a minimum valve-to-piston clearance of .120″. This operation should only be performed after the camshaft has been properly timed. Variations in cam timing or in the amount of valve lash can materially alter the piston/valve relationship. Clearance checking before the camshaft is timed is a waste of time. And it is a *must* operation whenever any engine modification is carried out which could alter the piston/valve relationship. Such modifications include: camshaft changing or regrinding; head and block milling; and increasing compression ratios by changing head gasket, crankshaft, rods or pistons.

A simpler method of measuring valve-to-piston clearance, although probably not quite so accurate can be carried out as follows:

1. Bring piston No. 1 to TDC on its compression stroke and adjust valve to correct clearance.
2. Rotate crank 360°. This will place both valves in their overlap position (lightly off their seats).
3. Place a dial indicator on top of valve spring retainer, or use a mechanic's scale beside the spring.

4. Wedge a small pry bar under the rocker arm and depress the valve until it touches the piston. Read the dial indicator or mechanic's scale and compare to net valve lift specification. Repeat the procedure for each cylinder.

Finding Top Dead Center (TDC)

Timing a camshaft has long been considered a simple matter of aligning the markings on the cam and crank sprockets. However, when setting up a high performance engine, which involves camshaft replacement, this method permits the crankshaft-to-camshaft relationship to be several degrees off, either advanced or retarded. Therefore it is important that a new camshaft be "degreed in" to insure maximum performance. To do this it is first necessary to make an accurate check for top dead center.

There are several methods of doing this, and the two most common involve the use of either a positive piston stop or a dial indicator and a degree wheel. The first of these is probably the most popular and is carried out as follows:

1. The crankshaft, pistons, camshaft, cam gearing and lifter should be in place. Rotate the crank until No. 1 piston appears to be at TDC. The lifters for No. 1 cylinder should be on the base circle of the cam.
2. Mount a degree wheel (360° protractor) on the nose of the crankshaft and fasten a suitable pointer to the block. Align the degree wheel so that the pointer indicates TDC and tighten the degree wheel to the crankshaft.
3. Rotate the crank opposite to its normal direction to lower No. 1 piston enough to allow for the installation of a piston stop. (See Fig. 176).
4. Install the stop on the cylinder block and slowly rotate the crank in its normal direction until the piston is against the stop. Record the reading on the degree wheel.
5. Rotate the crank opposite its normal direction until the piston again touches the stop. Record the reading on the degree wheel.

Note that reversing the crankshaft rotation takes up any clearance in the connecting rod bearings which could otherwise affect the second reading of the degree wheel.

6. If the degree readings on either side of TDC are the same, the pointer is reading the correct position of TDC, and no correction is needed. If the readings are not identical, it

will be necessary to remove the piston stop and rotate the crank until the pointer is aligned exactly half way between the two degree readings. This point is TDC. Make the necessary correction on the damper by making a scribe mark to indicate exact TDC. This will insure that future checks with a timing light will be properly oriented. Without disturbing the crank, carefully loosen the degree wheel and adjust it so the pointer aligns with TDC. Then tighten the degree wheel in place.

Timing the Camshaft

With TDC established and the degree wheel showing the correct location of TDC, the camshaft is now ready to be "degreed in".

The following procedure for accurately timing the camshaft involves the determination of the cam lobe center line or point of maximum lift in relation to TDC and degrees of crankshaft rotation.

1. Install the camshaft, sprocket and chain. With the camshaft in place, insert the tappet for the intake valve of No. 1 cylinder.
2. Place a dial indicator on the tappet with the cam at maximum lift. Make certain that you are working with the intake cam lobe and tappet for No. 1 cylinder. The degree wheel with the correct location of TDC should be installed on the crank.
3. The next step involves use of the timing card received with the camshaft from its manufacturer.
   Let us assume the card reads as follows:

Intake opens	60° BTDC
Intake closes	90° ABDC
Exhaust opens	94° BBDC
Exhaust closes	56° ATDC
Total intake duration	330°
Total exhaust duration	330°

Using this information, it is possible to compute the cam centerline or point of maximum lift. The cam centerline comes at half of total intake duration or 165°. To find when maximum lift occurs in relation to TDC, subtract the point of intake opening from the 165° or 165° − 60° = 105°. In this example, the point of maximum lift occurs 105° after top dead center.

Effect of Valve Lash on Cam Timing

A final point should be covered before leaving the subject of camshafts. Total valve duration and timing can be materially altered by changes in valve lash setting. Increased valve lash will shorten duration and make the valve open later and close earlier, since more crankshaft rotation is required to close the gap between the rocker arm and valve stem and open the valve. Tightening the lash will effectively lengthen duration and may increase engine output at higher speeds. Generally speaking, .010″ of valve lash difference will be about one degree of cam advance or retard. Varying the valve lash could be a way to increase power, but it will affect valve timing and can be extremely hard on the valve train.

Review Questions

Page

1. Define the terms "valve head," "valve face," "margin" and "valve seat." 38

2. What part of the engine contains the valve seats in an L-head engine? 38

3. Explain the difference between valve stem clearance and operating clearance. 39

4. What is the function of a valve spring? 39

5. Why is the term valve lifter preferable to valve tappet? 40

6. How does a mushroom-type valve lifter differ from the barrel type? 40

7. Which part of the cam actually opens the valve? 40

8. Why is it so important to have the correct valve operating clearance? 41, 44

9. Name the two common causes of valve leakage. 42

10. What would be the result of too little valve stem-to-guide clearance? Too much clearance? 42

11. What effect would weak valve springs have on engine performance? 43

12. How would valve springs that are too strong effect the valve train? 43

13. Explain how a rocker arm can cause valve guide wear. 42

14. How can an improperly tightened cylinder head cause valve seat distortion? 44

15. What sort of valve damage can occur as a result of a defective cooling system? 45

16. Explain how a lean air-fuel mixture can cause premature valve failure. 45

17. What effect does poor valve seat contact have on valve temperature? 46

18. How does sluggish valve closing cause valve trouble? 48

19. Some late model engines have the combustion chamber formed in the cylinder block; what is the principal advantage of this design? 50

20. Why is it almost impossible to have uniform compression pressure in an engine with cylinder head combustion chambers? 50

21. What is the procedure for making a compression test with a compression gauge? 50

22. What information does a compression test disclose? 49

23. What condition would be indicated by a low compression reading between two adjacent cylinders? 49

24. Describe the procedure for positioning the valve lifter correctly prior to a cold valve adjustment. 52

25. What is the purpose of the adjusting screws on some engines with hydraulic lifters? 53

26. Why is it inadvisable to remove a cylinder head when the engine is hot? 59

27. Why is it important to use a torque wrench when tightening cylinder heads? 61

28. Explain the meaning of the term "pounds-feet." 61

29. Explain the purpose of valve stem seals on overhead valve engines. 70

30. How is excessive valve stem clearance eliminated on engines without replaceable valve guides? 77

31. How does a valve spring tester measure spring tension? 78

32. When using a valve spring tester, what precaution should be taken to avoid weakening the spring? 78

33. When grinding valves, why is it important to have the correct valve seat width? 82

34. Explain the difference between valve face overlap and rim margin. 82

35. Aside from shortened valve life, what other effect would insufficient rim margin have on engine performance? 82

36. In valve work, what is meant by topping? throating? 83

37. The angle between the valve face and the seat is called the interference angle. What advantage is gained by the use of this angle? 83

38. After the valves have been ground, describe two methods of determining how tight the valves fit into the seats? 83

39. How is dry ice (carbon dioxide) used in valve work? 84

40. Why is it a mistake to assume that the oil in a hydraulic valve lifter gives a cushioning effect? 85

41. When repairing valve lifters, what precaution must be taken to maintain the original "selective" fit of the lifter components? 87

42. What are the most frequent causes of hydraulic lifter troubles? 87

43. What effect would too much oil in the crankcase have on hydraulic valve lifter operation? 88

44. How is a compression test made on a unit-type hydraulic valve lifter? 88

45. Describe the two methods commonly used to mark valve timing on chain drive valve gear. 93

46. What is meant by the taut side of a timing chain? 93

47. What types of camshaft drives are currently being used on most American passenger car engines? 93

48. Why is it practically impossible to have valve timing a mere two, three or five degrees early or late? 95

49. What is the purpose of a camshaft thrust plate? 96

50. On engines using timing gears, what is the purpose of not using a steel gear at the camshaft? 101

51. What type of valve lifter can be removed without removing the camshaft from the engine? 106

52. How does the oil pump or distributor drive gear interfere with camshaft removal? 106

53. Why do camshaft bearings so seldom require replacement? 111

54. What is the function of a welch or expansion plug? 112

Engine Piston System

Review Questions for This Chapter on Page 165

INDEX

INTRODUCTION

Bore wear, cylinder 118
Cold engines wear fast, why ... 119
Combustion chamber, oil loss
   into 121
Cylinder bore wear 118
Distorted cylinder bores 119
Oil consumption, excessive 121
Oil loss into combustion
   chamber 121
Oil loss by leakage 121
Piston ring wear 120
Piston slap 120
Seating rings 121

CYLINDER BORE SERVICE

Block inspection, cylinder 123
Blocks, repairing cracked 123
Bores, check cylinder 141
Bore measuring instruments .. 142
Bore, out-of-round cylinder 141
Bore, tapered cylinder 141
Bore, wavy cylinder 141
Boring bar construction 151
Boring bars, use of 151
Boring, preparation for 151
Boring procedure 152
Bore, welded 152
Cracked blocks, repairing 123
Cylinder block inspection 123
Cylinder bores, check 141
Cylinder bore measuring instru-
   ments 142
Cylinder bore taper 141
Cylinder sleeves or liners 152
External oil loss, checking for 122
Flexible hone 145
Glaze buster or breaker 144
High performance, preparing
   cylinder block for 152
Hone, flexible 145
Hone, rigid 147
Instruments, cylinder bore
   measuring 142

Internal oil loss, checking for 122
Oil loss, checking for external 122
Oil loss, checking for internal 122
Preparing cylinder block for
   high performance 152
Reboring 150
Rigid hone 147
Sleeves or liners, cylinder 152
Taper, cylinder bore 141

PISTON SERVICE

Cleaning pistons 133
Expanders, piston skirt 137
Groove depth for ring ex-
   panders 135
Pistons, cleaning 133
Piston inspection 133
Piston skirts, check for col-
   lapsed 136
Piston skirt expanders 137
Ring expanders, groove depth
   for 135
Ring groove clearance 134
Ring grooves widening 134
Ring land diameter 133
Widening ring grooves 134

PISTON RING SERVICE

Compressors, ring 132
Engine, disassemble for ring job 124
Fitting piston rings 129
Install rings on piston 131
Oil ring, U-flex 130
Piston rings, fitting 129
Piston rings, removing 128
Pistons and rods, remove 127
Reamer, use of ridge 125
Ridge reamer, use of 125
Ridge, remove ring 124
Ring compressors 132
Ring job, disassemble engine for 124
Rings on piston, install 131
Ring ridge, remove 124
Rings, fitting piston 129

Rings, removing piston 128
Rings to use, what 127
Rods and pistons, remove 127
Ring, U-flex oil 130
U-flex oil ring 130

PISTON RING & ROD SERVICE

Bushings, connecting rod 159
Connecting rod alignment,
   checking 162
Connect rod bushings 159
Connecting rod bushings, install 159
Connecting rod bushings, re-
   move. 159
Connecting rod bushings,
   swaging 160
Hand honing piston pin holes 160
Honing piston pin holes, hand 160
Pins, check for loose piston .. 154
Pin fits, piston 157
Pin holes, hand honing piston 160
Pin holes, reaming piston 158
Pin honing machine, piston ... 161
Pins, reaming fixtures for piston 158
Pins, remove piston 155
Pin sizes, piston 157
Pins too tight, piston 155
Piston pins, check for loose ... 154
Piston pin fits 157
Piston pin holes, hand honing 160
Piston pin holes, reaming 158
Piston pin honing machine ... 161
Piston pins, reaming fixtures for 158
Piston pins, remove 155
Piston pin sizes 157
Piston pins too tight 155
Piston to rod, assemble ...164, 165
Reamer fixtures for piston pins 158
Reaming piston pin holes 158
Rod alignment, checking 162
Rod bushings 159
Rod bushings, install 159
Rod bushings, remove 159
Rod bushings, swaging 160
Rod to piston, assemble ...164, 165

Introduction

The piston system involves all those parts which transmit the power of combustion to the flywheel. These parts are the piston assemblies, connecting rods, main and rod bearings, crankshaft, vibration damper, flywheel, cylinder block and cylinder head.

It should be realized that most repairs to the piston system are made for the purpose of reducing oil consumption or to eliminate objectionable noise. When parts become very loose and noisy, new parts may be necessary to forestall breakage. Worn parts are not likely to greatly affect fuel economy and power.

Obviously, to do repair work on the piston system, it is necessary to remove the cylinder head and oil pan. Some jobs require the removal of the engine—Corvair, for example.

After the cylinder head has been taken off, it should be cleaned and then the head and block should be examined for cracks, warpage, leakage and other defects.

If the complaint is high oil consumption or if loose main, rod or camshaft bearings are suspected, it is a good plan to check the bearings with an oil leak detector before removing the pistons. Excess oil leaking out of loose bearings is thrown on the cylinder walls by the whirling crankshaft, increasing the oil consumption. New rings will not bring oil consumption down to normal if main, rod or camshaft bearings leak too much.

The procedure for making a bearing leak test is given in the *Crankshaft and Bearings Chapter*.

In a new engine, the bores are straight, circular holes. Variation in straightness of the bore from top to bottom is only a fraction of a thousandth of an inch. And likewise the bores are circular within a fraction of a thousandth. Therefore the pistons slide up and down over almost perfectly straight surfaces and the rings press out against almost perfectly circular bores.

Each piston moves up and down about 3,000 times per mile in a typical car or 30 million times in 10,000 miles. So in spite of good lubrication, wear is bound to occur. The piston rings cause more wear than the pistons. They are springy and are designed so that they press outward against the bore in order to seal off the hot, high pressure gases in the combustion chamber as well as to prevent oil from flowing past the rings into the combustion chamber.

Most of the cylinder bore wear occurs at its upper end where the rings scrape up and down on the bore surface. By contrast, the smooth piston skirt causes very slight wear in the lower end of the bore. Hence, the bore gradually becomes tapered, being larger in diameter at the upper end than at the lower.

CYLINDER BORE WEAR

Bores wear more rapidly crosswise of the engine than fore and aft. This is because there are powerful side thrusts by the pistons as they move up and down. Fig. 1 shows a piston moving downward on its power stroke. The total force (due to gas pressure) pushing down on the piston is a ton or two and this force jams the piston hard against the bore as shown by the horizontal arrow.

Side thrusts, obviously, are due to the angularity of the connecting rod with respect to the piston. Side thrust is absent only when the piston is at top and bottom dead center. There is no side thrust lengthwise of the engine because there is no connecting-rod angularity. The inertia forces acting on the piston also produce strong side thrusts.

The side thrust causes the bore to wear into an oval shape, Fig. 2. Since the heaviest side thrust occurs during the power stroke, the side of the bore that takes this thrust wears most. This is called the major thrust side of the bore—the opposite side of the bore being called the minor thrust side. The major thrust side is always the right side of the bore as viewed from the driver's seat. Of course on the Corvair, one must stand in back of the car to face the front of the engine. But, by

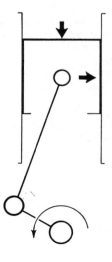

Fig. 1 On the power stroke the gas pressure places a heavy downward load on the piston, causing a powerful side thrust on it (shown by the horizontal arrow) because the connecting rod tries to wedge the piston against the cylinder bore

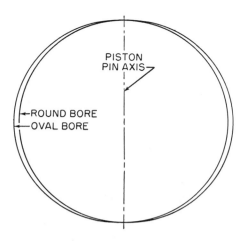

Fig. 2 *Because of piston side thrust, the cylinder bore wears oval, crosswise of the engine*

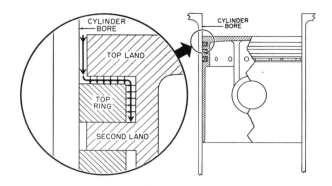

Fig. 3 *During the power stroke, gas pressure forces the top ring against the bore and also downward against the second land. Gas leakage past the ring is slight when the surfaces are unworn*

custom, American passenger car engines rotate counterclockwise when viewed from the driver's seat.

DISTORTED CYLINDER BORES

In addition to these two normal types of wear (which simultaneously cause the cylinders to grow tapered as well as to become out-of-round) the bores may go out of shape because of strains imposed upon them by unevenly tightened cylinder head nuts, distortion caused by abnormal block temperatures due to general overheating of the engine, or local overheating caused by clogged water jackets. These effects may cause low and high spots in the bore or the bores may wear to a wavy contour instead of a comparatively even taper.

Bores may be scratched (scored) or gouged due to various things including inadequate lubrication, tight pistons, tight rings, broken rings, broken piston, piston pin rubbing against bore, foreign particles in bore, and so on.

Pistons and rings wear as they travel up and down in their bores for any of the causes mentioned above.

Road dirt is a major cause of wear, especially in dusty territories.

WHY COLD ENGINES WEAR FAST

Wear is at maximum while the engine is warming up. It is negligible once the engine is warm unless the car is driven at high speed. Wear must obviously increase with speed.

Numerous road tests of 100,000 miles or more have proved that wear is negligible provided the

engine is kept running at moderate speeds and never allowed to grow cold. The explanation is that lubrication is ideal when the engine is warm but not adequate during the warming up period. When an engine has been stopped for a few hours it takes a number of minutes for it to reach normal temperature. And during this short period the cylinder bores, pistons and rings are not properly lubricated. The choke also interferes with lubrication. It supplies a copious wet stream of liquid gasoline to the cylinders which dilutes the oil on pistons, rings and bores to such a degree that the effectiveness of lubrication is very low and therefore pistons and rings rub against the bores and wear is comparatively rapid during the period that the choke is on. The oil is so thinned that it is unable to perform its function of keeping the wearing surfaces apart. Pistons and rings no longer ride on a tough oil film but may rub

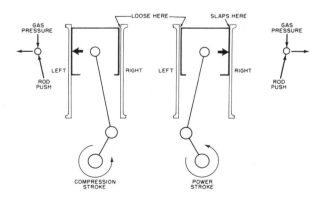

Fig. 4 *The diagram shows how the side thrust on the piston causes it to slap. It slaps against the right side of the bore shortly after it passes dead center to begin its power stroke*

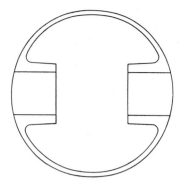

Fig. 5 *Most pistons have slightly oval skirts which are a few thousandths of an inch less along the pin diameter than at right angles to it*

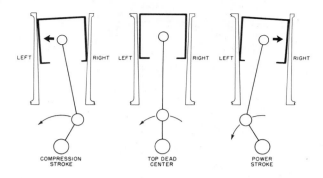

Fig. 6 *Showing how piston slap increases when cylinder bores wear. The bores wear fastest at the top and least at the bottom, tapering them. When pistons and bores were new the clearance was, say .002". But as the bores wear, the clearance increases and may eventually increase to .010", .020" or more. The greater the clearance, the more severe the slapping*

together because the diluted oil film is too thin and weak to keep them apart.

Because of the fact that most engine wear occurs during the starting period, an engine that is subject to short runs will wear much more rapidly than one operated mostly on long runs.

PISTON RING WEAR

Piston ring pressure should be sufficient to form a tight seal to prevent leakage of gases downward and oil upward. More pressure than this is undesirable because it needlessly increases ring wear and bore wear, and augments engine friction which reduces power and increases fuel consumption.

The sides of the rings and ring lands also wear. While the pistons are pulling the rings up and down, the rings are in constant motion sideways (radially) to accommodate themselves to irregularities in the contour of the bore. Even though the motion is only a thousandth of an inch or so, side wear may be considerable as the mileage piles up.

The top ring and its upper and lower lands wear fastest for several reasons: The ring and land wearing surfaces are exposed to the hot, high-pressure gases of combustion and therefore lubrication is imperfect. The outer face of the top ring is forced hard against the cylinder bore by the high pressure gases getting behind the ring, Fig. 3.

PISTON SLAP

On the compression stroke, Fig. 4, the compression pressure pushes downward on the piston while the rod resists this pressure by pushing upward on the piston pin. The combined action of

these two forces pushes (thrusts) the piston against the left side of the bore—because the rod slopes upward to the left.

During the power stroke, the rod slopes upward to the right. Combustion pressure pushes downward on the piston and the rod resists this pressure by pushing upward on the piston pin. The combination of these two forces thrusts the piston against the right side of the bore.

The direction of side thrust acting on the piston changes from left to right when the piston passes top dead center. This change in direction of thrust pulls the piston away from the left side of the bore and slaps it against the right side. If the clearance between piston and bore is too large a noise is heard which is called piston slap.

When the pistons and bores are unworn no piston slap will be heard when the engine is warm because the clearance is small. If the engine has aluminum pistons, slight slap may be heard when a cold engine is first started. However, these

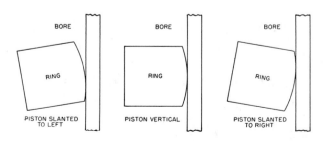

Fig. 7 *The rocking action of a loose piston reduces the effectiveness of the rings by rounding their outer faces*

pistons quickly warm up and expand, reducing the clearance and eliminating the noise.

Modern pistons, instead of being ground perfectly circular are ground slightly oval (cam ground) being about .009″ less in diameter across the piston pin hole than at right angles to the pin hole, Fig. 5.

Usually, aluminum skirts are ground about half a thousandth larger in diameter at the bottom of the skirt. In other words the skirt flares or tapers outward by .0005″.

However, after thousands of miles of service, the thrust forces acting on the skirt gradually reduce its diameter so that the skirt now tapers inward instead of outward. The skirt is said to have collapsed. This reduction in diameter is in addition to any surface wear of the skirt. The collapse increases the clearance between bore and skirt and results in more piston slap.

Piston slap may be thought of as a rocking action of the piston in the bore as shown in Fig. 6. As bore wear and skirt collapse increase, the rocking action increases. The rocking of the piston rounds off the outer faces of the rings and reduces their effectiveness, Fig. 7. Gas pressure leaks downward past these rounded surfaces and oil leaks upward into the combustion chamber and so oil consumption is increased.

SEATING RINGS

When new rings are installed they must seat themselves into the bore. To do this the bore must be slightly rough, so that in the first few hundred miles, rings and bores wear in together—or seat. If the rings do not seat, oil consumption will be high and in addition ring wear may be rapid because of excessive blow-by.

Therefore when rings are to be installed in worn bores, if there is a hard glaze on the bore surface, the glaze should be removed with a glaze buster or a spring-loaded hone.

The glaze appears to be caused by the deposit of a film of varnish or lacquer formed from some of the ingredients of fuel or oil.

EXCESSIVE OIL CONSUMPTION

There are only two ways in which an abnormal amount of oil can leave the crankcase of an en-

gine: through external leaks, and by getting into the combustion chambers where it is burned.

OIL LOSS BY LEAKAGE

External leaks may be classified as occurring under three different conditions:

1. When the pressure in the crankcase is normal, oil may be leaking from the engine past some gasket or through some loose metal-to-metal joint that is in direct contact with the oil in the oil pan.

2. When the crankcase pressure is above normal because of an excessive amount of blow-by or because the outlet and inlet of the crankcase ventilating system are partially or completely clogged, oil may be forced out of the crankcase through the front or rear bearings. When the crankcase pressure is too high, the underside of the engine generally has the appearance of leaking at every joint.

3. Sometimes the crankcase pressure is below normal because the ventilator outlet or the splash pan around it has become bent in such a way that the velocity of the air passing under the car at high speeds produces enough vacuum at the lower end of the ventilator outlet to suck enough oil vapor out of the crankcase to result in excessive oil consumption.

OIL LOSS INTO COMBUSTION CHAMBER

Oil can pass from the crankcase into the combustion chambers in only three ways:

1. Through a broken or porous vacuum diaphragm on the vacuum side of combination fuel pumps.

2. Through worn intake and exhaust valve guides on overhead valve engines and through worn intake valve guides on L-head engines.

3. Past the piston rings.

Piston System Service

CHECKING FOR EXTERNAL OIL LOSS

Since an abnormal amount of oil may be leaving the engine in any one or any combination of ways, it is good practice to make a thorough examination of the engine before it is disassembled for repairs.

Dense, blue smoke coming from the exhaust while the engine is runing at a fast idle speed indicates that too much oil is being used.

Leaky gaskets, leaking metal-to-metal joints, leaky oil seals, or any other sources of oil leaving the crankcase should be located as definitely as possible, because after disassembly is once started, it is more difficult to locate such sources of trouble.

Fresh oil on the clutch housing, oil pan, fuel pump, edges of valve covers, external oil lines, distributor shaft housing, base of crankcase filler tube, or on the bottom of the timing case cover is a fairly good indication of an external leak close to that point.

The hole in a cylinder block at the rear end of the camshaft is sealed with either an expansion plug or a cover plate. Quite often oil leaking downward from this opening is found on the rear end of the oil pan or on the bottom of the clutch housing and is mistaken for a rear main bearing leak.

Any area underneath an engine that is washed clean with oil is an indication that there may be an external leak from the crankcase somewhere in front of the washed area. Oil blown or sucked from the crankcase through the crankcase ventilator outlet pipe generally produces washed areas on the ventilator side of the chassis.

If washed areas are found on the chassis behind the crankcase ventilator outlet, the ventilator inlet should be examined to see if it is clogged. Possibly the sheet metal surrounding the lower end of the ventilator outlet tube has been bent or the outlet tube has been moved out of line in such a way that the velocity of the air passing under the car at high speeds produces enough vacuum at the lower end of the ventilator outlet to suck the oil from the crankcase.

The crankcase ventilator outlet and inlet should be examined to see if they are partially or completely clogged, and are thus causing high crankcase pressure to force oil out through joints that normally would not leak. The return passages in the valve cover plate between the crankcase venti-

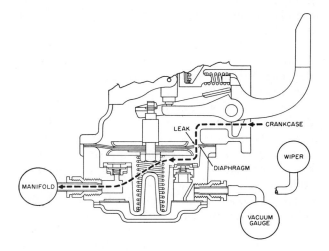

Fig. 8 *Diagram showing how oil is sucked into the engine through leaky diaphragm on vacuum side of combination fuel pump*

lator outlet and the crankcase should be thoroughly cleaned.

CHECKING FOR INTERNAL OIL LOSS

If the symptoms point to leaky rings, remove the spark plugs. If they are oily, the combustion chambers are receiving too much oil but do not forget that this may be due to: (1) Stuck rings which perhaps can be freed with a solvent. (2) A leaking vacuum pump diaphragm. (3) Leaky intake valve guides, or leaky intake and exhaust valve guides on overhead valve engines.

If blue smoke is seen at the exhaust, especially when the throttle is abruptly closed after decelerating from speed the rings are either worn or stuck.

If spark plugs are wet after engine has been idling for ten minutes but are dry after a few miles of driving, the cause is probably leaky intake valve guides.

If spark plugs are wet after road driving, either the vacuum pump leaks or the rings leak due to wear or sticking. Be sure the screws on the vacuum diaphragm are tight before making this test.

A leaky vacuum pump diaphragm causes high oil consumption by permitting engine suction to draw oil vapor from the crankcase into the intake manifold, and thus it flows to the cylinders where it is burned. The arrow, Fig. 8, indicates the path of the oil particles through the leaky diaphragm. The suction of the engine pulls the oil through the holes in the diaphragm, through the open discharge valve, then through the pipe leading to the intake manifold.

Whether the diaphragm leaks or not can readily be determined with a vacuum gauge. Disconnect the windshield wiper line at the vacuum pump and attach the vacuum gauge.

Start the engine and allow it to run for a few revolutions until the vacuum reading is a maximum. Shut off the engine. If the vacuum gauge needle remains stationary, there is no leakage. But if the needle starts moving toward zero, either the diaphragm leaks or the discharge valve leaks. To check for discharge valve leakage, start the engine. When vacuum becomes a maximum disconnect the line to the intake manifold at the vacuum pump. Put your thumb over the hole in the pump. If the vacuum gauge needle falls, the diaphragm leaks. If it does not fall, the discharge valve leaks.

Worn main, rod or camshaft bearings cause increased oil consumption. The oil which escapes from these bearings is thrown in all directions by the whirling crankshaft and therefore an excess amount is tossed onto the cylinder walls. Some of this excess oil makes its way to the combustion chamber where it is burned.

Leakage of oil from the bearings causes a drop in oil pressure. (Chevrolet engines up to 1953 have splash feed to connecting rods.)

Therefore, when deciding whether or not a ring job is necessary it is important to know whether worn bearings are contributing to the high oil consumption. So be sure to check the oil pressure —with the engine warm. If the oil pressure is 5 to 10 pounds below normal there is a good chance that bearings are worn. But do not forget that reduced oil pressure may be caused by various other troubles.

Whether or not the oil pressure is low, it is a good plan to check bearing leakage with an oil leak detector, as explained in the *Engine Bearing System Chapter*.

CYLINDER BLOCK INSPECTION

Before doing any work on rings, pistons, etc., it is advisable to inspect the block for defects.

Examine cylinder bores, top of block, valve seats and water jackets for cracks. Do not overlook the inside of the valve lifter chamber. If cracks are not large enough to be seen they may be revealed by wet or rusted areas.

A minute crack not visible to the naked eye may be detected by coating the suspected area with a mixture of 25% kerosene and 75% light engine oil. Mix up some zinc oxide in wood alcohol to form a thin, white "paint." Apply a coat to the surface. If there is a crack, the zinc oxide will be discolored. Note that there are several liquids on the market for detecting cracks.

Pressure testing the block will usually indicate the presence of a crack. Use 30 to 60 pounds water pressure or air with water. This test must be made with cylinder head installed unless dummy heads are available.

REPAIRING CRACKED BLOCKS

Repairing external cracks in cylinder blocks is readily accomplished by welding. But repairing internal cracks is quite an uncommon operation nowadays. It is discussed here merely to acquaint the reader with the methods used, especially during World War II when new blocks were hard to come by.

Three methods of repairing internal cracks were used: by applying a metallic sealing compound; by the insertion of screw plugs, and by electric arc welding.

Metallic Sealing Compound—A can of this compound is poured into the radiator. It will successfully seal cracks which are not too large. Run the engine at a fast idle until thoroughly warm. This will require 15 to 30 minutes depending on how cold the engine was to begin with— and whether the air temperature is high or low. It is a good idea to clean the cooling system with a cleaning compound before starting this work because the metallic sealer adheres better to clean cracks.

Screw Plugs—Specially made taper plugs of cast iron are available in various diameters. Plugs are installed at the ends of the crack. If the crack is more than ¾″ long, install additional plugs approximately ¾″ apart. The tension created by the plugs prevent the opening and closing of the crack under heat. The cracks remaining between the plugs can be closed by peening or the use of a metallic sealing compound or both. There are pneumatic peening hammers made for this work. Cut off the plugs with a cold chisel and smooth with a file or use the rotary files which are designed for use with a ¼″ electric drill. All types of cracks including cracked valve seats can be repaired by this method. The plug manufacturer's instructions should be carefully followed in all cases.

Fig. 9 shows how these plugs are used. Holes are first drilled at the ends of the crack to prevent

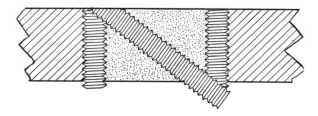

Fig. 9 *The vertical screw plugs are placed at the ends of the crack. Then the diagonal plug blocks off the crack*

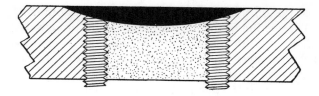

Fig. 10 *Screw plugs are installed at the ends of the crack. A groove is cut into the crack and filled with a special metallic compound (shown in black)*

it from increasing in length. The holes then are threaded, plugs inserted and filed off flush with the surface. Then a diagonal hole is drilled, a plug inserted and filed off flush. This plug completely blocks the crack.

In Fig. 10, the crack is sealed with a special metallic compound. Holes are drilled at the ends to the crack and screw plugs inserted. A groove is cut into the crack as shown in black. Then the groove is filled with a special metallic compound.

Figs. 9 and 10 merely show the principles used in mending cracks with screw plugs. There is an infinite variety of ways that plugs can be used for fixing various types of cracks including cracked valve seats. Therefore an instruction book supplied by a manufacturer of the equipment needed for the work is essential.

Electric Arc Welding—Cracks may be welded after being suitably grooved. An arc welding machine is required.

Piston Ring Service

DISASSEMBLE ENGINE FOR RING JOB

In order to remove the piston and rod assemblies, it is necessary that both the cylinder head and oil pan be removed. The head must be taken off because the piston assemblies must be removed through the top of the block because there is not enough room in the crankshaft area to permit the removal of pistons from below. (The procedure for removing cylinder heads is covered in detail in the *Engine Valve System* chapter.)

RING RIDGE, REMOVE

Before removing the pistons, examine the bores for ridges, Fig. 11. The top ring on the piston is usually poorly lubricated because it is exposed to the extremely hot, high-pressure combustion gases which blow or boil the oil off of it. Therefore cylinder bore wear is greatest at the upper end of the top ring travel. The worn bore leaves a "ridge" above the ring. Unless this ridge is negligible, it should be cut away before the piston can be pushed out. If a ridge can be felt by rubbing the

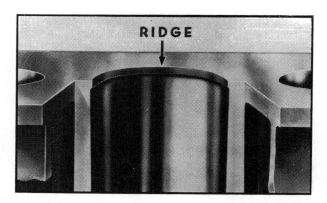

Fig. 11 *The top piston ring undercuts the cylinder bore, leaving a ridge which must be removed before the piston is taken out*

finger nail over the bore, it must be removed to avoid damage to new rings.

An attempt to force the piston up out of the bore may cause the ridge to bend or break the ring land just below the top ring.

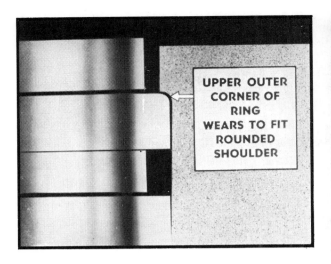

Fig. 12 *The old ring has a rounded shoulder*

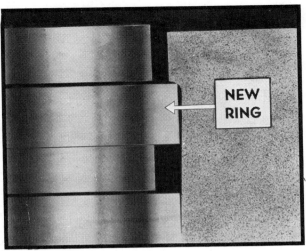

Fig. 13 *The new ring has a sharp edge or shoulder*

Note also that even if the piston is coaxed past the ridge without suffering damage, it is still necessary to cut away the ridge if new rings are installed. In explanation, note that the old top ring has a rounded upper edge, Fig. 12, which has worn itself into the ridge. When a new ring is installed with a sharp edge, Fig. 13, the edge will strike the ridge and may bend the land, Fig. 14, so that the second ring is clamped tight in its groove and therefore cannot function. A clicking sound will be heard when the ring strikes the ridge. The ridge is cut away by means of a special tool called a ridge reamer.

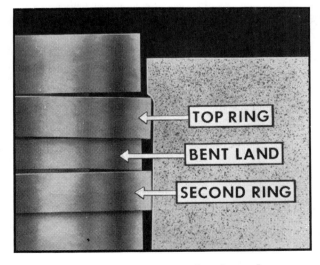

Fig. 14 *The new ring has bent the land below it. The land clamps against the second ring and prevents it from functioning properly*

Use of Ridge Reamer

There are a number of ridge reamers commercially available and the manufacturer's instructions should be carefully followed. But regardless of what make is used, it must be correctly installed in the bore and the cutting edge must be adjusted to the surface of the ridge. Care must be exercised not to allow the reamer to cut down into the bore below ridge more than 1/32″, as shown in Fig. 15. Cutting too far down into the bore will prevent the top ring from functioning properly and will greatly shorten its life.

Fig. 16 shows a typical ridge reamer installed in a bore, ready to remove the ridge. Fig. 17 explains its construction and principle of operation and tells you the names of its parts.

The edge of the cutter on all ridge reamers stands at a small angle to the bore, usually about two degrees, Fig. 18. Therefore it cuts a slightly tapered hole which blends smoothly with the bore.

To remove the ridge, proceed as follows:

1. Clean the top of the block.
2. Rotate the crankshaft to move the piston down to lower dead center. Wipe the bore with a clean rag wet with oil. Place a cloth over the piston to catch the cuttings from the ridge reamer. (After the ridge has been removed, turn the crankshaft half a revolution to bring the piston up to the top of the block, and remove the cloth with the cuttings.)
3. Lubricate the cutter with a few drops of oil. The maker recommends using a special oil.
4. Install the reamer in the bore as shown.

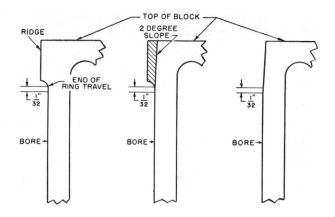

Fig. 15 *Ridge shown at left. Portion to be removed at center. Ridge removed at right. Ridge reamer should not be allowed to cut into bore below ridge more than 1/32"*

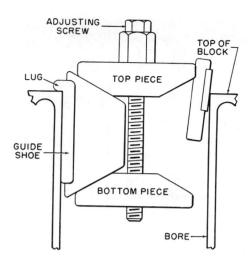

Fig. 16 *Ridge reamer installed in bore*

5. Expand the guide shoes by turning the hex head screw at the top until it is finger tight.

6. Then turn the hex head ½ turn to adjust the tension of the cutter.

7. Use a socket wrench to turn the cutter—clockwise only. If cylinder head studs are in the way use a deep socket. Press down firmly on the wrench with the left hand while turning it with the right. While removing the ridge inspect the progress frequently. This can be done without removing the reamer from the bore.

8. Turn the hex head screw counter-clock-wise to remove the pressure on the shoes and lift the reamer out of the bore.

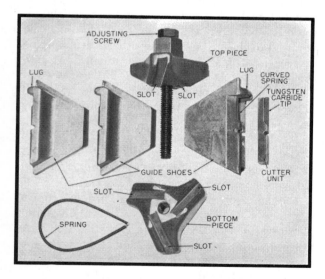

Fig. 17 *This ridge reamer consists of eight parts. Three guide shoes fit into slots in the top and bottom pieces. The latter are held together by the set screw which is threaded into the hole in the bottom piece. Turning the screw draws the bottom piece toward the top ·piece, thus forcing the guide shoes outward against the cylinder bore. The guide shoes are held in place by a coil spring. The cutter is mounted in a slot in one of the guide shoes. There is a flat spring in the back of the slot to hold the cutter against the cylinder bore. The cutter edge is a piece of tungsten carbide*

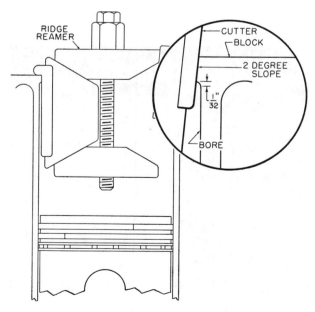

Fig. 18 *The cutter of the ridge reamer has a two-degree slope. The slope shown is purposely exaggerated*

126

Fig. 19   Connecting rod numbering

Remove Pistons and Rods

Before removing the connecting rod caps, note whether the rods and caps are numbered to indicate the cylinder to which they are fitted, Fig. 19. Also note on which side of the engine the numbers are placed. On In-Line engines, the numbers are usually on the camshaft side of the engine. On V8 engines there is no set rule as to which side the numbers are placed so be sure to note the location of the numbers before taking out the pistons and rods.

If the rods are not numbered, identify them with punch marks. Used pistons and rod bearings work best in their original cylinders because pistons conform to the wear pattern of the bores and rod bearings do likewise on the crankshaft.

Never number a rod with file marks. A file mark may start a crack which will eventually cause the rod to break.

In all engines with four or more cylinders, the rods move up and down in pairs. Therefore it is convenient to remove them in pairs. Rotate the crankshaft until a pair of rod caps are in the most convenient position for removal of rod bolt nuts. Remove the nuts and take off the caps.

Push the two pistons up through and out of the bores. Be careful the rods do not scratch or nick the crankshaft journals. To avoid this, plastic sleeves may be used over the rod bolts to protect the crankshaft as the rod is pushed up into the cylinder.

Rotate the crankshaft to bring two more rods in

position. Remove them, and continue in this manner until all pistons and rods are removed.

On Corvair engines, the complete assembly of cylinder, piston and rod is removed as a unit; this is done when the engine is out of the car.

WHAT RINGS TO USE

Each piston ring maker offers a variety of ring sets for both new and worn bores. Each set is designed to best serve a particular situation as regards the condition of the bores.

However, ring sets may be divided roughly into three kinds, as follows:

1. For perfectly reconditioned bores.
2. For bores not perfectly reconditioned or bores wtih moderate taper—say up to .007″ or thereabouts. These sets are also preferred by some repairmen for perfect bores on the theory that they do a better job when bores have become worn.
3. Badly worn bores with taper up to .015″ or even .020″.

Some piston ring makers ordinarily recommend sets 1 and 3.

Rings for these three kinds of bores for a three-ring piston might be:

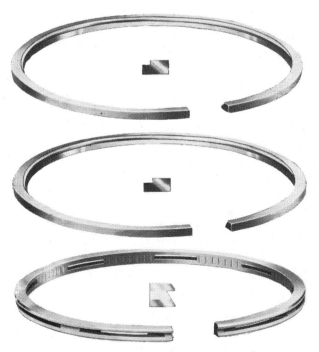

Fig. 20   Ring set for a new engine or for perfectly reconditioned bores. Two top rings are a modern standard compression type. Oil ring does not need a spring expander

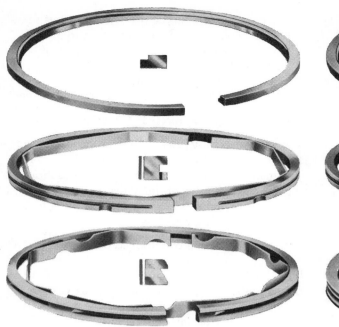

Fig. 21 *Ring set for less than perfect bores. Top ring is standard compression type. Second compression ring is more flexible and has a spring expander to help control oil. Oil ring has a spring expander to increase oil control*

Fig. 22 *Ring set for badly worn bores. Top ring is standard compression type. Second ring is same as in Fig. 21. Oil ring at bottom is designed to give maximum oil control. Steel rings above and below the oil ring assist in scraping oil from bore. All three rings in this groove are pushed against the bore by an expander spring*

1. Two regular compression rings and one regular oil control ring—same as used in a new engine, Fig. 20.

2. One regular compression ring, one more flexible compression ring, and a more flexible oil ring, Fig. 21.

3. Same compression rings as No. 2 but a very flexible oil ring, Fig. 22.

For engines with four rings above the piston pin, the rings shown in Figs. 20, 21 and 22 would be applied as shown in Fig. 23.

Ring combinations recommended by another leading ring manufacturer are given in Fig. 24.

The foregoing is offered to give the reader some idea as to the variety of sets available to meet the different conditions of bore wear.

The situation can be briefly summed up by stating that the greater the taper the more flexible the rings (except the top one) must be in order to follow the bore. The rings are contracted when at the bottom of the bore and must have the ability to expand on the up-stroke as the bore grows larger. And if the bore is out-of-round, the rings must do their best to shape themselves to the contour. This is especially important in connection with the bottom oil control ring.

The bores in a new engine may not be standard size. A car maker may produce a small percentage of engines with oversize bores—say .010″ oversize. Therefore, if the specified bore is 3.250″, some of the bores in some engines, or all the bores in others, may be 3.260″. Consequently, if a re-ring job is done on any of these engines, .010″ oversize rings should be used in every oversize bore.

Cylinders may also be oversize because they have been rebored one or more times in the past. If so, if new rings are to be installed in worn bores, select the ring oversize for the job.

REMOVING PISTON RINGS

Hold the piston upright by clamping the rod in a vise. Remove the top ring, then second, third, and fourth if there is one. All piston rings should be removed upward. Never try to slide a ring down over the piston skirt as there is danger of scratching the skirt. When replacing rings, install the bottom one first, and the top ring last. There is one exception: A U-Flex oil ring, Fig. 34, should be removed first and replaced last.

A piston ring removing and replacing tool is almost a necessity. There are numerous designs on

Fig. 23 (below) For four-ring pistons use the following ring numbers:

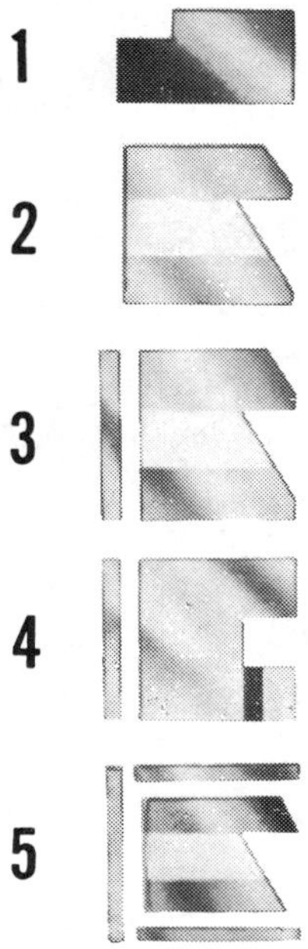

1

2

3

4

5

	Top Groove	Second Groove	Third Groove	Fourth Groove
Worn bore	1	4	5	2
Fair bore	1	4	3	3
Perfect bore	1	1	2	2

Fig. 24 (right) The numbered rings illustrated are applied in the following ways:

	Top Groove	Second Groove	Third Groove	Fourth Groove
3 rings above pin				
Worn bore	1	4	5	
Rebore or new bore	1	1	3	
4 rings above pin				
Worn bore	1	4	2	5
Rebore or new bore	1	1	2	3
3 rings—2 above pin and 1 below				
Worn bore	1	4	5	
New bore	1	1	3	
4 rings—3 above pin and 1 below				
Worn bore	1	4	5	3
New bore	1	1	3	2

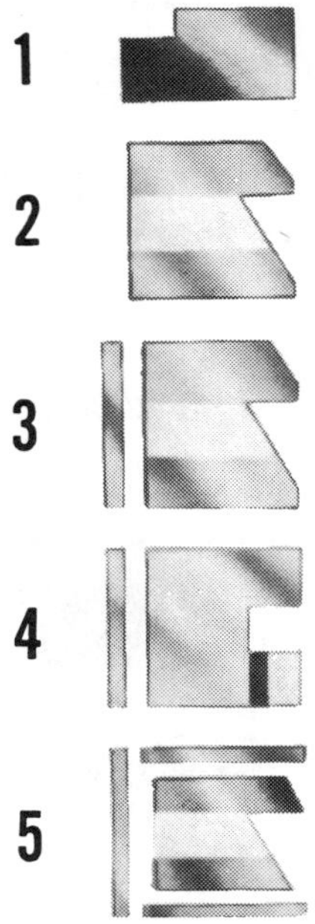

1

2

3

4

5

the market. One type is shown in Fig. 35. The rings are likely to be distorted if removed and replaced by hand.

FITTING PISTON RINGS

Some of the rings in a set may have a top and bottom. That is, one side of the ring must be installed toward the head of the piston. Follow the directions included with the ring package.

When a compression ring must be installed "right side up" it will be found that the upper inner edge is beveled or notched, Fig. 25. In other words, whenever the inner corner is modified in any way it should be on the upside of the piston.

Similarly if either a compression ring or a cast iron oil ring has its lower, outer corner modified, as in Fig. 26, for example, this corner should be on the down side of the piston.

When flexible oil rings have two flat steel rings, Fig. 27, the steel rings are usually installed above and below, as shown.

Before actually fitting the rings, of course, the pistons must be cleaned and otherwise prepared to receive them. But rather than go into the procedure for doing this work at this time, we leave it for later.

Roll each ring around in its groove, Fig. 28. If it binds, check the groove for dirt and burrs. If it still binds, reduce the ring width slightly by rub-

bing with fine emery cloth placed on a smooth surface.

Check the rings for side clearance if this has not already been done. Also check the ring groove depth, if necessary.

The ring end gap must be checked to be sure it is sufficient. Usually it is but it is not safe to install a ring without checking the gap.

If the bores have been fully reconditioned either by a rigid hone or reboring, place the ring in the bore near the top as shown in Fig. 29. Be sure that the ring is straight across the bore by turning a piston upside down, inserting it in the bore and pushing on the ring.

If the rings are being installed in worn bores, the gap should be checked with the ring about

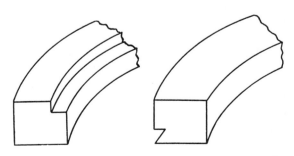

Fig. 25 Grooved inner edge of ring is the top side

Fig. 26 Grooved lower edge of ring is bottom side

Fig. 27　Steel rings are placed above
and below this flexible oil ring

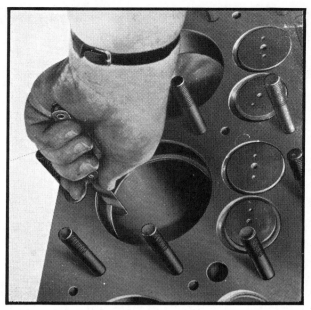

Fig. 29　Check the gap clearance
of each ring with a feeler gauge

Fig. 28　Roll the new piston ring around
in its groove to tell whether it binds

Fig. 30　These fixtures enable accurate filing
of piston ring gaps. This fixture is for straight
gaps. The four slots accommodate four different ring widths

half way down in the bore, that is, at the lower
end of the ring travel where wear is least.

Check the gap with a feeler. Use the gap recommended by the piston ring manufacturer. But
in the absence of these instructions the following
table may be used:

BORE	3″	3¼″	3½″	3¾″	4″
Top ring	.012″	.013″	.014″	.015″	.016″
Other rings	.010″	.011″	.012″	.013″	.014″

The purpose of the gap is to allow for ring expansion when rings are hot. The gap must be
large enough so that there is still some gap clearance when the rings are at their hottest. The gap
must never be less than the specified minimum.

But if any rings are found with less than the
minimum gap the ends should be filed. Filing
should be done carefully because gaps that are
much too large may permit too much leakage of
oil and gas pressure through the gap.

Fixtures for filing the rings are shown in Figs.
30, 31 and 32. If such a device is not available,
clamp a file in a vise as shown in Fig. 33. Press the
ends of the ring against the file while moving the

ring slowly back and forth. Try to keep the ends
of the ring flat against the file. Check the ring in
the bore occasionally until gap is correct.

U-FLEX OIL RING

A U-Flex oil ring, Fig. 34, does not require a
gap. The ends of the ring are designed to be in
contact at all times. This ring is so constructed
that it is resilient around its circumference and
therefore its expansion when heated is automatically absorbed. Before installing this ring be sure
it is the correct size for the bore.

130

Fig. 31 *This fixture is for rings with diagonal gaps*

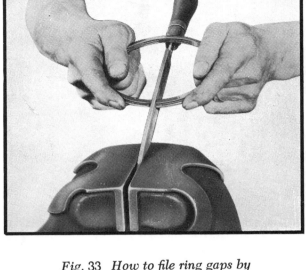

Fig. 33 *How to file ring gaps by means of file mounted in vise*

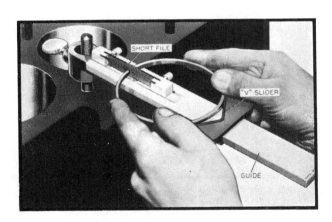

Fig. 32 *Ring gap is filed by mounting on V-shaped slider and moved back and forth on guide*

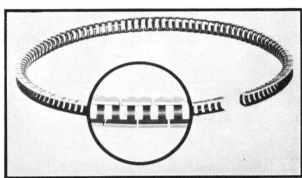

Fig. 34 *The U-flex oil ring should be removed first and replaced last*

INSTALL RINGS ON PISTONS

Mount the rod in a vise with the piston upright to support the piston while the rings are being put in their grooves.

A piston ring replacing tool, Fig. 35, should be used if ring distortion and breakage is to be avoided. All rings above the piston pin should be installed from bottom to top, that is, the lowest ring first and the top ring last.

If there is a ring below the piston pin, it may be installed either before or after the others.

After the rings are in place be sure they are free and do not bind. Check them by moving them laterally and also by rotating them in the grooves.

If binding exists, correct the condition before proceeding with the work.

Before installing pistons and rods in the block be sure to stagger the ring gaps. Rotate until adjacent rings are a half circumference apart. Or if preferred, space the gaps one-third revolution apart on three-ring pistons and one-fourth revolution apart on four-ring pistons. If the piston pin is pressure lubricated by a passage drilled in the rod, be careful that the oil ring gaps do not line up with the piston pin.

Lubricate the rings, ring grooves, piston skirts and bores with engine oil.

The last thing to do before installing a piston is to check the marks on piston and rod, so that it is correctly installed. Be careful, also, to install the piston in its correct cylinder bore.

Fig. 35 *Tool for removing and replacing piston rings. The tool expands the ends of the ring when the handles are compressed*

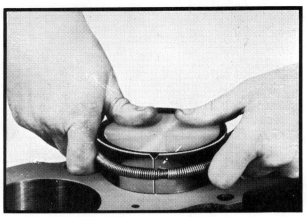

Fig. 36 *Using a piston ring compressor for inserting piston in bore*

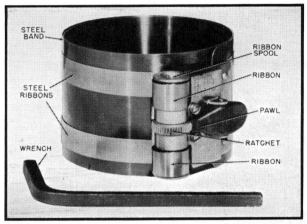

Fig. 37 *Ratchet compressor can be adjusted to any size of piston. Spool is turned by inserting wrench in square hole at top. Circular ratchet is attached to spool*

RING COMPRESSORS

A ring compressor is essential when installing pistons in the block. A simple design is shown in Fig. 36. It consists of a steel band flared at the upper end and contracted by the circular coil spring. It can be used only for a given bore size. The compressor is placed over the bore and the piston pushed down through it as shown.

Fig. 37 shows a ring compressor that can be adjusted to the piston diameter. It consists of an overlapping spring steel band placed within two steel ribbons whose ends are attached to the clamp spool. The device is adjusted to the approximate diameter of the piston. Then it is placed over the upper end of the piston with its top flush with the top of the piston. Tighten the band by inserting the wrench in the square hole in the top of the clamp shaft. When rings are fully compressed, back the ratchet off one notch. Insert the piston in the bore and tap the piston into place gently with the handle of a hammer.

Piston Service

CLEANING PISTONS

When cleaning aluminum pistons, do not use a caustic solvent such as lye, washing soda, etc., as they dissolve aluminum. A caustic solvent is readily recognized because it makes the fingers feel soapy—and if strong enough will smart and burn.

Use a stiff brush. If a wire brush is used be careful not to scratch the piston skirt or the ring grooves.

Hard carbon is best removed from the ring grooves by a special tool made for the purpose. There are a number on the market. One design is shown in Fig. 38. Be careful not to allow the tool to cut into the bottom of the groove.

Lacking such a tool, a satisfactory ring groove cleaner, Fig. 39, can be made by breaking a piston ring in two. Square off the broken end by filing or grinding. Be sure the edge is sharp and not rounded. Insert the end of the ring in the groove and push the ring forward to clean out the groove.

On some pistons, there is a narrow groove above the top ring which acts as a heat dam to prevent too much heat flowing from head to skirt. Carbon in this groove reduces the effectiveness of the dam, so be sure to clean this groove also.

Clean out the oil drain holes in the lower grooves by running a drill through them by hand. The drill should be the same size as the holes. It is best to mount the drill in a small hand reamer chunk to use as a handle.

PISTON INSPECTION

After the pistons have been thoroughly cleaned they should be carefully examined. Look for

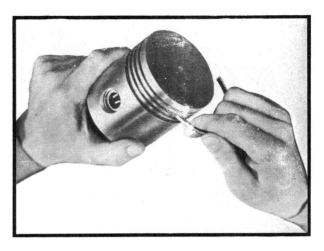

Fig. 39 A broken piston ring with a sharp, square end makes a good tool for cleaning ring grooves

cracks in the piston head, the ring lands and the skirts. Look for a small hole burned in the top of the piston, caused by excessive pinging. Needless to say, a piston with cracks or holes in it must be discarded. Pistons with badly scored skirts should be scrapped.

RING LAND DIAMETER

This paragraph applies to new pistons as well as old. The ring lands should not contact the cylinder

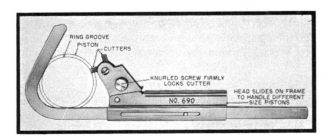

Fig. 38 This tool may be used to clean ring grooves but be careful not to cut the bottom of the groove. There are several widths of cutters for different ring grooves

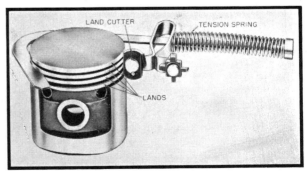

Fig. 40 Tool for reducing the diameter of the piston ring lands. The tool shown is rotated by hand. The spring presses the cutter against the piston. The same tool can be used for widening or deepening the ring grooves. The cutter head is simply slipped off of the handle, turned around and put back on

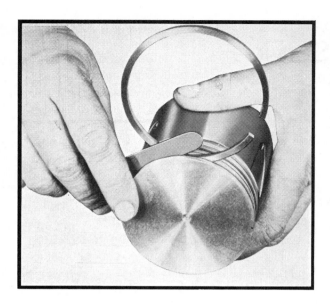

Fig. 41   *Checking ring side clearance with feeler gauge*

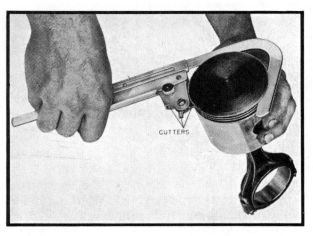

Fig. 43   *Cutting tool for widening ring grooves*

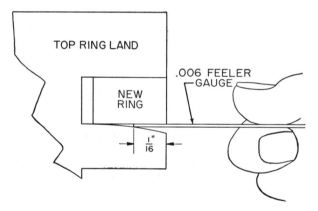

Fig. 42   *If tip of .006" feeler gauge can be inserted 1/16" or more, either discard the piston or enlarge the groove*

bore. If the faces of the ring lands show that they have been rubbing against the cylinder bore, the land diameters should be reduced. Land diameters should be less than the skirt diameter by .005" per inch of piston diameter for cast iron and .007" for aluminum. These figures are for the two top lands. Somewhat less clearance is needed on the lower lands.

The required reduction in land diameter can be made on a lathe, a piston grinding machine, or by means of a hand tool made just for the purpose, one of which is shown in Fig. 40.

RING GROOVE CLEARANCE

Piston rings must be free in their grooves so that they can follow the contour of the cylinder bore. Therefore a little clearance is required.

For compression rings use .001" per inch of bore diameter. For oil rings, .0007" per inch of bore diameter. For a 3-inch bore, for example, the clearances are: Compression ring, .003" (3 × .001"); oil ring, .0021" (3 × .0007")—use .002".

Ring manufacturers make a distinction between cast iron and aluminum pistons. Ring side clearances recommended are: Aluminum piston: Top groove, .0015" to .002"; other grooves, .001" to .0015". Cast iron piston: Top groove, .002" to .0025"; other grooves, .0015" to .002".

If ring clearances are insufficient, the ring should be rubbed lightly on a piece of fine emery cloth laid on a surface plate (or piece of plate glass).

To check the clearance on worn pistons, place a new ring in the groove and measure the clearance with a .006" feeler gauge, Fig. 41. Measure the land wear both above and below the ring. If the feeler can be inserted 1/16" or more. Fig. 42, the piston should either be discarded or the width of the groove enlarged to fit a wider ring (if available) or the same size ring can be used with a steel spacer made for the purpose.

WIDENING RING GROOVES

This operation is seldom if ever done on passenger car engines nowadays. But inasmuch as it is sometimes performed on large, expensive truck engines, it does no harm to acquaint the reader of its existence.

Fig. 44 *The tool shown in Fig. 40 can also be used for widening or deepening grooves*

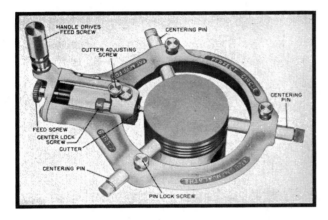

Fig. 45 *Cutting tool for widening ring grooves*

end of the tool is inserted in the groove and the tool rotated by hand. Be careful to remove metal only on the sides of the groove since the tool can also be used for increasing the depth of the groove. Fig. 44 shows a somewhat similar tool.

Fig. 45 illustrates another type of tool. The tongue on the three centering pins are inserted in the ring groove. The pins guide the piston. A cutter of the desired width is mounted in the cutter head and adjusted to contact the piston. Mount the piston in a piston vise or clamp the connecting rod in a vise. To cut the groove, turn the handle clockwise. The tool is fed into the groove automatically since a worm pinion on the handle meshes with the feed screw shown. Be careful not to increase the depth of the groove unless increased groove depth is desired.

In another design, the tool not only widens the groove 1/16″ but at the same time cuts a 1/32″ slot which anchors the spacer so that it won't rotate or float, Fig. 46.

GROOVE DEPTH FOR EXPANDER RINGS

These instructions are mainly for the installation of new rings on old pistons but note that occasionally groove depth in new pistons will be inadequate for the rings which are to be installed. In all cases abide by the ring makers instructions on this point.

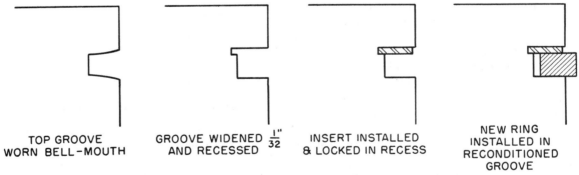

Fig. 46 *One design of regrooving tool not only widens the ring groove but also cuts a 1/32″ slot to anchor the steel space ring*

The width of the grooves may be enlarged on a lathe; or a hand tool may be used. There are several on the market. The tool shown in Fig. 43 is equipped with four widths of cutters. The cutter of the desired width is rotated into position and locked by the knurled screw. The hook-shaped

When a spring is used to expand the ring against the bore, Fig. 47, it is important that it have some working space. This space is determined by the depth of the groove left over after the ring is installed and the piston is placed in the cylinder bore.

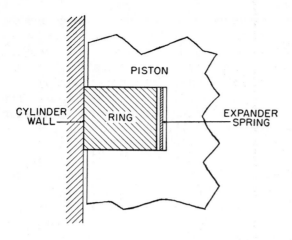

Fig. 47 *Ring groove depth must be such that the expander spring has correct clearance space in which to operate (3/16" to 5/64")*

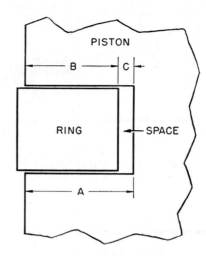

Fig. 48 *Diagram showing how to measure the clearance for an expander spring*

If spring space is too small, the pressure on the ring may be objectionably high, forcing the ring too hard against the cylinder bore. And there is also the possibility that the overcompressed spring will break. On the other hand, if the spring space is too great, the pressure of the ring will be inadequate.

Carefully follow the instructions accompanying the rings. Generally speaking, the clearance for the expander should not be less than about .050" (about 3/64") nor more than .080" (about 5/64").

The easiest way to check this is to measure the ring groove depth and then subtract the thickness of the ring, Fig. 48. The difference approximately equals the clearance available for the expander. Place a straight edge along the skirt at right angles to the pin hole, Fig. 49, and measure the depth of the groove. Then measure the depth of the ring. The difference is the space for the spring expander.

Fig. 50 shows a simple gauge for measuring ring groove depth. It is tapered from less than ⅛" to more than 7/32" as shown. Use a steel scale as a straight edge and push the gauge in as far as it will readily go. Then read the gauge marking nearest the straight edge.

If the ring groove depth is insufficient, it can be increased by means of a ring groove tool, Fig. 51, although the piston ring maker may suggest a lighter expander spring instead. Expander springs are available in three strengths: Standard groove, shallow groove and deep groove. A deep groove spring, of course, should be used when the groove is more than .080".

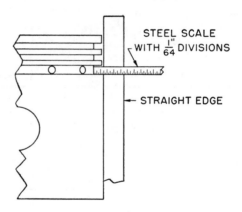

Fig. 49 *A steel scale and straight edge can be used to measure ring groove depth*

CHECK PISTON SKIRTS FOR COLLAPSE

After considerable service, piston skirts, especially those with vertical or nearly vertical slots in them, are likely to collapse. This means that the diameter of the skirt at the bottom is slightly smaller than it was when new.

To check for this condition, measure the diameter of the skirt at top and bottom at right angles to the piston pin, Fig. 52. If the skirt is not more than .003" smaller at the bottom than at the top, the piston may be re-used as is, but if the collapse is greater, the skirt should be expanded in a machine made for the purpose. The reason is that collapsed skirts are likely to produce objectionable piston slap. Expansion should bring the skirt back to its original diameter or perhaps .0005" larger.

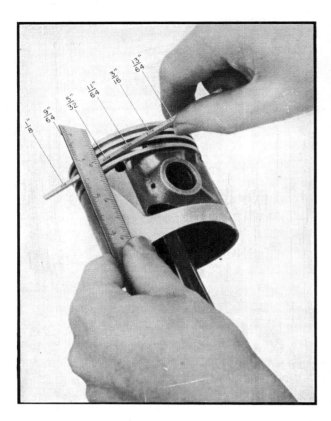

Fig. 50 *This simple ring groove depth gauge is marked off in divisions of 1/64".* The steel scale serves as a straight edge

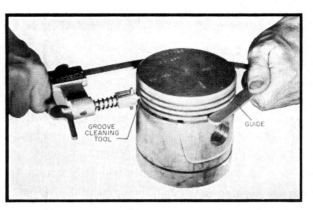

Fig. 51 *A chisel-like tool is used to clean the ring grooves*

Fig. 52 *Measure the diameter at the top of the skirt at the point shown*

PISTON SKIRT EXPANSION

There are a number of machines on the market for expanding piston skirts. However, it should be pointed out that the cost of expanding a set of piston skirts is often very close to the cost of a new set of pistons, so it may not be worth doing. If it is decided to go ahead with expansion, there are three methods of doing so: 1, by knurling the outside of the skirt; 2, by peening the inside of the skirt; and 3, by stretching the skirt after it has been heated to the proper temperature. All of these methods require special equipment of one sort or another.

Perfect Circle Nurlizer

The Perfect Circle Co. expands the skirt by knurling it on its thrust faces, Fig. 53. The process is called Nurlizing and the machine used is a Nurlizer, Fig. 54. This raises the metal on the surface as many thousandths of an inch as desired—up to an increase in skirt diameter of .015".

Knurling is accomplished by a small wheel with hard, sharp edges. The wheel is pressed against the piston skirt and then the piston is oscillated. The depth of the knurling is proportional to the pressure applied to the wheel. Knurling improves the wearing quality because the diagonal channels act as oil reservoirs. Some large fleet operators knurl the pistons of new trucks (before they are put into service) to prolong the life of pistons and bores.

The skirt can be enlarged to within .0005" of the desired size.

Kopper's Shot Peening

Kopper's makes a machine, Fig. 55, in which the piston skirt is bombarded by 1/16" steel balls accelerated to high velocity by air pressure. This is known as the Koetherizing process.

Fig. 53 *Piston knurled by Perfect Circle Nurlizer*

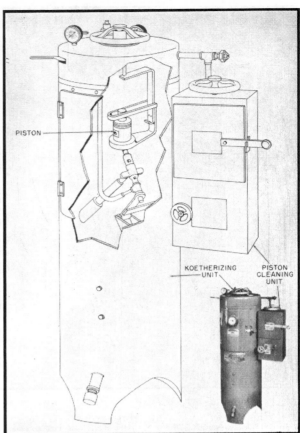

Fig. 55 *The piston is seen resting on a bracket at the center. Below it is the air nozzle which sends a blast of small steel shot against the inside of the piston skirt, expanding it by peening. The skirt is readily brought to the desired size within a tolerance of .0005"*

Fig. 54 *Nurlizer. Piston is mounted in chuck which is slowly oscillated by right hand. Left hand controls pressure of Nurlizer wheel which is mounted on piston rod. Dial gauge provides constant check on skirt expansion*

Ramco Ramconizer

The Ramconizer, Fig. 56, is a power-driven peening hammer which expands the skirt by peening it on the inside, Fig. 57. The piston is mounted on V guides so that it can be moved up and down and also rotated. An abrasive disc is used for smoothing up any high spots on the skirt.

Hastings Micro-Knurler

The outer surface of the piston skirt is expanded by knurling it. The machine used is shown in Fig. 58.

Pedrick Thermo-Resizer

At 750 degrees aluminum pistons are just pliable enough so that the pistons can be expanded

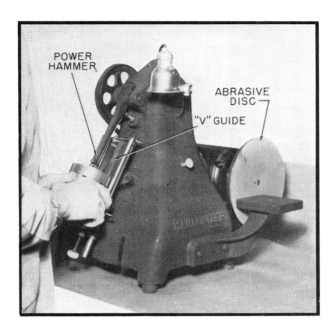

Fig. 56 *Ramconizer. The power hammer peens the inside of the skirt*

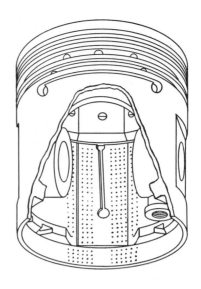

Fig. 57 *Sample pattern of peening by Ramconizer*

Fig. 58 *Hastings Micro-Knurler. The knurl pattern recommended by Hastings varies according to the piston type*

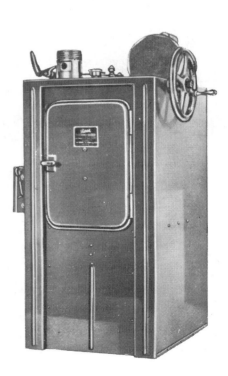

Fig. 59 *Pedrick Thermo-Sizer*

by force. After this, the piston is cooled by submerging in water.

The Pedrick Thermo-Sizer is illustrated in Fig. 59. The piston is heated in a thermostatically controlled electric oven to 750 degrees, Fig. 60. Then it is placed on a flat plate, Fig. 61, and the skirt is expanded by two lugs which are spread apart by turning the crank wheel at the right, Fig. 62.

HEAT.

Fig. 60
Pairs of pistons are heated in electric oven

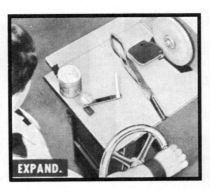

EXPAND.

Fig. 62 Turning the hand wheel expands piston by moving right lug to right

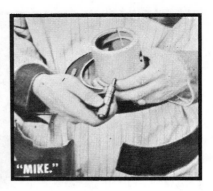

"MIKE."

Fig. 64
After cooling piston in water, amount of expansion is checked with micrometer

PLACE.   LUGS   ...O-SIZER

Fig. 61
Tongs are used to place hot piston over the expansion lugs. Left lug is stationary but right lug is slidably mounted in slot

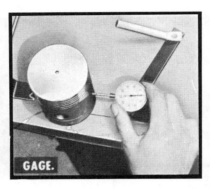

GAGE.

Fig. 63 Dial gauge in contact with piston skirt tells amount of expansion

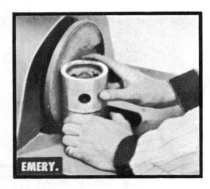

EMERY.

Fig. 65 Any slight high spots on skirt are smoothed up by emery wheel

Before the piston is expanded, a dial gauge mounted on an arm is swung into contact with the piston skirt, Fig. 63, and the dial set to zero. Then the skirt is expanded to the desired number of thousandths up to .015″ or perhaps .020″—to an accuracy of .0005″.

The piston is quenched in water and then checked with a micrometer, Fig. 64.

Finally, any high spots on the skirt are smoothed off by the motor-driven emery disc on the machine, Fig. 65.

Sealed Power Super-Sizer

This machine, Fig. 66, expands the piston skirt by a special peening process. A small power-driven hammer makes star-shaped indentions in-

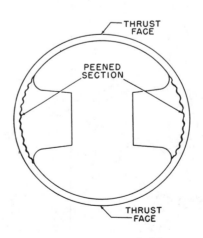

THRUST FACE

PEENED SECTION

THRUST FACE

Fig. 67 Piston is expanded by peening at points indicated. Advantage is that worn-in contours on thrust faces remain unchanged

Fig. 66 Sealed Power Super-Sizer

side the skirt, Fig. 67. A chart is supplied to show the number of stars required for each type of piston. The amount of skirt expansion depends on the peening time. For example, a Ford T-slot aluminum piston requires 10 seconds for .002″ skirt expansion and 30 seconds for .004″ expansion.

Cylinder Bore Service

CHECK CYLINDER BORES

Before cleaning or inspecting the worn pistons it is advisable to measure the cylinder bores. If bore taper is more than .015″, the cylinders should be rebored and new pistons and rings installed. Hence it is a waste of time to inspect the worn pistons if new pistons must be installed.

There are three sets of measurements to determine *taper, out-of-round* and *wavy* bore. Usually it is sufficient to measure the bore for taper and forget the other two. If taper is more than .015″, the cylinders should be rebored. If less than .015″ install new rings.

CYLINDER BORE TAPER

This measurement is taken crosswise of the engine because greatest wear can be expected to be crosswise. The bore is measured: (1) At the top of the ring travel just below the point where the ridge was removed. (2) At the bottom of the bore. Subtract measurement 2 from 1 to get the taper. Example: Diameter at top, 3.019″; at bottom, 3.002″; taper is .017″.

OUT-OF-ROUND BORES

Worn bores are no longer circular but oval. The diameter crosswise of the engine is larger than the diameter fore and aft. This measurement should be taken at top and bottom of bore—and at intermediate points if deemed necessary. The difference between the crosswise diameter and fore-and-aft diameter at the upper end of the ring travel indicates the amount the bore is out-of-round at this point. Similarly for bottom of bore and intermediate points.

If the cylinders are out-of-round more than .005″ reboring is recommended. However, one of the more conservative ring manufacturers states that rings installed in cylinders that were out-of-round .010″ have satisfactorily controlled oil consumption, adding that cylinders are more nearly round when warm. This company points out that if the rings are able to follow the taper, the out-of-roundness of cold cylinders can usually be ignored.

WAVY BORE

Measure bore at several points from top to bottom crosswise of engine to determine whether

141

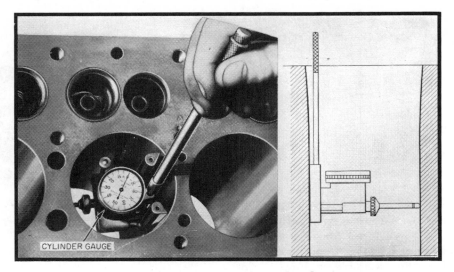

Fig. 68   A dial gauge is easy to read and fast to use

the surface has waves or high and low spots. Also measure bore waviness lengthwise of engine.

Look for the possibility of a ridge at the lower end of the bore or the lower end of the ring travel. The ridge can be honed out.

If the cylinder has waves, holes or pockets more than .001″ deep, hone them out completely or at

least until the depth is .001″ or less. If too deep for economical honing, rebore.

Look for scratches or gouges—score marks—in the bore. They can be honed out if not too deep. Otherwise, reboring may be required. It is not necessary to rebore the other cylinders unless they need it.

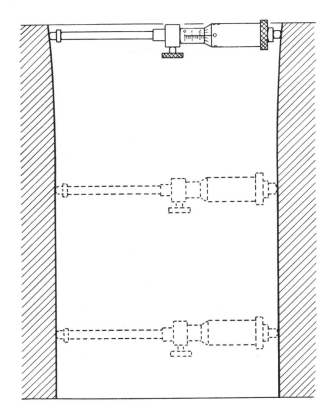

Fig. 69   Cylinder bore wear can be measured with an inside micrometer

CYLINDER BORE MEASURING INSTRUMENTS

Cylinder bores can be measured by (1) a dial gauge, (2) an inside micrometer, (3) a telescope gauge and outside micrometer. Note that micrometers may go out of adjustment, therefore, always check the zero point before taking a measurement, and adjust if necessary. Cylinder bore taper can also be measured with a set of long feeler gauges.

Dial Gauge Method

Adjust the thrust pin of the dial gauge, Fig. 68, so that it must be pushed in about ¼″ when it is inserted in the bore. The best way to use the dial gauge is to adjust an outside micrometer to the exact size of the original bore. Then fit the dial gauge between the contacts of the micrometer and adjust the dial carefully to read zero. Or dispense with the outside micrometer and set the dial gauge to read zero when close to the bottom of the bore where wear is slight.

Check for Taper, Out-of-Round and Wavy bore as already explained. For example, the reading of the dial gauge at the top of the bore crosswise,

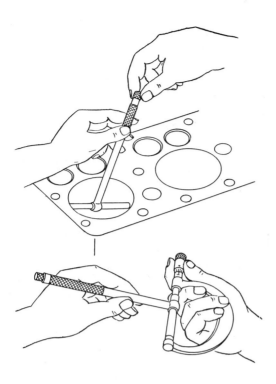

minus the reading at the bottom of the bore, gives the amount of taper in thousandths of an inch.

Inside Micrometer Method

With the inside micrometer, Fig. 69, measure the bore at the various points mentioned above and note the readings.

Telescope Gauge and Outside Micrometer Method

The telescope gauge, Fig. 70, consists of a rod slidably mounted in a tube. Insert the gauge in the bore and adjust it until it just fits the bore at the place selected. Then remove the gauge and measure the diameter with an outside micrometer.

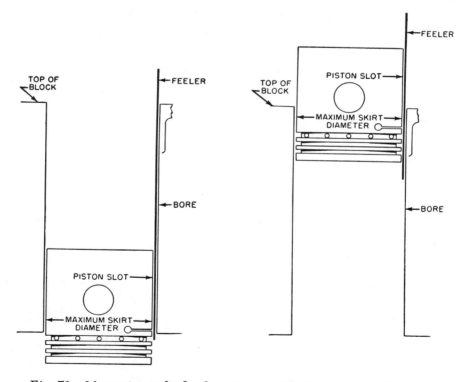

Fig. 71 Measuring cylinder bore taper with feeler gauge. Note position of piston in bore. For upper measurement, the maximum skirt diameter should be ½″ down from the top of the block. For lower measurement, the maximum skirt diameter should be close to the bottom of the bore

Feeler Gauge Method

The taper can also be measured with a set of long feeler gauges or by various thicknesses of shim stock. This method is not as accurate as the micrometer or dial gauge methods. Be sure the surfaces involved are free of oil.

One of the old pistons may be used for this job. Feelers of various thickness are used to measure the approximate clearance at top and bottom of bore. The difference in feeler thicknesses gives the taper.

The feeler clearance is measured crosswise of the engine because this is the direction of greatest wear. Fig. 71 shows how the feeler is used. The piston should be upside down with the piston pin fore and aft of the engine. Insert the feeler and the piston together, with the lower end of the feeler extending down beyond the piston skirt. The piston skirt is largest near the top of the skirt. Therefore this part of the skirt should be opposite the top and bottom of the bore when the feeler measurements are taken. At top and bottom, try feelers of various thicknesses until one is found that can be moved with a very gentle pull—while the piston is held stationary in the bore. Consider this the clearance at top or bottom as the case may be. The difference in feeler thickness required at top and bottom indicates the amount of taper.

Feeler gauges come in various thicknesses from .001″ up by thousandths or half thousandths. In checking piston clearances usually a ½″ wide gauge is used but occasionally the one inch wide gauge is specified. The length should be about 12 inches for most cylinder bores. For accurate work the gauge should not be kinked, bent or ragged.

Cylinder Bore Service Equipment

GLAZE BUSTER OR BREAKER

A glaze buster is designed solely to remove the glaze on the cylinder bore. The glaze buster shown in Fig. 72 consists of numerous pieces of special abrasive cloth which are clamped in slots in the hub. The pieces may be reversed or replaced by loosening the nut.

The buster is operated by an electric drill. Dip the abrasive in kerosene. Install the drill and move up and down in bore for 30 seconds. Then give it another coating of kerosene and operate it for another 30 seconds. Wipe bore clean.

The glaze breaker, Figs. 73 and 74, has three "pads" (made of laminations of felt and rubber) covered with abrasive sheets furnished in both medium and fine grit sizes. The three "tension" plates at the base of the pads can be adjusted to give the correct pad stiffness for all cylinders from 3″ to 4¼″ or even larger. The plates are correctly adjusted when they are readily compressed by hand, Fig. 74, so that they can be inserted in the cylinder. For cylinders smaller than 3 inches (down to 2.6″ bore) the tension plates are removed.

The glaze breaker is mounted in an electric drill and 8 or 10 complete strokes (up and down) are usually sufficient to remove the glaze. To reduce flying abrasive to a minimum the pads should be well soaked in oil.

Fig. 72 Strips of abrasive cloth are clamped into the hub in this glaze buster

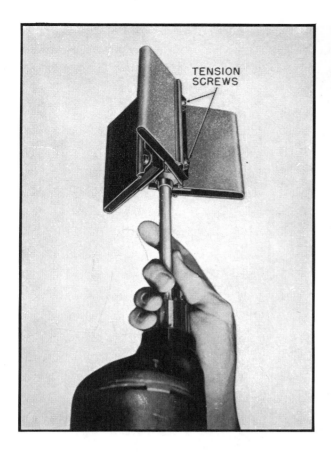

Fig. 73 *Three felt and rubber pads are covered with abrasive cloth*

Fig. 74 *Correctness of tension is tested by compressing pads by hand*

FLEXIBLE HONE

A flexible hone can be used for removing a small amount of metal from the bore—up to .003".

A typical flexible hone is shown in Figs. 75 and 76. It has two stones and two guides made of a felt composition. The hone is rotated by an electric drill. The stones and guides are pressed against the bore by both spring tension and centrifugal force. The hone shown is suitable for bore diameters from 2-9/16" to 4-3/4".

Insert hone in cylinder and start motor. Move hone up and down at the rate of about 30 times per minute. Allow hone to protrude slightly at top and bottom of strokes. Examine bore after 20 or 30 strokes and repeat if necessary—until required amount of metal has been removed.

Coarse and fine stones are available. Use coarse stone for fast cutting and fine stones for finishing. If only a slight amount of metal is to be removed, use fine stones.

Carefully observe the manufacturer's instructions as to care and use of hone. If felt guides are impregnated, the hone may be used dry. Dry honing gives faster cutting. Wet honing (with

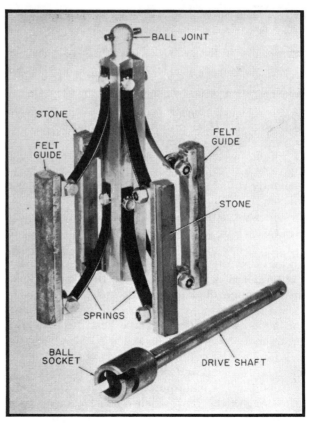

Fig. 75 *This flexible hone has two stones and two felt guides. The ball and socket drive permits honing a bore that is recessed into the dash*

145

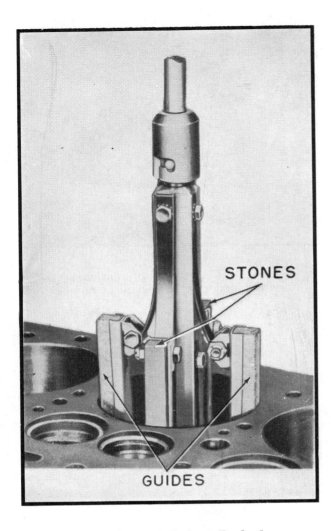

Fig. 76 *Hone installed in cylinder bore*

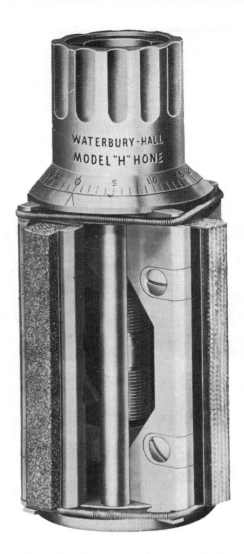

Fig. 77 *Expanding hone. Stone shown at left and felt guide at right*

kerosene) gives smoother finish. Kerosene is conveniently squirted on stones by means of ordinary oil can.

Stones must be kept clean. If dirty or gummed, clean with wire brush and kerosene.

For glaze removal, use 150 grit stones with rapid up and down motion of hone—fast enough so that the scratches the hone makes on the bore surface are at an angle of about 45 degrees.

The honing stones are expanded by springs plus the action of centrifugal force acting on the rotating tool.

When used for removing glaze about 10 complete strokes (up and down) should be sufficient.

However, the main purpose of the flexible hone is to smooth the bore when it is necessary to remove only a small amount of metal—not more than .003″. If more, use a rigid hone or rebore the cylinders.

If the bore is slightly scored, the flexible hone may be used to remove the marks if they are not more than .003″ deep. If deeper, use a rigid hone or rebore.

The flexible hone has the advantage that it requires no adjustment. Its disadvantage is that it removes the metal very slowly. Therefore, if somewhat more than .003″ is to be removed it will save time to use a rigid hone. Another objection is that it follows the taper of the bore.

If the bore has high or low spots (waves or pockets) with a variation of .001″ or less they can be ignored but if the high and low spots are from .001″ to .003″, use a flexible hone to remove the high spots. It is not absolutely necessary to remove the low spots completely. A few low spots will do no harm, especially if they are less than .001″ deep.

The flexible hone may be used to remove a

Fig. 78 Expanding hone installed in cylinder bore

ridge in the lower bore if not more than .002″ or .003″ deep. If the ridge is less than .001″ it can be ignored. If more than .003″ use a rigid hone.

RIGID HONE

The stones in a rigid expanding hone are held firmly in place to whatever diameter they are adjusted. They are adjusted to the desired size by turning the micrometer dial which reads in thousandths of an inch.

This hone may be used instead of a flexible hone for all work done by the flexible hone except glaze busting for which the rigid hone is not suited because it does not follow the taper of the bore.

Use a rigid hone to remove a ridge at the bottom of the bore if more than .003″. If less, use a flexible hone.

When bores are reconditioned by a rigid hone, they should be as good as new. That is, taper and out-of-round should be as little as .0007″ or even .0005″.

The rigid hone can be used for refinishing cylinders which are tapered up to .020″ but reboring may be faster and therefore more economical if much metal is to be removed. Opinions differ as to where the dividing line lies, that is, when does reboring become more economical than honing.

Fig. 79 Expanding hone with metal guides

For the sake of argument, let us say that .007″ taper is usually the economical limit. If taper is .007″ or less, honing may be cheaper but if more than .007″ it may be more economical to rebore. But no two experienced men are likely to agree as to whether the .007″ figure is just right, or too high or perhaps much too low. The hardness of the cylinder block, the power of the electric motor and the skill of the operator all have a bearing on what the figure should be.

This hone usually has two stones and two guide blocks, Figs. 77, 78 and 79, although four stones and no guide blocks are sometimes recommended with wet honing (wet honing means the stones are wetted with kerosene). The guide blocks may be made of fabric composition as in Figs. 77 and 78 or may be of metal as in Fig. 79.

The stones are positively expanded by turning the micrometer dial. For quick expansion to the

Fig. 80 Honing an engine block. The weight of the tool is supported by a long coil spring. Grit is removed by a "vacuum cleaner"

Fig. 81 Checking piston fit with feeler gauge

bore size, pull the micrometer nut upward and turn it counterclockwise until it fits the bore snugly (when hone is flush with top of bore). Then push the nut downward to bring the micrometer into use. If the adjustment is too tight in the bore pull micrometer upward and back off slightly.

Carefully follow the manufacturer's instructions both as to operation and care of the make and model of hone used. Be sure stones are clean. If not scrub them with a stiff brush and kerosene.

Note that most rigid hones have a micrometer nut with a scale reading from .000″ to .030″, Figs. 77 and 78. The scale is so large that tenths (of thousandths) are readily estimated.

The guide blocks should have from .010″ to .030″ clearance in the bore. If too tight, file, or dress with a rough honing stone. Tight guide blocks cause stones to chatter.

Stones must be clean or speed of cutting will be reduced. If coated with grease, oil or carbon, scrub with a stiff wire brush and kerosene. If they cannot be cleaned they should be discarded.

Tapered stones and guide blocks are caused by too little tension on the stones. This is avoided by frequent adjustment of the micrometer to keep the stones tight against the bore.

Before starting to hone, be sure the bores have been cleaned of oil and grease.

Crankshaft and camshaft should be completely covered with oily rags or wet newspaper so that no grit can reach any of the bearing wearing surfaces. This precaution is necessary not only for dry honing but also for wet.

If the bores are to be dry honed, a grit remover is recommended, Fig. 80. It works like a vacuum cleaner. Its suction hood fits over the top of the hone and sucks away loose grit from the stones, delivering it to the grit bag.

If the bores are to be wet honed, the hone should be dipped into a pail of kerosene occasionally.

The hone manufacturer's instructions should be carefully followed as to what stones to use, etc. However, on the average, coarse stones (about 150 grit) should be used for roughing and finer (about 250 grit) stones for finishing. If .003″ or less is to be removed, the job may be done with the finishing stones.

But if more than .003″ must be taken off, use the coarse stones to remove all but about the last .001″. Then switch to the finishing stones.

Place the hone well down in the bore and turn the micrometer nut until the hone just fits the bore. Turn the motor on and move the hone up and down in the lower bore with a steady motion,

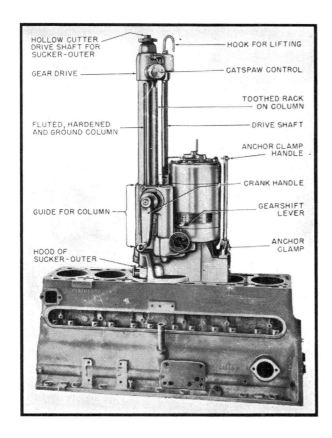

Fig. 82 Boring bar installed on cylinder block

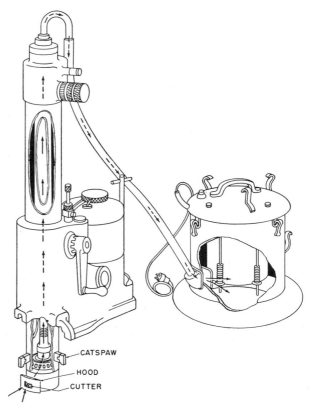

Fig. 83 A vacuum device removes chips from cutting tool. A small metal "hood," located just above the cutting tool, is the air intake. A mixture of air and chips flows up through the hollow boring bar to the top where a 9-foot hose connects with the vacuum cleaner

at the rate of about 30 complete strokes per minute.

The lower end of the bore should be honed first, because it is smallest here. On the down stroke the hone should extend out of the bore about an inch. Initial up-strokes should be only about 2″.

Immediately after starting the hone, if the tension on the stones is not enough, stop the hone and expand the stones slightly by turning the micrometer but be sure not to increase the tension enough to slow the motor down. As honing proceeds, the gradually increasing bore size requires frequent adjustment of the tension by turning the micrometer. At the same time the length of the upstrokes is gradually increased, until finally full strokes are used with the stones protruding about an inch at both top and bottom.

Let us assume, for example, that .010″ oversize pistons are to be fitted and that the specified piston clearance is .002″.

Honing is begun with coarse stones and finished with fine stones. Continue honing one bore with coarse stones until it is within one or two thousandths of finished size as determined by a mi-

crometer or by trying a new piston in the bore. When the piston is a tight fit in the bore it is close to size.

Note the micrometer reading, and then rough hone all the other bores to this size.

Then change to the fine stones and fit each bore to the piston selected for it. Check the final piston fit with a .002″ feeler (since we assumed the specified piston clearance was .002″). The fit is correct when a 5 to 7 pound pull is obtained on the feeler, Fig. 81. Some shops dispense with the spring scale and simply "feel" the clearance by pulling on the feeler.

As soon as the fit is correct, the lower inner edge of the skirt should be marked with scratch lines, the number of lines corresponding to the cylinder number—three lines for No. 3 cylinder, for example.

When honing is finished, be careful to wash away all grit. Use plenty of soap and water and a stiff bristle brush.

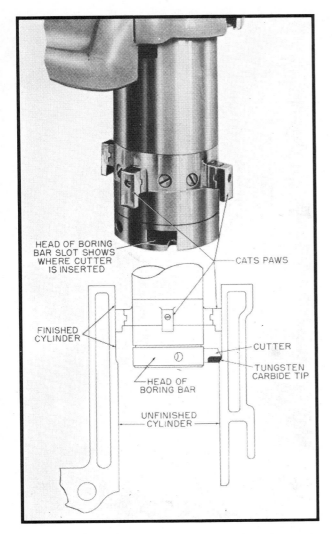

HEAD OF BORING
BAR SLOT SHOWS
WHERE CUTTER
IS INSERTED

CATS PAWS

FINISHED
CYLINDER

CUTTER

TUNGSTEN
CARBIDE TIP

HEAD OF
BORING BAR

UNFINISHED
CYLINDER

Fig. 84   The four cats-paws center the boring bar in the cylinder bore. The cutter is shown in the lower picture

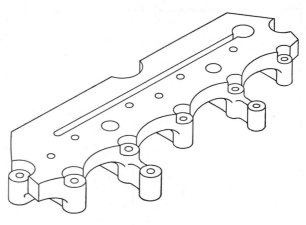

Fig. 85   The cast iron plate shown is used as a spacer on which the boring bar is mounted. It avoids the removal of cylinder head studs. The plate illustrated is for a Ford Six. Plates are also made for other widely used engines

REBORING

The cylinders should be rebored to fit the nearest oversize piston. Therefore if taper is .017″, for example, it is rebored for .020″ oversize pistons.

Pistons are usually available in the following oversizes: .010″, .020″, .030″, .040″, .050″—and perhaps .060″. There may also be oversizes less than .010″ such as .003″ and .005″.

The cylinder walls on some makes of engines are thicker than on others. Therefore some makes can be rebored to a slightly larger diameter than others.

If desired, the jobber shop can grind semifinished pistons to any non-standard oversize, such as .017″ or .025″, etc.

The cylinders are rebored to within about .001″ of the desired size and then finished to size by honing.

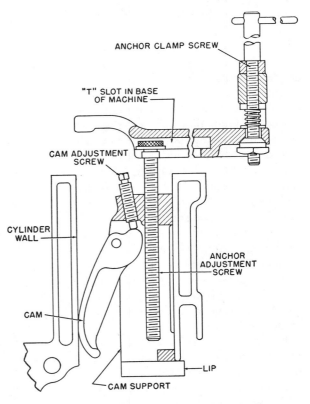

ANCHOR CLAMP SCREW

"T" SLOT IN BASE
OF MACHINE

CAM ADJUSTMENT
SCREW

CYLINDER
WALL

ANCHOR
ADJUSTMENT
SCREW

CAM

LIP

CAM SUPPORT

Fig. 86   Details of construction of the anchoring device which holds the boring machine tightly to the block

When the engine is rebored in the car, it sometimes happens that the rear cylinder is so close to the dash that the boring tool cannot be installed. In this case, the rear cylinder may be honed to size while the others are rebored.

Use of Boring Bars

The following information is intended to give a general idea as to how a rebore job is done. Fig. 82 shows a boring bar mounted on a cylinder block.

Boring bars all use the same principle of a rotating cutter driven by an electric motor to cut the metal from the bore. Naturally the details of the different makes of boring bars vary.

A Sucker-Outer, Fig. 83, is optional equipment on some models. It disposes of the chips. There is an electrically-driven fan which applies suction to a hood surrounding the cutter, picking up the chips that are carried up through the hollow cutter shaft and delivering them to a container surrounding the suction fan and its motor.

The torque required to rotate the tool depends on the bore diameter, the depth of cut and the hardness of the bore. A 50 per cent increase in bore diameter calls for a 50 per cent increase in torque; a deep cut requires more torque than a shallow cut; a hard bore demands more torque than a soft one. This explains why some of these tools have two to four speeds. The slower the speed the greater the torque.

One model has four speeds. The two higher speeds are for cast iron and the two lower speeds for hard steel sleeves. This model is described in some detail below.

Boring Bar Construction

The complete tool installed on the block is shown in Fig. 82. A rotating cutter with a tungsten carbide tip, Fig. 84, rebores the cylinder. The cutter shaft rotates within the heavy column which is mounted in the body of the machine, Fig. 82. The column does not rotate but it is fed downward by means of the toothed rack shown, which is operated by a gear in the base of the machine. The cutter, of course, moves down with the column.

The cutter shaft is rotated by the drive shaft which is rotated by the transmission gears which are driven by the motor.

Preparation for Boring

If the block has cylinder head studs instead of bolts, the studs must be removed, or a boring plate should be installed, Fig. 85. The thickness of the plate is greater than the length of the studs.

Before beginning a rebore job, the top of the block must be clean and free of burrs so that the base of the machine can rest squarely on the block. If dirt tilts the bar the finished bore will also be tilted.

Before a cylinder can be rebored, the cutting head must be accurately centered in the bore and then the machine must be firmly clamped to the block.

Before placing the machine on the block, raise the column all the way to the top by turning the large crank handle slowly.

The cutter head is centered in the bore by the four cats-paws, Fig. 84. Retract the cats-paws by turning the outer knurled wheel on the cats-paw control located near the top of the columns, Fig. 82.

Then run the boring bar down into the bore until the cats-paws are near the bottom where wear is least. Expand the cats-paws until they are tight in the bore by turning the outer knurled wheel just mentioned.

The anchor clamp, Fig. 86, holds the boring bar in place after it has been centered in the bore. The clamp is located in an adjacent bore. The device is inserted in the bore from underneath. The head of the long screw is slipped into the T-slot in the base of the boring bar and the screw is adjusted until the lip on the lower end of the clamp body is snug against the lower edge of the bore. Then turn down the cam adjustment screw until the cam is tight against the bore—yet not tight enough to distort the bore.

Turn down the anchor clamp screw until it is just snug against the top of the block. Do not tighten it enough to mar the block. This screw is seen to right of motor in Fig. 82.

After the tool is centered, retract the cats-paws and turn the large crank slowly to raise the cutter head up to the top of the bore.

Cylinders are rebored to within about .001″ of finished size and then finished to size by honing. Each piston is individually fitted to its bore during the honing process and checked for fit with a feeler. The procedure is the same as described for finish honing.

The next step in reboring is to set the cutter to the desired oversize. For example, assume that the original cylinder bore was 3⅛″ or 3.125″ and that .020″ oversize pistons are to be fitted. Assume also that piston clearance for this make and model of engine is .002″. Therefore, the refinished bore diameter will be 3.125″ plus .020″ plus .002″ clearance or 3.147″. However, .001″ should be left for honing. Consequently, the cutter should be set to 3.146". All cylinders should be bored to this size.

Boring Procedure

Boring is started by engaging the clutch which operates the feed gearing. This is done by pulling out on the ring on the crank handle shaft, Fig. 82. Immediately start expanding the cats-paws slowly and evenly until they touch the finished bore. Continue to exert pressure on the cats-paws until they have completely entered the bore. Continue to hold the outer knurled wheel while you engage the pawl and ratchet on the inner knurled wheel until a definite resistance is felt. The paws are now set and need no further attention while the boring of the cylinder is in progress. The cats-paws steady the cutter head and prevent chattering of the tool.

The cutter, with its tungsten carbide tip is shown in the lower half of Fig. 84. It fits into a slot in the cutter head. The slot is clearly illustrated in the upper picture.

To adjust the cutter, it is removed and placed in a special micrometer fixture which is supplied with the boring bar.

After the cutter is adjusted, it should be inserted in the cutter head.

Measure the depth of the cylinder bore and set the automatic stop to this depth so that the motor will shut off when the bore is finished. Attach the Sucker-Outer if used and turn on its motor.

Welded Bore

If a cracked bore is repaired by welding, the weld should be finished by boring. If only enough metal is removed to smooth up the weld and little or no metal is removed from the rest of the bore it should be possible to use the original piston, if undamaged. The bore should be smoothed up with a hone after boring.

CYLINDER SLEEVES OR LINERS

If a cylinder bore is cracked or so badly worn or scored that more than .060" must be removed by reboring, a cylinder sleeve may be installed, Fig. 87. The sleeve is usually made of a high grade of alloy cast iron. Sleeves are supplied in various thicknesses such as 1/16", 3/32", 1/8" and 5/32", the 3/32" being the most popular. The inside diameter comes finished to standard bore size for a particular make and model of engine. The engine is rebored to the outside diameter of the

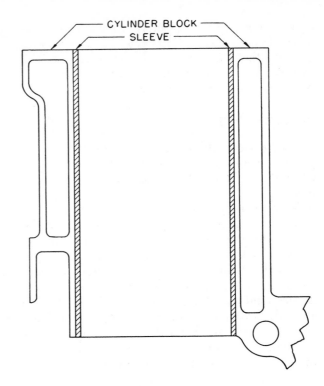

Fig. 87 *The sleeve is a tight fit in the bore*

sleeve and the sleeve is pulled into position by a suitable tool, Fig. 88. Fig. 89 illustrates the method of removing a worn sleeve.

"PREPARING CYLINDER BLOCK FOR HIGH PERFORMANCE"

Before disassembly of components from the block, check the deck height. This is the distance from the top of the piston at TDC to the top or deck surface of the cylinder block, Fig. 90. This is a critical dimension as it relates directly to the compression ratio. It must be held constant to ensure a proper balance of combustion chamber volumes when the head is installed.

Completely disassemble cylinder block, removing everything from it that can be removed—crankshaft, camshaft, water jacket plugs, etc., and give it a Magnaflux test to make sure that there are no cracks. Next check the main bearing bores for specified diameter and alignment with each other.

To check main bearing alignment, place an arbor of the correct size in the bearings and tighten the bearing caps to the correct torque, Fig. 91. If the arbor can be freely turned under this

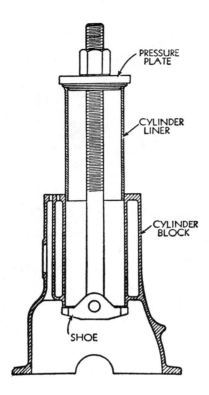

Fig. 88 *A shoe which is larger than the bore is attached to the bore. Then the sleeve (liner) is forced into the bore by turning the nut at the top*

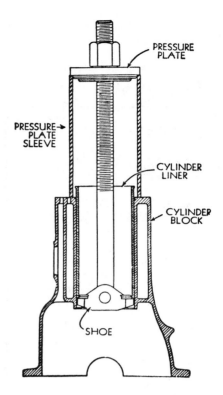

Fig. 89 *A special puller is used to remove a cylinder sleeve (liner). A shoe slightly smaller than the hole in the block is pinned to the screw. The sleeve is pulled out (upward) by turning the nut at the top*

condition, the bearings are in alignment. If the arbor binds, the bearings must be line bored.

Check the deck surfaces of the block to see that they are parallel with the crankshaft axis. This must be done after the bearings are line bored. The distance from the centerline of the crank to the deck surface of the block must be the same at front and rear. If they are not the same, the head surfaces on the block will have to be milled or ground until they are parallel to the crank. They should be cut the least amount necessary to produce a true flat surface.

Remove casting burrs and slag with a small hand grinder. This helps to ensure that these protrusions, or flashings, don't break away and close off a critical oil passage or scuff moving surfaces. The grinding operation also helps to reduce stress areas where cracks are most likely to start.

Run a bottoming tap down all bolt holes to clean out any deposits. A small chamfer on the cylinder head bolt or stud holes will prevent the top thread from pulling up when the bolts or studs are tightened down. This would prevent the cylinder head from sealing properly to the block.

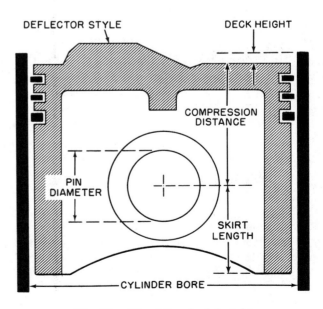

Fig. 90 *Checking deck height*

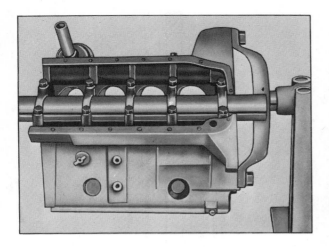

Fig. 91   Checking main bearing bore alignment

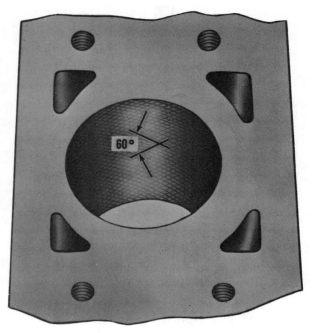

Fig. 92   Cross-hatch pattern

Check all dowel pin heights. If any are longer than their mating holes a proper seal cannot be attained.

Polish tappet bores with fine grit sandpaper to remove rough spots.

Rough-up rear main bearing oil seal machined surface with a center punch. This helps to prevent the seal from slipping in the groove at high speed.

Run a small bottle brush or drill through all oil passages.

Check and refinish cylinder bores as required.

Visually inspect the walls for scores, cracks, spotty wear, etc. Measure the bores for taper or out-of-round condition, and if required, bore the cylinders. Leave approximately .002″ stock for the honing operation.

Hone the cylinder walls to their final dimension. Honing is recommended both to remove glaze when installing new rings—or a glaze breaker can be used—and as the final step in reboring. Wet honing, with a cutting fluid, is preferable to dry honing.

The recommended surface finish to produce the 60° pattern shown in Fig. 92, is best achieved by frequent cleaning of the stones. The rate of vertical strokes of the hone should be increased as the speed of the hone is increased.

Break all sharp edges with a small file. (Around cylinder bore chamfers, main bearing bore chamfers, etc.)

Clean the block surface thoroughly and coat its machined surfaces with a light oil film until it is ready for assembly.

Piston Pin & Rod Service

CHECK FOR LOOSE PISTON PINS

Before installing new rings, check the piston pins for looseness. If wear is considerable it can be felt. Hold the piston in one hand and grasp the rod with the other hand. Push and pull on the rod and also rock it back and forth in the plane of the piston pin. If there is noticeable play, the pin is too loose.

A more delicate test for wear in pin holes is to squirt some light oil on the end of the pin (after removing lock rings), wipe off the surplus and then note whether there is any motion in the oil film between pin and hole when the rod is rocked.

Instead of holding the piston in the hand it may be mounted in a piston vise, Fig. 93. Or the pin may be wedged onto an easy-out clamped vertically in a vise, Fig. 94.

If the piston pins are pressure lubricated through a rifle-drilled hole in the connecting rod,

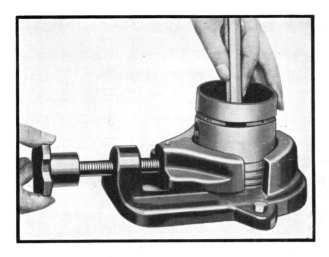

Fig. 93   Vise for holding a piston

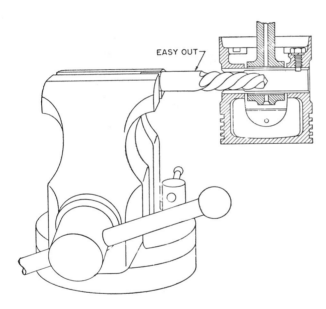

EASY OUT

Fig. 94   How to support a piston with an "Easy Out"

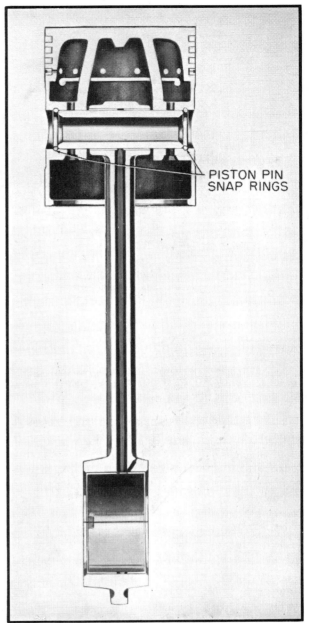

PISTON PIN
SNAP RINGS

Fig. 95   The full floating piston pin is retained by snap rings

excessive oil consumption may result if the pins are loose in their piston holes, since the extra clearance between pins and piston permit extra oil to flow onto the cylinder bores.

PISTON PINS TOO TIGHT

Sometimes when pistons are removed, the pins will be found to be too tight, as indicated by the fact that some effort is required to rock the piston on the rod. Usually this is caused by a gum or varnish deposit on the wearing surfaces. Remove the pins and clean the surfaces with a suitable

solvent. If difficulty is experienced in removing the pins, heat the pistons in boiling water to soften the deposit.

REMOVE PISTON PINS

Before removing the piston pins, the pistons should be numbered by making light file marks on the inner edge of the bottom of the piston skirt. One mark for No. 1 piston, two marks for No. 2 piston, etc. Put these marks on the same side as

155

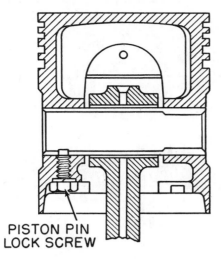

PISTON PIN
LOCK SCREW

*Fig. 96   The piston pin is locked
to the piston by a set screw*

*Fig. 97   A screw clamps
the rod to the piston pin*

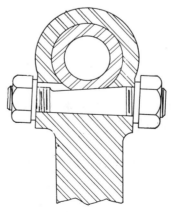

*Fig. 98   The nut on the
right pulls the taper pin
out when the nut on
the   left   is   removed*

the rod number and be careful to reassemble the
rods and pistons accordingly.

The procedure in removing piston pins depends
on whether the pin floats, or is locked to the piston
or to the rod, or is a pressed fit.

Full-Floating Pin—Remove the snap rings at
either end of the pin, Fig. 95, by means of a pair
of pliers with pointed jaws; or pry the snap rings
loose with a screw driver. When the pin is reassembled to the piston always use new snap rings.

Push out the pin with the fingers. If the pin is
too tight to be removed by hand use a brass or
copper drift. Insert the drift in the piston pin hole.
Hold the piston in one hand and the drift in the
other. Then gently pound the free end of the drift
on a metal plate. If the piston is aluminum, pin
removal is facilitated by heating the piston in hot
water.

Set Screw Pin—Remove lock screw, Fig. 96, and
then remove pin. A piston vise, Fig. 93, is recommended when removing the lock screw, not only
for convenience but also because it prevents possible distortion of the piston.

Oscillating Pin—Two types are shown in Figs.
97 and 98. In Fig. 97, the rod is clamped tight to
the pin. Loosening the screw permits the removal
of the pin.

In Fig. 98 the lock screw has a tapered flat surface which mates against a flat surface on the
piston pin. To free the screw, loosen the nut on
the left and tighten the one on the right until the
pin is free. Then push or drive pin out of piston.

When reinstalling pin, be sure the flat spot is in
the position shown. Loosen the nut on the right

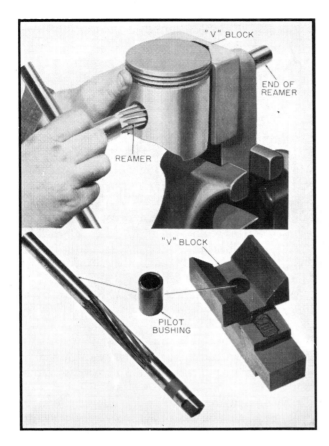

Fig. 99 *The V-block is mounted in the vise. The right size pilot bushing is inserted in the hole in the V-block. The piston is put into position and the reamer inserted*

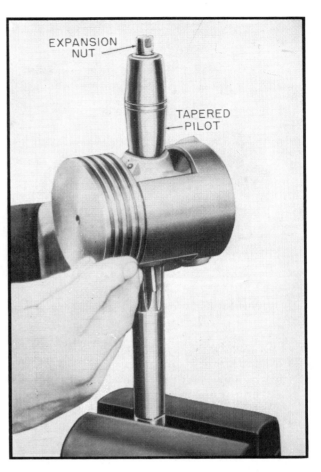

Fig. 100 *The reamer is mounted in the vise. The piston is installed on the reamer. Then the tapered pilot is slipped over the end of the reamer*

and tighten the left nut until lock screw is drawn up tight. Then tighten nut at right securely.

Press Fit Pin—Mount the piston and rod on a press and, using a suitable arbor, press the pin out of the rod. The press and arbor are used to install the pin in the rod. Any fit less than that which requires the use of a press is too loose.

PISTON PIN SIZES

Piston pins supplied by car makers are made in standard size and two or three oversizes. Oversizes of .003" and .005" are the most common. But some car makers offer oversizes of .001" and .003" and .008", or .003", .005" and .010".

Piston pin manufacturers offer pins in standard size and oversizes of .003" and .005" with perhaps some additional oversizes for some of all makes of cars—.0015" for example.

PISTON PIN FITS

Full Floating Pin—Alloy and cast iron pistons, light tap.
Oscillating Pin—Alloy piston, palm push.
  Cast iron piston with bronze bushings, thumb push.
  Cast iron piston without bushings, light thumb push.
Set Screw Pin—Alloy Piston: Set screw end, light tap; other end, thumb push.
  Cast iron piston: Set screw end, light tap; other end, palm push.

Pin fits should be checked with the piston and pin at the same temperature. If the piston has been heated to remove the pin, allow it to cool off before checking the pin fit.

Diamond boring of pin holes is used by some manufacturers. Note that for each type of pin, the "push effort" is less with diamond boring than with honing or reaming.

157

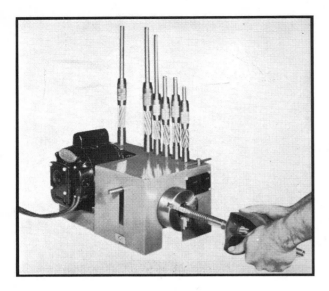

Fig. 101 *Electrically driven reamer tool*

Full Floating Pin—Alloy piston, palm push.
    Cast iron piston, thumb push.
Oscillating Pin—Alloy piston, thumb push.
    Cast iron piston with bronze bushings, thumb push.
    Cast iron piston without bushings, light thumb push.
Set Screw Pin—Alloy and cast iron pistons:
    Set screw end, light tap; other end, thumb push.

Piston pins that are press-fitted—and these are almost universal today—are approximately .001″ larger in diameter than the rod bore, and no bushing is used in the rod.

REAMING PIN HOLES

An expansion reamer should be used. The reamer is expanded or contracted by turning a nut on the end of the reamer.

Expansion reamers come in a range of sizes from 1/2″ diameter upwards by 1/16″ graduations.

A piston can be reamed simply by clamping the reamer in a vise and rotating the piston on it. However, in order to insure a perfectly true hole the reamer should be piloted. Fig. 99 shows one design and Fig. 100 another.

When reaming a piston, be sure that both the reamer and the pin holes are clean. The reamer blades should be sharp. When a reamer becomes dull it should be returned to the factory for sharpening. There is, of course, a limit to the number of times it can be sharpened.

The reamer should be kept amply lubricated with engine oil throughout the reaming process.

Measure the new pins with a micrometer to be sure they are the correct size.

Reaming the piston hole to size requires a gradual and cautious enlargement of the hole until it is just the right diameter.

The first step is to make a light cut for the purpose of "fitting" the reamer to the hole. Make the cut through both piston bosses and then try the piston pin. It won't go into the hole but at least it will give you some idea as to how much metal is still to be removed.

Expand the reamer a few tenths (ten-thousandths), make another cut and try the pin again —and so on until the pin is the correct push fit. With each new cut expand the reamer less and less. The last cut might be only a fraction of a thousandth of an inch.

While fitting the pin to the piston, do not try to force it all the way into the boss. Instead, just try to insert the pin about 1/16″ into the boss. Turn the pin while pushing it in.

After the pin is fitted to the first piston, measure the reamer diameter with a micrometer. Then adjust the micrometer to read .0005″ less. Adjust the reamer to the micrometer and ream the remaining pistons to this size. After this is done, fit each pin to its own piston by taking very light cuts until the fit is correct.

Be careful to keep each pin and its piston together. Either insert the pin in the piston or turn the piston upside down and put the pin inside it.

REAMER FIXTURES FOR PISTON PINS

In Fig. 99, the reamer is piloted in a V-block. There is a pilot bushing for each reamer size. To set up the fixture, select the correct size pilot bushing, insert it in the hole in the V-block, and mount the V-block in the vise as shown. The V-block insures that the piston is accurately lined up at right angles to the pilot hole. The piston is held by one hand while the reamer is turned with the other.

In Fig. 100, the reamer is stationary and the piston is turned with both hands. The piston is placed on the reamer as shown and then the tapered pilot is slipped down over the top of the reamer. The nut for expanding the reamer is shown at the top. Every effort should be made by the hands to provide a purely rotary motion without side thrust.

Fig. 101 illustrates an electrically-driven reamer tool. The reamer is mounted in the chuck. The piston is slipped over the reamer and stroked back and forth until the reamer runs free. Then the reamer is expanded slightly, etc. In other words

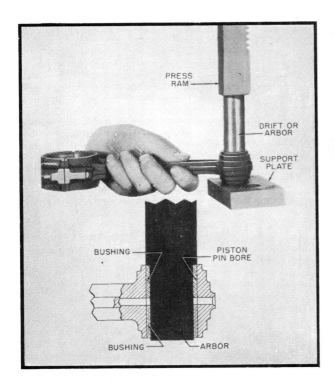

Fig. 102   The bushing is removed by pressing it out with an arbor operated by an arbor press

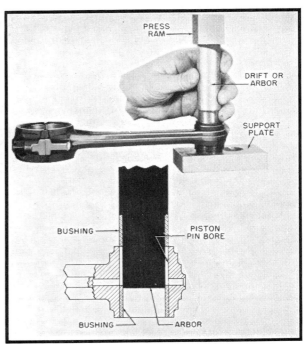

Fig. 103   Pressing new bushing into position

the method is similar in principle to hand reaming but much faster.

ROD BUSHINGS

New bushings may be required if standard size pins are to be installed. But if oversize pins are to be used, the rod bushings should be refinished oversize unless they are loose in their rod bores or badly scored, in which case new bushings should be installed.

The bushings in the rod may be in one piece, running the length of the rod bore, or two half-length bushings may be used. Bushing walls may be of normal thickness or may have thin walls.

ROD BUSHINGS, REMOVE

To remove a bushing, Fig. 102, select a drift of suitable diameter. The diameter of the shank of the drift should be slightly smaller than the rod bore. The diameter of the end of the drift should be a little less than the bushing bore.

Insert the drift in the rod and put the rod on the arbor press. The drift should be directly under the press ram and the side face of the rod should rest squarely on the surface of the support plate. Operate the ram lever to push the bushing out.

If an arbor press is not available the bushing may easily be removed as follows: Take the blade out of a hacksaw. Slip the blade through the bushing and connect the blade to the hacksaw frame. Then saw a slot all the way through the bushing, after which it is a simple matter to drive it out of the rod.

ROD BUSHINGS, INSTALL

If the rod has an oil hole in it be sure to line up the oil hole in the bushing with the oil hole in the rod. Then press the bushing into the rod until it is flush with the sides of the rod. Check to be sure that the oil holes in bushing and rod are lined up.

It is always good practice to use a special replacing drift, Fig. 103, when installing a bushing. The end of the drift should fit the bore of the bushing snugly so that when the bushing is pressed into the rod the snug fit will prevent distortion of the surface of the bushing bore.

If an arbor press is not available the bushing is easily installed as follows: File a slight bevel on the leading edge of the bushing so that it can be started in the rod hole. Then place the rod and bushing between the jaws of a vise. Before closing the vise, make sure the bushing is square with the rod hole. Then push the bushing in the rod by turning the handle of the vise.

Fig. 104 *Swaging new thin-wall bushing to secure it to rod bore*

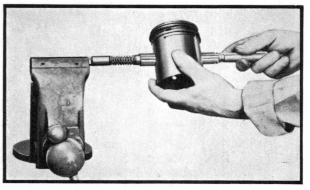

Fig. 105 *Hand hone mounted in vise. Piston is rotated by hand*

Fig. 106 *Foot switch for controlling electric drill when honing*

SWAGING ROD BUSHINGS

A special replacing drift is particularly desirable when installing thin-walled bushings.

After a thin-walled bushing is installed it is likely to come loose in service unless it is forced tightly against the rod bore. This is done by a swaging tool, Fig. 104. The tool is slightly tapered, the diameter at the upper end being slightly larger than the bore of the bushing. The tool is forced down through the bushing, causing the metal to flow until it presses so tightly against the rod bore that there is no danger of its coming loose. It also burnishes the bushing bore. Burnishing, by definition, is the rubbing of a hard tool over a surface to make it smooth. The burnishing, in this case, is a by-product of the all-important swaging operation. The tool is often called a burnishing tool although it is primarily a swaging tool.

After the thin-walled bushing is swaged, it should be finished to the required size by honing or reaming—preferably by a piston pin hole honing machine as the walls are considered too thin for reaming or hand honing tools.

HAND HONING PISTON PIN HOLES

A hand hone is used in the same manner as a reamer. It can be mounted in a vise, Fig. 105, or the piston can be mounted in a fixture and the hone turned with a wrench. Use a fine abrasive cloth recommended by the maker of the tool. Like reamers, piston pin hones come in a range of sizes.

The tool is expanded in the same manner as a reamer. As with a reamer, the tool is expanded gradually during the honing process until the piston pin just fits the hole. Both piston holes and rod bushings can be honed.

If the piston pin holes have to be enlarged more than .002″, it will save time to put the hone in an electric drill which has a speed of about 300 to 500 rpm. Mount the drill in a vise with the hone horizontal. The drill is conveniently started and stopped by a foot switch, Fig. 106, made for the purpose.

Clean the piston to be honed with gasoline. Select the proper oversize pin for each piston before honing (usually .003″ oversize). Select the largest hone possible to insert in the hole.

If only .001″ to .002″ is to be removed, a drill is hardly necessary because you may remove more metal than desired before you are aware of it. This is especially true on alloy pistons.

There are several effective methods of removing only a small amount of stock.

1. Clamp the driving end of piston hone in a bench vise, then revolve piston by hand, traversing the piston back and forth.

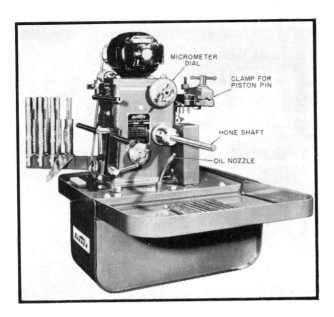

Fig. 107 The hone is mounted on the shaft indicated. The piston pin is placed in the clamp shown so that it will be handy for testing the fit

2. Place the piston in a piston vise and rotate hone in the hole by means of any standard tap or reamer wrench.

3. Any hand brace can be used very handily by using same as in drilling with bit and brace.

For removing more stock in piston holes any portable electric drill with an approximate speed of 300 to 500 rpm. is recommended, or use a drill press.

Clamp portable drill in bench vise or use piston hone in drill press at the recommended speed. Insert piston hone in drill chuck, collapse sufficiently to slide over piston hone. Expand hone snugly in piston hole (piston hone expanded into piston too tight will have a tendency to tear the abrasive cloth and wedge the hone in hole and will not cut). Start the drill and traverse piston back and forth ¼″ beyond the edge of abrasives on either end of piston hone. After honing for a few seconds the drill will run free. Stop the drill, remove piston from hone without moving the graduated adjusting screw and check the hole in the piston.

After each run of the piston hone (which is usually 30 to 45 seconds) reverse the piston end for end and hone as before: When the piston is honed to size so that the piston pin just enters snugly and you notice a slight difference in one end of piston, enter piston hone into the small end of piston, and traverse back and forth with only a short stroke. Hone this end of piston to the same size as the other end, using the piston pin as a

plug gauge. This method of honing will produce a straight round hole in both ends of the piston.

IMPORTANT: Use a dry cloth or bristle brush to clean the abrasives after each run. Do not use a wire brush or sharp abrasive for cleaning, as they reduce the life of abrasives. After each run when abrasives have been cleaned, slide piston on hone, release lock screw and expand graduated screw two graduations. Proceed as before. The expanding of two graduations is very important: more than two graduations at each run will not remove any more stock, but will shorten the life of abrasives.

PISTON PIN HONING MACHINE

Fig. 107 illustrates a machine for honing piston pin holes and connecting rod bushing holes. It is more accurately called a grinder. It can be used for other small holes—kingpins, for example.

The piston is placed on the hone and the micrometer dial is turned until the mandrel is expanded to fit the hole. The hone should be copiously lubricated throughout the honing process. This is accomplished by a built-in oil pump which delivers a steady stream of oil.

Start the machine and stroke the piston back and forth. Slowly turn the dial to the right until it begins to cut. Stroke the piston back and forth on the hone while holding it in both hands. When the piston feels "free", retract the hone by pressing the floor pedal. Remove the piston, turn the dial to expand the hone a few thousandths. Put the piston back on the hone. Remove foot from pedal to allow the hone to expand. Stroke the piston back and forth at the rate of about 20 complete strokes per minute. Every few strokes remove the piston and reverse it. Whenever removing or replacing piston always depress pedal to retract hone.

When the piston feels free, remove it and try the piston pin for fit.

Expand the hone a little more and repeat the process and so on until the fit is perfect.

Note that as the pin hole approaches the desired size, the micrometer dial should be advanced with caution. When the fit is almost perfect the dial should be advanced only a fraction of a ten thousandth of an inch. When the pin fits perfectly, place it inside the upturned piston.

After sizing the pin hole in the first piston, turn the micrometer dial back about 5 ten thousands (.0005″) and hone the other piston holes to this diameter. Then hone the holes in each piston to the exact fit required for the pin selected for it.

The procedure for honing connecting rods is the same as for pistons. The rod should be mounted

Fig. 108 *Honing a rod bushing. The rod is mounted in a suitable holder so that it can be grasped by both hands*

MICROMETER DIAL

HONING STONE

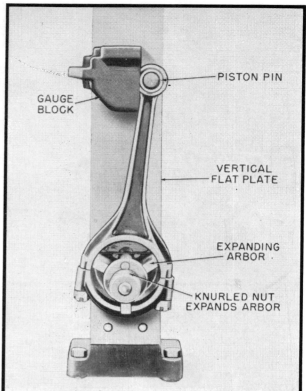

PISTON PIN

GAUGE BLOCK

VERTICAL FLAT PLATE

EXPANDING ARBOR

KNURLED NUT EXPANDS ARBOR

Fig. 110 *Testing connecting rod for twist*

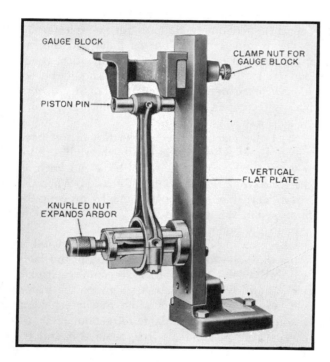

GAUGE BLOCK

CLAMP NUT FOR GAUGE BLOCK

PISTON PIN

VERTICAL FLAT PLATE

KNURLED NUT EXPANDS ARBOR

Fig. 109 *Testing connecting rod for bend*

in a holder, Fig. 108, so that it can be properly grasped in both hands.

The instructions given above are intended to give the reader a general idea as to how a honing machine is used. However, it is strongly recommended that when using this or any other machine the manufacturer's instructions should be carefully followed.

CHECK ROD ALIGNMENT

The alignment of both old and new rods should be checked before they are installed. There are a

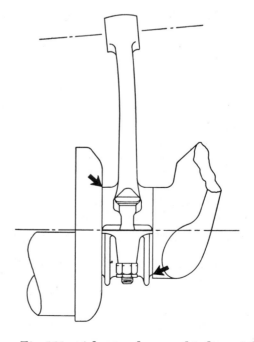

Fig. 111 *A bent rod causes binding of the rod bearing at the points indicated by the arrows. There is similar binding and wear at the piston pin*

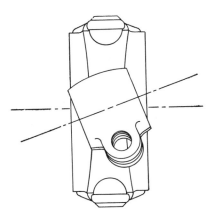

Fig. 112 *A twisted rod causes binding and increased wear on rod bearing, piston pin and piston*

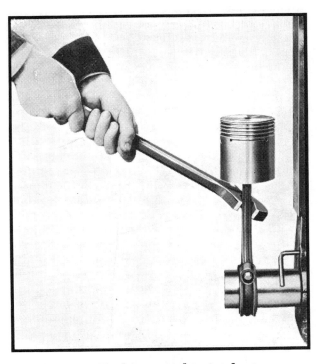

Fig. 113 *Tool for straightening bent connecting rod. The same wrench-like tool can be used for correcting twist. The rod is shown mounted on the aligning tool*

number of aligning tools in use including those illustrated in Figs. 109 and 110.

The rod should be checked for bend and twist. If the rod is bent its piston pin will not be parallel with the crankpin, Fig. 111. This cocks the rod bearing on the crankpin, causing extra wear where the bearing is pinched—as shown by the arrows. The same thing happens at the piston pin.

In a twisted rod, Fig. 112, the piston pin bore is not in fore-and-aft alignment with the crankshaft axis and may cause increased wear on rod bearings, piston pin and piston.

When checking a rod, the bearing cap should be tight but do not install the rod bearings.

The rod aligning tool, Fig. 109, is permanently mounted on the bench. To the base is bolted a vertical flat plate and the aligning gauge block is mounted on it. The base of the gauge block fits around the left edge of the plate, Fig. 110. There is a clamp nut at the rear so that the block can be positioned for any length of rod.

To check a rod, install the piston pin and the bearing cap but not the bearings. Tighten the rod nuts to torque wrench specifications.

Expand the arbor by turning the nut until the arbor is firmly seated in the rod bore.

To check the rod for bend, place the gauge block in the position shown in Fig. 109 with the legs of the gauge resting on the pin. If only one leg makes contact, check the clearance with a feeler. If the clearance is more than .002″ the rod should be straightened with the bending iron shown in Fig. 113 until clearance is less than .002″ or until no light can be seen between pin and block.

To check the rod for twist, place the aligning gauge in the position shown in Fig. 110 and clamp

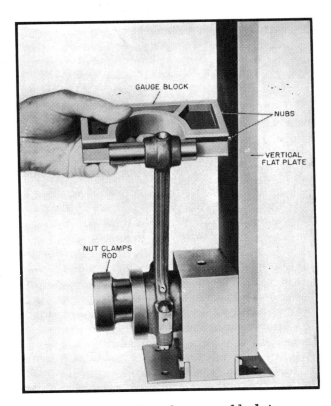

Fig. 114 *When the gauge block is vertical the nubs indicate bend—and when horizontal they show twist*

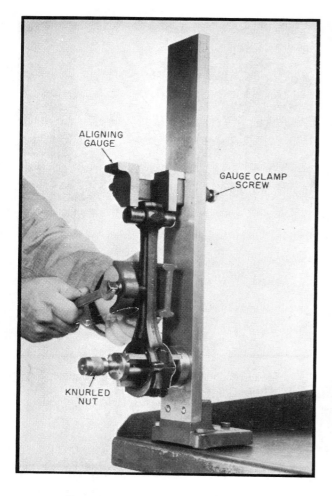

Fig. 115 *A special clamp type fixture is used to straighten the rod*

Fig. 116 *The aligning gauge is placed under the piston to check the correctness of its alignment by means of a feeler gauge*

it in place. Then bring the pin in contact with the gauge. If one end of the pin fails to contact the gauge, check the clearance with a feeler. If more than .002″, the rod should be untwisted until clearance is less than .002″. Then recheck for both bend and twist.

After the piston and rod are assembled, check the alignment of the piston. Its top surface should be parallel to the gauge within .002″. Use the set up shown in Fig. 109.

Fig. 114 shows another make of tool for testing connecting rod alignment. The rod is clamped in position vertically as shown. The gauge block has a V groove to center it on the piston pin. The gauge block is placed horizontally to determine twist and vertically for bend.

Two round nubs contact the vertical flat plate if the rod is in alignment. If one of the nubs has a clearance of .002″ the rod should be straightened.

Another design of aligning tool is seen in Figs. 115 and 116. The knurled nut expands the arbor until it fits the rod bore. A special straightening tool is used, Fig. 115. Turning the screw straightens the rod. Fig. 116 illustrates how the tool is used for checking piston alignment.

ASSEMBLE PISTON TO ROD

Match up piston and rod according to their marks, so that they are in correct relationship.

Wipe off pin with clean cloth and wipe out bores in piston. Coat pin with engine oil.

Hold the piston in one hand. Start the pin into the piston and push it about ⅛″ beyond the inner edge of the boss. Then install the rod over the pin and push the pin through the second boss.

If the pin is a light tap fit, start the pin into the hole by hand. Then, while holding the piston in one hand, tap the pin lightly with a soft hammer. Or install a soft drift in the pin and pound the drift lightly on the bench or tap the drift with a hammer.

If the pin is locked to the rod, be sure the flat or groove in the pin lines up with the lock screw in the rod.

If the pin is locked to the piston, line up the hole in the pin with the lock screw. If one end of this pin is split insert this end first.

If the new pistons with pins already fitted happen to be tighter than a light tap fit, as they are in some engines, heat the pistons in boiling water or on an electric hot plate before installing pins. This recommendation applies to very tightly fitted pins with a clearance of .0001″ or less, down to .0003″ negative clearance—that is the pin is .0003″ larger than the pin holes in the piston.

PISTON AND CONNECTING ROD ASSEMBLY FOR HIGH PERFORMANCE

Pistons that are typically used in high performance engines combine strength and durability and lightweight characteristics with controlled heat expansion features, along with dome and skirt configurations that permit proper working clearances at all temperatures that are likely to be experienced.

They are selected to provide the desired compression ratio.

Chrome piston rings are highly desirable in high performance engines. One authority claims an additional 19 horsepower through careful selection of piston rings.

Connecting rods, in addition to being Magnafluxed, checked for alignment and de-burred, must be balanced with great accuracy. Special high performance rods have weight milling pads at both ends for balancing purposes.

When installing rods to the crankshaft, torque the caps to specification, rotate the shaft, remove the bearing cap and inspect it for burrs on the bearing surface. This is an extra precaution before final assembly.

Review Questions

Page

1. What is the function of the piston system? 118

2. Name the parts of the piston system. 118

3. Most repairs to the piston system are made for what purpose? 118

4. How do worn engine bearings affect oil consumption? 118

5. Which causes more wear, pistons or rings? 118

6. Viewing the engine from the driver's seat, which side of the engine is the major thrust side? 118

7. Under what operating condition does the most wear take place in an engine? 119

8. Why does the top piston ring wear faster than the other rings? 120

9. Explain what condition would result in piston slap. 120

10. Why is it necessary to break the glaze on cylinder walls in the process of doing a ring job? 121

11. What are the two common causes of pressure build-up in the crankcase? 121

12. What might be the result of excessive crankcase pressure? 121

13. What three ways can oil pass into the combustion chambers? 121

14. Blue smoke from the exhaust indicates what condition? 122

15. How does a leaky vacuum pump diaphragm cause high oil consumption? 122

16. How may an "invisible" crack in a cylinder block be detected? 123

17. What three methods may be used in repairing cracked blocks? 123

165

18. After the cylinder head and oil pan have been removed, what important operation must be performed before the pistons are removed for the purpose of installing new rings? 124

19. What might be the result if new piston rings were installed without removing the ring ridges? 124

20. Why is it so important to install the same pistons and rods in the cylinders from which they were removed? 127

21. Why should connecting rods never be identified by file marks? 127

22. In what portion of a worn cylinder bore should a piston ring be placed to measure the end gap? 129, 130

23. What could be the result of installing piston rings with insufficent end gap? 130

24. What type of piston ring does not require an end gap? 130

25. What tool should be used to install rings on pistons? 131

26. What is the last thing to do before installing piston and rod assembly in the cylinder? 131

27. What tool should be used when installing pistons in cylinders? 132

28. When cleaning aluminum pistons, what type of solution should be avoided? 133

29. What is the purpose of the narrow groove above the top ring on some pistons? 133

30. Why is it necessary for piston rings to move freely in their grooves? 134

31. What three methods are used to expand piston skirts? 137, 138

32. What three measurements should be taken of cylinder bores in order to determine if a rebore job is needed? 141

33. Describe three methods of measuring cylinder bores. 142

34. What is the purpose of a glaze buster? 144

35. When should a flexible hone be used? 145

36. Under what circumstances would the use of a rigid hone be preferable to a flexible hone? 147

37. Under what circumstances would cylinder sleeves be used in an engine that had no sleeves originally? 152

38. How many methods are currently being used to retain piston pins? 157

39. Why is a "full-floating" piston pin so called? 155, 156

40. What is an "oscillating" piston pin? 156

41. If a press is not available, describe a simple method of removing a piston pin bushing from a connecting rod. 156

42. What precaution should be taken when installing a connecting rod bushing having an oil hole? 159

43. What is the purpose of "swaging" a connecting rod bushing? 160

44. What is the purpose of checking connecting rod alignment? 162

45. In what two ways might a connecting rod be out of alignment? 163

Crankshaft and Bearings

Review Questions for This Chapter on Page 188

INDEX

INTRODUCTION

Main bearings 167
Rod bearings 167

CRANKSHAFT & BEARING SERVICE

Bearing bores out-of-round, main 186
Bearing caps, warped main 186
Bearing clearance, methods of checking 175
Bearing crush fit 179
Bearings, doweled main 181
Bearing installation, check 180
Bearings, install rod 179

Bearing journals, "miking" main 173
Bearing knocks 172
Bearing oil leak detector 170
Bearing oil seal, rear main 182
Bearings replace, main 180
Bearing spread 179
Bearings to use, what size 174
Bearing troubles 172
Clearance, rod side 179
Crankcase, warped 186
Crankshaft, checking for wear 173
Crankshaft end play 181
Crankshaft grinder 185
Crankshaft, hand polishing 185
Crankshaft, preparing for high performance 187
Crankshaft removal 184
Crankshaft straightness, check 186
End play, crankshaft 181

Engine oil passages, clean 178
Flywheel ring gear, replace 186
Flywheel service 187
Grinder, crankshaft 185
Main bearing bores out-of-round 186
Main bearing caps, warped 186
Main bearings, doweled 181
Main bearing journals, "miking" 173
Main bearing oil seal, rear 182
Main bearings, replace 180
Oil leak detector, bearing 170
Oil passages, clean engine 178
Oil seal, rear main bearing 182
Rod bearings, install 179
Rod journals, "miking" 173
Rod side clearance 179
Taper shims for worn bearings 178
Troubles, bearing 172
Worn bearings, taper shims for 178

Introduction

MAIN AND ROD BEARINGS

Main and connecting rod bearings used in all modern engines consist of two semi-circular halves usually referred to as bearing inserts or shells, Figs. 1 and 2. Each shell is made up of a steel back coated on its inner surface with a thin layer of babbitt. Babbitt is a bearing alloy containing about 90% tin, 5% antimony, 4% copper and small amounts of other metals. Other babbitt alloys may have somewhat different composition. There are other bearing alloys in use which contain cadmium, lead and copper; aluminum is also used as a bearing surface in some modern engines. These bearings have a total thickness of approximately 1/16″ with the bearing surface material approximating .003″.

Bearings should last upwards of 100,000 miles provided that: (1) Engine is run at correct operating temperature, (2) there is unfailing lubrication by clean oil of good quality, (3) not much driving is done at excessive speed, (4) the bearings were in perfect condition at the beginning with respect to materials, fit and workmanship.

As long as the bearings and journals are smooth and true and constantly separated by an oil film, the rate of wear is very slow. But if the oil film fails to separate the bearing surfaces even for a few seconds, damage will be done. Failure of the oil film may be due to lack of oil, low or no oil pressure, or excessively thin oil.

If the lack of an oil film permits the journal and its bearing to rub together, rapid wear results and in fact the heat caused by the high friction may melt the babbitt almost instantly, thus ruining the bearing and causing a knock.

Rapid bearing wear may be caused by fine, hard gritty particles such as road dust floating in the air. These particles are hard enough to wear away the bearing journals as well as the bearings. The particles get into the oil in one of two ways. They may enter the engine by way of the carburetor and make their way past the pistons to the crankcase. Or they may enter by way of the crankcase ventilating system.

Bearings wear faster than their journals because they are made of much softer material.

Rod bearings wear faster than main bearings

Fig. 1 Thin shell connecting rod bearing

because they are the most heavily loaded. They have to withstand the high combustion pressure as well as the high inertia pressures resulting from the pistons moving up and down through the cylinders at 2,000, 3,000 or even 4,500 revolutions per minute.

Bearing trouble can be avoided by keeping the air cleaner, and the cleaner on the crankcase ventilating system clean, and by changing the oil filter cartridge and the oil before it becomes dirty or badly diluted or otherwise contaminated.

Because bearings are so frequently forced to operate under adverse conditions, they should always be checked for wear when new piston rings are installed.

There are three principal reasons for bearing work. (1) Excessive oil leakage from worn bearings which causes increased oil consumption. (2) Bearings have damaged surfaces. (3) Bearings are so badly worn that they cause a knock. Various kinds of bearing damage are illustrated in Figs. 3 to 11.

If oil pressure is 5 or 10 pounds below normal when the engine is warm, it may indicate that an excessive amount of oil is escaping from the bearings because they are too loose. Normal oil pressure for all modern engines is 30 to 50 pounds per square inch. But remember that there are other causes of low oil pressure.

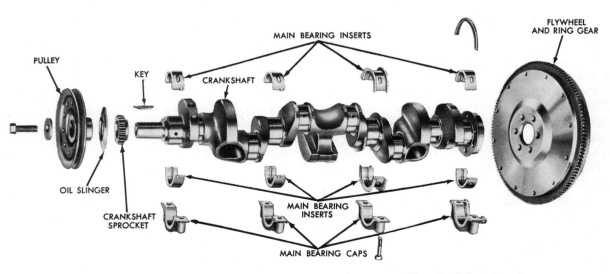

Fig. 2 Six-cylinder crankshaft with four main bearings. The third bearing from the front has flanges which take the end thrust of the crankshaft

Fig. 3 *Bearing ruined by minute particles of cast iron which became embedded in the soft bearing metal. The particles came from an engine that was not cleaned after a cylinder bore reconditioning job*

Fig. 4 *Medium sized particles of dirt or hard carbon scored this bearing*

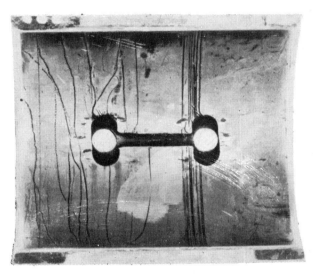

Fig. 5 *Bearing scored by large foreign particles*

Fig. 7 *Scored and melted bearing caused by lack of clearance*

Fig. 6 *Not enough clearance caused this bearing to fail. Much of the bearing surface is melted away*

Fig. 8 *Another bearing failure caused by lack of clearance. About two-thirds of the surface was liquefied and washed away by the oil*

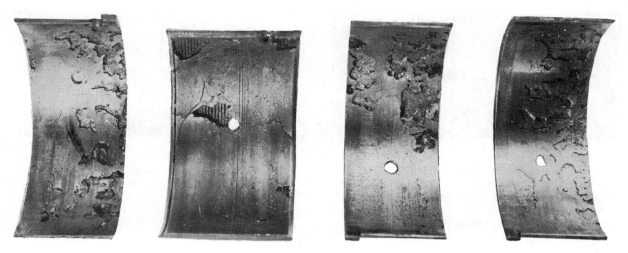

Fig. 9   *Overspeeding the engine forced these badly overloaded bearings to break down*

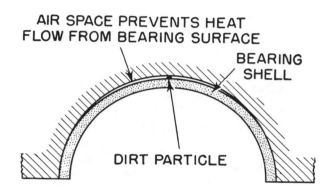

Fig. 10   *The backs of the bearings must fit tight against the bearing bore. Consequently, these surfaces must be clean. The particle of dirt shown, even if less than .001″, may cause bearing failure for two reasons: The particle prevents good contact between bearing and bore, therefore, heat conductivity is prevented at this point and the bearing metal melts in this region. The bearing may also fatigue in this area because the bearing will be bent in this area*

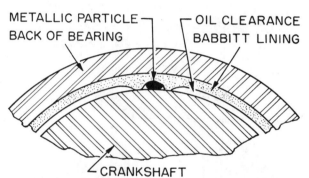

Fig. 11   *This diagram shows what happens when a hard particle embeds itself in the babbitt lining. The bearing metal is forced up around the high spot. The friction melts the babbitt away, spoiling the bearing at this point. At the same time the hard particle scores the crankshaft. A number of such particles will soon ruin the bearing and perhaps score the journal sufficiently to require regrinding*

Crankshaft and Bearing Service

BEARING OIL LEAK DETECTOR

Leakage is assumed to be excessive if oil pressure is below normal or if bearing clearances are somewhat greater than they should be.

Oil leakage increases very rapidly with increase in bearing clearance. For example, in a certain engine where specified bearing clearance is .0015″ the leakage is at a satisfactory minimum. If we say this leakage rate is equal to 1, a clearance of .003″ gives 5 times the leakage while .006″ clearance permits 25 times the leakage, Fig. 12.

The only sure way to determine the amount of leakage is to use a bearing leak detector. With it,

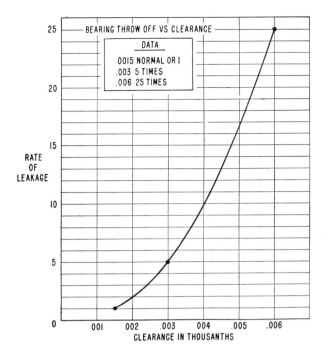

Fig. 12 *How oil leakage increases with bearing clearance*

If the leakage at a given bearing is 15 to 20 drops per minute (at 25 pounds pressure) the bearing is in good condition. The bearing may still be considered serviceable if the leakage is from 30 to 120 drops per minute. If leakage is less than 15 drops, either the oil passage to this bearing is clogged or the bearing is too tight. Leakage of more than 120 drops per minute indicates a worn bearing, which should be replaced with a new one.

A typical oil leak detector is shown in Fig. 13. It is a compact, easily handled unit with an oil tank a little more than a foot long. It is filled with SAE 20 oil to the level of the filler plug. The hose at the right is connected to the engine oil pressure system. The left hose is attached to the shop air line. A pressure reducing valve reduces the air pressure to 20 to 25 pounds. To use the detector, remove the oil pan so that the leakage can be observed. Connect the leak detector oil hose to the engine lubricating system. It may be attached at the front where the oil gauge pipe connects with the block or at the oil pressure switch.

Open the shut-off cock which admits air into the oil tank. Adjust the pressure reducing valve until the gauge on the tank reads 20 to 25 pounds. Open the oil cock and check the pressure gauge. Readjust the pressure reducing valve to 20–25 pounds, if necessary. As testing proceeds, check gauge pressure and readjust as required.

To catch the oil that leaks from the bearings, use a suitable pan, the length and width of the crankcase, or cover the floor underneath the crankcase with an oil-absorbing compound or fine sawdust.

Leakage of main bearings and camshaft bearings can be determined with the crankshaft in any position. Full oil pressure is supplied to these

the amount of leakage at each bearing can be seen. Guesswork is eliminated.

Without the aid of a leak detector, objectionably leaky bearings may be overlooked and on the other hand bearings in satisfactory condition may be discarded because the repairman does not want to risk the possibility of doing a ring or piston job only to find that oil consumption is still too high because of leaky bearings.

The leak detector can also be used to determine whether newly installed bearings are too tight or too loose.

Fig. 13 *Bearing oil leak detector is used to apply oil pressure to bearings when oil pan is off. Amount of leakage indicates whether or not bearings should be replaced*

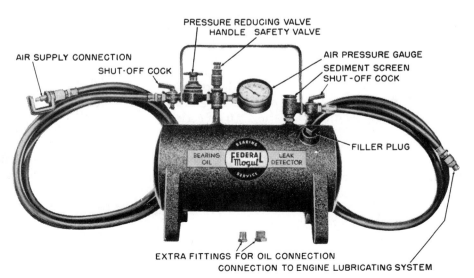

Fig. 14   Rough crankpin journal

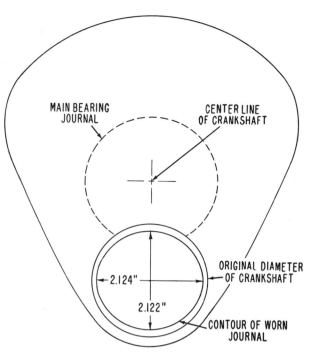

Fig. 15   This rod journal is .002″ out-of-round (2.124″ minus 2.122″)

bearings continuously throughout each revolution regardless of crankshaft position. But this is not true of rod bearings. Remember that the rod bearing is fed by a drilled passage in the crankshaft which registers once every revolution with the oil channel which supplies the adjacent main bearing. Therefore oil pressure to the rod bearing is only at maximum when the crankshaft oil passage lines up with its supply channel.

If the leak detector indicates loose bearings, new bearings should be installed.

When a bearing job is finished, it is a wise precaution to use a leak detector to determine whether bearing leakage is within reason.

BEARING TROUBLES

If one rod bearing knocks, it may be a sign of failure of additional bearings because of faulty lubrication. On the other hand, the lack of lubrication may be due to clogged oil passages leading to this bearing, in which case all or most of the other bearings will be undamaged. Other causes of failure of one rod bearing are: (1) Binding between bearing and journal due to too tight a fit or to a misaligned rod. (2) A slug of abrasive particles which by chance are delivered only to this bearing —and score the bearing. (3) Defect in manufacture of this bearing. (4) Out-of-round rod bore. (5) Dirt between rod bore and bearing shell prevents proper seating of bearing. (6) Bearing improperly installed. (7) Slightly loose rod cap. (8) New bearing installed on rough crankpin, Fig. 14, crankpin out-of-round or tapered too much.

Similar considerations apply to main bearings.

BEARING KNOCKS

If rod bearings knock, the upper and lower halves should be replaced. Sometimes just one rod

will fail—often because of a clogged oil passage leading to this bearing. If so, the other rod bearings may be okay—but be sure to inspect them and check for wear. Clean out the oil passages in the crankshaft as described further on.

If rod bearings are worn, the lower halves of the main bearings should also be inspected. If the lower halves are okay the upper halves can be assumed to be in good condition. However, if the lower halves must be replaced the upper halves must also.

If one main bearing develops a knock it may be caused by a clogged oil passage to this bearing, in which case all other bearings may be okay. However, to be on the safe side all main and rod bearings should be inspected and their clearances checked.

If rod bearing wear is suspected because of low oil pressure and high oil consumption—and if an oil leak detector is not available—looseness of rod bearings can often be determined by shaking the rod. But even if looseness is not evident, bearing clearance should be checked by methods which will be described shortly.

Always check rod bearing clearances when doing a ring job. If the bearing clearance is satisfactory—as explained further on—inspect the bearings for faults. The bearing surfaces should be smooth and free of blemishes such as deep scratches, pits, melted or broken metal, as indicated in Figs. 3 to 11. If any of these faults are noted the bearings should be discarded.

CHECKING CRANKSHAFT FOR WEAR

Before checking bearings for wear, the crankshaft journals should be inspected for surface smoothness, nicks, and for out-of-round and taper. If this checking indicates that the journals must be reground, there is no need to check bearing clearances for wear since the reground shaft requires new undersize bearings, Fig. 14.

Rod journals can be reground without removing the crankshaft by portable equipment, whereas regrinding of the main bearing journals requires removal of the crankshaft because the work must be done in a special crankshaft grinding machine.

The rod and main bearing journals should also be inspected for smoothness. If the journal surface is scored or pitted the journal should be reground unless the rod journals can be smoothed up with a lapping tool if the indentations are not too deep.

A good test for satisfactory journal smoothness is to rub the tips of the fingers over the surface. If the roughness can be felt, reduced bearing life will result and lapping or grinding of the crankshaft is called for.

After a crankshaft has been reground, check the smoothness by rubbing a penny lengthwise of the journal. If a copper line shows on the journal, it should be lapped smooth.

In the case of a rod bearing journal, note that the loads imposed on the journal are greatest in the region of upper and lower dead center. Therefore with the rod at dead center, journal wear will be greatest in the up-and-down direction and least in the crosswise direction, Fig. 15.

To measure the journal, bring it to the lower dead center position and mike it up-and-down and crosswise at the middle of the journal. The difference between the two measurements is the amount of out-of-round. For example, if the crosswise diameter is 2.124″ and the vertical diameter is 2.122″, the out-of-round is .002″, Fig. 15.

The journal should also be miked at both ends on the vertical diameter to determine taper.

For maximum bearing life, obviously out-of-round and taper should be negligible. Also the greater the out-of-round and taper, the more the bearing life is reduced.

For excellent bearing life, out-of-round should not exceed .001″ and taper should be less than .001″—preferably much less. This means that the

crankshaft should be reground if taper exceeds .001″.

Practical experience proves that very good bearing life is obtained when the out-of-round does not exceed .0015″ and taper is not more than .001″.

"MIKING" ROD JOURNALS

The first thing to do is use a micrometer to measure the amount of out-of-round and taper of the rod journals (also called crankpin journals).

Taper is likely to be greater at the vertical diameter, Fig. 15, therefore measure the vertical diameter at both ends of the journal. On a given engine with rod journals a nominal size of 2⅛″, the vertical diameter of one end of the journal might be 2.1235″ and the other end 2.1230″. The taper is the difference between these two dimensions, or .0005″.

A rod journal wears more rapidly on its vertical diameter. Therefore the horizontal diameter of a worn journal is always greater than the vertical diameter. Measure the horizontal and vertical diameters near the middle of the journal. The horizontal diameter might be 2.124″ and the vertical diameter 2.123″. The difference between the two diameters is the out-of-round .001″ in this case. Therefore the journals do not need regrinding provided their surfaces are smooth because .0015″ is usually considered the out-of-round limit.

However, assume that the horizontal diameter is 2.1237″ and the vertical diameter is 2.1216″. The difference (out-of-round) is .0021″. The journal is .0006″ over the usual out-of-round limit of .0015″. The crankshaft should be reground.

"MIKING" MAIN BEARING JOURNALS

If the crankshaft has been removed from the engine, an outside micrometer can be used.

However, it is often desirable to mike the main journals with the crankshaft in place. The V-block gauge, Figs. 16, 17, 18, permits miking all crankshafts whether in place or removed from engine.

The crankshaft caliper, Figs. 19 and 20 can be used if the upper halves of the main bearings can be removed.

Whenever the crankshaft is miked in place by either of these methods, be sure that it is properly supported by its bearings. Either remove and replace the caps one by one, or remove all the intermediate caps and mike their journals. Then put these caps back and remove the end caps.

The V-block gauge, Figs. 16, 17, 18, consists of two pads which make a 30-degree angle with its

Fig. 16   *V-block crankshaft gauge*

Fig. 17   *V-block gauge is pressed against main bearing journal*

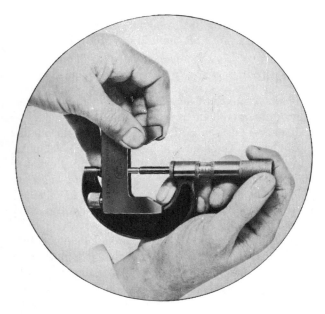

Fig. 18   *Distance between upper and lower buttons equals the radius of the main bearing journal*

center line. The upper button, as it is called, is really a spring backed plunger locked by tightening the knurled nut.

To use the gauge, loosen the knurled nut, press the plunger all the way down and lock it by tightening the knurled nut. Place the V pads in firm contact with the journal, Fig. 17. Hold it there with one hand and loosen the nut with the

other until the plunger clicks against the journal. While still holding the gauge tight, lock the plunger by tightening the nut. Measure the distance between plunger and button with a micrometer, Fig. 18. This distance equals the radius of the crankshaft. Multiply the reading by two to obtain the diameter.

Be sure the micrometer adjustment is accurate because any error in measurement is doubled. All surfaces must be perfectly clean before using the gauge to avoid error in measurement.

WHAT SIZE BEARINGS TO USE

Replaceable main and rod bearings are generally available in the standard size and under sizes of .001″, .002″, .010″, .020″, and up in some cases.

Automotive machine shops have semi-finished bearings which can be machined to any desired undersize up to .060″.

If a crankshaft is reground it is customary to remove .010″ of metal and fit .010″ under size bearings unless a greater undersize is necessary to smooth up badly damaged journals. In some cases where too great a cut is required to clean up a journal, it is advisable to build up the journal with metal and regrind it to the standard size, a job that is often done by operators of large fleets of vehicles.

The standard size bearing and the .001″ and .002 undersize bearings are intended for installation on crankshafts which do not require regrinding.

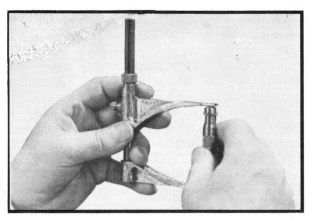

Fig. 19 *The main bearing journal diameter is determined with a crankshaft caliper and then the distance between the jaws on the caliper is measured with an inside micrometer, Fig. 20. The jaws can be accurately adjusted to the journal diameter*

METHODS OF CHECKING BEARING CLEARANCE

There are three methods of checking rod and main bearing clearances: (1) Shim method, (2) Plastigage method, (3) Micrometer method.

If main bearings are checked by the Plastigage method with the engine in the chassis, the crankshaft must be supported in order to take up clearance between the upper bearing insert and crankshaft journal. This can be done by tightening bearing caps of adjacent bearings with .005″ to .015″ cardboard (such as a calling card) between lower bearing shell and journal. Use extreme care when this is done to avoid unnecessary strain on the crankshaft and bearings or a false measurement may be obtained. *Be sure to remove cardboard.*

Checking Clearance by Shim Method

We know that a standard size bearing will fit but if the journal is worn it may give more clearance than necessary. So the preference is for a .002″ undersize bearing if it will fit, or a .001″ undersize bearing, assuming that the .002″ undersize is too tight and the standard size allows more clearance than needed.

A piece of brass or copper shim stock is used,

Fig. 21. Its width should be ½″ and its length should be slightly less than the length of the bearing. Most shops use a .002″ shim thickness. The edges of the shim should be smooth and free from burrs. Install a new .002″ undersize bearing in the rod and cap and coat it with clean engine oil. Lubricate the shim with engine oil and place it lengthwise in the bottom of the rod bearing cap. Install the rod on the crankshaft and draw the nuts up to specified torque wrench tightness. Turn the crankshaft by hand through one revolution. If there is a slight drag, the bearing fit is okay. If there is no drag, the bearing clearance may be less than ideal but still fairly satisfactory.

Fig. 21 *Checking bearing clearance with piece of shim stock*

175

The "loose" bearing, however, should be reasonably satisfactory provided the clearance is not more than .004″. Therefore, try a piece of .004″ shim for drag. Both car and bearing makers are reluctant to state positively what the maximum permissible bearing clearance is. But one of the best of them gives a maximum worn clearance of .0045″ for rods and mains. Maximum clearance for new bearings on this make of car is .002″. Also note that many car makers specify a maximum clearance for new bearings of .003″. Therefore the .004″ figure does not seem excessive.

Instead of turning the crankshaft to determine rod bearing drag, the rod bearing may be checked for fit by tapping it lightly with a hammer to move it lengthwise of the rod journal. If only a light tap is required the fit is not too tight.

When the fit of the .002″ undersize bearing is checked with a .002″ shim, if the drag is heavy instead of light, the bearing is too tight.

Therefore, try a .001″ undersize bearing. Use a .002″ shim as before. If drag is heavy instead of light a standard size bearing must be used.

Regardless of the size of bearing used it is a good plan to be sure that the new bearing is perfectly free on its journal when the shim is removed and the rod installed. Move it back and forth lengthwise by hand. If it is not perfectly free, remove the rod bearing and find out why. It may be binding at the edges or it may be the wrong size.

If brass or copper shim stock is not handy, paper may be used instead. Paper also has the advantage that there is no danger of marring the bearing. Mike up various thicknesses of paper which are handy. Use a strip of paper ½″ wide and slightly shorter than the length of the bearing. Wet it with oil before using.

The .002″ shim stock which is generally used is conservative. It insures plenty of clearance. There isn't much danger of fitting the bearings too tight. It conforms to the machinist's rule of thumb that bearing clearance ought to be .001″ for each inch of bearing diameter.

However, a .0015″ shim is likely to do a more accurate job than a .002″ shim. A rod or main bearing clearance of .0015″ is within the clearance range of most car engines.

The virtue in using a .0015″ shim is simply that a .002″ shim may tell you that a .002″ undersize bearing is too tight whereas the .0015″ shim may indicate the .002″ undersize is okay. Therefore if you are guided by the .002″ shim you may decide to use a standard size bearing when you would be better off with .002″ undersize. To a lesser degree this also applies to .001″ undersize bearings.

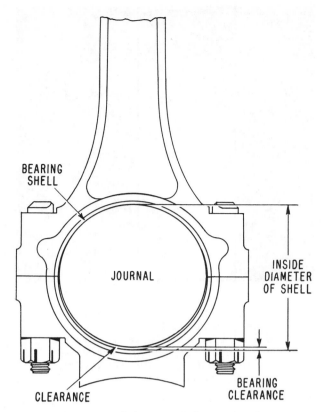

Fig. 22 *The journal diameter plus the bearing clearance equals the inside diameter of the bearing*

Checking Clearance by Micrometer Method

Before explaining the micrometer method of deciding what size bearing to use, it is necessary to discuss bearing diameters, journal diameters and specified bearing clearances, Fig. 22.

Standard size bearing shells are made so that their inside diameter equals the maximum journal diameter plus the minimum specified bearing clearance. For example, if the maximum journal diameter is 2.1250″ and the minimum specified bearing clearance is .0005″, then the inside diameter of the bearing is 2.1255″.

If a standard size bearing shell has an inside diameter of 2.1255″ as mentioned above, the undersize bearings are so many thousandths less than this figure. Therefore the .002″ undersize would be 2.1235″ and the .010″ undersize would be 2.1155″, and so on.

Likewise if the minimum specified bearing clearance is .001″ and the maximum diameter of a new journal is 2.125″, the inside diameter of the bearing is the sum of the two figures or 2.126″. The .002″ undersize would be 2.124″ and so on.

It is impossible to make any part of any machine—a bearing journal for example—to perfect

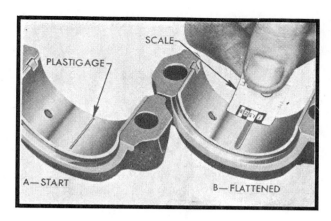

Fig. 23 Left—A round strip of Plastigage is placed in the bearing. Right— The width of the flattened strip is measured with the printed scale on the paper gauge which shows that the bearing clearance is slightly less than .0015" (1.5)

accuracy such as 2.125"—no more and no less. Therefore most car makers usually allow a tolerance of .001" although a few use a tolerance of .0005". Thus the journal diameter might vary from 2.125" to 2.124".

The dimensions just given for a 2⅛" rod journal, namely 2.125" to 2.124" mean that in a new engine of this particular make the journals may have a diameter of 2.125" or 2.124" or anywhere in between these limits. In other words the journal tolerance is said to be .001".

Likewise bearing clearance has a tolerance range, for example, .0005" to .0015". This means that a clearance of less than .0005" is considered too little and a clearance of more than .002" is too much.

Whether to install a standard size rod bearing or a .001" to .002" undersized bearing can be readily determined by a micrometer—without the use of shim stock or Plastigage except perhaps as a final check when the new bearings are installed. The check is insurance against a bearing whose size is not what it is supposed to be. The check also shows up any errors in reading the micrometer. This method has the advantage that the correct bearing size can be selected with the certainty that it will fit. No time is wasted trying first one bearing size and then another.

The first step is to measure the rod journal for out-of-round and taper, as already described, to decide whether or not the journal should be reground.

If journals do not need regrinding, the next step is to decide whether standard size bearings or .001" or .002" undersize will provide the best fit.

Which of the three sizes of bearings to install depends on the maximum diameter of the worn rod journal. If the rod journal when new had a nominal diameter of 2.125", use a standard size bearing if the journal diameter measures between 2.125" and 2.124".

If the rod journal diameter is between 2.124" and 2.123", use a .001" undersize bearing, if available.

If the diameter is between 2.123" and 2.122", use a .002" undersize bearing.

Fit a standard size bearing if the diameter of the worn journal is not more than .001" under par.

Checking Clearance by Plastigage Method

Plastigage is placed on the lower half of the bearing, Fig. 23, and the cap is installed and tightened to the specified torque. This flattens the Plastigage to whatever the bearing clearance is. The cap and bearing are removed.

The greater the width of the strip the less the bearing clearance and vice versa. The paper scale measures the clearance in thousandths of an inch. Actually the scale can be read to ½ thousandth or less.

Plastigage is intended to be used for checking worn bearings and also for checking newly installed bearings to make sure that the clearance is within the required limits.

The Plastigage comes in a sealed envelope. It is made for three ranges of bearing clearances each with a different color of wax, as follows: Green, .001" to .003"; red, .002" to .006"; blue, .004" to .009". The blue is intended for very large crankshafts. For automobiles, the red range is usually used for checking worn bearings. But if new undersize bearings are installed, it may be necessary to use the green range to determine whether or not the new bearing is too tight. Bear in mind that the green range measures as little as .001" clearance while the minimum for the red range is .002".

Before using Plastigage, clean the journals and the bearings so that all oil is removed. The Plastigage is soluble in oil and therefore oil will affect the accuracy of the reading.

Do not rotate the crankshaft when using Plastigage.

As explained previously, a rod journal wears more vertically than horizontally. Consequently if Plastigage is placed at the center of the bearing with the rod at lower dead center, maximum clearance will be registered. Likewise, clearance is a minimum at 90 degrees from this point. Best results are obtained if bearing clearance is measured at the halfway point. The rod journal is placed 30 degrees from lower dead center and the Plastigage is placed ¼″ away from the vertical axis.

To decide whether to install a standard size bearing or .001″ or .002″ undersize compare the clearance indicated by the flattened Plastigage with the specified clearance. For example, let us assume that this clearance is .0005″ to .0025″.

If the Plastigage reading is about .003″ or less, try a standard size bearing. Install the bearing and check with Plastigage. If the clearance is less than the specified .0025″, the clearance fit is okay.

If the Plastigage reading is .0035″ more or less, try a .002″ undersize bearing which reduces the clearance to .0015″.

If the clearance is .0025″ to .0030″, try a .001″ undersize bearing if available.

If an undersize bearing is installed be sure that it is a free fit on its journal. If not, the bearing is too small.

For main bearings, the procedure is the same as for rod bearings except the crankshaft can be checked in any position.

When checking main bearings with Plastigage, all caps should be tight.

If the bearings are badly worn, a paper shim should be installed in each main bearing except the one being checked. Otherwise the weight of the crankshaft will press down on the Plastigage and therefore the Plastigage reading will be too small.

TAPER SHIMS FOR WORN BEARINGS

Shimming Worn Bearings

Efforts have been made to economize by re-using worn bearings with thin shims between the bearing back and the housing to compensate for wear. Very often this method is not successful, and it is not recommended.

One of the problems in attempting to make this kind of adjustment is that bearings do not wear round. The heaviest loads occur near the area of the crown of each bearing half (at right angles to the parting plane of the bearing). Therefore wear is greatest in this area. As a result, adjustment with shims of uniform thickness reduces the bearing inside diameter all the way around. In some sections the diameter is reduced too much, causing insufficient oil clearance and consequent oil starvation.

Tapered shims, that are thickest at the crown area and thinner close to the bearing parting faces, have also been used. But because bearings

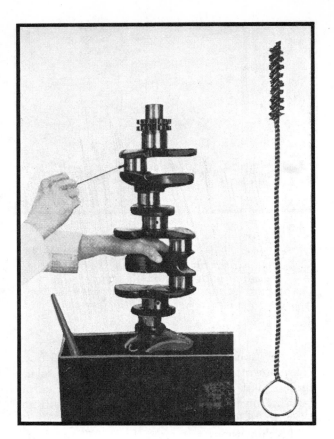

Fig. 24 *The crankshaft oil passages are cleaned with a rifle brush*

do not wear exactly oval either, the use of these shims often results in non-uniform oil clearance.

Any shim between the bearing back and housing has another effect. That is, they increase the bearing crush drastically, often resulting in permanent distortion of the bearing shell.

CLEAN ENGINE OIL PASSAGES

Before installing new bearings it is desirable to clean out the engine oil passages in both cylinder block and crankshaft. Also clean the inside of the crankcase and the oil pan.

If the engine is in the car, use kerosene or a special solvent in conjunction with compressed air to blow out the passages. Automotive jobbers can supply you with a suitable nozzle—and a special solvent.

If the engine is out of the car, the cylinder block and crankshaft should be soaked in a special cleaning tank before any work is done on either the block, the crankshaft or any other engine parts. When dirt has been loosened up, blow out all oil passages in the block and crankcase with compressed air. If this cleaning equipment is not

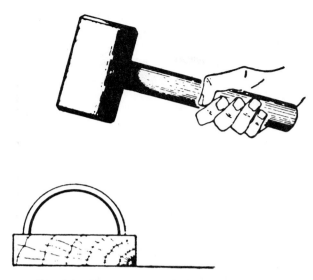

Fig. 25   How to expand thick shell bearing

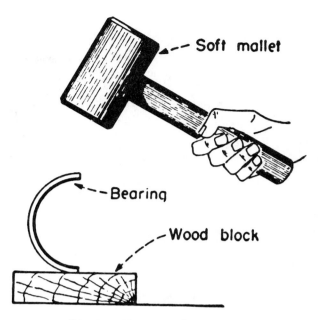

*Fig. 26   How to reduce spread
of thick shell bearing*

available, use the method described in the preceding paragraph.

Whenever there are oil passages to be cleaned, a rifle brush, Fig. 24, can be used in conjunction with any of the methods described above. Or, if necessary, the rifle brush can be used alone. Wet the brush with a suitable solvent or, if possible, submerge the shaft in solvent and allow it to soak for a few minutes. Then use the rifle brush.

ROD BEARINGS, INSTALL

Upper and lower bearings may be identical or the upper and lower bearings may have different oil grooving. If the rod has an oil spurt hole and a drilled passage to lubricate the piston pin there will be holes in the upper bearing half to allow the delivery of oil to the spurt hole and the drilled passage. Therefore, when assembling the upper half of the bearing to the rod be sure that the holes are properly aligned.

When installing each bearing half, be sure that the lip on one edge of the bearing is accurately seated in its notch in the rod or cap.

BEARING SPREAD

Both main and rod bearings are made with a slight spread so that they will stay in place in the bearing bore or bearing cap when inserted. This makes for ease of installation. If a thin shell bearing does not have enough spread to stay in place, expand it by pulling slightly on the ends with the fingers.

If the bearing is a thick-shell type it can be expanded by placing on a flat surface, Fig. 25, and

tapping it with a soft mallet. If thick shell bearing has too much spread, reduce spread by tapping it as shown in Fig. 26.

BEARING CRUSH FIT

The lower halves of thin shell rod or main bearings should project slightly (.001″ to .003″) above the flat face of the cap, Fig. 27. The purpose of the projection is to compress the bearing into the bore to seat the outside of the shell tightly against the cap. Intimate contact is necessary not only to conduct heat away from the bearing but also to provide a firm support for the heavy loads imposed on the bearing, Fig. 28.

ROD SIDE CLEARANCE

Be sure that the rod does not bind endwise on

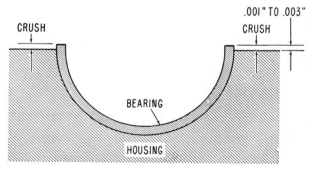

*Fig. 27   Bearing shell should extend
.001″ to .003″ beyond face of cap*

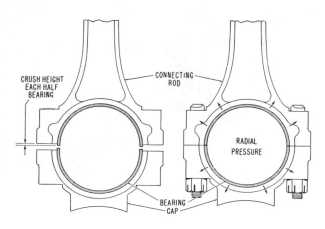

Fig. 28 *Result of bearing crush*

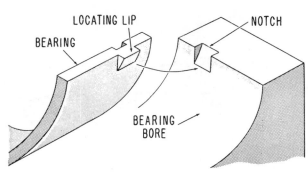

Fig. 30 *Lip on bearing should fit into the notch shown*

the crankshaft. For safety there should be a few thousandths clearance, say .004".

MAIN BEARINGS, REPLACE

To install main bearing shells, remove and replace the bearing caps one at a time. It may facilitate the removal and replacement of the upper shells if all main bearing cap nuts are first loosened ½ turn.

The upper shell is pushed out of its seat bore by rotating the crankshaft after inserting a flat-headed tool (providing bearing shells are not doweled), Fig. 29, in the oil hole in the crankshaft. The thickness of the head must be less than the thickness of the bearing shell. Lacking this tool, the head of a common nail may be used instead. The oil hole is at an angle and the end of the nail head or tool should be parallel to the

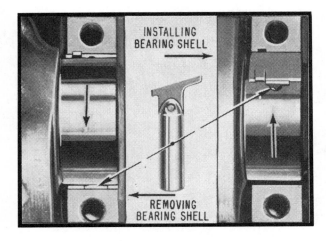

Fig. 29 *Tool for removing and replacing upper half of thin shell bearing*

journal. One end of the shell has a lip on it which fits into a notch in the bearing bore, Fig. 30. It should be obvious that the shell should be removed by pushing on the opposite end. But when installing a new shell, push on the end which has the lip. When the bearing is in place be sure that the lip is accurately seated in the notch. To install the shell, first dip it in engine oil. Then insert the unnotched end in the notched side of the bearing bore. Push it in as far as it will go by hand and then finish the job by rotating the crankshaft. If a heavy push is required, find out why. Probably the shell was not started squarely.

CHECK BEARING INSTALLATION

When installing new bearings, as described above, check each rod and main bearing after it is installed to be sure it is not too tight. If a bearing binds, it is often due to sharp edges on the sides of the bearing shell in which case, dull the edges very slightly with a scraping tool. There is, of course, also the possibility that the bearings are the wrong size.

If the crankshaft is reground and undersize bearings are installed, it is to be expected that bearings and journals will fit together perfectly—and usually they do. But be sure that the crankshaft can be rotated freely when all main bearings have been installed and the caps drawn up to the correct torque wrench readings. Likewise when each rod is installed, see that it can be moved by hand lengthwise on the journal.

If any bearing binds, remove it to determine cause. Always tighten bearing cap nuts to torque wrench specifications. Then check bearings with shim or Plastigage to be sure that the clearance is not too much or too little.

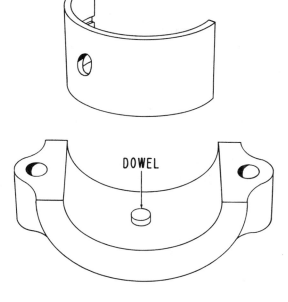

Fig. 31 *On some 4-cyl. Willys engines, the main bearing halves are retained by dowels*

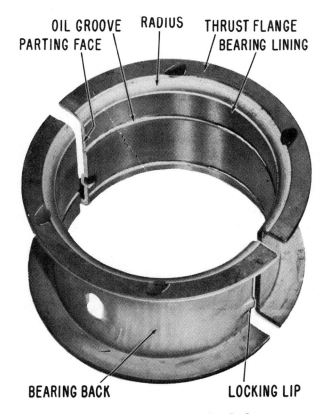

Fig. 32 *Typical crankshank thrust bearing. Note that the ends of the bearing halves have flanges to take the thrust of the crankshaft*

Fig. 33 *Checking thrust clearance with feeler gauge*

DOWELED MAIN BEARINGS

On some 4-cyl. Willys engines, the main bearing halves are held in place by dowels, Fig. 31. Consequently the upper halves of the main bearings cannot be removed without removing the crankshaft. Once the crankshaft is out of the way the bearing halves can readily be lifted out.

CRANKSHAFT END PLAY

There is a certain amount of endwise thrust on the crankshaft and this thrust is controlled to a specified limit by various means, usually by flanges on one of the main bearings. On cars with manual shift transmission, when the clutch pedal is depressed the crankshaft and flywheel assembly is pushed forward. On automatic transmissions with a torque converter, the thrust is exerted rearward.

However, even if there never was any thrust on a crankshaft a thrust bearing would be desirable in order to confine the movement of the crankshaft lengthwise. Any main bearing may be selected to function as a thrust bearing by the engine designer.

The endwise motion or clearance at the thrust bearing must be less than the endwise clearance at the other main bearings (and at the rod bearings) because the crankshaft must be allowed to rub against the edges of these bearings or bearing trouble will ensue.

Fig. 32 illustrates a typical thrust bearing. The ends of the bearing inserts are flanged and coated with bearing metal. The adjacent crankshaft faces, of course, are ground smooth the same as is the journal itself.

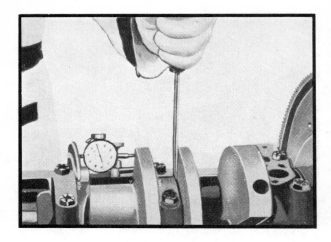

Fig. 34   Checking thrust clearance with dial gauge

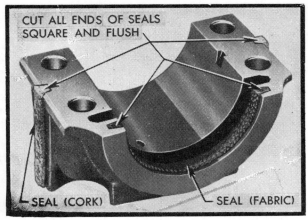

Fig. 35   The picture of the bearing cap illustrates the fabric oil seal and also the cork seals at the sides of the cap

To determine the clearance at the thrust bearing, pry the crankshaft forward at an adjacent main bearing by means of a screw driver and then check the clearance with a feeler gauge, Fig. 33, or with a dial gauge, Fig. 34.

If new thrust bearing shells are installed, check the clearance. If less than the specified minimum, the bearing flanges may be dressed down to get the desired clearance or a new bearing should be tried. Measure the distance between the thrust faces on the crankshaft with a micrometer. Then subtract the minimum thrust clearance from this length. The difference is the length that the bearing should be reduced. If the clearance is too great, a new thrust bearing should be installed and clearance checked.

When a crankshaft is reground, it is often necessary to remove a little metal from the thrust faces on the crankshaft. This may require the installation of a thrust bearing with flanges thicker than standard. The automotive machine shop which grinds the crankshaft should supply a thrust bearing to give the correct clearance.

REAR MAIN BEARING OIL SEAL

It is important that oil be prevented from leaking out of the rear main bearing, not only for the sake of oil economy but also because it may ruin the clutch facings if the car has a clutch.

Some engines have a single piece rubber seal which is fit to the groove in the flange at the rear of the block.

However, the most widely used type of rear main bearing oil seal is shown in Fig. 35. It consists of two lengths of wicking or fabric, the upper and lower halves being carried in suitable grooves

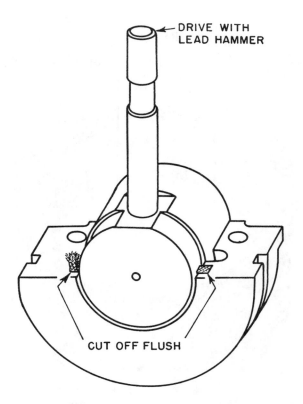

Fig. 36   The tool shown is being used to hold the oil seal in place while the ends are trimmed off. The tool is also used to install the seal firmly in the groove. A razor blade or sharp knife is used for trimming

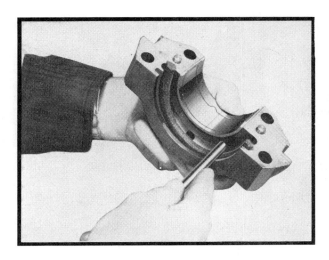

Fig. 37 *Method of pressing oil seal in place with a round rod*

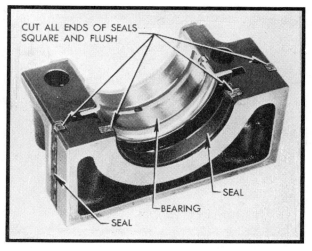

Fig. 38 *Correct assembly of oil seals in bearing cap*

in the bearing structure. The illustration shows the lower seal in the bearing cap.

To renew any type of rear main bearing oil seal, it is obviously necessary to remove the oil pan. Renewing the lower oil seal only will very often stop the leak. However, whenever the crankshaft is removed for other work, it is good practice to install new seals, both upper and lower, regardless of how effective they appear to be.

In servicing the most common type of oil seal, Fig. 35, bear in mind that the materials used for side seals of this type include synthetic rubber or cork, fibre composition or wood blocks manufactured to the correct size. In replacing these side seals, use the same material as removed unless another material has been substituted by the car maker.

To replace the seal, remove the rear main bearing cap and bearing insert. Take out the old seal, clean the groove and install a length of the new seal with the ends sticking up above the face of the cap. Force the seal firmly in place by means of the tool shown in Fig. 36. Lacking this tool or its equivalent, at least use a round rod to force the seal in the groove, Fig. 37. With the side seals installed, the cap assembly should be as shown in Fig. 38.

Although the usual practice is to remove the crankshaft when the upper half of the seal is to be replaced it is possible to do a satisfactory job without removing the crankshaft as follows:

To remove the seal, use needle-nosed pliers to grasp the end of the seal most accessible. Pull the seal downward while rotating the crankshaft slowly in the direction that the seal is being removed.

To install the new seal, fasten a length of wire or strong string such as fishing line securely to one end of the new seal. See that the point of fastening is not bulky and that it is not over ⅜″ from the end of the seal. Coat the seal with Lubriplate. Pass the free end of the wire or string up over the crankshaft at the point where the seal is to be installed. Then exert a firm, steady pull on the wire or string and at the same time rotate the crankshaft slowly in the direction of the pull. This will help to move the seal in position. When the installation is completed, trim the ends of the seal flush with the engine block.

One-Piece Rubber Seal

Fig. 39 illustrates the one-piece seal used on a number of engines. The seal does not fit in a

Fig. 39 *One-piece rubber seal*

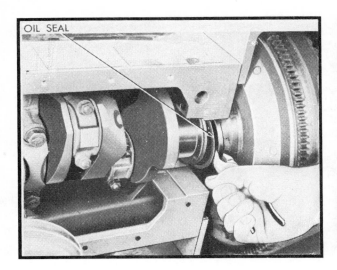

Fig. 40 *The old one-piece rubber seal is pulled out with pliers*

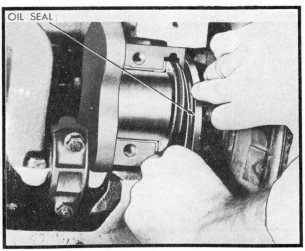

Fig. 41 *The one-piece oil seal is pushed onto the rear cylinder block flange. Be sure to install the lip of the seal against the oil slinger on the crankshaft*

groove. Instead it fits on the flange which forms the rear of the bearing bore. The lip of the seal presses against the oil slinger on the crankshaft. Leakage is thus prevented.

The oil seal is pulled out with a pair of pliers after the bearing cap has been removed, Fig. 40. Then the new seal is slid into place, Fig. 41. However, the details of the operation are as follows:

Remove engine oil pan and clutch housing pan (if used).

Loosen all main bearing cap screws three turns to provide adequate clearance for inserting new seal.

To install a new seal, first lubricate its ends with cup grease, tire soap or glycerine. Wipe off any lubricant on the end faces. Start one end onto the flange in the block and push it up until it is near the top. Then work the other end into place. Spread some of the lubricant on the groove in the seal so that the bearing cap will slip readily into place.

The lip of the seal that contacts the crankshaft should be coated with oil before installing the cap to prevent scuffing the seal.

When installing the seal, be sure that the sealing lip is adjacent to the oil slinger and not backwards. Oil leakage can be expected if lip faces rear.

CRANKSHAFT REMOVAL

Although it is possible to remove the crankshaft on many cars without removing the engine from the chassis it is not practical because the usual reasons for taking out the crankshaft are for major service operations that can be done best with the engine out of the car.

With the engine out of the car, mount it on an engine stand or on a low workbench. Remove cylinder head or heads, manifolds, etc. with engine right side up. Then turn engine upside down and remove oil pan, vibration damper, fan pulley, flywheel, timing case cover, etc. Take out the pistons and rods. Be sure the main bearing caps are suitably marked as to their location before removing them. Then lift out the crankshaft.

Fig. 42 *Tool for grinding crankshaft in a lathe. The electric motor rotates the grinding wheel while the crankshaft is slowly rotated by the lathe*

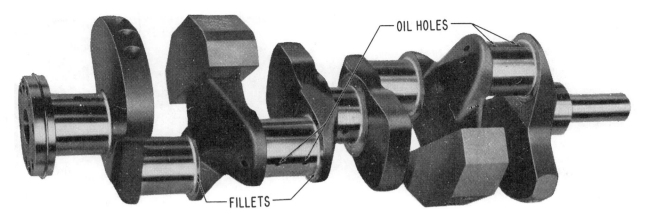

OIL HOLES

FILLETS

Fig. 43 On a reground crankshaft, the fillets must have the correct radius. Sharp edges should be removed from oil holes

HAND POLISHING CRANKSHAFT

If crankshaft journals are slightly rough or burred they can be smoothed up by hand with No. 320 grit emery cloth or a fine honing stone. In either case, wet the journal with engine oil.

CRANKSHAFT GRINDER

Grinding a crankshaft is a job for an automotive machine shop. The work requires a skilled operator and an expensive machine made for the purpose. However, the portable fixture shown in Fig. 42 is designed for regrinding crankshaft journals when the crankshaft is mounted in a lathe. The fixture is clamped to the lathe carriage.

The adjustable hook is placed under the journal. Its purpose is to hold the grinding wheel in contact with the journal surface. When a rod journal is ground the hook and the grinding wheel reciprocate up and down because they are linked to the swinging arm.

The same grinding wheel unit can be used for grinding rod journals when the engine is in the car—in conjunction with a fixture that is clamped to the underside of the crankcase.

On reground crankshafts it is most important for the oil holes and the fillets, Fig. 43, to blend smoothly into the journal surface. The fillet radius on the reground shaft should be the same as for a new shaft. If the radius is too large the edges of the bearing may wear rapidly. If too small, the crankshaft may crack in time and break at one of the fillets.

Other causes of crankshaft breakage include:

1. Bending due to misalignment of bearings.
2. Bending due to improper main journal grinding.
3. Bending due to the weight of an overhung flywheel.
4. Torsional vibration.

The last two, of course are faults of design, and there is nothing the mechanic can do about it. The first two, though, can be avoided by taking sufficient care in the measurement and assembly of the engine.

Practically all crankshafts are hardened to a

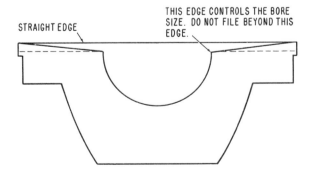

STRAIGHT EDGE

THIS EDGE CONTROLS THE BORE SIZE. DO NOT FILE BEYOND THIS EDGE.

Fig. 44 Warped main bearing cap. File the surfaces flat as indicated by dotted lines

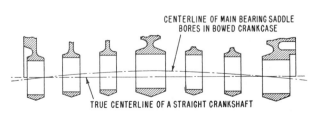

CENTERLINE OF MAIN BEARING SADDLE BORES IN BOWED CRANKCASE

TRUE CENTERLINE OF A STRAIGHT CRANKSHAFT

Fig. 45 Main bearing bores out of alignment

Fig. 46 Checking alignment of crankshaft with dial gauge

Fig. 47 How to measure flywheel runout with dial gauge

certain extent by either of two basic processes. One is called Tocco hardening, and is a process by which the main journal and crankpin areas are hardened to an appreciable depth. These crankshafts can be reground because the hardening reaches a depth of at least .050″.

The other, called Tufftriding, hardens the crankshaft all over. This a low temperature salt-bath nitriding process used to increase wear resistance and fatigue life of steel parts subject to high operating stresses. It can be applied to any steel crankshafts or other parts regardless of whether they were Tufftrided in original manufacture or not.

The hardened skin produced by Tufftriding is very shallow, only .0003″ to .0005″ thick. An intermediate layer of .015″ to .018″ thickness below the thin hardened surface will not give good crankshaft life. Therefore Tufftrided shafts are not suitable for grinding unless they are re-hardened, and this is a specialized process for which few shops are equipped.

To check any shaft to determine if it has been Tufftrided, either of two methods may be used.

First, a chemical method. A drop of 10% aqueous solution of copper ammonium chloride (which can usually be obtained from your local pharmacy) is placed on the shaft. If the solution turns from its light blue color to a reddish brown (copper) color in less than 10 seconds, the shaft has not been Tufftrided. The time element is important. If the color change takes longer than 10 seconds, the shaft has probably been Tufftrided.

Second, a file test method. A sharp medium-fine mill file is applied to any crankshaft surface other than the journals, such as a crank cheek. If metal can be removed under light pressure, the shaft has not been Tufftrided, and may be reground. If the file glides over the surface and no metal is removed, the shaft has been Tufftrided and cannot be reground without re-hardening.

Quite a large number of crankshafts are Tuff-trided in original manufacture, by almost all makers, and it is therefore highly desirable that all shafts be checked by either of the above methods before sending them out for grinding.

MAIN BEARING BORES OUT-OF-ROUND

Crankshaft main bearing bores may go out-of-round in use. If they are out-of-round more than .002″, main bearing life will be reduced. The remedy is to install semi-finished main bearings and line bore them to size. Check bores with inside micrometer.

WARPED MAIN BEARING CAPS

It is a good plan to check main bearing caps for flatness whenever they are removed, Fig. 44. A warped cap may cause reduced oil pressure and increased oil consumption. Warped caps should either be discarded or trued up by grinding or filing the cap as indicated in Fig. 44 but be careful not to go below the edges of the bearing bore. If the material is removed with a file, finish the surface smooth with emery cloth. Lay the cloth on a flat surface, wet it with engine oil and rub the cap back and forth until it is smooth.

WARPED CRANKCASE

If the main bearing bores are out of alignment more than .002″, Fig. 45, bearing life will be reduced. The greater this misalignment the shorter the bearing life. The cure is to install semi-finished main bearings and line bore them to size.

CHECK CRANKSHAFT STRAIGHTNESS

Whenever the crankshaft is removed it should be checked for straightness. Mount it in V blocks supporting the end main bearing journals, Fig. 46, and place a dial gauge in contact with one of the intermediate or center journals. Rotate the crankshaft through one revolution. If the dial gauge hand varies more than .003″, the crankshaft should be straightened in an arbor press.

Some authorities claim that it is not good practice to straighten a crankshaft because it will go out of alignment again. They prefer to true the crankshaft up by regrinding the journals. Obviously, if a crankshaft is a few thousandths out of line, regrinding the journals to .010″ undersize will correct the condition—or regrind the .020″ undersize, etc. The point is that by the very nature of the regrinding process, the finished journals are automatically brought into line.

FLYWHEEL SERVICE

If the surface of the flywheel is scored, cracked or burned, it should be refinished in a lathe. If more than a slight amount of metal must be removed, a new flywheel should be installed. It is important that the rear surface of the flywheel be smooth to avoid undue wear of the clutch facing.

Also examine the flywheel bolt holes. If these are worn (due to running with a loose flywheel) the flywheel should be renewed. In other words, the bolts should fit into the bolt holes rather snugly.

Before installing the flywheel on the crankshaft, check the clutch shaft pilot bearing in the crankshaft for wear, and be sure that the mating surfaces are clean and free from burrs. Put the flywheel in place on the crankshaft and install and tighten the bolts securely.

Check Flywheel Run-Out

Mount a dial indicator on the flywheel as shown in Fig. 47. If the run-out is more than .008″, remove the flywheel from the crankshaft and examine all contacting surfaces for burrs and dirt. If any are found, remove them. Then install the flywheel on the crankshaft and check the run-out again. If the run-out is caused by an untrue crankshaft or flywheel, the condition can be corrected by an automotive machine shop.

Flywheel Ring Gear, Replace

If the starter ring gear has some broken teeth or if the teeth are badly worn or chipped, the gear should be replaced with a new one. If a new gear is to be installed, remove the flywheel from the crankshaft. Then cut the ring gear part way through with a hacksaw and use a cold chisel to cut the gear all the way through. The gear usually falls off when split.

While removing the gear, support the flywheel on flat blocks of wood, with the front face of the flywheel upward. The blocks should be located within the circumference of the gear so as not to interfere with its removal.

Before installing the new gear, be sure to remove any dirt or rough spots from the flywheel surface that contacts the ring gear and do likewise for the inner circumference of the gear.

Note that the chamfered teeth on the gear should be on the "same side" as the starter pinion. In other words, if the pinion is forward of the flywheel the chamfered teeth should be forward when installed on the flywheel—and should point rearward if the starter pinion is to the rear of the flywheel. Remember that the teeth are chamfered to make it easier for the pinion to engage the ring gear.

To install the ring gear on the flywheel, the gear is expanded by heating and then the flywheel is slipped into it.

To do this, place the ring gear on a flat plate on the floor. Take a piece of soldering wire the length of the ring gear circumference and lay it carefully on the upper face of the gear. Press it down against the gear so that it makes good contact at all points. Warm the gear slowly with a welding torch, running the flame around and around the gear. Do not direct the flame at the solder. When the gear is warm enough to melt the solder all around, brush off the solder and insert the flywheel into the gear. Be sure that the gear is tight against the flange on the flywheel. When the gear has cooled, it should be perfectly tight on the flywheel.

PREPARING CRANKSHAFT FOR HIGH PERFORMANCE

One of the most important operations in preparing a high performance engine is balancing. This is highly specialized work and is only performed in specially equipped shops. It involves removal of metal at heavy spots and the addition of metal at light spots.

Not only must the crankshaft itself be balanced, but the entire rotating and reciprocating assembly must be balanced together. This includes connecting rods, pistons, rings, and bearings, together with flywheel, clutch, vibration damper, fan pulley, and timing gears.

Review Questions

Page

1. What types of main and connecting rod bearing are most widely used? 167

2. What is the approximate total thickness of a modern bearing? 167

3. What is the predominant metal in a babbitt bearing? 167

4. What metal, in addition to babbitt is becoming widely used as bearing surface material? 167

5. Why do bearing inserts wear faster than their journals? 167

6. What four conditions would guarantee long bearing life? 167

7. Explain how the carburetor air cleaner and crankcase ventilating system affect the bearings. 167

8. What set of bearings in an engine is subjected to the most wear? 167

9. What symptoms other than a bearing knock would indicate the need for bearing replacement in an engine? 168

10. What is the only sure way to determine the amount of oil leakage at each bearing? 170

11. When testing the oil leak rate at the main bearings, would the position of the crankshaft be of any importance? 171, 172

12. Give two reasons why a bearing would fail prematurely if a dirt particle were imbedded between the bearing and the cap. 170

13. In case of a single bearing failure, what would be the most logical source of trouble? 172

14. What are the names of the bearing surfaces on a crankshaft? 173

15. When bearing service is indicated, under what condition is it unnecessary to check bearing clearances for wear? 172

16. The object of measuring a crankshaft journal is to establish what two dimensions? 173

17. When "miking" a crankshaft, the difference in what two measurements would indicate an out-of-round journal? 173

18. Why is it impossible to use an ordinary outside micrometer to measure main bearing journals without removing the crankshaft? 173, 174

19. Explain the operation of two types of gauges that can measure main bearing journals without removing the crankshaft. 175

20. Explain why the measurement of a main bearing journal taken with a V-Block gauge must be multiplied by two. 174

21. In what sizes are main and rod bearings available? 174

22. Describe three methods of checking bearing clearance. 175

23. Describe the positioning of the crankshaft before checking main bearing clearance by the Plastigage method. 177

24. Explain briefly how the micrometer is used to determine the correct size of bearing to be installed. 176

25. Why is it necessary to remove all traces of oil from a bearing surface before using Plastigage? 177

26. When using Plastigage for checking bearing clearance, what would a well-flattened strip of this material indicate? 177

27. What would be the result of a Plastigage test if the crankshaft were accidentally moved? 177

28. What should be the position of the crankshaft journal for any rod bearing that is being measured by the shim or Plastigage method? 177

29. What parts of an engine should be cleaned before new bearings are installed? 178

30. Why are bearings made so that they project slightly above the bearing cap before assembly? 179

31. Explain how "bearing spread" aids in bearing installation. 179

32. What two methods might be used to keep main bearings from turning with the crankshaft? 180, 181

33. On most engines the upper main bearing can be removed with the crankshaft in place. How is this accomplished without damaging the bearing surfaces? 180

34. One of the main bearings in any engine is of a special type. What is this bearing called? 181

35. How is crankshaft end play determined? 181

36. What reason other than oil economy is there for an effective rear main bearing oil seal? .. 182

37. Explain the difference between a rear main bearing oil seal and a rear main bearing cap side seal. 183

38. Describe the most common type of rear main bearing oil seal. 182

39. Is it always best to remove an engine for crankshaft removal even though this operation might be performed with the engine in the chassis? 184

40. How is the starter ring fastened on the flywheel when it is not welded? 187

Engine Oiling Systems

Review Questions for This Chapter on Page 218

INDEX

INTRODUCTION

Engine, how lubricated 198
Engine oil 191
Engine oil contaminants 199
Engine oil pan 199
Filler pipe, oil 199
Friction 197
Leakage preventives, oil 199
Lubrication 190
Multi-viscosity oils 192
Oil classifications 196
Oil contaminants 199
Oil, engine 191
Oil filler pipe 199
Oil leakage preventives 199
Oil pan, engine 199
Oils, multi-viscosity 192
Oil viscosity 192

OILING SYSTEM SERVICE

By-pass oil filters 214
Crankcase ventilation 216

Draining procedure, oil 201
Engine wear, general 203
Filter elements, replace 214
Filter, oil, installing new 214
Filter, oil, service 214
Filters, oil 214
Filters, oil by-pass 214
Filters, oil, full flow 199, 214
Filters, sealed container, re-
place 214
Full flow oil filters 199, 214
Gear oil pumps 209
Oil change intervals 201
Oil consumption, analysis of
high 203
Oil consumption, causes of ex-
cessive 201
Oil draining procedure 201
Oil filter, installing new 214
Oil filters 214
Oil filters, by-pass 214
Oil filter service 214
Oil filters, full flow 214
Oil leaks, external 203
Oil level, checking 200

Oil lines 204
Oil pan, removing 205
Oil pan, replacing 205
Oil pressure 203
Oil pressure, high 204
Oil pressure, low 204
Oil pressure, no 203
Oil pressure relief valves 213
Oil pressure warning light 204
Oil pumps, gear 209
Oil pumps, installing 213
Oil pumps, removing 207
Oil pumps, rotor 211
Oil screen clogging 207
Oil screens 206
PCV system tests 217
Relief valves, oil pressure 213
Rotor oil pumps 211
Screen, oil, clogging 207
Screens, oil 206
Test PCV systems 217
Vacuum diaphragm, check for
leaky 203
Valves, oil pressure relief 213
Ventilation, crankcase 216

Introduction

LUBRICATION

The thousands of parts that go to make up the complete automobile are either fastened tightly to one or more other parts or they move against other parts. If they move against other parts they require lubrication in order to keep wear at a minimum, to provide for smooth and even movement of the parts and to keep out water and dirt. In some places, as in the engine, rubbing speeds are high and pressures are great so that without proper lubrication there would be so much friction that heat would melt some of the parts, expand metals, close up clearances and cause the parts to stick.

Moving parts operate under a great variety of conditions including heat, cold, low or high speed, heavy or light loads, exposure to water, sand, dirt, air and other things that would tend to cause rust, corrosion, expansion, shrinkage, and wear.

To secure the best lubrication, these various moving parts require a variety of types of lubricants such as engine oils, gear lubricants and greases, each of which is available in a variety of light and heavy grades and with other varying characteristics to make them suitable each for a particular need.

Practically all lubricants are made from petroleum and are therefore called petroleum products. While petroleum in some form or other forms the main body of most lubricants, other chemicals, substances or soaps are added to provide required qualities that the petroleum itself does not possess.

Petroleum is a dark colored, slightly greenish

liquid that comes from the ground. It does not mix with water and will burn when ignited. It becomes thicker when cold and thins out when heated. It has many other natural characteristics of which these are the principal ones.

Petroleum is a mixture of thousands of different chemical compounds, all made up of hydrogen and carbon and usually some impurities such as slight amounts of sulfur and other substances. Petroleum is called a hydrocarbon because its basic ingredients are carbon and hydrogen. Thousands of its different compounds have been separated or isolated or combined in various ways and uses found for them. Apparently only the surface has been scratched because research laboratories are discovering more and more new compounds all the time and finding uses for many of them.

ENGINE OILS

Modern engine oils are designed as carefully as any engine part. The requirements of the engine are carefully measured under a wide variety of operating conditions. The oil designer can then determine the qualities wihch must be built into the motor oil.

Modern engines are built with extremely small clearances, and their precision parts are much more sensitive to rust, corrosion and the presence of deposits than older types of engines. Oils for the modern engine must be able to prevent such deposits from occurring.

By-products of incomplete combustion are the primary source of the substances which cause corrosion and other engine deposits. When gasoline is burned in an engine, more than a gallon of water is formed for each gallon of fuel burned. Although most of the water is in a gaseous form and goes out the exhaust pipe, some condenses on the cylinder walls (especially in cold weather, before the engine has warmed up) and escapes past the piston rings and is trapped in the crankcase. Other corrosive combustion gases also escape past the piston rings and are condensed or dissolved in the oil. The life of almost every engine part depends on the ability of the oil to neutralize the effect of such corrosive substances. To provide this protection, oil-soluble chemical compounds are added to the oil during manufacture.

Under ideal conditions, gasoline burns (in the presence of an adequate supply of air) to form carbon dioxide and water. For a variety of reasons, a gasoline engine cannot burn all the fuel completely. Some of the partially burned gasoline is converted into soot or carbon. The black smoke emitted from the exhaust when the gasoline-air mixture is extremely rich in gasoline is largely soot. Some of this soot and partially burned fuel escapes past the rings and is picked up in the oil. Such materials together with water tend to form sludge and varnish deposits on critical engine parts. If sludge is allowed to build up, oil passages may become restricted or clogged, resulting in insufficient oil flow to engine parts and rapid failure of these parts.

Some of the additives used in modern motor oils help the oil to resist thinning when hot or thickening when cold, help form lubricating films strong enough to withstand the extreme pressures found in some parts of the engine, and prevent entrapped air from foaming. Others serve as dispersing agents to keep sludge and varnish-forming products of combustion suspended in so fine a form that they do not cause damage or deposits. Still others help prevent rusting and corrosion by neutralizing acids and/or retard the deterioration (oxidation) of the oil which takes place at abnormally high temperatures.

As the engine is operated, the level of contaminants in the oil is constantly increased, since combustion products are continually being formed and picked up by the oil. This increasing level of contaminants makes it more and more difficult for the oil to protect and lubricate the engine.

For example, the additives which disperse the sludge-forming materials and prevent rust and corrosion are used up in performing their function. When this occurs, the oil can no longer do its job effectively. It must then be drained and replaced with a fresh supply.

This explains the commonly heard statement that oil does not wear out. Strictly speaking, that is perfectly true. The base oil does not wear out. But the additives which are essential to the proper functioning of the oil do become depleted, and it is not practical, in the field, to clean the oil and replenish the additives.

A properly serviced air cleaner is a big help in preventing most of the road dirt from reaching the oil, and periodic oil filter changes help keep the larger solid contaminants from circulating with the oil and causing wear. But neither of these two accessories is the complete answer to keeping the oil completely effective indefinitely. The engine crankcase must be drained regularly to remove the accumulations of contaminants and refilled with clean new oil.

The rate of contamination depends to a large extent on the kind of driving conditions under which the engine is operated. The more severe the conditions, the more frequently the oil should be changed. However, it is impractical for the individual driver to determine when the contaminant level is too high.

To overcome this problem, each automobile manufacturer recommends oil change intervals for his cars, but these recommendations change from year to year, because of changes in engine design and construction, and vary from manufacturer to manufacturer. Some recommend different intervals for different kinds of driving conditions. While this makes sense, it is difficult for the average motorist to remember. Therefore, the American Petroleum Institute (API) has come up with a universal recommendation: Change oil at least every three months or 3,000 miles, whichever comes first. But never exceed the car maker's warranty requirements for oil change.

What A Motor Oil Must Do

It must:

1. Permit starting and circulate promptly.
2. Lubricate and prevent wear.
3. Reduce friction.
4. Protect against rust and corrosion.
5. Keep engine interior parts clean.
6. Cool engine parts.
7. Seal combustion pressures.
8. Be non-foaming.

These requirements are simple to state but changes in engine design and in driving conditions have made them increasingly difficult to attain. The following sections describe in some detail how these requirements are met.

Oil Permits Starting

The relative ease with which an engine may be cranked and started depends not only on the condition of the battery, ignition system, proper fuel volatility and mixture ratio, but also on the motor oil. If the oil is too viscous or heavy at starting temperatures, it will impose such a drag on the moving parts that the engine cannot be cranked fast enough to start promptly and keep on running. This is especially true of modern high compression engines which require more power to crank.

Cold thickens all oils. So an oil for winter use must be thin enough to allow the battery and starter motor to crank the engine properly at the lowest temperature at which it will be used. In addition, the oil must be fluid enough to flow to the bearings the instant the engine starts in order to reduce wear. But it must also be thick enough to provide adequate protection when the engine reaches normal operating temperatures.

The characteristic of oil which determines the ease of cranking is its viscosity. Viscosity is a measure of the resistance of oil to motion of flow. This resistance, or fluid friction of the oil, keeps it from being squeezed out from between engine surfaces when they are moving under load or pressure. This resistance is due to the adhesive and cohesive forces of the oil molecules in the oil itself. But this same resistance to motion or flow is responsible for the major amount of drag imposed on the starter during cranking. Therefore it is important to use oil having suitable viscosity characteristics to insure satisfactory cranking, proper oil circulation and high temperature protection.

The Society of Automotive Engineers (SAE) has established a viscosity-range classification system for crankcase lubricating oils which is used world-wide. All motor oils are classified according to this system and each is assigned as SAE number or numbers signifying the viscosity range or ranges into which each falls. The numbers in use today are SAE 5W, 10W, 20W, 20, 30, 40 and 50. Thick, slow-flowing oils have high numbers and thin oils have low numbers. The "W" denotes oils that are suitable for winter service under specific temperature conditions from the freezing point and below. To make sure that oils classified with the "W" have proper flow characteristics at low temperatures, their viscosities are determined at 0° F. SAE numbered oils which do not include the "W" have their viscosities measured at 210° F. to be certain that they have adequate viscosity.

Because the effect of temperature changes upon viscosity varies widely with different types of oil, a standard of measurement of the amount of viscosity change with the change of temperature has been established and is known as Viscosity Index (V.I.). A high viscosity index oil is one that shows a relatively small change in viscosity over a wide range of temperatures.

Today, through the selective choice of crude oil base stocks, newer refining methods, and the addition of special chemicals, there are many high viscosity index oils that are light enough for easy cranking at low temperatures and heavy enough to perform satisfactorily at high temperatures. These oils will meet the viscosity requirements and do the work of two or more SAE grades and are labelled SAE 5W-20, SAE 5W-30, SAE 10W-20W-30, SAE 10W-30, SAE 10W-40 and SAE 20W-40.

These are known as all-season or all-weather oils, as well as multi-viscosity grade, multi-grade, or multi-viscosity oils. The term "multi-viscosity" is a misnomer because oils vary in viscosity with temperatures, whether single or multi SAE grade. Multi-grade (as well as single grade) oils are recommended by all car manufacturers.

There is not complete agreement between car manufacturers in regard to SAE grade recommendations for different temperature conditions and oil companies set limits for their own oils. To avoid starting problems due to use of the wrong SAE grade, car owners should change, before cold weather, to the SAE grade which is suitable for the lowest temperature to be expected wherever the car may be used.

If specific information is not available, the following is a composite of car manufacturer and oil company advice:

Lowest atmospheric temperature expected	Single grade oils	Multi-grade oils
32° F.	20–20W, 30	10W–30, 10W–40
0° F.	10W	10W–30, 10W–40
Below 0° F.	5W*	5W–20, 5W–30

* SAE 5W single grade oils should not be used for sustained high speed driving (above 60 mph).

Oil Distribution

Once an engine is started, the oil must circulate promptly and lubricate all moving surfaces to prevent harmful metal-to-metal contact and consequent wear, scoring or seizure of engine parts. Oil films on bearings and cylinder walls are displaced rather quickly with movement and pressure. Therefore these oil films must be replenished by adequate flow and proper distribution of oil through the lubrication system.

Prompt oil circulation at low temperatures—freezing and below—is dependent on the pour point and viscosity of the oil at the lowest temperature to be expected.

The pour point of an oil is the lowest temperature at which the oil will just flow or pour under the force of gravity. Normally, the pour point of an oil has little or nothing to do with the ease of cranking. If the pour point of the oil is not several degrees lower than the starting temperature, however, prompt oil circulation cannot be achieved.

The viscosity of the oil must be low enough at the starting temperature for the engine to be cranked rapidly enough to enable the engine to start. Once the engine starts, however the viscosity of the oil must be high enough at top operating temperatures to assure adequate protection of the working parts.

Oil Must Lubricate and Prevent Wear

The most important function of the oil is to lubricate and prevent wear of moving surfaces. In many parts of the engine the oil can establish a complete unbroken oil film between the surfaces, which is constantly replenished. This is known as full film lubrication, or hydrodynamic lubrication.

Full film lubrication occurs when the working surfaces are in motion and are completely separated by a relatively thick oil film. The most important quality of the oil in keeping parts separated is its viscosity at the temperature existing. The viscosity must remain high enough to maintain complete metal-to-metal separation. Since the metals do not make contact in full film lubrication, wear cannot occur unless a solid particle large enough to exceed the thickness of the oil film causes abrasion or scratching of the moving surfaces. The crankshaft bearings, the connecting rods, camshaft and piston pins normally operate under full film conditions.

Under some conditions, it is impossible to establish a complete oil film between the moving parts and there is more or less intermittent metal-to-metal contact between the high spots on the sliding surfaces. This is called boundary lubrication, where the load is only partly supported by the oil film. Boundary conditions always exist during starting and stopping of the engine and during the break-in period of a new or reconditioned engine, when surfaces are relatively rough and irregular. Boundary lubrication conditions also exist in the area of the top piston ring because the oil supply is limited and temperatures are relatively high.

On some engine parts, the loads are high enough to squeeze out or to rupture the oil film and to permit appreciable metal-to-metal contact. When this occurs, the friction generated between the surfaces produces sufficient heat to cause one or both of the metals in contact to melt and to weld to the other. This results either in immediate seizure, or in the surfaces being torn apart and roughened, which progressively worsens the situation.

Extreme pressure conditions can develop from lack of lubrication, inadequate clearance, extreme heat, heavily loaded parts, and sometimes from the use of the wrong type or grade of oil for the operating conditions. In modern engines, the valve operating system, cams, valve lifters, push rods, valve stem tips and parts of the rocker arms operate under extreme pressure conditions because they carry heavy loads on very small contact areas. The unit loading may be very high, up to as much as 200,000 psi. This is many times greater than loads on the connecting rod bearings or piston pins.

Modern oils contain additives which react with the metals to form surface coatings that are highly adhesive. These coatings reduce the friction between the metals and prevent welding.

Oil Reduces Friction

Where full film lubrication is present, no metal-to-metal friction exists. Nevertheless, force is required to move the parts relative to one another. This is the force necessary to overcome the viscosity effect of the lubricant. The viscosity must be high enough to maintain an unbroken film but should not be greater than necessary, since this will increase the force necessary to overcome it. This is the reason manufacturers specify the recommended viscosity by indicating the SAE number of the oil to be used at various atmospheric temperatures. It is an attempt to be certain that the lubricant will have adequate but not excessive viscosity at normal operating temperatures. When an oil becomes contaminated with solids (soot, dirt, etc.) or with sludge, the viscosity is increased and the engine becomes less efficient—that is, it must generate more power and burn more fuel to do the same amount of work. Contaminant levels in the oil must be kept low for this reason, and this can be accomplished only by regular drain intervals.

Metal-to-metal contact does occur during boundary and extreme pressure lubrication conditions which appear in various parts of the engine. The viscosity of the oil has only a small effect on reducing friction under these conditions. However, the amount and type of chemical additives blended into the oil during manufacture do have a very marked beneficial effect on reducing friction in these circumstances.

Oil Can Keep Engine Interiors Clean

It was pointed out earlier that modern oils can prevent dangerous accumulations of sludge and varnish on engine parts and in the crankcase, timing gear covers and valve covers. The basic objective here is not simply to maintain the engine in a clean condition, but to prevent such deposits of sludge and varnish from interfering with the proper operation of the engine.

The varnish and sludge-forming materials do little or no harm in the oil but they may accumulate to too high a level and/or clump together to form large masses which can restrict flow to various parts of the engine. Or they can combine with oxygen from the air and be baked by engine heat to a sticky or hard binder which keeps the engine parts from moving freely. What we see as sludge in the engine is merely large amounts of such materials clumped together to form a large viscous mass. The water vapor which condenses in the crankcase in cold engine operation aggravates this condition. Therefore it can be seen that sludge formation is generally a problem of low engine temperature operation. The materials which form sludge are a combination of water from condensation, dirt and products of oil deterioration and incomplete combustion. The rate at which these materials accumulate in the crankcase oil is related to factors of engine operation. Rich mixtures which occur on starting or from a sticking choke, restricted air cleaner, or poor ignition, increase the rate.

Straight mineral oils have only a limited ability to keep these sludge-forming materials from clumping and forming a large mass of sludge in the interior of the engine. Modern oils have chemical additives known as detergent/dispersants blended into them during manufacture. These work mainly by keeping the sludge-forming materials from clumping together. In other words, the sludge-forming materials are dispersed in tiny particles throughout the oil and kept suspended in such form that they cannot settle out on engine parts.

The sludge-forming materials are so small in size when first formed that they cannot be seen even with a high powered optical microscope. In such form and size even the finest oil filters remove them and they are much smaller than the thickness of the oil film on engine parts. Therefore, they will not cause wear or damage as long as they remain in a small size and are well dispersed. The function of the detergent/dispersant in a modern oil is to suspend these contaminants in such fine form within the oil so that they can be removed when the oil is drained.

In service the detergent/dispersant is used up doing its job of suspending contaminants. While in theory, more detergent/dispersant could be added to a used oil after the original additive is exhausted, this is not practical or economical in practice.

By contrast with sludge formation, the varnish-forming materials react differently in an engine. These materials combine with oxygen from the air in the crankcase. At first the resulting compounds are only slightly more viscous than the oil but they continue to react with the oxygen and themselves, and are baked by engine heat into a hard, tenacious coating on the hotter parts of the engine.

The compression rings and pistons are particularly sensitive to varnish deposits, and if allowed to accumulate, the piston rings may become frozen or stuck in their grooves. They can no longer perform their jobs efficiently and vehicle operation suffers.

Detergent/dispersants are very effective in preventing varnish deposits. Because the materials are dispersed in the oil rather than being allowed to come in contact with one another, the chemical reaction of combining together is prevented. Because this is fundamentally a process which depends on oxidation, the detergent also works by interrupting this oxidation process. Certain of the additives used to prevent corrosion also prevent the oxidation process from occurring.

As these additives are used up, the oil becomes less and less effective in preventing sludge and varnish deposits, and should be replaced.

Detergent/dispersants used in motor oil have only a very limited ability to clean up deposits of sludge and varnish already formed in an engine. Once formed, sludge and varnish can only really be removed during mechanical overhaul. The function of a detergent/dispersant in a modern oil is to prevent the formation of sludge and varnish, not to remove such deposits after they have formed.

Modern engines cannot tolerate excessive amounts of sludge and varnish. So-called non-detergent motor oils (mineral oils without additives) do not provide a modern engine with adequate protection against the accumulation of sludge and varnish and do not contain the necessary chemical to prevent damage to the engine's valve operating mechanisms. Detergent/dispersant type oils for modern engines are recommended today by all U.S. and foreign car manufacturers.

Combustion Chamber Deposits

All engines must allow some oil to reach the area of the top piston ring in order to lubricate both the rings and cylinder walls. This oil is then exposed to the heat and flame of the burning fuel. Part of this oil then actually burns off. Modern oils burn cleanly under these conditions, leaving little or no residue (carbon). Detergent/dispersants keep the piston rings free in their grooves in order to minimize the amount of oil that reaches the combustion chamber. This not only reduces oil consumption, but more importantly, keeps the amount of combustion chamber deposits at a minimum.

Some deposits in the combustion chamber are unavoidable, since they result from incomplete combustion of the fuel. However, a modern oil can minimize such deposits by keeping rings operating freely, keeping combustion pressures high and promoting more complete combustion of the fuel.

Excessive combustion chamber deposits adversely effect engine operation. They may deposit on the spark plugs causing them to short out, cause pinging or knock and other combustion irregularities, and generally cause loss of efficiency and economy of operation. Because they act as heat barriers, pistons, rings, spark plugs and valves are not properly cooled and are damaged or actually fail.

The function of a motor oil in preventing excessive combustion chamber deposits is limited to two areas:

1. That portion of the oil which reaches the combustion chamber should burn cleanly.
2. The oil must keep the rings free to do their job of minimizing the amount of oil which reaches the combustion chamber.

Motor oil is often blamed for excessive combustion chamber deposits. However, it is worth remembering that if an engine is not using an excessive amount of oil, the oil cannot cause excessive deposits.

Multi-grade oils (SAE 5W-20, 5W-30, 10W-30, 10W-40 and 20W-40) are manufactured from lubricating oil stocks which burn cleanly and contain additives which leave little or no deposits in the combustion chamber.

Oil Cools Engine Parts

Water is usually considered to be the sole cooling agent in an automobile engine. Actually, only about 60% of the cooling job is done by the water, and only the upper part of the engine is so cooled. Water cools the cylinder walls, the cylinder heads and valves, but the crankshaft, main and connecting rod bearings, the camshaft and its bearings, the timing gears, the pistons and many other parts in the lower section of the engine depend almost entirely on lubricating oil for necessary cooling. All of these parts have definite temperature limits which must not be exceeded or failure will result. Some parts can tolerate fairly high temperatures, but others, such as the main and connecting rod bearings, must run relatively cool or they will fail rapidly during operation.

These parts must be supplied with plenty of cool oil, which picks up the heat and carries it to the crankcase where it is ejected to the surrounding air. Some idea of the temperatures involved may help in understanding this important function of a motor oil.

Combustion temperatures are 2000° to 3000° F. Certain parts of the valves may be at temperatures of 1000° to 1200° F. Pistons sometimes reach 1000° F. and this heat is conducted down the connecting rod to the rod bearings. Tin and lead, which are common metals used in bearings, become very soft at around 350° F. and melt at 450° and 620° F. respectively. After warm up, crankcase oil temperatures may reach 180°–225° F. and oil is supplied to the bearings at these temperatures. At the bearing, the oil picks up heat and in many engines leaves the bearing at temperatures of about 225° to 250° F. This keeps the bearing from exceeding these safe temperatures. The hot oil then drains back to the crankcase and splashes onto the bottom and sides where heat is removed and carried away by the surrounding air.

To keep this cooling process going, large volumes of oil must be constantly circulated to bearings and other engine parts. If the supply of oil is interrupted, parts heat up rapidly both as a result of increased friction and from combustion temperatures.

While only a small quantity of oil is needed at any one time to provide lubrication, the oil pump must circulate many gallons of oil per minute to provide cooling of engine parts. Chemical additives and the physical properties of the oil have little effect on its power to provide cooling. The secret is to get large quantities of oil constantly circulating through the engine and over hot engine parts. The engine designer has provided for this with large capacity oil pumps, and lines and oil passages large enough to handle the volume of oil necessary.

However, if these oil lines or passages become partially or completely plugged with deposits, oil does not circulate properly, the vital job of cooling engine parts may not be done and early failure may result. This is another reason for changing oil before the contaminant level becomes too high. The level of oil in the crankcase must be maintained and should never indicate more than one quart low on the dipstick in order to be certain that sufficient volume of oil is present in the engine to provide for proper cooling.

Oil Seals in Combustion Pressures

The surfaces of the piston rings, ring grooves and cylinder walls are not completely smooth. If examined under a microscope, these surfaces con-

sist of minute hills and valleys. The rings therefore cannot completely prevent high compression and combustion pressures from escaping into the low pressure area of the crankcase. This costs power and reduces efficiency. The motor oil fills in these hills and valleys on ring and cylinder wall surfaces and helps seal compression and combustion pressures. Because the oil film at these points is rather thin—generally less than .001″ thick—the oil cannot compensate for excessive wear of rings, grooves and cylinder walls.

Oil Must Be Non-Foaming

Foam is simply a lot of air bubbles in an oil, which do not collapse readily. Because of the many rapidly moving engine parts, air in the crankcase is constantly being whipped into the oil. These air bubbles normally float to the surface and break or collapse. However, water and certain other contaminants slow up the rate at which such air bubbles collapse and large amounts of foam are the result.

Modern oils contain a small amount of a chemical additive called a foam inhibitor, which speeds up the foam collapse and causes air bubbles to break almost as soon as they are formed.

Foam is a poor heat conductor, so engine cooling is impaired if the amount of foam is excessive. Foam does not have much ability to carry the load and prevent wear in hydraulic valve lifters and bearings, since air is easily compressible. Oil on the other hand is virtually incompressible.

Oil Classification

The SAE classification system mentioned earlier only identifies viscosity and does not indicate anything about the type of the oil, its quality or the service for which it is intended. So a system of classification has been developed for the API by a joint industry committee composed of representatives of engine builders and oil companies. Letter designations are used which indicate the kind and type of service for which the oil is suitable. Both gasoline and diesel engine oils are so classified, although aviation oils are not included. All the factors which affect oil and engine performance have been considered, including engine design, fuel characteristics, operating conditions, the type of lubricating oil, and maintenance. There are two classifications for gasoline engine oils and three for diesel engine oils. The gasoline engine oil classifications are:

Service MS. This represents the most severe service for modern gasoline engines which have

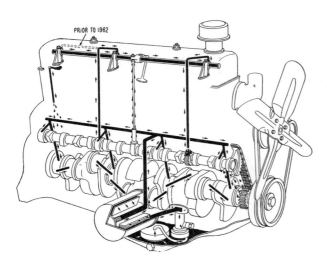

Fig. 1 *Engine oiling
system on Ford engine*

deposit, wear and corrosion control problems, and which operate at extremely high or low temperatures, including stop-and-go service, short trip operation, or which experience long periods of idling. Also, it includes high speed expressway operation.

This classification demands the highest quality in a motor oil. Because of the service conditions under which such oils must produce satisfactory lubrication, these oils are compounded with a wide variety of additives, including detergent/dispersants. Service MS oils meet all the requirements for engines operating under the car manufacturers' warranty, including those equipped with emission control devices. Service MS oils are recommended for all cars and all driving conditions.

Service MM. This represents a more moderate service requirement than Service MS. It includes moderate to high speeds, loads and temperatures. These engines generally have deposit and bearing corrosion control problems when crankcase oil temperatures are high. This classification defines oils which are inhibited against oxidation and bearing corrosion but contain less detergent/dispersant additive than oils for Service MS.

Service MM oils give reasonable performance in moderate service in non-critical engines such as in older model cars, but are neither suitable nor recommended if stop-and-go, idling or short trip operation is involved. Oils of this classification are not recommended for modern engines operating under manufacturers' warranty.

In addition to the above service classifications, there was originally a Service ML classification, but this was discontinued in May, 1969. Oil for Service ML is not recommended for use in passenger car engines, although it is recommended by the manufacturers of some two-cycle engines and some older model farm tractors.

FRICTION

Friction is the resistance set up between two surfaces that move against each other. In the automobile there is some friction between parts due to the load or weight of one part against another as in the wheel bearings which carry the load of the car and the body to the wheels. If the wheels operated without lubrication and without ball or roller bearings the friction would be so great that the parts would heat up, shred the metal and eventually either destroy the bearings or melt the metal.

The pistons move up and down in the cylinders, often at speeds in excess of 40 mph. Without a lubricant to separate the metal surfaces the friction would generate enough heat to melt the metal pistons, rings, cylinders and other parts. This situation is further complicated by the fact that the piston rings press strongly against the cylinder walls as they expand outward. Furthermore, there is considerable side pressure of the rings and pistons against the cylinder walls when the connecting rod is at an angle or is off dead center. This is caused either by the pressure of the explosion against the top of the piston or to a lesser extent to the drag of the rod on the intake stroke followed by side pressure on the compression stroke.

There is friction between the cams and the valve lifters, both at the rubbing points between the cams and the lifters and the side pressure of the valve lifters against the guides. There is friction between the gear teeth of the timing gears of the engine, at all the main and connecting rod bearings, the water pump shaft, the oil pump, the distributor shaft, rotating parts of the transmission no matter of what type, the rear axle gears, universal joints and in fact every moving part on the entire car.

At some of these points the pressure is extremely heavy as between the teeth of the hypoid rear axle gears where pressure may run up to many tons during periods of fast driving, hill climbing or pulling through soft sand or mud in lower transmission gears.

Friction cannot be eliminated but it can be greatly reduced by the use of lubricants.

Lubrication of the various parts is accomplished in a number of different ways.

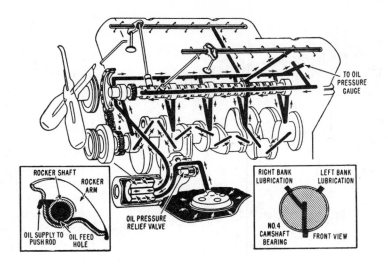

Fig. 2 *Engine oiling system on Plymouth engine*

HOW ENGINE IS LUBRICATED

The engine is lubricated from the oil in the pan, Figs. 1 to 4. This oil is circulated to the various parts by a pump that is located in the lower part of the engine, or sometimes in the front cover.

The oil, some 4 to 6 quarts, is picked up by the oil pump after being drawn through a screen which is intended to prevent the passage of larger pieces of dirt. From the oil pump, the oil is forced under pressure to the crankshaft main bearings and the camshaft bearings.

Oil is distributed evenly over the surfaces of the crankshaft main bearings by grooves in the bearings. A portion of this oil is forced into holes in the crankshaft which carry a supply to each connecting rod bearing. The oil that is thrown off the connecting rods is splashed onto the lower parts of the cylinder walls, from where it is carried up and distributed around the entire cylinder wall. Some is also splashed onto the underside of the pistons, where it performs two functions—cooling the pistons and lubricating the piston pins. The surplus from the cylinder walls, main bearings and connecting rod bearings is splashed over the camshaft, valve lifters, gears and any other internal working parts.

Hydraulic valve lifters are lubricated by a direct pressure line at the same time as the camshaft bearings. This oil, from the hydraulic lifters, is also used sometimes to lubricate the overhead valve gear through hollow push rods. In other engines, there is an oil pipe that carries oil to the valve rocker arm shaft and to other parts that need oil.

The oil line is connected to a mechanism which

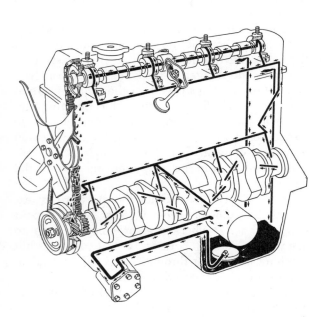

Fig. 3 *Engine oiling system on Jeep overhead camshaft engine*

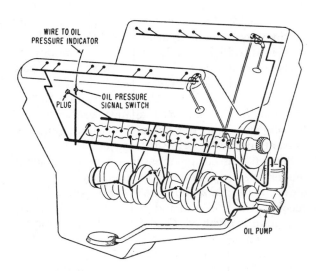

Fig. 4 *Engine oiling system on 1963 Cadillac engine*

carries the pressure indication to the oil gauge or indicator lamp on the instrument panel. This mechanism can be mechanical or electrical.

All modern cars have a full flow oil filter in the engine lubrication system. It is connected into the system right after the oil pump and before the crankshaft main bearings, so that all oil is filtered before going to any bearings.

OIL LEAKAGE PREVENTIVES

Oil is prevented from leaking out of the engine in several ways. The front and rear main bearings are provided with oil seals. At the rear bearing, there is an oil slinger which is simply a disc on the crankshaft. This being of greater diameter than the shaft, throws off any oil that creeps along the shaft and drains down into the pan. In addition there is a main bearing seal which prevents oil from oozing out along the shaft. As long as the seals are in good condition and in place, there should be little or no oil leakage from the shaft ends. If the bearing is badly worn or if the seal leaks or is out of position, there may be considerable leakage of oil, usually at the rear end and this may get into the clutch housing or leak out onto the ground depending on the design of the parts.

ENGINE OIL PAN

The engine oil pan which holds the engine oil supply of about 4 to 6 quarts is held to the crankcase by a number of cap screws or bolts and a gasket, usually cork, is placed between the pan and the crankcase so that when the cap screws or bolts are snugged up, there will be an oil-tight joint. Under conditions of normal use, time and for other reasons, these screws may loosen and may require tightening to keep oil from leaking out along the gasket.

The only other openings in the crankcase are the crankcase drain plug which is usually to the rear of the oil pan and the filler pipe which is at the side, the front or the top of the engine. The oil dipstick or level gauge is always on the same side or near the oil filler pipe so that the oil level can be checked and refilled from the same spot.

OIL FILLER

Modern engines are filled with oil through a filler on top of the engine, and nearly all have a cap which seals the opening. Older engines have an oil filler pipe which extends from the crankcase. This is closed with a filler cap so constructed as to prevent rain or splashed water from enter-

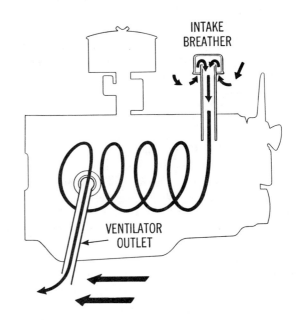

Fig. 5 Road draft system of crankcase ventilation

ing. The filler cap often contains an air cleaner made of metal wool and a strainer to keep out dust and dirt. This is necessary because in the course of heating and cooling the crankcase breathes air in and out. In addition, the ventilating arrangement draws its air into the crankcase through this cap.

Other ventilating arrangements on older engines include the provision of an opening toward the rear of the engine with a scoop which pulls the air out of the crankcase. Some engines also used to have an intake scoop through which fresh air was blown by the fan and by the forward motion of the car. These openings were usually provided with elementary air cleaners, which needed to be cleaned and serviced from time to time, Fig. 5.

ENGINE OIL CONTAMINANTS

The greatest source of engine oil contamination comes from unburned liquid fuel which may pass by the piston rings from the combustion chamber. To this liquid fuel may also be added water that is condensed from the process of combustion and harmful and corrosive chemicals which are also formed during the process of combustion.

Many of the oil contaminants, as they are called, form easily when the engine is cold. If the engine is run long enough at full operating temperature, they can be evaporated and may be drawn off through the ventilating system.

Water vapor is one of the products of combustion. That is, it is chemically necessary to form some water in the form of vapor in order for the oxygen of the air to combine with the carbon in the gasoline. The hydrogen that is released combines with the oxygen of the air to form water. If some is present this combines with some of the oxygen and hydrogen to form one of the acids, sulfuric or sulfurous. Either one is corrosive.

With the cold engine, the water vapor, acid vapor and some of the fuel vapor condense on the cold cylinder walls. Whenever the cylinder walls are colder than about 140 deg., condensation will take place just as it does on the outside of a glass of water or on the bathroom window or mirror when you take a warm shower and release water vapor which condenses onto anything cooler.

This condensed water and fuel are carried down past the piston rings and into the crankcase where, being heavier than oil, they sink to the bottom. Later they may form an emulsion of the consistency of mayonnaise, may combine with dirt or worn metal or carbon particles or, if the crankcase remains hot enough, they may reduce by evaporation or may be strained out by the oil filter.

The condensed water does not mix with the oil but forms sludge, paste or other material which is neither solid nor liquid and which often clogs up oil passages, oil screens, and in severe cases, occupies considerable bulk in the crankcase. This is always liable to cause trouble.

The fuel that works down into the crankcase will mix with the oil and it tends to thin it out or dilute it. This not only reduces the carrying capacity of the oil and the lubrication value but it also tends to deposit inside the engine and to leak out. When the engine has run long enough to warm up, this condensed fuel evaporates and is carried out through the ventilating system. However, it takes time and mileage to bring the oil up to full operating temperature and to keep it there. The amount of water and fuel that is condensed in a few minutes on a cold day may take hours of engine running to work off.

Much, if not all, of the fuel that is condensed in the few seconds or minutes of starting comes from the use of the choke. It is not possible to start an automobile engine cold with the same mixture as that which it later runs on when heated up. Further, the fuel itself is more difficult to evaporate when cold so the mixture has to be made extremely heavy with the choke. If of the automatic type, the choke will usually be set a little on the rich side to make sure of a start. As the engine heats up, the automatic choke releases and gradually returns mixture to its proper value.

If the choke is of the hand-operated type, there is almost always a tendency for the driver to overchoke, to keep the mixture too rich till the engine warms up and then, most unfortunately of all, many times the driver forgets all about the choke and drives merrily along with a stream of black smoke coming out of the exhaust and with unseen harm being done to the inside of the engine in the form of oil dilution and carbon formation from the over-rich mixture. Not only this, but the soot mixes with the oil and in time this becomes abrasive and causes wear.

In order for the lubricating oil to perform the way it is planned, it is necsseary that the proper grade or body of oil be used for the temperatures and that the entire cooling system of the engine be maintained so that it operates at the proper temperature. The oil level must be maintained between certain limits, not too much and not too little. If some of the inside parts of the engine wear too much and open up the clearances, too much oil will be used and where oil passages are allowed to clog up, too little oil may pass. If the level gets too low, the oil may overheat under excessive load conditions and if it is carried too high, too much will be thrown on the cylinder walls as a result of which carbon will form rapidly in the cylinder heads and around the valves, the spark plugs will become fouled and other operational difficulties will be likely to occur.

Oiling System Service

CHECKING OIL LEVEL

The level of the oil in the engine crankcase is checked from time to time with a gauge rod or dipstick usually on the right or left side of the engine. It is recommended that the oil level be checked whenever gasoline is put into the fuel tank.

Oil is consumed more rapidly when the car is operated at high speeds, when the oil is thin and during very hot weather conditions. Oil consumption due to worn cylinders, pistons, piston rings or other parts is usually evidenced by a bluish smoke from the exhaust. Leakage will lower the oil level.

The correct level is at the FULL mark on the dipstick or slightly below. It is not necessary to add oil until the level of the oil has dropped about 1 qt. below full.

The level should never be carried higher than the FULL mark. Excess oil will be thrown on the cylinder walls, will likely bypass the piston rings and cause excessive carbon deposits as well as wasting the oil.

The level should never be allowed to drop below the lower mark on the dipstick. This mark is usually about one quart below the FULL mark. An average engine with a crankcase capacity of 5 qts. would be operating on less than 4 qts. with oil at the low level. Less than this amount is not safe for adequate lubrication as the oil does not get a chance to lose its heat after cooling the parts it is supposed to cool and the remaining oil will have to do a bigger job in lubricating and cleaning. If a long trip is contemplated, if the weather is excessively hot, or if fast driving is to be done or there are tough hills to climb, it is best to carry the level right up to but not higher than the FULL mark.

CAUSES OF EXCESSIVE OIL CONSUMPTION

Excessive oil consumption may be due to any of the following:

1. Loose or leaky drain plug or gasket.
2. Defective oil pan gasket or loose pan bolts.
3. Cracked or broken crankcase oil pan.
4. Leakage of oil through faulty rear main bearing oil seal.
5. Leaky valves.
6. Carrying oil level too high.
7. Oil leaks around valve cover gaskets or covers, loose oil line connections, loose oil filter cover, leaks around overhead valve covers or seals.
8. Leaks in any part of the oil system where the crankcase oil also serves as an automatic transmission fluid.
9. Loose bearings which permit too much oil to leak out of the ends of the bearings to be thrown onto the cylinder walls.
10. Poor quality oil or oil that has been diluted beyond the point of safe lubrication.
11. Oil that is thinner or of a lower viscosity than that recommended for the existing temperature or driving conditions.

OIL DRAINING PROCEDURE

1. For best results, drain crankcase oil when the oil is hot or at normal operating temperature. If the draining can be done immediately after the car has been on a long or fast run, a maximum amount of dirt and sediment will be in suspension in the oil and will drain out when the plug is removed. The hot oil will also be thin and will drain rapidly.

2. Allow the crankcase to drain for at least 3 minutes, preferably longer, especially if the oil is cold. Ideal conditions are to allow draining for 3 minutes after the stream of oil has dwindled to a succession of drops.

3. Examine any deposit of sediment on drain plug for possible worn metal, sludge or foreign matter in the discarded oil. Excessive deposit of sediment indicates the desirability of using a cleaning or highly detergent oil to "digest" foreign material. Evidence of excessive thinning of oil, water in the oil or other irregularities calls for appropriate action.

OIL CHANGE INTERVALS

When to drain the crankcase and replace with fresh oil depends largely on how the car is driven. Car manufacturers recommend that the oil be drained at mileages of from 2,000 to 6,000 under ideal driving conditions but at lesser intervals when subjected to abnormal conditions.

The old statement that oil never wears out is probably somewhere near true. Oil wearing out is not the cause for changing oil. It is the contamination of the oil from outside sources that makes it necessary.

Stop-and-go driving and running the engine cold can add enough contaminants to the oil to require an oil change at as low an interval as a few hundred miles. The contaminants help to form cold weather sludge and some types of fuels may contain sulfur which combines to form corrosive acids.

Cars operated under very dusty conditions will also need more frequent oil changes. A car that has been driven through a dust storm may take enough gritty dust into the engine in a few minutes to require a change of oil.

These are operating conditions at their worst. Between the best of conditions that may require engine oil changes only every 3,000 or more miles and the worst of conditions which may require a change every few hundred miles lie a great variety of driving conditions. For proper lubrication, the oil change interval should fit the kind of driving.

Ideal driving conditions for engine oil are when the car is driven for long distances at moderately high speeds with the engine at top operating temperature and when the driving is through air that has little dust, when the air cleaner is serviced regularly and when the oil filter is changed at proper intervals.

Under these conditions, with the aid of the air cleaner and oil filter, the crankcase keeps itself clean of foreign contaminants, dirt, fuel dilution and water and the oil will stay in serviceable condition for a long time and for many miles. This is engine lubrication at its best and under these conditions the crankcase oil might last indefinitely except for a small amount of worn metal.

At the other end of the scale of driving conditions is the car that is driven at low speeds, low mileages and with many stops during which the engine cools off. This is generally referred to as stop-and-go driving. It is at its worst in cold weather.

In stop-and-go driving, the engine warms up a little, then gets cold again and this is repeated time after time. Typical of such driving is that of doctors who drive a little, then stop for a patient or at a hospital; salesmen who go short distances and make stops; commuters who drive short dis-

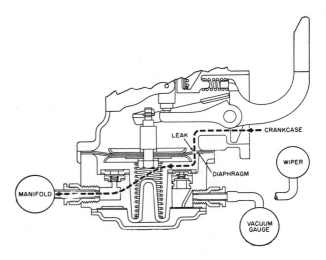

Fig. 6   The arrow indicates the path of oil particles through the leaky vacuum diaphragm. Suction of the engine pulls the oil through the holes in the diaphragm, through the open discharge valve, then through the pipe leading to the intake manifold

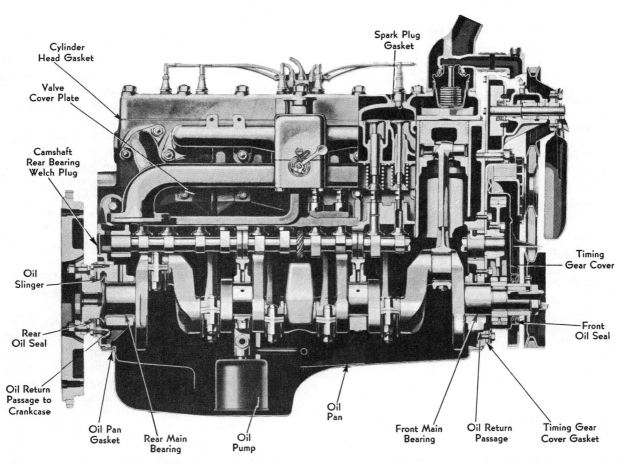

Fig. 7   Sectional view of engine showing parts where oil leakage is likely to occur

ANALYSIS OF HIGH OIL CONSUMPTION

Before tearing down an engine for the purpose of eliminating the cause of excessive oil consumption, it should first be checked to see whether or not it is actually burning too much oil. A black oily deposit in the end of the muffler tail pipe, coupled with the fact that blue smoke comes out of the tail pipe when the engine is running at moderate speeds, are definite indications that too much oil is being burned in the engine. Blue smoke burned should not be confused with black smoke caused by too rich a carburetor mixture or with white steam coming from the tail pipe during engine warm-up period, especially on a cool but humid day.

Many cars have a combination fuel and vacuum pump, the vacuum section being used to operate windshield wipers. A broken or porous diaphragm in the vacuum section of these pumps can cause plenty of blue smoke when the rest of the engine is in perfect condition. Such a diaphragm allows oil to be drawn from the crankcase to mix with the fuel in the intake manifold and be injected directly into the combustion chamber where it is burned and wasted. Fig. 6 shows how this is accomplished.

CHECK FOR LEAKY VACUUM DIAPHRAGM

One way to check for this condition is to disconnect the vacuum line running to the intake manifold and blow through it to see if any oil comes out. Another way is to stick a wooden match in the vacuum hole in the manifold to get an oil smear. If there is any trace of oil on the match the diaphragm is defective and the pump should be repaired. However, the fact that a new diaphragm is installed does not necessarily mean that the vacuum pump is in good condition as it is possible for a new diaphragm to be sufficiently porous to allow oil to pass through it. Therefore, after the engine has been operated, it is a good idea to check again for the presence of oil in the manifold vacuum line.

GENERAL ENGINE WEAR

If the engine is actually burning too much oil, inspection may show that it is caused by a general wear of parts in addition to piston rings. It may be that valve guides and bearings are worn and if these parts are not replaced, it will do no good merely to install a new set of piston rings because they will not prevent oil from getting into the combustion chambers. Therefore, since there is a close relationship between rings, pistons, cylinders, piston pins, valves and bearings, all these parts should be checked and replaced as necessary.

EXTERNAL OIL LEAKS

Fig. 7 indicates points to be checked in looking for an external oil leak. Fresh oil on the clutch housing, oil pan, oil pump, edges of valve covers, external oil lines, distributor shaft, base of oil filler tube, or on the bottom of the timing case cover are points to look for external oil leaks.

Any area under the engine that is washed clean with oil is conclusive proof that there is an external oil leak somewhere in front of the washed area.

Sometimes the camshaft rear welch plug is not tight. Oil leaking downward from this opening is often thought to be a rear main bearing oil leak because the oil is found on the bottom of the clutch housing or on the rear of the oil pan.

OIL PRESSURE

On all modern automobile engines the normal oil pressure at cruising speeds is upwards of 30 pounds per square inch.

The gradual increase in engine bearing clearances as a result of normal wear produces a proportional drop in oil pressure. An engine having seen much service will register generally lower oil pressure than a new or rebuilt one. It may be justifiable on a worn engine to use a heavier oil to cut down oil pumping and a smoky exhaust and thus, incidentally, raise the oil pressure to that when new. But it would be altogether wrong to raise the oil pressure by changing the adjustment of the oil pressure relief valve. The outcome of the latter procedure would only aggravate the trouble by allowing less oil to be by-passed and forcing a greater quantity to the cylinders and bearings.

Bear in mind that the registered oil pressure is merely an indication of the proper functioning of the oiling system. Sudden changes in oil pressure are sure signs of trouble.

NO OIL PRESSURE

If the oil pressure drops to zero, stop the engine and check the oil level. If the dipstick indicates that the proper quantity of oil is in the engine, check the following:

1. Oil gauge may be broken.
2. Line leading from engine to gauge may be clogged.
3. There may be trouble in the engine itself.

To locate the cause of the trouble, disconnect the oil gauge line at the engine end and start the engine. If no oil flows from the open connection, stop the engine immediately because the trouble is most likely due to a broken oil line or a clogged oil pump or oil strainer screen.

However, if oil flows from the connection when the engine is running, either the oil gauge is defective or the line to the gauge is clogged. To determine which it is, connect the line to the engine again and disconnect it at the gauge end. Start the engine and if oil comes out of the open end of the line the gauge is at fault. If no oil comes out the line is clogged.

Sometimes an accumulation of dirt or sludge plugs the small hole in the oil gauge connection— either at the tip of the line or the hole in the gauge. This may be reopened with a fine wire and the gauge action will be restored. If this is not the case, however, the defective part should be replaced.

It sometimes happens that the oil gauge registers normal for a brief period and then suddenly drops to zero. If this occurs, stop the engine and, after about a minute, start it up again to see if this action recurs. If it does, very likely the oil screen is clogged. Oil screens in modern engines usually are provided with some means of prevention for such emergencies as will be explained further on.

OIL PRESSURE WARNING LIGHT

Some cars utilize a warning light on the instrument panel in place of the conventional oil gauge to warn the driver when the oil pressure is dangerously low. The warning light is wired in series with the ignition switch and the engine unit—which is an oil pressure switch.

The oil pressure switch contains a diaphragm and a set of contacts. When the ignition switch is turned on the warning light circuit is energized and the circuit is completed through the closed contacts in the pressure switch. When the engine is started, build-up of oil pressure compresses the diaphragm, opening the contacts, thereby breaking the circuit and putting out the light.

The oil pressure warning light should go on when the ignition is turned on. If it does not light,

disconnect the wire from the engine unit and ground the wire to the frame or engine block. Then if the warning light still does not light when the ignition switch is on, replace the light bulb.

If the warning light goes on when the wire is grounded to the frame or cylinder block, the engine unit should be checked for being loose or poorly grounded. If the unit is found to be tight and properly grounded, it should be removed and a new one installed. (The presence of sealing compound on the threads of the engine unit will cause a poor ground.)

If the warning light remains lit when it normally should be out, replace the engine unit before proceeding further to determine the cause for a low oil pressure indication.

The warning light sometimes will light up or will flicker when the engine is idling even though the oil pressure is adequate. However, the light should go out when the engine is speeded up. There is no cause for alarm in such cases; it simply means that the pressure switch is not calibrated precisely correct.

HIGH OR LOW OIL PRESSURE

Oil pressure should not be judged too high or too low until it is known whether the proper oil for the season is being used and the engine is warmed up to normal operating temperature. If the correct oil is being used and the pressure remains high after the engine is warmed up, a sludged condition is indicated. This calls for a good flushing job and a thorough cleaning of the engine.

Low oil pressure is a sign of insufficient oil or badly diluted oil. If the oil level dipstick indicates the oil to be "rusty," is too thin, or if it feels gritty when rubbed between the fingers, it should be replaced with clean oil, preferably after first flushing out the engine.

Low oil pressure may also be caused by worn engine bearings, worn oil pump gears, leaky connection in oil lines, a sticking oil relief valve plunger, or an oil pump cover gasket that is too thick, allowing too much clearance between gears and cover.

OIL LINES

Any investigation involving low or no oil pressure should include an inspection of external oil lines, such as the line leading to the oil gauge and those to the oil filter. A leak at any of these lines may result from a crack or from a loose or defective connection. A leaky connection or cracked line will generally be indicated by oil leaking at the defective point. But, leaving nothing to chance, tighten all connections with a wrench. Then start the engine to create pressure in the lines and look closely for cracks all along the lines. A cracked oil line must be replaced.

REMOVING OIL PAN

To remove an oil pan, lift out the oil level dipstick. Unscrew the drain plug and allow the oil to drain in a suitable receptacle. After the oil has drained, install and tighten the drain plug, using a new plug gasket as a precaution against future leakage. Then, after making a careful inspection of the front suspension and oil pan attaching points to see what (if any) preliminary work is necessary before the pan can be removed, perform this work and then remove all screws or bolts holding the pan in position.

However, before removing the two last screws completely, steady the pan to prevent its dropping. After taking out the two remaining screws, lower the pan. While being lowered, if the crankshaft interferes with the pan's removal, turn the engine over by means of the flywheel or the starter until the pan clears the crankshaft throws or counter-weights.

Removing the oil pan on some cars involves no more work other than raising the car to provide working space and removing the attaching screws or bolts. On other cars it may be necessary to disconnect a steering tie rod or drag link to provide room to release the pan. Sometimes this can be avoided by jacking up the engine. In this way, the engine is raised while the front end remains on the floor, thus increasing the area between the steering rods and oil pan. In some instances, such as some 6-cylinder Chevrolet cars, the removal procedure is so complex and time-consuming that it is recommended that the engine be removed from the chassis.

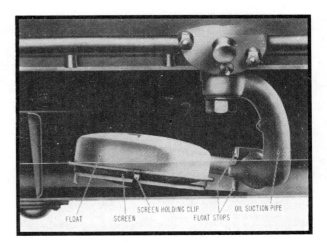

Fig. 8 Floating type oil screen

REPLACING OIL PAN

Be sure the oil pan and strainer are thoroughly clean. Scrape all traces of old gasket material and gasket cement from the oil pan flanges and crankcase surfaces. If the flanges on the pan around the bolt holes are bellied as a result of a previous tightening, they should be flattened with a hammer as a precaution against oil leakage. Be sure the oil pan flanges are clean and dry before applying the new gasket.

Spread a thin film of gasket cement on the oil pan flanges and stick on the gasket, being sure to line up the gasket holes with those on the pan. Allow the cement to become "tacky" (sticky) before installing the pan. If gasket cement is not available, stick on the gasket with heavy grease.

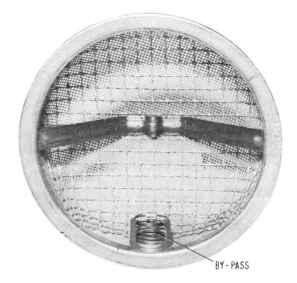

Fig. 10 Fixed oil screen is fitted with a spring-loaded by-pass which opens when the screen becomes clogged

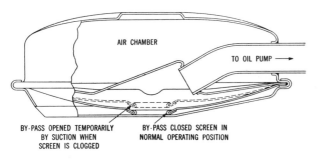

Fig. 9 How floating oil screen works

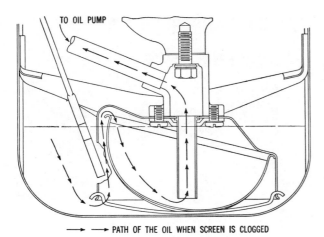

TO OIL PUMP

→ → PATH OF THE OIL WHEN SCREEN IS CLOGGED

Fig. 11 The oil screen has an edge above the normal oil level. Oil is drawn over this edge when screen becomes clogged

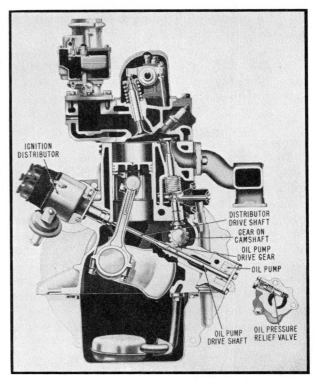

IGNITION DISTRIBUTOR

DISTRIBUTOR DRIVE SHAFT
GEAR ON CAMSHAFT
OIL PUMP DRIVE GEAR
OIL PUMP
OIL PRESSURE RELIEF VALVE
OIL PUMP DRIVE SHAFT

Fig. 12 Cross section of an engine with an externally-mounted oil pump

The general procedure for replacing an oil pan after the gaskets have been applied is to raise it until it contacts the crankcase. Line up the holes and insert two cap screws, one on each side of the pan. Before going any further, be sure the gaskets were not damaged or disturbed as the pan was set in place. Install the remaining cap screws or bolts and tighten them securely. As the gaskets are soft and compress with tension, it is necessary to test the tightness of the cap screws two or three times. Tightening should stop when the edge of the gasket near the cap screw begins to bulge, as further tightening may crush the gasket.

Complete the job by connecting or replacing any parts that were removed. Test the tightness of the oil pan drain plug. Fill the pan with the proper grade and amount of oil. Start the engine and inspect the oil gauge for pressure. After the engine has warmed up, inspect the oil pan for possible leaks.

OIL SCREENS

The possibility of oil screen clogging, resulting in loss of oil pressure, is remote because most car engines incorporate provisions to by-pass the oil in case the screen becomes clogged.

If the screen becomes clogged, and the oil pump draws oil directly from the crankcase, the benefit of the screen is lost and the engine bearings may be damaged by large abrasive particles. This is less serious, however, than engine failure due to loss of oil pressure.

In the majority of present designs, an automatic

by-pass is used in the oil screen. Should screen clogging occur due to sludge, varnish or ice formation, the increased suction opens this by-pass.

This construction is part of the floating type screen shown in Fig. 8. Clogging is unlikely with the floating screen due to the position of the screen away from the bottom where ice, sludge and other objectionable material usually accumulate.

The by-pass in this design, Fig. 9 is formed as an integral part of the screen. A passageway in the center of the screen is normally closed by pressure of the screen against the bottom plate of the assembly. Increased suction on the screen pulls it away from this plate, thereby opening the by-pass.

To clean floating type screens, unfasten the screen holding clip and detach the screen. Wash it in kerosene or other suitable solvent and blow all openings clear with compressed air. If the engine is badly sludged, unfasten the oil suction pipe and give it a good cleaning by swabbing the inside of the pipe with a rag. After connecting the pipe and replacing the screen, be sure the hinge action is free and that the assembly is positioned so that the crankshaft throws will not interfere with it as it floats upward against the stop.

Fig. 10 shows a fixed screen design in which a separate spring-loaded bypass is used. When the

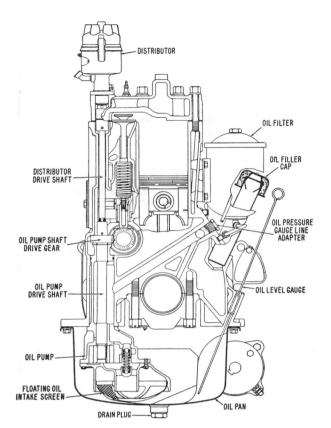

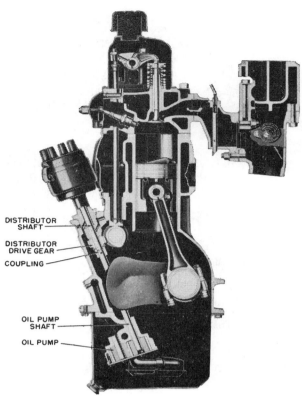

Fig. 13 *Cross section of an engine with an internally-mounted oil pump*

Fig. 14 *Cross section of an engine showing oil pump drive gear attached to distributor shaft*

screen is clean this bypass is closed but when suction increases, the valve opens. Be sure to check the spring action before replacing the pan.

Several engine makers employ a fixed bypass and depend upon its position above the normal oil level to prevent continuous flow of oil through the opening, Fig. 11. The screen is inclined at an angle with a portion extending above the oil level. An open-bottom shroud or housing around the screen takes oil from a relatively large area at the bottom of the oil pan and insures oil supply even if the level becomes low. This shroud also reduces the possibility of the pump drawing air if the oil becomes too stiff in cold weather for free flow to the intake. Increased suction due to screen clogging raises the liquid level inside the housing and oil spills over the top of the screen. If ice is the cause of clogging the level drops as soon as the ice melts. The screen then functions normally.

OIL SCREEN CLOGGING

Oil screen clogging is more likely in winter when ice formation and sludge are prevalent. Water freezes in small particles which collect on the oil screen. Sludge is the result of a conglomer-

ation of water, carbon and partly burned fuel particles. Water is the chief offender as cold weather sludge emulsion does not form when water is absent. Higher engine temperatures, efficient crankcase ventilation and less short-run driving are among the means possible to minimize the detrimental effects of water in the oil.

Water seldom is troublesome in summer because it collects less rapidly and the higher operating temperatures more effectively remove any that does accumulate. Oil screen clogging still is possible in summer, however, but the most likely causes are varnish due to oil break-down at high temperatures.

Whenever the oil pan is removed for any reason, the screen should be cleaned. The by-pass, if fitted, should be checked to insure it is in proper working condition. If the screen is collapsed or if there is an opening through which oil can pass without going through the screen, a replacement is necessary.

REMOVING OIL PUMPS

In most cars the gear on the camshaft drives both the oil pump and ignition distributor. And

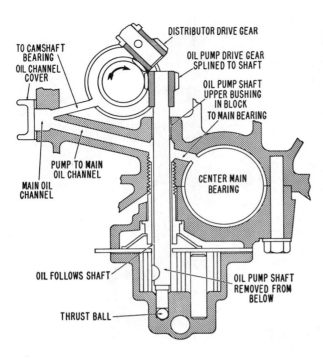

*Fig. 15 Oil pump and drive details
on a Nash 6-cylinder engine*

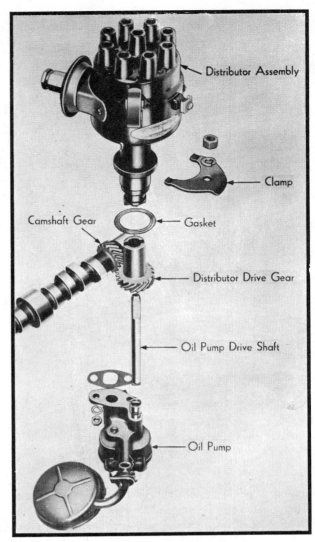

*Fig. 16 Layout of a Cadillac (to 1962)
oil pump and distributor drive*

when the drive gear for these units is attached to the oil pump shaft and is coupled directly to the distributor drive shaft, Fig. 12, it is essential that the pistons are positioned properly with relation to ignition timing before the oil pump is removed or installed.

Note that the oil pump in this case, Fig. 12, is attached to the outside of the engine and, therefore, it can be removed without dropping the oil pan. Before removing the pump, however, first turn over the engine to bring No. 1 piston on its compression stroke and continue cranking until it reaches its firing position. Lift off the distributor cap and make a chalk mark on the distributor housing adjacent to the rotor, which should be in the firing position for No. 1 cylinder. Then remove the capscrews which attach the pump to the engine and lift it off. Unless it cannot be avoided, do not disturb the position of the distributor while the pump is off the engine.

Fig. 13 shows an engine in which the same arrangement is employed. This means that the same precautions regarding ignition timing must be taken. But since the oil pump is located inside the engine, the oil pan must be removed. After removing the pan, unfasten the pump from its mounting and take it out.

The design shown in Fig. 14 is similar to the above except that the drive gear is attached to the distributor drive shaft. No special precautions

relative to ignition timing are necessary in this case since the pump can be taken down without disturbing the timing.

Another example is shown in Fig. 15. A separate gear is used to drive the distributor and oil pump shafts, both meshing with the gear on the camshaft. In removing the pump from these engines, no attention need be paid to ignition timing. Note, however, that the oil pump drive gear is splined to the drive shaft. And after removing the oil pan and taking out the pump from below, the drive gear is removed separately.

Fig. 16 shows the oil pump and distributor drive on Cadillac engines to 1962. The top of the distributor drive gear is slotted to receive the tangs at the lower end of the distributor drive shaft. The oil pump drive shaft has a flatted side

Fig. 17 Disassembling Studebaker Champion oil pump

lows: Take off the pump cover and pick out the "C" washer from the pump shaft, Fig. 17. Then remove the pump body and pull out the shaft and gears.

On some cars, a gear type oil pump is housed in the engine front cover, Figs. 18 and 19. When servicing these pumps, because they are mounted above the level of the oil in the crankcase, special precautions must be taken before starting the engine.

On the Cadillac type, before installing the oil filter, fill the oil passages in the filter support bracket with engine oil. This will allow the pump to prime itself when the engine is started.

On the Buick Special and Olds F-85 type, the pump gear pocket should be packed with vaseline (not chassis lube). Reinstall the gears so the vaseline is forced into every cavity of the gear pocket and between the gear teeth. Unless the pump is packed as directed, it may not prime itself when the engine is started.

which fits into a similarly-shaped hole in the distributor drive gear. With this construction the usual precautions for ignition timing outlined for Fig. 12 must be observed.

On Studebaker Champion engines, the same arrangement as that shown in Fig. 12 is used to drive the oil pump and distributor and the same precautions relative to ignition timing apply in this case. However, removing the pump is a piece-by-piece proposition and is accomplished as fol-

GEAR OIL PUMPS

The most commonly used oil pumps are those of the positive gear type, Fig. 20, which are driven directly from the gear on the camshaft. It consists of two gears enclosed in a close-fitting housing. As the gears turn, oil is taken from the inlet pipe and is directed to the spaces between the teeth and is

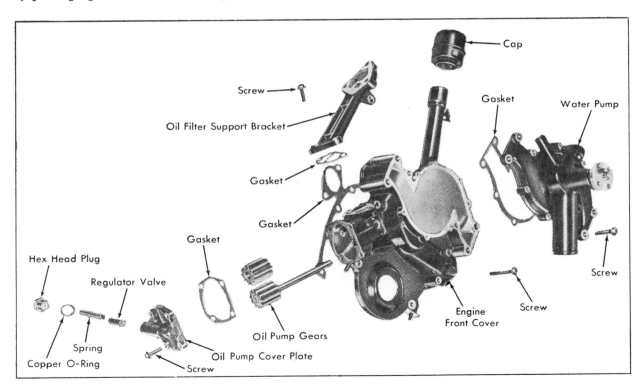

Fig. 18   Engine front cover and oil pump disassembled. Cadillac 1963 and later

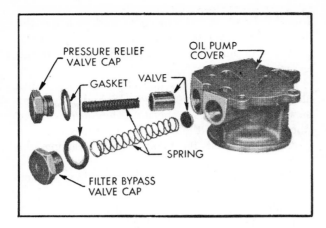

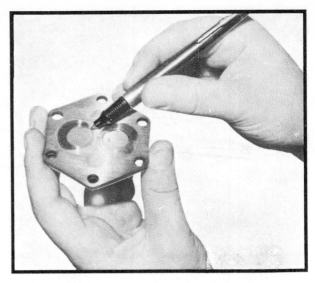

Fig. 19 *Engine front cover mounted oil pump. Buick Special and Olds F-85*

Fig. 21 *Example of a worn oil pump cover*

carried around to the outlet where the action of the teeth meshing together squeezes it out of the spaces and forces it from the pump. In many pumps of this type, the oil pressure relief valve is built into the pump body. To inspect the pump, remove the cover and wash all parts in kerosene or other suitable solvent. Examine the gears for damage, excessive backlash between the teeth, and looseness of the drive gear on the shaft.

If a .010″ feeler gauge can be inserted between any mating teeth, consider the backlash excessive and replace the gears. Move the shaft from side to side in the housing and if more than just a perceptible movement is felt, replace the pump housing or if the housing is provided with bushings, replace the bushings.

Fig. 21 shows a pump cover that has been worn

out by too much end play in the pump shaft which permitted the pump gear to ride against the cover. This cover must be replaced because one of the controlling factors of oil pump pressure is the clearance between the bottom of the gears and the pump cover. Always check the shaft for end play either by moving the gear up and down or by inserting a feeler gauge between the gear and pump body, Fig. 22. If there is space enough to insert a .005″ feeler the drive gear should be replaced. Some pumps have replaceable drive gears; in others the gear is integral with the shaft.

Check the clearance between the face of the

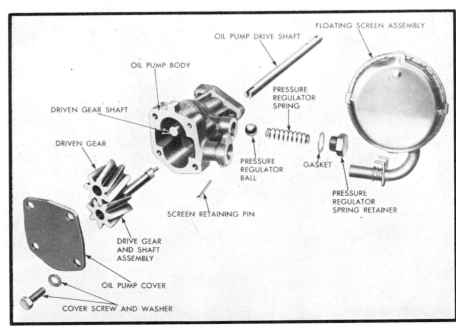

Fig. 20
Exploded view of a typical gear oil pump

Fig. 22 *Checking pump shaft end play with a feeler gauge inserted between gear and housing. In this particular pump, the body casting provides the bearing surface for the oil pump shaft*

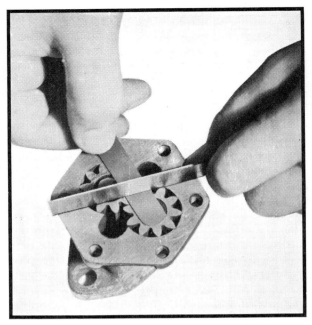

Fig. 23 *Checking clearance between face of gears and pump body with steel scale and feeler gauge*

gears and pump body, using a machinist's scale and feeler gauge, Fig. 23. If the clearance exceeds .005″ the gears should be replaced.

After disassembling the pump as explained below, check the bearing surface of the shaft for any evidence of roughness and replace the shaft if it is not smooth.

Examine the condition of the idler gear pin. In most cases the pin is integral with the pump body and, when worn, requires a new pump body.

Some idler gears have a replaceable bushing. If the gear is satisfactory from the standpoint of gear lash but loose on the idler pin, install a new bushing.

To disassemble the pump, remove the cover and pick out the idler gear. If the oil pressure relief valve is built into the pump, unscrew the spring retainer and take out the parts, being sure to note their arrangement so they may be reinstalled without difficulty.

The next step is to remove the upper drive gear if one is used. Usually it is pressed on the shaft and kept from rotating by a straight pin driven in the hole drilled through the gear and shaft. File off one end of the pin and, using a small drift or punch, drive out the pin and take off the gear. In a few pumps of this type, the gear is held by a Woodruff (semicircular) key. And before removing the gear, note carefully whether the gear is positioned flush with the end of the shaft or how far it extends beyond the end so that the new gear may be installed accordingly.

After removing the upper drive gear, withdraw the shaft and lower gear. Then inspect all parts in the manner previously described, replacing those that are worn or otherwise damaged.

When assembling the pump, dip the gears in engine oil to provide for initial lubrication and priming. There must be a perfect gasket seal between the pump body and cover in order to assure suction. A few pumps do not use a gasket between these parts but where gaskets are used, the thickness is very important—the gasket determining the clearance between the face of the pump gears and cover which in turn affects oil pressure.

ROTOR OIL PUMPS

A typical rotor pump, Fig. 24 is driven by the gear on the camshaft in the same manner as conventional gear pumps. It differs, however, in that it has an inner and outer rotor in the pump body instead of gears. In operation, oil is drawn from the crankcase through the oil screen and passes through a drilled passage to the pump from which it is forced through drilled passages to the engine bearings.

To disassemble and inspect the pump, remove the cover and gasket. Hold a hand over the cover opening and, with the pump upside down, turn the drive shaft until the outer rotor slips out. Slide the shaft and inner rotor assembly out of the pump body. Wash all parts in kerosene and dry with compressed air, if available.

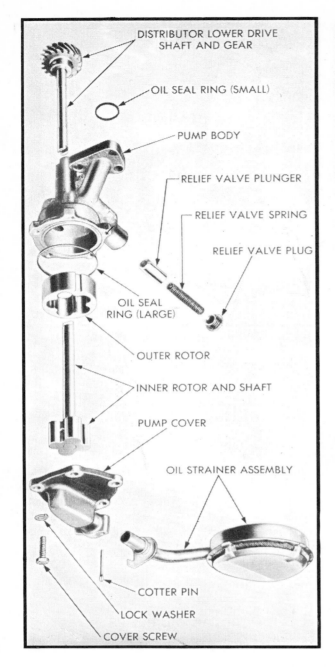

DISTRIBUTOR LOWER DRIVE
SHAFT AND GEAR

OIL SEAL RING (SMALL)

PUMP BODY

RELIEF VALVE PLUNGER

RELIEF VALVE SPRING

RELIEF VALVE PLUG

OIL SEAL
RING (LARGE)

OUTER ROTOR

INNER ROTOR AND SHAFT

PUMP COVER

OIL STRAINER ASSEMBLY

COTTER PIN

LOCK WASHER

COVER SCREW

Fig. 24   Typical rotor type oil pump

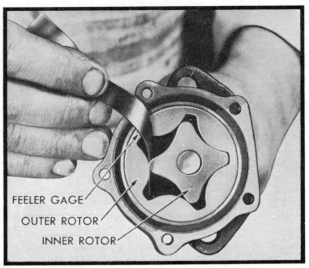

FEELER GAGE
OUTER ROTOR
INNER ROTOR

*Fig. 25   Clearance between
rotors should be .010″ or less*

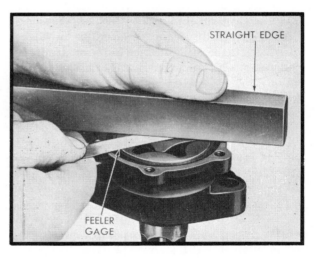

STRAIGHT EDGE

FEELER
GAGE

*Fig. 26   Clearance between rotors and
straight edge should be .004″ or less*

Match the rotors together, Fig. 25, and measure the clearance as illustrated. It should be .010″ or less; if more replace both rotors.

Place a straight edge across the pump body between the screw holes, Fig. 26, and, with a feeler gauge, measure the clearance between the top of the rotors and the straight edge. This measurement should be .004″ or less. Press the outer rotor to one side and measure the clearance, Fig. 27, between the rotor and body. This measurement should be .008″ or less. Should either of

the above measurements be more than specified, replace the pump body.

Place the straight edge across the cover, Fig. 28, and try to insert a .001″ feeler gauge between the cover and straight edge. If the feeler can be inserted or if the cover is scratched or grooved, replace it with a new cover.

When installing a new rotor on the shaft, press it on until the end of the shaft is flush with the face of the rotor. Drill the pin hole and install the pin. Slide the rotor and shaft into the pump body.

Press the drive gear on the shaft until end play on the shaft is .003″ to .010″. Press the rotor down

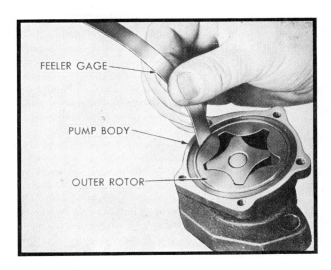

Fig. 27 *Clearance between outer rotor and body should be .008" or less*

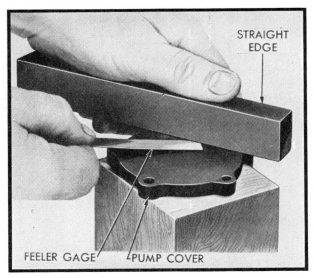

Fig. 28 *Clearance between pump cover and straight edge should be less than .001"*

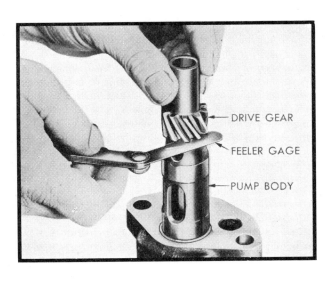

Fig. 29 *Clearance between drive gear and body should be from .003" to .010"*

into the body and measure the clearance as shown in Fig. 29. Install the pin and peen over both ends. If the pin holes do not line up, drill a new hole between the gear and shaft, making the hole at right angles to the other hole. Slide the outer rotor into the pump body. Install a new gasket and tighten the cover screws down evenly.

INSTALLING OIL PUMPS

Replacing an oil pump is a matter of reversing the work done when it was removed. However, if the pump is of the type having the drive gear fastened to the oil pump shaft, Fig. 11, with the ignition distributor shaft coupled directly to it, first turn over the engine to bring No. 1 piston to its firing position. Then turn the distributor shaft so the rotor is in the position to fire No. 1 cylinder and note the angle at which the tongue or slot on the lower end of the distributor shaft faces when installed in the engine.

Having carefully noted this angle, install the oil pump and mesh the drive gear with the gear on the camshaft so the tongue or slot in the pump shaft will engage the end of the distributor shaft. Fasten the pump temporarily and install the distributor. If the distributor rotor is in the No. 1 firing position, the installation is correct and the pump may be fastened permanently. However, if the rotor must be moved in order to engage the pump shaft to the distributor shaft, remove the pump and mesh the gears correctly.

OIL PRESSURE RELIEF VALVES

Since the output of any positive displacement type pump, such as those described, is many times

the normal requirement of the engine oiling system, an oil pressure relief valve is used to by-pass the excess oil back into the oil pan when the pressure exceeds the rated capacity of the pump. This is shown schematically in Fig. 30.

Whenever an oil pump is removed for cleaning or repairs, the oil relief valve should be taken apart and cleaned. This must not be overlooked because if the valve is held open with sludge or carbon it will not seat properly and will therefore allow oil to leak through it before the pressure reaches the proper maximum. This will decrease

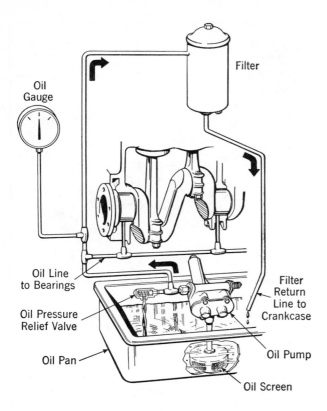

Oil Gauge

Filter

Oil Line to Bearings

Oil Pressure Relief Valve

Oil Pan

Filter Return Line to Crankcase

Oil Pump

Oil Screen

Fig. 30 Schematic drawing of an engine oiling system

Fig. 31 Example of an oil pressure relief valve which is accessible from the outside of the engine

the amount of oil supposed to be delivered to the bearings.

Where the relief valve is not built as part of the oil pump, it is connected in the oil line but is usually accessible from the outside of the engine for inspection, Fig. 31. This construction is used on L-head Chrysler-built engines and a few others.

OIL FILTERS

All cars are now fitted with full flow oil filters as standard equipment. As stated earlier, these are located in the lubrication system immediately after the oil pump, and before the main bearings, so that all oil is filtered before "going into action." Practically all of these filters are of the "spin-on" type, which means that the whole filter unit is a throw-away unit. It is screwed onto the engine block by *hand pressure only*—it must not be overtightened or the rather thin metal case might be distorted, causing leaks.

The filter should be changed at the intervals recommended by the car manufacturer, and to remove it, it may be necessary to use a special filter wrench, as the gasket may have caused it to

stick. But never use the wrench to tighten the filter. A useful precaution that helps to prevent the gasket from sticking is to wipe a little engine oil on it with the finger before installing it.

There are two basic types of full flow filters— the pleated paper type and the depth filtration type. The first makes use of a very large area of specially treated paper as the filtering element. It is pleated in order to get the large area into the limited space of the filter can. This is known as a surface type filter because the oil only has a very short travel through the filtering medium.

The depth type uses a cartridge with filter media so arranged that the oil must pass through a relatively large depth of material before leaving the filter. This type is made of any of a number of different materials—soft white cotton thread blended with wood wool; cotton thread and sisal; paper and wood chips; paper and barley hulls. The object is to trap the very fine particles of dirt and prevent them from getting into the engine lubricating system.

It is not possible to tell from the outside what materials a filter is made of, so it is important that only filters made by reputable manufacturers be used.

Good filtration of the oil is extremely important, especially in modern engines which are built with relatively small clearances, and which can be quickly ruined by dirty oil. This cannot be too strongly emphasized, Fig. 32.

Older cars have used several different types of filtration. For example, some have used a by-pass type filter system, others a shunt type.

The by-pass system, as its name implies, by-

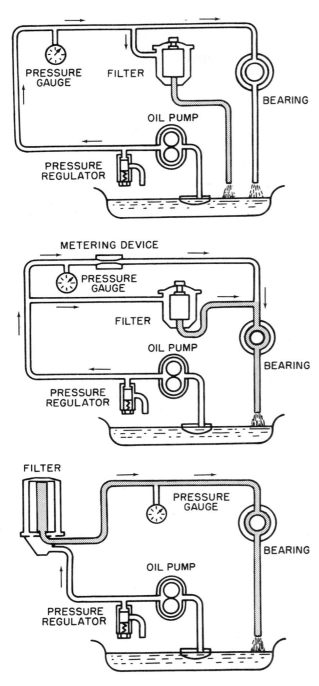

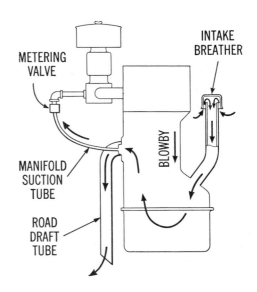

Fig. 33 Schematic drawing showing how either the road draft or manifold suction systems of crankcase ventilation operate

Fig. 32 Oil filter systems bypass, shunt and full flow

passes a small quantity of the oil pumped by the engine oil pump through a very fine filter which removes almost all impurities from the oil and then returns the oil to the oil pan. The amount of oil that passes through the filter varies from 5% to 20% of the total, according to the engine. The remaining 80% to 95% of the oil passes unfiltered either to the engine bearings or through the pressure relief valve if the bearings cannot take it all.

Engine protection with this system is uncertain, but oil cleansing is good and service life of the filter is good.

The shunt system—which the by-pass system is sometimes called—is neither full flow nor by-pass, but a combination of both. In this system, oil from the pump is passed into the filter on the way to the bearings, but after entering the filter housing, it has the option of either passing through the filter cartridge or through an opening around the cartridge. This shunt opening is large enough to handle the entire oil supply to the bearings after the filter becomes completely plugged, but with a pressure drop at the bearings. In actual operation, when the filter cartridge is new, more oil will go through the cartridge than through the shunt passage. This means that the bearings receive full flow protection according to the percentage of oil passing through the cartridge. As the cartridge collects dirt, this percentage will drop until ultimately the full flow protection becomes zero. Since the cartridge handles only a percentage of the bearing supply, it functions somewhat like a by-pass filter even though the initial flow through the cartridge starts out considerably higher than with a by-pass type cartridge.

Another way of describing the shunt system is to say that some of the oil is sidetracked, filtered, and then fed back into the main oil stream. As in the case of the by-pass system, the shunt filter cleans the oil a little at a time, but what it cleans, it cleans very thoroughly. Variations of this system provide the shunt either at the oil pump itself or somewhere between the pump and filter.

There is a fourth type of filter system that is becoming popular on large diesel engines, where engine failure can be very expensive. This is the dual filtration system, in which both full flow and by-pass systems are used together, either by using two separate filters, or by combining the two functions in one filter housing. This system provides what appears to be the ultimate in filtration with the facilities and equipment available today.

Older types of filter have consisted of a throw-away type of housing, but instead of being a spin-on type, the housing is held to the engine block or the firewall by a clamp, and the oil connections are made by pipe unions. Another type made use of a housing like that described above, but the top is removable so that the old cartridge can be removed, the housing cleaned and a new cartridge installed.

Filters for High Performance Engines

Competition filters differ from conventional filters in that they are constructed to withstand more severe service. Street type filters use thin wall steel casings which are adequate for average pressure surges. High performance filters are constructed of heavy gauge (one third heavier than conventional) metal.

The contour of the can is such that no weak points occur during the stamping process, and the can is made to tolerate pressures up to 400 psi. The base plates are similarly constructed of heavier gauge steel to prevent warpage and oil leakage caused by high pressure surges. The ports in the base plates are designed for maximum oil flow. Thus oil pressure remains high at high engine speeds.

CRANKCASE VENTILATION

Crankcase ventilation has an important function in controlling sludge and keeping the engine lubricating system in good condition. Ineffective or inoperative crankcase ventilators are responsible for lubricating troubles serious enough in some cases to cause engine failure.

Two methods of crankcase ventilation are in use, the road-draft system and manifold vacuum system, both shown schematically in the same drawing in Fig. 33.

Fig. 5 shows a typical road draft type. As indicated by the direction of the arrows, air is driven into the crankcase by fan draft through a copper gauze filter in the oil filler cap. It then circulates around the inside of the engine and is discharged through the ventilator outlet pipe, carrying with it the water vapor which collects in the crankcase, particularly in cold weather. Whenever the oiling system is cleaned, the filler and outlet pipes should be removed and flushed out thoroughly. The gauze filters should also be washed in clean gasoline or kerosene.

In order to control smog, the manifold vacuum system of crankcase ventilation is standard equipment on 1961–62 cars to be used in California and is mandatory on all cars starting with 1963 models. The correct operation of this system depends upon a free flow of air from the carburetor air cleaner through the oil filler tube and engine to the control valve mounted on the intake manifold, Fig. 33. The arrows indicate the direction of the flow of air.

The system sucks crankcase vapors into the intake manifold to be burned in the combustion chamber. The flow of the vapors is controlled by the ventilator valve. The valve is actuated by engine vacuum working against spring tension. The high vacuum at engine idle provides minimum ventilation; low vacuum at road speeds provides maximum ventilation.

Servicing the system consists of checking the valve, the tubing and the air intake. The valve should be removed and checked for proper operation and for harmful deposits. If necessary, disassemble the valve, clean the parts in solvent and dry with compressed air. Inspect each component carefully. If the spring is distorted or the valve is worn the valve should be replaced. When reassembling the valve, make sure the end coil of the spring is snapped in the groove which is usually just under the head of the valve.

Remove the tubing and blow out any deposits with compressed air. When reinstalling the valve and tubing, make sure all connections are tight to prevent air leaks.

On some engines, the system air intake is a hose from the air cleaner to the oil filler tube. Make sure the hose is in good condition, and all connections are air tight. Check the oil filler tube cap to be sure it makes an air tight seal on the tube. Air leaks can easily be checked for by squirting kerosene at the connections.

On other engines, the system air intake is a filter type oil filler tube cap. Check the cap filter to be sure it is free of dirt. Clean it in solvent if necessary.

If the control valve becomes clogged with carbon or other foreign matter the ventilation system will not operate and a slight pressure will build up in the crankcase which may cause oil leakage at the rear main bearing or by the piston rings. And

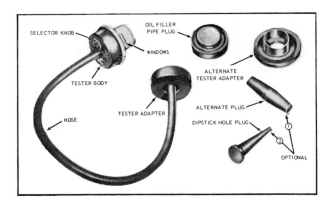

Fig. 34   PCV system tester

should the valve fail to seat it will be impossible to make the engine idle satisfactorily.

If idle speed is slow, unstable, rolling, with frequent stalling; if there is back flow through the crankcase breather (it may be necessary to remove the oil filler cap or the air cleaner to see this condition) and if the engine compartment is covered with an oily mist, the ventilator valve, or PCV (Positive Crankcase Ventilation) valve may be completely clogged or stuck in the open position.

A valve stuck in the closed position is also indicated by breather back flow and an oily engine compartment.

If the valve is stuck in the intermediate position, the indication will be a rough fast idle and stalling.

The PCV valve should be cleaned every six months or 6000 miles, whichever comes first, and more frequently when there is a lot of engine idling in cold weather.

When the valve assembly is removed for cleaning, place a finger over the open end of the hose or tube and have the engine started. If the hose and carburetor passages are open and operating normally, a strong suction will be felt and there will be a large change in engine idling quality when the end of the hose is uncovered. If these conditions are not found, the carburetor passages and/or ventilator hose are clogged and must be cleaned or replaced. The carburetor should be removed from the engine and the ventilator passages cleaned by dipping the lower part of the carburetor in the cleaning fluid. A pipe cleaner or wire can be used to clean the passages.

PCV SYSTEM TESTS

If a condition of rough or loping engine idle speed is evident, do not attempt to compensate for this idle condition by disconnecting the crank-

case ventilation system and making carburetor adjustments. The removal of the system from the engine will adversely affect the fuel economy and engine ventilation with resultant shortening of engine life.

To determine whether the loping or rough idle condition is caused by a malfunctioning crankcase ventilation system, perform either of the following tests.

Regulator Valve Test

1. Install a regulator valve known to be good in the crankcase ventilation system.
2. Start engine and compare engine idle condition to the prior idle condition.
3. If the loping or rough idle condition remains when the good regulator valve is installed, the crankcase ventilation system is not at fault. Further engine component diagnosis will have to be made to find the cause of the malfunction.
4. If the idle condition proves satisfactory, replace the regulator valve and clean hoses, fittings, etc.

Air Intake Test

This test uses the AC positive crankcase ventilation tester, Fig. 34, which is operated by the engine vacuum through the oil filler opening.

1. With engine at normal operating temperature, remove oil filler cap and dipstick.
2. Connect one end of the hose to the tester body and connect the other end of the hose to the tester adapter.
3. Use the dipstick hole plug to plug the opening in the dipstick tube.
4. Insert the tester adapter in the filler cap opening and turn the selector knob to No. 2 (Fig. 34).
5. If the vehicle has a system with the tube from the air cleaner going into the oil filler cap, disconnect the tube at the filler cap and plug the tube.
6. Start engine and let it idle.
7. With plugs secure and tube free of kinks, hold tester body upright and note color in the tester windows. Following lists the various colors and probable cause or related condition of the system.

Green: System operating properly.

Green & Yellow

1. Regulator valve or system partially plugged.
2. Slight kink in tester hose.
3. Slight engine blow-by.

4. Plugs from kit or engine vacuum lines are not properly sealed.
5. Tester knob improperly set.

Yellow

1. Regulator valve or system fully plugged.
2. Tester hose kinked or blocked.
3. Blow-by at maximum capacity of regulator valve.
4. Plugs from kit or engine vacuum lines are not properly sealed.
5. Tester knob improperly set.

Yellow & Red

1. Regulator valve or system partially or fully plugged.
2. More engine blow-by than regulator valve can handle.
3. Vent hose plugged or collapsed.

Red

1. Regulator valve or system fully plugged or stuck.
2. Vent hose plugged or collapsed.
3. Extreme blow-by.

Review Questions

Page

1. At what points in an engine is lubrication necessary? 190

2. What is the raw material from which almost all lubricants are made? 190

3. Besides lubricating moving parts, what other important function does engine oil perform?
   194, 195, 196

4. What is the name of the residue which tends to collect on the inside of an engine? 194

5. What are clean-up additives in engine oil called? 194

6. A.P.I. is the abbreviation for what agency in the oil industry? 192

7. Define each of the following classifications of engine oil: Service MS, Service MM and Service ML. 196

8. Under what operating conditions would oil classified as Service ML be unsuitable? 197

9. Which characteristic of lubricating oil is known as viscosity? 192

10. What is the name of the numbers which grade viscosity? 192

11. Explain how the term "multi-viscosity" applies to engine oil. 192

12. What is friction? 194

13. What component of the engine supplies the force necessary to circulate the oil through the lubricating system? 198

14. What is the difference between a full flow type oil filter and a partial flow or by-pass type? 214

15. What is the greatest source of oil contamination? 199

16. During what phase of engine operation is the oil exposed to the most contamination? 199

17. Name the common substance which is one of the products of combustion. 199

18. For best results should the crankcase oil be drained when it is hot or cold? 201

19. What is the prime reason for periodic oil changes? 201

20. What attachment on an engine can lengthen the interval between oil changes? 202

21. Under what operating conditions is it necessary to change oil most frequently? 201

22. What factors, other than mechanical wear inside of an engine, might account for excessive oil consumption? 201

23. What should be done immediately when an oil gauge fails to register? 201

24. What is the first step in checking for no oil pressure? 204

25. Describe the operation of an oil pressure warning, light system. 204

26. Why is it unlikely that a completely clogged oil intake screen will shut off crankcase oil from the oiling system? 206

27. What is the main advantage of having a floating type oil intake screen? 206

28. What component of an engine drives the oil pump and distributor? 207

29. What precaution, if any, must be taken regarding ignition timing when the oil pump is removed? 208

30. How many different arrangements are there for driving the oil pump? 208, 209

31. What type of oil pump is most commonly used? 209

32. Name the points of measurement that determine whether or not a gear type pump is serviceable. 210

33. What effect does a worn pump cover have on the operation of the pump? 210

34. Describe the basic difference between a rotor type oil pump and a gear type oil pump. .. 211

35. What must be done to an oil pump prior to installation to insure proper priming? 211

36. There are four points of measurement that determine whether or not a rotor type oil pump is serviceable, what are they?... 212

37. Why is the thickness of the oil pump body-to-cover gasket important? 211

38. What is the purpose of an oil pressure relief valve in an oiling system? 213

39. Is the oil pressure regulator always located in the oil pump body? 214

40. How is oil flow to the oiling system maintained with a completely clogged shunt filter? 215

41. What is the difference between a full-flow and by-pass type of oil filter? 214

42. What should be done to the container gasket of an oil filter cartridge before a new element is installed? 214

43. How do competition oil filters differ from conventional units? 216

44. How tight should "spin-on" type filters be installed? 214

45. What is the best location of an oil filter on an engine? 214

46. Explain the purpose of crankcase ventilation. 216

47. Name two types of crankcase ventilating systems. 216

48. What might be the result of a clogged crankcase breather pipe? 216

49. At what manifold vacuum does the vacuum operated crankcase ventilation system provide maximum ventilation? 216

50. What might be indicated by a rough idle on an engine with a manifold vacuum type crankcase ventilation system? 216

Emission Control Systems

INDEX

Air Pump Systems	225	Dual Area Diaphragm 266
American Motors Exhaust Gas		Electric Assist Choke........ 221
Recirculation (EGR) 236		Ford CTAV System......... 268
American Motors Engine		Ford Decel Valve............ 266
"MOD" System 236		Ford DVB System........... 267
Aspirator Air System........ 236		Ford ESC System........... 261
Catalytic Converters 269		Ford EGR System........... 262
Choke Hot Air Modulator..... 224		Ford Electronic Distributor
Chrysler CAP 250		Modulator 260
Chrysler CAS 253		Ford HCV System........... 269
Chrysler Exhaust Gas		Ford High Speed EGR Modulator
Recirculation (EGR) 257		Sub-System 261
Chrysler Ignition 257		Ford IMCO System......... 260
Chrysler Lean Burn Engine		Ford Spark Delay Valve...... 262
Electronic Spark Advance... 249		Ford TAV System........... 267
Chrysler NOx 255		Ford TRS System........... 261
Chrysler Orifice Spark Advance		Ford TRS +1............... 262
Control (OSAC) 257		

Fuel Evaporative Emission Controls 271
General Motors CCS........ 242
General Motors CEC........ 245
General Motors EFE........ 248
General Motors EGR........ 246
General Motors SCS........ 246
Non-Air Pump Systems...... 235
Pulse Air Injection Reactor (PAIR) 235
Temperature Operated Vacuum By-Pass Valve 257
Thermostatic Controlled Air Cleaner, TAC & Auto Therm Air Cleaner Systems...... 239
Transmission Controlled Spark (TCS) 243

Before talking about how various emissions are controlled, it would be well first to describe what these emissions are and where they come from.

First, there are unburned hydrocarbons. These are simply unburned gasoline. They appear in four areas, each of which has to be controlled separately. These are the engine crankcase, the exhaust, the carburetor and the gasoline tank. Emissions from the crankcase are handled by the crankcase ventilation system, already described.

Hydrocarbons normally are harmless and invisible, but when held in suspension and subjected to sunlight, can combine with oxides of nitrogen to form photochemical smog. In heavy concentration this smog will cause irritation to the eyes and nose. Control of emissions from the carburetor and gasoline tank will be described later.

Second, there is the actual exhaust gas, which consists mainly of carbon dioxide and water, which are harmless; carbon monoxide, which is deadly; oxides of nitrogen, which, if not dangerous, do appear to be objectionable, as they cause the smog that brings eye smarting. And there is also a certain amount of unburned hydrocarbons in the exhaust.

Carbon dioxide (CO_2) and water are the normal result of burning a mixture of gasoline and air. Carbon monoxide (CO) and the unburned hydrocarbons in the exhaust are the result of incomplete combustion of gasoline and air. They can be removed, or transformed into CO_2 and water by additional burning—by adding extra air while they are still very hot; by slowing down the burning in the cylinder so that they have more time to burn before being expelled; or by passing

them over a catalyst in a special muffler where they are persuaded to combine with more air and become harmless. This latter method is at this time, being studied extensively by vehicle manufacturers.

Oxides of nitrogen can be reduced but not eliminated, by reducing the temperature of combustion in the cylinder, but that increases carbon monoxide and hydrocarbon emissions. And they can be reduced by lowering the compression ratio (which also reduces cylinder temperature). But all this reduces power output. In fact the methods of controlling exhaust emissions are changing very rapidly.

The sealed crankcase, which was the first step in reducing automotive emissions, came in 1961-62. Since then new steps have been added each year, and it seems probable that further changes will appear regularly for some time to come until the emissions are reduced to acceptable amounts.

Basically two methods are being used to reduce unburned hydrocarbons and carbon monoxide in the exhaust. The first to be developed and now gradually dropping from favor, is the air pump method. It is used by all makers and involves the use of a special air pump which forces air through jets into each exhaust port to provide extra oxygen so that the noxious gases can be burned up. In addition to the air pump, there are several valves to control and direct the air and a number of hoses to conduct the air to the desired locations. This system seems to be the most effective.

The second method involves burning the noxious gases in the cylinders. It was initiated by Chrysler and refinements of it are now used by all makers

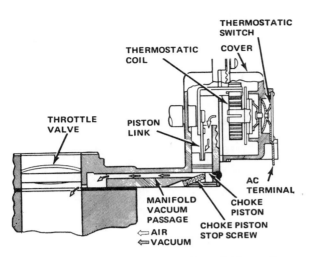

Fig. 1 *American Motors and Ford Motor Co. electric assist choke*

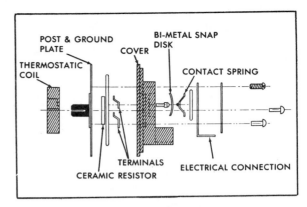

Fig. 2 *General Motors electric assist electric choke assembly*

on most models. It accomplishes its purpose by using a much weaker than normal idle mixture and a retarded spark at idle. (It's at idle speed and during deceleration that emissions are greatest.) As far as it goes neither method is 100% effective. This, of course, is a great over-simplification of the method, but it will be described more fully later on.

Some of the refinements that have been added to the second method include special valves to control vacuum to the distributor advance mechanism, specially calibrated carburetors with idle mixture adjustment stops to prevent over-enrichment, thermostatically controlled engine air intakes, and higher temperature coolant thermostats. In addition, as new engines are designed especially to reduce emissions, the bore to stroke and compression ratios are reduced. To overcome the loss of efficiency resulting from these changes, engine displacement is being increased in turn increasing fuel consumption.

ELECTRIC ASSIST CHOKE

Most 1973-77 American Motors, Chrysler, Ford and General Motors vehicles are equipped with an electric assist choke. Cadillac models with fuel injection incorporate a fast idle valve which is electrically controlled and serves the same purpose as the choke. This device aids in reducing the emissions of hydrocarbon (HC) and carbon monoxide (CO) during starting and warmup (choke on) period. The electric assist choke is designed to give a more rapid choke opening at temperatures of about 60° to 65° F. or greater and a slower choke opening at temperatures of about 60° to 65° F. or below.

The electric assist choke system does not change

any carburetor service procedures and cannot be adjusted. If system is found out of calibration the heater control switch and/or choke unit must be replaced.

American Motors & Ford

The electric choke system, Fig. 1, consists of a choke cap, thermostatic spring, a bimetal temperature sensing disc (switch), and a ceramic positive temperature coefficient (PTC) heater. The choke is powered from terminal or tap of the alternator. Current is constantly supplied to the ambient temperature switch. The system is grounded through a ground strap connected to the carburetor body. At temperatures below approximately 60 degrees, the switch opens and no current is supplied to the ceramic heater located within the thermostatic spring. Normal thermostatic spring choking action then occurs. At temperatures above approximately 60-65 degrees, the temperature sensing switch closes and current is supplied to the ceramic heater. As the heater warms, it causes the thermostatic spring to pull the choke plates open within 1-1½ minutes.

General Motors

The main components of the electric assist choke system, Fig. 2, consist of the thermostatic coil, ceramic resistor, cover, bi-metal snap disc and contact spring. The ceramic resistor is divided into a small center section for gradual heating and a large outer section for rapid heating of the thermostatic coil. The electric actuated ceramic resistor heats the thermostatic coil, gradually relaxing coil tension and allowing the choke valve to open.

At air temperatures below 50° F., electric current applied to the small section of the ceramic

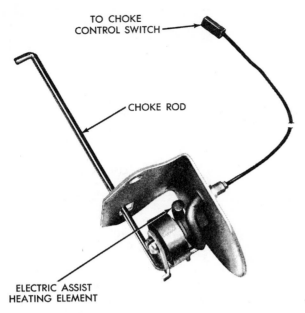

Fig. 3   Chrysler Corp. electric assist choke

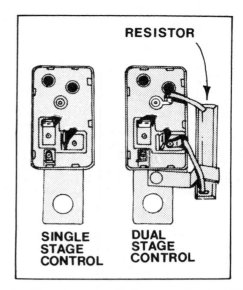

Fig. 4   Electric assist choke control unit. 1975-76
Chrysler Corp.

resistor, allows slow opening of the choke valve for good engine warm-up. As the small section of the ceramic resistor continues to heat, a bi-metal disc causes the spring loaded contact to close and apply electric current to the large section of the ceramic resistor which increases the heat flow to the thermostatic coil for more rapid opening of the choke valve.

At air temperatures between 50-70° F., electric current applied to the small section, or both the small and large sections of the ceramic resistor, will produce the amount of heat required to control the choke valve position for good engine operation in these temperature ranges.

At air temperatures above 70° F., electric current applied to the small section of the ceramic resistor and through the spring contact to the large section of the ceramic resistor, provides rapid heating of the thermostatic coil for quicker choke valve opening when leaner air/fuel mixtures are required at warmer temperatures.

For operation and servicing of the fast idle valve, refer to the "Fuel Injection Section".

Chrysler

NOTE: The wattage of the choke heater is part of the choke calibration and may change from year to year.

The control switch, Fig. 3, is connected to the ignition switch from which electrical power is obtained and transferred through electrical con-

nection to the control switch. The 1973 switch serves two purposes:
1. Above 63° F. the control switch will energize the choke heater.
2. After a period of time, the control switch will de-energize the choke heater. The shut-down will result after the control switch warms to about 110° F. by engine heat and a small electrical heater within the switch. Since the heater control switch is mounted to the engine and near the carburetor, some winter operation may energize the choke heater. This could happen after the choke has opened without benefit of electric heat. If this happens it will have no adverse effect on engine operation, and will soon be turned off.

The 1974 control switch serves three purposes:
1. Below 58° F. the control switch will partially energize the choke heater.
2. Above 58° F. the control switch will fully energize the choke heater.
3. The control switch will de-energize the choke heater at approximately 110° F. During winter operations, engines will experience three stages of choke heat: partial heat during engine warm-up, full heat after engine warm-up and no heat well after engine warm-up. Engine starts during summer temperatures will not experience the partial heat stage.

Two different control switches are used on 1975-77 vehicles: a single stage unit and a dual stage unit, Figs. 4 and 5.

The 1975-76 single stage control switch serves

two purposes:

1. Above 68° F., the control switch will energize the choke heater.
2. At about 130° F., the control switch will de-energize the choke heater.

Since the control switch is engine mounted, some cold engine operation may energize the choke heater, especially if the choke valve has been opened without electric heat. This condition which has no adverse effect on engine operations, will return to normal.

The 1977 single stage control switch shortens choke duration only above 80° F. Below 55° F., electric heat is not available until the engine is at normal operating temperature. Normal engine heat will then warm the control and energize the choke heater, but only after the choke has opened by engine heat.

The 1975-76 dual stage control switch serves four purposes:

1. Below 58° F., the control switch will partially

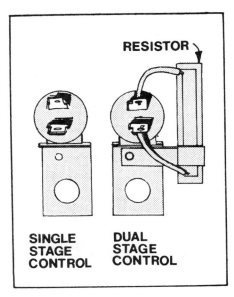

Fig. 5 Electric assist choke control unit. 1977 Chrysler Corp.

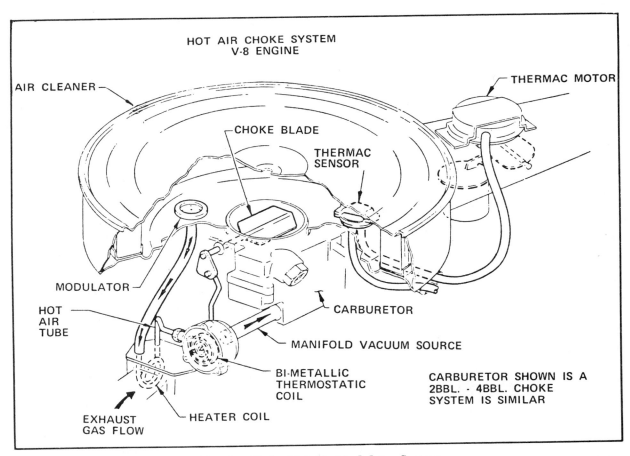

Fig. 6 Choke Hot Air Modulator System

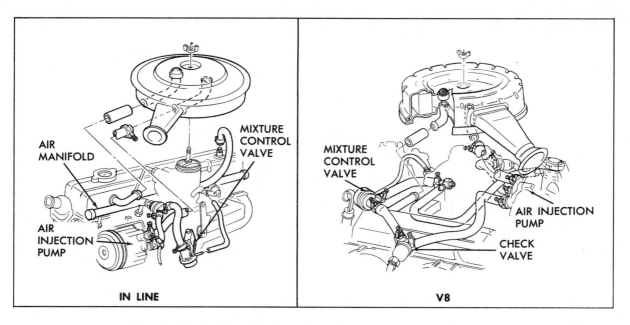

Fig. 7 Typical installation of an air pump system with a mixture control valve, otherwise known as a backfire by-pass valve

energize the choke heater.
2. Between 58-68° F., the control switch will either partially energize or fully energize the choke heater.
3. Above 68° F., the control switch will fully energize the choke heater.
4. At about 130° F., the control switch will de-energize the choke heater.

The 1977 dual stage control switch shortens choke duration above 80° F. and stabilizes choke duration during cold weather operation. During hot weather operation, electric assist heat is hotter than during cold weather operation assist level.

Cold weather heat levels are regulated by an electrical resistor connected to both terminals of the control. Below 55° F., electrical power is reduced by the resistor. Above 80° F., the resistor is bypassed by a switch inside the control to supply full power.

Engines started during cold weather conditions will experience two levels of choke heat: low during engine warm-up and high after engine warm-up. High heat levels occur after the choke is open to insure an open choke condition under all driving conditions and to minimize choking action which can occur after short stops during cold weather operation.

Engines started in hot weather conditions will not experience low choke heat levels. Engines started hot will only experience high heat because the switch is normally warmer than 80° F.

NOTE: The heating element should not be exposed to or immersed in any fluid for any purpose. An electric short in the wiring to the heater or within the heater will be a short of the ignition system.

CHOKE HOT AIR MODULATOR

General Motors

All 1975-77 Buick V6-231, V8-350 and V8-455, Oldsmobile Omega and Starfire V6-231 and V8-350 and Pontiac Ventura and Sunbird V6-231 and V8-350, incorporate a choke hot air modulator located in the bottom of the air cleaner housing to provide heated filtered air to the choke housing, Fig. 6. When air cleaner temperature is below 68° F., the air heated by the heater coil will pass through a .005 inch orifice in the modulator, allowing very little hot air flow over the bi-metallic thermostatic coil and providing slower choke warm-up. Above 68° F. air cleaner temperature, the modulator opens, allowing more air flow and faster choke warm-up.

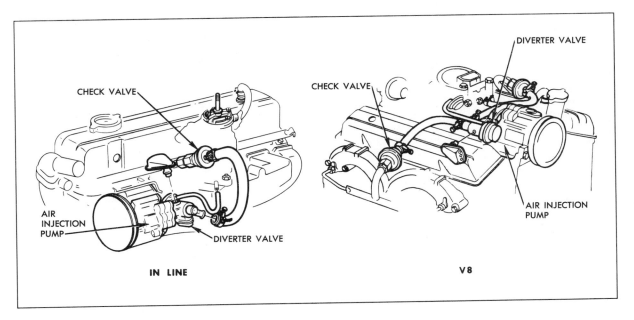

Fig. 8   Typical installation of an air pump system with a diverter valve, otherwise known as an air by-pass valve. 1968-77

AIR PUMP SYSTEMS

American Motors Air Guard, Chrysler Air Injection, Ford Thermactor & GM Air Injection Reactor (A.I.R.)

This system is used to reduce carbon monoxide and hydrocarbon emissions by adding a controlled amount of air to the exhaust gases through the exhaust ports. This causes oxidation of gases and an appreciable reduction of carbon monoxide and hydrocarbon emissions.

All air pump systems, Figs. 7 and 8, consist of an air injection pump, air injection tubes (one for each cylinder), a mixture control or backfire by-pass valve (added in 1966, '67), a diverter or air by-pass valve (added in 1968), check valves (one for in-line engines, two for V-8 engines), air manifolds, pipes and hoses necessary to connect the various components.

Carburetors and distributors for engines with an air pump system are designed especially for these engines, and they should not be interchanged with

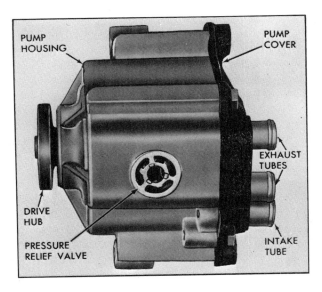

Fig. 9   Air injection pump with separate air filter. 1966-67

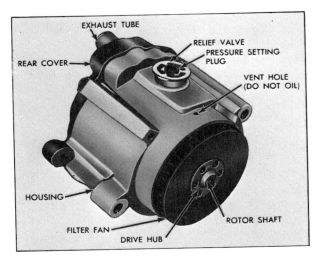

Fig. 10   Air injection pump with integral centrifugal air filter. Starting 1968

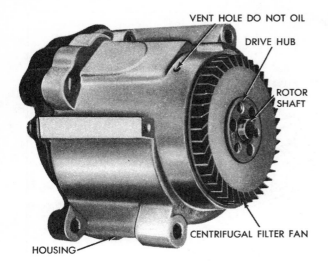

Fig. 11   *Air injection pump with integral centrifugal air filter without pressure valve*

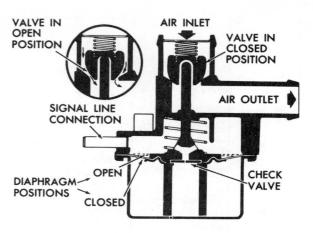

Fig. 12   *Typical mixture control or backfire by-pass valve*

or replaced by carburetors or distributors for engines without the air pumps.

The air injection pump, Figs. 9, 10 and 11, compresses the air and injects it through the air manifolds, hoses and injection tubes into the exhaust system in the area of the exhaust valves. The fresh air ignites and burns the unburned portion of the exhaust gases, thus minimizing the noxious content of the exhaust.

The mixture control or backfire by-pass valve, Fig. 12, when triggered by a sharp increase in manifold vacuum, as when the throttle is suddenly closed, supplies the intake manifold with fresh filtered air to lean out the fuel-air mixture and prevent exhaust system backfire.

The diverter or air by-pass valve, Figs. 13, 14, 15 and 16, when similarly triggered by a sharp increase in manifold vacuum, shuts off the injected air to the exhaust ports and helps to prevent backfiring during this period when the mixture is exceptionally rich. During engine overrun, all the air from the pump is dumped through the muffler on the diverter or air by-pass valve. At high engine speeds, the pump produces more air than the en-

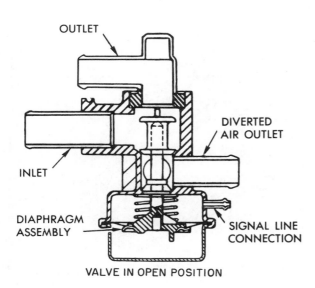

Fig. 13   *Typical diverter or air by-pass valve without pressure relief valve. General Motors*

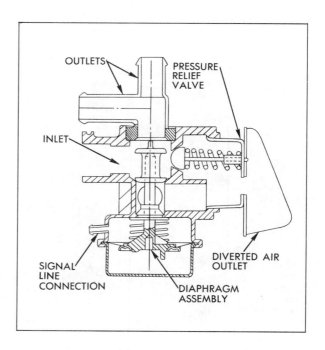

Fig. 14   *Typical diverter or air by-pass valve with integral pressure relief valve. General Motors*

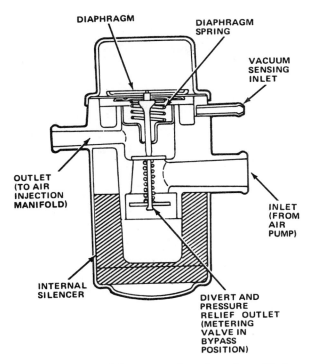

Fig. 15   *Typical diverter valve. American Motors*

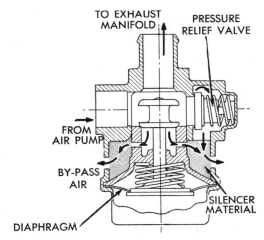

Fig. 16   *Typical diverter valve. Chrysler Corp.*

gine can use, and the excess is dumped through the pressure relief valve when that valve is part of the air pump, Figs. 9 and 10, or through the diverter or air by-pass valve when the pressure relief valve is part of that valve, Fig. 14.

The check valve or valves prevent exhaust gases from entering and damaging the air injection pump, as back flow can occur even under normal operating conditions.

When properly installed and maintained, the system will effectively reduce exhaust emissions. However, if any system components or any engine component that operates in conjunction with the air pump system should malfunction, exhaust emissions might increase.

Because of the relationship between engine operating condition and unburned exhaust gases, the condition of the engine and tune-up should be checked whenever the air pump system seems to be malfunctioning. Particular care should be taken in checking items that affect fuel-air ratio, such as crankcase ventilation system (PCV), the carburetor and carburetor air cleaner.

Air Switching Valve

This valve, Fig. 17, is used on 1977 Chrysler Corp. vehicles to switch air injection from the exhaust ports to an injection point further downstream in the exhaust system after engine warm-up so that air injection will not offset the effect on NO_x emissions by the EGR.

When the engine is warming up and NO_x emissions are low, air injection to the exhaust ports is acceptable, however, as the gases become hotter they tend to oxidize more rapidly by the addition of secondary air causing an increase of NO_x emissions. At this time, the air switching valve switches the air injection.

A vacuum signal from the CCEVS causes the switching valve to open allowing all air injection into the exhaust ports. When CCEVS shuts off vacuum to the switching valve, the valve closes and bypasses most of the air injection to the injection point downstream in the exhaust system. A bleed hole in the switching valve permits a small amount of air pump air to be injected into the exhaust ports at all times to assist in CO and HC reduction.

Power Heat Control Valve

This valve is used on 1977 Chrysler Corp. ve-

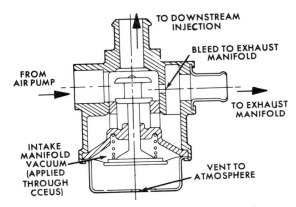

Fig. 17   *Air switching valve. Chrysler Corp.*

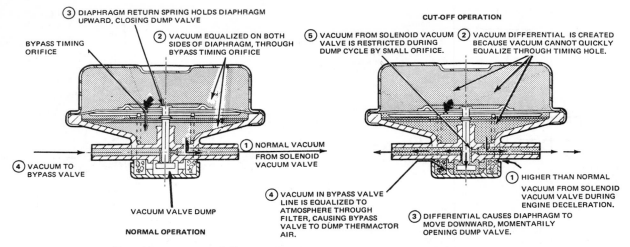

③ DIAPHRAGM RETURN SPRING HOLDS DIAPHRAGM UPWARD, CLOSING DUMP VALVE

BYPASS TIMING ORIFICE

② VACUUM EQUALIZED ON BOTH SIDES OF DIAPHRAGM, THROUGH BYPASS TIMING ORIFICE

④ VACUUM TO BYPASS VALVE

① NORMAL VACUUM FROM SOLENOID VACUUM VALVE

VACUUM VALVE DUMP

NORMAL OPERATION

CUT-OFF OPERATION

⑤ VACUUM FROM SOLENOID VACUUM VALVE IS RESTRICTED DURING DUMP CYCLE BY SMALL ORIFICE.

② VACUUM DIFFERENTIAL IS CREATED BECAUSE VACUUM CANNOT QUICKLY EQUALIZE THROUGH TIMING HOLE.

① HIGHER THAN NORMAL VACUUM FROM SOLENOID VACUUM VALVE DURING ENGINE DECELERATION.

④ VACUUM IN BYPASS VALVE LINE IS EQUALIZED TO ATMOSPHERE THROUGH FILTER, CAUSING BYPASS VALVE TO DUMP THERMACTOR AIR.

③ DIFFERENTIAL CAUSES DIAPHRAGM TO MOVE DOWNWARD, MOMENTARILY OPENING DUMP VALVE.

Fig. 18   Vaccum differential valve. 1975-77 Ford with Thermactor and catalytic converter

hicles to increase the flow of hot exhaust gases through the left-hand exhaust manifold to rapidly bring the mini-catalyst up to operating temperature. The CCEVS controls the manifold vacuum signal necessary to activate the valve. At coolant temperatures below a predetermined level, manifold vacuum is applied to the valve and all exhaust gases flow through the left hand exhaust manifold. Above this temperature no vacuum is applied to the valve and exhaust gas flows through both the right and left exhaust manifolds.

Vacuum Differential Valve (VDV)

On 1975-77 Ford vehicles with the Thermactor system and catalytic converters, a VDV Fig. 18, is used to control the operation of the air bypass valve. Under normal operation, vacuum applied through the VDV holds the valve upward, blocking the vent port and allowing Thermactor air flow. During acceleration or deceleration or in case of system failure, the VDV momentarily cuts off vacuum flow to the bypass valve, diverting the Thermactor air flow to atmosphere. In case of excessive pressure or system restriction, the excess pressure will unseat the valve in the lower part of the bypass valve, allowing a partial flow of air to atmosphere. At the same time, the valve in the upper part of the valve remains unseated allowing a partial flow of air to the exhaust manifold.

The VDV, Fig. 19, used on Chevrolet Chevette models is located in the air by-pass valve vacuum signal line and is used to prevent backfire in the exhaust system by directing the air bypass valve to divert the air pump output to atmosphere under high vacuum conditions. A sudden rise in intake manifold vacuum during deceleration periods cre-

ates a vacuum condition under the VDV diaphragm to pull the vacuum dump valve disc down and interrupt the vacuum signal to air bypass valve by diverting the vacuum source to atmosphere. Elimination of vacuum to the air bypass valve causes diversion of air pump output to atmosphere for about 2 seconds. The timing orifice hole in the VDV equalizes the vacuum pressure on both sides of the diaphragm allowing the valve to return to its normal position and permitting vacuum continuity in the air bypass valve signal line.

Differential Vacuum Delay & Vacuum Separator Valve

This valve, used on Chevrolet Chevette models, is located in series with the VDV in the vacuum

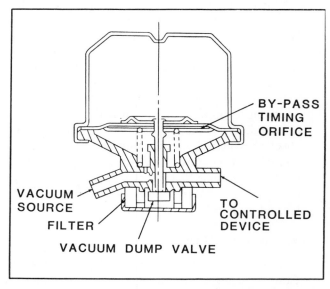

BY-PASS TIMING ORIFICE

VACUUM SOURCE

FILTER

VACUUM DUMP VALVE

TO CONTROLLED DEVICE

Fig. 19   Vacuum differential valve (VDV). 1976 Chevrolet Chevette

228

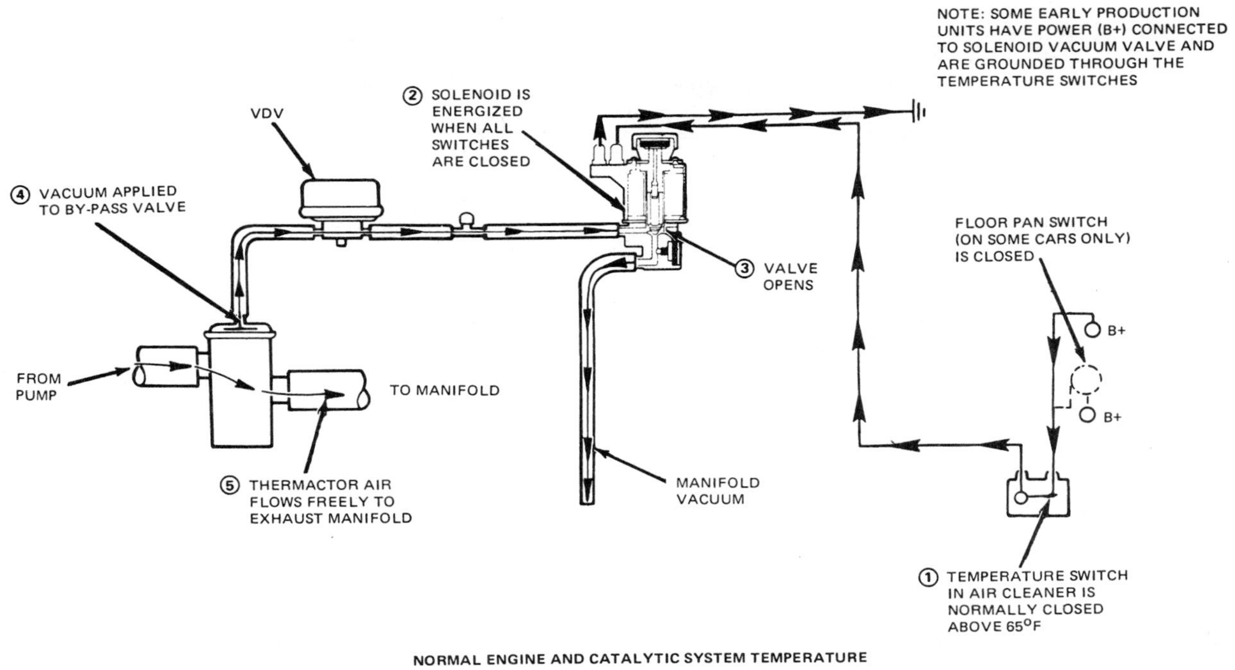

NOTE: SOME EARLY PRODUCTION UNITS HAVE POWER (B+) CONNECTED TO SOLENOID VACUUM VALVE AND ARE GROUNDED THROUGH THE TEMPERATURE SWITCHES

VDV

④ VACUUM APPLIED TO BY-PASS VALVE

② SOLENOID IS ENERGIZED WHEN ALL SWITCHES ARE CLOSED

③ VALVE OPENS

FLOOR PAN SWITCH (ON SOME CARS ONLY) IS CLOSED

FROM PUMP

TO MANIFOLD

⑤ THERMACTOR AIR FLOWS FREELY TO EXHAUST MANIFOLD

MANIFOLD VACUUM

① TEMPERATURE SWITCH IN AIR CLEANER IS NORMALLY CLOSED ABOVE 65°F

NORMAL ENGINE AND CATALYTIC SYSTEM TEMPERATURE

Fig. 20   Cold engine lockout system. Normally closed switches

signal line to the air bypass valve. The differential vacuum delay and separator valve aids in control of vehicle exhaust emissions by delaying diversion of the air from the air bypass valve during hard accelerations from 16-30 seconds. This is accomplished by the air bleed restrictions which allow only gradual reduction in the vacuum trapped in the signal line by the check valve.

During steady engine operation where there is no vacuum change to close the check valve, a constant .01 cubic foot per minute of filtered underhood air is drawn through the restrictors to purge the signal line below the check valve of all fuel and contaminants.

During deceleration when manifold vacuum increases, the pressure differential across the check valve causes it to open, thus immediately allowing the vacuum differential and air bypass valve

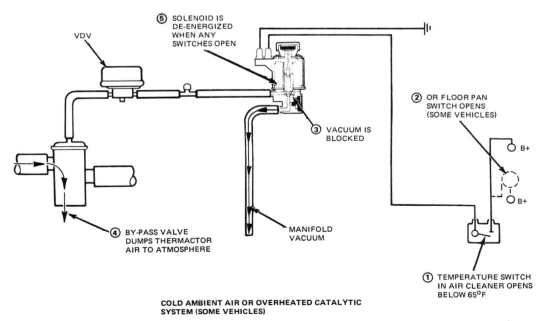

⑤ SOLENOID IS DE-ENERGIZED WHEN ANY SWITCHES OPEN

VDV

③ VACUUM IS BLOCKED

② OR FLOOR PAN SWITCH OPENS (SOME VEHICLES)

④ BY-PASS VALVE DUMPS THERMACTOR AIR TO ATMOSPHERE

MANIFOLD VACUUM

① TEMPERATURE SWITCH IN AIR CLEANER OPENS BELOW 65°F

COLD AMBIENT AIR OR OVERHEATED CATALYTIC SYSTEM (SOME VEHICLES)

Fig. 21   Cold engine lockout system. Normally closed switches

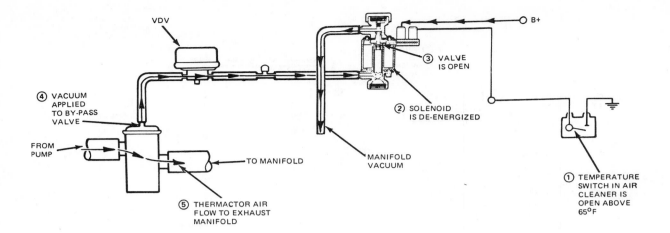

VDV

④ VACUUM
APPLIED
TO BY-PASS
VALVE

FROM
PUMP

TO MANIFOLD

③ VALVE
IS OPEN

② SOLENOID
IS DE-ENERGIZED

B+

MANIFOLD
VACUUM

⑤ THERMACTOR AIR
FLOW TO EXHAUST
MANIFOLD

① TEMPERATURE
SWITCH IN AIR
CLEANER IS
OPEN ABOVE
65°F

NORMAL ENGINE TEMPERATURE
(AND NORMAL CATALYST TEMPERATURE
ON SOME VEHICLES)

Fig. 22   Cold engine lockout system. Normally open switches

to be controlled by the increasing vacuum.

Differential Valve Delay Valve (DVDV)

This delay valve, used on some 1975-77 vehicles, delays air by-pass during periods of low engine manifold vacuum. During sudden drops in manifold vacuum, DVDV delays operation of vacuum differential valve (VDV). A sudden raise in manifold vacuum will open check in delay valve.

Exhaust Check Valve

This valve, used on some 1975-77 vehicles, al-

lows Thermactor air to enter exhaust port drillings, but prevents reverse flow of exhaust gases in event of improper operation of system components. This valve is located between by-pass valve and exhaust port drillings, either on air manifold or engine.

Cold Engine Lockout System

This system is used on 1975-77 Ford vehicles with the Thermactor system and catalytic converters. It consists of an electrically operated vacuum solenoid in the vacuum circuit to the by-pass valve and temperature sensors in the air cleaner housing and floor pan (on some models).

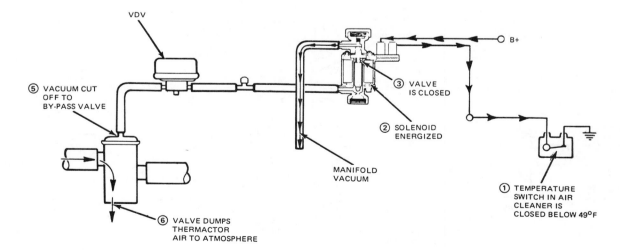

VDV

⑤ VACUUM CUT
OFF TO
BY-PASS VALVE

③ VALVE
IS CLOSED

② SOLENOID
ENERGIZED

B+

MANIFOLD
VACUUM

⑥ VALVE DUMPS
THERMACTOR
AIR TO ATMOSPHERE

① TEMPERATURE
SWITCH IN AIR
CLEANER IS
CLOSED BELOW 49°F

COLD AMBIENT AIR
OR OVERHEATED
CATALYST SYSTEM
(SOME VEHICLES)

Fig. 23   Cold engine lockout system. Normally open switches

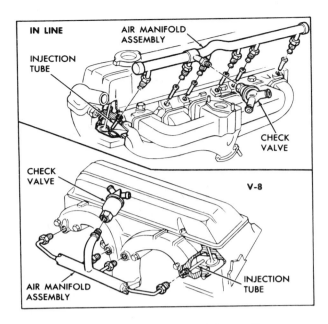

Fig. 24   Typical air manifold installations

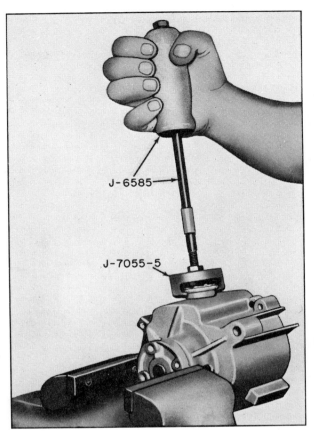

Fig. 25   Removing pressure relief valve with slide hammer tool shown

Two different type systems are used. The first type consists of normally closed switches that provide current to the solenoid only when the engine is warm. The second type consists of normally open switches that provide current to the solenoid only when the engine is cold. Operation of both systems is as follows:

Normally Closed Switches, Type 1, Figs. 20 & 21

With engine at normal operating temperature and all components operating properly, electrical path is provided through the solenoid vacuum valve, air cleaner sensor, floor pan temperature sensor (on some models) and to ground. This provides vacuum flow through the vacuum solenoid valve, VDV and to the bypass valve, allowing Thermactor air flow.

If any of the switches should operate due to low intake air temperature or overheated floor pan (on some models), the vacuum solenoid will stop vacuum flow, causing the bypass valve to divert Thermactor air to atmosphere.

Normally Open Switches, Type 2, Figs. 22 & 23

With engine at normal operating temperature and all components operating properly, the solenoid vacuum valve is inoperative, thereby providing vacuum flow through the solenoid vacuum valve, VDV and to the bypass valve, allowing Thermactor air flow. With the engine cold, the air cleaner sensor operates, providing current flow

through the solenoid vacuum to ground. If the floor pan sensor operates (on some models), it will provide current flow through the solenoid to ground.

If the solenoid vacuum valve operates, it stops the vacuum flow to the bypass valve, causing the Thermactor air flow to be diverted to atmosphere.

Maintenance

Engine tune-up should be checked whenever the air pump system seems to be malfunctioning, especially items affecting air/fuel ratio.

Because of the similarity of many parts, typical illustrations and procedures are given in the following text.

Air Manifold, Hose and Tube, Fig. 24

1. Inspect all hoses for deterioration or holes.
2. Inspect all tubes for cracks or holes.
3. Check all tube and hose routing for interference that may cause wear.
4. Check all hose and tube connections.
5. If a leak is suspected on the pressure side of

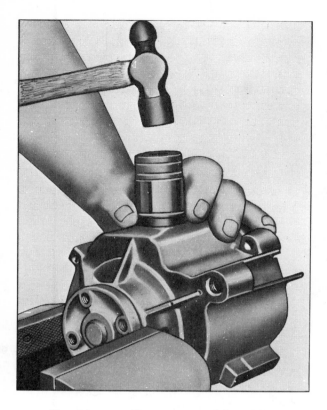

Fig. 26   *Installing pressure relief valve*

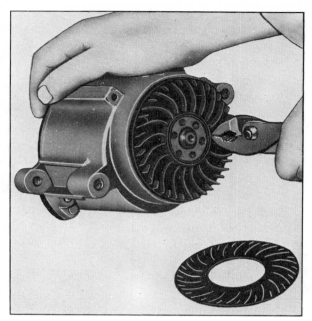

Fig. 27   *Removing centrifugal type pump air filter*

the system, and any tubes and/or hoses have been disconnected on the pressure side, the connections should be checked for leaks with a soapy water solution. With the pump running, bubbles will form at any leak.

6. To replace any hose or tube, note the routing, then remove the hoses or tubes as required.

Note that the hoses used with this system are made of special material to withstand high temperature. No other type of hose should be used.

Check Valves

1. Check valves should be inspected whenever the hose is disconnected from the valve or whenever valve failure is suspected.
   Note that if an air pump shows any signs of the presence of exhaust gas, this is an indication that the check valve has failed.
2. Orally blow through the check valve (toward the air manifold) then attempt to suck back through the check valve. Flow should be in one direction only, toward the air manifold.
3. To replace a check valve, disconnect the pump outlet hose at the check valve. Remove the valve from the air manifold, taking care not to bend or twist the manifold.

Diverters or Air By-Pass Valve

1. Check condition and routing of all lines, especially the signal line (the small diameter hose between intake manifold and the diverter valve). All lines must be secure, without crimps, and not leaking.
2. Disconnect signal line at valve. A vacuum signal must be available at the valve with the engine running.
3. With engine warmed up to operating temperature and throttle at curb idle, no air should be escaping through the valve's muffler. Open and quickly close the throttle. A momentary blast of air should discharge through the valve's muffler for at least one second. If it does not, and there is a good vacuum signal at the valve, then the valve is defective and should be replaced.

CAUTION: Although sometimes similar in appearance, these valves are designed to meet the particular requirements of individual engines. Therefore be sure to install the right valve for the engine being worked on.

4. To replace a valve, disconnect the vacuum signal line and valve exhaust hose or hoses.
5. Remove diverter or air by-pass valve from pump, and the muffler from the valve assembly, noting the angle of attachment.
6. Install muffler to new valve at previously noted angle.

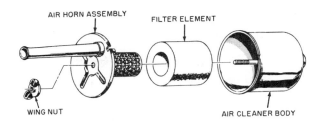

Fig. 28 *Conventional type pump air filter*

7. Install diverter or air by-pass valve to pump or bracket with new gasket.
8. Install outlet and vacuum lines and check system for leaks.

Mixture Control or Backfire By-Pass Valve

1. Check condition of all lines, especially the signal line. A defective signal or outlet line will cause malfunctioning of the mixture control or backfire by-pass valve.
2. Disconnect pump-to-valve inlet hose at pump.
3. A leaking valve will be indicated by an air gushing noise coming from the hose. Place palm of hand over hose. Little or no pull with a gradual increase is normal. If an immediate strong pull is felt, or air noise is heard, the valve is defective and must be replaced.
4. Open and close the throttle rapidly. Air noise should be evident and then gradually decrease. If air noise is not evident, or if the valve is noisy, it should be replaced.
5. To replace this valve, disconnect the signal line, air inlet and air outlet hoses, and remove valve.

6. Install new valve and connect hoses.

CAUTION: Like the diverter or air by-pass valve, these valves are designed to meet particular requirements of individual engines. Therefore be sure to install the correct valve for the engine being worked on.

Air Injection Tubes

There is no periodic service or inspection for these parts. However, whenever the cylinder head is removed from in-line engines, or exhaust manifolds from V-8 engines, inspect the tubes for carbon build-up and warped or burned tubes. Remove any carbon with a wire brush. Warped or burned tubes must be replaced.

1. To replace a tube, remove carbon, and using penetrating oil, work tube out of cylinder head or exhaust manifold. Then install new tube.

Air Injection Pump

1. Accelerate engine to about 1500 rpm and observe air flow from hose or hoses. If flow increases as engine speed rises, the pump is operating properly. If flow does not increase with engine speed, or if there is no air flow, proceed as follows:
2. Check for proper drive belt tension.
3. Check for leaky pressure relief valve. Air may be heard leaking with the pump running.

It is important to remember that the air pump system is not completely noiseless. Under normal conditions noise rises in pitch as engine speed increases. To determine if excessive noise is the fault of the system, operate the engine with the pump belt removed. If the excessive noise dis-

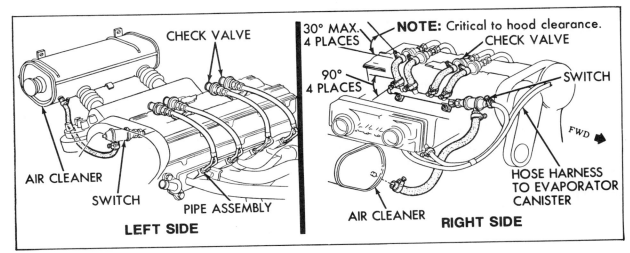

Fig. 29 *Pulse Air Injection Reactor System. Chevrolet Cosworth Vega*

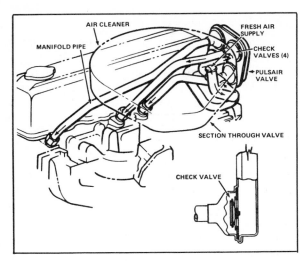

Fig. 30   Pulse Air Injection Reactor System. 1977
General Motors with 4 cylinder engine

appears with the belt removed, proceed as
follows:

4. Check for proper installation of the relief
   valve silencer (if equipped).
5. Check for seized air pump.
6. Check hoses, tubes, air manifolds and all connections for leaks and proper routing.
7. Check carburetor air cleaner for proper installation.
8. Check air pump for proper mounting.
9. If none of the above conditions exist and the
   air pump is excessively noisy, remove and replace the pump unit.
10. To replace the pump, disconnect hoses at
    pump and remove pump pulley, unfasten and
    remove pump.
11. Install new pump with mounting bolts loose.
    Then install pump pulley and drive belt.
12. Adjust drive belt tension and connect hoses at
    pump.

Pressure Relief Valve

When the pressure relief valve is incorporated
in the diverter or air by-pass valve the complete
valve must be replaced. If the relief valve is in the
pump, proceed as follows:

1. Pull relief valve from pump, Fig. 25.
2. Using a 15/16″ socket, Fig. 26, tap relief valve
   into housing until valve shoulders on housing.
   Use extreme care to avoid distorting housing.

NOTE: Various length pressure setting plugs designed for the particular requirements of the vehicle being worked on determine the pressure
required to open the relief valve. Usually the
pressure setting plugs are color-coded. To remove
the pressure setting plug, carefully unlock the legs

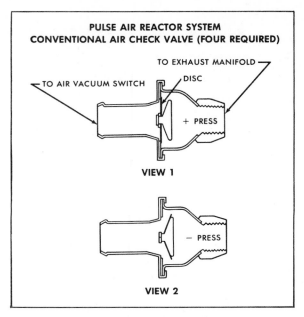

Fig. 31   PAIR System check valve. Chevrolet
Cosworth Vega

from the inside surface of the relief valve with a
small screwdriver. To install the plug, carefully
push it into the relief valve until the legs lock. If
a pressure setting plug is to be re-used, be sure
the leg angles are sufficient for the plug to lock in
place.

Centrifugal Pump Filter, Fig. 27

1. To replace the centrifugal type filter, remove

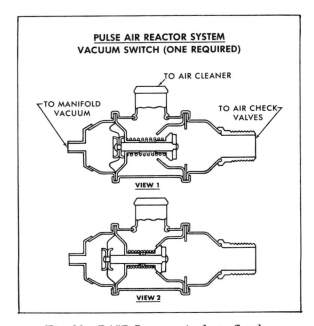

Fig. 32   PAIR System air shut off valve.
Chevrolet Cosworth Vega

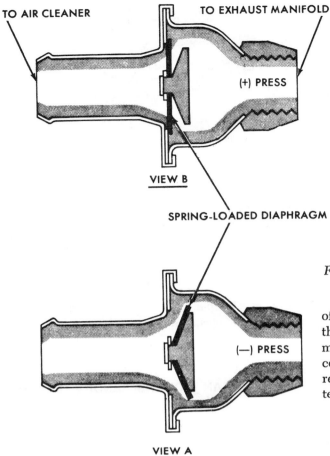

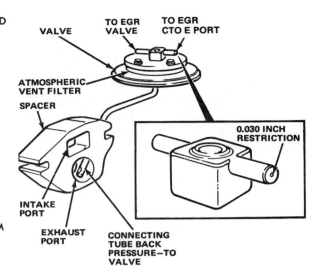

TO AIR CLEANER

TO EXHAUST MANIFOLD

(+) PRESS

VIEW B

SPRING-LOADED DIAPHRAGM

(—) PRESS

VIEW A

Fig. 33   *Aspirator valve. 1977 Chrysler Corp.*

VALVE

TO EGR VALVE

TO EGR CTO E PORT

ATMOSPHERIC VENT FILTER

SPACER

0.030 INCH RESTRICTION

INTAKE PORT

EXHAUST PORT

CONNECTING TUBE BACK PRESSURE—TO VALVE

Fig. 34   *Exhaust back pressure transducer sensor.*
(Typical)

drive belt and pump pulley. Pry loose outer disc of filter fan, being careful to prevent fragments from entering the air intake hole.

2. Install the new filter by drawing it on with the pulley and pulley bolts. Do not attempt to install a filter by hammering it or pressing it on.

3. Draw the filter down evenly by alternately tightening the bolts. Make certain that the outer edge of the filter slips into the housing. The slight amount of interference with the housing bore is normal.

Conventional Type Filter, Fig. 28

Remove and replace the air filter element as suggested by the illustration. The element is not cleanable. It must be replaced. Position the assembled air horn and filter element in the air cleaner body, being sure the tang is fitted into the slot.

NON-AIR-PUMP SYSTEMS

These are the systems which improve the quality of the exhaust gases by burning a proportion of them in the cylinders. This is accomplished by modifications to the carburetor, distributor and combustion system. Also by the addition of extra refinements which improve the efficiency of systems and improve driveability.

PULSE AIR INJECTION REACTOR (PAIR)

General Motors

This system, Figs. 29 and 30, utilizes exhaust pressure pulsations to draw fresh air from the air cleaner into the exhaust system.

The Cosworth Vega system consists of one air "check valve," Fig. 31, located on each of the four exhaust manifold legs adjacent to the cylinder head, and a pulse air shut off valve through which filtered air flows to the check valves.

On all other models, this system uses a series of pipes and check valves in the pulse air valve to route fresh air into the exhaust manifold, Fig. 30.

Engine operation causes a pulsating flow of exhaust gasses which are of positive or negative pressure, depending on whether the exhaust valve is seated or not. If pressure is positive, the disc is forced to the closed position, allowing no air flow into the exhaust system. If pressure is negative, the disc will open allowing fresh air flow into the exhaust system. Due to the inertia at high engine rpm, the disc fails to follow pressure pulsations. Therefore, the disc will remain closed preventing air flow into the exhaust system.

Throttle closure during deceleration, momentarily causes air fuel mixtures too rich to burn.

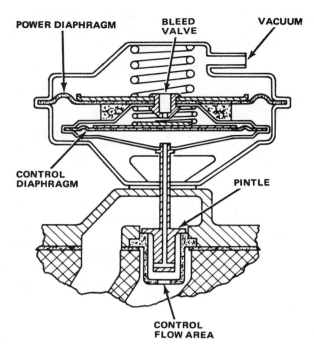

Fig. 35   EGR Back-Pressure Valve and Back Pressure Sentor unit. 1977 American Motors

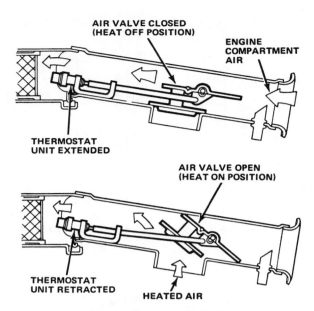

Fig. 36   Mechanically controlled thermostatic air cleaner. American Motors

These mixtures, when combined with air will become combustible, causing after fire. On Cosworth Vega models, a Pulse Air Shut Off Valve, Fig. 32, is used to momentarily divert air from the exhaust system to prevent after fire. This valve is calibrated to close only at vacuums exceeding idle vacuum which occur only during engine overrun. A spring is used to hold the valve open during idle, part throttle and wide open throttle.

ASPIRATOR AIR SYSTEM

Chrysler Corp.

This system is used on 1977 Chrysler Corp. catalyst equipped vehicles except 6-225, V8-360 (4 barrel carb. with high altitude calibration) and California. The valve in this system, Fig. 33, uses exhaust pressure pulsations to draw air into the exhaust system to reduce CO and HC emissions. It draws fresh air from the "clean" side of the air cleaner and past a one-way spring loaded diaphragm. The diaphragm opens to permit fresh air to mix with the exhaust gases during negative pressure (vacuum) pulses. When the pressure is positive, the diaphragm closes and no exhaust gases are allowed to flow past the valve. The aspirator valve works most efficiently at idle and slightly above idle when the negative pulses are maximum. At higher engine speeds, the aspirator valve remains closed.

AMERICAN MOTORS "ENGINE MOD" SYSTEM

This system controls exhaust emission levels by using composition cylinder head gaskets instead of steel gaskets and a special carburetor and distributor calibration. The carburetor incorporates idle limiters. The distributor centrifugal advance is calibrated to provide best performance and economy in the driving range and ignition timing is retarded only at idle speed (T.D.C.) to reduce exhaust emission levels at this slow engine speed. These engine modifications will result in a more complete combustion. Thermostatically controlled air cleaners are also used on most units to speed up engine warm-up.

AMERICAN MOTORS EXHAUST GAS RECIRCULATION (EGR)

The EGR system consists of a diaphragm actuated flow control valve (EGR valve), coolant temperature override switch, low temperature vacuum signal modulator, high temperature vacuum signal modulator and connecting hoses.

The purpose of the EGR system is to limit the formation of oxides of nitrogen (NOx) by diluting the fresh intake charge with a metered amount of exhaust gas, thereby reducing the peak temperatures of the burning gases in the engine combustion chambers.

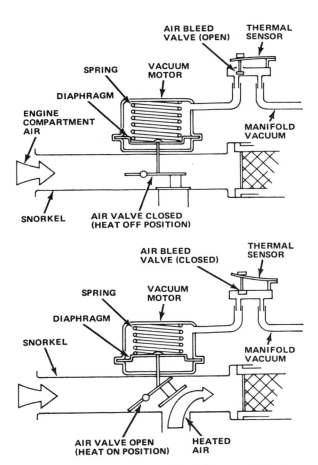

Fig. 37 *Vacuum controlled thermostatic air cleaner. American Motors*

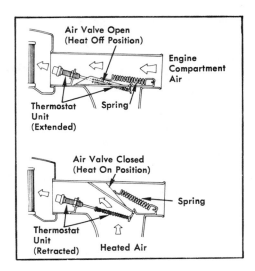

Fig. 38 *Temperature operated duct and valve assembly*

EGR Valve

The EGR valve is mounted on an machined surface at the rear of the intake manifold on V-8 engines and on the side of the intake manifold on six cylinder engines. The valve used with an automatic transmission is calibrated differently than the valve used with a manual transmission.

The valve is held in a normally closed position by a coiled spring located above the diaphragm. A special fitting is provided at the carburetor to route ported (above throttle) vacuum through hose connections to a fitting on the valve which is located above the diaphragm. A passage in the intake manifold directs exhaust gas from the exhaust crossover passage (V8 engine) or from below the riser area (six cylinder engine) to the EGR valve. When the diaphragm is actuated by vacuum, the valve opens and meters exhaust gas through another passage in the intake manifold to the floor of the manifold below the carburetor.

Coolant Temperature Override Switch

This switch is located at the coolant passage of the intake manifold (adjacent to oil filler tube) on a V8 engine or at the left side of the engine block (formerly the drain plug location) on a six cylinder engine. The outer port of the switch is open and not used. The inner port is connected by a hose to the EGR fitting at the carburetor. The center port is connected to the EGR valve.

When the coolant temperature is below 115° F. (160° F. on some models), the center port of the switch is closed and no vacuum signal is applied to the EGR valve, therefore, no exhaust gas will flow through the valve. When the coolant temperature reaches 115° F. (160° F. on some models), both the center and inner port of the switch are open and a vacuum signal is applied to the EGR valve. However, the vacuum signal to the EGR valve is subject to regulation by low and high temperature signal modulators.

Low Temperature Vacuum Signal Modulator

This unit is located at the left side of the front upper crossmember, just ahead of the radiator, on the same mounting bracket as the TCS ambient temperature override switch and is connected to the EGR vacuum signal hose. The modulator is open when ambient temperatures are below 60° F. This causes a weakened vacuum signal to the EGR valve and a resultant decrease in the amount of exhaust gas being recirculated.

High Temperature Vacuum Signal Modulator

This unit is located at the rear of the engine compartment and is connected to the EGR vac-

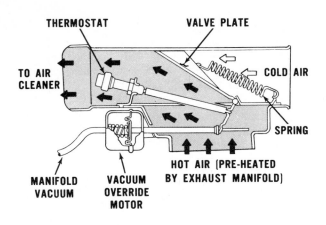

Fig. 39   Duct and valve assembly with
vacuum override motor

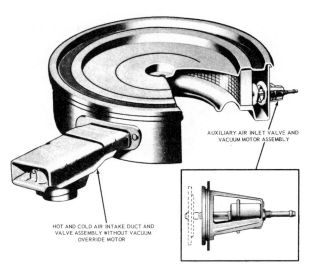

Fig. 40   Air cleaner with auxiliary
air inlet valve and vacuum motor

uum signal hose. The modulator opens when the underhood air temperature reaches 115° F. and causes a weakened vacuum signal to the EGR valve. As a result, the amount of exhaust gas being recirculated is decreased.

Exhaust Back-Pressure Sensor

Some 1974, all 1975 California vehicles and all 1976 models are required to have an exhaust back pressure transducer sensor, Fig. 34. This device consists of a diaphragm valve, a spacer and a metal tube. The EGR valve is mounted to the sensor spacer and is modulated by the sensor.

The EGR system, when equipped with a back-pressure sensor, obtains a vacuum signal at the carburetor spark port and not the EGR port. The vacuum signal passes through the EGR CTO (Coolant Temperature Override) switch (when coolant temperature exceeds 115° or 160° F.) to the valve portion of the sensor where it is modulated by exhaust back-pressure.

NOTE: The inlet nipple of the exhaust back-pressure sensor has a .030 inch restriction. The vacuum line from the EGR CTO must be connected to this nipple.

When exhaust back-pressure is relatively high, as during acceleration and some cruising conditions, exhaust back-pressure traveling through the metal tube overcomes spring tension on the diaphragm within the back-pressure sensor valve, and closes the valve atmospheric vent.

With the back-pressure sensor valve no longer vented to atmosphere, the vacuum signal now passes through the back-pressure sensor valve, and the EGR valve. When vacuum signals the EGR valve, exhaust gas recirculation commences.

When exhaust back-pressure is too low to overcome diaphragm spring tension, the vacuum signal is vented to atmosphere and does not pass through to the EGR valve. With no vacuum signal applied to the EGR valve, exhaust gas does not recirculate.

All six cylinder and some V8-304, 360 engines incorporate a steel restrictor plate under the exhaust back-pressure sensor. The restrictor plate limits the rate of EGR flow, thereby improving driveability. Note that gaskets are used on both sides of the plate.

The back-pressure sensor is not serviceable and must be replaced if defective.

EGR Back-Pressure Valve & Sensor Unit

On 1977 vehicles, the exhaust back pressure sensor is an integral part of the EGR valve Fig. 35 thereby combining the functions of the EGR valve and exhaust back pressure sensor into one unit. Calibration of this unit is accomplished by using different diaphragm spring loads and flow control orifices.

Exhaust gas recirculation is controlled by the movable pintle. In the closed position, spring pressure holds the pintle against its seat, confining the exhaust gases in the exhaust manifold. While the vacuum bleed valve in the power diaphragm is open, carburetor vacuum cannot pull the pintle off its seat. When the exhaust back pressure conducted through the hollow pintle is enough to

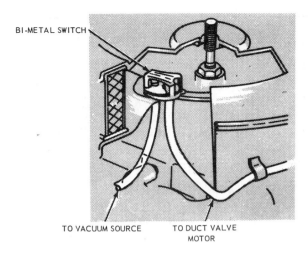

BI-METAL SWITCH

TO VACUUM SOURCE   TO DUCT VALVE MOTOR

Fig. 41 Vacuum operated duct and valve assembly

overcome control spring pressure, full vacuum is applied to the power diaphragm and the pintle moves to start EGR. When back pressure drops sufficiently, the control diaphragm moves away from the bleed valve to stop EGR.

THERMOSTATIC CONTROLLED AIR CLEANER, TAC & AUTO-THERM AIR CLEANER SYSTEMS

American Motors

Mechanically Controlled

This system consists of a heat shroud placed over the exhaust manifold, a hot air hose, and an air duct and valve assembly, Fig. 36.

The air duct and valve assembly regulates the temperature of air entering the air cleaner by selecting either air from the engine compartment, or heated air from the shrouded exhaust manifold.

During engine warmup when the air temperature entering the air duct is less than the calibrated temperature, the thermostat is in the retracted position and the air valve is held in the open position (heat on) by the spring, thus allowing hot air flow from the shrouded exhaust manifold.

As air temperature passing around the thermostat unit rises, the thermostat starts to open, pulling the air valve down and allowing cooler air from the engine compartment to enter the air cleaner. When the incoming air reaches the calibrated operating temperature, the air valve is in the closed position (heat off) so that only engine compartment air enters the air cleaner.

Vacuum Controlled

This system consists of a heat shroud, a hot air hose, a thermal sensor and a vacuum motor and air valve assembly, Fig. 37.

The thermal sensor incorporates a bleed valve which regulates the amount of vacuum applied to the vacuum motor which controls the air valve position to supply either heated air from the exhaust manifold or air from the engine compartment.

During engine warmup when underhood air temperatures are low, the air bleed valve is closed allowing sufficient vacuum to be applied to the vacuum motor, holding air valve in the open position.

As the air temperature entering the air cleaner reaches the calibrated temperature, the bleed valve opens, decreasing the amount of vacuum applied to the vacuum motor. The diaphragm spring in the vacuum motor moves the air valve into the closed position (heat off), thus allowing only underhood air to enter the air cleaner.

During hard acceleration, manifold vacuum drops causing the valve to close regardless of temperature to obtain maximum airflow through the air cleaner.

Ford Motor Co.

Temperature Controlled

Carburetor air temperature is thermostatically controlled by the air duct and valve assembly. Air from the engine compartment, or heated air from the shrouded exhaust manifold is supplied to the engine, Fig. 38.

During the engine warm-up period when the air temperature entering the air duct is less than 105° F., the thermostat is in the retracted position and the air valve is held in the closed position by the air valve spring, thus shutting off the air from the engine compartment. Air is then drawn from the shroud at the exhaust manifold.

As the temperature of the air passing the thermostat unit rises, the thermostat starts to open and pulls the air valve down. This allows cooler air from the engine compartment to enter the air cleaner. When the temperature of the air reaches 130° F, the air valve is in the open position so that only engine compartment air is allowed to enter the air cleaner.

Vacuum Controlled

Some Ford systems incorporate a vacuum override motor, Fig. 39. This motor during cold ac-

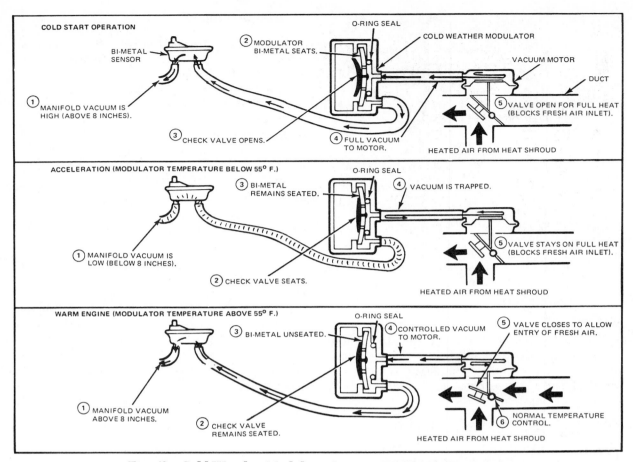

Fig. 42 Cold Weather Modulator System. 1975-77 Ford Motor Co.

celeration periods provides additional air to carburetor. The decrease in intake manifold vacuum during acceleration causes the vacuum override motor to override the thermostat control, opening the system to both engine compartment air and heated air from the exhaust manifold shroud.

Yet another Ford application of same system uses a vacuum motor, Fig. 40, installed on the perimeter of the air cleaner to take the place of the vacuum override motor. When the manifold vacuum is low, during heavy engine loading or high speed operation, a spring in the vacuum motor opens the motor valve plate into the air cleaner. This provides the maximum air supply for greater volumetric efficiency.

Thermostatically Controlled Vacuum Operated

A vacuum operated duct valve with a thermostatic bi-metal control, Fig. 41, is used on some installations. The valve in the duct assembly is in an open position when the engine is not operating. When the engine is operating at below normal operating temperature, manifold vacuum is routed through the bi-metal switch to the vacuum motor

to close the duct valve allowing only heated air to enter the air cleaner. When the engine reaches normal operating temperature the bi-metal switch opens an air bleed which eliminates the vacuum, and the duct valve opens allowing only cold air to enter the air cleaner. During periods of acceleration the duct valve will open regardless of temperature due to the loss of manifold vacuum.

Cold Weather Modulator

This modulator, Fig. 42, located in the air cleaner housing of some 1975-76 Ford vehicles, prevents the air cleaner door from opening to non-heated outside air at ambient temperatures below 55° F. At ambient temperatures above 55° F., the modulator is inoperative. During acceleration at ambient temperatures below 55° F., the modulator located in-line between the bi-metal sensor and vacuum duct motor, will close off vacuum to the motor and hold the duct open.

Air Cleaner Temperature Switch

1975-76 vehicles with CTAV systems and those

240

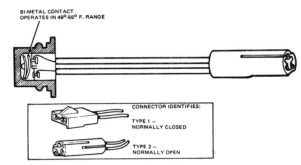

CONNECTOR IDENTIFIES:

TYPE 1 —
NORMALLY CLOSED

TYPE 2 —
NORMALLY OPEN

*Fig. 43   Air cleaner temperature switch. 1975-76
Ford Motor Co.*

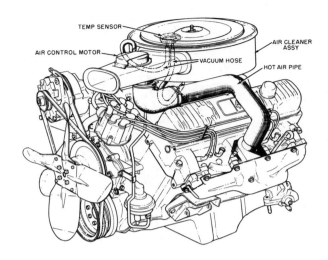

*Fig. 44   Controlled Combustion System
(CCS) installed on a V8 engine*

TEMP SENSOR

AIR CONTROL MOTOR

VACUUM HOSE

AIR CLEANER ASSY

HOT AIR PIPE

with Thermactor systems used with catalytic converters, are equipped with a bi-metallic temperature switch mounted in the air cleaner housing. The two type switches used may be identified by the connector, Fig. 43. The first type has normally closed contacts, while the second type has normally open contacts. Refer to "Cold Engine Lockout" under "Air Pump Systems", for operation of these switches.

All General Motors

Figs. 44 thru 47. Carburetor air temperature is controlled by a pair of doors, located in the air cleaner snorkel, which channel either pre-heated or under hood air to the carburetor.

Preheated air is obtained by passing under hood air through ducts surrounding the exhaust manifold, causing it to pick up heat from the manifold surface. The heated air is then drawn up through a pipe to the air cleaner snorkel.

Under hood air is picked up at the air cleaner

snorkel in the conventional manner.

The two air mixing doors work together so that as one opens, the other closes and vice versa. When under hood temperature is below approximately 90 deg. F., the cold air door closes, causing the hot air door to open. Hot air from the exhaust manifold stove is then drawn into the carburetor. As the under hood temperature increases, the cold air door begins to open until the temperature reaches approximately 115 to 130 deg. F., at which time the cold air door is fully open and the hot air door is fully closed.

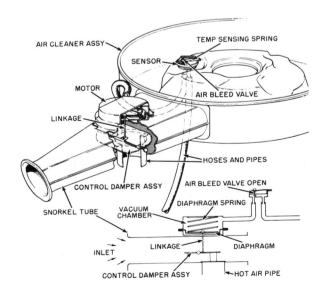

Fig. 45   Cold air door open

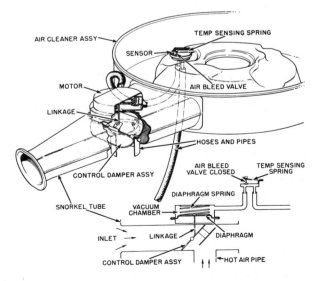

Fig. 46   Hot air door open

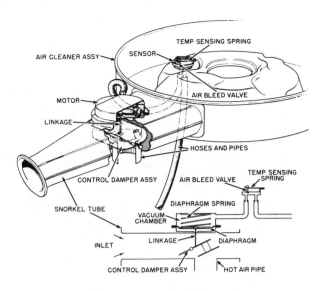

Fig. 48   Thermo vacuum switch. Typical of General Motors CCS, Ford IMCO and Chrysler NOx, OSAC (temperature sensing valve)

Fig. 47   Cold and hot air doors both partially open

The doors are controlled by a vacuum motor mounted on the air cleaner snorkel. This motor, in turn, is controlled by a sensor inside the air cleaner which regulates the amount of vacuum present in the vacuum motor according to air cleaner temperature. Whenever manifold vacuum drops below 5-8 inches, depending on the unit, the diaphragm spring in the motor will open the cold air door wide in order to provide maximum air flow.

The vacuum motor and control door assembly in the left snorkel on outside air induction units does not have a sensor and is controlled only by manifold vacuum. This snorkel remains closed until full throttle is obtained. With manifold vacuum at 6-8 inches, the door will open, allowing maximum air flow.

GENERAL MOTORS CCS

Controlled Combustion System

The C.C.S. system, Fig. 44, is designed to keep the air entering the carburetor at approximately 100 deg. F. or above so that the carburetor can be calibrated to operate leaner without affecting engine performance, provide improved fuel economy, eliminate carburetor icing and improve engine warm-up.

Because of the leaner carburetor calibration, the ignition timing at idle must be retarded. This is done by means of a "ported" vacuum advance, with the vacuum take-off just above the throttle

plate(s), so that there is no vacuum advance at closed throttle but there is advance as soon as the throttle is cracked slightly. Ignition timing is set at or near TDC and centrifugal advance does not start until approximately 1000 rpm.

Thermo Vacuum Switch

Because of the increased possibility of engine overheating at idle with the C.C.S. calibration, the thermo vacuum switch, Fig. 48, is added to the system on some engines. This switch senses engine coolant temperature and, if temperature reaches 220 deg. F., the switch valve moves to allow manifold vacuum to reach the distributor, to advance the timing and allow the engine to run cooler.

Idle Stop Solenoid

Most cars equipped with the Controlled Combustion System have an idle stop solenoid mounted on the carburetor, Fig. 49, to prevent engine operation after the engine is shut off (dieseling). Dieseling is prevalent with the CCS system, especially with automatic transmission because of the retarded ignition timing and higher operating temperature of the engine. Idle speed is higher with automatic transmissions.

The solenoid eliminates this dieseling by fully closing the throttle valve(s) when the ignition is turned off.

When the ignition switch is turned on, the solenoid coil is activated and the plunger is driven to its full extended position. The plunger acts on the throttle valve lever and sets the throttle valve(s) in a position to achieve specified idle rpm.

Fig. 49   Idle stop solenoid

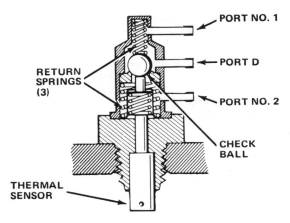

Fig. 50   Coolant temperature override switch.
American Motors

When the ignition is turned off, the solenoid is de-energized and the plunger retracts into the solenoid, causing the throttle valve(s) to close to a position controlled by the low idle adjusting screw. At this point the throttle valve(s) is open only enough to allow the engine to run at a much lower rpm. This lower setting keeps the throttle valve(s) from completely closing and scuffing the throttle bore(s).

The lower idle speed setting is made with a set screw when the solenoid is electrically disconnected. The setting for normal idle speed is adjusted, with the solenoid energized, through the hex screw in the plunger and/or by repositioning the solenoid in its mounting clamp.

NOTE: To set the solenoid while starting a hot engine, the accelerator pedal must be depressed approximately one third its travel. When starting a cold engine, however, the accelerator must be fully depressed to set both the choke and the solenoid.

CAUTION: Unburned gases emit undesirable amounts of hydro-carbons, predominantly during idle operation. One would naturally think that if the idle mixture is made as lean as possible, exhaust emission would be reduced. This is true up to a point. However, if the idle mixture is made too lean, the hydro-carbon content may increase above acceptable limits. The amount of carbon monoxide exceeds acceptable limits when idle mixtures are too rich. Conversely, the hydro-carbon content exceeds acceptable limits if the idle mixture is too "lean." Consequently, proper idle adjustment is essential.

TRANSMISSION CONTROLLED SPARK (TCS)

American Motors

This system is designed to provide vacuum spark advance during high gear operation and certain engine conditions. The resultant is a lower peak combustion pressure and temperature during the power stroke, significantly reducing exhaust emissions.

Basically the system incorporates an ambient temperature override switch, solenoid vacuum valve, solenoid control switch and in some systems a coolant temperature override switch.

The ambient temperature override switch, senses ambient temperatures and completes the electrical circuit from the battery to the solenoid vacuum valve when ambient temperatures are above 63° F.

The solenoid vacuum valve is attached to the intake manifold. When the valve is energized, carburetor ported vacuum is blocked and the distributor vacuum line is vented to atmosphere through a port in the valve, resulting in no vacuum advance. When the valve is de-energized, ported vacuum is applied to the distributor resulting in normal vacuum advance.

The solenoid control switch, located at the transmission, opens or closes in relation to car speed for automatic transmission equipped cars or gear range manual transmission equipped. At speeds above 34 MPH on automatic transmission equipped cars, high gear on manual transmission equipped cars, the switch opens and breaks the ground circuit to the solenoid vacuum valve. At speeds under 25 MPH on 1971-73 models or 35 MPH on 1974-76 model automatic transmission equipped cars, lower gear ranges manual transmission equipped, the

243

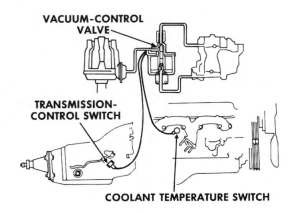

Fig. 51   Typical TCS emission control

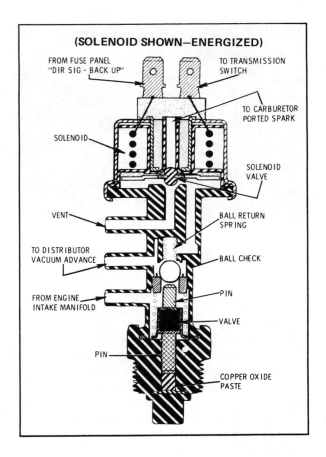

Fig. 52   TVS-TCS combination valve

switch closes and completes the ground circuit to the solenoid vacuum valve.

On automatic transmission equipped vehicles, the switch is automatically operated by the speedometer gear speed on 1971-73 models or transmission governor oil pressure on 1974-77 models. On manual transmission vehicles, the switch is manually operated by the shifter shaft.

The coolant temperature override switch, Fig. 50, is threaded into the thermostat housing on V8 engines or into the left rear side of the cylinder block on six cylinder engines. Its purpose is to improve driveability during the warm-up period by providing full distributor vacuum advance until the engine coolant has reached 160° F. on V8 engines and 1975-76 six cylinder engines or 115° F. on 1974 California six cylinder engines with manual transmission and 1977 Matador with V8-360. The switch incorporates a thermal unit which reacts to coolant temperatures to route either intake manifold or carburetor ported vacuum to the distributor vacuum advance diaphragm.

The CTO system may be used in place of or in conjunction with the TCS system depending upon application.

General Motors

This system is designed to provide vacuum spark advance during high gear operation only. The resulting ignition timing in the lower gears significantly reduces exhaust emissions.

Basically, the TCS system consists of a vacuum control valve, transmission control switch and a coolant temperature switch. The valve controls the vacuum signal to the distributor vacuum advance unit in response to a signal from either switch, Fig. 51.

The vacuum control valve is of solenoid design and is installed in the vacuum line between the carburetor and the distributor. When energized by the transmission control switch, the valve blocks the vacuum source and vents the advance unit through the carburetor air horn.

When the engine is cold, and in some units when the engine temperature exceeds 210° F, the coolant temperature switch overrides the system and provides full manifold vacuum to the advance unit, advancing timing and in turn lowering engine operating temperature. This switch operates through a relay mounted on the firewall. Some systems use combination TVS-TCS valve, Fig. 52, to provide system over-riding vacuum to the advance unit.

A TCS system change on mid-year 1973 and all 1974 Pontiac models (vehicles manufactured after March 15, 1973) will incorporate the following revisions, Fig. 39: A Start-up Relay Switch supplies full advance (ported advance on manual transmission vehicles) in any gear for 20 seconds after engine start.

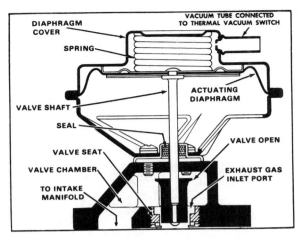

Fig. 53 Single diaphragm EGR valve cross section

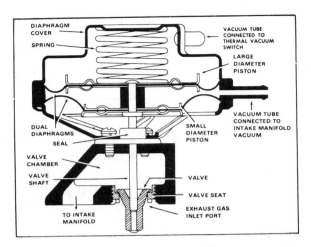

Fig. 54 Dual diaphragm EGR valve cross section

A Distributor Vacuum Spark Thermal Valve on 1973 models or Distributor Spark-EGR Thermal Vacuum Switch on 1974 models, sensing air-fuel mixture temperature, provides full advance in any gear when the mixture temperature is below 62° F. When air-fuel mixture temperature rises above 62° F., the thermal valve closes, shutting off vacuum. In turn to provide vacuum advance, the distributor must be energized, depending on condition, by the TCS Switch, Cold Feed Switch or the Hot Coolant Switch.

The TCS Switch grounds the distributor solenoid in high gear on all applications and in reverse gear on 1974 vehicles equipped with automatic transmissions only when the Cold Feed Switch is closed. The Cold Feed Switch, depending on application, closes when cylinder head temperature reaches 125°, 140° or 155° F. Regardless of TCS Switch position, the Hot Coolant Switch grounds the distributor solenoid when coolant temperature is over 240° F.

On some models, a Distributor Vacuum Spark Delay Valve is installed between the distributor solenoid and distributor. This valve restricts the rate of initial vacuum supplied to the distributor, full vacuum will be supplied gradually.

The Time Relay is an electrical on-off type switch. When the coil is energized, it starts heating the bi-metal strip and opening the normally closed relay points within 20 seconds after the ignition switch is turned on. If for some reason the ignition key is left in the "ON" position and vehicle is not started within 20 seconds and the relay completes its "countdown," vacuum advance will be blocked off until the relay has cooled off.

NOTE: Once the relay has run one cycle (about 20 seconds), after the key has been turned on, the relay must cool off before it will reactivate, even if the key is switched "OFF" and then turned "ON" again.

GENERAL MOTORS CEC

Combined Emission Control System

This system is designed to provide vacuum spark advance during high gear operation only, as does the TCS system, but the CEC solenoid vacuum switch also regulates curb idle and high gear deceleration throttle positions, further reducing emissions.

The CEC system consists of a vacuum control solenoid valve, transmission control switch, coolant temperature switch and a time-delay relay.

When the solenoid is in the non-energized position, vacuum to the distributor vacuum advance unit is shut off and the distributor is vented to the atmosphere through a filter at the opposite end of the solenoid. When the solenoid is energized by one of the switches or the relay, the vacuum port is uncovered and the plunger is seated at the opposite end, shutting off the clean air vent. This routes vacuum to the distributor. The solenoid is energized in high gear by the transmission switch. The coolant temperature switch overrides the transmission switch to energize the solenoid and provide vacuum advance below 82° F., and the time-delay relay is incorporated into the circuit to energize the solenoid for approximately 15 seconds after the ignition key is turned on.

When the solenoid plunger is in the non-energized position, it allows the throttle to close to

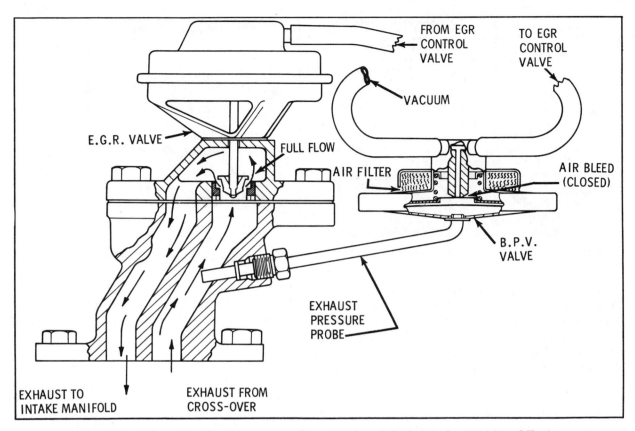

Fig. 55   Exhaust back pressure transducer. Oldsmobile (typical of 1977 Cadillac)

the curb idle setting. When the solenoid plunger is in the energized position, it keeps the throttle open to the high gear deceleration position which controls hydrocarbon emission.

GENERAL MOTORS SCS

Speed Controlled Spark

This system is essentially the same as the TCS system except that here the vacuum spark advance is controlled by the vehicle's speed rather than the gear position of the transmission.

The SCS system consists of a vacuum advance solenoid valve, a speed sensing switch and a temperature switch. The vacuum control valve is installed in the vacuum line between the carburetor and the distributor. When energized, this valve cuts off vacuum to the distributor. The speed sensing switch de-energizes the vacuum control valve at speeds above 38 mph allowing normal vacuum spark advance. Abnormal operating temperatures (below 85 degrees F. and above 220 degrees F.) will cause the temperature switch to de-energize the vacuum control valve, over-riding the speed sensing switch.

GENERAL MOTORS EGR

Exhaust Gas Recirculation

This system, Figs. 53 and 54, is used to reduce oxides of nitrogen emissions at the engine's exhaust. This is accomplished by introducing exhaust gases into the intake manifold at throttle positions other than idle. It consists of an E.G.R. valve mounted on a special intake manifold. The exhaust gas intake port of the E.G.R. valve is connected to the intake manifold exhaust crossover channels where it can pick up exhaust gases.

As the throttle valves are opened and the engine speeds up, a vacuum is applied to a vacuum diaphragm in the EGR valve through a connecting tube. When the vacuum reaches approximately 3″ Hg., the diaphragm moves upward against spring tension and is in the full-up position at approximately 7″ to 8″ Hg. of vacuum. This diaphragm is connected by a shaft to a valve which closes off the exhaust gas port. As the diaphragm moves up, it opens the valve in the exhaust gas port which allows exhaust gas to be pulled into the intake manifold and enter the cylinders. The exhaust gas port must be closed during idle as the mixing of exhaust gases with the fuel air mixture

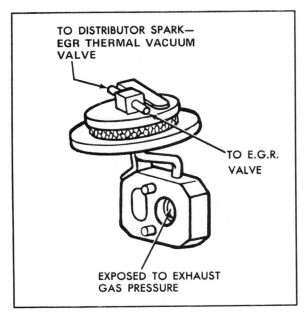

Fig. 56   Exhaust back pressure transducer.
Pontiac

at this point would cause rough running.

The dual diaphragm EGR valve, Fig. 54, is designed to provide increased exhaust gas recirculation rates when engine loads increase.

NOTE: Manifold vacuum is used as the signal to indicate the engine load.

The valve is similar to the single diaphragm valve except that a second diaphragm has been added to the valve and is connected to the upper diaphragm, with a spacer, thus both diaphragms move together. A manifold vacuum signal is applied to the volume between the two diaphragms. The upper diaphragm has a larger diameter piston than the lower diaphragm, therefore the load caused by the manifold vacuum between the two diaphragms aids the spring load. Thus as the engine load increases and manifold vacuum decreases the combined load of the spring and the vacuum chamber are reduced allowing the valve to open further for a given EGR vacuum signal.

Therefore, for high intake manifold vacuums (such as cruising), the opening is less than for low manifold vacuums obtained during accelerations. The valve now is capable of providing more recirculation on accelerations where loads are higher and the tendency to produce NOx is greater.

Exhaust Back Pressure Transducer Valve

Figs. 55 and 56.—This valve, used on some 1974-

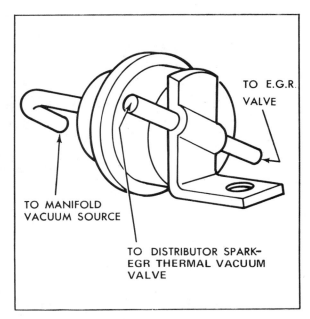

Fig. 57   Vacuum bias valve. Pontiac

77 vehicles, modulates EGR flow according to engine load. The device consists of a diaphragm valve, a spacer and a metal tube.

The EGR system, when equipped with a back pressure transducer valve, obtains a vacuum signal at the carburetor spark port and not at the EGR port. This vacuum is modulated by the transducer and, in turn activates the EGR valve.

When exhaust back-pressure is relatively high, as during acceleration and some cruising conditions, exhaust back-pressure traveling through the metal tube overcomes spring tension on the diaphragm within the back-pressure transducer valve, and closes the valve atmospheric vent.

With the back-pressure transducer valve no longer vented to atmosphere, the vacuum signal now passes through the back-pressure transducer valve, and the EGR valve. When vacuum signals the EGR valve, exhaust gas recirculation commences.

When exhaust back-pressure is too low to overcome diaphragm spring tension, the vacuum signal is vented to atmosphere and does not pass through to the EGR valve. With no vacuum signal applied to the EGR valve, exhaust gas does not recirculate.

Vacuum Bias Valve

This valve, Fig. 44, used on some California 1974-75 Pontiac vehicles, is located between the EGR valve and the distributor spark-EGR thermal vacuum valve. At high manifold vacuum conditions (such as cruising), the VBV decreases EGR flow and in turn, acts to reduce surge. When NOx formation is high and manifold vacuum is low as dur-

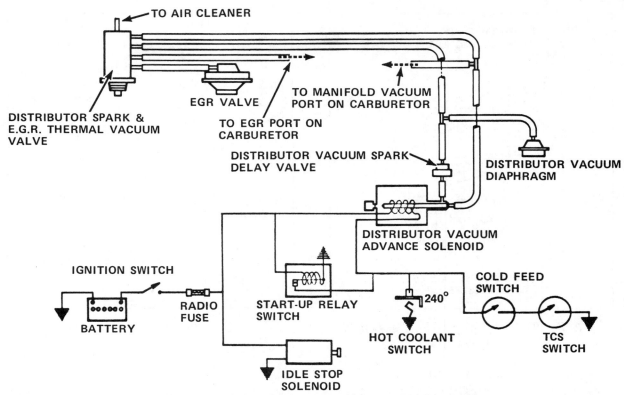

Fig. 58   1974-75 Pontiac TCS-EGR system. V8 (Typical)

ing acceleration, the VBV does not reduce EGR flow.

An EGR system change on all mid-year 1973 V8 and 1974-75 six cylinder Pontiac models (vehicles manufactured after March 15, 1973) will incorporate a new EGR Thermal Vacuum Valve which senses engine coolant temperature. This thermal valve, located between the carburetor and EGR valve, controls vacuum to the EGR valve. When coolant temperature is below 95° F. on 1973 models or 100° F. on 1974-75 models, the thermal valve isolates the EGR valve from the vacuum source, therefore eliminating exhaust gas recirculation. When coolant temperature is above 95° F. on 1973 models or 100° F. on 1974-75 models, the thermal valve opens and allows vacuum to actuate the EGR valve, allowing ported recirculation. On 1974-75 Pontiac vehicles equipped with V8 engines, the EGR Thermal Vacuum Valve and the Distributor Spark Thermal Vacuum Valve is incorporated into one assembly, Fig. 58. This Distributor Spark-EGR Thermal Vacuum Valve senses air-fuel mixture temperature. This thermal valve is located between the EGR Valve and its vacuum source. The valve prevents actuation of the EGR valve when air-fuel mixture temperature is below 62° F. When air-fuel mixture temperature exceeds 62° F., the valve opens and allows vacuum to activate the EGR valve.

GENERAL MOTORS EFE

Early Fuel Evaporation

Valve Controlled

This system is used on 1975-76 vehicles to provide a source of rapid heat for quick induction system warm-up during cold engine operation. Rapid heat is more desirable because it provides for better fuel evaporation and a more uniform mixture.

The vacuum operated EFE heat control valve mounted between the exhaust manifold and pipe, Fig. 59, directs a portion of the exhaust gasses through the intake manifold and to the base of the carburetor during engine warm-up. A thermal vacuum switch (TVS) is used to regulate the EFE valve at or below a predetermined temperature. Refer to chart.

Buick . . . . .120° F.
Cadillac
    1975-76 Exc. Seville . . . . .150° F.
    1977 Exc. Seville . . . . .120° F.

Chevrolet
    Six Cylinder . . . . .150° F.
    4-140, V8-350, 400 . . . . .180° F.

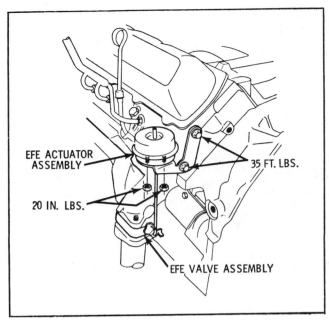

Fig. 59   *Typical EFE valve installation.*
General Motors

V8-454
   1975 ....100° F.
   1976 ....150° F.
  Oldsmobile ....120° F.
  Pontiac ....150° F.

When engine coolant (oil on 6-250) is at or below the specified temperature, the TVS will cause vacuum to flow to the EFE control valve, diverting the exhaust gasses through the intake manifold to the base of the carburetor. At temperatures above those specified, the TVS closes off vacuum to the EFE control valve, stopping the flow of exhaust gasses through the intake manifold.

Orifice Controlled

The orifice controlled EFE system is used on some 1977 Pontiac models. This system consists of an orifice restriction in one end of the exhaust crossover pipe which increases the flow of exhaust gases through the exhaust crossover under the intake manifold. This type of EFE is in effect whenever the engine is running.

CHRYSLER LEAN BURN ENGINE ELECTRONIC SPARK ADVANCE

This system, Fig. 60, used on some 1976-77 engines, allows the engine to run at an air/fuel ratio of 18 to 1. Since the engine runs lean, NOx, HC and CO emissions are reduced without the use of EGR, OSAC, air pumps, catalytic converters, vac-

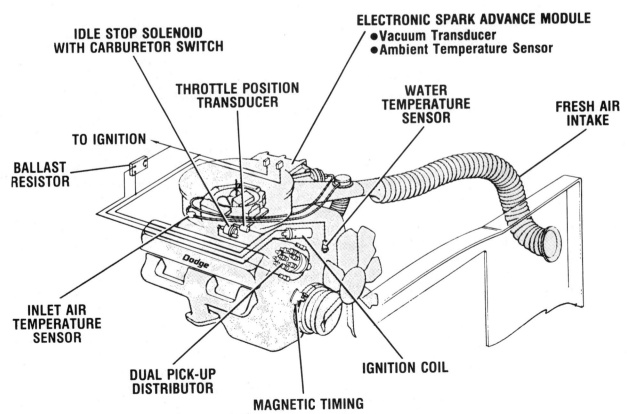

Fig. 60   *Chrysler Electronic Spark Advance System*

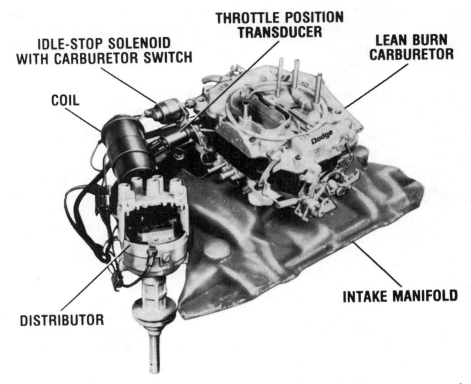

Fig. 61 Chrysler Electronic Spark Advance System solenoid locations

uum advance units, some special heat shields and all interconnecting hoses. However, for California applications, the use of catalytic converters and heat shields is retained.

Six sensors are mounted on the engine which monitor the following operating conditions:

1. Engine condition when started (Hot or cold start).
2. Engine RPM and piston position in relation to TDC.
3. Intake manifold vacuum.
4. Throttle position, Fig. 61.
5. Rate of change in throttle position.
6. Carburetor air temperature.
7. Coolant temperature.

The information from the sensors is sent to the spark control computer which calculates the precise instant the fuel charge should be fired in each cylinder to provide the best combination of driveability, fuel economy and emission control.

The computer, which is the brain of the electronic spark advance system, sends signals for each cylinder to the electronic distributor. The distributor, in turn, controls the firing of the spark plugs. The distributor incorporates two magnetic pick-ups. One pick-up is used during engine start and operation for sixty seconds after which the second pick-up is activated and the "Start" pick-

up is deactivated, regardless of operating conditions. Also, the system incorporates a "Limp-In" mode whereby, if ignition failure occurs, vehicle can still be driven into a service station.

CHRYSLER CAP

Cleaner Air Package

The Cleaner Air Package is in addition to the Positive Crankcase Ventilation (PCV). The PCV device is designed to control the emission of hydrocarbon vapors from the crankcase, whereas the Cleaner Air Package controls the emission of hydrocarbon vapors (unburned gasoline) and carbon monoxide in the vehicle exhaust.

The Cleaner Air Package is engineered to continuously control carburetion and ignition at the best settings for performance and combustion during all driving conditions. These adjustments keep unburned gasoline and carbon monoxide in the exhaust at a minimum concentration.

Only three special components are involved in the CAP installation, Fig. 62. The carburetor and distributor have been re-designed and a new component—the Vacuum Advance Control Valve —has been added. In addition, some cooling system components have been changed to handle the increased heat rejection at idle and low speed.

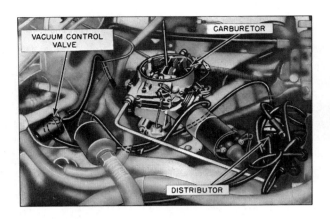

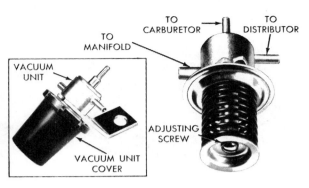

Fig. 62 Chrysler Cleaner Air Package (CAP). Unlike the others this system uses no air pump

Fig. 63 Distributor vacuum advanced control valve, deceleration valve

Special Carburetor and Ignition

The carburetor is specially calibrated to provide leaner mixtures at idle and low speed operation. The distributor is designed to give retarded timing at idle. The vacuum advance control valve, in conjunction with the distributor, provides advanced timing during deceleration.

CAP idle timing for all engines is retarded instead of advanced. Exhaust emission is reduced at idle by using leaner air/fuel mixtures, increased engine speed, and retarded ignition timing. The higher air flow at this idle condition approximates the desirable conditions of cruise. CAP, therefore, is designed to operate with late timing during idle, and with conventional spark advance during acceleration and cruise.

Vacuum Advance Control Valve, 1967-68 All; 1969 6-170 with Std. Trans. & 1969 V8-426 Hemi

Early ignition of the air/fuel mixture is needed during deceleration to provide the most efficient combustion and reduced exhaust emissions. The vacuum advance control valve provides that additional spark advance during deceleration. The vacuum advance control valve is connected by vacuum hoses to the carburetor, to the intake manifold, and to the distributor vacuum chamber, Fig. 63.

Carburetor vacuum and manifold vacuum act on the vacuum advance control valve. From these two signals, the vacuum advance control valve senses engine speed and load conditions, and relays a vacuum signal to the distributor to vary spark timing when necessary to reduce emissions to an acceptable level.

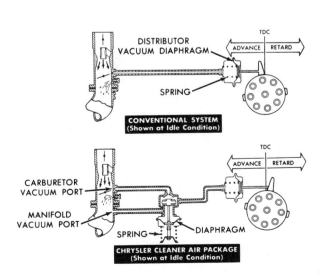

Fig. 64 Engine idle condition

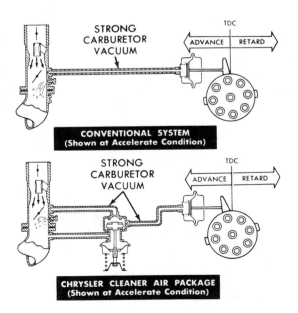

Fig. 65 During acceleration

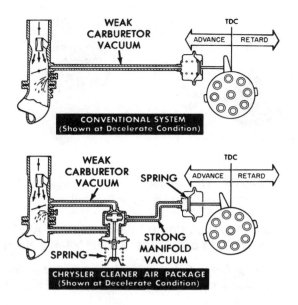

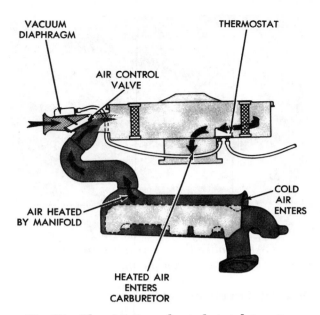

Fig. 66   During deceleration

Fig. 67   Chrysler Corp. heated air inlet system

Engine Idle, Fig. 64

The initial idle timing is retarded as much as 15 degrees from conventional timing. The vacuum advance control valve does not affect timing at idle because the distributor vacuum chamber receives the same vacuum signal as in the conventional system, namely, carburetor vacuum. At idle it is not strong enough to overcome the distributor vacuum diaphragm spring.

Manifold vacuum acts on the vacuum control valve diaphragm, but is not strong enough to overcome the vacuum control valve spring. So the spring holds the vacuum control valve closed to manifold vacuum and opens the low carburetor vacuum.

Acceleration, Fig. 65

During acceleration and cruise, manifold vacuum is not strong enough to actuate the CAP

control valve. Thus the CAP system operates in the same manner as the conventional system. The throttle blade opens enough to permit the distributor vacuum advance to function, and spark timing is advanced according to the amount of vacuum created by the pumping action of the pistons.

Deceleration, Fig. 66

The conventional system provides the highest emissions under deceleration conditions. Carburetor vacuum is too weak to overcome the distributor advance diaphragm spring.

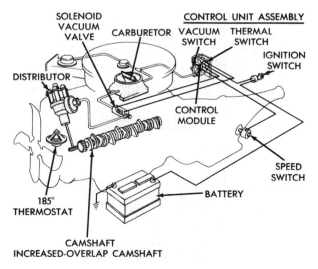

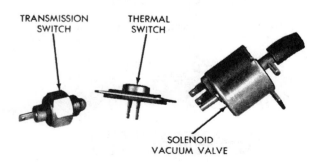

Fig. 68   NOx system components. Manual trans.

Fig. 69   NOx system. Automatic trans.

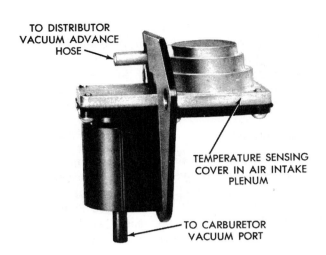

Fig. 70 Chrysler OSAC valve

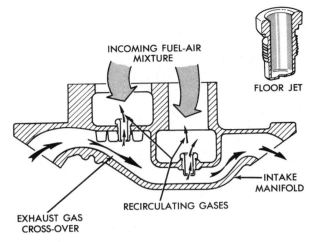

Fig. 71 EGR floor jets V8 configuration

Manifold vacuum is at its strongest under deceleration conditions. Therefore, the CAP system uses manifold vacuum rather than carburetor vacuum to control spark timing.

Manifold vacuum is strong enough to overcome the vacuum advance control valve spring and the distributor vacuum diaphragm spring, moving spark timing to the maximum advance condition.

CHRYSLER CAS

Cleaner Air System

This system uses higher inlet air temperature,

higher idle speeds, retarded ignition timing, leaner carburetor mixtures and built in modifications such as lower compression ratios, increased overlap camshafts and redesigned intake manifolds and combustion chambers.

Heated Air System, Fig. 67

This system uses a thermostatically controlled air cleaner to maintain a pre-determined air temperature entering the carburetor when underhood temperatures are less than 100° F. By maintaining this temperature, the carburetor can be calibrated leaner, improve engine warm-up and minimize carburetor icing. Temperature is controlled by

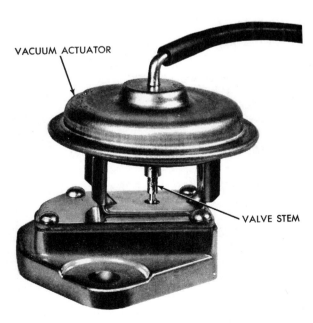

Fig. 72 Chrysler EGR control valve

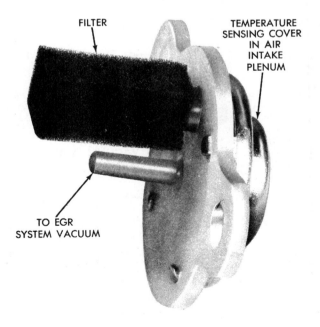

Fig. 73 Chrysler EGR temperature control valve

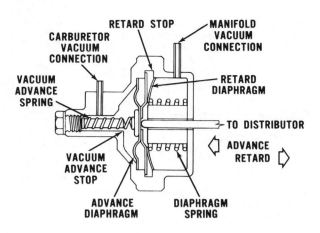

Fig. 74 Dual diaphragm vacuum advance valve

intake manifold vacuum, a temperature sensor and a vacuum diaphragm which operates the heat control door in the air cleaner snorkel.

During engine warm-up, air is heated by a shroud surrounding the exhaust manifold, then the air is piped to the air cleaner snorkel and into the carburetor. The vacuum diaphragm controls the air control valve which is closed to outside air. Therefore all air entering the carburetor is heated.

During normal operation, as the air entering the air cleaner increases, the air control valve opens to allow heated air to mix with cold air to keep the air entering the carburetor at about 100° F.

During wide-open throttle operation or at any time engine vacuum is below 4-6 inches Hg., the hot air duct is closed off allowing only cold air to enter the carburetor.

Dual Snorkel

The dual snorkel air cleaner performs at low temperature and above 105 deg. F. basically like a single snorkel air cleaner except that on deep throttle acceleration, both snorkels are open (when manifold vacuum drops between 5″ Hg).

The "non-heat" air snorkel is connected to manifold vacuum through a tee in the vacuum hose between the carburetor and the sensor.

Lower Compression Ratios

Compression ratios have been lowered by various modifications in the piston head design and in the quench height. This reduction in compression ratios permits the engine to operate satisfactorily on lower octane fuel, thereby achieving a slight reduction in HC and NOx emission levels.

Combustion Chamber Design

The combustion chamber is designed to elimi-

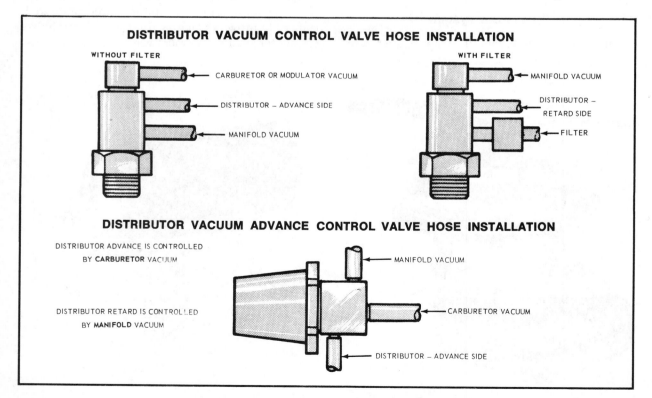

Fig. 75 Distributor vacuum control valve and distributor vacuum advance control valve

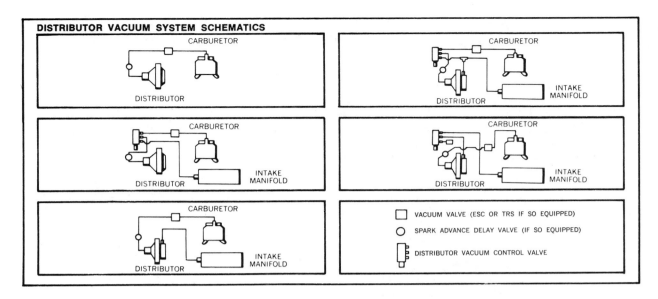

Fig. 76   Ford distributor vacuum system schematics

nate pockets and close clearance spaces which have a tendency to quench the flame before all of the air-fuel mixture is burned. By increasing the quench heights, more complete burning of the air-fuel mixture is achieved, thereby substantially reducing the HC emission levels.

Increased Overlap Camshaft

The increased valve overlap produced by the camshaft causes some dilution of the incoming air-fuel mixture. This dilution lowers the peak combustion temperature which results in lowered NOx emission levels.

Redesigned Intake Manifold

Intake manifolds have been redesigned to promote rapid fuel vaporization during engine warm-up. The exhaust crossover floor of the intake manifold between the inlet gases and exhaust gases has been thinned out with improved thickness control, thereby reducing the time required to get the heat from the exhaust manifold gases into the inlet gases. By adding this additional heat, fuel vaporizes quicker, and leaner air-fuel mixtures can be used, resulting in lower CO emission levels.

Idle Speed Solenoid

Because of the high idle speeds used on some high performance engines, these engines have an electrical solenoid throttle stop which holds the throttle at the correct idle position when energized but de-energizes when the ignition is turned off, allowing the throttle blades to close more com-

pletely, thereby eliminating the possibility of "after-run" or "dieseling."

Distributor Solenoid

Some engines have a solenoid incorporated in the distributor vacuum advance mechanism to retard the ignition timing when the throttle is closed. At closed throttle, and with the idle adjusting screw in the closed position, electrical contacts on the carburetor throttle stop cause the distributor solenoid to energize. This retards the ignition timing to reduce emissions during hot idle conditions. Cold or part throttle starting is not penalized because the distributor solenoid is not energized unless the hot idle adjusting screw is against the throttle stop contact.

NOTE: Ignition timing must be set at closed throttle to give accurate setting.

CHRYSLER NOx

Oxides of Nitrogen

The NOx system controls oxides of nitrogen emissions by allowing vacuum spark advance only in high gear (manual transmission), or above 30 mph (automatic transmission), and with the use of an increased overlap camshaft and a 185 degree F. coolant thermostat. Vacuum to the distributor is controlled by a solenoid vacuum valve mounted in the line between the carburetor vacuum port and the distributor. When the solenoid is energized the

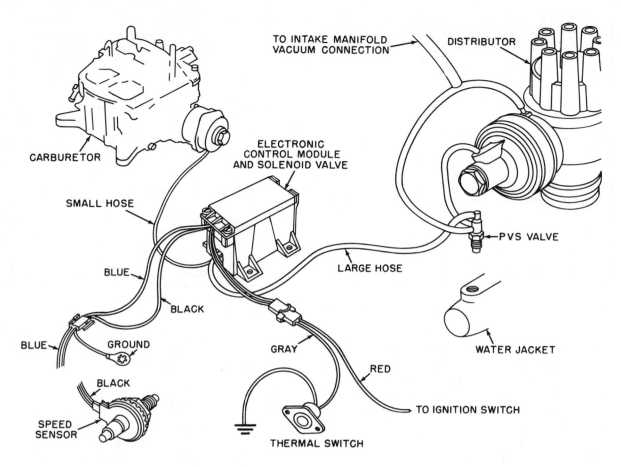

Fig. 77 Electronic distributor modulator system schematic

plunger shuts off vacuum to the distributor and vents it to the atmosphere. When it is de-energized the plunger opens allowing normal vacuum spark advance. There are two separate systems employed to control the solenoid vacuum valve, one for vehicles equipped with manual transmissions and the other for those having automatic transmissions.

Manual Transmission

The NOx system for manual transmissions consists of a solenoid vacuum valve, transmission switch and in the 1971 system a thermal switch, Fig. 68. The solenoid vacuum valve is mounted and operates as explained above. The transmission switch is mounted on the transmission housing and is used to sense the transmission gear position. It remains closed ("on") in any gear below high which energizes the solenoid thereby preventing vacuum spark advance. It opens when the top gear is selected permitting normal vacuum spark advance. The thermal switch is mounted on the firewall and senses ambient air temperature. If the temperature is below 70 degrees F., this switch will be open. This breaks the circuit be-tween the transmission switch and the solenoid valve leaving the NOx system inoperative and allowing normal vacuum spark advance in all gears.

Automatic Transmission

The NOx system for automatic transmission equipped vehicles consists of a solenoid vacuum valve, a speed switch and a control unit assembly, Fig. 69. The solenoid vacuum valve is mounted and operates as explained above. The speed switch senses the vehicle speed and is mounted in line with the speedometer cable. The control unit assembly mounts on the fire wall. It contains three parts, the control module, the thermal switch and the vacuum switch. It senses ambient temperature and manifold vacuum. These components work together for one purpose, to prevent vacuum spark advance under following conditions:

1. Temperature above 70 degrees F.
2. Speeds below 30 mph.
3. Acceleration necessary on 1971 vehicles only.

Whenever all conditions are present, the solenoid vacuum valve will be energized shutting off vacuum to the distributor.

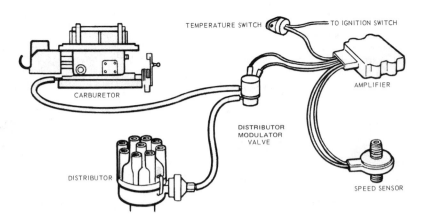

Fig. 78   Ford Electronic Spark Control system

CHRYSLER IGNITION

Electronic Ignition

A better control of exhaust emissions is achieved through the use of the "Electronic Ignition." By eliminating the breaker points, engine misfiring and increased emissions caused by worn or misadjusted breaker points is eliminated.

Distributor Solenoid, 1972-73

A start only solenoid is used on some distributors to provide additional spark advance during engine starting. The solenoid is located in the vacuum unit attached to the distributor housing and operates only while the ignition switch is in start position.

Use of the solenoid provides improved starting characteristics while maintaining a low level of hydrocarbon and carbon monoxide emissions at idle.

CHRYSLER ORIFICE SPARK ADVANCE CONTROL (OSAC)

The OSAC system, Fig. 70, is used on all 1973-77 engines, to aid in the control of NOx (Oxides of Nitrogen). The system controls the vacuum to the vacuum advance actuator of the distributor.

A tiny orifice is incorporated in the OSAC valve which delays the change in ported vacuum to the distributor by about 17 seconds when going from idle to part throttle. When going from part throttle to idle, the change in ported vacuum to the distributor will be instantaneous. The valve will only delay the ported vacuum signal when the ambient temperature is about 60° F. or above. Vacuum is obtained by a vacuum tap just above the throttle plates of the carburetor. This type of tap provides no vacuum at idle, but provides manifold vacuum as soon as the throttle plates are opened slightly. Proper operation of this valve depends on air tight fittings and hoses and on freedom from sticking or plugging due to deposits.

TEMPERATURE OPERATED VACUUM BY-PASS VALVE

This vacuum by-pass or Thermal Ignition Control (TIC) valve, Fig. 48, is used on some engine applications to reduce the possibility of engine overheating under extremely high temperature operating conditions. When engine coolant temperature at idle reaches 225° F., the valve opens automatically and applies manifold vacuum directly to the distributor for normal vacuum spark advance. This will by-pass the NOx or OSAC system. This increases engine idle speed and provides additional engine cooling. When the engine has cooled to normal operating temperature, the NOx or OSAC system is restored to normal operation.

CHRYSLER EXHAUST GAS RECIRCULATION (EGR)

In this system, exhaust gases are circulated to dilute the incoming fuel air mixture. Dilution of the incoming mixture lowers peak flame temperatures during combustion and thus limits the formation of NOx.

Floor Jet Exhaust Gas System 1972-73

In this system the exhaust gases are introduced into the intake manifold through jets in the floor below the carburetor, Fig. 71. An orifice in each jet allows a controlled amount of exhaust gas to be drawn through by engine vacuum to dilute

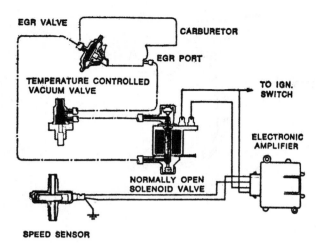

Fig. 79 Ford high speed EGR modulator
sub-system components

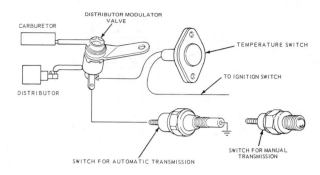

Fig. 80 Ford TRS system

incoming fuel and air. In eight cylinder engines, exhaust gases are taken from the intake manifold exhaust crossover passage. While in six cylinder engines, gases are taken from the exhaust manifold "plenum" chamber located at the "hot spot" below the carburetor riser.

NOTE: In 1973 two additional systems are used to control the rate of exhaust gas recirculation, depending on engine model. These systems are: Ported Vacuum Control System and Venturi Vacuum Control System.

Both systems use the same type exhaust gas recirculation (EGR) control valve, Fig. 72, only the method of controlling the valve is different. The valve is a vacuum actuated, poppet type unit used to modulate exhaust gas flow from the exhaust gas crossover into the incoming air fuel mixture.

Venturi Vacuum Control System

The venturi vacuum control system utilizes a vacuum tap at the throat of the carburetor venturi to provide a control signal. This vacuum signal is amplified to the level required to operate the EGR control valve. Elimination of recycle at wide open throttle is accomplished by a dump diaphragm which compares venturi and manifold vacuum to determine when wide open throttle is achieved. At wide open throttle, the internal reservoir is "dumped," limiting output to the EGR valve to manifold vacuum. The valve opening point is set above the manifold vacuums available at wide open throttle.

NOTE: This system is dependent primarily on

engine intake airflow as indicated by the venturi signal, and is also affected by intake vacuum and exhaust gas back pressure.

Ported Vacuum Control System

The ported vacuum control system utilizes a slot type port in the carburetor throttle body which is exposed to an increasing ratio of manifold vacuum as the throttle blade opens. This throttle bore port is connected through an external nipple directly to the EGR valve. The flow rate is dependent on three variables, 1) manifold vacuum, 2) throttle position, and 3) exhaust gas back pressure. Recycle at wide open throttle is eliminated by calibrating the valve opening point above manifold vacuums available at wide open throttle as port vacuum cannot exceed manifold vacuum. Elimination of wide open throttle recycle provides maximum performance.

Temperature Control Valve

The plenum mounted temperature control valve, Fig. 73, is utilized on the ported vacuum control system and the venturi vacuum control system. The valve reduces the recycle rate at low ambient temperature for improved driveability. The unit contains a temperature sensitive bimetal disc which senses plenum air temperature. The snap action of the disc unplugs a calibrated orifice to provide the bleed air. Calibration is protected by an air filter unit.

Coolant Control Exhaust Gas Recirculation (CCEGR)

1974-77 engines equipped with EGR use a CCEGR valve mounted in the radiator top tank, engine block or water pump housing. The purpose of this valve is to allow exhaust gas recirculation

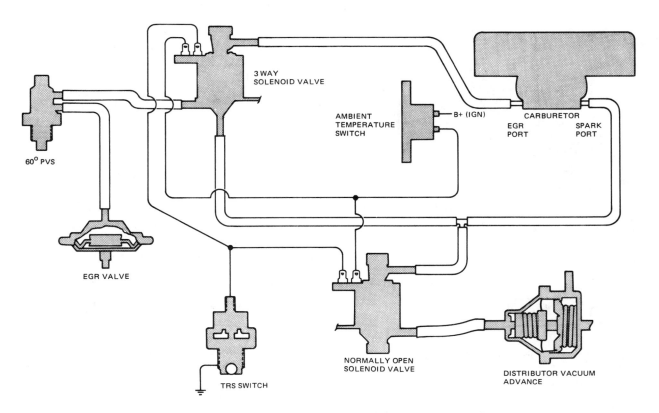

Fig. 81   Ford transmission regulated spark one system—TRS + 1

only after the engine has reached a predetermined temperature.

Coolant Controlled Engine Vacuum Switch (CCEVS)

This switch is used on 1977 vehicles to improve hot driveability by preventing operation of the Idle Enrichment System, Power Heat Valve System and Air Injection Switching System after the coolant temperature has reached a predetermined temperature. When the engine coolant temperature is below 86° F. (green), 108° F. (orange) or 138° F. (neutral), the valve opens allowing manifold vacuum to operate one or more of the systems previously mentioned.

EGR Delay System

Some 1974 vehicles are equipped with an EGR Delay System, which has an electrical timer mounted on the dash panel in the engine compartment controlling an engine mounted solenoid. This solenoid which is connected with vacuum hoses to the carburetor venturi and vacuum amplifier, prevents EGR operation for about 35 seconds after engine start up.

EGR Time Delay & Idle Enrichment System

On most 1975-77 vehicles, a time delay device is used in the EGR system which works in conjunction with the idle enrichment system to improve engine starting and performance. This system consists of an electronic timer and either one or two vacuum solenoids. The electronic timer energizes the solenoid(s) during starting and for 35 seconds thereafter. When the solenoid(s) energizes, vacuum to the EGR amplifier is cut off, thereby stopping exhaust gas recirculation. At the same time, vacuum is applied to the idle enrichment system through the solenoid and/or a block coolant valve when the engine block temperature is below 98° F. for California vehicles or 150° F. for all other vehicles. On 1975-76 Calif. and all 1977 except U8-360, idle enrichment is controlled only by engine coolant temperature. The system used on V8-318 engines except California, functions basically in the same manner except that it uses two vacuum solenoids, one controlling the EGR system, the other controlling the idle enrichment system. Thirty-five seconds after engine start-up, the solenoid(s) de-energizes, allowing normal EGR operation and stopping idle enrichment.

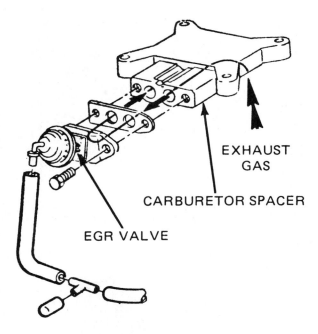

Fig. 82   Ford V8 EGR valve hook up

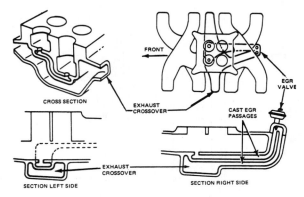

Fig. 83   Ford floor entry EGR system

EGR Maintenance Reminder System

This reminder system on 1975-77 vehicles uses a warning light in the instrument cluster, to alert the driver of the vehicle at 15,000 mile intervals to have the EGR system checked for the following:

1. Vacuum hose operation
2. EGR valve operation
3. CCEGR valve operation
4. Timer and solenoid operation
5. Intake manifold and EGR valve flow passages

Once the EGR system has been properly serviced, the reminder system must be reset for the next 15,000 mile interval and to turn off the warning light. If the EGR switch has to be replaced before a 15,000 mile interval, the EGR system maintenance check must be performed and the new switch installed.

FORD IMCO SYSTEM

Improved Exhaust Emission Control System

This system combines a thermostatically controlled air cleaner and higher engine operating temperature with leaner carburetor calibration and later ignition timing under closed throttle operating conditions.

Dual Diaphragm Distributor

In addition to the conventional centrifugal and vacuum advance control units, the unit, Fig. 74, uses a separate diaphragm to retard the spark timing under closed throttle conditions. The advance diaphragm is connected to the carburetor above the throttle plate(s) so that when the throttle is opened, the timing is advanced. The retard diaphragm is connected to the intake manifold so that during closed throttle operation, when manifold vacuum is high, the timing is retarded to provide more complete combustion.

Distributor Vacuum Control Valve

This valve, Fig. 75, is exposed to cooling system temperature and when coolant temperature exceeds normal limits during long idle periods, the valve opens a vacuum passage to the advance diaphragm of the distributor which speeds up the engine idle lowering the temperature.

Distributor Vacuum Advance Control Valve

This valve, Figs. 63 and 75, provides the necessary ignition advance during acceleration periods to provide the most efficient combustion and reduce emissions.

NOTE: Application of the distributor vacuum control valve and the dual-diaphragm vacuum advance mechanism will vary from vehicle to vehicle, Fig. 76.

FORD ELECTRONIC DISTRIBUTOR MODULATOR

Description

The system operates to prevent spark advance below 23 mph on acceleration and below approximately 18 mph on deceleration. Control by the modulator is canceled out if the outside air temperature is below 58 deg. F., allowing the distributor to operate through the standard vacuum control system, Fig. 77.

The modular system consists of four components: speed sensor, thermal switch, and electrical control amplifier-solenoid valve. The control amplifier and solenoid valve are combined in one assembly and mounted in the passenger compartment on the dash panel. The speed sensor is connected to the speedometer cable. The thermal switch is mounted near the front door hinge pillar on the outside of the cowl panel. It may be mounted on either the right or left side.

FORD ESC SYSTEM

Electronic Spark Control System

This system, Fig. 78, reduces the exhaust emissions of an engine by providing vacuum spark advance only at speeds above 24 to 33 mph (depending on the engine application). It consists of a speed sensor, an electronic amplifier, an outside air temperature switch and a vacuum control valve. The vacuum control valve is inserted between the carburetor vacuum advance port and the distributor primary advance connection. This valve is normally open, but when energized electrically by the electronic amplifier it closes to cut off vacuum to the primary vacuum advance unit on the distributor thus preventing vacuum spark advance. The temperature switch, which is mounted in either the right or left A-pillar, senses outside air temperature. A temperature below 49 degrees F. will cause the switch contacts to open, thereby de-energizing the vacuum valve and allowing normal vacuum advance at all speeds. A temperature of 60 degrees plus or minus 5 degrees F. causes the contacts to close, thereby cutting off vacuum to the advance side of the distributor at speeds below 24 to 33 mph. On deceleration the vacuum advance cut-out speed is approximately 18 mph.

On some applications the vacuum hose connections between the carburetor and distributor may route through a PVS valve. This valve serves as a by-pass or safety override switch. When the coolant temperature reaches 230 degrees F., manifold vacuum is applied directly to the primary (advance) side of the distributor advancing the timing and thereby lowering operating temperature.

FORD HIGH SPEED EGR MODULATOR SUB-SYSTEM

The high speed EGR modulator sub-system

used on some V-8 engines, Fig. 79, is basically the same in operation as the ESC system described previously. This system cuts off exhaust gas recirculation flow by stopping vacuum flow from the EGR port to the EGR valve at speeds above 64 mph, in turn improving driveability.

The vacuum solenoid valve installed in the vacuum line is normally open (not energized), allowing vacuum flow from the EGR port to the EGR valve. The EGR system remains functional when the valve is not energized.

The speed sensor driven by the speedometer cable, produces an electric signal directly proportional to vehicle road speed, signalling the amplifier to energize the vacuum solenoid valve at which time the electronic module receives the signal from the speed sensor and amplifies it to provide a usable signal to the vacuum solenoid valve.

When the vehicle speed exceeds approximately 64 mph (trigger speed of the amplifier) the circuit to the ignition switch is completed and the normally open vacuum solenoid valve is energized. The plunger moves upwards and shuts off the EGR port vacuum and the vent at the bottom of the vacuum valve is opened, bleeding vacuum from the EGR valve and hose. Spring force closes the EGR valve which remains non functional until the vacuum solenoid valve is de-energized, at speeds below approximately 64 mph.

NOTE: There is a continuous internal vacuum bleed provided by the vent at the top of EGR valve. Whether the valve is in a closed or open position, this vent purges the vacuum supply hose from carburetor of any gasoline vapor.

FORD TRS SYSTEM

Transmission Regulated Spark Control System

This system, Fig. 80, reduces the exhaust emissions of an engine by providing vacuum spark advance only in high gear. It consists of a vacuum control valve, an outside air temperature switch, and a transmission switch. The vacuum control valve is inserted between the carburetor vacuum advance port and the distributor primary advance connection. This valve is normally open, but when energized electrically by the transmission switch it closes to cut off vacuum to the primary vacuum advance unit on the distributor thus preventing vacuum spark advance. The temperature switch,

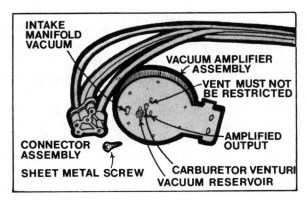

Fig. 84   Venturi vacuum amplifier
(single connector type)

which is mounted in either the right or left A-pillar, senses outside air temperature. A temperature below 49 degrees F. will cause the switch contacts to open, thereby de-energizing the vacuum valve and allowing normal vacuum advance in all gears. A temperature of 60 degrees plus or minus 5 degrees F. causes the contacts to close, thereby cutting off vacuum to the advance side of the distributor in all but high gear.

FORD TRS +1

Transmission Regulated Spark +1

The TRS +1 system, Fig. 81, consists of two separate vacuum control systems, that are electrically controlled by input information from a manual transmission gear selector switch, and an outside ambient air temperature switch. The TRS function of the TRS +1 system is identical to the function performed by the 1972 TRS system. The plus 1 system of the TRS +1, controls the selection of the carburetor vacuum source for the vehicle EGR system. The EGR vacuum supply source can be either carburetor spark port, or carburetor EGR port depending upon the manual transmission gear selected, and the outside ambient air temperature.

FORD SPARK DELAY VALVE

This unit is used in conjunction with some of the other Ford systems. Its purpose is to further reduce emissions by delaying the spark advance during rapid acceleration and by cutting off advance immediately upon deceleration.

This plastic disc-shaped valve is installed in the carburetor vacuum line at the distributor advance diaphragm. It is a one way valve and will not operate if installed backwards. The back side of the valve must be toward the carburetor. This

AMPLIFIER CODED FOR PORT CONNECTIONS:

O – OUTPUT TO EGR VALVE   V – VENTURI VACUUM
R – FROM RESERVOIR   A – ATMOSPHERE (VENT)
S – VACUUM SOURCE
    (SPARK OR EGR PORT)

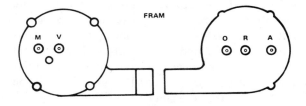

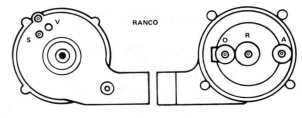

Fig. 85   Venturi vacuum amplifier
(dual connector type)

valve cannot be repaired or checked for proper operation. However, this valve can be checked for proper operation on 1975 vehicles.

NOTE: On all systems which employ the dual diaphragm distributor the line which has high vacuum at idle (normal operating temperature) is connected to the secondary (retard) side of the distributor vacuum advance unit. This is the connection closest to the distributor cap.

FORD EGR SYSTEM

Exhaust Gas Recirculation

In this system the exhaust gases are metered through the EGR valve to a passage in the carburetor spacer to dilute the air fuel mixture entering the combustion chambers. Dilution of the incoming mixture lowers peak flame temperatures during combustion and thus limits the formation of nitrogen oxides (NOx).

Most eight cylinder engines use the "Spacer

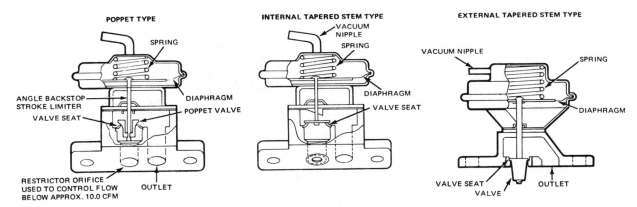

Fig. 86   Ford EGR valves

Entry" EGR System which has the EGR valve mounted on the rear of the carburetor spacer, Fig. 82. The exhaust gases are taken from a drilled passage in the exhaust crossover of the intake manifold. The exhaust gas is then routed through a metered EGR valve to a passage in the carburetor spacer and fed into the primary bore. Some 1974 eight cylinder engines use the "Floor Entry" EGR system, which has the EGR valve mounted on the rear of the intake manifold. The EGR valve controls the exhaust gases that enter specially cast passages in the manifold from the exhaust crossover passage. When the valve opens, the exhaust crossover is then opened to the two drilled passages in the floor of the intake manifold riser under the carburetor, Fig. 83.

On six cylinder engines, the EGR system is basically the same as the Spacer Entry EGR Sys-

tem except that exhaust gas is routed directly from the exhaust manifold.

Two variables control the operation of the EGR system, 1) engine coolant temperature and 2) carburetor vacuum. When engine coolant temperature is below the specified level the EGR system is locked out by a temperature controlled vacuum switch. This vacuum switch is installed in series with the EGR valve. This valve receives vacuum from a port in the carburetor body. When the valve is closed due to lower coolant temperature, no vacuum is applied to the EGR valve and no exhaust gas is fed to the air-fuel mixture. When the engine coolant temperature reaches the specified level, the valve opens allowing vacuum to be applied to the EGR valve. Exhaust gas is then fed to the air-fuel mixture.

The second factor controlling EGR operation is

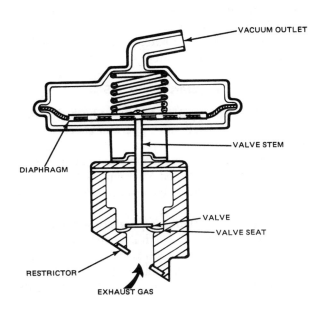

Fig. 87   Ford poppet-type EGR valve

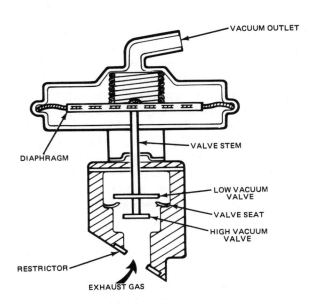

Fig. 88   Ford modulating type EGR valve

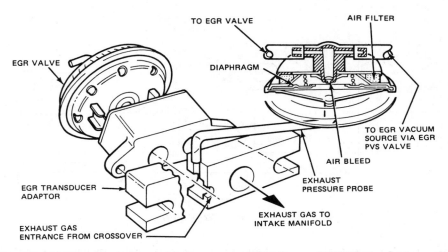

Fig. 89   EGR valve exhaust back pressure transducer. 1976-77 Ford Motor Co.

carburetor vacuum. The location of the EGR port in the carburetor determines at what point vacuum is sent to the EGR valve. Vacuum should be fed to the EGR vacuum control valve when the primary throttle plate reaches a position corresponding to a road speed of approximately 20 mph under light acceleration.

A Venturi Vacuum Amplifier, Figs. 84 and 85 used in 1974-77, uses a weak venturi-vacuum signal to produce a strong intake manifold vacuum to operate the EGR valve, thereby achieving an accurate, repeatable and almost exact proportion between venturi airflow and EGR flow. This assists in controlling oxides of nitrogen with minimal sacrifice in driveability.

There are three types of EGR valves, the poppet type, modulating type and the tapered stem type, Fig. 86.

The poppet type valve, Fig. 87, consists of springloaded diaphragm, and a valve stem and valve operating in an enclosed valve body. At approximately 3 inch Hg of vacuum, the valve begins to open. The valve stem is pulled forward unseating the valve and allowing the exhaust gas to flow into the valve chamber. Venturi vacuum will then pull the gas from the chamber into the air-fuel flow and then into the combustion chambers. Once the valve has been unseated the only means of limiting exhaust gas flow is the size of the flow restrictor placed in the inlet port of the valve body. The size of the restrictor will vary according to engine application.

On the modulating type valve, Fig. 88, an additional disc has been added to the valve stem below the main valve. The modulating valve operates exactly like the poppet valve when vacuum is between approximately 3 in. Hg. and 10.5 in. Hg.

When vacuum reaches approximately 10.5 inches, the lower disc (high vacuum flow restrictor) approaches the shoulders of the valve seat and restricts the flow of exhaust gas. The purpose of the modulation of gas flow is to improve driveability on certain engine models.

NOTE: The EGR valve and vacuum control valve cannot be repaired and must be replaced if damaged.

EGR Back Pressure Transducer

Used on some 1976 EGR systems, the back pressure transducer is connected to an adapter between the EGR valve and intake manifold. The transducer modulates EGR flow by varying vacuum signal to EGR valve according to exhaust back pressure. Exhaust back pressure is sensed in pressure cavity of transducer spacer, Fig. 89.

Cold Start Cycle (CSC)

The EGR/CSC system regulates both distributor advance and EGR valve operation according to coolant temperature by sequentially switching vacuum signals. The major system components are: a 95° F. EGR-PVS (Ported Vacuum Switch) valve, a SDV (Spark Delay Valve) and a vacuum check valve, Fig. 90.

When engine coolant temperature is below 82° F., the EGR-PVS valve admits carburetor EGR port vacuum (at about 2500 RPM) directly to the distributor advance diaphragm, through the one way check valve. At the same time, the EGR-PVS valve shuts off carburetor EGR vacuum to the EGR valve and transmission diaphragm.

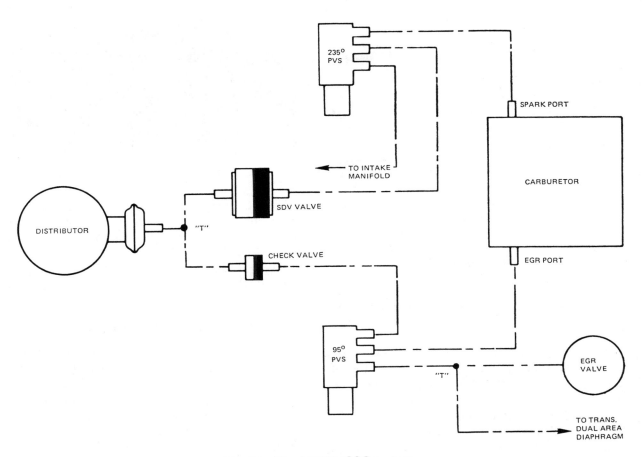

Fig. 90   Ford EGR/CSC system

When engine coolant temperature is 95° F. and above, the EGR-PVS valve is actuated and directs carburetor EGR vacuum to the EGR valve and transmission diaphragm instead of the distributor. At temperatures between 82° and 95° F., the EGR-PVS valve may be open, closed or in mid position.

The Spark Delay Valve (SDV) delays carburetor vacuum to the distributor advance by restricting the vacuum signal through the SDV for a predetermined time. During normal acceleration, little or no vacuum is admitted to the distributor advance diaphragm until acceleration is completed and engine coolant temperature is 95° F. or higher.

The check valve blocks off vacuum signal from the SDV to the EGR-PVS valve so that carburetor spark vacuum will not be dissipated when the EGR-PVS valve is actuated above 95° F.

The 235° F. PVS valve which is not part of the EGR-PVS system is connected to the distributor vacuum advance to prevent engine overheating while idling.

Cold Start Spark Advance (CSSA)

Used on some 1975-77 engines, this system, Fig. 91, is added to the distributor control to provide

manifold vacuum when engine coolant is below 125° F. by providing vacuum from the intake manifold, through the Distributor Retard Control Valve (DRCV), the CSSA PVS and to the distributor. At 125° F. and above, the vacuum flows from the carburetor spark port, through the cooling PVS, the SDV, the CSSA PVS and to the distributor. Above 235° F., the vacuum flows through the SDV and CSSA PVS to the distributor. If engine overheats at idle, increased vacuum will flow to the distributor to increase engine speed. When engine coolant temperature decreases, spark advance will be controlled by the carburetor spark port.

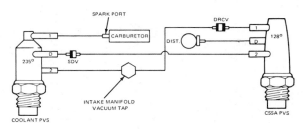

Fig. 91   Cold Start Spark Advance System. 1975-77 Ford Motor Co.

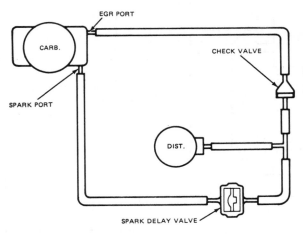

Fig. 92   Dual Signal Spark Advance (DSSA)
system

Dual Signal Spark Advance System (DSSA)

1975-76

This system uses a spark delay valve (SDV) and a one way check valve to provide improved spark and EGR performance during mild acceleration, Fig. 92. The check valve prevents spark port vacuum from reaching EGR and causing excessive EGR flow. The valve also prevents EGR port vacuum from diluting spark port vacuum which could result in improper spark advance. The SDV permits application of full EGR vacuum to distributor vacuum advance during mild acceleration. During cruise conditions, EGR port vacuum is applied to EGR valve and spark port vacuum is applied to distributor vacuum advance.

DUAL-AREA DIAPHRAGM

On 1973-77 vehicles, new dual-area diaphragms are used, Fig. 93. These diaphragms offset effects of engines using the EGR system and equipped with automatic transmissions. The new diaphragms permit vehicles to function with satisfactory shift spacing and shift feel.

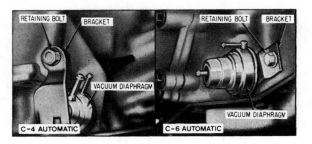

Fig. 93   Dual diaphragm vacuum modulator

To test, remove the vacuum diaphragm and test unit using an outside vacuum source. Set regulator on tester to 18 in. Hg. with end of vacuum hose blocked off then connect vacuum hose to vacuum diaphragm unit. If unit does not hold 18 in. Hg. reading, the diaphragm is leaking and must be replaced.

FORD DECEL VALVE

This valve, Fig. 94, used on the 1600cc, 2000cc, 2300cc and 2800cc engines, is mounted on the intake manifold adjacent to the carburetor and meters an additional amount of fuel and air during engine deceleration periods. This additional fuel and air, together with engine modifications, permits more complete combustion with the resultant being lower levels of exhaust emissions. During engine deceleration, manifold vacuum forces the diaphragm assembly against the spring in the decel valve, which in turn raises the decel valve (open position). With the valve open, existing manifold vacuum pulls a metered amount of fuel and air from the carburetor, which travels through the decel valve body assembly into the intake manifold. The decel valve remains open

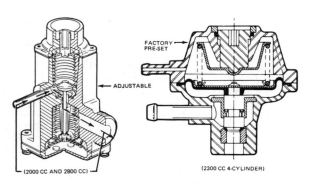

Fig. 94   Ford Decel Valve

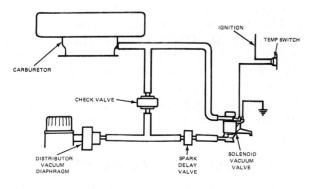

Fig. 95   Ford Delay Vacuum By-Pass (DVB) System

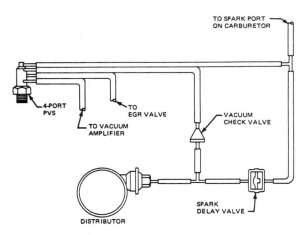

*Fig. 96   Delay vacuum by-pass (DVB) system.
1975-76 Mord Motor Co.*

FORD DVB SYSTEM

Delay Vacuum By-Pass System

1973

This system is used to by-pass spark delay valve (SDV) at temperatures below 49° F. The system incorporates two components of TRS system (ambient temperature switch and electrically operated vacuum control solenoid valve) and a check valve. The DVB system does not operate at temperatures above 65° F., thus the normal function of SDV will not be effected in warm weather, Fig. 95.

1975-76

This system allows full spark port vacuum to distributor vacuum advance below PVS opening temperature. The system uses a check valve, spark delay valve (SDV) and 4 port PVS, Fig. 96. Above PVS opening temperature top two ports of PVS close, routing spark port vacuum signal through spark delay valve.

and continues to feed additional air and fuel for a specified time.

NOTE: On 1975-76 V6 Mustang models, the fuel decel system is controlled by vehicle speed through a transmission mounted switch. This switch prevents the system from operating during deceleration at speeds below 11 mph. On 1977 Mustang models with air conditioning, the fuel decel system is designed to operate only when the air conditioner is operating.

FORD TAV SYSTEM

Temperature Activated Vacuum

This system selects either the carburetor spark

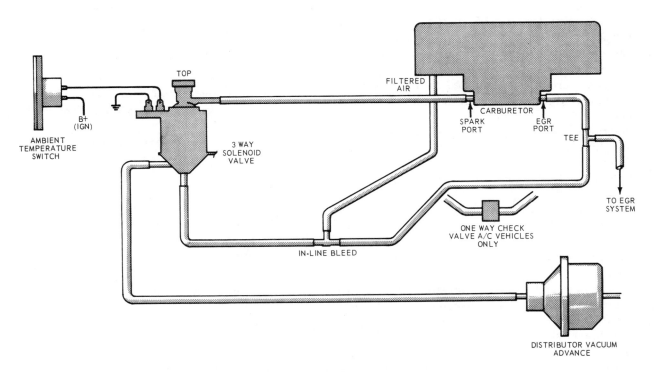

Fig. 97   Ford Temperature Activated Vacuum system—TAV

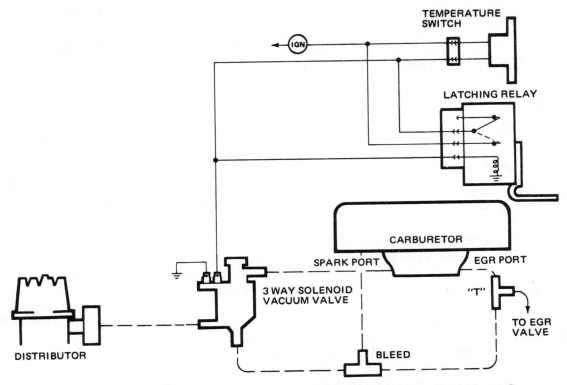

Fig. 98   *Cold Temperature Activated Vacuum (CTAV) System. Ford Motor Co.*

port vacuum, or the carburetor EGR port vacuum as a function of outside ambient air temperature.

The EGR system can be used in addition to the TAV system, although systems work independently of each other.

The TAV system, Fig. 97, consists of an ambient temperature switch, a three-way vacuum valve, and an external inline vacuum bleed. The three-way vacuum valve is used to select the carburetor vacuum source that is supplied to the distributor vacuum advance mechanism. The ambient temperature switch provides the switching circuit to determine which vacuum source will be selected as a function of outside air temperature. The inline vacuum bleed function is to purge the vacuum line in the TAV system of any excessive gasoline vapors.

The basic difference between a TAV system and the standard IMCO system is the selective control feature provided by TAV system for distributor vacuum advance as a function of outside air temperature.

When the ambient air temperature is above 60 degrees F. the three-way vacuum valve is energized, therefore the EGR vacuum is controlling the distributor advance. When the ambient air temperature is below 49 degrees F. the three-way vacuum valve is de-energized, therefore the spark port vacuum is controlling the distributor advance.

NOTE: The TAV system controls spark advance below 49 degrees F. while the EGR system controls spark advance above 60 degrees F.

FORD CTAV SYSTEM

Cold Temperature Activated Vacuum

This system operates basically the same as the TAV system previously discussed except that a latching relay, Fig. 98, has been added. The latch-

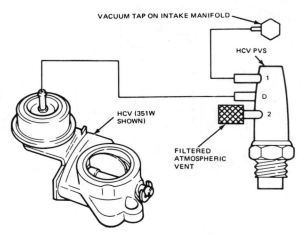

Fig. 99   *Heat Control Valve (HCV) System. 1975-76 Ford Motor Co.*

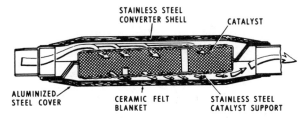

Fig. 100   *American Motors & General Motors catalytic converter*

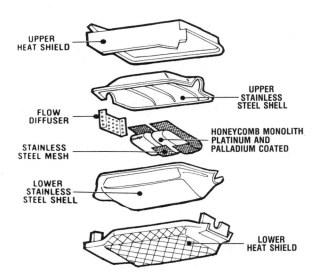

Fig. 101   *Chrysler Corp. catalytic converter*

ing relay, activated by temperature switch closing remains energized regardless of temperature switch position which prevents system cycling due to minor ambient temperature changes.

The temperature switch energizes the three-way vacuum valve and latching relay when ambient temperature is above 65° F. When ambient temperature is below 49° F., the system is inoperative and the distributor diaphragm and EGR valve receives vacuum directly from its respective carburetor ports.

FORD HCV SYSTEM
Vacuum Operated Exhaust Heat Control Valve

Used on all 1975 vehicles except those equipped with the 2300cc and V8-460 engines and some 1976 models this system, Fig. 99, provides quick induction system warm-up for better cold engine fuel vaporization.

A vacuum operated heat control valve mounted between the exhaust manifold and pipe, directs a portion of the exhaust gases through the intake manifold during engine warm-up. On cold starts, manifold vacuum is directed to the heat control valve (HCV) through the top two ports in HCV PVS (ported vacuum switch), closing the HCV. When engine coolant temperature reaches a predetermined value, the PVS closes off vacuum and

vents the PVS allowing the HCV to close under spring tension. The three PVS valves used may be identified as follows:

PVS Body Color	Opening Temp. (° F.)
Black	92-98
Blue	125-131
Purple	157-163

CATALYTIC CONVERTERS

The catalytic converter serves two purposes: it permits a faster chemical reaction to take place and although it enters into the chemical reaction, it remains unchanged, ready to repeat the process. The catalytic converter combines hydrocarbons (HC) and carbon monoxide (CO) with oxygen to form water (H_2O) and carbon dioxide (CO_2).

The catalyst is structured in the form of pellets, Fig. 100 (American Motors and General Motors),

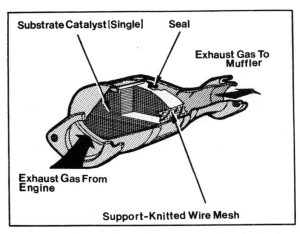

Fig. 102   *Ford catalytic converter with single substrate catalyst*

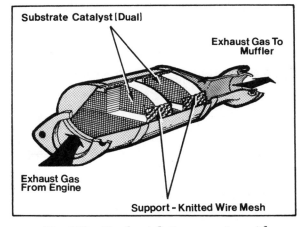

Fig. 103   *Ford catalytic converter with dual substrate catalyst*

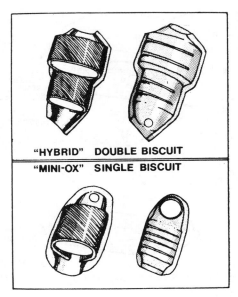

"HYBRID" DOUBLE BISCUIT

"MINI-OX" SINGLE BISCUIT

*Fig. 104   Chrysler Corp. "Hybrid" & "Mini-Ox"
catalytic converters*

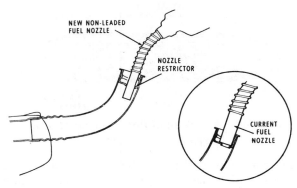

NEW NON-LEADED
FUEL NOZZLE

NOZZLE
RESTRICTOR

CURRENT
FUEL
NOZZLE

*Fig. 105   Fuel tank filler safety neck for all
vehicles equipped with catalytic converters
(typical)*

or a honeycomb monolithic composition, Fig. 101 (Chrysler Corp.) and Figs. 102 and 103 (Ford). The catalyst consists of a porous substrate of an inert material, coated with platinum and other noble metals, the catalytically active materials.

1977 Chrysler Corp. vehicles use two mini-oxidation catalytic converters in conjunction with the main underfloor converter. Their main purpose is to initiate exhaust gas oxidation before the gases reach the underfloor converter. The "mini-ox", Fig. 104, is a single biscuit catalytic converter and is shaped so that only one small biscuit can fit inside the can. The "hybrid" converter, Fig. 104, utilizes a large biscuit and a small biscuit. A Power Heat Control Valve is used to increase the flow of exhaust gases through the left-hand exhaust manifold to rapidly bring the mini-catalyst up to operating temperature.

1977 American Motors California models use monolithic type warm up converters located ahead of the pellet type converter. Vehicles with 6 cylinder engines use one warm up converter and vehicles with 8 cylinder engines use two warm up converters.

This device, located in the exhaust system between the exhaust manifold and muffler, requires the use of heat shields, in some cases, due to its high operating temperatures. The heat shields are necessary to protect chassis components, passenger compartment and other areas from heat related damage.

A smaller fuel filler tube neck is incorporated to prevent the larger service station pump nozzle, used for leaded fuels, from being inserted into the filler tube, thereby preventing system contamination, Fig. 105.

CAUTION: Since the use of leaded fuels contaminates the catalysts, deteriorating its effectiveness, the use of unleaded fuels is mandatory in vehicles equipped with catalytic converters. The catalytic converter can tolerate small amounts of leaded fuels without permanently reducing the catalyst's effectiveness.

Most Chrysler Corp. vehicles equipped with catalytic converters have a catalyst protection system, Fig. 106. This system maintains engine idle above 2000 rpm during certain conditions to protect the catalyst from high temperatures. Also this system prevents complete throttle closure at high engine rpm, in turn reducing the catalyst temperature under deceleration conditions.

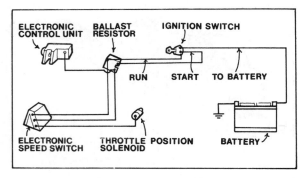

ELECTRONIC
CONTROL UNIT

BALLAST
RESISTOR

IGNITION SWITCH

RUN   START   TO BATTERY

ELECTRONIC
SPEED SWITCH

THROTTLE POSITION
SOLENOID

BATTERY

*Fig. 106   Chrysler Corp. catalyst protection
system*

Fuel Evaporative Emission Controls

CAUTION: The utmost care should be exercised when using a torch in the area of the fuel evaporation system as an open flame near these hoses may cause a fire and ultimate explosion.

Also installation of a fill cap from a non-emission fuel tank will render the system inoperative, since the non-emission fill cap is vented and the system must be sealed to function properly. Also if a non-vented fill cap is installed on a conventional tank, the result will be a serious deformation or a total collapse of the fuel tank.

NOTE: Vapor line hoses used in these systems are made from a special rubber material. Bulk service hoses are available for service and will be marked "EVAP." Ordinary fuel hoses should not be used as they are subject to deterioration and may clog system.

Two different systems are used to accomplish a reduction of evaporative losses to less than six gallons a day. 1971-77 American Motors, 1972-77 Chrysler and all General Motors store the fuel vapors in a carbon-filled canister before burning them in the engine. 1967-71 Chrysler and 1967-70 American Motors store the vapors in the crankcase. Ford fits refined versions of both systems.

On pre 1971 American Motors vent system, fuel vapors are routed from the tank, through a check valve and connecting lines to the valve cover of the engine, Fig. 1. On V8 engines, the line is routed to the left valve cover. Vapors are then drawn into the PCV system and burned along with the normal air-fuel mixture. Most fuel tanks incorporate an integral fuel expansion tank to provide an air displacement area for normal fuel expansion, unless the tank itself is designed to provide an adequate air displacement area for fuel expansion.

A check valve is included to prevent the flow of liquid fuel to the valve cover under all operating conditions.

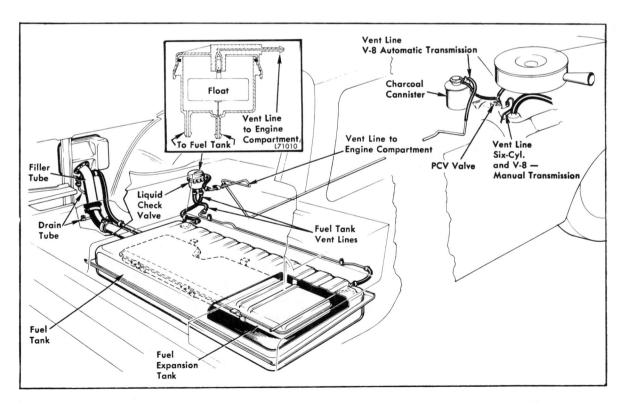

Fig. 1   Evaporative emission control system with charcoal canister. Typical American Motors (up to 1976)

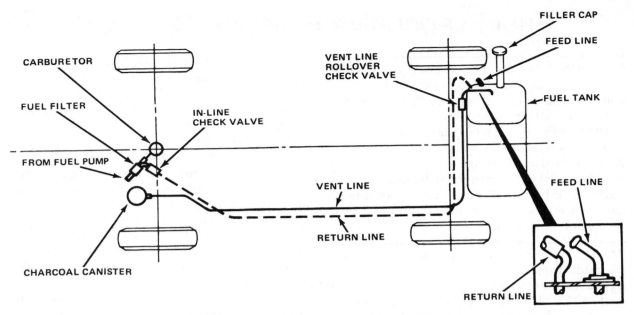

Fig. 2   Evaporative emission control system. Typical 1977 American Motors

The 1971-77 engines incorporate a "Fuel Vapor Storage (Charcoal) Canister," Figs. 1 and 2. This canister contains activated charcoal granules which absorb and store the fuel tank vapors until they are drawn into the intake manifold through the PCV system on six cylinder engines or the carburetor air cleaner on V8 engines.

On fuel tanks without a fuel expansion tank, a liquid check valve, Fig. 3, is used to prevent liquid fuel from entering the vapor lines.

Two rollover check valves are used on 1977 vehicles to prevent fuel spillage in case of a rollover accident. The check valve located in the vent line, Fig. 4, has a steel ball loose within its guides which drops to seat a plunger when the unit is inverted. The check valve located in the fuel filler cap has a steel ball in a plastic housing which drops in its orifice when the vehicle is tipped sufficiently.

The filler cap includes a two-way relief valve which is closed to atmosphere under normal op-

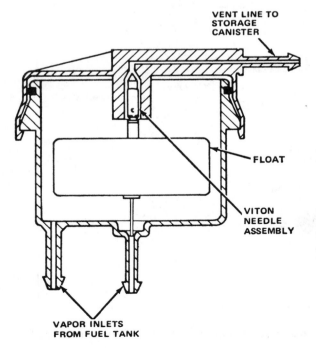

Fig. 3   Liquid check valve. American Motors

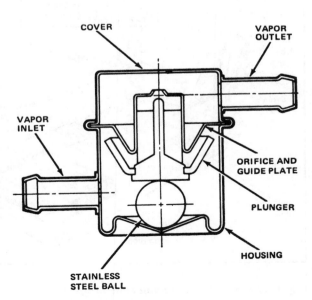

Fig. 4   Rollover check valve. American Motors

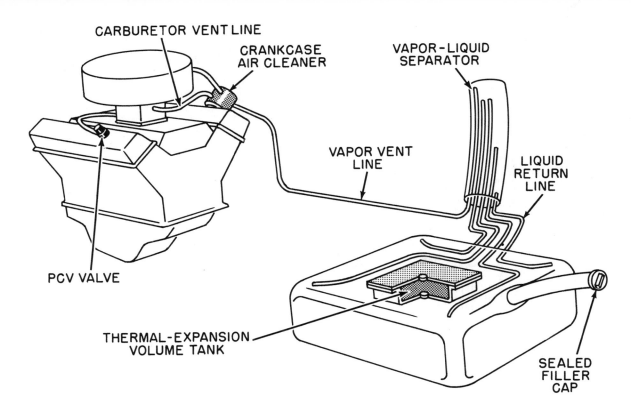

Fig. 5   Evaporative emission control system. Chrysler Corp., typical 1967-71

erating conditions and opens only when an abnormal pressure or vacuum develops within the tank. It is normal to occasionally encounter an air pressure release when removing the filler cap.

In the 1967-71 Chrysler system, fuel vapors from both the sealed gasoline tank and the carburetor are fed to the crankcase rather than the canister, Fig. 5. When the engine is running, the vapors are purged to the intake manifold through the existing crankcase ventilation system.

Chrysler's liquid-vapor separator is a vertically mounted 2-in. diameter steel tube containing four tank vent lines and a line to the crankcase. Line heights are different, with the crankcase vent being the highest. This vent also has a small orifice to further inhibit liquid fuel transfer.

In the 1972-77 Chrysler system, Figs. 6 thru 10, when fuel tank is filled to the base of the filler tube, vapors can no longer escape, they become trapped above the fuel. Vapor flow through the vent line is blocked by the limiting valve; and the filler tube is blocked by fuel preventing more fuel to enter the tank. At any time pressures in the tank rise above operating pressure of the limiting valve, about ½ psi, the valve opens and allows vapors to flow forward to the charcoal canister. Due to the configuration of the fuel tank on some models and all station wagons, vapor separator tanks are not required. The charcoal canister is a feature on all

models for the storage of fuel vapors from the fuel tank and carburetor bowl. A vacuum port located in the base of the carburetor governs vapor flow to the engine. On some models, each corner of the fuel tank is vented and each of the hoses from these vents is connected to a vapor separator. A tube from the separator leads to the charcoal canister. Evaporated fuel vapor from the fuel tank, flows through the separator to the canister. The canister used in 1973-77 vehicles will have three hoses and no purge valve. The purge valve previously located on top of the canister has been eliminated by using an additional ported vacuum connection on the carburetor for purging the canister. This utilizes the throttle plates of the carburetor as purge valve. This system will improve hot idle quality by eliminating canister purging during idle. Some limited production, High Performance vehicles will continue to use the earlier type two stage canister which utilizes an integral purge valve. This canister can be identified by four hose connections while the new type canister uses only three. In order to meet Federal Safety Standards, a rollover check valve is used on 1976-77 vehicles to prevent fuel spillage in the event of an accidental rollover.

Ford refines the ECS designs by fitting a 3-way valve between the separator and the vapor storage, Figs. 11 and 12. Fuel vapor is transported either to

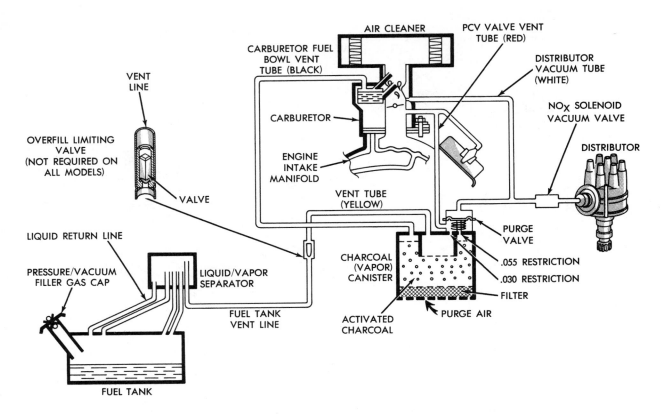

Fig. 6   Evaporative emission control system. 1972 Chrysler Corp., (typical)

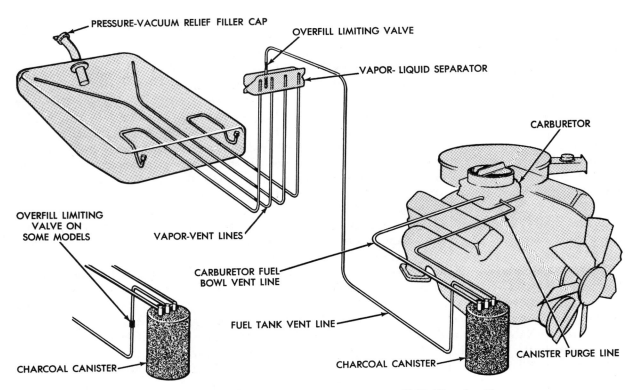

Fig. 7   Evaporative emission control system. 1973 Chrysler Corp.

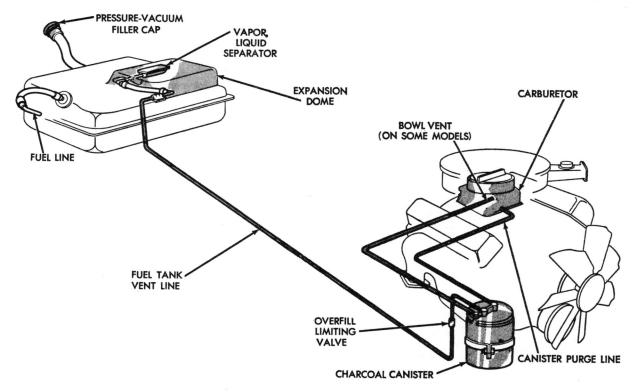

Fig. 8 Evaporative emission control system. 1974-75 Chrysler Corp.

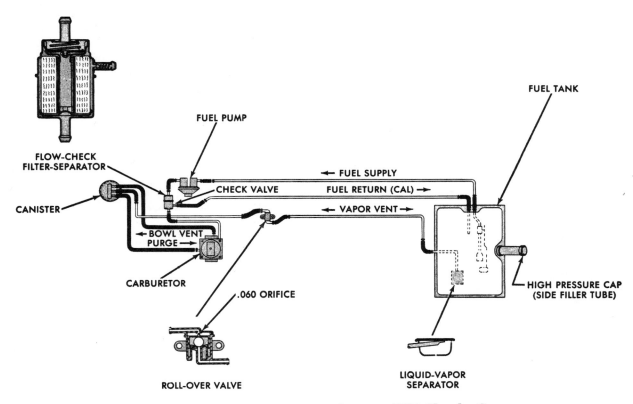

Fig. 9 Evaporative emission control system. 1976 Chrysler Corp.

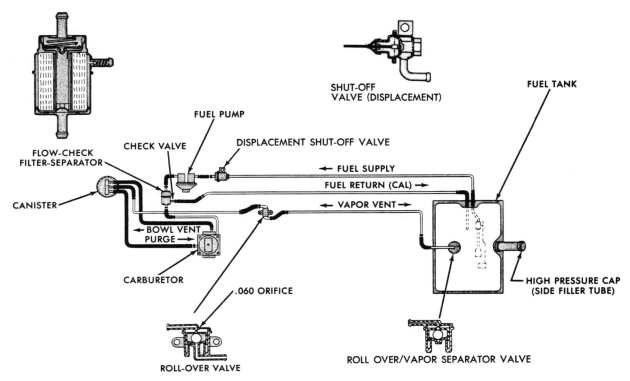

Fig. 10 Evaporative emission control system. 1977 Chrysler Corp.

a carbon canister or the engine crankcase, depending on the year of the vehicle and/or engine application. It is then drawn into the engine, from the carbon canister to the air cleaner, or from the crankcase through the PCV system.

The fuel tank is designed to limit the fill capacity to provide space for normal fuel expansion. A separate tank is located above the fuel tank to separate vapor from liquid and prevent any liquid from entering the system.

A combination valve on the forward side of the fuel tank isolates the fuel tank from engine pressures and allows vapor to escape from the separator to the carbon canister. It also relieves excessive fuel tank pressure and allows fresh air to

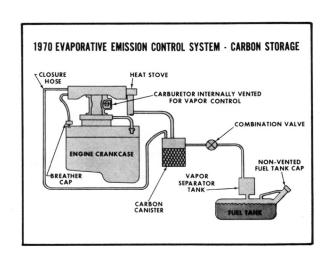

Fig. 11 Evaporative emission control system.
Ford carbon storage type

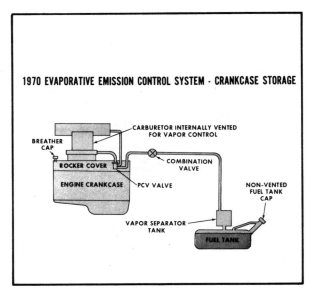

Fig. 12 Evaporative emission control system.
Ford crankcase storage type

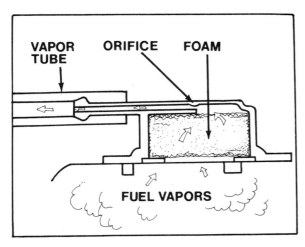

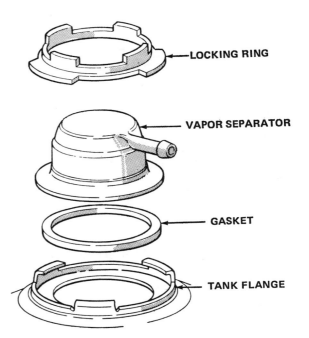

Fig. 13 *Vapor separator cross section*

Fig. 14 *Tank mounted vapor separator*

be drawn into the tank as fuel is used.

The fill control vent system which provides positive control of fuel height during fill operations is made possible by the design of the filler pipe and by vent lines within the filler neck or fuel tank. This system is designed so that about 10% of tank capacity will remain empty when the tank is filled. This space allows for thermal fuel expansion and temporary storage of fuel vapors. The pressure and vacuum relief valve system operates through the use of a sealed fill cap with a built-in pressure and vacuum relief valve. Under normal operating conditions, the valve opens to relieve pressure when it exceeds ¾ to 1¼ psi. When fuel tank vacuum reaches ½ inch mercury maximum, the valve opens allowing air to enter the

system. The vapor vent and storage system used on vertically mounted fuel tanks consists of a vapor separator, Fig. 13, mounted on the uppermost surface of the tank. The empty space at the top of the tank provides adequate breathing space for the vapor separator. Horizontally mounted fuel tanks use a raised mounting section for the vapor separator. This raised section provides additional breathing space for the vapor separator since the space allowed for thermal expansion of fuel is not

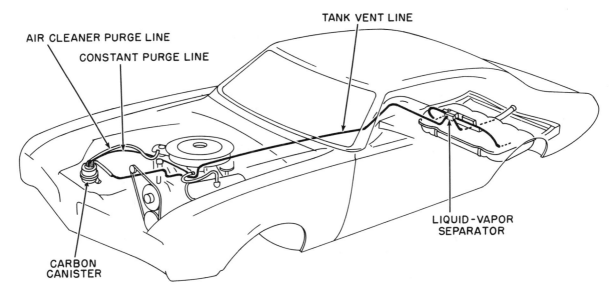

Fig. 15 *GM Evaporative Control (ECS) System (typical)*

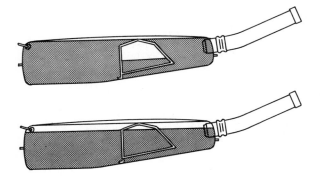

Fig. 16   Sealed fuel tank expansion

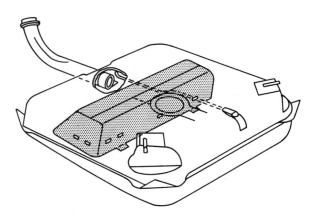

Fig. 17   Sealed fuel tank expansion

as deep as it is on vertically mounted tanks.

The vapor separator which acts as a baffle to prevent fuel from entering the charcoal canister, Fig. 14, consists of a small hole in the outlet connected to the vapor tube plus open cell foam to separate liquid fuel and fuel vapors. The fuel vapors in the tank go through the opening in the vapor separator and into the vapor tube. A fuel vapor return system is used on some engines to reduce the amount of fuel vapor entering the carburetor. It consists of a fuel vapor separator installed in the fuel supply line between the pump and carburetor and a one piece vapor return line from the separator to the fuel tank. Fuel vapors are collected in the separator and routed to the fuel tank where they recondense or are contained by the evaporative emission control system.

In the GM Evaporative Control System (ECS), vaporized fuel is fed from the sealed gas tank through a liquid-vapor separator to the storage canister, Fig. 15. There, when the engine is not running, the fuel is absorbed by the activated charcoal. When the engine is running, purge air flows through the canister carrying the fuel to the engine where it is burned. Constant purging of

vapors from the canister is accomplished through a calibrated orifice in the canister center connection to the PCV hose and/or the air cleaner snorkel. All GM carburetor float bowls are internally vented to trap bowl losses in the intake manifold, air cleaner and canister lines. Some carburetors have a vent switch in the bowl to divert the vapors to the canister. Other carburetors are insulated to reduce bowl temperatures.

All fuel tanks include a fill limiting baffle which allows space for normal fuel expansion which can be as much as three gallons on a hot day, Figs. 16 and 17. To prevent damage to the tank from excessive internal or external pressure resulting from

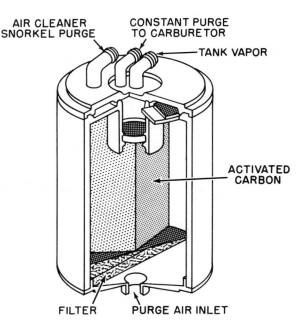

Fig. 19   Charcoal canister typical

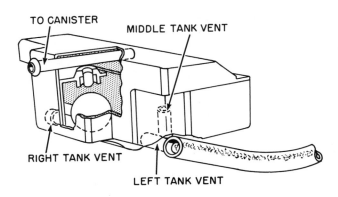

Fig. 18   Vapor separator typical

the closed system, a filler cap is used, which includes a two-way relief valve which opens only when an abnormal pressure or vacuum develops within the tank.

A liquid vapor separator is provided to prevent liquid fuel from entering the system. Liquid fuel entering the separator is spilled back into the tank while raw fuel vapor is passed into the lines.

Three expansion vents connect the tank with the liquid-vapor separator. The vents are positioned so that one is always above the liquid whatever the vehicle's attitude. The separator, Fig. 18 contains a ball float that rises, closing a needle check valve to prevent liquid fuel from passing to and overloading the carbon.

Fuel vapor enters the canister at a small cylindrical chamber above the charcoal. If the engine is running, most vapors are immediately drawn off to the carburetor. The excess is trapped by the activated charcoal. When engine speed is above idle, air is drawn through the carbon, where it picks up the trapped fuel, and is fed to the air cleaner snorkel, Fig. 19.

That then is the position regarding exhaust and evaporative emissions. Changes are occurring every day, and undoubtedly will continue to occur frequently for some time to come.

It is important, therefore, for the mechanic to watch for changes and to follow absolutely the specifications for adjustment and tune up which are plainly displayed under the hood of every car made since about 1968. Really accurate settings are absolutely essential for correct operation of engine and exhaust emission equipment.

The 4-Cycle Diesel and How It Works

INDEX

Diesel Engine Compared To The
   Gasoline Engine 280
DIESEL FUEL SYSTEM
Characteristics of Diesel
   Combustion 282
Distributor Type System 295

Fuel Filters 304
Fuel Supply Units 298
Governors 300
Injection Systems 286
Unit Injection System 296
Wobble Plate Pump System 292

**RECENT DEVELOPMENTS IN
DIESEL ENGINEERING**

Mercedes-Benz 5 Cylinder
   Diesel 305
Oldsmobile 350 V8 306

The diesel engine bears the name of Dr. Rudolph Diesel, a German engineer. He is credited with constructing, in 1897, the first successful diesel engine using liquid fuel. His objective was an engine with greater fuel economy than the steam engine, which used only a small percentage of the energy contained in the coal burned under its boilers. Dr. Diesel originally planned to use pulverized coal as fuel, but his first experimental engine in 1893 was a failure. After a second engine also failed, he changed his plan and used liquid fuel. The engine then proved successful.

DIESEL ENGINE COMPARED TO THE GASOLINE ENGINE

General Mechanical Construction

The diesel engine is mechanically similar to the gasoline engine but is somewhat heavier in construction. Both engine types utilize air, fuel, compression, and ignition. Intake, compression, power and exhaust occur in the same sequence; arrangements of pistons, connecting rods, and crankshafts are similar. Both are internal combustion engines; that is, they extract energy from a fuel-air mixture by burning the mixture inside the engine.

Fuel Intake and Ignition of Fuel-Air Mixture

In principles of operation, the main difference between gasoline and diesel engines, Figs. 1 and 2, is the two different methods of introducing the fuel into the cylinder and of igniting the fuel-air mixture. Fuel and air are mixed together before they enter the cylinder of a gasoline engine, then the mixture is compressed by the upstroke of the piston and is ignited within the cylinder by a spark plug. (Devices other than spark plugs, such as "firing tubes," are sometimes utilized.) Air alone enters the cylinder of a diesel engine; the air is compressed by the upstroke of the piston and the diesel fuel is injected into the combustion chamber near the top of the upstroke (compression stroke). The air becomes greatly heated during compression and the diesel fuel ignites and burns as it is injected into the heated air. No spark plug is used in the diesel engine; ignition is by contact of the fuel with the heated air, although

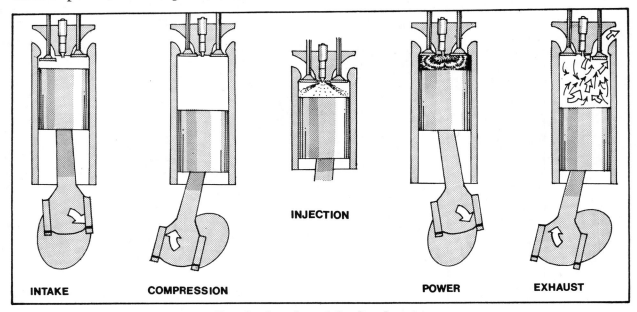

INJECTION

INTAKE   COMPRESSION   POWER   EXHAUST

Fig. 1   4-strokes of the diesel engine

GASOLINE ENGINE

ON DOWNWARD STROKE OF PISTON, INTAKE VALVE OPENS AND ATMOSPHERIC PRESSURE FORCES AIR THROUGH CARBURETOR WHERE IT PICKS UP A METERED COMBUSTIBLE CHARGE OF FUEL. THE MIXTURE GOES PAST THE THROTTLE VALVE INTO CYLINDER SPACE VACATED BY THE PISTON

DIESEL ENGINE

ON DOWNWARD STROKE OF PISTON, INTAKE VALVE OPENS AND ATMOSPHERIC PRESSURE FORCES PURE AIR INTO THE CYLINDER SPACE VACATED BY THE PISTON, THERE BEING NO CARBURETOR OR THROTTLE VALVE. CYLINDER FILLS WITH SAME QUANTITY OF AIR, REGARDLESS OF LOAD ON THE ENGINE.

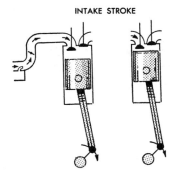

INTAKE STROKE

ON UPSTROKE OF PISTON, VALVES ARE CLOSED AND MIXTURE IS COMPRESSED, USUALLY FROM 70 TO 125 PSI, DEPENDING ON COMPRESSION RATIO OF ENGINE.

ON UPSTROKE OF PISTON, VALVES ARE CLOSED AND AIR IS COMPRESSED TO APPROXIMATELY 500 PSI.

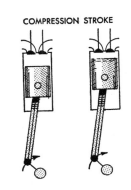

COMPRESSION STROKE

COMPRESSED FUEL-AIR MIXTURE IS IGNITED BY ELECTRIC SPARK. HEAT OF COMBUSTION CAUSES FORCEFUL EXPANSION OF CYLINDER GASES AGAINST PISTON, RESULTING IN POWER STROKE.

HIGH COMPRESSION PRODUCES HIGH TEMPERATURE FOR SPONTANEOUS IGNITION OF FUEL INJECTED NEAR END OF COMPRESSION STROKE. HEAT OF COMBUSTION EXPANDS CYLINDER GASES AGAINST PISTON, RESULTING IN POWER STROKE.

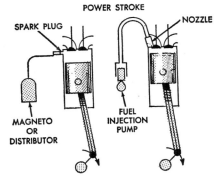

POWER STROKE

SPARK PLUG — MAGNETO OR DISTRIBUTOR — NOZZLE — FUEL INJECTION PUMP

UPSTROKE OF PISTON WITH EXHAUST VALVE OPEN FORCES BURNED GASES OUT, MAKING READY FOR ANOTHER INTAKE STROKE.

UPSTROKE OF PISTON WITH EXHAUST VALVE OPEN FORCES BURNED GASES OUT, MAKING READY FOR ANOTHER INTAKE STROKE.

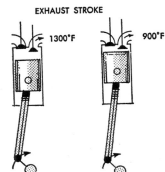

EXHAUST STROKE

1300°F   900°F

Fig. 2   Sequence of diesel and gasoline engine events

"glow plugs" are used in some models of diesel engines to assist only in starting the engine. Pressure developed by the compression stroke is much greater in the diesel engine, in which pressures as high as 500 psi are common. For each pound of pressure exerted on the air, there will be a temperature increase of about 2° F. At the top of the compression stroke (when pressure is highest), the temperature in the chamber will be about 1,000° F. This heat ignites the fuel almost as soon as it is injected into the cylinder, and the piston, actuated by the expansion of burning gases, then moves down on the power stroke. In a gasoline engine, the heat from compression is not enough to ignite the fuel-air mixture and a spark plug is therefore necessary.

Control of Speed and Power

The speed and the power output of diesel engines are controlled by the quantity of fuel injected into the cylinder. This is opposed to the common gasoline engine, which controls speed and power output by limiting the amount of air admitted into the carburetor. Therefore the difference between the diesel and the gasoline engine is that the diesel engine controls the quantity of fuel, whereas the gasoline engine regulates the quantity of air. In the diesel engine, a varying amount of fuel is mixed with a constant amount of compressed air inside the cylinder. A full charge of air enters the cylinder on each intake stroke. Because the quantity of air is constant, the amount of fuel

injected determines power output and speed of engine. As long as the amount of fuel injected is below the maximum established by the manufacturer in designing the engine, there is always enough air in the cylinder for complete combustion. The gasoline engine air intake is controlled by the carburetor. The amount of air and its velocity, in turn, control the quantity of fuel that is picked up and mixed with air to be admitted into the cylinder. The amount of mixture available for combustion determines power output and speed of engine. It is apparent, therefore, that the controlling factor is the quantity and velocity of air passing through the carburetor.

Combustion Processes

In the diesel engine, there is continuous combustion during the entire length of the power stroke; pressure resulting from combustion remains approximately constant throughout the stroke. In the gasoline engine, however, combustion is completed while the piston is at the upper part of its travel. This means that the volume of the mixture stays about the same during most of the combustion process. When the piston moves down and the volume increases, there is little additional combustion to maintain pressure. Because of these facts, the cycle of the gasoline engine is often referred to as having *constant-volume* combustion while the diesel cycle is said to have *constant-pressure* combustion.

Diesel Fuel System

CHARACTERISTICS OF DIESEL COMBUSTION

Diesel Fuels

The fuels used in modern high-speed diesel engines are a product of the petroleum refining process. They are heavier than gasoline because they are obtained from the leftovers, or residue, of the crude oil after the more volatile fuels such as gasoline and kerosene have been removed. The large, slow-running diesel engines used in stationary or marine installations will burn almost any grade of heavy fuel oil, but the high-speed diesel engines used in automotive installations

require a fuel oil as light as kerosene. Although diesel fuel is different from gasoline, its specification requirements are just as exacting as those of gasoline. Of the various properties to be considered in selecting a fuel for diesel engines, the most important are cleanliness, viscosity, and ignition quality.

Cleanliness

Probably the most necessary property of a diesel fuel is cleanliness. The fuel should not contain more than a trace of foreign substance; otherwise, fuel pump and injector difficulties will occur. Diesel fuel, because it is heavier and more viscous than gasoline, will hold dirt in suspension for longer periods of time. Therefore, every precaution must be taken to keep dirt out of the fuel system or to eliminate it before it reaches the pumps. Water is more objectionable in diesel fuels than it is in gasoline because it will cause ragged operation and corrode the fuel system. The least amount of corrosion of the accurately machined surfaces in the injection equipment will cause it to become inoperative.

Viscosity

The viscosity of an oil is an indication of its resistance to flow. The higher the viscosity, the greater the resistance to flow. The viscosity of a diesel fuel must be sufficiently low to flow freely at the lowest temperatures encountered, but it must also be high enough to properly lubricate the closely fitted pump and injector plungers. It must also be sufficiently viscous to avoid leakage at the pump plungers and dribbling at the injectors. The viscosity of a fuel also determines the size of the fuel-spray droplets which, in turn, govern the atomization and penetration qualities of the spray.

Ignition Quality

The ignition quality of a diesel fuel is its ability to ignite spontaneously under the conditions existing in the engine cylinder. The spontaneous-ignition point of a fuel is a function of temperature, pressure, and time. Since it would be difficult to artificially reproduce these factors as they exist in an engine cylinder, the best apparatus for measuring the ignition quality of a fuel is an actual diesel cylinder running under controlled operating conditions. The yardstick used for measuring the ignition quality of diesel fuels is the cetane-number scale. The cetane number of a fuel is obtained by comparing the operation of the unknown fuel in a special test engine with the operation of a known reference fuel in the same engine. The reference fuel is a mixture of alpha-methyl-naphthalene, which will hardly ignite when used alone, and cetane, which will readily ignite at temperatures and pressures obtained in a diesel cylinder. The cetane number indicates the percent of cetane in a reference fuel which will just match the ignition properties of the fuel being tested.

Knocking

It has been observed that diesel engines knock, particularly at light loads. This knock is believed to be due to the rapid burning of the charge of fuel accumulated during the delay period between the time of injection and ignition. When the fuel is injected, it must first vaporize, then superheat until it finally reaches the spontaneous-ignition temperature under the proper conditions to start combustion. Time is required for sufficient fuel molecules to go through this cycle to permit ignition. This time is called *ignition lag* or *ignition delay*. During this same time, other portions of the fuel are being injected and are going through the same phases but behind the igniting portion; therefore, as the flame spreads from the point of ignition, appreciable portions of the charge reach their spontaneous-ignition temperatures at practically the same instant. This rapid burning causes a very rapid increase in pressure, which is accompanied by a distinct and audible knock. Increasing the compression ratio in the diesel engine will decrease the ignition lag and thereby decrease the tendency to knock. The reverse is true in a gasoline engine. Increasing the compression ratio in a gasoline engine leads to pre-ignition and, in addition, tends to make detonation worse. Knocking in the diesel engine is affected by a large number of factors besides compression ratio. The type of combustion chamber, air flow within the chamber, the type of nozzle, the injection pressure conditions, the fuel temperature, and the air temperature are all factors, as are the characteristics of the fuel itself. For these reasons, more can be done in the design of a diesel engine to make it operate smoothly without detonation than is possible with the gasoline engine.

Combustion Chamber Design

The fuel injected into the combustion space of a diesel engine must be thoroughly mixed with the compressed air and distributed as evenly as possible throughout the chamber if the engine is to function under the principles discussed previously. None of the liquid fuel should strike the chamber walls. It is essential that the shape of the combustion chamber and the characteristics of the injected fuel spray be closely related. There are many types of combustion chambers in use today, but they are all designed to produce one effect, to bring sufficient air into contact with the injected fuel to provide complete combustion at a constant rate. All modern combustion chamber designs may be classified under one of the following: 1. Open, 2. Precombustion, 3. Turbulence, 4. Divided chambers. Designs which fall under two or more headings will be covered under the heading which is the most applicable.

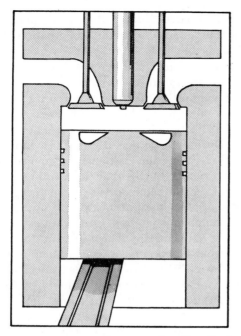

Fig. 3   Open combustion chamber design

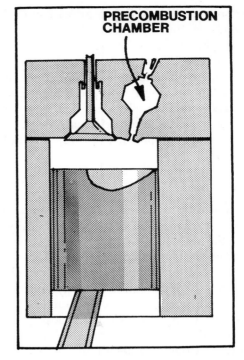

Fig. 4   Precombustion chamber design

Open Chamber

The open chamber, Fig. 3, is the simplest form of chamber, but its use is limited to slow-speed engines. The fuel is injected directly into the combustion space at the top of the cylinder. The combustion space, formed by the top of the piston and the cylinder head, is shaped to provide a swirling action of the air as the piston comes up on the compression stroke. There are no special cells, pockets, or passages to aid the mixing of fuel and air. This type of chamber requires higher injection pressures and a greater degree of fuel atomization than is required by the other types to obtain the same degree of mixing.

Precombustion Chamber

The precombustion chamber, Fig. 4, is an auxiliary chamber at the top of the cylinder. It is connected to the clearance volume above the piston through a restricted throat or passage. The precombustion chamber conditions the fuel for final combustion in the cylinder and distributes the fuel throughout the air in the cylinder in such a way that complete, clean burning of all the fuel is assured. On the compression stroke of the engine, air is forced into the precombustion chamber and, since the air is compressed, it becomes very hot. Thus, at the beginning of injection this small chamber contains a definite volume of air. Con-

sequently, combustion of the fuel actually starts in the precombustion chamber, since the fuel is injected into the chamber. Only a small part of the fuel is burned in this chamber because there is only a limited amount of oxygen present in this small chamber with which it can unite. The small

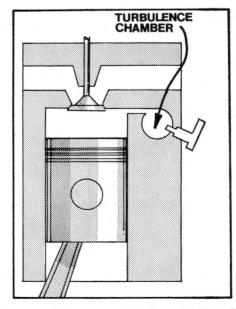

Fig. 5   Turbulence combustion chamber design

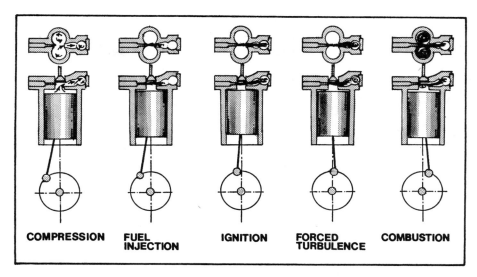

Fig. 6   Lanova divided combustion chamber design

predetermined amount that burns creates heat which, in turn, creates high pressure within the precombustion chamber; as injection continues, this high pressure forces the fuel at great velocity into the cylinder. There is ample oxygen present in the cylinder to burn all the fuel completely, regardless of the speed or load under which the engine is operating. Fuel injection pressures need not be as high with this type of chamber as in the open type. A coarser spray is satisfactory because the function of the chamber is to vaporize the fuel further before it enters the cylinder.

Turbulence Chamber

The turbulence chamber, Fig. 5, is similar in appearance to the precombustion chamber, but its function is different. There is very little clearance between the top of the piston and the head, so that a high percentage of the air between the piston and the cylinder head is forced into the turbulence chamber during the compression stroke. The chamber is usually spherical, and the opening through which the air must pass becomes smaller as the piston reaches the top of the stroke, thereby increasing the velocity of the air in the chamber. This turbulence speed is approximately 50 times crankshaft speed. The fuel injection is timed to occur when the turbulence in the chamber is the greatest. This insures a thorough mixing of the fuel and the air, with the result that the greater part of combustion takes place in the turbulence chamber itself. The pressure created by the expansion of the burning gases is the force that drives the piston downward on the power stroke.

Divided Chamber

The divided chamber, or combination precombustion chamber, probably is better known by the trade name, Lanova combustion chamber. Like the open chamber combustion system, the main volume of air remains and the principal combustion takes place in the main combustion chamber; but unlike the open chamber combustion system, the combustion is controlled. Like the turbulence-chamber type, the Lanova system depends on a high degree of turbulence to promote thorough mixing and distribution of the fuel and air; but unlike the turbulence-chamber this entails no increase in pumping losses. Ninety percent of the combustion chamber is directly in the path of the in-and-out movement of the valves. The turbulence in the Lanova system is dependent upon thermal expansion and not on engine speed, unlike the other systems which depend on engine speed.

Primarily, the Lanova system involves the combination of the figure-8-shaped combustion chamber situated centrally over the piston, and a small air chamber known as the *energy cell*, Figs. 6 and 7. In its latest development, this energy cell is comprised of two separate chambers, an inner and an outer. The inner chamber, which is the smaller of the two, opens into the narrow throat between the two lobes of the main combustion chamber through a funnel-shaped venturi passage. The larger outer chamber communicates with the inner one through a second venturi. Directly opposite the energy cell is the injection nozzle.

During the compression stroke, about 10 percent of the total compressed volume passes into the energy cell, the remainder staying in the fig-

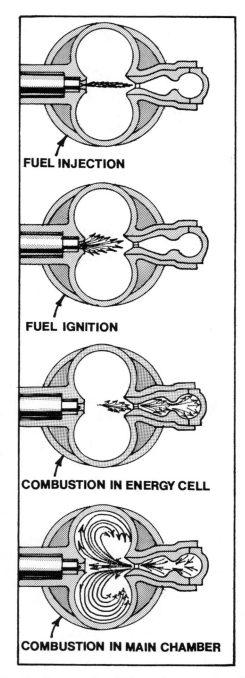

FUEL INJECTION

FUEL IGNITION

COMBUSTION IN ENERGY CELL

COMBUSTION IN MAIN CHAMBER

*Fig. 7   Lanova divided combustion chamber
fuel flow*

ure-8-shaped combustion chamber. The fuel is in-jected in the form of a pencil stream which passes directly across the narrow throat of the combustion chamber, most of it penetrating the energy cell. A small portion of the boundary layer follows the curvature of the combustion chamber lobes and swirls into vortexes within them, thus, indicating a weak combustion. The fuel entering

the energy cell is trapped, for the most part, in the small outer cell, but a small part passes into the larger outer cell where it meets a sufficient quantity of superheated air to explode violently. This explosion produces an extremely rapid rise to high pressure within the steel energy cell, which blows the main body of the fuel lying in the inner cell back into the main combustion chamber, where it meets the main body of air. Here, because of the shape of the chamber, it swirls around at an exceedingly high rate of turbulence, thus burning continuously as it issues from the energy cell. The blow-back of fuel into the combustion chamber is controlled by the restriction of the two venturi connecting the energy cells, consuming an appreciable period of time and producing a prolonged and smooth combustion in which the rate of pressure rise on the piston is gradual.

INJECTION SYSTEMS

Fuel Injection Principles

Methods of Injection

There are two methods of injecting the fuel against the air pressure in the cylinder of a diesel engine: 1. air injection, where a blast of air from an external source forces a measured amount of fuel into the cylinder 2. solid injection, where the fuel is forced into the cylinder by a direct pressure on the fuel itself. The following coverage will be limited to those systems utilizing solid injection, since the air injection system has been proved impractical for automotive installations.

Fuel Atomization and Penetration

The fuel spray entering the combustion chamber must conform to the shape of the chamber so that the fuel particles will be well distributed and thoroughly mixed with the air. The shape of the spray is determined by the degree of atomization and penetration produced by the orifice through which the fuel enters the chamber. *Atomization* is the term used to denote the size of the drops into which the fuel is broken, whereas *Penetration* is the distance from the orifice which an oil drop attains at a given phase in the injection period. Basically, the penetration of a spray depends on the length of the nozzle orifice, the diameter of the orifice outlet, the viscosity of the fuel, and the pressure on the fuel. Penetration increases with the increasing ratio of the length of the orifice to

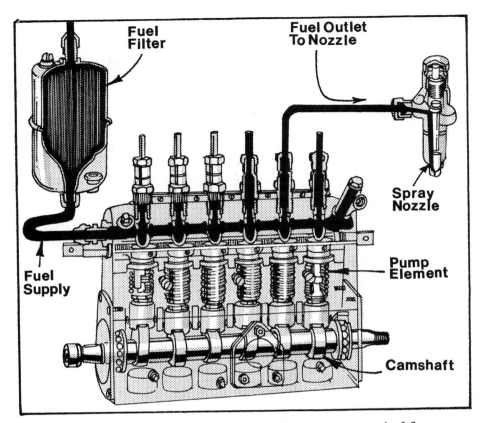

Fig. 8   Multiple unit diesel engine fuel injection pump fuel flow

its diameter; atomization, however, (inversely proportional) is increased by decreasing the ratio of the length of an orifice to its diameter. Because of the relationship between penetration and atomization, a compromise is necessary if uniform fuel distribution is to be obtained. The amount of pressure required for efficient injection is dependent on the pressure of the air in the combustion chamber, the size of the orifice, the shape of the combustion chamber, and the amount of turbulence produced in the combustion space.

Function of Injection System

It is impossible to cover the operation and construction of the many types of modern injection systems in this text. However, the operation of a few of the more common systems will be discussed. Emphasis will be placed on functional operation rather than mechanical details. The function of each system is to meter the fuel accurately, deliver equal amounts of fuel to all cylinders at a pressure high enough to insure atomization, and control the start, rate and duration of injection. If this three-fold function is kept in mind, the operation of the various systems will be more easily understood.

Multiple-Unit Injection Pump System

General Operation

In this system each cylinder has an individual injection pump which meters the fuel and delivers it under high pressure to the spray nozzles which lead into the combustion chambers. The pumps are mounted in a common housing, are operated by a common camshaft, and utilize the same control mechanism to insure an equal amount of fuel in each cylinder at the proper time. The flow of fuel in a typical multiple-unit injection pump system is shown in Fig. 8. Diesel fuel oil flows from the supply tank through a fuel filter to the fuel supply pump. The fuel supply pump forces the fuel through an additional filter to the injection pumps. The fuel injection pumps force a measured amount of fuel through high-pressure lines to the spray nozzles in the combustion chambers. Surplus fuel flows from the injection pumps through a check valve on the common housing and is returned to the fuel supply tank.

Fuel Injection Pumps

A phantom cross-sectional view of a typical multiple-unit injection pump, is shown in Fig. 8.

287

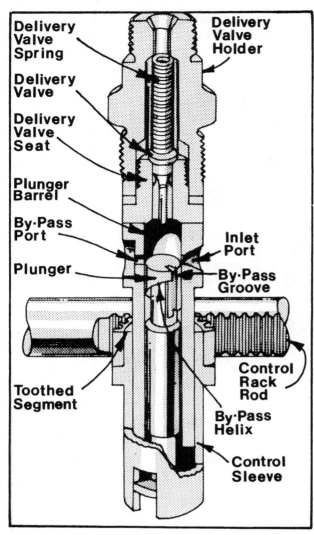

Delivery Valve Spring

Delivery Valve

Delivery Valve Seat

Plunger Barrel

By-Pass Port

Plunger

Toothed Segment

Delivery Valve Holder

Inlet Port

By-Pass Groove

Control Rack Rod

By-Pass Helix

Control Sleeve

Fig. 9   Fuel injection pump element and control rack

This pump is mounted on the engine in such a manner as to permit it to be driven by the engine. The pump camshaft, near the bottom of the housing, is carried on ball bearings. The cam lobes cause the upward movement of the plungers, and springs produce the downward motion. The cams are arranged to actuate the individual pumps in the same sequence as the firing order of the engine to eliminate the necessity for crossing lines leading from the pumps to the spray nozzles. On the 4-stroke-cycle engine, the injection pump is driven at one-half engine speed.

The individual pumps are the lapped-plunger constant-stroke metering bypass type. The quantity of fuel delivered to the spray nozzle is regulated by the time the plunger covers the bypass port. The plunger stroke remains constant at all loads. The injection must be timed to occur as demanded by the requirements of the engine. Volumetric control is effected by rotating the plunger. Two ports lead to the plunger barrel, Figs. 9 and 10, one of the ports is the inlet port and the other is the bypass port. The plunger has a groove around its circumference, which has a circular lower edge and helical upper edge. The space thus formed with this alignment is interconnected with the top face of the plunger through a vertical slot. Hence, any fuel above the plunger will flow down the vertical slot and fill the helical space. At the lower end of the plunger are two lugs which fit in corresponding slots in the bottom of an outer sleeve fitted around the pump barrel. The upper portion of the sleeve is fastened to a gear segment which meshes with a horizontal toothed rack. Any move-

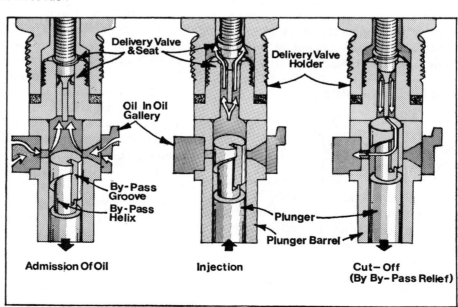

Fig. 10   Fuel injection pump plunger

Delivery Valve & Seat

Delivery Valve Holder

Oil In Oil Gallery

By-Pass Groove

By-Pass Helix

Plunger

Plunger Barrel

Admission Of Oil

Injection

Cut-Off (By By-Pass Relief)

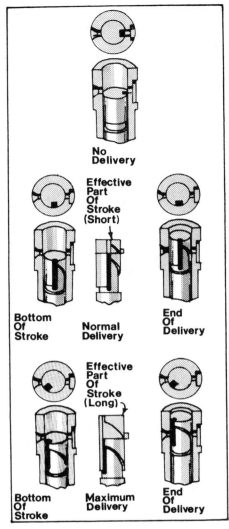

Fig. 11   Injection pump plunger positions

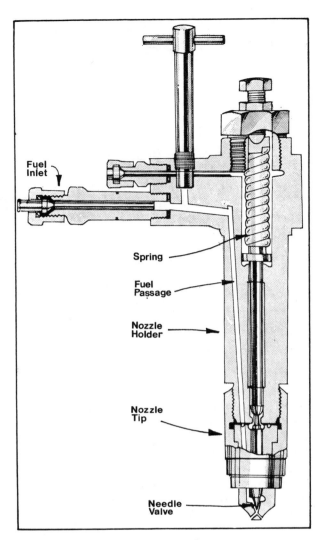

Fig. 12   Multiple unit fuel injection pump spray nozzle

ment of the rack rotates the outer sleeve and plunger relative to the bypass port in the stationary pump barrel.

Fuel from the sump rushes into the barrel as soon as the upper edge of the plunger opens the two opposite ports in the barrel. This action begins during the downward stroke of the plunger, and the ports remain open until the plunger starts moving upward. After the plunger covers the ports on its upward stroke, the pressure exerted on the fuel causes the spring-loaded delivery valve to lift off its seat, thereby, permitting the fuel to discharge into the tubing which leads to the spray nozzle. The delivery of the fuel ceases as soon as the helix on the plunger uncovers the bypass port in the barrel. At this instant, the pressure chamber is interconnected with the sump by way of the vertical groove and the helix on the plunger. This alignment thus relieves the pressure in the barrel.

The delivery valve is quickly returned to its seat by the combined action of its spring and the great difference in pressure that exists between the barrel and the high-pressure line. In returning to its seat, the delivery valve performs a double function: 1. It prevents excessive draining of the fuel from the high-pressure line, 2. It relieves the pressure in the high-pressure line. This pressure relief is accomplished by an accurately lapped displacement piston on the delivery valve. Before the delivery valve actually reseats, it reduces the pressure in the high-pressure line by increasing the volume by a quantity equal to the volume of the displacement piston.

For maximum delivery, the plunger is positioned in the barrel so that it will nearly complete its stroke before the helix indexes with the bypass port, Fig. 11. For zero delivery, the plunger is turned in its barrel until the vertical slot registers

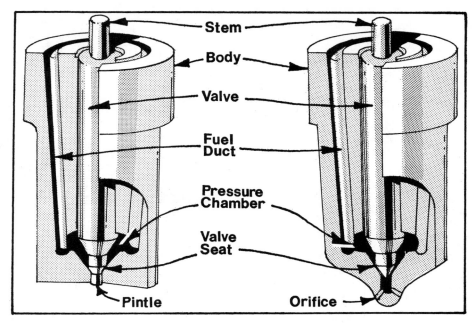

Fig. 13   Pintle and hole fuel nozzles

with the bypass port, Fig. 11. In this position, the pressure chamber is connected with the sump during the entire stroke of the plunger. Any position between no fuel to maximum fuel delivery can be obtained by moving the control rack in or out, as the movement of the rack causes the plunger to rotate a proportionate amount. The same rack controls the position of all the plungers simultaneously, thereby insuring the injection of equal amounts of fuel in each cylinder of the engine.

Spray Nozzles and Nozzle Holders

General

For proper engine performance, the fuel oil must be injected into the combustion space in a definite spray form. This is accomplished by the spray nozzle, which is held in the correct position in the cylinder head by the nozzle holder. The type of unit, Fig. 12, is most commonly used with the multiple-unit injection pump.

Operation

The fuel delivered by the injection pump flows through the high-pressure line and enters the nozzle-holder inlet stud. Then it passes through the edge filter, flows through the ducts in the holder and nozzle body, and flows down into the pressure chamber of the spray nozzle above the valve seat. There, the pressure of the fuel oil acts

on the differential area of the nozzle valve. At the moment when the pressure of the fuel exceeds the pressure exerted by the adjusting spring, the nozzle valve is lifted off its seat and the fuel is forced through the orifices and sprayed into the combustion chamber of the engine. The nozzle valve returns to its seat after the injection pump has ceased to deliver fuel. The hydraulic opening pressure of the spray nozzle may vary from 1,000 to 4,000 psi, depending on engine combustion-chamber requirements. A certain amount of seepage of fuel between the lapped surfaces of the nozzle valve and its body is necessary for lubrication. This leakage oil accumulates around the spindle and in the spring compartment, from which it drains through the leakoff connection provided for that purpose.

Spray Nozzles

Because of the widely differing requirements in the shapes of the fuel spray for various combustion chamber designs, and because of the wide range in engine power demands, there is a large variety of nozzles used with multiple-unit injection pump systems. Essentially, there are two basic groups: pintle nozzles and hole nozzles, Fig. 13. Pintle nozzles are generally used in engines having precombustion, turbulence, or divided chambers, whereas the hole nozzles are generally used with open combustion chamber designs.

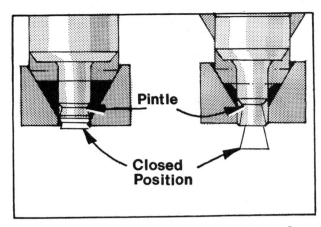

Fig. 14 Pintle and the throttling pintle nozzles

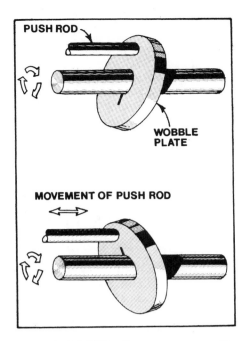

Fig. 15 Wobble plate pump operation

Pintle Nozzles

In pintle nozzles, Fig. 14, the nozzle valve carries an extension at its lower end in the form of a pin (pintle) which protrudes through the hole in the nozzle bottom. This requires the injected fuel to pass through an annular orifice, producing a hollow cone-shaped spray, the nominal included angle of which may be from 0° to 60°, depending on the combustion system requirement. The projection of the pintle through the nozzle orifice induces a self-cleaning effect, reducing the accumulation of carbon at this point.

A specific type of pintle nozzle extensively used in small-bore, high-speed diesel engines is the *throttling nozzle*. It differs from the standard pintle nozzle in that the pintle projects from the nozzle body for a much greater distance, and the orifice in the bottom of the nozzle body is much longer, Fig. 14. The outstanding feature of the *throttling nozzle* is its control of the rate at which fuel is injected into the combustion chamber. The pintle extends through the nozzle orifice when no fuel is being injected, and hydraulic pressure from the injection pump causes the pintle to rise for fuel delivery. At the beginning of the injection period, only a small quantity of fuel is injected into the chamber because the straight section of the pintle is in the nozzle orifice. The volume of the fuel spray is then progressively increased as the pintle is lifted higher. When the straight section of the pintle moves out of the orifice the tapered tip of the pintle in the orifice provides a larger opening for the flow of fuel. Another type of throttling nozzle has its pintle flush with the nozzle-body tip for no fuel delivery and extended through the body for maximum fuel delivery. In this type of nozzle, fuel under high pressure from the injection pump acts on the seat area of the pintle, forcing it outward against a preloaded spring. This spring, through its action on a spring hanger, also returns the pintle to its seat, sealing the nozzle against further injections or dribble when the line pressure is relieved at the pump. When the pintle moves outward, due to fuel pressure, an increasingly larger orifice area is opened around the flow angle of the pintle.

Hole Nozzles

The hole nozzles have no pintle but are basically similar in construction to the pintle type. They have one or more drilled round passages through the tip of the nozzle body beneath the valve seat, Fig. 13. The spray from each orifice is relatively dense and compact, and the general spray pattern is determined by the number and arrangement of the holes. As many as 18 spray holes are provided in the larger nozzles, and the diameter of these drilled orifices may be as small as 0.006 inch. The spray pattern may or may not be symmetrical, depending on the engine combustion chamber design and fuel distribution requirements.

The size of the hole determines the degree of atomization attained. The smaller the hole the greater the atomization, but if the hole is too small, it will be impossible to get enough fuel into the chamber during the short time allowed for in-

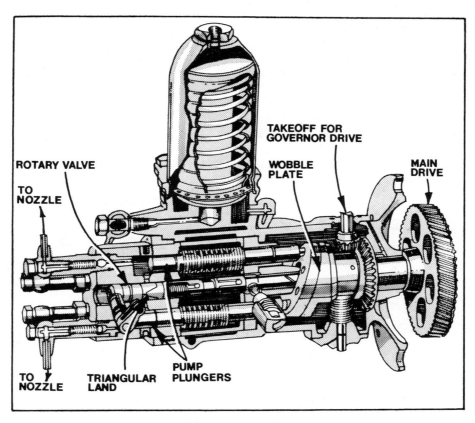

ROTARY VALVE

TO NOZZLE

TAKEOFF FOR GOVERNOR DRIVE

WOBBLE PLATE

MAIN DRIVE

TO NOZZLE

TRIANGULAR LAND

PUMP PLUNGERS

Fig. 16   Wobble plate fuel pump

jection. If the hole is too large, there will be an overrich mixture near the nozzle tip and mixture will become leaner as the distance increases from nozzle tip. Multiple-hole nozzles overcome this difficulty, because the holes can be drilled small enough to provide proper atomization and a sufficient number can be provided to allow the proper amount of fuel to enter during the injection period.

Nozzle Holders

The nozzle holder holds the spray nozzle in its correct position in the engine cylinder, provides a means of conducting fuel oil to the nozzle, and conducts heat away from the nozzle. The holder also contains the necessary spring and a means of pressure adjustment to provide proper action of the nozzle valve, Fig. 12. The body has drilled passages for conducting the fuel from the inlet connection to the nozzle, and its lower end is provided with an accurately ground and lapped surface which makes a leakproof and pressure-tight seal with the corresponding lapped surface at the upper end of the nozzle. The nozzle is secured by means of the cap nut, Fig. 12. At its upper end, the nozzle valve has an extension of reduced diameter (referred to as the *stem*) which makes

contact with the lower end of the spring-loaded spindle. Adjustment of nozzle-valve opening pressure is accomplished by means of the spring pressure-adjusting screw. The adjustment in other types of holders, which are not illustrated, is accomplished by means of spacers between the top of the spring and the upper spring seat.

WOBBLE-PLATE PUMP SYSTEM

The major difference between the wobble-plate pump system and the injection system previously described, is in the injection-pump unit itself. In the wobble-plate pump, all the pump plungers are actuated by a wobble plate instead of the customary camshaft having an individual cam for each pump plunger. Also, the metering of the fuel is accomplished by a single axially located rotary valve in the wobble plate unit, whereas the rotary movement of the individual plungers controls the amount of fuel in the multiple-unit injection-pump system.

Wobble Plate Pump Principles

A plate is mounted at an angle on a shaft so

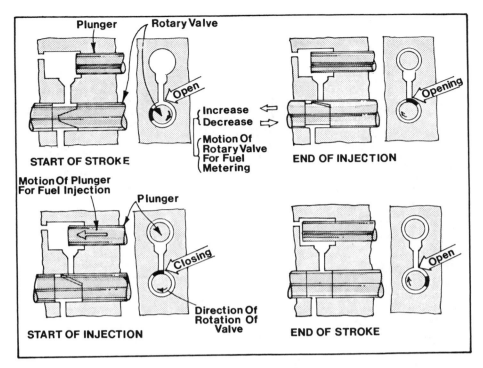

Fig. 17   *Four strokes of rotary valve and the single pump plunger*

that, as the shaft rotates, the plate also rotates and any particular spot on the plate will not only rotate but will also move laterally, Fig. 15. The pump derives its name from the fact that the plate appears to wobble back and forth as it rotates. The end of a push rod is placed in a guide plate which lays against the wobble plate. The push rod is held in a bore in the pump body so that it can move only in a direction parallel to the wobble-plate shaft. The rotation of the wobble plate then causes the guide plate to wobble, thus moving the push rod back and forth. The push rod is connected to a pump plunger, Fig. 16, so that the movement to the left actuates the pump on its delivery stroke and a spring returns it on the suction stroke. The injection pump unit contains one pumping plunger for each engine cylinder. Half the number of plungers are always moving to the right on their filling stroke, the other half are moving to the left on the delivery or injection stroke.

The rotary metering valve is driven by the same shaft that drives the wobble plate. The rotary valve consists of a lapped cylindrical shaft closely fitted in a barrel to prevent fuel from escaping at its ends. Fuel is admitted to the barrel at the center of the valve, which contains a spool-like reduction in diameter. This reduction in diameter acts as a fuel reservoir. The recess portion of the valve is in the shape of a band broken by a triangular land of the same diameter as the ends of the valve, Fig. 17. A separate port leads from the recess to the end of each plunger bore. Through this port the plunger cavity is supplied with fuel. The angular relation of the valve and the wobble plate is such that the valve land will cover each port at the time the respective plunger is at approximately its maximum speed of travel in the direction of discharge. Prior to port closing and after port opening (caused by the movement of the triangular land across the port), fuel displaced by the plunger in its pressure stroke flows back through the port into the recess. The fuel trapped in the plunger cavity when the port is closed is forced through a check valve into the high-pressure lines and then to the spray nozzle.

To obtain zero delivery, the valve is moved to a position where the ports are never closed by the land, Fig. 16. The movement of the plungers then merely causes fuel to move in and out through the ports without building up a pressure sufficient to open the delivery valves and cause injection. To cause the pump to deliver fuel, the valve is moved so that, during rotation, the triangular-shaped land closes the port when the plunger is moving in the discharge direction. Further endwise movement of the valve causes a wider portion of the land to

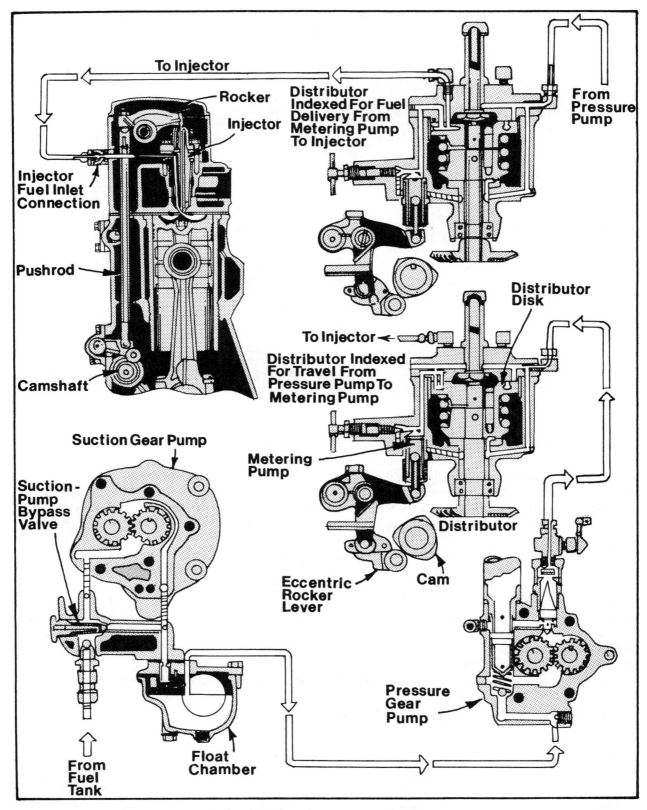

To Injector

Rocker

Injector

Distributor Indexed For Fuel Delivery From Metering Pump To Injector

From Pressure Pump

Injector Fuel Inlet Connection

Pushrod

Camshaft

To Injector

Distributor Disk

Distributor Indexed For Travel From Pressure Pump To Metering Pump

Suction Gear Pump

Metering Pump

Suction-Pump Bypass Valve

Distributor

Eccentric Rocker Lever

Cam

Pressure Gear Pump

From Fuel Tank

Float Chamber

Fig. 18   Distributor type injection system fuel flow

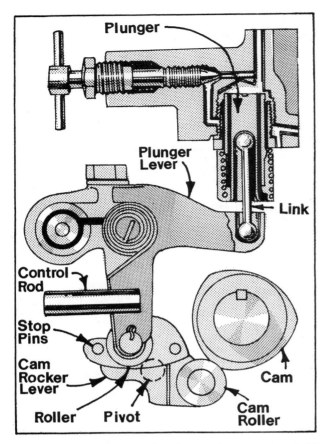

Fig. 19 *Variable stroke plunger for metering fuel*

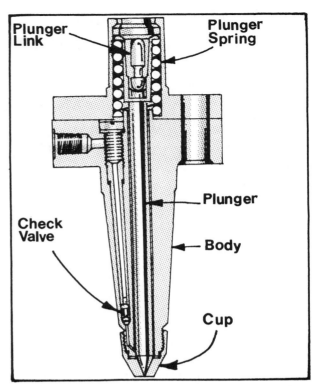

Fig. 20 *Distributor type fuel system injector*

pass across the ports, thus, increasing the duration and the quantity of fuel injected. Each port is closed for exactly the same duration of plunger travel and therefore exactly the same quantity of fuel is delivered to each of the engine cylinders.

DISTRIBUTOR-TYPE SYSTEM

The distributor injection system used in automotive diesel engines is classed as a low-pressure system in that pumping, metering, and distribution operations take place at a relative low pressure. The high pressure required for injection is built up in the injector at each cylinder, Fig. 18. A suction gear pump lifts fuel from the tank and delivers it to the float chamber, from which a second gear pump delivers it, at low pressure, to the distributor. Fuel passes through the distributor to the metering pump, then through the distributor again, and then to the injector where it is injected into the cylinder.

Distributor

The distributor consists of a rotating disk and a stationary cover to which are connected fuel lines running to individual injectors. The disk and cover have a series of holes, which, when properly aligned, form passages from the fuel supply pump to the metering pump. This occurs when the metering plunger moves down on its suction stroke and thus permits the barrel to fill with oil. The disk continues to rotate and lines up with the correct discharge hole in the cover just as the metering plunger rises by action of the main fuel-pump cam. This forces the fuel into the proper injector line.

Metering Units

The metering pump is a closely fitted reciprocating type pump, obtaining its motion through a link from the plunger lever. The plunger lever is operated by a vertical lever controlled in turn by an eccentric rocker lever running directly off a cam on the fuel pump main shaft. The position of the vertical lever in the eccentric part of the rocker lever determines the travel of the plunger lever and therefore the travel of the metering plunger, Fig. 19. As the metering plunger starts upward on its controlled stroke, it pushes fuel to the injector through passages formed by the rotating distributor disk. The stroke of the metering plunger, which

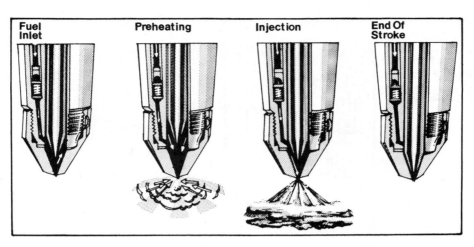

Fuel Inlet | Preheating | Injection | End Of Stroke

Fig. 21   Four fuel injection cycle stages

determines the amount of fuel going to each injector, is varied by changing the position of the vertical lever between the stop pins in the cam rocker lever. When the roller is against the inside stop pin, the vertical lever and the metering plunger are at the end of their travel consequently no fuel is being delivered to the engine. If the vertical lever is moved over to the outside pin, the stroke of the metering plunger is sufficient to deliver enough fuel to the engine for full speed and load.

Injector

The injector consists of a forged body with a properly fitted plunger, Fig. 20. This plunger is forced down against spring action by a rocker arm actuated by push rods from the engine camshaft. Mounted on the end of the body is a cup which contains the nozzle tip. On the intake stroke of the engine, the fuel metering pump forces a charge of fuel of the exact amount for the load and speed of the engine into this cup. The operation of the metering pump requires that the fuel line and passage leading to the cup be *filled with fuel*, Fig. 21. It naturally follows that any fuel added at the fuel metering pump end will push the same amount of fuel into the injector cup. The fuel lies in the cup during the compression stroke of the engine, and the compressed air is forced through the small spray holes in the cup. The fuel oil in the tip of the cup is exposed to the intense heat and blasting of the compression, and, thus, is *preheated* and broken up, Fig. 21. A few degrees before top center, the plunger is forced down and the preheated fuel charge is *driven out or injected* into the cylinder, Fig. 21; now the plunger *ends its stroke*. A small check valve is located in the lower

end of the fuel passage in the injector body. This check valve prevents the compression pressures from blowing the fuel back and filling the lines with air.

UNIT INJECTION SYSTEM

The unit injection system consists of injectors, fuel supply pump, fuel oil filters, and the fuel oil manifolds. Fuel is drawn from the supply tank through the primary filter by the fuel supply pump. From the pump, fuel is forced through the secondary filter and to the fuel intake manifold which supplies the injectors. Surplus fuel, flowing through the injectors, is returned through the fuel outlet manifold to the supply tank.

Injector Mounting

Unit injectors, Fig. 22, combine the injection pump, the fuel valves, and the nozzle in a single housing to eliminate the high-pressure lines. These units provide a complete and independent injection system for each cylinder and are mounted in the cylinder head, with their spray tips slightly below the top of the inside surface of the combustion chambers. A clamp, bolted to the cylinder head and fitting into a machined recess in each side of the injector body, holds the injector in place in a water-cooled copper tube which passes through the cylinder head. The tapered lower end of the injector seats in the copper tube, forming a tight seal to withstand the high pressures inside the cylinder.

Injector Operation

Fuel oil is supplied to the injector, Fig. 23, at a pressure of about 20 psi and enters the body at the

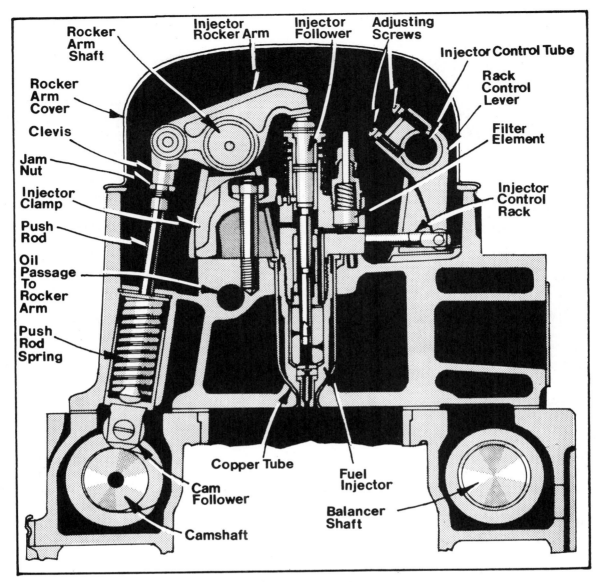

Fig. 22   Single unit fuel injector, cylinder head mounted

top through the filter cap. After passing through the fine-grained filter element in the inlet passage, the fuel oil fills the annular (ring-shaped) supply chamber between the bushing and the spill deflector. The plunger moves up and down in this bushing, the bore of which is connected to the fuel supply in the annular chamber by two funnel-shaped ports, one on each side at different heights.

The injector rocker arms, Fig. 22, are actuated through push rods from the engine camshaft. The motion of the injector rocker arm is transmitted to the plunger by the follower, which bears against the return spring. In addition to this reciprocating motion, the plunger can be rotated in operation around its axis by the gear, which is in mesh with

the control rack. Each injector control rack is connected by an easily detachable joint to a lever on a common control tube which, in turn, is linked to the governor and throttle. For metering purposes, a recess with an upper helix and a lower helix or a straight cutoff is machined into the lower end of the plunger. The relation of this upper helix and lower cutoff to the two ports changes with the rotation of the plunger. As the plunger moves downward, the fuel oil in the high-pressure cylinder or bushing is first displaced through the ports back into the supply chamber until the lowest edge of the plunger closes the lower port. The remaining oil is then forced upward through the center passage in the plunger into the recess be-

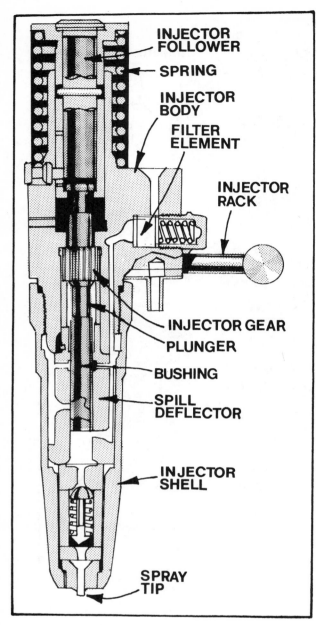

Fig. 23   Fuel injector, typical

injection), the upper port is not closed by the helix until after the lower post is uncovered. With the control rack in this position, all the fuel charge is forced back into the supply chamber and no injection of fuel takes place. With the control rack pushed in completely (full-load), the upper port is closed shortly after the lower port has been covered, thus producing a full effective stroke and maximum injection. From the no-injection position to full-load position (full rack movement), the contour of the helix advances the closing of the ports and the beginning of injection.

The various positions for downward travel of the plunger with the rack in a fixed position, are shown in Fig. 25. On the downward travel of the plunger, the metered amount of fuel is forced through the center passage of the valve assembly, through the check valve, and against the spray-tip valve. When sufficient fuel pressure is built up, the spray-tip valve is forced off its seat and fuel is discharged through several small orifices in the spraytip and atomized in the combustion chamber. The check valve prevents any leakage from the combustion chamber into the fuel system if the spray-tip valve is accidentally held open by a small particle of dirt, allowing the injector to operate until the particle works through the valve.

On the return upward movement of the plunger, the high-pressure cylinder is again filled with oil through the ports. The constant circulation of fresh coal fuel through the injectors, which renews the surplus fuel supply in the chamber, helps to maintain even operating temperature of the injectors, and also effectively removes all traces of air which might otherwise accumulate in the system. The amount of fuel circulated through the injector is in excess of maximum needs, thus insuring sufficient fuel for all conditions.

FUEL SUPPLY UNITS

Fuel Supply Pumps

Fuel injection pumps must be supplied with fuel oil under pressure because they have insufficient suction ability. Therefore, all injection systems require supply pumps to transfer fuel from the supply tanks to the injection pumps. Pumps used for this purpose have a positive suction lift, and their performance is largely independent of any reasonable variations in viscosity, pressure, or temperature of the fuel. The pumps in use today are of the gear, plunger, or vane types. The gear type is similar in operation to the pump used in the gasoline engine lubrication system, covered

tween the upper helix and the lower cutoff, from which it can flow back into the supply chamber until the helix closes the upper port. The rotation of the plunger, by changing the position of the helix, retards or advances the closing of the ports and the beginning and ending of the injection period. At the same time, it increases or decreases the desired amount of fuel which remains under the plunger for injection into the cylinder.

The various plunger positions from "NO-INJECTION" to "FULL-LOAD" are shown in Fig. 24. With the control rack pulled out completely (no-

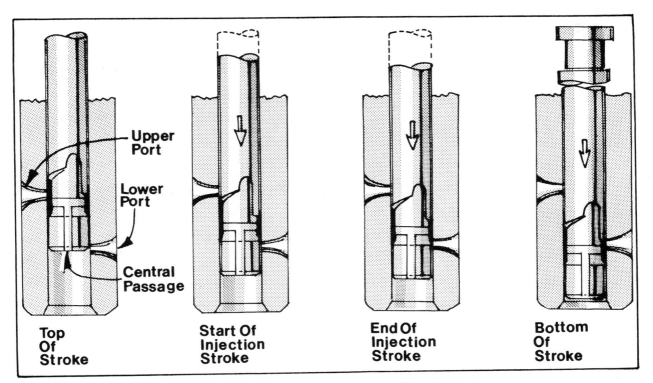

Upper Port
Lower Port
Central Passage

Top Of Stroke

Start Of Injection Stroke

End Of Injection Stroke

Bottom Of Stroke

Fig. 24   Plunger movement stages to full load injection

elsewhere in this textbook, hence the discussion of supply pumps in this paragraph will be limited to the vane and plunger types.

Vane Type

The vane type fuel oil pump, Figs. 26 and 27, has an integral steel rotor and shaft, one end is supported in the pump flange and the other end in the cover; it revolves in the body, the bore of which is eccentric to the rotor. Two sliding vanes are placed 180° apart in slots in the rotor and are pressed against the body bore by springs in the slots. Two oil seals on the pump shaft prevent leakage of fuel or lubricating oil. A drain hole between the two seals leads to the atmosphere.

When the shaft is rotated, the vanes pick up fuel at the inlet port and carry it around the body to the outlet side, where the fuel is discharged. Pressure is produced by the wedging action of the fuel as it is forced toward the outlet port by the vane. A spring-loaded horizontal relief valve is provided in the cover of the pump, connecting the inlet and outlet ports, and opens at a pressure of approximately 55 psi. This valve does not normally open, since its purpose is to relieve excessive pump pressure, if any of the fuel lines or filters become plugged, and build up an extremely high pressure in the pump. When the valve opens, fuel passes

from the discharge side (pressure side) to the suction side of the pump, relieving pressure.

Plunger Type

This type of pump, Fig. 28, is usually mounted directly on the housing of the injection pump and is driven by the injection pump camshaft. It is a variable-stroke self-regulating plunger-type pump that will build pressure only up to a predetermined point.

As the injection pump cam allows the plunger to be forced by its spring toward the camshaft, the suction effect created opens the inlet valve and permits the fuel to enter the plunger spring chamber. As the cam lobe drives the plunger against its spring, the fuel is forced by the plunger through the outlet valve and around into the chamber created in back of the plunger by its forward movement. As the injection pump cam continues to rotate, it allows the plunger spring (which is now under compression) to press the plunger backward again, forcing the fuel oil behind the plunger out into the fuel line leading to the filters and injection pump. At the same time, the plunger is again creating a suction effect, which allows additional fuel to flow through the inlet valve into the spring chamber. This pumping action continues as long as the fuel is being used by the

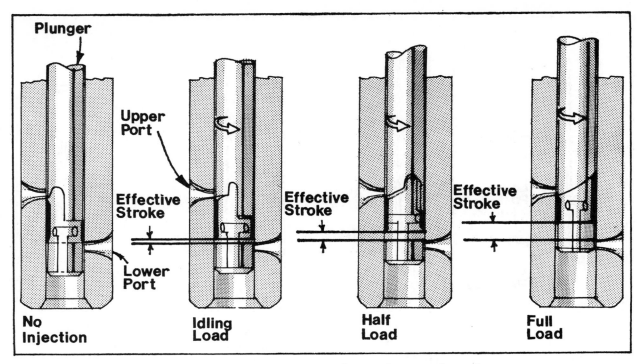

Plunger

Upper Port

Effective Stroke

Lower Port

No Injection

Idling Load

Effective Stroke

Half Load

Effective Stroke

Full Load

Fig. 25   Positions of injector during downward travel

injection pump, fast enough to keep the supply pressure from rising to the point where it equals the force exerted by the spring on the plunger. The pressure between the supply pump and the injection pump holds the plunger stationary against the spring and away from the rod. This prevents further pumping action until the pressure drops enough to permit the plunger to resume operation. This entire cycle is automatic and continues as long as the engine is running.

GOVERNORS

All diesel engines require governors to prevent over-speeding of the engines under light loads. Automotive diesel engines also demand control of the idling speed. Any of the installations provides a variable-speed control which, in addition to controlling minimum and maximum speeds, will maintain any intermediate speed desired by the operator. Engine speed in a diesel engine is controlled by the amount of fuel injected; consequently, the injection system is designed to supply the maximum amount of fuel which will enable the engine to operate at full load and reach a predetermined maximum speed (rpm). However, if the maximum fuel charge was supplied to the cylinders with the engine operating under "PARTIAL LOAD" or "NO LOAD," the engine speed would increase beyond the critical range and soon cause a failure. Thus, it can be seen that the governor must con-

trol the amount of fuel injected in order to control the engine speed.

Governors may be actuated through the movement of centrifugal flyweights or by the air-pressure differential produced by a governor valve and venturi assembly. The centrifugal flyweight type may incorporate a mechanical linkage system to control the injection pump, or it may include a hydraulic system to transmit the action of the weights to the pump. Where the rate of acceleration must be high, the governor-controlling weights must be small to obtain the required speed of response from the governor. These small weights may not exert sufficient force to control the injection equipment; instead, the injection pump will be controlled by a servo piston utilizing pressure from a pump within the governor. The centrifugal weights actuate a valve which controls the amount of oil going to the servo piston.

Mechanical (Centrifugal)

The operation of the mechanical governor is based on the centrifugal force of rotating weights counterbalanced by springs. When the speed of the engine increases, the weights fly outward, pulling with them suitable linkage to change the setting of the injection pump control rod. The governor linkage is connected to the injection pump in such a manner that the spring moves the control mechanism toward the full-fuel position, and

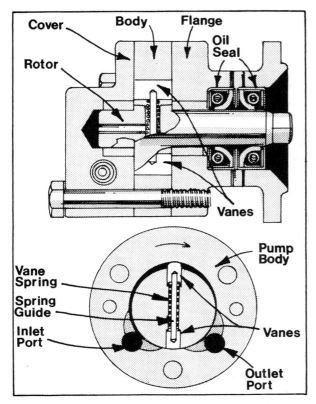

Fig. 26   Vane type fuel supply pump

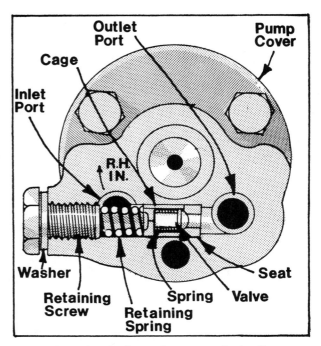

Fig. 27   Fuel supply pump relief valve

the action of the fly-weights reduces the amount of fuel delivered.

Utilizing a variable speed governor, Figs. 29 and 30, the operator varies the governor spring tension to control the quantity of fuel and does not at any time move the injector control rack directly. The control rack of the injection pump is connected to the yoke of the governor in such a manner that any movement of the yoke will directly affect the quantity of fuel injected. The spring tension is controlled by the operating lever, the movement of which is determined by the position of the foot throttle. The travel of the operating lever is limited by the idle and maximum-speed screws. With the weights fully collapsed (engine stopped), the spring moves the sliding sleeve and yoke so the fuel injection pump is in the full-fuel position. When the weights are fully extended, the sliding sleeve and yoke move to the right and decrease the amount of fuel delivered.

If the load on the engine is decreased, the engine tends to accelerate. However, when the engine accelerates, the governor flyweights move outward as a result of increased centrifugal force. Since the flyweights are in contact with the sliding sleeve assembly, this movement causes a longitudinal movement of the sleeve to the right. This

movement continues until an equilibrium is established between the governor spring force and the centrifugal force exerted by the flyweights. This occurs when the engine returns to the original speed, as determined by the position of the foot throttle and its effect on the governor spring.

If the load on the engine increases, the engine tends to slow down, thereby causing an inward movement of the flyweights. As the weights move inward, resulting in reduced force on the sliding sleeve, the compressed governor spring shifts the sleeve to the left until the spring force and the centrifugal force exerted by the flyweights are again balanced. In this way, the yoke, following the movement of a sliding sleeve, moves the control rack of the fuel injection pump toward the more-fuel position thereby returning the engine to the preset speed.

To accelerate the vehicle, depress the foot throttle, which, in turn, increases the spring tension. This causes the yoke to pivot to the left, thereby increasing the supply of fuel. The flyweights move outward as a result of increased engine speed and prevent the control rack from reaching the full-fuel position, unless the foot throttle is fully depressed. Deceleration is accomplished in the reverse manner. Spring pressure is decreased, the engine slows down, the flyweights move inward, and a balanced condition between the flyweights and the spring is obtained at a lower engine speed.

The adjustable bumper spring prevents rapid

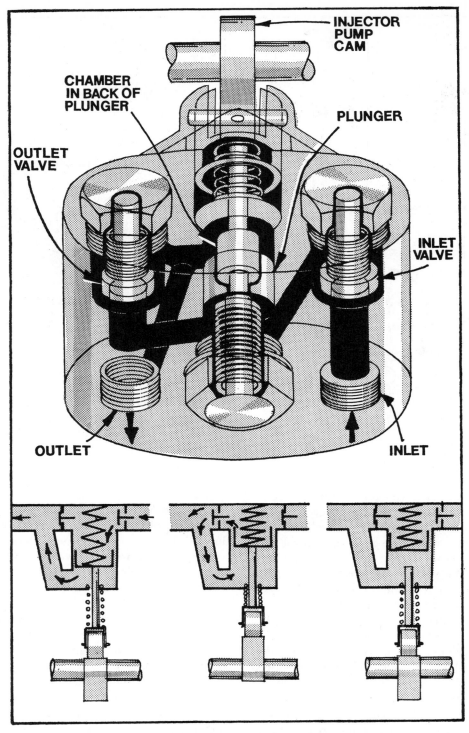

INJECTOR PUMP CAM

CHAMBER IN BACK OF PLUNGER

PLUNGER

OUTLET VALVE

INLET VALVE

OUTLET

INLET

Fig. 28   Plunger type fuel supply pump

oscillations of the control rack at low no-load engine speeds. The spring contacts the yoke at idling speed and insures steady operation of the governor. The bumper spring also assists in preventing stalling of the engine on sudden deceleration to idle speed, as it prevents the control rack of the injection pump from moving into the full-stop position when this speed change occurs.

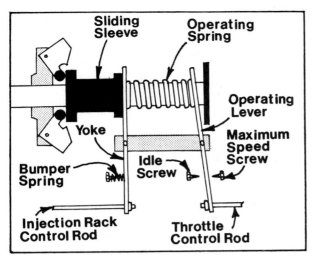

Fig. 29   *Variable speed governor operation*

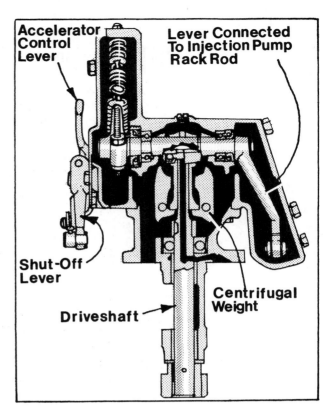

Fig. 30   *Variable speed governor*

Pneumatic

The actuating force for the pneumatic or vacuum-controlled governor is the pressure drop caused by the velocity of air passing through a venturi located in the air intake manifold. The governor, Fig. 31, consists essentially of an atmospheric-suspended diaphragm connected by linkage to the control rack of the fuel injection pump. The chamber on one side of the diaphragm is open to the atmosphere, and the chamber on the other side is sealed and connected to the venturi in the manifold. In addition, there is a spring acting on the sealed side of the chamber, which moves the diaphragm and control rack to the full-fuel position when the engine is not operating and both sides of the diaphragm are at atmospheric pressure. When the engine is running, however, the pressure in the sealed chamber is reduced below the atmospheric pressure existing in the other chamber. The amount of pressure reduction depends on the position of the governor valve and the speed of the engine. It is this pressure differential that positions the diaphragm and consequently, the control rack. The governor valve is controlled by a lever which is connected by suitable linkage to the foot throttle. There is no actual connection between the foot throttle and the governor or fuel injection pump.

If the engine is operating under load and the speed (rpm) is below governed speed, the velocity of air passing through the venturi is comparatively low and only a slight pressure differential is present. The spring moves the diaphragm and control rack toward the full-fuel position and the engine

speed approaches that of governed speed. The same principle prevents the engine from overspeeding at light loads. As the engine speeds up, the velocity of air through the venturi increases, with the result that the pressure differential at the diaphragm is increased. This differential is suf-

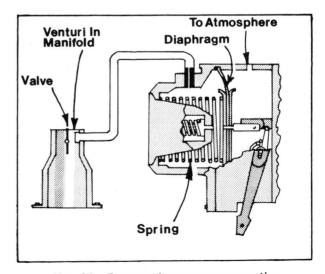

Fig. 31   *Pneumatic governor operation*

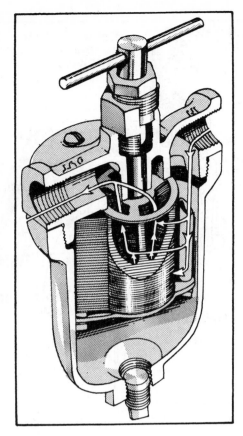

Fig. 32　Cleanable type metal fuel filter

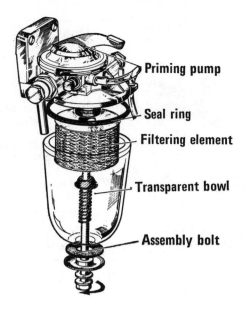

Priming pump
Seal ring
Filtering element
Transparent bowl
Assembly bolt

Fig. 33　Water trap type filter

ficient to overcome the spring force and to cause the diaphragm and control rack to move toward the STOP position. When the engine is operating at governed speed with the valve wide open, the pressure differential is just slightly below that of the spring force, and the diaphragm remains in the full-delivery position.

For any position of the governor valve between idling and full load of the engine, the diaphragm finds its relative position. Since any movement of the diaphragm is also transmitted to the control rack, the amount of fuel delivery is definitely controlled at all engine speeds. As the pressure drop between the chambers is increased, the diaphragm is moved in the direction calling for reduced fuel delivery. As the pressure drop is decreased, the spring can move the control rack in the direction of greater fuel delivery. Therefore, in order to increase the speed of the engine, the governor valve is opened; to decrease the engine speed, the valve is closed.

FUEL FILTERS

Thorough and careful filtration is especially necessary to keep diesel engines efficient. Diesel fuels are more viscous than gasoline. They contain more gums and more abrasive particles, which may cause premature wear of the injection equipment. The abrasives may consist of material difficult to eliminate during the process of relining, or they may enter the fuel tank through careless refueling. Whatever the source, it is imperative that means be provided to protect the system from these abrasives.

Most diesel engine designs include two filters in the fuel supply system to protect the closely fitted parts in the pumps and nozzles. The primary (coarser) filter is usually located between the supply tank and the fuel supply pump. The secondary (finer) filter is found between the fuel supply pump and the injection pump. Additional filtering elements are frequently installed between the injection pump and the nozzle.

Diesel fuel oil filters are referred to as *full-flow filters,* since all the fuel must pass through the filters before reaching the injection pumps. Filters must be inspected regularly and cleaned or replaced if maximum efficiency is to be maintained. Some filter elements are cleanable, Fig. 32, but most filtering elements must be replaced when they become dirty. Since water is an extremely dangerous polutant of diesel fuel, water trap filters are used very extensively, Fig. 33. A diesel oil filter usually incorporates an air vent to release any air which might accumulate in the filter during operation.

Recent Developments in Diesel Engineering

Traditionally, automobile diesel engines have been 4-cylinder powerplants. For various reasons, the latest diesel engines have broken with that tradition. We've recently seen the introduction of a 5-cylinder diesel from Mercedes-Benz and a V8 diesel from the Oldsmobile Division of General Motors.

MERCEDES-BENZ 5-CYLINDER

The goal in designing the 300D 5-cylinder engine, Fig. 34, was to gain more power for better acceleration, without sacrificing fuel economy. Within present technology and cost parameters, the best way to increase light-duty passenger car diesel engine output is through an increase in piston displacement. Since the individual cylinder volume of Mercedes-Benz's 240D 4-cylinder diesel engine, at 600cc (36.6 cu. in.), was considered the maximum for optimum combustion efficiency with the Mercedes-Benz pre-chamber system, the building of a larger displacement 4-cylinder engine was ruled out. On the other hand, an inline 6-cylinder diesel was considered too large and too heavy because of structural requirements. The solution was the addition of another cylinder to the 240D engine.

Engine bore, stroke, combustion chamber design, valve sizes and injection timing was the same for both the 240D and the 300D. Many parts, such as pistons, connecting rods, valves, rocker arms, bearings and injector nozzles were interchangeable. The block, oil pan, cylinder head, crankshaft (with six main bearings and vibration damper), camshaft (overhead) and oil pump were designed for the addition of the fifth cylinder. The dimensions of the cylinder head pre-chamber were changed slightly.

At 515 pounds, the weight of the 5-cylinder engine roughly corresponded to that of the 2.8-liter, 6-cylinder gas engine. It was 68 pounds heavier than the 240D engine.

The new 3.0-liter engine (183.4 cu. in.) developed 77 HP (net) at 4000 RPM and 115 lbs. ft. of torque (net) at 2400 RPM. Max. engine speed was governed at 4350 RPM. Despite the 24% increase in net horsepower over the 240D, fuel consumption was nearly identical for both.

For further reduction of hydrocarbon emissions and a small improvement in mileage, the 300D injection pump was fitted with reverse flow damping valves. These valves are one way valves that prevent a small "afterspray" of fuel following the

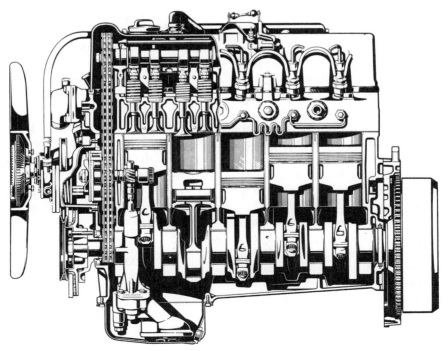

Fig. 34 Five cylinder Mercedes-Benz diesel engine

main injection pulse due to the reflected pressure wave in the line between the pump and the nozzle. Later versions of the 240D also were fitted with the reverse flow damping valves.

300D emission values were as follows: 0.23 grams/mile HC, 1.43 grams/mile CO and 1.55 grams/mile NOx. And, of course, no catalysts were used on either the 300D or 240D.

The 300D injector pump was a new development, utilizing a mechanical governor and vacuum-controlled engine shutoff. This was the exact opposite of the previous arrangement, which had a vacuum governor and mechanical shutoff. The mechanical governor resulted in smoother driving characteristics and better engine control during shifting with an automatic transmission. Mechanical governing enabled the throttle butterfly to be eliminated. The accelerator pedal was connected directly to the governor.

Mechanical governing also made it possible to use a paper air filter element instead of an oil-bath type. This was due to the fact that vacuum changes in the intake manifold due to clogging of the filter did not affect governing.

The most noticeable change, especially for experienced diesel drivers, was the new key start-and-stop system which combined pre-glowing, starting and stopping, eliminating the pull-push knob. When the key was turned to the driving position, an orange control lamp in the instrument cluster lit up and pre-glowing took place automatically. This system was then applied to future 240D models.

When the light went out, the engine could be started by turning the key in the normal manner. The duration of the pre-glowing, which depends on the temperature of the coolant water in the cylinder head, was timed automatically. A safety device limited it to a maximum of 2¼ minutes.

For stopping, moving the key to the "off" position operated a vacuum control valve via a cam in the steering lock. The valve connects the stop mechanism with the brake system vacuum booster. Thus, the control rod was pulled to the stop position.

To achieve steady temperatures of the engine, especially in short-distance operation and city traffic and to further improve engine durability, the capacities of the thermostat and the water pump were increased. The thermostat, as in the 240D, was located at the water inlet to the engine.

The radiator was also larger and was combined with an oil cooler, which was bypassed until the oil temperature was over 200° F. The bypass was a thermostatic valve in the oil filter head.

In order to guarantee adequate cooling in extreme conditions and to save horsepower, a temperature-controlled viscous-drive fan was installed.

The engine was attached to the subframe and floor unit by three rubber mounts. To further reduce engine vibrations in the passenger compartment, two special shock absorbers were installed between the engine mounts and the subframe.

Vibrations were a particular problem, not only because the new 3-liter engine was a diesel but because it was a 5-cylinder engine. An inline 5-cylinder engine is inherently plagued with imbalanced moments of both the first and second order, whereas even in a 4-cylinder engine there is no first order imbalance. The only refinement advantage that a 5-cylinder can claim for itself is that of additional torque output.

Consequently, the development of the engine and its mountings required a lot of work. The crankshaft itself, with its throws at 72°, had four balance weights and a straightforward torsional vibration damper, the results of more or less routine development work.

The weight and position of the crankshaft counterweights had to be chosen so that the first and second order imbalance could be absorbed in all driving conditions by appropriately designed rubber engine mountings. The mountings also required extensive development, and the two located at either side of the cylinder block incorporated a small hydraulic tube damper.

Another important point in obtaining the required running refinement was to increase the rigidity of the bellhousing and gearbox in the vertical plane running along the centerline of the car, a requirement that necessitated a modification in the shape of the gearbox housing. This was not specific to the 5-cylinder engine, however, and had already been carried out on the 240D. Before it was dropped into the car, the engine was balanced as a whole. It's also worth noting that, as in all Mercedes-Benz cars, the two-piece driveshaft and its central bearing mounting were specifically tuned to suit the particular type of engine and reduce resonance periods.

OLDSMOBILE 350 V8

Beginning with the 1978 model year, the Oldsmobile Division of General Motors offered a 350 cubic inch (5.7 liter) diesel engine, Fig. 35, for use in the Division's B-body and C-body (full size) models. The engine was also offered as an

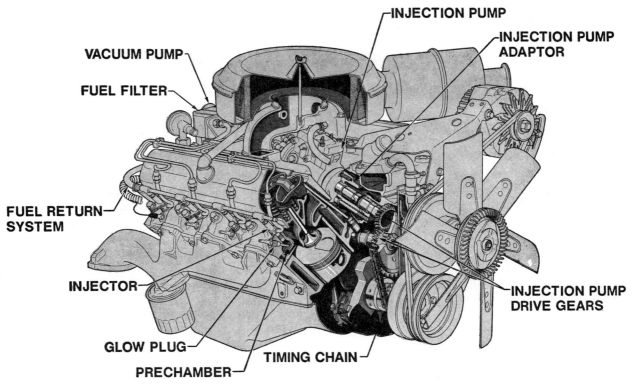

Fig. 35   V8-350 Oldsmobile diesel engine

Labels: VACUUM PUMP, FUEL FILTER, INJECTION PUMP, INJECTION PUMP ADAPTOR, FUEL RETURN SYSTEM, INJECTOR, GLOW PLUG, PRECHAMBER, TIMING CHAIN, INJECTION PUMP DRIVE GEARS

option in some Chevrolet and GMC light duty trucks.

The new diesel engine shared many dimensions with the 350 cid gasoline V8, which provided the starting point for the diesel conversion.

The compression ratio in the diesel was nearly three times that of the gasoline engine at 22.3 to 1. This compression ratio was also higher than other production automotive diesel engines, most of which had compression ratios of 21 or 22 to 1. So most of the major components of the diesel had to be strengthened, including block, crank, camshaft, rods and lifters.

The major difference between the gasoline and diesel 350 V8 engines was the diesel's fuel injection system. The fuel injection pump, of course, controlled ignition by timing and metering the amount of fuel delivered to each cylinder. The Olds diesel used a rotary-type injection pump which performed the functions of both the distributor and carburetor in the gasoline engine. The timing of the injection pump was permanently set at the factory.

The fuel was delivered to the pre-combustion chamber located in the cylinder head. Ignition started in this pre-chamber, which was added to minimize emissions, noise, and smoke. A glow plug

was used to assist in cold starting. Dual 12-volt batteries provided the heat source for the glow plugs and the additional power necessary for cranking this high compression engine.

The diesel glow period, or "wait-to-start," ranged from a few seconds on a warm day to 60 seconds or more at subzero temperatures. The engine used Number Two diesel fuel above 20° F and Number One below 20° F.

The "wait-to-start" period was monitored by an electronic device that signaled the driver when the engine was ready to start. The "Wait" light in the instrument cluster was shut off and the "Start" light turned on.

A small diaphragm vacuum pump powered by the camshaft was added because diesel engines produce minimal manifold vacuum. This vacuum pump supplies the necessary vacuum to operate the servos for heater, air conditioning, and cruise control.

A hydraulic brake booster used in conjunction with the power steering pump powered the hydraulic assisted brakes.

In summary, except for the fact that the engine was based on a 90-degree V8 cylinder block, the rest of the engine was quite conventional in design.

The Wankel and How It Works

Review Questions for This Chapter on Page 319

INDEX

Compression Phase 314
Development of Engine 318
Engine Construction 317
Engine Performance 317
Engine Specifications 319
Exhaust Phase 315
Intake Phase 313
Major Engine Components............. 308
Power Phase 314
Rotary Combustion Process............. 309

The Wankel Rotary Combustion Engine represents a complete departure from ordinary internal combustion engine design. Despite this fact, its operation is quite easy to understand; provided one understands conventional four-stroke cycle automotive engines to which it will be compared.

There is nothing mysterious about the principles on which the Wankel RC (rotary combustion) engine is based. It, like all internal combustion engines, is a machine that converts the heat energy of the fuel burned into useful mechanical work. To accomplish this, the Wankel RC utilizes the familiar Otto-cycle as does its reciprocating, piston engine counterpart. That is, a mixture of fuel and air is introduced into a confined chamber in an Intake phase. This mixture then has pressure put on it in a Compression phase, following this it is ignited and expands in a Power phase and, finally, burned gases are discharged from the chamber in an Exhaust phase.

In a piston engine these phases are called strokes and the whole sequence is described as a four-stroke cycle, after the strokes made by the piston to complete one entire cycle. In the Wankel Rotary Combustion engine the four phases cannot be described as strokes as no piston is involved so the whole conversion of heat engery into work is called by the general term, Otto-cycle process, not by the term four-stroke cycle process.

Thus, it is apparent that a major difference between the conventional four-stroke cycle automotive engine and the Wankel RC lies in the matter of the piston or pistons. The latter engine substitutes a roughly triangular shaped rotor as a power harnessing device that, as the name of the engine implies, rotates within the working chamber. This system dispenses with the piston and connecting rod which converts the reciprocating motion of the piston to the circular motion of the crankshaft. In short, the Wankel RC translates power directly into a circular path. This circular motion is utilized, as is the identical circular motion of the crankshaft in a reciprocating engine, to propel the car.

In addition to eliminating intermediate devices such as the piston, piston pin and connecting rod, the rotary engine does not need a complicated valve system. It uses ports in the working chamber, like the ports in the cylinder of a two-stroke cycle engine, that dispense with the complicated valve arrangement of ordinary automobile engines. It is this simplicity, this fewer number of working parts that give the Wankel RC engine its initial potential advantage over other internal combustion engines.

To appreciate this simplicity look at Fig. 2 which shows both a Wankel and a modern V8 piston engine disassembled. The Wankel has slightly over half as many parts. Furthermore, assembled, the RC is considerably smaller and lighter, sometimes less than half the weight of a V8 of equivalent output.

Major Engine Components

Instead of a cylinder and piston, the rotary combustion engine uses a rotor and rotor housing. The rotor housing and a cylinder are comparable in that they both provide a working chamber for the energy conversion process to take place. In the same way, the rotor and a piston are similar because they both do work on the fuel air mixture in compressing it and, in turn, have work done on them when the fuel air mixture is ignited during the power phase. Because the rotor and the rotor housing entertain a circular motion during the cycle while the piston and cylinder accomodate a linear, up-and-down motion, the comparison is not exact. It is the step-by-step sequence of events that differs between the two types of engines, not the essential function of the basic components.

In a piston engine the crankshaft and flywheel furnish the power take-off point. Their exact counterparts in a rotary combustion engine are the mainshaft and flywheel. The flywheels on both engines are approximately the same and the crankshaft and mainshaft differ only in that the U-shaped crankshaft throw on a crankshaft becomes a solid lobe on the mainshaft. Since there are presently only one or two rotors used on auto-

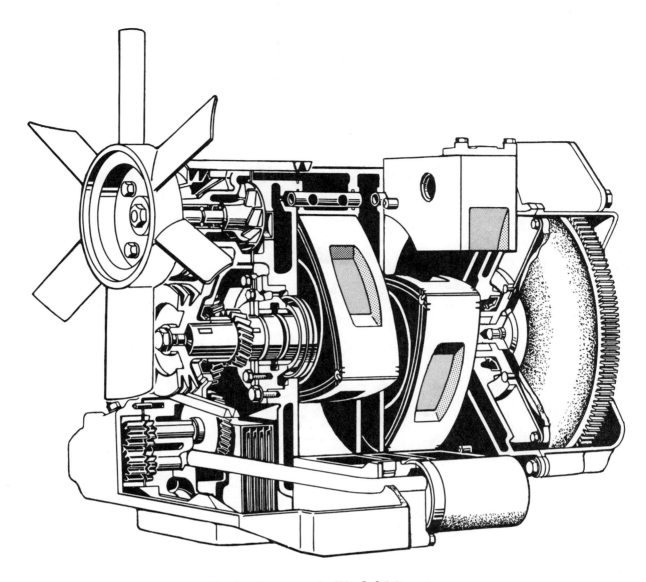

Fig. 1 *Cutaway of a Wankel RC engine*

motive Wankel RC engines the output shaft has only one or two lobes, this is in contrast to the four throws on a V8 crankshaft.

Both kinds of engines utilize closely similar fuel, cooling, lubrication, and ignition systems. Wankel RC's have been equipped with both fuel injection devices and carburetors and either unit could be interchanged on a piston engine of the equivalent cubic inch displacement with a minimum of modification. The rotary engine is currently being produced in both water and air cooled versions (though the last named is not being used in automobiles) and the heat transfer requirements for a four-stroke engine and a Wankel of the same power output are roughly equal. Any modern ignition system is applicable to the rotary engine and lubrication requirements remain the same

with one slight difference. Current Wankel engines must have oil added to the gasoline, like a two-stroke cycle engine, or use some other method such as a metering device feeding oil into the intake to assure lubrication of the rotor seals.

Rotary Combustion Process

It is difficult initially to follow the sequence of events that occur in a rotary combustion engine for two reasons. First, several phases are taking place simultaneously within the rotor housing. Secondly, we are used to picturing in our minds the Otto-cycle as it takes place in a piston engine: in an up-and-down, easy to follow sequence. Adjusting to the overlapping phases and the complicated double rotary motions of the rotor and the mainshaft requires some practice.

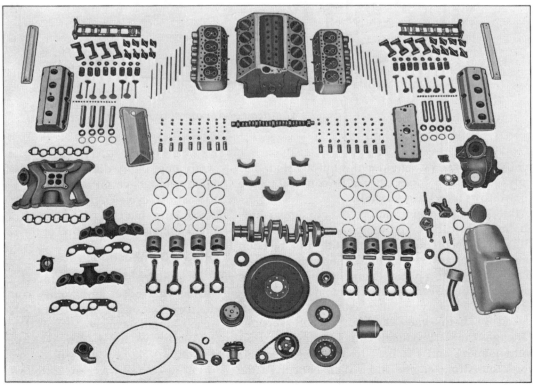

Fig. 2   Comparison of Wankel exploded (top view)
to a modern V8 engine exploded (bottom view)

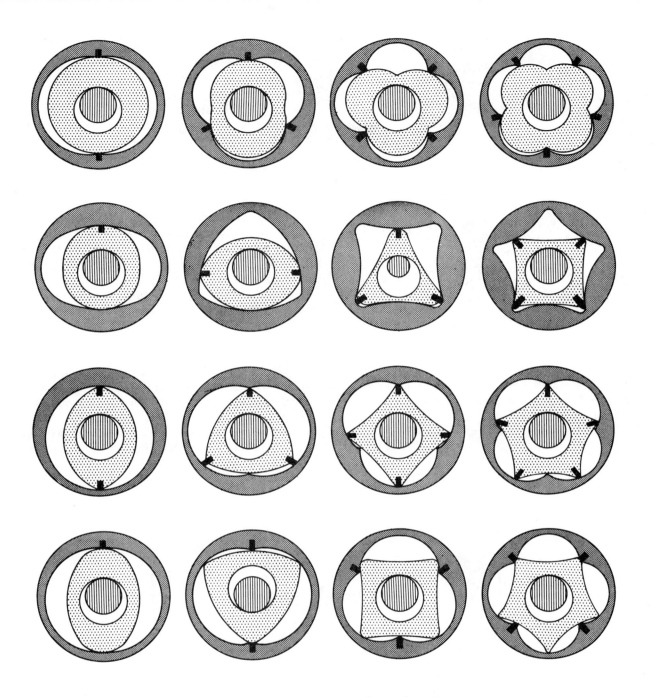

*Fig. 3 Some of the rotor and rotor housing
configurations that are possible*

To simplify it, consider the rotor housing as being two intersecting shallow cylinders. The three sided, triangular, rotor turns in this cavity so that its path describes a circle at the junction of the two cylinders while it rotates around its own center at the same time. Called a planetary motion, the effect is the same as picturing the world moving in a circular path around the sun while it also turns on its own axis.

These two distinct rotations of the rotor are timed so that each of the rotor's three corners, called apexes, are always in contact with the walls of the rotor housing, Fig. 4.

The mainshaft is centered at the geometric midpoint of the rotor housing and its lobe, or eccentric as it is known, follows the rotor as it traces its circular path around that midpoint. The output shaft makes three revolutions about its center while the rotor is making one complete turn about its center on the eccentric of the mainshaft. While

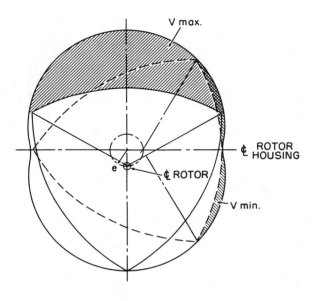

Fig. 4   Rotor positions with rotor housing

the rotor makes one revolution, of course, each apex will have completely traversed the wall of the rotor housing.

To assure that there is correct motion between rotor and shaft, a gearing arrangement is pro-vided. One gear is placed around the center of rotation of the rotor while another, smaller gear meshing with it is fixed to the mainshaft or engine housing. If these timing gears were not present the rotor could simply flop about or jam in the rotor housing without traversing its walls. Putting it another way, the output shaft eccentric only makes certain that the rotor describes the right motion about the shaft and does nothing to make it turn while describing that motion; this function is taken over by the timing gears.

Having three sides, the rotor allows three com-plete combustion processes composed of four phases each, to take place while it makes a single traverse of the rotor housing wall, Fig. 5. There-fore, each of the rotor sides acts as a movable combustion chamber providing, successively, in-take, compression, power and exhaust phases. By studying the size of this "combustion chamber" space as the rotor turns it is easy to see that its volume changes just as its equal number in a cylinder does as the piston moves up and down. But, it will be noticed that by watching all three "combustion chambers" various different phases of the cycle are occurring in each of the three. Never, of course, are the same phases taking place in more than one of the three chambers at the same

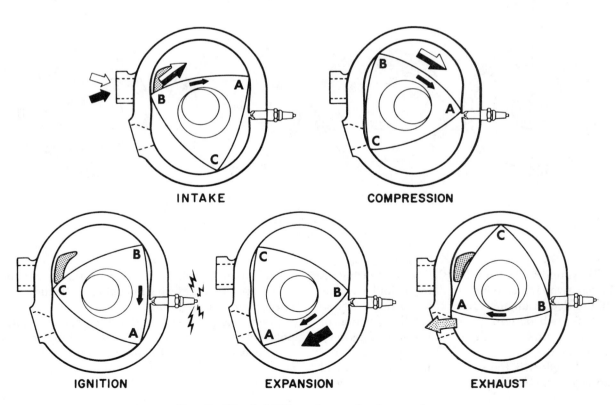

Fig. 5   Wankel RC engine cycle of operation

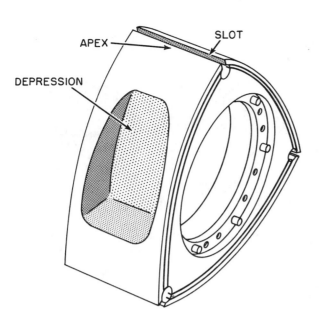

Fig. 6 *Rotor apex, slot end depression*

moment. At this point, the individual phases can be studied as they occur.

Intake Phase

Direction of rotation of the rotor shown in Fig. 5 is clockwise. Looking at one flank of the rotor, indicated by the black arrow, it can be seen that apex A began the intake phase by uncovering the intake port in the side of the rotor housing. The chamber under consideration is that delineated by the rotor flank AB and it is almost at its maximum volume when it is in the position shown over the heading INTAKE. In this position the chamber can be compared to a cylinder with the piston approaching bottom dead center on the intake stroke.

Notice that the fuel-air mixture entered the

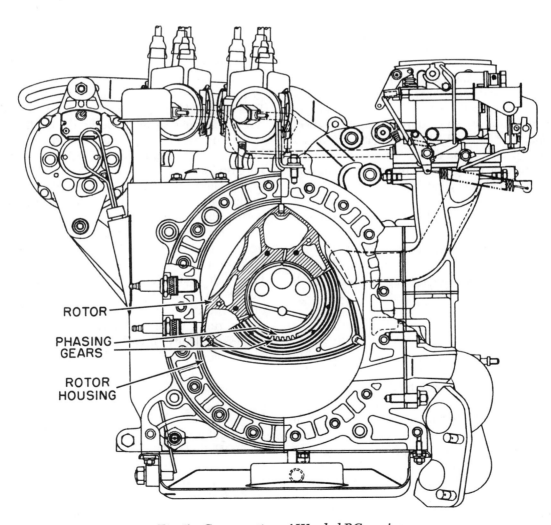

Fig. 7 *Cross section of Wankel RC engine*

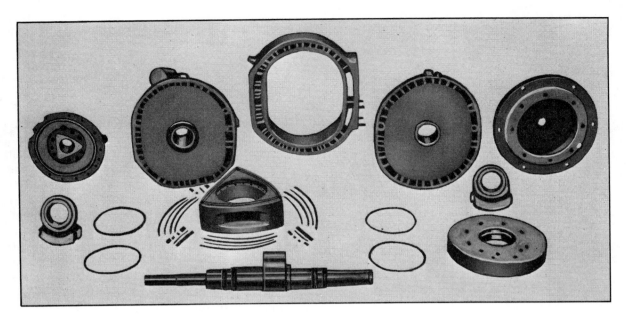

Fig. 8   Primary components of Wankel engine

chamber through the port in the side of the rotor housing. This type is called a side port and contrasts with the peripheral type which is set in the rotor housing wall, as per the lower port shown over the heading EXHAUST. Some engines use both a side and peripheral port or a pair of peripheral ports, with one or the other opened by means of a butterfly valve only during high speed operation. The port which is open at all times is called the primary, the other, the secondary. It is characteristic of the Wankel RC engine that inlet (volumetric) efficiency can be made almost ideal. This is so because practically any port size and shape can be accomodated in the rotor housing of a particular engine. Timing of the intake port, its location with respect to the center line of the rotor housing on its short axis, unfortunately can not be varied greatly without sacrificing performance in certain speed ranges.

Compression Phase

During compression, chamber size is gradually decreased as can be seen by comparing chamber volume over the headings, INTAKE and COMPRESSION in Fig. 5. Comparison of the chamber volume between its maximum, slightly more than shown over INTAKE, and minimum, that shown over IGNITION, is the compression ratio. It is exactly comparable to the compression ratio of a four-stroke cycle engine as obtained when cylinder volume with the piston at bottom dead center is divided by cylinder volume with the piston at top dead center.

Compression ratio of an RC engine is controlled by engine geometry and a great variety of different ratios have been experimented with. In general, compression ratios on production Wankel engines are about the same as those used on piston engines, usually between 8.5 and 10.5 to 1. But, it should be noted in this respect that Wankel engines do not seem to be as demanding in the matter of octane requirements and engines with high compression ratios (10:1) run well on regular gasoline.

Not shown in Fig. 5 but present on an actual engine are the seals that contain gas pressure within the chambers. Apex seals run across the edges of the three apexes, A, B & C and ride against the rotor housing wall to prevent gas leakage at those points between adjacent chambers. Side seals are set in the flanks of the rotor, just in from the edges, and ride against the sides of the rotor housing. Serving the same function as piston rings, these seals have been the source of endless difficulty in earlier Wankel engines. Presently, most problems with them seem to have been overcome and good service life is predicted.

Power Phase

With the rotor as shown over the heading IGNITION in Fig. 5, the spark plug fires and the fuel-air charge is ignited. This position of the rotor, by direct analogy with the piston of a reciprocating engine, is known as top dead center. During subsequent movement of the rotor, from IGNITION to EXPANSION in Fig. 5, work is

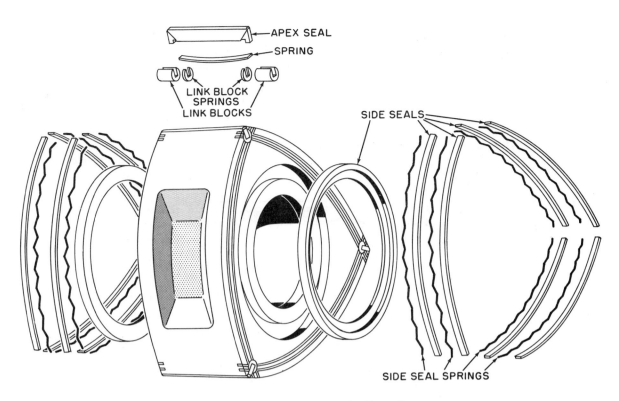

APEX SEAL

SPRING

LINK BLOCK
SPRINGS

LINK BLOCKS

SIDE SEALS

SIDE SEAL SPRINGS

Fig. 9   *Rotor apex and side seals*

done on the rotor by the expanding gas, the entire process being called the power phase.

It will be appreciated that pressure applied by the expanding gas is transmitted through the rotor to the eccentric and then to the mainshaft. In this, the eccentric on the mainshaft, not shown in Fig. 5, but represented by the crescent shaped space between the rotor and mainshaft, acts in every way like the throw on a piston engine crankshaft in transmitting power. The phasing gears, one of which surrounds the rotor center opening and the other attached to the mainshaft or engine housing and meshing with it, do not transmit power. They are there simply to assure correct relative motion between mainshaft and rotor. To transmit power from the rotor to the mainshaft, either a roller or plain bearing is fitted to the eccentric where it mates with the rotor, Fig. 7.

The spark plug shown in Fig. 5, can be positioned in any of several locations. Unlike a piston engine, spark plug location is not over critical and it is generally placed either slightly before or slightly after the top dead center point of the rotor. That is, it is just above or just below the short axis of the rotor housing. In some cases two spark plugs are fitted, one in each of the positions noted. The plug or plugs usually fire at or near top dead center at idle and a spark advance is provided, just as on a piston engine, to fire them

earlier and earlier as engine speed increases.

The spark plug can be set almost flush with the rotor housing wall or inset into a small pocket, as shown in Fig. 5. Usually the rotor flanks have depressions, as shown in Fig. 6, that increase chamber volume over that provided between the rotor and rotor housing wall.

Up to this point we have been concentrating our attention on one single chamber, that one confined by the rotor flank between apexes A and B. If you consider the other two flanks BC and CA, it can be seen that their chambers have also been going through separate phases of the Otto-cycle. Chamber BC being one phase behind AB and chamber CA being one phase ahead. Therefore, the spark plug must fire three times for every revolution of the rotor.

Exhaust Phase

As soon as rotor apex A passes over the exhaust port the exhaust phase begins. Because of the relatively high gas pressure in the chamber, as in the cylinder of a four-stroke or two-stroke cycle engine with the piston well down on the power stroke, a fairly small port is adequate to exhaust the burned charge. For this reason double ports, as often used on the intake, are unnecessary.

Timing of the exhaust phase in relation to the

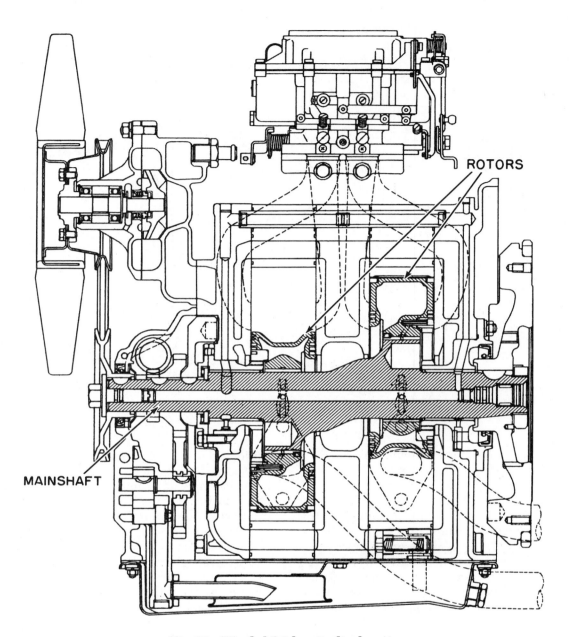

ROTORS

MAINSHAFT

Fig. 10   Wankel RC longitudinal section

power phase is determined by the position of the port in the rotor housing wall. With a clockwise rotor motion, as in Fig. 5, placing the port further down, counterclockwise, on the housing wall allows earlier beginning of the exhaust phase and vice versa. Likewise, varying the position of the intake port or ports has the same effect on the timing of the intake phase. Four-stroke cycle engine intake and exhaust timing, of course, are determined by the design of the camshaft while two-stroke cycle engines accomplish this timing in very much the same way as the Wankel, by locating the ports at certain points along the cylinder wall with respect to piston travel.

Here the similarities with other positive displacement internal combustion engines cease and it can be readily seen that both intake and exhaust ports of a Wankel RC engine are open at all times. Passage of an apex over either port simply confines gas flow to one particular chamber; it does not shut off a port set in the rotor housing wall (side inlet ports are momentarily shut off by the rotor apexes). This feature is considered to be very advantageous and means that exhaust flow out of and intake flow into a rotary combustion engine are different from the same processes in a piston engine. Despite this, a normal automotive exhaust silencer (muffler) and intake silencer (air cleaner) are sufficient to keep Wankel engines quiet.

Before leaving the subject of the energy conversion process, it would be wise to note another unique feature of the Wankel engine. By comparing the positions of the mainshaft index mark in Fig. 5, it is obvious that each phase of the Otto-cycle extends over 270 degrees of mainshaft rotation. This could also be determined by noting that the mainshaft makes three revolutions to one turn of the rotor and during this period the four phases of the Otto-cycle have been completed:

$$\frac{3 \text{ (turns of the mainshaft)} \times 360 \text{ degrees}}{4 \text{ (phases)}} = 270 \text{ degrees}$$

The strokes of a four-stroke cycle piston engine, on the other hand, occupy 180 degrees of crankshaft rotation, from top dead center of piston travel to bottom dead center or vice versa.

Engine Construction

Wankel RC engines are made of common materials such as those used in ordinary automobile engines and they do not require exotic or unusual alloys. While many different materials have been tried for the different components, aluminum alloys, cast iron, and various grades of steel are generally preferred.

The engine consists, in a one rotor version, of six or more primary components, Fig. 8. A mainshaft mounts the rotor, which is in turn contained within a separate rotor housing. The rotor housing is bolted between two end covers and the alignment between them, which is very critical, is maintained by dowel pins. The mainshaft projects out of both end covers where it is supported by ball, roller or plain bearings. One or two flywheels are fitted to the projecting stubs and these flywheels are not symetrically weighted as on a four-stroke cycle engine. Since the mainshaft does not use counterbalance weights to offset the mass of the eccentric or rotor, these must be provided for in the disc shaped but unevenly weighted flywheel. These flywheels have a ring gear to accommodate the starter motor.

Additional rotors are accommodated by adding the rotor and its appropriate rotor housing plus an intervening spacer that separates the two housings. This spacer projects down to the mainshaft but does not have to support it after the fashion of an intermediate web with its main bearing in a piston engine. The mainshafts are very short and stiff so they do not require support at this point. When multi-rotor engines are built, units with up to four rotors have been built though not for automotive use, they are assembled out of components which are used for single rotor construction. This saves design and production time and cost and provides for a family of engines of different displacements and power outputs but with the same basic components.

The critical parts of the Wankel engine design are the bore of the rotor housing, the rotor itself, and the rotor seals. From design and manufacturing viewpoints it can be appreciated that the bore configuration, its shape called an epitrochoid, and the shape of the rotor present problems.

From the performance viewpoint, it is the rotor housing bore and the rotor seals that give the most trouble. Rapid seal wear, the wear problem with seals is much more complex than it is with the corresponding piston rings of a reciprocating engine, plagued the acceptance of the engine for years. When harder apex seals were used to lessen wear, they left chatter marks on the wall of the rotor housing.

Presently, a number of different solutions have been proposed and several adopted to overcome these twin problems. Some manufacturers utilize a hard coating on the rotor housing wall while others vary the materials out of which the seals are made as well as the design of the seals themselves. All engines in production are claimed to have largely eliminated the two problems or at least brought them under control to the point that seal wear and bore chatter can be kept to safe limits.

Engine Performance

Positive displacement engines such as reciprocating piston or rotary combustion types are compared on the volume of air they displace. This figure is considered to represent the potential each engine possesses for power development, although it does not enter into power development calculations themselves. The idea is that the higher an engines displacement is, the more horsepower it is theoretically capable of producing.

To obtain this figure for a reciprocating engine, the volume contained in one cylinder with the piston at bottom dead center is multiplied by the number of cylinders. Both two and four stroke cycle engines are calculated in the same way, but, because a two-stroke has twice as many power strokes it should ideally produce twice the power of a four-stroke with the same displacement. This has been the source of endless controversy among engineers because of this fact and many therefore feel that displacement is not a good basis of comparison between the potentials of various engines. Nevertheless, it is still used. Enter the Wankel RC. It further compounds the problem because its potential power output is seemingly even less related to displacement than is the power output of a two-stroke cycle piston engine. Furthermore, not many engineers agree on how the comparative displacement of Wankel should be calculated.

One widely used method of obtaining this figure follows. It is based on the notion that since a single rotor Wankel engine completes two energy conversion cycles during two rotations of the mainshaft, it has the power potential of a two-stroke cycle engine. Thus, a double rotor Wankel RC has a displacement equivalent, in terms of crankshaft rotation, to a four-stroke cycle, four cylinder piston engine if the volume of one Wankel chamber is the same as the volume of one cylinder of the piston engine.

The International Automobile Federation (the FIA, the sanctioning body for international auto racing), therefore, determines rotary combustion engine displacement by doubling the volume of one chamber and multiplying that figure by the number of rotors. For example, if a certain Wankel RC engine has a chamber volume of 60 cubic inches, its displacement would be:

$$2 \times 60 \text{ cu. in.} \times 2 \text{ (number of rotors)} = 240 \text{ cu. in.}$$

Of course, not all companies producing Wankel engines subscribe to this method, so when displacement figures are compared it is necessary to find out how the Wankel determination was made.

Whether displacement or some other means of comparison is used, Wankel RC engine efficiency and performance at present are roughly equivalent to four-stroke cycle engines.

Potential performance to be realized with further development of the Wankel is, however, another matter. Consideration of the three basic efficiencies of an internal combustion engine, namely, thermal, volumetric, and mechanical, indicates where this potential might lead.

The thermal efficiency of either a Wankel or a four-stroke cycle engine is to a large extent determined by the compression ratio. It seems that future rotary combustion engines will tolerate higher compression ratios than will their opposite number piston engines because of lower fuel octane requirements. Another factor of performance related to thermal efficiency is engine speed.

Without inertia effects of a complicated valve arrangement or reciprocating parts, the Wankel engine should tolerate considerably higher rotating speeds than presently employed.

Volumetric efficiency, or "breathing" of the Wankel is already high because of their porting layout and other considerations. So is mechanical efficiency because of the Wankels fewer number of moving parts.

Presently, the RC engines strongest claim is large output from a small, simple and consequently light package. This, coupled with low manufacturing cost, one estimate lists manufacturing costs for an RC at $1.00 per pound of engine weight versus $2.00 per pound for a modern V8, are the principal attractions of this new engine.

Development of the Engine

Considering their recent invention, rotary combustion engines have been developed to an amazing degree. While the piston type internal combustion engine has had about ninety years in which to develop, the first working model was demonstrated by Dr. Nicholas Otto in 1878, the rotary combustion engine has had less than fifteen. While it is true that the concept of a rotary combustion engine has been around since the early part of this century, it remained for Dr. Felix Wankel of Lindau, Germany, to first successfully test such a machine in 1957. It is not surprising, then, that the Wankel RC engine has not yet realized as much of its potential as has the reciprocating engine.

Other kinds of rotary machines such as windmills no doubt furnished the inspiration for the idea of a rotary combustion engine. Dr. Wankel, taking that idea from early experimenters, systematically investigated most of the possible configurations, Fig. 3 shows a few of the hundreds of possible designs, over a long period before he conceived and began laying out his first engine in 1954. Early collaborating with the German car company, NSU, he had built several engines by the end of 1957. The final version, with a modest displacement of 7.6 cu. in., developed an astonishing 29 hp at 17,000 rpm in a 100 hour test.

The early engine was decidedly different from current examples in that the rotor housing also turned and was contained within a larger stationary housing. Called the SIM type, this engine's rotor turned but did not transmit power. The turning rotor housing accomplished this through gearing to an output shaft. The complexity of the engine can be appreciated when it is realized that the turning rotor also held the three spark plugs required together with their leads and that the hollow rotor shaft also served as the fuel-air intake pipe.

A machine similar in most respects to the SIM, but driven as a compressor by outside means, was used as a supercharger by NSU. Fitted to a motorcycle, the supercharger helped the engine produce the equivalent of 260 hp per 61 cu. in. If this kind of performance could be coaxed from a modern V8, a 283 cu. in. version would produce over 1200 hp.

NSU also contributed much in helping to simplify the original Wankel RC engines. It was with their help that Dr. Wankel redesigned the engine to the less complicated form we know today, called the PLM or planetary rotation engine. The PLM type Wankel engines are being produced under license or tested in most of the automobile making countries of the world. Cars made by NSU in Germany and Toyo Kogyo in Japan are equipped with Wankel R's and Curtiss-Wright Corp., the licensee in the U.S.A., is far along in the development of several models for automobile, truck, aircraft and industrial use.

Engine Specifications

Manufacturer .... Toyo Kogyo (Japan)
Car .... Mazda
Engine Model .... 110S, Wankel RC
Rotors 2
Brake Horsepower . 123 @ 7000 RPM (mainshaft)
Torque ... 96.2 lb. ft. @ 3500 RPM (mainshaft)
Equivalent Displacement .... 120.5 cu. in.
Compression Ratio 9.4:1
Mainshaft Eccentricity .... 0.59 in.
Cooling .... Water, rotors oil cooled
Carburetor .... 1 Stromberg four-barrel
Ignition .... Coil with 2 spark plugs per rotor
Weight .... 225 lbs w/water & oil
Power: Weight Ratio .... 0.48 HP per lb.
Length .... 20.08 in.
Width .... 23.43 in.
Height .... 21.65 in.
Ports .... 1 exhaust, 2 intake per rotor

Review Questions

Page

1. Why can't the four phases of operation in the Wankel engine be described as strokes? 308

2. What is the proper name for the operating cycle used in both the Wankel and the common reciprocating engine? 308

3. What is the main difference between the conventional engine and the Wankel? 308

4. What is the Wankel's initial advantage over other internal combustion engines? 308

5. What are the major components of the Wankel? 308

6. How does the Wankel work? 309

7. How are valves eliminated in the Wankel? 308

8. How is it that, in the Wankel, inlet efficiency can be almost ideal? 314

9. How is the compression ratio controlled in the Wankel? 314

10. In the Wankel, what parts serve the same function as the piston rings in a conventional engine? 314

11. Is spark plug location very critical in the Wankel? 315

12. How is the exhaust phase timing accomplished in the Wankel? 315

13. What materials are preferred in the construction of the Wankel? 317

14. At what point in the Wankel is wear a major problem? 317

15. How does the performance of the Wankel compare with that of other types of engines? .. 318

The Turbine and How It Works

Review Questions for This Chapter on Page 325

INDEX

Cold Weather Operation 324
Fuel Consumption 324
How The Turbine Works 320
Introduction 320
Reduced Air Pollution 324
Regenerative Turbines 322
The Burner 325
The Fuel System 325
The Ignition System 325

INTRODUCTION

In its basic concept, the gas turbine is the simplest of all engines. There are no reciprocating parts. Combustion is continuous, eliminating the need for precise timing of each event in the cycle. The ignition system is needed only when starting, and a cooling system is unnecessary. There is no sliding metal-to-metal contact, as in a piston engine; only a few bearings and shafts require lubrication.

The gas turbine offers several other advantages for motor vehicle use. It is much lighter and smaller than gasoline or diesel engines of comparable power. The lack of reciprocating parts in the engine makes it inherently smoother. It runs on a wide variety of liquid and gaseous fuels. Because most turbines have built-in "torque converters," there is no need for a clutch, and the number of gear ratios in the transmission is usually fewer than required for a reciprocating engine.

Although turbines have been installed in a number of experimental passenger cars, trucks and buses, no standard production motor vehicle in the United States is yet available with turbine power.

The term, "gas turbine," is misleading. The engine gets its name from the fact that power comes from the force of burning gases striking the blades of a turbine. However, the hot gases may be created by the combustion of either liquid or gaseous fuel—or even finely powdered solid fuel. Turbines have been operated on gasoline, natural gas, diesel fuel, aircraft jet fuel, powdered coal and perfume. Most vehicular turbines burn diesel fuel.

HOW THE TURBINE WORKS

In its simplest form, the turbine consists of a centrifugal compressor and a turbine wheel, mounted on opposite ends of the same shaft; a burner; and a means of providing an electrical spark. Air enters the compressor inlet, then is compressed to about four times atmospheric pressure, increasing its temperature by several hundred degrees. The compressed and heated air enters the burner, or combustor, where fuel is introduced and the mixture ignited by an electric spark. Once combustion begins, it is continuous; therefore, the ignition system is used only when starting. The burning gases expand rapidly, striking the blades of the turbine, spinning the shaft and providing power for the compressor.

The earliest turbines developed barely enough power to keep themselves running. Virtually all of the power produced by the turbine was consumed by the compressor. The remaining energy in the burning gases escaped to the atmosphere.

To capture more of the energy of the hot gases, engineers added a second stage, or power, turbine. The second stage provides power to produce useful work—to drive a vehicle, turn a generator or operate a pump.

The second-stage turbine can be mounted on the same shaft as the first-stage turbine and compressor or on a separate shaft. Each design has its advantages. The single-shaft version is simpler mechanically. It's good for an application such as a generator, which must maintain a relatively constant speed, regardless of fluctuations in load. The high rotating mass develops the inertia to compensate for sudden changes in electrical load.

Gas turbines designed for vehicular use are of the split-shaft type. The compressor and first-stage turbine are mounted on one shaft; the power turbine is fastened to a second shaft, on the same axis but not connected to the first.

The portion of the turbine containing the compressor and compressor turbine is called the gasifier section, while the second-stage turbine, reduction gears and output shaft make up the power section.

The split-shaft turbine develops very high starting torque, making it ideal for vehicular applications. When the vehicle is stationary, the power turbine, which is geared to the rear wheels, does not rotate, but the first-stage turbine idles at approximately 20,000 rpm. As the vehicle accelerates from a stop, first-stage turbine speed rises, reaching a maximum of 45,000–50,000 rpm. The burn-

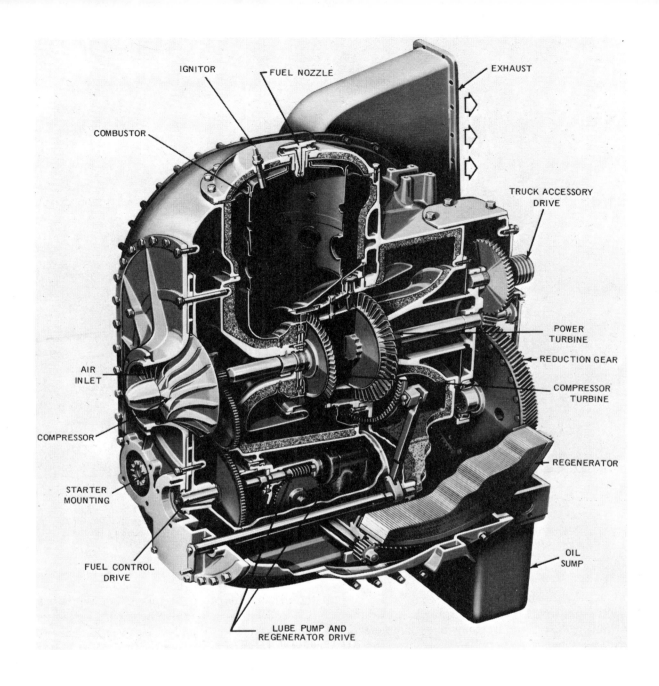

IGNITOR — FUEL NOZZLE — EXHAUST

COMBUSTOR

TRUCK ACCESSORY DRIVE

POWER TURBINE

REDUCTION GEAR

AIR INLET

COMPRESSOR TURBINE

COMPRESSOR

REGENERATOR

STARTER MOUNTING

OIL SUMP

FUEL CONTROL DRIVE

LUBE PUMP AND REGENERATOR DRIVE

Fig. 1   Cutaway of a Ford Motor Co. turbine

ing gases, after passing through the first-stage turbine, strike the blades of the second-stage turbine, turning it and the driving wheels of the vehicle. At maximum speed, the second-stage turbine is turning at approximately the same speed as the first stage. The effect is similar to that of the hydraulic torque converter used in automatic transmissions. The engine can't stall, and maximum performance can be achieved with a trans-

mission of fewer gear ratios than required with a piston engine.

Because of the high operating speed of the second-stage turbine, reduction gearing is necessary for turbine-powered vehicles. The shaft on which the second-stage turbine is mounted drives the input gear, which meshes with an output gear mounted on the output shaft. Gear ratio is usually about 10 to 1, reducing output shaft speed to 1/

Fig. 2   A prototype Ford 707 gas turbine installed in Ford tractor, one of a fleet of turbine-powered trucks running in regular service between Ford plants in Michigan and Ohio. A similar turbine has been installed in a Continental Trailways cross-country bus. The gas turbine is designed to replace diesel engines in the 300- to 450-hp range

10th of power turbine speed.

Both the single-shaft and split-shaft turbines previously described are known as simple-cycle turbines. Once the burning gases have passed through the two turbine stages, they escape to the atmosphere, still at temperatures of 1,000–1,200 degrees F. Exhausting burned gases at such high temperatures has two big drawbacks. First, much of the energy in the expanding gases is wasted, resulting in high fuel consumption. Second, the emission of hot gases from a motor vehicle could be hazardous.

REGENERATIVE TURBINES

To extract more energy from the burning gases and reduce exhaust temperature, designers added a regenerator. This device is simply a heat exchanger, which transfers some of the heat from the exhaust gases to the intake air. While a stationary heat exchanger can be used, most automotive gas turbines employ rotating regenerators, geared to turn very slowly. In one turbine designed for vehicular use, the two regenerators turn at 9 rpm while the gasifier turbine and compressor are spinning at 18,000 rpm.

The regenerator can be made in the form of a disc or a drum. It is of honeycomb construction, similar to an automobile radiator, but made of material such as stainless steel, capable of withstanding high temperature. As the regenerator rotates, air at relatively low temperature and pressure passes through one portion, while exhaust gases at high temperature and pressure flow through another area. The incoming air picks up heat from the hot surfaces of the regenerator, which have been exposed to the exhaust gases. As a result, temperature of the air entering the combustor is raised by several hundred degrees.

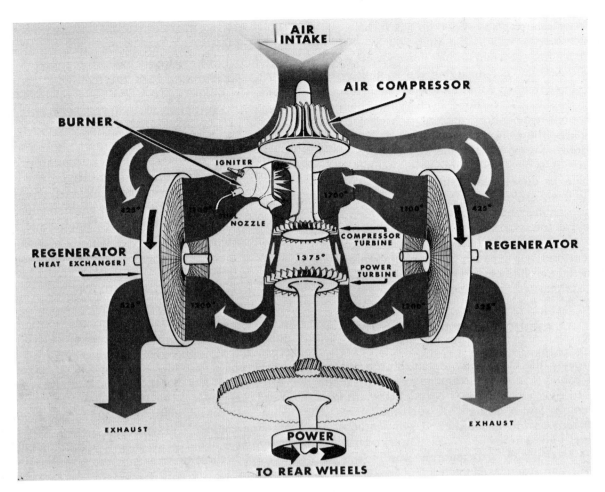

Fig. 3   Chrysler turbine operational flow

Without a regenerator, it would be necessary to burn additional fuel to achieve the same increase in temperature. Because heat is transferred from exhaust gases to incoming air, exhaust temperature is reduced to a safe level.

Tracing the flow of air, fuel and burning gases in a turbine aids in the understanding of the engine's cycle. The example used is the Chrysler turbine installed in experimental passenger cars loaned to selected drivers as part of Chrysler's field testing program, Fig. 3. The turbine was a split-shaft design with dual regenerators, mounted on opposite sides of the engine to ensure even heat distribution and prevent the development of thermal stresses.

Air enters the compressor, where it is compressed to approximately four times atmospheric pressure, increasing its temperature to about 425 degrees F. It then flows through the slowly revolving regenerators, raising the temperature to 1,100 degrees. The hot air enters the burner, where fuel is added. When the engine is first started, an electric igniter, similar to an oversized automotive spark plug, produces the spark to ignite the mixture of fuel and air. Once combustion begins, it is continuous; the electric spark is no longer needed.

The burning gases reach a temperature of approximately 1,700 degrees, as they leave the burner and flow to the first-stage turbine. A portion of their energy is expended in driving this turbine, which furnishes power for the compressor and all accessories. In the process, the gases give up some of their heat, leaving the first-stage turbine at a temperature of about 1,375 deg. They then strike the blades of the second-stage, or power, turbine, turning the output shaft, which is geared to the driving wheels of the vehicle. After passing through the blades of the second-stage turbine, the gases cool to about 1,200 deg. They then flow through the regenerators, reducing their temperature to approximately 525 deg. as they escape from the exhaust stacks.

All of the engine's useful power comes from a power turbine that is smaller in diameter than an ordinary dinner plate. A 200-hp gas turbine, with all accessories, weighs less than 500 lb. A diesel truck engine of comparable horsepower weighs 1,600–2,000 lb., not including the cooling system and clutch that are not needed with the turbine.

The split-shaft turbine, even when equipped with dual regenerators, has only four major moving parts—the compressor and compressor turbine assembly, the power turbine and the two regenerators. A typical V8 reciprocating engine, either gasoline or diesel, has 83 major moving parts, including a crankshaft, camshaft, flywheel, 8 pistons, 8 connecting rods, 16 valves, 16 push rods, 16 valve lifters and 16 rocker arms. The turbine also requires fewer bearings, needs no cooling system and has a much simpler ignition system than the gasoline engine and a fuel system that's less complicated than that of a diesel.

REDUCED AIR POLLUTION

Because the gas turbine operates on a mixture that contains a high percentage of air in proportion to fuel, combustion is complete, resulting in extremely low emissions of such pollutants as carbon monoxide and unburned hydrocarbons. As federal and state exhaust emission standards become increasingly stringent, the turbine's inherently clean exhaust characteristics make it more attractive to engineers. Unlike the reciprocating engine, it does not require extensive modifications to conform to the new regulations.

COLD-WEATHER OPERATION

The turbine is particularly well-suited to cold-weather operation. It spins almost as freely at 0 deg. as it does at 100 deg. because there's no film of congealed oil between sliding surfaces to put an additional load on the starter motor. The igniter produces repeated sparks until the fuel mixture starts to burn. There's no need for precise timing of the spark.

A turbine engine requires no warm-up. Once combustion starts and the gasifier turbine reaches idle speed, the engine is capable of picking up full load, without stalling, regardless of outside temperature. For the comfort of operator and passengers, heat is available a few seconds after the engine starts.

Despite its many basic advantages, the turbine has not yet achieved commercial success in vehicular use. Most applications have been in specialized vehicles, such as oil-field servicing trucks, mobile log chippers and other machines in

which the turbine's favorable power-to-weight ratio outweighs its two major disadvantages of high initial cost and heavy fuel consumption.

The gas turbine's high fuel consumption and expensive selling price are closely related. Efficiency of the engine goes up sharply with increases in operating temperature, but the maximum permissible operating temperature is limited by the ability of material in the turbine wheels to withstand the tremendous heat. High-priced alloys are capable of operating at elevated temperatures, but they increase the cost of the engine. At present, temperature at the first-stage turbine inlet is limited to about 1,700–1,750 deg.

FUEL CONSUMPTION

Under the conditions where it works best (heavy load and relatively constant speed), a regenerative turbine can match or exceed the fuel mileage of a gasoline engine and even approach the economy of a diesel. At light load and varying speeds, however, even the most efficient turbine consumes considerably more fuel than does a gasoline engine. Without a regenerator, fuel consumption is about twice that of a gasoline engine and nearly three times as much as a well-designed diesel.

For vehicular use, the turbine has two other serious handicaps—lag in throttle response and little or no engine braking. The delay in throttle response is due to the necessity of increasing first-stage turbine speed from 20,000 to about 45,000 rpm before maximum torque can be applied to the second-stage turbine. There is no engine braking in a split-shaft turbine because there is no mechanical connection between the gasifier section, in which compression takes place, and the power section, which drives the wheels. When a vehicle powered by a piston engine descends a grade with closed throttle, the flow of power is reversed, the driving wheels turning the engine, instead of the engine turning the wheels. Engine compression acts as a brake to help slow the vehicle.

One method of solving both of these problems is a variable nozzle arrangement. In most turbine engines, burning gases are directed against both the first-stage and second-stage turbines through fixed nozzles, each in the form of a ring of airfoil-shaped vanes. The vanes are positioned to ensure that the gases strike the turbine blades at the angle that will provide maximum efficiency through most of the engine's operating range.

To increase efficiency and provide engine braking, one form of variable nozzle system employs a ring gear meshing with sectors attached to each vane. Positioning of the vanes is accomplished by hydraulic pressure, controlled by signals from the accelerator pedal and the transmission. At idle, the vanes are positioned to direct the hot gases straight into the power turbine blades—90 degrees from the direction of rotation. As the accelerator pedal is depressed, the vanes turn, changing the flow of gas to the direction of the turbine's rotation.

For engine braking, the variable nozzle responds to signals from both accelerator pedal and transmission. A signal from the transmission governor tells the hydraulic control system if the vehicle is coasting at sufficient speed to require engine braking. If it is, releasing the accelerator pedal causes the vanes to shift to reverse position, directing the hot gases against the power turbine opposite to the direction of rotation.

THE FUEL SYSTEM

The gas turbine's fuel system is basically simple, but it needs several controls to protect it against itself—devices that prevent overheating, overspeeding and hung starts. Because the turbine always operates with an excess of air, there is no need to control the proportions of fuel and air, as in a gasoline engine. In most systems, a transfer pump, usually driven electrically, pumps fuel from the supply tank to the high-pressure pump, which is driven by the gasifier section of the turbine. The rate of fuel flow is controlled jointly by accelerator pressure and signals from the governor, also driven by the first-stage, or gasifier, turbine. When the vehicle is coasting, with accelerator released, fuel flow is shut off.

Should first-stage inlet temperature exceed a safe level, temperature probes signal the pump to reduce the flow of fuel. Some turbines are also protected against hung starts. This condition oc-

curs when the charge of fuel and air ignites but the first-stage turbine does not turn the compressor fast enough to produce the excess air required to cool the turbine. If the first-stage turbine does not reach the required speed within a specified time, fuel is shut off to prevent overheating of the turbine blades.

THE BURNER

The burner, or combustor, looks like a tin can with a series of holes drilled around its circumference. Compressed and heated air enters the burner and is mixed with fuel injected from a nozzle mounted in the burner cover. The igniter, which works in the same manner as an ordinary spark plug, is also installed in the burner cover.

THE IGNITION SYSTEM

In an aircraft or industrial turbine, an electric spark is needed to ignite the fuel charge only when the engine is started. Turbines for land-based vehicles often incorporate a fuel shut-off, which operates when the vehicle is decelerating, reducing exhaust emissions and conserving fuel. To prevent the engine from stalling, the flame must be re-lit when the engine returns to idle speed or is accelerated. The simplest method of accomplishing this task is to use an ignition system that fires continuously. A typical unit employs a set of cam-driven points, similar to those in a conventional automotive distributor; a coil; a condenser; and a shielded igniter. The igniter fires 80 to 200 times per second, depending upon engine speed. The system is much simpler than that of a gasoline engine because the spark is applied at only one point, instead of six or eight, and it need not be timed to occur at a specific stage in the engine's cycle. The spark merely starts the fire; if it's early or late, the engine doesn't know the difference. The turbine won't knock, lose power or overheat because of faulty timing.

Review Questions

Page

1. What are the major advantages of the gas turbine for motor vehicle use?   320

2. What fuels are suitable for gas turbines?   320

325

3. What are the four basic parts of a simple turbine? 320

4. What is the function of the second-stage turbine? 320

5. Why is the split-shaft turbine ideal for vehicular use? 320

6. Why is reduction gearing necessary for turbine-powered vehicles? 321

7. What is the basic difference between the simple-cycle and regenerative turbines? 322

8. How much does the regenerator increase inlet air temperature? 322

9. How many major parts are there in a split-shaft turbine? 324

10. What are the two major disadvantages of the gas turbine? 324

11. How does fuel consumption of the gas turbine compare with that of gasoline and diesel engines? 324

12. What is the function of a variable nozzle? 324

13. What is a hung start? 325

14. Where is the fuel nozzle mounted? 325

15. Why is it necessary for the ignition system of a vehicular turbine to fire continuously? 325

Cooling System

Review Questions for This Chapter on Page 306

INDEX

INTRODUCTION

Anti-freeze compounds 340
Anti-freeze data 340
Caps, radiator 332
Cap, radiator pressure....... 332
Capacity, extra cooling...... 340
Cooling capacity, extra...... 340
Cooling system corrosion..... 341
Cooling system, Corvair...... 338
Cooling system damage...... 340
Cooling system, how it works.. 327
Corrosion, cooling system.... 341
Corrosion damage 342
Corvair cooling system...... 338
Drain cocks and plugs....... 332
Distributing tube, water...... 331
Fan 330
Hose 337
Overcooling 341
Overheating 341
Overheating in traffic........ 339
Pressure cap, radiator....... 332
Pump, water 329
Radiator 331
Radiator caps 332
Radiator pressure cap....... 332
Rust formation, effects of.... 342
Rust prevention, importance of 342
Temperature control, water... 336
Thermostats 334
Tube, water distributing...... 331
Water boils, when........... 338
Water distributing tube 331

Water pump 329
Water temperature control.... 336

PREVENTIVE MAINTENANCE

Belts, fan blades and........ 348
Blades, and fan belts....... 348
Cap, radiator pressure....... 348
Cylinder head joint leakage... 346
Fan blades and belts........ 348
Hose leakage 345
Leakage 344
Pressure cap, radiator....... 348
Radiator leakage 344
Radiator pressure cap....... 348
Thermostat 347
Water jacket leakage........ 346
Water level 343
Water pump leakage........ 346

SERVICE OPERATIONS

Anti-freeze protection 356
Anti-freeze, testing 356
Belt adjustments 359
Belts, drive 358
Cleaning the system........ 354
Compounds, stop-leak 360
Cooling system, clean....... 354
Cooling system, flushing..... 355
Fan belt 358
Flow tester, radiator........ 359
Flushing the system........ 355

Hose inspection 357
Hose, replace 358
Hydrometer, use of........ 357
Plug, welch, renew........ 359
Pump service, water........ 360
Radiator flow tester........ 359
Radiator removal 363
Stop-leak compounds 360
Testing anti-freeze 356
Water distributing tube 331
Water pump service 360
Welch plug, renew 359

TROUBLE SHOOTING

Air suction test............ 351
Cap, test, pressure........ 352
Combustion leakage test.... 351
Cooling troubles, air cooled
    engine 353
Coolant, frozen 349
Foaming 350
Frozen coolant 349
Gauge, test, temperature.... 353
Pressure cap test.......... 352
Radiator clogging 353
Temperature gauge test...... 353
Test, air suction 351
Test, combustion leakage.... 351
Test in emergencies........ 348
Test, pressure cap......... 352
Test, temperature gauge..... 353
Test, thermostat 352
Thermostat test 352

Introduction

HOW COOLING SYSTEM WORKS

The purpose of the cooling system is to prevent temperatures of cylinders, pistons, valves and other engine parts from rising high enough to destroy them or to destroy the oil which lubricates them.

The cylinder bores and cylinder head are surrounded by water passages called water jackets. The water pump draws water from the bottom of the radiator and sends it through the water jackets in the cylinder block, then up into the cylinder head from whence it is delivered to the top tank of the radiator. The water is cooled as it flows down through the radiator passages by air flowing through the core, impelled by both the fan and the forward motion of the car.

Fig. 1 shows schematically how the coolant circulates through the system in a V8 engine. The coolant is drawn from the bottom of the radiator by the water pump and delivered to both cylinder blocks simultaneously. The coolant circulates

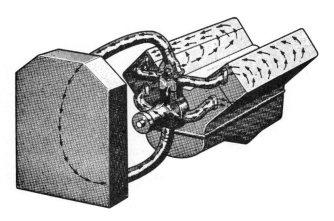

Fig. 1 Schematic drawing showing flow of coolant through a V8 engine

The liquid in the cooling system may be either water or a mixture of water and anti-freeze. The term "coolant" is commonly used to describe both of them.

New cars are equipped at the factory with a thermostat that is designed to open at 180 to 200 degrees. Therefore, maximum water temperature will not exceed 5 degrees above the designed opening temperature of the thermostat.

In other words, cooling temperature never exceeds these figures unless the cooling system is below par or unless the car is driven at high speed and/or the atmospheric air temperature is more than 100 degrees.

Fig. 2 illustrates a typical cooling system. Arrows show the direction of water flow. The water pump and fan are mounted on opposite ends of the same shaft. A pulley attached to the fan hub is driven by a fan belt connected to a pulley on the front end of the crankshaft.

The water inlet passage at the top of the radiator is connected to the cylinder head by rubber hose. Likewise the water outlet passage from the lower radiator tank is connected to the pump by rubber hose.

A thermostat valve in the cylinder head is auto-

around the cylinders and up through drilled holes to the cylinder heads. After circulating through the heads, the coolant flows through the thermostat housing (also called water manifold) which is located at the top of the water pump.

To sum it up: Water cools the engine and air cools the water. Anything which prevents this water-air system from working properly will cause overheating.

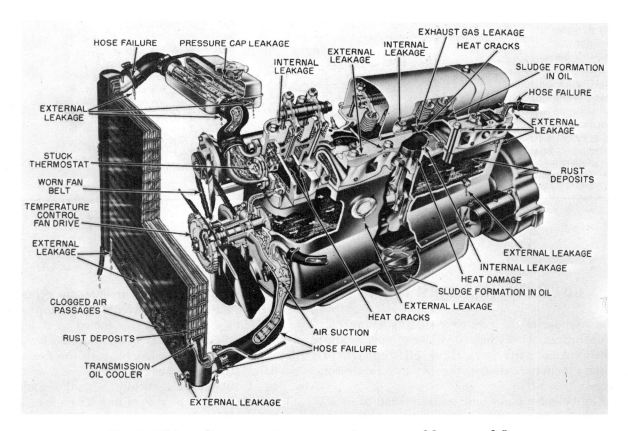

Fig. 2 This cooling system incorporates features used by many different car makers. Legends indicate operation and service check points

Fig. 3 *Cutaway view of engine showing how water passages communicate between cylinder block and cylinder head (arrow) and how water distributing tube (when used) directs stream of water to exhaust valve passages through which the hot burned gases pass*

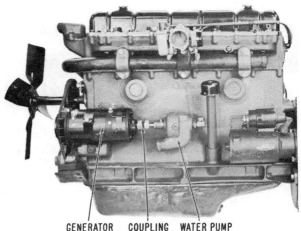

GENERATOR   COUPLING   WATER PUMP

Fig. 4 *Side-mounted water pump driven by generator*

matically closed when the engine is cold and open when the engine is warm.

When closed, it prevents flow of water through the radiator. Therefore, the heat of combustion quickly warms the water in the water jackets to normal operating temperature. Not until then does the thermostat start to open and permit flow of water to the radiator. Therefore, the engine warms up in minimum time. The thermostat keeps the cooling system at normal temperature in cold weather, as will be explained later.

The cylinder head is bolted to the cylinder block. The joint between head and block is sealed by a cylinder head gasket which prevents leakage of water and combustion gases. There is a series of holes which permit flow of water from block to head, Figs. 2 and 3.

WATER PUMP

The water pump is mounted on the front end of the cylinder block, Fig. 2, except on Rambler In-

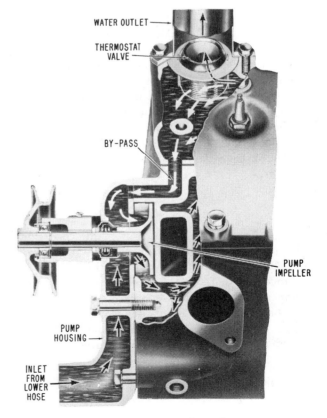

WATER OUTLET

THERMOSTAT VALVE

BY-PASS

PUMP IMPELLER

PUMP HOUSING

INLET FROM LOWER HOSE

Fig. 5 *Arrows show flow of water through pump*

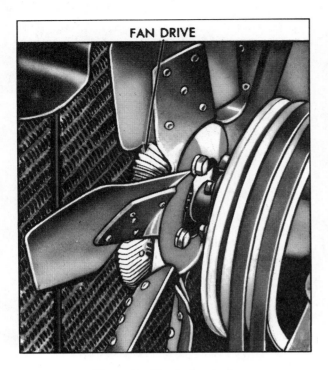

Fig. 6 Typical variable-speed fan installed

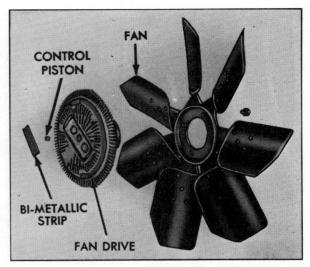

Fig. 7 Variable-speed fan with flat bi-metal thermostatic spring

Line engines up to 1956, which have the pump on the side, Fig. 4. When the pump is placed at the front of the cylinder block, the front end of the pump shaft is used to drive the fan on most cars.

The water pump is always a centrifugal type. The rotation of the impeller blades, Fig. 5, throws the water radially outward by the action of centrifugal force, delivering the water to the pump discharge openings at considerable pressure. At the same time, suction is created at the pump intake to draw water from the bottom of the radiator.

FAN

A fan usually consists of four blades, occasionally more, bolted together with a drive pulley to a hub. Ordinarily, this hub is mounted on the water pump drive shaft, the fan belt driving both the pump and the fan. Drive pulley and fan blades can be removed without disturbing the water pump. On some models a variable speed fan drive system is incorporated with the fan drive.

Variable Speed Fans

The fan drive clutch, Fig. 6, is a fluid coupling containing silicone oil. Fan speed is regulated by the torque-carrying capacity of the silicone oil.

The more silicone oil in the coupling the greater the fan speed, and the less silicone oil the slower the fan speed.

Two types of fan drive clutches are in use. On one, Fig. 7, a bi-metallic strip and control piston on the front of the fluid coupling regulates the amount of silicone oil entering the coupling. The bi-metallic strip bows outward with a decrease in surrounding temperature and allows a piston to move outward. The piston opens a valve regulating the flow of silicone oil into the coupling from a reserve chamber. The silicone oil is returned to the reserve chamber through a bleed hole when the valve is cloed.

On the other type of fan drive clutch, Fig. 8, a heat-sensitive, bi-metal spring connected to an opening plate brings about a similar result. Both units cause the fan speed to increase with a rise in temperature and to decrease as the temperature goes down.

In some cases a Flex-Fan is used instead of a Fan Drive Clutch. Flexible blades vary the volume of air being drawn through the radiator, automatically increasing the pitch at low engine speeds.

Fan Drive Clutch Test

Run the engine at a fast idle speed (1000 rpm) until normal operating temperature is reached. This process can be speeded up by blocking off the front of the radiator with cardboard. Regardless of temperatures, the unit must be operated

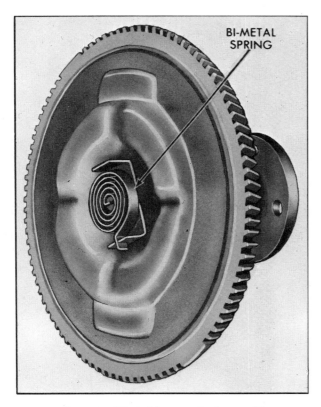

Fig. 8 *Variable-speed fan with coiled bi-metal thermostatic spring*

BI-METAL SPRING

together and separate the fan from the drive clutch. Next remove the metal strip on the front by pushing one end of it toward the fan clutch body so it clears the retaining bracket. Then push the strip to the side so that its opposite end will spring out of place. Now remove the small control piston underneath it.

Check the piston for free movement of the coupling device. If the piston sticks, clean it with emery cloth. If the bi-metal strip is damaged, replace the entire unit. These strips are not interchangeable.

When reassembling, install the control piston so that the projection on the end of it will contact the metal strip. Then install the metal strip with any identification numerals or letters facing the clutch. After reassembly, clean the clutch drive with a cloth soaked in solvent. Avoid dipping the clutch assembly in any type of liquid. Install the assembly in the reverse order of removal.

The coil spring type of fan clutch cannot be disassembled, serviced or repaired. If it does not function properly it must be replaced with a new unit.

WATER DISTRIBUTING TUBE

On some engines a water distributing tube is inserted in the water jacket (see Fig. 3). The front end of the tube receives the water directly from the pump. Holes in the tube discharge jets of water to provide extra cooling for the exhaust valve seats and valve stems.

RADIATOR

There are two types of radiators presently in use. The first is the downflow variety in which the coolant moves vertically from top to bottom. The other type is the crossflow in which liquid movement is from one side of the radiator to the other. Heat transfer from the coolant to the outside air takes place in the radiator core which consists of many vertical or horizontal passages which terminate on either end in tanks. Downflow systems have two tanks, at the top and bottom of the core while the crossflow radiator has its tanks mounted on the sides, at left and right. On the crossflow, water inlet is still at the top of one of the side tanks and it exits at the bottom of the other side tank.

The various parts of the radiator are soldered together. The components are made of copper or brass, in preference to other metals, because of their high heat conductivity, ease of soldering, and resistance to corrosion. In recent years alumi-

for at least five minutes immediately before being tested.

Stop the engine and, using a glove or a cloth to protect the hand, immediately check the effort required to turn the fan. If considerable effort is required, it can be assumed that the coupling is operating satisfactorily. If very little effort is required to turn the fan, it is an indication that the coupling is not operating properly and should be replaced.

Service Procedure

CAUTION: When it becomes necessary to remove a fan clutch of the silicone fluid type, Fig. 7, the assembly must be supported in the vertical (on car) position to prevent leaks of silicone fluid from the clutch mechanism. This loss of fluid will render the fan clutch inoperative.

The removal procedure for either type of fan clutch assembly is generally the same for all cars. Merely unfasten the unit from the water pump and remove the assembly from the car.

The type of unit shown in Fig. 7 may be partially disassembled for inspection and cleaning. Take off the capscrews that hold the assembly

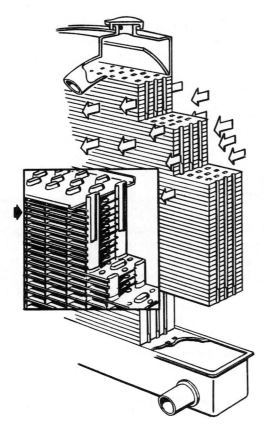

Fig. 9 Details of a
tubular radiator core

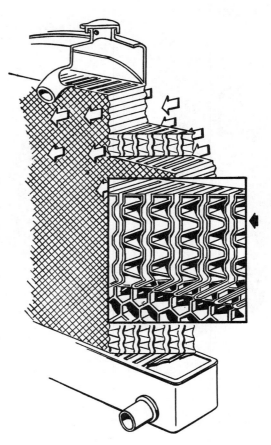

Fig. 10 Details of a
cellular radiator core

num has been used to some extent for this purpose. Regardless of the metals used, radiator cores are of two general types, tubular and cellular.

The radiator core, Fig. 9, consists of tubes to which thin copper or brass fins are soldered in order greatly to increase the radiating surface. The tubes may be round, or flattened as shown.

In the cellular core, Fig. 10, the water passages and air fins are made by soldering preformed sheets of thin copper or brass as shown. The water passages are about 1/16″ wide.

The top tank of the radiator is equipped with a baffle which distributes the water over the top of the radiator core so that all the radiator passages are fully supplied.

DRAIN COCKS & PLUGS

There is always a drain cock at the bottom of the radiator and there may be one or more drain cocks in the cylinder block to drain the water out of the water jackets. Screw plugs are sometimes used instead of drain cocks, especially in the cylinder block.

RADIATOR CAPS

In the left view of Fig. 11 is illustrated a radiator cap which permits the cooling system to operate under atmospheric air pressure since the overflow pipe is open to the atmosphere. There is slow but constant evaporation of water. Water may also be lost through the overflow pipe whenever the brakes are applied with full force. This results in a high surge of water in the top tank—when some of it escapes through the overflow pipe. The overflow pipe, in spite of its faults, is an essential part of the cooling system. Without it, the radiator or hose connections would burst whenever the cooling system boils. Bursting would occur whenever the radiator were completely filled with water. Later, when the water was heated, expansion would create an irresistible pressure.

RADIATOR PRESSURE CAP

Pressure caps are standard equipment on all cars made in the past several years. The cap, Fig.

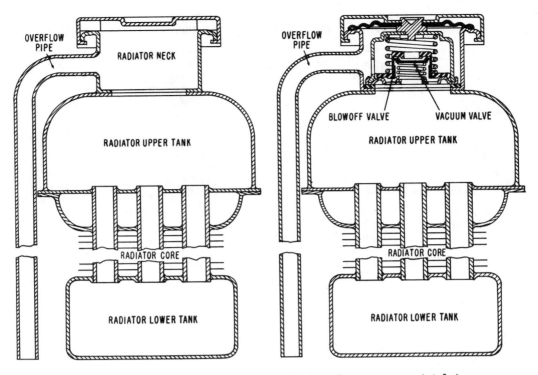

Fig. 11 *Atmospheric radiator cap* (*left*) *and pressure cap* (*right*)

12, has a spring closed valve which opens at some designated point above atmospheric pressure. Each make and model car have cooling systems designed to operate at a certain pressure above atmospheric and caps are supplied for this specific pressure. There are, therefore, caps available with pressure ratings ranging in steps from 4 to 19 pounds. Thus equipped, water coolant in these systems has its boiling temperature raised from 225 for the 4 pound caps to 257 degrees for the 19 pound systems. In the range under consideration, every pound of pressure put on the cooling system by means of the pressure cap raises the boiling temperature about 2.5 degrees.

Actual pressure required to open a certain rated cap may vary slightly. So if a pressure cap designed to open at 14 pounds in fact opens at anything between 13 and 15 pounds it is acceptable.

Fig. 13 shows a pressure cap and filler neck while Fig. 14 illustrates the details of a typical cap. In the left picture the large coil spring seals the pressure valve and its rubber gasket tight against its seat in the filler neck. The picture at the

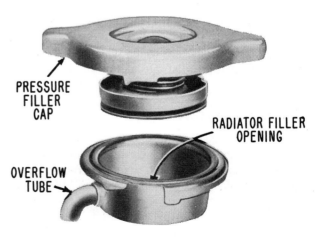

Fig. 13 *Radiator pressure cap and filler neck*

Fig. 12 *Radiator pressure cap*

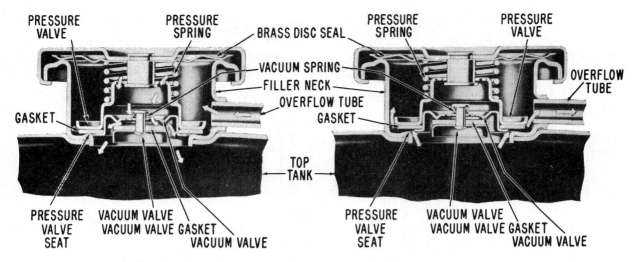

Fig. 14 *Details of pressure cap showing vacuum relief (left) and pressure relief (right)*

right shows that pressure in the cooling system has opened the pressure valve, allowing discharge of steam and water through the overflow pipe as indicated by the white arrows. Note that the top edge of the filler neck is sealed by a springy brass disc attached to the cap.

If the coolant boils, there will be a rapid condensation of steam when the engine is shut off and, in consequence, several inches of vacuum may be created within the cooling system. As a result, the end tanks could be crushed inward. To avoid this possibility, all pressure caps are equipped with a small vacuum valve, one type is shown in Fig. 14. The picture at left shows this vacuum valve open with atmospheric air flowing into the system by way of the overflow pipe as indicated by the white arrows. When there is no vacuum in the radiator, the valve is closed by the small coil spring shown. Another type of vacuum relief valve has a weighted end that keeps the cap, and thus the system, vented to the atmosphere most of the time. It is only when the pressure builds up in the radiator that the valve is forced closed.

Closed Cooling Systems

So-called closed cooling systems are arranged like conventional systems except that they have an additional expansion tank. Usually connected to the inlet tank of the radiator, the expansion tank provides reserve cooling capacity and reserve expansion space. Generally attached to the radiator by means of a conventional radiator hose, Fig. 2, it is located at the highest point in the system. When such a closed system is installed, the pressure cap is placed on the expansion tank and the radiator itself has no cap. The place where it would ordinarily be is taken up by the tube leading to the expansion tank.

If the coolant boils, instead of simply leaving the system via the radiator cap and overflow pipe it can expand in the extra tank and, if the boiling is not severe, it will be cooled there and no liquid will be lost from the system. It is thus especially useful in cases of "afterboiling" such as occur when the engine is turned off immediately after a hard run. Under such circumstances the cooling system can not dissipate the accumulated heat rapidly enough and it will boil slightly for a short time. If severe boiling occurs in a closed system, due to some malfunction, the pressure cap will open just as it would in an ordinary system and coolant will be lost.

In another variation of the closed cooling system, the expansion tank, called in this case an overflow tank, is connected to a conventional radiator having a plain cap by means of a vent tube. Serving exactly the same function as the previous tank, the overflow unit can be located at any level in the system except the lowest point and is equipped, as before, with a pressure cap. Coolant is added through the regular filler neck and if it boils the radiator simply overflows into the extra tank. On cooling down, the system syphons the liquid back from the overflow tank.

THERMOSTATS

There are two kinds of thermostat valves, the poppet type, Fig. 16, and the butterfly type, Fig.

Fig. 16   Poppet valve type thermostat

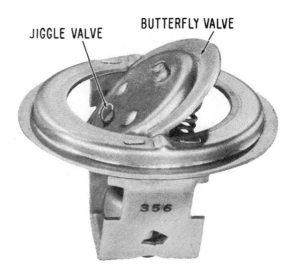

Fig. 17   Butterfly valve type thermostat

17. Both valves are shown fully open. There are three methods of opening a thermostat valve:

1. By a bellows which contains a volatile liquid whose vapor pressure expands the bellows and pushes the poppet valve open.

2. By the use of fine crystals (inside a small metal container called a pellet) which expand when heated and thus open the valve with the aid of suitable linkage.

3. By a bi-metal coil which expands when heated and thus opens the valve.

Bellows Type Thermostats

The bellows is partly filled with a liquid which boils at a low temperature. Before it is filled, the air is pumped out of it. Then the liquid is put in and the bellows is sealed. Consequently, at air temperature there is a vacuum in the bellows which holds the poppet valve closed against its seat. The liquid starts to boil at its opening temperature—160 degrees for example. As the temperature rises from this point, the vigor of boiling increases and the vapor pressure goes up until the thermostat is fully open when the temperature has risen 20 to 25 degrees or more.

Pellet Type Thermostat

The powder-like crystals are tightly packed inside the metal pellet, Fig. 18. When they expand they force the rod upward. The rod acts on a crank which rotates the butterfly shaft to open the valve. When they contract the valve is closed by the springs shown, which pull on the valve.

A different butterfly construction is shown in Fig. 19. The valve has no shaft but extensions on the butterfly are pivoted in holes in the housing. The spring holds the valve closed by the pull of its upper end on a slotted jaw which is riveted to the butterfly. The valve is opened by the pellet rod pushing upward on a piece which is attached to the slotted jaw.

Thermostat Bleed Hole

All thermostats have a small bleed hole (about 1/16″) in the valve to permit the escape of air and bubbles of steam from the water in the system to the top tank of the radiator when the thermostat is closed. These gases are carried through the hole along with a small stream of water. Some thermostats have a "jiggle valve," Fig. 17, consisting of a rod in the escape hole. A small weight is hung from the rod. The jiggling of the weight and rod prevents clogging of the hole. Sometimes the bleed hole is located in the radiator outlet.

Thermostat Temperature for Alcohol

If alcohol anti-freeze is used, a thermostat which opens at 160 degrees should be used because a thermostat which opens at a higher tem-

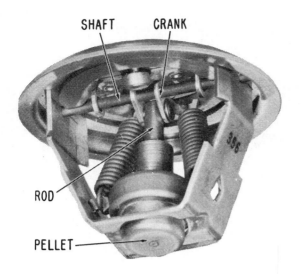

Fig. 18 *Details of pellet type thermostat*

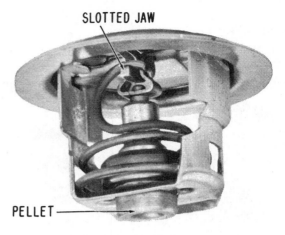

Fig. 19 *Another type of pellet type thermostat*

perature will permit the alcohol to boil, especially if the car is driven hard in mild winter weather.

Thermostat Temperature for Heater

The effectiveness of the car heater is proportional to the coolant temperature. Therefore, a thermostat that opens at a high temperature will keep the car warmer. If the thermostat opens at a higher temperature than 160 degrees, a permanent type antifreeze must be used.

WATER TEMPERATURE CONTROL

In order to explain how the thermostat controls water temperature, let us assume that the cooling system is in normal condition. We also assume that it is designed with enough capacity to operate without over-heating when the air temperature is 100 degrees.

Assume, for example, that the thermostat starts to open at 160 degrees and is fully open at 180 degrees. When the engine is started from cold, the water temperature quickly rises to 160 degrees and the thermostat starts to open. If the weather is very cold, the thermostat will open a little, say to 165 degrees. If it tries to open more, the increased flow of water will cool it off and the thermostat will drop back to the 165-degree position.

Actually, however, a thermostat starting to open at 160 degrees will seek to maintain minimum temperature control subject to the influence of engine heat load and ambient (under hood) air temperature.

If the car is driven hard on a 100-degree day, the thermostat is wide open when 180-degree water temperature is reached. Recall that we assumed that the cooling system was designed for proper cooling at 100 degrees air temperature. If the air temperature is higher than 100 degrees and/or the cooling system is not in perfect condition, the water temperature will rise higher than 180 degrees. The thermostat has no longer any control over the water temperature.

There are two ways of using a thermostat to control water temperature: the choke type and the by-pass type.

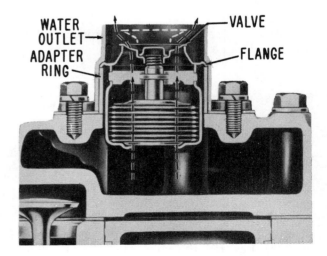

Fig. 20 *Choke type of temperature control. When thermostat is closed it blocks flow of water to radiator*

Fig. 21   A curved molded hose

Choke Type of Temperature Control

The choke type of temperature control is shown in Fig. 20. When the thermostat is closed, it blocks the flow of water to the radiator. Arrows show flow of water when the thermostat is open. In this picture, it would seem that the body of the thermostat blocks the flow of water even when it is open. But this is not true as Fig. 16 clearly shows.

By-Pass Type of Temperature Control

The by-pass type is illustrated in Fig. 5. When the thermostat is open, water flows as indicated by the black arrows. It enters at the pump inlet, flows to the pump impeller and from thence flows through the cylinder block, and up into the cylinder head. Then it flows forward through the head to the front where it goes upward through the open thermostat valve to the radiator outlet which is connected to the top of the radiator.

When the thermostat is closed, no water can flow through the radiator and, therefore, the pump takes its supply from the cylinder head as shown by the white arrows and discharges it into the block as indicated by the black arrows, Fig. 5. Thus the pump simply circulates the water through the engine until the thermostat opens. When the thermostat opens a little, water still flows to the pump through the small entry passage but the bulk of the water flows out of the very much larger outlet passage and from thence to the radiator.

The by-pass water passage is clearly shown in Fig. 5. When the thermostat is closed, the water pump draws water from the cylinder head (see arrow) and pumps it into the cylinder block through passages not shown; the direction of water flow is indicated by an arrow.

When the thermostat valve is open, the pump draws water from the lower part of the radiator and delivers it to the cylinder block as indicated by the lower arrow. A small amount of water still is circulated through the by-pass passage.

Thermostat Replacement

To remove a thermostat, first remove the water outlet housing (also called the thermostat housing) after taking out the flange screws. Remove the gasket. Inspect the thermostat. Clean gasket surfaces.

Be careful to install the thermostat valve right side up. The bellows or capsule should be on the underside in order to be in contact with the hot water in the cylinder head. Some thermostats are stamped front or top. Be sure to install them this way.

On some by-pass types, there is a rubber gasket above the valve—usually a rubber washer which fits between the top of the thermostat and the housing. If this gasket is not installed, the thermostat will give no temperature control because the water flows unhindered through the space where the gasket should be. If an upper gasket is required, it is packaged with the new thermostat.

When installing a thermostat, coat the gasket surfaces with gasket cement. Use a new gasket. After installing, be sure to tighten the screws snug. Then after the engine has been brought up to normal operating temperature test for leakage.

Selecting a Thermostat

There are several manufacturers of thermostats and all of them make a line of thermostats to fit all cars. The opening temperature of the thermostat is marked on it.

Note that if there is alcohol anti-freeze in the cooling system it is not advisable to use a thermostat that opens at a higher temperature than 160 degrees because alcohol is likely to be lost, especially during warm spells in winter. This may happen even with a pressure cap because of "after boil," which means that boiling occurs after the engine is stopped.

When a hot engine is shut off, the coolant in the water jackets must cool off the hot engine. The absorption of this heat may cause the alcohol in the water to boil and the pressure created by the boiling may open the pressure cap and alcohol will be lost.

HOSE

Hose for cooling systems and heaters is made with inside diameters of 3/8″, ½″, 5/8″, ¾″, 1″, 1⅛″, 1¼″, 1½″, 1¾″, 2″, 2¼″, 2½″, 2¾″, 3″. The smaller diameter hoses listed are for heater connections.

Fig. 22 A flexible hose

There are various types of hose to meet a variety of requirements as follows:

1. Straight hose cut to correct length.

2. Straight hose in three-foot lengths or longer. It may be used for slight bends.

3. Curved molded hose which is made to exact length, Fig. 21.

4. Curved molded hose. A section is cut from the hose to match the old hose removed from the car. This hose is suitable for both upper and lower radiator connections.

5. Flexible hose, Fig. 22. This type of hose is used to connect the lower tank of the radiator to the water pump. The water pump causes a suction in this type hose and, when necessary, it is equipped with a wire spring to prevent its collapse. It can be cut to various lengths so that 12 to 20 sizes will fit most cars.

CORVAIR COOLING SYSTEM

The engine is entirely shrouded with sheet metal pieces that attach directly to the engine and form a plenum chamber—a condition in which the pressure of the air in an enclosed space is greater than that of the outside atmosphere.

A centrifugal blower, mounted to the top of the crankcase cover, spins on a vertical shaft to deliver cooling air outward and downward over the cylinders and heads. The air then enters a duct under each bank from where it travels rearward to be exhausted at an opening at the rear of the engine.

The rate of engine cooling is regulated by a bellows-type thermostat at the lower part of the plenum. This thermostat operates a cooling air valve which moves in and out of the eye of the blower to control the air flow. The ring closes the blower air intake until the engine has reached its correct operating temperature. In the event of a failed thermostat bellows, the ring will remain in the open position to prevent overheating of the engine.

The blower, which runs on a sealed, permanently lubricated ball bearing, is belt-driven by a pulley mounted at the extreme rear end of the crankshaft. A generator drive pulley at the left rear of the engine and an idler pulley at the right rear provide a means of changing belt direction from a vertical plane at the crankshaft pulley to a horizontal plane at the blower pulley.

An oil cooler through which a portion of the air passes before discharge, is mounted above the air exhaust duct near the left rear corner of the engine.

WHEN WATER BOILS

Water boils at 212 degrees at sea level where the atmospheric pressure is 14.7 pounds per square inch. This pressure represents the total weight of air resting on each square inch of the earth's surface. Think of it as a one-square-inch column more than 100 miles high.

The air becomes progressively thinner as we go upward. At 10,000 feet there are only 10 pounds of air pressure on each square inch of the earth's surface and water boils at 193 degrees. At 15,000 feet, the air pressure is only 8.5 pounds and water boils at 185 degrees.

We may sum up these facts by stating that as pressure is reduced, the boiling point temperature is reduced because it is easier for steam bubbles to form as the pressure is lowered.

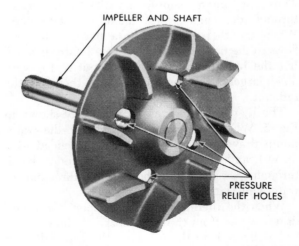

Fig. 23 Water pump impeller with curved blades for high-pressure operation

On the other hand, if pressure is increased beyond 14.7 pounds per square inch, it becomes more difficult for steam bubbles to form and, therefore, it requires a higher temperature to cause boiling.

The table below gives the temperature at which water boils with various radiator pressure caps. If the cap leaks, the boiling point is 212 degrees.

Pressure	Boiling Temperature
0	212
4	224
7	232
9	237
13	246
14	249

The effect of altitude on atmospheric pressure and boiling temperature is shown in the table below.

Altitude	Atmospheric Pressure	Boiling Temperature
0	14.7	212
2500	13.4	207
5000	12.1	202
7500	11.0	198
10000	10.0	193
12500	9.1	189
15000	8.5	184

What Pressure Cap Does

Under normal conditions the pressure in the cooling system will be maintained at the pressure which is stamped on the pressure cap. However, when overheating occurs, instead of the water boiling at 212 degrees as it would without a pressure cap, it does not boil until it reaches the temperature given in the table above.

This is a very desirable advantage because it often happens that faults in the cooling system might be just enough to cause boiling at 212 degrees but not enough to boil the water at some higher temperature. And it should also be realized that the higher the water temperature, the greater is the quantity of heat that the air absorbs, since the rate of absorption is proportional to the difference in temperature between the surrounding air and the water.

For example, if the air temperature is 100 degrees and the water temperature is 212 degrees, the difference is 112 degrees. With a 13-pounds pressure cap the water boils at 246 degrees; the temperature difference is 146 degrees, and the gain in cooling ability is 146/112 or nearly 30 per cent.

A pressure cap is of great value if something goes wrong with the cooling system at high altitude. For example, at 5000 feet, water boils at 202 degrees but with a 4-pound cap it cannot boil until the cooling water reaches 224 degrees.

The preceding remarks may have implied that pressure caps make cars less prone to overheating than cars made years ago. This is not quite true. Radiators today are smaller than they used to be because a new model is styled without much regard as to how much space is needed under the hood for the machinery. Consequently, in many cases the engineering department has been up against it to find enough room for a radiator of adequate size. At the same time, fancy grilles have interfered with the flow of air to the radiator. Pressure caps are the answer to the problem because they permit the water to rise beyond 212 degrees without boiling.

Because of modern styling, today's cars are likely to overheat in slow stop-and-go traffic driving in hot weather even though they will not overheat when driven at top speed when the atmospheric temperature is 100 degrees. The explanation is that one way to obtain satisfactory cooling capacity on the open road is to circulate the water much faster. This means that water pump pressure must be greatly increased.

Overheating in Traffic

The principal causes of overheating in traffic are (1) the inefficiency of centrifugal water pumps at low rpm; (2) smaller radiator frontal areas; (3) more underhood obstructions due to engine shape and accessory installations which create higher underhood temperatures and back pressures, and (4) torque converter heat loads which appreciably contribute to low speed stop and go overheating in hot weather.

The older type "low speed" engine water pump had an impeller with straight radial blades which had ample pumping capacity for idling and very low speed, as well as for higher speeds. To make a high volume pump for the modern high speed engine the impeller blades are curved, Fig. 23. Such an impeller can be designed for maximum efficiency at any desired pump speed or engine speed and at this speed it is much more efficient than the straight-blade pump. However, since its efficiency falls off rapidly when the pump is run at lower (or higher) speeds, modern pumps have greater cross-sectional area both at the intake and outlet, and thereby compensate for the curved shape of the impeller.

EXTRA COOLING CAPACITY

In excessively hot climates and/or if the engine is overloaded as by pulling a heavy trailer up long grades, extra cooling system capacity may be required. Methods of increasing the cooling capacity are described below. Most car makers offer one of these methods:

1. A slightly thicker radiator core.
2. A fin and tube radiator core with more fins, or a cellular core with more and finer water passages.
3. A radiator core made of copper instead of brass because copper is a better conductor of heat.
4. A higher speed fan obtained by reducing the diameter of the fan pulley or increasing the size of the crankshaft pulley.
5. A fan of greater capacity with five blades instead of four, wider blades and/or blades with greater curvature (pitch), or the fan diameter may be increased.
6. Closed cooling systems add capacity by using an auxiliary tank in addition to the usual two radiator tanks. Called expansion, reserve, surge and overflow tanks these devices serve several functions that in effect increase the heat rejection ability of the cooling system.

ANTI-FREEZE COMPOUNDS

The purpose of an anti-freeze compound is to depress the freezing point of the coolant so that it will remain fluid in freezing weather. There are only two anti-freeze compounds in common use, namely, methanol and ethylene glycol. (All anti-freeze base compounds should be properly inhibited to prevent rust and corrosion in the cooling system.)

Methanol is an abbreviation of methyl alcohol which is the chemical name for wood alcohol.

When adding anti-freeze to a car, it is necessary to know what kind of anti-freeze is already in the system. Methanol has a characteristic odor. Ethylene glycol has no noticeable odor.

When the word alcohol is used in this discussion it means methanol.

ANTI-FREEZE DATA

The addition of methanol alcohol or ethylene glycol base anti-freezes to water depresses the freezing point of the coolant. For example, pure ethylene glycol freezes at $-9°$ F. However, as water is added the freezing point is lowered. The maximum freezing protection of about 92° F below zero is given by a 66% pure ethylene glycol to 34% water solution.

The number of quarts of anti-freeze to use in a cooling system for whatever freezing point is desired is listed on charts supplied to service stations by the anti-freeze maker. The quantities of anti-freeze specified in these charts for prevention of freezing at a given temperature mean that no ice crystals will form if the temperature does not go any lower.

For example, a methanol chart will show that three quarts of methanol in an 18-quart cooling system prevents the formation of ice crystals down to 15 degrees. As the temperature drops below 15 degrees, more and more ice crystals are produced, until finally the radiator core and/or lower hose become clogged with slush ice which either greatly reduces the water circulation or stops it entirely; then the water in the water jackets boils. This explains why you see steaming cars in very cold weather.

To elaborate this point, when the engine is idling or the car is driven slowly on a cold day, the ice crystals will accumulate in the lower hose connection and when enough of them have formed the hose will be completely clogged. On the other hand, if the car is allowed to stand with the engine shut off, the ice crystals will form in the passages of the core and will eventually block it completely.

Slush cannot form if the cooling system contains enough anti-freeze to give full protection down to the lowest cold weather temperature anticipated. It is interesting to note, however, that if the cooling system contains 10 per cent or more of anti-freeze neither the cylinder block nor the radiator will be damaged because you have a mixture which is composed of 90 per cent ice crystals surrounded by 10 per cent liquid anti-freeze. This thick but semi-liquid mixture is unable to produce enough pressure to damage either the engine or the radiator.

COOLING SYSTEM DAMAGE

The regulation of engine operating temperature by the cooling system is indispensable for dependable performance, maximum power and economy of operation.

When a car is new all radiator passages, water jackets, hose and other parts of the cooling system are clean and are able to carry away waste heat efficiently. After awhile, however, if the system is not kept leak-proof and protected against corrosion, rust will form in the water jackets of the cylinder block and head and will be carried to the radiator during normal water circulation. This rust

Fig. 24 *Burned valves caused by overheating*

acts as a blanket and prevents the water from being properly cooled and restricts its circulation. If allowed to remain, these deposits will eventually shut off circulation entirely, resulting in engine overheating.

Overheating

Overheating not only causes engine knock and loss of power, but also will result in damage to bearings and other moving parts. Cylinder heads and engine blocks are often warped and cracked by terrific strains set up in the overheated metal, especially when cold water is added afterward without allowing the engine to cool.

If a car is operated with boiling water, steam pressure forces large quantities of water out of the cooling system through the radiator overflow pipe. More violent boiling occurs, and still more water is lost. Finally, water circulation is stopped and cooling fails completely. This means that operating an engine with boiling water for even a short length of time may be actually driving that engine to destruction. Fig. 24 shows a pair of burned valves caused by overheating.

Overcooling

Although less sudden in effect than overheating, overcooling may be equally dangerous to the engine. Low engine operating temperature, especially during freezing weather, results in excessive fuel consumption, dilution of engine oil by unburned fuel, and formation of sludge from condensation of water in cylinder and crankcase.

Lubrication failure may follow sludge formation and lead to serious engine damage. Burned fuel vapors also mix with water in the crankcase and form corrosive acids which attack engine parts. Fig. 25 are examples of piston and pin damage due to overcooling.

Fig. 25 *Piston and pin damage due to overcooling*

Cooling System Corrosion

A chemical combination of iron, water and air produces rust. The water jacket of the engine has a large mass of iron exposed to the cooling water, and no cooling system is free of air. Thus all elements of rust formation are found in the cooling system. Over 90 per cent of the solid matter that clogs radiators is rust.

The rate of rust formation in the cooling system is influenced by many conditions of service and operation. Air mixed with water can increase corrosion of iron as much as 30 times. The normal source of air in the cooling system is the radiator top tank. At high engine speeds, the rush of water into the radiator is great enough to drive air into the liquid and carry air bubbles down through the water tubes in the radiator. If the water level is allowed to drop as low as the top of the tubes, suction of the water pump will draw air into the overflow pipe and down through the tubes.

Heat speeds up corrosion and unfortunately the rate of corrosion with iron appears greatest at water temperature corresponding to best engine performance. Iron, solder and copper will corrode more than twice as fast at 175 degrees than at 70 degrees.

Some waters are less corrosive to iron than distilled or rain water. But others that contain dissolved mineral salt impurities are particularly

Fig. 26 Example of extreme rust formation in cylinder block

harmful to cooling system metals. Any acid condition in water will increase iron corrosion and rust formation. Hard water containing large amounts of lime and certain other minerals will deposit scale at "hot spots" in the engine water jacket if large quantities of such water are added to the cooling system over a period of time.

The cooling water may become contaminated as a result of extended service, a faulty condition within the system, or from improper maintenance. Excessive aeration from a neglected suction leak at the water pump or at any point between the pump and radiator speeds up corrosion and shortens the rust-free life of the coolant. Combustion gas dissolved in water from a leak at the cylinder head joint has a similar effect to aeration. Corrosive contamination of the water can also result from failure to neutralize and flush out cleaning solution.

Effects of Rust Formation

If rust deposits are allowed to build up in the water passages of the block or head, Fig. 26, they may hold enough heat in the metal to create local "hot spots", especially around the valve seats. Steam pressure from local boiling at such places is a hidden although common cause of overflow loss. The metal may get so hot as to cause sticking, warping, or burned valves, or even a cracked block or head.

Rust deposits have their most harmful effects in the radiator. Even a small amount of fine rust particles continually circulating through the radiator has a tendency to spread out in the form of a thin, hard scale on the inside of the narrow water tubes, Fig. 27. This scale first reduces cooling efficiency of the radiator by insulating the tubes from the water. As more rust becomes lodged in the tubes, circulation is restricted and boiling starts. When boiling starts, large amounts of rust are stirred up in the water jacket and carried over into the radiator. Further operation of the car will result in overheating, loss of power and engine damage.

Corrosion Damage

Although a less common cause of trouble than rust clogging, corrosion damage to metal parts can be equally serious. For example, when a water distributing tube in the block becomes perforated by corrosion, Fig. 28, water distribution in the water jacket is completely upset. Some valves and cylinders will be robbed of proper circulation and cooling, and hot spots, overheating, sticking valves, and even heat-cracking may follow.

Corrosion prevention is especially important for such parts that are so completely hidden within the engine that preventive inspection is impractical and detection of failure is difficult. Among other metal parts sometimes damaged by corrosion are radiators, water pumps, cylinder heads, Fig. 29, and core hole plugs. Thin metal parts are weakened by corrosion and crack more easily when subjected to vibration and strain.

Importance of Rust Prevention

A rusty cooling system may seem to function satisfactorily under moderate operating conditions

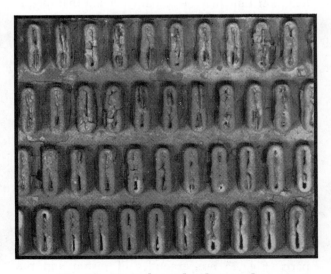

Fig. 27 Rust clogged radiator tubes

342

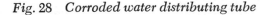

Fig. 28   Corroded water distributing tube

but will fail to cool the engine under more severe conditions, often just at the time when full power output is more urgently needed. The system can be kept practically rust-free, and loss of cooling efficiency from rust formation can be avoided by periodic corrosion preventive services.

Protection of the cooling system against rust corrosion is accomplished by adding an inhibitor to the cooling water. These inhibitors are commercially available, and laboratory tests have shown that a corrosion inhibitor in water reduces the normal rusting of iron at least 95 per cent.

Corrosion protection is particularly important during warm weather driving when water alone is used as a coolant, since there is more air in water and more rusting in the system. In very cold weather, control of coolant circulation by the thermostat may reduce the flow into the radiator to only a few gallons per minute and very little air is driven into the coolant.

In hot weather with thermostat wide open, the flow into the radiator at high engine speeds may increase to 100 gallons per minute or more in some engines. The resulting increase in coolant

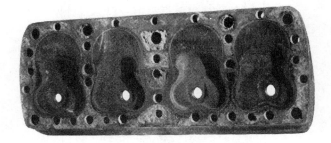

Fig. 29   Corroded cylinder head

aeration, together with a higher metal and coolant temperature, greatly speeds up the rate of rusting.

Many anti-freeze compounds contain a corrosion inhibitor. If one of these compounds is installed, no inhibitor of any kind should be added. If a cooling system is dirty enough to require cleaning, anti-freeze solutions would surely be contaminated and probably inhibitors depleted. The system should have a fresh fill of anti-freeze or inhibited water.

Preventive Maintenance

WATER LEVEL

The level of the water in the radiator is the starting point for proper cooling system preventive maintenance. Water level should be checked accurately and frequently for three reasons: (1) To make sure the system contains enough water; (2) as a guide to the condition of the cooling system; (3) to avoid overfilling.

If rust is to be avoided, it is important to make sure that the system contains a sufficient quantity of water at all times. Low water level may prevent proper circulation, especially at low engine speeds. At higher engine speeds, low water level allows a large volume of air to become mixed with the liquid. Air bubbles in the water not only reduce the capacity of the water to carry away heat, but also promote rapid rust formation and corrosion, and may cause excessive foaming and

water loss out of the overflow pipe. In any case, water shortage leads to overheating, operating difficulties and engine damage.

If possible, the water level should be checked at approximately the minimum safe operating temperature, since the level rises as the engine warms up and falls as the engine cools down. And if the cooling system is clean, leak-proof and in proper working order, very little water should be lost through evaporation or any other cause. Therefore, any unusual amount of water needed to bring the level up to the standard height should be regarded as a possible indication of trouble developing in the cooling system.

If the radiator is continually filled above the proper level, any changes in the level or the quantity of water additions will be of little value as an indication of the condition of the cooling system. Both water and anti-freeze expand when

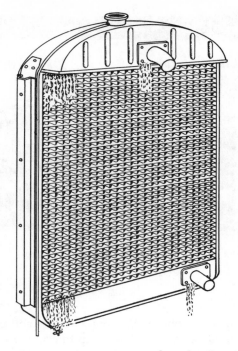

Fig. 30   *Common leakage points in radiator*

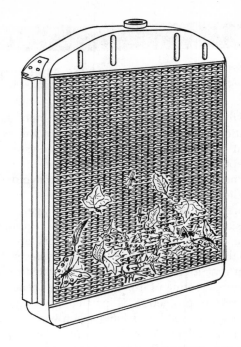

Fig. 31   *Air circulation through radiator impeded by leaves, insects, etc.*

heated, and if there is not enough air space left in the radiator for this expansion, some liquid will be lost through overflow. If the system contains a corrosion inhibitor, it will become diluted and weakened if overfilling is practiced.

Unnecessary additions of water increase water scale deposits which interfere with removal of heat from the engine. Overfilling also wastes anti-freeze, and when a system containing anti-freeze is overfilled with water, it may lead to a freeze-up.

When necessary to add water when the engine is cold, it is enough to cover the top of the tubes by ½ inch.

LEAKAGE

Engine vibration, road shocks and deterioration of gaskets, as well as wear, breakage or corrosion of metal parts, may create leakage. If the car is equipped with a radiator pressure cap, it creates additional pressure in the system, thereby increasing the leakage tendency at hose connections and other water joints.

Small leaks which show dampness or even dripping when the water is cold may not be noticed when the engine is hot, due to rapid evaporation of the liquid. Rusty or grayish-white stains at joints in the radiator or engine water jackets are usually indications of leakage, even though there

appears to be no dampness. Even small leaks should not be neglected since they often become larger, sometimes suddenly, and generally while the car is being driven.

Radiator Leakage

Engine vibration and road shocks put a strain on all radiator seams and joints that may lead to breakage and leakage, particularly in the water tubes, the tanks and the inlet and outlet fittings, Fig. 30. Additional strain is set up by extreme changes in metal temperatures, especially during cold weather operation. Neglect of small leaks may result in excessive leakage, rust clogging and overheating difficulties. Thus it is extremely important to keep the radiator mounting properly adjusted and tight at all times, and to detect and correct promptly even the smallest leaks.

The primary function of the radiator is to transfer heat efficiently from the water to the air. This is not possible without clean, straight air fins and unobstructed air passages. Air fins are easily bent and damaged by impact of small stones and other accidental causes. Flying mud, dust, sand, grass, leaves, large insects, paper and other debris, Fig. 31, may clog air passages in a very short time.

The problem of maintaining sufficient air flow through the radiator is often complicated by fog lamps, license plates, bumper guards and other

Fig. 32 *Example of hose deterioration*

Fig. 33 *Leakage at metal water joints*

assessories placed in front of the radiator. If an accessory is mounted so that it is causing overheating, it should be placed elsewhere or removed entirely.

Hose Leakage

Leaks are more common at radiator hose connections than anywhere else in the cooling system.

Engine vibration has a tendency to wear and loosen hose connections. Clamps may buckle the hose and threads on the clamp bolts are often stripped.

The hose itself has a limited service life. Heat and water cause hose swelling, hardening, cracking and rotting. Deterioration of hose usually takes place more rapidly from the inside, Fig. 32, so that outside inspection is not dependable. Hardening of old hose increases the difficulty of keeping connections leakproof. Hose failures not only result in leakage but may also cause restriction of water circulation through clogging or col-

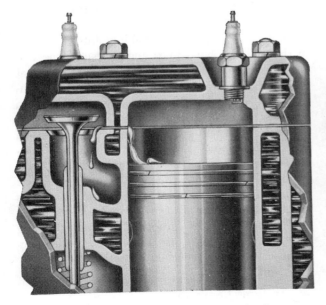

Fig. 34 *Water leakage into cylinder bore and valve port due to broken cylinder head gasket*

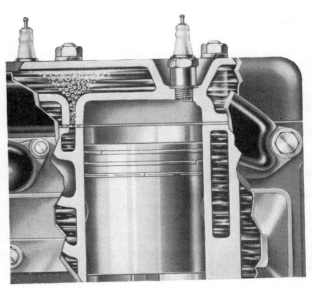

Fig. 35 *Combustion gases blown into cooling system due to loose cylinder head*

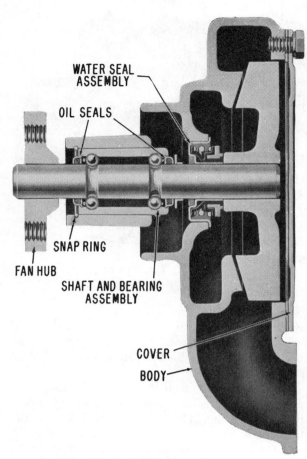

Fig. 36   Sealed type water pump

lapsing. Rubber particles from rotted hose linings will clog radiator water tubes and are very difficult to remove. Rotted hose may break open without warning and cause sudden large water losses. Frequent outside examination of all hose and connections and careful inside inspection of rubber hose whenever the connections are opened require little time and can save much trouble.

Water Jacket Leakage

The engine water jacket has gasketed water joints and a number of metal water joints in both the cylinder head and block where preventive maintenance neglect may result in leakage. Vibration, pressure, and wide changes in engine temperature impose strains on all these joints. Gaskets deteriorate from the effects of heat, water and pressure. Gasket joints at the thermostat housing and water pump mounting are common points of leakage. Metal joints, Fig. 33, such as core hole plugs, drain plugs, drain cocks, temperature gauge fittings, and connections at water bypass tubes, are all subject to leakage.

Corrosion leakage occasionally develops in metal water joints. Any leakage at water jacket joints or casting cracks is aggravated by pump pressures, which may run as high as 35 pounds per square inch. Pump pressures are naturally greater at higher engine speeds and while the thermostat is closed. The radiator pressure cap (if fitted) also allows additional pressure to build up to prevent the coolant from boiling.

Cylinder Head Joint Leakage

The joint between the cylinder head and engine block actually consists of a large number of individual water joints at water transfer ports, which are all sealed by the cylinder head gasket. All of these joints are subjected to the strain of extreme temperature changes within the engine, and also to combustion pressures as high as 600 pounds per square inch. Internal leakage at the cylinder head gasket cannot be detected from outside inspection.

Leakage of water into the engine interior, Fig. 34, can cause serious damage, especially in cold weather. Either water or anti-freeze solution, when mixed in large quantities with engine oil will form sludge which may cause lubrication difficulties. If internal leakage is not promptly discovered and corrected, serious engine damage can result. Even though the joint is tight enough to prevent liquid leakage, the slightest looseness will allow combustion gases to be blown into the cooling system, Fig. 35. This can force water out of the overflow pipe. Burned gases dissolve in the water to form acids which cause rapid rust formation and attack other metal parts.

Considering the many possible points of leakage in the cylinder head joint and the seriousness of water leakage into the engine, it is imperative that the cylinder head always be kept leakproof. Cylinder head bolts cannot be evenly tightened with an ordinary wrench. The use of a torque wrench is necessary to obtain proper uniform pressure on all bolts and to avoid warpage of the head or distortion of the block at valve seats and cylinder bores from overtightening. The extreme importance of maintaining cylinder head joint tightness demands careful attention to all instructions on the installation of new gaskets, proper order of tightening bolts, correct torque to apply, and rechecking torque following a new gasket installation.

Water Pump Leakage

The water pump is the only power-driven unit in the cooling system. Maximum pump speed of

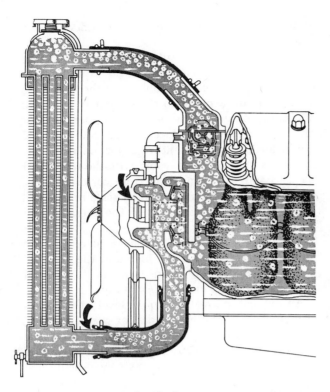

Fig. 37 *Slight leak at water pump seal or hose connection (arrows) allows air to be sucked into cooling system at high engine speeds, causing water loss and overheating*

cially with a high pressure gun, forces grease into the cooling system, which contributes to clogging and overheating.

Forced water circulation is so necessary in the modern cooling system that any reduction in pumping capacity causes a loss of cooling effectiveness. Complete pumping failure is invariably followed by sudden overheating and operating difficulties.

Loss of water is not the only trouble that can result from a water pump leak. Water leakage at the shaft, if not properly corrected, will destroy lubrication and cause corrosion and wear of the shaft and bearings. Even a slight leak at the pump seal or in the connections between the pump and radiator will allow air to be sucked into the cooling system at high engine speeds, Fig. 37. Air suction into the system through a perforated rubber shaft seal can force enough liquid out of the overflow pipe to cause serious water shortage in a short period of high speed engine operation.

Mixing of air with water (aeration) reduces heat transfer and may raise engine temperature high enough to cause overheating at high engine output. Furthermore, the introduction of air into the system may speed up rusting as much as 30 times and also greatly increase corrosion of all cooling system metals. Clogging and corrosion go hand in hand with neglected water pump leakage and air suction.

5,000 rpm is not uncommon. Some pumps circulate more than 7,500 gallons of water an hour. Pumping failures are often caused by broken or loose drive belts, but edge wear of impeller blades and wear of the pump housing also reduce pumping capacity.

Sand, rust and other abrasive foreign matter in the coolant have a tendency to wear away impeller blades. Corrosion of the impeller and housing may result from failure to use corrosion inhibitor with water or to discard rusty anti-freeze solution.

Leakage of the pump is a more common trouble than pumping failure. The pump housing joint is under strain from the pump drive and may work loose and leak if the mounting bolts are not kept tight.

In the sealed-type (packless) pump, Fig. 36, the self-adjusting seals are subject to wear, deterioration and leakage. Thrust seal washers and seats are prematurely worn by abrasive action of sand, dirt and rust in the cooling water and by operation with engine overheating. Bearing and shaft damage, which leads to leakage and pump failure, can result from neglect of lubrication in pumps that require it. But over-lubrication, espe-

THERMOSTAT

The function and operation of the thermostat is such that this unit does not have an indefinite service life and can fail with little or no advance warning. The valve and operating mechanism is subject to extreme temperature changes, corrosion and also to wear and bending movement. Rust or foreign matter in the cooling water interferes with proper thermostat operation, and overheating from any cause may damage it.

The automatic control of engine operating temperatures provided by the thermostat is absolutely necessary—winter and summer—for efficient engine performance. If the valve fails to close properly the engine will run too cool; then sludge formation and other harmful effects of overcooling can take place. If the valve fails to open properly, engine temperature will rise and overheating difficulties may follow. Engines should not be operated with the thermostat removed except in cases of emergency, and then for only as short a time as necessary to obtain a new one.

The temperature gauge should be observed during engine warm-up and on road tests in order

to be sure the thermostat is functioning properly. Whenever the gauge continually indicates unsafe low or high temperatures, the thermostat should be removed and tested, as described later on.

FAN BLADES AND BELTS

Bent fan blades or a loose, bent, misaligned or damaged fan shroud, interferes with proper air flow and reduces cooling. Periodic inspection and servicing of the fan and shroud (if equipped) is essential to proper engine cooling.

Preventive maintenance of the fan drive belt is also of great importance because this belt usually drives the fan, water pump and generator. Continuous flexing, friction and heat cause fan belt cracking, fraying, wear and deterioration.

Loose adjustment may result in slippage, rapid belt wear and an overheated engine. Overtight adjustment also wears the belt and causes early failure of shafts and bearings in the fan, water pump or generator.

A neglected fan belt may break without warning and cause sudden overheating and operating difficulties. Therefore, inspection of the fan belt condition and adjustment should never be neglected. Close examination is necessary to discover small flaws, particularly since belts usually begin to crack through from the inside. Immediate replacement of a doubtful belt is good insurance against car failure during operation.

RADIATOR PRESSURE CAP

A radiator pressure cap has more effect on cooling system operation than is generally realized. A properly operating pressure cap increases the normal margin of safety between water operating temperature and boiling point five degrees or more. This additional margin of safety helps to prevent boiling during operation in hot weather, at high altitudes and when driving under heavy load.

The pressure cap is subjected to high cooling temperatures which cause relatively rapid deterioration of the gasket. The valves and the underside of the cap are exposed to the extremely corrosive effects of hot steam and air in the upper radiator tank. Since the cap is located above the normal liquid level, it receives little protection from rust inhibitors in the cooling water, with the result that the cap and valves may fail from corrosion damage. Even a small amount of rust scale or dirt will interfere with the operation of the pressure and vacuum valves.

Frequent removal and replacement of the cap for water level observation increases the possibility of leakage and pressure loss, due to wear of the gasket and cap locking mechanism.

An air leak above the liquid level in the radiator, such as at the cap gasket or pressure valve, will prevent pressure from building up, and the benefits of the pressure cap will be lost. Water may boil in some cooling systems even at normal operating temperatures if the cap is not pressure-tight. If the pressure valve fails to open, sufficient pressure may build up in the system to break radiator seams or blow off hose connections. Failure of the vacuum valve to open when the system cools may cause collapse of hose and other parts which have no internal support.

To avoid damage to the cap gasket and gasket seat on the filler neck, care should be exercised in removing and replacing the cap. The cap should be turned to the "vent" position before removing to allow escape of hot steam that might cause personal injury.

When filling the radiator, metal filling spouts or nozzles should not be allowed to come in contact with the filler neck gasket seat. Proper maintenance consists of inspection of the cap, seat and gasket, periodic cleaning of the cap and valves, checking of valve operation, and testing for tightness of valves and cap seal.

Trouble Shooting

TESTS IN EMERGENCIES

When an engine overheats, the trouble is indicated in several ways. In addition to high reading on the temperature gauge, there is usually a noticeable "ping" and loss of power when the engine is accelerated. In extreme cases of overheating, the engine will continue to rotate for several seconds after the ignition switch has been turned off. Should the engine overheat when the ignition timing is known to be correct, valves properly adjusted and valves in good condition, then look for cooling system failure, a clogged exhaust system, or a mechanical failure in the engine.

When overheating occurs, turn off the ignition switch and allow the engine to cool. At three- or four-minute intervals, rotate the engine a few times with the starter to prevent the piston rings from sticking in the ring grooves. When the engine has cooled to 180 degrees or less, start it up and *slowly* fill the cooling system. Then examine the system to determine whether the coolant has been lost because of boiling or because of leaking.

To see whether or not the thermostat is operating, start the engine and allow it to run at a fast idle. If the thermostat is closed no water activity will be apparent when looking into the filler neck. When the engine reaches operating temperature, if the thermostat has opened, there will be some stirring around in the water which, of course, is caused by the operation of the water pump. No stirring around of the water means that the thermostat has not opened.

Another way of checking this is to run the engine at a fast idle until operating temperature is reached. Then compress the upper hose with your hand, Fig. 38, and accelerate the engine. If the water is forced through the restricted hose as the engine is accelerated, the thermostat is operating and, therefore, open.

If neither of these methods produce the results desired, replace the thermostat or remove it and operate the car until convenient to install a new one.

If the possibility of thermostat trouble has been eliminated and overheating continues, test the pump by squeezing the upper hose again, Fig. 38. There should be an increased flow of coolant through the hose as the engine is accelerated. If the flow does not increase, the pump is at fault.

If the coolant is leaking from a hose connection, tighten the hose clamp screw. If the leak continues, replace the hose or tightly wrap the connection with tape until a permanent repair can be made. Do the same thing if the leakage is due to a defective pipe.

If the leak is due to a defective radiator core, remove the core and have it repaired. In an emergency, whittle a plug of soft wood slightly larger than the hole and insert it in the opening. Allow the plug to remain until a permanent repair can be made.

If the leak is coming from the water pump shaft it indicates the seal is defective. The water pump must be replaced. If, on the other hand, the leak is

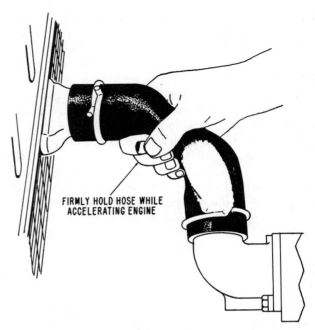

Fig. 38 Squeezing hose to feel if water is flowing to radiator

coming from the joint between the water pump and the engine a broken gasket is indicated. This can be replaced by removing the water pump.

If inspection shows that the overheating is not caused by leakage, it indicates that the coolant is not circulating properly. Adjust the fan belt if necessary, or if the belt is broken install a new one.

Should tests and corrective adjustments of the pump and thermostat fail to relieve the overheating, test the radiator for clogging. To do this, start the engine and allow it to run until it reaches operating temperature. Turn the ignition off and feel each part of the core surface with the hand, beginning with the top center section. If some sections are cool to the touch while others are warm, the radiator core is probably clogged and must be cleaned.

The foregoing instructions are suggestions for locating the cause of cooling system troubles in emergencies. Special trouble shooting procedure for locating and testing for relatively uncommon cooling system complaints are as follows:

FROZEN COOLANT

In freezing weather, a sudden rise in engine temperature when the engine is first started may indicate frozen coolant. A freeze-up can be checked by examining the coolant in the upper radiator tank or by attempting to draw it into a hydrometer, or by opening a drain cock. Squeezing the radiator hose is not a dependable test because the hose may feel hard when cold, even though the coolant is not frozen. Feeling the radiator core with the bare hand for cold spots after the engine has been running a short time may reveal the temporary clogging condition caused by a slush freeze in cooling system, Fig. 39.

If water is allowed to freeze solid in the system, the practical way to thaw it out, without causing further damage to engine and cooling system, is to allow the car to stand in a warm place until all the ice is melted. Under no circumstances should operation of the engine be attempted when solid freezing occurs as this may cause freeze-cracking of the cylinder head, Fig. 40, cylinder block, radiator, water pump and connections. After the ice has thawed, examine the cooling system carefully to be sure this has not occurred.

At temperatures below the freezing point of anti-freeze solution, there is no solid freezing but the mass of small ice crystals (slush ice), Fig. 39, is formed in the solution which may stop circulation through the radiator. The safest way to thaw out this slush ice is to stand the car in a warm place without running the engine. But depending upon the severity of the freeze-up, it may be safely thawed by running the engine if certain precau-

tions are taken: Cover the radiator, start the engine and allow it to idle for about a minute. Then stop the engine and allow it to remain at rest for the same period of time. Repeat this procedure several times, each time lengthening the running time interval. The heat radiated from the engine in this manner will slowly thaw the ice.

FOAMING

Some water has a greater tendency to foam than others, due to the minerals or impurities contained in them. Contamination of anti-freeze solution or aging of solution from extended service increases foaming tendencies. Foaming of the coolant does not mean a head of foam on the surface of the coolant in the radiator, but refers to the small air bubbles that are caught and held in the body of the coolant, giving the coolant a milky appearance and increasing its volume. Foaming, therefore, may indicate combustion gas leakage into the coolant, air suction into the system, or abnormal foaming tendencies in the coolant itself.

If foaming is found, drain the cooling system and replace with fresh water. Then operate the engine under the same conditions of speed and temperature as before. If there is no foaming,

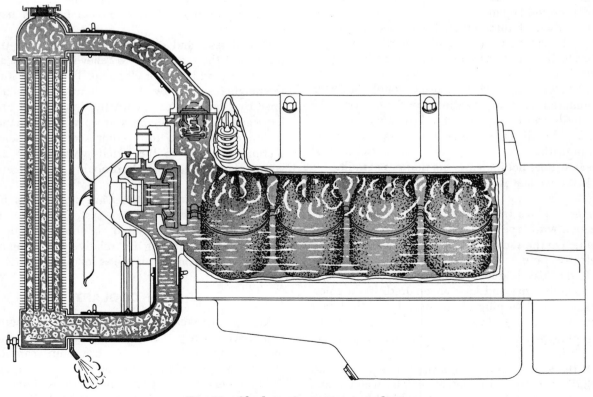

Fig. 39   Slush ice formation in radiator

Fig. 40   Cylinder head with valve port cracks

discard the drained coolant and add corrosion inhibitor to the water in the system, or drain the water and install fresh anti-freeze solution, depending on the season. If foaming continues with plain water, test for air suction, and for combustion gas leakage into the system, if found necessary after the air suction test.

AIR SUCTION TEST

First be sure the coolant level is low enough to prevent it being lost through the overflow pipe

Fig. 41   Air suction test of cooling system, using glass jar and hose to test for air bubbles

during the test. Then, if the car is equipped with a radiator pressure cap, block open the pressure valve and put the cap on tight. Attach a suitable length of hose to the overflow pipe, allowing the hose to stand in a vessel of water, Fig. 41. Run the engine until the temperature gauge stops rising and remains stationary. With the engine running at maximum safe speed for several minutes, watch for air bubbles in the water. In the absence of combustion gas leakage, a continuous stream of bubbles indicates that air is being sucked into the system.

To be sure of this, however, carefully examine the water pump shaft seal, lower hose connections, lower part of radiator, and all other possible points of leakage on the suction side of the pump. Correct the leaks and repeat the test. If air bubbles continue, or if no leaks are located, make a combustion leakage test. If combustion leakage is found, correct the condition and repeat the air suction test to be sure the condition is remedied. If no combustion leakage is found or if the air suction test still shows bubbles, repair the water pump.

COMBUSTION LEAKAGE TEST

Combustion leakage is most commonly caused by a defective cylinder head gasket or loose head, but it may also be caused by warpage in the head joint or small cracks in the cylinder head or block.

To make the test, start with the engine cold. Remove the drive belt entirely or at least from the crankshaft pulley, Fig. 42. Drain the system until the coolant is level with the top of the cylinder block. Then remove the upper hose and thermostat and pour water in the radiator until it flows out of the thermostat opening (engine outlet). If necessary, block the upper hose connection at the radiator with a hand to avoid spillage.

Start the engine. Accelerate it six or eight times and watch the engine outlet opening for bubbles

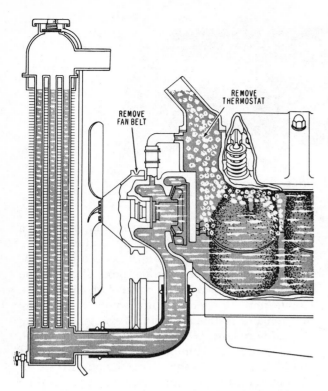

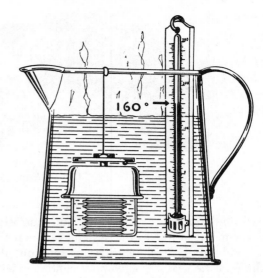

Fig. 43 *Checking operation of thermostat. Valve should open when water temperature reaches opening temperature of thermostat valve. Drawing shows valve is open at 160 degrees, this being a 160-degree thermostat*

Fig. 42 *Combustion leakage test of cooling system. Air bubbles in engine outlet with engine running confirms this*

or sudden rise of liquid while accelerating, and also when the engine drops back to idling. Either bubbles or liquid rise indicates combustion leakage into the coolant.

To detect very small leaks, jack up the rear wheels, run the engine at maximum safe speed in high gear and load it gradually and intermittently by applying the foot brake. Make this test quickly before boiling starts, since steam bubbles give misleading results.

If leakage is apparent, remove the cylinder head and inspect the gasket, head and block gasket surfaces, cylinder bores, combustion chambers and water jackets for evidence of leaks, blowby, coolant obstructions and any other defects.

THERMOSTAT TEST

Hang the thermostat by its frame in a vessel of water, Fig. 43, so the thermostat unit does not touch the bottom of the container. Heat the water and measure the temperature (an oven thermometer will do). If the valve opens at a temperature more than 10 degrees below the specified opening or does not open at a temperature 10 to 15 degrees above the specified opening, the thermostat should be replaced with a new one.

If the valve can be pulled or pushed off its seat

with a slight effort when the thermostat is cold, the unit may be considered defective.

Whenever the thermostat is removed for any reason, it should be cleaned, tested and examined carefully.

PRESSURE CAP TEST

To test a radiator pressure cap for airtightness, attach a rubber tube to the bottom of the radiator overflow pipe and block open the pressure valve. With the cooling system cold, suck on the tube and apply your tongue to the tube opening. If your tongue adheres to the tube the system is reasonably airtight. If not, the point of air leakage can usually be determined by blowing tobacco smoke into the tube and watching to see where it comes out. These tests will generally reveal air leaks above the liquid level, including leakage at the valve cap gasket, but will not show whether or not the valves leak.

Pressure and vacuum valves may be checked for leakage by putting the bottom side of the cap in your mouth with the lips over the valve cage opening. If it is possible to blow through the opening, one or both valves are leaking. Collapsing of hose or other cooling system parts indicates that the vacuum valve is not opening when the engine cools down, and the cap should be replaced with a new one.

The cap and valves should be flushed by spraying a stream of water, preferably hot, through the holes in the cap valve cage while moving the pressure valve up and down with a suitably-shaped blunt pin or pencil. The valve should work freely and seat properly, otherwise, the cap should be replaced. The gasket and seats in the cap and filler neck should be examined for damage that may cause leakage.

TEMPERATURE GAUGE TEST

With the engine started and running at fast idle, watch the action of the gauge to see that it indicates gradual temperature rise. No movement of the gauge after a reasonable warm-up period may indicate that the thermostat is stuck open. Suddenly rising or unusually high temperature may indicate coolant shortage or freeze-up or other serious defect in the cooling system.

The possibility of a false temperature indication from a defective gauge should not be overlooked when checking for overheating or overcooling. An approximate check for gauge accuracy can be made by inserting a thermometer into the coolant in the upper radiator tank. Stop the engine when the thermometer indicates rated full open thermostat temperature. Wait until the thermometer stops rising and compare its reading with the temperature gauge on the instrument panel. The gauge should normally read within 10 degrees of the thermometer reading.

For a more accurate test, remove the temperature gauge thermal unit from the engine and suspend it in water which is heated to at least 120 degrees. Then suspend a thermometer in the water and compare the thermometer readings at several temperatures with readings of the temperature gauge on the instrument panel taken at the same time.

RADIATOR CLOGGING

Severe clogging may be detected by feeling the radiator core for cold spots immediately after engine operation. Another approximate method of checking for radiator clogging is to remove the lower hose. Then plug the radiator outlet, block the inlet openings and fill the radiator with water. Remove the radiator cap and radiator outlet plug and check the time it takes for draining. The radiator may be considered clogged if the draining time is noticeably longer than that required for a clean radiator of the same type.

AIR COOLED ENGINE COOLING TROUBLES

Overheating of the cooling system on an air-cooled car may be due to any of the following:

1. Oil cooler dirty.
2. Blower belt loose.
3. Blower belt worn or oil soaked.
4. Thermostat sticking closed.
5. Incorrect cooling air valve opening.
6. Engine cooling fins plugged.
7. Incorrect ignition or valve timing.
8. Brakes dragging.
9. Improper grade and viscosity oil being used.
10. Fuel mixture too lean.
11. Valves improperly adjusted.
12. Defective ignition system.
13. Exhaust system partly restricted.
14. Loose engine shield seals.
15. Spark plug boots loose.
16. Engine sheet metal loose.

When tuning up an air cooled engine, be certain that all metal shrouds and shields are in place and properly fitted to prevent air leaks. Inspect the blower belt to be sure it has the proper tension and that the blower assembly is in good condition.

Examine the openings in the rear lid to see that they are free of dirt, twigs, leaves, paper, and so on, as this is the source of engine cooling air. Lubricate hinges of air outlet baffle and see that they move freely.

Service Operations

CLEANING THE SYSTEM

An examination of the cooling system for cleanliness should be made periodically. The quickest way to do this is to scrape a finger around the inside of the radiator filler neck and if an accumulation of rust or oily muck is brought forth, the system should be cleaned.

The color of the cooling liquid can be conveniently checked by drawing a sample into a suitable hydrometer or anti-freeze tester. The appearance of rust in the sample is an indication that either the corrosion inhibitor has lost its effectiveness, or the system was not treated with an inhibitor. Rusty water or anti-freeze solution should be drained and discarded. Following the draining, the system should always be cleaned before fresh coolant is installed.

To clean the system, remove the radiator cap to let in air so that the system will drain faster. Then open all drain cocks, leaving them open until the system is completely drained. Draining points are at the bottom of the radiator, on the side of the cylinder block and, in some cases, under the water pump.

If the car is equipped with a hot water heater, drain it also. If the heater is mounted on the dash, disconnect the hose at the lower heater connection. If the heater is under the seat, remove one hose at the heater and disconnect the other at the engine block. To empty the heater completely, apply a little compressed air to the connection at the engine.

After emptying the system, close all drains and reconnect the heater hose. Add the correct amount of good cleaning compound (instructions printed on container) and fill the system with water. Be sure the water level is low enough so that the solution will not be lost out of the overflow pipe.

Run the engine at a fast idle for at least 30 minutes after the cleaning solution has heated up. Then drain the solution and, if the cleaning compound used is one that requires a neutralizing agent after its use, add the neutralizer and fill the system with water. Run the engine for at least five minutes after the neutralizing agent has heated up, then drain it out and refill with fresh water. Again run the engine for a few minutes to wash out the neutralizing agent and drain the system. Finally, refill with fresh water and corrosion inhibitor or with the proper anti-freeze solution, depending on the season.

Fig. 44 Cleaning radiator openings with cleaning gun

During the engine idling periods, it is important to cover the radiator and keep the cover adjusted so that a temperature of 180 to 200 degrees is maintained. The engine develops so little heat while running without load that the thermostat valve remains partially or fully closed. Covering the radiator opens the valve quickly; but if the cover is removed the valve will close again, even though the temperature shows little change. It is a good idea to remove the thermostat during the cleaning operation to be sure the system is not being restricted.

As a part of the cleaning operation, sediment and foreign matter should be removed from the radiator pressure cap valves (if fitted), the overflow pipe and radiator core air passages.

To check for a clogged radiator overflow pipe, fill the radiator to the top of the filler neck and if it fails to flow through the overflow pipe, it is clogged. Usually the obstruction can be cleared out by pushing a length of iron wire down through the pipe. Any deposits embedded in the lower end of the pipe may be reached by poking the wire through the bottom opening. Of course, this should be checked before the cleaning operation, but in the event that it was forgotten, be sure to drain off the excess water from the radiator to bring it down to the proper level.

Dirt and insects may be cleaned out of the radiator air passages by applying a radiator core cleaner gun to the back of the core, Fig. 44. If such equipment is not available, wet the outside

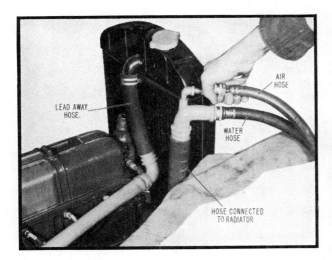

Fig. 45 Reverse flushing radiator by forcing water through it in opposite direction to that when operating normally

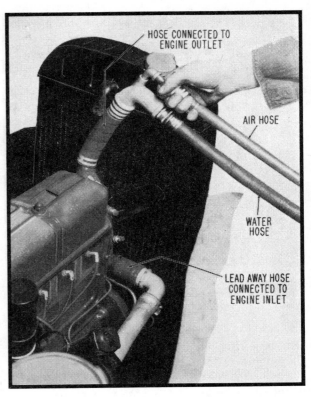

Fig. 46 Reverse flushing engine by forcing water through engine outlet and out through inlet pipe

surfaces of the radiator core until all foreign matter is thoroughly saturated. Then direct a stream of water through the air passages *from the engine side of the core* until all passages are free.

FLUSHING THE SYSTEM

Cooling systems so badly clogged that they do not respond to the above treatment should be reverse flushed, using air pressure to force the water through the system. Reverse flushing is just what the name implies—flushing in the direction opposite to the normal flow of water through the system. Reverse flushing is used to get behind the corrosion deposits to force them out.

First remove the thermostat, as cold water will cause it to close and result in building up pressure which might cause damage. Then remove the upper and lower hose connections and replace the radiator cap. Attach a lead-away hose at the top of the radiator and a piece of new hose to the radiator outlet connection. Insert the flushing gun in this hose, Fig. 45. Connect the water hose of the flushing gun to a water outlet and the air hose to an air line. Turn on the water and when the radiator is full, turn on the air in short blasts, allowing the radiator to fill between blasts of air. Continue in this manner until the water from the lead-away hose runs clear. Be sure to apply the air gradually as a clogged radiator will stand only a limited pressure.

After flushing the radiator, attach the lead-away hose to the water pump inlet and a length of new hose to the water outlet connection at the top of the engine. Insert the flushing gun into the

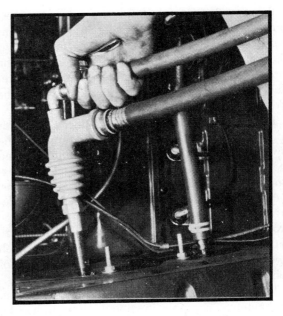

Fig. 47 Flushing out hot water heater

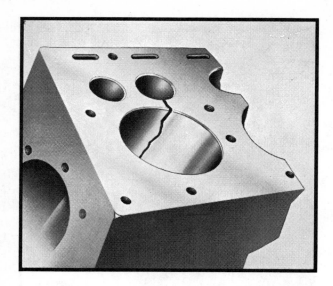

Fig. 48 *Crack caused by cooling system freeze-up*

new hose, Fig. 46. Turn on the water and when the engine water jacket is full, apply the air in short blasts. Continue this sequence until the water from the lead-away hose runs clear.

If the car has a hot water heater, it should be flushed out separately, Fig. 47, as rust deposits will built up in the heater core just as they do in the radiator. If flushing is done in the spring and the heater will not be used until the fall, it is wise to discard the heater hose. If the flushing is being done in the fall, replace the hose.

ANTI-FREEZE PROTECTION

When water freezes it forms solid ice and expands approximately 9 per cent in volume. This expansion takes place with a terrific force and if allowed to take place in the cooling system, the force of the expansion will crack the engine water jacket, Fig. 48, and cause serious damage to the radiator and other parts.

The amount of anti-freeze required depends upon the capacity of the cooling system, the lowest temperature likely to be encountered, and the type of anti-freeze to be installed. For example, if the cooling system capacity is 20 quarts, 6 quarts of methanol (alcohol) protects to 5 degrees below zero; but 6 quarts of permanent type protects to 4 degrees above zero. Anti-freeze makers supply their dealers with charts which give full information for all makes and models of vehicles.

To install anti-freeze, first drain the system completely. Then pour in enough water to half fill the cooling system. Add the required amount of

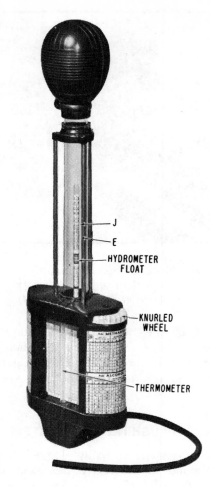

Fig. 49 *Hydrometer for determining amount of anti-freeze in coolant*

anti-freeze and finish filling with water. Start the engine and run it for a few minutes. This allows the solution to mix thoroughly and permits the thermostat to open, which will release any trapped air that may be in the engine block. Be sure the level of the solution is low enough to allow for expansion, and check the system for leaks and recheck level before putting the car in service.

TESTING ANTI-FREEZE

To be sure the anti-freeze will give protection at the coldest temperature to which the engine is likely to be exposed, and to avoid overheating difficulties from slush freeze-up, the freezing protection of the solution should be tested at least weekly and more often if the need is indicated because of water additions or weather conditions.

Always use a good hydrometer for checking the strength of anti-freeze solution. The accuracy of the hydrometer should be checked occasionally by taking readings on prepared solutions of known freezing point. Before reading the hydrometer, the barrel should be filled and emptied several times in order to equalize the temperature of all parts. Accurate readings cannot be obtained if the float and inside of the glass barrel are dirty.

When taking a sample of the solution, if the level has dropped to a point where the hydrometer cannot reach it, drain a small quantity of the solution into a clean glass jar and make the test from the jar. After testing, pour the solution back into the radiator and add the necessary amount of anti-freeze to give the proper protection. If necessary, add water to the proper level and then check the system for leaks with the engine running and when not running.

It is well to note that alcohol type anti-freeze solutions will boil away if the engine overheats. If the car has a pressure cap and if there is no leakage in the cooling system, add alcohol to bring the coolant up to its proper level. On the other hand, if ethylene glycol is used and the engine overheats, only water will boil away. Therefore, water should be added to bring the level up to normal.

USE OF HYDROMETER

An hydrometer is used to determine at what temperature the coolant will freeze. There are a number of them on the market, one of which is shown in Fig. 49.

The hydrometer float consists of a glass tube which is weighted at the bottom so that it will float upright as shown. Obviously, the lighter the liquid, the lower it will sink and the heavier the liquid the higher it will ride.

Methanol is 60 per cent as heavy as water, while ethylene glycol is 11.5 per cent heavier than water. Consequently, the scale on the hydrometer float indicates the amount of anti-freeze in the coolant.

Liquids expand and become lighter as their temperature rises. Consequently, it is necessary to know the temperature of the coolant when reading an hydrometer. Therefore, the device is equipped with a thermometer.

For example, with ethylene glycol, if the level of the liquid is at E, Fig. 49, and the temperature is 160 degrees, the table on the hydrometer shows that there is 50 per cent ethylene glycol in the coolant. However, if the level of the liquid is at E and the coolant temperature is 50 degrees, there is only 20 per cent ethylene glycol in the coolant.

Similarly, when using this hydrometer to determine the amount of methanol in the coolant, if the level of the liquid is at J, Fig. 49, it means that 15 per cent of the coolant is methanol if its temperature is 160 degrees but that nearly 40 per cent is methanol if its temperature is 50 degrees. All this may sound very complicated. However, the directions on the hydrometer make the temperature problem quite simple.

To use the hydrometer shown, you first look up the capacity of the cooling system of the car. This information is given in a compact table on the back of the device. The capacity figures are for cars without heater. The directions recommend that 2 quarts be added if the car has a heater. Let us assume that the capacity with heater is 20 quarts and that zero protection is desired.

There are four anti-freeze tables inside the housing and any one of them can be brought into position by turning the knurled wheel.

The next step is to fill the hydrometer with coolant. Insert the rubber tube in the filler neck. Squeeze the bulb hard and release it to draw liquid up into the chamber containing the hydrometer float. Adjust the liquid level by operating the bulb so that the hydrometer float floats freely and does not touch either at the top or the bottom. Be sure to hold the hydrometer vertical when reading it.

Let us assume that the coolant level is at E, Fig. 49, and that the coolant temperature is 120 degrees. Now assume that protection is wanted down to 20 degrees below zero. A table on the housing of the hydrometer shows that 3 quarts of anti-freeze should be added. Draw 4 or 5 quarts from the radiator into a pail. Pour 3 quarts out of the pail and throw away. Add 3 quarts of anti-freeze directly into the cooling system and pour the contents of the pail back into the radiator.

With this hydrometer the coolant can be checked at any temperature between 55 and 160 degrees.

HOSE INSPECTION

If the outside of the hose is hard, cracked or rotted, install new hose. When cylinder head or radiator has been removed, inspect inside of hose for cracks and loose rubber which has separated from the fabric. If there are any loose flaps of rubber or if portions of the rubber surface have been torn off and carried away, hose should be renewed. When in doubt as to hose condition, install new hose. If one hose connection does not pass inspection, renew all hose to avoid future trouble.

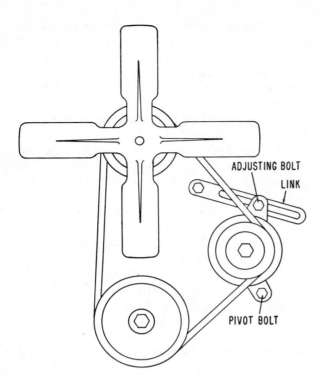

Fig. 50 Generator supported by pivot bolt below and link above. Loosen pivot bolt and link nut. Move generator outward. When fan belt tension is correct, tighten link nut and pivot bolt

HOSE, REPLACE

Apply a coat of water-proof gasket cement to the metal connections. Install the hose. Install the hose clamps and tighten them securely. The gasket cement not only is insurance against leakage but also acts as a lubricant which makes hose installation easier. When installing a hose, never use grease as a lubricant as it will rot the rubber.

DRIVE BELTS

Fan belts and other driving belts should be replaced when they begin to show wear. Belts are designed to contact the sides of their pulleys and should not touch the bottom of the pulley groove.

The sides of the pulley should have a polished appearance because of belt contact but if the bottom of the groove is polished it indicates that the felt does not fit properly or is much too tight. To be on the safe side a new belt should be installed.

A belt should be replaced when it becomes frayed. Before installing the new belt examine the sides of the pulleys for rough spots. If any are found, replace the pulley or smooth it up with fine emery cloth. A pulley with a broken flange should be discarded.

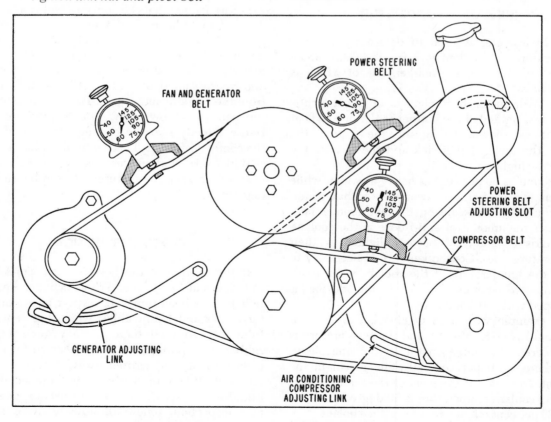

Fig. 51 Showing use of belt strand tension gauge

Fig. 52 Cup type of welch plug. These plugs are made necessary because the holes they enclose are used to empty the casting of sand when the cylinder block or cylinder head is cast

Fig. 53 Disc type of welch plug

Pulleys at all times should be kept free of oil and grease because lubricants may cause the belt to slip, may pick up dirt that will soon fray the belt or may ruin the belt by softening the rubber used in its construction.

Belt Adjustments

Fig. 50 illustrates the most common fan-generator-water pump belt arrangement, while Fig. 51 shows a typical belt set-up when air conditioning and power steering are involved. The adjustment links and slots shown are in general use on most cars.

Precise adjustment of drive belts is a must. Loose adjustment of fan-generator belts may cause overheating and low generator output; slippage of the power steering pump belt may not provide adequate oil pressure, and slippage of the air conditioner compressor belt may not provide enough pressure for proper operation of the system. Too tight adjustments wear belts prematurely and causes early failure of shafts and bearings in the driven units.

Car manufacturers stress the importance of proper belt adjustments and almost all of them recommend the use of a belt strand tension gauge, Fig. 51. The gauge is placed midway on the belt as shown. Then the belt is adjusted until the proper tension is indicated on the gauge. In the example shown, the fan-generator belt is adjusted to 60 lbs, the power steering belt to 80 lbs, and the compressor belt to 72 lbs.

If such a gauge is not available, general practice is to adjust a belt of ordinary length so it has a deflection of ½″ at its midway point; longer belts would require a greater deflection, say 3/4″ to 1″.

RADIATOR FLOW TESTER

A radiator flow tester measures the maximum flow of water through the radiator in gallons per minute. The device consists of an open test tank filled with water, a variable speed pump driven by an electric motor, and a meter which registers the flow in gallons per minute. The radiator can be tested either on the car or off.

A hose from the pump is connected to the lower radiator outlet. Water is pumped upward through the radiator and flows out of the top radiator inlet. If the radiator is off the car, water spills out and falls into the test tank. If the radiator is on the car, a hose is connected to the top radiator inlet so that the water issuing from it is returned to the tank.

With the radiator cap removed, the pump speed is increased until the water level in the radiator rises until it is even with the top of the overflow tube. Then the meter is read to determine the flow in gallons per minute. The operator then refers to a table which gives the flow rate for all makes and models of cars. If the flow rate is much less than normal the radiator should be cleaned. The radiator flow is tested again after cleaning to make sure that flow is normal.

WELCH PLUG, RENEW

Welch plugs are used to seal holes in cylinder block or cylinder head which were necessary in connection with making the casting.

A leaky welch plug, Figs. 52 and 53, should be replaced by a new one. To remove it, punch or drill a hole (about 3/8″) in its center. Insert a punch or rod in the hole and pry out the plug. Clean away any rust or dirt from the sides of the hole. Apply a waterproof gasket cement to the sides of the hole and to the edges of the plug. Insert the plug in the hole with the curved top toward you. Press or tap it in as far as it will go. To lock it in place, expand it by hitting its center with a hammer and drift. The drift should be about half the diameter of the hole.

If a welch plug is located at the rear of the cylinder block, it may be necessary to remove the engine to get at it. Some mechanics cut a large hole in the dash panel in order to reach the plug. Special "hole saws" are used for this purpose although their primary purpose is to cut holes in instrument panels to mount instruments.

359

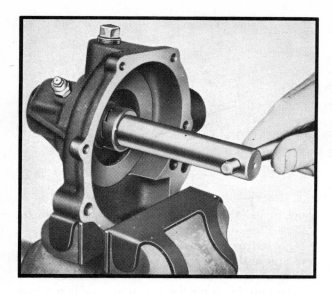

Fig. 54 *Smoothing up water pump seal surface with cutting tool. The seat is then finished by polishing with crocus cloth*

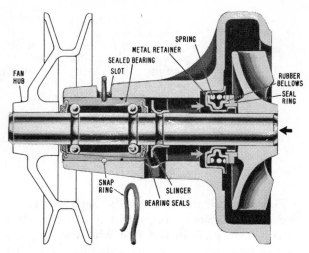

Fig. 55 *This water pump does not have a cover. Both the impeller hub and fan hub are flush with the ends of the shaft Oil seal assembly is stationary. Arrow at right shows where force is applied to push shaft out of impeller. Two arrows to left of seal show direction seal is forced out of housing*

If a hole saw is not available and if it is not feasible to remove the engine for economic or other reasons, the leak may be stopped by pouring a stop-leak compound in the cooling system.

STOP-LEAK COMPOUNDS

Although stop-leak compounds are usually effective in curing small leaks at least temporarily, they cannot be depended upon to correct leakage which involves combustion pressures at the cylinder head and cylinder head gasket.

There are a number of brands of stop-leak compound in either liquid or powder form. Bring the engine up to normal operating temperature and pour a specified amount in the filler neck opening, according to the directions on the can or package.

Continue to run the engine 15 minutes or so until the leaks are sealed. Stop-leak compound remains liquid when in the coolant but as soon as it is exposed to air at the point of leakage it congeals.

In the case of a large leak, two doses may be required. If this fails to stop the leak, the leak is too large and should be repaired.

If the cooling system is subsequently drained and if there is stop-leak in the system, it should be given a new dose of stop-leak when it is refilled.

If a cooling system with a pressure cap boils there may be slight leakage through the cracks as long as the system is under pressure. But the leakage should stop as soon as boiling ceases and pressure falls to zero.

WATER PUMP SERVICE

Water pumps rarely give trouble and, therefore, it is not often that a pump has to be overhauled or replaced with a new or rebuilt one.

The principal trouble is leakage of water. If the leakage is at the gasket between the pump and the cylinder block, the remedy, of course, is to install a new gasket after cleaning the metal surfaces which mate with the gasket. Many pumps also have a gasket and cover, and leakage may require a new gasket. However, in both the cases just mentioned the leakage may be due to loose bolts or nuts; therefore, tightening them may cure the leakage, thus making the removal of the pump unnecessary. When renewing water pump gaskets it is recommended that both sides of the gasket be coated with a waterproof gasket cement before gasket installation. Note that the use of a stop-leak compound may make renewal of the gaskets unnecessary.

If there is leakage at the seal it is necessary to overhaul or replace the pump. The leakage may be due to worn seal parts or to worn bushings or bearings. (A badly worn ball bearing will be noisy.) Not only does water leak out of the seal but also when the pump is operating at high speed its suction may draw air into the cooling system. The air bubbles in the water may reduce the effectiveness of the system and produce overheating.

Leakage at the pump may also be due to blow holes in the pump body or to a crack caused by a severe freezing. In most cases this leakage can be cured by putting a stop-leak compound into the radiator.

Water Pump Removal

Drain radiator. Remove hose connections to water pump. Loosen fan belt adjustment and remove belt. Remove fan blades. Remove bolts or nuts holding pump to cylinder block and take off pump and gasket. If the pump has a cover remove it and its gasket. If car has a fan shroud it may be necessary to loosen or remove it to get the fan out.

It seems hardly necessary to remind the reader that if there are any other parts in the way of pump removal they must be taken off. For example, it may be necessary to remove oil filter pipes. And on cars with power steering the hydraulic pump may have to be removed. And the engine support bolts removed if there is insufficient space between the pump and radiator to permit removal of the water pump. In some cases, the engine may have to be jacked up to provide clearance between the water pump and the engine support bolts before the pump can be removed. In any event, a careful inspection of the area around the water pump should disclose what additional work has to be done.

Water Pump Repairs

For water pumps, suitable pullers and drifts are required. An arbor press is practically a must to push parts off and on. On some pumps with bushings, a burnishing tool is recommended to give the bushings the desired inside diameter and surface finish although a reamer may be used instead.

Most water pumps have the following features in common as shown by the accompanying pictures:

The front end of the shaft drives the fan hub.

The rear end of the shaft drives the pump impeller.

To prevent leakage of water at the shaft, there is a seal assembly between the pump body and the impeller hub.

The seal assembly includes a spring-backed seal ring which prevents leakage by either pressing against the impeller hub or the pump body, depending on the construction. On some pumps, the seal ring rotates with the impeller. On others it is stationary.

Some pumps have covers while others do not.

The pump shaft is mounted on three types of bearing arrangements as shown in the pictures. These are: a sealed bearing with two rows of balls; two bushings; one bushing with a sealed bearing with one row of balls. In a sealed bearing the lubricant is sealed in at the time of manufacture and should last for the life of the pump. Bushings require occasional lubrication and, therefore, are provided with a grease fitting. The sealed bearing and the shaft are sold as a unit, since the inner races for the balls consist of grooves in the shaft.

Before taking the pump apart, note that when the pump is put back together again the impeller and fan hub must be correctly positioned on the shaft—exactly as they were before they were disassembled. Therefore, before removing the fan hub, measure the distance from the fan hub to the front end of the shaft. Likewise, measure the distance from the impeller hub to the rear end of the shaft. These measurements should be written down so they won't be forgotten. Of course, if the fan hub and impeller hub are flush with their respective shaft ends, there is no problem. The impeller hub is usually flush with the end of the shaft but not always. The fan hub may or may not be flush. In some cases one or both hubs may be almost flush. It is particularly important to have the impeller accurately positioned on its shaft as otherwise it may rub against the pump body or its cover. After installing the impeller, check the clearance between the impeller and pump body with a feeler gauge. It should be at least .010″. If the pump has a cover, be sure the impeller does not rub against the cover. If it does, more clearance is required.

After the pump has been disassembled, its parts should be thoroughly cleaned. However, do not allow any solvent to touch a sealed bearing as the solvent may get into the bearing and render the lubricant therein useless. Rubber seal parts may also be damaged by solvent. A complete new seal assembly should be installed, regardless of the condition of the old seal.

Rotate the sealed bearing on its shaft. If there is any indication of sticking or grinding or if there is excessive play, install a new shaft and bearing assembly.

If the shaft rotates in bushings, renew the bushings and shaft if there is noticeable side play. Renew the shaft if it is scored or corroded where

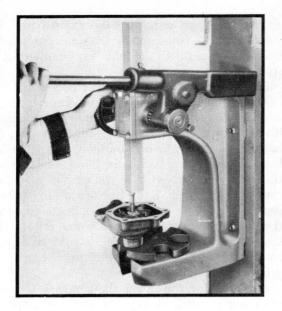

Fig. 56 Pressing shaft out of impeller with arbor press, using a bolt as a drift

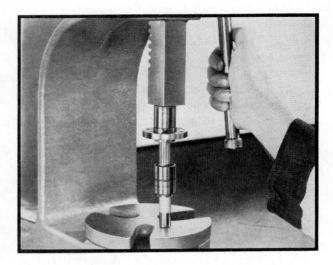

Fig. 57 Pressing fan hub onto pump shaft, using a large socket wrench as a hollow drift

it rotates in the bushings. The shaft should be dipped in light engine oil before installing. Be sure to fill the lubricant chamber with water pump grease.

Before installing the pump shaft, examine the surface against which the seal ring presses. This surface is either the front of the impeller hub or on the adjacent pump body. If the surface is rough it should be smoothed up with a facing tool of the type shown in Fig. 54, and finished with crocus cloth. If the seal surface is so badly damaged that a good refacing job cannot be obtained, renew either the impeller or the pump body, whichever is necessary.

Inasmuch as water pumps are usually replaced with new or rebuilt units, the following paragraphs will deal with the overhaul of one typical unit.

Disassemble Fig. 55 Pump

Remove snap ring.

Press shaft out of impeller by applying force as shown by the arrow at right, Fig. 55. Fig. 56 shows how an arbor press is used to press the shaft out.

Press the fan hub from the shaft, if necessary.

With shaft and bearing and impeller out of the way, press seal assembly out of pump body to the right as shown by the arrows, Fig. 55.

Assemble

Press fan hub on shaft, Fig. 57, until it is flush with end of shaft. Install new seal assembly by pressing it in place, Fig. 58. Install shaft and sealed bearing by means of a hollow drift which presses on the outer race of the bearing. Align snap ring groove in bearing with groove in pump body and install snap ring. Press impeller onto end of shaft, Fig. 59, until it is flush with end of shaft.

Fig. 58 Pressing seal assembly into housing, using a large socket wrench as a hollow drift

RADIATOR REMOVAL

The bottom of the radiator is usually bolted to the front cross member of the frame or to brackets attached to the frame. If there are spacer washers between the bottom of the radiator and its support be careful to install the same number of washers on each bolt when putting the radiator back on. In some cases there are brackets at the sides of the radiator to support it.

If the battery is at the side of the engine, disconnect the battery ground strap at the battery. Otherwise, if the radiator slips while removing or replacing it and touches the live battery terminal, the radiator may be seriously damaged.

Before starting to remove the radiator, look the engine compartment over carefully to determine what is in the way. Remove fan blades if they are in the way although on some cars it is sufficient to rotate the fan blades by hand so that they will not interfere with the radiator outlet while lifting the radiator out. When the radiator is disconnected and all is clear otherwise, lift up the radiator and swing it back over the top of the engine.

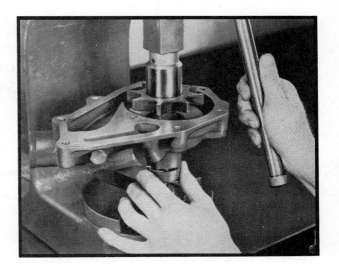

Fig. 59 Pressing impeller on shaft, using a large socket wrench as a hollow drift

Review Questions

		Page
1.	What is the purpose of a cooling system?	327
2.	What is the proper name of the liquid in a cooling system?	328
3.	Name the major components of a cooling system.	328
4.	Most thermostats operate in what temperature range?	328
5.	What is the most common location of the water pump on an automobile engine?	329
6.	What type of water pump is always used in the cooling system?	330
7.	Describe the general direction of flow in a cooling system.	327
8.	What is the main advantage of using a pressure type radiator cap?	333
9.	What is the purpose of the vacuum valve on a radiator pressure cap?	334
10.	Name two types of thermostats.	334
11.	What are three methods which might be used to operate the valve on a thermostat?	335
12.	What is a thermostat bleed hole?	335
13.	What is ambient air temperature?	336
14.	What type of anti-freeze must be used with a thermostat that opens at a temperature higher than 160° F?	336
15.	How does thermostat operation govern the effectiveness of a hot water type heater?	336

16. Explain the difference between a choke type and a by-pass type temperature control system. 337

17. How can it be determined whether or not the thermostat is being installed right side up? 337

18. Either of what two types of radiator hose should be used when the outlets to be connected are not in a straight line? 338

19. What is the principal difference between the cooling system of the Corvair and that of conventional cars? 338

20. How does altitude affect the boiling point of water? 338, 339

21. Explain why it is especially important to have a pressurized cooling system on cars that operate in mountainous country? 339

22. Name the main reason why modern water pumps have curved impeller blades. 339

23. What particular additive is common to all types of anti-freeze? 340

24. Ethylene glycol is the main ingredient in what type of anti-freeze? 340

25. Would pure ethylene glycol in the cooling system give the most anti-freeze protection in cold weather? 340

26. Name two adverse effects of overheating. 341

27. Name two adverse effects of overcooling. 341

28. Most of the solid matter that clogs up radiators originates in what part of the cooling system? 341

29. Besides inadequate cooling, what harm does low coolant level in the radiator do to the engine? 341

30. In what respect does rust formation in the radiator affect cooling system operation? 342

31. What is the prime function of the radiator? 344

32. At what points is a leak in the cooling system most likely to occur? 344

33. Name two adverse results of exhaust gases being blown into the cooling system. 346

34. How does a leaking water pump increase corrosion in the cooling system? 347

35. What is the most common type of malfunction in a water pump? 347

36. Is it good practice to remove the thermostat for Summer driving? 347

37. What could be expected to happen if a fan belt were over-tightened? 348

38. Describe two simple tests for checking thermostat operation. 349

39. Describe a simple test for a clogged radiator core. 349

40. What is the best way to thaw out a frozen cooling system? 350

41. What are the two most likely causes of foaming of the coolant? 350

42. What is the most common cause of combustion leaks into the cooling system? 351

43. List the items that would be needed to test a thermostat. 352

44. Why is the term "reverse flush" used to denote this particular type of cooling system service? 355

45. Why does water burst cylinder blocks and radiators when it turns to ice? 356

46. What is the necessity for a thermometer on an anti-freeze hydrometer? 357

47. What substance must never be used to facilitate hose installation? 358

48. Why do some cylinder heads and all cylinder blocks have holes which make welch plugs necessary? 359

49. How is a leaking welch plug replaced? 359

50. On most water pumps, what parts of the cooling system are attached to each end of the pump shaft? 361

Fuel System

Review Questions for This Chapter on Page 478

INDEX

COMPONENTS OF FUEL SYSTEM

Air cleaners 372
Carburetor 373
Dual exhaust system 377
Exhaust system 376
Fuel filters 368
Fuel gauges 373
Fuel lines 373
Fuel pump 368
Fuel tanks 367
Intake manifolds 373
Manifold heat control 375
Ram induction manifolds 374
Vacuum pump operation 371

ENGINE FUEL

Combustion 378
Compression pressure 378
Detonation 379
Octane rating 379

PHYSICS OF CARBURETION

Vaporization 380
Vaporization by heat 381
Vaporization by spraying 381
Vaporization by vacuum 381
Weight of air 380

HOW A CARBURETOR WORKS

Accelerating system 384
Air bleed principles 387
Balanced carburetor 386
Carburetor accessories 388
Choke system 385
Float system 382
How altitude affects
    carburetion 388
Idle and low speed system . . . 383
Part throttle system 383
Power system 384

Throttle valve 382
Types of carburetors 388
Venturi action 385

OPERATION OF CARTER CARBURETORS

ABD 395
AFB 393
AVS 394
BBD, BBS 396
RBS 397
YF 389
YH, WCFB 391

OPERATION OF FORD CARBURETORS

One-barrel 398
Two-barrel 398
Four-barrel 399

HOLLEY CARBURETORS

1904, 1908, 1960 399
1909 401
1920 403
1931 404
2300, 4150, 4160 404

OPERATION OF ROCHESTER CARBURETORS

B, BC, BV 404
2GC, 2GV 408
4GC 411
4MV, 4MC 414
Model H 406
Model HV 407
Model M, MV 408

OPERATION OF STROMBERG CARBURETORS

WA, WW 417

FUEL SYSTEM SERVICE

Air cleaner 421
Carburetor 421
Flexible fuel lines 428
Fuel filters 427
Fuel pipes 425
Fuel pump 427
Fuel tank 429
Intake manifold 429
Manifold heat control valve . . . 430
Muffler and tail pipe 430
Windshield wiper tubing 430

TROUBLE SHOOTING PROCEDURE

Excessive fuel consumption . . 434
When engine won't start 431
When engine won't idle 432
When engine loses power 433

CARBURETOR SERVICE

Carter BBS overhaul 439
Cleaning parts 439
Ford dual carburetor overhaul . 442
Holley 2300 overhaul 448
Rochester 2GC overhaul 452
Service tips 437
Stromberg WW overhaul 457

FUEL PUMP SURVICE

Fuel pump overhaul 466
Fuel pump tests 464
Performance tips 469
When capacity is low 465
When pressure is high 465
When pressure is low 464
Vacuum pump test 466
Vacuum pump troubles 466

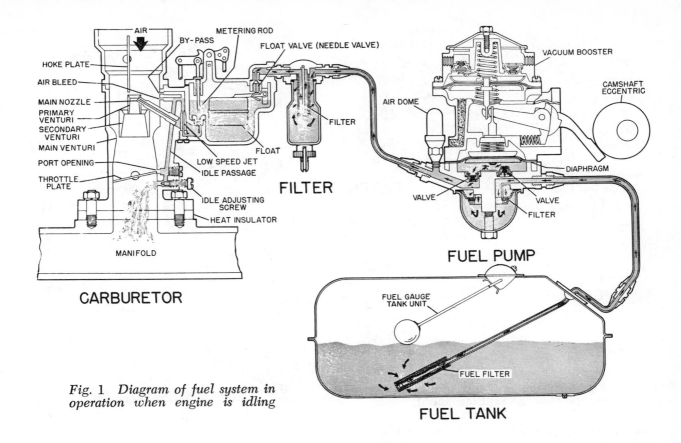

CARBURETOR

AIR
BY-PASS
HOKE PLATE
AIR BLEED
MAIN NOZZLE
PRIMARY VENTURI
SECONDARY VENTURI
MAIN VENTURI
PORT OPENING
THROTTLE PLATE
MANIFOLD
METERING ROD
FLOAT VALVE (NEEDLE VALVE)
FLOAT
LOW SPEED JET
IDLE PASSAGE
IDLE ADJUSTING SCREW
HEAT INSULATOR

FILTER
FILTER

FUEL PUMP
VACUUM BOOSTER
CAMSHAFT ECCENTRIC
AIR DOME
DIAPHRAGM
VALVE
VALVE
FILTER

FUEL TANK
FUEL GAUGE TANK UNIT
FUEL FILTER

Fig. 1 Diagram of fuel system in operation when engine is idling

Components of Fuel System

The fuel system consists of:

1. A tank for storing the fuel.
2. A fuel gauge on the instrument panel which keeps the driver informed as to the amount of fuel in the tank.
3. Pipes for conveying the fuel from the tank to the engine.
4. Filters and cleaners for straining and cleaning the air and fuel.
5. A fuel pump for transferring the fuel from the tank to the carburetor.
6. A carburetor which feeds the engine with a vaporized mixture of fuel and air.
7. An intake manifold which forms a path for fuel mixture to flow to the engine's combustion chambers.
8. The exhaust system consisting of the exhaust manifolds, which directs the burned gases through exhaust pipe, muffler and tailpipe to the atmosphere. Frequently a smaller muffler, called a resonator, is installed toward the end of the tailpipe.

Figs. 1 and 2 show how the fuel travels through the system when the engine is idling and when cruising along.

FUEL TANKS

Since all modern fuel systems are pump fed, the fuel tank can be placed at the most convenient point in the vehicle—which is at the rear of the chassis under the trunk compartment of the car. The exception of this is the rear engined Corvair which has the fuel tank in the forward compartment.

The tank has an inlet or filler pipe, and an outlet. The outlet, with a fitting for fuel line connection, may be in the top or side of the tank. The lower end of the outlet pipe is placed about one-half inch from the bottom of the tank so that any sediment which collects in the tank will not be carried to the carburetor. Baffle plates may be placed in the tank to reinforce the sides and bottom and to prevent the fuel from surging or splashing. The baffles are welded to the sides and

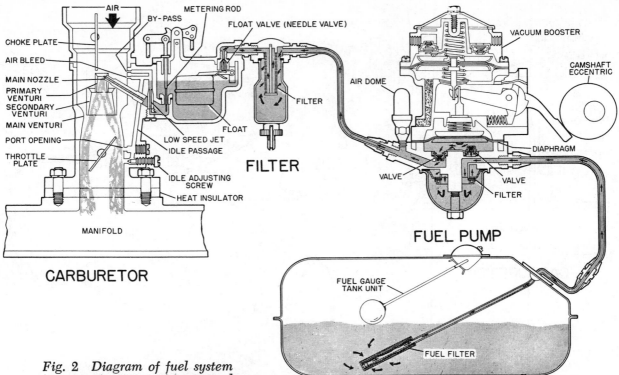

Fig. 2 *Diagram of fuel system in operation at cruising speed*

bottom and are notched or perforated so that the fuel can flow freely from one section to the other. A drain plug is placed in the bottom so the tank can be drained and cleaned.

FUEL FILTERS

These may be located any place between the fuel tank and carburetor, Fig. 3. One may be in the tank itself, in the fuel pump or in the carburetor.

The most common type of fuel filter, shown in Figs. 5 and 6, is placed between the fuel tank and mechanical fuel pump. In this type, the fuel enters the glass bowl and passes up through the filter screen and out through the outlet. Any water or solid matter caught by the filter fall to the bottom of the glass filter bowl where they can be readily seen and removed. Dirt in fuel generally comes from rust scale in tank cars, storage tanks or drums. Water comes from condensation of moisture in the fuel tank.

FUEL PUMPS

Mechanical fuel pumps are of the diaphragm type shown in Figs. 4 and 5. Fig. 6 differs in that the pump has a vacuum booster section. This booster section has nothing to do with the fuel system, except that it is operated by the fuel pump arm. In Fig. 4, the inlet valve and filter are

built in the body of the pump, and in the other two types, Figs. 5 and 6, the inlet valve and filter are separate from the pump body but are part of the assembly.

Mechanical Fuel Pump Operation

During the first or suction stroke, the rotation of the eccentric on the camshaft actuates the pump operating arm, which pulls the lever and diaphragm downward against the pressure of the

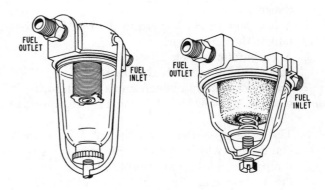

Fig. 3 *Types of fuel filters used in fuel line*

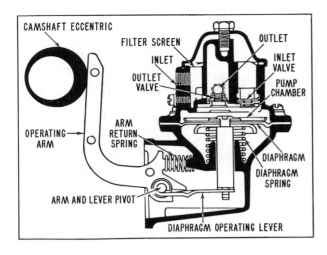

Fig. 4 *Mechanical fuel pump with built-in fuel filter*

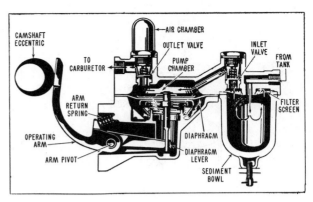

Fig. 5 *Mechanical fuel pump with a separate fuel filter built integrally with the pump*

diaphragm spring, producing a suction (vacuum) in the pump chamber. The suction holds the outlet valve closed and pulls the inlet valve open, making the fuel flow from the supply tank through the inlet up through the filter screen and down through the inlet valve into the pump chamber. During the return stroke, the diaphragm is forced up by the diaphragm spring, the inlet valve closes and the outlet valve is forced open, allowing the fuel to flow through the outlet to the carburetor.

The operating lever is hinged to the pump arm so that it can be moved down but cannot be raised by the pump arm. The pump arm spring makes the arm follow the cam without moving the lever. The lever is only moved upward by the diaphragm spring. The pump, therefore, only delivers fuel to the carburetor when the fuel pressure in the outlet is less than the pressure maintained by the diaphragm spring. This condition arises when the fuel passage from the pump into the carburetor float chamber is open and the float needle valve is not seated. The foregoing description can be followed by referring to the fuel pump shown in Fig. 5.

Electric Fuel Pump Operation

While electrically operated fuel pumps have been used for many years on trucks, buses, etc., and as replacements for mechanically operated fuel pumps on automobiles, they have only recently become original equipment on cars. The replacement types, of which Fig. 7 is an example, typically utilize a diaphragm arrangement similar to mechanical pumps except that the actuating mechanism is an electrical solenoid.

A great departure from usual fuel pump design is the electrically driven turbine type used on the Buick Riviera, Fig. 8. It utilizes a small turbine wheel driven by constant speed electric motor, the entire unit being located in the fuel tank, submerged in the fuel itself. Operating continuously whenever the engine is running, the pump maintains a constant pressure capable of supplying the maximum fuel demands of the engine.

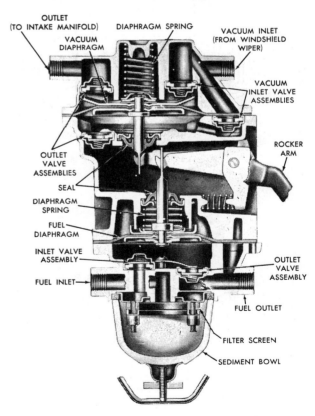

Fig. 6 *Mechanical fuel pump with fuel filter and vacuum section for windshield wiper operation*

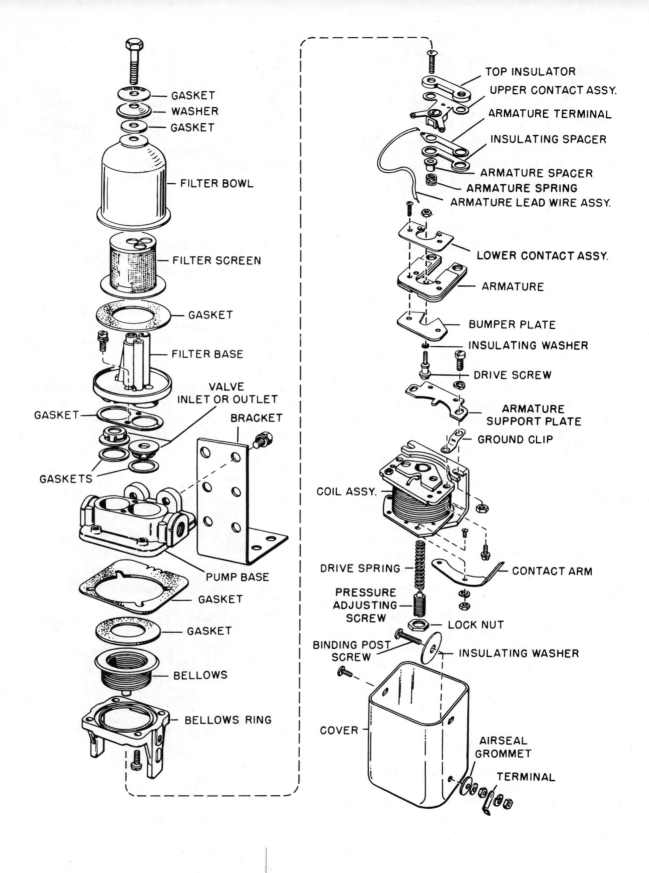

GASKET
WASHER
GASKET
FILTER BOWL
FILTER SCREEN
GASKET
FILTER BASE
VALVE
INLET OR OUTLET
BRACKET
GASKET
GASKETS
PUMP BASE
GASKET
GASKET
BELLOWS
BELLOWS RING

TOP INSULATOR
UPPER CONTACT ASSY.
ARMATURE TERMINAL
INSULATING SPACER
ARMATURE SPACER
ARMATURE SPRING
ARMATURE LEAD WIRE ASSY.
LOWER CONTACT ASSY.
ARMATURE
BUMPER PLATE
INSULATING WASHER
DRIVE SCREW
ARMATURE
SUPPORT PLATE
GROUND CLIP
COIL ASSY.
CONTACT ARM
DRIVE SPRING
PRESSURE
ADJUSTING
SCREW
LOCK NUT
BINDING POST
SCREW
INSULATING WASHER
COVER
AIRSEAL
GROMMET
TERMINAL

Fig. 7   Autopulse electric fuel pump

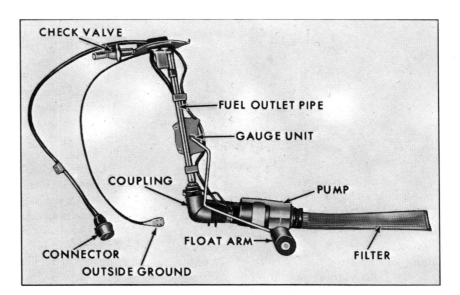

Fig. 8 Buick Riviera electric fuel pump, 1969–70

CHECK VALVE

FUEL OUTLET PIPE

GAUGE UNIT

COUPLING

PUMP

FLOAT ARM

FILTER

CONNECTOR

OUTSIDE GROUND

At lower engine speeds, when less fuel is required, the pump does not deliver its full potential because the turbine is not a positive displacement type like the mechanical fuel pump. That is, the turbine will run without pumping fuel and hence, needs no means of varying fuel delivery rate like its mechanical counterpart must have. The fact that the fuel can slip past the spinning turbine blades eliminates the need for pump inlet and outlet valves and the need for varying its speed. A check valve above the fuel pump in the fuel line prevents the fuel from flowing back into the tank when the engine is shut off. If this valve were not present fuel starvation on start-up might occur since it takes a somewhat longer time for an electric turbine fuel pump to get up pressure in the fuel line than it does for a positive displacement mechanical pump.

Vacuum Pump Operation

Many fuel pumps have a vacuum booster section which operates the windshield wipers at an almost constant speed, Fig. 6. The fuel section (lower) functions in the same manner as that in ordinary fuel pumps. However, the rotation of the camshaft eccentric in this type pump also operates the vacuum booster section by actuating the pump arm which pushes a link and the bellows diaphragm assembly upward, expelling the air in the upper chamber through its exhaust valve out into the intake manifold. On the return stroke of the pump arm, the diaphragm spring moves the bellows diaphragm downward, producing a suction in the vacuum chamber. This suction opens the intake valve of the vacuum section and draws air through the inlet passage from the windshield wiper.

When the windshield wiper is not being used, the intake manifold suction (vacuum) holds the diaphragm upward against the diaphragm spring pressure so that the diaphragm does not function with every stroke of the pump arm. When the manifold suction (vacuum) is greater than the suction produced by the pump, the air flows from the windshield wiper through the inlet valve and the vacuum chamber of the pump and out the exhaust valve outlet to the manifold, leaving the vacuum section inoperative. With a high suction (vacuum) in the intake manifold the operation of the windshield wiper will be the same as if the pump were not installed. However, when the intake manifold suction is low, as it is when the engine is accelerated or operated at high speeds, the suction of the pump is greater than that in the manifold and the vacuum section operates the windshield wiper at nearly constant speed.

Some pumps have the vacuum section placed in the bottom of the pump instead of in the top as shown in Fig. 6, but the operation is basically the same.

Fuel Pump Performance

It is essential that the fuel pump deliver sufficient fuel to supply the requirements of the engine under all operating conditions and that it maintain sufficient pressure in the line between the fuel pump and carburetor to keep the fuel from boiling and to prevent vapor lock.

Excessive fuel pump pressure holds the carburetor float needle valve off its seat, causing high gasoline level in the float chamber which in turn increases gasoline consumption.

The pump usually delivers a minimum of ten gallons of gasoline per hour at top engine speeds,

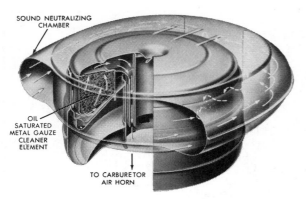

*Fig. 9   Air cleaner of
the oil saturated type*

under an operating pressure of from about 2½ to 7 pounds. The highest operating pressure will be attained at idling speed and the lowest at top speed.

AIR CLEANERS

These devices, Figs. 9, 10, 11, are designed to separate dust and other foreign matter from the incoming air before it enters the carburetor. Since thousands of cubic feet of air are drawn from within the hood of the car and passed through the engine cylinders, it is important that this air be clean. When driving along dusty roads, dust is drawn through the radiator and ultimately finds its way into the engine unless the air is filtered and cleaned. Foreign matter in the engine causes excessive wear and operating troubles.

In the type filter shown in Fig. 9, the filtering element is usually made from copper gauze or copper wool saturated with oil and packed into position between wire or other open-work sections of the cleaner body. Air entering the engine passes through the filtering element which traps the dust and air. The filter element should be washed in gasoline or kerosene at regular lubrication periods, or oftener if operating conditions warrant. After washing out the dirt and old oil, dip the filter element in clean engine oil, allowing the excess oil to drain before reassembling the filter.

The gauze or wool in the filter also acts as a flame arrester during a severe flashback or back-fire. The filter shown in Fig. 9 contains a silencing

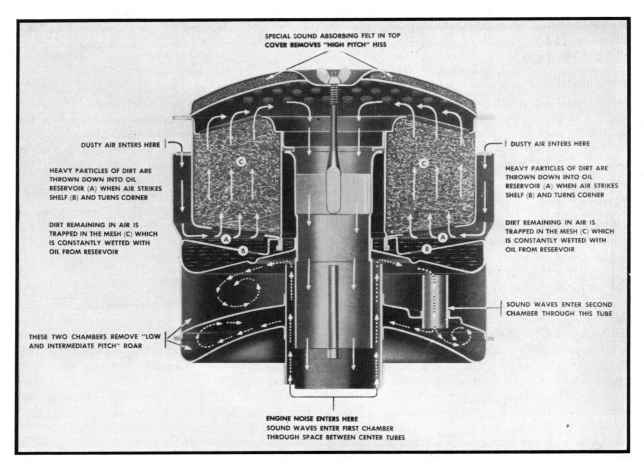

Fig. 10   Air cleaner of the oil bath type

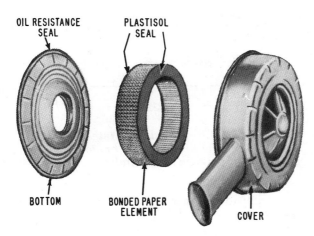

OIL RESISTANCE SEAL   PLASTISOL SEAL

BOTTOM   BONDED PAPER ELEMENT   COVER

Fig. 11 Air cleaner with accordion-pleated cellulose fiber filtering element

unit consisting of intake passages designed to muffle air noises.

The air cleaner shown in Fig. 10 is of the oil bath type. A cavity in the base of this cleaner contains a quantity of oil. The incoming air is forced to reverse its direction of flow immediately above the surface of the oil and the heavier particles of dust are carried into the oil by their momentum. The air then passes through an oil saturated copper gauze where the remaining dust is removed.

Fig. 11 illustrates a paper-element type of air cleaner first introduced on many 1957 cars. The air cleaner element is of the disposable dry type and consists of a cylindrical cellulose fiber material, pleated to permit maximum filter area. On each end of this cylinder, the fiber is embedded in end plates to provide an efficient dust seal. On both sides of the fiber, rust resistant wire screen furnishes compressive strength. The fine mesh of the inner screen also acts as a flame arrester in case of backfire.

The fiber of the element passes air through with low restriction, but any dust or dirt in the air deposits on the pleated outer surface. The fiber is flame-proofed and will retain its filtering efficiency under normal concentrations of gasoline vapor, engine oil and water vapor.

FUEL GAUGES

Automobiles are equipped with fuel gauges which are operated in conjunction with the vehicle's electrical system. The two types are the electric meter and rheostat type, and the bimetal and resistance wire type. The details of their operation are discussed in the *Dash Gauges* chapter.

FUEL LINES

Fuel lines connecting the various units of the fuel system are usually made of rolled steel and sometimes drawn copper. Steel tubing, when used for fuel lines, is generally rust proofed by being copper or zinc plated.

Fuel lines are placed away from exhaust pipes, mufflers and manifolds so that excessive heat will not cause vapor lock. They are attached to the frame, engine and other units so that the effect of vibration is minimized and so they are free of contact with sharp edges which might cause wear. In places of excessive movement, as between the car's frame and rubber-mounted engine, short lengths of gasoline resistant flexible tubing are used.

CARBURETOR

The carburetor is an instrument for metering fuel and air, and for atomizing and vaporizing the fuel charge. The *Physics of Carburetion* and *How a Carburetor Works* are explained further on.

INTAKE MANIFOLDS

An intake manifold is a system of passages which conduct the fuel mixture from the carburetor to the intake valves of the engine. Manifold design has a great deal to do with the efficient operation of an engine. For smooth and even operation, the fuel charge taken into each cylinder should be of the same strength and quality. The distribution of the fuel should, therefore, be as even as possible. This depends greatly on the design of the intake manifold. Dry fuel vapor is the ideal form of fuel charge but present-day fuel prevents this unless the mixture is subjected to high temperature. If the fuel charge is heated too highly, the power of the engine is reduced because the heat expands the fuel charge. Therefore, it is better to have some of the fuel deposited on the walls of the cylinders and manifold passages. Manifolds in modern engines are designed so that the amount of fuel condensing on the intake manifold walls is reduced to a minimum.

The inside of the intake manifold is smooth and the passages are large enough so as not to obstruct the flow of the fuel mixture. The bends or curves in the manifold are designed so that a minimum amount of fuel will be condensed on the walls of the manifold.

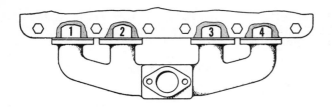

Fig. 12 *Four-cylinder intake manifold with one branch for each cylinder*

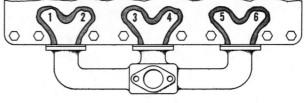

Fig. 13 *Six-cylinder intake manifold with one branch for each two cylinders*

Figs. 12 to 15 illustrate the arrangement of intake manifolds for various type engines.

RAM INDUCTION MANIFOLDS

Available on some 1958–61 Chrysler engines, the system consists of twin air cleaners, twin four-barrel carburetors and two manifolds containing eight long tubes of equal length (four for each manifold), Fig. 16.

The system is designed to increase power output by 10 per cent in the middle speed range (1800–3600 rpm). Each manifold supplies one bank of cylinders and is carefully calculated to harness the natural supercharging effect of a ram induction system. By taking advantage of the pulsations in the air intake column caused by the valves opening and closing, sonic impulses help pack more mixture into the combustion chambers. A brief explanation of how this is accomplished follows:

In an engine with conventional manifolds, several cylinders are supplied by one manifold with branches of varying lengths. The incoming fuel-air mixture does not have a smooth, steady flow into the combustion chambers; rather its behavior is erratic due to the inertia of the moving mixture and the opening and closing of the valves.

Chrysler engineers, after much experimenting with the ram effect on engine induction systems, knew they could calculate effective intake tube lengths for various engine speeds, and the tubes could be placed between the carburetor and the intake ports, or extended from the conventionally-mounted carburetor.

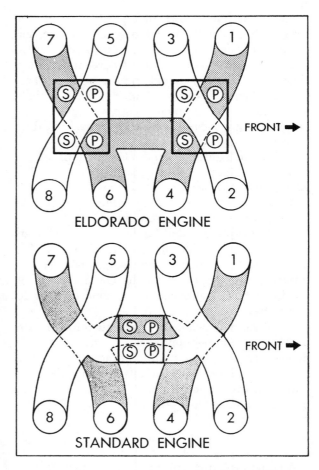

ELDORADO ENGINE

STANDARD ENGINE

Fig. 15 *Eight-cylinder Cadillac intake manifolds for two four barrel carburetors (upper) and one four barrel carburetor (lower). S is for secondary and P is for primary side of carburetor*

Fig. 14 *Eight-cylinder intake manifold for two-barrel carburetor*

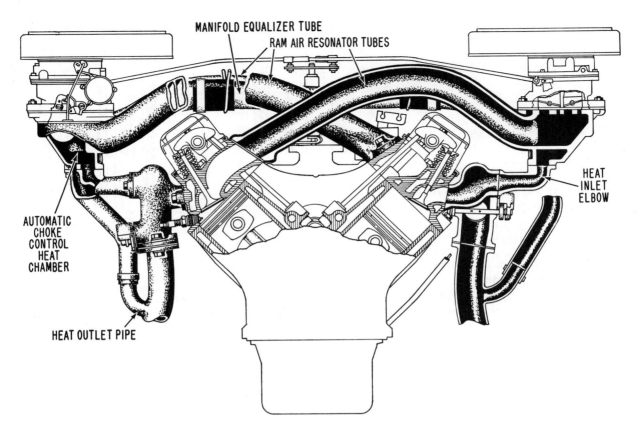

MANIFOLD EQUALIZER TUBE

RAM AIR RESONATOR TUBES

HEAT INLET ELBOW

AUTOMATIC CHOKE CONTROL HEAT CHAMBER

HEAT OUTLET PIPE

Fig. 16   Chrysler ram induction manifolds

By using telescoping tubes and an engine dynamometer they found that with the carburetor mounted on the engine and the tubes extending from it the power increase obtained was in the 2400–4500 rpm range. But since most driving is done in the middle speed range (1800–3600 rpm) this would not do for passenger cars. Therefore, the carburetors were mounted to the sides of the engine with the tubes extending to the intake ports. Then, by trial and error, it was found that eight tubes, 30 inches long, would produce a steady power increase in the middle speed range. In other words, all the tubes were tuned to the same frequency which eliminated the power-robbing pulsations in the power curve. An added advantage with this set-up is that the carburetors can be kept cooler while the engine is running, and can pick up less conducted heat from the engine when it is not running.

In the Chrysler system, the air-fuel mixture from each carburetor flows into a chamber directly below the carburetor, then passes through the long individual intake branches to the opposite cylinder bank. The right-hand carburetor supplies the air-fuel mixture for the left-hand cylinder bank, whereas the left-hand carburetor supplies the right cylinder bank. The passages be-

tween the right- and left-hand manifolds are interconnected with a pressure equalizer tube to maintain balance of the engine pulsations.

MANIFOLD HEAT CONTROL

Most engines have automatically operated heat controls which utilize the exhaust gases of the engine to heat the incoming fuel-air charge during starting and warm-up so as to improve vaporization and mixture distribution. The heat control is regulated by a coiled thermostatic spring mounted on the exhaust manifold, Fig. 17. A counterweight is mounted on the other end of the heat control valve shaft and this counterweight, in conjunction with the thermostatic spring, operates to close and open the heat control valve.

When the engine is cold, all of the exhaust gas is deflected to and around the intake manifold "hot spot," Fig. 18. As the engine warms up, the thermostatic spring is heated and loses tension, thereby permitting the counterweight to gradually change the position of the heat control valve so that at higher driving speeds with a thoroughly warmed engine, the exhaust gases are passed direct to the exhaust pipe and muffler as shown by the dotted arrows.

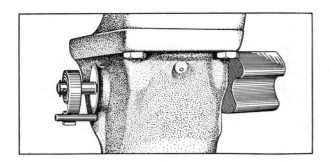

Fig. 17 Manifold heat control thermostatic spring and counterweight

Ram Induction Manifold

In the Chrysler ram induction system, there is a heat control chamber in each manifold to operate the automatic choke and to heat the fuel mixture after warm-up. A hear control valve in each exhaust manifold by-passes exhaust gas through an elbow to the intake manifold heat control chamber. Heat outlet pipes then carry the gas down to a "Y" connector under the heat control valve.

EXHAUST SYSTEM

This system, when used on an "In-Line" engine, such as in Fig. 19, consists of an exhaust manifold, exhaust pipe, muffler, and muffler outlet (or tail) pipe. The burned gases from the engine are conducted to the muffler, and from there they are led from the muffler outlet pipe to the atmosphere.

A muffler quiets the noise of the exhaust by damping the sound waves created by the opening and closing of the exhaust valves. When an exhaust valve opens it discharges the burned gases

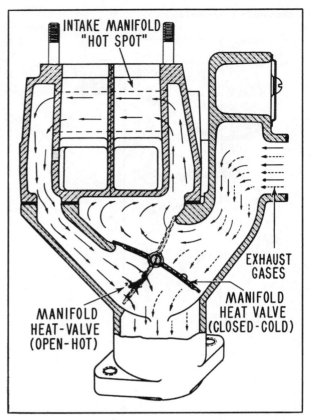

Fig. 18 Manifold heat control operation

at high pressure (around 75 pounds) into the exhaust pipe which is at low pressure (a little above atmospheric, 15 or 16 pounds). This action creates sound waves that travel through the flowing gas, moving much faster than the gas itself, up to about 1400 m.p.h., that the muffler must silence. It does this principally by converting the

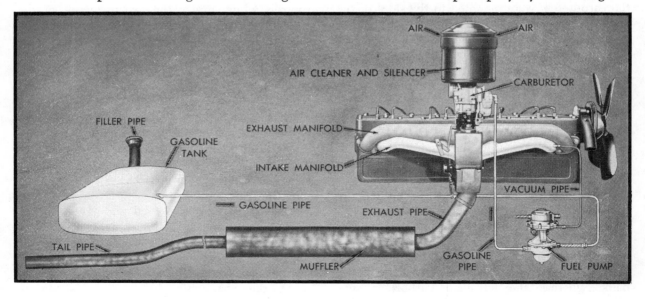

Fig. 19 Fuel and exhaust system

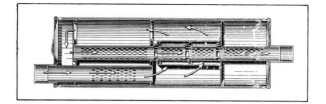

Fig. 20 *Muffler which oper-*
ates on reverse flow principle

Fig. 21 *Straight through*
type of exhaust muffler

sound wave energy to heat. This is usually accomplished by passing the exhaust gas, with its accompanying wave pattern, throughout perforated tubes into chambers of various sizes. Passage through the perforations and subsequent reflection within the chamber forces the sound waves to dissipate their energy. Hereafter, the sound wave pattern and gas flow will be treated as one in describing the various types of mufflers.

The muffler shown in Fig. 20 operates on the reverse flow principle. It is oval-shaped and has multiple pipes. Four chambers and a double jacket are utilized to accomplish muffling of exhaust noise. Exhaust gases are directed to the third chamber, forced forward to the first chamber, from where they travel the length of the muffler and are exhausted into the tail pipe. Arrows in picture show direction of gas flow.

Fig. 21 shows a straight through type muffler. This muffler is provided with a central tube, perforated with many openings which lead into an outside chamber packed with a sound absorbing (insulating) material. As the exhaust gases expand through the perforated inner pipe into the outer chamber, they come in contact with the insulator and escape to the atmosphere under a constant pressure. In this manner the expanding chamber tends to equalize or spread out the pressure peaks of the exhaust from each individual cylinder of the engine.

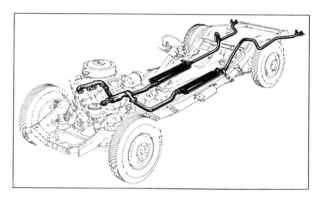

Fig. 23 *Dual exhaust system*

Engines of the V-8 type require two exhaust manifolds and either one or two mufflers and often accompanying resonators. If one muffler is used, the exhaust pipe from one manifold meets the other one in the form of a "Y", Fig. 22.

DUAL EXHAUST SYSTEM

Two types of dual exhaust systems are shown in Figs. 23 and 24. Each exhaust manifold connects to a separate exhaust pipe, muffler and tail pipe. The resonators shown in Fig. 24 can be considered as being additional mufflers as they are used to reduce exhaust noise further.

The advantage of a dual exhaust system is that the engine exhausts more freely, thereby lowering the back pressure which is inherent in an exhaust system. With a dual exhaust system a sizeable in-

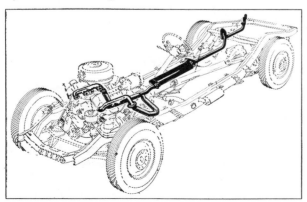

Fig. 22 *V-8 engine exhaust sys-*
tem with one Y-shaped exhaust
pipe and one muffler

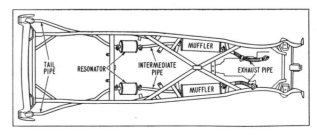

Fig. 24 *Dual exhaust*
system with resonators

crease in engine horsepower can be obtained because the "breathing" capacity of the engine is improved, thereby leaving less exhaust gases in the engine at the end of each exhaust stroke, which leaves more room for an extra intake of the air-fuel mixture.

Engine Fuel

A fuel is a substance composed principally of hydrogen and carbon in such proportions that it will burn in the presence of oxygen and liberate heat energy. By burning fuel in an internal combustion engine this heat energy can be transformed into mechanical energy. Liquid fuels are ideal for internal combustion engines because they can be economically produced, have a high heat value per pound, an ideal rate of burning and can be easily handled and stored.

The most common internal combustion engine fuels are gasoline, kerosene and Diesel fuel oil. Gasoline, because of its many advantages, is used to a much greater extent than any other fuel in internal combustion engines having spark ignition. It has a better rate of burning than other fuels, and due to the ease with which it vaporizes, it gives quick starting in the coldest weather, smooth acceleration and maximum power.

Diesel fuel oil ranks next to gasoline in quantity used. It can be produced as economically as gasoline but its use is limited to Diesel type engines.

The use of kerosene as a fuel for internal combustion engines is usually limited to farm tractors, marine and stationary engines which operate at a fairly constant speed. Its characteristics are such that it cannot be properly mixed with air and controlled in variable speed engines.

COMBUSTION

The burning of a fuel in the presence of oxygen is called combustion.

A gallon of gasoline weighs about 6½ pounds of which 5 pounds is carbon and the remainder hydrogen. The air necessary to burn a gallon of gasoline is composed of about 19 pounds of oxygen and 72 pounds of nitrogen and other gases. In burning a gallon of gasoline, about 11 of the 19 pounds of oxygen from the air combine with the 5 pounds of carbon in the fuel to form carbon monoxide and carbon dioxide gases. Complete combustion converts all the carbon to carbon dioxide gas. However, complete combustion is never attained in the engine and, as a result, carbon monoxide is also formed.

While the carbon and oxygen are uniting to liberate heat, which is converted into energy to run the engine, the 1 pound of hydrogen in the fuel unites with the remaining 8 pounds of oxygen in the air to form about 9 pounds of water; that is, over 1 gallon of water is formed for every gallon of gasoline burned. The water passes off with the burned exhaust gases in vapor form. In hot weather it is not of great importance, but in cold weather the water vapor partially condenses, which usually causes rust and corrosion in the muffler.

The nitrogen in the air is not affected. It remains stable and acts as a cooling agent, reducing the maximum burning temperature that could be obtained if pure oxygen were supplied instead of air.

A Diesel engine differs from a gasoline engine. Operating on an excess amount of air, a Diesel engine burns the fuel more completely and the exhaust is ordinarily free from carbon monoxide gas. It is, however, given off when the combustion in a Diesel engine is incomplete, usually due to faulty fuel injection.

When fuel burns inside the cylinders of an internal combustion engine, the temperature of the mixture is raised by the heat given off. Actual burning temperatures of upwards of 4,000 degrees have been recorded in operating engines.

COMPRESSION PRESSURE

Within practical limits, the more fuel that is compressed in the combustion chamber (compression pressure) the more efficient it is and the more power it produces. As a fuel burns the pressure it creates (combustion pressure) is about four times greater than the compression pressure. A fuel subjected to a compression pressure of 100 pounds per square inch will develop about 400 pounds per square inch combustion pressure as it burns. Should the compression pressure be raised to 150 pounds per square inch, the power produced by the engine will be greatly increased because the combustion pressure will have been raised to about 600 pounds per square inch.

DETONATION

When the compression pressure is very high, the fuel mixture tends to explode instead of burning uniformly and slowly, causing detonation, knock or ping. Fuel knock, besides being an annoying sound, results in loss of power, overheating, increased fuel consumption and severe shock to spark plugs, pistons, connecting rods, bearings and crankshaft. It has been known in extreme cases to chip porcelain from spark plugs and crack cylinder and valve heads.

Improvements in combustion chamber design have helped to reduce the detonation. When the temperature of an unburned fuel can be kept below the detonation temperatures by engine cooling, the increased compression pressures result in increased combustion pressures. Hot areas, like the exhaust valve, radiating heat to the unburned charge, may cause detonation. By placing the spark plugs close to these hot areas so that the fuel starts to burn near them and then moves away from them, the tendency toward engine knocking can be minimized.

Carbon formation in a combustion chamber increases the tendency of a fuel to knock because, being an excellent heat insulator, it reduces the effective cooling surface in the region of the last-to-burn portion of the fuel charge.

OCTANE RATING

The ability of a gasoline to resist detonation is called its octane or anti-knock rating. A gasoline from asphaltic base crude oil produces less knock than one from paraffinic base crude. Cracked gasoline has less tendency to knock than straight run gasoline. All marketed gasolines are a blend of straight run and cracked gasolines, so unless their blending is controlled, their anti-knock qualities will vary.

Engineers and refiners have devised a method of determining and comparing the anti-knock qualities of gasolines by using a special one-cylinder engine, known as the C.F.R. (Co-operative Fuel Research Committee) fuel testing engine, in which the compression pressure can be raised or lowered. A device records and measures the knocking effect of the fuel being tested.

A mixture of iso-octane, which has a very high anti-knock rating, and heptane, which produces a pronounced knock, is used as a reference fuel to establish an anti-knock standard. The anti-knock value or octane number of a gasoline being tested is represented by the percentage of volume of iso-octane that must be mixed with normal heptane in order to duplicate the knocking of the gasoline being tested. Octane numbers range from 50 in third grade gasoline to 110 in aviation gasolines. Since an octane number of 100 indicates a fuel having an anti-knock value equal to that of iso-octane, a number higher than 100 indicates that the anti-knock value is that much greater than that of iso-octane.

If the octane rating of a gasoline is naturally low, the fuel will detonate as it burns and power will be applied to the pistons in hammer-like blows. The ideal power is that which pushes steadily rather than hammers against the piston.

The octane rating of a gasoline can be raised by treating it with a chemical.

A treated fuel is one which contains a chemical that is not a fuel. The most satisfactory chemical known is tetra-ethyl lead compound, which is added to the gasoline in the proportion of about 1 to 1200 by volume, depending on the fuel and the anti-knock value desired.

Tetra-ethyl lead is a liquid which mixes thoroughly with gasoline and vaporizes completely. Ethylene dibromide prevents the tetra-ethyl lead from forming lead oxide deposits on spark plugs and on valve seats and stems. Red dye is added to identify an ethyl treated gasoline and to warn against its being used as anything but an engine fuel.

An engine that does not knock on a low octane fuel does not increase in efficiency when operated on fuel with a higher octane rating. If the knock does not stop, some mechanical adjustments are probably necessary. By adjusting the spark timing of an engine using a low anti-knock gasoline so that it will fire later (retarding the spark), the knocking will be eliminated, but fuel consumption will be increased and the engine will overheat. It may be less expensive to use a higher priced, high octane gasoline with an advanced spark than to use a cheap, low octane gasoline with a retarded spark.

Physics of Carburetion

All automobile engines require a fuel charge in the engine cylinders. This fuel charge is a mixture of air and a vapor obtained from gasoline. The gasoline is atomized (broken up), partially vaporized and mixed with the correct proportion of air in the carburetor. To understand the mechanics of a carburetor it is first necessary to understand the principles of carburetion.

WEIGHT OF AIR

It is not generally appreciated that air has weight and that this weight decreases as the height above sea level increases. The weight of air (atmospheric pressure) has a definite bearing on carburetor design, construction and adjustment. Mechanics are most interested in adjustment, since it varies with each grade of gasoline, with various load conditions and with different altitudes.

To show that air has weight, two identical air-tight vessels, each containing 1 cubic foot of air are placed on a balance as shown in Fig. 25. If all the air is extracted from one vessel with a vacuum pump, it will weigh 1¼ ounces less than the other.

A cubic foot of air weighs 1¼ ounces, but a cubic foot of gasoline weighs about 775 ounces (48½ pounds). This means that a cubic foot of air will have to be multiplied by 620 to equal the weight of a cubic foot of gasoline.

The ideal air-fuel ratio for an automobile engine is 15 parts of air to 1 part of gasoline. Accordingly, to burn 1 cubic foot of gasoline, 15 times 620 or 9,300 cubic feet of air is needed. Converting this ratio to gallons: 9,000 gallons of air is needed to burn 1 gallon of gasoline the air-fuel ratio is 9000:1 by volume. To put it another way, ½ teaspoonful of gasoline has to be mixed with 1 cubic foot of air to obtain the ideal 15 to 1 ratio.

VAPORIZATION

In order to secure a good air-fuel mixture, gasoline must first be atomized (broken up) and vaporized. If ½ teaspoonful of gasoline is poured into a cubic foot of air the gasoline simply drops to the bottom of the vessel and remains there without mixing with the air except by the slow process of evaporation.

Vapor is a gasified liquid. When a liquid vaporizes it occupies more space, and it will also float in air. Evaporation occurs at all temperatures, but it is more rapid in hot weather. All volatile liquids placed in contact with the air will start to evaporate and in this vapor form may present extremely dangerous fire and explosion hazards. The loss from evaporation is considerable, especially in extremely warm weather, and must be considered when handling fuels in bulk. As a rule, the evaporated particles are not visible, but in most cases the vapors from an evaporated gasoline give off an odor which may aid in detecting abnormal evaporating conditions.

It is possible to accelerate vaporization of a fuel by a number of methods. The vaporization of a gasoline in an automobile engine is carried out progressively. The breaking up of a fuel is started by the action of the needle valve and the venturi tubes. It is partially vaporized and atomized by the suction produced in the top of the venturi tube. It is further vaporized as it passes through the intake manifold, by the heat transferred to the manifold from the engine, and it is almost completely vaporized during the compression stroke of the piston by the heat of compression and the heat left by the previously burned fuel. In carburetor design, the practice is to start vaporization as soon as possible.

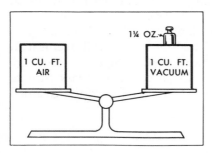

Fig. 25 One cubic foot of air is balanced by 1¼ ounces, demonstrating that air has weight

Fig. 26 Spraying ½ teaspoonful of gasoline into a cubic foot of air mixes them fairly well

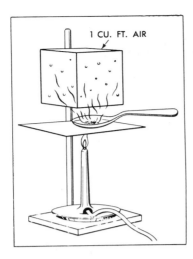

Fig. 27 *Demonstrating use of heat to vaporize fuel and mix it with air*

Vaporization by Spraying

If an atomizer is used for vaporizing a fuel, as shown in Fig. 26, it is possible to spray ½ teaspoonful of gasoline into a cubic foot of air very quickly and have it mix fairly well, since it enters the air in the form of a mist. The use of an atomizer represents one of the most important principles of carburetion. When the bulb is pressed, the suction created at the nozzle by the expelled air draws portions of the gasoline up through the tube. As it draws the gasoline out, the air from the bulb strikes it and breaks it up into small particles, or in other words, atomizes it. "Atom" is a term applied to very small parts of any element and atomization means breaking into small particles.

Vaporization by Heating

A liquid can be vaporized by applying heat to it. The greater the heat, the more rapid the vaporization. If heat is applied to ½ teaspoonful of gasoline, as shown in Fig. 27, the gasoline will be driven off very quickly in the form of gasoline vapors. This vapor rises and if the vessel containing the cubic foot of air is immediately above it the air and vapor mix fairly well.

This heating principle is made use of in carburetors by several methods. The older method was to make use of a "stove" built around the exhaust manifold at a point close to the carburetor so that the air being drawn into the carburetor was first passed around the manifold and heated. While stoves are still used to a limited extent, the modern practice is to use the "hot spots" around the manifold passages (see Fig. 18). The atomized gasoline, as it passes from the carburetor into the

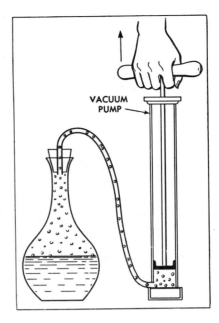

Fig. 28 *Vaporization of fuel by vacuum*

manifold, comes in direct contact with a heated surface and vaporizes rapidly.

Vaporization by Vacuum

The earth is surrounded by an ocean of air, many miles in height, that presses upon the earth with a pressure of 14.7 pounds per square inch at sea level. Atmospheric pressure is exerted in all directions. Air is continually forcing its way into any place that contains no air; that is, air will always try to fill a vacuum. Of course, the term "vacuum" means a lower air pressure than atmospheric pressure.

Air can be forced from a bottle half filled with water by connecting it to a vacuum pump, as shown in Fig. 28. As pressure is lifted from the water, it begins to vaporize and it will then boil at 80 to 90 degrees instead of the normal 212 degrees F. If half the air is removed from a sealed container, the air left will expand as the pressure is removed, leaving the air pressure within the container only about half that of the atmospheric pressure on the outside of the container. This unequal pressure speeds up the vaporization action.

Vaporization by vacuum has a very important place in carburetor operation. A vacuum is the principal means of drawing fuel into the mixing chamber of a carburetor. A venturi tube, which is located just ahead of the mixing chamber, is used in combination with the suction of the engine to increase the vacuum. The lowered pressures due to the vacuum in the venturi allow the atomized fuel to expand and vaporize more quickly.

How a Carburetor Works

In order for a carburetor to deliver fuel to an automobile engine under all operating conditions it must have the following features:

1. A float system which provides a means of storing fuel to be used as needed.
2. A throttle valve to control the speed of the engine.
3. An idle and low speed system for engine operation while standing still and for speeds up to about 20 miles per hour.
4. A part throttle system for operation at cruising speeds.
5. A power system for operating at high speeds.
6. An accelerating pump system to furnish an extra charge of fuel for quick bursts of speed as in passing a car.
7. A choke system to supply a richer air-fuel mixture for starting a cold engine.

FLOAT SYSTEM

Fig. 29. The fuel pump delivers gasoline to the float bowl. As the fuel enters the float bowl through the open needle valve, it raises the float until the needle valve, which is attached to the float, closes in its seat, thus preventing any more fuel from entering the float bowl.

As the fuel is used up by the engine, the float gradually drops, the needle valve opens, allowing more fuel to be pumped into the float bowl by the fuel pump.

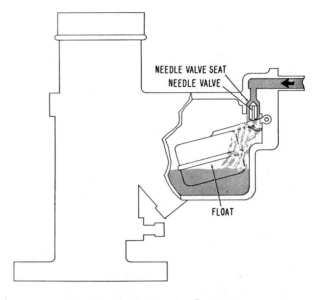

Fig. 29   Diagram of a float system

THROTTLE VALVE

The throttle valve controls the amount of air-fuel mixture passing through the carburetor, thereby controlling the speed of the engine. The throttle valve is a round, flat metal plate attached to a throttle shaft. The assembled plate and shaft is positioned in the carburetor, Fig. 30. When closed it prevents air from passing through the carburetor; when open it permits air to flow through. The quantity of air flowing through the

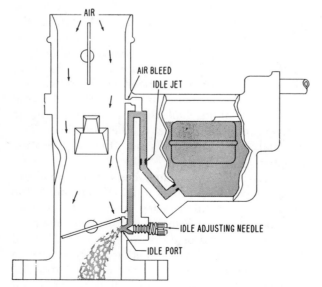

Fig. 30   Diagram of an idle system

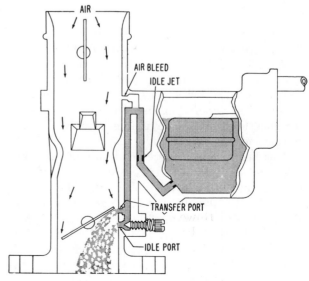

Fig. 31   Diagram of low-speed system

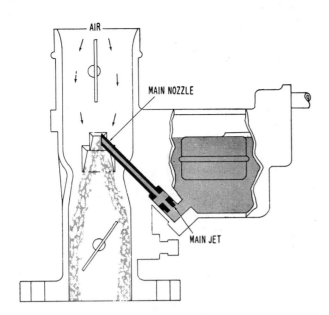

Fig. 32   *Diagram of part-throttle system*

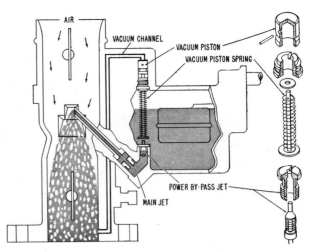

Fig. 33   *Diagram of power system using power valve*

carburetor when the throttle valve is open is determined by the degree of opening. This is controlled by the accelerator pedal which is linked to the throttle shaft.

IDLE AND LOW-SPEED SYSTEM

Let's assume that you are waiting for a red traffic light to change to green. The engine is warm and running at idle speed and your foot is on the brake pedal. Under these conditions the throttle valve is closed, Fig. 30.

Since the throttle valve is closed, we have to provide a means of admitting air-fuel mixture under the throttle valve, otherwise the engine would not run. To accomplish this, two openings are provided in the air passage, one below the throttle valve (idle port) and the other above it (transfer port). These two openings are connected by a passage, the function of which is to allow air to by-pass the closed throttle valve. A second passage is also provided, one end of which communicates with the fuel in the float bowl and the other end to the top (air) opening, Fig. 30.

Now, with the engine idling, the engine pistons are creating a suction (vacuum or low air pressure), but with the throttle valve closed, the main air passage through the carburetor is blocked so that the low pressure area is under the closed throttle valve. However, since the lower opening (idle port) is under the throttle valve, air naturally rushes down the vertical passage. At the same time the low pressure in the vertical passage also lowers the pressure in the passage which communicates with the float bowl. As a result fuel

(which is under atmospheric pressure) is forced out of the float bowl and through the passage where it is mixed with the air stream rushing down the vertical passage to the idle port opening, from whence it is delivered to the engine. As shown, an adjusting needle at the idle port can be turned to provide a leaner or richer air-fuel mixture.

For low speed operation (15 to 20 mph) the throttle valve is opened a little, gradually exposing the transfer port, causing a discharge of air-fuel mixture from this port as well as the idle port, Fig. 31.

The quantity of gasoline fed to the idle and low speed system is controlled by the idle or low speed jet, Fig. 31. A jet is a fitting with a hole drilled in it to a precise dimension so only a predetermined quantity of fuel can pass through it.

PART THROTTLE SYSTEM

When the traffic light turns green you step on the accelerator pedal lightly. This opens the throttle valve, which no longer blocks the passage of air through the carburetor. Thus, the low pressure at the idle passage no longer exists. In consequence, the idle and low speed system stops functioning and the main metering or part throttle system automatically comes into operation, Fig. 32.

Air now rushes down through the carburetor and past the main nozzle sucking a quantity of fuel out of the nozzle to mix with the air stream. The resultant air-fuel mixture is delivered to the engine.

The amount of gasoline fed to the main nozzle

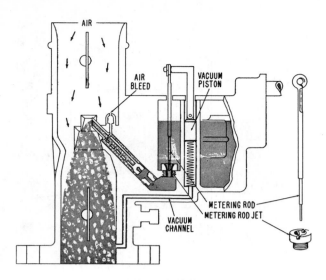

Fig. 34   *Diagram of power system using metering rod*

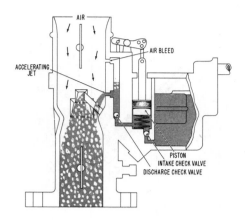

Fig. 35   *Diagram of accelerating system*

is controlled by the size of the hole drilled in the main jet.

The part throttle system operates only at cruising speeds depending on carburetor design.

POWER SYSTEM

It should be obvious that it requires more fuel to run a car at 80 miles per hour than at 50. For this purpose it is necessary to supply additional fuel for the richer "power" mixtures during wide open throttle and high speed operation. Two methods are used to supply this richer mixture, the power valve and the metering rod.

Power Valve: The power valve provides additional fuel for high speed operation by permitting fuel flow through an auxiliary power jet, Fig. 33. Normally controlled by a vacuum-operated piston or diaphragm, the power valve is held closed by intake manifold vacuum until the throttle valve is opened and the manifold vacuum falls below 4 to 6 inches of mercury. A spring then forces the valve open, permitting fuel to flow through the power jet and to join that discharged by the main jet.

Metering Rod: In this design, Fig. 34, all fuel is metered through the main jet (metering rod jet) during both part throttle and full throttle operation. Like most other jets, the metering rod jet is just a fitting with a precisely drilled hole through which fuel can pass from the float bowl to the main nozzle. The metering rod fits into this jet, which is a metal rod with several steps on it, each successive step somewhat larger in diameter than the preceding one. The metering rod is raised out

of its jet and lowered into it in direct relation to the pressure applied to the accelerator pedal.

Like the power valve discussed above, the metering rod is normally controlled by a vacuum-operated piston. During part throttle operation, the larger section of the rod is in the jet, restricting the flow of fuel. As the throttle valve is opened wider by depressing the accelerator pedal, the metering rod is lifted correspondingly out of the jet, so that a smaller diameter section of the rod is now within the jet and more gasoline can enter the main nozzle. When the accelerator is depressed to its limit, the thinnest diameter section of the metering rod permits the maximum quantity of fuel to flow into the main nozzle.

ACCELERATING SYSTEM

Let's say you are driving at 35 miles per hour and want to pass a car in a hurry. Of course, you step on the accelerator pedal and thus open the throttle valve wide. By doing this, one would think that this should increase the amount of air flowing through the carburetor and with it the amount of fuel sucked out of the float bowl. Actually, however, the air responds to this sudden demand quickly, but gasoline, being much heavier than air, lags behind, so that the result would be just a lot of air but not enough fuel to push the engine into high speed. To overcome this deficiency, we use the accelerating system.

As shown in Fig. 35, a typical accelerating pump system consists of a piston linked to the throttle, two check valves, an air bleed and a jet. When the throttle is closed, the piston draws fuel into its cylinder through the intake check valve. As

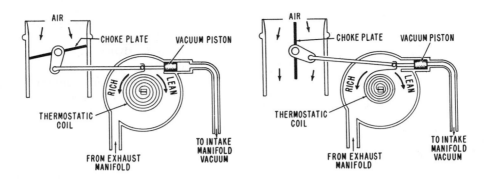

Fig. 36 Operation of an automatic choke

the throttle is opened suddenly, the piston forces fuel through the discharge check valve and the accelerating jet into the carburetor air stream, just enough to mix with the air until fuel starts coming out of the main nozzle. As a result, there is no flat spot in the operation of the engine when fast acceleration is desired, and the acceleration is smooth and fast.

Since a low pressure (vacuum) occurs at the accelerating jet during normal operation, an air bleed is incorporated to counteract this vacuum, thereby preventing syphoning of fuel when the accelerating pump is not operating. This is accomplished by a higher air pressure being applied to the air bleed.

CHOKE SYSTEM

When a cold engine is being started it requires a richer air-fuel mixture, that is, one that contains a higher proportion of gasoline than when the engine is warm. To accomplish this, an additional valve is placed in the air inlet of the carburetor (air horn) above the main nozzle so that when this valve is closed, it greatly reduces the normal air flow in the air horn. This results in extreme unbalance of air pressures. In the air horn, the vacuum produced by the pistons of the engine reduce the pressure to about 5 pounds per square inch, while a pressure of nearly 15 pounds (at sea level) acts upon the gasoline in the float bowl. This difference in pressure forces a lot of gasoline from the float bowl into the air passage and into the cylinders, producing the rich mixture necessary to start a cold engine. As soon as the engine is warm, the choke valve must be opened in order to restore the normal air-fuel ratio.

Although on some cars the choke is operated manually by the driver, most modern carburetors are equipped with a built-in automatic choke control which is a combination of a thermostatic bimetal coil spring and a vacuum-operated piston,

Fig. 36. Both the spring and piston are linked to the choke valve.

During cold starting the coil spring holds the choke valve closed. As soon as the engine starts, intake manifold vacuum acting on the piston opposes the spring action, tending to open the choke. As the engine warms up, heat drawn from the exhaust manifold through a tube causes the thermostatic spring to lose its tension, thereby permitting the choke valve to fully open.

VENTURI ACTION

If a constriction (venturi) is placed in a pipe and water is forced through the pipe, the water in the constriction will flow faster than the water in other parts of the pipe. This principal can easily be demonstrated by allowing water to flow from a garden hose with a tapered nozzle and without a nozzle. The taper in the nozzle forms the restriction and since its diameter is smaller than the diameter of the hose, the water will flow faster out of the nozzle than it will if no nozzle is attached to the hose.

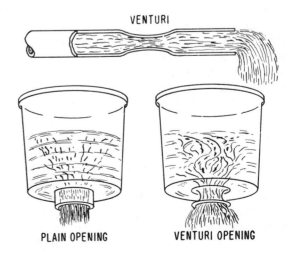

Fig. 37 Principle of venturi tube

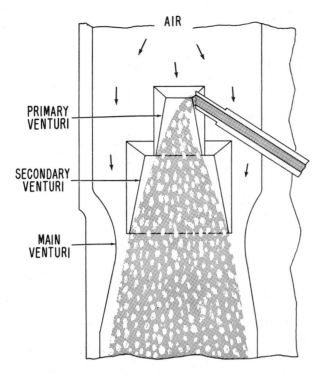

AIR

PRIMARY
VENTURI

SECONDARY
VENTURI

MAIN
VENTURI

Fig. 38   Triple venturi

Similarly, if two containers filled with water are placed side by side, one with a one inch plain opening and the other with a one inch venturi or bell-mouth opening, Fig. 37, the water will flow out of the venturi opening at a faster rate than out of the plain opening.

This same accelerating action takes place as air flows through the venturi of a carburetor.

A venturi tube in a carburetor will develop increased air flow and a higher vacuum within the venturi throat. A problem in carburetion is to secure the correct amount of suction around the needle valve at slow engine speeds and still allow enough air to enter at high engine speeds to maintain the desired ratio of air and fuel. The venturi tube lends itself to these extremes by increasing the vacuum at low speeds without restricting the flow of air at high speeds.

As the piston within the engine cylinder moves downward on the intake stroke, a suction is created around the fuel nozzle in the narrowest part of the venturi and air enters and picks up the gasoline as it passes. After passing the venturi, the partially atomized air and gasoline enters a mixing chamber below the venturi and completes atomization. A partial vacuum exists in the mixing chamber and just as water boils more readily in a vacuum than it does under normal air pressure, so gasoline tends to vaporize more readily in this partial vacuum.

Venturi size must of necessity be a compromise for both high and low speed operation. Because the maximum power an engine can develop is limited by the amount of air it can breathe in, the venturi size should offer minimum resistance to the larger volume of air flowing at high engine speed. On the other hand, a small venturi is desirable at low engine speeds to provide sufficient air velocity for controllable fuel metering and good fuel atomization.

The modern approach to this problem is the use of two or more venturis arranged in series, Fig. 38. Commonly used in modern carburetors, the multiple venturi design serves two purposes: First, the added venturis build up air velocity in the smaller primary venturi, thereby increasing the force available at the main nozzle for drawing and atomizing fuel. Second, air by-passing the primary venturi forms an air cushion around the rich mixture discharged by the venturi, tending to improve mixture distribution by preventing fuel from contacting the carburetor walls.

BALANCED CARBURETOR

The modern carburetor not only must incorporate all the features discussed previously but it must eliminate outside affects on air-fuel ratios—such as a dirty air cleaner.

Regardless of the excellence of design, air cleaners offer some resistance to the flow of air through their filtering elements and passages. Moreover, the air-fuel mixture varies with air

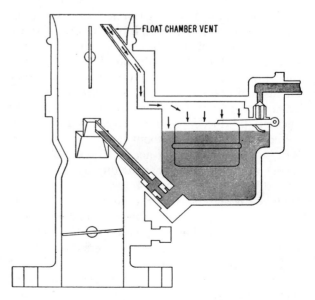

FLOAT CHAMBER VENT

Fig. 39   Carburetor vented to air horn, allowing atmospheric pressure on fuel in float bowl

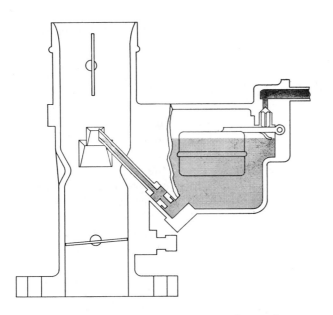

Fig. 40 *Fuel level in float bowl in relation to level in main nozzle. Level in nozzle is always slightly below nozzle tip*

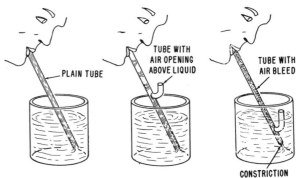

Fig. 41 *Demonstrating the air bleed principle*

cleaner cleanliness, increasing as the filter becomes clogged. This variable effect is essentially overcome in modern carburetors by venting the fuel bowl to the carburetor air horn above the choke valve, as shown in Fig. 39. By this means, the only pressure difference causing the fuel to flow during cruising and high speed operation is that created by air velocity through the venturi. This construction is known as a balanced carburetor.

AIR BLEED PRINCIPLES

Various means of mixture compensation have been employed to correct the enrichening effect of the main jet when air flow increases. Most modern automobile carburetors, however, employ the "air bleed" principle to insure uniform air-fuel mixture ratio during part throttle operation. Before describing its operation, the principles of the air bleed must be understood.

In a plain tube elementary carburetor, the correct fuel supply is maintained by the main jet. A venturi is placed over the nozzle of the plain tube (jet) that receives fuel from a simple carburetor float bowl. When the engine is in operation, the action of its pistons produces suction on the nozzle. Particles of fuel are drawn from this nozzle into the venturi tube. Such an arrangement gives satisfactory operation at cruising speeds. However, the mixture becomes so lean at low and idle speeds that the engine slows down and stops due to lack of suction in the venturi.

If it is desired to have the engine operate at a speed of, let's say, 10 miles per hour, the size of the fuel jet must be materially increased. However, this enlarged jet would increase the fuel delivered at cruising speeds, giving too rich a mixture for satisfactory operation. In addition, the engine would not idle properly no matter what size opening was used, because the suction at the nozzle at extremely low speed would be so weak that fuel would not flow into the mixing chamber of the carburetor.

To prevent the fuel in any carburetor float bowl from overflowing at the nozzle, the float and needle valve is designed so that the fuel level will always be slightly below the nozzle tip, Fig. 40. Consequently, considerable suction (vacuum) is required to lift the fuel from the nozzle into the mixing chamber. At low engine speeds, the fuel is likely to come out of the nozzle in large drops, and not atomize properly.

To help break up the fuel into smaller particles as they leave the nozzle, the air bleed principle is used in modern carburetors. The principle is demonstrated in Fig. 41. As the liquid is sucked into the tube (left view) it is drawn up in a solid column free of air. If a small opening is placed in the tube above the level of the liquid (center) air bubbles will enter the tube and the liquid will be drawn upward in a continuous series of large drops with air spaces between them. At right, which illustrates the air bleed principle, an air bleed tube with one end open to the air has been set into the main tube at a point below the level of the liquid. Suction on the main tube now introduces a considerable quantity of air along with the liquid. The constriction in the lower end of the main tube makes the air bleed opening larger, so that the liquid will be more completely broken up and mixed with the air as it is drawn upward. A

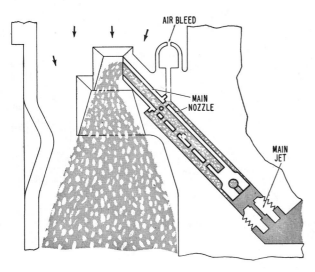

Fig. 42   *Diagram showing air bleed principle applied to carburetor*

very fine balance of the two openings must be maintained when the air bleed principle is applied to a carburetor.

Although design details differ, Fig. 42 shows the general principle of air bleed compensation as applied to a carburetor. Air supplied through a bleed hole in the carburetor throat enters the space between the main nozzle and its surrounding tube. With the engine running at medium speed, the difference in the pressure between the air bleed and the nozzle tip forces the fuel level down in the space, tending to uncover the upper bleed holes in the nozzle. Increasing air flow drops the surrounding fuel level still further, uncovering additional nozzle bleed holes. This lowers the suction in the main jet, thereby reducing fuel delivery at the nozzle and preventing the mixture from becoming too rich at high air flows. In normal practice, the sizes of the air bleed, main jet and nozzle are experimentally determined by the engine manufacturer so that the resulting mixture curve supplies the desired fuel delivery characteristics.

The air bleed serves another important function. Since it provides an emulsion of fuel and air in the main nozzle, the response is quicker upon acceleration than would be the case if only gasoline flowed out of the nozzle. Because gasoline is much heavier than air, it would, therefore, lag behind due to its greater inertia.

HOW ALTITUDE AFFECTS CARBURETION

The gasoline in the float bowl is subjected to atmospheric pressure by means of the vent which communicates with the air horn of the carburetor.

But the air rushing through the venturi when the engine is running is reduced in pressure to about half that of atmospheric pressure. This pressure differential which causes fuel flow is dependent on air velocity at the tip of the nozzle, which varies with the volume of air passing through the venturi. Thus, the weight of the fuel delivered by the fuel metering system is proportional to the volume of air—*not its weight*.

At sea level, where the air pressure is 14.7 pounds per square inch, one pound of air occupies approximately 13.3 cubic feet. At an elevation of 5,000 feet above sea level, however, the air pressure is only 12.2 pounds per square inch, so that one pound of air occupies about 15.4 cubic feet. As a result, there is 14 per cent less oxygen by weight in each cubic foot of air supplied to the engine when operating at an altitude of 5,000 feet, and consequently the mixture ratio is richer. Smaller or leaner jets must therefore be used at high altitudes to maintain the desired air-fuel ratios.

CARBURETOR ACCESSORIES

In addition to its primary function of supplying the required air-fuel mixtures for varying engine demands, a modern carburetor usually provides assistance in controlling other features of engine and vehicle operation. These include:

1. A vacuum supply passage for automatic spark advance control (discussed in the *Ignition Chapter*).
2. Electric switches to operate the starter circuit (described in the *Starter Switch Chapter*).
3. A dashpot which prevents engine stalling with automatic transmissions by slowing throttle closure.
4. A vacuum-controlled kickdown switch to limit automatic transmission downshift speeds.

TYPES OF CARBURETORS

For the sake of simplicity, the preceding discussion has been based on a single barrel downdraft carburetor. The same principles apply, however, to updraft and horizontal carburetors.

As the name implies, air flows up through the updraft carburetor, picking up fuel from the nozzle, idle ports, or accelerating jet before being delivered to the manifold. Although this type carburetor was supplanted by the downdraft design for passenger cars many years ago, it still is used on trucks, farm tractors and marine engines.

Like the updraft carburetor, the horizontal carburetor is mounted on the side of the engine. However, air passes horizontally through the carburetor air horn and venturi in a horizontal carburetor before entering the manifold.

Modern eight-cylinder engines carry dual or four-barrel carburetors, the dual carburetor also being used on some six-cylinder engines.

Dual carburetors consist of two primary venturi tubes, each containing the necessary parts for a complete carburetor except for a common float bowl. With this arrangement on V-8 engines, each tube supplies air-fuel mixtures to half of the engine's cylinders. As a result, a V-8 dual carburetor insures more equal air-fuel distribution.

The four-barrel carburetor is in essence composed of two dual carburetors in a single housing. The primary section forms a complete dual carburetor, while the secondary section normally incorporates only the throttle valves and fixed idle and main metering systems. During normal cruising, only the primary section supplies air-fuel mixtures. However, when the primary throttles are opened wide, the secondary throttles, controlled by either delayed-action linkage from the primary throttles or air flow through the primary venturis, also open in response to the engine's demand for more air. Thus, the four-barrel carburetor provides more power through better breathing at high speeds, with no sacrifice of fuel economy at low and cruising speeds.

Carter Carburetors

YF SERIES

The YF Series is a single barrel carburetor combining the basic features of the fundamental carburetor previously described with several new features. These include a diaphragm type accelerating pump, a bowl vent tube to maintain the correct air-fuel ratio at all times regardless of restriction in the air cleaner due to dirt. It also has a diaphragm operated metering rod, both vacuum and mechanically controlled. Figs. 43 to 47 show the construction of the various circuits.

Mechanical Metering Rod Action

Fig. 45. During part throttle operation, manifold vacuum pulls the diaphragm assembly down, holding the metering rod arm against the pump lifter link. Movement of the metering rod will then be controlled by the pump lifter link, which is connected to the throttle shaft. This is true at all times when the vacuum under the diaphragm is strong enough to overcome the tension of the lower pump diaphragm spring. The upper pump spring serves as a bumper upon deceleration and a delayed action spring on acceleration.

Vacuum Metering Rod Action

Under any operating condition, when the tension of the lower pump diaphragm spring overcomes the pull of vacuum under the diaphragm, the metering rod will move toward the wide open throttle or power position.

The restriction and air bleed in the vacuum passage provide a lower and more uniform

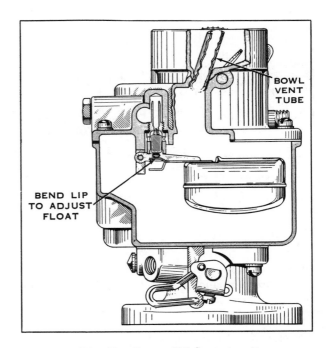

Fig. 43 Carter YF float circuit

vacuum condition in the chamber below the diaphragm.

Anti-Percolator

To prevent vapor bubbles in the nozzle passage and low-speed well caused by heat from forcing fuel out of the nozzle, anti-percolator passages, calibrated plugs and bushings are used, Fig. 45.

Their purpose is to vent the vapors and relieve the pressure before it is sufficient to push the fuel out of the nozzle and into the intake manifold.

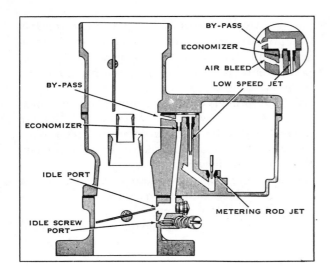

Fig. 44 *Carter YF low speed circuit*

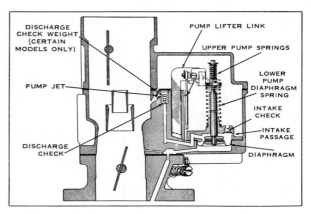

Fig. 46 *Carter YF pump circuit*

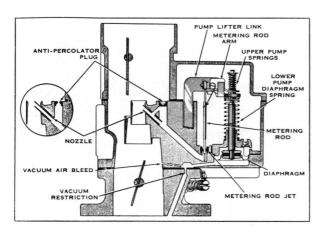

Fig. 45 *Carter YF high speed circuit*

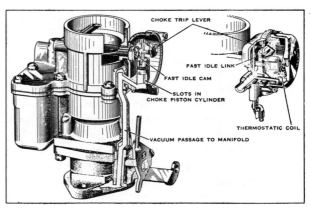

Fig. 47 *Carter YF choke circuit*

Anti-percolator plugs, bushings and the main nozzle are permanently installed and must not be removed in service.

Accelerating Pump Circuit

Fig. 46. The accelerating pump circuit provides a measured amount of fuel, which is necessary to insure smooth engine operation for acceleration at speeds below about 30 miles per hour.

Pump action is controlled both mechanically and by manifold vacuum in the same manner as the metering rod. When the throttle is closed, the diaphragm moves downward and fuel is drawn into the fuel pump chamber.

There are three types of intake systems used on YF carburetors: intake passage, intake check valve, intake passage plus an intake check valve.

When the diaphragm moves downward, the discharge check is seated. When the throttle is

opened, the diaphragm moves upward, forcing fuel out through the discharge passage, past the discharge check, and out of the pump jet. The pump discharges through the main nozzle on models where no pump jet is used. On models with an intake passage, a measured amount of fuel is returned to the bowl through this passage. Carburetors that have only an intake check valve do not discharge fuel back into the bowl. When the diaphragm moves upward, the intake check (where used) must seat.

If the throttle is opened suddenly, the upper pump spring will be compressed, resulting in a smoother pump discharge of longer duration.

Manifold vacuum is applied to the underside of the diaphragm at all times the engine is in operation. When manifold vacuum decreases to the point where the lower pump diaphragm spring overcomes the pull of vacuum, the diaphragm moves upward and a pump discharge results.

The pump jet is pressed into the casting during manufacture and must not be removed in service.

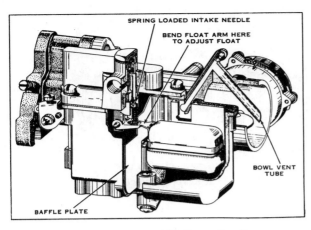

Fig. 48 *Carter YH float circuit*

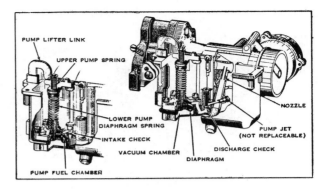

Fig. 51 *Carter YH pump circuit*

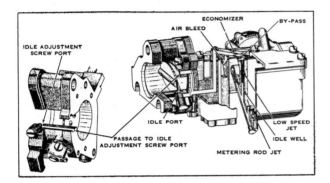

Fig. 49 *Carter YH low speed circuit*

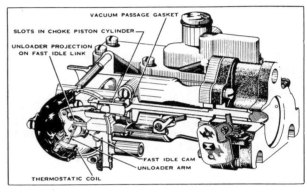

Fig. 52 *Carter YH choke circuit*

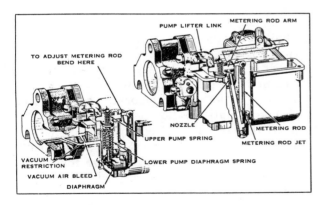

Fig. 50 *Carter YH high speed circuit*

YH SERIES

The YH carburetor may be compared with the YF unit with the circuits rearranged to operate in a horizontal or sidedraft position. It has the five conventional circuits as are used in YF carburetors. Figs. 48 through 52 show the construction of the various circuits.

WCFB SERIES

The WCFB carburetor is basically two dual carburetors contained in one assembly. The section containing the metering rods, accelerating pump and choke is termed the primary side of the carburetor while the other section is called the secondary side. It has five conventional circuits, the same as other Carter carburetors. They are:

Two float circuits.
Two low speed circuits.
Two high speed circuits.
One pump circuit.
One choke circuit.

Float Circuits

Fig. 53. The primary and secondary bowls are separated by a partition. The fuel line connection is above the secondary needle valve and seat. Fuel is supplied to the primary needle and seat through the passage in the bowl cover.

The bowls are vented to the inside of the air horn, and (on certain models) also to atmosphere. Bowl vents are calibrated to provide proper air pressure above the fuel at all times.

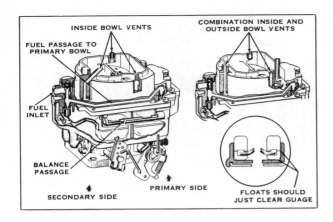

Fig. 53   Carter WCFB float circuits

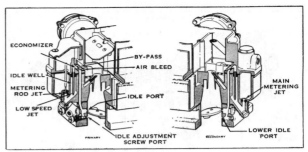

*Fig. 54   Carter WCFB
low speed circuits*

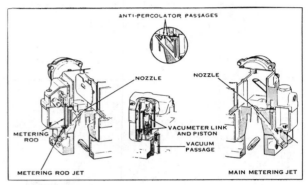

*Fig. 55   Carter WCFB
high speed circuits*

A connecting passage along the outside of the body effects a balance of the fuel levels and air pressures between the two bowls.

Low Speed Circuits

Fuel for idle and early part throttle operation is metered through the low speed circuits.

Gasoline enters the idle wells through the metering rod jets on the primary side of the carburetor and through the main metering jets on the secondary side, Fig. 54.

The low speed jets measure the amount of fuel for idle and early part throttle operation. The air by-pass passages, economizers and idle air bleeds are carefully calibrated and serve to break up the liquid fuel and mix it with air as it moves through the passages to the idle ports and idle adjustment screw ports. There are no idle adjustment screws on the secondary side of the carburetor.

The idle ports are slot shaped. As the throttle valves are opened, more of the idle ports are uncovered, allowing a greater quantity of fuel-air mixture to enter the carburetor bores. The secondary throttle valves remain closed at idle.

The vapor vent ball check, operated by the arm on the countershaft, provides a vent for fuel vapors to escape from the carburetor bowls to the outside at idle and when the engine is not in operation.

High Speed Circuits

Fuel for part throttle and full throttle operation is supplied through the high speed circuits, Fig. 55.

Primary Side—The position of the metering rods in the metering rod jets controls the amount of fuel flowing in the high speed circuit of the primary side of the carburetor. The position of the

metering rods is dual controlled—mechanically by movement of the throttle, and by manifold vacuum applied to the vacuum piston on the vacumeter link.

Secondary Side—Fuel for the high speed circuit of the secondary side is metered at the main metering jets (no metering rods are used).

Throttle values in the secondary side remain closed until the primary throttle valves have been opened a pre-determined amount. They arrive at wide open throttle position at the same time as the primary throttles. This is accomplished by linkage between the throttle levers. Certain models have offset throttle valves mounted above the secondary throttle valves, which are called "auxiliary throttle valves." Air velocity through the carburetor controls the position of the auxiliary throttle valves. When the accelerator is fully depressed, only the primary high speed circuit will function until there is sufficient air velocity to open the auxiliary throttle valves. When this occurs, fuel will also be supplied through the secondary high speed circuit.

Auxiliary throttle valves, or the secondary throttle valves on models without auxiliary valves, are locked closed during choke operation to insure faster cold engine starting.

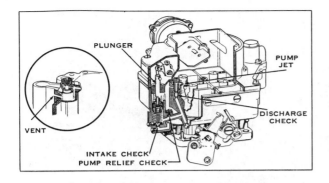

Fig. 56 *Carter WCFB pump circuit*

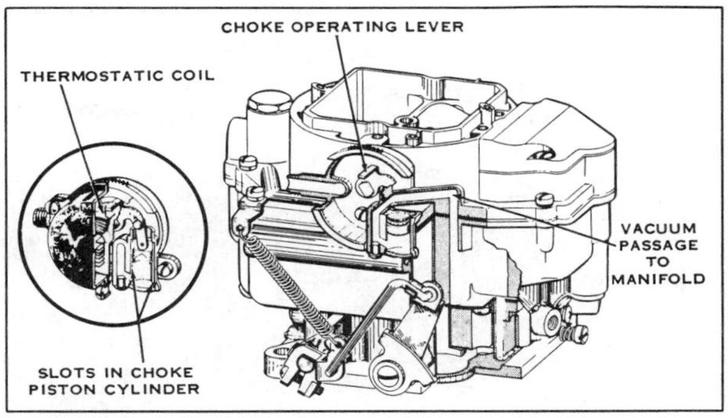

Fig. 57 *Carter WCFB choke circuit*

Anti-Percolator

Fig. 55. To prevent the vapor bubbles in the nozzle passages and low speed wells, caused by heat forcing fuel out of the nozzles, anti-percolator passages and calibrated plugs and bushings are used. Their purpose is to vent the vapors and relieve the pressure before it is sufficient to push the fuel out of the nozzles and into the intake manifold. Anti-percolator plugs, bushings and main nozzles are permanently installed and must not be removed in service.

Pump Circuit

Fig. 56. This circuit is found only on the primary side of the carburetor. It provides a measured amount of fuel, which is necessary to insure smooth engine operation for acceleration at speeds below approximately 30 miles per hour.

When the throttle is closed, the pump plunger moves upward in its cylinder and fuel is drawn into the pump cylinder through the intake check. The discharge check is seated at this time to prevent air being drawn into the cylinder. When the throttle is opened, the pump plunger moves downward, forcing fuel out through the discharge passage, past the discharge check, and out of the pump jets. When the plunger moves downward the intake check is closed, preventing fuel from being forced back into the bowl.

If the throttle is opened suddenly, the upper pump spring will be compressed by the plunger shaft telescoping, resulting in a smooth pump discharge of longer duration.

At speeds above approximately 30 miles per hour, pump discharge is no longer necessary to insure smooth acceleration. When the throttle valves are opened a pre-determined amount, the pump plunger bottoms in the pump cylinder, eliminating pump discharge due to pump plunger movement at high speeds.

During high speed operation, a vacuum exists

at the pump jets. To prevent fuel from being drawn through the pump circuit, the passage to the pump jets is vented by a cross passage to the carburetor bowl above the fuel level. This allows air instead of fuel to be drawn off the pump jets.

On certain models a pump relief check prevents excessive pressure in the discharge passage during acceleration.

Choke Circuit

Fig. 57. This circuit operates in the same manner as on other Carter carburetors. And like single and two barrel carburetors, it is provided with a fast idle device and unloader mechanism, the operation of which has previously been described.

AFB SERIES

This carburetor, Fig. 58, while essentially the same as the WCFB, contains many new features, some of which are a new location for the step-up rods and pistons. The step-up rods, pistons and

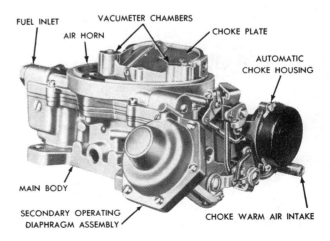

Fig. 58 *Carter AFB four-barrel carburetor*

393

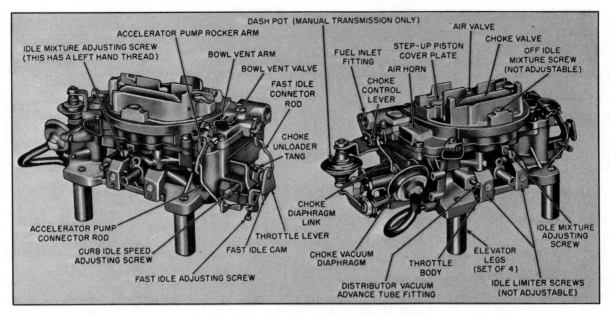

Fig. 59   Carter AVS four-barrel carburetor

springs are accessible for service without removing the air horn, or the carburetor from the engine. The venturi assemblies (primary and secondary) are replaceable and contain many of the calibration points for both the high and low speed systems.

All the major castings of the carburetor are aluminum, with the throttle body cast integral with the main body. This allows an overall height reduction in the carburetor. The section containing the accelerator pump and integral choke is termed the primary side of the carburetor. The other side is the secondary. The five conventional systems used in the WCFB carburetor are also used in this unit.

AVS SERIES

Similar to the AFB, the four barrel AVS carburetors, Fig. 59, have several novel features that make them different from the former. The AVS designation means, "Air Valve Secondary" and they have a spring loaded air valve located above the fuel nozzles on the secondary side of the carburetor that provides very smooth response when the secondary throttles are opened. Instead of the built-in automatic choke of the AFB's, this new series has a remotely located temperature sensing choke coil mounted on the intake manifold over the exhaust crossover passage.

The primary side of the carburetor uses venturi

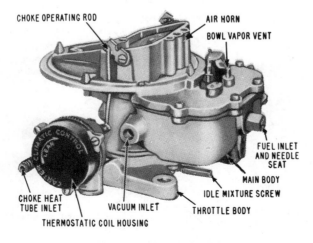

Fig. 60   Carter ABD two-barrel   carburetor

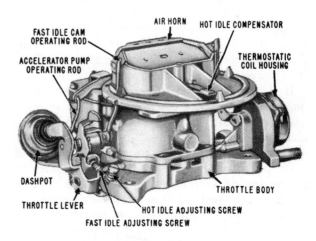

Fig. 61   Carter ABD two-barrel   carburetor

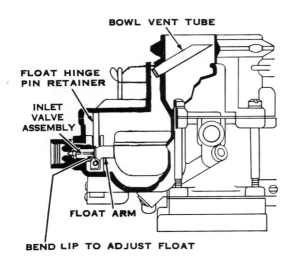

Fig. 62 *Carter BBD and BBS float system*

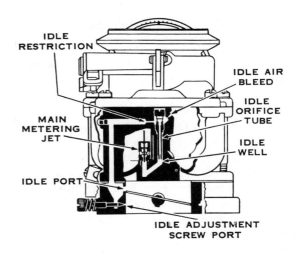

Fig. 63 *Carter BBD and BBS low speed system*

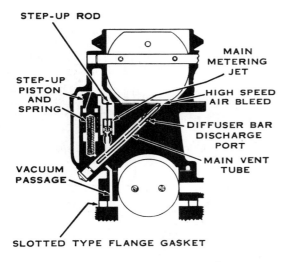

Fig. 64 *Carter BBD and BBS high speed system*

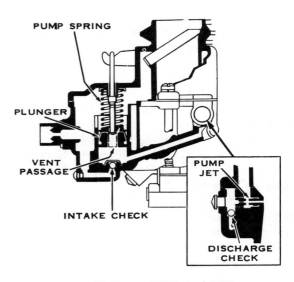

Fig. 65 *Carter BBD and BBS accelerator pump system*

clusters for fine fuel control in the idle and economy operating ranges. Fuel nozzles are pressed into the secondary side of the fuel bowl, eliminating protuberances in the secondary bores. This gives the carburetor exceptionally high air capacity in the power ranges.

ABD SERIES

This carburetor, introduced in 1960, is a two-barrel unit that contains the five conventional circuits. Like the AFB series, which it resembles, the air horn, float bowl and body, and throttle body are made of aluminum castings, Fig. 60 and 61.

A hot idle compensator, Fig. 61, is included in the low speed system. During long periods of idling with an extremely hot engine, the fuel in the carburetor bowl becomes hot enough to form vapors. These vapors enter the carburetor bores by way of the inside bowl vents. The vapors mix with the idle air and are drawn into the engine causing an excessively rich mixture and a loss in rpm or engine stalling. Also, the decrease in the density of the air caused by extreme high underhood temperatures can reduce the idle speed.

The hot idle compensator is calibrated to open under these temperature conditions, permitting additional air to enter the manifold below the throttle plates and mix with the fuel vapors to provide a more combustible mixture.

BBD & BBS SERIES

These carburetors are similar in design, the chief difference being that the BBS is a single barrel unit whereas the BBD is a dual barrel carburetor.

Each throat of the BBD carburetor supplies an air-fuel mixture to four specific cylinders. Thus it is essentially two carburetors in one housing. Each throat contains its own idle air bleed, high speed air bleed, idle orifice tube, main vent tube, main metering jet, metering port, idle port, idle mixture adjustment and throttle valve.

Metering of the fuel in the accelerating pump system on the BBD carburetor is accomplished by two accurately drilled orifices, one for each throat, in the discharge cluster. In the BBS carburetor, a replaceable accelerating pump jet is used.

Figs. 62 and 65 illustrate the various systems. The float, low speed and pump systems operate in the conventional manner, but inasmuch as the high speed system is somewhat different, herewith is a description of its operation.

High Speed System

Fig. 64. When the engine is under a heavy load, suddenly accelerated, or operated at very high engine speeds, the step-up system supplies additional fuel through the diffuser bar discharge port. Fuel flow through the fuel passage of the main metering jet is controlled by the movement of the step-up rod which in turn is moved by a spring and a vacuum-controlled piston. A vacuum passage to the intake manifold is provided for by a drilled passage in the carburetor body and throttle body, and a slotted flange gasket.

Under normal driving conditions, manifold vacuum exerts a strong pull on the vacuum piston. This holds the piston down, keeping the step-up

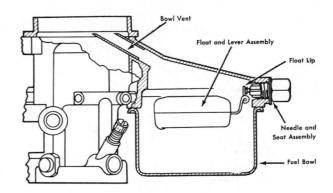

Fig. 66   *Carter RBS float system*

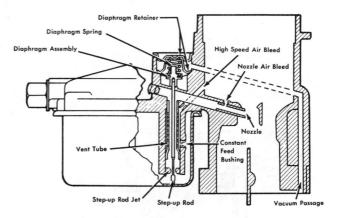

Fig. 68   *Carter RBS high speed system*

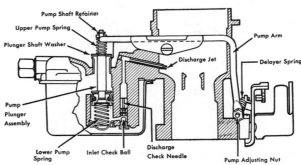

Fig. 69   *Carter RBS accelerator pump system*

rod in the fuel passage of the main metering jet. Fuel then flows around the rod, through the jet, and through the diffuser bar discharge port.

When manifold vacuum falls off, due to a heavy load, sudden acceleration, or very high engine speed, the spring moves the piston up, moving the step-up rod out of the main metering jet fuel passage. Additional fuel is then supplied to the engine.

Air is drawn through the high speed air bleed

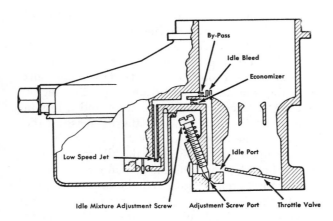

Fig. 67   *Carter RBS low speed system*

396

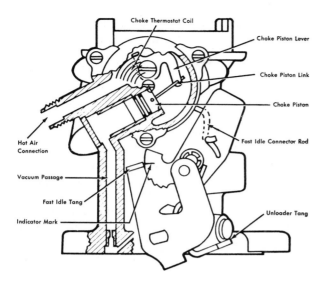

Fig. 70   Carter RBS choke system

and mixes with the fuel surrounding the main vent tube. The mixture is then drawn from the diffuser discharge ports.

RBS SERIES

This carburetor uses a single lightweight aluminum casting with a pressed steel bowl. Adjustments are readily accessible and most calibration points are located in the casting. Fuel pickups are located near the centerline of the carburetor bore to gain the benefits of a concentric bowl carburetor, yet so located that engine heat being radiated through the bore and conducted through the casting is not readily conducted to the fuel in the bowl. Vapor vents allow rapid dissipation of vapors to assure smooth idle and to minimize hard starting while engine is hot. A diaphragm controlled step-up provides instantaneous response to engine demands. Figs. 66 to 70 illustrate the various systems.

Float System

The float assembly extends around the metering portion of the casting to produce the effect of twin floats yet remains a single unit. It assures little or no change in float setting due to vibration or heat. The resilient needle seat has the ability to digest small foreign particles in the fuel to minimize leakage of fuel or flooding.

The bowl is vented to the inside of the air horn through a single tube which is calibrated to pro-

vide the correct air pressure above the fuel under all operating conditions, Fig. 66. An outside vent is provided for the bowl (around the pump plunger shaft) during idle and off idle operation to prevent fuel vapors under extreme heat conditions being fed to the engine which might result in uneven idle. It also assures fast hot restarts.

Low Speed System

Fuel is metered as it enters the lower end of the low speed jet, Fig. 67, and flows up through the tube where air metered through the by-pass mixes with the fuel. The fuel and air then passes through the economizer to the idle bleed where a second metered amount of air is introduced. This mixture is discharged into the manifold through the idle port hole and the idle adjustment screw port.

During curb idle only a small amount of the idle port is exposed to intake manifold vacuum. As the throttle is opened slowly, more of the port is exposed to allow a calibrated increase in the amount of mixture fed to the engine.

High Speed System

During part throttle operation the relatively high vacuum from the intake manifold is transferred through a passage to the upper surface of the spring loaded diaphragm assembly. Manifold vacuum opposing the calibrated spring, provides an economical mixture control to the engine at all times except when full power is required. When the diaphragm is up, the larger diameter (lower end) of the step-up rod is in the jet to provide the economy mixture.

As the throttle is opened, manifold vacuum is decreased. The calibrated diaphragm spring then moves the diaphragm downward and the step-up rod, attached to the diaphragm, is lowered in the jet. The smaller diameter of the rod permits the metered increase in fuel flow to provide for the additional power needs.

During acceleration the same action takes place to richen the mixture. However, the rod is raised in the jet as soon as terminal acceleration is reached when manifold vacuum indicates the need for a less rich mixture.

A vent tube with calibrated side holes is pressed into the high speed well around the step-up rod. A metered amount of air is fed from the bore of the carburetor to the annulus around the vent tube. This air passes through the side holes in the vent tube to mix with the fuel before it flows through the nozzle to the air stream. This aerated fuel permits immediate fuel atomization as it emerges from the tip of the nozzle. This action is further helped by a second air bleed near the tip of the nozzle, Fig. 68.

Accelerator Pump System

As the throttle is closed, the plunger is raised in the cylinder and fuel from the bowl flows into the cylinder through the check ball, Fig. 69. No air enters the cylinder due to the sealing action of the pump discharge needle in its seat.

As the throttle is opened, the spring on the connector link is compressed, which in turn pushes the plunger down in the cylinder. Fuel is forced past the discharge needle and out through the pump jet and into the air stream. During the discharge stroke the intake ball is on its seat to prevent fuel from flowing back into the bowl.

Choke System

When the engine is cold, tension of the thermostatic coil holds the choke valve closed, Fig. 70. As the engine is cranked, air pressure against the offset choke valve causes the valve to open slightly against the coil tension. Manifold vacuum applied to the choke piston also tends to pull the choke valve open. When the engine starts the choke valve assumes a position where tension of the coil is balanced by the pull of vacuum on the piston and force of the air stream against the offset choke valve. A fast idle cam, controlled by choke valve position, holds the throttle open the correct amount to provide a speed commensurate with engine temperature.

Ford Carburetors

ONE-BARREL CARBURETOR

Introduced in 1963 the Ford one-barrel carburetor, Figs. 71 and 72, consists of two main assemblies, the main (upper) body and the throttle (lower) body.

The upper body contains the major metering components of the carburetor which are: main and idle fuel, power valve, float chamber vent, and fuel inlet systems.

The lower body contains the fuel bowl, accelerating pump, idle mixture adjusting screw, and spark valve. A hydraulic dashpot is also included in the lower body for use on cars equipped with automatic transmission. The carburetor is available with a manual choke or with an automatic choke.

TWO-BARREL CARBURETOR

The Ford two-barrel carburetor, introduced in 1957, consists of two main assemblies: the air horn and the main body, Fig. 73.

The air horn assembly, which serves as the main

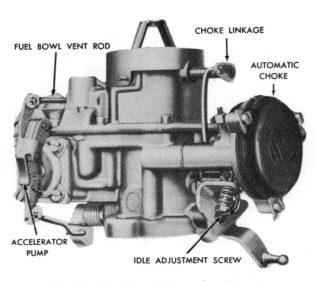

Fig. 71 Ford single barrel carburetor

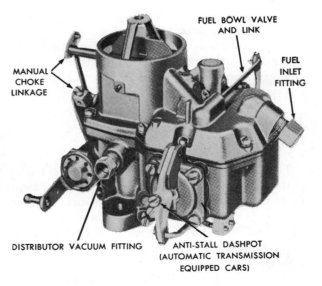

Fig. 72 Ford single barrel carburetor

398

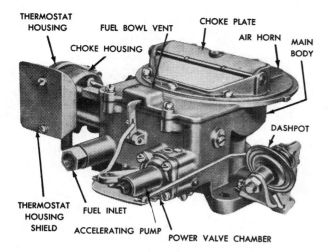

Fig. 73 *Ford two-barrel carburetor*

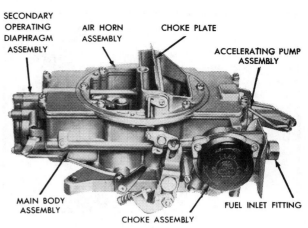

Fig. 74 *Ford four-barrel carburetor*

body cover, contains the choke plate and the fuel reservoir vent.

The main body houses the throttle plates, the accelerating pump, the power valve and the fuel reservoir. The automatic choke housing is attached to the main body.

Each barrel contains a main and booster venturi, main fuel discharge, accelerating pump discharge, idle fuel discharge and throttle plate.

The carburetor has four basic fuel metering systems. They are the idle fuel system, the accelerating system, the main fuel system and the power fuel system. A fuel inlet system provides the various fuel metering systems with a constant supply of fuel. In addition, an automatic choke system provides a means of temporarily enriching the mixture to aid in starting and operating a cold engine.

FOUR-BARREL CARBURETOR

This carburetor, Fig. 74, is basically two dual carburetors built into one housing. It consists of two main assemblies: the air horn and the main body.

The air horn assembly, which serves as the main

body cover, contains the choke plate and the primary fuel reservoir vent.

The main body houses the primary and secondary throttle plates, the accelerating pump, the power valve, the secondary operating diaphragm, and the fuel reservoirs. The automatic choke housing is attached to the main body.

The two front (primary) barrels each contain a main and booster venturi, main fuel discharge, accelerating pump discharge, idle fuel discharge, and the primary throttle plate.

The rear (secondary) barrels each have a main and booster venturi, secondary fuel discharge, and a vacuum-operated throttle plate.

The carburetor has a primary fuel circuit and vacuum-operated secondary fuel circuit.

The primary circuit has four basic fuel metering systems. They are the idle fuel system, the accelerating system, the main fuel system and the power fuel system. A fuel inlet system for both the primary and secondary circuits provides the various fuel metering systems with a constant supply of fuel. In addition, an automatic choke system provides a means of temporarily enriching the mixture to aid in starting and operating a cold engine.

Holley Carburetors

ONE-BARREL CARBURETORS

Models 1904, 1908, 1960

These carburetors, Fig. 75, are single barrel units with a single venturi. It is a compact unit

less than two-thirds as high as carburetors of standard design having a comparable capacity. In reducing its height the conventional air horn has been eliminated and the choke plate is placed in the venturi. In addition to its normal function, the

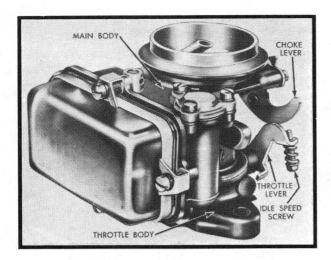

Fig. 75 *Holley carburetor with glass fuel bowl. Models* 1904, 1908, 1960

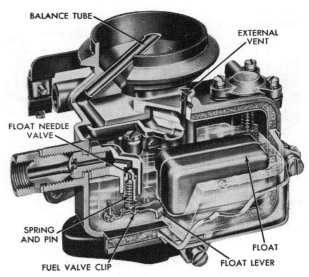

Fig. 76 *Holley* 1904, 1908, 1960 *float system*

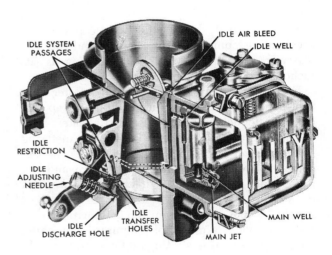

Fig. 77 *Holley* 1904, 1908, 1960 *idle system*

Fig. 78 *Holley* 1904, 1908, 1960 *main metering and high speed systems*

choke plate, when open, aids in the distribution and vaporization of the fuel discharged by the main nozzle.

Fig. 76 illustrates the float system. The fuel, under pressure from the engine's fuel pump, enters the carburetor through the float needle valve.

Fig. 77 shows the idle system. The air and fuel travel down a passage under the air bleed and through the idle passages to the idle discharge holes.

Fig. 78 pictures the main metering and power systems. During part throttle operation (main metering system) a supply of air is introduced into the main fuel well by the high speed air bleed. The fuel and air travel up the main well and are then discharged into the venturi where the fuel vaporizes and mixes with the air flowing through the carburetor.

The power system vacuum diaphragm and spring are actuated by the vacuum below the throttle plate. The diaphragm and spring are held in the "up" position which allows the valve to remain closed until the vacuum drops to approximately 6 to 7 inches of mercury. Under load, as when climbing hills, the vacuum drops because it is necessary to open the throttle wider in order to maintain speed. When the vacuum drops below 6.5 inches of mercury, the power valve is opened by the spring and held open by the diaphragm rod. Additional fuel then flows through the power valve, into the main well and out the main discharge nozzle.

The accelerating pump system is shown in Fig. 79. Fuel is drawn into the pump chamber, through the pump inlet passage and the pump inlet ball check valve on the upward stroke of the

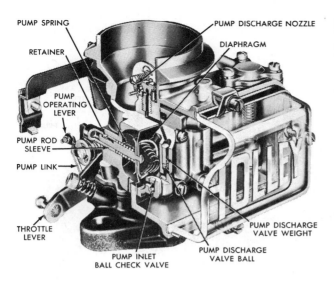

Fig. 79 *Holley* 1904, 1908, 1960
accelerating pump system

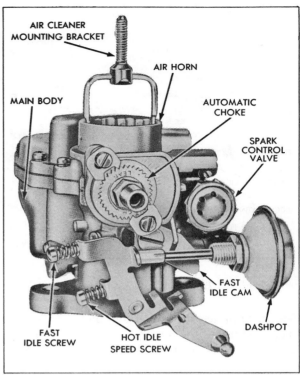

Fig. 80 *Holley Model* 1909 *carburetor*

pump diaphragm. When the throttle is opened, the diaphragm is moved on its discharge stroke, thus closing the pump check valve and overcoming the weight of the pump discharge needle valve. The accelerating fuel then goes around this valve and out the pump discharge nozzle.

The manually-operated choke functions in the same manner as previous Holley carburetors without automatic choke.

Model 1909

This carburetor, Fig. 80, consists of two main sub-assemblies: the air horn and the main body.

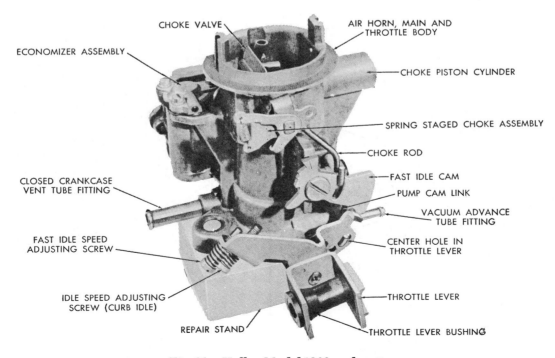

Fig. 81 *Holley Model* 1920 *carburetor*

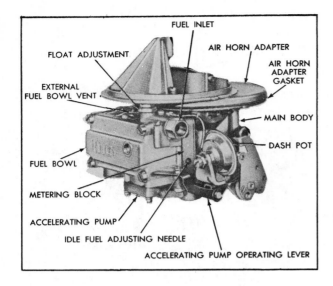

Fig. 82 *Holley Model* 2300 *carburetor*

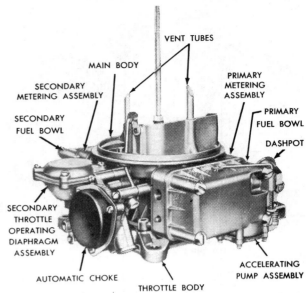

Fig. 83 *Holley* 4150, 4160
four-barrel carburetors

The air horn acts as a cover and includes the fuel inlet system and the automatic choke mechanism. The main body includes the fuel bowl, the fuel metering systems, the throttle shaft and plate, and the spark control valve.

The carburetor has the conventional five fuel metering systems to provide the correct fuel-air

mixture for all phases of engine operation. An anti-stall dashpot, for cars equipped with automatic transmission, is attached to the carburetor and controls the closing rate of the throttle plate.

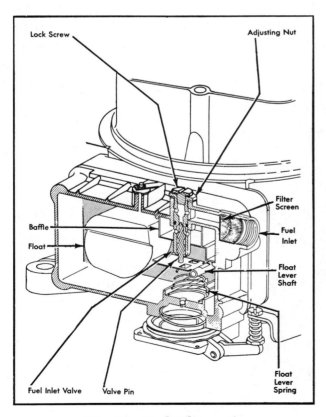

Fig. 84 *Fuel inlet system.*
Holley 2300, 4150, 4160

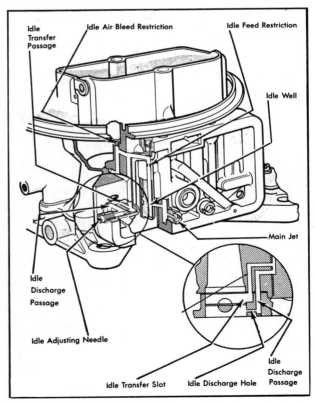

Fig. 85 *Idle   system.*
Holley 2300, 4150, 4160

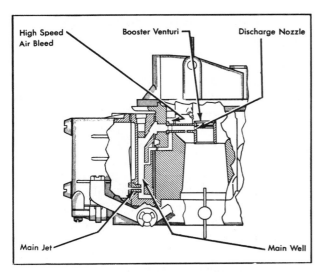

Fig. 86   Main metering system. Holley 2300, 4150, 4160

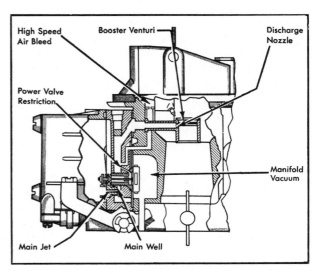

Fig. 87   Power system. Holley 2300, 4150, 4160

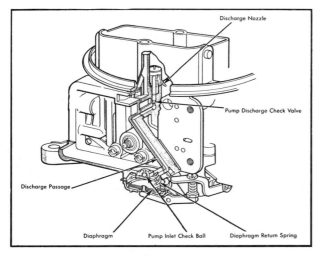

Fig. 88   Accelerating pump system. Holley 2300, 4150, 4160

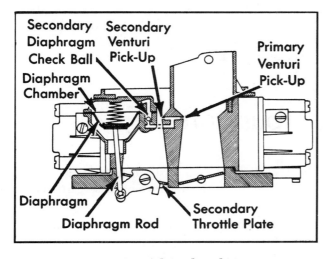

Fig. 89   Secondary throttle operating system. Holley 4150, 4160

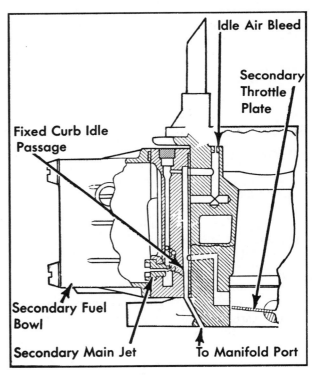

Fig. 90   Idle transfer system. Holley 4150, 4160

Model 1920

In this carburetor, Fig. 81, fuel from the bowl flows into four basic fuel metering systems which are: the idle system, the main metering system, the power system and the accelerating pump system. The choke valve, located in the bore of the carburetor, is connected to a well-type automatic choke.

Additional fuel for acceleration is supplied by a diaphragm type, mechanically operated pump. The pump is operated by a cam connected by linkage to the throttle shaft. An override spring on the pump insures a prolonged discharge of fuel for smoother operation.

A power valve, mounted in the metering body and actuated by manifold vacuum, delivers the additional fuel necessary for full power and high speed operation.

The fuel bowl is vented by an external vent valve located in the top of the bowl. The vent valve is connected by linkage to the throttle shaft so that the valve is opened a prescribed distance when the vehicle is at idle or completely shut down.

Some models of this carburetor are equipped with a spring staged choke. This choke, Fig. 81, is a device incorporated in the choke mechanism which reduces the choke plate closing torque when cranking the engine at temperatures below zero. Thus the spring staging of the choke is a better match for the engine's starting mixture requirements at both low and moderate temperatures.

Model 1931

This carburetor features a one piece main body casting along with a large capacity fuel bowl. The fuel inlet and float assembly located in the center of the fuel bowl cover maintains a stable fuel level for best performance on turns. The large capacity fuel bowl is designed to efficiently handle fuel vapors.

The automatic choke is contained in a housing which is an integral part of the carburetor body.

The carburetor metering systems are similar to the Model 1904 and 1908 discussed previously.

TWO & FOUR BARREL CARBS
Models 2300, 4150, 4160

Model 2300 is a two-barrel carburetor, Fig. 82, while models 4150 and 4160 are four-barrel units of the same basic design as the two-barrel unit, Fig. 83. The two primary bores of both type carburetors supply the fuel-air mixture throughout the entire range of operation for all four basic fuel metering systems as well as the fuel inlet system, Figs. 84 to 88. The two secondary bores of the four-barrel carburetors function only when speed or load requires them. Each barrel has its own venturi, idle system, main metering system, booster venturi and throttle plate.

In the secondary system of four-barrel units, at lower speeds the secondary throttle plates remain nearly closed. When engine speed increases to a point where additional breathing capacity is needed, the vacuum controlled secondary throttle plates open automatically.

The secondary fuel bowl is equipped with a fuel inlet valve which regulates the flow of fuel into the bowl—the same as the primary fuel bowl. The three secondary fuel systems are the transfer system, main metering system and by-pass system, Figs. 89 and 90.

The transfer system begins to function when the secondary throttle plates begin to open. At this point the fuel flows through the secondary main jets into the idle passages which are similar to those in the primary metering body.

When the secondary throttle plates are opened further the pressure differential causes the secondary main metering system to begin functioning.

When manifold vacuum drops to a pre-determined value, the secondary power valve opens, thus allowing a full mixture to be discharged in the secondary booster venturi as the secondary throttle plates are opened. These three secondary systems provide a smooth transition of power instead of a sudden surge.

Rochester Carburetors

ONE BARREL CARBURETORS

Models B, BC & BV

These carburetors are of the same design with Model B equipped with a manually operated choke while Models BC and BV have automatic chokes, Figs. 91, 92 and 93. In addition to the fuel inlet system, the four basic metering systems are used. The distinctive features of these carburetors are:

1. Concentric fuel bowl: Regardless of any shift of fuel level in the bowl the main metering jet is always immersed in fuel.
2. Centrally located main discharge nozzle: A shift

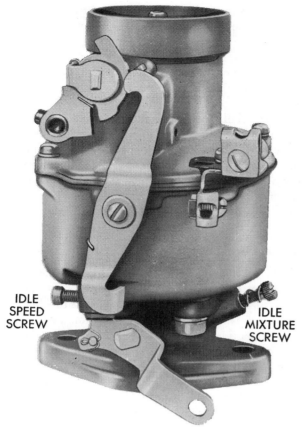

Fig. 91   *Rochester Model B carburetor*

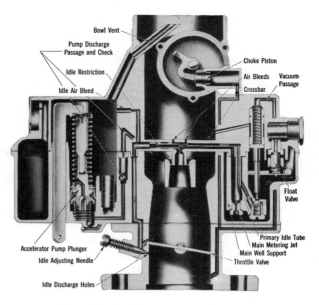

Fig. 92   *Rochester BC carburetor*

in fuel level has little or no effect on the rate of discharge from the nozzle.

3. The main well and support assembly, which contains the main metering jet and the power

valve, is attached to the cover and suspended in the float bowl.

4. Twin floats operate the needle valve to control the fuel level.

The description which follows applies to both models except where reference is made to the automatic choke which refers to the BC and BV models.

On the Model BV carburetor, Fig. 93, the choke thermostatic coil is located on the exhaust manifold and is connected to the choke valve shaft by a connecting rod. The exhaust mounted choke coil provides excellent choke response to supply the

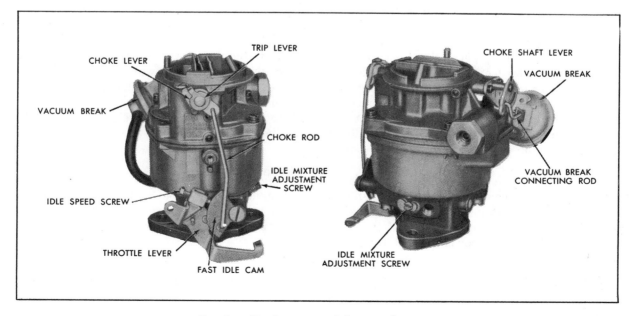

Fig. 93   *Rochester Model BV carburetor*

correct fuel mixtures to the engine during the warm-up period. A vacuum diaphragm unit mounted on the carburetor air horn opens the choke valve, just after starting, to a point where the cold engine will run without loading or stalling.

Idle System

The idle fuel first passes through the float bowl through the main metering jet attached to the bottom of the main well assembly. This fuel is then drawn up the main well by manifold vacuum to the crossbar of the air horn. Air joins the fuel through the air bleeds in the center of the crossbar. This fuel-air mixture is then calibrated as it passes through the idle restriction and is drawn down the passage in the float bowl to the throttle body. The idle fuel is then metered to the engine by the idle adjusting needle below the throttle valve.

Part Throttle System

As the throttle valve is opened further (from idle position) air at a higher velocity is drawn down the carburetor throat, and the fuel and air begin to pass from the main discharge nozzle to meet the increased engine demand. Further throttle opening will result in higher fuel flow from the nozzle and decreased flow from the idle system until it eventually cuts out altogether.

Power System

A vacuum-operated power system provides the additional fuel needed for sustained high speed operation or for increased road load power.

A direct manifold passage within the carburetor to the engine intake manifold operates this system. At any manifold vacuum above approximately 5 inches of mercury the power actuating piston is held in the "up" position against the compression of the power spring, by manifold vacuum. Thus no fuel passes through the power valve.

In accordance with the principle that any sudden acceleration causes a drop in manifold vacuum, the power spring has been calibrated so that any vacuum below approximately 5 inches of mercury, forces the power actuating piston "down".

The end of this piston then unseats the spring-loaded ball in the power valve. Fuel passes readily around the ball into the base of the main well support. The calibrated power restriction meters

the fuel prior to its joining the fuel from the main metering jet. Conversely, as the manifold vacuum rises above approximately 5 inches of mercury, the power piston is drawn immediately to the "up" position, and the spring-loaded ball of the power valve closes, returning the carburetor to the economical part throttle mixtures.

The relief passage which is drilled from the bore of the air horn into the power piston passage serves to relieve any vacuum built up around the piston diameter. This vacuum, if unrelieved, would draw fuel past the piston and down the vacuum passage into the manifold, resulting in an overly rich mixture.

Accelerating Pump System

To provide fuel for smooth, quick acceleration, a double spring pump plunger is used. The rate of compression of the top spring versus the bottom spring is calibrated to insure a sustained charge of fuel for acceleration.

To exclude dirt, all fuel for the pump system first passes through the pump screen in the bottom of the float bowl. It is then drawn past the ball check into the pump well on the intake stroke of the plunger. Upon acceleration, the force of the pump plunger seats the ball check and forces fuel up the discharge passage. The pressure of the fuel lifts the pump outlet ball check from its seat. The fuel is then sprayed on the edge of the venturi by the pump jet and delivered to the engine.

The pump plunger head has been designed to eliminate fuel percolation in the pump system. When the engine is not operating, any build-up of fuel vapors in the pump will rise and by-pass the vent ball. This allows the hot fuel and vapors to circulate up the passage in the plunger head and return to the float bowl. Without this feature, any vapor pressure build-up would evacuate the fuel in the pump system into the intake manifold, causing poor initial acceleration due to lack of fuel in the pump system as well as difficult hot weather starting.

Choke System

The automatic choke is the conventional bi-metal thermostatic spring type and functions in the same manner as described for Carter carburetors.

Model H, 1960–61 Corvair

The Corvair engine for 1960–61 uses two of these carburetors, one on each intake manifold,

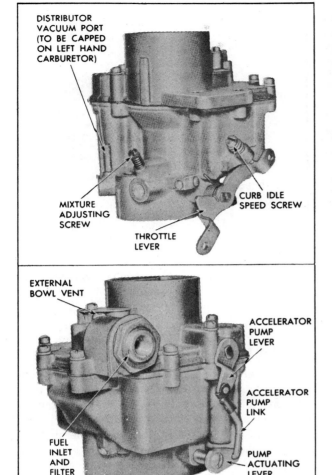

Fig. 94 Rochester Model H carburetor

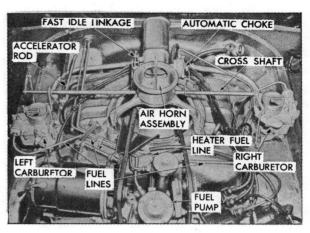

Fig. 95 Corvair carburetion
system. 1960–61

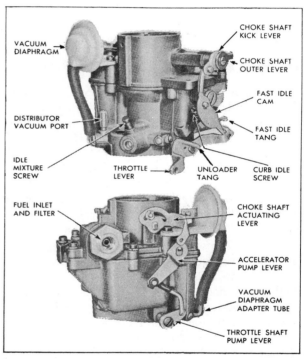

Fig. 96 Rochester Model HV carburetor with built-in automatic choke

Figs. 94 and 95. The choke is positioned in the centrally located air horn assembly at the top of the engine. From the air cleaner, two air tubes direct filtered air to the top of each carburetor. The two carburetor throttle shaft levers are connected by a cross shaft which is actuated by the accelerator linkage. The air horn assembly, while separated from the carburetors, is nevertheless a definite part of the carburetion system. Outside air passes through the air horn, through the air cleaners and arrives as clean filtered air at each carburetor. The automatic choke, choke modifier and fast idle linkage are located on the air horn assembly and are controlled by the same cross shaft which operates the carburetors.

This carburetor is fundamentally the same design as the Rochester Model B, utilizing the same basic fuel metering systems.

Model HV, 1962–1966 Corvair

These engines use two identical carburetors of the type shown in Fig. 96, with built-in automatic choke. The set-up differs from earlier models in that the choke no longer is mounted in the centrally located air horn. The automatic choke mechanism consists of a thermostatic control coil mounted to the lower side of the cylinder head, linked directly to the carburetor choke valve shaft, and a vacuum diaphragm mounted on the air horn.

407

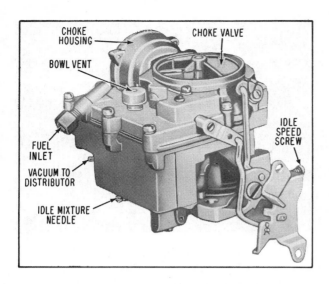

Fig. 97   *Rochester Model 2GC carburetor*

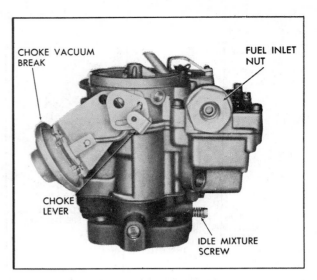

Fig. 98   *Rochester Model 2GV carburetor*

Models M & MV

Called the Monojet, this carburetor is a single-bore downdraft instrument which employs triple concentric venturis in conjunction with a refined metering system that improves mixture control and performance.

A plain tube nozzle is used with the multiple venturi arrangement. Fuel flow to the nozzle from the main metering system is controlled by a mechanically and vacuum operated variable orifice jet. On the mechanical side, a specially tapered rod operates in the fixed orifice of the main metering jet and is connected directly by linkage to the main throttle shaft. A vacuum operated enrichment system is used in addition to provide good performance during moderate to heavy acceleration. The main metering system also has an adjustable flow feature, provided to correct the slight variations that occur during normal production of the components.

Idle System

A separate and adjustable idle system is used in conjunction with the main metering system to meet fuel requirements during idle and low speed operation. It incorporates a vertical slot design off-idle discharge port to provide a good transition between idle and main metering system operation. On some Monojet carburetors a hot idle compensator is added to maintain smooth engine idle during periods of prolonged hot engine operation.

Choke System

These carburetors are designed so that either an automatic or manual choke can be fitted. A conventional choke valve is located in the air horn bore. On automatic choke models, the vacuum diaphragm unit is an integral part of the air horn and the choke coil is remotely mounted on the intake manifold. The conventional bi-metallic spring coil is connected to the choke shaft by linkage. Extra enrichment is provided on automatic choke models during cold start by a special feature which allows use of low torque thermostatic coils for increased fuel economy.

TWO BARREL CARBURETORS

Models 2GC, 2GV

Model 2GC, Fig. 97 and 2GV, Fig. 98 are two-barrel units of the calibrated cluster design, Fig. 99. The cluster casting embodies the small or secondary venturi, the high speed passages, the main well tubes and nozzles, the idle tubes, and the calibrated air bleeds for both the low and high speed metering systems, as well as the accelerating pump jets.

The cluster fits on a platform provided in the body casting of the carburetor so that the main well and idle tubes are suspended in the fuel.

This method of design and assembly serves to insulate the main well tubes and idle tubes from engine heat, thus preventing heat expansion and percolation spill-over during hot idle periods of operation and during the time the hot engine is not operating.

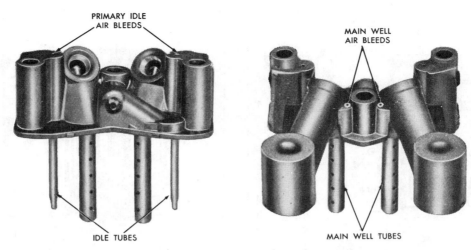

IDLE TUBES

MAIN WELL TUBES

Fig. 99   Rochester 2GC, 2GV cluster casting

An external vent with a protective cover to keep out dirt is located in the center of the bowl cover to provide adequate venting of the unit under all types of operating conditions. No internal tube venting methods are used.

Float System

Fig. 100. Entering fuel first travels through the inlet strainer to remove foreign particles which might block jets or passages. Then the fuel passes through the needle and seat into the carburetor bowl. As in all carburetors, the fuel flows into the bowl until the rising liquid level raises the float to a position where the valve is closed. The float tang prevents the float from traveling too far downward.

Idle System

Fig. 101. In the curb idle speed position, the throttle valve is opened slightly, allowing a small amount of air to pass through between the wall of the carburetor and the edge of the throttle valve.

Fuel is drawn from the fuel bowl through the main metering jets into the main well. It is metered by the idle fuel metering orifice at the lower tip of the idle tube and travels up the idle tube. When the fuel reaches the top of the idle tube, it mixes with air drawn through the primary idle air bleed (the larger of the two holes on top of the cluster, Fig. 99) and the mixture moves through the horizontal idle passage.

The smaller of the two holes on top of the cluster is the secondary idle air bleed, through which more air enters and combines with the mixture.

The fuel-air mixture next moves down the vertical idle passage to the three idle discharge holes

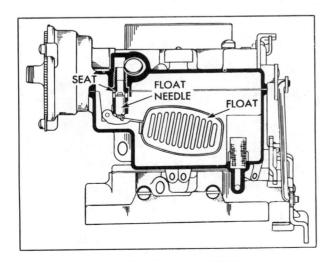

SEAT

FLOAT
NEEDLE

FLOAT

Fig. 100   Rochester 2GC, 2GV float system

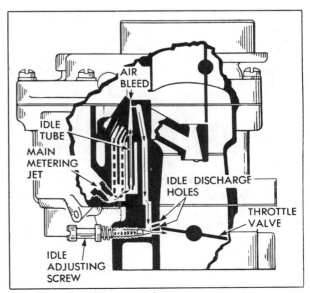

AIR
BLEED

IDLE
TUBE

MAIN
METERING
JET

IDLE DISCHARGE
HOLES

THROTTLE
VALVE

IDLE
ADJUSTING
SCREW

Fig. 101   Rochester 2GC, 2GV idle system

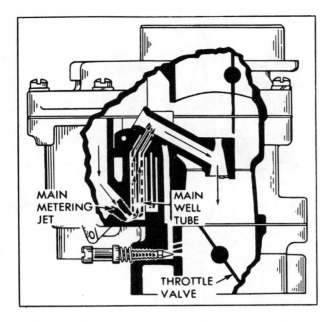

Fig. 102   *Rochester 2GC, 2GV*
part throttle system

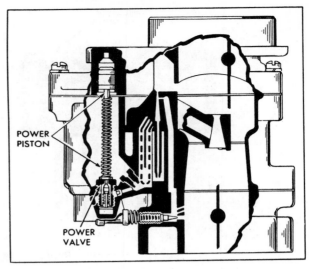

Fig. 103   *Rochester 2GC, 2GV power system*

located just above the throttle valve. Through these holes additional air is added to the mixture, which then passes through the idle needle hole.

In addition to this mixture of fuel and air, there is air entering the bore through the slightly open throttle valve. The position of the idle adjustment needle governs the amount of fuel-air mixture admitted to the carburetor bore, the same as in all carburetors under discussion.

Part Throttle System

Fig. 102. Opening of the throttle valve progressively exposes the three idle discharge holes to manifold vacuum and the air stream with the result that they deliver additional fuel-air mixture for fast idle engine requirements.

As the throttle is opened still further, a greater air velocity passes through the carburetor, the fuel flows through the main metering system and decreased flow from the idle system until it eventually cuts out altogether.

Power System

Fig. 103. A spring-loaded power piston, controlled by vacuum, regulates the power valve to

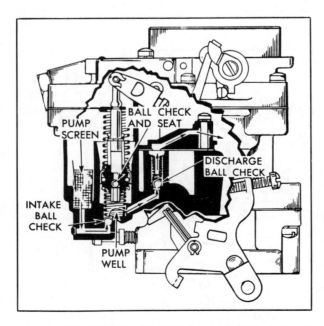

Fig. 104   *Rochester 2GC, 2GV accelerating pump system*

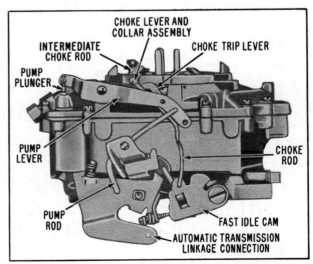

Fig. 105   *Rochester Model 4GC carburetor*

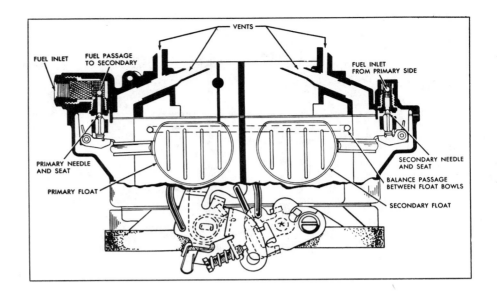

Fig. 106 Rochester 4GC float system

VENTS

FUEL INLET

FUEL PASSAGE TO SECONDARY

FUEL INLET FROM PRIMARY SIDE

PRIMARY NEEDLE AND SEAT

PRIMARY FLOAT

SECONDARY NEEDLE AND SEAT

BALANCE PASSAGE BETWEEN FLOAT BOWLS

SECONDARY FLOAT

supply additional fuel according to engine speed and load. The power piston vacuum channel is open to manifold vacuum beneath the throttle valves; thus the vacuum in the channel rises and falls with manifold vacuum.

During idle and part throttle operation, the manifold vacuum in the channel is normally high enough to hold the power piston in the fully raised position against the tension of the spring. As the manifold vacuum drops with load, the calibrated spring forces the piston down against the power valve, to open and allow additional fuel flow through the calibrated power restrictions into the main wells.

A two-step valve allows a gradual increase in fuel flow as the power valve is opened. At full throttle position, the power valve is fully opened to permit maximum calibrated fuel flow from the power system.

As the load decreases, manifold vacuum increases. The increasing vacuum pull on the piston gradually overcomes the spring tension and returns the power piston to its original raised position, with the valve fully closed.

Accelerating Pump System

Fig. 104. This system functions in the same manner as described for Model B and BC carburetors.

Choke System

As with conventional built-in automatic chokes, the system used with this carburetor includes a thermostatic coil, housing, choke piston, choke valve, and a fast idle cam and linkage. It is controlled by a combination of intake manifold vacuum, air velocity against the offset choke valve, atmospheric pressure and hot air from the choke stove. It functions in much the same manner as Carter carburetor chokes.

The Model 2GV carburetor is also an automatic choke model but the thermostatic coil is located on the manifold and is connected to the choke valve by linkage. A vacuum break is used to open the choke valve to a point where the engine will operate without stalling or loading up after the engine is started.

FOUR BARREL CARBURETORS

Model 4GC

Model 4GC, Fig. 105, is essentially two 2GC carburetors in a single casting. The "Primary Side" of the carburetor contains all the conventional systems (float, idle, part throttle, power, pump and choke). The "Secondary Side" supplements the primary side with separate float and power systems.

Float System

Fig. 106. Both the primary and secondary floats function identically and in the same manner as outlined for Model 2GC carburetor. However, a passage in the float bowl, slightly above the normal fuel level, connects the primary and secondary float bowls. In this way, any abnormal rise in level on one side will be absorbed by the other without disrupting engine operation.

Both sides of the carburetor are externally and internally vented to allow even pressure of fuel and air at all times and to allow the escape of fuel vapors during hot idle operation.

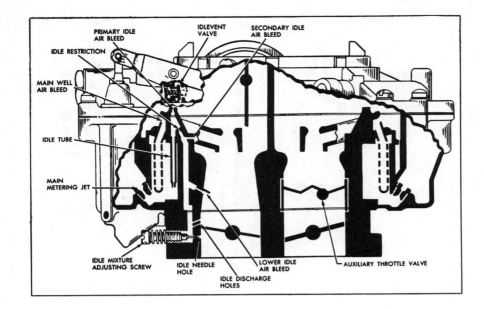

Fig. 107 *Rochester 4GC idle system*

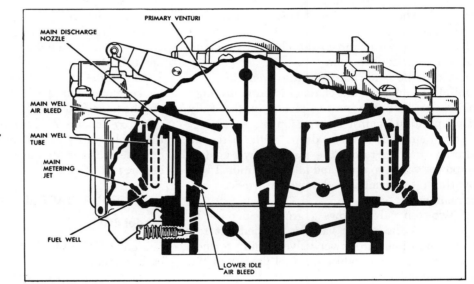

Fig. 108 *Rochester 4GC part throttle system*

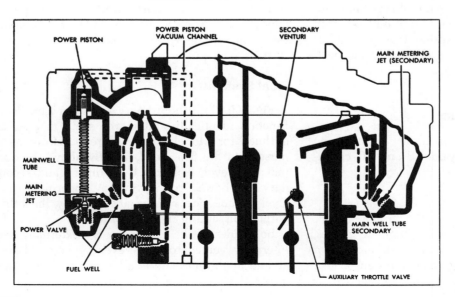

Fig. 109 *Rochester 4GC power system*

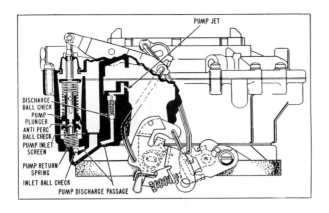

Fig. 110 *Rochester 4GC accelerating pump system*

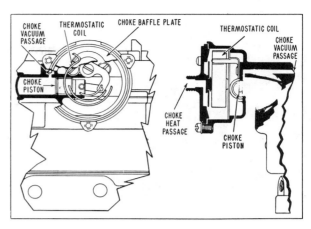

Fig. 111 *Rochester 4GC choke system*

To aid in the venting of the carburetor bowl, an idle vent valve and spring assembly is installed in the bowl cover and air horn assembly, Fig. 107. The function of this valve assembly is to improve idle when the engine is warm by venting fumes outside the carburetor rather than into the air cleaner area.

Idle System

Fig. 107. An adjustable idle system on the primary side only on some units and on both sides on others. On units having the secondary side inoperative the idle passages are blocked by gaskets. Except for this note the idle system functions the same as described for the 2GC carburetor.

Part Throttle System

Fig. 108. This system also functions in the same manner outlined for Model 2GC carburetor. On some units the primary side functions alone up to about 40° of throttle opening, whereas on other carburetors some fuel is used on the secondary side.

Power System

Fig. 109. As the primary throttle valves are opened past 40 degrees, mechanical linkage between the secondary and primary throttle valves start to open the secondary valves. The ratio of motion is such that by the time the primary valves have reached wide open position, the secondary

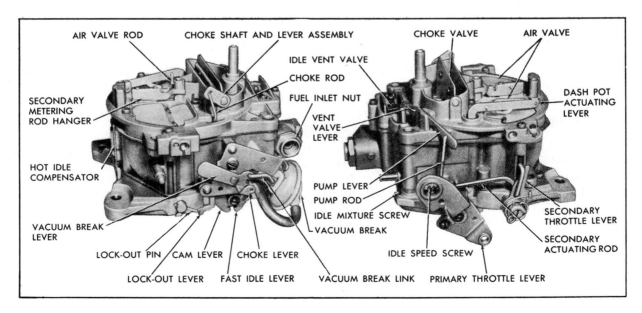

Fig. 112 *Rochester Model 4MV carburetor*

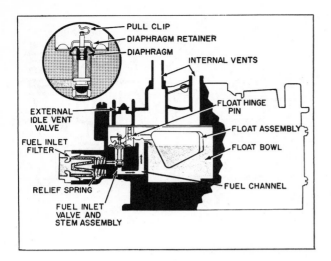

Fig. 113   Rochester Models 4MV,
4MC *float system*

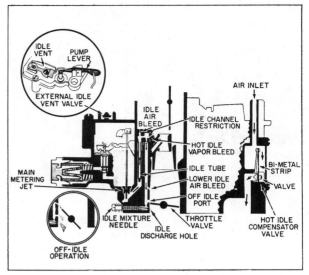

Fig. 114   Rochester Models 4MV,
4MC *idle system*

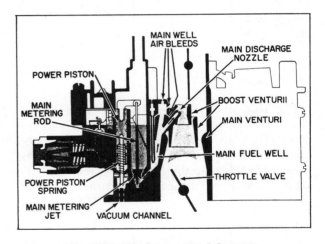

Fig. 115   Rochester Models 4MV,
4MC *main system*

valves are also wide open. With all throttle valves wide open, the venturi systems in both sides feed fuel-air mixture through their respective main metering systems.

A pair of spring-loaded, air velocity operated, auxiliary throttle valves are located in the secondary bores above the regular throttle valves. When the throttle valves are moved to their wide open position and engine speed is low, there is sufficient air flow through the secondary bores to force the auxiliary valves to open. This will concentrate all air flow through the primary throttle bores with better metering of fuel and air. In this condition the carburetor is functioning as a two-barrel unit. As engine speed increases, the force of the air acting on the auxiliary valves increases to the point where the auxiliary valves are forced to open.

In addition, fuel flow is supplemented through a vacuum-controlled power valve on the primary side. This valve functions in a manner similar to that described for Model 2GC carburetor.

Accelerating Pump System

Fig. 110. Since the secondary throttle valves remain closed during part throttle operation, only the primary side needs the extra boost of fuel for acceleration. Hence the primary side only contains the pump system. It functions in the same manner as Model 2GC carburetor.

Choke System

Fig. 111. Since the secondary throttle valves remain closed for idle and part throttle operation, only the primary side requires a choke system.

When the choke is closed, the fast idle cam is raised. The raised position of the fast idle cam "locks out" any opening of the secondary throttle valves by means of a lockout lever, which is free to move only when the cam is fully lowered.

Model 4MV, 4MC

The "Quadrajet" carburetor, Fig. 112, is a 4-barrel unit with versatility and principles of operation that make it adaptable for small to large engines without design changes.

The fuel bowl is centrally located to avoid problems of fuel slosh, causing engine turn cutout, and delayed fuel flow to the carburetor bores.

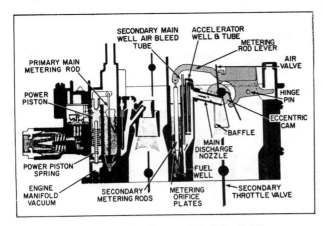

*Fig. 116 Rochester Models 4MV,
4MC power system*

The float needle valve is pressure balanced to permit use of a single small float.

The primary side has small bores and a triple venturi for fine fuel control in the idle and economy ranges. The secondary side has large bores and an air valve for high air capacity.

A hot idle compensator, consisting of a bi-metal strip, a valve and a mounting bracket located at the rear of the carburetor adjacent to the secondary bores supplies additional air to the idle mixture during prolonged hot idle periods.

The primary side has six systems of operation; float, idle, main metering, power, pump and choke. The secondary side has one metering system which supplements the primary main metering system and receives fuel from a common float chamber.

Float System

Fig. 113. Fuel enters the float chamber at the top to prevent incoming fuel vapors from mixing with solid fuel in the bottom of the float bowl. A plastic filler block is located in the top of the float chamber just above the float valve. This block prevents fuel slosh on severe braking and maintains a more constant level to prevent stalling.

The float chamber is internally and externally vented to allow an even pressure of fuel and air at all times and to permit fuel vapors to escape during hot engine operation. The external vent prevents rough engine idle and long hot engine starting.

Idle System

Fig. 114. A hot idle compensator valve is used which permits more air to enter the manifold during hot engine operation. This offsets the richer mixtures and maintains a smoother idle. Except for this the idle system operates the same as outlined previously for the 2GC carburetor.

Main Metering System

Fig. 115. This system consists of main metering jets, vacuum operated metering rods, main fuel well, main well air bleeds, discharge nozzles and triple venturi. The system operates as follows:

During cruising speeds and light engine loads, manifold vacuum is high and holds the metering rods down in the main metering jets against spring tension. Fuel flow from the float bowl is metered between the metering rods and the main jet orifice.

As the primary throttles are opened beyond the off-idle range allowing more air to enter the engine manifold, air velocity increases in the venturi. This causes a drop in pressure in the large venturi which is increased many times in the double boost venturi. Since the low pressure (vacuum) is now in the smallest boost venturi, fuel flows from the main nozzles as follows:

Fuel flows from the float bowl through the main metering jets into the main fuel well and is bled with air from the vent at the top of the main well and side bleeds. The fuel in the main well is mixed with air from the main well air bleeds and then passes through the main discharge nozzle into the boost venturi. At the boost venturi the mixture then combines with the air entering the engine through the carburetor and passes through the intake manifold and on into the cylinders.

Power System

Fig. 116. On part throttle and cruising ranges, manifold vacuums are sufficient to hold the power piston down against spring tension so that the larger diameter of the metering rod tip is held in the main metering jet. Mixture enrichment is needed at this point. However, as engine load is increased to a point where mixture enrichment is required, the spring tension overcomes the vacuum pull on the power piston and the tapered primary metering rod tip moves upward in the main metering jet. The smaller diameter of the rod tip allows more fuel to pass through the main metering jet and enrich the mixture flowing into the primary main wells and out the main discharge nozzles. As engine speed continues to increase, the primary side can no longer meet the engine requirements and then the secondary side is used as follows:

When the engine reaches a point where primary

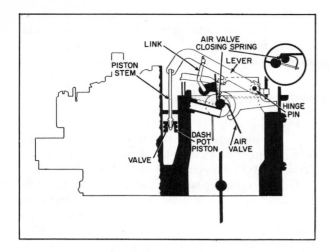

Fig. 117 *Rochester Models 4MV,*
4MC air valve dashpot

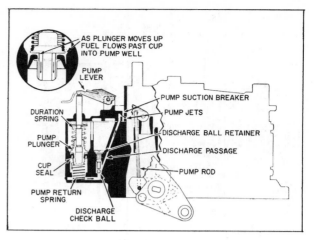

Fig. 118 *Rochester Models 4MV,*
4MC accelerating pump system

bores cannot meet engine demands, the primary throttle lever through connecting linkage to the secondary throttle shaft lever, begins to open the secondary throttle valves. Air flowing through these bores creates a low pressure (vacuum) beneath the air valve, atmospheric pressure on top of the air valve forces the valve open against spring tension. This allows the required air for increased engine speed to flow past the air valve.

When the air valve begins to open, the upper edge of the valve passes the accelerating well port. The port is then exposed to manifold vacuum and will start to feed fuel from the accelerating wells.

The secondary main discharge nozzles are located below the air valve and above the secondary throttle plates. Being in the area of lowest pressure, they begin to feed fuel as follows:

When the air valve begins to open it rotates a plastic cam attached to the main air valve shaft. The cam pushes on a lever attached to the secondary main metering rods. The cam pushes the lever up, raising the metering rods out of the secondary orifice plates. Fuel flows from the float chamber through the secondary orifice plates into the secondary main wells, where it is mixed with air from the main well tubes. The mixture next travels from the main wells to the secondary discharge nozzles and into the secondary bores. Here the fuel is mixed with air traveling through the secondary bores to supplement the mixture delivered from the primary bores, and then goes on into the engine.

Air Valve Dashpot

Fig. 117. The secondary air valve has an attached piston assembly which acts as a dampener to prevent oscillation of the valve due to engine

pulsations. The dampener piston operates in a well which is filled with fuel from the float bowl. The motion of the piston is retarded by fuel which must by-pass the piston when it moves up in the fuel well. The piston is loosely attached to a plunger rod. The rod has a rubber seal which retains the dampener piston to the plunger rod and also acts as a valve. The purpose of the valve is to seat on the piston when the air valve opens and the piston rod moves upward. This closes off the area through the center of the piston and slows down the air valve opening to prevent secondary discharge nozzle lag.

Accelerating Pump System

Fig. 118. This system is located on the primary side of the carburetor. It consists of a spring loaded plunger and return spring operating in a fuel well. The plunger is operated by a lever on the air horn which is connected to the throttle lever by a rod.

During throttle closing the plunger moves up in the well and allows fuel to enter the well through a slot in the top of the well. It flows past the cup seal into the bottom of the well.

When the throttle valves are opened the linkage forces the plunger down. The pump cup seats instantly and fuel is forced through the discharge passage where it unseats the pump discharge check ball and passes on to the pump jets where it sprays into the venturi area.

The pump discharge check ball seats in the pump discharge passage during upward motion of the plunger so that air will not be drawn in; otherwise a momentary acceleration lag could result.

During high speed operation, a vacuum exists

at the pump jets. A cavity just beyond the jets is vented to the top of the air horn, outside the bores. This acts as a suction breaker so that when the pump is not in operation fuel will not be pulled out of the pump jets into the venturi area. This insures a full pump stream when needed and prevents any fuel "pull over" from the pump discharge passage.

4MV Choke System

A thermostatic coil is located in the engine manifold and is calibrated to hold the choke valve closed when the engine is cold. A vacuum break is used to overcome the thermostatic coil and open the choke valve to a point where the engine will run without loading or stalling.

4MC Choke System

The system in this model carburetor is basically the same as the one discussed for the 4MV type except that the thermostatic coil is mounted on the carburetor.

Stromberg Carburetors

ONE & TWO BARREL CARBS

WA & WW Series

These carburetors are fundamentally the same, the WA model being a one barrel unit whereas the WW Model is a two barrel carburetor, Figs. 119 and 120. Both models consist of two main assemblies, namely the air horn and the main body. The air horn serves as a fuel bowl cover and includes parts of the idle system, choke system, accelerating and power systems. The main body includes the fuel inlet, fuel bowl, fuel metering systems and throttle mechanism.

Float System

Fig. 121. Fuel enters the carburetor at the gasoline inlet, flowing through the float needle valve and seat and into the float chamber, where it is maintained at a constant level by the float.

The float chamber can be vented by either an external or an internal vent. The internal vent is in the air horn and, because of its position, the air pressure on the gasoline in the float chamber is balanced with the pressure in the air horn. This results in the mixture remaining correct, regardless of the fact that dirt gradually accumulates in the air cleaner.

Idle System (1st Stage)

Fig. 121. When an engine idles at its slowest speed, the throttle is held open slightly by the throttle stop screw. The idle speed may be varied by turning this screw. Any throttle valve position thus obtained controls the amount of air entering the engine and thereby regulates the engine's idling speed.

At closed throttle or slow engine speeds, the fuel is delivered through the idle system. The fuel is taken from the base of the main discharge jet, flowing into the bottom of the idle tube, where it is metered. From the idle tube it flows through a connecting channel where air from the idle air bleed is mixed with it so that a mixture of fuel and air passes down the channel and is discharged from the idle discharge holes. The idle needle valve controls the quantity of fuel discharged from the primary hole, thereby controlling the mixture ratio.

Idle System (2nd Stage)

Fig. 122. When the engine runs at no load, slightly faster than idle speed, it operates in the second stage of the idle system. While in this range of operation the fuel is delivered into the barrel from the upper discharge hole as well as from the needle valve hole.

Main Metering System

Fig. 123. The main metering system controls the flow of fuel during part throttle operation. Fuel flows from the float chamber into the main metering jet and then into the base of the main discharge jet. Air is bled through the high speed bleed into the main discharge jet so that a mixture of fuel and air is discharged from the main discharge jet into the carburetor barrel.

The main discharge jet is designed so that if any vapor bubbles are formed in the hot gasoline,

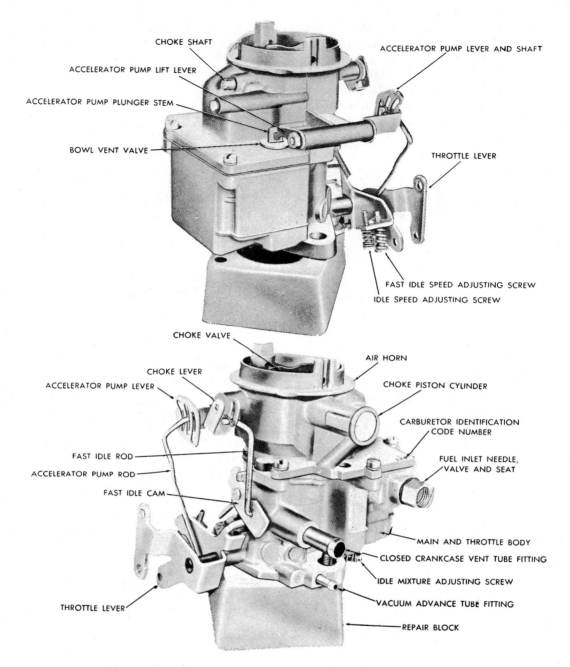

CHOKE SHAFT

ACCELERATOR PUMP LIFT LEVER

ACCELERATOR PUMP PLUNGER STEM

BOWL VENT VALVE

ACCELERATOR PUMP LEVER AND SHAFT

THROTTLE LEVER

FAST IDLE SPEED ADJUSTING SCREW

IDLE SPEED ADJUSTING SCREW

CHOKE VALVE

CHOKE LEVER

ACCELERATOR PUMP LEVER

AIR HORN

CHOKE PISTON CYLINDER

CARBURETOR IDENTIFICATION CODE NUMBER

FAST IDLE ROD

ACCELERATOR PUMP ROD

FAST IDLE CAM

FUEL INLET NEEDLE, VALVE AND SEAT

MAIN AND THROTTLE BODY

CLOSED CRANKCASE VENT TUBE FITTING

IDLE MIXTURE ADJUSTING SCREW

VACUUM ADVANCE TUBE FITTING

THROTTLE LEVER

REPAIR BLOCK

Fig. 119   Stromberg Model WA carburetor

the vapors follow the outside channel *around* the main discharge jet instead of passing through the center of the jet. These vapor bubbles collect and condense in the dome-shaped high speed bleed, and thereby eliminate percolating troubles.

Power System

Fig. 124. For a maximum power or high speed operation, a richer mixture is required than that necessary for lesser throttle opening.

A vacuum controlled piston automatically operates the power jet in accordance with the throttle opening. When the throttle is closed, a high manifold vacuum is present and the power piston is moved to its "up" position against the tension of the spring. When the throttle is opened to a point where additional fuel is required, the manifold vacuum has decreased sufficiently so that the spring on the power piston moves the piston down, thereby opening the power jet to feed additional fuel into the main metering system.

418

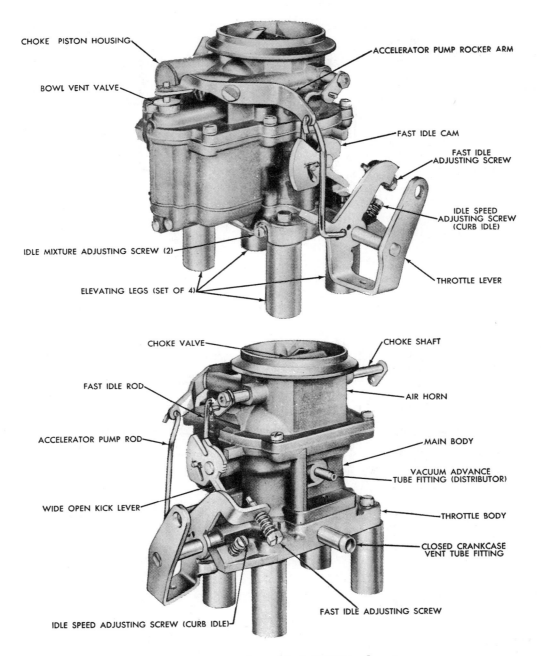

CHOKE PISTON HOUSING

ACCELERATOR PUMP ROCKER ARM

BOWL VENT VALVE

FAST IDLE CAM

FAST IDLE ADJUSTING SCREW

IDLE SPEED ADJUSTING SCREW (CURB IDLE)

IDLE MIXTURE ADJUSTING SCREW (2)

THROTTLE LEVER

ELEVATING LEGS (SET OF 4)

CHOKE VALVE

CHOKE SHAFT

FAST IDLE ROD

AIR HORN

ACCELERATOR PUMP ROD

MAIN BODY

VACUUM ADVANCE TUBE FITTING (DISTRIBUTOR)

WIDE OPEN KICK LEVER

THROTTLE BODY

CLOSED CRANKCASE VENT TUBE FITTING

IDLE SPEED ADJUSTING SCREW (CURB IDLE)

FAST IDLE ADJUSTING SCREW

Fig. 120   *Stromberg Model WW carburetor*

Accelerating Pump System

Fig. 125. In most designs the accelerating pump is directly connected to the throttle so that when the throttle is closed, the pump piston moves up, taking in a supply of fuel from the float chamber through the inlet check valve in the pump cylinder.

When the throttle is opened suddenly, the piston on its down stroke creates a pressure that closes the inlet check valve, forcing open the bypass jet, and discharging a metered quantity of fuel through the pump discharge nozzle. The pump duration spring provides a follow-up action so that the discharge carries over a period of time.

419

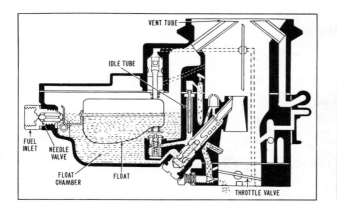

Fig. 121 *Stromberg WW float system*

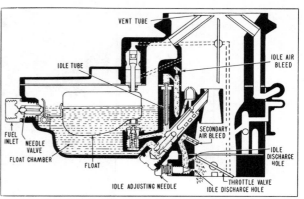

Fig. 122 *Stromberg WW idle system (2nd stage)*

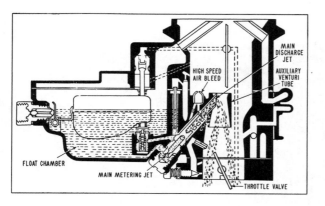

Fig. 123 *Stromberg WW main metering system*

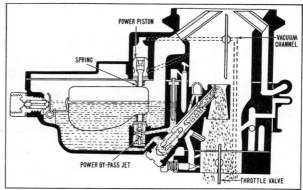

Fig. 124 *Stromberg WW power system*

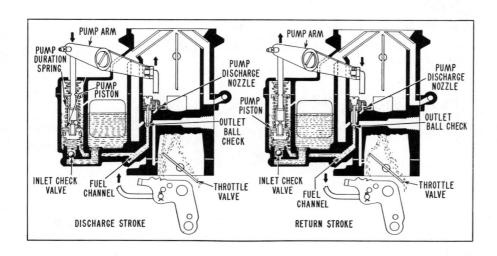

Fig. 125 *Stromberg WW accelerating pump system*

Fuel System Service

AIR CLEANER

An excessive amount of dust, dirt and oil on the cleaner body and at the points of air entry are definite indications of a dirty air cleaner. When such conditions are found, remove the air cleaner and clean it thoroughly.

Several types and many shapes of air cleaners are used, two of the more common types being shown in Figs. 126 and 127. Each type requires a certain amount of disassembly to inspect the oil level and to determine the condition of the oil within the reservoir. Some types do not have the oil bath feature, in which case the filtering element is merely oil-wetted.

The paper element type filter, introduced on many cars in 1957, should be cleaned and inspected every 5000 miles under normal operating conditions (every 1000 miles under dusty operating conditions). Every 12,000 miles the element should be replaced with a new one. To clean the element, repeatedly drop it squarely on a flat surface from a height of two inches until all loose dust has been removed. Turn the element over and repeat the process. *Do not clean by any other method or the element will be damaged. Do not wash, air blast or oil element.*

On the oil-wetted and oil-bath type cleaners, Figs. 126 and 127, scrape the dirt from the air cleaner reservoir. Wash the filter element and, after drying it, re-oil with engine oil. Set it aside to allow the surplus oil to drain from the element. Then on oil bath type cleaners, wash the cleaner reservoir and wipe it dry. Add oil of special grade to indicated level.

CARBURETOR

Examine the carburetor for fuel leaks at a point where the cover fastens to the float bowl and at the points where threaded caps or lead plugs cover the fuel passages. Inspect all caps covering check valves and jets. Tighten any of these joints that leak.

Clean the fuel strainer at the carburetor. Tighten the flange nuts or cap screws holding the carburetor gasket by noting the condition of the gasket edge showing at the edge of the flange.

Test the action of the accelerator. It should operate the throttle from closed to wide open position.

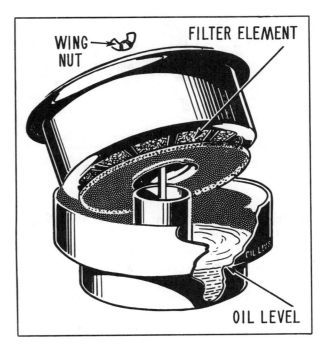

Fig. 126   One type of oil bath air cleaner

Fig. 127   Another type of oil bath air cleaner

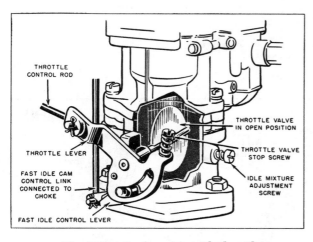

Fig. 128 *Carburetor with throttle in wide open position*

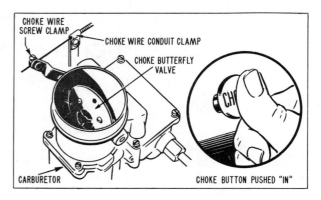

Fig. 129 *Manual operated choke control*

When the fuel used contains a dye for coloring, a fuel leak is sometimes indicated by an accumulation of the dye on the carburetor, fuel lines or fuel tank. It should be remembered, however, that certain castings such as the fuel pump or carburetor may permit a very slight amount of gasoline seepage which will be indicated by a layer of the dye. This seepage is inconsequential and should be ignored.

Fuel Strainer at Carburetor

Remove the fuel strainer from the carburetor, or remove the cover from the strainer. Wash the strainer in cleaning fluid or clean it with compressed air. Examine the strainer gasket. Replace it if it is compressed or damaged.

Throttle Control

Examine the throttle control to be certain that the throttle will range from idle to full open position. Examine the carburetor control levers and the rod leading from the accelerator to the controls. These must be free at all connections and should not bind or rub against the engine or floorboards.

The range of throttle opening can be determined by operating the accelerator pedal. The opening range should extend from the point where the idle screw contacts the stop to where the throttle lever stop contacts the full open position lug on the carburetor body.

The range of throttle opening is determined in the following manner: Remove the air cleaner. Set the choke valve to wide open position. Should the choke be automatic, the valve must be held open unless the engine has been thoroughly warmed.

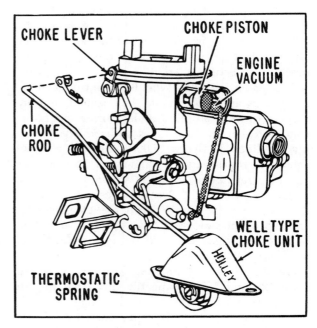

Fig. 130 *Well-type choke mounts in exhaust manifold*

Open and close the throttle. Observe the action of the throttle valve at the carburetor. The valve should be in a vertical position, Fig. 128, when the accelerator pedal is fully depressed. If not, adjust the length of the accelerator control rod at the threaded ends to obtain the full operating range.

Manual Choke Control

Examine the action of the choke valve to be certain that it operates at full range from open to closed position. When the choke valve disc has a spring-loaded poppet valve as a part of the disc, be sure that the valve is free on the supporting stem and that the spring holding the valve in place is not damaged.

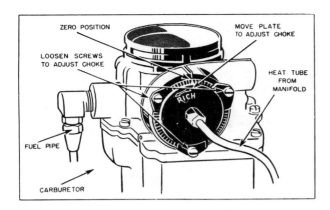

Fig. 131 Exterior view of a built-in automatic choke control

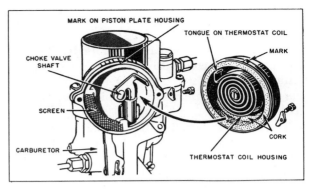

Fig. 132 Interior view of a built-in automatic choke control

To adjust this type of choke control, Fig. 129, tighten the nut holding the choke control to the instrument panel. Push the choke button in. Tighten the stationary conduit clamp at the carburetor. Loosen the set screw on the choke wire screw clamp. Hold the choke valve fully open and tighten the screw clamp against the wire. Test for extremes of travel as the choke button is moved from open to closed position.

Manifold Mounted Automatic Choke

The operation of this type of choke control, Fig. 130, is based upon the combination of intake manifold vacuum, a vacuum operated piston, an offset choke valve and a thermostatic coil spring located in a cavity in the exhaust manifold and connected to the choke lever by a rod.

Heat from the manifold governs the tension of the thermostat coil spring. The fast idle cam operates in conjunction with the automatic choke mechanism to provide the correct throttle opening to prevent engine from stalling during warm-up.

To function properly, it is important that all parts be clean and work freely. Other than an occasional cleaning the choke requires no servicing. It is very important, however, that the choke control unit work freely in the well and at the choke shaft.

Move the choke rod up and down to check for free movement at the pivot. If the unit binds a new choke unit should be installed as the unit is serviced as an assembly only. Do not attempt to repair or change the index setting.

Built-in Automatic Choke Control

This type automatic choke control, Fig. 131, is attached to the upper part of the carburetor. The

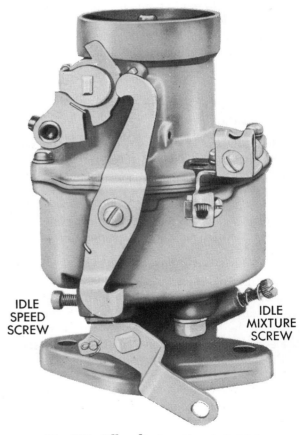

Fig. 133 Idle adjustment screws on a single barrel Rochester carburetor

cover is marked with a scale and pointer, and has a mark at the center for a neutral (or "Index") setting. An arrow on the face of the cover indicates the direction in which the cover should be turned from the center position to obtain a leaner or richer mixture.

To adjust a choke of this type, Fig. 132, remove the housing attaching screws. Take off the cover with care so that the tongue on the thermostat coil releases from the choke valve shaft as the cover is moved away from the carburetor body. Remove and clean the screen. Replace the screen. Then install the thermostat coil housing or cover with the marks up. Insert the attaching screws and retainers.

There is a specified setting for all carburetors having this type choke control. Lacking this specific information, turn the housing counter-clockwise until the pointer and the coil housing marks show neutral ("O" or "Index"), then adjust two notches rich. When in this position the choke should close and seat against the carburetor air horn (engine cold) and then move to full open position as the engine warms up to operating temperature. Should it fail to do so, loosen the housing attaching screws and rotate the housing one notch at a time according to the arrow and symbol on the housing, to obtain a leaner or richer setting.

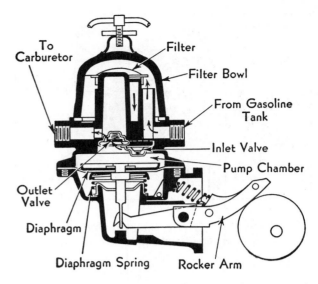

Fig. 134   *Fuel pump*

Carburetor Idle Speed and Mixture Adjustment

The initial setting of the idle adjustment screw ranges from ½ to about 2 turns from the seated position of the screw tip.

Turn the adjustment screw clockwise by hand until it seats lightly, Fig. 133. Then open it about one turn. Start the engine and run it until it reaches normal operating temperature (choke valve wide open).

Slowly turn the adjustment screw to the right a little at a time until the engine shows a tendency to hesitate and stop. Then turn it to the left until the engine runs smoothly. Continue turning the screw to the left until the fuel mixture is rich and the engine lopes or gallops. Again turn the screw to the right until the engine runs evenly.

On two-barrel carburetors there is an idle adjustment screw for each barrel. On four-barrel carburetors there are also two idle adjustment screws on the primary side of the carburetor. When idle ports are provided on the secondary side they are non-adjustable or they are rendered non-functional by being blocked with gaskets.

When adjusting the idle mixture on two- and four-barrel carburetors, adjust one screw at a time until the engine runs smoothly as directed above. Then adjust the other screw in like manner.

After the idle mixture has been adjusted, reduce the engine speed by backing out the throttle adjusting screw until the engine idle speed is satisfactory. Each car manufacturer specifies a definite idle speed, usually between 350 and 500 engine

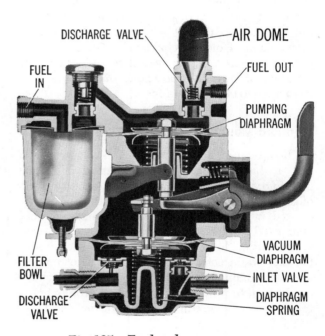

Fig. 135   *Fuel and vacuum pump*

rpm. This data is important on cars with automatic transmission because if the speed is too great the car will creep forward when standing with engine running and shift lever or button in Drive position.

Carburetor Removal

The general procedure for removing a carburetor is to remove the air cleaner. Disconnect choke and throttle controls, fuel pipe and vacuum ad-

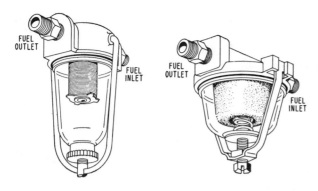

Fig. 136 *Two types of fuel filters*

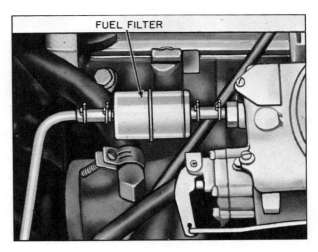

Fig. 137 *Disposable fuel filter*

vance pipe. Unfasten carburetor from manifold flange and lift it off the engine.

Handle the carburetor carefully and hold it in an upright position. Do not allow it to tilt, for if there is an accumulation of sediment in the fuel bowl it will become agitated, pass into the jet openings and clog them. Never place the carburetor in a location where dirt can enter any opening. It is a good idea to mask or tape the carburetor air horn and fuel inlet openings.

Carburetor Installation

When the carburetor used is a replacement, be certain that it is similar in all respects to the one removed. The attaching flange must be the same, the fuel connections must match; the throttle lever, the choke control and all connections must be identical.

Be certain that the gaskets are in good condition. If the carburetor has a governor, any port holes cut through the gasket must align with the governor flange and also match the holes in the carburetor. Tighten the carburetor flange nuts evenly and securely. Connect the throttle and choke control, the fuel pipe and vacuum advance pipe. Assemble the air cleaner to the carburetor and adjust the carburetor idle mixture and speed as outlined above.

Before attempting to start the engine, pump the carburetor full of fuel by cranking the engine with the starter with the ignition off.

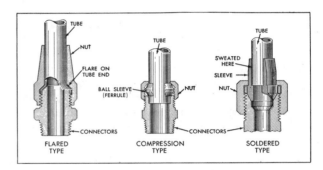

Fig. 138 *Various types of pipe connections*

Fig. 139 *Two piece fuel connection*

FUEL PIPES

Examine all fuel pipes carefully for cracks, dents or wear. The pipe must be securely fastened and all loose clips and clamps securely tightened. If the pipe rubs against the car frame or strikes any sharp metal edges, move it slightly to clear the object.

Never bend or change the shape of a sweater steel fuel pipe as this will cause the seam to open, starting a leak. Inspect and carefully tighten the fuel pipe connections at units such as the tank,

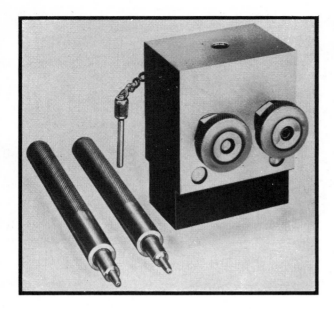

Fig. 140 Flaring tool for fuel pipe ends

fuel filter, fuel pump and carburetor.

When a flexible connection is used to connect a fuel unit to the fuel pipe, examine it for leaks and cracks and replace it if damaged. The new flexible hose must be long enough to sag when connected. This allows it to flex with movement between the units, instead of breaking, when the car is operated on rough roads.

Fuel Pipe Connections

Fig. 138 pictures various types of fuel pipe connections. When the connections are of two pieces (a nipple that screws into the unit and a nut that is screwed into the nipple) use two wrenches for loosening or removing the nut, Fig. 139. When installing a union nut to a fuel connection, start it by hand and turn finger tight; then tighten it further with a wrench. This will prevent cross threading.

When tightening such a fuel pipe connection, use two wrenches, Fig. 139. First loosen the union nut, tighten the nipple, and then retighten the union nut. Tightening in this manner prevents twisting off.

Fuel Pipe Replacement

It is not always necessary to replace a cracked or broken fuel pipe, nor is it always necessary to remove the pipe to complete a repair. However, if the pipe is broken beyond the immediate range of the coupling point it should be replaced.

First disconnect the unions coupling the pipe to the fuel units. Next, loosen or remove the clamps holding the pipe in place. Remove the pipe.

The new pipe must be the same size, have the same form and the same length as the old. Loosely clamp the pipe in place. The connections should match those of the fuel unit to which it couples. Start the union nuts by hand and screw them finger tight. If there is a strain or tension as the pipe is coupled, it will cause the pipe to break or leak later on. It is permissible, with this coupling operation, to slightly bend a copper pipe to conform to the fuel unit. However, never bend a sweated steel fuel pipe beyond a slight degree, as bending breaks the seam. Tighten the clamps holding the pipe in place and tighten the unit or coupling nuts.

Fuel Pipe Repair

If the pipe is broken at a point next to the connection, the pipe can be repaired. Slide the union nut along the pipe, away from the end. File the end of the pipe squarely. Clean the filings from the pipe and apply a new sealing member. This seal may be a ferrule, or a nut and ferrule combined, or a nut that compresses against the pipe when under tension, Fig. 138. When such a repair is made, be sure that a strain on the pipe is not created because of its shorter length.

When the pipe has a flared end, it too at times

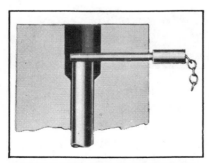

INSTALL TUBING TO PROPER DEPTH

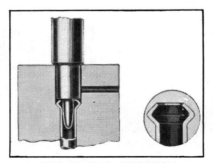

FIRST FLARING OPERATION

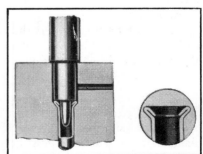

FINISH FLARING OPERATION

Fig. 141 Flaring tool in operation

can be repaired without removing the pipe, providing there is sufficient space to attach and operate the flaring tool, Fig. 140. When such a repair is made, first cut off the damaged flare, clean the filings from the pipe, and place the pipe in the flaring tool so that the end of the pipe contacts the stop pin (left view, Fig. 141). Tighten the nuts holding the pipe in the flaring tool and remove the stop pin. Apply lubricating oil on the end of the pipe. Then flare the end of the pipe with the die as shown in the center and right views, Fig. 141, using a hammer on the tool to form the double lap flare shown.

If the pipe should crack while it is being flared, it does so because it has become hardened with use, vibration, or compression. If the cracked pipe is made of copper, cut off the end and soften the metal by annealing. This is accomplished by holding the end of the pipe in the flame of a blow torch. This should be done after removing the pipe from the vehicle and being certain that it has been drained completely. Heat the pipe until it becomes a light red in color. Then plunge it immediately in cold water and allow it to cool. This procedure softens the pipe so that it can be flared without cracking.

FUEL FILTERS

Fig. 136 illustrates two types of fuel filters. Remove the drain plug or bowl to empty the filter. Should a noticeable amount of water or dirt be present, the filter should be cleaned.

As filter elements are fragile, wash them carefully in a cleaning fluid. Gently brush it or clean it with compressed air of low pressure. Insert it carefully into the filter body. Examine gaskets on the filter bowl and (if so equipped) on the filter element as well.

If the filter employs a knurled nut to secure the element, assemble the element with care and fasten it finger tight. Never use pliers or a wrench to tighten the element. Install the cover and tighten the screw holding it in place.

When installing the filter to the carburetor, or at any other point in the fuel system, be sure the outlet, stamped "outlet" or "out", is upstream, in terms of fuel flow, from the inlet.

Start the engine and allow it to run until the fuel pipe and filter are filled with fuel. Inspect the gasket and drain plugs for leaks.

Disposable fuel filters, one of which is shown in Fig. 137, are now widely used. A sealed unit designed for a specific service life (usually 12,000 miles), they cannot be disassembled for cleaning and must be replaced periodically when no longer

effective in removing foreign matter from the fuel. They are generally equipped with a male connector at either end for use with flexible gasoline resistant rubber or plastic fuel tubing. When installing this type filter care should be taken that it be inserted in the line pointing in the right direction. An arrow or an appropriate legend is usually stamped on the case to indicate the proper direction of fuel flow.

ELECTRIC FUEL PUMP

Electric fuel pumps cannot be serviced on the car except to test pump output and/or pressure and if these are nil, to check to be certain that the malfunction is not caused by a current supply failure. When the engine is idling, fuel pressure at the carburetor should be 4½ pounds. Fuel flow under the same conditions should be 1 pint or more in 30 seconds.

If fuel flow or pressure is inadequate or nonexistent, the voltage supply should be checked at the pump electrical connections on top of the fuel tank. If 12 volts or more are available at that point, the pump is defective and should be replaced. If, no current is flowing to the pump further electrical tests should be made.

These electrical pumps utilize a multiple current supply system. Electricity for the pump is obtained from the starter circuit on engine start-up and from the ignition switch under control of the engine oil pressure when the engine is running. Oil pressure must be 3 pounds or more when the engine is running to energize the fuel pump. Trouble in any of these electrical circuits will cause fuel pumping troubles.

Electric fuel pumps of the turbine type are easily removed and installed. The electrical connectors must be removed before the pump is removed from the fuel tank; these connections are reached from under the car. Access to the pump is gained by pulling back the trunk floor mat and then removing the screws that hold the gas gauge tank unit and pump cover plate to the tank. After disconnecting the fuel hose from the cover plate, remove the entire unit from the tank by unscrewing the retaining cam ring. The pump is taken off the tank unit by first releasing it from the fuel pipe coupling and then removing the two electrical connections before pulling it free. Reverse the procedure to install the pump.

MECHANICAL FUEL PUMP

In most passenger car engines a cam on the camshaft engages with, and directly actuates, the face of the rocker arm of the fuel pump, Figs. 134 and 135.

Fig. 142   *Fuel line with raised ring*

Fig. 143   *Plain tube connection*

Examine the inside of the sediment bowl for dirt or water. A glass bowl can be inspected without removal. A metal bowl or metal cover, however, must be removed for inspection. If necessary, remove the sediment bowl and thoroughly clean it. Wash the screen in cleaning fluid or clean with compressed air. Use a new gasket when installing the bowl, as the pump will not operate with an air leak at this point. This operation must be performed oftener during freezing weather as ice in the bowl will stop the fuel flow.

Inspect the fuel connections for leaks, as indicated by fuel dripping or seeping from joints, and tighten if necessary. Tighten the cover screws—these hold the pump together and also hold the diaphragm in place.

Fuel Pump Testing

Although special testing gauges are available for testing fuel pumps, as will be explained under *Fuel Pump Service,* an emergency method is as follows:

First be sure there is fuel in the supply tank. Disconnect the fuel pipe at the carburetor. Test the fuel pump by cranking the engine with the starter (ignition off). If no fuel is forced from the open end of the fuel pipe, the pump is defective and should be repaired or replaced. Be certain, however, that the pump has been properly inspected and tightened as previously outlined.

Fuel Pump Removal

Disconnect the fuel pipes at the pump. Clean the dirt and grease from the base of the pump at the point where it is fastened to the engine. Unfasten the pump from the engine and lift it out, together with the gasket.

Fuel Pump Installation

If the pump is a replacement compare it with the one removed. Be certain that it is the same type and size. Identify it further by the numbers and letters stamped on it, particularly the number

on the pump arm. Examine the replacement pump to be sure the connections are clean. Remove any dust caps or plugs from the connections. Transfer the connections from the old pump before the new pump is installed and securely tighten them. Place a new gasket on the studs of the fuel pump mounting or over the fuel pump opening in the crankcase, holding it in position with gasket seal or grease.

Insert the pump rocker arm through the opening in the crankcase. Slide the pump into position and tighten the nuts or cap screws. Start the union nuts by hand and then tighten them to the connections with a wrench. Prime the pump by cranking the engine with the ignition off. Then turn on the ignition to start the engine and examine all connections for fuel leaks.

FLEXIBLE FUEL LINES

Tubing under this category is meant to include only plain, unreinforced rubber and plastic fuel lines. Other flexible piping, such as the reinforced length connecting the fuel pump with the gas line, have standard connectors of the type described in the preceding section.

Widely used for fuel, vacuum, windshield washer and other low pressure purposes, this type line comes in a complete range of inside and outside diameters and is very easy to install. When used in fuel service, the rubber or plastic tubing is secured over rigid connectors as shown in Fig. 142. Notice that these male connectors have a raised ring, like the ball sleeve or ferrule shown with the compression fitting in Fig. 138, over which the tube must pass. A ring clamp or threaded clamp is placed on the far side of the raised ring to secure a leak-proof seal. In vacuum and other less critical applications, a plain tube rigid connector, without the raised ring, can be used. This type connection, shown in Fig. 143, is not recommended for fuel system use. Another precaution is that only certain compositions of rubber and plastic can be exposed to gasoline without disintegrating, thus, the tubing must be of one of these compositions. Further, even some of these have a tendency to "age" and become brittle over a period of time and should be inspected periodically.

FUEL TANK

Examine the anchor bands or straps securing the tank to the chassis. These should be free of cracks or bends. Tighten the straps. If they are lined with web strips, rubber or some such material, be sure that the strips are in place between the bands and tank before tightening.

Inspect the tank for wear and for leaks, particularly at the points where the tank seats in the bracket or carrier. Inspect the filler neck at the point where it is soldered and riveted to the tank, or be sure it is not loose or leaking.

Check to see that the tank vent is open. If the vent is located in the filler cap, poke through the opening with a piece of wire. If the vent consists of a tube anchored inside the filler neck, it must extend to the top, and must be so placed that both fuel and air will flow freely. The vent may also be located in the fuel gauge housing. Regardless of its location, the vent hole must be open; if not the fuel will not flow, and the tank may even collapse due to the vacuum created inside the tank.

Method of Tightening Fuel Tank

If the tank is found to be loose, tighten as follows: First, oil the threaded ends of the bands and attaching nuts with light engine oil or with a special penetrating oil. As the oil penetrates the threads, it will allow the nuts to turn easily and will prevent the bands or straps from twisting when the wrench is applied.

Tighten the screws holding the fuel gauge to the tank, and should there be a fuel leak at this point which tightening will not stop, replace the gasket. While you are in the area, examine the fuel gauge wire for damage and the terminals for tightness. This should be done at the instrument panel as well.

When Water Is in System

When water accumulates in the fuel tank, it must be drained and all fuel units must be inspected to determine whether or not water has lodged in them. Remove the drain plug and drain the tank, raising first one end of the tank and then the other, to be sure all the water runs out.

If alcohol in any form is available, pour one quart in the fuel tank. The alcohol will gather and mix with the scattered drops of water at the bottom of the tank. When the mixture of alcohol and water has been well agitated, remove the plug and allow the mixture to drain.

When water has accumulated to any degree in the tank, undoubtedly it is also present in other fuel units, and it may be necessary to drain them also.

Fuel Tank Removal

To remove a fuel tank, first drain the fuel into clean containers of a type to comply with all safety requirements. Then disconnect the fuel pipe. If the fuel filler neck is soldered and riveted to the tank, it cannot be removed. Other types can be removed when necessary. If the filler neck is coupled to the tank with a short section of hose, it can be removed by loosening the attaching clamps. If it is attached by a spanner nut, it can be taken off easily by use of a spanner wrench.

Disconnect the fuel gauge. The connections to the gauge are readily accessible, either directly or through a hand hole provided for the purpose. Oil the threads on the bands or straps and remove the nuts. It may be necessary to remove a shield or guard before lifting the tank out of the vehicle.

Fuel Tank Installation

Before installing a replacement tank, inspect it carefully to make sure that it is a duplicate of the one removed. Using the light of a flash lamp or a well protected electric lamp on an extension cord, see that the inside of the tank is perfectly clean.

If the new tank is not equipped with a fuel gauge, transfer the old gauge, installing it with a new gasket and tightening securely. Cement the webbing or rubber strips on the mounting brackets and set the tank in position. Oil the threads on the tank hold-down straps and be sure the nuts are free on the threads. Set the bands or straps on the tank with the threaded ends projecting through the mounting brackets. Tighten the nuts. Couple the fuel pipe. Attach the filler neck if it is the detachable type. Connect the fuel gauge.

Fill the tank and, at the same time, observe the action of the gauge. It should register a rise from "empty" to "full" as the tank is filled. Since the gauge float does not rest on the bottom of the tank, some fuel will be added before the gauge starts to read above "empty".

INTAKE MANIFOLD

Test for Air Leaks

With the engine idling, test the intake manifold gasket for air leaks by spreading oil from the spout of an oil can along the gasket edges. An air leak is indicated when oil is drawn past the gaskets by the suction of the engine. Tighten the

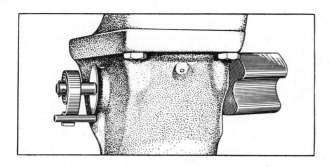

Fig. 144   Manifold heat control valve

nuts or cap screws holding the intake manifold to the engine, and retest for leaks. If tightening fails to stop the air leaks, replace the gaskets.

When making a test for leaky manifold gasket, examine the pipe or tube leading to the windshield wiper for air leaks (if the wiper is electric, there will be no such tube). Test with oil in a manner similar to the gasket test. An air leak in this pipe, or one at a joint, can be detected by a shrill, hissing sound. A short section of rubber tubing is usually employed to provide a flexible connection between the intake pipe and the windshield wiper tube. Replace any broken or rotted rubber tubing and any broken metal tubing.

WINDSHIELD WIPER TUBING REPLACEMENT

Unless an electric wiper is fitted, the windshield wiper operates from the vacuum created in the intake manifold of the engine. As the speed of the engine increases, or when the engine operates at full throttle, the vacuum in the manifold decreases. Unless a vacuum booster is used in connection with it, Fig. 135, the wiper will cease to operate with the engine operating at full throttle. Therefore, when testing the action of the wiper, be sure that the test is made with the engine running at slow speed.

Two pieces of metal tubing, connected by a rubber tube, usually serve as a vacuum line to the manifold. The rubber tube is used because movement of the engine, independent to that of the car body, would break a continuous metal tube. The rubber tube is easily pulled off the metal tube and replaced. However, the metal tube may be installed behind the dash pad and through the windshield post; if so, it can be reached for replacement only by removing some interior trim.

MANIFOLD HEAT CONTROL VALVE

The exhaust manifold conveys the exhaust gases away from the engine. However, before the hot gas is completely dissipated, it is utilized to further the vaporization of the fuel. This is accomplished by diverting a portion of the exhaust gas to a "hot spot" in the intake manifold or by conveying it through a passage surrounding the point where fuel enters into the intake manifold.

The heat is usually directed to the hot spot by a bimetal thermostatically controlled valve, attached to a shaft having supporting bearings in the manifold, Fig. 144.

The shaft must turn freely in the bearings in order that the thermostatic spring can move the valve to closed position as the manifold temperature increases. Heat will scale the shaft and cause it to bind in the bearings. If the valve does not move freely, it may be loosened by applying penetrating oil to the shaft and bearings. A solution of washing soda and kerosene may be substituted for the oil. If this method fails, remove the unit and clean the shaft with fine sandpaper or emery cloth.

MUFFLER AND TAIL PIPE

The muffler and tail pipe have no connection with the fuel system but they do have a direct bearing on power developed by the engine and on engine performance in general. These two units must always be inspected when a power loss develops. An accumulation of carbon may clog the muffler, or the tail pipe may be crushed or otherwise clogged, thus restricting the escape of exhaust gas and affecting engine power.

Momentarily accelerate the engine to high speed. When the exhaust gas builds up a high pressure in the exhaust system, reduce the engine speed to normal idle. A hissing sound indicates that some part of the system is restricted. The system should then be examined carefully and the obstruction removed.

Inspect the tail pipe and remove any obstruction present. If the end of the pipe is clogged, use a screwdriver, sharp stick or other pointed instrument to open the pipe. If the pipe is kinked, insert a bar or an old axle shaft in the pipe. If necessary, strike the pipe lightly with a hammer as the bar is inserted.

Trouble Shooting Procedure

WHEN ENGINE WON'T START

When the engine fails to start, when the ignition system is giving adequate spark and properly timed, the fuel system or the engine itself is at fault. Causes of fuel system failure are as follows:

1. No fuel in tank.
2. Choke inoperative or flooding.
3. Water in fuel system.
4. Clogged vent in fuel tank filler cap, filler pipe or tank gauge housing, whichever is provided.
5. Fuel pump failure.
6. Vapor lock.
7. Clogged fuel pipe.
8. Leaking fuel pipe or connection.

No Fuel in Tank

Determine by the fuel gauge or with a stick whether there is sufficient fuel in the tank. The fuel must cover the end of the suction pipe leading to the fuel pump. This is usually about one inch. If the fuel level is low, tilting the vehicle will cause the fuel to flow away from or toward the end of the suction pipe.

Carburetor Fuel Test

After determining that the fuel tank has sufficient fuel, find out whether the fuel reaches or enters the carburetor float bowl. Do this by disconnecting the fuel line at the carburetor. If the carburetor fitting is of the elbow type, hold the elbow with pliers to prevent it from breaking while unscrewing the union nut.

Crank the engine with the starter (ignition off). If fuel is ejected from the open end of the fuel pipe during this test, clean the carburetor screen at the carburetor. Be sure the opening of the fitting at the carburetor is not restricted and that the carburetor needle valve is not sticking.

Inspect the fuel ejected from the pipe to be sure there is no water present. Connect the fuel pipe to the carburetor. If the fuel fails to flow from the end of the fuel pipe on this test, inspect the fuel pump.

Choke Inoperative or Flooding

Examine the choke for free operation. Be sure the choke will fully close and that it will fully open. If the choke is stuck in the open position, remove the air cleaner and restrict the air passage to the carburetor with your hand while the engine is being cranked. This will enable the engine to start by providing a richer fuel mixture.

If the automatic choke is stuck in the closed position, or if the manual choke is used to excess while attempting to start the engine, or if the carburetor float level is too high, the carburetor and intake manifold will be flooded with fuel and the engine will not start. This condition can be detected by fuel seeping from between the carburetor float bowl and the bowl cover, or around the throttle shaft end in the carburetor body, or it may be evident by seeping past the gasket between the carburetor flange and intake manifold.

If the engine will not start due to overchoking, set the choke and throttle valves to full open position. Then as the engine is cranked with the starter, the excess fuel will be drawn from the intake manifold and cylinders, restoring normal fuel mixture in the combustion chambers.

A condition similar to flooding by overchoking is sometimes encountered when trying to start a hot engine. This is caused by "percolating" within the carburetor when the engine is stopped. This action is caused by the warming of fuel in the carburetor float bowl, causing it to expand and escape through the carburetor jets into the intake manifold. All modern carburetors have provision to prevent this condition by venting the float bowl to the atmosphere or by the use of mechanical anti-percolator valves. However, if this condition should occur, the engine can be started in the same manner as one flooded by excessive choking after the engine has been allowed to cool for a few minutes.

Should the inspection show that the fuel is reaching the carburetor and that the choke is functioning normally (not overchoking) and still the engine fails to start, examine the fuel system for water as outlined below.

Water in Fuel System

Water can cause engine failure by clogging the carburetor passages and jets. Allow a small amount of fuel to flow into a small container and examine it for water, which will take the form of a bubble or globule in the bottom of the container.

Another test for water in the fuel can be made by removing the fuel pump bowl or the fuel strainer bowl and pouring the fuel on the ground. The fuel will spread rapidly, but the water will remain concentrated in a bubble or globular form for a time. If there is only a small amount of water in the pump bowl, wipe the bowl clean with a rag. Reinstall the pump bowl and fill it by rotating the engine with the starter and again test the engine for starting. If water again appears in the fuel bowl, the entire system must be drained and cleaned as discussed under *Fuel System Service*.

In an emergency, remove the drain plug from the fuel tank and allow a gallon or more of fuel to drain from it. Install the drain plug and examine drained fuel for water. If a considerable amount of water is drained off with the fuel, repeat the operation until the drained fuel is free of water. Install and tighten the drain plug.

Disconnect the fuel pipe at the fuel tank and at the fuel pump and blow through the pipe from the fuel pump end to remove the water. Connect the fuel pipe to the tank and filter or fuel pump. Again drain the filter, or filter and fuel pump bowls. It is sometimes necessary to drain the carburetor.

Clogged Fuel Tank Vent

If the fuel test indicates that the fuel does not reach the carburetor, examine the air vent in the fuel tank filler cap or in the tank proper, as outlined under *Fuel System Service*. If the vent is clogged, remove the obstruction, otherwise the fuel will not flow from the tank to the pump. If this is not possible, leave the filler cap off until a repair can be made. However, if the vent is clear, test the fuel pump as follows:

Fuel Pump Failure

To test a mechanically-operated fuel pump—the type used on all passenger cars—disconnect the inlet pipe and operate the pump by cranking the engine with the starter. If fuel is ejected from the fuel pump outlet, the pump is working.

Vapor Lock

What appears to be an out-of-fuel condition is sometimes encountered in hot weather operation. It is apparent by a miss or skip in the engine, gradually slowing down or complete stoppage as if the throttle were being closed, or by failure to start after the engine is stopped.

This condition, known as vapor lock, is caused by fuel vaporizing in the fuel system and restricting the flow of fuel. When this condition occurs,

the engine should be allowed to cool before attempting to start it. Cold water may be poured over the fuel pump and fuel pipes to hasten the cooling of the fuel.

In exceptional cases, it may be necessary to insulate the fuel pipes and pump from engine heat. This may be done by inserting a metal shield, or wrapping the pipe with asbestos or tape at points where the heat may be picked up by the pipe or pump.

Clogged Fuel Pipe

To check for a clogged fuel pipe, disconnect the pipe at the pump inlet and remove the tank filler cap. Apply air under low pressure to the open end of the fuel pipe. If compressed air is not available, use an empty oil gun as an air pump, or blow into the pipe. Air bubbles venting through the fuel or a gurgling sound within the fuel tank indicates the fuel line is open and that the filter is not clogged.

If this procedure relieves the cause of engine failure, the obstruction has been blown from the fuel pipe back into the tank, and the tank must be cleaned as soon as possible as outlined under *Fuel System Service*.

Test for Leaks

When testing any fuel pump, bear in mind that air entering the fuel pipe will prevent the fuel pump from working. All fuel pipes and connections between the pump and tank must be air tight.

Examine the fuel pipes and connections for leaks. Test each pipe union nut with a wrench to be certain it is tight. Move the pipe back and forth by hand near each connection to determine whether it is loose at the connection. A pipe that is loose at the union nut or connection must be repaired as outlined under *Fuel System Service*.

WHEN ENGINE WON'T IDLE

When the ignition system has been thoroughly examined and found satisfactory and the engine fails to idle or fails to run smoothly at idle speed, the failure is probably in the fuel system or in the engine itself. The following checks should be made:

1. Carburetor out of adjustment.
2. Carburetor flooded.
3. Air leaks at intake manifold gaskets or in windshield wiper tube.
4. Engine mechanical failure.

Carburetor Out of Adjustment

Start the engine and turn the throttle adjusting screw until the engine runs at a fast idle speed. Examine the choke to be sure it is fully open when the engine is warm. Then adjust the idle mixture and idle speed as outlined under *Fuel System Service*. If an idle mixture adjustment cannot be obtained, the carburetor should be removed and the idle system inspected for clogged jets or passages.

Carburetor Flooded

There are several causes for this condition. The float level may be adjusted too high, allowing the fuel to rise and run from the main discharge nozzle when the engine is idling. The float may be punctured, filled with gasoline and sunk, failing to close the needle valve. Dirt may lodge on the needle valve seat and allow an excess passage of fuel. The fuel pump may supply the fuel at too high a pressure and not allow the needle valve to close. The choke valve may be defective or (in extreme cases) stuck.

If one or more of these conditions are present, they can generally be observed because fuel will leak around the carburetor body. In extreme cases, fuel may even drip from the muffler tail pipe.

First, examine the choke to be sure it is in the full open position when the engine is warm. Lightly tap the carburetor body near the needle valve with a screw driver handle to dislodge any particles which may have lodged on the needle valve seat.

If the carburetor still floods it indicates that the carburetor is excessively dirty, or that the carburetor float is punctured. It may be that the fuel pump is building up too much pressure. For tips on carburetor and fuel pump service, consult the remarks under their respective headings.

Air Leaks at Intake Manifold or Windshield Wiper Pipe

Air leaking past the intake manifold gaskets will mix with the fuel and air in the manifold, allowing the mixture to become too lean. As a result the engine will not idle smoothly or will not idle at all.

Test for air leaks as outlined under *Fuel System Service*. If leaks are found, tighten the carburetor flange nuts and the nuts holding the manifold to the engine. Tighten the carburetor body screws.

If the test indicates that leaks are still present at the intake manifold after the nuts are tightened, the gasket is probably broken or compressed and must be replaced.

Excess air may also enter the intake manifold through a leaking windshield wiper pipe or a power brake vacuum pipe.

To overcome the effect of a leaky intake manifold gasket in an emergency, set the throttle stop screw so that the engine will run at the slowest idle speed possible without stalling. The gaskets should be replaced as soon as possible.

Engine Mechanical Failure

If the foregoing tests fail to correct the idling condition, the failure is probably due to a stuck engine valve, which is usually accompanied by a loud engine knock. It may also be due to a miss on one cylinder or excessive vibration.

WHEN ENGINE LOSES POWER

If the engine shows a marked loss of power when the ignition system is in proper adjustment, the failure is probably in the fuel or exhaust system, or in the engine itself. The following items should be checked:

1. Carburetor mixture too lean on acceleration.
2. Carburetor mixture too rich.
3. Fuel pump defective or vapor locked.
4. Muffler tail pipe clogged.

Mixture Too Lean on Acceleration

If loss of power is caused by a lean fuel mixture, it usually is accompanied by a "popping back" in the carburetor when the throttle is opened rapidly. When this happens, test as follows:

With ignition off, remove the air cleaner and set the choke wide open. If the choke is automatic, it may have to be held open. Inspect the volume of discharge of accelerator jet by looking inside the carburetor past the choke valve. At the same time, open the throttle valve by moving the throttle rod rapidly. Fuel ejected from the accelerator jet should now be visible as a spray in the carburetor throat. (Do not open the throttle more than once or twice as flooding will occur and it will be difficult to start the engine.) If the spray is observed, and the "popping back" still occurs, then the indication is that the accelerating pump has been furnishing insufficient fuel for quick getaway, and the pump stroke must be adjusted to furnish an added amount of fuel.

If no spray is observed in the above test, the

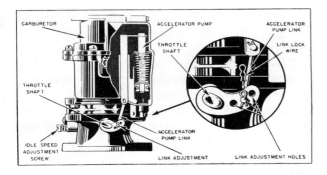

Fig. 145 Accelerating pump control with adjustment holes in pump arm

accelerator jet may be clogged or the accelerating pump may not be operating because it is disconnected or its valves are faulty.

A partly clogged main jet is indicated when the engine operates at low speed but fails to respond at full throttle. The main jet may be cleared by speeding the engine and closing the choke until the engine nearly stops. Repeat this procedure several times. If the main jet is still not cleared by this process, remove the carburetor and effect a repair.

Accelerator pumps are classified as adjustable and non-adjustable. When adjusting the accelerator pump, bear in mind that more fuel (longer stroke adjustment) is needed during cold weather at sea level than during hot weather or at a high elevation. If the car is to be permanently transferred from one area to another, change the accelerator pump stroke, if necessary, to correspond to the altitude and the prevailing temperature in the new area of operation.

One type of carburetor has an accelerating pump with a series of holes (usually three) drilled through the pump arm, Fig. 145. The adjustment is made by moving the connecting link from one locating hole to another. The second type has an accelerator link rod with a right-angle bend at the lower end. Adjustment is accomplished by bending the link.

If a carburetor with a non-adjustable pump provides insufficient fuel for acceleration, the carburetor must be repaired.

Mixture Too Rich

Loss of engine power caused by a rich fuel mixture can be detected by the tendency of the engine to "surge" at high speeds or on a heavy pull. This condition is usually caused by an excessive amount of fuel entering the carburetor. This may be due to a punctured carburetor float or a defective needle valve, or by an improper adjust-

ment of the accelerator pump. It can be detected by the presence of fuel seeping around the end of the throttle shaft and by dense black smoke ejected through the muffler tail pipe.

Fuel Pump Defective or Vapor Locked

If the engine still has a marked loss of power after checking the above, indications are that the mixture is too lean, or that vapor lock has blocked the flow of fuel. Examine the fuel pump output and check for vapor lock as previously directed.

If the fuel pump is operating efficiently and if the mixture is still too lean for effective power, clean the fuel pipe as already explained.

Muffler Tail Pipe Clogged

If the tail pipe is sharply kinked, or plugged with mud or clogged to a degree that the exhaust gas will not fully vent from the pipe, the engine power will be reduced. This condition usually can be detected by accelerating the engine to high speed, then closing the throttle. The exhaust gases passing through the restricted passage will produce a hissing sound as the engine loses speed. Unkink or unclog the pipe as described under *Fuel System Service*.

Mechanical Failure

If all the foregoing conditions have been eliminated as causes of loss of power, the indication is that the failure is in the engine, cooling system or brake system.

EXCESSIVE FUEL CONSUMPTION

Complaints of excessive fuel consumption require a careful investigation of the owner's driving habits and operating conditions as well as the mechanical condition of the engine and fuel system; otherwise much needless work may be done in an attempt to increase fuel economy.

Driving habits which seriously affect fuel economy are: high speed driving, frequent and rapid acceleration, driving too long in low and second speed when getting under way, excessive idling while standing.

Operating conditions which adversely affect fuel economy are: excessive acceleration, frequent starts and stops, congested traffic, poor roads, hills and mountains, high winds, low tire pressures.

High speed is the greatest contributor to low gas mileage. Air resistance increases as the square of the speed. For instance, a car going 60 miles an hour must overcome air resistance four times as great as when going 30 miles an hour. At 80 miles an hour the resistance is over seven times as great as when going 30 miles an hour. Over 75 per cent of the power required to drive a car 80 miles an hour is used in overcoming air resistance, while at 30 miles an hour only 30 per cent of the power required is used to overcome air resistance.

Gas mileage records made by car owners never give a true picture of the efficiency of the engine fuel system since they include the effects of driving habits and operating conditions. Because of the wide variation in these conditions, it is impossible to give average mileage figures for cars in general use. Therefore, any investigation of a mileage complaint must be based on an accurate measurement of gasoline consumption per mile under proper test conditions.

Gasoline Mileage Test

There are a number of gasoline mileage testers commercially available that measure fuel consumption precisely. Manufacturers of these devices furnish full instructions for their use. However, an inexpensive tester can be made with a quart oil can, suitable fittings and a tube to connect the can to the carburetor.

Drill or punch a hole in the bottom of the can and solder a fitting to the hole. The fitting at the carburetor end should be the same as the existing fitting at the carburetor inlet. Arrange a suitable handle or wire hook to the can so it can be mounted either under the hood or in the driver's compartment.

Before making the test, disconnect the fuel pipe at the carburetor and plug the pipe opening with a small cork so that the fuel will not spurt out during the test. Run the engine until all the fuel in the carburetor is used up. Then connect the tester tube to the carburetor.

With the can mounted at a higher level than the carburetor so fuel will flow into the carburetor by gravity, pour exactly one quart of gasoline into the can.

Make the test on a reasonably level road, at fixed speeds, without acceleration or deceleration. Test runs should be made in both directions over the same stretch of road to average the effect of grades and wind resistance. Test runs made at 30, 50, and 70 miles an hour will indicate the approximate efficiency of the low speed, high speed and power systems of the carburetor and show whether fuel consumption is actually abnormal. Under the conditions given, the fuel consumption in miles per gallon, based on the normal economy of a car capable of giving 20 miles per gallon at 20

miles per hour, should be approximately as follows:

Constant Speed	Miles Per Gal.
20	20.0
30	19.7
50	15.9
70	8.0

If it takes 5 miles to empty the can, it means that the fuel consumption is 20 miles per gallon, since there are 4 quarts to the gallon.

If the test indicates that the fuel consumption is above normal, check the following before deciding to take the carburetor apart:

1. Check all gasoline pipe connections, fuel pump bowl gasket, gasoline filter gasket, and carburetor bowl gasket.
2. Check for low tire pressures.
3. Check for dragging brakes.
4. Late ignition timing causes loss of power and increases fuel consumption. Dirty or worn out spark plugs are wasteful of fuel.
5. Use of gasoline of such low grade that ignition timing must be retarded to avoid excessive detonation will give very poor fuel economy.
6. Check for sticking manifold heater valve or improper setting of the thermostat.
7. Check for dirty or clogged air cleaner element and for excessive oil in the crankcase.
8. Check for sticking choke valve and improper setting of the automatic choke thermostat.
9. Check for insufficient valve operating clearance or sticking valve.
10. Check for excessive fuel pump pressure.
11. Check for carburetor idle adjustment. On Carter carburetors, the metering rod setting may be checked without removing the carburetor. For all other corrections to high speed and power systems on all carburetors, the carburetor must be removed and disassembled.

Changing Carburetor Jets

Under no circumstances should leaner than standard jet sizes, metering rods and other calibrations of a carburetor be changed from factory specifications. The specified calibrations must be adhered to unless these are later changed by a bulletin issued by the carburetor manufacturer.

Carburetor calibrations have been determined by exhaustive tests with laboratory equipment and instruments which accurately measure overall performance and economy. Besides, the leanest possible mixture obtainable by the use of smaller jets, etc., will not increase mileage as much as 10 per cent, and will often impair engine performance.

Carburetor Service

It is beyond the scope of this book to provide an exhaustive treatise on the overhaul procedures of the hundreds of carburetors in use. Although there are perhaps two dozen basic carburetor designs in use in recent years, each one has many models and each model requires individual adjustment specifications. Therefore, to provide the reader with a working knowledge of what it takes to be a carburetor specialist, we shall outline in a general way the requirements of carburetor servicing, including the overhaul procedure on a popular model of a Carter, Ford, Holley, Rochester and Stromberg carburetor.

There are a number of special tools which one must have available for accurate and efficient carburetor rebuilding. Figs. 146 and 147 show the tools a carburetor specialist uses in servicing Stromberg and Carter carburetors. If the reader doubts the value of these tools, let him pick up an old junk carburetor and take it apart to see how far one can go before being stopped for the want of one of those special tools. Even if most of the work could be done without a number of these tools, very likely a lot of time would be lost and more important, an accurate and efficient job would not be produced.

Therefore, in the following pages, we shall confine our discussion to carburetor service tips, the procedure for rebuilding a few popular carburetors, and the adjustments which must be made precisely in order for the carburetor to function as it was intended.

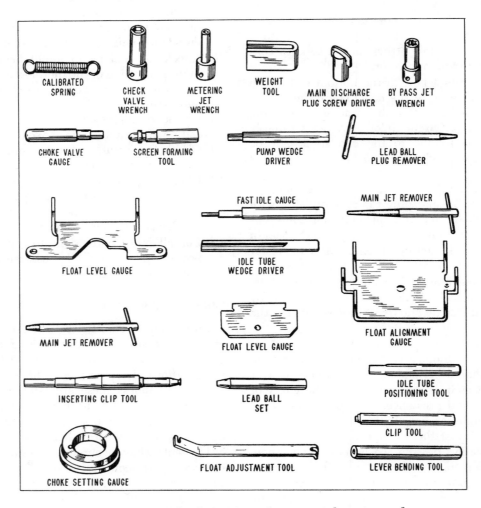

Fig. 146 Partial display of Stromberg special service tools

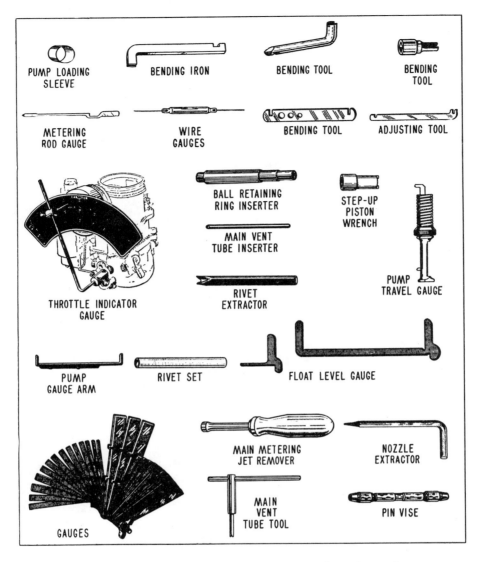

Fig. 147   Partial display of Carter special service tools

SERVICE TIPS

1. Dirt is the great enemy of good carburetion. It not only fills up air and fuel passages, but it also accelerates the wear of delicate parts. Therefore, make sure that the work bench, hands, tools and all parts are absolutely clean.

2. To prevent the entrance of dirt while the carburetor is off the engine, tape the opening in the manifold.

3. When disassembling a carburetor, it is a good idea to group the parts as they come out of each assembly. An inexpensive muffin pan which can accommodate six muffins is an ideal container to keep the parts of the various systems of the carburetor together.

4. Rap screws and jets with a light hammer and screwdriver to loosen them. Forcing tight fit-ting parts with a screwdriver will burr the slots and make it difficult to install them.

5. After all parts have been disassembled, clean the castings and all parts thoroughly with an approved carburetor cleaner and remove any gaskets that may have been overlooked in the disassembly.

6. Use cleaning solutions and compressed air for cleaning holes in carburetors. Acids, wires, sandpaper or drills may do damage beyond repair.

7. Before ordering parts, make a careful list of the parts needed. On the average, a carburetor repair kit contains all the parts required for rebuilding a carburetor. It includes a full set of gaskets, needle valve and seat assembly, bowl cover fuel strainer, all jets, metering rods (Carter), throttle connector rod, accelerating

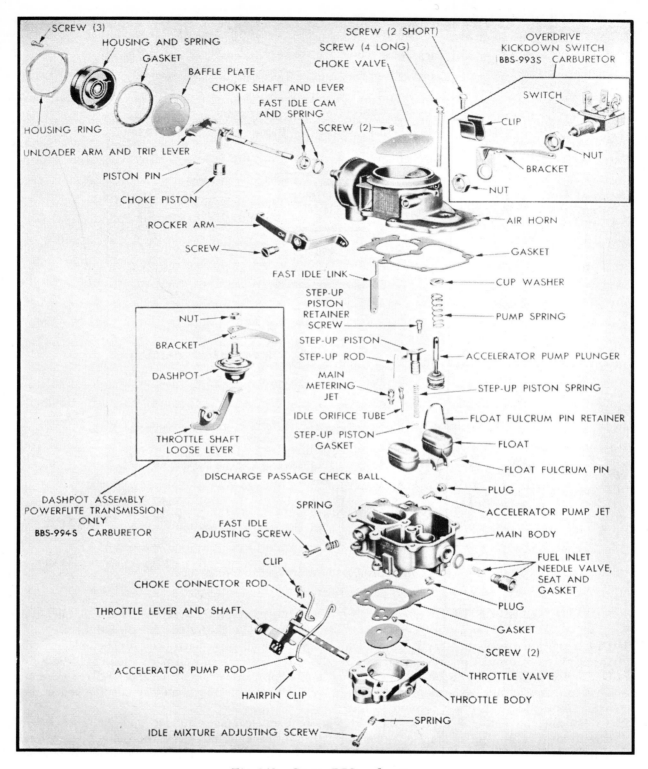

SCREW (3)

HOUSING AND SPRING

GASKET

BAFFLE PLATE

CHOKE SHAFT AND LEVER

FAST IDLE CAM AND SPRING

HOUSING RING

UNLOADER ARM AND TRIP LEVER

PISTON PIN

CHOKE PISTON

ROCKER ARM

SCREW

FAST IDLE LINK

SCREW (2 SHORT)

SCREW (4 LONG)

CHOKE VALVE

SCREW (2)

OVERDRIVE KICKDOWN SWITCH IBBS-993S CARBURETOR

SWITCH

CLIP

NUT

BRACKET

NUT

AIR HORN

GASKET

CUP WASHER

STEP-UP PISTON RETAINER SCREW

PUMP SPRING

NUT

BRACKET

DASHPOT

STEP-UP PISTON

STEP-UP ROD

MAIN METERING JET

IDLE ORIFICE TUBE

STEP-UP PISTON GASKET

ACCELERATOR PUMP PLUNGER

STEP-UP PISTON SPRING

FLOAT FULCRUM PIN RETAINER

FLOAT

FLOAT FULCRUM PIN

THROTTLE SHAFT LOOSE LEVER

DISCHARGE PASSAGE CHECK BALL

PLUG

ACCELERATOR PUMP JET

DASHPOT ASSEMBLY POWERFLITE TRANSMISSION ONLY BBS-994S CARBURETOR

SPRING

FAST IDLE ADJUSTING SCREW

MAIN BODY

CLIP

FUEL INLET NEEDLE VALVE, SEAT AND GASKET

CHOKE CONNECTOR ROD

THROTTLE LEVER AND SHAFT

PLUG

GASKET

ACCELERATOR PUMP ROD

SCREW (2)

THROTTLE VALVE

HAIRPIN CLIP

THROTTLE BODY

SPRING

IDLE MIXTURE ADJUSTING SCREW

Fig. 148   *Carter BBS carburetor*

pump plunger, pump check valves, and various other items which experience dictates should be replaced on a particular carburetor model. There may be a part or two of improved design to correct some small fault.

8. After obtaining the correct repair kit, empty

the package and substitute each new part for the old one which is in the cups of the muffin pan. As each part is substituted, discard the old one to avoid the possibility of mixing the new with the old.

9. Soak new gaskets in kerosene (never plain

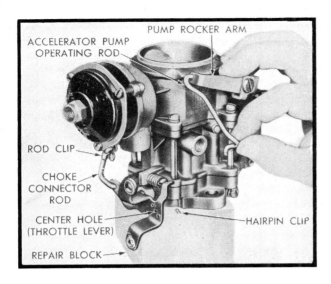

Fig. 149 *Removing accelerator pump rod. Carter BBS*

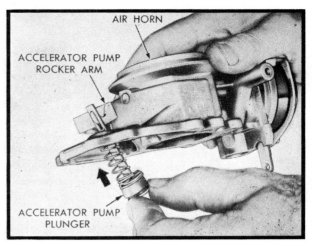

Fig. 150 *Removing accelerator pump plunger. Carter BBS*

water). Dry gaskets may not seal or they may crack.

10. Soak pump pistons in kerosene before using. Never use a screwdriver or a knife to insert the piston in the pump cylinder as damage to the plunger leather may result. Instead, use a suitable loading sleeve.

11. Make sure all jets, screws, and seats are firmly tightened, and that gaskets (when called for) are under these parts.

12. Assemble parts so they fit freely and operate smoothly. Do not force as the parts may not work afterward.

13. All carburetor linkages such as connector rods, fast idle linkages, etc., should not be lubricated as the increased accumulation of dirt and grit this will cause will promote wear.

CLEANING PARTS

Regardless of the number of new parts that are used in rebuilding a carburetor, the job in the end will not be satisfactory unless all metal parts are thoroughly cleaned. Because of the nature of carburetor parts, with numerous small passages subject to fouling with tenacious carbon and gum deposits, ordinary cleaning processes are entirely inadequate. The correct procedure is to use a cleaning bath in which metal parts can be immersed and "soaked" for sufficient time after disassembly to thoroughly clean all surfaces and passages.

There are several preparations which have been developed especially for cleaning carburetors. Regardless of the cleaning material used, however, be sure to rinse the parts thoroughly in kerosene

or white gasoline to remove all gummy deposits that have been softened by the cleaner. Blow out all passages in castings with compressed air and blow off all parts so they are free of solvent.

Do not soak cork, plastic or leather parts in the cleaner. Wipe such parts with a clean cloth.

Remove all carbon from barrels of the body flange so that the throttle valves may close properly. Be sure to clean all carbon out of the idle ports.

CARTER BBS OVERHAUL

This type carburetor is used on Chrysler Corp. six-cylinder engines. The BBS denotes that the unit is a Ball and Ball single barrel carburetor. The two barrel counterpart is the BBD carburetor.

The procedure which follows applies in general to all models of this type carburetor but to avoid confusion in the reader's mind, specifically it refers to a single model. Bear in mind also that the adjustment specifications given apply only to the particular model carburetor under discussion. Fig. 148 shows an exploded view of the BBS version.

Carburetor Disassemble

1. Remove choke connector rod and accelerator pump operating rod, Fig. 149.
2. Remove air horn attaching screws and carefully lift straight up to remove air horn assembly. Discard the gasket. *Long screws attach throttle body to main body. Use care to prevent accidental damage to throttle body.*
3. Disengage accelerator pump plunger from rocker arm by pushing up on bottom of plunger and sliding it off the hook, Fig. 150.

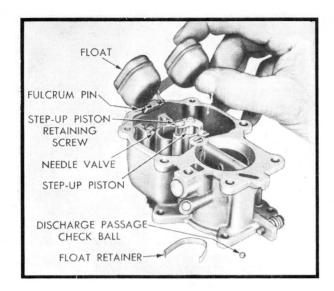

Fig. 151 *Removing float assembly. Carter BBS*

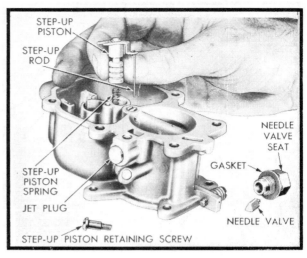

Fig. 152 *Removing or installing step-up piston. Carter BBS*

4. Remove float pin retainer, float pin and float, Fig. 151.
5. Remove step-up piston, spring, step-up rod, main metering jet and gasket, and idle orifice tube, Fig. 152.
6. Remove accelerator pump discharge check ball. To remove the pump jet, first remove jet plug.
7. Remove idle adjustment screw and spring from throttle body. Fig. 153 shows details of throttle body.

Carburetor Assemble

Testing Accelerator Pump System—Install the plunger in the cylinder and the discharge check ball on its seat. Pour a small quantity of gasoline in the bowl. Move the plunger up and down

slowly several times to expel all air from the pump passage. Hold the ball down firmly with a brass rod and raise the plunger, Fig. 154. Press the plunger down. No fuel should flow from the pump inlet or discharge passage. If gasoline is evident from either point, clean the passages again and repeat tests. If leakage is still evident, replace the check ball.

Main Body—Install the accelerator pump jet and plug. Install idle orifice tube and tighten securely. Install main metering jet and gasket. Tighten securely. Install step-up piston spring and step-up rod. See Fig. 152 and carefully guide step-

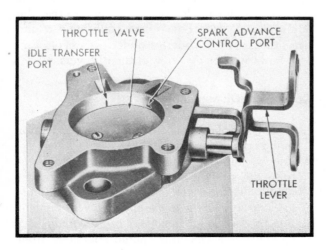

Fig. 153 *Throttle body details. Carter BBS*

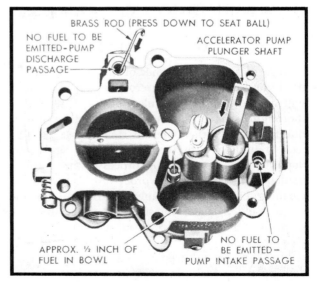

Fig. 154 *Testing accelerator pump system. Carter BBS*

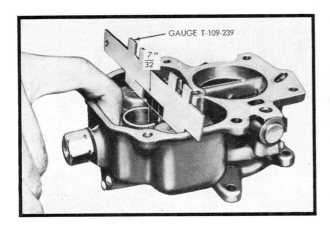

Fig. 155 *Checking float level height. Carter BBS*

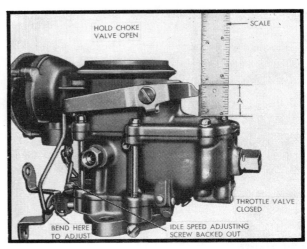

Fig. 156 *Accelerator pump adjustment. Carter BBS*

up rod into main metering jet. Place float assembly and inlet needle valve in position.

Checking Float Level—Seat needle with finger pressed against float lip. There should be 7/32″ from top of crown of each float to the top of the main body, Fig. 155. Each float must be adjusted for this setting. It is important that the floats do not touch the sides of the bowl.

Air Horn and Throttle Body—Place a new gasket on throttle body and position main body, making sure they are aligned. Assemble pump plunger, spring and cup washer and insert through air horn, engaging pump arm. Place a new gasket on main body and position air horn. Install attaching screws and tighten securely. Attach choke connector rod and accelerator pump operating rod.

Adjustments

Accelerator Pump—Back out throttle adjusting screw and open the choke valve so that the throttle valve can be completely seated in the carburetor bore. The adjustment is made with the pump connector rod in the center hole of the throttle lever. With the throttle valve closed, measure the distance between the top of the float bowl cover to the end of the plunger shaft, marked "A" in Fig. 156. This dimension varies according to the particular carburetor. If necessary, carefully bend the connector rod at the lower angle to obtain the desired result.

Fast Idle and Unloader—Remove thermostatic coil housing, gasket and baffle plate. Back out

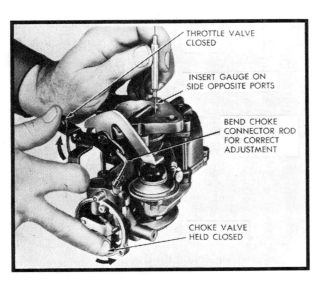

Fig. 157 *Fast idle adjustment. Carter BBS*

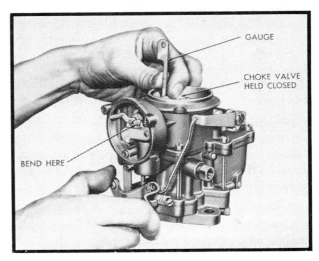

Fig. 158 *Unloader adjustment. Carter BBS*

Fig. 159 *Overdrive kickdown switch adjustment. Carter BBS*

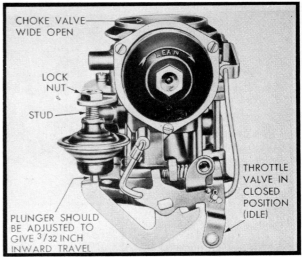

Fig. 160 *Dashpot adjustment. Carter BBS*

throttle adjusting screw. Open throttle valve partially and hold choke valve in fully closed position. Then close throttle valve. This procedure will allow fast idle cam to revolve to fast idle position. Measure the clearance between the throttle valve and bore of carburetor on the side opposite the idle port, Fig. 157. The clearance varies between one model carburetor and another. This clearance can be adjusted by bending the choke connector rod at the lower angle.

The unloader adjustment must be made after the fast idle adjustment has been performed. Hold the throttle valve wide open and close the choke valve as far as possible without forcing. Clearance is measured between the upper edge of the choke valve and the inner wall of the carburetor air horn, Fig. 158. This clearance dimension also varies according to carburetor model. To obtain the correct adjustment, bend the arm on the choke trip lever.

Overdrive Kickdown Switch—Open the throttle valve to wide open position and adjust the hex nuts on the switch to have 1/64 to 3/64″ clearance between the kickdown switch lever and the switch stem guide, Fig. 159.

Powerflite Dashpot—Maximum dashpot action is obtained by loosening the lock nut and adjusting the unit so that the dashpot plunger shaft can be moved inwardly approximately 3/32″ when the throttle valve is tightly closed, Fig. 160. After adjustment is made, tighten lock nut.

FORD DUAL CARBURETOR OVERHAUL

Disassemble

To facilitate working on the carburetor and to avoid damage to the throttle plates, use bolts through the carburetor retaining stud holes with a nut above and below the flange. Be sure that the bolt diameter and thread size match the hole diameter. Bolt length should be approximately 2¼″.

Air Horn

1. Referring to Fig. 161, remove air cleaner anchor stud. Disconnect choke plate operating rod at choke housing lever by loosening set screw.
2. Remove retaining screws and remove air horn and gasket.
3. Remove hairpin retainer that attaches choke operating rod to choke shaft lever. Remove rod.
4. Slide choke rod felt seal and two washers from between retainer and air horn.
5. If seal retainer bracket is damaged, remove it by driving out the retaining rivet.
6. If choke plate or shaft is damaged, file off peened end of choke plate retaining screws and remove screws, plate and shaft.

Main Body

1. Remove fuel inlet fitting and gasket.
2. Use a hook and disconnect float shaft retainer. Then remove float and shaft, and fuel inlet needle and clip.

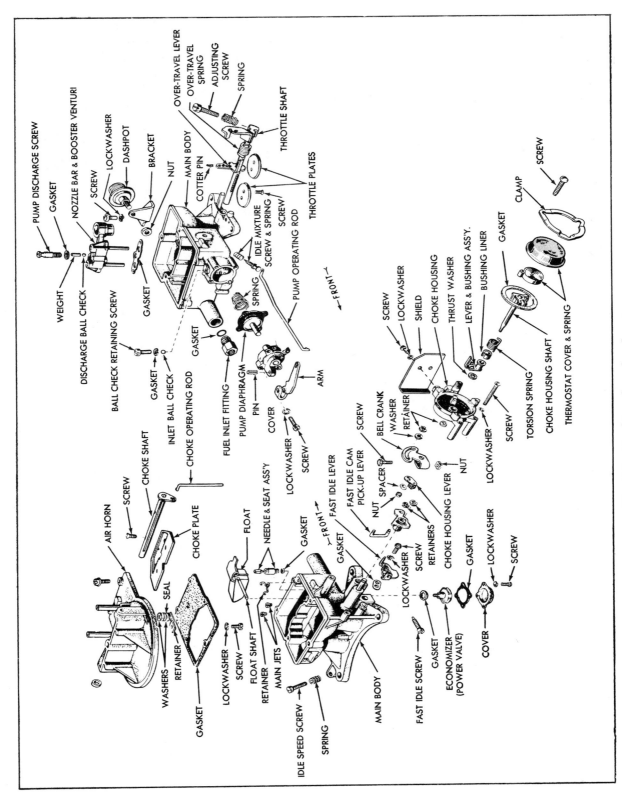

Fig. 161 Ford two-barrel carburetor

443

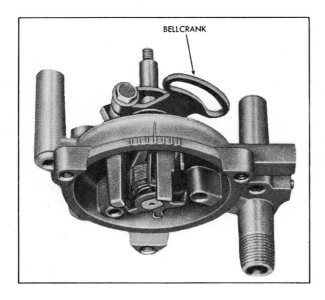

*Fig. 162 Bellcrank instal-
lation. Ford two-barrel*

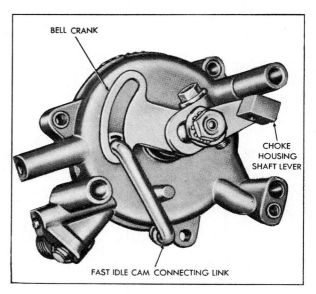

*Fig. 163 Choke housing linkage
installation. Ford two-barrel*

3. Remove fuel inlet needle seat, gasket, and float shaft retainer, and remove main jets.
4. Remove booster venturi attaching screw gasket, then lift nozzle bar and booster venturi assembly and gasket out of main body.
5. Invert main body and let accelerating pump discharge ball and weight fall into your hand.
6. Invert main body and remove power valve cover and gasket, then remove power valve and gasket.
7. Remove accelerating pump operating rod, then remove pump cover, diaphragm and spring. To release the clip, it is necessary to press on the tab portion of the clip while withdrawing the rod. Do not pry clip from rod.
8. Remove acclerating pump inlet ball check retainer screw and gasket. Remove ball check.
9. Remove choke shield.
10. Disconnect fast idle cam pickup lever at fast idle cam.
11. Remove thermostatic coil housing retaining screws and clamp, then remove housing and gasket.
12. Remove choke housing and gasket.
13. Remove choke housing lever attaching nut, spacer and lever. Loosen screw on bellcrank clamp and slide bellcrank off choke housing shaft and lever assembly. Remove fast idle cam rod.
14. Remove retainer from choke housing shaft and lever and slide shaft out of choke housing.
15. Slide thermostatic coil lever and bushing off choke housing shaft, taking care to allow for the twisting action of the torsion spring when

the parts are separated. Remove torsion spring and sleeve bearing from thermostatic coil lever and bushing.
16. Remove fast idle cam retainer and slide fast idle cam off boss on main body.
17. Remove nut and washer securing fast idle adjusting lever to throttle shaft and slide lever off shaft.
18. Remove distributor vacuum line fitting, the anti-stall dashpot (if equipped), idle fuel adjusting needles, and idle speed adjusting screw and spring.
19. If it is necessary to remove throttle plates, lightly scribe both plates along throttle shaft and mark each plate and corresponding bore with a number or letter for proper installation.
20. Remove throttle plate screws and plates, then slide throttle shaft out of body.
21. Remove accelerating pump overtravel lever retainer. Slide overtravel spring and lever off shaft.

Assembly

Make sure all holes in the new gaskets have been properly punched and that no foreign material has adhered to them. Make sure also that the accelerating pump diaphragm is not torn or cut. Refer to Fig. 161 while performing the following operations.

Air Horn

1. Position choke shaft and plate in air horn.
2. Close choke plate to position it in shaft, then

install and tighten screws. Stake screws with duck bill pliers or other suitable staking tool.

3. If choke seal retainer was removed, enlarge rivet hole with a No. 28 drill (.140″ dia.) and thread the hole with a No. 8-32 tap. Secure seal retainer with a No. 8-32 screw.

4. Position rod seal between the two brass washers and slide them into position between retainer and air horn.

5. Install choke operating rod and hairpin type retainer.

Main Body

1. If throttle plates were removed, place the accelerating pump overtravel spring, with shortest tang end first, over boss on pump overtravel lever. Place short tang of spring under lug on lever.

2. Slide overtravel lever and spring, spring end first, on throttle shaft.

3. Hook longest tang of spring over throttle lug of throttle lever and install pump overtravel lever retainer.

4. Slide throttle shaft into main body.

5. Referring to line scribed on throttle plates during disassembly, install plates in their proper location with the screws just snug (not tight). Hold throttle plates closed and hold main body up to light. Little or no light should show between throttle plates and throttle bores. Tap plates lightly with a screwdriver handle to seat them, then tighten the screws. Stake screws with duck bill pliers or other suitable staking tool.

6. Install idle speed screw and spring.

7. Position anti-stall dashpot.

8. Install distributor vacuum passage fitting.

9. Place fast idle adjusting lever on throttle shaft and install washer and nut.

10. Slide fast idle cam on boss of main body and install retainer.

11. Install white sleeve bearing and torsion spring on thermostat lever and bushing. Rolling sleeve bearing around itself will assist it in maintaining conformity around brass bushing.

12. Slide thermostat lever, bushing and torsion spring on choke housing shaft and engage spring.

13. Install complete choke housing shaft and lever in choke housing so that thermostat lever is at top of choke housing. Install choke housing shaft retainer.

14. Install bellcrank on choke housing shaft and position it so that lever is against shaft re-

tainer and the slot portion is to the front of choke housing, Fig. 162. (Refer to "Adjustments.")

15. Install fast idle cam pickup rod on bellcrank.

16. Position choke housing lever on choke housing shaft and install spacer, washer and nut, Fig. 163.

17. Place new choke housing gasket on main body. Position choke housing on main body, engaging fast idle cam pickup lever in hole in fast idle cam.

18. Install choke housing and screws.

19. Position thermostatic spring housing gasket and housing on choke housing and align index mark on spring housing with middle index mark on choke housing.

20. Install clamp screws, and choke shield.

21. Drop accelerating pump inlet ball in fuel inlet passage of the pump chamber and install washer and retaining screw.

22. Install the diaphragm return spring on the boss in the chamber. Insert the diaphragm assembly in the cover and place the cover and diaphragm assembly in position on the main body.

23. Install the cover screws finger tight, then push the accelerating pump plunger the full distance of its travel and tighten the cover screws.

24. Position the accelerator pump operating rod in the inner hole of the link arm of the accelerator pump. Install the other end of the rod in the No. 3 hole (third hole from the throttle shaft). This position is for average and warm temperature operation. If necessary, the rod may be installed in the No. 4 hole position (farthest from throttle shaft for cold weather operation. The No. 1 and No. 2 hole positions may be used for unusually hot operating conditions.

25. Invert the main body and install the power valve and gasket, then install the cover and gasket.

26. Install the idle adjusting needles and springs. *Turn the needles in gently with the fingers until they just touch the seat, then back them off 1½ turns for a preliminary idle adjustment.*

27. Insert the main jets.

28. Position the float shaft retainer in the groove on the fuel inlet needle seat, then install the seat and gasket.

29. Slide the float shaft in the float lever.

30. Install the fuel inlet needle clip in the groove on the fuel inlet needle and hook the assembly on the float tab.

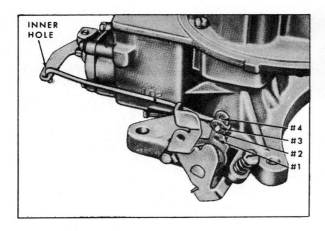

Fig. 164 *Accelerating pump stroke adjustment. Ford two-barrel*

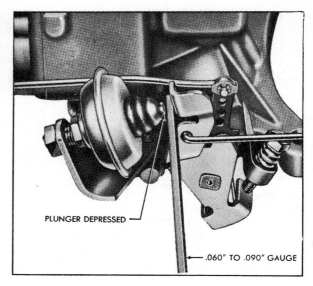

Fig. 165 *Anti-stall dashpot adjustment. Ford two-barrel*

ADJUSTMENTS

Accelerator Pump Stroke

Fig. 164. Install pump operating rod in the inner hole of the pump operating link. The other end of the pump rod should be installed in the third hole of the overtravel lever from the throttle shaft for average weather conditions. The fourth or outer hole is for extreme cold weather operation. The first and second holes are for any required variations from the standard positions.

Anti-Stall Dashpot

Fig. 165. To adjust the dashpot operating clearance, adjust the engine idle speed.

Hold the throttle in the closed position, and then, fully depress the dashpot plunger stem using a thin blade screwdriver.

Check the available clearance between the plunger tip and the throttle lever (or the dashpot operating lever on a single-barrel carburetor).

To adjust the clearance on all eight cylinder carburetors, loosen the locknut and rotate the dashpot in a direction to provide .060–.090 inch clearance. Tighten the locknut to secure the adjustment.

To adjust the clearance on the six cylinder carburetor, loosen the locknut and rotate the dashpot in a direction to provide .120–150 inch clearance.

Automatic Choke Adjustment

The normal setting for the thermostatic housing is two marks lean. The final setting may be varied but not to exceed two marks from the normal setting.

Choke Housing Torsion Spring Adjustment

Fig. 166. The initial choke plate pull-open torsion spring tension is adjustable. The outer end of the choke housing shaft lever has three spider-like fingers. These fingers provide for an adjustment of the torsion spring tension that governs the initial opening of the choke plate after a cold start.

The normal position for the tang of the torsion spring is on the center finger of the spider-like lever. The other positions on the lever are intended for increasing or decreasing the tension of

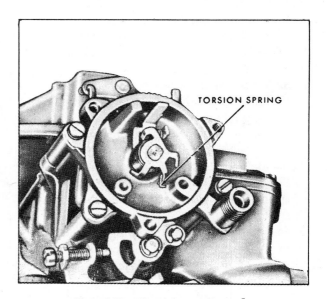

Fig. 166 *Torsion spring adjustment. Ford two-barrel*

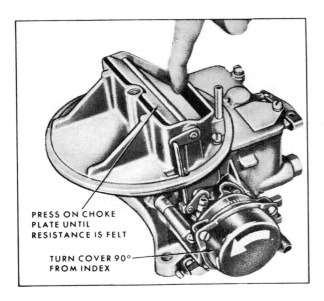

PRESS ON CHOKE
PLATE UNTIL
RESISTANCE IS FELT

TURN COVER 90°
FROM INDEX

Fig. 167 *Checking pull-down
opening. Ford two-barrel*

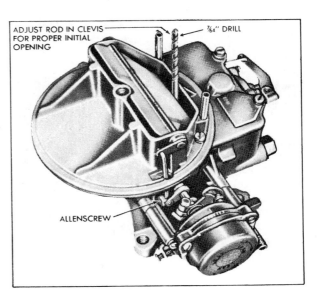

ADJUST ROD IN CLEVIS
FOR PROPER INITIAL
OPENING

7/64" DRILL

ALLENSCREW

Fig. 168 *Measuring pull-down
clearance. Ford two-barrel*

the spring to correct for leanness or richness in the automatic choke during drive-away, immediately after a cold start.

If there is an apparent leanness in the carburetor during drive-away after a cold start in low ambient temperatures, tension of the choke housing spring should be *increased* by moving the tang of the spring from the center finger to the finger that is to the left of the center position.

If there is an apparent richness and tendency to load-up during drive-away after a cold start in low ambient temperatures, the tension of the choke housing spring should be *decreased* by

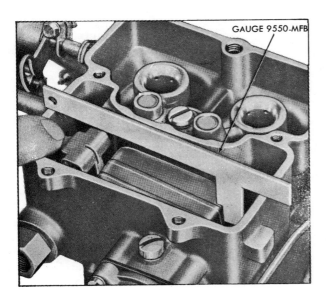

GAUGE 9550-MFB

Fig. 169 *Checking float
level. Ford two-barrel*

moving the tang of the spring from the center finger to the finger that is to the right of the center position.

Initial Choke Plate Pull-Down Setting

1. Turn the choke coil housing 90 degrees counterclockwise to assure positioning of the choke linkage in the fully closed choke position, Fig. 167.
2. Press the choke plate toward the open position until resistance to movement is felt.
   CAUTION: *Do not use force or the tang on the lever in the choke housing will be bent or broken.*
3. Measure the clearance between the lower edge of the choke plate and the inside vertical surface of the air horn. The clearance should be 7/64 inch. To adjust the clearance, loosen the set screw on the choke plate operating lever clevis with a 1/16 inch Allen wrench, and move the choke plate oprating rod up or down within the clevis until a clearance is achieved, Fig. 168.
4. Recheck the fast idle cam to choke housing clearance and adjust it, if necessary. (This clearance should be 0.050 inch.)
5. Reset the choke coil housing to the specified setting.

Float Level

Hold the float in the uppermost position by pressing downward on the float tab. Place Gauge

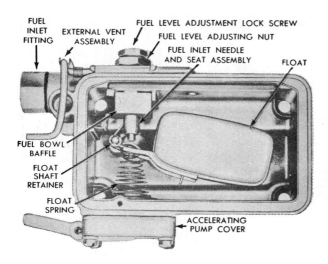

Fig. 170 *Fuel bowl assembly. Holley 2300*

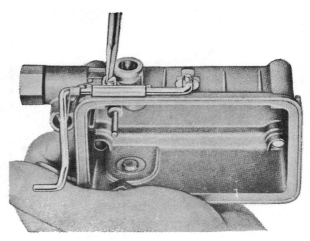

Fig. 171 *Removing external vent. Holley 2300*

9550-MFB on the bowl lengthwise to the float. The gauge should just contact the float. Position the gauge as shown in Fig. 169.

HOLLEY 2300 CARBURETOR OVERHAUL

Disassembly

Fuel Bowl & Metering Block

1. Referring to Fig. 170, unfasten screws and slide fuel bowl and metering block off main body. Separate metering block from fuel bowl and discard gaskets. Remove baffle from metering block.
2. Remove idle adjusting screws from metering block. Using a socket wrench, remove power valve and gasket. Remove main jets with a suitable wrench.
3. Remove fuel level adjusting lock screw and gasket. Turn adjusting nut counterclockwise and remove locknut and gasket. Remove fuel inlet needle and seat.
4. Using needle nose pliers, remove float retainer. Slide float off shaft. Remove spring from float.
5. Remove baffle plate from fuel bowl.
6. Remove external vent, Fig. 171.
7. Remove fuel level sight plug and gasket, fuel inlet fitting, gasket and filter screen.
8. Invert fuel bowl and remove accelerating pump cover, diaphragm and spring. The pump inlet ball check is not removable.

Main Body

1. Remove air cleaner anchor stud. Invert carburetor and remove throttle body and discard throttle body gasket.

2. Remove rod from choke housing shaft and lever assembly. Remove thermostat spring housing and gasket, then remove choke housing and gasket from body.
3. Remove choke housing shaft and fast idle cam assembly. Remove choke piston and lever.
4. Remove retaining pin from choke plate shaft lever, then remove choke rod. Remove choke plate from its shaft, then slide shaft and lever out of choke plate housing. Remove choke rod seal.
5. Remove accelerating pump discharge nozzle screw, then lift pump discharge nozzle and

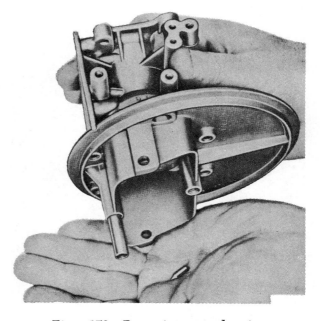

Fig. 172 *Removing accelerating pump discharge needle. Holley 2300*

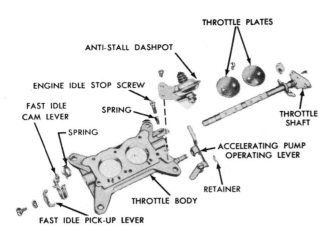

Fig. 173 *Throttle body assembly. Holley 2300*

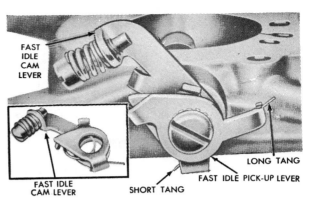

Fig. 174 *Fast idle cam lever installation. Holley 2300*

gasket out of main body. Invert main body and let pump discharge needle fall out, Fig. 172.

Throttle Body

1. Remove anti-stall dashpot (if equipped). Remove accelerating pump operating lever. Remove fast idle pick-up lever and fast idle cam lever and spring.
2. If it is necessary to remove throttle plates, lightly scribe both plates along throttle shaft and mark each plate and its corresponding bore with a number or letter for proper replacement. Remove screws and throttle plates. Slide throttle shaft out of throttle body and remove accelerating pump cam.

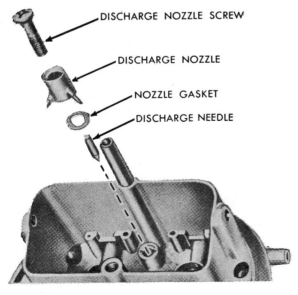

Fig. 175. *Installing pump discharge needle. Holley 2300*

Assembly

Make sure all holes in the new gaskets are properly punched and that no foreign material has adhered to the gaskets. Make sure the accelerating pump diaphragm is not torn or cut.

Throttle Body

1. If throttle plates were removed, refer to Fig. 173. Place accelerating pump cam on throttle shaft, then slide throttle shaft into throttle body.
2. Position throttle return spring so that small tang fits into slot in throttle lever and long tang rests against wide open throttle stop.
3. Referring to marks made on throttle plates when disassembled, install plates in proper position with screws snug (not tight).
4. Hold throttle body up to the light. Little or no light should show between throttle plates and throttle bores. If plates are properly installed and there is no binding when the throttle shaft is rotated, tighten screws and stake them.
5. Place fast idle cam pick-up lever spring inside fast idle cam pick-up lever and position lever on throttle shaft so that longest end of spring rests on longest arm of pick-up lever, Fig. 174. Install screw and lockwasher.
6. Install anti-stall dashpot (if equipped) and accelerating pump operating lever and retainer.

Main Body

1. Drop accelerating pump discharge needle into its well, Fig. 175. Seat needle with brass drift and a light hammer. Make sure it is free. Position pump nozzle gasket and nozzle in main body, then install retaining screw.
2. Referring to Fig. 176, place choke plate in air

449

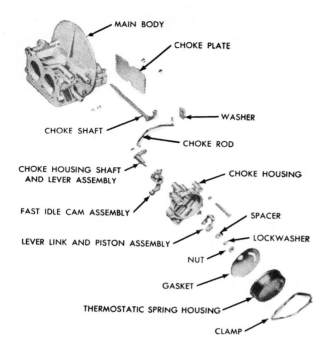

Fig. 176 *Choke plate and housing. Holley* 2300

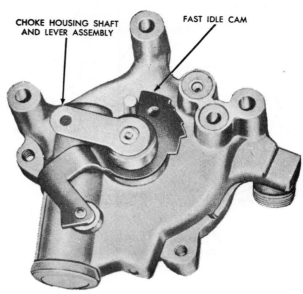

Fig. 177 *Choke housing linkage installed. Holley* 2300

horn, then position choke plate in shaft. Install and stake retaining screws.

3. Install choke plate rod seal in retaining grooves on main body. Slide choke plate rod through opening in main body and secure it to choke lever with cotter pin.

4. Install fast idle cam assembly on brass bushing on back of choke housing with bushing on cam facing outward, Fig. 177.

5. Position piston and lever in choke housing, then install choke housing shaft and lever and secure lever and piston to it with spacer, lockwasher and nut.

6. Lay main body on its side and position choke gasket on passages on main body, Fig. 178.

7. Position screws in choke housing. Place choke housing on main body, inserting choke rod in housing lever as housing is placed in position. Be sure projection on choke rod is placed under fast idle cam so that cam will be lifted when choke plate is closed.

8. Using needle nose pliers, install choke rod retaining pin.

9. Place thermostat spring housing and gasket in position on choke housing, engaging thermostat spring on spring lever. Position spring housing to one notch in rich direction from the mid-position mark, then install clamp and screws.

10. Invert main body and position throttle body gasket on main body. Place throttle body on

main body and secure with screws and lockwashers. Install air cleaner anchor stud.

Fuel Bowl & Metering Block

1. Referring to Fig. 179, place pump diaphragm spring and diaphragm in pump chamber. Install cover with screws finger tight. Make sure diaphragm is centered, then compress diaphragm with pump operating lever and tighten cover screws.

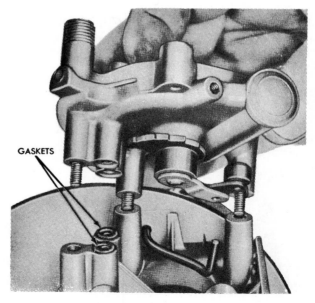

Fig. 178 *Choke housing installation. Holley* 2300

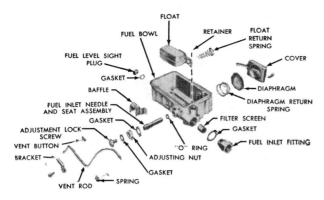

Fig. 179   Fuel bowl assembly. Holley 2300

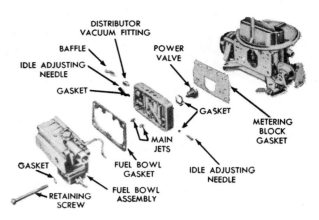

Fig. 180   Fuel bowl and meter-ing block assembly. Holley 2300

2. Install fuel level sight plug and gasket, fuel inlet filter screen, gasket and fitting.

3. Install fuel bowl vent, Fig. 171.

4. Slide baffle plate on ridges in fuel bowl.

5. Install spring on float and slide float on shaft. Using needle nose pliers, install float retainer.

6. Apply vaseline to a new "O" ring seal and slide it on fuel inlet needle and seat assembly.

7. Position fuel inlet needle and seat through top of bowl. Align flat in I.D. of nut with flat on O.D. of fuel inlet needle and seat. Adjust needle so that there is ⅞" clearance between floor of fuel bowl and bottom of float with fuel bowl inverted.

8. Install fuel level adjusting lock screw and gasket.

9. Install power valve and gasket, main jets, idle fuel adjusting needles and gaskets. Turn idle adjusting needles in gently until they just touch seat, then back them off ¾ to 1¾ turns for a preliminary idle adjustment.

10. Referring to Fig. 180, position metering block gasket on dowels on back of metering block. Lay metering block in place on main body. Position baffle on metering block, then place fuel bowl gasket on metering block.

11. Place retaining screws and new compression gaskets in fuel bowl. Lay bowl on metering block and tighten retaining screws.

ADJUSTMENTS

See Fig. 181 for location of idle, accelerating pump stroke and anti-stall dashpot adjustments. Adjust the idle mixture and engine speed as outlined previously in the Fuel System Service section of this chapter. Adjust the anti-stall dashpot as directed to establish a clearance of .045".

Fuel Level

1. Position car on a level floor.

2. Operate engine until normal temperature is reached. Remove air cleaner.

3. Place a suitable container below fuel level sight plug to collect any spill-over of fuel, Fig. 182.

4. With engine stopped, remove fuel level sight plug and check fuel level. Fuel level within bowl should be at the lower edge of sight plug opening within plus or minus 1/16".

5. If the level is too high, drain the fuel bowl and refill it before altering the float setting. This will eliminate the possibility of foreign material causing a temporary flooding condition. To drain the bowl, remove one lower retaining screw and allow the bowl to drain. Install the screw, start the engine to fill the bowl. After the fuel level has stabilized, stop the engine and check the fuel level.

6. To adjust the level, loosen lock screw on top of

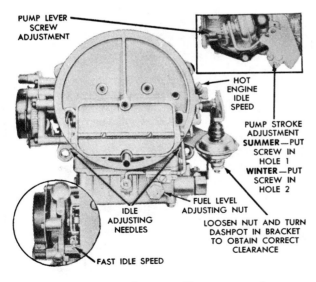

Fig. 181   Carburetor idle, pump and dashpot adjustments. Holley 2300

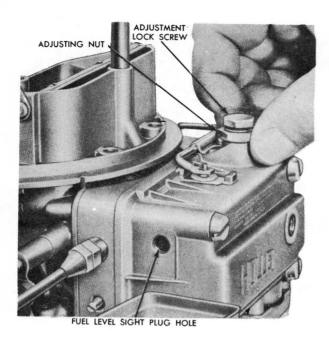

Fig. 182 *Fuel level adjustment. Holley 2300*

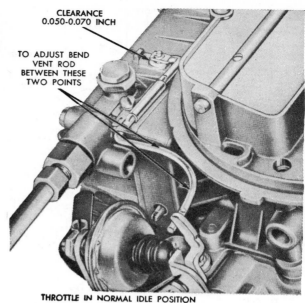

Fig. 183 *Fuel bowl vent adjustment. Holley 2300*

fuel bowl just enough to allow rotation of the adjusting nut underneath. Turn adjusting nut as required to obtain proper fuel level. A 1/6 turn of the adjusting nut will change fuel level at the sight plug opening by 3/64".

Fuel Bowl Vent

Refer to Fig. 183 and adjust the bowl vent valve as directed. The clearance shown should be established between the vent button and the machined surface of the fuel bowl with the throttle in the normal idle position.

ROCHESTER 2GC OVERHAUL

Adjustments on Car

Automatic Choke—Choke cover setting is indicated by the index markings on the housing and air horn casting. With choke cover set in this position, choke valve should be just closed at 75 degrees F. (engine and carburetor must be cooled down to room temperature).

Fast Idle Cam Index—No adjustment of fast idle speed is provided since the steps on the fast idle cam are correctly proportioned to give the correct speed stops above normal idle speed. It is necessary, however, to have the correct relationship between fast idle cam position and choke valve position. To check and adjust the setting, proceed as follows:

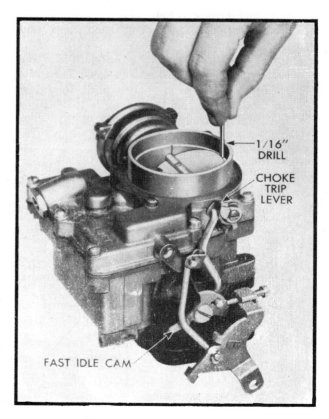

Fig. 184 *Checking fast idle adjustment. Rochester 2GC*

452

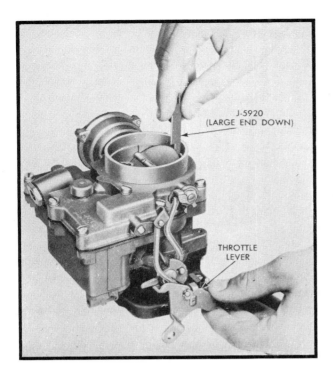

Fig. 185 *Checking unloader adjustment. Rochester 2GC*

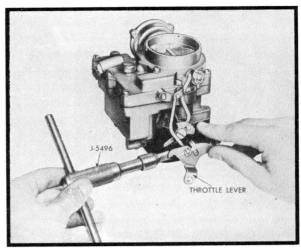

Fig. 186 *Adjusting unloader. Rochester 2GC*

Place end of idle screw on next to highest step on fast idle cam, Fig. 184. A 1/16″ drill should slide easily between upper edge of choke valve and bore of carburetor as shown. If necessary, bend choke trip lever tang until prescribed clearance is obtained.

Unloader—Place throttle in wide open position. Large end of tool shown in Fig. 185 should slide freely between upper edge of choke valve and

bore of carburetor. If necessary, bend tang of throttle lever with tool shown in Fig. 186 to obtain the necessary clearance.

Carburetor Disassemble

Flooding, stumbling on acceleration and other performance complaints are in many instances caused by the presence of dirt, water or other foreign matter in the carburetor. To aid in diagnosing the cause of the complaint, the carburetor should be carefully removed from the engine without draining the fuel from the bowl. The contents of the fuel bowl may then be examined for contamination as the carburetor is disassembled.

Bowl Cover—Remove the three attaching screws and remove the choke cover and coil as-

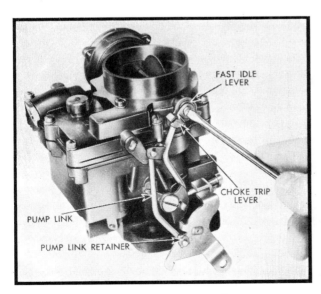

Fig. 187 *Removing or replacing choke trip lever. Rochester 2GC*

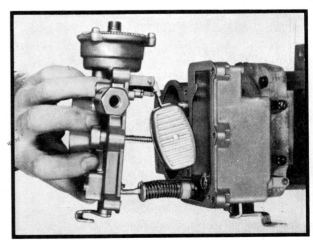

Fig. 188 *Removing or replacing bowl cover. Rochester 2GC*

453

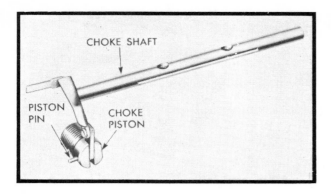

Fig. 189 *Choke shaft and piston assembly. Rochester 2GC*

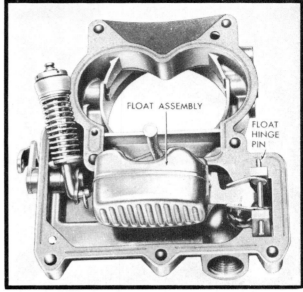

Fig. 190 *Float-to-bowl cover construction. Rochester 2GC*

sembly. Remove choke cover gasket and baffle plate. Remove retaining screw at end of choke shaft, Fig. 187, and carefully pry on choke trip lever, fast idle link and lever. Lever can be removed from link by turning until slot in lever will pass over tang in link. Rotate link until it will slip out through slot in fast idle cam.

Remove fuel filter. Remove filter screen retainer nut and gasket with ¾" wrench and remove screen. Disconnect pump link from throttle lever by removing retainer. Link can be removed completely by rotating until it clears pump lever, Fig. 187. Remove eight screws and lift cover from bowl, Fig. 188.

Remove two screws and take out choke valve. Rotate choke shaft counterclockwise to free choke piston from housing, then pull piston and choke

shaft from carburetor, Fig. 189. Remove two screws and take off choke housing and gasket.

Place up-ended cover on flat surface. Remove float hinge pin and lift float assembly from cover, Fig. 190. Float needle may now be removed from float. Remove float needle seat and gasket with wide blade screwdriver. Remove power piston, Fig. 191. Remove retainer on pump plunger shaft and plunger assembly from pump arm, Fig. 191. The pump lever and shaft may be removed by loosening set screw on inner arm and removing

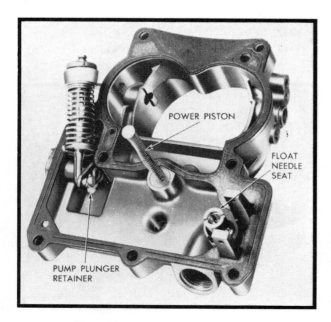

Fig. 191 *Power piston and pump plunger. Rochester 2GC*

Fig. 192 *Pump plunger return spring and inlet filter screen. Rochester 2GC*

Fig. 193 *Main metering jets and power valve. Rochester 2GC*

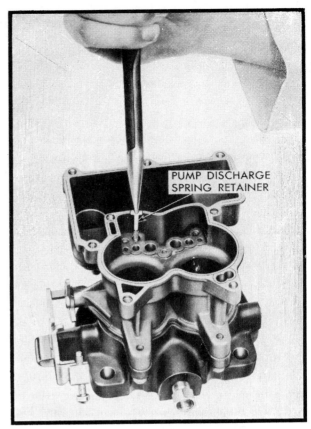

Fig. 194 *Removing or replacing pump discharge spring retainer. Rochester 2GC*

outer lever and shaft. The pump plunger may be further disassembled by compressing the spring and removing the retainer.

Disassemble Bowl—Remove pump inlet filter screen and pump plunger return spring. Remove check ball from bottom of pump well, Fig. 192. Remove main metering jets and power valve, Fig. 193. Remove three screws on top of cluster and lift out cluster and gasket.

Using long-nose pliers, Fig. 194, remove pump discharge spring retainer. Then remove the spring and check ball.

Invert carburetor and remove three large bowl-to-throttle body attaching screws, and remove throttle body and gasket. Remove fast idle cam.

Remove idle adjusting needles and springs from throttle body, and the idle screw from throttle lever.

Cleaning and Inspection

1. Thoroughly clean carburetor castings and metal parts in cleaning solvent. *Choke coil and housing, and pump plunger should not be immersed in solvent. Use clean gasoline only on pump plunger.*
2. To avoid damage to gasket between choke housing and air horn do not soak air horn in cleaner or solvent if choke piston housing has not been removed.
3. Check float needle and seat for wear. If wear is noted the assembly must be replaced.

4. Check float lip for wear and float for dents. Check floats for leaks by shaking.
5. Check throttle and choke shaft bores in throttle body and cover castings for wear or out-of-round condition.
6. Inspect idle adjusting needles for burrs or ridges. Such a condition requires replacement.
7. If wear is noted on steps of fast idle cam, it should be replaced as it may upset engine idle speed during the warm up.
8. Inspect pump plunger leather. Replace plunger if leather is damaged.
9. Check both filter screens for dirt or lint. Clean them and if they are distorted or remain plugged, replace.
10. If for any reason parts have become loose or damaged in the cluster casting, it must be replaced.

Carburetor Assemble

Throttle Body—Install idle screw in throttle lever. Screw idle adjusting needles and springs into throttle body until finger tight. Back out

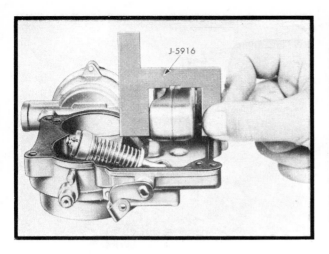

Fig. 195   Checking float level. Rochester 2GC

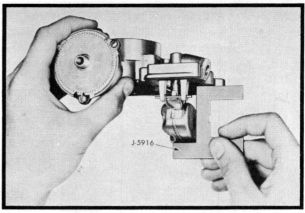

Fig. 196   Checking float drop. Rochester 2GC

screws 1½ turns as a preliminary idle adjustment. Place new throttle body gasket in position and attach throttle body. Tighten screws evenly and securely.

Bowl Assemble—Install fast idle cam. Drop pump discharge check ball (steel) into discharge hole and replace spring and retainer, Fig. 194. Replace cluster and gasket, tightening screws evenly and securely. Make certain center screw is fitted with gasket to prevent pump discharge leakage.

Replace main metering jets and power valve, Fig. 193. Drop pump inlet ball check into hole in pump well (ball is aluminum). Install pump return spring, pressing it with finger to center it in pump well. Replace pump inlet strainer, pressing carefully into position.

Bowl Cover Assemble—Place new gasket into position and attach choke housing to air horn. Tighten screws securely. Assemble choke piston to shaft with pin, with flat cut side of piston toward air horn, Fig. 189. Place shaft in choke housing and rotate clockwise so that piston slides in housing. Install choke valve on choke shaft with letters "RP" facing upward. Center choke valve before tightening screws. (While centering choke valve, install fast idle lever, choke trip lever and tighten attaching screws temporarily.)

Replace pump lever assembly. Install float needle seat and gasket, using wide blade screwdriver. Place power piston in vacuum cavity, being sure piston travels freely in cavity, Fig. 191. Attach plunger shaft with retainer, with shaft end pointing inward. Install cover gasket. Attach needle to float, carefully position float and insert hinge pin, Fig. 190.

Float Adjustments—To adjust float level, place float gauge, Fig. 195, in position over float and resting on gasket surface. Highest point of float should just touch gauge. Adjust by bending float arm. Measurement from gasket surface to high point of float is 1 15/64."

Check and adjust float drop. With air horn right side up so that float can hang free, the distance from the gasket surface to the lowest point of the float should be 1 29/32" and can be measured with float gauge shown in Fig. 196. To adjust, bend float tang.

Final Assembly—Place cover on bowl, making certain that accelerator pump plunger is correctly

Fig. 197   Checking pump link. Rochester 2GC

positioned and will move freely, Fig. 188. Install and tighten 8 cover screws evenly and securely. Install filter screen with closed end toward air horn. Install strainer nut and gasket in cover.

Install pump link and retainer, Fig. 186. To adjust pump link, place float and pump gauge on air horn next to air intake with single leg of gauge downward toward pump link, Fig. 197. With throttle stop screw backed off so that throttle valves are tightly closed, the top surface of the pump rod should just touch the end of the gauge. The measurement is .922."

Place baffle plate and choke housing gasket into position and install choke coil and cover. Rotate cover counterclockwise until index marks on cover and housing are aligned. Attach three retainers and screws to choke housing and tighten securely.

Place link on fast idle cam and choke lever. Place choke lever on cover with tang facing outward and toward pump lever. Install spacer washer and trip lever so that tang of trip lever is under tang of choke lever, and install retaining screw, Fig. 187.

Complete the job by adjusting the fast idle and choke unloader as directed under *Adjustments*.

STROMBERG WW OVERHAUL

Carburetor Disassemble

Air Horn—Remove the fast idle rod and pump operating rod. Remove all air horn attaching screws and carefully lift off vertically the air horn assembly. Disengage the pump plunger rod by tilting slightly and remove pump plunger, Figs. 198 and 199.

To remove the vacuum power piston, use an open end wrench and a wood block as shown in Fig. 200. Use care as pressure is applied as the assembly is staked in position. The choke plate and shaft can be removed if necessary. Care should be exercised when removing screws to prevent breaking them off in the shaft.

Main Body—Remove idle tubes as shown in Fig. 201. Invert carburetor and remove accelerator pump inlet check ball. Do not remove the two dome-shaped high speed bleeders in the main discharge strut of the main body. Remove accelerator pump discharge cluster and invert the body to allow the accelerator pump discharge check ball to drop out, Fig. 202.

Remove float inlet needle and seat. Inspect for grooving. Use a small screwdriver to pry out the float fulcrum retaining spring. Cover the float chamber to prevent spring flying out. Then lift out float.

Remove power by-pass jet and gasket, Fig. 203. Test plunger action of by-pass jet. Invert main body and remove main metering jet plugs. A special tool is needed to remove main metering jets (see Fig. 146). Then use the tool shown in Fig. 204 to remove the main discharge jets or tubes. As shown in Fig. 207, this tool has a tapered right-hand thread and should be screwed into the jet. The threads formed by the tool will not damage the jets.

The main body is attached to the throttle body by four screws in the bottom of the throttle body. If separated, always use a new gasket.

Throttle Body—A new throttle shaft can be installed but if clearance between shaft and bore is excessive enough to cause poor idling, the assembly should be replaced.

To remove the shaft, first remove the lock nut located in the choke housing. Then remove the throttle valve retaining screws. These should be removed with care to prevent screws being broken in the shaft. Mark the valves so they are installed in the correct bores, Fig. 205. When valves are installed, make sure that the small notch in the valve is toward the idle port.

Carburetor Assemble

Assemble throttle body to main body using a new gasket. Place main discharge jets on the tool shown in Fig. 206 and install into position. Make sure the opening in end of tube (diagonal cut end) is facing opposite side of small venturi. Insert main metering jets over discharge jets and tighten. Then install gasket and plug. Install power by-pass jet and gasket.

Testing Accelerator Pump—Install the accelerator pump inlet check ball (3/16" diameter) in check ball seat at the bottom of the pump cylinder. Install the accelerating pump discharge check ball (⅛" diameter) in the orifice in the center passage of the discharge strut section of the main body.

Pour clean gasoline into the carburetor bowl, approximately ½" deep. Raise the plunger and press lightly on the plunger shaft to expel the air from the pump passage. Using a small, clean brass rod, hold the discharge check ball firmly down on its seat, Fig. 207. Again raise the plunger and press downward. No fuel should be ejected from either the intake or discharge passages. Install the discharge cluster gasket, cluster and screw. Tighten securely.

Float Level—Check the float for leaks or damage. If satisfactory for further service, install in

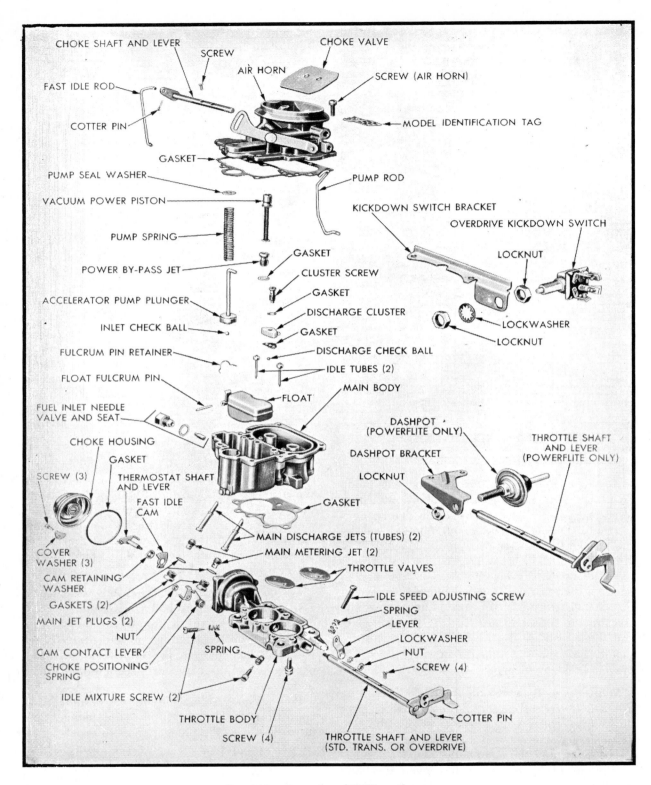

Fig. 198 Stromberg WW carburetor

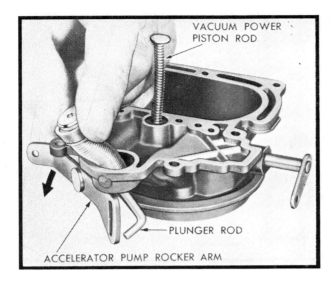

Fig. 199 *Removing or installing accelerator pump assembly. Stromberg WW*

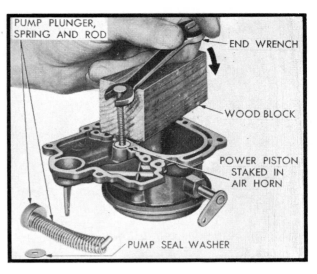

Fig. 200 *Removing power piston. Stromberg WW*

position in carburetor bowl. Assemble fuel inlet needle valve, seat and gasket, then insert in position in main body. Tighten securely. Install float fulcrum pin retaining spring and force under lip of boss to keep fulcrum pin in position.

Using a "T" scale or the tool shown in Fig. 208, check the float setting. The top of the float must be 3/16″ from the top of the main body (gasket removed) with the gauge at the center of the flat and the float lip held firmly against the fuel inlet needle. To change the float setting, bend the float lip toward the needle to lower, and away from the needle to raise the float, Fig. 209.

Main Body Assemble—Install idle tubes in main body. Install vacuum power piston and

plunger in air horn. Lock in position by prick punching on the retaining rim. Compress the piston plunger to be sure no binding exists. If the piston sticks or binds enough to hinder smooth operation, install a new piston assembly.

Slide a new air horn gasket over the accelerator pump plunger and down against the air horn. Lower the air horn slightly down on the main body with the accelerator pump plunger sliding into its well, Fig. 210. Be sure plunger leather does not fold or curl back. Install air horn retaining screws and lock washers and tighten securely. Work the pump plunger several times to be sure it operates freely.

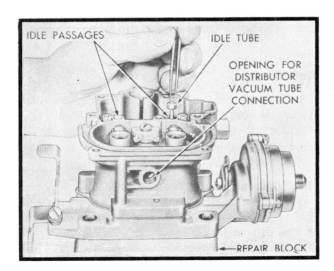

Fig. 201 *Removing or installing idle tubes. Stromberg WW*

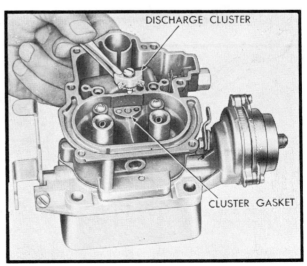

Fig. 202 *Removing or installing accelerator pump discharge cluster. Stromberg WW*

Fig. 203 *Removing or installing power by-pass jet. Stromberg WW*

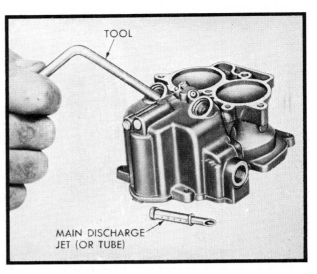

Fig. 204 *Removing main discharge tubes. Stromberg WW*

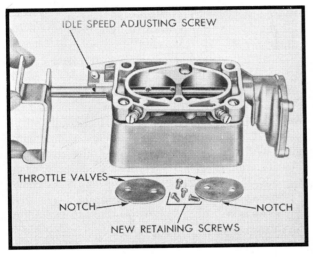

Fig. 205 *Throttle valves and shafts. Stromberg WW*

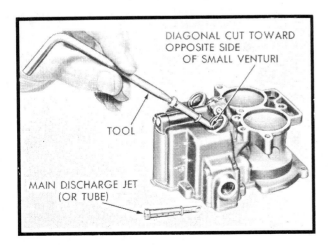

Fig. 206 *Installing main metering discharge tubes. Stromberg WW*

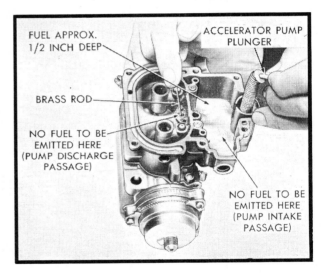

Fig. 207 *Testing accelerator pump intake and discharge check balls. Stromberg WW*

Adjustments

In making some of these adjustments, a special calibrated coil spring (Stromberg Tool No. T-25906) is essential. This calibrated spring is shown installed in Figs. 213, 215 and 217.

Fast Idle—To make this adjustment, remove the thermostat coil housing and install the special calibrated spring with a thermostat cover screw at one end and hook the other end to the long tang of the thermostat lever. Remove the spring from the fast idle adjusting screw, Fig. 211, and reinstall the screw without the spring.

With the throttle valves fully closed, turn the fast idle screw in so it just contacts the throttle

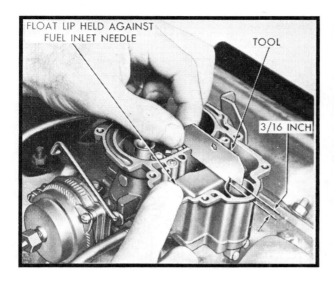

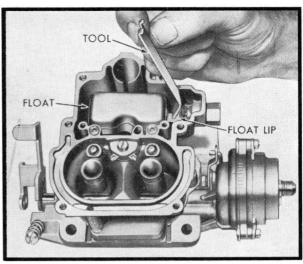

Fig. 208 *Checking float level. Stromberg WW*

Fig. 209 *Bending float lip to obtain correct setting. Stromberg WW*

lever. Then turn the adjusting screw in 4½ turns for 8-cylinder carburetors or 5⅛ turns for 6-cylinder carburetors. This will open the throttle valves a predetermined amount.

With the choke valve closed and the throttle level held against the fast idle screw, insert the proper size drill between the tang on the contact lever and the first step of the fast idle cam, Fig. 211. To adjust, slide the tool shown in Fig. 212 over tang on cam contact lever. Bend tang away from or toward fast idle cam until correct clearance has been obtained. After adjusting, make sure the tang is parallel to the throttle shaft. Remove the fast idle speed adjusting screw and install the spring.

Fast Idle Cam—To position the fast idle cam, open the throttle valves sufficiently to clear the fast idle cam. Close the choke valve tightly by applying pressure on the choke valve. Now, force the throttle valves closed against the tension of the cam positioning spring as shown in Fig. 213. In this position, the tang on the cam contact lever should just clear the high step of the fast idle cam, Fig. 213. The clearance should not exceed .020″. Adjust, if necessary, by bending the fast idle rod (at angle) until the correct clearance has been obtained, Fig. 214.

Choke Positioning Spring—To make this adjustment, slide the special weight tool shown in Fig.

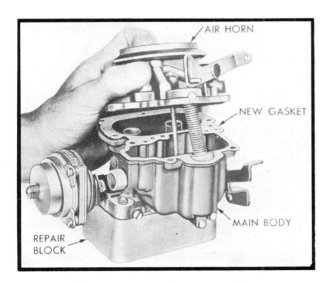

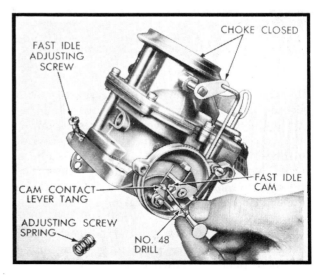

Fig. 210 *Installing air horn on main body. Stromberg WW*

Fig. 211 *Checking clearance for fast idle. Stromberg WW*

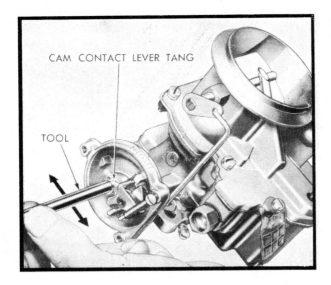

Fig. 212 *Bending cam contact lever to obtain fast idle setting. Stromberg WW*

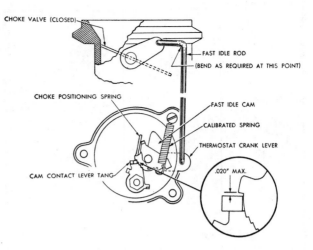

Fig. 213 *Fast idle cam position adjustment. Stromberg WW*

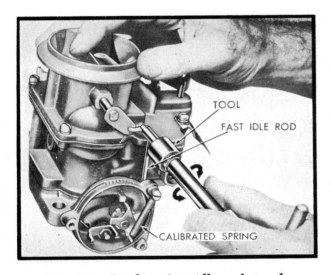

Fig. 214 *Bending fast idle rod to obtain fast idle cam setting. Stromberg WW*

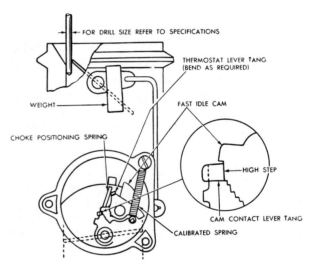

Fig. 215 *Choke positioning spring adjustment. Stromberg WW*

215 over the choke lever. Hold carburetor in a vertical position. Close throttle valves, making certain the cam contact lever tang is resting on the high step of the fast idle cam (it may be necessary to rotate the fast idle cam to obtain this adjustment). Now, lightly close the throttle valves. The choke valve should open just enough to insert the proper drill between choke valve and wall of air horn, Fig. 215.

To adjust, bend the tang of the thermostat lever that contacts the choke positioning spring, using long nose pliers as shown in Fig. 216. After adjustment has been made, remove weight from choke lever.

Unloader Adjustment—To make this adjustment, lightly hold the choke valve closed, then open the throttle valves wide. The choke valve should open sufficiently to allow the proper size drill to be inserted between the choke valve and wall of air horn, Fig. 217. To adjust, bend the tang of the cam contact lever with long nose pliers, Fig. 218.

Remove the special calibrated spring and screw, then install the thermostat coil housing and gasket, being sure to align index marks.

Hold the choke open and then open and close the throttle valves. Failure to obtain full throttle operation indicates improper assembly or adjust-

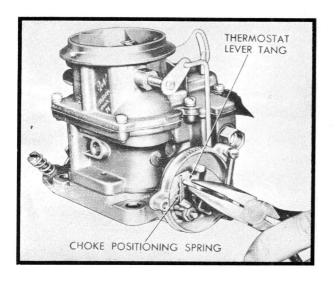

Fig. 216 *Bending thermostat lever tang to obtain choke positioning spring setting. Stromberg WW*

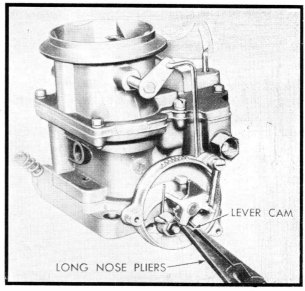

Fig. 218 *Bending cam contact lever tang for correct unloader setting. Stromberg WW*

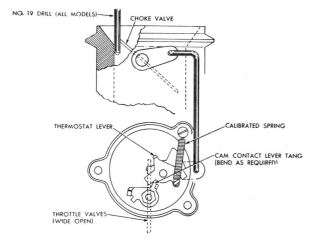

Fig. 217 *Unloader adjustment. Stromberg WW*

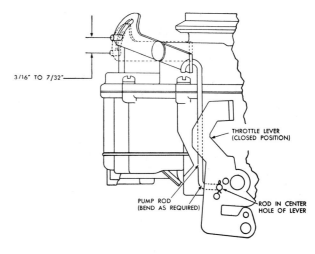

Fig. 219 *Accelerator pump adjustment. Stromberg WW*

ment of the choke mechanism. With the throttle valves held in open position, open the choke valve slowly to wide open position. There should be no bind throughout the entire travel of the choke mechanism.

Accelerator Pump—The following adjustment is made with the accelerator pump rod in the center hole of the throttle lever. To check the pump travel, hold the carburetor in a vertical position. Then operate the pump to permit the check ball at the bottom of the well to take its normal position on the seat. With the choke valve held open,

measure the travel of the pump as the throttle valves are moved from open to fully closed position. The pump travel should measure 3/16″ to 7/32″ on 6-cylinder cars, Fig. 219.

To adjust, remove the pump rod from the center hole of the throttle lever and bend it at the angle as shown in Fig. 220 up or down until correct travel has been obtained. Reinstall rod.

Dashpot—With the idle speed adjusting screw set for normal idle speed, there should be 1/16″ to 3/32″ clearance between the end of the dashpot plunger on the throttle lever when the plunger is pushed to the end of its travel. To adjust, loosen the locknut and turn the dashpot in the bracket to obtain correct clearance.

Kickdown Switch—With the throttle valves wide open, there should be 1/64″ to 1/32″ clearance between ear on throttle lever and the threaded portion of the switch housing. To adjust, loosen the adjusting lock nuts (see Fig. 198) and move the switch in or out until correct clearance has been obtained.

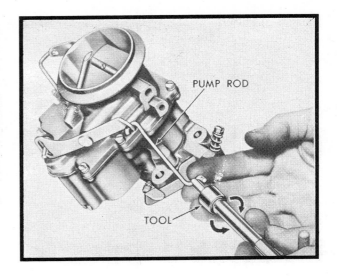

Fig. 220   Bending pump rod at lower angle for correct pump setting. Stromberg WW

Fuel Pump Service

FUEL PUMP TESTS

The fuel pump can be tested on the car with a pressure gauge, a hose and a pint measuring can. With this equipment, it is possible to check the fuel pump to see if it is delivering the proper amount of gasoline at the correct pressure.

Pressure Test—To make the pressure test, disconnect the fuel pipe at the carburetor inlet and attach the pressure gauge and hose between the carburetor inlet and the disconnected fuel pipe, Fig. 221. Take the pressure reading with the engine running. The pressure should be between 2 and 6 pounds, depending on the pump model and the car on which it is installed. The pressure should remain constant or return very slowly to zero when the engine is stopped.

Capacity Test—To make this test, connect hose so the pump will deliver gasoline into the pint measure held at carburetor level. Run the engine at idle speed and note the time it takes to fill the measure. On the average it should take from 20 to 30 seconds, depending on the pump being tested.

WHEN PRESSURE IS LOW

Low pressure indicates extreme wear on one part, small wear on all parts, ruptured diaphragm, dirty valve or gummy valve seat.

Wear in the pump usually occurs at the rocker

arm pivot pin, Figs. 222 and 223, and on the contacting surfaces of the rocker arm and links. Due to the leverage design, wear at these points is multiplied five times in the movement of the diaphragm. It is apparent therefore, that very little wear will materially reduce the stroke of the

Fig. 221   Testing fuel pump pressure

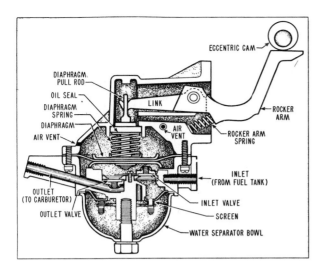

Fig. 222 Schematic cutaway view of fuel pump

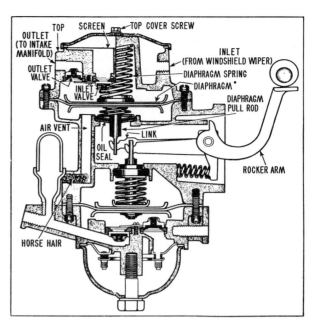

Fig. 223 Schematic cutaway view of fuel and vacuum pump

diaphragm. The worn parts must be replaced for a satisfactory correction.

The diaphragm pull rod has an oil seal around it which prevents the hot oil vapors from the crankcase coming in contact with the diaphragm. If this seal is damaged, the oil vapors have a tendency to shorten the life of the diaphragm.

The first three conditions—extreme wear on one part, small wear on all parts, and ruptured diaphragm—are brought about by usage, while dirty and/or poor fuel is usually the cause of valve trouble.

WHEN PRESSURE IS HIGH

High pressure is caused by a tight diaphragm, fuel between diaphragm layers, diaphragm spring too strong, pump link frozen to rocker arm.

A tight diaphragm will stretch slightly on the down stroke. As the pump operates, the diaphragm will rebound on the up stroke beyond its normal position, much as a stretched rubber band when it is suddenly released. This rebound will cause a higher than normal pressure in the pump chamber.

A loose diaphragm retainer nut or poor riveting on the diaphragm assembly may allow fuel to seep between the diaphragm layers. This will cause a bulge in the diaphragm and have the same effect as a diaphragm that is too tight.

A diaphragm spring that is too strong also causes a high pressure, for the diaphragm will operate longer before pressure of the fuel on the diaphragm will overcome the diaphragm spring.

On a combination pump, Fig. 223, there are

times when the operating parts may become badly corroded and the links freeze to the rocker arm. In this condition the pump operates continually, resulting in a very high pressure and a flooding carburetor.

The remedy for all these conditions is to remove the pump for replacement or repair, using a repair kit.

WHEN CAPACITY IS LOW

A pump is extra efficient and will never starve the engine when it supplies fuel equal to or above the capacity of the pump. The pressure, of course, must be within specifications.

Low capacity is usually caused by an air leak in the intake pipe at these points: fuel pipe fitting at pump, bowl flange or diaphragm flange, fuel bowl. (It is assumed that the conditions of too little fuel have already been checked and the fuel pump is the cause of the difficulty.)

An air leak at fuel pipe fittings indicates either poor installation of pump or a defective fitting. The fitting should be tightened or replaced.

A leak at the diaphragm flange may be caused by a warped cover casting, loose diaphragm cover screws or foreign material between cover casting and diaphragm.

A leak at the bowl flange of the cover casting can usually be corrected by the installation of an extra gasket. A warped top cover indicates that the pump must be replaced.

A chipped glass or bent metal bowl may cause a leak at the bowl flange as may a defective gasket or foreign material between gasket and bowl or cover casting. A chipped glass bowl must be replaced while a dented metal bowl can be straightened.

VACUUM PUMP TROUBLES

To assist the manifold vacuum to operate the windshield wiper at a uniform rate under any engine load is the only function of the vacuum pump, Fig. 223. Of course, "uniform rate" infers a wet windshield glass and not one covered with snow or ice. Failure to do the above indicates difficulty in either the vacuum system of the pump, windshield wiper motor, or tubes and connections.

Symptoms of trouble in the vacuum section show up in four ways: oil consumption, slow windshield wiper action, poor idle, noise.

In some cases it has been found that an engine which has given very good oil mileage suddenly appears to be using oil. Upon investigation, it will often be found that the vacuum booster has a ruptured diaphragm and is drawing oil fumes from the crankcase into the intake manifold. This can be checked by removing the cover of the vacuum section.

When the windshield wiper slows down excessively under engine load, it usually is an indication of a ruptured vacuum diaphragm or defective valve action in the vacuum pump. This condition may not be discovered immediately as the windshield wipers may not be used for long intervals.

Oil will be evidenced in the cover casting recesses if the diaphragm is ruptured. The pump should be removed and the diaphragm replaced.

On some cars the engine will idle very poorly when the vacuum diaphragm is ruptured. This is true when the tube from the vacuum pump is connected to one end of the intake manifold. The air leak through the vacuum section will give those cylinders on the end a lean mixture which results in a miss or poor idle. In many cases, this leads one to believe the valves of the engine are sticking; but if they are ground, the miss or poor idle remains.

This condition can be checked by removing the vacuum pump tube at the manifold and plugging the hole. If the miss or rough idle disappears, the trouble is an air leak through the pump or tube connections. The pump should be removed and repaired or the connections tightened to eliminate this trouble.

VACUUM PUMP TEST

With a combination fuel and vacuum pump the windshield wiper should operate at 80 to 100 strokes per minute through all ranges of car speed and load. The windshield should be wet when the test is made, otherwise the action will be slow.

Checking with Vacuum Gauge

To check the vacuum section, disconnect both inlet and outlet tubes and attach a vacuum gauge to the inlet (side that goes to windshield wiper). It is assumed that the engine, windshield wiper motor and blade, and connecting tubing have been checked and are in satisfactory condition.

Read the vacuum gauge when the engine is running at 1000 rpm (about 20 mph). It should read from 7 to 12 inches of vacuum on a normal pump. If the reading is below this figure, the pump should be removed and repaired or replaced. When making this test, the tube to the manifold should be plugged and the pump outlet should always be open or damage may result to the mechanism.

Checking Without Gauge

Disconnect the outlet tube (to manifold) from the pump and plug the end. Then operate the engine from an idle through slow acceleration to about 40 mph. If the wiper starts operating at about 15 mph and reaches full speed at about 40 mph, the vacuum section is okay. If it does not operate, it may be the windshield wiper motor. This can be checked by connecting the intake manifold directly to the windshield wiper tube. Then slowly accelerate the engine from idle to about 25 mph. The wiper should operate at full speed. If it does not it can be assumed the wiper motor or tubing is defective.

FUEL PUMP OVERHAUL

Although there have been many types of fuel pumps produced, it is not necessary to know all of the details of every type in order to service them. Therefore, we will describe the procedure for overhauling a popular type fuel pump which pumps gasoline only, and a combination fuel and vacuum pump, which, in addition to pumping fuel, also assists in the operation of the windshield wiper as already mentioned.

For each make and model of pump, a repair kit is available which contains all the parts needed to obtain a thoroughly satisfactory job. The parts include everything except the body and cover castings.

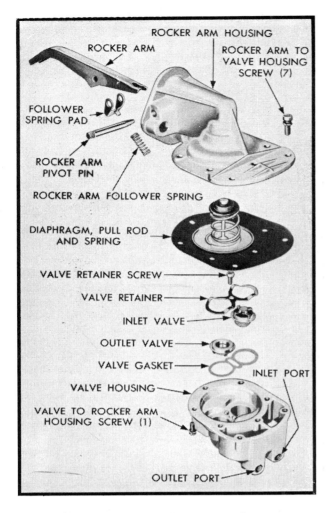

Fig. 224 *Exploded view of fuel pump*

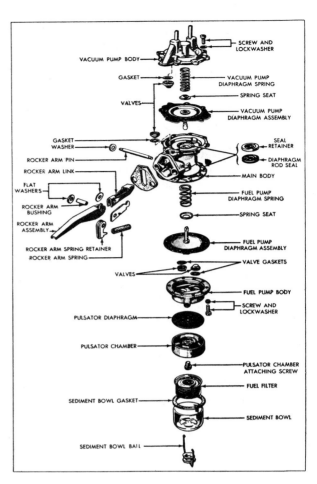

Fig. 225 *Exploded view of fuel and vacuum pump*

Servicing Fuel Pump

An exploded view of a typical fuel pump is illustrated in Fig. 224. All parts of the pump except the diaphragm should be thoroughly cleaned in a suitable solvent. Gum deposits can be removed with denatured alcohol. Examine diaphragm for cracks, breaks or torn screw holes. Inspect push rod oil seal for wear, breaks or tears and examine the valves for proper sealing.

To disassemble, remove the pivot pin in order to disengage the rocker arm from the push rod. Remove the body screws and separate the rocker arm housing from the valve housing. Valves are held in position by a spring retainer.

When installing valves, always use a new gasket. Tighten retainer securely. Install the diaphragm into the rocker arm housing. Tie the follower spring down on the arm and install the arm into the housing, engaging the push rod. Install pivot pin and remove the material used to tie down the follower spring.

Assemble the rocker arm housing to the valve housing but do not tighten the screws. Hold the diaphragm in fully compressed position and then tighten the screws. This will prevent tearing the diaphragm.

Combination Pump Disassemble

Fig. 225 shows an exploded view of a late model fuel and vacuum pump, which is fairly representative of all types of combination pumps. To disassemble, proceed as follows:

1. Loosen retainer nut on bail to release sediment bowl. Remove bowl, gasket, filter, and bail.
2. Scribe a mark with a file on the main body, fuel pump body and pulsator chamber so that these parts may be reassembled in their original position.
3. Remove one screw and lock washer retaining the pulsator chamber to the fuel pump body.

Then remove the pulsator chamber and diaphragm.

4. Remove the screws and washers that hold the fuel pump body to the main body. Remove fuel pump body.

5. Remove staking marks around valves and pry valves out of fuel pump body with screwdriver.

6. Press the diaphragm in and twist to unhook the rod from the rocker arm link. If the rod is not readily removed by this method, it may be unhooked by pushing against it with a blunt punch inserted through the rocker arm opening.

7. Remove one attaching screw on each side of the vacuum pump body and replace each with two long (1½″) screws. These screws will serve as pilots and also relieve spring tension.

8. Remove remaining attaching screws and lockwashers, then back off alternately on each pilot screw until all spring tension is relieved. Then remove pilot screws and vacuum body.

9. Remove diaphragm return spring and spring seat.

10. Remove vacuum pump diaphragm, using same procedure as when removing fuel pump diaphragm.

11. Remove staking marks around valves. Pry valves from main body and vacuum body with screwdriver.

12. File or grind peened end of rocker arm pin flush with washer.

13. Drive pin from main body with blunt punch.

14. Remove rocker arm assembly.

15. Remove staking marks and pry vacuum and fuel diaphragm rod seals and retainers from main body.

16. Smooth all seal and valve counterbores in main body, fuel pump body, and vacuum pump body with scraper. This will remove metal projections.

Combination Pump Assemble

1. After cleaning all parts and obtaining the repair kit, install the fuel and vacuum diaphragm rod seals and retainers with a suitable driver and stake the body to secure retainers and seals in place.

2. Place gaskets in position and install both fuel pump outlet and inlet valves. Stake valves securely in place.

3. Install new valves and gaskets in main body and vacuum booster pump body. Stake all valves in vacuum pump body. Stake all valves except inlet valve in main body. Inlet valve is located adjacent to pump mounting flange. A very thin coat of sealing compound applied to the edges of the valve cage and rod seal in the vacuum booster section will help prevent oil leakage.

4. Assemble rocker arm.

5. Place rocker arm spring in position and insert arm assembly into main body. Hold in place with a thin punch.

6. Hold pump body so that fuel side is facing down. Jar the body so that diaphragm operating link will fall toward the fuel pump side.

7. Insert fuel diaphragm rod, with spring and seat in place, through the main body seal. Hook rod over end of rocker arm link. Lubricate diaphragm rod lightly with cup grease before installing.

8. Using a new rocker arm pin, drive the holding punch (mentioned above) out of the pump body. Install pin retaining washer and peen pin.

9. Place fuel pump body in position on main body, aligning the scribe marks made prior to disassembly. Make sure that the diaphragm is in the proper position. Start all screws, making sure they pass through the holes in the diaphragm without tearing the fabric. Turn all screws in evenly.

10. Hold main body so that vacuum side faces down. Jar the body to cause rocker arm diaphragm operating links to fall toward vacuum pump side.

11. Insert vacuum diaphragm rod through the main body seal, tilting rod slightly away from rocker arm links. Hook rod over both link arms. Lubricate rod lightly with cup grease before installing.

12. Place spring seat and diaphragm return spring on diaphragm.

13. Drop vacuum booster pump body over the spring, and align the scribe marks made prior to disassembly.

14. Install the long (1½″) pilot screws and turn down alternately until the shorter attaching screws can be engaged. Remove pilot screws and install two remaining attaching screws.

15. Tighten the attaching screws alternately across the booster pump body.

16. Align scribe marks and install pulsator diaphragm and chamber. Secure with screw and lockwasher.

17. Install fuel strainer and new bowl gasket in pulsator chamber.

18. Replace bail, position bowl and tighten bail nut finger tight.

19. Install pump on engine, using a new gasket,

TYPICAL V8 ENGINE HEAT BALANCE

Heat Energy	Wide Open Throttle	Cruising
Exhaust loss	35% to 50%	30% to 50%
Cooling system loss	20% to 25%	25% to 30%
Friction loss	5% to 8%	10% to 15%
Useful work (Brake HP)	25% to 30%	15% to 25%

Fig. 226   Typical "Heat Balance" for a V8 engine

and alternately tighten attaching screws to prevent distortion.

20. Connect fuel and vacuum pipes.
21. Test fuel and vacuum booster pump as outlined previously.

PERFORMANCE TIPS

Obtaining added performance by modifying the fuel and exhaust systems usually involves other changes to the engine as well. Simply changing fuel or exhaust system components without simultaneously making other engine alterations severely restricts the performance increase available. Thus, it is common that in installing multiple carburetors the engine will also be equipped with cylinder heads with a higher compression ratio as well as a special camshaft. It is the combination of the three that provides a performance increase out of proportion to the individual contribution of multiple carburetion, a racing cam or boosted compression.

The art of engine modification to increase output is so advanced that literally tens, if not a hundred, of different combinations are available for every production engine. It would be impossible to list all such combinations here so basics will be discussed instead.

Before going into details, several precautions should be mentioned. If an engine is to be modified it must be in excellent mechanical condition and be of a good design. Most American production engines are of such design and can be safely modified provided they are sound mechanically. Bad rings, sloppy pistons, burned valves, etc., not only detract greatly from any performance gains that might be achieved they also physically endanger the safe running of the engine. The other warning is made to conserve your own time and energy. It is far easier, faster and safer to install high performance equipment obtained from the car manufacturers themselves or from a specialty producer of speed equipment than it is to try and build it yourself.

Basic Concepts

A look at a typical "heat balance" chart, Fig. 226, shows what an inefficient machine an internal combustion engine is. The heat balance simply shows what percentage of the energy available from the burning of the fuel is turned into useful work, how much is lost as heat to the exhaust and cooling systems, what quantity is lost to friction in turning the engine over, etc. If any of these losses can be reduced, engine performance will be increased by a corresponding amount. It is as simple as that.

Reducing the losses, however, is not a simple matter. All production engines represent compromises in design in order to make them reliable, quiet, tractible, and reasonably economical under a wide variety of driving conditions. When modifying such an engine, a small sacrifice in reliability, for example, will be tolerated by the builder in return for a performance increase. Similarly, a somewhat louder car is not objectionable to the driver in return for the livelier performance available with straight-through mufflers. One is, in effect, almost invariably trading off—compromising—some element of engine operation in order to get the increased performance.

Provided one is willing to sacrifice something for it, reducing heat losses becomes another matter. How actually is it done? Engineers speak of the three basic elements of internal combustion engine performance as: 1. Thermal efficiency; 2. Mechanical efficiency and; 3. Volumetric efficiency. Increasing any one of them improves performance and is accompanied by a corresponding rise in useful work output on the heat balance (there are exceptions to this rule and they will be remarked as they come up).

Thermal efficiency is concerned with the actual burning of the fuel and is traditionally improved by a variety of techniques including changing valve timing and lift, increasing the compression ratio and other innovations. Few changes in the fuel and exhaust systems will affect thermal efficiency so we are not concerned with it except as noted later. Mechanical efficiency is an expression of the amount of friction developed in the engine as a result of simply running it as a machine. It is generally quite high and can safely be ignored as a significant approach to higher outputs.

Volumetric efficiency is the direct result of how an engine "breathes" and can be much enhanced by fuel and exhaust system modification. Other factors enter into the question of engine breathing such as valve size, lift and timing, but many aspects of fuel and exhaust system design affect it profoundly. Supercharging, multiple carburetors, fuel injection, intake and exhaust "tuning", manifold and port design, are but several conventional routes to increased performance via increasing engine volumetric efficiency.

Ports

Intake and exhaust port size and shape are very important in letting an engine breathe properly. While much has been written on the subject, there is still much confusion between theory and practice.

In the forefront of any discussion is the question of port shape. Many people labor under the illusion that round ports are superior to rectangular ones. This is decidedly not the case and since many modern V8 engines have square or rectangular ports, long, weary hours with a rotary file will be saved if this fact is remembered. Actually, it is not whether a flat sided port is better or worse than a circular one but, rather, which of them is better in a given situation? Even this point is the subject of much research effort by the car companies and a definitive answer is not available.

The safest course, then, is to leave the ports the same shape they were originally. Apart from any other consideration, there is always the point that there may not be enough metal in the port area of the cylinder head casting to permit changing their shape.

This must also be taken into account before attempting to enlarge the ports substantially. Ideally, port area on both intake and exhaust should be equal to the area between, respectively, intake and exhaust valves and their seats when the valves are fully open. This presumes a separate port for each valve. Modern engine design is good in this respect and only when high lift camshafts are used are the ports usually enlarged.

The most important use of port alteration is in the matter of matching the manifold ports with the cylinder head ports. They should mate as close to perfectly as possible. Checking such alignment requires an extremely accurate template or one of the machinists' techniques such as the use of prussian blue on one of the surfaces. Intake and exhaust manifold gaskets are generally not suitable to use as templates because they are deliberately made somewhat oversize to take care of some mismatching in port size and location.

A final point and the most hotly debated of all, How much good does polishing the ports do? The theory is that polishing provides better gas flow. In fact, flow velocity is very much slower at the sides of the port than it is even a little way in

toward the center so it doesn't make that much difference if the port walls are a trifle rough. What is very important, though, is to remove any large projections in the ports—thicker, say, than 0.010 in.—such as casting marks because they do obstruct gas flow.

Carburetors

It is almost as bad for performance to have too many carburetors as too few. Improving volumetric efficiency by providing more carburetor throat are—either by adding carburetors or using ones with larger throats—is a good idea but can easily be overdone.

Two opposing factors must be considered when engine carburetion is changed. First, it is desirable to provide as much carburetor throat area as possible to reduce restriction in the intake system and improve engine breathing. But, at the same time throat area must not be so large that poor atomization of the fuel results which cuts down on the engine's thermal efficiency. It is the question of air velocity through the carburetors that is critical.

The larger the carburetor throat area, the lower will be the air velocity. Air requirements change with engine speed so at low RPM, when much less air is needed, velocity is less to begin with. Velocity is about 10 times higher at maximum RPM than it is at low engine speeds. Therefore, it is obvious that a compromise must be made to assure good velocity at low speed while simultaneously providing enough throat area to prevent strangling the engine at high speed.

One method of obtaining these conflicting goals is to use carburetors with primary and secondary barrels. The primaries open at all RPM's, but the secondaries only do so during periods of peak air demands.

A truly staggering variety of special, multiple carburetion systems are available from the car and specialty manufacturers. When buying one—making your own for a V8 is almost unthinkable because of the design and construction complexities involved—the above principles should be kept in mind.

Fuel Injection

Injecting fuel directly into the cylinder in diesel engine fashion or injecting it into the port overcomes the compromise necessary with carbureted engines, described in the previous section. A separate injection pump and suitable nozzles or injectors assure good atomization of the fuel regard-

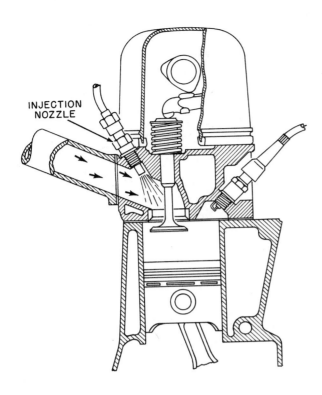

Fig. 227   Port injection system

less of the air velocity through the intake system. An engine with fuel injection, therefore, can use "wide open" intake pipes that offer no restriction to air flow into the engine. Result, higher (close to maximum) volumetric efficiency.

Of the two systems, the direct cylinder injection method, Fig. 228, is preferred, though generally more expensive, than port injection, Fig. 227. This is because it dispenses with even the slight restriction of port fuel nozzles and because better atomization of the fuel is obtained under all engine speeds. Injection of fuel directly into the combustion chamber has other advantages as well. Both systems, while appearing simple, are quite complex and require very accurate pumps and metering devices. On a well designed cylinder injection engine, output can go 20% higher than that available with carburetors.

Supercharging

Supercharging can easily raise volumetric efficiency over 100%. It is not difficult to see how. If an engine cylinder at a certain speed—volumetric efficiency decreases with increasing RPM on an unaspirated (unsupercharged) engine—is only

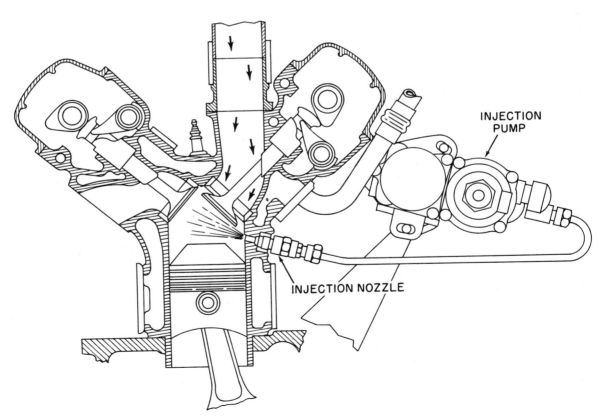

Fig. 228   Direct cylinder fuel injection system

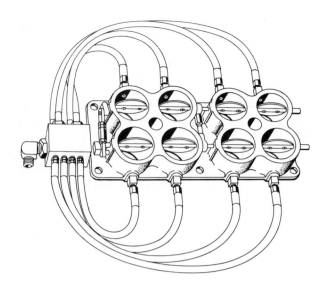

Fig. 229   Bolt on port injection

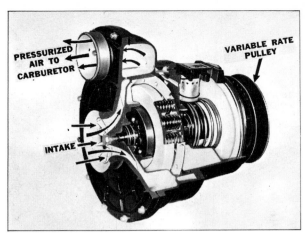

Fig. 230   Centrifugal blower

¾ filled by the air fuel-charge, the engine has a volumetric efficiency of 75% at that speed. Being an air pump, the supercharger can produce pressures above those resulting from the weight of the atmosphere and can completely fill the cylinder regardless of engine speed. In fact, as long as supercharger pressure exceeds atmospheric pressure—14.7 lb. at sea level—the cylinder will receive more charge than an engine relying on normal breathing possibly could and volumetric efficiency goes over 100%.

The amount of over-pressure a supercharger produces is called the "boost" and an engine in top mechanical condition will handily sustain a 5 to 6 lb. boost with very little sacrifice in engine life. With this amount of boost such an engine should put out 30% to 40% more power without other modifications. Engines have been built that will accommodate boost pressures of up to around 50 lb. and in return have had horsepower increased to 3 and 4 times that which a corresponding engine without a supercharger would produce.

For high boost pressures, the engine must be modified extensively. Because supercharging is such a relatively easy route to high performance there is a tendency to overdo it. Supercharging does put greater stresses on bearings and reciprocating parts that increase rapidly with boost pressure and the sad end of many a "blown" engine has been a blown engine.

There are two general types of blowers and three methods of driving them. Centrifugal, Fig. 230, and radial superchargers are, in effect, fans; not being, therefore, positive displacement pumps, air can slip past the moving vanes at lower speeds. Positive displacement superchargers include both

Roots, Fig. 231, and vane types and they do pump a fixed volume of air which is dependent on their rotating speed. In general, Roots blowers—the GMC diesel engine variety are great favorites for automotive adaption—will provide positive boosts from low engine RPM on up while centrifugal types must be revved up to produce pressures above atmospheric.

The most common method of driving a supercharger is to run it off the engine by using V-belts and pulleys. In so doing, power is lost in driving the blower and positive displacement blowers, matched in air output capacity to the engine's displacement, can easily eat up 30 to 40 HP on a big V8. Power consumption by such a super-

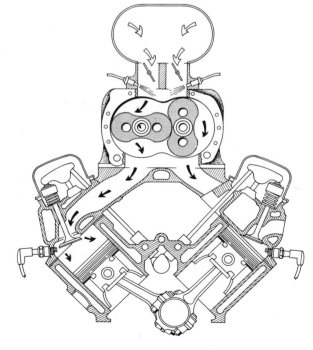

Fig. 231   Roots supercharger

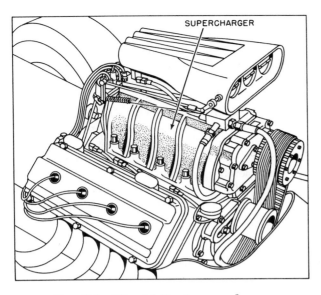

Fig. 232 *V8 with Roots supercharger*

charger is also, of course, a matter of boost pressure; the higher the boost, the higher the power requirements.

An alternate method is to drive the blower by an outside power source such as a separate small engine or an electric motor. Such installations are obviously unsuited for ordinary street application and find use in all-out, short time competitive events.

Most intriguing drive system of all is the turbo-blower. Adopted from diesel engine work, the exhaust-driven blower requires no mechanical drive from the engine and is run by the engine's own expanding exhaust gases. With a turbine wheel, like a gas turbine engine, the blower does not consume engine power during operation. Referring back to the heat balance, it utilizes some of the heat normally lost to the atmosphere through the exhaust. In this sense it provides a "free" supercharge.

A centrifugal blower is always linked to the turbine wheel of a turbo-blower because the exhaust gases do not provide enough energy to drive a positive displacement pump at low RPM. Consequently, positive boost pressures are only available at higher engine RPM's.

Exhaust System

There is a major misconception about high performance exhaust system design. It is that great performance gains can be realized simply by relieving the back pressure caused by conventional mufflers, resonators and connecting plumbing. This is not true. Relieving back pressure completely by substituting carefully made headers

terminating in wide open pipes without mufflers will only increase maximum output by 5% to 7%.

This presumes that a "tuned" exhaust system is not used. Such a system, see next section, will produce appreciably higher gains.

If, however, you decide to go for the smaller gain alone, the rules are simple. The easiest method of all is to simply install "straight-through" mufflers, described earlier in the chapter, in place of the conventional mufflers and resonators. If a V8 is being modified, there should be separate exhaust systems for each bank of cylinders . . . i.e., a dual exhaust system.

The next step is to replace the cast iron, stock exhaust manifolds with headers—specially fabricated manifolds with longer, more gently curved pipes leading to the individual cylinders. A V8 normally uses a single header for each bank of four cylinders and these are usually adequate. In general, the fewer the number of cylinders discharging into a common exhaust system the better. Hence, 6 cylinder engines often have two separate systems of 3 cylinders each or even three systems of two cylinders each. When multiple manifolds are used, it is best to design them so that two successive cylinders in the firing order do not discharge into the same system. On a V8 or a Corvair 6 this is automatically taken care of by having individual headers for each bank since on either engine the firing order jumps from one bank to the other alternately.

The important things to remember in buying or building an exhaust system are: 1. Be sure header ports match the cylinder head ports in size and location (see section, *Ports*); 2. There are no abrupt changes of tubing diameter leading from one part of the system to the other and; 3. All bends in the system should be gradual, not abrupt.

Intake and Exhaust System Tuning

Intake and exhaust tuning exploit a strange phenomenon that is still imperfectly understood. In so doing, they can provide performance gains that are appreciable, rivaling those of a super-charger supplying a modest boost. But, in providing this gain, intake and exhaust tuning operate in the realm of conventional thermodynamic processes. Horsepower increases are strictly attributable to reducing energy losses out the exhaust and by increasing volumetric efficiency (see Heat Balance, Fig. 226), not by some kind of "black magic".

One is apt to think of a moving column of gas, be it the air-fuel mixture in the intake or the burnt gases in the exhaust, as a simple flow situation like water moving in a garden hose. There is more to it than that and gas flow is one of the most complicated studies one can get involved in. Fortunately, we are concerned here with only one other element of gas movement beside its simple flow—the wave motion.

When the exhaust valve opens at the end of the power stroke the pressure in the cylinder is very much greater than it is in the exhaust pipe. Releasing that pressure into the exhaust pipe creates a wave motion that is superimposed on the actual flow of gas leaving the cylinder. Similarly, when the piston starts down on the intake stroke, cylinder pressure is below that of the intake system. This, in turn, creates a wave pattern in the moving intake gas; because the pressure differential is less on the intake side, the energy of its waves are less than those in the exhaust.

In both intake and exhaust, the waves generated are a variety of sound wave. That is, waves are created that move through the flowing gas by alternately compressing and expanding successive layers of gas like a fast acting accordion. Called sound waves of finite amplitude, they are unlike ordinary sound waves in that they carry a great deal more energy. In comparison to gas flow they travel at high velocities: while the exhaust gas at high RPM in a racing engine may be traveling at speeds to 500 MPH, its wave will be moving at rates around 1900 MPH through it. Similarly, the intake wave will be traveling at speeds of about 750 MPH through the air-fuel mixture which will be moving at about 150 MPH.

It should be obvious that the exhaust wave moving outward from the cylinder has an average pressure higher than the exhaust gas itself, and, hence, is a compression wave. When it reaches the end of an exhaust pipe—without mufflers or abrupt changes in diameter—it is reflected by the atmosphere back up the pipe as a rarefaction, or suction, wave. Fig. 235, shows this in diagrammatic form—notice that the wave continues to be reflected, first by the atmosphere then by the cylinder.

The intake wave, on the other hand, starts as a pulse of rarefaction and is reflected by the atmosphere as a pulse of compression.

At high engine speeds (wave amplitude—energy—increases with engine speed) a reflected wave in the intake, pulse of compression, can be at 2 pounds above atmospheric pressure. On the exhaust side, the reflected wave, a pulse of rarefaction, may be at 6 to 8 pounds below atmospheric.

How can these waves be utilized to increase engine performance? This is quite easy to understand, although there are a great many elements that enter into the calculations in actual practice. If the pulse of compression in the intake arrives at the cylinder while the intake valve is still open on the intake stroke, it will force extra fuel-air mixture into the combustion chamber and act as a "free" supercharge. That is, no engine power is lost in providing this boost and it is, therefore, free in the same sense that the boost from a turbo-blower supercharger is free.

If the pulse of rarefaction reflected in the exhaust can be made to arrive at the cylinder at the end of the exhaust stroke when *both* intake and exhaust valves are open during the overlap period, two benefits result. First, the pulse of suction helps remove the last traces of burnt gas from the cylinder. Next, it will induce a suction in the intake system via the slightly open intake valve and help draw extra fuel-air mixture into the cylinder, again providing a free supercharge.

Many factors modify the strength of the waves as well as their affect. Tending to give better results because the pulses are stronger are: Fast opening valves (steep cam profile); early opening exhaust valves (combustion pressure is still high); high compression ratios (higher combustion pressures); unrestricted ports; high RPM; intake and exhaust pipes of substantially constant diameter (sharp diameter changes reflect the pulses); proper intake and exhaust pipe diameter and length; and as few cylinders as possible using common manifolds (waves from other cylinders in the manifold cause disturbances that change and often negate the original wave pattern). Because the waves in the exhaust system carry more energy than do those in the intake, higher performance gains are had by tuning the exhaust. It is, of course, better to tune both systems than one alone.

It can be seen that arranging things so that the correct pulses in the intake and exhaust systems arrive at the cylinder at the right moments to act beneficially is a question of timing. Timing is accomplished by making the two systems the right lengths so as to assure correct pulse arrival at a certain RPM.

The above statement exposes the major drawback of intake and exhaust tuning. Tuned systems only work at certain engine RPM's. At other engine speeds they either do not work or work detrimentally. Detrimental affects are reduced by tuning the system for high RPM's so that negative results are confined to low speeds where pulse strength is much lower.

It is readily apparent why tuning only produces

PROPERTIES OF FUELS

Fuel	Octane Rating	Calorific Value CHU/Lb.	Heat Release CHU/Lb. air	Air/Fuel Ratio Max. power
Gasoline	80	10,500	875	12.5
Methyl Alcohol	100	4,760	865	5.5
Ethyl Alcohol	100	6,400	855	7.5
Ether	Low	8,160	815	10.0
Acetone	100	6,830	718	9.5
Nitrobenzene	—	6,000	750	8.0
Nitromethane	—	1,890	945	2.0

Fig. 233   Properties of fuels

gains over a certain band of engine RPM's. Wave speed remains fairly constant while engine speed varies greatly. Thus, a tuned exhaust system of a calculated length will assure correct pulse arrival at one engine speed but if the engine is running slower the pulse will arrive too early to be of benefit. While various methods have been suggested to

$$L = \frac{TV}{N}$$

eliminate this drawback—adjustable length systems, varying pulse speed by changing the density of the gas, etc.—none have proved too successful.

Use of a megaphone (diverging nozzle), Fig. 236, at the end of the exhaust system increases pulse amplitude and widens the band of engine speeds at which a tuned exhaust system produces performance gains. Use of bell-mouth intake pipes, Fig. 237, accomplishes the same result on the intake side. With both installed, tuning can be made to provide performance increases over a 1000 to 1500 band of engine RPM's.

How are the length calculations made? On an engine having a separate exhaust pipe and intake pipe for each cylinder, of constant diameter, the

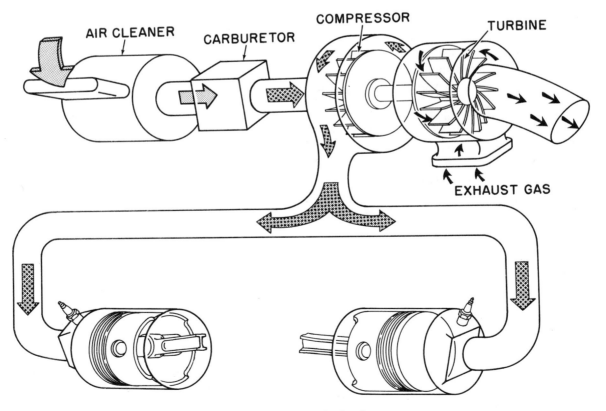

Fig. 234   Corvair Turbocharger

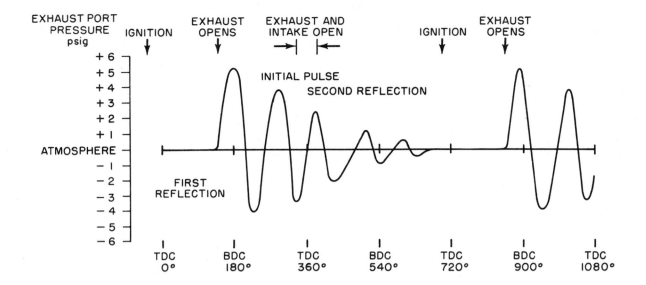

Fig. 235   Intake and exhaust tuning wave diagram

following formula will assure the correctly timed arrival of a beneficial wave at the desired engine RPM.

$$L = \frac{TV}{N}$$

L = Length of intake or exhaust system in inches from the valve to the pipe end (if a megaphone is used the measurement is made to the beginning of the megaphone).

N = Engine speed in revolutions per minute at which performance increase is desired. The specific RPM chosen should lie in the middle of the RPM band over which tuning benefits will apply.

V = Average velocity of the wave in feet per second. For the intake this is the speed of sound in air—1088 Ft. per second at sea level. For the exhaust this varies from about 1300 to 1800 Ft. per second—depending on engine speed, throttle opening and load on engine, being at a maximum at high RPM, with a load and wide open throttle.

T = Time expressed in degrees of crankshaft ro-

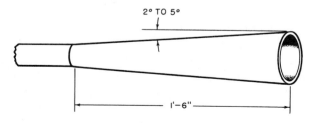

Fig. 236   Exhaust megaphone

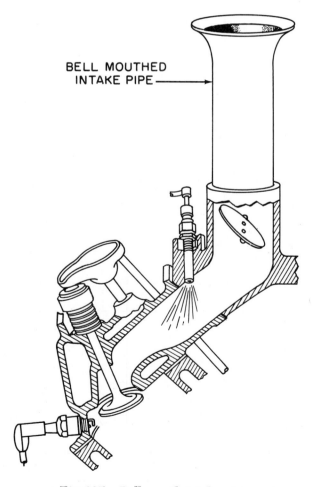

Fig. 237   Bell mouth intake pipe

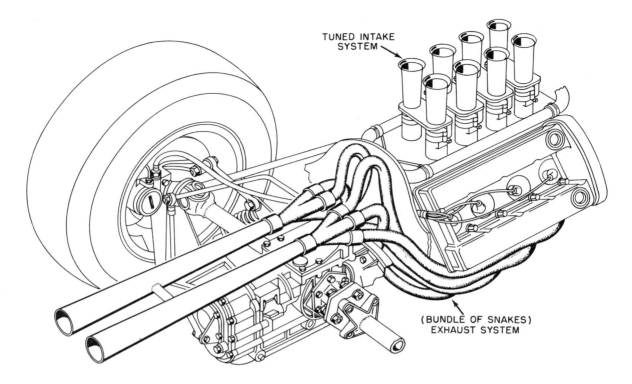

TUNED INTAKE
SYSTEM

(BUNDLE OF SNAKES)
EXHAUST SYSTEM

Fig. 238   Tuned intake and exhaust system

tation it takes for a wave to travel down the pipe, to be reflected at the open end and return to the cylinder. This is the point in the cycle that it is desired to have the wave arrive. On the exhaust, T is usually chosen as the number of degrees between the opening of the exhaust valve and the middle of the overlap period when intake and exhaust valves are open together. On the intake, T is usually chosen as the number of degrees between the beginning of the intake stroke and a point just before the intake valve closes, around 180 degrees.

It is obvious that tuning will work on systems having more than a single cylinder firing into the tuned exhaust pipe, as witness the "bundle of snakes" exhaust systems used on Grands Prix and other V8 race car engines. These engines have four cylinders in each tuned system and the calculations differ from those given above. Tuned intake systems always have one intake pipe for each cylinder, as for example on a fuel injection engine where such a provision is readily available. A word of caution, so called "tuned systems" available from some speed equipment manufacturers are, in fact, not tuned systems at all. Measuring and judging commercially available systems by the information given in this section should help

guide you in determining whether the system really is a tuned one or not.

Fuels

Changing the fuel the engine burns, of course, does not directly affect, and is little effected by, volumetric efficiency. Many different fuels have been experimented with and a few—either in the pure state or mixed with other fuels—have very great performance potential.

Basically, there are only a few relationships that must be understood to appreciate the potential of different fuels. At the outset, the calorific content—amount of heat released in burning—is of primary importance. A fuel of high calorific content will provide more power from a given engine than will a fuel with a lower value . . . if other things are equal.

They never are equal, so other things have to be taken into consideration. One is, How much air must be combined with the fuel for burning? If the calorific content of the fuel is low, such as it is with alcohol, but less air is required to burn it, a given engine may do better with such a fuel. Fig. 233 gives the calorific content of several possible fuels together with the weight of air required to combust them. The significant comparative figures up to this point are shown in the column, Heat Release per Lb. of air.

It can be seen that both ethyl and methyl alcohol compare favorably with gasoline on this basis although both have much lower calorific values. The only major difference between them being that much more alcohol would be required than gasoline and this is, of course, a minor consideration if all-out performance is of primary importance. Because much more alcohol is needed, carburetors set up for it as a fuel would have to have larger jets.

Several other factors have to be taken into consideration in selecting a fuel for maximum performance. One is, How high a compression ratio can be tolerated with a certain fuel? In general, the higher the octane rating of the fuel, shown in the first column in Fig. 233, the higher will be the permissible compression ratio. This is a

factor that makes alcohol so attractive for racing— higher ratios can be used with it than with most gasolines.

A final element in fuel selection is what makes the "nitros" the fuels of choice in some situations. If the fuel carries its own oxygen with it—within its own molecular structure—the engine builder can get higher outputs per pound of fuel burned. This is exactly what nitromethane and nitrobenzene do.

This explanation is necessarily over simplified and before anyone begins to experiment with fuels other than gasoline he must inform himself thoroughly on the subject. Forgetting the time and expense which might be lost by setting the engine up incorrectly, there is the matter of personal safety. In addition to other dangers, several of the fuels listed are both highly volatile and extremely poisonous.

Review Questions

		Page
1.	Name the components of the fuel system.	367
2.	How is fuel fed to the carburetor on all moden engines?	367
3.	What is the purpose of baffle plates in a fuel tank?	367
4.	Where is the fuel filter most commonly located?	368
5.	How does water accumulate in a fuel tank?	368
6.	What part of the engine drives a mechanical fuel pump?	368
7.	Name the part of a fuel pump which causes fuel to flow in the fuel lines.	369
8.	On what part of the fuel pump does fuel pressure depend?	369
9.	What type of fuel pump assists in operating the windshield wipers?	371
10.	What two conditions must a mechanical fuel pump maintain in order to satisfy the fuel requirements of an engine under all operating conditions?	371
11.	Besides filtering air, what two important functions does an air cleaner fulfill?	372
12.	Where is a flexible fuel line usually located?	373
13.	What is a carburetor?	373
14.	What is an intake manifold?	373
15.	Explain how the manifold heat control operates.	375
16.	If an engine were run with the exhaust pipe disconnected, the exhaust would be very noisy. Why is this so?	376
17.	How does a muffler quiet the noise of exhaust?	376
18.	What is one undesirable feature of any exhaust system?	376

19. What is the main advantage of having dual exhausts on a V8 engine? 377

20. Engine fuel is composed principally of what two elements? 378

21. What are the main advantages of gasoline as an engine fuel? 378

22. The burning of a fuel in the presence of oxygen is a process referred to by what name? .. 378

23. What element in the air-fuel mixture is not affected chemically by combustion? 378

24. Octane rating refers to the ability of gasoline to resist what objectionable process? 379

25. What is the ideal air-fuel ratio for an automobile engine? 380

26. What three conditions do the intake manifold and carburetor create that result in good vaporization of the fuel? 380

27. In order for a carburetor to deliver fuel to an engine under all operating conditions, it must have what seven features? 382

28. What does the throttle valve control? 382

29. Explain why it is necessary to have an accelerating system for satisfactory engine operation. 384

30. Explain how venturi action feeds gas from the carburetor bowl into the intake manifold. 385

31. What kind of spring operates any type of automatic choke? 385

32. What type of carburetor design overcomes the effect of air resistance in an air cleaner? .. 386

33. What is the function of an air bleed in a carburetor? 387

34. How does altitude affect carburetion with respect to jet size? 388

35. For what reason does a four barrel carburetor give better engine performance than a two barrel carburetor? 389

36. What is an anti-percolator device on a carburetor? 389

37. What is the main difference between the primary and secondary side of a Carter WCFB Carburetor? 391

38. Name three types of air cleaners. 421

39. Explain how a paper element type air cleaner may be cleaned. 421

40. What two basic types of automatic chokes are currently being used? 423

41. In the process of adjusting the carburetor idle, usually two separate adjustments have to be made. What are they? 424

42. When adjusting the idle on cars with automatic drive, which adjustment requires special attention? 424

43. While loosening or tightening a fuel fitting, why is it necessary to use two wrenches? 426

44. What type of fuel line has to be handled with special care while bending? 426

45. What might be the two results of a clogged tank vent? 429

46. What points should be checked when an engine won't start due to a possible defect in the fuel system? 431

47. Describe a simple test for a restriction in the exhaust system. 434

48. Before deciding that a car is giving poor mileage due to a mechanical condition, what important factor should be considered? 434

49. What two tests are needed to determine whether or not a fuel pump is serviceable? 464

50. What equipment is needed to perform a fuel pump test? 464

Fuel Injection

INDEX

ROCHESTER SYSTEM

System Components 482
Air Meter 483
Fuel Meter 483
Operating Principles 485
Starting System 485
Cold Enrichment System 486
Idle System 486
Acceleration System 486
Power System 486
Hot Idle Compensator 487

BENDIX ELECTRONIC SYSTEM

System Description 487

Fuel Delivery Subsystem 488
Air Induction Subsystem 489
Engine Sensors 491

**BOSCH D-JETRONIC &
L-JETRONIC SYSTEMS**

Operating Principles 494
Components 495
Fuel System 495
Induction System 496
Pressure Sensor With Full
    Load Enrichment 497
Temperature Sensor (Intake Air). 498

Control System 498
Electronic Control Unit 498
Correction Factors 498

**BOSCH K-JETRONIC
SYSTEM (CIS)**

Description 500
Fuel Supply 503
Compensation Units 507
Auxiliary Starting Assembly . . . 508
Construction of System
    Components 509
Electrical Circuit 510

The internal combustion engine's purpose is to produce horsepower to drive the vehicle. It does so by producing heat which, in turn, is the by-product of the engine's internal combustion of an air-fuel mixture. All gasoline-powered automotive engines must get this air and fuel (usually gasoline) into their combustion chambers.

The conventionally carbureted automotive engine depends upon the vacuum created by the downstroke of its pistons to suck the air-fuel mixture into the combustion chambers.

Generally speaking, this is a good method of getting the job done. It is a relatively simple system that works under most conditions in a satisfactory manner. However, the carburetor is not a very sophisticated device for metering fuel and while it does do the job, it does not do it in the most efficient manner possible. In fact, the imprecise nature of the carbureted intake system leads to wasted fuel and an overabundance of exhaust emissions. In fact, in the areas of exhaust emission control and high performance, the carburetor has clearly been eclipsed by a more precise fuel metering system which is fuel injection.

The fuel injected engine utilizes that same vacuum condition to suck in ambient air. But the fuel is squirted either into a modified intake "manifold," or directly into each cylinder head intake port, or through a fuel injection nozzle directly into the combustion chamber.

Thus, though fuel injected engines do depend on engine vacuum to suck air into the cylinders, the fuel is delivered via a pump through the injection system and is non-dependent on engine vacuum. This allows precise fuel metering to each cylinder. With fuel injection, the fuel is delivered to each cylinder. This permits precise fuel distribution, vastly improved fuel use efficiency, and therefore, much better economy plus a lessened output of harmful emissions.

Fuel injection is interrelated with reduced harmful emissions because a fuel injected engine can run leaner. That is, the air-fuel ratio contains more air and less fuel than "normal." Furthermore, the air-fuel ratio can be more tightly controlled and precisely tailored to the engine's needs than can the ratio on a carbureted engine. An engine's "optimum" air-fuel ratio is the same whether the engine is carbureted or injected. The ratio requirement changes according to whether the engine is idling, accelerating, or cruising. The optimum ratio is about 9.5-to-1 at idle (for pre-1968 automobiles), whereas emission-controlled late-model engines require a 14-to-1 ratio at idle. Cruising mixtures as lean as 16-to-1 are required on emissions engines operating near sea level. Suffice it to say that, generally, 11.5-to-1 is rich (that is, less air in the ratio than "normal"), and that 15-to-1 is a lean mixture because there's more air than "normal" in the ratio, and thus less fuel.

A further advantage of fuel injection is its more even distribution of fuel which, in turn, enables fuel injected engines to meet stringent emission regulations. This is because fuel injected engines discharge their fuel right at or quite close to the intake valve and do not depend upon air velocity for air/fuel mixing in the intake manifold or combustion chamber.

By the way, these same properties are what make fuel injection systems so popular in racing engines. However, there the injection systems are tuned for rich mixtures and more engine power, rather than lean mixtures, for better emission control.

In short, fuel injection helps control the efficient burning of the air-fuel mixture. And high efficiency can work for you whether you are after performance, economy or lower emissions.

A Short History

There are presently various types of mechanical and electronically controlled fuel injection systems utilized by domestic and foreign automakers. And you may think of fuel injection as a relatively new development, yet, various manufacturers have been trying to circumvent the carburetor for more than 50 years.

Bosch began in 1912 in its experiments with fuel injection concepts primarily for increased engine power output, which, until the recent and overriding emissions requirements, was the major thrust of fuel injection development. By 1932 Bosch was ready to series-produce piston operated aircraft engines equipped with gasoline injection systems. Bosch's first automotive injection setups for series production were introduced in 1952.

The history of American fuel injected engines is integrally connected to racing, units having appeared as early as the 1940s on Indianapolis and other racing cars.

Enderle and Hilborn fuel injection units first appeared in the 1940s and became virtual necessities on modern drag racing engines, both naturally aspirated and supercharged. Today's turbocharged Indianapolis, Formula I, and drag racing engines utilize fuel injection that works in conjunction with superchargers and turbochargers.

Cadillac's use of a Bendix electronic fuel injection system in 1975 was the first employment of domestic fuel injection in an American production automobile since the famous fuel injected Corvettes passed from the scene in 1965.

The Chevrolet Corvettes employed Rochester fuel injection as an option from 1957 until 1965. The Rochester unit was unique, and originally designed for extreme cornering situations during racing and slalom events, because carburetors just could not function properly under such conditions. The unit, however, not only served diligently in racing for nine years, but also delivered excellent fuel economy and performance on the street.

With Chevrolet leading the way with production line installation of the Rochester fuel injection units on Corvettes and even high performance versions of their passenger cars, other American manufacturers began looking at fuel injection as a performance and an image booster.

Pontiac built 1500 special 1957 Bonneville convertibles with bucket seats and fuel injected engines in late 1957 as a test. Although the test seemed successful, and Pontiac announced the injection option as available for 1958, they never built any 1958 cars with injection, instead going to three 2-barrel carburetion for their top performance models.

Also, in 1958, the Rambler Rebel and the Chrysler 300D, both special high performance models, were available on a limited basis with an electronically controlled Bendix Electrojector system. However, as production line and field problems developed with these early units, they disappeared, leaving only the Corvette Rochester system, which would continue as a regular production option until 1965.

In contrast to American auto manufacturers, the imported carmakers have continued on the fuel injection development road in partnership with the Robert Bosch Corporation.

Bosch and Volkswagen cooperated in the development and 1967 introduction of Bosch's first real electronically controlled injection system, the D-Jetronic. Various model VWs have used this system ever since.

As more and more automakers came to Bosch and needed different performance characteristics from an electronic fuel injection system, Bosch evolved the L-Jetronic system that is more common now than the earlier D-Jetronic.

And when manufacturers asked for a simpler system that could accomplish the same results as the L-Jetronic but without the inherent complexities of an electronic brain controlling the system, Bosch developed the K-Jetronic which is a mechanically controlled continuous flow injection system.

At present, among U.S. manufacturers, only Cadillac utilizes a fuel injection system. It is manufactured by Bendix and is electronically controlled. However, it is expected that this situation will change in the future as more and more manufacturers seek a precise fuel metering system to control engine efficiency and emissions.

Since each system has its own operating characteristics and components, it will be helpful for you to understand how each system operates. We will not discuss the Bendix Electrojector system since it was only installed on a few cars in 1958 and is not even remembered by most students of automotive technology.

Rochester System

The Rochester fuel injection system was improved throughout its nine year life cycle. However, the basic components and operating principles remained the same throughout the nine years, Fig. 1.

Some of the improvements, Fig. 1A, are as follows:

1. Vent screen and baffle added to the fuel meter providing for a more stable fuel mixture.
2. Choke calibrations improving engine warm-up performance.
3. Solenoid controlled by-pass fuel circuit (for cold engine starting) replacing the manifold vacuum controlled, cranking signal valve circuit.

Note that some additional improvements to this system are explained elsewhere in this chapter.

Vent and Baffle

The vent screen is located behind the welsh plug in the discharge port on the spill side of the spill valve. The screen reduces vapors or bubbles being discharged into the bowl, thus stabilizing fuel mixture by providing a more solid fuel to the pump.

The vent baffle is attached to the float side of the fuel meter cover to protect the external vent screen from fuel wetting and provide a more stable idle.

Choke

The addition of a choke piston and the relocation of the choke valve stop (in the diffuser cone) to allow 10° open position allows more initial choke for initial cold start. The choke piston will open the choke valve to the 30° open position after the initial cold start.

By-Pass Fuel Circuit

When the by-pass fuel circuit is in operation, the entire output of the *engine* fuel pump is delivered through a by-pass line directly to the fuel distributor (spider) where the fuel then passes through a ball-check valve and is routed to the nozzle in each cylinder. The control solenoid for the by-pass fuel circuit will be energized to open a fuel valve and provide fuel delivery through this circuit whenever the ignition key is held in the "Start" position and the accelerator pedal is not depressed more than ⅓ of its total travel distance. A microswitch, mounted on the injector throttle linkage (tripped by depressing the accelerator pedal beyond the first third of its travel), will stop all fuel delivery through the by-pass fuel circuit.

SYSTEM COMPONENTS

(1) the *Air Meter,* which supplies a vacuum

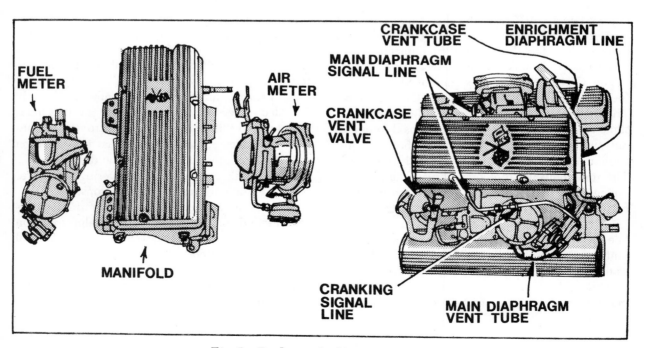

Fig. 1   Rochester fuel injection system

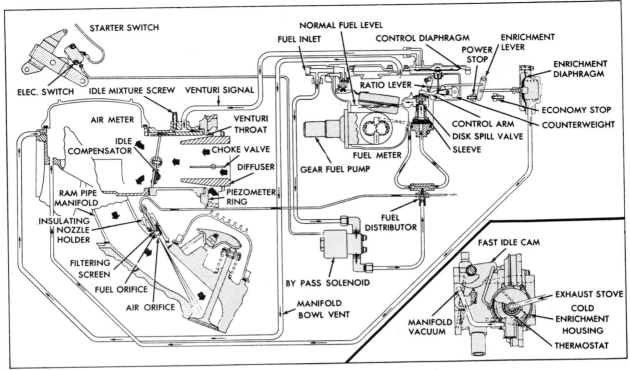

Fig. 1A *Rochester fuel injection system, fuel flow including improvements*

control signal to the fuel meter in response to engine load demand;

(2) the *Fuel Meter,* which interprets the vacuum control signal from the air meter and regulates the fuel flow to the nozzles accordingly; and

(3) the *Intake Manifold,* which provides the distribution system with the rammed air flow to the engine cylinders.

Air Meter

The air meter consists of three main parts—the throttle valve, the cold enrichment valve and diffuser cone assembly and the air meter body, Fig. 2.

The throttle valve controls the flow of air into the system and is connected mechanically to the accelerator pedal.

The diffuser cone is suspended in the bore of the air meter inlet. Its design provides a highly efficient, annular venturi between the air meter body and the cone, Fig. 3. This type of venturi produces the minimum restriction to air flow that is a vital factor in engine breathing capacity.

The air meter body houses the foregoing components plus idle and main venturi signal systems.

Main Venturi Signal

Main venturi vacuum signals are generated at

the venturi by air flowing past an annular opening formed between the air meter body and the machined piezometer ring. The signal is then transmitted thru a tube to the main control diaphragm in the fuel meter, Fig. 4.

The venturi vacuum signal (except at idle speeds) will always be a direct measure of air flowing into the engine; therefore the signal can be used to automatically control fuel air ratios to the engine cylinders.

Idle Air

Approximately 40% of the air requirement at idle enters the engine by way of the nozzle blocks from an air connection tapped into the air meter body, Fig. 5. The remaining air is controlled by adjusting throttle valve position with the positive idle stop screw located externally on the air meter. Turning the screw in increases speed and turning the screw out decreases idle speed.

Fuel Meter

The fuel meter contains a float controlled fuel reservoir very similar to that used in conventional carburetion, Fig. 6. Fuel is supplied to the fuel meter by the existing engine fuel pump. Fuel enters the meter thru a 10 micron filter, passes thru the fuel inlet valve, and spills directly into the main reservoir of the fuel meter where the high pressure wobble pump picks it up.

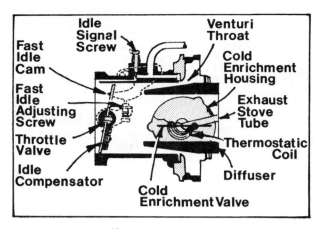

Fig. 2   Air meter

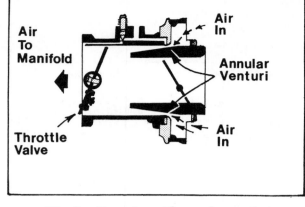

Fig. 3   Air meter with annular venturi

High Pressure Wobble Pump

The spur gear type pump has been replaced in the 1963 model with a high pressure wobble pump. The pump is located in the lower part of the fuel meter main reservoir and is completely submerged at all times, Fig. 7. The pump is powered by a flexible shaft driven by the distributor and rotates at ½ engine speed. Nominal fuel pressures vary up to 200 psi, depending on engine speed. Fuel supply not used by the engine is spilled back into the fuel meter by means of the fuel control system.

Fuel Control System

Fuel pressure (flow) from the positive displacement wobble pump must be regulated to provide the correct flow to the nozzles.

The Rochester Injection System is regulated by the amount of fuel spilled or bypassed away from the nozzle circuits.

The 1963-65 model was designed with a recirculating fuel flow from the high pressure pump to the nozzle distributor block and back to the fuel meter spill ports. A three piece valve is located in series between the wobble pump, nozzle distributor block and spill ports, Fig. 8. When a high fuel flow is required, the spill plunger or disc is moved downward, closing off spill ports to the fuel meter reservoir. The back pressure now created forces the fuel through a spring loaded check valve in the nozzle distributor block and into the nozzle feed lines. According to engine requirements, variation in back pressure is achieved by the positioning of the spill disc. With increased pressure less fuel can return to the bowl through the restricted spill ports. Correspondingly, on lower fuel requirements the spill disc is progressively raised, decreasing back pressure and allowing more fuel to return through the spill ports.

The accelerator pedal is not directly connected to the spill plunger. Fuel control is accomplished by a very precise linkage system. This linkage system is carefully counterbalanced so that the only forces acting on the system are fuel pressure

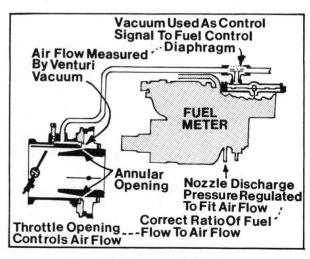

Fig. 4   Main control diaphragm

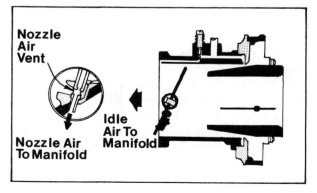

Fig. 5   Idle air system

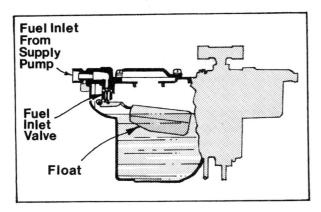

Fig. 6 Fuel meter

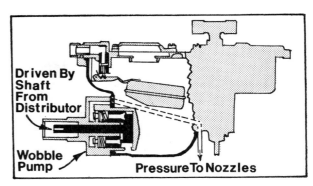

Fig. 7 High pressure wobble type fuel pump

and diaphragm vacuum. This precision balancing of the linkage makes the unit extremely sensitive to the slightest change in venturi vacuum signal on the main control diaphragm.

The control linkage is located in the fuel meter, Fig. 9. One end of the fuel control lever rests directly on the spill plunger head and controls spill plunger or disc position. The other end of the control lever is connected by a link to the main control diaphragm. The control lever pivots on another lever called the ratio lever. When the diaphragm pulls the control lever upward, the roller end pushes the spill plunger or disc downward, closing off the spill ports and thus increasing fuel flow to the nozzles. When the diaphragm allows the control lever to fall, fuel pressure forces the spill plunger or disc upward and opens the spill ports to by-pass fuel into the reservoir, reducing fuel flow to the nozzles.

Operating Principles

Starting System

Cold engine starting conditions require extra

fuel to compensate for poor fuel evaporation. Pumping the accelerator will not provide this fuel because there is no accelerator pump as in conventional carburetion. The accelerator should be depressed once and then released, allowing the throttle to be preset for starting by the fast idle cam. During cranking rpm, the signal generated at the idle needle and air meter venturi is very low and has to be boosted. This boost comes from a normally spring loaded open cranking signal valve located at the enrichment diaphragm housing, Fig. 10. The open cranking signal valve allows direct manifold *cranking vacuum to* react on and lift the main control diaphragm closing the spill valve. In addition, the enrichment diaphragm is spring loaded to hold the ratio lever at the rich or "Power" stop, Fig. 10, thus maximum fuel flow available at cranking speeds is directed to the nozzles. Immediately upon starting, or when manifold vacuum reaches 1″ Hg, the engine manifold vacuum overcomes the springs in the cranking signal valve and enrichment diaphragm, and the Fuel Injector operates on the normal idle system.

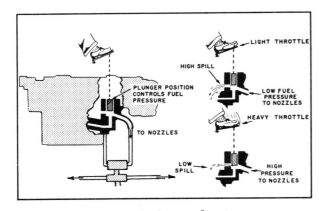

Fig. 8 Fuel control system

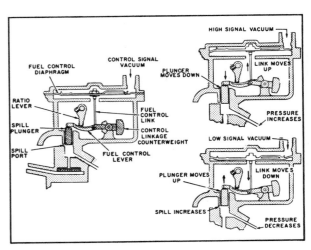

Fig. 9 Fuel control linkage

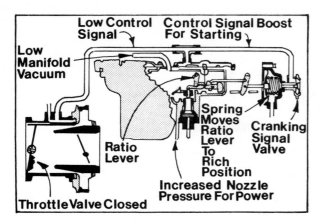

Fig. 10   Engine starting system

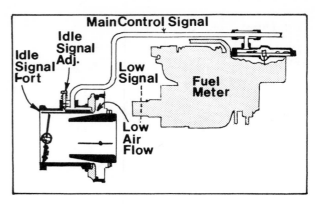

Fig. 11   Idle signal system

Cold Enrichment System

The idle system used on the latest design Fuel Injector has been greatly simplified. The newly designed diffuser cone incorporates a "choking" valve that simplifies the cold enrichment system found on early models.

To achieve a maximum signal at the control diaphragm during cold engine operation, the choke valve is held in a closed position by a thermostatic coil. This forces all air flow through the venturi thereby creating a high signal, even at relatively slow engine speeds.

As engine heat relaxes the thermostat and allows the choke valve to open, less air passes thru the venturi decreasing the generated signal.

Idle System

Fuel control during warm-engine-idle, is accomplished thru the idle circuit signal acting on the main control diaphragm, Fig. 11. The ratio lever is at the "Economy" stop. Air for combustion now enters as previously described, through the idle circuit and the vented nozzle blocks.

Acceleration

At normal driving speeds, acceleration is instantaneous. As the throttle is opened, three operations take place to provide the necessary added fuel during acceleration, Fig. 12:

1. opening the throttle valve causes an increase in air flow which increases the venturi signal at the main diaphragm,

2. the momentary reduction of manifold vacuum causes the ratio lever to move to the power stop,

3. the calibrated restriction in the main control signal circuit retains any signal left prior to the acceleration demands, thus adding to the total signal before it bleeds thru the restriction.

Power

The fuel/air mixtures necessary during power demands are similar to acceleration demands. The quickly obtained wide-open-throttle causes a reduction in manifold vacuum and the ratio lever moves to the power stop. More important is the radically increased air flow and hence venturi signal directed to the main diaphragm.

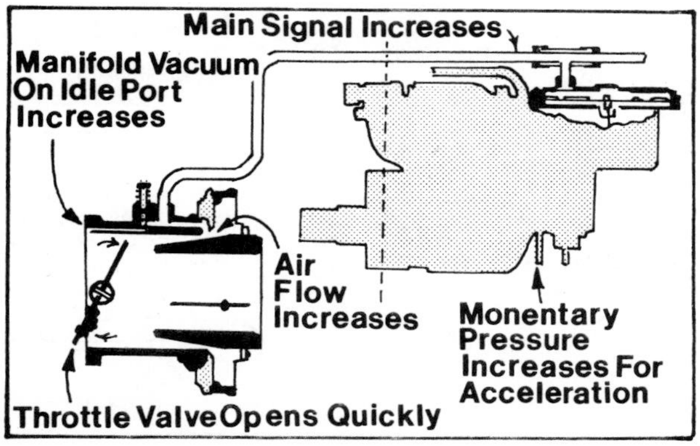

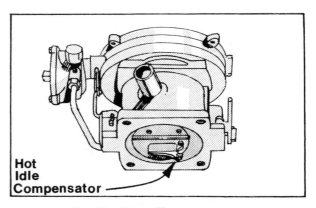

Fig. 12   Acceleration system

Fig. 13   Hot idle compensator

Hot Starting and Unloading

Hot starting and unloading operations definitely require no "extra fuel."

For hot starting and unloading, the throttle valve should be held from half to wide open.

NOTE: If start does not occur in a few seconds, do not continue cranking. Stop and repeat starting procedures.

Hot Idle Compensator

During extreme heat conditions, fuel vapors can cause rich idling which results in engine roughness and stalling.

A thermostatically operated valve is located on the top side of the air meter throttle valve. As idling temperatures increase to the activity range of the thermostatic spring, the valve opens exposing a calibrated orifice thru the throttle valve. A predetermined amount of additional air can now bleed into the over-rich manifold and restore idle mixtures to a correct ratio. Since the compensator is factory calibrated, no adjustment is necessary and should be replaced if defective, Fig. 13.

Bendix Electronic Injection (EFI)

Bendix Electronic Fuel Injection (EFI) basically involves electrically actuated fuel metering valves which, when actuated, spray a predetermined quantity of fuel into the engine. These valves or injectors are mounted in the intake manifold with the metering tip pointed toward the head of the intake valve. This arrangement is commonly known as port injection. The injector opening is timed in accordance with the engine frequency so that the fuel change is in place prior to the intake stroke for the cylinder.

Gasoline is supplied to the inlet of the injectors through the fuel rail at high enough pressure to obtain good fuel atomization and to prevent vapor formation in the fuel system during extended hot operation. When the solenoid operated valves are energized the injector metering valve (pintle) moves to the full open position and since the pressure differential across the valve is constant, the fuel quantity is changed by varying the time that the injector is held open.

The injectors could be energized all at one time (called continuous injection) or one after the other in phase with the opening of each intake valve (called sequential injection) or they can be energized in groups.

The Cadillac EFI System, Fig. 14, is a two-group system. In a two-group system, the eight injectors are divided into two groups of four each. Cylinders 1, 2, 7 and 8 form group 1 while group 2 consists of cylinders 3, 4, 5 and 6. All four injectors in a group are opened and closed simultaneously while the groups operate alternately.

The amount of air entering the engine is measured by monitoring intake manifold absolute pressure, the inlet air temperature, and the engine speed (in rpm). This information allows the Electronic Control Unit to compute the flow rate of air being inducted into the engine, and consequently, the flow rate of fuel required to achieve the desired air/fuel ratio for the particular engine operating condition. Each of the groups is activated once for every revolution of the camshaft and two revolutions of the crankshaft.

The input/output block diagram, Fig. 15, represents Cadillac's two group EFI system. In this system, the prevailing engine conditions are monitored with sensors and provide information to the Electronic Control Unit (ECU). The ECU converts the multi-variable input information into an injector pulse width, which opens the injectors for the proper duration and at the proper time with respect to the cylinder firing sequence.

The object of the ECU is to calculate fuel requirements for the engine for various combinations of inputs from the sensors, to determine an injector pulse width and provide accurate control of the air/fuel ratio.

SYSTEM DESCRIPTION

The EFI System is comprised of four major subsystems:

1. Fuel Delivery,
2. Air Induction,
3. Sensors, and
4. Electronic Control Unit. The function of these separate subsystems is integrated into the system operation.

Fuel Delivery

The fuel delivery system, Fig. 16, includes an in-tank boost pump, a chassis-mounted constant-displacement fuel pump, a fuel filter, the fuel rails,

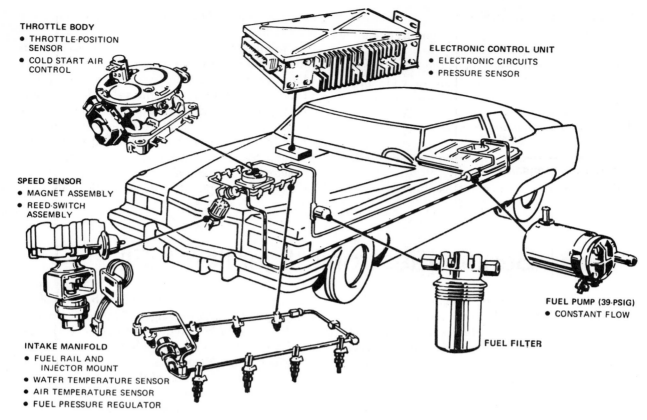

THROTTLE BODY
• THROTTLE-POSITION SENSOR
• COLD START AIR CONTROL

ELECTRONIC CONTROL UNIT
• ELECTRONIC CIRCUITS
• PRESSURE SENSOR

SPEED SENSOR
• MAGNET ASSEMBLY
• REED-SWITCH ASSEMBLY

FUEL PUMP (39-PSIG)
• CONSTANT FLOW

FUEL FILTER

INTAKE MANIFOLD
• FUEL RAIL AND INJECTOR MOUNT
• WATER TEMPERATURE SENSOR
• AIR TEMPERATURE SENSOR
• FUEL PRESSURE REGULATOR

Fig. 14   Bendix Electronic Fuel Injection (EFI). Cadillac

one injector for each cylinder, a fuel pressure regulator, and supply and return lines. The fuel pumps are activated by the electronic control unit (ECU) when the ignition is turned on and the engine is cranking or operating. (If the engine stalls or if the starter is not engaged, the fuel pumps will deactivate in approximately one second.) Fuel is pumped from the fuel tank through the supply line and the filter to the fuel rails. The injectors supply fuel to the engine cylinders in precisely timed bursts controlled by the electrical signals from the ECU. Excess fuel is returned to the fuel tank.

Fuel Delivery System Components

Fuel Tank

The fuel tank used on cars equipped with EFI incorporates a reservoir directly below the sending unit-in-tank pump assembly. The "bath tub" shaped reservoir is used to insure a constant supply of fuel for the in-tank pump, even at low fuel level and severe maneuvering conditions. The fuel returned to the tank by the fuel pressure regulator is dumped directly into the reservoir as an additional means of keeping the pump intake below the fuel level.

In-Tank Boost Pump

Two electric fuel pumps are used with the EFI system. Located inside the fuel tank and an integral part of the fuel gage tank unit, the in-tank pump is used to supply fuel to the chassis-mounted pump and to prevent vapor lock on the suction side of the system.

The chassis-mounted and in-tank fuel pumps are electrically connected parallel to the ECU and are protected by an AGC-10 fuse located in the harness near the ECU.

Chassis-Mounted Fuel Pump

The Chassis-mounted fuel pump is a constant-displacement, roller-vane pump driven by a 12 volt motor. The pump incorporates a check valve to prevent back-flow. This maintains fuel pressure when the pump is off. The pump has a flow rate of 33 gallons per hour under normal operating conditions (39 PSI). An internal relief valve provides over-pressure protection by opening at an excessive pressure (55-95 PSI). The pump is mounted under the vehicle, forward of the left rear wheel on all except Eldorado vehicles, and forward of the right rear wheel on the Eldorado.

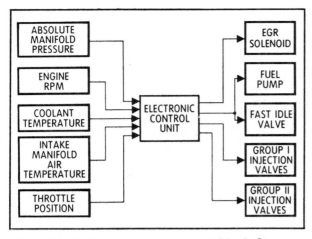

Fig. 15   EFI system input/output block diagram

Fuel Filter

The Fuel Filter consists of a casing with an internal paper filter element capable of filtering foreign particles down to the 20 micron size. The filter element is a throwaway type and should be serviced regularly.

Fuel Pressure Regulator

The Fuel Pressure Regulator contains an air chamber and fuel chamber separated by a spring-loaded diaphragm. The air chamber is connected by a rubber hose to the throttle body assembly.

The pressure in this chamber is identical to the pressure of the intake manifold. The changing manifold pressure and the spring control the action of the diaphragm valve, opening or closing an orifice in the fuel chamber. (Excess fuel is returned to the fuel tank.) This regulator, being connected to the fuel rail and intake manifold, maintains a constant 39 psi differential across the injectors.

The fuel pressure regulator is mounted on the fuel rail toward the front of the engine.

Injection Valve

The Injection Valve is a solenoid operated pintle valve that meters fuel to each cylinder. Based upon a pulsed signal from the electronic control unit, the valve opens for the proper time interval (pulse-width) to deliver the exact amount of fuel required. When energized, the valve sprays the fuel in fine droplets. When de-energized, it prevents further fuel flow to the engine. Handling of the injection valve requires special care to avoid possible damage to the metering tip. The injection valves are located on the intake manifold above the intake valve for each cylinder.

Air Induction

The Air Induction subsystem, Fig. 17, consists of the throttle body assembly, fast idle valve assembly, and intake manifold.

Air for combustion enters the throttle body and

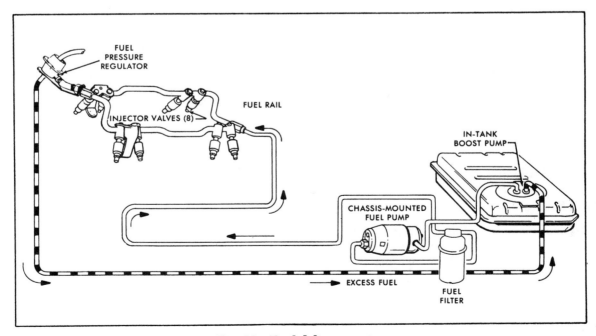

Fig. 16   Fuel delivery system

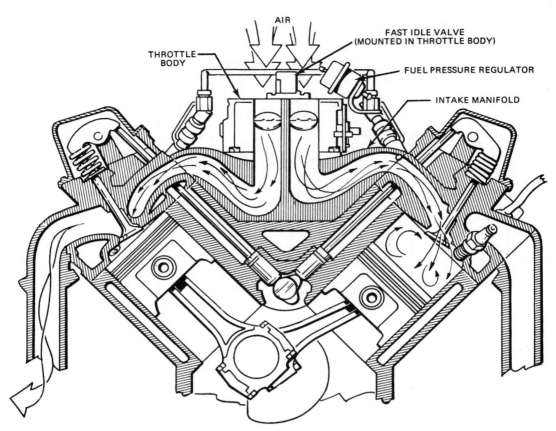

Fig. 17 Air induction subsystem

is distributed to each cylinder through the intake manifold. The primary air flow rate is controlled by the throttle valves which are connected to the accelerator pedal linkage. An adjustable idle by-pass air passage is incorporated within the throttle body that allows a regulated amount of air to by-pass the throttle valves. The throttle blades are also pre-set slightly open when the throttle lever is resting against the idle stop position. This adjustment is not to be altered.

Additional air for cold starts and warm-up is provided through an electrically controlled fast idle valve, which is incorporated in the top of the throttle body.

Air Induction System Components

Intake Manifold

The dual plane intake manifold is basically the same for EFI equipped vehicles as for carburetor equipped vehicles. The EFI intake manifold differs from the carburetor version in the following ways:

1. Only air is distributed to the intake port through the intake manifold.

2. The manifold contains a port above each cyl-

inder for injection valve installation. The fuel is injected directly toward the top of the cylinder intake valve.

3. A port is available to receive the air temperature sensor.

4. No exhaust heat passage is used. The exhaust passages from the right cylinder head are used only for EGR.

5. The EGR passage delivers exhaust gas from the right cylinder head to an area just ahead of the manifold bores under the throttle body and then into the manifold.

Throttle Body

The throttle body, Fig. 17, consists of a housing with two bores and two shaft mounted throttle valves connected to the vehicle accelerator pedal by mechanical linkage. Fittings are incorporated to accommodate vacuum connections. The end of the throttle shaft opposite from the accelerator lever, controls the throttle position switch (described below under Engine Sensors). An adjustable set screw on the front of the throttle body enables adjustment of warm engine idle speed Turning the screw clockwise restricts the air flow passing through the idle air channel, lowering the

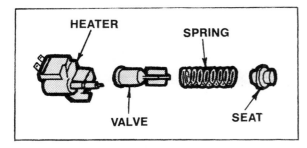

Fig. 18   Fast idle valve

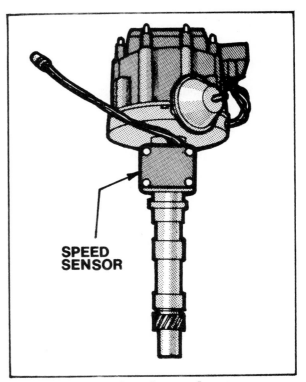

Fig. 19   Speed sensor location

engine speed (rpm). Turning the screw counterclockwise permits a greater air flow, thereby increasing the engine idle speed.

A large post on top of the throttle body contains the fast idle valve.

Fast Idle Valve

The fast idle valve, Fig. 18, consists of a plastic body that houses an electric heater, a spring and plunger and a thermal sensitive element.

The device is installed in the top of the throttle body. Electrically, it is connected to the fuel pump circuit through the electronic control unit. When the cold engine is started, the open valve allows supplemental air to bypass the throttle. The heater warms the thermal element which expands, forcing the spring and plunger toward the orifice, reducing the supplemental air flow and thus reducing the engine speed to the normal level. The fast idle valve will have no further effect after the thermal element reaches approximately 140° F. (60° C.). The rate of valve closure is a function of time and temperature. The warmer the ambient temperature, the faster the valve will close, reducing the engine idle speed. At 68° F. (20° C.), the valve will close in approximately 90 seconds; at —20° F. (—29° C.), the valve will require approximately 5 minutes to close.

Engine Sensors

The sensors are electrically connected to the Electronic Control Unit and all operate independently of each other. Each sensor transmits a signal to the ECU, relating a specific engine operating condition. The ECU analyzes all the signals and transmits the appropriate commands.

Manifold Absolute Pressure Sensor (MAP)

Manifold Absolute Pressure (MAP) Sensor monitors the changes in intake manifold pressure controlled by engine load, speed and barometric pressure variations. These pressure changes are supplied to the electronic control unit circuitry in the form of electrical signals. The sensor also monitors the changes in the intake manifold pressure due to changes in altitude. As intake manifold pressure increases, additional fuel is required. The MAP sensor sends this information to the ECU so that the pulse width will be increased. Conversely, as manifold pressure decreases the pulse width will be shortened.

The sensor is mounted within the electronic control unit. A manifold pressure line, routed with the engine harness, connects it to the front of the throttle body.

Throttle Position Switch

The Throttle Position Switch is mounted on the throttle body and connected to the throttle valve shaft. Movement of the accelerator causes the throttle shaft to rotate (varying the opening of the throttle blades). The switch senses the shaft movement and position (closed throttle, wide open throttle, or position changes), and transmits appropriate electrical signals to the electronic control unit. The electronic control unit processes these signals to determine the fuel requirement for the particular situation.

Temperature Sensors

The Temperature Sensors (Coolant and Air)

491

are comprised of a coil of high temperature nickel wire sealed into an epoxy case, and molded into a brass housing with two wires and a connector extending from the body. The resistance of the wire changes as a function of temperature. Low temperatures provide low resistance and as temperatures increase, so does resistance. The voltage drop across each sensor is monitored by the ECU.

The Air Temperature Sensor is located on the rear side of the intake manifold and is connected to the engine harness. The Coolant Temperature Sensor is located on the heater hose fittings at the rear of the right cylinder head.

Both temperature sensors are identical and completely interchangeable.

Speed Sensor

The Speed Sensor is incorporated within the special ignition distributor assembly (HEI), Fig. 19. It consists of two components. The first has two reed switches mounted to a plastic housing. The housing is affixed to the distributor shaft housing. The second is a rotor with two magnets, attached to and rotating with the distributor shaft.

The rotor rotates past the reed switches causing them to open and close. This provides two types of information: synchronization of the ECU and the proper injector group with the intake valve timing (phasing); and the engine rpm for fuel scheduling.

Electronic Control Unit (ECU)

The electronic control unit is a preprogrammed analog computer consisting of custom electronic circuits housed in a steel case. The unit is installed above the glove box within the passenger compartment. It is electrically connected to the vehicle power supply and the other EFI components by a harness that is routed through the firewall.

The ECU receives power from the vehicle battery when the ignition is set to the ON or CRANK position. During cranking and engine operation, the following events occur.

Information Received from EFI Sensors

1. Engine Coolant Temperature
2. Intake Manifold Air Temperature
3. Intake Manifold Absolute Pressure
4. Engine Speed and Firing Position
5. Throttle Position and Change of Position

Commands Transmitted by ECU

1. Electric Fuel Pump Activation
2. Fast Idle Valve Activation
3. Injection Valve Activation
4. EGR Solenoid Activation

The desired air/fuel ratios for various driving and atmospheric conditions are designed into the ECU. As the above signals are received from the sensors, the ECU processes the signals and computes the engine's fuel requirements. The ECU issues commands to the injection valves to open for a specified time duration. The duration of the command pulses varies as the operating conditions change. All injection valves in each group open simultaneously upon command.

The EFI System is activated when the ignition switch is turned to the ON position. The following events occur at that moment:

1. The Electronic Control Unit (ECU) receives vehicle battery voltage.
2. The Fuel Pumps are activated by the ECU. (The pumps will operate for approximately one second only, unless the engine is cranking or running.)
3. All engine sensors are activated and begin transmitting signals to the ECU.
4. The electrically heated Fast Idle Valve is activated. (The Fast Idle Valve is connected to the Fuel Pump electrical circuit and like the fuel pumps, will only be activated for approximately one second unless the engine is cranking or running. This action is controlled by the ECU.)
5. The EGR solenoid is activated, blocking the vacuum signal to the EGR valve at coolant temperatures below 130° F. (60° C.).

The following events occur when the engine is started:

1. The Fuel Pumps are activated for continuous operation.
2. The heater element of the Fast Idle Valve is activated for continuous operation.
3. The Throttle Body air-bypass controls the air flow to the intake manifold with the throttle valves closed. When the vehicle accelerator is depressed, the throttle valves open, creating an air flow similar to a carburetor.
4. The Fuel Pressure Regulator maintains the fuel pressure in the Fuel Rail (Injector inlet) at 39 psi higher than the intake manifold air pressure by returning excess fuel to the fuel tank.
5. The following signals from the primary sensors are continuously received and processed by the Electronic Control Unit.
   a. Engine Coolant Temperature
   b. Intake Manifold Air Temperature

c. Intake Manifold Absolute Air Pressure

d. Engine speed and firing position

e. Throttle Position changes or wide open throttle (W.O.T.) position

6. The ECU emits electrical signals alternately to each injector group, precisely controlling the opening and closing time (pulse width) to deliver fuel to the engine.

Bosch D-Jetronic and L-Jetronic (ECGI)

Both of these Electronically Controlled Gasoline Injection systems (ECGI) are almost identical except for one feature. On the earlier D-Jetronic system, fuel delivery was regulated by an intake manifold pressure sensor. On the newest L-Jetronic system, the fuel delivery is regulated by the amount of air flow into the engine.

The air-flow sensor has several advantages over the intake-manifold pressure sensor used with Bosch D-Jetronic system: all changes within the

engine (wear, carbon deposits, valve lash, etc.) are automatically compensated for because they all affect air flow; exhaust gas can be recirculated without any special provisions; idle stability is better; and both acceleration enrichment and altitude compensation are virtually automatic.

Naturally, all the normal advantages of fuel injection over carburetion accompany this system. This includes more power output because long tuned intake passages are possible with no fuel

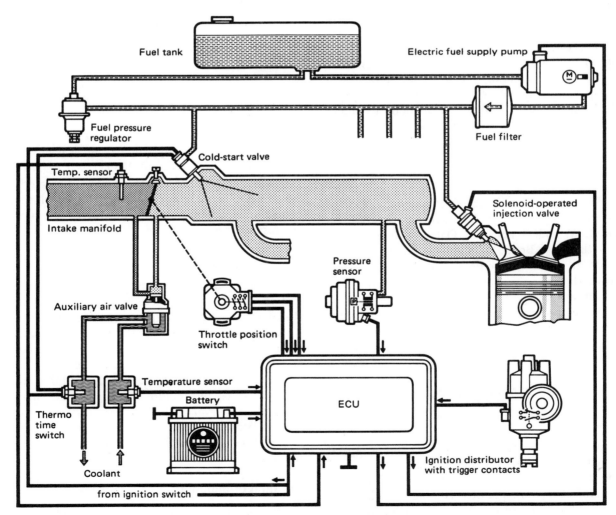

Fig. 20   Bosch Electronically Controlled Gasoline Injection (ECGI)

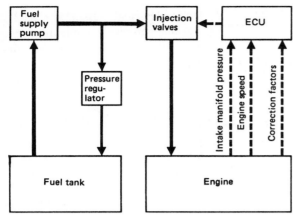

Fig. 20A   ECGI operation

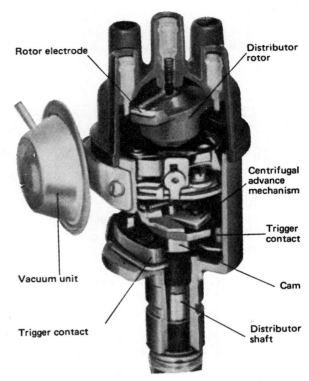

Fig. 21   Trigger contact type ignition distributor

separation, and no heat need be added to vaporize the fuel. Lower emissions are easy to achieve because of the tight control over both air/fuel ratio and cylinder-to-cylinder distribution.

Ten years of electronic fuel-injection production has brought many improvements to the L-Jetronic system. Integrated circuits have reduced the number of individual electronic components from 300 to 80. Also, detecting elements are now much cheaper to manufacture and significantly more reliable in service. There is tremendous flexibility inherent in any electronic control, so this system will quite handily accept the Lambda probe for three-way catalyst control of exhaust emissions. That alone should buy it a place in many future cars.

There is really only one factor that has limited L-Jetronic's popularity (it's currently used only on the BMW 630CSi and 530i, Datsun 810 and 280-Z, and the VW Beetle). It suffers from a stigma, as do all other electronic components: if it can't be fixed with a wrench, mechanics would rather not know about it.

We will concentrate here on the L-Jetronic, since it is the more commonly used system and because more and more manufacturers will be using this system in the future.

OPERATING PRINCIPLES

The Bosch Electronic Gasoline Injection System (ECGI), Figs. 20 and 20A, is controlled by the intake manifold pressure and the engine speed which injects the fuel into the intake manifold about 10-15 cm before each cylinder inlet valve. The injected fuel quantity is regulated by controlling the time during which the solenoid-operated injection valves open. Since the orifice area

of the injection valves is precisely calibrated and the fuel pressure is held constant, the injected fuel quantity depends solely on the length of time the injection valve is open. The system is controlled by the Electronic Control Unit (ECU), which processes information received from the engine concerning its operating condition and which then transmits electrical pulses to the valves to control the instant and duration of each injection.

Instant of Injection

The beginning of the pulse which causes the injection valve to open is determined by the ignition distributor contacts (trigger contacts), depending on the position of the camshaft. These contacts are located in the distributor under the centrifugal advance mechanism and are activated by a cam on the distributor shaft, Fig. 21. In addition, the ECU receives information on the engine speed from the intervals between the trigger pulses (which come from the trigger contacts), and this information is one of the inputs used to calculate the duration of injection.

Duration of Injection

The open time of the injection valves, and therefore the quantity of fuel which is injected, is de-

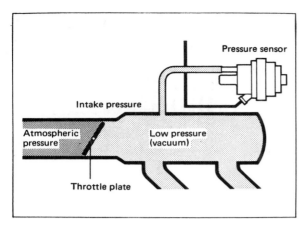

Fig. 22 Intake manifold pressure conditions and location of the pressure sensor

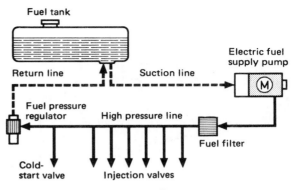

Fig. 23 Fuel system

termined for the most part by two factors: the speed of the engine and the load on it. As mentioned above, information on the engine speed is transmitted to the ECU from the distributor trigger contacts. The engine load (part or full load) can be determined from the pressure conditions in the intake manifold, Fig. 22. The pressure prevailing here at any time is converted into an electrical quantity by the pressure sensor and is also fed to the ECU, Fig. 22. The ECU then commands the injection valves, by means of current pulses, to inject more or less fuel. In this way the "basic fuel quantity" is determined.

In addition to this "basic fuel quantity," under certain operating conditions an additional definite, precisely metered quantity of fuel must be injected. These conditions require *correction factors* covered further on in this chapter.

Components

The system consists essentially of the fuel system, the induction system, and the control system.

Fuel System

In the fuel system, a fuel supply pump draws the fuel from the fuel tank and forces it through a filter and through the distribution line and its branches to the solenoid-operated injection valves. The fuel pressure at the valves is held constant by a fuel pressure regulator. Excess fuel flows back pressure-less to the tank. To prevent foreign particles from reaching the injection valves there is a fuel filter fitted between them and the pump, Fig. 23.

Each cylinder has a solenoid-operated injection valve which opens once for each working cycle (rotation of the camshaft). Depending on engine design, the injection valve is located either in the intake manifold or in the cylinder head. In every case, however, injection takes place into the intake manifold before the inlet valve.

In order to reduce the complexity of the ECU, in 4-cyl. engines two groups of two injection valves each are formed. The valves in a given group are connected electrically in parallel and open simultaneously, Fig. 24. The same applies for 6- and 8-cyl. engines (two groups of three valves each, four groups of two valves each respectively).

The 6-cyl. engine injection timing chart shows the sequence of start injection, opening of the inlet valve, and ignition point, referred to the crankshaft position in degrees, Fig. 25. The air-fuel mix-

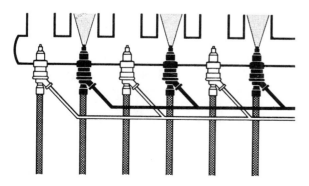

Fig. 24 Injector valve operation

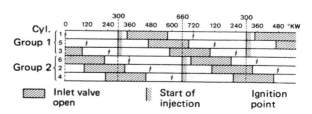

Fig. 25 Six cylinder engine injector timing chart

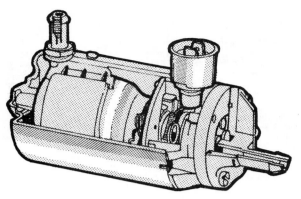

Fig. 26　Electric fuel supply pump

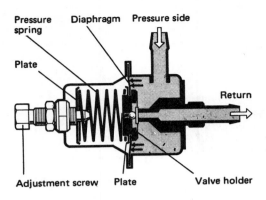

Fig. 27　Pressure regulator

ture is stored for each cylinder before its inlet valve and is drawn into the cylinder when the inlet valve opens. Ignition of the mixture takes place in each cylinder shortly after its inlet valve closes.

Electric Fuel Pump

The fuel supply pump in this system is a roller-type pump, Fig. 26, driven by a permanent-magnet electric motor. An eccentric rotor mounted on the motor shaft has metal rollers housed in pockets around its circumference which are forced outward by centrifugal force and which act as seals. In the gaps formed between the rollers the fuel is forced through the pump and into the high-pressure line. The pump is designed as a so-called "wet pump," which means that the electric motor is filled with fuel. There is no danger of explosion, however, because the pump housing never contains an explosive mixture. The pump delivers fuel at about 21 and 31 gallons/hour (about 80 and 120 liters/hour).

When the ignition is turned on at the ignition switch, the pump runs for only about one second and then stops. Only after the engine has started is the pump switched on again by the ECU through a relay. This protective system (called flooding protection) prevents flooding of the particular cylinder with fuel, in event of a possible defective injection valve when the ignition has been switched on.

Pressure Regulator

The fuel pressure is held at a constant value by the pressure regulator. This regulator consists of a metal housing in which a spring-loaded diaphragm frees the opening to an overflow duct if the pressure setting is exceeded. The pressure is adjusted by changing the pretension of the diaphragm spring, Fig. 27.

Solenoid-Operated Injection Valve

The solenoid-operated injection valve, Fig. 28, consists essentially of a valve body and the nozzle needle to which the solenoid plunger is attached. The movable plunger is attached rigidly to the nozzle needle, which is pressed against the valve seat in the nozzle body by a helical spring. The solenoid coil is mounted in the rear section of the valve body, with the guide for the nozzle needle in the front section.

The current pulses received from the ECU build up a magnetic field in the coil. As a result, the plunger is attracted and lifts the nozzle needle from its seat. This opens the path for the pressurized fuel. Valve lift is about 0.15 mm, and lift period about 1/1000 second. Depending on the amount of fuel required to be injected, the injection valve is held open from 2/1000 to 10/100 seconds.

Induction System

The volume of air required for combustion of the fuel is drawn through the air cleaner and past the air throttle valve into the intake manifold. From here individual intake tubes of equal length branch out to each cylinder. This design ensures that absolutely equal amounts of air are fed to the individual cylinders. Generation of the air-fuel mixture starts with the injection of the fuel charge into the charge of air which has been drawn into the cylinder by the reciprocating movement of the piston.

In the intake manifold, atmospheric pressure prevails in front of the throttle plate (air cleaner side), whereas after the throttle plate there is a lower pressure which is constantly changing depending on the throttle plate position, Fig. 22. This changing absolute pressure in the manifold is used

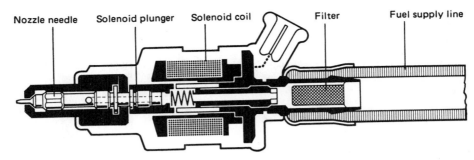

Fig. 28   Solenoid operated injection valve

as a parameter for determining the most important item of information, the engine load.

Pressure Sensor with Full-Load Enrichment

The pressure in the intake manifold is provided by the pressure sensor, Fig. 29. For this reason we also refer to this sensor as an "intake manifold pressure sensor," Fig. 22.

The pressure sensor measuring system is installed in a pressure die-cast housing which is connected to the intake manifold of the engine by a hose. The pressure sensor contains two evacuated aneroids which shift the position of a plunger inside a coil. As the load increases, that is, as the absolute pressure in the intake manifold increases, the aneroids are compressed, moving the plunger more deeply into the iron core of the coil, thus increasing the inductance of the coil. This results in a longer opening pulse to the injection valve and thus a larger amount of fuel is injected.

When the air throttle valve is closed the intake manifold pressure is low, the aneroids are less compressed, and they move the plunger out of the iron core. This lowers the inductance of the coil,

and as a result, the opening pulse is shorter and the valves inject a smaller amount of fuel.

The inductive data transmitter (the coil) in the pressure sensor is connected to an electronic timer in the ECU which determines the duration of the opening pulse to the injection valve. During the pulse the valve is open. In this way the intake manifold pressure is converted directly into a corresponding pulse duration (injection duration).

It should basically be kept in mind that the start of injection is activated by the trigger contacts in the distributor, while the duration of injection, and thus the injected fuel quantity, is determined by the pressure sensor through the electronic timer in the ECU.

Pressure Sensor with Altitude Correction

In newer gasoline injection systems, the pressure sensor contains two aneroids, but only one of these is evacuated—the other is open to the atmosphere. By this arrangement not only the absolute pressure in the intake manifold is taken into account, but also the difference between the atmospheric pressure and the pressure in the manifold. Practically speaking, this means that in the part-load range the engine is matched far better to changes in altitude, Fig. 30.

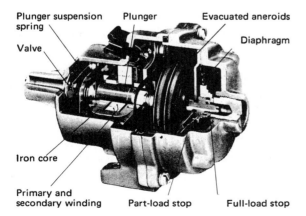

Fig. 29   Pressure sensor incorporating the full load enrichment switch

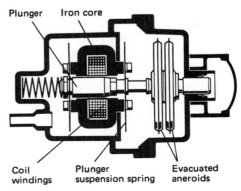

Fig. 30   Pressure sensor without enrichment switch

497

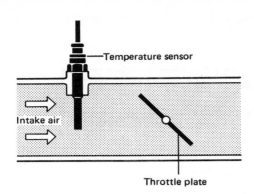

Fig. 31   *Intake air temperature sensor location*

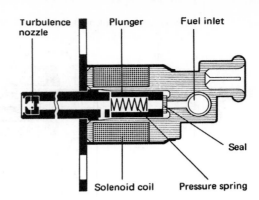

Fig. 32   *Cold start valve*

Temperature Sensor (Intake Air)

The injected fuel quantity is controlled primarily by the absolute pressure in the intake manifold, with a correction made for the engine speed. However, this control of the injected fuel quantity is only exact at constant temperature, because at low ambient temperatures the weight of the air drawn in is greater, so the air-fuel mixture becomes leaner if the control system does not take the air temperature into account. This becomes of greatest importance at ambient temperatures of $-20°$ C. to $0°$ C. ($-4°$ F. to $32°$ F.) when Combustion Miss can result. In order to avoid this disadvantage, a temperature sensor is installed in the intake manifold. This sensor reacts to decreasing air temperatures by increasing the injected fuel quantity corresponding to the increase in the air density, Fig. 31.

The temperature sensor consists of a metal body shaped like a hexagonal-head screw inside which a so-called NTC resistor is installed. NTC stands for "Negative Temperature Coefficient" and means that the resistance of this type of resistor decreases sharply as its temperature increases.

Control System

The purpose of the control system in the ECGI is to meter a precise volume of fuel, corresponding to engine operating conditions, and to ensure its delivery to the charge of air drawn in by the engine during each operating cycle.

Electronic Control Unit (ECU)

The most important unit in the ECGI system is the Electronic Control Unit (ECU) which is responsible for metering the fuel quantity. The items of information needed for this are collected by transducers (pressure sensors, temperature sensors, etc.) and are fed as electrical quantities to the ECU. After processing these quantities, the ECU generates the current pulses (opening pulses) which excite the solenoid coil of the solenoid-operated injection valves.

Depending on its design, the ECU is constructed of 250-300 components, including about 30 transistors and 40 diodes. The unit operates at battery voltage but voltage variations occurring during driving do not affect its functioning. The electrical units of the injection system are connected to the ECU through a cable harness which is terminated in a 25-pole plug. When the ignition is turned on, the injection system is made operational through a relay.

The ECU is designed on the basis of printed-circuit engineering. We already foresee, however, that as a result of the introduction of integrated circuits in the unit, its construction costs can be reduced and its reliability increased still further.

Correction Factors

In addition to the exact metering of fuel for full- or part-load under normal (warm) operating conditions of the engine, the following series of additional correction factors are required for perfect engine performance:

1. Enrichment of the mixture for cold starting.
2. Matching of the mixture during engine warm-up.
3. Enrichment of the mixture when accelerating.
4. Cut-off of fuel during engine braking (now superseded in some cases by combustion during this mode of operation).

Cold Starting

When the engine is cold some fuel condenses in the intake manifold and on the cylinder walls.

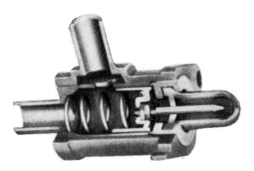

Fig. 33 Auxiliary air valve for water-cooled engines

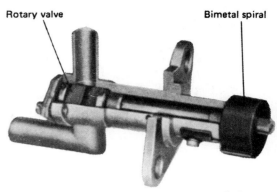

Rotary valve   Bimetal spiral

Fig. 34 Auxiliary air valve for air-cooled engines

As a result, less fuel is mixed with the air drawn in to form a combustible mixture than in a warm engine; that is, a homogenous, combustible mixture is not formed. If the mixture is too lean, ignition combustion miss can take place. The function of the solenoid-operated cold-start valve, Fig. 32, near the air throttle valve is to enrich the air in the intake manifold with finely atomized fuel. It injects fuel only when the starting motor is activated and when at the same time a thermoswitch or thermo time switch immersed in the coolant (or fitted in the engine compartment) is closed.

Depending on the temperature it senses, the thermoswitch interrupts or completes the electric circuit leading to the cold-start valve. The thermo time switch performs the same function as the thermoswitch, but also limits the time that the valve is switched on.

Solenoid-Operated Cold-Start Valve

Great importance is placed on fine atomization of the fuel in the solenoid-operated cold-start valve, Fig. 32. A helical spring presses the movable plunger of the magnetic circuit with its seal against the valve seat and blocks the incoming flow of fuel. When the plunger is drawn back the valve seat is freed. The fuel then flows laterally past the plunger and reaches the turbulence nozzle. In this nozzle the fuel is given a rotating movement and is discharged from the nozzle in a finely atomized condition.

Warm-Up

The warm-up phase of engine operation follows the cold start. During this phase the engine requires considerable fuel enrichment because a portion of the fuel condenses on the cold cylinder walls. Moreover, without additional fuel enrichment a considerable drop in engine speed would be noticeable after fuel injected from the cold-start valve is cut off.

As the temperature increases, the enrichment steadily decreases, and it stops when the operating temperature is reached. This process is controlled in water-cooled engines by a temperature sensor immersed in the coolant and in air-cooled engines by a sensor in the cylinder head. Enrichment also takes place during starting operations, so that at low temperatures, additional fuel to that sprayed into the manifold by the cold-start valve is provided. Immediately after starting at −20° C. (−4° F.), two to three times as much fuel must be injected, depending on engine type, than when the engine has reached operating temperature.

In addition to a richer air-fuel mixture during cold-start and the following warm-up phases of operation, a larger volume of air is also required during idling. In order to overcome the higher frictional losses the engine must produce a greater torque during idling. This is made possible by the so-called auxiliary charge of air. The auxiliary charge of air is controlled by a bypass valve, the auxiliary air valve, which bypasses the air throttle valve.

Auxiliary Air Valve

In the auxiliary air valve, Fig. 33, a temperature-dependent expansion element moves a control piston and changes the flow area of the air duct. The area of the variable opening in the duct is changed dependent on the temperature so that at every starting temperature the desired idle speed can be attained. As the engine temperature increases, the area of the air duct is continually reduced and at a coolant temperature of about +60° C. to +70° C. (141° F. to 159° F.) the duct is completely closed.

The auxiliary air valve can also be designed as a rotary valve, which is activated by a bimetal spiral. In air-cooled engines the bimetal spiral is inserted in the crankcase and surrounded by the engine oil vapors, Fig. 34.

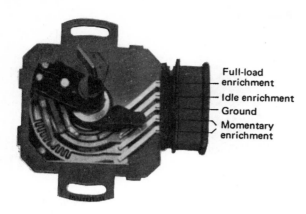

Full-load
 enrichment
Idle enrichment
Ground
Momentary
 enrichment

Fig. 35 Throttle position switch

Acceleration

When the air throttle valve is opened quickly, for example, during acceleration, the pressure sensor reports the pressure rise with a slight delay to the ECU. This slight response delay is present because the pressure build-up in the intake manifold (and thus in the pressure sensor) requires a certain time after the change in position of the throttle plate. In order to bridge this slight delay in pressure sensor response, the throttle position switch is fitted with auxiliary contacts which cause additional injection pulses to be generated by the ECU when the throttle valve is opening.

Throttle Position Switch

The sliding contacts in the throttle position switch slide over contact tracks and are actuated directly by the throttle plate shaft depending on the movement of the throttle plate. They bridge the contact tracks, thereby causing additional opening pulses to be generated by the ECU for acceleration enrichment, or fuel cut-off, when the vehicle is operating in an engine braking mode. In newer systems the throttle position switch, Fig. 35, performs the function of full-load enrichment.

Full-Load Operation

During part-load operation the fuel is metered

so that the fuel consumption and the proportion of unburned components in the exhaust gas are kept as low as possible. When operating under full-load, however, the injected fuel quantity is determined on the basis of attaining maximum engine power; for example, additional fuel must be injected at such a time. Information required for this full-load enrichment is provided from the pressure switch, or the diaphragm section of the pressure sensor.

In newer systems the full-load enrichment is controlled by an additional contact in the throttle position switch. As a result, in these systems the pressure switch, or the diaphragm section of the pressure sensor, is not used for full-load enrichment.

Engine Braking

When the engine is driven by the rolling vehicle (for example, on a downgrade), the air throttle valve is fully closed. In order to save fuel and to reduce the exhaust emission, the fuel supply is completely cut off during this mode of operation. In order to keep the engine from stalling if the clutch is disengaged, however, the supply of fuel is restored at engine speeds between 1,000 and 1,500 rpm. In the same way, injection is resumed if the engine speed approaches the idle speed when the vehicle is braked. The "air throttle valve closed" information is signalled by the throttle position switch to the ECU, which calculates the engine speed on the basis of the intervals between the triggering pulses received from the distributor.

In some other systems, combustion is continued during engine braking and the engine finds its own speed on the basis of this mode of operation. The reason combustion is continued in these systems is that fewer harmful exhaust components are emitted from an engine which is running, than from an engine in which the combustion chamber has cooled off following the cessation of combustion. When the combustion chamber has cooled off, this results in poor mixing of the air and fuel which, when combustion starts again, leads to the emission of an impermissible level of unburned, and therefore harmful, exhaust components.

Bosch K-Jetronic (CIS)

DESCRIPTION

The Bosch Continuous Injection System, Fig. 36, is a mechanically and continuously operating fuel injection system which does not have to be driven

by the vehicle engine. The fuel is pumped by an electrically-driven roller cell fuel pump. During operation the intake air quantity is metered by an air-flow sensor installed in front of the throttle plate. Depending on the position of the accelerator

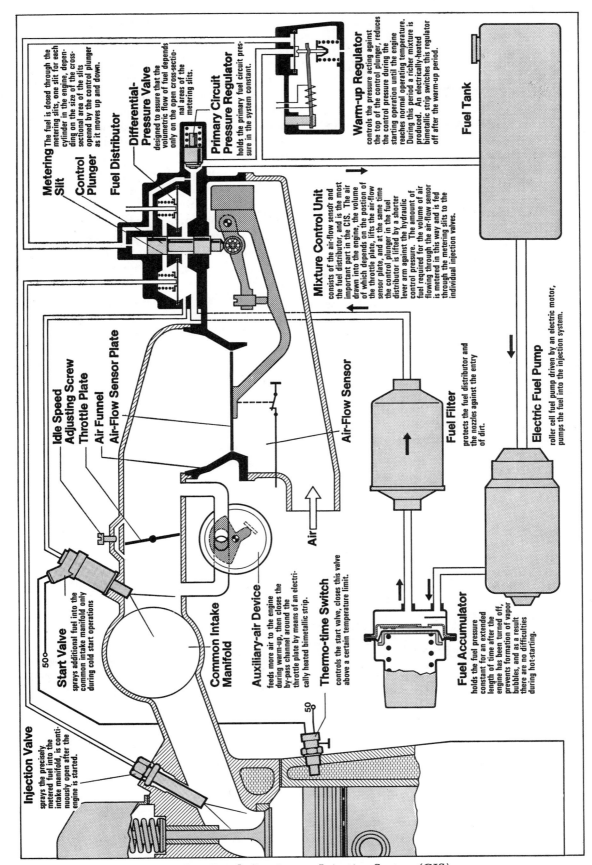

Metering The fuel is dosed through the metering slits, one slit for each cylinder in the engine, depending on the size of the cross-sectional area of the slits opened by the control plunger as it moves up and down.

Differential-Pressure Valve designed to assure that the volumetric flow of fuel depends only on the open cross-sectional areas of the metering slits.

Primary Circuit Pressure Regulator holds the primary fuel circuit pressure in the system constant.

Warm-up Regulator controls the pressure acting against the top of the control plunger, reduces the control pressure during the starting operation until the engine reaches normal operating temperature. During this period a richer mixture is produced. An electrically-heated bimetallic strip switches this regulator off after the warm-up period.

Fuel Tank

Metering Slit

Control Plunger

Fuel Distributor

Mixture Control Unit consists of the air-flow sensor and the fuel distributor, and is the most important part in the CIS. The air drawn into the engine, the volume of which depends on the position of the throttle plate, lifts the air-flow sensor plate, and at the same time the control plunger in the fuel distributor is lifted by a hydraulic control pressure. The amount of fuel required for the volume of air flowing through the air-flow sensor is metered in this way and is fed through the metering slits to the individual injection valves.

Idle Speed Adjusting Screw

Throttle Plate

Air Funnel

Air-Flow Sensor Plate

Air-Flow Sensor

Air

Fuel Filter protects the fuel distributor and the nozzles against the entry of dirt.

Electric Fuel Pump roller cell fuel pump driven by an electric motor, pumps the fuel into the injection system.

Start Valve sprays additional fuel into the common intake manifold only during cold start operations.

Common Intake Manifold

Auxiliary-air Device feeds more air to the engine during warm-up, then closes the by-pass channel around the throttle plate by means of an electrically heated bimetallic strip.

Thermo-time Switch controls the start valve, closes this valve above a certain temperature limit.

Fuel Accumulator holds the fuel pressure constant for an extended length of time after the engine has been turned off, prevents formation of vapor bubbles, and as a result there are no difficulties during hot-starting.

Injection Valve sprays the precisely metered fuel into the intake manifold, is continuously open after the engine is started.

Fig. 36   Bosch Continuous Injection System (CIS)

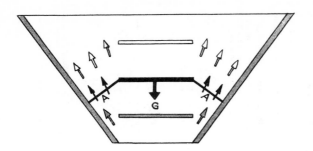

Fig. 37 Air flow sensor plate

Description

The principle of air-flow sensing in the CIS is comparable to the well-known suspended body principle: a round disc (the air-flow sensor plate) rises in a conical air tunnel until its weight and the force of the air flow against the bottom of the disc are in equilibrium, Fig. 37. If the volume of air drawn into the engine increases, the rate of air flow also increases through the original annular cross-sectional area in the air-flow sensor. As a result, the flow force increases, and the air-flow sensor plate is forced farther upward until the original flow rate (= flow force) exists at the new and larger cross-sectional area in the funnel. Here, the air-flow sensor plate again comes to rest. The position of the air-flow sensor plate in the air funnel thus represents a measure of the volumetric rate of air flow through the air funnel and is therefore a measure of the quantity of fuel required. The air-flow sensor plate rises a distance approximately proportional to the volumetric rate of air flow, Fig. 38.

Construction

The air-flow sensor, Fig. 39, consists of the air funnel and the air-flow sensor plate mounted on the lever which is supported at its fulcrum. The weights of the lever and air-flow sensor plate are

pedal, which controls the position of the throttle plate, a varied quantity of air is drawn into the engine. Therefore depending on the volume of air metered, a fuel distributor meters a quantity of fuel to the individual cylinders through the associated injection valves which produces an optimum air-fuel mixture with regard to engine power, fuel consumption, and exhaust-gas emission.

The air-flow sensor and the fuel distributor are combined into one assembly, called the mixture control unit. The precisely metered quantity of fuel is fed to the injection valves which continuously spray the fuel in finely atomized form into the intake manifold in front of the cylinder intake valves. From there the fuel is drawn into the engine cylinders combined with air when the intake valves open.

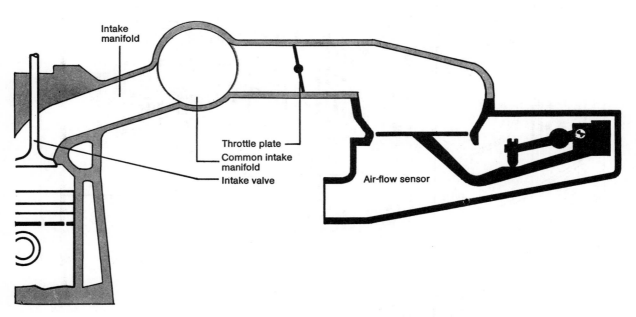

Fig. 38 Engine cylinder incorporating the air flow sensor

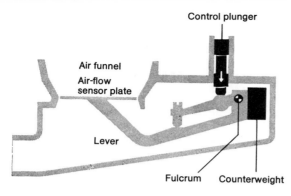

Fig. 39   Air flow sensor

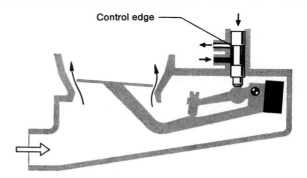

Fig. 40   Mixture control unit fuel control

balanced by the counterweight. A plunger (the control plunger) operated by the hydraulic pressure transmits a force opposing the air force through the lever and then to the air-flow sensor plate. The intake air flowing through the air funnel lifts the air-flow sensor plate until the force of this air and the opposing force of the plunger reach equilibrium.

In this position of equilibrium, which is a measure of the volumetric rate of air flow, the control plunger is positioned at a certain point in the fuel distributor, and its horizontal control edge opens the rectangular metering slits by a certain uniform amount, Fig. 40. The fuel which flows through these openings is then fed to the injection valves.

As a result of the linear action of the air-flow sensor and fuel distributor, and because these components are joined by a lever to form a single operating assembly, an initially exact and stable basic matching of air and fuel results for a constant air factor. The exhaust gas regulations requirements are achieved in the basic design of the CIS by a modification in the shape of the funnel in the air sensor, Fig. 41. Specifically, in order to match the air-fuel ratios to the various load levels —(idle, part load, full load) the air funnel becomes wider in stages, Fig. 41. In those parts of the funnel where the wall is steeper than in the basic shape, the air-flow sensor plate must be forced higher in order to develop a state of equilibrium than would be the case in an air funnel with straight walls. As a result, a richer mixture is developed during idle and full-load. Conversely, if the walls of the funnel rise less steeply (flatter funnel) than in the basic shape, a leaner mixture is developed.

FUEL SUPPLY

When discussing the fuel supply system in the

CIS, we must differentiate between the primary fuel circuit and the control circuit. In addition, the pressure in the fuel lines leading to the injection valves, must also be considered.

Primary Fuel Circuit

In this circuit, a roller cell pump combined with an electric motor draws the fuel from the fuel tank and pumps it through a fuel accumulator and a fine fuel filter to the fuel distributor, Fig. 36.

The electric fuel pump is set in operation when the vehicle engine is started, and the power to the pump motor is cut off when the engine stops turning. This process is actuated by the air-flow sensor contact through the safety relay, Fig. 42.

The fuel accumulator must fulfill the following three functions:

1. dampen supply pump noise by a built-in fuel spinner;

2. delay the build-up of pressure in the primary fuel circuit when the engine is started—this ensures that the control plunger is in the zero position at this time; and

3. retain pressure in the system after the engine has been turned off in order to ensure good hot-restarting performance.

In the fuel distributor, the fuel initially enters a

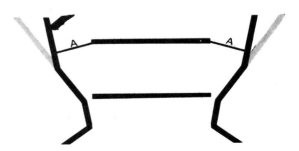

Fig. 41   Cone compensation in the air flow sensor

503

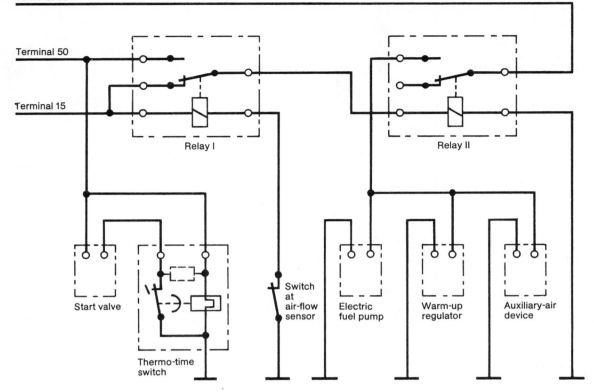

Terminal 30

Terminal 50

Terminal 15

Relay I

Relay II

Start valve

Thermo-time switch

Switch at air-flow sensor

Electric fuel pump

Warm-up regulator

Auxiliary-air device

Fig. 42   CIS wiring diagram

passage which joins the lower chambers of the differential-pressure valves. As a result, the same fuel pressure, held constant by the primary circuit pressure regulator built into the fuel distributor, prevails in each of the lower chambers. This primary circuit pressure regulator is designed as a plunger-type regulator and regulates the system overpressure. Excess fuel flows back through a return line under no pressure back to the fuel tank.

When the engine is operating, the fuel passes through the metering slits, to the upper side of the diaphragm of the differential-pressure valves and then flows through the fuel injection lines to the injection valves.

The injection valves are designed so that they finely atomize the fuel even at low fuel flow rates.

A supply line leads from the primary fuel circuit to the start valve.

When the engine is turned off, the primary circuit pressure regulator lets the pressure in the system fall rapidly to the opening pressure of the injection valves, then it holds the pressure at this level by means of a rubber valve seat which takes effect at that time, Fig. 43. As a result of the rapid drop of pressure in the fuel supply system, the "dieseling" or "running on" phenomena after the ignition has been turned off which occurs in many

vehicles is prevented without additional switching devices.

Control Circuit

The control circuit branches off from the primary fuel circuit through a restriction bore in the fuel distributor. A connection line leads from the fuel distributor to the regulator for the control pressure which is called the warm-up regulator, Fig. 44. At normal operating temperature, the warm-up regulator holds the control pressure constant, it lowers the control pressure only when the engine is cold and during the warm-up period. The control pressure acts through a damping restriction on the control plunger and thereby develops the force which opposes the air force in the air-flow sensor, Fig. 45.

Excess fuel flows from the warm-up regulator under no pressure back to the fuel tank.

The warm-up regulator is mounted on the vehicle engine in such a way that it can absorb heat from the engine block and thus assume the temperature of, the engine block. As a result, when the engine is started in the semi-warm condition unnecessary enrichment (causing an over rich condition) of the fuel is avoided.

504

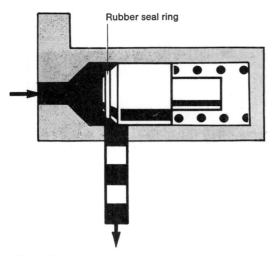

Rubber seal ring

Fig. 43   *Primary circuit pressure regulator*

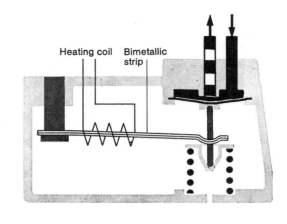

Heating coil   Bimetallic strip

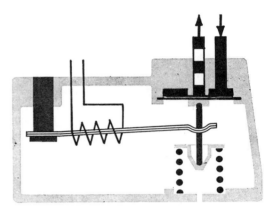

Fig. 44   *Warm-up regulator. Upper diagram with engine cold, lower diagram with engine warm*

The damping restriction over the control plunger performs a special function. Under conditions of pulsating air flow which occur, for example, at a low engine speed and high load, the air-flow sensor plate in the air-flow sensor is deflected too far from its base position. As a result, the mixture is automatically enriched even though the basic setting is for a lean mixture. The function of the damping restriction is to dampen the movement of the air-flow sensor plate during pulsating flow. At the same time, this restriction regulates the distance by which the air-flow sensor plate will swing too far, determining the momentary transitional enrichment of the fuel during acceleration; this restriction therefore creates the conditions necessary for optimum driving performance.

Fuel Distributor

The fuel must be distributed uniformly to the different cylinders in the engine. In the CIS, the assembly designed to do this operates on the basis of simultaneous control of the open cross-sectional

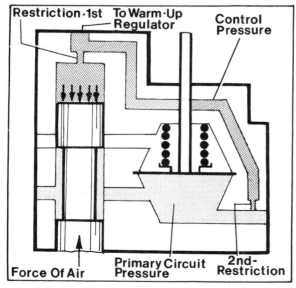

Restriction·1st   To Warm·Up Regulator   Control Pressure

Force Of Air   Primary Circuit Pressure   2nd-Restriction

Fig. 45   *Relationship between the primary circuit and control pressures*

Fig. 46   *Fuel distributor barrel metering slits*

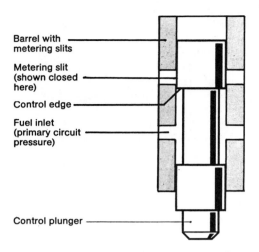

Barrel with metering slits

Metering slit (shown closed here)

Control edge

Fuel inlet (primary circuit pressure)

Control plunger

Fig. 47   Fuel distributor barrel with control plunger

area of metering slits machined into the barrel in the fuel distributor, Figs. 46, 47 and 48. The barrel has as many slot-shaped (rectangular) openings (metering slits) as there are cylinders in the engine. A differential-pressure valve follows each metering slit for purposes of holding the drop in pressure at the metering slits constant at various flow rates. As a result, possible effects of variations in the primary system pressure and differences in the opening pressure of the injection valves are eliminated. With a constant drop in pressure at the metering slits, the amount of fuel flowing to the injection valves depends only on the open cross-section of these slits. The fuel distributor operating characteristic line is therefore linear.

The differential-pressure valve, Fig. 49, is a diaphragm valve consisting of a lower and an upper chamber with a steel diaphragm between them. In the lower chamber the so-called primary system overpressure prevails, while in the upper chamber there is a slightly lower overpressure. The pressure differential is produced by the heli-

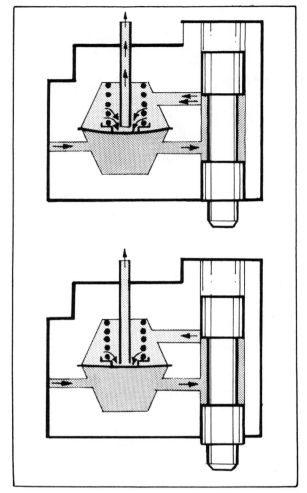

Fig. 48   Fuel distributor barrel fuel flow. Upper diagram with high flow rate, lower diagram with low flow rate

cal spring built into the upper chamber.

Under these conditions, equilibrium of forces exists at the diaphragm of the differential-pressure valve.

If more fuel flows through the metering slit into the upper chamber, the pressure there rises tem-

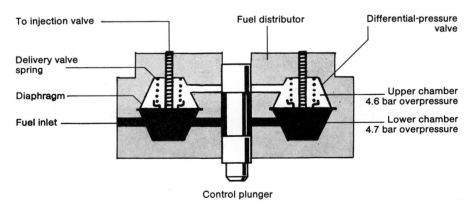

To injection valve

Delivery valve spring

Diaphragm

Fuel inlet

Fuel distributor

Differential-pressure valve

Upper chamber 4.6 bar overpressure

Lower chamber 4.7 bar overpressure

Control plunger

Fig. 49   Fuel distributor differential pressure valve

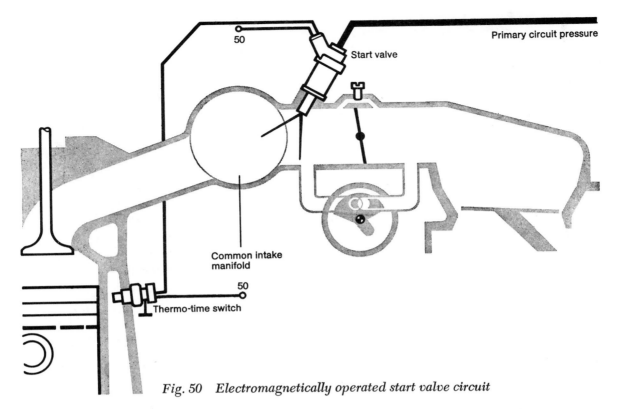

Fig. 50   Electromagnetically operated start valve circuit

porarily. The steel diaphragm is forced downward and enlarges the cross-section of the outlet leading to the injection valve until a slight differential pressure again prevails at the metering slit.

At higher rates of fuel flow, the diaphragm opens a larger annular cross-section, so the pressure differential remains constant. If the rate of fuel flow decreases, the diaphragm decreases the discharge cross-section. The total travel of the diaphragm is only a few hundredths of a millimeter.

COMPENSATION UNITS

If needed, compensation units are available for the CIS for cold starting, warm-up, and full-load enrichment, Fig. 44. In order to meet the more exacting requirements arising from stricter exhaust gas regulations, for instance during catalytic operation etc., additional compensation and control modifications are possible.

Warmup Compensation

During the warmup period of an engine, two compensations are basically required compared with conditions at normal operating temperature:

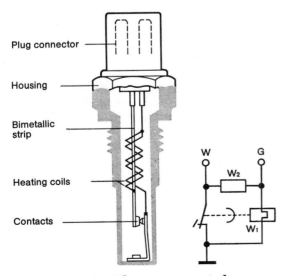

Fig. 51   Thermo-time switch

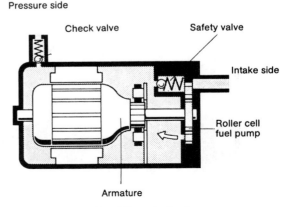

Fig. 52   Electric fuel pump

Fig. 53   Roller cell fuel pump

1. Compensation for condensation losses on the cold walls of the combustion chamber and intake manifold.
2. Compensation for power lost as a result of greater friction.

Compensation is made for condensation losses by a richer mixture. This function is performed by the warm-up regulator. It lowers the pressure applied to the control plunger during the warm-up period, as a result of which the air-flow sensor plate is lifted a greater distance from its rest position by the same rate of air flow, and the open cross-sectional area of the metering slits in the barrel assembly is therefore enlarged.

The change in the control pressure during the warm-up period takes place as follows:

As long as the engine is cold, a bimetallic strip presses against the delivery valve spring. As a result, the pressure on the diaphragm is reduced, the discharge cross-section is enlarged and the control pressure is thus lowered. When the engine is started, the electrical system designed to heat the bimetallic strip is switched on. This strip warms up, relaxing the pressure on the delivery valve spring, and at a certain temperature it rises completely up from the spring plate. This means that after the warm-up period the delivery valve spring is fully effective. A fairly high control pressure must develop before the discharge cross-section opens.

Compensation is made for the power lost as a result of greater friction by feeding a larger volume of the air-fuel mixture to the engine than

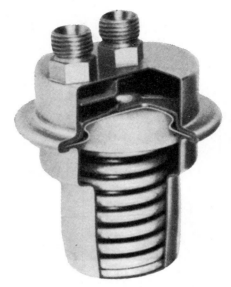

Fig. 54   Fuel accumulator

corresponds to the position of the throttle plate. This is done by bypassing the throttle plate with an auxiliary-air device, Fig. 36, in which the cross-sectional area of the channel which is open to flow of air is controlled by a pivoted blocking plate with a specially shaped hole; the movement of this plate is dependent on an electrically heated bimetallic strip. At normal operating temperature this extra channel is closed.

Auxiliary Starting Assembly

The auxiliary starting assembly consists essentially of an electromagnetically-operated start valve, Fig. 50, which is switched on at the beginning of the starting process, and a thermo-time switch which limits the duration of time that the valve is open, or at higher temperatures, prevents this valve from opening at all. The start valve sprays additional fuel into the common intake manifold.

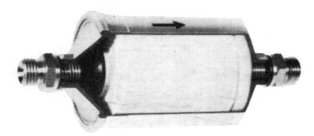

Fig. 55   Fuel filter

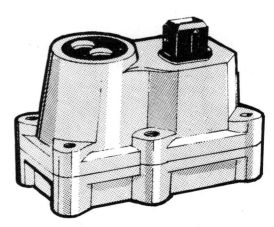

Fig. 56   Warm-up regulator

Fig. 57   Injection valve

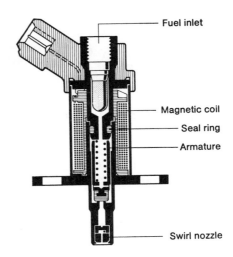

Fuel inlet
Magnetic coil
Seal ring
Armature
Swirl nozzle

Fig. 58   Start valve

Because of the relatively high primary circuit pressure (overpressure), the fuel is well atomized by the swirl nozzle.

The thermo-time switch, Fig. 51, closes or opens the power circuit leading to the start valve depending on the engine temperature. During cold starting, the power circuit is interrupted depending on the temperature of an electrically-heated bimetallic strip. The thermo-time switch has either a single heating coil W_1, or—if rapid heating is required—two heating coils, W_1 and W_2. When the switching temperature is reached, the contacts are opened, cutting off power to the start valve. This switches heating coil W_2 off. Heating coil W_1 holds the contacts open until the end of the starting operation.

Hot or Warm Starting

As a result of suitable design of the various valves and of the fuel accumulator in the system, sufficiently high pressurization is assured even after the engine has been turned off for an extended length of time, so that formation of vapor bubbles in the lines is prevented.

CONSTRUCTION OF SYSTEM COMPONENTS

Electric Fuel Pump

The fuel pump in the CIS is a roller cell pump driven by a permanent-magnet electric motor, Fig. 52. The rotor disc, mounted on the motor shaft, is fitted with metal rollers in notches around its circumference which are pressed against the excentrically designed pump housing by centrifugal force and act as seals. The fuel is carried in the gaps between the rollers and is then forced into the fuel injection tubing, Fig. 53.

In this system, the fuel flows directly around the electric motor. There is no danger of explosion, however, because there is never a combustible mixture inside the pump housing. This pump delivers several times the quantity of fuel actually required, so the excess fuel is diverted at the primary circuit pressure regulator and it flows under no pressure back to the fuel tank.

Fuel Accumulator

The fuel accumulator is divided by a diaphragm into the accumulator chamber and the spring chamber, Fig. 54. In front of the diaphragm is a steel partition with a plate valve for fuel intake and a restriction bore for return of the fuel. The diaphragm chamber is filled through the plate valve and a large hole. The diaphragm itself bulges downward against the pressure of the spring until it reaches the spring plate limit stop on the accumulator housing under the diaphragm; it remains in this position as long as the vehicle engine is operating. When the engine is turned off, the space inside the fuel accumulator becomes available for the system, and the fuel flows back through the restriction.

Fuel Filter

The fuel filter is a line filter with a paper cartridge, Fig. 55. Following the paper cartridge is a fine-mesh filter designed to catch any particles which may be released by the paper cartridge. For this reason, the direction of flow printed on the housing must be strictly adhered to. If necessary, the entire filter should be replaced. The filter is connected at both ends into the fuel system by means of threaded connectors.

Mixture Control Unit

The main components in the mixture control unit are the air-flow sensor and the fuel distributor, Fig. 36.

The relationship between the air-flow sensor and the fuel distributor is set by means of the idle mixture adjusting screw. When this screw is turned to the right a richer mixture is produced,

Fig. 59   Thermo-time switch

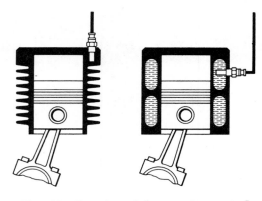

Fig. 60   Position of thermo-time switch

when turned to the left, a leaner mixture is produced. The idle mixture is adjusted with this screw.

Warmup Regulator

A two-hole flange is used to mount the warmup regulator on the engine, Fig. 56; the electrical connection is made by means of a plug connector. The control pressure is low with a very cold engine. As the engine warms up, this pressure rises continuously to the normal overpressure. The return line is connected to the smaller threaded connection.

Injection Valve

The injection valve, Fig. 57, opens automatically by fuel pressure and has no metering function. It is supported in a specially shaped molded rubber part. This valve is pressed, not screwed, into place. The hexagonal section on the valve shaft is provided to hold the valve stationary while the fuel line is attached. The fuel is finely atomized, even at a low rate of flow.

Start Valve

In the electrically-operated start valve, Fig. 58, a helical spring presses the movable armature in the magnetic circuit together with the seal against the valve seat and closes the fuel inlet. When the armature is drawn back, however, the fuel inlet is opened. The fuel then flows along the sides of the armature to the swirl nozzle. There, a swirling motion finely atomizes the fuel.

Thermo-Time Switch

The thermo-time switch, Fig. 59, limits the length of time that the start valve remains open, and at higher temperatures it prevents the start valve from opening at all. This switch is therefore mounted on the engine in such a way that it can absorb heat from the engine, Fig. 60.

Auxiliary-Air Device

The auxiliary-air device, Fig. 61, is mounted by

means of a two-hole flange at some point on the engine where the temperature is characteristic of the engine's operating condition. A plug is provided for the electrical connection.

ELECTRICAL CIRCUIT

The air-flow sensor plate in the air-flow sensor activates a switch which is closed when the engine is turned off. When the air-flow sensor plate rises from its seat, however, the ground line from Relay I is opened.

When the ignition is switched on (Terminal 15), Relay I is energized. Relay II remains at rest, however, and the electric fuel pump remains switched off.

When the engine is started (Terminal 50), the control current for Relay II flows through the working contact of Relay I, and the electric fuel pump is switched on through the working contact of Relay II. At the same time, current starts to flow to the warm-up regulator and to the auxiliary-air device. The start valve is also switched on at this time through a thermo-time switch.

As soon as the engine draws in air, the air-flow sensor plate in the air-flow sensor rises from its seat and opens the ground line from Relay I. Relay I is then de-energized and returns to its off-position. Relay II remains energized, however, and the electric fuel pump continues to operate.

If the engine comes to a stop as a result of exceptional conditions, the electric fuel pump is also automatically stopped even though the ignition is still turned on. This results from the switch at the air-flow sensor closing. This switches Relay I to the working position and interrupts the control line leading to Relay II. At this point there is no longer a connection from Terminal 30 to the electric fuel pump.

Electricity

Review Questions for This Chapter on Page 517

INDEX

Electrical Conductors. 516
Electrical Measurements 513
Types of Circuits. 512

If electricity seems more mysterious than some of our more prosaic tools like hammers and pliers, it is mainly because electricity is invisible. There is no way of seeing an electric current flow through a wire.

But that does not mean that electricity is hard to understand. Nobody ever saw the wind, either. We see trees blowing or dust moving, and we take it for granted that there is an invisible force—a moving air *current*—causing these reactions. We can easily learn how to use this invisible current to sail a boat or fly a kite without bothering too much why the air current happened to be there. In the same way, we can learn to understand and use the invisible current we call electricity.

An electric current is simply a gale of electrons moving at the rate of thousands of miles per second. They flow readily through a wire because the individual atoms are spaced at relatively vast distances.

There are multitudes of free or unattached electrons in the vast spaces between the atoms in the wire. When no current is flowing these electrons wander around haphazardly.

However, when the wire is connected into an electrical circuit, the voltage acts as an invisible force to drive these electrons through the wire at fantastic speeds, just as the invisible wind blows a cloud of leaves between the trees in a forest.

It is sufficient for our purpose to picture voltage as an invisible force which urges electrons to flow through a wire. Thus voltage may be thought of as electrical pressure just as a wind is caused by air pressure. Note that the atoms in the wire offer resistance to the flow of electrons just the same as a forest of trees offers resistance to flow of wind.

Whenever an electron is in motion it somehow surrounds itself with a proportionately small magnetic "field". Therefore, whenever current flows through a wire, the billions of moving electrons produce a magnetic "field" surrounding the wire, the strength of the field being proportional to the amperage (that is, proportional to the number of electrons involved). This phenomenon of magnetism is of importance in connection with the various electro-magnetic units on a car.

It was Benjamin Franklin who first realized that current flow through a circuit had to be in one direction or the other and so he guessed that flow was from positive to negative. Ever since that day it has been customary to speak of current flowing from positive to negative—and in practical electrical work it is still customary to think of current flowing from positive to negative even though just the opposite is true.

Positive and negative terminals are frequently called plus (+) and minus (−) terminals. Battery and generator terminals are marked with (+) and (−) signs to denote their polarity.

In an automobile, current is supplied to the various electrical units by either the storage battery or the generator. The generator is driven by the engine. When the engine is stopped, the battery supplies current to operate the starter, ignition system, lights, horn and other units.

When the engine is running above a minimum speed corresponding to about 10 m.p.h. in high gear, current for the various electrical units is supplied by the generator. Equipped with an alternator, another type of generator, the engine at idle will supply sufficient current to more than meet its own ignition requirements. Hereafter, the term generator is meant to include both ordinary generators and alternators.

Usually the generator produces more current than is required for the electrical units and the excess flows to the battery where it is stored as chemical energy.

Some time later when current is needed when the engine is stopped, the chemical energy may be converted back into electrical energy and allowed to flow to the various units in the electrical system.

A generator or battery does not create electricity. The electricity is always present in the form of the very small unit of electricity carried by each wandering electron. All a battery or generator does is to create the pressure that moves these hordes of electrons against the resistance of the circuit.

The action of a generator might be compared with an automobile fan. As the fan rotates, it creates a current of air by piling up air to the rear of the fan, creating a pressure, and sucking air away from the front of the fan, creating a suction.

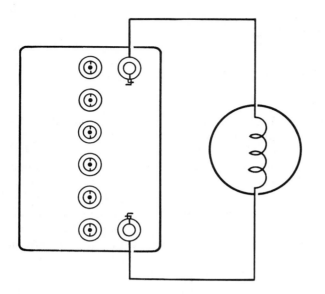

Fig. 1   A simple electrical circuit con-sisting of a lamp connected to a battery

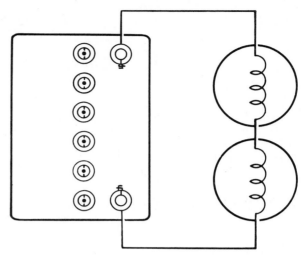

Fig. 2   Series circuit. Current must flow through one lamp after the other before returning to the battery

We don't have to know everything there is to know about the air or the fan to understand that these whirling blades will suck air in on one side of the fan and blow it out on the other.

Similarly, as the auto's generator rotates, it creates a current of electrons, pushing them out at one terminal and sucking them in at the other. We don't have to know all the physicist's theories about electricity in order to picture that action.

The force or pressure which "blows" the current along its way is commonly expressed in volts. Volts are comparable to wind pressure.

Note that electricity, unlike a stream of water or air, will not flow unless there is a complete circuit from the generator or the battery out through the various electrical devices and then back to the generator or battery again. Break the circuit any-where and the current will stop flowing.

Suppose we connect up a lamp bulb and a car storage battery, as in Fig. 1. This forms a *circuit*, for the current of electrons flows in a complete circuit from the battery along one wire to the lamp, through the lamp filament, and back along the other wire to the battery again.

TYPES OF CIRCUITS

There are three types of circuits: series, parallel and series-parallel. For example, if you wanted to hook up two lamps with an automobile battery, there are two ways in which you might do it, namely, in series and in parallel.

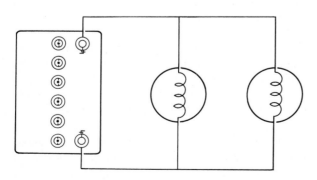

Fig. 3   Parallel circuit. Current from battery is divided between two lamps

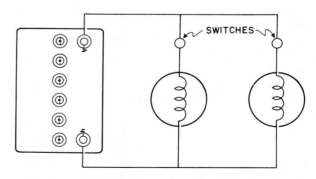

Fig. 4   Series-Parallel circuit. Lamps are in parallel with cir-cuit to battery but switches are in series with their lamps

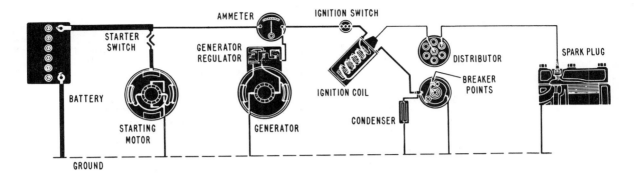

Fig. 5. *The major units in the electrical system are connected in parallel while the minor units, especially switches, are connected in series with the units they control. Note that breaker points, cutout relay and generator regulator points come under the head of switches. The broken line indicates that this part of the circuit consists of the frame, engine and other metal parts of the car*

Series Circuit—In Fig. 2 the current must flow through one lamp, then through the other before returning to the battery. The two lamps are said to be connected in series.

Parallel Circuit—In Fig. 3 the current divides. Part of the current flows through one lamp, while the remainder flows through the other one. This is called a parallel circuit since there are two parallel paths for the current.

Series-Parallel Circuit—Fig. 4 shows a series-parallel circuit. The lamps are in parallel but each switch is in series with its lamp.

All three circuits are of great importance with respect to the electrical system of the automobile since all major units are in parallel with the battery and the generator, while the switch or relay controlling each unit is in series with the unit which it controls, Fig. 5.

ELECTRICAL MEASUREMENTS

A current flow of a certain number of amperes may be compared with the flow of so many gallons of water per minute through a pipe. Both amperes and gallons indicate the quantity flowing.

Pressure in pounds per square inch is used to force the water through the pipe while pressure in volts is required to force a current through a wire.

When water flows through a pipe some of the pressure is used to overcome the frictional resistance offered by the pipe. A wire also offers resistance to the flow of electric current. The unit of resistance is the ohm.

Different metals vary greatly as to their resistance. Copper has an exceptionally low resistance. Steel has 10 times the resistance of copper and cast iron 50 times. The resistance of copper cable used in an automobile is very small. For example, the resistance of No. 14 cable (about 1/16 inch in diameter) is only 1/1000 ohm per foot of length.

The resistance of a wire is proportional to its length. For example, the resistance of six feet of cable will obviously be six times the resistance of just one foot. It should also be obvious that resistance varies according to the cross-sectional area of the wire or cable. In other words if the area of a wire is half what it should be then the resistance will be doubled for any given length of wire.

When renewing cables on a car be sure that the diameter of the metal cable is adequate. If the diameter is too small the flow of current will be throttled and the electrical unit to which the wire connects will not work properly. For example, a lamp will be less bright than it should be.

In a given circuit with a given resistance the amperes of current flowing are proportional to the voltage. For example, if the voltage is doubled the amperage is also doubled.

Likewise, for example, if the battery provides 12 volts to deliver 2 amperes to a certain lamp cir-

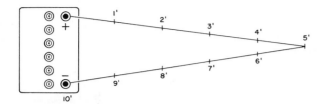

Fig. 6 A wire 10 feet long is connected to a 12-volt battery

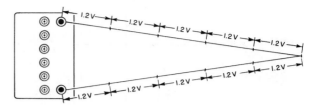

Fig. 7 The voltage drop along the whole wire is 12 volts. Therefore, the drop along each foot of wire is 12/10 volt, 6/5 volt

cuit, a generator supplying 15 volts to this circuit will deliver 15/12 x 2 amperes or 2½ amperes, making the light glow brighter than before.

The volt, ampere and ohm take their names from three early electrical researchers named Allesandro Volta, Andre Ampere and George Ohm. The "size" of the volt, ampere and ohm have been so chosen that a pressure of one volt applied to a resistance of one ohm causes a current of one ampere to flow.

This relationship between volts, amperes and ohms may be simply expressed by the equation:

$$\text{Volts} = \text{Amperes} \times \text{Ohms}$$

This equation is known as Ohm's law. It may be used to find out how many volts are required to produce a current of so many amperes against a resistance of so many ohms. For example: How many volts are needed to force 6 amperes against a resistance of 2 ohms? Using the equation above we have:

$$\text{Volts} = 6 \times 2 = 12$$

The equation may be transposed to permit the ohms resistance to be found when amperes and volts are known. As follows:

$$\text{Ohms} = \frac{\text{Volts}}{\text{Amperes}}$$

If we want to know the amperes when volts and ohms are known, the equation becomes:

$$\text{Amperes} = \frac{\text{Volts}}{\text{Ohms}}$$

A fully charged battery provides approximately 12 volts at its terminals, considering only the new 12 volt automobile electrical systems and not the older 6 volt systems. In other words, a voltmeter connected across its terminals will read 12 volts. If we connect a wire to the two terminals, the voltage drops gradually from one end of the wire to the other, the total drop, of course, being 12 volts. For example, if the wire is 10 feet long, the voltage drop will be 12/10 volts per foot, as in Fig. 6; if 100 feet long, 12/100 volts per foot, and so on.

Therefore, let us assume that the 10 foot wire has a total resistance two ohms (2/10 or 1/5 ohm per foot), Fig. 6. Ohm's law states that:

$$\text{Amperes} = \frac{\text{Volts}}{\text{Ohms}}$$

Battery voltage is 12 and the resistance of the wire is two ohms. Therefore, the current flowing through the wire is:

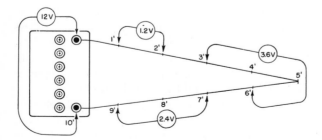

Fig. 8 Voltmeters show the voltage drop along various portions of the 10-foot wire

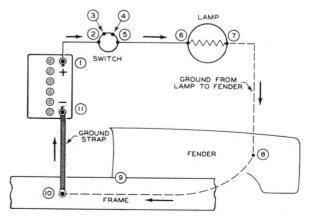

Fig. 9 Simplified diagram of a lamp circuit on a car

$$\frac{12 \text{ Volts}}{2 \text{ Ohms}} = 6 \text{ Amperes}$$

Remembering that the resistance of the wire per foot is 2/10, or 1/5, the voltage drop per foot is:

$$6 \text{ Amperes} \times 1/5 \text{ Ohm} = 6/5 \text{ Volt}$$

The voltage drop per foot is shown in Fig. 7.

The same wire is shown in Fig. 8 with voltmeters connected in various ways to show the voltage drop between various points on the wire. The voltmeters are represented by circles and the voltage drop is indicated within the circle.

Fig. 9 shows a simplified lamp circuit with arrows indicating the flow of current. There are cable conductors from battery to switch, from switch to lamp and from the lamp to a ground on the front fender. From thence, current flows through the fender to the frame and along the frame to the ground strap, then through the ground strap back to the battery. Path of current through fender and frame is shown by a broken line.

There are 11 connections in this simplified circuit, numbered 1 to 11. No. 8 represents the metal-to-metal contact between fender and frame. There are two connections within the switch, designated by numbers 3 and 4. When all these connections are clean and tight the voltage drop at each connection is negligible. Likewise the voltage drop in the conductors is negligible (including fender and frame). Therefore practically the entire 6 volts at the battery is available at the lamp and therefore the lamp glows with satisfactory brightness.

However, even slight looseness at any connection will increase the voltage drop at this point and thus rob the lamp of some of the voltage it needs.

A loose connection may also permit dust, lubricant and/or water to enter. Water permits corrosion, including rust. Dust, lubricant and corroded surfaces offer increased resistance to flow of current. Voltage drop at these points may become sizable and proportionately diminish the brightness of the lamp. If the surfaces at a connection are completely coated with dirt, lubricant and/or corrosion the resistance at such a point may be infinite with the result that no current whatever flows. In this case almost the entire voltage drop of the battery will occur across this connection.

A voltmeter is a most useful instrument to locate a loose or dirty connection—not forgetting a rusty fender-to-frame connection.

It should be remembered that 6 to 8 volts is such a low electrical pressure that even a slight amount of dirt may completely stop the flow of current.

What has just been said concerning the simplified lamp circuit, Fig. 9, is equally true of all the circuits in the car.

When current is drawn from the battery there is a drop of voltage across the battery terminals, the amount of drop being proportional to the amperes flowing. The drop is slight for moderate amounts of current—say 25 amperes or so. However, when the starter switch is closed, a large current (300 amperes or more) flows through the starter circuit and voltage across the battery drops to about 4½ —a loss of 7½ volts (still considering a 12-volt system). This drop is due to the fact that the current flow is so heavy that the battery is unable to maintain its voltage. In other words, its chemical processes are not fast enough to keep the voltage up where it should be.

In any electrical device it is the quantity of amperes which produces the result we are after. This remark holds true regardless of the type of device, whether it is a lamp, ignition coil, starting motor, cigarette lighter or what not. In every case, a certain amperage is required to do the job in hand. The necessary amperage is that called for by the size of the unit. So it should be obvious that a 100 watt bulb should require about twice the amperage of a 50 watt bulb if both are to glow with the same brilliancy. Likewise a 2 horsepower starting motor requires about twice the amperage of a 1 horsepower motor. An ignition coil which needs 6 amperes will be much more powerful than a 4-ampere coil. A 1 horsepower 6-volt starting motor can be made to produce 2 horsepower by applying 12 volts to it because doubling the voltage doubles the power. (The figures given in this paragraph are true in principle. For simplicity any complicating factors which may alter our round number values are ignored, and it is not recommended that the power of a starting motor be doubled by doubling the voltage.)

The preceding paragraph shows that electrical power depends on both volts and amperes. To be exact, electrical power equals:

$$\text{Volts} \times \text{Amperes} = \text{Watts}$$

The watt is the unit of electrical energy and 746 watts are the equivalent of one horsepower.

The ampere and ohm are actually measured according to internationally accepted methods, while the volt is derived from Ohm's Law.

One ampere is defined as the (constant) current required to electrolytically deposit .00025 pounds of silver per second when the current flows through a silver nitrate solution.

One ohm is the resistance offered at 32 degrees Fahrenheit by a column of mercury approximately 40 inches long having a uniform cross-sectional area of .0016 square inches.

One volt is the electrical pressure (electromotive force) required to cause a current of one ampere to flow against the one ohm resistance of the mercury column just mentioned.

The value of the volt is calculated from Ohm's Law. In other words, after having determined the value of the ampere and the ohm as stated above, the value for the volt equals ampere times ohm, or:

$$1 \text{ Volt} = 1 \text{ Ampere} \times 1 \text{ Ohm}$$

ELECTRICAL CONDUCTORS

In an automobile, five principal types of conductors of electricity are used. These are wires, cables, straps, engine and chassis parts, and air (between the spark plug points). The term wire needs no definition. A cable consists of numerous strands of wire twisted together. A cable is obviously more flexible than a wire of the same diameter. Therefore, it is easier to handle and not so likely to break. A strap, as its name implies, is flat. It is woven of strands of wire for the sake of flexibility. On some cars a strap instead of a cable is used to connect the battery to the frame or engine. If the battery is grounded on the frame a ground strap may also be used between engine and frame in order to make a good electrical connection between the two. Except for the ground strap (or straps), all the conductors on the car are insulated. Generally speaking the conductor is enclosed in natural or synthetic rubber and then one or two layers of braided cotton or silk are added. Often in addition, the outer layer of braiding is lacquered. On high-tension cable, instead of covering the rubber with braiding and lacquer, the rubber may be enclosed in a rubber-like synthetic material.

An electric current produces heat in the conductors through which it moves. The larger the number of amperes flowing through a wire the hotter it gets. Also the greater the resistance in ohms the more the temperature of the wire rises.

The electric lamp is based on the principle just described. The lamp contains a slim filament of tungsten which is heated white hot by the amperes flowing through it. Tungsten is used because of its unusually high melting point (5000° F) which is twice that of steel.

The filament is designed to operate at a white hot temperature close to its melting point. The filament metal is so hot that it gradually vaporizes and in this way "wears out" until the filament becomes so thin it melts and breaks and then the lamp must be replaced. Meanwhile the interior surface of the bulb is blackened by the condensation of the vaporized metal. The hotter the filament the more efficient it is as a light producer— that is, the more light it emits for a given amperage. Hence it is good economics to operate the filament at such a very high temperature rather than reduce the temperature to increase the life of the filament. In short, the longer the filament life the more current is required for a given brilliancy of a lamp.

A slight reduction in voltage (and therefore amperage) delivered to the lamp results in a marked decrease in brilliance. Even a half volt reduction at the lamp will cause a large decrease in the amount of light produced. Hence the importance of keeping the lamp circuit connections clean and tight.

On the other hand, because the filament operates at a temperature close to its melting point, just a small increase in voltage above normal will vaporize the filament at a rapid rate and thus reduce lamp life to just a few hours or even to zero. For if there is sufficient increase in voltage the filament will melt instantly.

A cigarette lighter is based on the same principle as the bulb. A coil of wire is heated red hot by an electric current.

In designing electrical units for an automobile the problem of temperature rise caused by current flow must be kept in mind. Wiring within the units must not become hot enough to damage the insulation. Hence the capacity of a unit, such as a generator, is limited by the practical question: will the production of more amperes singe the insulation? And the same remark holds true for external wiring connecting the various units in the electrical system. Overheating of insulation is mainly avoided by using wires of adequate size.

However, the generator is equipped with a fan which blows cooling air through its interior. In this way the generator is able to deliver a higher amperage without damage to the insulation than would be possible without the air cooling feature.

Review Questions

Page

1. The movement of what kind of particles in a conductor results in an electric current? 511

2. When a wire is connected into a circuit, what electrical unit must be present to cause a flow of current? 511

3. What effect does an electron moving in a conductor have on its immediate sphere of influence? 511

4. What is polarity in an electrical circuit? 511

5. What is the true direction of current flow in an electrical circuit? 511

6. Does the true direction of current flow agree with Benjamin Franklin's observations? 511

7. What two units in an automobile's electrical system supply electric power? 511

8. Do the sources of electric power in an automobile create electricity? 511

9. What particular condition must exist in a circuit before current can flow through it? 512

10. Name three types of circuits. 513

11. Which particular unit of electrical measurement might be compared to pressure in a water pipe? 513

12. Which particular unit of electrical measurement might be compared to resistance in a water pipe? 513

13. Which particular unit of electrical measurement might be compared to rate of flow in a water pipe? 513

14. Why is copper used almost exclusively in wiring? 513

15. What is meant by the cross-sectional area of a wire? 513

16. How does the cross-sectional area of a wire affect current flow? 513

17. What is OHM's law? 514

18. Write the equation which expresses OHM's law. 514

19. Write the OHM's law equation in the proper form to determine resistance when volts and amperes are known. 514

20. Write the OHM's law equation in the proper form to determine amperes when volts and resistance are known. 514

21. What is voltage drop in a circuit? 514, 515

22. What factor in a circuit is responsible for voltage drop? 514

23. Which instrument is most useful in locating loose or corroded electrical connections? 515

24. What unit in an automobile's electrical system causes the greatest voltage drop when in operation? 515

25. What two factors are used to calculate wattage? 515

26. How many watts are equivalent to one horsepower? 515

27. What is the difference between a cable and a wire? 516

28. What is a strap as used in an automobile's electrical system? 516

29. What characteristics of a bulb filament causes it to glow when current is passed through it? 516

30. Why is the generator on an automobile air cooled? 516

Magnetism

Review Questions for This Chapter on Page 529

INDEX

Electrical Motors............................527
How Generator Works.......................522

Most of the electrical units in a car employ not only electricity but also magnetism (produced by electricity) to make them work. The list of electro-magnetic units includes the starting motor (and any other electric motors), generator, ignition coil, horn, horn relay, solenoids, cutout relay, current regulator, voltage regulator, electric windshield wiper and so forth.

A permanent magnet, which requires no electricity to keep it magnetized, is used in all speedometers, in many ammeters and in the dash unit of the gasoline gauge. The best known permanent magnet is the horseshoe type often used as a toy.

One of the main purposes of this chapter is to explain briefly the relationship between magnetism and electricity so that the reader will understand how the various electro-magnetic units work. A few of the statements concerning magnetism will have to be taken at their face value because a complete explanation would be likely to confuse the average reader and thus do more harm than good.

For some reason, whenever an electron is in motion, a magnetic field surrounds that electron. (The word "field" means the field of magnetic influence which surrounds the electron.) Likewise, whenever an electric current flows through a conductor, a magnetic field surrounds that conductor. This total field is the sum of the fields surrounding all the electrons flowing through the wire. Therefore, the intensity of the field is proportional to the amperage. For example, doubling the amperage doubles the strength of the magnetic field because twice the number of electrons are flowing.

It is well known that a "magnetic" field attracts iron and steel. In Fig. 1 we show a wire which has been dipped in fine powder-like iron filings. In Fig. 2 we have doubled the amperage (by doubling the voltage by using two dry cells in series) and it will be seen that the quantity of iron filings clinging to the wire has approximately doubled.

Before continuing our explanation of electro-magnetism it is desirable to discuss permanent magnets.

Everybody knows that a magnet attracts iron, steel and cast iron. Obviously there is a magnetic

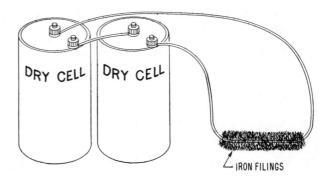

Fig. 2 *As compared with Fig. 1, two dry cells in series give twice the voltage, twice the current, and magnetic field strength is doubled, thus attracting approximately twice the quantity of iron filings*

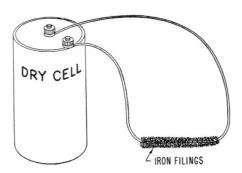

Fig. 1 *The magnetic field surrounding the wire attracts iron filings*

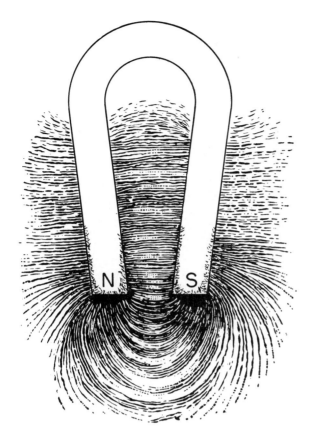

Fig. 3   Lines of magnetic force produced by a horseshoe magnet

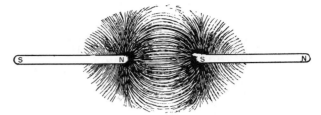

Fig. 4   Unlike poles attract each other

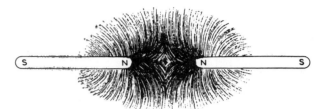

Fig. 5   Like poles repel each other

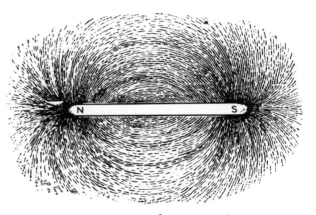

Fig. 6   Complete magnetic field of a bar magnet

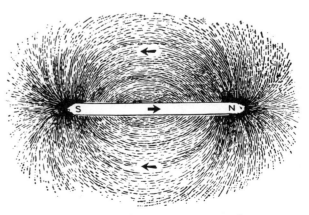

Fig. 7   Magnetic lines of force flow through the air from north pole to south pole and then back through the steel bar from south to north, thus completing the magnetic circuit

field of attraction extending from the magnet to the metal it attracts.

When a piece of hard steel is magnetized it retains its magnetism almost indefinitely and therefore is called a permanent magnet. On the other hand, soft steel and cast iron lose almost all their magnetism as soon as the magnetizing influence is withdrawn. The magnetizing influence may be another magnet or a conductor through which a current flows.

The size and shape of the field may be shown by placing the magnet under a sheet of cellophane and then sprinkling iron filings on the sheet, Fig. 3. The filings arrange themselves in definite lines which are called lines of magnetic force, or lines of force for short. These lines exert a force when they attract a piece of iron. Note that the lines are concentrated at the ends or poles of the horseshoe.

Experiment shows that the fields at the two poles are different. One is distinguished from the other by calling one the north pole and the other the south pole. Like poles repel each other and unlike poles attract each other. This statement can readily be proved by bringing the ends of two bar magnets close to each other as shown in Figs. 4 and 5. (A complete bar magnet is shown in Fig.

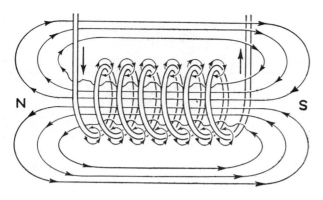

Fig. 8 *A coil of wire is called a solenoid*

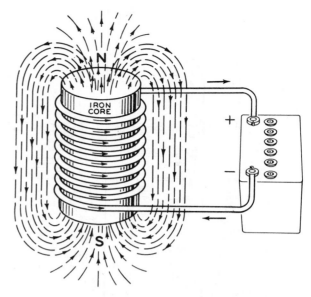

Fig. 9 *An iron core inserted within the solenoid coil concentrates the strength of the magnetic field*

6.) In Fig. 4 the unlike poles attract each other, the lines of force acting like the pull of so many rubber bands. In Fig. 5 the like poles are repelled, the lines of force acting as though they were stiff rubber bristles. In an electric motor it is magnetic repulsion which rotates the armature and thus changes electrical power into rotating mechanical power. On the other hand, in an electric generator, it is magnetic attraction by which mechanical power is changed into electrical power. These statements will be amplified later in this chapter.

A piece of iron or steel may be magnetized by placing it within the field of a permanent magnet. Fig. 4 will serve to illustrate the process. Assume that only the left bar is a permanent magnet and that the right bar is unmagnetized when it is placed in the position shown. The magnetic field around the north pole of the permanent magnet at the left somehow transforms the piece at the right into a magnet, creating a south pole. Thus the magnet at the left attracts the formerly unmagnetized piece at the right. It should be noted that whenever a piece of iron is magnetized at one end, an opposite pole appears at the right end. Fig. 6 illustrates the idea.

Magnetic lines of force "flow" through a circuit the same as does electricity. They flow from north to south pole, and then from south pole through the iron or steel magnet back to north pole, Fig. 7. There is a law for the flow of magnetism which corresponds to Ohm's Law. Ohm's Law states that: Volts (pressure or electromotive force) *equals* amperes (intensity or quantity of current) *times* resistance (to flow).

The corresponding magnetic law states that: Magnetomotive force (pressure) *equals* flux (intensity of flow) *times* reluctance (resistance to flow). Reluctance of air is 2500 times that of iron, which means that for maximum flux, any air gaps in the circuit of iron or steel should be a minimum

as to thickness or clearance.

The force of an electro-magnet is concentrated and intensified by making the wire into the form of a coil, Fig. 8, called a solenoid. The strength of the field is proportional to the number of amperes flowing through the coil times the number of turns in the coil.

The lines of force in the magnetic field are still further concentrated by the insertion of an iron core, Fig. 9. Solenoids with iron cores are used to open and close relay switches. Generator and motor armatures and field poles, and ignition coils are solenoids with iron cores. All solenoid cores are made of soft iron or soft steel so they will lose practically all their magnetism as soon as the solenoid circuit is broken. It is the solenoid coil which supplies the magneto-motive force just mentioned to create the magnetic circuit.

A magnetic field represents stored magnetic energy just the same as a compressed spring represents stored mechanical energy. The mechanical energy in a compressed spring is released when the spring is allowed to expand to its normal position.

The strength of the magnetic field in a solenoid increases or decreases as the current rises or falls. When current flow is constant the strength of the magnetic field is constant.

The magnetic field represents electric energy which is stored in magnetic form. When the solenoid circuit is first closed, the current rises to a maximum. During this time some of the energy in the current is converted into magnetic energy and

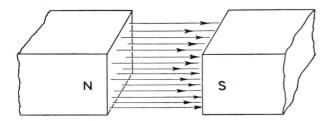

Fig. 10 *Close up sketch of the adjacent poles of two bar magnets. Arrows indicate lines of force flowing from north pole to south*

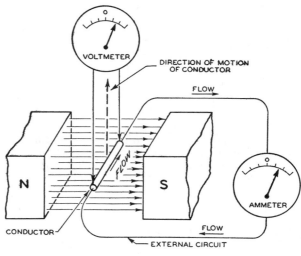

Fig. 13 *Reversing the motion of the conductor reverses the voltage and current*

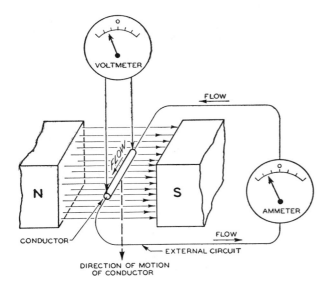

Fig. 11 *When a conductor moves at right angles to the lines of force, the conductor generates a voltage which causes a current to flow*

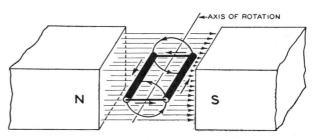

Fig. 14 *This rotating coil produces maximum current (and therefore voltage) in the position shown because its conductors are momentarily moving at right angles to the lines of force*

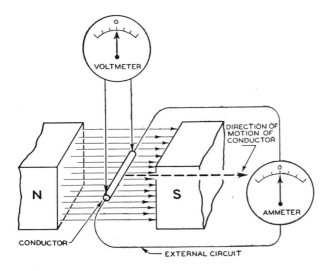

Fig. 12 *No voltage is generated in the conductor when it moves parallel to the lines of force. Therefore, no current flows*

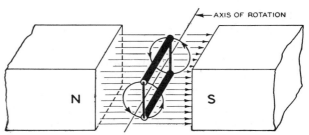

Fig. 15 *In this diagram the rotating coil is moving parallel to the lines of force and, therefore, produces neither voltage nor current. The coils in Figs. 14 and 15 are one-quarter revolution apart*

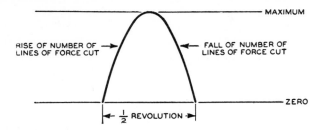

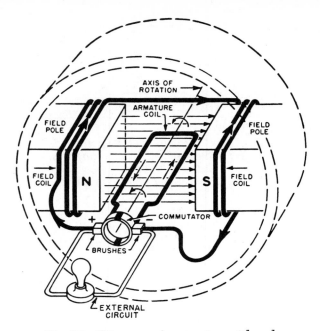

Fig. 16 This diagram illustrates how the number of lines of force cut and, therefore, voltage and current rise and fall during one-half revolution of the conductors shown in Figs. 14 and 15

stored as such in the field surrounding the solenoid. When the solenoid circuit is broken, the magnetic energy is transformed back into electric energy in the solenoid and a spark may jump across the switch points while they are being opened. The statements in this paragraph are of special interest in connection with ignition coils, as will be explained in the *Ignition Chapter.*

HOW GENERATOR WORKS

Fig. 10 is a diagram which shows the poles of two bar magnets, with arrows running from north to south pole. The arrows represent lines of force. (This diagram is nothing but a simplified drawing based on Fig. 4.)

If a conductor is moved at right angles to the lines of force, Fig. 11, a voltage will be generated in the wire, causing a current to flow through the wire and through the external circuit.

No voltage is produced in the conductor when it moves parallel to the lines of force, Fig. 12, because no lines of force are cut.

With north and south magnetic poles at left and right as shown in Fig. 11 and with the conductor moving downward, current flow is in the direction shown by the arrows. If the wire is moved upward instead of downward, current flow is in the opposite direction, Fig. 13.

Fig. 14 shows a single armature coil placed in a magnetic field. Assume that the coil was made by connecting the ends of the conductors shown in Figs. 11 and 13. As the coil is rotated, current flow is as shown by the arrows. This illustration explains the principle of the generator. Current flows toward you in the left conductor and away from you in the right conductor.

If the coil in Fig. 14 is rotated through one-quarter revolution, the conductors are then moving parallel to the lines of force. No voltage is generated and therefore no current flows, Fig. 15.

One-quarter revolution after Fig. 15 the coil is

Fig. 17 Diagram of generator with only one armature coil and only two commutator segments. Armature coil is moving at right angles to lines of force

again in the same situation shown in Fig. 14.

To summarize: When a conductor is rotating in a field, Fig. 14, it cuts a maximum number of lines of force when moving at right angles to them, and cuts no lines when moving parallel to them, Fig. 15. During the rotation of a conductor from the parallel to right angle position (from Fig. 15 to 14) the number of lines of force cut is gradually increased from nothing to the maximum. Likewise when the conductor rotates from the right angle position to parallel, the number of lines cut is decreased from a maximum to nothing.

Fig. 16 shows how the lines cut, rise from zero to maximum and then fall to zero again during each half-revolution. Voltage and current rise and fall in the same manner. This rise and fall occurs endlessly during all the half revolutions through which the conductors are rotated.

The principles discussed in connection with Figs. 1 to 16 will now be applied to a simple direct current generator with only one armature coil, Fig. 17. In this diagram there are the same magnetic poles and the same armature conductors as shown in the preceding pictures but some necessary details have been added. Voltage, and the resulting current, are produced by the rotation of the armature coil in the magnetic field.

The construction of the simple generator shown in Fig. 17 is as follows: The single armature coil is located in slots in a cylindrical piece of soft steel called the armature core (not shown). The ends

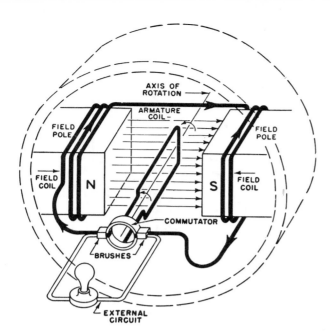

Fig. 18 *Same as Fig. 17 except armature coil is moving parallel to lines of force*

series and are connected to the brushes. The external circuit is also connected to the brushes.

Even though the pole pieces and the armature core are made of soft iron or soft steel, a slight amount of magnetism (residual magnetism) remains in them from the last time the generator was used. If the generator is a new one it must be initially magnetized by connecting it to a battery for a moment. Therefore, a generator has a slight but not negligible magnetic field when it is not in use.

When the armature coil is first rotated, a slight voltage is produced, causing a slight current to flow through the field coils. This current increases the strength of the magnetism in the field coils and in this way, in a few seconds, the voltage in the generator is built up to its normal value and the generator is ready for use.

Note that the field coil circuit and the external circuit are in parallel. Therefore, the current produced by the armature coil divides at the brushes. In an actual generator, about 1/15 of the current produced in the armature goes to the fields and the balance to the external circuit.

When the field circuit is in parallel with the external circuit, the device is called a "shunt" generator because some of the current produced is shunted into the field circuit.

When the generator is running, the magnetic lines of force produced by the armature and the magnetic lines of force produced by the field coils attract each other and try to prevent armature rotation. Therefore, engine power is required to pull apart or cut these lines of force in order to drive the generator. This attraction is the same in principle as the attraction of the two bar magnets shown in Fig. 4.

Current flows out into the external circuit through the positive (+) brush and returns by way of the negative (−) brush to the armature coil. One half revolution later, the right and left hand sides of the armature coil have changed places but so have the two segments; therefore,

of the coil are soldered to semicircular segments of copper to form the "commutator." The two segments are insulated from each other. Two copper composition "brushes" are pressed against the commutator, thus providing a sliding electrical contact when commutator and armature coil are rotated. The brushes, of course, are stationary. That is, they do not rotate. The brushes are supported in suitable holders which, for simplicity, are not shown. For the same reason the brush springs which hold the brushes against the commutator are not shown.

The magnetic field is produced electrically by coils of wire which are wound around the iron "field poles" (these are solenoids with iron cores but they are never called that. The terms used are field coils and field poles). The two coils are in

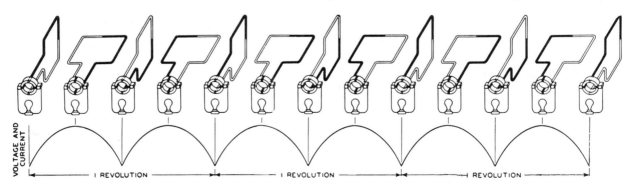

Fig. 19 *Diagram showing rise and fall of voltage and current during six half-revolutions*

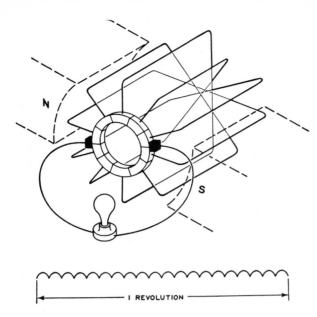

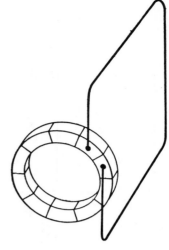

Fig. 20 *This armature, with 10 coils and 10 commutator bars, provides a comparatively smooth flow of voltage and current. There are 20 small pulsations during one revolution*

Fig. 22 *In an actual generator the ends of the coil are connected to adjacent commutator bars. For simplicity, this sketch shows an armature with only 10 commutator bars instead of the customary 20 or more*

Fig. 21 *When 20 armature coils and 20 commutator bars are used there is almost a perfectly smooth flow of current and voltage*

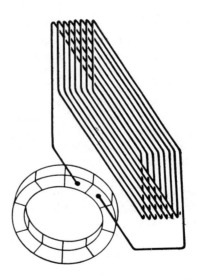

Fig. 23 *A 6-volt generator has about 8 turns per coil as shown. A 12-volt alternator, in contrast, would have more turns per coil*

current still continues to flow as shown—and so on for an infinite number of half revolutions. In other words, this is called a direct current generator because current and voltage always flow in the same direction.

Fig. 18 represents any quarter revolution before or after Fig. 17. At this time the armature coil is moving parallel to the lines of force and, therefore, no voltage or current are generated.

When the armature coil is moving from Fig. 18 to Fig. 17, there is a rise in voltage and current from zero to a maximum, and then when the coil moves from Fig. 17 to Fig. 18, there is a drop in current and voltage from a maximum to zero. (Within limits, the faster the armature runs the greater is the voltage and current generated.) Fig. 19 shows how voltage and current rise and fall during three complete revolutions of the armature coil.

Instead of the pulsating current produced by a single armature coil, a smooth current is desirable

and this is obtained by increasing the number of coils on the armature. An actual generator has upwards of 20 coils.

Fig. 20 shows a 10-coil armature as well as a diagram representing the voltage and current

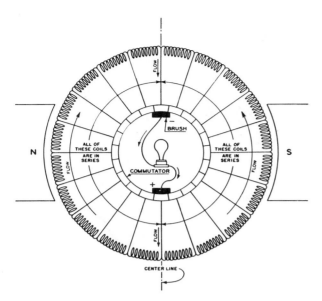

Fig. 24 *Wiring diagram of generator armature showing 20 coils and 20 commutator bars. Each coil has 8 loops which are intended to indicate that each coil has 8 turns. The brushes are located on the inner circumference of the commutator instead of on the outer circumference where they should be. This change in brush location was made simply so that the brushes would not interfere with the wiring diagram*

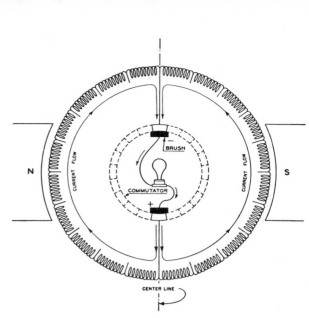

Fig. 25 *Wiring diagram showing that right and left sides of armature windings are in parallel with the two brushes*

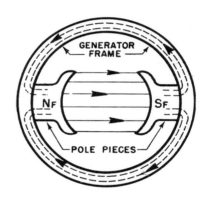

Fig. 26 *Diagram of magnetic field in two-pole generator showing lines of force*

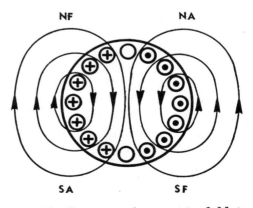

Fig. 27 *Diagram of magnetic field in generator armature showing lines of force*

pulsations during one revolution. As the armature rotates, each coil in turn is connected to the brushes to deliver current to the field coils and to the external circuit. Fig. 21 shows the current and voltage pulsations in a 20-coil armature. They give practically a smooth flow of current.

There are two serious practical objections to the 10- or 20-coil armature shown in Figs. 20 and 21. First, only the one or two coils whose commutator bars are in contact with the brushes are able to deliver current. All of the other coils which are cutting lines of force are unable to deliver current because they are not in contact with the brushes.

Second, when the commutator bars of a particular coil break contact with the brushes there is objectionable arcing between these bars and the brushes when the circuit is broken by the commutator bar moving out of contact with the brush. The arcing is due to the fact that the coil is generating its full voltage when the circuit is broken. Excessive commutator burning results.

As will be presently explained, both these difficulties are solved by connecting the coils to adjacent commutator bars, Fig. 22. In an actual generator, each coil, instead of having one turn,

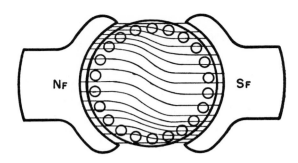

Fig. 28 Resultant magnetic field caused by interaction of the fields illustrated in Figs. 26 and 27

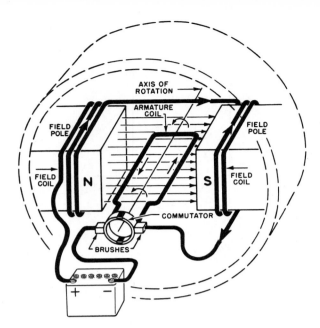

Fig. 29 Wiring diagram of series wound motor. The field coils are in series with the armature

has 8 or more, Fig. 23, in order to increase the voltage and current output. That is, 8 turns will give 8 times the voltage of one turn because 8 times as many lines of force are cut, and therefore provides 8 times the amperage of the single coil.

A typical generator armature with 20 coils of 8 turns each cannot be represented by the type of sketch shown in Fig. 23. The result would be nothing but a meaningless confusion of lines.

Therefore, we are forced to resort to the wiring diagram, Fig. 24, which shows that the ends of adjacent coils are connected to each other and also that each pair of coil ends is connected to a commutator bar. The commutator bars, of course, are insulated from each other. There are 8 loops in each coil in Fig. 24 to represent the 8 turns in the coil. Remember, however, that each of the 20 coils actually is arranged as shown in Fig. 23.

With the brushes located as shown in Fig. 24, the armature circuit is divided in two halves, left and right. These two halves are in parallel with the brushes. This statement will be clearer if we eliminate the commutator bars not in use, Fig. 25. It should be perfectly plain now that the left and right sides may be considered as separate circuits which are in parallel with the brushes.

It should be noted also that the coils in each half are in series. In either half, voltage is generated in all those coils which are cutting lines of force and the individual voltages add up to, let's say, 15 volts across the brushes. Thus, in Fig. 25, current flows up through the two halves of the armature from the vicinity of the positive brush to the negative brush under a total pressure of 15 volts. Also this 15-volt pressure "across" the brushes drives the armature current out through the positive brush, thence through both external circuit and field circuits back to the negative brush.

As stated previously, the armature is a sort of fan which blows electrons through the circuit wires, the number of electrons which pass a given point each second being proportional to the amperes of current.

The brushes are thicker than the width of the commutator bars, therefore, a brush makes contact with a new bar before it breaks contact with the preceding bar. Consequently the circuit remains unbroken and sparking at the brushes is avoided—unless contact between brushes and commutator is interrupted by dirt, oil, commutator roughness, weak brush springs, and/or sticking brushes.

The magnetic field produced by the pole pieces alone is shown in Fig. 26 and the magnetic field produced by the armature alone is shown in Fig. 27. NF and SF denote north and south magnetic poles (F stands for field). Likewise NA and SA indicates the fields produced by the armature. Remembering that like magnetic poles repel and unlike poles attract, NF attracts SA but repels NA with equal force. Similarly SF attracts NA but repels SA. Note that repulsion and attraction exactly balance.

However, when the magnetic fields in Figs. 26 and 27 are combined, the lines of force in the resulting field is as shown in Fig. 28. The combined field is distorted in such a way that the forces of attraction are greater than the forces of repulsion, therefore, engine power is required to rotate the armature against the excess force of attraction (in a generator).

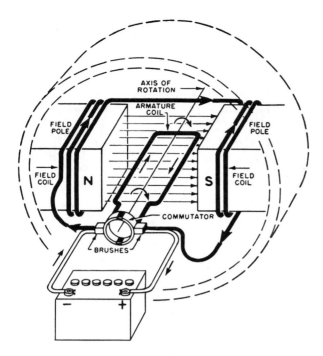

Fig. 30 *Wiring diagram of a shunt wound motor. The field coils are in parallel with the armature windings*

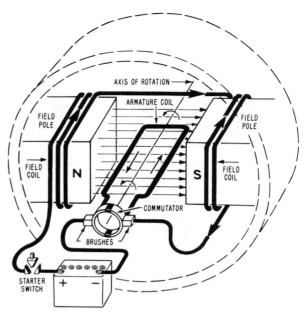

Fig. 31 *Wiring diagram of starter circuit*

Just the opposite is true in an electric motor. In this case, distortion makes the forces of repulsion greater than the forces of attraction and it is this excess of repulsive magnetic force which rotates the armature. A generator will run as a motor if connected to a battery.

ELECTRIC MOTORS

There are two common types of direct current electric motors, series wound and shunt wound. In a series wound motor, Fig. 29, the field coil windings and the armature windings are in series with each other whereas in a shunt wound motor the field coils are in parallel with the armature windings, Fig. 30.

The fan motor in an automobile heater, for example, is shunt wound because this type of winding gives a constant maximum speed because the field strength has a constant maximum value for a given voltage—say 15 volts. The speed of the fan may be reduced by introducing resistance in the field to reduce the number of amperes flowing through the field and thus cut down the field strength. The weaker the field, the slower the fan runs.

A starting motor is series wound because this type of winding provides maximum armature torque or turning effort at zero speed and this is

just what is wanted to start the engine moving. When the starter switch is first closed, Fig. 31, the battery supplies a maximum of about 300 amperes (12-volt system) and this heavy current produces extremely powerful fields in both the field poles and the armature core. But after just a few seconds the armature is up to its normal cranking speed and the amperage has dropped to about 150.

With a fully charged battery registering 12 volts, the battery voltage drops to 9 while the starter is in operation because the battery is unable to change its chemical energy fast enough to maintain 12 volts. However, 9 volts is adequate to force 150 amperes through the starter to drive the device at normal speed.

Automobile starters always have four poles in order to provide sufficiently powerful magnetic fields to produce adequate power for cranking the engine. On a few of the less expensive cars only two of the four poles have windings. The magnetic circuit, Fig. 32, is the same in both cases but naturally the fields are not quite as strong with only two poles wound as with four poles wound.

In a four-pole motor (or generator) there are two pairs of armature windings (instead of one pair for the two-pole machine illustrated in Figs. 29 and 30). Therefore, four brushes are required.

It is interesting to note that with 9 volts across the armature and a 150 ampere current flowing through the starter, the number of watts of power

527

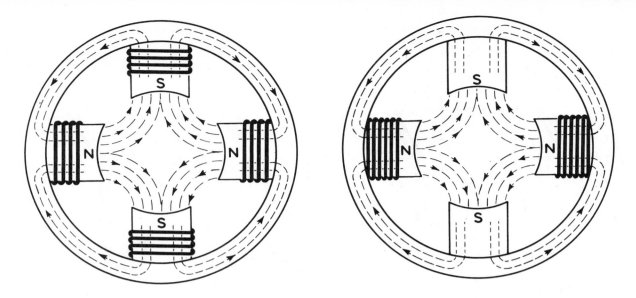

Fig. 32 *Diagram of starters with four field coils (left) and two field coils (right). The magnetic field circuits are the same in both cases*

(watts equals voltage times amperage) absorbed by the starter is 9 times 150 or 1350 watts. One horsepower equals 746 watts. Therefore, the starter absorbs 1.8 hp of electrical energy. Deducing electrical and mechanical losses in the starter, the net horsepower available for cranking the engine is about one and one half horsepower on the average car.

As explained in the chapter on *Electricity*, when current flows through a conductor, the conductor is heated. Therefore, all electrical apparatus must be designed with conductors of such size that the temperature rise is not only insufficient to melt the wire but not even sufficient to char the insulation. Therefore, all wires, whether within electrical units or whether connecting the various units in a circuit must be of sufficient size so that there is no danger of charring the insulation.

The size or capacity of an ignition coil, a voltage regulator, a starter or a generator, etc., are all designed with this temperature problem in view. For example, if it were not for the temperature

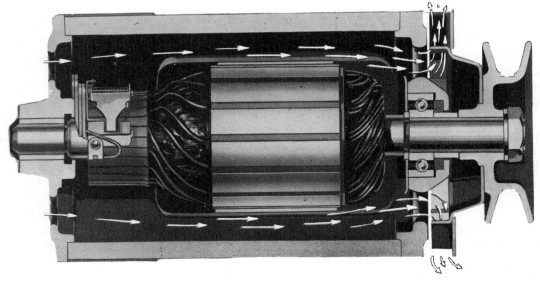

Fig. 33 *This typical generator is air-cooled by a fan attached to the armature. Arrows show direction of air flow*

rise, a much smaller generator could be used for a 12-volt, 35-ampere output and likewise for starters and other electrical units.

In the case of the modern generator, its output is increased by including an air fan (at its driven end) which cools the field and armature coils with the result that increased amperes may be produced without harm to the insulation, Fig. 33.

In the 12-volt generator, field strength and number of armature conductors are both increased without increasing the size of the generator. This is the most important factor in increasing generator output, and is possible because finer wires can be used to carry 12-volt currents; more armature conductors can be put in the same space; more powerful field coils, wound with more turns of finer wire, can be placed in the same size frame as used for the old 6-volt generator.

In the case of 12-volt starters, the field coils are wound with finer wire, which increases field strength. More armature conductors, carrying slightly lower currents than its 6-volt counterpart, react with stronger magnetic fields to produce extra cranking power. In spite of the greater power output the 12-volt starters are about the same size as 6-volt units.

If by chance the current in a circuit becomes excessive because of a short circuit, a ground or excessive voltage, the circuit is often protected by a fuse or circuit breaker. The fuse is made of a low melting point metal which melts and breaks the circuit before damage is done to the insulation. A circuit breaker is an electro-magnetic device whose points separate to break the circuit before amperage becomes high enough to injure the insulation.

Review Questions

Page

1. What produces magnetism? 518

2. What is the difference between an electro magnet and a permanent magnet? 518

3. Name four electromagnetic units in an automobile. 518

4. What governs the strength of a magnetic field? 518

5. Magnetic lines of force can be made visible by what simple device? 519

6. Does a magnet have a definite polarity? 519

7. In what direction do the lines of force move in the air gap between the poles of a horse-shoe magnet? 519

8. In which way do the conductors in a generator armature move in relation to the lines of force of the field. 522

9. Describe the construction of a simple generator. 522, 523

10. How are the magnetic lines of force produced in a generator field? 523

11. What does a commutator brush do in a generator? 523

12. Field poles and armature cores are made out of what kind of material? 523

13. What is residual magnetism? 523

14. What tends to resist armature rotation in a generator? 523

15. What is a shunt generator? 523

16. How does the number of individual coils and the number of windings in each coil on a generator armature affect power output? 524

17. To which two commutator bars is each end of any one armature coil fastened? 526

18. How are the ends of each adjacent coil connected? 526

19. What are two common types of direct current electric motors? 527

20. What step had to be taken to increase generator capacity and starter torque on late model cars without increasing the size of these units? 529

Electrical System

Review Questions for This Chapter on Page 559

INDEX

12 Volt electrical systems 539	Circuit, lighting 538
Cable, how to determine size of 544	Circuit, starting 535
Cable, selecting proper size... 551	Circuit, wipers 537
Cables, high-tension 545	Diagrams, wiring 531
Cables, ignition 545	Electrical systems, 12 volt ... 539
Cable size requirements 544	Electrical system service 540
Circuit, charging 535	High-tension cables 545
Circuit, horn 537	How to determine size of cable 550
Circuit, ignition 537	Ignition cables 545
Circuit, instrument 538	Lighting circuit 538

Locating trouble with voltmeter 540
Selecting proper size cable ... 551
Service, electrical system 540
Size, cable requirements 544
Systems, 12 volt electrical ... 539
Trouble, locating with voltmeter 540
Voltmeter, locating trouble with 540
Wiring diagrams 531

In dealing with the electrical system of the modern automobile, perhaps the most important point to remember is that the *Single Wire* system is used exclusively. Instead of running two wires from the battery to a lamp, for example, one wire is used to connect the battery to the lamp, and the frame of the vehicle is used in lieu of the other wire to carry the current to or from the battery, depending upon which post of the battery is grounded.

Where the ground is on the positive post of the battery, all operating electrical units are connected to the negative side of the battery, and current flows from the ungrounded (live) post of the battery through the chassis of the car (the ground) which is connected to the positive battery post.

Where the negative post of the battery is grounded, the current flow is in the opposite direction, through the chassis, then through the electrical units and back into the battery through the wires connecting them to the battery.

Thus, current flow is always from negative to positive regardless of which terminal is grounded. Confusion occurs in the minds of some people because they associate negative (−) with ground and, in fact, this convention is sometimes used. Nevertheless, in automotive practice the actual battery post that is grounded, whether positive or negative, is designated the "ground".

While all cars presently being produced have 12 volt systems, the text will frequently refer to 6 volt examples as well. This is done to furnish a contrast between modern 12 volt electrical systems and the older 6 volt systems or to better illustrate basic principles.

The principle on which the circuit is based is quite simple: All the main units are connected in parallel, Fig. 1, while the switches that control these units are naturally connected in series with the units they control.

The term switches usually mean those devices for closing and opening circuits which are operated by hand or foot at the will of the driver or his passengers. Switches that are automatically operated include the distributor breaker points, and

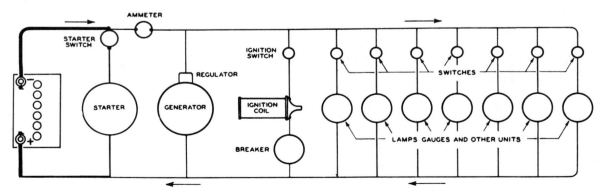

Fig. 1   All the main units in the electrical system are connected in parallel

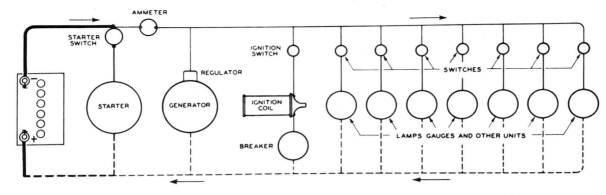

Fig. 2   This diagram is the same as Fig. 1 except the broken line indicates that this part of the circuit is through the metal parts of the car

various starter, generator and horn relays. For simplicity, only hand operated switches are shown in Fig. 1 except for the breaker.

WIRING DIAGRAMS

In Fig. 1, two parallel wires are shown and the various units are shown between them. Current flows out from the battery by way of the top wire and back to the battery by way of the lower wire. However, since the car is mostly made of metals, two wires are unnecessary, since the metal of the car itself can be used as the return conductor, Fig. 2, where the broken line is intended to indicate that the return circuit is through the car itself. Sometimes a heavy solid line is used instead of a broken line.

The same scheme is used in telegraph circuits, Fig. 3. The sender's instrument sends impulses out over a single wire. The impulses operate various receiving units along the way and the return circuit to the sender's instrument is through the ground (earth itself)—hence the name ground. We will call this a grounded return circuit. The earth is a satisfactory conductor for the purpose.

The sending instrument is simply a handy switch connected to a battery or other source of direct current. The receiving instruments are noisy, magnetically-operated relays which give audible "dots" and "dashes."

Fig. 4 is the same as Fig. 1 except that each ground connection is represented by an inverted triangle made of parallel lines.

Because of the large cross-sectional area of the various metal parts in the car the resistance is small even though iron and steel have a much higher resistance than copper. Therefore, the volt-

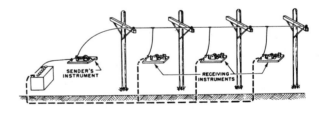

Fig. 3   Diagram of simple telegraph circuit

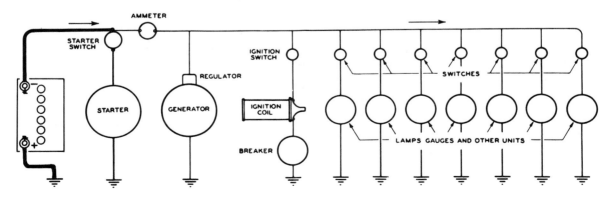

Fig. 4   This diagram is the same as Fig. 1 except "ground" connections are indicated by inverted triangles made of parallel lines

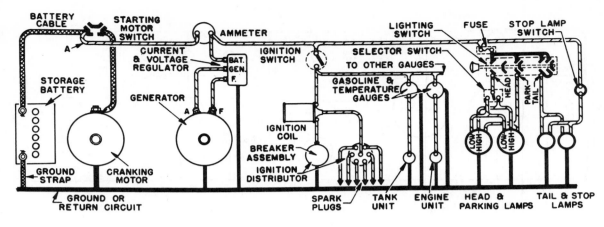

Fig. 5 *Typical car wiring diagram. The heavy black line indicates the grounded portions of the electrical system*

age drop in this part of the circuit is negligible provided that clean and tight metal connections exist between the various metal parts involved.

For example, provided headlamp units, stop and tail lamp units, etc. are held in place with a single bolt or screw having a clean, tight connection with the body, current will flow satisfactorily and voltage drop will be negligible. However, if the unit is loose and good metal-to-metal contact is not made, there will be a voltage drop and that particular light will not be as bright as it should be.

Fig. 5 is a wiring diagram which gives more details than Fig. 1 although the arrangement of units is similar. The ground or return circuit is shown as a heavy black line because this method seems clearer than the conventional methods shown in Figs. 2 and 4.

A car wiring diagram which includes all units

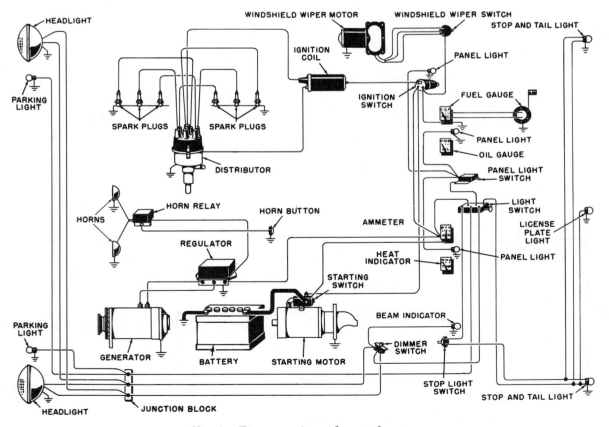

Fig. 6 *Diagram of car electrical system*

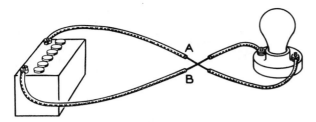

Fig. 7 *Chafed insulation causes a short circuit between A and B*

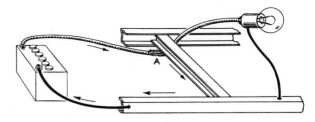

Fig. 8 *Chafed insulation at A allows bare wire to touch some metal part of car, short-circuiting the lamp as shown by the arrows*

Fig. 9 *Symbol for a 3-cell battery*

+

Fig. 10 *Symbol for positive side of circuit*

—

Fig. 11 *Symbol for negative side of circuit*

Fig. 12 *Symbol for a simple switch*

Fig. 13 *Symbol for a spring-type switch such as horn button or stop light switch*

Fig. 14 *Symbol for a ground connection. Circuits are usually completed through the chassis rather than through wires. Connections to various units not insulated from the chassis are called ground connections*

Fig. 15 *Symbol for a non-inductive resistance such as resistances in a generator regulator*

Fig. 16 *Symbol for an inductive resistance such as the field coils in a generator and starting motor*

Fig. 17 *Symbol for an ignition coil (induction coil). The primary winding is shown smaller than the secondary winding and in this case the secondary winding is connected to the primary winding. On some ignition coils the one end of the secondary winding would be shown grounded*

Fig. 18 *Symbol for a fuse*

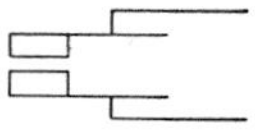

Fig. 19 *Symbol for breaker points*

Fig. 20 *Symbol for a condenser*

Fig. 21 *Symbol for a lamp bulb*

Fig. 22 *Left—Symbol for two connected wires. Right—Wires not connected*

except extras such as heater and radio is shown in Fig. 6. Because of its complication it is necessary to show the ground by triangles of parallel lines as in Fig. 4. All the units are in parallel although they may not seem to be at first glance.

The various units are usually grounded by electrically connecting one side of the unit to its housing. Then the housing is attached to some metal portion of the car. Therefore, the simple act of installing the unit automatically completes the ground connection.

The term "short circuit" means that the circuit has been accidentally shortened. For example, in Fig. 7, if the insulation of the two wires should rot or chafe away at A-B, so that the bare wires touch each other, current would flow from the battery, through point A-B instead of through the lamp and, therefore, the lamp would fail to light.

It is not uncommon for the insulation on a wire to be chafed away by rubbing on some metal part of the car. As soon as the bared wire touches the metal of the car a short circuit occurs. This type of short circuit is often called a "ground" or we say that the wire is grounded since current flows from wire to ground instead of to the unit to which it is supposed to go, Fig. 8. Chafed insulation at A causes current from the battery to flow through the frame cross member instead of to the lamp,

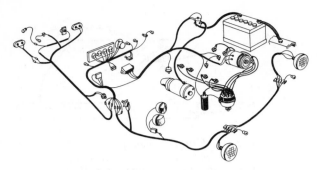

Fig. 23 *Layout of a typical wiring harness connected to various units*

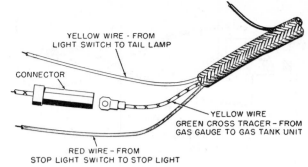

YELLOW WIRE - FROM
LIGHT SWITCH TO TAIL LAMP

CONNECTOR

YELLOW WIRE
GREEN CROSS TRACER - FROM
GAS GAUGE TO GAS TANK UNIT

RED WIRE - FROM
STOP LIGHT SWITCH TO STOP LIGHT

Fig. 24 *A color code to identify the various leads is used. Only three examples are shown*

and from thence through the frame back to the battery.

The usual wiring diagram is not always presented in the manner shown in Fig. 6. Instead of picturing the units in their realistic form, various symbols are used to indicate them. The most common symbols used are shown in Figs. 9 to 22, inclusive.

At this time it might be well to take the confusion out of the wiring diagram shown in Fig. 6. There are a great many circuits to follow. But inasmuch as only one circuit usually gives trouble at a time we will break them down into their seven basic circuits, tell what each includes, and what each one does. (In the illustrations, each circuit is emphasized by the heavy black portions.) The seven basic circuits are:

1. The starting circuit.
2. The charging circuit.
3. The ignition circuit.

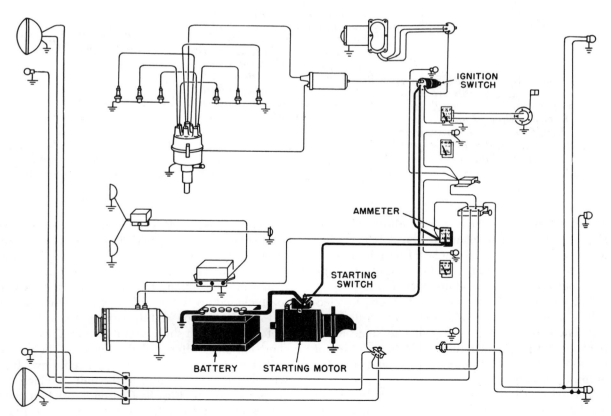

IGNITION
SWITCH

AMMETER

STARTING
SWITCH

BATTERY   STARTING MOTOR

Fig. 25 *Starting circuit diagram (heavy black)*

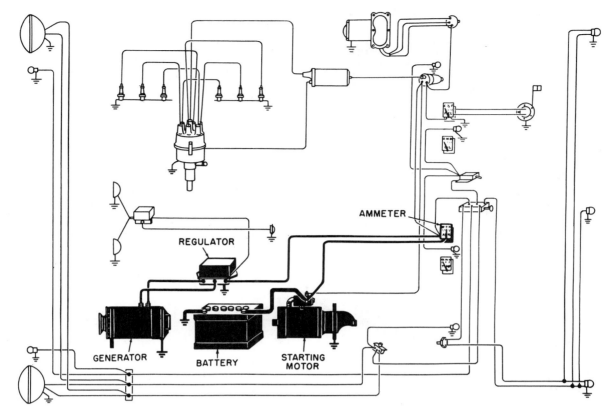

Fig. 26   Charging circuit diagram (heavy black)

4. The horn circuit.
5. The electric windshield wiper circuit.
6. The instrument circuit.
7. The lighting circuit.

Bear in mind that wiring diagrams are invariably schematic, which means that the position of the wires and units in the diagram are not necessarily the position they assume on the vehicle. Fig. 23 is an example of a typical wiring harness connected to various units. Fig. 24 shows a few examples of how a color code is used to identify wires and the units to which they are connected.

At the end of this chapter some car manufacturer wiring diagrams have been included so the reader will be acquainted with the various circuits and styles.

The Starting Circuit

As shown in Fig. 25, this circuit includes the battery, starting switch, ignition switch, starting motor, connecting wires and car frame.

The starting circuit is designed to carry high current with a minimum loss of voltage. The starting motor cranks the engine for starting while the battery supplies the power and the switch controls the operation.

The starting motor must supply a high cranking power for a short time. And it must engage its pinion with the engine flywheel when cranking, and disengage the pinion as soon as the engine starts.

The Charging Circuit

Fig. 26 illustrates the component parts of this circuit. They include the battery, generator, generator tell-tale light, regulator, connecting wires and car frame.

The generator, this term is used to include both ordinary generators as well as alternators, supplies all the electricity both for recharging the battery and for use by the other electrical equipment. Notice that the generator tell-tale light is used on this and subsequent diagrams to indicate when the generator is charging. Some cars come equipped with an ammeter instead of a tell-tale light that shows the actual amount of current the generator is providing at any given moment or the amount of circuit discharge if it is not charging.

The regulator controls the output of the generator to conform to the requirements of the various circuits, prevents high voltage and resultant damage to the generator and electrical system, and keeps the battery fully charged unless the

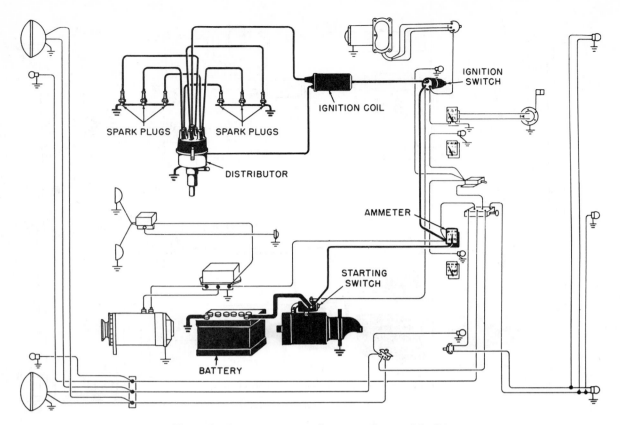

Fig. 27   Ignition circuit diagram (heavy black)

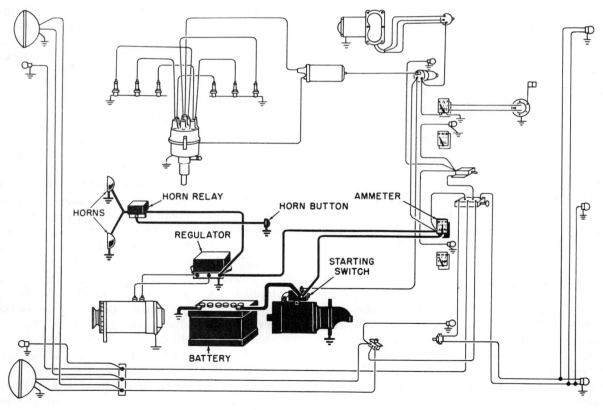

Fig. 28   Horn circuit diagram (heavy black)

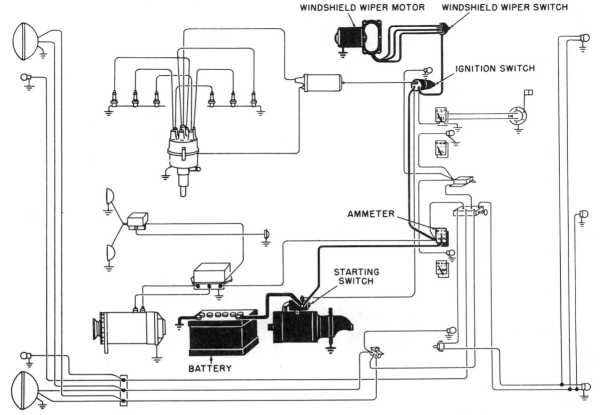

WINDSHIELD WIPER MOTOR   WINDSHIELD WIPER SWITCH

IGNITION SWITCH

AMMETER

STARTING
SWITCH

BATTERY

Fig. 29   Electric windshield wiper circuit diagram (heavy black)

demands have exceeded the capacity of the generator.

The Ignition Circuit

The circuit, Fig. 27, includes the battery, generator tell-tale light, ignition switch, coil, distributor, spark plugs, connecting wires and car frame.

The purpose of the ignition system is to ignite the fuel in each cylinder at the proper time and to vary the timing to suit the driving or operating conditions.

The ignition coil transforms the low battery voltage into high voltage.

The distributor times the spark and distributes the high voltage to the correct spark plug.

The spark plug is a device for making a spark to ignite the fuel inside the cylinder.

All battery ignition systems have the main components illustrated in Fig. 27. However, the design and location of the coil, distributor and spark plugs vary widely for different engines. The connections to the coil, distributor and spark plugs do not vary greatly, except for the location of the terminals and the order in which the cap towers are connected to the spark plugs.

The Horn Circuit

The dual horns illustrated in Fig. 28 have a circuit which includes the battery, horns, horn relay, horn button, connecting wires and car frame.

The horns are designed to make a warning sound when the horn button is operated.

The relay eliminates the necessity of the button carrying all of the current and reduces burning of the horn contacts.

When only one horn is used, the relay is usually eliminated and the horn is not grounded. The circuit is simply from the battery to horn and to ground and to button.

Often another type relay is used with dual horns. On this type, the relay winding is not connected to the battery terminal within the relay but is connected externally at some other point, usually the ignition switch. This makes the horns inoperative except when the ignition is turned on.

The Electric Windshield Wiper Circuit

The circuit shown in Fig. 29 is used for a single motor which operates two wiper arms and pro-

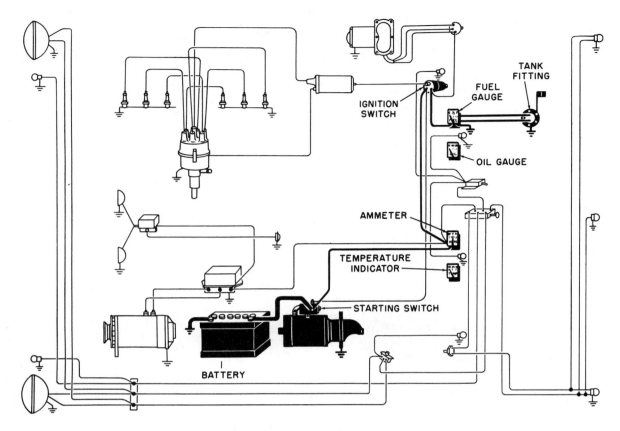

Fig. 30 Dash gauge circuit diagram (heavy black)

vides a fixed parking position. It includes the battery, wiper motor, wiper switch, ammeter, connecting wires and car frame.

Through crank action the motor operates two wiper arms and blades whenever the control switch is turned on. Usually the control switch includes a resistor which is connected in the circuit to give a second speed to the wiper motor.

Some windshield wiper motors are designed to operate only one arm. These are usually connected with a single wire from the motor through the switch and ammeter to the battery.

The Instrument Circuit

This circuit, Fig. 30, includes the generator, generator tell-tale light, fuel gauge, temperature gauge, and oil pressure tell-tale light. In actual practice, the generator dash unit may be either a tell-tale light or ammeter, the oil pressure dash unit either a tell-tale light or a pressure gauge, and the engine temperature dash unit either a tell-tale light or a temperature gauge. The gasoline supply dash unit is, of course, invariably a gauge rather than a warning light.

The arrangement in Fig. 30 with warning lights

for generator and oil pressure and a gauge for temperature was chosen to show typical wiring for both type dash units. In rare cases, engine temperature and oil pressure indicators are gauges that read the appropriate information directly and are connected to the engine independent of, and without connection to, the electrical system.

No matter whether a gauge or a warning light is used, electrical oil pressure and temperature indicators are similar in operation and circuitry to the fuel gauge. These circuits include the battery, an electrical sending unit mounted at the appropriate point on the engine or in the fuel tank and the dash mounted warning light or gauge. They are activated by the ignition switch so that when it is turned off the circuits are inoperative.

Electrical fuel, oil and temperature gauges may have one or two wires connecting the dash unit to the sending unit but all have a circuit which begins at the battery and is grounded in the send $\jmath$ unit.

The Lighting Circuit

The lighting circuit, Fig. 31, consists of the battery, generator tell-tale light, many lights, various switches, connecting wires and car frame.

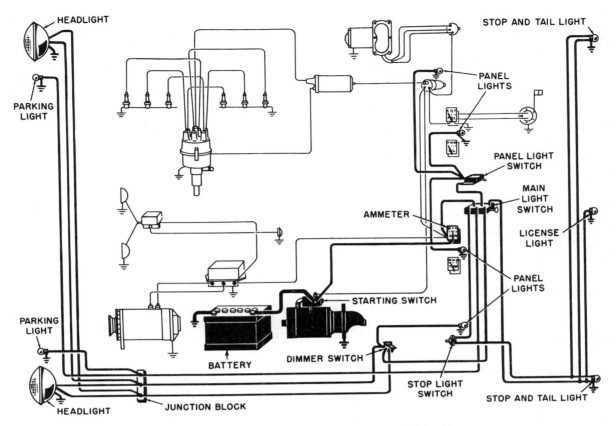

Fig. 31 *Lighting circuit diagram* (*heavy black*)

The lights may be for road lighting, panel lighting, signaling or for some other purpose. Each light or set of lights has one or more switches to control its use.

Some installations may not have as many lights as illustrated in Fig. 31, while other applications may have more. However, any additional lamps usually have a simple circuit from the battery through the ammeter (or indicator light) to the control switch or switches where it divides and goes to each lamp. The lamp socket is grounded to complete the circuit.

Protective devices consisting of fuses, fusible links and thermal cutouts are used in various combinations in the lighting and other circuits to prevent damage to electrical units and the fire hazard of circuit overloading. Such devices function by interrupting the circuit when current flow exceeds a certain figure, such as occurs when a wire short circuits.

There are other electrical circuits designed to control automatic transmissions, overdrives, power tops and other devices. Such circuits are widely varied, apply only to one particular application and do not have the universal usage of those already described.

12 VOLT ELECTRICAL SYSTEMS

The 12 volt electrical system, adopted on most cars starting in 1953, was chosen over the old 6 volt system for several reasons:

1. To provide adequate power to feed the ever growing number of electrical accessories used on present day cars.
2. To provide greater starting torque with the same size starters.
3. To provide greater available voltage in the ignition system for higher speed, higher compression engines.
4. To permit smaller diameter wiring to be used in the electrical system.

Generator Requirements

In 1935 or thereabouts, passenger car generators were designed to produce about 20 amperes. In 1952, generator capacity was increased to over 50 amperes in some cases and even this high output was found to be insufficient to supply the electrical needs of all the accessories used on some cars.

For all practical purposes, the limit of generator output was reached in 1952 with a 6-volt system as generators had become bulky and wiring heavy and clumsy. With a 12-volt system, the same amount of electrical power can be delivered with half the amperage, because power is measured in watts, which are simply amperes multiplied by volts. In other words, when a generator is required to deliver 600 watts, this amount of power can be delivered in two ways: 100 amperes at 6 volts or 50 amperes at 12 volts. In each case the amperes multiplied by the volts produces the same number of watts (600).

As it is the current in amperes flowing through a conductor that determines the required size or gauge of the wire, it is obvious that with a 12-volt system, smaller size wires can be used to deliver the same amount of power as the larger sizes would deliver in a 6-volt system.

Ignition System Requirements

Turning to the ignition system, the adoption of high speed, high compression 8-cylinder engines by most of the automotive industry has made it more difficult to maintain adequate voltage in the ignition system for all engine operating conditions.

Compression in the combustion chamber of the gasoline engine bears the same relation to the firing of a spark plug as does resistance in an electrical conductor to the passage of current through it. The higher the resistance of the conductor, the less current can flow through it. The higher the compression ratio, the weaker the spark. More

current can be forced through a conductor of a certain resistance value by increasing the voltage across the conductor. By the same token, the strength of the spark can be increased by increasing the voltage across the spark plug. This is one reason for increasing the system voltage from 6 to 12 volts.

Another factor is the higher speeds developed by the new type engines. With higher speeds, the old 6-volt ignition coil does not get sufficient time to become thoroughly saturated and therefore cannot produce enough energy to create a good spark. The 12-volt system is one way of putting more energy into the coil at high speeds without overloading and burning the breaker points at the lower speeds.

Comparing the available outputs of 6 and 12-volt coils at 3600 rpm engine speed, the 6-volt coil will produce 16,000 volts whereas the 12-volt coil will produce 19,000 volts. Thus, at high engine speeds the 12-volt coil produces adequate voltages for this type of operation where the 6-volt coil frequently fails.

Electrical System Changes

With a 12-volt system, many changes had to be made in the electrical units of the car. The battery, starter, generator, ignition coil, the various electrical motors, the light bulbs—all had to be redesigned for 12-volt use but their basic construction remains the same as the 6-volt units. *However, the ignition distributor, breaker points and condenser remain the same as in the 6-volt system.*

Electrical System Service

The following text will deal only with such service that pertains to the electrical system in general. Subsequent chapters describe the functions and service requirements of the various components such as starting motors and switches, generators and regulators, ignition systems, batteries and instruments.

LOCATING TROUBLE WITH VOLTMETER

A voltmeter is preferable to an ammeter in locating trouble in the electrical system because it

isn't necessary to disconnect any wires when making voltage tests. Besides, sometimes when an ammeter is connected in a circuit, the cause of the trouble is removed until the wire is reconnected or, if the trouble was due to a loose connection that wasn't apparent, the reconnected wire would eliminate the cause with the result that you would wonder what was the trouble in the first place.

It might be well to point out here that the difference between an ammeter and a voltmeter is that the ammeter measures the quantity of current flowing (amperes) in a circuit whereas a voltmeter indicates the pressure (volts) that pushes the current through the circuit. In view of this, therefore, an ammeter must always be connected in a circuit in *series*, Fig. 32, while a voltmeter can be connected in *parallel* with a circuit or *across* the circuit—never in series.

When a voltmeter is connected *across* a circuit, Fig. 33, it indicates the total voltage in the circuit. But when it is in *parallel* with a circuit, Fig. 34, any reading on the voltmeter is an indication of excessive resistance in the circuit and this reading is the voltage drop caused by the resistance. Of course, with the parallel hook-up, if the voltmeter reads zero, it indicates that all connections are clean and tight, and the cable or wire is the correct size and is in good condition.

Bear in mind that the flow of electrical current (amperage) depends upon the pressure (voltage). This voltage is used up as the current flows through resistances.

Frequently, this loss in voltage is extremely small. Therefore, a low-reading voltmeter, Fig. 35, is necessary to reveal it. The full-scale reading of a low-reading voltmeter for this purpose should be three volts, each volt divided into tenths. These divisions are important because, even though the loss of one-tenth of a volt seems unimportant, several such losses in the starting circuit, for example, may add up to each to keep the engine from starting.

The reason for such starting trouble is obvious when it is known that under cold start conditions, the voltage at the battery may drop to four volts or less, which is barely enough to crank the engine fast enough to start it.

NOTE—In subsequent chapters, the use of a low-reading voltmeter will be described as well as the permissible voltage drop limits recommended.

ELECTRICAL WIRING

The wiring supplied at the factory for starting, lighting, ignition and other circuits when the car is assembled is usually adequate. But when electrical accessories are added, it is important that the circuit in which they are connected be of sufficient capacity to take care of the increased electrical load. It may be necessary to replace the old cable with cable of a larger gauge or to add a separate cable to carry the additional load.

There are two things to remember in connection with cables: (1) A small cable offers more resistance to the flow of electrical current than a larger cable. Resistance goes *up* as cable capacity goes *down*. (2) A long cable offers more resistance than a shorter cable. *Resistance* goes up as *length* goes up.

Gauge of Cable

Cable size is expressed in terms of gauge. Gauge indicates the cross-sectional *area* of the copper in the core, Fig. 36. For example, a No. 2 gauge cable has 79% as much copper as a No. 1 gauge cable, Fig. 37, and, therefore, has only 79% as much current carrying capacity. Likewise, a No. 4 gauge cable, Fig. 38, has 50% as much copper as a No. 1 gauge cable and, therefore, has only 50% of the current carrying capacity.

The amount of current that a cable will carry increases as the area of copper increases (with minimum voltage drop). Likewise, the amount of voltage drop that will occur within a cable decreases as the copper area increases (current remaining constant), Fig. 39.

As stated previously, a long cable offers more resistance to current flow than a short cable.

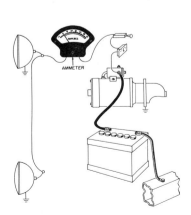

Fig. 32 An ammeter must always be connected in a circuit in series

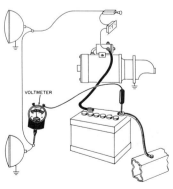

Fig. 33 A voltmeter connected across a circuit measures the total voltage in a circuit

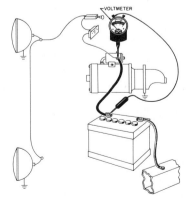

Fig. 34 A voltmeter connected in parallel with a circuit measures the resistance in the circuit expressed in terms of voltage drop

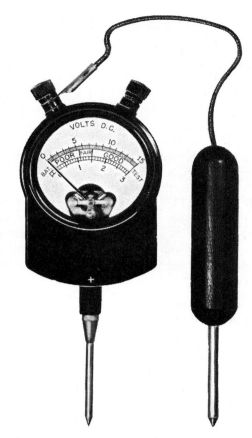

cable is the surest way to supply the starting motor with enough current at a high enough voltage.

In designing motor vehicles, engineers generally specify cables of No. 1 or No. 0 gauge because they realize that cable of ample size is necessary for satisfactory performance.

In warm weather with good connections and no extra current draw from lights or accessories, an engine will start with a smaller cable. But a lot of starting must be done under conditions that are far from ideal. With an undersize cable, the starting motor will not develop its greatest turning effort, and a fully charged battery may be unable to start the engine.

When the starting switch is closed in a normal starting circuit, the resistance in the circuit determines the amount of current. This current is of high amperage, which drops as the starting motor comes up to speed. With an abnormal circuit (undersize cables, corroded terminals and poor connections), closing the starting switch will show a lower current draw because of the loss in voltage with corresponding lower cranking speed.

We know that the load, or resistance, of the engine determines the speed of starting. However, a point to bear in mind is that all direct-current motors are also generators, and that when a direct-current motor comes up to speed it generates a

Fig. 35 *Low reading voltmeter with two scales. The 0–15 volt scale is used to check the total voltage across the circuit. The 0–3 volt scale is used in parallel with the circuit to check voltage drop. Wire is connected to 3-volt scale terminal. To use 15-volt scale connect wire to other terminal*

Therefore, in order to carry the same loads without excessive voltage drop, long cables must have a larger copper core than short ones, Fig. 40.

Starting Motor Cables

The starting motor must be supplied with enough current to crank the engine, and there must be a high enough voltage applied at the terminals of the starting motor to crank the engine fast enough to start it. Using No. 1 or No. 0 gauge

Fig. 37 *The greater the area of copper in a cable the greater its capacity*

Fig. 38 *The heavier cables carry the lower gauge numbers. No. 1 gauge has eight times the area as No. 10*

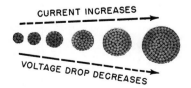

Fig. 39 *As the area of copper becomes greater, resistance decreases and current carrying capacity increases*

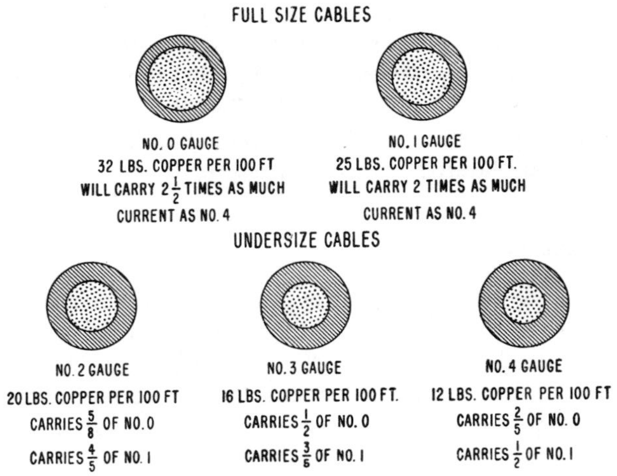
FULL SIZE CABLES
NO. 0 GAUGE
32 LBS. COPPER PER 100 FT
WILL CARRY $2\frac{1}{2}$ TIMES AS MUCH
CURRENT AS NO. 4
NO. 1 GAUGE
25 LBS. COPPER PER 100 FT.
WILL CARRY 2 TIMES AS MUCH
CURRENT AS NO. 4
UNDERSIZE CABLES
NO. 2 GAUGE
20 LBS. COPPER PER 100 FT
CARRIES $\frac{5}{8}$ OF NO. 0
CARRIES $\frac{4}{5}$ OF NO. 1
NO. 3 GAUGE
16 LBS. COPPER PER 100 FT.
CARRIES $\frac{1}{2}$ OF NO. 0
CARRIES $\frac{3}{8}$ OF NO. 1
NO. 4 GAUGE
12 LBS. COPPER PER 100 FT
CARRIES $\frac{2}{5}$ OF NO. 0
CARRIES $\frac{1}{2}$ OF NO. 1

Fig. 40 Doubling the length of a cable doubles its resistance. Therefore, a long cable must have a larger copper core than a short cable to carry the same current without excessive voltage drop

voltage opposite to that applied to its terminals. It is this voltage which governs the amount of current draw. That is why the speed of the starting motor causes the amount of current to decrease from the maximum as determined by the resistance of the circuit.

Voltage loss is the result of resistance. Since a rapid cranking will start an engine much more quickly than a slow cranking speed, current must be delivered to the starting motor with the least loss in voltage. In cold weather the voltage of a battery under load is less than in mild weather. Moreover, the oil in the engine is more viscous or heavy-bodied, making the engine more difficult to turn over. There is also a tendency for the valves to be sluggish in action and to stick. Starting in cold weather thus raises the current requirements, since a greater turning effort is needed to spin the engine. With undersize cables, it may not be possible to turn the engine over fast enough to start it.

Reserve voltage is important to the successful operation of a motor vehicle under all conditions. There must be reserve for cold weather, for the primary and secondary ignition circuits, and for the lighting circuits, all of which are fed through the battery cables.

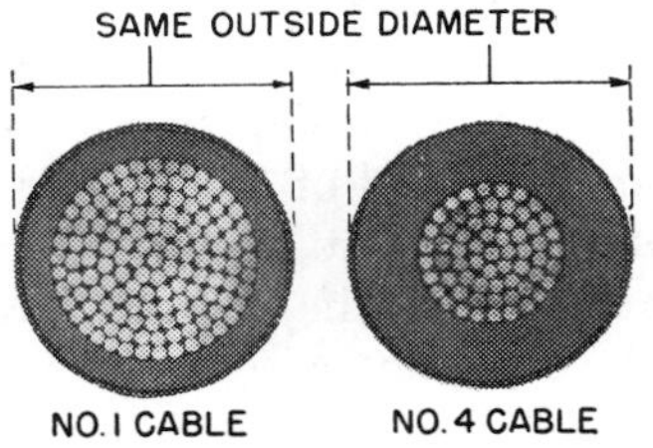
SAME OUTSIDE DIAMETER
NO. 1 CABLE
NO. 4 CABLE

Fig. 41 Showing two cables with the same outside diameter. As shown by the copper area, the outside diameter means nothing

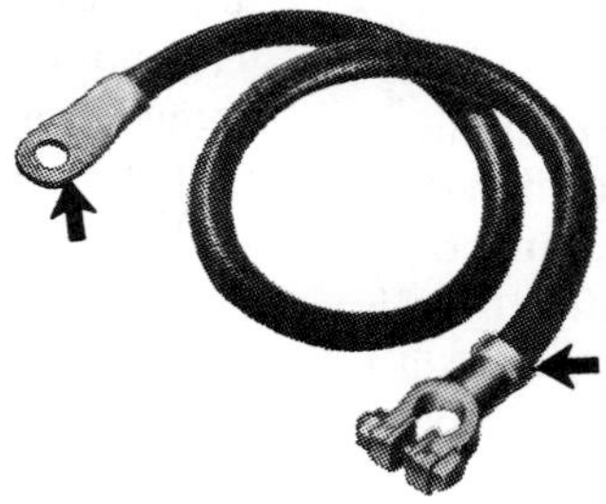

Fig. 43 All the current carried by a cable must flow through its terminal. Therefore, good design and construction of terminals are essential to efficiency

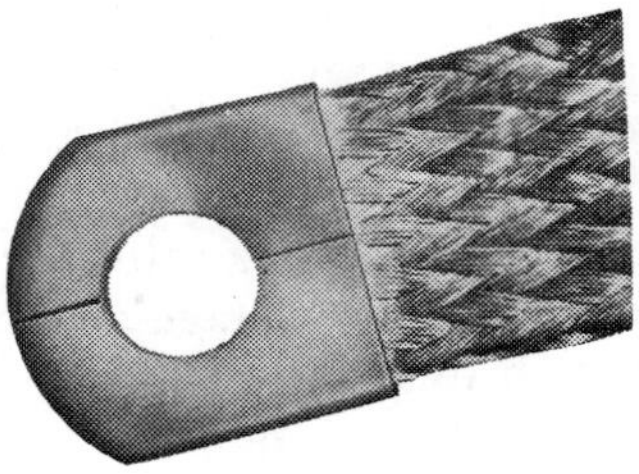

Fig. 45 Ground strap terminals are formed and soldered around the ends of the braided copper to assure good electrical contact and mechanical strength

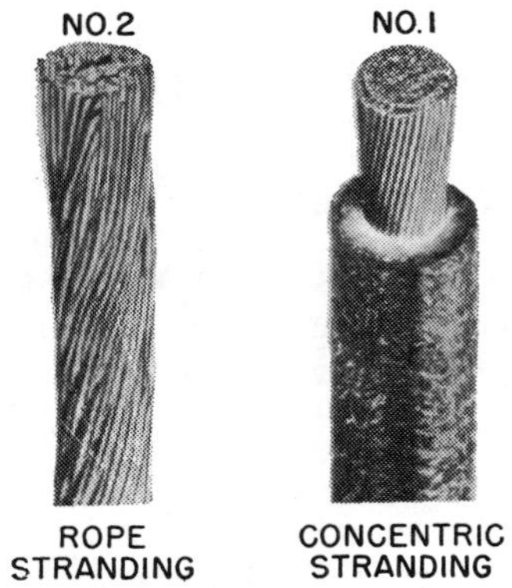
NO. 2
NO. 1
ROPE
STRANDING
CONCENTRIC
STRANDING

Fig. 42 The cores of the two cables shown are about equal but the compact concentric stranding contains 26% more copper

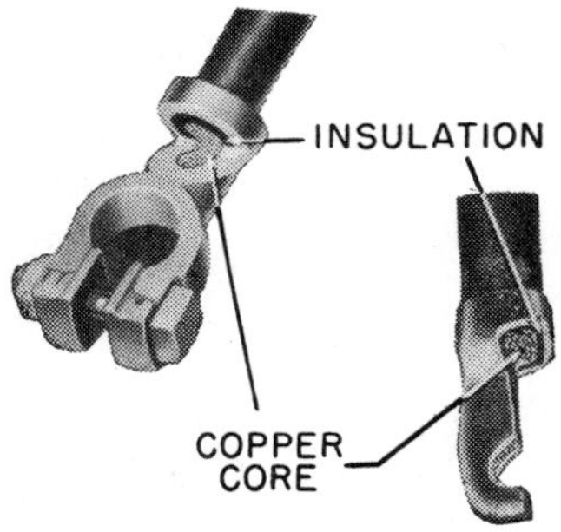
INSULATION
COPPER
CORE

Fig. 44 Terminals must be strong and must make good contact with the cable core to maintain high conductivity

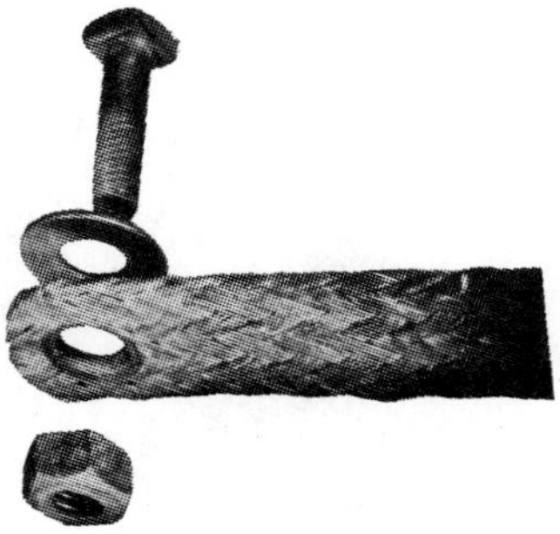

Fig. 46 In attaching a ground strap that lacks a formed terminal clip, use a large flat washer to increase the area of contact and to prevent the braid from curling up at the edges

Fig. 47 *Improved battery terminals include a copper band formed over the braid before the terminals are attached. This construction increases conductivity by 30%*

Cable Capacity

Bear in mind that the outside diameter of a cable may have nothing to do with its gauge, Fig. 41. A thicker insulation can make any smaller gauge cable look like a No. 1 gauge cable, though it has less current carrying capacity.

To illustrate the difference in core capacities, a No. 1 gauge battery cable has 26% more copper in it than a No. 2 gauge cable, and, therefore, 26% greater current-carrying capacity. When it is realized that there are a little over six volts at the battery to start with, the importance of proper current-carrying capacity is apparent.

Since the primary current for ignition usually is taken from a connection at the starting switch, it is possible that undersize cables or poor connections may so reduce the voltage at starting that there is not enough for ignition. That is why a push start will often cause an engine to fire when no amount of turning with the starting motor will do the work.

Another thing to bear in mind about cable is that rope-stranding is less compact than the concentric type, Fig. 42. A No. 2 gauge rope-strand

core has approximately the same diameter as a No. 1 gauge concentric-strand core, although the No. 1 gauge concentric type has 26% more copper.

The surest way to avoid installing undersize cable by mistake is to be certain that the cable is marked plainly as to size.

Cable Terminals

There is more to battery cables than their size. There is the question of the design and construction of the terminals, Fig. 43, and the methods used in attaching the terminals to the copper core.

The two most necessary things for satisfactory terminals are strength and high conductivity. Terminals must be attached to the copper core so that they make good contact with it, and they must be able to remain tight in spite of the vibration of the motor vehicle. As shown in Fig. 44, there must be a large area of contact, providing high conductivity, and the braided insulation must be covered to eliminate fraying.

Like the cable, the ground strap must have as much current-carrying capacity at the terminals as there is in the stranded wire. The construction shown in Fig. 45, employs a clip that is formed and soldered around the end of the ground strap. This improves conductivity and provides a rugged connection when bolted to the engine, transmission or frame, as the case may be.

When there is no clip, contact is made only where the bolt head forces the rough braid to tough the ground. It is necessary, therefore, to place a large flat washer between the bolt head and ground strap, Fig. 46, to increase conductivity and prevent the soldered braid from curling up at the edges.

At the terminal end, the best method is to form a copper band over the braided group strap before it is attached to the terminal, Fig. 47. This increases conductivity at the joint by 30%.

Why Cables Need Replacement

From the time a battery cable is installed, there are several factors working against it, and even

Fig. 48 *In checking the secondary ignition circuit, look for hard, brittle or cracked insulation on high tension cables*

Fig. 49 *A recessed cable core indicates the probability that the entire cable has become worthless*

the best cable is involved in a fight that eventually becomes hopeless.

The greatest enemy of battery cables is acid. Sulphuric acid from the battery finds its way to the battery terminals and the cable insulation, attacking the terminals and the insulation around the cable, and even eating away the copper core itself. A battery cable or ground strap that is acid-eaten is extremely brittle, and may break at any time.

Corrosion

Corrosion on battery cables can be prevented to a considerable extent by frequently inspecting the cable and removing the corrosion with a solution of soda and water, with ammonia or with plain hot water. The terminal should be removed and the battery post scraped to make a good, clean connection. When the cable terminal is replaced and properly tightened, a coat of vaseline should be applied.

During warm-weather operation, there is a possibility that the battery will be overcharged. This causes gassing, which spreads acid over the top of the battery, resulting in corrosion. Reducing the charging rate helps to correct this condition on old cars. Later models are equipped with a voltage and current regulator by which the charging rate is controlled.

Some mechanics, when confronted with a corroded cable, cut the cable back of the terminals, strip the insulation, and apply a new terminal to the old cable. This practice is not good, because shortening the cable may impose a strain on the battery post and cause it to loosen. Moreover, it is difficult to get as good a connection when attaching the terminal as the manufacturer can with his specialized equipment.

Abrasion

Another cause of battery cable deterioration is abrasion. Often the insulation is rubbed through where the cable comes in contact with the frame of the car or some other metal surface. This causes a dead short which may seriously damage some of the electrical units of the vehicle. A *short to ground* is the cause of many dead batteries, and may result in a fire. When being removed, battery cables should always be detached from the battery post first to prevent accidental shorting.

Battery Terminals

A universal battery clamp terminal has a post bore approximately half-way between the sizes of negative and positive posts. To attach to the larger post, it should be spread sufficiently to slip over the post easily. In attaching to the grounded post, it is merely drawn up by the bolt far enough to make proper contact.

A terminal should never be driven onto the battery post. Whenever more than a light tapping is required to force the terminal into position, the bolt should be loosened and the jaws spread farther apart.

A condition frequently run into is one where the terminals come loose from the core, usually because they were not attached by the proper method. This condition is hard to detect because the cable appears to be good. Yet loose terminals may cause high voltage in the charging and ignition circuits of older model cars without voltage regulators or in cars with regulators which may be defective or inoperative, resulting in burned-out lamps, pitted and burned breaker points, damaged ignition coils and other troubles.

IGNITION CABLES

Primary Cables

Primary cables are of the proper gauge when originally installed on a new car and are sufficient for carrying the desired current. But they are sometimes replaced by cables too small to carry the current required for primary ignition without excessive voltage drop. Added accessories may also cause an overload. (See the *Ignition* chapter for procedure for making voltage drop tests in this circuit.)

To avoid installing undersize cables by mistake, make sure before removing an old cable that the right gauge replacement cable is on hand and that its size is marked plainly.

Other mistakes made in replacing worn or broken cables include mismatching of cables with a specific resistance. For example, some ignition coils use special primary cables to provide a variable resistance that is vital to correct ignition operation. Replacing this cable with ordinary wiring alters the resistance and can lead to short contact point life and other difficulties. Such special cables are clearly indicated on wiring diagrams and must be replaced with a cable of the same core material and of the same length.

Secondary or High Tension Cables

Two types of high tension ignition cables are now in use. Both use heavy insulation of rubber and plastic in various combinations to carry the

Fig. 50 *Corroded distributor sockets cause poor engine performance. Clean every socket with a wire brush*

Fig. 51 *An air gap between the bottom of the terminal and the bottom of the socket causes spark jumps that result in corrosion and burned contacts*

high voltage involved. The first type has a stranded copper or aluminum conducting core and while no longer supplied as original equipment, since the early 1960's, it is often used as replacement wiring. The other kind has a linen or fiberglass core impregnated with carbon that eliminates radio and television interference, hence, it is supplied as factory installed wiring on new cars. While the fiberglass core has considerable strength, the linen core can be broken if roughly handled and such cable is not always available in reel lengths to be cut to size for replacement.

Unlike the primary ignition circuit the secondary circuit does not lend itself to testing with a voltmeter because of its high voltage and high frequency current. Therefore, the condition of the cable is best determined by inspection.

A sharp snapping sound, which can be heard quite distinctly when the engine is running, a decided miss, excessive vibration at idling speed, and uneven idling performance may be caused by faulty insulation on the spark plug cables or the coil-to-distributor cable.

Even though engine operation seems to be normal under light load or idling conditions, faulty high tension cable can be the cause of intermittent or regular misfiring under load.

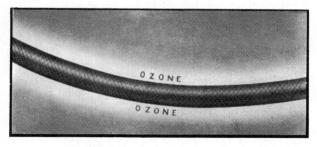

Fig. 53 *Ozone, a powerful oxidizing agent, is formed from the air by the electrical field around a high tension cable. Rubber insulation quickly goes bad under the action of ozone, resulting in an electrical loss, weakening the spark*

Causes of Misfiring

When an engine is under heavy load, the higher compression in the cylinder increases the resistance, and a greater voltage is required to jump the spark plug gap. As a result, any weakness in the cable insulation near enough to a ground becomes the path of least resistance, and the spark jumps through this weak spot to the ground rather than across the spark plug gap. But when the throttle is more nearly closed, and compression is lower, the path of least resistance may be across the spark plug gap, in which case the engine will fire regularly. (The short-to-ground will take place much more regularly in moist air than when the air is dry.)

When an engine has been completely overhauled, the new pistons, rings, valves, etc., bring the compression up much above what it was before the overhaul job. This higher compression often shows up weaknesses in the ignition system that were not noticeable before the overhaul and, unless the ignition system also is overhauled, the engine may perform much worse than before being overhauled. Frequently the fault is in the cable. Therefore, every engine overhaul should include the installation of a new set of high tension cable, unless the cable was comparatively new before the overhaul.

Checking Insulation

In examining metallic core spark plug cables, first look at the insulation and the terminals, Fig. 48, of the high tension cable from the coil to the distributor. If the insulation is hard, brittle or cracked, the cable should be replaced.

If the end of the metallic core appears to have

shrunk back into the insulation, Fig. 49, and is not making good contact with the terminal, it is likely that the entire cable has so deteriorated that it should be replaced.

Carbon impregnated core cables suffer the same insulation problems as noted above, but it is cable core continuity that is frequently at fault and this is hard to detect by simple visual inspection. Without the tensile strength of a metallic core, the linen inner material can separate if jerked abruptly which introduces a high resistance break in the core. This, in turn, means that the spark plug may fire under some conditions but not under others, result, mysterious misfiring. Continuity must be checked in a suspected non-metallic core cable by electrical test before it can be given a clean bill of health.

Remove all spark plug cables from the distributor cap sockets and examine the condition of each terminal and socket. Every socket should be thoroughly cleaned, Fig. 50, as it is not uncommon to find enough resistance in a corroded socket to definitely affect engine performance.

When the spark plug cables are replaced in the distributor cap sockets, care must be taken to see that the cables are pushed all the way into the sockets. If this is not done, the sparks jump the air gap in the socket, Fig. 51, and cause corrosion and burned contacts.

Sometimes spark plug cables are covered by metal plates on the side of the engine. These cables should be examined frequently for faulty insulation, since they are subjected to more heat than are exposed cables.

HIGH TENSION CABLE

The value of high tension cables depends on their ability to carry surges of high tension electrical current from the ignition coil to the spark plugs without insulation failure. High tension cables must withstand heat, cold, oil, grease, chafing, corona.

Of these, corona is perhaps the most baffling problem because, unlike the other effects mentioned, corona, a strange electrical phenomenon, is not readily visible. Yet it is probably the worst enemy of spark plug cables because it deteriorates rubber very rapidly.

Effects of Corona

The surge of high tension current through the circuit builds up an electrical field around each cable, Fig. 53. This electrical field liberates oxygen

Fig. 54 *Some protection from corona was obtained by covering the rubber with braided cotton and an outer coating of lacquer*

in the surrounding air to form ozone, which attacks the organic rubber of the insulation if it is not properly protected. This ozone causes the rubber to deteriorate, thereby losing its insulating qualities. An electrical loss results, depriving the spark plugs of their full charge, and causing a decrease in engine power.

For years the lacquered cable was the highest quality high tension cable known, and yet the life of this cable was none too long. In its construction, Fig. 54, the copper core is enclosed in a rubber covering to give the cable the proper dielectric (insulating) strength. The rubber is then covered with cotton braid and a coating of lacquer to protect it from deterioration.

In the newer cable, however, the braid and lacquer have been replaced with an inorganic material, Fig. 55, that is far more efficient in protecting the rubber against heat, cold, oil, grease, abrasion and corona.

This inorganic sheath, when subjected to standard engineering tests proved to have far longer life than lacquered cables.

Fig. 55 *Cable covered with inorganic material*

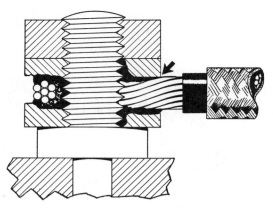

Fig. 56 *In a dirty or corroded connection, current flow is decreased because the area of metal-to-metal contact is reduced. Arrow shows the slight remaining path for current in a corroded connection*

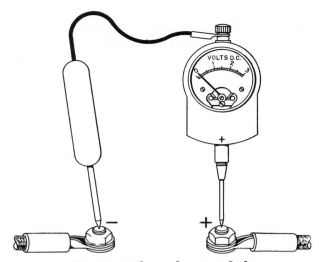

Fig. 57 *Voltage drop in a light-switch is tested across its terminals with the switch closed*

Installing High Tension Cable

Sharp bends in high tension ignition cable should be avoided. Cable carriers or manifolds, if metal, should be carefully grounded. Heat should be avoided as much as possible. See that vibration is reduced to a minimum and that sharp edges do not contact the cable since every precaution must be taken to avoid abrasion. Oil should be prevented from getting on rubber covered or lacquer cable. Only the best-engineered inorganic-sheathed cable will withstand the effects of oil.

It is important to have good ventilation around these cables, for lack of ventilation greatly shortens cable life.

Rubber nipples should be used to prevent moisture from getting into the sockets in the distributor cap, since electricity passing through moisture will form nitric acid, which attacks the core and causes corrosion.

How to Replace Cables

The quickest way to replace high tension cables that are carried in a manifold is to cut off the old cable close to the spark plug terminal and strip the insulation from the end. Then strip the distributor end of the new spark plug cable and twist the two cores together so that, as the old cable is pulled out, the new one is pulled into position. Then the new cable may be cut to the proper length, the rubber nipple and distributor clip applied, and the cable inserted into the distributor cap.

Occasionally, after the installation of a new set of ignition cables, detonation or pinging occurs in the engine, particularly on acceleration. This sim-ply means that the ignition timing had previously been advanced beyond normal adjustment in an attempt to get good engine performance with the old, worn-out cable. Therefore, after installing new cable, it may be necessary to reset the ignition timing. (See *Ignition* chapter for details.)

LIGHTING CIRCUIT

From the standpoint of safety alone, it is essential that the lighting circuit be at its best. Manufacturers have contributed to safety by improving lights and circuitry, providing better protection for terminals and connections to retard corrosion, and so on, and these improvements indicate the efforts constantly being made to improve motor vehicle lighting.

Adequate candlepower is possible only when everything in the system is correct. Lamps must have sufficient output and be focused correctly, all terminals must be tight and clean, switches good, and cables in good condition and of the correct gauge.

Besides the more common and easily found lighting circuit troubles, there are other faults not so easily found, faults that can seriously impair the efficiency of the lighting circuit, such as voltage drop in the fuse clips and in the foot switch. In case of trouble, look also for loose terminals of the type having the tips swedged on without being soldered.

The most common causes of inefficient lighting circuits are high voltage and low voltage.

High Voltage

Considering high voltage first, we find in the case of older cars not having voltage regulators,

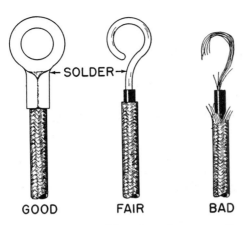

 is placed at top left. Below it:

Fig. 58 A soldered-on lug, or solidly soldered (dipped) strands must be used to form a good connection

GOOD   FAIR   BAD

SIZE AND AREA OF WIRE		
(A) WIRE DIAMETER (INCHES)	(B) AMERICAN WIRE GAUGE	(C) CIRCULAR MIL AREA
.4600	0000	211600
.4096	000	167800
.3648	00	133100
.3249	0	105500
.2893	1	83690
.2576	2	66370
.2294	3	52640
.2043	4	41740
.1620	6	26250
.1285	8	16510
.1019	10	10380
.0808	12	6530
.0640	14	4107
.0508	16	2583
.0403	18	1624
.0319	20	1022
.0284	21	810.1
.0253	22	642.4
.0225	23	509.5
.0201	24	404.0
.0179	25	320.4
.0159	26	254.1
.0142	27	201.5
.0126	28	159.8
.0112	29	126.7
.0100	30	100.5
.0089	31	79.70
.0079	32	63.21
.0070	33	50.13
0063	34	39.75
0056	35	31.52
0050	36	25.00

Fig. 59 The circular mil area of a single strand, multiplied by the number of strands, gives the cross-sectional area of the cable. The middle column shows the gauge of cables of different areas

that the causes may be: Loose connections at the ammeter; loose connections between generator, ammeter and battery; undersize leads between generator, ammeter and battery. With modern cars equipped with properly functioning regulators, high voltage is not a big problem.

Nevertheless, low or weak electrolyte in the battery, improperly adjusted or inoperative voltage regulator or a shorted generator field circuit will produce high voltages in cars equipped with voltage regulators. Such high voltages cause headlamp flare-up when engine speed is increased.

Improperly applied terminals produce a high resistance, as do terminals in which breaking of strands or corrosion has greatly reduced the area of contact, for corrosion or dirt prevents metal-to-metal contact. Notice, in Fig. 56, that current can pass from the wire to the terminal only at the point having an area that is only a small part of the total cross-sectional area of the cable. A connection of this kind may look good, but no connection of any kind should be less in contact area than the cross-section area of the cable carrying the current.

Low Voltage

Low voltage is trickiest of all enemies of the lighting circuit. Possible causes are (1) low water in battery, (2) corroded battery cables, (3) poor ground connections at frame or engine, (4) undersize cable.

Low voltage due to any of these causes may be checked with a voltmeter. Voltage drop in the lighting circuit must be kept at a minimum. All connections must be clean and tight.

In checking voltage drop across lighting switches, Fig. 57, the switch must be on to allow current to pass through it. Voltage drop of more than 0.5 volt is considered excessive.

Terminals Connections

All connections should terminate in soldered or carefully crimped ends. Proper terminals should be used and wire strands should never be twisted around a screw and fastened.

Wires that are not properly terminated cause voltage drop, because corrosion affects the exposed portions of the metal. Since improper connections make only a small surface contact, the corrosion easily gets between the contacting surfaces. A bundle of loose strands is flattened when compressed between a binding post and nut and good contact is impossible. Besides, the sharp corners of the nut may cut some of the strands, further increasing the resistance. If a soldering lug is not at hand, at least coat the end of the wire with solder before forming a loop, Fig. 58.

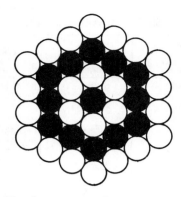

Fig. 60 *In a uniformly stranded cable, each layer contains six more strands than the previous layer. The total of the strands in all layers, plus the center strand, gives cables of 7, 19, 37, 61, 91, 127, 169, etc., strands*

Effect of High and Low Voltage

Ordinarily, the lamps used in motor vehicle headlamps operate on voltages that vary. But the effect of excess voltage on the life of a lamp is out of proportion to the increase in voltage. For example, the ordinary life of a lamp working at the design voltage of 7 volts may be around 200 hours, but when operating at 7.7 volts, its useful life may drop to as low as 60 hours.

Since tests show that a 10% loss in voltage results in a 30% loss in candlepower, we can see how important it is to use the proper size cables in re-wiring the lighting circuit. To do this, it is necessary to have on hand a reliable re-wiring guide which gives the gauge and type of wiring required in all the main points in the electrical system of a motor vehicle.

HOW TO DETERMINE SIZE OF CABLE

Automotive cable is made up of numerous separate wires in order that it be flexible. These wires vary in number and diameter to form different sizes of cables.

A cable having a single thick wire, and a cable having many wires whose total cross-sectional area equals that of a single large wire would be of the same gauge. Therefore, to determine the gauge of any cable, it is necessary to find the total cross-sectional area of all its separate wires, then ascertain what is the gauge of a single wire having an equal area. The accompanying table, Fig. 59, listing wires according to their diameters, gives the gauge and the area (in circular mils) for each size of wire.

Cables composed of numerous wires seldom have exactly the same area as a single wire cable, because of limitations imposed by the number of strands that can be used to form a uniformly round cable. Fig. 60 shows, in enlarged scale, that it requires at least seven strands to make a uniform conductor, and that each additional layer requires six more strands than the preceding layer. Therefore, flexible cables usually have 7, 19, 37,

6-VOLT SYSTEM		12-VOLT SYSTEM		Total Length of Cable in Circuit from Battery to most Distant Electrical Unit									
AMPERES (APPROX.)	CANDLE POWER	AMPERES (APPROX.)	CANDLE POWER	10 Feet	20 Feet	30 Feet	40 Feet	50 Feet	60 Feet	70 Feet	80 Feet	90 Feet	100 Feet
				Gauge	Gauge	Gauge	Gauge	Gauge	Gauge	Gauge	Gauge	Gauge	Gauge
0.5	3	1.0	6	18	18	18	18	18	18	18	18	18	18
0.75	5	1.5	10	18	18	18	18	18	18	18	18	18	18
1.0	8	2	16	18	18	18	18	18	18	18	16	16	16
1.5	12	3	24	18	18	18	18	18	16	16	16	14	14
2.0	15	4	30	18	18	18	16	16	16	14	14	14	12
2.5	20	5	40	18	18	18	16	14	14	14	12	12	12
3.0	25	6	50	18	18	16	16	14	14	12	12	12	12
3.5	30	7	60	18	18	16	14	14	12	12	12	10	10
4.0	35	8	70	18	16	16	14	12	12	12	10	10	10
5.0	40	10	80	18	16	14	12	12	12	10	10	10	10
5.5	45	11	90	18	16	14	12	12	10	10	10	10	8
6.0	50	12	100	18	16	14	12	12	10	10	10	8	8
7.5	60	15	120	18	14	12	12	10	10	10	8	8	8
9.0	70	18	140	16	14	12	10	10	8	8	8	8	8
10	80	20	160	16	12	12	10	10	8	8	8	8	6
11	90	22	180	16	12	10	10	8	8	8	8	6	6
12	100	24	200	16	12	10	10	8	8	8	6	6	6
18	—	36	—	14	10	8	8	8	6	6	6	4	4
25	—	50	—	12	10	8	6	6	4	4	4	2	2
50	—	100	—	10	6	4	4	2	2	1	1	0	0
75	—	150	—	8	4	2	2	1	0	0	00	00	00
100	—	200	—	6	4	2	1	0	00	000	000	000	0000

Fig. 61 *Required cable sizes for single wire six and twelve volt battery systems*

61, 91, 127, 169, etc., strands. (Exceptions are rope-stranded, or extra-flexible, cables, and cables using extremely fine stranding, since these cables can be made reasonably uniform in roundness regardless of the number of separate strands.)

To determine the size of a cable, using the table shown in Fig. 59, proceed as follows:

1. Count the number of strands.

2. Measure the diameter of a single strand in thousandths of an inch, using a micrometer.

3. In column "A" of the table, find the diameter of the wire you have measured, and on the same line in column "C", find its area.

4. Multiply the area of a single wire by the number of strands to get the total area.

5. In column "C", find the figure that is closest to the total area obtained by Step 4, and on the same line in Column "B", note the gauge number of a single wire having that area. This number is the gauge of the cable.

SELECTING PROPER SIZE CABLE

Cable too small in gauge size for the load (lamps, starter and other equipment) will cause a drop in the battery voltage delivered to such equipment. A 10% drop in battery voltage to the lamps will result in a 30% drop in candlepower. This means that if there are twelve volts at the battery and, because of undersize cable, or additional equipment, or a long run, the voltage at the headlamps is 10.8 volts, a 32-candlepower lamp will deliver only 22.4 actual candlepower.

To determine the proper size of cable for a circuit to be used for an electrical system, using the chart in Fig. 61, proceed as follows:

1. Determine the length of cable needed to reach from the battery to the most distant lamp in the circuit (return circuit has been allowed for in the chart).

2. *By Candlepower:*—Add up the total maximum candlepower ever to be used in the circuit. (If you now use a 3-C.P. bulb where you might later change to a 6-C.P., use 6 in adding up C.P.) Read down to the closest corresponding figure in the *Candlepower* column of the chart.

3. *By Amperes:*—If the approximate amperes load in the circuit is known, read down the *Amperes* column of the chart to the closest corresponding figure.

4. Read across on the line representing the C.P. or the amperes of the circuit to the correct foot-length column. The figure there found indicates the proper gauge of cable to be used throughout the circuit.

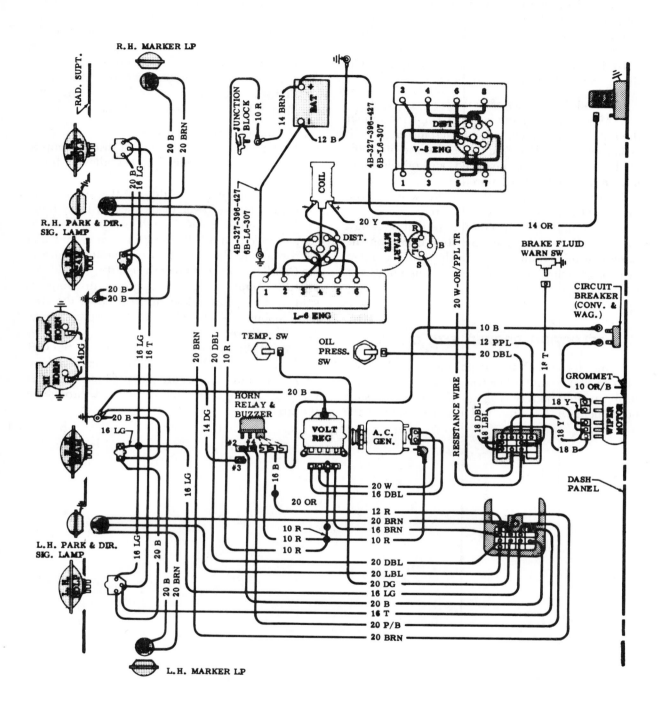

Front Lighting and Engine Compartment—1969 Chevrolet

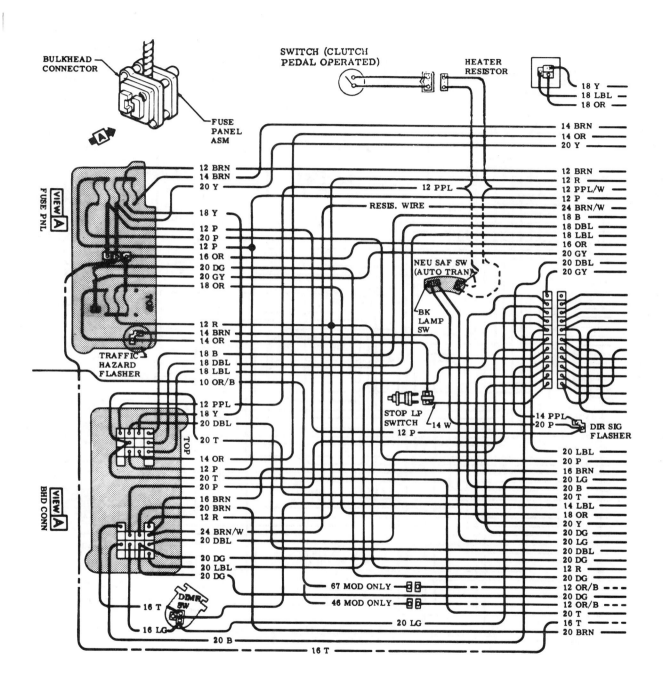

BULKHEAD
CONNECTOR

FUSE
PANEL
ASM

SWITCH (CLUTCH
PEDAL OPERATED)

HEATER
RESISTOR

18 Y
18 LBL
18 OR

14 BRN
14 OR
20 Y

VIEW A
FUSE PNL

12 BRN
14 BRN
20 Y

12 BRN
12 R
12 PPL/W
12 P
24 BRN/W
18 B
18 DBL
18 LBL
16 OR
20 GY
20 DBL
20 GY

12 PPL

RESIS. WIRE

18 Y
12 P
20 P
12 P
16 OR
20 DG
20 GY
18 OR

NEU SAF SW
(AUTO TRAN)

BK
LAMP
SW

12 R
14 BRN
14 OR

TRAFFIC
HAZARD
FLASHER

18 B
18 DBL
18 LBL
10 OR/B

12 PPL
18 Y
20 DBL

STOP LP
SWITCH

14 W
12 P

14 PPL
20 P

DIR SIG
FLASHER

TOP

20 T
14 OR
12 P
20 T
20 P

20 LBL
20 P
16 BRN
20 LG
20 B
20 T
14 LBL
18 OR
20 Y
20 DG
20 LG
20 DBL
20 DG
12 R
20 DG
12 OR/B
20 DG
12 OR/B
20 T
16 T
20 BRN

VIEW A
BHD CONN

16 BRN
20 BRN
12 R
24 BRN/W
20 DBL

20 DG
20 LBL
20 DG

DIMR
SW

67 MOD ONLY

46 MOD ONLY

16 T

16 LG

20 B

16 T

20 LG

Fuse Panel—1969 Chevrolet

553

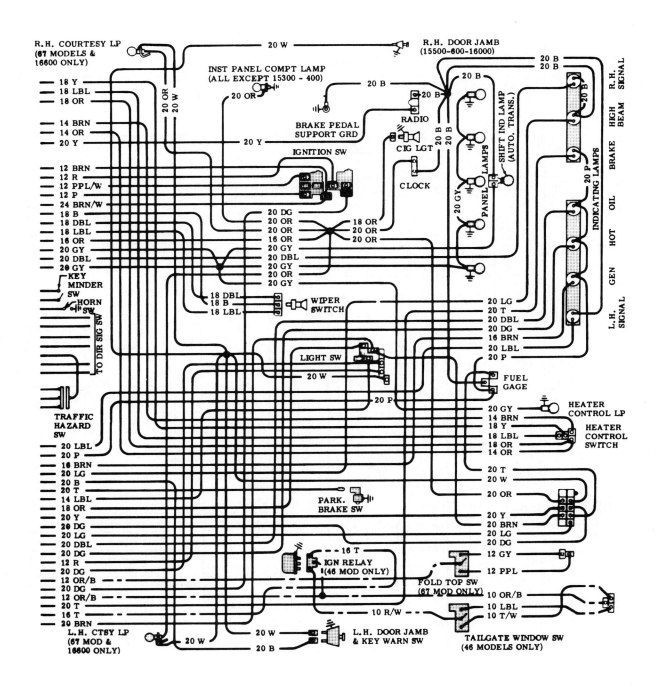

Instrument Panel—1969 Chevrolet

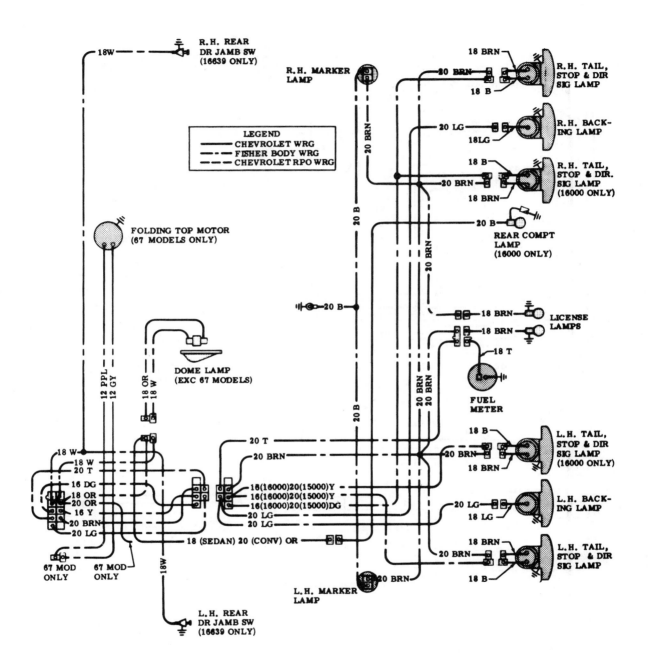

Body and Rear Lighting—1969 Chevrolet (exc. Wagons)

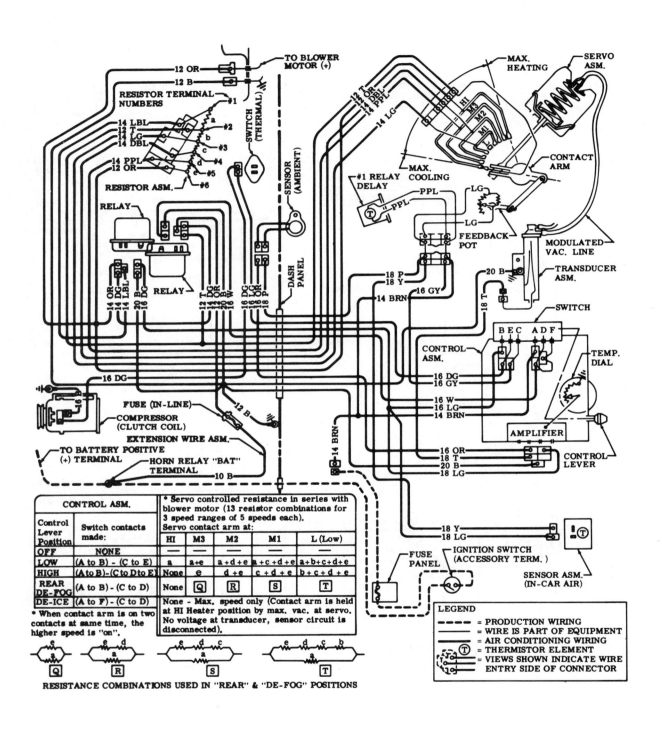

1969 *Chevrolet Comfortron Circuit*

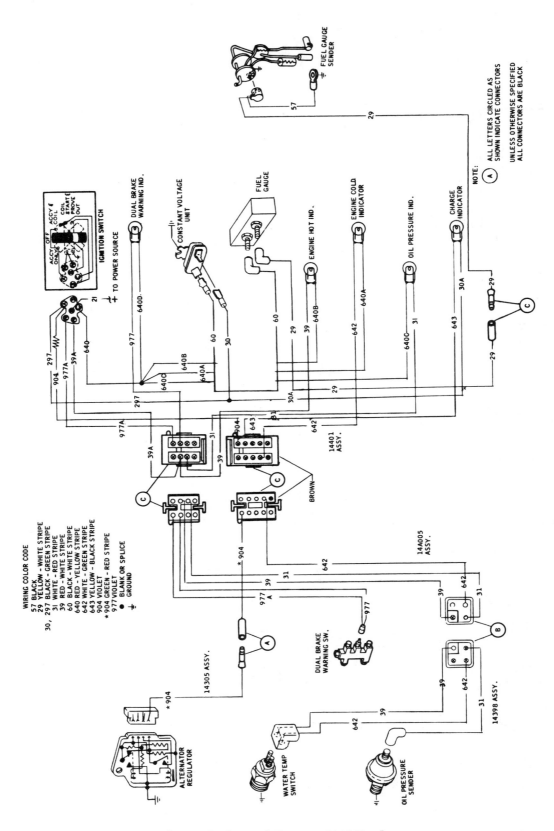

Indicator Lights and Gauges—1968 Ford

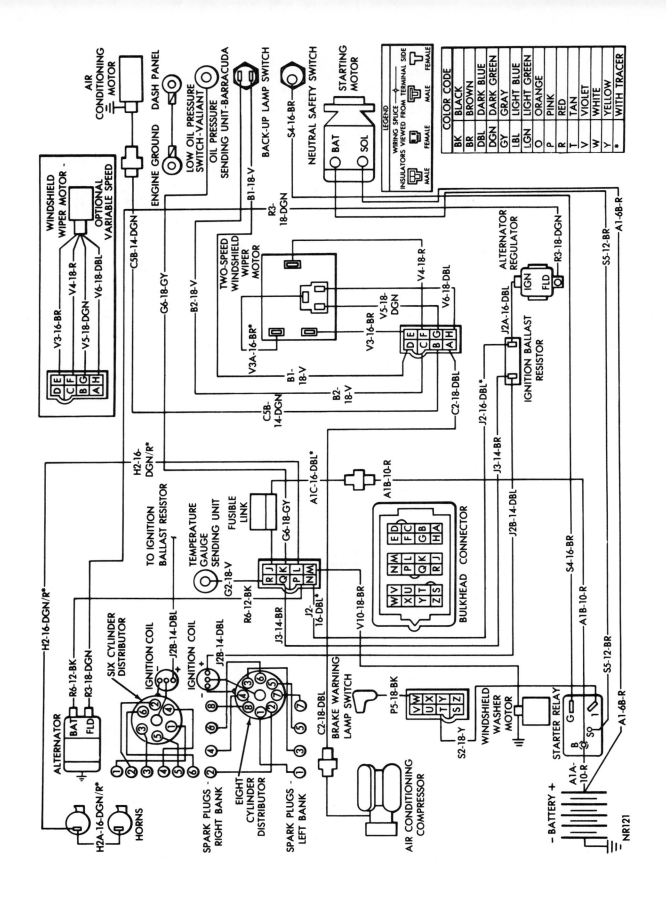

Engine Compartment—1968 Valiant and Barracuda

Review Questions

		Page
1.	The electrical circuit in an automobile is known as what kind of system?	530
2.	In the type of wiring system used in an automobile, what part of the construction acts as a path for current?	530
3.	How are all main electrical units connected into the electrical system?	530
4.	How are switches connected in relation to the units they control?	530
5.	What is the name of that part of the circuit that is made up of the body and chassis?	530
6.	Why do the chassis and body parts offer little electrical resistance even though iron would be a poor conductor if used in wiring?	531
7.	Why is good metal-to-metal contact important in the mounting of a light?	532
8.	What is meant by "short circuit"?	533
9.	What is meant by a "grounded" wire?	533
10.	What would be likely to cause a wire to become "grounded"?	533
11.	What are the seven basic electrical circuits in an automobile?	534, 535
12.	What is a schematic wiring diagram?	535
13.	What are the major components of the starter circuit?	535
14.	What are the major components of the charging circuit?	535
15.	What are the major components of the ignition system?	537
16.	What major component is common to the starter, charging and ignition circuits?	537
17.	What is the purpose of an ignition system?	537
18.	What is the function of the ignition coil?	537
19.	Why is a relay usually included in a horn circuit?	537
20.	What two methods might be used to inform the driver whether or not the generating system is in working order?	535
21.	How is damage to the lighting system prevented in case of a short?	539
22.	Give two main reasons why auto manufacturers have changed to 12 volt electrical systems.	539
23.	How does the power output of a 12 volt generator at 20 amps, compare with the power output of a 6 volt generator at 20 amps?	539
24.	How does the size of the wires in a 12 volt system compare with the size of the wires in a 6 volt system?	539
25.	How does compression pressure affect the firing of a spark plug?	540
26.	In what way would an ignition coil operating on 6 volts fall short of ignition requirements for modern V8 engines?	540
27.	What parts in a distributor are the same for both 6 volt and 12 volt systems?	540
28.	What type of meter is it best to use for locating electrical troubles?	540
29.	How must an ammeter always be connected in a circuit?	540

30. How must a voltmeter always be connected in a circuit? 541

31. What would be indicated by a voltage drop between any two points in a circuit? 541

32. Does the voltage drop have to be large in order to cause trouble in a circuit? 541

33. What two things should be remembered concerning cables? 541

34. What is the term which expresses cable size? 541

35. The term which expresses cable size indicates what dimension of a cable? 541

36. Does the gauge number of a cable increase or decrease as the size of the cable increases? 542

37. How would increasing the length of a cable affect its resistance? 541

38. How does the starter cable compare in size with the other cables in an automobile's electrical system? 542

39. What is the best way to avoid using the wrong size cable? 544

40. What is the greatest enemy of battery cables? 545

41. What preventive maintenance will lengthen battery cable life? 545

42. What should never be done when installing a cable terminal on a battery post? 545

43. What might be the cause of burned out bulbs and breaker points? 545

44. What part of the electrical system cannot be tested with a voltmeter? 546

45. Why is it that a defect in the high tension system might cause misfiring under heavy engine loads, yet operate perfectly under moderate loads? 546

46. What is ozone? 547

47. What generates ozone? 547

48. What must take place in a circuit while testing for voltage drop? 549

49. What unit of measurement is used to give the size of each wire in a cable? 550

Ignition System

Review Questions for This Chapter on Page 619

INDEX

Advance spark 578
Analysis of spark plug condition 588
Breaker point service 596
Causes of coil failure 584
Chrysler distributor overhaul.. 598
Chrysler distributor service .. 598
Coil, causes of failure 584
Coil, ignition 561
Coil polarity, importance of
    correct 586
Coil, service ignition 583
Coil testers, types of 587
Coil, test ignition 586
Condenser 565
Condenser service 592
Current, paths of flow in
    ignition circuit 567
Delco-Remy Corvair distributor 615
Delco-Remy distributor
    overhaul 598
Delco-Remy distributor service 598
Distributor 568
Distributor inspection & tests. 596
Distributor, overhaul Prestolite 598
Distributor, overhaul Chrysler . 598
Distributor, overhaul Delco-
    Remy 598
Distributor, overhaul Holley .. 611
Distributor, service Prestolite . 598
Distributor, service Chrysler . 598

Distributor, service Delco-Remy 598
Distributor, service Holley ... 611
Distributor, tests 609
Holley distributor overhaul .. 614
Holley distributor service 611
Ignition circuit, paths of current
    flow in 567
Ignition coil 561
Ignition coil causes of failure .. 584
Ignition coil service 583
Ignition coil tests 586
Ignition miss 583
Ignition system service 581
Ignition system trouble
    shooting 581
Ignition timing 611
Miss, ignition 583
Overhauling Prestolite
    distributors 598
Overhauling Chrysler
    distributors 598
Overhauling Delco-Remy
    distributors 598
Overhauling Holley distributors 614
Paths of current flow in ignition
    circuit 567
Performance tips 617
Plug, condition analysis of
    spark 588
Plug, resistor 591

Plugs, service spark 587
Plug, spark 565
Points, service breaker 596
Polarity, importance of correct
    coil 586
Prestolite distributor overhaul. 598
Prestolite distributor service .. 598
Resistor spark plug 591
Service Prestolite distributor . 598
Service, breaker points 596
Service, condensers 592
Service Delco-Remy distributor 598
Service Holley distributor 611
Service ignition coil 583
Service ignition system 581
Service spark plugs 587
Spark advance 578
Spark plugs 565
Spark plug, resistor 591
Spark plug condition,
    analysis of 588
Spark plug service 587
System, trouble shooting
    ignition 581
Test & inspect distributor 596
Tests, ignition coil 586
Timing, ignition 611
Trouble shooting ignition
    system 581
Types of coil testers 587

Introduction

The purpose of the ignition system is to supply sparks across the points of the spark plugs to ignite the combustible mixture in the cylinders. The principal units in the ignition circuit, Fig. 1, are the battery, ammeter or tell-tale light, ignition switch, ignition coil, distributor, spark plugs and the necessary wires which connect these parts. The distributor shaft is rotated by the engine camshaft.

Note that there are two circuits, called primary and secondary, or low tension and high tension, respectively. The primary circuit carries low voltage and the secondary circuit high voltage. Ordinarily the primary circuit operates on 12–14 volts while the high-voltage circuit may produce up to 20,000 volts.

As explained in the chapter on *Electricity*, voltage should be thought of as electrical pressure which forces current through the circuit against the resistance of the circuit. Thousands of volts are required in the secondary circuit in order to force the current through the extremely high resistance of the air gap between the spark plug points.

The low-voltage current in the primary circuit is changed to high-voltage current in the secondary circuit by means of an induction coil which is more commonly called an ignition coil.

IGNITION COIL

The ignition coil contains two coils of wire, Fig. 2. One coil is part of the low-tension circuit and the other is part of the high-tension circuit. The

561

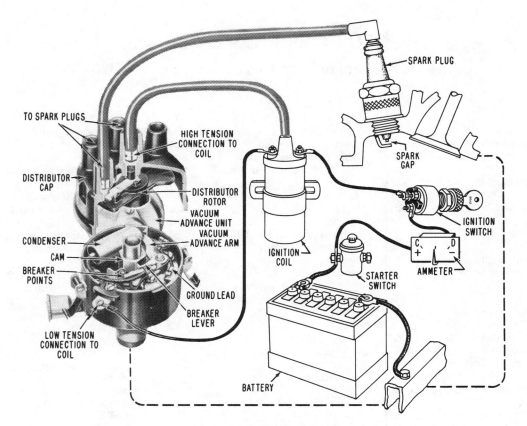

Fig. 1   *Typical ignition system*

two coils are called low- and high-tension windings or primary and secondary windings, respectively.

The diagram, Fig. 2, has been purposely made as simple as possible and therefore contains only the elements required to produce a spark—that is, to transform low-voltage current from the battery or generator into high-voltage current which has pressure enough to jump the spark plug gap.

With the breaker points closed, primary current flows as indicated by the lines. And as explained in the chapter on *Electricity*, whenever a current flows, a magnetic field surrounds the current. The magnetic field is concentrated by the use of numerous turns of wire which forms the primary coil. The magnetic field is further concentrated by the presence of the iron core. In consequence, the magnetic field created by current in the low-tension winding completely permeates the high-tension winding. The magnetic field represents magnetic energy in storage obtained from the electric current in the low-tension circuit. This energy can be compared to the stored mechanical energy in a compressed spring.

The stored magnetic energy which completely surrounds both the high and low-tension wind-

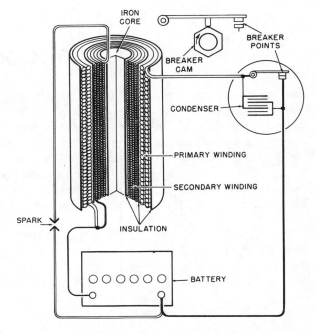

Fig. 2   *Basic principle of ignition system*

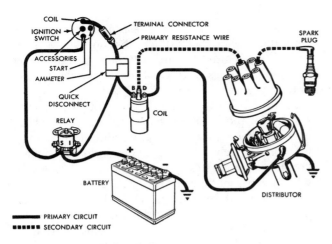

Fig. 3   Typical ignition circuit

PRIMARY CIRCUIT
SECONDARY CIRCUIT

ings remains in storage just as long as current continues to flow through the low-tension circuit. However, the instant the breaker points are snapped open, the current stops flowing and the magnetic field collapses. But while collapsing, its energy is transformed back into electrical energy in both the low and high-tension windings, producing a very high voltage in the secondary winding and a moderate voltage in the primary winding. It is as though the magnetic energy surrounding the wires of the windings "drains" into the windings and somehow turns its energy into an electric current.

Another way to explain this phenomenon is to note that the magnetic field produced by the primary winding consists of magnetic lines of force. As explained in the chapter on *Electricity,* when a wire moves at right angles to magnetic lines of force it "cuts" them and in so doing a voltage is produced in the wire. These lines of force can be compared to phantom rubber bands which are cut in two as the wire moves through them but immediately join themselves together again once the wire has passed.

When the primary circuit is broken by opening the breaker points and the magnetic field collapses, the magnetic lines of force shrink inward to nothing. During this motion they are "cut" by all the wires in both the low- and high-tension windings. Therefore, a voltage is *induced* in *both* the primary and secondary windings.

The voltage induced in the primary winding may be anywhere from 100 to 300 volts depending on the number of turns in the primary winding and other factors. A typical coil has a primary winding consisting of 240 turns of No. 20 (coarse) copper wire while the secondary winding has 21,000 turns of No. 38 (fine) copper wire. The ratio of turns in secondary to primary is 87.5.

Tests show that the maximum voltage the secondary is capable of producing is approximately equal to the ratio of turns multiplied by the induced voltage in the primary. Therefore, if the induced voltage in the primary is 100, the secondary voltage is 8750. If the induced voltage in the primary is 200, the secondary voltage can be 17,500.

The actual voltage required for the spark to jump the spark plug gap may be less than 20,000 volts, or even less than 10,000. The voltage required increases with the breadth of the spark plug gap, being considerably more for a .040″ gap than for a .025″ gap. The voltage required also increases with compression pressure. With throttle open the compression pressure may be 150 psi (pounds per square inch) whereas with throttle closed and engine idling the compression pressure can go as low as 35 pounds. Therefore, with the throttle closed and a moderate spark plug gap, the voltage required for the spark to jump the gap may be only 5,000 or even less.

Regardless of throttle opening and point gap, the voltage induced in the secondary increases until it is sufficient to cause the current to jump the gap. The induced voltage does not increase beyond this point. Therefore, if 5,000 volts is enough, that is all that the secondary winding produces. Ten to 20,000 volts may be required on open throttle. Voltage required increases with the size of the plug gap and also with compression pressure.

Some excess voltage capacity is always built into the coil so that sparks will still be delivered when the ignition system is not in perfect condition. For example, a slightly dirty connection anywhere in the low-tension circuit might reduce the current flowing through this circuit enough to reduce the maximum secondary voltage by a few thousand volts. But even so, the voltage would be adequate to operate the engine. However, a still dirtier connection may prevent ignition on open throttle, and in an extreme case will prevent any current at all from flowing through the primary circuit. Some excess voltage will also enable somewhat dirty spark plugs to fire.

Returning to Fig. 2, note the details of coil construction. In the center is a laminated iron core which concentrates the magnetic field. Surrounding it are the secondary windings which in turn are surrounded by the primary windings. There are iron laminations between the primary windings and the sheet steel housing which is called a can. The wires in the windings are covered with insulation. Three thin insulating cylinders separate (1) the secondary from the core, (2) the

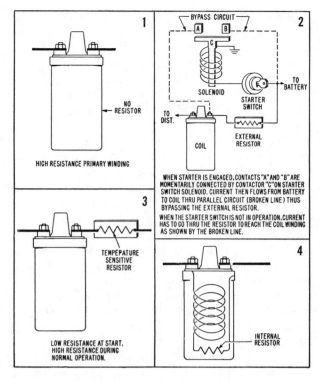

Fig. 4 Twelve volt coil and resistor circuits

primary from the secondary and (3) the primary from the iron laminations just inside the can.

The iron core in the coil is laminated because a solid chunk of iron would generate stray currents within itself caused by the rapid build up and collapse of the magnetic field while the coil is in operation. Objectionable overheating of the core would result. This effect is largely eliminated by making the core out of many laminations. The same remark applies to the laminated iron just inside the case.

The primary winding is connected between the ignition switch and the breaker points, Fig. 3. One end of the secondary winding is connected to the primary wire running to the breaker points and the other end of the secondary winding is attached to the main high-tension terminal in the coil tower.

12 Volt Coils

The ignition coils used with 12-volt systems are specially designed *6-volt coils* which operate with a resistor connected in series with the primary ignition circuit. The purpose of the resistor is to prolong the service life of the distributor breaker points.

In the Prestolite system the resistor consists of an ordinary resistance wire that is sensitive to temperature. It has a lower resistance value when cold than when hot. When the ignition is first turned on, more current will flow through the primary windings of the coil for a very short time until the resistor heats up.

In the Delco-Remy and Ford systems, the resistor is of the constant temperature type; that is, it is not affected by temperature and its resistance is approximately the same when cold and when hot. However, in this system a feature is employed which shorts out the resistor while the engine is being cranked by the starter and automatically puts the resistor back into the coil circuit as soon as the starter switch is released. This is accomplished by by-passing the resistor through the starter solenoid. The solenoid has an additional terminal from which a wire runs directly to the coil. The resistor is by-passed by means of a "finger" inside the switch housing which is attached to the additional switch terminal.

Under no circumstances should a 12-volt system be operated with the resistor shorted out as the breaker points will burn up in short order. What happens is that 12 volts will be placed across the 6-volt coil and this will double the normal current through the breaker points.

The resistance used on some cars starting with 1960 consists of a high resistance wire incorporated in the instrument panel wiring harness. The wire used with Delco-Remy systems is stainless wire, plastic coated and covered with a glass braid. A similar wire is used on Ford Company cars.

The four 12-volt coil and resistor setups in use are shown in the wiring diagrams, Fig. 4.

No. 1 circuit shows a 12-volt coil without any resistor in the primary circuit. The primary winding, however, has a high resistance value.

No. 2 circuit shows the most common Delco-Remy type. It has an external resistor which does not change with temperature. The primary winding of the coil is similar to that of a 6-volt coil but it has a higher resistance value. Note that in this set-up the ignition circuit is wired so that the external resistor is shorted out while the starter switch is operating.

No. 3 circuit is the most common Prestolite type. It uses an external resistor whose resistance value changes with temperature. When the car is being started, the cold resistor permits a higher current through the coil primary resulting in easy starting. As the resistor warms up, its resistance increases to cut down the primary current through the coil for normal operation.

No. 4 circuit shows a Prestolite coil with the resistor incorporated within the coil housing.

Fig. 5 *External view of condenser. The metal housing is grounded on the breaker assembly on which it is mounted*

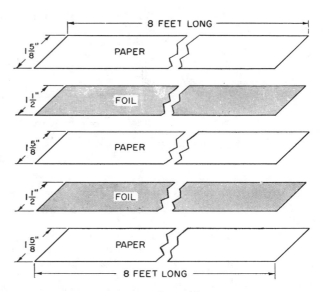

Fig. 6 *A typical condenser consists of two strips of tin foil sandwiched between three strips of paper. The strips shown are 8 feet long. The width of the paper strips is 1⅝" and the strips of foil are 1½"*

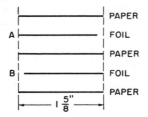

Fig. 7 *One strip of foil (A) is lined up with the left edge of the paper. The other strip (B) is lined up with the right edge of the paper*

Fig. 8 *Condenser after the ends have been dipped in solder. It is then inserted in a pasteboard tube*

CONDENSER

The condenser is connected across the breaker points, Fig. 2. A condenser is shown in Fig. 5. The condenser itself is encased in a metal housing about ⅝" in diameter. A typical condenser consists of two strips of tin foil 1 to 1½ inches wide and eight feet long, Fig. 6. They are sandwiched between three strips of tissue-thin translucent paper which is specially impregnated with an insulating compound. The paper strips are about ⅛" wider than the strips of foil. One strip of foil, Fig. 7, is lined up with the left edge of the paper and the other strip is lined up with the right edge. The five strips are rolled up into a cylinder and

both ends are dipped in solder, Fig. 8. Thus the solder on each end forms a good metallic connection for one of the strips of metal foil. The condenser is inserted in an insulating pasteboard tube. Fig. 9 illustrates the assembled condenser while Fig. 10 shows a diagram which is often used to indicate a condenser.

The capacity (capacitance) of the condenser should match the capacity of the coil. Some coils require larger condensers than others. The capacitance of condenser is given in microfarads and the three typical ranges are .20 to .25 mfd (microfarads), .25 to .28 mfd and .28 to .32 mfd.

The condenser not only helps build up voltage in the secondary windings of the coil but it also reduces arcing across the breaker points to a minimum when the points are snapped open. The arcing cannot be completely eliminated. Excessive arcing indicates condenser trouble or dirty or loose condenser connections.

SPARK PLUG

Fig. 11 illustrates typical spark plugs. The center electrode is insulated from the shell by a ceramic insulator such as porcelain. The side electrode is fastened to the spark plug shell which is grounded on the cylinder head. A high-tension wire from the distributor is attached to the top of the center electrode. The current flows down the

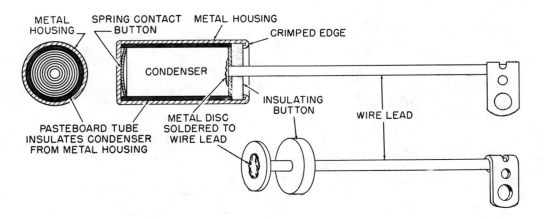

Fig. 9 *This drawing shows the construction of the complete condenser assembly. The edge of the steel housing is crimped over so that it makes a water-tight contact with the thick insulating button. The crimping holds the metal disc on the wire lead tight against the soldered surface at the right end of the condenser. The crimping also compresses the spring contact button at the left so that good contact is maintained between the end of the steel housing and the adjacent solder surface on the condenser*

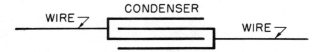

Fig. 10 *This type of diagram is often used in a wiring diagram to indicate a condenser*

center electrode, jumps the gap to the side electrode, and thus produces a spark which ignites the fuel-air mixture in the combustion chamber. Most of the spark plugs in use today have a 14 millimeter thread diameter although some have 18 millimeter and a very few have 10 millimeter. These sizes when expressed in inches are respectively 0.55″, 0.71″ and 0.40″.

Spark plugs are made in several heat ranges so that a plug with a suitable heat range for a given engine and a given set of operating conditions is available. A very cold plug may be desirable for continuous high-speed driving whereas a very hot one may be required if the car is always operated at low speed.

Note that a spark plug is exposed to the intense heat of combustion which may exceed 4,000 degrees at the beginning of each power stroke on open throttle.

The heat the plug absorbs is conducted to the cylinder head as shown in Fig. 12. In this way the nose of the plug is kept cool enough so that the center electrode is not melted or eroded by the hot flame nor the insulator damaged by excessive temperature. On the other hand, if the plug is too

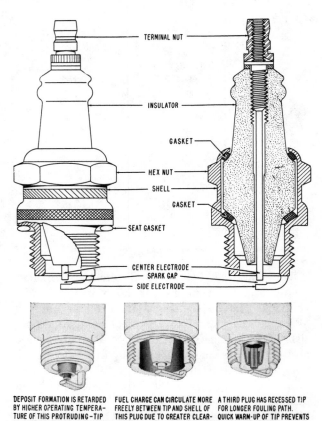

DEPOSIT FORMATION IS RETARDED BY HIGHER OPERATING TEMPERATURE OF THIS PROTRUDING-TIP PLUG. COOL INCOMING MIXTURE PREVENTS OVERHEATING AT HIGH SPEED

FUEL CHARGE CAN CIRCULATE MORE FREELY BETWEEN TIP AND SHELL OF THIS PLUG DUE TO GREATER CLEARANCE. DEPOSITS MUST COLLECT OVER WIDER AREA TO CAUSE FOULING

A THIRD PLUG HAS RECESSED TIP FOR LONGER FOULING PATH. QUICK WARM-UP OF TIP PREVENTS BUILD-UP OF DEPOSITS DURING LOW SPEED OPERATION

Fig. 11 *Typical spark plugs*

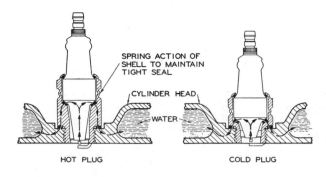

Fig. 12   *Hot and cold spark plugs*

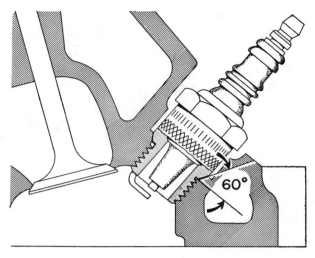

Fig. 13   *Spark plug with tapered seat which requires no gasket*

cold, the insulator will become fouled with carbon particles bound together with baked oil.

The hottest part of the plug is the tip of the nose. Heat flows up through the insulator and down through the shell, as shown by the arrows. The longer the heat path through the insulator and shell, the hotter the plug (and vice-versa) because the amount of heat that flows decreases as the length of the path is increased. Therefore, the cold plug, when used in an engine, has a tip which is considerably cooler than the hot plug. Thus the cold plug is recommended when the car is driven on open throttle most of the time. But this plug will start to foul up when operated at low speed. The fouling may be subsequently burned off if the period of low-speed driving is short. But if the period is long, the steady collection of carbon particles and baked oil will permit some current to leak across the fouled insulator and some missing will occur, especially on open throttle. Eventually the deposit will be heavy enough so that all the current flows through the deposit and none jumps the gap.

On the other hand, the hot plug will not foul when driven continuously at low speed but if it is driven for some length of time at high speed the tip of the insulator and its electrode will be ruined.

The spark plug gasket is an important factor in conducting the heat from the plug to the cylinder head. It may seriously obstruct the flow of heat unless the plug is properly tightened in place. This is essential for good contact of the gasket with the plug and cylinder head. The rate of heat conduction will be reduced if the contact is poor. (Spark plugs used on some cars of the Ford Group do not use gaskets but rather the plug seat is tapered so that when installed an air-tight seat is provided, Fig. 13.)

A cold plug may easily be ruined by excessive heat if driven hard when insufficiently tight and if hot plugs are driven hard when not properly tightened their ruination is certain.

PATHS OF CURRENT FLOW IN IGNITION CIRCUIT

Having described the coil, condenser and spark plug, we are now in a position to explain how the current makes its rounds through the ignition circuits.

Refer to Fig. 3 which illustrates a typical ignition system. With engine stopped and ignition switch turned on the flow of current in the primary circuit is from battery to starter switch terminal, to ammeter (or indicator light), to ignition switch, to primary winding in coil, to insulated breaker point, to grounded breaker point attached to metal distributor housing, from whence current flows through engine, to ground strap from engine to frame (if used), and through frame through ground strap connected to battery.

If the engine is running and the generator is charging the battery, the generator supplies the ignition current, delivering it to the ammeter (or indicator light). From this point the primary current flow is the same as described in the previous paragraph.

The secondary current originates in the high-tension winding in the coil. It flows out through the secondary coil terminal to a terminal in the center of the distributor cap. The cap also contains the terminals for the wires connected to the spark plugs.

Within the cap is an engine-driven rotating switch called a rotor which switches the high-

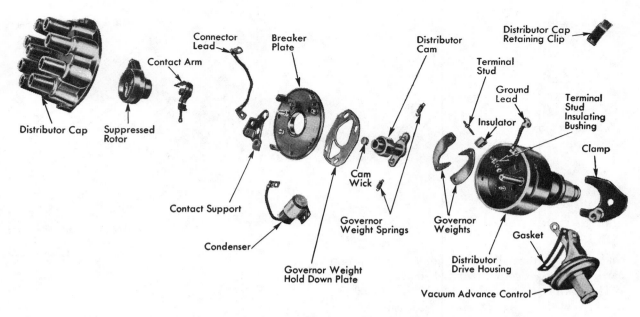

Fig. 14   Exploded view of a Delco-Remy distributor

tension current to one spark plug after another according to the engine's firing order. The current jumps the gap within the spark plug (igniting the mixture) and from thence flows through the cylinder block and frame to the battery ground strap, then through battery, to starter switch terminal, to ammeter (or indicator light), to ignition switch, to coil, to secondary winding.

The coil automatically provides only enough voltage to produce a spark at the spark plug electrodes. This might be as little as 5000 volts on closed throttle and up to 20,000 on wide open throttle. Most of this high voltage is used up in pushing the current across the spark plug gap. Therefore, the voltage used in the return circuit— all the way from spark plug shell to engine, to frame, to battery, to coil—is slight.

DISTRIBUTOR

The principal parts of the distributor unit include the cap, rotor, housing, breaker assembly, breaker cam, advance weights and drive shaft. (Distributors used on some cars do not use the centrifugal advance weights.) Fig. 14 illustrates a typical Delco-Remy distributor.

The distributor cap and rotor are molded from a resinous compound which is an excellent insulator. Brass terminals are installed in the cap when it is being molded. The high tension cable from the coil is inserted in the center terminal in the cap. Surrounding this terminal are terminals into which the spark plug cables are inserted.

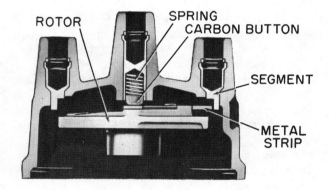

Fig. 15   This picture shows how a spring-backed carbon button is used to make contact with the metal strip on the rotor

The rotor is mounted on top of the distributor shaft. There is a hole in the hub of the rotor which fits onto the top of the breaker cam unit. There is a flat spot in the hole in the rotor hub which mates with a corresponding flat spot on the cam unit. This construction provides a positive drive (without slippage) for the rotor. A typical rotor is shown in Fig. 15. The clearance between the outer end of the metal strip and segments on the cap is very slight and, therefore, this clearance gap offers relatively small resistance to the flow of the high-voltage current.

Fig. 16 shows a typical breaker mechanism as well as the centrifugal advance assembly. The breaker plate has a pivot on which is mounted the

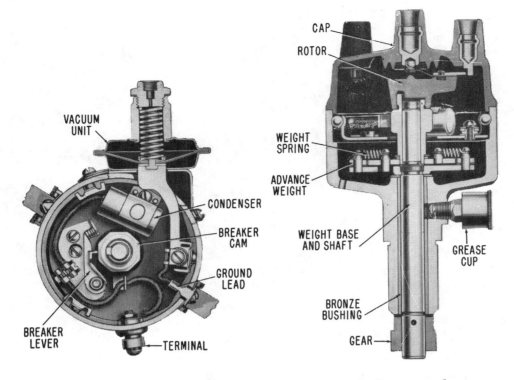

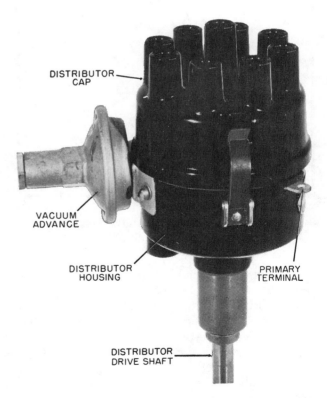

Fig. 16 *Picture of breaker mechanism and centrifugal advance mechanism*

contact support plate which carries the stationary breaker point. Also mounted on this point is the breaker lever arm to which the movable breaker point is attached.

The breaker cam unit is hollow and fits over the top of the distributor shaft. The base plate for the centrifugal weights is attached to this shaft. The weights are pivoted on the pins on the base plate. The cam is driven by the shaft by means of the centrifugal advance weights, as explained later on. Note that the shaft is carried in a long bronze bushing and provision is made for lubrication.

At this point it is desirable to discuss a typical distributor in detail so that the reader may obtain a thorough knowledge of the construction of the various parts and how they function. Other distributors may vary as to minor details but all of them have the same essential parts (except Ford distributors mentioned above).

Figs. 17 to 19 illustrate a typical distributor. Fig. 17 is a picture of the complete unit. Fig. 18 is an exploded view and Fig. 19 a sectional view. Figs. 18 and 19 should be studied together. Fig. 18 shows what the individual parts look like while Fig. 19 shows their relationship when assembled. The exploded view will also be helpful in understanding the construction shown in succeeding illustrations.

Fig. 20 shows the rotor and the inside of the

Fig. 17 *External view of an 8-cylinder Prestolite distributor. Primary terminal is connected to ignition coil by a wire*

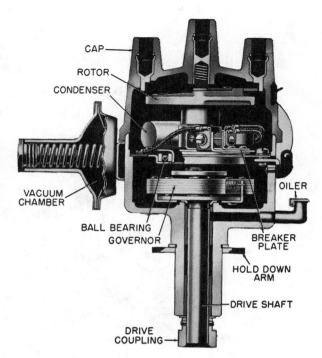

Fig. 19 *Sectional view of Fig. 17*

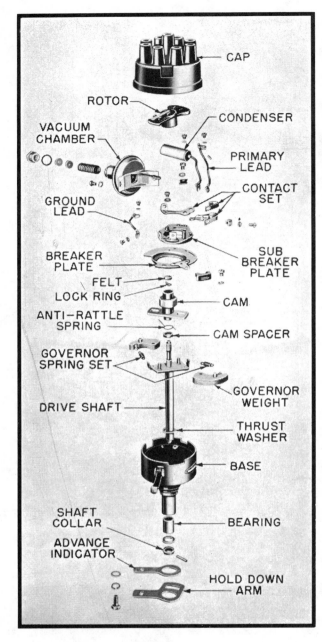

Fig. 18. *Exploded view of Fig. 17*

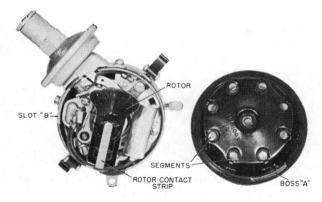

Fig. 20 *Picture of rotor and inside of distributor cap*

distributor cap. Secondary current from the coil is delivered to the metal contact strip on the rotor which distributes it to the metal segments to which the spark plug cables are connected. The curved boss "A" on the inside of the cap fits into slot "B" in the housing so that it is impossible to install cap incorrectly.

Fig. 21 shows the rotor removed. The flat spot in the rotor hub fits on the flat spot on the upper end of the breaker cam assembly. Consequently the rotor is always in the correct position when installed.

Fig. 22 illustrates the breaker mechanism. The shaft, driven from the engine camshaft, rotates in two bronze bushings. The breaker cam unit is hollow and fits over the top of the shaft. The breaker assembly is attached to the housing by two screws. An oil cup lubricates the bushings.

Fig. 23 is a disassembled view of Fig. 22.

Fig. 24 shows the centrifugal advance mechanism. The base plate is permanently attached to the shaft. The two centrifugal weights are pivoted on pins which are riveted to the base plate. The weights are held inward by small coil springs.

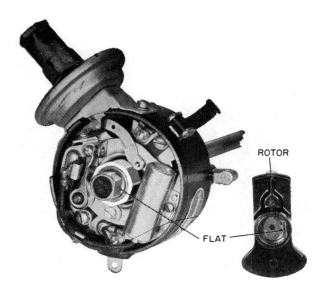

Fig. 21   *Rotor is removed*
by pulling it off

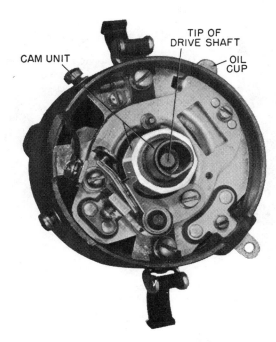

Fig. 22   *Breaker and advance assembly*

There is a pin extending upward from each weight. These pins engage slots in the plate welded to the base of the cam. Spring tension is sufficient to hold the weights inward when the engine is idling (or not running). But as engine speed is increased, centrifugal force pulls them outward and thus advances the cam which advances the spark. The cam is driven in the following manner: The distributor drive shaft is rotated by the camshaft. The drive shaft rotates the advance weights which in turn rotate the cam.

Fig. 25 shows the weights pulled inward by springs (left view). Note that the pins are at the inner ends of the slots in the cam plate. The right view shows maximum advance. Centrifugal force has pulled the weights outward. Note that the pins are at the outer ends of the slots. The cam advance is 12 degrees with respect to the distributor shaft, or 24 degrees on the crankshaft—which makes two revolutions to the camshaft's one.

Fig. 26 shows the breaker base plate which is attached to the housing by two screws. The flat springs under the screws retain the ball bearing. The primary terminal is insulated from the base plate.

Fig. 27 shows the advance plate. The ball bearing fits into a depression in the breaker plate. The ball bearing permits the advance plate to rotate through an 8-degree arc for vacuum advance. The pivot is for mounting the contact support arm and breaker arm. The pin is for attaching the link which is operated by the vacuum diaphragm. The eccentric screw is for adjusting the breaker points.

Fig. 28 shows the advance plate installed on the

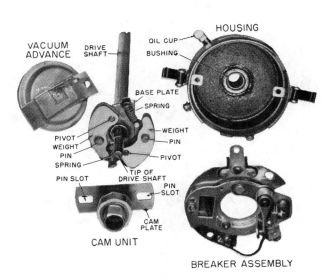

Fig. 23   *Disassembled view of Fig. 22*

breaker plate. The slot limits the motion of the lug on the advance plate.

Fig. 29 shows the contact support arm installed on the pivot on the advance plate. This arm carries the stationary breaker point. The breaker points are adjusted by turning the eccentric screw which swings the breaker arm through the small angle indicated. The lock screw must be loosened before the eccentric screw can be turned.

Fig. 30 illustrates the breaker arm and spring. A copper strip is riveted to the arm. This strip

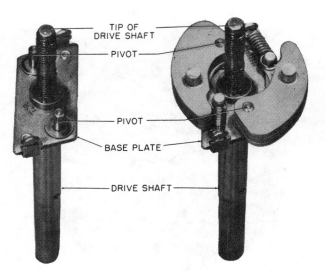

Fig. 24 *Centrifugal advancing mechanism*

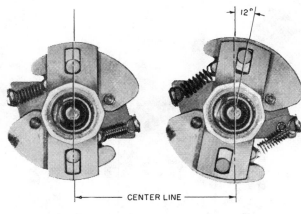

Fig. 25 *Centrifugal advance mechanism showing zero advance at left and maximum advance at right*

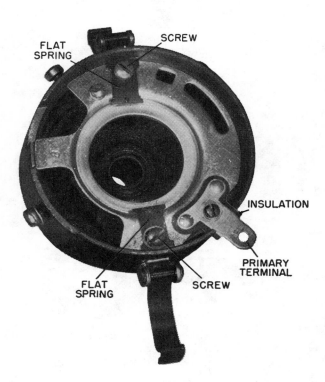

Fig. 26 *Showing breaker base plate attached to housing by two screws*

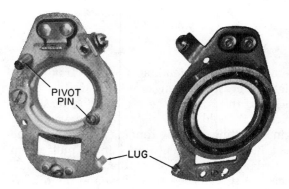

Fig. 27 *Advance plate. Top view shown at left, bottom view at right*

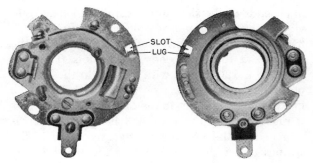

Fig. 28 *Advance plate installed on breaker plate. Top view shown at left, bottom view at right*

carries most of the current although some flows through the thin spring. Some breaker arms do not have a copper strip in which case the steel spring is heavy enough to carry the current. The arm has a composition bushing which pivots on a pin.

Fig. 31 pictures the breaker arm installed on the contact arm. The breaker arm is insulated from the contact arm by the composition bushing.

Fig. 32 shows the condenser installed on the advance plate. There is a metal-to-metal contact between the condenser housing and the plate. The wire leading from the condenser is attached to the terminal (No. 2) on the advance plate. One side of the condenser is grounded on the advance plate

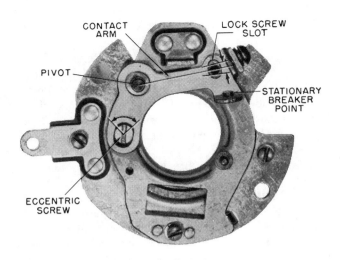

Fig. 29   Contact support arm installed
on pivot on advance plate

Fig. 31   Breaker arm installed
on contact arm

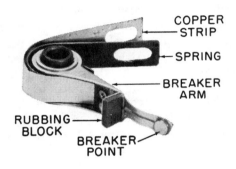

Fig. 30   Breaker arm and spring

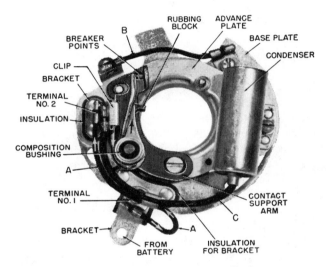

Fig. 32   Condenser installed
on advance plate

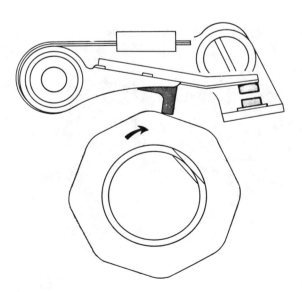

Fig. 33   Tip of rubbing block is
closer to cam than in Fig. 34; therefore, point opening is greater

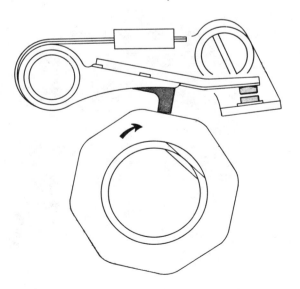

Fig. 34   In this view point opening
is less than shown in Fig. 33

573

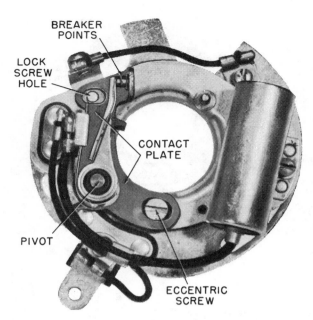

Fig. 35   Breaker point adjustment details

bracket is insulated from the breaker base plate on which it is mounted. A rubber grommet on the terminal prevents entry of dust. Current flows through wire "A" to terminal No. 2, which consists of an insulated bracket mounted on the advance plate. The breaker arm is completely insulated. It has a non-conducting composition bushing. The rubbing block which contacts the cam is made of the same material. There is a clip on terminal No. 2 which holds the breaker arm spring and its copper conductor in place. Current flows from wire "A" to the movable breaker point at the end of the breaker arm. From here current flows through the stationary breaker point on the contact support arm and from thence to the advance plate on which it is mounted. A ground wire "B" carries the current from the advance plate to the base plate which in turn is grounded on the distributor housing to which it is attached.

The breaker point gap depends upon how far away the rubbing block is from the cam. In Fig. 33 the tip of the rubbing block is closer to the cam than in Fig. 34; therefore, the point gap is greater. Obviously, the closer the rubbing block is to the flank of the cam the larger the point gap.

Fig. 35 (also see Fig. 29) illustrates the adjustment details. The contact plate swings on the same pivot as the breaker arm. It is adjusted by turning the eccentric screw whose head is about 1/16″ off center, permitting the head to move the arm 1/16″ to left or right, making a total travel of 2/16″ or 1/8″. This action moves the breaker point in or out a total of about 1/10″ which is more adjustment than is necessary. Fig. 36 (right

and the other side is connected by wire "C" to terminal No. 2. For proper functioning of the breaker mechanism, all electrical connections must be clean and tight. Insulation on wires must be good. Insulation elsewhere must be clean and in good condition.

Current flow through the breaker assembly is as follows: Current from the battery is delivered to terminal No. 1 which is on a flat bracket which extends out through the distributor housing. This

Fig. 36   Position of rubbing block when points are open and closed

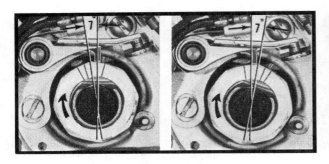

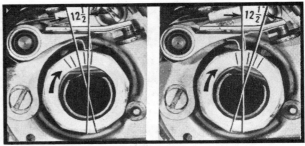

Fig. 37 *When points on a typical eight are set to a .015" gap the points are open while the cam rotates through an arc of 14 degrees. The left picture shows the cam position at the moment of point opening. The right view shows the cam at point of closing. The cam angle is 31 degrees (45 minus 14)*

Fig. 38 *When the gap on the same eight-lobe cam distributor shown in Fig. 37 is set at .020", the cam must rotate through an arc of 25 degrees to open and close the points. The cam angle or dwell is 20 degrees (45 minus 25)*

view) shows the rubbing block touching the flank of the cam, at which position the point gap is at a maximum.

The breaker gap must be wide enough so that the current flow is stopped abruptly when the points open. If the points are too close together there will be some arcing across them and this will reduce the available voltage at the spark plugs. On the other hand, the larger the gap the smaller the time that the points are closed. Therefore, too large a gap will cause missing at high speed.

The recommended gap varies according to the make and model of the engine. However, the gap for four-cylinder engines and most sixes is between .018" and .022", and between .015" and .018" for most eights.

The distributor cam makes one revolution for each two engine revolutions. The cam has as many lobes as there are cylinders: four lobes for a four, six for a six and eight for an eight. The breaker points open and close four, six or eight times per cam revolution for fours, sixes and eights, respectively.

Since 360 degrees represent one revolution of the breaker cam, a six-cylinder cam opens and closes the points for each cylinder while rotating through a 60-degree arc (360/6). On an eight the cam must open and close the points while moving through a 45-degree arc (360/8). On the average six, the points are closed 37 degrees and open 23 degrees (37 + 23 = 60). On the average·eight the points are closed for 31 degrees and open 14 degrees (31 + 14 = 45). The number of degrees that the points are closed is called the cam angle or dwell—31 degree dwell, for example.

Fig. 37 shows an eight-cylinder distributor with points adjusted to .015". In the left view the

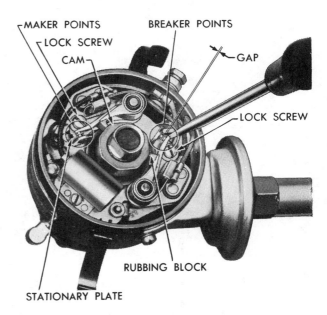

Fig. 39 *Prestolite double breaker distributor*

points have just opened. In the right view the points are just ready to close again. The cam angle or dwell is 31 degrees.

In Fig. 38 the points have been adjusted to .020". This results in a cam angle of 20 degrees. With this cam angle the points are not closed long enough at high speed to permit the coil to become fully saturated with magnetism and, therefore, the engine misfires.

Fig. 39 pictures the double breaker distributor used on some cars of the Chrysler group. Two sets of points are used for the purpose of building up the primary current so that when operating at high speed, efficient operation is assured.

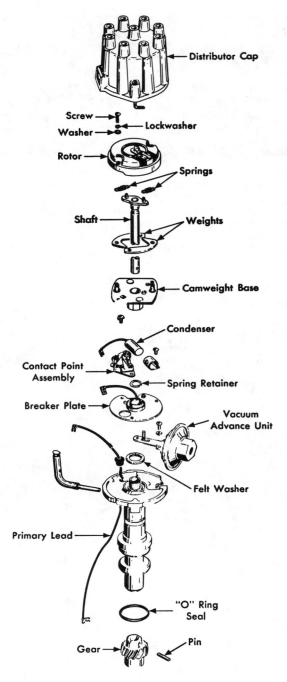

Fig. 40 *Delco-Remy external adjustment type distributor*

Distributor Cap

Screw
Lockwasher
Washer
Rotor
Springs
Shaft
Weights
Camweight Base
Condenser
Contact Point Assembly
Spring Retainer
Breaker Plate
Vacuum Advance Unit
Felt Washer
Primary Lead
"O" Ring Seal
Gear
Pin

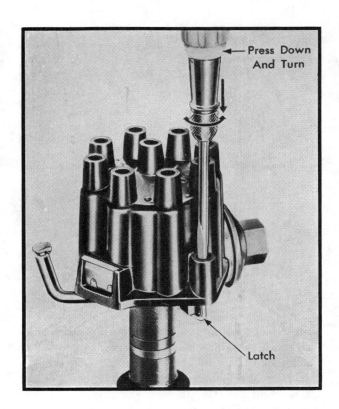

Fig. 41 *Distributor cap removal*

Press Down And Turn

Latch

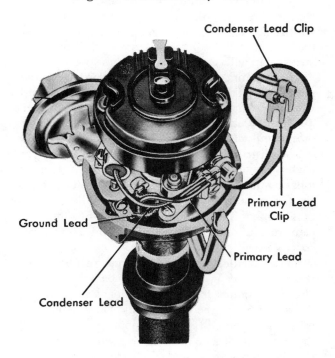

Fig. 42 *Primary and condenser lead positions*

Condenser Lead Clip

Primary Lead Clip

Primary Lead

Condenser Lead

Ground Lead

The points are connected in parallel between coil and ground and are staggered in relation to the eight-lobe cam. The overlapped contacts result in longer coil saturation, and as they are in a parallel circuit, no ignition occurs until both sets of points are open.

As the cam rotates, the first set of points closes the primary circuit. As it rotates a little further, the second set of points closes, but since they are connected in parallel, the circuit is not changed. Further rotation of the cam causes the first or "circuit maker" points to open. But, again, the

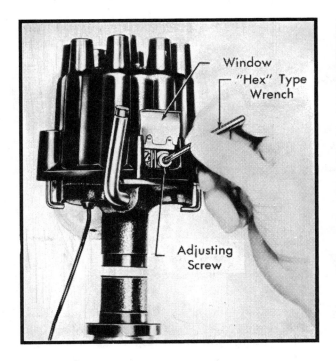

Fig. 43 *Adjusting breaker points through window*

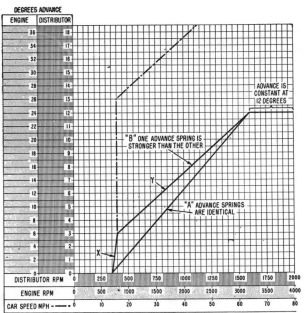

Fig. 44 *Centrifugal spark advance diagram*

circuit is not interrupted because the second or "circuit breaker" points are still closed. Later, the "circuit breaker" points open and break the primary circuit, causing a spark at the plug.

Since the "make" and "break" points are timed to close and open at the exact instant necessary for efficient engine operation, adjustment of points is an important factor in correct distributor operation.

New points can be adjusted with a feeler gauge in the conventional manner as follows: Rotate the distributor shaft until the rubbing block of one set of points is on the high spot of the cam. Then, with a screwdriver blade in the triangular opening, close or open the points to a clearance of .015 to .018″ by turning the screwdriver blade against the stationary point plate. Repeat this procedure on the other set of points.

Fig. 40 is an exploded view of a Delco-Remy external adjustment type distributor. This unit features a rotor with a built-in radio suppressor, and a window in the side of the distributor cap through which the breaker points can be adjusted, Fig. 41.

The breaker point set is replaced as a complete assembly. The replacement set has the breaker lever spring tension and point alignment pre-adjusted at the factory. Only the point gap requires adjustment after replacement. To facilitate the removal and installation of the distributor cap,

the conventional spring latches or clips have been replaced with two spring-loaded latches with slotted heads. To remove the cap, Fig. 41, insert a screwdriver in the upper slotted end of the cap retainers, press down and turn until latch is disengaged.

When installing new points, have the primary and condenser leads in the position shown in Fig. 42.

To adjust the points, operate the engine and lift the adjustment window provided in the distributor cap. Insert an Allen wrench, Fig. 43, into the head of the adjusting screw. Turn the adjusting screw clockwise until the engine begins to misfire. Then turn the wrench a half-turn (180 degrees) in the opposite direction (counterclockwise) which will give the proper point gap.

On all distributors, once the breaker gap is in correct adjustment, it gradually goes out of adjustment because of wear at the tip of the rubbing block. This wear obviously reduces the gap and invites misfiring caused by arcing across the points. This trouble may occur at all speeds but is most likely to be noticed at low speed.

If the car maker's specifications, instead of calling for .015″ gap, give a range of .015″ to .018″, it is desirable to set the gap at .015″ if the car is operated mostly at high speed although this setting may require rather frequent resetting of the points because of rubbing block wear. On the other hand, if the car is not driven at high speed,

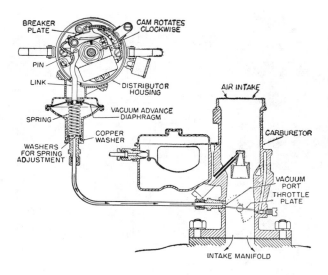

Fig. 45 *The spark is advanced by rotating the advance plate on which the breaker plate is mounted*

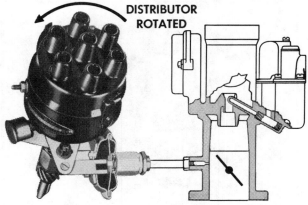

Fig. 46 *In this design vacuum advance is obtained by rotating the entire distributor*

the points should be set at .018". This allows for an extra .003" rubbing block wear and makes for infrequent point adjustment.

If the distributor shaft bushings are worn a few thousandths of an inch, the shaft and cam will wobble and therefore the breaker point gap will vary in random fashion from too much to too little. When the gap is small, misfiring (because of arcing) will occur at all speeds but especially at low speed. If the gap is too large, missing will occur at high speed because the cam angle or dwell is too small. Large and small gaps may occur at random during each revolution of the distributor shaft.

SPARK ADVANCE

In this section, spark advance is discussed in terms of crankshaft degrees. It should be realized, however, that when testing a distributor, that it rotates at half the speed of the crankshaft. Therefore, if the maximum centrifugal advance is 25 crankshaft degrees and the maximum vacuum advance is 16 crankshaft degrees, the advance when measured in degrees rotation of the distributor shaft is one half—or 12½ degrees centrifugal advance and 8 degrees vacuum advance.

For best power on open throttle, the pressure resulting from combustion should reach a maximum shortly after the piston has passed top dead center. In a modern engine, this pressure may be upwards of 600 pounds per square inch. Maximum pressure occurs about 15 degrees after top dead center, at which moment the piston has moved about 1/16" on its power stroke.

Inasmuch as it takes time for the mixture to burn, it is necessary to ignite it in advance of the arrival of the piston at top dead center. For example, on a certain engine operating at 1200 rpm, the time required to complete the combustion process is 1/200 second (.005 second). At first glance this may seem to be a negligible interval but it really is a large one because 1200 rpm is 20 revolutions per second. Therefore, the crankshaft makes one revolution (360 degrees) in 1/20 second (.05 second). Therefore, combustion requires 1/10 of a revolution and the crankshaft rotates 36 degrees during combustion. Thus, in order to have combustion completed 15 degrees after top dead center, ignition must occur 21 degrees before top dead center (21 + 15 = 36). At higher engine speeds, additional spark advance is required in order to obtain maximum torque at all engine speeds.

Centrifugal Advance

The spark is advanced by centrifugal weights within the distributor housing. The maximum advance varies considerably on different makes, ranging roughly from 18 to 40 crankshaft degrees. The engine speed at which maximum advance occurs may be anywhere from 2400 to 4200 rpm, depending on make and model of engine.

Centrifugal advance is zero when the engine is idling. Usually, advance begins at about 800 engine rpm and from this point increases steadily with engine speed until the maximum is reached.

Many engines in use are timed so that when idling, the spark occurs at top dead center. Others are timed so that ignition takes place 2 to 8 degrees before the piston arrives at top dead center. So, if for example, the spark is timed for 5 degrees

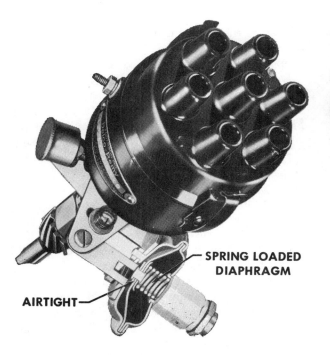

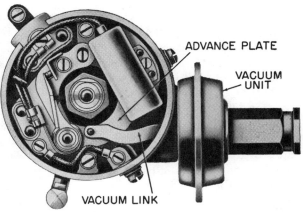

Fig. 49 *This picture shows link which operates advance plate*

Fig. 47 *The distributor in this picture is in no-advance position because the spring has pushed the diaphragm to the left as far as it will go*

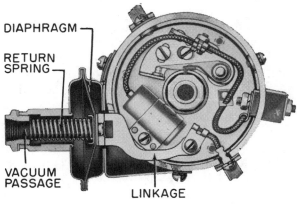

Fig. 50 *Diaphragm is in no-advance position. The entire breaker mechanism is mounted on three balls which allow it to rotate clockwise when the vacuum pulls the diaphragm to the left*

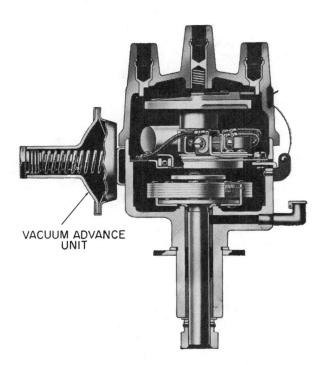

Fig. 48 *Vacuum advance unit in no-advance position. The spring has the diaphragm all the way to the right*

before top dead center and maximum centrifugal advance is 25 degrees, the total is 30 degrees.

If the two advance weight springs have the same strength, the advance weights will start to work at some low speed, such as 350 rpm, and the spark will be advanced at a steady rate up to some maximum speed, as shown at "A", Fig. 44. It will be seen that the advance is zero up to 350 rpm of the distributor shaft, which is 700 rpm of the crankshaft.

As engine speed increases, the spark is advanced steadily until it reaches a maximum of 12 degrees as measured with respect to the distributor shaft or 24 degrees with respect to the crankshaft. The maximum advance indicated in this example is obtained when the distributor shaft is rotating 1600 rpm, at which time the engine is rotating 3200 rpm.

Above 1600 distributor rpm (or 3200 crankshaft rpm) the advance remains constant at 12 distributor degrees (or 24 crankshaft degrees). The number of degrees advance is limited by the length of the slots in the advance plate, which means that in this case the advance plate cannot advance more than 12 degrees. With longer slots the advance could be higher, 15 degrees, for example. With shorter slots, the advance would be lower, say 10 degrees—or whatever is desired by the car maker.

If a maximum advance of 12 degrees were wanted at 2400 crankshaft rpm, weaker springs would be required. Or if a car maker did not want maximum advance until 4000 rpm, stiffer springs would be used. In other words, the strength of the springs determines the slope of "A", Fig. 44—the weaker the springs the steeper the slope and vice-versa.

Better engine performance is frequently obtained by using a weak spring on one advance weight and a strong spring on the other, as shown at "B", Fig. 44. When the engine is idling, the strong spring is slack. Therefore, only the weak spring opposes the centrifugal force of the advance weights.

The weak spring permits an extremely rapid advance as indicated by the steep line "X". The diagram shows that the spark advances from 0 to 3 degrees while the engine speed is increasing from only 700 to 800 rpm. At this speed all the slack has been taken out of the strong spring. Therefore, from this speed upward, both springs resist the outward motion of the advance weights and therefore, the rate of advance is reduced as indicated by the much reduced steepness of the line "Y" (as compared with "X"). As in the case of "A", the length of the slots stops the advance at 12 distributor degrees or 3200 engine rpm. Any increase in engine speed beyond this point cannot increase the advance beyond 12 degrees.

Vacuum Advance

Maximum engine torque at any given engine speed is obtained when the spark is advanced close to the point where pinging occurs. Pinging, of course, only occurs on open throttle. On part throttle, improved economy can be obtained by advancing the spark further than that provided by the centrifugal advance. This extra advance is supplied by a vacuum-operated device.

Details of vacuum advance units are given in Figs. 45 to 50. It should be noted that there are two ways of advancing the spark. In Fig. 45 the breaker assembly is rotated and in Fig. 46 the whole distributor is rotated. In either case, there is a spring-backed vacuum diaphragm. The spring holds the mechanism in the no-advance position unless or until there is sufficient vacuum acting on the diaphragm to move it by overcoming spring pressure. Note that the vacuum port in the carburetor is located just above the throttle when it is closed with the engine idling. However, just as soon as the throttle plate is opened a little, the edge of the plate is above the vacuum port and therefore engine vacuum acts on the diaphragm. In a typical case, the vacuum diaphragm starts to advance when there is 6 inches of vacuum and reaches full advance with 14 inches. Fig. 47 shows the same unit with full advance.

Figs. 48 and 49 show a Prestolite distributor wherein vacuum advance is obtained by rotating the advance plate. Fig. 48 illustrates the vacuum diaphragm and its spring as well as the ball bearing on which the advance plate is mounted. Fig. 49 shows that the diaphragm link is attached to a pin on the advance plate.

In the Delco-Remy distributor, Fig. 50, the complete breaker plate assembly is rotated by the vacuum diaphragm. This assembly is carried in three balls which run in a circular groove in the metal housing.

The amount of advance used varies from 15 to 25 degrees on the crankshaft on different models of engines. When an engine is running on open throttle, vacuum is not sufficient to operate the vacuum advance unit. On a typical engine six inches of vacuum is required to cause the vacuum unit to start advancing the spark and the advance reaches a maximum of 16 degrees when the vacuum is increased to 14 inches. Therefore, vacuum advance starts when the throttle is closed far enough to produce six inches of vacuum, and upon further closing the advance increases until it reaches a maximum when the vacuum is 14 inches.

On a given car running 40 mph, the throttle opening might be such that a vacuum advance of 10 degrees is obtained. At this speed the centrifugal advance might be 20 degrees. Therefore, the total advance is 30 degrees (10 + 20). The vacuum advance is designed to give no advance when the throttle is closed and the engine is idling.

Ignition System Service

TROUBLE SHOOTING

One of the most frequent troubles found in the ignition system is the burning of distributor breaker points. They turn dark gray or a bluish color and, in most cases, they are removed and replaced with new points which also burn up in a short time. This is bad practice because points never burn of themselves. It is simply an indication of trouble elsewhere in the electrical system which causes an abnormal amount of current to pass through the breaker points.

There are various reasons why points burn but high voltage probably accounts for well over half of them. And this condition shortens the life of every electrical unit on the car.

It must be remembered that all electrical units currently used on passenger cars are designed to operate on 12–14 volts. Under normal conditions the generator supplies the current to the battery and the electrical units about 14.5 volts. But if any obstruction to the passage of the generator current develops, the generator voltage will immediately increase, and excessive voltage, will be impressed upon all the electrical units, shortening their useful life. In an extreme case, such as the breaking of a battery cable, when the generator charging circuit is wide open, the voltage may run upwards of 30 volts. This will not only burn up the generator but all the bulbs will blow out and the ignition coil will burn out. In less severe cases, such as a loose connection or a corroded battery cable terminal, the results will not be as noticeable but bulbs will have to be replaced more often and the breaker points will burn.

Other causes of point burning are described below:

Bad Condenser

A condenser may be over or under-capacity. It may leak, it may be short-circuited, or it may have defective internal connections which cause high resistance. There are a number of good quality condenser testers commercially available and each manufacturer furnishes detailed instructions for its use. How a typical condenser tester is used is described further on.

Causes of High Voltage

1. Loose or corroded battery terminal which may look perfect on the surface yet have a high resistance contact at the battery post. In other words, it may be mechanically tight but it may not be electrically tight.
2. Generator charging rate too high.
3. Battery overcharged.
4. Engine mountings loose, producing an imperfect ground.
5. Loose ground strap on rubber-mounted engines.
6. A sulphated battery or one with defective cells will cause generator voltage to build up.

Grease or Oil in Distributor

Too much oil on the felt wick in the distributor cam well will be splashed all over the interior of the distributor by centrifugal force. Too much or the wrong kind of grease (vaseline, for instance) is sometimes used on the distributor cam. In either case, this oil or grease gets on the points in the form of an insulating black film through the action of the spark. This condition can be distinguished from a high-voltage burn by the fact that the film can be scraped off with a knife.

Loose Connection

A loose connection anywhere will cause trouble. If the connection is very loose, it can usually be detected by feel, since it will be warm or hot to the touch. The most reliable method of finding loose connections is by the use of a low-reading voltmeter—which is described further on.

Incorrect Breaker Point Gap

Too large or too small a gap will upset the balance between the coil action and the condenser effect, and this will cause points to burn.

Worn Distributor Shaft Bushings

Too much play between the distributor shaft and bushing in the housing will cause the breaker cam to be pushed to one side when it strikes the breaker arm rubbing block. Thus the points do not open sufficiently and they burn.

Also, if the lobes of the distributor cam are not uniform, erratic engine idling and shorter breaker point life will result.

Fig. 51 *What to look for when checking for high resistance in the primary circuit of distributor. In addition to the points indicated, look for external circuit high resistance at ignition switch terminals, ammeter terminals, coil terminals, and broken or poorly insulated wires in this circuit*

Worn Breaker Arm Pivot Pin Bushing

This will cause the points to open slowly instead of breaking cleanly. The result is that the points will burn.

Improper Breaker Arm Spring Tension

This tension should be within recommended limits as too low a pressure will cause misfiring at high speeds and too high a pressure will cause excessive wear on the cam and rubbing block.

Checking Distributor with Voltmeter

Any abnormal resistance in the primary circuit may be checked with a low-reading voltmeter. To use such a voltmeter, proceed as follows:

1. Turn on the ignition but do not start the engine.
2. Make sure that the breaker points are open.
3. Place the voltmeter prod on the distributor side of the ignition coil and place the handle prod on any convenient ground. (This assumes that the battery is grounded on the negative post; if the ground is on the positive post of the battery reverse the connections.)
4. The voltmeter reading should be 6 or 12 volts, depending on the system being tested. A read-

ing of less than 6 or 12 volts indicates a loose connection inside the ignition switch, or between the switch and distributor.

5. Repeat this test between the distributor stud to which the coil wire is connected, and on the insulated breaker arm. A reading of less than 6 or 12 volts indicates an imperfect connection. No reading indicates a complete open or a ground in the insulated bushing that goes through the distributor housing.
6. Now close the breaker points and place the two voltmeter prods across the points. There should be no reading on the voltmeter. Any reading whatsoever indicates the presence of dirt or oil between the contacts.
7. With the points still closed, test between the insulated breaker arm and a clean ground on the engine. Any reading on the voltmeter indicates that the distributor is not making a good electrical contact with the engine, which will result in poor performance. A remedy for this condition is to connect a length of wire between the distributor body and the nearest convenient spot on the engine to establish a good, artificial ground to replace the defective ground.

Hard Starting

Under mild climatic conditions, if the starter cranks the engine freely the battery voltage will be approximately 10 volts. But if the engine and oil are cold, maximum cranking load is put on the battery and starter and the battery voltage may be less. At these cranking voltages, if there is high resistance in the ignition primary circuit, the ignition coil output may be affected and cause hard starting. Therefore, when hard starting is encountered on cars where the starter cranks the engine freely, check the primary circuit to determine whether or not it has abnormal resistance, which will be indicated by the conditions shown in Fig. 51.

The ignition secondary circuit is another source of hard starting. If the cable from the coil to the distributor, or spark plug to distributor cables are not making good electrical contact at their terminals, the ignition coil secondary voltage may not be sufficient to force current across the spark plug gaps. A common cause of poor electrical contact at the distributor cap terminal sockets is corrosion.

Moisture settling in the distributor terminal sockets—in the presence of high-voltage electricity and nitrogen from the air—will form an acid. This condition will corrode and etch the cable strands, terminals and the imbedded brass sockets in the distributor head.

If spark plug gaps or the clearance between distributor segments and the end of the rotor is too wide, the resistance of the wide gaps will cause hard starting. Of course, a defective ignition coil will cause hard starting. (More on the subject of coils, condensers and spark plugs later on.)

IGNITION MISS

Although engine misfiring can be caused by some mechanical failure or by trouble in the fuel system, the most frequent cause is due to the ignition system. When the trouble is electrical, it can be classified as: (1) miss on acceleration and hard pull, (2) miss at high speed, (3) miss at low speed.

Miss on Acceleration and Hard Pull

Ignition miss that occurs under these conditions is usually caused by one or more of the following:

1. Spark plug gaps too wide.
2. Spark plug insulator cracked.
3. Spark plug insulator coated with excessive carbon deposits.
4. Secondary cable from coil to distributor having poor electrical contact at terminals.
5. Spark plug cables having poor electrical contact at terminals.
6. Ignition coil defective.
7. Distributor cap, coil tower or rotor having breakdown in insulation as a result of a crack or a carbon streak.
8. Breakdown in insulation of high tension cables—fractured or oil soaked.
9. Gap between distributor cap and rotor segment burned too wide.
10. High tension current jumping from rotor to primary wire in the distributor.

Ignition Miss at High Speed Only

This condition as a rule can be traced to one of the following:

1. Improper spark plug gap.
2. Deposit on spark plug insulator.
3. Insufficient breaker arm spring tension.
4. Breaker arm binding on its pivot pin.
5. Breaker point gap too wide.
6. Dirty or oxidized breaker points.
7. Worn bearings at movable breaker plate assembly.
8. Fractured primary lead between distribution primary terminal and breaker arm.

Ignition Miss at Low Speed

This condition as a rule can be traced to one of the following:

1. Improper spark plug gap.
2. Carbon deposit on spark plug insulator.
3. Breaker gap set too close.
4. Worn bearings at movable breaker plate assembly.
5. Fractured primary lead between distributor primary terminal and breaker arm.

NOTE—When the engine is running on part throttle at a speed of 20 to 35 mph, and the throttle valve is suddenly opened wide, compression is at its peak. This high pressure increases the electrical resistance at the spark plug gaps with the result that the ignition coil secondary builds up a considerably higher voltage than that required for part throttle operation. Unless the insulation in the secondary circuit is reasonably good, the high voltage brought about by peak compression will break down the insulation and cause the engine to misfire.

The gaps at spark plugs and at the distributor cap-to-rotor are the only gaps that should be in the secondary circuit. High tension cable to ignition coil tower, or high tension cables at distributor and at spark plug terminals should be mechanically tight and make good electrical contact. If these connections are not free from corrosion and making good electrical contact, an abnormal resistance will be built up in the secondary circuit. This condition may prevent spark at the spark plugs when the engine is accelerated.

IGNITION COIL SERVICE

Ignition coils do not normally require any service except to keep all terminals and connections clean and tight. In addition, the coil should be kept reasonably clean but it must not be subjected to steam cleaning or similar high pressure cleaning methods which may cause moisture to enter the coil. Moisture inside of any ignition coil is probably the principal factor that determines coil life. Ignition coils are subjected to rain, snow and road splash. Also, some moisture gets inside the coil from condensation and the normal breathing of the coil caused by temperature change.

The high tension terminal socket may be corroded as a result of arcing by previous failure to properly insert the end of the cable into the socket. Corrosion may also develop in sea coast areas due to the salt air.

Any corrosion will cause resistance to the flow

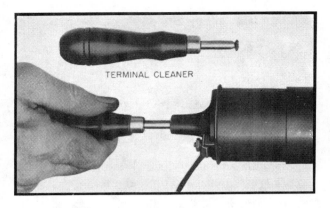

Fig. 52 Cleaning high tension terminal socket of coil with special cleaner made for the purpose. This tool is also used for cleaning out distributor cap cable sockets. Sandpaper wrapped around a pencil will also do an adequate job

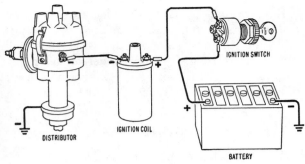

Fig. 53 Wiring connections for coil with negative grounded system

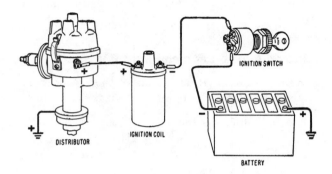

Fig. 54 Wiring connections for coil with positive grounded system

of current. Therefore, the socket should be thoroughly cleaned with a special terminal cleaner, Fig. 52, with sandpaper wrapped around a pencil or with a suitable stiff wire brush. Clean the cable terminal with sandpaper.

CAUSES OF COIL FAILURE

As mentioned above, moisture inside of any ignition coil is an important factor determining its life. But premature coil failure is caused by several other conditions, the chief reason being the same high voltage which burns breaker points.

The ignition coil, like the rest of the car's electrical system, is designed to operate efficiently at 6–8 or 12–14 volts depending on the system. Even with this normal voltage, a certain amount of heat develops in the coil, as heat is the natural result of current flowing through any conductor.

With a high-voltage condition, however, the increased voltage naturally causes an increase in the current flowing through the coil and, consequently, an increase in heat. But the really bad part of it is that the heat developed is *proportional to the square of the current*—not to the current itself. In other words, an increase of one-half in current will produce more than twice the heat.

This is exactly what happens to a coil on a car with high voltage. Considering a car with a 6-volt system as an example it is not unusual to find that the voltage is 10 volts, which is one-third increase over the normal 7½ volts. In such cases, the heat developed in the coil is about twice the normal, and the coil must overheat.

Any coil that is subjected to twice the normal heat for any length of time will gradually weaken,

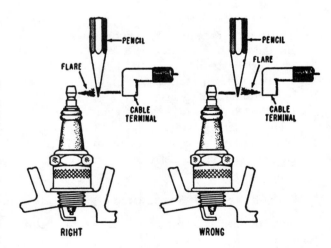

Fig. 55 Checking coil polarity

as the high heat will char the insulation in the secondary winding and cause internal shorts. Eventually the coil will fail completely.

Therefore, in most cases where you find burnt breaker points on a car, the coil is also liable to go bad as both conditions are caused mostly by high voltage. If the coil does go bad under these condi-

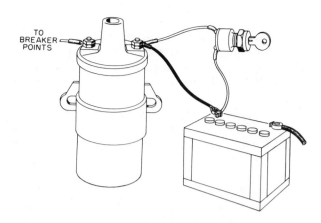

Fig. 56 Temporary connection to cut ignition switch out of circuit when checking operation of ignition coils

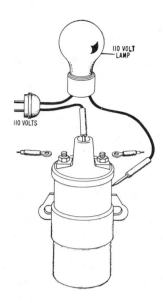

Fig. 58 Using test points and lamp to check for grounded windings. If coil is defective, bulb will either light or tiny sparks will be noted as test point is rubbed on outside shell

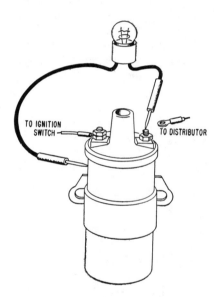

Fig. 57 Using test points and light to check for an open circuit in the coil primary winding. If bulb lights, coil is okay

The above conditions may even cause the spark to jump between the high tension and the primary terminals of the coil, over and along the bakelite top. It is a well known electrical characteristic that all electrical currents travel in the path of least resistance. Therefore, whenever the resistance of the spark plug gap under compression plus the resistance of the defective unit becomes greater than the resistance of the insulation between the high tension terminal and the primary terminal, the spark will jump there instead of the spark plug gap. This is especially true in damp weather, due to moisture on the bakelite top which reduces the electrical resistance there, or when grease or dirt are present on the coil top. The metallic particles in the dirt also provide a path of low resistance for the spark. Besides, any moisture that would dry up on a clean coil top will keep the dirty top damp.

Sometimes the spark jumps right through the bakelite of a high tension terminal at a point opposite a primary terminal. You will find in most of such cases that the high tension cable is not all the way in the socket.

tions, it will do no lasting good to install a new coil unless the cause of the high voltage condition is found and corrected.

Another cause of coil failure is an abnormal strain put upon the secondary winding of the coil by resistance in the high tension circuit. This can be caused by a defective radio suppressor, wide gaps between spark plug electrodes (due either to wrong adjustment or natural burning away of the electrodes in service) or a burnt rotor segment. In any one of these cases, the coil is compelled to develop an abnormal amount of voltage in order to overcome the additional resistance, and this abnormal voltage often breaks down the insulation between the layers of the secondary winding.

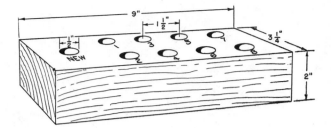

Fig. 59 Diagram of a spark plug holder with numbered holes so each spark plug can be identified as to the cylinder from which it was taken. The provision for the extra hole is for a new plug for purposes of comparison

Leaving the ignition key on overnight or for a long time is another frequent cause of coil breakdown. In this case, the uninterrupted current through the primary overheats the coil and damages the insulation. After a period of time, this will cause the coil to fail.

IMPORTANCE OF CORRECT COIL POLARITY

As you know if you have studied the chapters on *Magnetism* and *Electricity,* polarity means the quality of having opposite poles. In a magnet, for instance, there is a north pole and a south pole. A battery has a positive pole and a negative pole. Although it has been established that the flow of current in a wire is from the negative to the positive, the practice of considering that electric current flows from positive to negative has been established in textbooks and service station work for so many years that it is advisable to go along on that assumption. Therefore, so far as we are concerned with the automotive electrical system, the principle to remember is that current flows from the positive pole to the negative pole.

Let us apply this principle to ignition coil operation and performance. In the automotive electrical system, the battery is connected either by grounding the negative post or grounding the positive post. Where the ground is on the negative post of the battery, all operating electrical units are connected to the positive side of the battery, and the current flows from the live (ungrounded) post of the battery, through the units and back through the chassis of the car (the ground), which is connected to the negative post of the battery, Fig. 53.

Where the positive post of the battery is grounded, the current flows in the opposite direction—through the chassis, then through the units, and back into the battery, Fig. 54.

It has been conclusively established that the high tension spark in the combustion chamber jumps with greater ease if the live (center) electrode of the spark plug is of negative polarity. As the live electrode of the spark plug is connected directly to the high tension winding of the ignition coil, it follows that the ignition coil must be so constructed internally that the end of the high tension winding, which is connected to the spark plug, be of negative polarity.

From the above, it must *not* be deduced that an ignition coil will not operate unless the live end of the secondary winding is of negative polarity. What will happen if the polarity is wrong is that the efficiency of the coil will drop 15% or more, and while this may not have an apparent effect under normal operating conditions, it most certainly will be felt if conditions demand the maximum coil efficiency. For instance, in cold weather starting, wrong coil polarity has often resulted in failure to start, something that always haunted the winter driver. It is that 15% loss in efficiency, when conditions require peak performance, that is the decisive factor in such cases.

Checking Coil Polarity

Most coils are marked positive and negative at the primary terminals. When installing or connecting a coil be sure to make the connections as shown in Figs. 53 and 54.

If perchance the coil is not marked as to its polarity, it can be checked by holding any high tension wire about 1/4" away from its spark plug terminal with the engine running. Insert the point of a wooden lead pencil between the spark plug and the wire, Fig. 55. If the spark flares and has a slight orange tinge on the spark plug side of the pencil, polarity is correct. If the spark flares on the cable side, coil connections should be reversed.

IGNITION COIL TESTS

If poor ignition performance is obtained and the coil is suspected, it may be tested on the car or it may be removed for the test.

Ignition coils are often condemned when the trouble is actually in the ignition switch. A completely defective ignition switch will produce an open primary circuit, giving the same indication as if the coil were completely dead. A partly defective ignition switch will cause a weak spark.

Both of these conditions are often blamed on the coil.

By cutting the ignition switch out of the circuit, it can easily be determined whether or not the coil is defective or whether the fault lies with the ignition switch.

A wire should be connected to the terminal of the coil to which the battery wire is normally connected, Fig. 56. This temporary connection jumps the ignition switch. If the trouble is eliminated when the engine is started, it is obvious that the ignition switch was the offender—not the coil.

TYPES OF COIL TESTERS

Two types of testers are used in testing ignition coils. One of these makes use of an open or protected spark gap while the other reports the condition of the coil on a meter. The second type of tester is usually so designed as to permit testing of the coil without making any connections to the secondary terminal. This eliminates certain variables caused by altitude, atmospheric or spark gap electrode conditions which are usually present in the spark gap type of test.

The spark gap type of tester should always be used comparatively. That is, a coil known to be good should be compared with a questionable coil. Both coils should be at the same temperature and identical test leads must be used.

When using a meter type coil tester to test a coil without removing it from the car, be very careful to avoid touching the tester case to the car. Many such testers have a ground connection to the case. Touching the case to the car would produce a short circuit and possible serious damage to the equipment.

Details of testing procedures and the manner in which the various testers are used are furnished by the manufacturer of the equipment.

Simple Ignition Coil Check

In the absence of any testing equipment a simple check of the ignition coil can be made as follows: Turn on the ignition switch with breaker points closed. Remove the high tension cable from the center socket of the distributor cap and hold it about ⅜" away from a clean spot on the engine. Then open and close the breaker points with an insulated screwdriver. If the coil and other units connected to it are in good condition a spark should jump from the wire to the engine. If not, use a jumper wire from the primary terminal of the distributor to the engine; if the primary is in good condition a spark will occur.

Simple Test for Open or Grounded Circuit

To test for an open circuit in the primary winding, disconnect the distributor wire from the coil and use test points (or clips) as shown in Fig. 57. If the bulb lights the coil is okay.

To test for an open secondary winding, put one test point in the high tension terminal socket and the other one at one of the primary terminals. If the secondary winding is *not* open, the lamp will *not* light but tiny sparks will be noted as the test points are rubbed over the terminals. If the secondary winding is open, no sparks will occur and the lamp will not light.

All ignition coils with metal containers used on passenger cars can be tested for grounded windings with a 110-volt test lamp. Isolate coil by disconnecting wires from it. With one test point in coil tower and the other on paint-free outside shell, defective coil will be indicated if bulb lights or if tiny sparks appear when test point is rubbed on outside shell, Fig. 58.

SPARK PLUG SERVICE

In order to judge the condition of spark plugs accurately, they should be removed from the engine along with their gaskets (if used). Place each plug in a holder or in a position which will enable you to identify the cylinder from which it was taken. Fig. 59 shows a drawing of such a holder which can easily be made out of wood.

Cracked or broken insulators will tell their own story and such plugs should be discarded. Other conditions, some easily remedied, some of a more serious nature, can best be individually described. Remember that the threads and gaskets of the spark plugs tell an accurate story and various types of deposits on the firing end of the insulators, as well as the condition of the electrodes, clearly indicate the condition of each plug.

Removing Plugs from Engine

1. Remove wires from plugs.
2. Blow out any dirt accumulation from the spark plug recesses to prevent it from getting into the combustion chamber when the plug is removed. Use compressed air, a tire pump or an empty insecticide spray gun for the purpose.
3. Slide the spark plug wrench down over the spark plug as far as it will go. Be sure the wrench fits snugly over the hex portion of the shell, otherwise the wrench may slip and break

Fig. 60 *Example of excessive electrode wear due to the plug operating too hot*

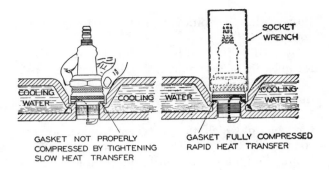

GASKET NOT PROPERLY COMPRESSED BY TIGHTENING SLOW HEAT TRANSFER

GASKET FULLY COMPRESSED RAPID HEAT TRANSFER

SOCKET WRENCH

COOLING WATER

COOLING WATER

COOLING WATER

Fig. 61 *Spark plug gasket installation*

Fig. 62 *Example of a cracked insulator. This type of break is always caused by an outside blow or a poorly fitting wrench when removing or installing spark plugs*

the insulator. Never use an open end wrench to remove spark plugs. Use deep socket wrenches only. The spark plug socket wrenches sold by some accessory stores fit too loosely around the hex and may result in damage to the plug. Therefore, always use deep socket wrenches manufactured by one of the companies specializing in high grade tools.

4. Unscrew the spark plug counterclockwise.

5. Take the gasket (if used) off the lower end of the plug, or remove it from the recess in the cylinder head if it did not come off with the plug.

ANALYSIS OF SPARK PLUG CONDITION

Original equipment spark plugs are designed for efficient operation for average driving conditions. Under unusual conditions, however, a different type of plug may give more satisfactory service. For example, if the car is driven most of the time at slow speed, such as door-to-door service, or is used for a greater part of the time in metropolitan areas where traffic is slow, the operating temperature would be low and a hotter plug may be required. The same car, if operated for a greater proportion of the time at high speed over long distances, may require a colder plug because the operating temperatures would be higher.

A practical guide to determine spark plug performance is to study the appearance of the firing tip of the insulators after they have been in operation (without idling) for a few hours under nor-

mal driving conditions. The color of the spark plug insulation will indicate whether the plug is too hot or too cold for the engine or whether the carburetor adjustment is correct.

Normal Spark Plugs

When a plug is operating normally for a relatively short time, the insulator nose will have a soft powdery deposit of grayish tan, and electrode corrosion will be slight.

When the same plug has operated normally for a long time, the insulator nose will have a heavier powdery deposit and still will retain its grayish tan color with slightly more electrode erosion.

Another indication of a normally operating plug is one in which the deposits on the insulator will have a dull, rusty appearance but without any signs of blisters. Electrode erosion will be **slight**.

Plugs Operating Too Hot

If the deposits on the insulator nose are badly blistered and fused, with excessive electrode erosion, it indicates prolonged operation under extremely severe conditions. Such cases require the substitution of a cooler plug of perhaps two heat ranges—certainly at least one, Fig. 60.

When the insulator is white with dark spots near the tip, it is a sign that the plug is in the early stages of being overheated. Reinstall the plug or plugs and after a few hours operation, remove and inspect again to see if the condition has worsened. If it has, a cooler plug is indicated. If the condition has not become worse, the indication is that the heat was not being transferred to the cooling water properly. Such a condition may be caused by a poor gasket or the plug not being tightened properly.

Plugs Operating Too Cool

When a plug is operating too cool for a comparatively short time, a dull black film of carbon develops on the insulator nose and exposed shell surfaces.

Later on, this "sooted" appearance builds up into a black accumulation of oil and carbon, partially or completely fouling the plug so that it misfires.

In severe cases, the deposits become a sludgy black accumulation of oil and carbon, almost filling the area between the insulator and shell.

A cold operating plug rarely shows electrode erosion.

Oil Fouling

Oil fouling is readily identified by a wet, shiny black deposit, generally distributed over all surfaces of the firing end of the plug. An oil-fouled plug nearly always indicates that the engine has badly worn or sticking piston rings, sloppy pistons, or sticking valves. While spark plugs cannot cure oil pumping, a hotter type spark plug will help engine performance.

Gas Fouling

A dry, fluffy black deposit indicates a gas-fouled spark plug. This deposit results either from too rich a mixture, poor choke adjustment or other carburetor faults. It can also be caused by a weakness in the ignition system, including spark plugs too cold for the service, or spark plugs that have been excessively cleaned and are worn out.

Spark Plug Gaskets

A properly seated gasket shows uniform flatness and performs its two-fold purpose of sealing the plug against compression loss and it provides the bridge or heat path over which the accumulated heat within the spark plug itself can travel into the cylinder head to be dissipated in the circulating water of the cooling system surrounding the spark plug hole.

The proper tightening of spark plugs is the most important item of all as 50% of troubles due to overheated spark plugs are caused by the plugs being too loose in the cylinder head. As shown in Fig. 61, if the spark plug is not tightened sufficiently to fully compress the gasket, there will not be adequate contact between the plug shell and the cylinder head. This would retard the flow of heat from the spark plug to the water jacket, and result in overheating of the spark plug, preignition, burning of insulator and electrodes and leakage.

Whenever spark plugs are removed, always use new gaskets upon installation to assure a good, tight seat. (Late model cars of the Ford Group employ spark plugs with tapered seats, thereby requiring no gaskets.)

Excessive tightening of spark plugs not only tends to increase the spark gap but it distorts the gasket so that compression blow-by occurs which is evident by carbon being present on the gasket.

Fig. 63 Example of a broken insulator at the lower end. This type of break is usually caused by bending or straining the center electrode when adjusting the gap. Abnormally hot operating conditions may also cause this type of break

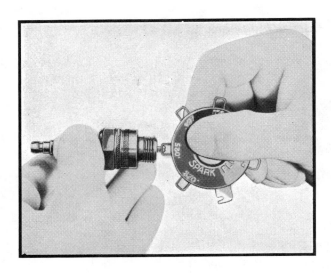

Fig. 64   *Checking spark plug gap
with a round wire gauge*

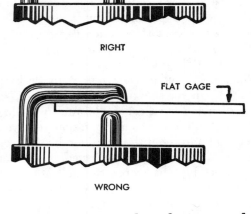

Fig. 65   *Showing right and wrong spark
plug gauges. Flat gauge will not indicate
true gap with worn electrode*

Broken Spark Plug Insulators

When the upper end of the insulator is broken, Fig. 62, it is never caused by conditions in the engine. It is always due to an outside blow or a poor fitting wrench when installing or removing plugs. A newly broken insulator may not cause misfiring immediately, but as soon as oil or moisture penetrates the fracture, the plug will misfire. This condition is often apparent to both eye and ear. The sound is a cracking snap and the appearance is a small flash to the metal shell's top. The actual crack may not be visible because it may occur just below the crimped part of the shell.

Fig. 63 shows a broken insulator at the lower end. In most cases this type of break is caused by bending or straining the center electrode when regapping. Always gap plugs by bending the side electrode only. This type of breakage may also occur when operating conditions are abnormally hot or when a plug too hot for the application is used.

Cleaning

Ideally, spark plugs should be cleaned every 3000 to 4000 miles. Spark plugs having an oily deposit must be degreased and dried thoroughly before sand blast cleaning to prevent gumming and packing of the cleaning compound. (Sand blasting equipment is essential for spark plug cleaning as there is no other way to clean a plug properly. There are a number of manufacturers of this equipment and full instructions for its use accompany each unit.)

Heavy deposits of carbon or oxide also must be scraped out to avoid excessive blasting which will cut away the insulator at the center electrode and cause the plug to run colder.

Before degreasing and scraping, place the spark plugs in carburetor cleaning fluid to dissolve the gum and grease. Soak the plugs for 15 to 30 minutes. Thoroughly dry the interior of the plugs with compressed air. Then scrape out all deposits from the shells and insulators with a pointed steel scraper. Blow out all scrapings with an air stream before applying the sand blaster.

Sand blast each plug until the interior of the shell is clean and the entire insulator is white. Examine the interior of the plug in a good light. Remove any cleaning compound with the scraper and air stream. If there are still traces of deposits, try to scrape it out rather than by excessive blasting. Finish the job with a light blasting operation.

Thoroughly clean the electrodes with fine sandpaper, paying particular attention to the inner side of the ground electrode. If a blue scale is present, it must be removed as it is an insulator. Use an ignition point file or steel scraper to break off this scale.

Adjusting Plug Gaps

Adjust the electrode gap according to the engine manufacturer's specifications, using a round wire gauge only, Figs. 64 and 65. Then set the gap with the tool shown in Fig. 66 or its equivalent, bending the ground electrode only. Never bend the center electrode as breakage of the insulator will result.

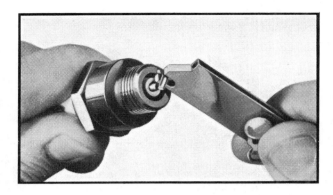

Fig. 66 Bending tool for adjusting spark plug gap. Always bend the side electrode, never the center one as the insulator may break

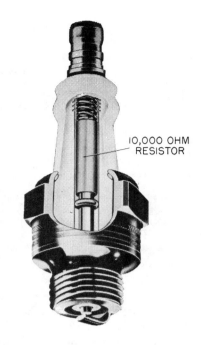

10,000 OHM RESISTOR

Fig. 67   Cutaway view of resistor spark plug

Installation of Plugs

After the plugs are properly gapped, the following procedure is recommended for their installation:

1. Thoroughly clean the gasket seating space.
2. Screw the spark plug in by hand as far as it will go.
3. Carefully fit a deep socket of the proper size over the plug and pull on the wrench very lightly until you feel contact with the gasket.
4. Slowly increase the pull on the wrench (you will feel the gasket compressing) until the resistance to pull suddenly becomes very great, indicating that the gasket has been fully compressed.
5. Clean the top of the insulator to obtain a clean, dry surface.

Precautions When Tightening Plugs

Always use a good quality deep-well socket wrench and a torque wrench when tightening plugs. When using a torque wrench, tighten the plugs to the values given below. These figures are maximum and are based on spark plug and engine threads being clean.

Plug Thread	Iron Heads	Aluminum Heads
10 MM	14 lbs. ft.	12 lbs. ft.
14 MM	30 lbs. ft.	28 lbs. ft.
18 MM	34 lbs. ft.	32 lbs. ft.
⅞″	37 lbs. ft.	35 lbs. ft.

If a torque wrench is not available, be guided by the following recommendations for, as stated previously, 50% of the troubles due to overheated spark plugs are caused by the plugs being too loose in the cylinder head.

The smaller the plug size the less pull on the wrench handle is required to fully compress the gasket. For instance, a 14 MM plug requires less pull on the wrench handle than a ⅞″ plug. This is particularly important in the case of 10 MM plugs, where only just enough pull should be exerted on the wrench to compress the gasket. And to avoid the possibility of damaging 10 MM plugs, the wrench handle should not be more than 4″ long.

The final tightening of spark plugs in cast iron heads should be made with the engine at normal operating temperature. For aluminum heads, the final tightening should be made with the engine cold (normal room temperature).

Installation of plugs in aluminum heads require particular care as there is danger of stripping the threads in the cylinder head. Never use graphite or other lubricating compounds on the threads as lubricants will retard heat transfer by separating the metal of the threads. It also reduces friction in the threads which may result in stripping the threads in the cylinder head.

RESISTOR PLUGS

The resistor spark plug, Fig. 67, is a specially designed plug with a built-in 10,000-ohm resistor. The resistor eliminates that part of the spark discharge which causes interference with radio and television. It also eliminates most of that part of the spark which causes heavy electrode erosion

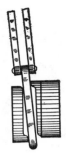

Fig. 68
Breaker points worn out of alignment. A flat .020″ gauge will space contacts .030″ to .040″. Use wire gauge only

Fig. 69
Breaker points worn unevenly. A flat .020″ gauge will space contacts .030″ to .050″. Use wire gauge only

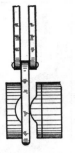

Fig. 70 *Breaker points with metal transfer from one to the other. Using a flat .020″ gauge to set these contacts actually will space them .040″ to .050″*

Fig. 71
Only correctly aligned breaker points can be properly spaced with a flat gauge. New contacts are usually made up of one convex surface operating against one with a flat surface

Fig. 72 *Showing advantage of employing one convex point as against two flat points when points are misaligned*

so that a wider gap setting can be used without the necessity of frequent gap adjustments.

When installing resistor spark plugs in a car not previously equipped with them, the question arises as to whether or not resistors already in the ignition system be removed. These resistors may be in the secondary cable between the coil and distributor, in the distributor tower, or at the spark plug.

In general, it is recommended that these resistors be removed when resistor plugs are installed. This recommendation is made not because the resistor is harmful, but because very few of the resistors used at these points have the required characteristics. In some cars, a resistor is built into the distributor. These are reliable units and seldom give trouble.

Presently, carbon impregnated core ignition cables (see Electrical System Chapter) are used as resistors to suppress radio interference. They are standard equipment on all passenger cars and require no supplementary resistor equipment.

Cleaning Resistor Plugs

Resistor plugs should be cleaned like regular plugs at regular intervals, especially when operating on highly leaded fuels. It is important that the insulator firing tips be cleaned of such deposits which tend to leak off current to ground and thus short out plugs.

CONDENSER SERVICE

Condensers used in ignition systems must be in perfect condition to be efficient. They must function under adverse conditions such as heat and moisture, and must withstand high-voltage surges of 350 volts and upward from the ignition system.

A condenser should not be condemned because the points are burned or oxidized. Oil vapor, or grease from the distributor cam or high resistance may be the cause of such a condition.

Condensers should be tested with a reliable condenser tester for leakage, breakdown, capacity, and resistance in series in the condenser circuit. Manufacturers of condenser testers furnish complete instructions as to their use. How a typical one is used is described further on.

Breakdown of a condenser is usually caused by the high-voltage surge of the primary circuit when the points open, which punctures the insulation of the condenser and causes it to fail. Leakage represents the lowering of the internal resistance of the insulation, and a small amount of current leaks through, preventing its effectiveness to stop arcing at the breaker points.

Low resistance of the condenser leads, both the ground connection and the flexible lead to the breaker point, is imperative for good results. When this resistance is raised trouble starts.

Capacity in microfarads must be correct for each type of ignition. If too low, point burning results; if too high the efficiency of the spark is reduced at high speeds.

Inspect the condenser for broken leads, frayed insulation and a loose or corroded terminal. Make sure the condenser is firmly mounted and makes a

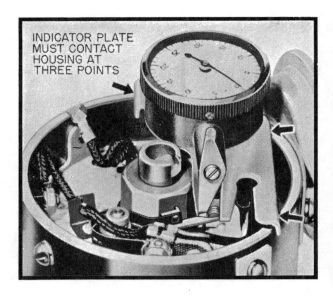

Fig. 73 *Dial indicator for measuring point opening on Delco-Remy internal adjustment distributors*

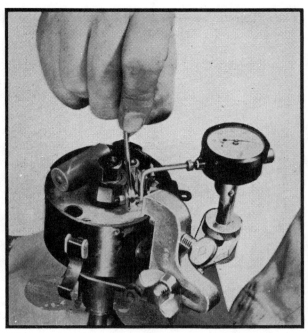

Fig. 74 *Dial indicator for measuring contact point opening on Prestolite distributors*

good ground contact. Be sure the terminal is tight.

BREAKER POINT SERVICE

The normal color of breaker points should be light gray. If the contact surfaces are black it is usually caused by oil vapor or grease from the cam.

Figs. 68, 69 and 70 show the condition of breaker points after they have been in operation for several thousand miles. These illustrations show why it is difficult to space contacts with a feeler gauge. Unfortunately, breaker points do not wear evenly, and with each thousand miles of operation, the surfaces deviate from being parallel with each other.

Fig. 68 shows what happens when contacts are not in correct alignment—they lap over each other. Fig. 69 shows uneven wear of the contact surfaces, while Fig. 70 pictures the development of a crater and projection, usually caused by a metal transfer from one contact to the other.

Fig. 71 shows an enlarged view of a new set of breaker points. The right-hand contact has a convex surface while the left-hand point has a flat surface. This convexity is scarcely visible to the naked eye as it amounts to approximately .002″ from the center of the contact to its outside extremity. The advantage claimed for this design is that the contact of surfaces which break the arc will be nearer to the mass of metal in the two

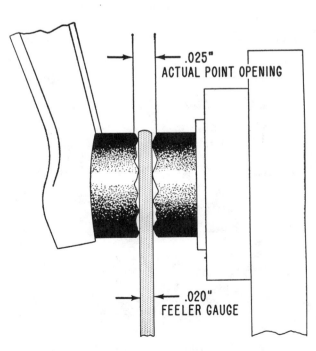

Fig. 75 *Why flat feeler gauge will not provide accurate point spacing if points are rough*

Fig. 76 Showing how metal from one contact transfers to the other

contacts, which gives better heat radiation. A further advantage is that should the points be misaligned, more metal will be in contact when a convex point is used than if both points were flat. Fig. 72 illustrates how this is accomplished.

Prestolite is strongly against filing their contact points because the cutting surface of the file produces high spots on the contact surfaces, which means concentration of current and heat in extremely small areas.

Delco-Remy recommends that points that are blackened or slightly burned or pitted should be cleaned with a special point dressing stone or a clean point file. In dressing the point, remove the high spots only as it is not necessary to remove all traces of build-up or pit.

Sandpaper or emery cloth should never be used to clean up points, since particles of sand or emery may imbed in the points and cause rapid burning and wear.

On Prestolite and Delco-Remy internal adjustment distributors contact points should be spaced either on a distributor test fixture or with a dial gauge, Fig. 73 and 74. Either of these methods not only eliminates the possibility of a wrong gap setting, but if the points are slightly rough but otherwise in alignment, there is the danger of an incorrect gap, Fig. 75. A further advantage of this equipment is that it uncovers irregularities between cam lobes. (Manufacturers of distributor test fixtures furnish complete instructions as to their use.)

If the contacts develop a crater or depression on one point and a high spot of metal on the other, the cause is an electrolytic action transferring metal from one contact to the other, Fig. 76. This can be the result of some operation of the car. A slow speed driver in city traffic or door-to-door delivery vehicles will be one extreme, and high speed long distance driving would be the other extreme. It may also be due to an unbalanced ignition system, which can sometimes be improved by a slight change in condenser capacity.

If the build-up is on the positive point, Fig. 77, install a condenser of higher capacity. If the build-up is on the negative point, a condenser of lower capacity may eliminate the condition, Fig. 78.

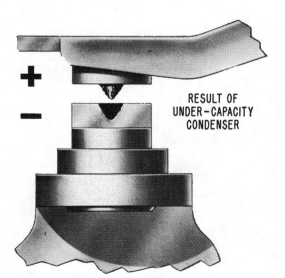

Fig. 77 Mound on positive point. If this condition is chronic, a condenser of somewhat higher capacity should be used

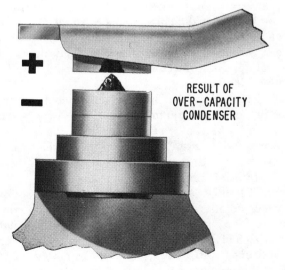

Fig. 78 Mound on negative point. If this condition is chronic, a condenser of somewhat lower capacity should be used

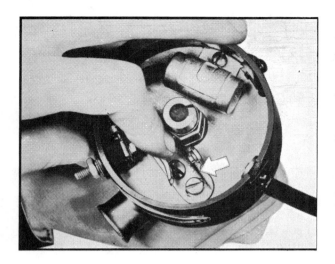

Fig. 79 *Oil on contact points shown by smudgy line on point support and breaker plate*

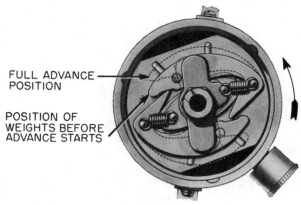

Fig. 80 *Operation of a typical automatic advance mechanism*

One of the most frequent causes of contact point failure is the presence of oil or grease on the contact surfaces, usually from overlubrication of the wick at the top of the cam or too much grease on the breaker arm rubbing block. This condition is indicated by a smudgy line on the point support and breaker plate, Fig. 79. If caught in time the contacts can be cleaned and the residue left on them can be wiped off by drawing a piece of lint-free tape between the contacts.

When new contacts are installed, the breaker arm should be free on the pivot pin, the contacts lined up with their outside diameters registering

perfectly, and contact made in the center of the contact surfaces.

Check the alignment of the rubbing block with the cam by using a thin strip of white paper and carbon paper, held between the rubbing block and cam. By rotating the cam against the paper, a carbon impression will be made. If a straight-line impression is obtained from the top to bottom of the rubbing block against the carbon paper, even though the top and bottom edges make contact, it will be unnecessary to "run in" the block to improve the contact. If only one end of the rubbing block is contacting the cam, look for a twisted breaker arm or a bent pivot pin.

Breaker arm spring tension is extremely important. This tension should be within specified limits as too low a pressure will cause missing at high speeds and too high pressure will cause excessive

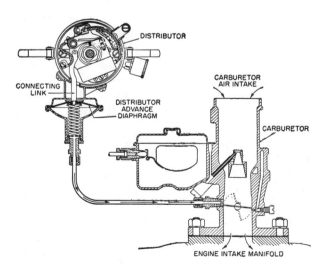

Fig. 81 *Vacuum advance mechanism of the type which is mounted on the side of the distributor and is connected to the breaker plate by a connecting link. The breaker plate rotates as vacuum conditions change*

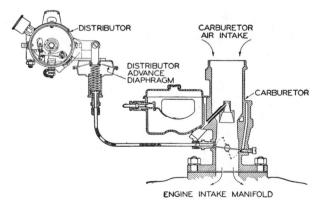

Fig. 82 *Vacuum advance mechanism of the type which is clamped around the distributor so that the entire distributor is rotated as vacuum conditions change*

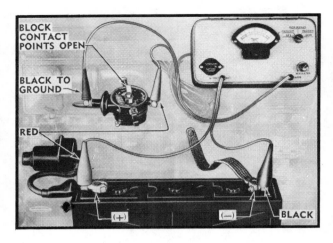

Fig. 83 Condenser tester connected into a negative-grounded system. For positive-grounded systems, reverse connections

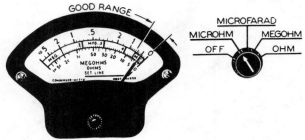

Fig. 84 Condenser resistance test reading

wear of the cam and rubbing block. How this tension is checked and adjusted will be explained further on.

DISTRIBUTOR INSPECTION & TESTS

All parts of the distributor must be inspected and tested, and worn or defective parts must be replaced to insure satisfactory ignition.

The contact points may be cleaned and adjusted without removing the distributor, but if the interior is dirty or saturated with oil or new parts are to be installed, the distributor must be removed from the engine.

To determine whether it is necessary to remove the distributor, as well as to determine what parts replacements are necessary, first make all the following tests and inspections:

Checking Automatic Advance

The automatic advance mechanism must operate freely and the springs must return the advance weights to the full retard position during idle speed operation.

Remove the distributor cap and rotor. Apply several drops of oil to the cam wick and replace the rotor. Turn the rotor in the direction of its normal rotation until the advance weights are fully extended, Fig. 80. Release the rotor and allow the springs to return the weights to the retard position, which will be indicated by a metallic click as the weights strike the stop. Repeat several times.

If the springs do not return the weights to the stop, or if there is free movement of the rotor and cam in the retard position, the indication is that

the weight springs are too weak and new ones must be installed.

Checking Vacuum Advance

The two types of vacuum advance mechanisms used on Prestolite and Delco-Remy distributors are shown in Figs. 81 and 82. In Fig. 81, the mechanism is attached to the distributor breaker plate so that only the breaker plate assembly rotates. In Fig. 82, the mechanism is connected to the distributor body so that the entire distributor moves.

With the type shown in Fig. 81, the breaker plate must move smoothly in the distributor housing and the return spring in the vacuum chamber unit must return the breaker plate to the full retard position when manifold suction is released. Rotate the breaker plate by pushing against the condenser and note whether movement of the plate is rough. Also note whether the spring in the vacuum chamber unit returns the breaker plate to the full retard position. If movement of the plate is rough, it indicates dirty or loose bearings in the breaker plate or distributor housing and the distributor must be disassembled for repairs. If the vacuum chamber unit fails to return the breaker plate to the full retard position, the indication is that the spring is broken.

With the type shown in Fig. 82, the distributor must move freely and the spring in the vacuum chamber unit must return the distributor to the full retard position when manifold suction is released. Test the action of the vacuum chamber spring as outlined above. Rotate the distributor and note whether there is any indication of roughness. If there is, the unit must be taken apart for repairs.

Checking Condenser Circuit

It is essential that a good condenser tester of the type shown in Fig. 83 be used to check (a)

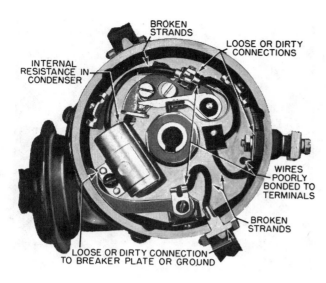

Fig. 85 *Points of resist-ance in condenser circuit*

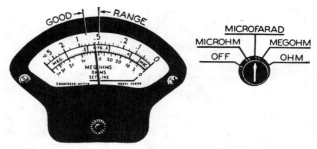

Fig. 86 *Condenser capacity test reading*

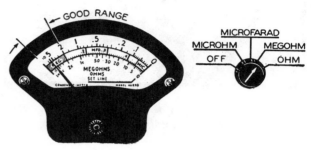

Fig. 87 *Condenser insulation test reading*

leakage in *megohms,* (b) lead resistance in *microhms* and (c) capacity in *microfarads.* Note that the tester shown has two battery leads and two condenser test leads, and that it is connected into a negative-grounded system. For a positive-grounded system, reverse the lead connections. *The condenser must be at normal operating temperature when it is being tested.*

With the tester hooked up as shown block open the breaker points with a piece of fibre or stiff paper such as a piece of a calling card.

Condenser Resistance (Microhm) Test

The resistance test, measured in microhms, is to determine if there is high series resistance in the condenser circuit caused by poor contacts or connections, or high resistance in the condenser itself. (A microhm is 1/1,000,000 ohm.)

Turn the tester switch to the "Microhm" position, Fig. 84. The reading should be within the area shown. If not, there is excessive resistance in the condenser circuit. To isolate the cause of the resistance, connect the positive test lead to the condenser pigtail terminal and the negative lead to the condenser shell. (For positive-grounded systems, reverse the connections.) If this brings the reading within the area shown in Fig. 84, the excessive resistance is at some point between the primary terminal and the condenser pigtail terminal, or the condenser shell is not properly grounded to the breaker plate and distributor housing. Examine each part of the circuit as indicated in Fig. 85 and eliminate the cause of the high resistance.

If the meter does not read within the area shown in Fig. 84 with the tester directly connected to the condenser as above, the condenser itself has excessive internal resistance and should be replaced.

Condenser Capacity (Microfarad) Test

The capacity test is to determine if the condenser capacity is actually within the proper specification limits. Although it is generally agreed that a condenser capacity of .24 microfarad is ideal for passenger car use, there are models of some car makes which specify capacities of .18–.23, .25.–28 and .28–32 microfarads. The actual specification limits for all other passenger cars is .20 to .25 microfarads. (A microfarad is one-millionth [.000001] of a farad.)

Turn the meter switch to the "microfarad" position, Fig. 86. The meter should read .20 to .25 microfarad in the middle scale (or within the limits specified for the particular condenser being tested). If the reading is not within the correct limits, the condenser does not have the proper capacity and should be replaced.

Condenser Insulation (Megohm) Test

The insulation test, measured in megohms, is to determine if the condenser insulation will hold a

Fig. 88 *Example of ignition timing mark on vibration damper*

Fig. 89 *Example of ignition timing mark on flywheel*

charge satisfactorily. (A megohm is 1,000,000 ohms.)

Turn the tester switch to the "Megohm" position. The meter should then read in the area marked "Meg", Fig. 87. If the reading is not within the area shown, the condenser does not have the proper internal insulation and it should be replaced.

DISTRIBUTOR SERVICE

Prestolite, Chrysler, Delco-Remy

If the foregoing inspections and tests indicate that the distributor requires cleaning or installation of new parts, remove the distributor from the engine so that the work can be done properly.

Recently the Ford Motor Co. made use of the name Auto-Lite and the older Auto-Lite name was changed to Prestolite and electrical equipment made by that firm has been given the new name throughout the text.

Removing Distributor from Engine

Each time a distributor is removed and reinstalled, or when a new one is installed, it is essential that it be properly timed with the engine. Therefore, first determine whether the ignition timing mark is on the flywheel or on the vibration damper, Figs. 88 and 89. Then, to make it easily visible, clean the mark and trace a narrow line on it with chalk or paint.

1. Turn over the engine by pulling on the fan belt until No. 1 piston is up on its *compression stroke* and as indicated by the ignition timing mark being adjacent to the indicator pointer on the flywheel housing or timing case cover. (If it is difficult to turn over the engine due to its compression, relieve the compression by taking out the spark plugs.)
2. Trace No. 1 spark plug wire to its terminal in the distributor cap.
3. Mark the distributor with a screwdriver or paint directly under the No. 1 terminal so that the rotor position for No. 1 cylinder will be known when the distributor is reinstalled.
4. Release the distributor cap clamps and raise the cap (with wires attached) and note the position of the rotor. Its metal strip should be directly over the mark made on the housing. If not, the distributor drive shaft gear or coupling is broken, the drive pin is sheared, or the timing chain has jumped or broken.
5. Disconnect the primary wire at the distributor terminal.
6. Remove the vacuum advance connection and the distributor clamp hold-down bolt.
7. Raise the distributor from its mounting.

DISTRIBUTOR OVERHAUL

Prestolite, Chrysler, Delco-Remy

The following procedure describes the general procedure when servicing a Prestolite, Chrysler and Delco-Remy distributor wherein the breaker points are adjusted internally.

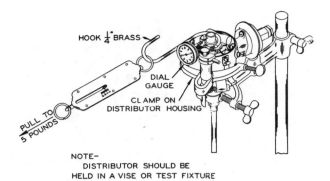

Fig. 90 *Checking wear of distributor shaft and bushings with dial gauge and spring scale*

HOOK ¼" BRASS

PULL TO 5 POUNDS

DIAL GAUGE

CLAMP ON DISTRIBUTOR HOUSING

NOTE—
DISTRIBUTOR SHOULD BE HELD IN A VISE OR TEST FIXTURE

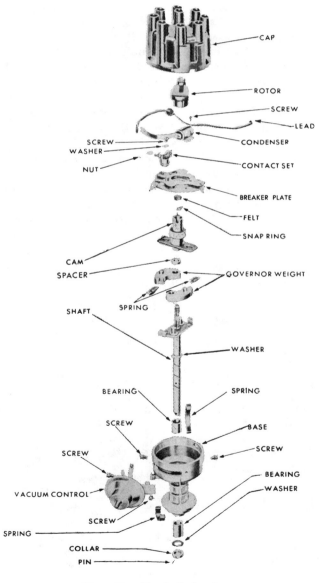

CAP
ROTOR
SCREW
LEAD
SCREW
WASHER
NUT
CONDENSER
CONTACT SET
BREAKER PLATE
FELT
SNAP RING
CAM
SPACER
GOVERNOR WEIGHT
SPRING
SHAFT
WASHER
BEARING
SPRING
SCREW
BASE
SCREW
SCREW
VACUUM CONTROL
BEARING
WASHER
SCREW
SPRING
COLLAR
PIN

Fig. 91 *Exploded view of a Chrysler distributor*

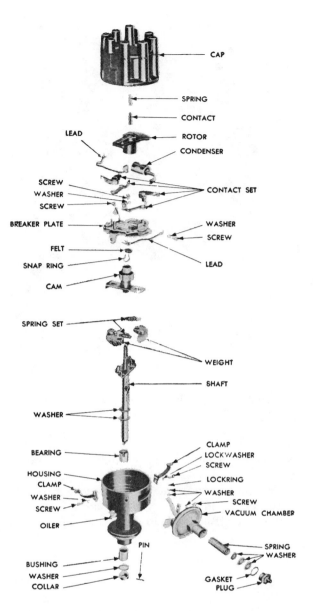

CAP
SPRING
CONTACT
LEAD
ROTOR
CONDENSER
SCREW
WASHER
SCREW
CONTACT SET
BREAKER PLATE
WASHER
SCREW
FELT
SNAP RING
LEAD
CAM
SPRING SET
WEIGHT
SHAFT
WASHER
BEARING
HOUSING
CLAMP
WASHER
SCREW
OILER
CLAMP
LOCKWASHER
SCREW
LOCKRING
WASHER
SCREW
VACUUM CHAMBER
SPRING
WASHER
BUSHING
WASHER
COLLAR
PIN
GASKET
PLUG

Fig. 92 *Exploded view of an Prestolite distributor*

The extent to which a distributor should be disassembled will depend upon the condition of the interior and the parts to be replaced. If the distributor is clean and only parts on or above the breaker plate are to be replaced, wipe away any dirt with a clean cloth.

If the interior is dirty or oily, the distributor should be completely disassembled for cleaning and inspection. Any attempt to wash out the distributor without disassembling it will result in getting dirt below the breaker plate where it cannot be blown out.

If the automatic advance is not functioning properly or if the breaker plate bearings are rough

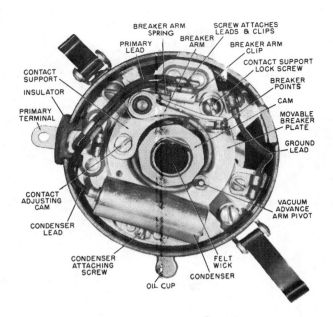

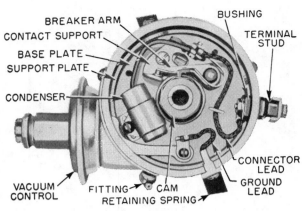

Fig. 93   *Parts of a vacuum advance distributor with cap removed*

Fig. 95   *Parts of a Delco-Remy internal adjustment distributor with cap removed*

as determined by the simple checks outlined previously, the distributor must be disassembled and the defects corrected. Likewise, if a check shows that the distributor shaft and bushings are worn excessively, complete disassembly is required in order to install new bushings.

The most accurate method of checking for wear in the shaft and bushings is with a dial gauge and spring scale, Fig. 90. With five pounds pull on the

scale, maximum wear should not exceed .005". More than .005" wear requires the replacement of the bushing in the distributor housing.

If a dial gauge and spring scale are not available, a fair estimate of the amount of wear may be determined by first turning the shaft so that the breaker points are open. Measure the breaker gap with a round wire feeler gauge. Then push the shaft in the direction which will open the points wider. Hold the shaft in this position and again check the breaker gap. The difference between the first check and the second is the amount of wear present.

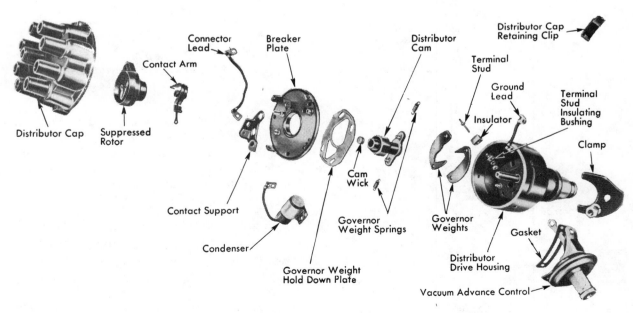

Fig. 94   *Exploded view of a Delco-Remy internal adjustment distributor*

600

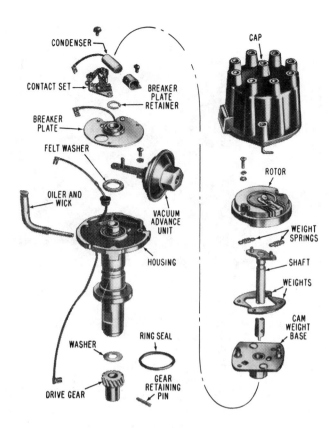

Fig. 96 *Exploded view of Delco-Remy external adjustment distributor*

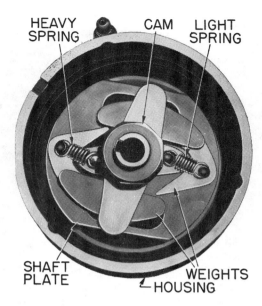

Fig. 97 *Top view of a Delco-Remy internal adjustment distributor with breaker plate removed to show centrifugal governor mechanism*

Disassembling Prestolite & Chrysler Distributor

Figs. 91, 92, 93. To remove the breaker points, remove the screw and clip which holds the primary lead wire, condenser lead wire and breaker arm spring. Lift the breaker arm off its pivot pin. Remove the retaining lock screw and lift off the contact support.

To remove the vacuum control unit, release the hairpin lock from the vacuum control arm pivot. Remove the screws holding the vacuum chamber to the distributor housing and withdraw the unit from the distributor.

To remove the breaker plate, take out the two screws and bearing retainer clips, which are adjacent to the distributor cap clips, and lift out the breaker plate. The breaker plate may be separated from the sub plate by pushing the two plates apart after disconnecting the attached wire leads.

To expose the governor weights, remove the felt wick from the well of the cam and pick out the lock ring, which fits in a groove on the shaft and holds the cam in position. Lift off the cam, which will expose the governor weights and springs.

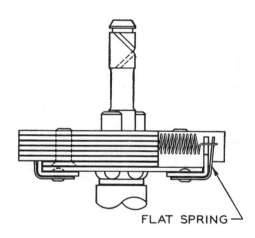

Fig. 98 *Flat spring used on some governors to provide a rapid spark advance*

Disassembling Delco-Remy Distributor

Internal Adjustment Type. Referring to Figs. 94 and 95, remove the breaker plate ground lead. Unfasten and remove the vacuum control unit. Disconnect the primary terminal connector lead from the terminal stud and remove the stud and bushing.

Remove the three screws which attach the support plate to the distributor housing and lift the breaker plate assembly from the housing. Do not disassemble the breaker plate at this time.

601

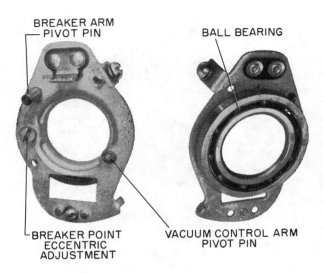

BREAKER ARM
PIVOT PIN

BALL BEARING

BREAKER POINT
ECCENTRIC
ADJUSTMENT

VACUUM CONTROL ARM
PIVOT PIN

Fig. 99 Top and bottom view of ball bearing mounted breaker plate on Presto-lite vacuum advance distributor

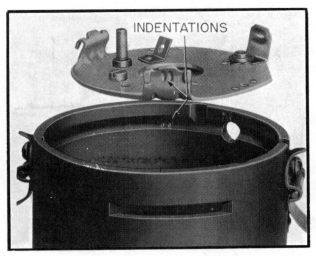

INDENTATIONS

Fig. 100 Delco-Remy housing and breaker plate showing worn places caused by the rolling of the three balls on which the breaker plate revolves

Remove the governor advance weight plate, springs and weights, and distributor cam.

External Adjustment Type, Fig. 96

1. Remove rotor by removing two attaching screws, lock washers and flat washers. Note that the rotor is doweled to the weight base so that it can be installed in only one position.
2. Remove both weight springs and advance weights.
3. Remove retaining pin from gear by filing off staking and driving it out of the gear with a drift and hammer. Support distributor in such a way that the shaft will not be damaged when driving the pin out.
4. Slide gear off the shaft.
5. Pull shaft and cam-weight base from housing.
6. Remove contact assembly.
7. Remove condenser hold-down screw and lift off condenser.
8. Remove spring retainer and raise plate from housing.
9. Remove two attaching screws, lock washers and plate ground lead, and remove vacuum advance unit.
10. Remove felt washer from around bushing in housing. *No attempt should be made to service the shaft bushings in the housing as the housing and bushings are serviced as an assembly.*

Governor Weights and Springs

In servicing the distributor, all weights should be removed from the hinge pins, cleaned and

checked for excessive wear, either in the weights or pins, or the plate which is slotted for the movement of the pins on top of the weights. Replacement should be made if there is any appreciable wear in the slots, as any wear at this point would change the characteristic of the automatic advance.

If these parts are in good condition, the hinge pins should be lubricated before being reassembled, by greasing the hinge pins and filling the pockets of the governor weights with grease. Do not use vaseline for this purpose as its melting point is comparatively low.

When installing new centrifugal governor assemblies, it is important that the spacer washers between the housing and shaft be installed correctly. If incorrectly installed, the governor assembly will be too high, causing it to rub against the bottom of the breaker plate.

Governor weight springs should always be replaced with new springs as there is no way of measuring the calibration of these springs in the field. Prestolite and Delco-Remy supply new springs in sets for individual units to insure the correct use.

On some distributors, both springs are alike, while on others there is one heavy and one light spring, as in Fig. 97. Another combination that may be found is an additional flat spring on the outside of the outer spring posts, Fig. 98. As the governor speed is increased, the flat springs are first pulled against the posts by the eyes of the coil springs to provide a rapid spark advance of a few degrees before the coil springs pull against the spring posts.

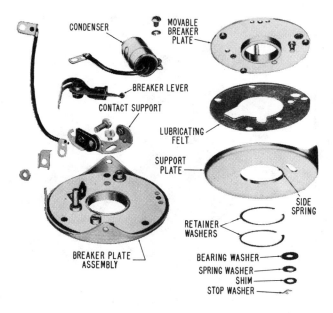

Fig. 101 *Construction of Delco-Remy internal adjustment breaker plate assembly*

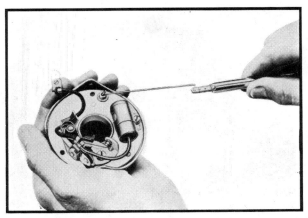

Fig. 102 *Checking Delco-Remy internal adjustment breaker plate friction with ounce spring scale. The correct pull to move the breaker plate while holding the support plate is usually 11 to 18 ounces*

Breaker Plate Bearings

On the Prestolite and Chrysler distributors under discussion here, the movable breaker plate is mounted on a ball bearing, Fig. 99. This bearing should be checked before being installed to see that there is no friction in any part of its rotation, due to worn balls or races. If the bearing inner race, which is part of the breaker plate, is loose, no attempt should be made to swedge it in place as this requires special tools to do the job properly. In such cases, replace the breaker plate and bearing assembly.

On some Delco-Remy distributors, the movable breaker plate rotates on three balls located between the outside diameter of the plate and distributor housing. The housing should be inspected for excessive wear in the groove where the balls roll. Fig. 100 shows the indentations in the housing and breaker plate caused by the balls pressing against them. The remedy is to replace the housing and breaker plate.

On other Delco-Remy internal adjustment distributors, a different method is used in mounting the movable breaker plate in the distributor which eliminates the three balls previously used. Fig. 101 illustrates this breaker plate construction. Note that the breaker plate to which the vacuum advance link is fastened has three bakelite buttons which rest on a support plate which is screwed to the distributor housing. An oil saturated felt be-

tween the breaker and support plates assures lubrication for the feet as they move in the support plate with changes in the amount of vacuum advance. A post fastened to the movable breaker plate extends through the slot in the support plate and carries a bearing washer, spring washer, shim washer and stop washer. These, together with one or two retaining washers which snap into a groove in the movable breaker plate hub, hold the two plates together as an assembly. Production tolerances determine if one or two retainer washers are used. The number of shim washers on the post is very important since they must furnish enough pressure against the spring washer to prevent tipping of the movable breaker plate, but not exert so much pressure as to cause excessive friction between the two plates.

Inspect the breaker plate assembly for looseness, which would permit tipping or rattling of the base plate. Normally it is unnecessary to disassemble this unit. However, if it is very dirty it may be disassembled for cleaning and inspection.

Remove the stop washer and shims from the base plate post, Fig. 101. Remove the retainer washer from the base plate hub and separate the parts.

When reassembling the breaker plate, saturate the felt washer with 10W engine oil, then install the base plate on the support plate. Place the small side spring in the recess in the support plate so that the two points will be between the ends of the retainer washer when installed. The spring helps to prevent side play of the base plate. Install

603

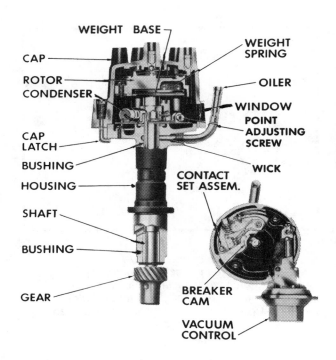

Fig. 103 Sectional view of Delco-Remy external adjustment distributor

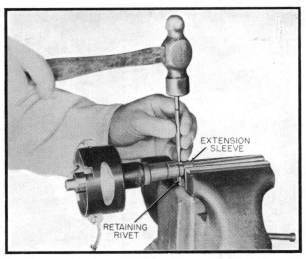

Fig. 104 Driving out extension sleeve rivet on Prestolite distributor

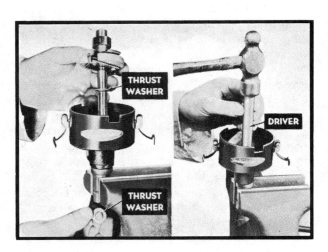

Fig. 105 Removing shaft and driving out bushing on Prestolite distributor

Fig. 106 Driving in distributor shaft bushing on Prestolite distributor

one retainer washer in the groove of the base plate hub. The hub is staked at four places to provide end play of approximately .015" between the base plate and the support plate. Place the fiber, spring, shim and stop washers on the base plate post in the order named as shown in Fig. 101.

Whether the breaker plate has been disassembled or not, attach an ounce spring scale, Fig. 102, and check the pull required to move the base plate on the support plate. The pull should be from 11 to 18 ounces. If the pull is less than 11 ounces, add shim washers to the base plate post between the stop and spring washers. If the pull exceeds 18 ounces, remove shim washers as required. This assembly tension is very important in preventing tipping of the base plate and controlling the amount of friction opposing rotation of the base plate on the support.

On the Delco-Remy external adjustment type distributor, Fig. 103, the centrifugal advance components are located above the breaker plate and cam. This arrangement allows the cam and

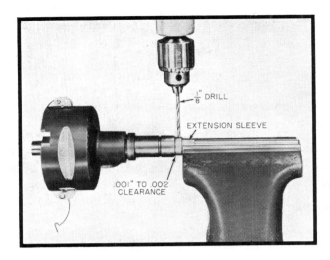

Fig. 107 *Drilling a hole in new drive shaft for rivet on Prestolite and Chrysler distributor*

Fig. 108 *When cam is assembled, be sure governor weight pins rest against inner ends of both slots as shown*

breaker lever to be located directly adjacent to the upper bushing for increased stability. The breaker plate is of one piece construction and rotates on the outer diameter of the upper bushing. The plate is held in position by a retainer clip in the shaft bushing.

Vacuum Advance Units

On Prestolite, Chrysler and Delco-Remy vacuum advance mechanisms, the basic principle of operation is that the spring retards the spark and vacuum advances it. The calibration of this unit is accomplished by changing the spring tension. On Prestolite and Chrysler units, a combination of flat washers of various thicknesses is used between the end of the spring and the brass retainer fitting which is screwed into the outer end of the vacuum diaphragm housing. Depending upon the result desired, any alteration is accomplished by adding or removing washers. Adding to the washer thickness increases spring tension and vice-versa. The same result is obtained on Delco-Remy distributors by changing the spring to one of greater or lesser tension, as required.

Distributor Shaft and Bushings (Prestolite & Chrysler)

If the previous inspection indicated that wear in the shaft and bushing is excessive, proceed as follows:

1. Grip the distributor shaft extension sleeve in a soft-jawed vise and drive out the extension sleeve rivet, Fig. 104.
2. Remove the sleeve from the shaft. Then drive out the extension shaft pin and remove the shaft.
3. Push the drive shaft from the housing, being careful that the thrust washers are not lost, Fig. 105.
4. The drive shaft bushing may be pressed out or driven out, Fig. 105, using a driver of slightly smaller diameter than the outside diameter of the bushing.
5. Carefully press or drive the new bushing into place, Fig. 106, being sure the driver is smooth and will not burr the bushing.
6. Drill a 3/16″ hole in the bushing through the grease hole in the housing.
7. Install the drive shaft into the housing, being sure the thrust washers are in place.
8. Install the extension shaft and attach it with the retaining pin.
9. Install the extension sleeve so that it rests snugly against the thrust washer.
10. Drill a ⅛″ hole through the extension sleeve and drive shaft if a new shaft is being installed, Fig. 107.
11. Drive the extension sleeve retaining rivet into place and peen it over on both ends.
12. Use a feeler gauge to see that there is .001″ to .002″ clearance between the extension sleeve and thrust washer, Fig. 105. If there is not enough clearance, a thinner thrust washer should be installed. If too great use a thicker thrust washer.

Distributor Shaft and Bushing (Delco-Remy)

Except for the external adjustment type distributor in which the shaft bushings are not serviced separately from the housing, removal and replacement of bushings in the internal adjustment type distributors are similar to that described for Prestolite units. To remove the shaft, file the peened end of the distributor shaft collar attaching pin, drive out the pin and remove the shaft from the distributor housing. Press or drive the bushing out of the housing and install the new one.

To assemble, coat the surface of the shaft running in the bushing with No. 2 cup grease and insert the shaft and weight base in the housing. Install washers and collar on the shaft against the housing and insert and peen over the collar retaining pin.

Assembling Prestolite & Chrysler Distributor, Figs. 91, 92, 93

1. See that the grease pockets in the governor weights for the hinge pin are packed with grease before governor weights are installed.
2. Install new governor weight springs (unless a complete governor is to be used, which includes new springs). See that the governor weight outer posts are approximately at right angles to the driving plate.
3. Install the grease cup or oiler in the housing and force a small amount of lubricant into the shaft and bushings. Then refill the cup or oiler.
4. Grease the slots for the cam stop plate.
5. Install the cam assembly, being sure that both governor weight pins rest against the inner end of both slots. Sometimes it is necessary to turn the cam 180 degrees so that the slots and weight pins will be in their proper relation, Fig. 108.
6. Put a few drops of oil on the upper end of the shaft so that the cam will be free to move.
7. Install the breaker base assembly and be sure that the bearing clamps are installed with the concave side down (tits up) so that when the screws are tightened, they will securely clamp the bearings in place.
8. Check the breaker arm rubbing block in relation to the cam to be sure that the full length of the rubbing block rests on the cam and no part of it projects above the cam. In some distributors it may be necessary to raise the cam by adding thrust washers underneath it.
9. Lock the cam in place by inserting the lock ring in the groove in the top of the shaft.
10. Apply three drops of oil in the cam well and insert the felt wick without further oiling.
11. Check all lead wire connections to be sure they are tight.
12. Install the vacuum unit on the distributor, being sure the connecting link moves freely on the base plate pin.
13. Be sure the breaker arm pivot bushing is properly fitted to the pivot pin. It should be a snug fit and free to move on the pin. Lubricate with one drop of oil.
14. The breaker arm rubbing block should be lined up with the cam by the use of a thin strip of white paper and carbon paper held between the rubbing block and cam. By rotating the cam against the paper, a carbon impression will be made, showing which way the arm should be bent. Any bending necessary should be done between the hinge pin and rubbing block.
15. Do not use a file or sandpaper on the rubbing block wearing surface as it is given a hard burnished surface during manufacture. The use of an abrasive would destroy this finish and cause rapid wear.
16. The breaker points should be lined up with their outside diameters registering perfectly and the contact made near the center of the contact surfaces. Contact alignment is accomplished by bending the stationary contact bracket. Never bend the breaker arm between the rubbing block and contact.
17. Adjust the breaker point gap and check and adjust breaker arm spring tension as outlined further on.

Assembling Delco-Remy Distributors. Internal Adjustment Type Figs. 94, 95, 97

1. Install advance weights and springs. Wipe the surfaces of the advance parts with a rag moistened with light engine oil to prevent rusting. Place one drop of light engine oil on each weight pivot pin.
2. Install the cam. Saturate the felt in the cam with light engine oil but do not oil beyond what the felt will readily absorb.
3. Install the breaker plate assembly in the distributor.
4. Install the vacuum assembly on the distributor. Check to see that the vacuum control linkage is fitted to the connector bearing on the breaker plate so there is no upward or downward thrust on the plate when the vacuum control is operated.
5. Install new breaker points and condenser. Make certain the contact points are well aligned. The movable contact must line up with

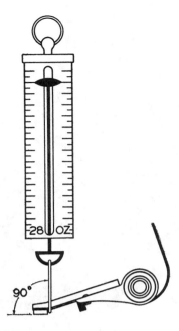

Fig. 109 *Method of hooking spring scale to breaker arm when checking breaker arm spring tension*

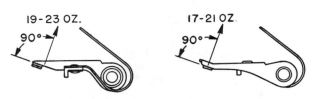

Fig. 110 *Spring tension specifications for two types of breaker arms*

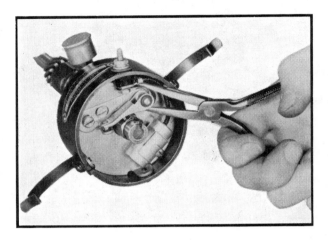

Fig. 111 *Using long nose pliers to decrease breaker arm spring tension by pinching spring*

the stationary contact or extensive pitting is apt to occur.

6. Check to see that the condenser connections are tight.

7. Place a small drop of engine oil on the breaker arm pivot and a small amount of petroleum jelly on the cam.

8. Check the breaker arm spring tension as described below.

Checking Breaker Arm Spring Tension

On all distributors, the breaker arm spring tension must be within specifications and is checked and adjusted as follows:

Turn the distributor shaft so that the breaker arm rubbing block is on a flat between two cam lobes. Measure the spring tension with an ounce spring scale hooked to the breaker arm as close as possible to the contact point on the end of the arm, Fig. 109. Pull the spring scale against the breaker arm at right angles to the contact surface, Fig. 110. Take the scale reading just as the points separate. An accurate way to determine just when the points separate is to hook up a small light bulb in series with a battery and connect it across the contacts. As soon as the light goes out the points are separated.

If the pressure is too high, it can be decreased by pinching the spring carefully, Fig. 111. If the pressure is too low, remove the breaker arm from

the distributor and bend the spring away from the lever.

Adjusting Breaker Point Gap

There are two types of breaker contact adjustments. One type, Fig. 112, has an adjusting screw with a lock nut, and the other, Fig. 113, utilizes an adjusting cam screw and lock screw.

If the contacts are new, an adequate job can be done with a *round* wire feeler gauge. But if the contacts are only slightly rough, a dial indicator, Figs. 114 and 115, or a cam angle meter is essential if the correct gap is to be obtained.

On six-cylinder engines, the breaker gap specification is usually from .018″ to .022″, and from .015″ to .018″ on eights. When adjusting new points, set the gap to the highest limit since the rubbing block on the breaker arm will wear during subsequent operation. This wear will naturally decrease the breaker gap but it will still be within the specified limits.

When making the adjustment with a feeler gauge, loosen the lock nut or lock screw, turn the cam so as to bring the rubbing block on its highest

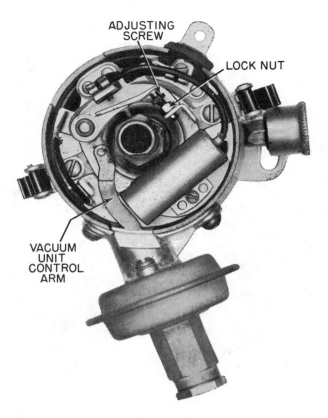

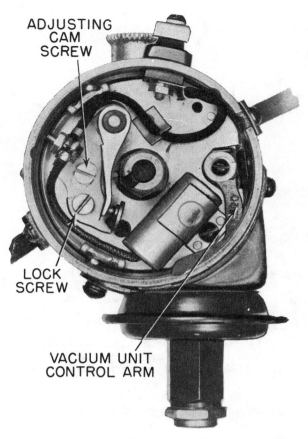

Fig. 112 *Adjusting screw type contacts. A contact wrench and a wire feeler gauge or dial indicator are used to adjust gap. After contacts are adjusted to the correct gap, tighten lock nut and recheck gap*

Fig. 113 *Adjusting cam type breaker contacts. Loosen lock screw and turn adjusting cam to give correct gap. Tighten lock screw and recheck gap*

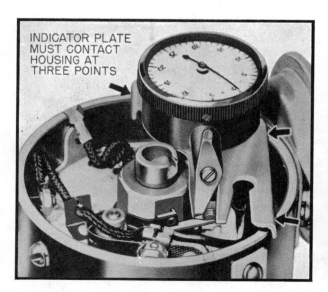

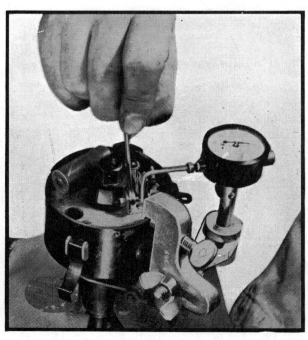

Fig. 114 *Checking Delco-Remy internal adjustment breaker contact gap with dial gauge*

Fig. 115 *Checking Prestolite breaker gap with dial gauge*

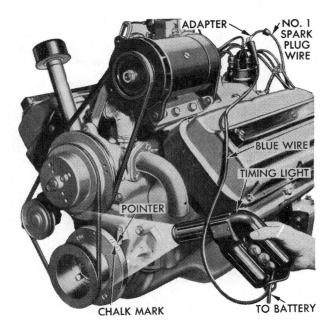

Fig. 116 *Checking ignition timing with timing light*

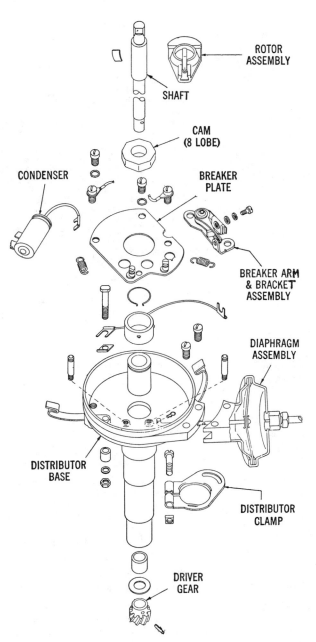

Fig. 117 *Exploded view of Holley distributor*

point and turn the adjusting screw as required to obtain the desired setting.

On the type shown in Fig. 112, hold the adjusting screw with one wrench while tightening the lock nut with another wrench. Recheck the adjustment.

On the type shown in Fig. 113, after establishing the correct gap, hold the adjusting cam screw stationary with a screwdriver and tighten the lock screw. Recheck the gap.

When using a dial gauge, Figs. 114 and 115, follow the instructions furnished by its manufacturer. The usual procedure is to mount the gauge on the distributor as directed by the manufacturer. Then turn the distributor shaft to bring the rubbing block on a flat between two cam lobes. Set the gauge face so the indicator needle is exactly at zero. Now turn the distributor shaft to bring the rubbing block on the high point of a cam lobe. Finally, adjust the contact gap as required to get the proper indication on the gauge.

Assembling Delco-Remy External Adjustment Type Distributor

Assembly of this distributor is the reverse of the disassembly procedure outlined previously. When installing a gear on the shaft, use a new retaining pin. Support the gear and shaft assembly and stake the pin securely. The pin must be tight in the hole to prevent any movement between the gear and the shaft.

TESTING DISTRIBUTOR

Before installing an overhauled distributor on the engine, its operation must be checked on a distributor testing machine to make sure the vacuum and centrifugal advances are working according to the specifications for the distributor being serviced. There are a number of reliable machines of this type on the market and the manufacturer of each furnishes complete instructions for its use.

609

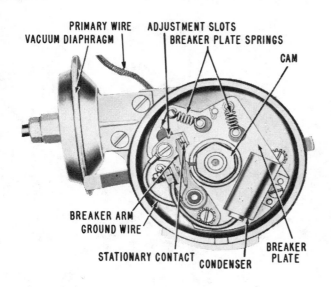

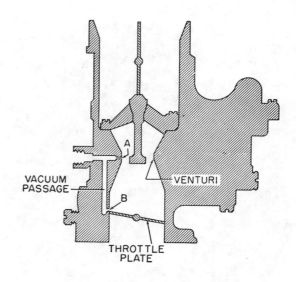

Fig. 118 *Details of Holley distributor with cap removed*

Fig. 119 *Carburetor vacuum passages for Holley distributor*

Checking Governor (Centrifugal) Advance

To check the centrifugal advance, mount the distributor on a testing machine that will show the distributor rpm and degrees of advance. Operate the distributor in the correct rotation and increase the speed until the spark begins to advance. Reduce the speed slightly and set the indicator at zero. Increase the speed to the value specified to give 1 degree advance. If the advance is not 1 degree, stop the distributor and bend the outer spring lug on the weak weight spring to change its tension. Check this point again, then operate the distributor at the specified speed to give an advance just below the maximum. If this advance is not as specified, stop the distributor and bend the outer spring lug on which the heavy spring is mounted (second spring on governors with two identical springs). Recheck the zero point and the above two points and make whatever readjustments are necessary. Then recheck the advance at all points specified again. When making this check, operate the distributor up and down the speed range. If there is a variation between the readings for increasing and decreasing speeds, it indicates that the governor action is sluggish and requires overhaul.

Checking Vacuum Advance

Vacuum advance should be checked on a testing machine that has a controlled source of vacuum and a vacuum gauge. Mount the distributor on the machine and connect the vacuum line.

Use two wrenches to tighten the vacuum connection, being careful not to apply a torque to the vacuum chamber as this could cause leakage where the diaphragm is clamped in the housing. Turn on the vacuum pump to give a reading of 10 to 20 inches of vacuum and shut off the pump. If

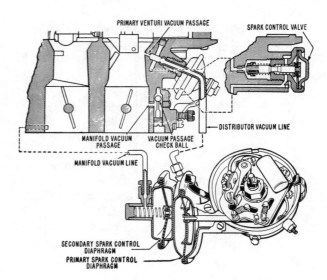

Fig. 120 *Distributor vacuum controls used on 1956 V-8 engines of the Ford group with four-barrel carburetors. Note the double vacuum diaphragm which is used for better control of breaker plate advance characteristics. The purpose of the second diaphragm is to provide a rapid spark retard at the acceleration tip-in point. The primary spark control diaphragm operates in the same manner as the single type diaphragm*

the gauge reading falls, it indicates leakage in the vacuum chamber, pump, gauge or connections which should be located and corrected before tests and adjustments are made.

Remove all vacuum from the distributor and operate it at a speed above the maximum governor advance speed to eliminate all spark variations due to the governor. Set the indicator at zero and apply vacuum to give one of the advance figures specified for the unit being tested. If the advance is incorrect, change the spacing washers between the vacuum chamber spring and nut on Prestolite and Chrysler distributors, or install a different spring on Delco-Remy distributors. When one point of the vacuum advance curve is adjusted, the others should be checked. If they are not within limits, it indicates either incorrect spring characteristics or leakage in the vacuum chamber and lines.

If the maximum advance is not correct, make sure the vacuum control linkage parts are correctly assembled and have not had an incorrect part installed.

Installation of Distributor

Before installing the distributor, check the timing mark on the flywheel, vibration damper or crankshaft pulley, Figs. 88, 89, to be certain that the engine has not been rotated while the distributor was off, and that it still remains set on the timing mark for No. 1 cylinder.

If a new distributor is being installed, scratch or chalk a mark on it to correspond to the mark made on the old distributor and use this mark as a guide for the initial position of the rotor as the new distributor is installed.

Temporarily set the distributor in its mounting with cap removed, being careful to see that the primary terminal and vacuum control (if used) are in position to connect to the wire and pipe, respectively. However, do not connect them at this time.

With the rotor in approximately the same position as when the distributor was installed (in line with mark on distributor housing) allow the distributor to settle down to its permanent position in the mounting, being certain that the screw hole for the clamp hold-down bolt is in the center of the clamp slot.

In the case of a gear-driven distributor, notice that the rotor will move from the position from which it was set as the distributor is moved into position. When this occurs, raise the distributor and turn the rotor just far enough beyond the desired position to allow for the change made by

the gear movement, and again set the distributor in place. Install the hold-down screw or bolt. Connect the primary wire to its terminal on the distributor, the vacuum pipe to the vacuum control unit and install the cap and wires. The distributor should now be properly timed.

Ignition Timing

Ignition timing should be checked with a timing light because this shows the actual timing with engine running. Such a light is hooked up to an engine in Fig. 116.

Connect the timing light to the No. 1 spark plug and to the battery according to the directions furnished by the manufacturer of the instrument. Disconnect vacuum tube or hose at distributor and plug opening in the tube or hose so idle speed will not be affected.

Start the engine and allow it to idle at its normal idling speed. Direct the timing light beam at the timing mark on the flywheel or vibration damper. To make the timing mark more visible, mark it with chalk or paint. Each time No. 1 spark plug fires, the light should strike the timing mark. If it does not, loosen the distributor attaching screws and turn the distributor housing very slowly in the direction required to bring the timing mark directly under the light beam. Tighten the distributor attaching screws and recheck the timing to be sure it has not changed during the tightening process. Connect vacuum tube or hose to distributor.

In the absence of a timing light, the timing may be altered from the initial setting by road-testing the car. For best performance and fuel economy, the setting should be one which will produce a slight spark knock or "ping" when accelerating from about 10 mph on wide open throttle.

HOLLEY DISTRIBUTOR SERVICE

"Full Advance" Type

Used on Ford, Mercury, Continental, Lincoln up to 1956, this type distributor, Figs. 117 and 118, bears a strong physical resemblance to a Prestolite and Delco-Remy unit, the fundamental difference being that no centrifugal governor advance is used.

The spark advance is regulated by the vacuum differential at the carburetor. Distributor advance is operated by a vacuum unit mounted on the distributor. One side of the vacuum unit is connected to the breaker plate by direct linkage and the other side is connected by a vacuum line to the carburetor.

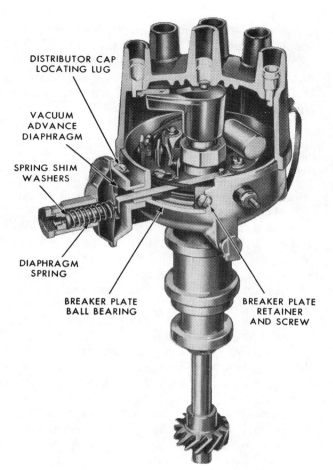

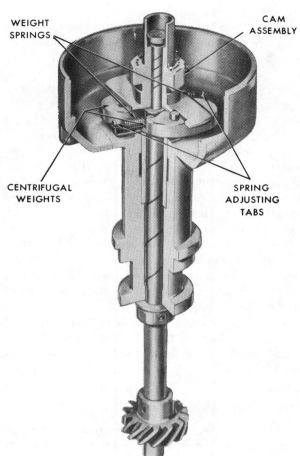

Fig. 121 *Holley centrif-ugal advance distributor*

Fig. 122 *Holley centrif-ugal advance mechanism*

The spark advance characteristics are controlled by two breaker plate springs working against the distributor vacuum control diaphragm, Fig. 118. The amount of spark advance is determined by the amount of vacuum supplied to the distributor and by adjustment of the breaker plate springs.

The carburetor has a vacuum passage with openings at both the venturi tube and a point just above the throttle plate, Fig. 119, so that the vacuum in the distributor line is a combination of the carburetor throat and venturi vacuums. The lower opening is above the edge of the throttle plate when the engine is idling, consequently the spark is retarded (also see Fig. 120).

Under normal road load or part throttle operation the vacuum is high, and the spark will become advanced at 18 to 35 mph.

When the engine is accelerating, the vacuum at the venturi increases as engine speed increases. However, the manifold vacuum (at throttle plate opening) decreases considerably from the normal

road load vacuum. The net result of these two changes is to lower the vacuum at the distributor diaphragm while the springs retard the spark advance from its road load setting. As the car speed increases, the venturi vacuum and the manifold vacuum continue to increase.

The breaker plate springs are precision set at the factory with special stroboscopic equipment. This equipment is available for adjustment purposes in the field. Shops having conventional type distributor testing machines can include a mercury column to take care of the settings of these springs as the ordinary vacuum gauge will not provide the necessary accuracy—which calls for tenths of an inch of mercury not readily measurable on a conventional vacuum gauge.

Centrifugal Advance Type

Used on V8 Fords after 1957, this distributor, Figs. 121 and 122, is basically the same as the

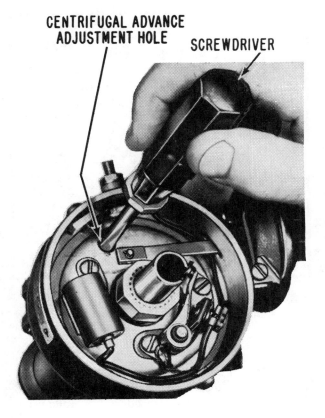

CENTRIFUGAL ADVANCE
ADJUSTMENT HOLE   SCREWDRIVER

*Fig. 123   Holley, centrif-
ugal advance adjustment*

conventional Prestolite and Delco-Remy internal adjustment distributors, the chief difference being that the centrifugal advance mechanism can be adjusted through a slot in the breaker plate as follows:

With the distributor mounted on a test machine, and with the vacuum hose disconnected, operate the distributor in the direction of normal rotation and increase the rpm until the mechanism begins to advance. Reduce the speed to where there is no advance and zero the advance scale. Increase the speed to the value specified for the initial advance of the distributor being serviced. If the correct advance is not indicated at this rpm, stop the distributor and bend the primary spring bracket with a screwdriver as shown in Fig. 123 to change its tension. Bend the bracket away from the distributor shaft to decrease advance and toward the shaft to increase it.

The primary spring is the spring that is under tension when the distributor shaft is not rotating. To determine which spring is under tension, insert a hook into the adjusting slot and move each spring. The secondary spring will be under less tension than the primary spring.

Check the minimum advance point again, then operate the distributor at the specified rpm to give an advance just below the maximum. If this advance is not up to specifications, stop the distributor and bend the secondary spring bracket to give the correct advance.

Breaker Points

The breaker point assembly consists of the stationary breaker point bracket, breaker arm and primary wire terminal. This assembly is mounted on the breaker plate as a unit and can be replaced without removing the distributor from the engine.

To remove the points, disconnect the primary and condenser leads. Remove the screws which secure the point assembly to the breaker plate and take off the point assembly.

To install, place the primary and condenser leads, lock washer and nut on the primary terminal and tighten the nut securely. Position the point assembly on the breaker plate. Install the holding screws. Be sure the ground wire terminal is on the screw nearest to the adjustment slot and the lock washer is used under the screw at the opposite end. Adjust the point gap by means of a screwdriver placed in the adjustment slots indicated in Fig. 118. When the correct gap is established, tighten the lock screw and recheck the spacing.

Distributor Removal

Before removing a distributor from an engine which is timed correctly, be sure to scribe a mark on the distributor housing indicating the position of the rotor, and another mark on the engine and housing to indicate the position of the housing.

The distributor can be reinstalled when the rotor is in line with the mark without rotating the engine crankshaft to obtain the proper timing.

After making the marks as recommended, take off the distributor cap with wires attached. Disconnect the primary wire and vacuum line. Remove the distributor hold-down bolts and lift the unit from the engine.

Installing Distributor

Align the rotor with the mark made previously on the distributor body. Install the distributor in the engine, with the housing mark aligned with the mark previously made on the engine. Tighten the hold-down screw or screws. Check and adjust the ignition timing, using a timing light if possible.

On 6-cylinder overhead valve engines the oil pump shaft has a tang which engages with a slot in the distributor shaft. On overhead valve V8's,

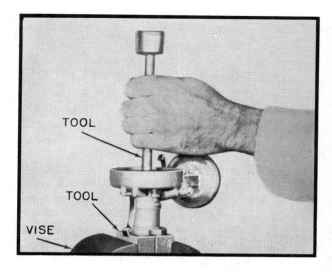

Fig. 124   *Removing bushing from distributor housing*

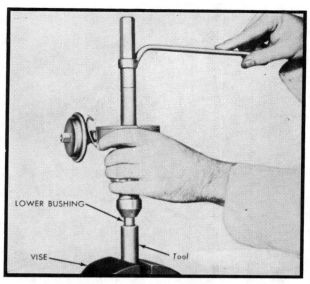

Fig. 125   *Installing lower bushing in distributor housing*

the tang is on the distributor shaft and engages with a slot on the oil pump shaft. If the tang is not engaged, the distributor body will not seat against the block, and possible breakage of the flange may result when the retaining screws are tightened.

HOLLEY DISTRIBUTOR OVERHAUL

Disassembly of Distributor

Although several distributors have been used since 1949, their construction differs only slightly so that the following description will serve for all. Referring to Figs. 117 and 118, the disassembly procedure is as follows:

1. File off the head of the pin which attaches the driver gear to the lower end of the distributor shaft.
2. Drive out the pin with a punch and slip off the gear.
3. Pull the shaft and assembled parts out of the housing.
4. To remove the breaker plate, first take out the points as previously described.
5. Disconnect and remove the condenser.
6. Remove the hairpin retainer and disconnect the vacuum control rod.
7. Release the tension on the breaker plate springs by turning the eccentric studs to which they are attached. Do not stretch these springs as it may make them useless for further adjustment when checking the vacuum advance.
8. Remove the lock ring attaching the breaker plate to the upper bushing and lift the breaker plate from the housing.
9. If it is necessary to replace the primary wire and ground wire, remove them at this time.

Bushing Installation—this operation calls for the use of the tools illustrated in Figs. 124 to 128 or their equivalent. The procedure is as follows:

1. Drive out the lower bushing, Fig. 124.
2. Invert the housing and drive out the upper bushing.
3. Place a new lower bushing in position on the bushing installation tool, then install the bushing in the housing as shown in Fig. 125. Turn the handle on the tool until the lower bushing is flush with the distributor housing.
4. Position the upper bushing in the housing with the lock ring end up. Then install the bushing as shown in Fig. 126.
5. Burnish both bushings to provide a free-turning shaft with barely perceptible side play (.0015″), Fig. 127.

Assembling Distributor—Assemble the breaker plate and related parts on the shaft and place the shaft in the housing. Place the spacer on the gear end of the shaft and install the gear. Press the gear on the shaft, Fig. 128, and check the end clearance with a feeler gauge as shown, which should be from .005″ to .008″. When the gear is pressed on the shaft far enough to establish this clearance, drill a hole in the shaft and insert a new pin, peening it over to be certain it cannot come out.

Ignition Timing—After the distributor is installed as outlined previously, adjust the ignition timing so that when the distributor is ready to fire No. 1 cylinder, the ignition timing indicator on the

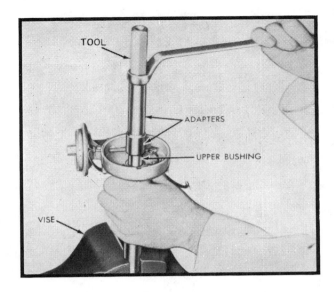

Fig. 126 *Installing upper bushing in distributor housing*

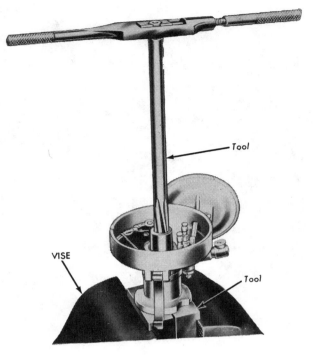

Fig. 127 *Burnishing bushings in distributor housing*

vibration damper is in line with the pointer on the front cover, Fig. 129.

For best performance and fuel economy, the timing should be set by means of a timing light, the use of which having already been described.

DELCO-REMY CORVAIR DISTRIBUTOR 1960–61

This distributor, Fig. 130, has the breaker plate located below the centrifugal advance mechanism and uses the outer diameter of the shaft bushing for its bearing surface. The position of the vacuum control unit when mounted on the distributor housing determines the location of the breaker plate. Note the calibrated scale stamped on top of the breaker plate, Fig. 131. To correctly position the breaker plate during assembly at the factory a mark on the calibrated scale is located opposite the edge of the slot in the distributor housing. This position should not be changed unless there is evidence that the plate has been moved from its original setting.

If the distributor has been disassembled, refer to Fig. 130 for guidance when reassembling.

1. Install felt washer around bushing in housing.
2. Install vacuum unit but do not tighten hold-down screws at this time.
3. Install breaker plate in housing and spring retainer on upper bushing.
4. Position breaker plate with 23° mark in line with scribe mark on housing, Fig. 131, and tighten vacuum unit hold-down screws.

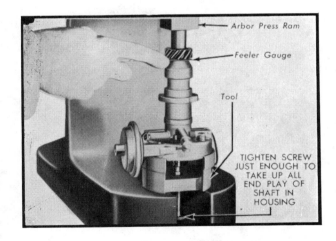

Fig. 128 *Install drive shaft gear so there is .005" to .008" clearance between gear and housing at point indicated by feeler gauge*

5. Install breaker points.
6. Install condenser.
7. Install cam and weight base assembly on shaft. If lubrication in grooves at top of shaft was removed during disassembly, replace with Plastilube #2 or equivalent.
8. Install shaft and cam weight assembly in housing.

615

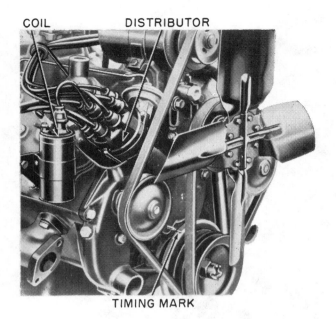

COIL   DISTRIBUTOR

TIMING MARK

*Fig. 129   Ignition timing
mark and pointer*

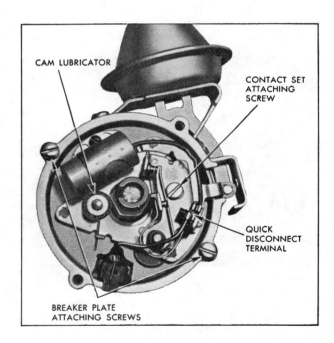

CAM LUBRICATOR

CONTACT SET
ATTACHING
SCREW

QUICK
DISCONNECT
TERMINAL

BREAKER PLATE
ATTACHING SCREWS

*Fig. 132   Breaker plate as-
sembly for later Corvair, Chevy
II and Tempest 4-cylinder*

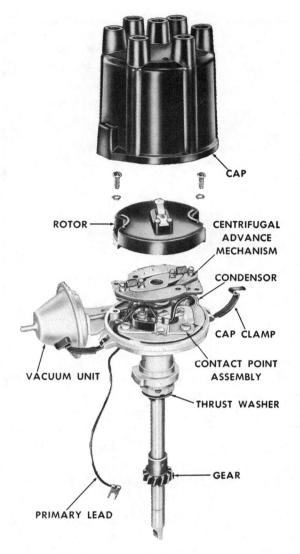

CAP

ROTOR

CENTRIFUGAL
ADVANCE
MECHANISM

CONDENSOR

CAP CLAMP

CONTACT POINT
ASSEMBLY

VACUUM UNIT

THRUST WASHER

GEAR

PRIMARY LEAD

*Fig.   130   Delco-Remy   dis-
tributor used on 1960–61 Corvair*

12. Adjust breaker arm spring tension and breaker points.

1962 and Later Distributor

If the distributor has been disassembled, re-assemble it as follows. Fig. 132 shows the details of the breaker plate and attaching parts.

1. Replace cam assembly to shaft. Lubricate top end of shaft with light engine oil prior to replacing.
2. Install weights on their pivot pins. Install weights, weight cover and stop plate.
3. Lubricate shaft and install in housing.

9. Slide thrust washers and gear on shaft and secure with pins.
10. Install advance weights and springs.
11. Install rotor.

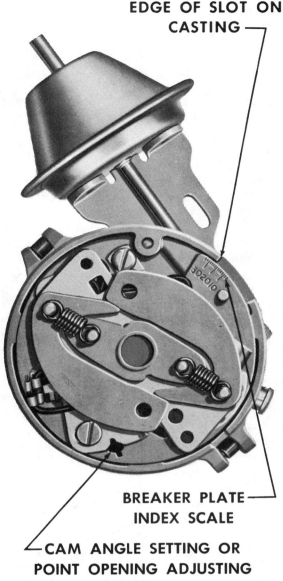

Fig. 131 Corvair 1960–61 distributor showing calibrated scale stamped on top of breaker plate

BREAKER PLATE INDEX SCALE

CAM ANGLE SETTING OR POINT OPENING ADJUSTING SLOT

4. Install thrust washers and driven gear to shaft and secure with roll pins. Check to see that shaft turns freely. Install driven gear with mark on hub in line with rotor segment.

5. Install breaker plate.

6. Attach condenser and breaker point set in proper location with appropriate attaching screws, Fig. 132. Connect primary and condenser leads to breaker point set quick disconnect terminal. *Contact point set pilot must engage matching hole in breaker plate.*

7. Attach vacuum control assembly to distributor housing, using upper mounting holes.

8. Adjust breaker arm spring tension and breaker gap.

9. Install rotor.

PERFORMANCE TIPS

The ignition system, while incapable of supplying performance increases by itself, is the most vital link in realizing the full potential of other performance modifications. No matter whether the route chosen to increase engine output was through improved volumetric or thermal efficiencies or simply through upping engine RPM's, the ignition system must be equal to its task or the performance gains will be limited.

Only a few years ago the answer to all high performance ignition needs was simple but expensive. Throw away the entire battery operated ignition system and replace it with a magneto. This is no longer necessary thanks to vast improvements in conventional ignition systems and they can now be modified to meet most high intensity spark and high RPM demands.

Starting with a standard ignition system. If compression pressures and engine speeds remain about standard or a little above, some simple modifications will produce quite startling results. Most factory distributor centrifugal advance curves are a compromise and they can be altered for higher performance with little sacrifice in reliability and fuel octane requirements.

Two things must be kept in mind when altering ignition advance characteristics. Initial advance should not be changed nor should total advance. Any changes made should not produce "ping" (detonation) anywhere in the RPM range and if they do they must be corrected *immediately*. Detonation is one of the most destructive of all malfunctions and must be avoided at all costs.

Compared to the stock advance curve, the new curve will rise more sharply yet still not exceed original total advance. This is achieved by removing metal from the centrifugal advance weights or by altering the tension of the springs that control their movement. The work must be done with a distributor testing facility at hand to check results. An altered advance curve might—bear in mind each engine will differ in this respect—look like this:

	Stock	Modified
Degrees advance	4 @ 1000 RPM	6 @ 1000 RPM
	7 @ 1500 RPM	10 @ 1500 RPM
	10 @ 2000 RPM	11 @ 2000 RPM
	12 @ 2500 RPM	12 @ 2500 RPM

Tailoring the advance curve is not restricted to nearly stock engines; highly modified power plants will benefit even more from such changes.

Before going into other modifications it will be necessary to review a few fundamentals. Remember that high engine speeds shorten the time primary current can flow in the coil. This reduces the intensity of the magnetic field created in the coil thereby reducing the strength of the secondary voltage delivered to the plugs when the magnetic field collapses. What we are trying to obtain is the strongest magnetic field possible so that secondary voltage will be at a maximum when we need it. To accomplish this it is necessary to replace some of the components of the stock ignition system with special parts.

The easiest change is the installation of a high voltage ignition coil. A greater number of windings provides a stronger magnetic field and, consequently, a heftier secondary voltage. There are a wide variety of these special coils available from the car companies and specialty manufacturers. Keep in mind that changing to such a high voltage coil requires that the condenser be changed as well. Usually the coil is supplied with the proper condenser but if it is not the system described earlier in the chapter for the selection of a suitable condenser can be followed.

Certainly the next most popular modification, one that will assure a hotter spark at even higher RPM's, is the addition of a second set of ignition points. Its purpose is to increase the period of time that the primary current flows. Dual point distributor plates are available for most distributors and they provide somewhat "cleaner" firings as well as the higher RPM potential at a minimal cost.

Next on the performance menu come the "double" ignition systems. These systems have dual coils, condensers, sets of points, and a double tipped rotor. Their purpose is also to increase field buildup time. Naturally, such systems also use distributor cams with only half as many lobes as the engine has cylinders—for example, a V8 will have a four-lobe cam. In essence, the engine is now equipped with two separate ignition systems which work off a common distributor. In so doing, the original single system providing 10,000 sparks per minute is replaced with two systems which need only supply 5,000 sparks each per minute. The result is that the time for field buildup is doubled.

For still higher RPM's and higher compression pressures, the builder can turn to a transistorized ignition system (described in the next chapter) or a capacitance discharge system. The latter works on a simple operating principle. Consider a condenser as a "storage tank" for electricity and install one of high capacity in the primary circuit. Under conventional practice the current flowing to the primary side of the coil does so at a fairly leisurely rate. The above mentioned condenser—capacitor is its proper name, hence the term capacitance discharge—"force feeds" current to the primary, shortening the time required to saturate the low voltage side of the coil. In practice, a capacitance discharge system achieves this build-up time up to eight times faster than ordinary system and will adequately satisfy ignition demands at engine speeds to 10,000 RPM. It is necessary that a complete capacitance discharge ignition system be substituted for the stock unit, forbidding the simple addition of a suitable condenser to the primary side of an ordinary ignition system.

Before any steps are taken to improve ignition performance by the modification or replacement of components, be sure that your stock system will not serve. Regardless of whether you use such a stock ignition or a highly modified one, the following checks will assure that it is operating at maximum efficiency: Keep the battery full of water and free of corrosion.

Keep battery and *all* other electrical connections clean and tight.

Ground the battery directly to the engine with suitable ground straps.

Solder both terminals on metal core secondary ignition cables and keep all secondary cables away from hot manifolding and away from each other.

Use proper ignition ballast resistors with 12-volt coils to prevent damage to them and the points.

Be sure the distributor is sufficiently but not over lubricated and that its action is tight and free of "play."

Review Questions

Page

1. Name two circuits within the ignition system. 561

2. How many separate coils are there in an ignition coil? 561

3. What does the magnetic field in the ignition coil represent? 562

4. Explain how induction takes place in an ignition coil. 563

5. Explain how the resistor in a 12 volt Prestolite coil controls input current. 564

6. Explain how the coil resistor is used in Delco-Remy 12 volt ignition systems. 564

7. What must never be done to the coil resistor in a 12 volt ignition system? 564

8. Describe the construction of a condenser. 565

9. What term is used to express the capacity of a condenser? 565

10. How is the condenser connected into an ignition system? 565

11. What does excessive arcing at the points indicate? 565

12. What is spark plug heat range? 566

13. Would it be best to use a hot or cold spark plug in a car that is used mostly for long trips? . 567

14. Why is it essential to maintain a good contact between the spark plug and the cylinder head? 567

15. What part in a distributor switches high tension current to each spark plug according to the engine's firing order? 567, 568

16. What part of the distributor does the high tension cable from the coil lead into? 568

17. Name the principal parts of a distributor. 568

18. Does the high tension side of the ignition coil always deliver the same voltage in proportion to the primary voltage? 568

19. Explain the operation of the centrifugal advance mechanism in a distributor. 570, 571

20. What is the ratio of rotation between the distributor cam and the crankshaft? 571

21. How many degrees of rotation represent one revolution of the distributor cam? 575

22. What is dwell angle? 575

23. What is a common result of insufficient dwell angle? 575

24. What is the main advantage in using a double breaker distributor? 575

25. What distinguishing feature makes it easy to recognize the Delco-Remy external adjustment type distributor? 577

26. Explain the necessity for spark advance. 578

27. On what factor of engine operation does centrifugal advance depend? 578

28. A 10 degree advance at the distributor cam results in how many degrees of advance at the crankshaft? 578

29. On what factor of engine operation does vacuum advance depend? 580

30. On most distributors, what is the position of the vacuum advance when the engine is idling? 580

31. Where is the vacuum port to the distributor usually located on a carburetor? 580

32. Under what operating condition would distributor advance be at the maximum? 580

33. What is the most common trouble to be found in the ignition system? 581

34. What precaution must be observed while lubricating a distributor? 581

35. Describe the effect of improper breaker arm spring tension on distributor operation. 582

36. What are three classifications of engine miss? 583

37. Name three factors that shorten the life of an ignition coil. 584

38. How does excessive resistance in the high tension circuit tend to injure the ignition coil? 585

39. Will an ignition coil operate at all if it is hooked into a circuit without regard to polarity? 586

40. What is a practical guide for appraising spark plug performance by inspection? 588

41. Why is the round wire gauge the only type suitable for adjusting spark plugs? 590

42. Which electrode of a spark plug should be bent when adjusting the gap? 590

43. Name two advantages in using a resistor type spark plug. 591

44. What is the advantage in using a distributor test fixture or dial indicator for adjusting breaker point gap? 594

45. The process which deposits metal from one breaker point member to the other is known as what kind of action? 594

46. What two types of vacuum advance mechanisms are used on Delco-Remy and Prestolite distributors? 596

47. What is the name of each of the three tests made on a condenser in the process of checking for possible defects? 596, 597

48. Describe how a timing light is used to adjust ignition timing. 611

49. What is the basic difference between a "full vacuum" type Holley distributor and all other distributors? 611

50. On the Holley "full vacuum" type distributor, vacuum to the distributor is a combination of vacuum at what two points in the carburetor? 612

620

Transistorized and Electronic Ignition Systems

INDEX

TRANSISTORIZED IGNITION
Delco-Remy Type 628
Operating Principles 628
Maintenance 631
Trouble Shooting 629
Ford Type 624
Circuit Components 625
Glossary 625
Operation 625

Service Recommendations &
  Specifications 626
Trouble Diagnosis 626

ELECTRONIC IGNITION
A Brief History 621
Advantages 621
Conclusion 641
Chrysler Corporation 633

Ford Motor Company 635
General Motors 636
High Energy Ignition System ... 637
Micro-Processed Sensing &
  Automatic Regulation
  (MISAR) 639
Imported Car Systems 640
Prestolite (American Motors) ... 632
The Future 640

It was almost inevitable that, as this country progressed through its space age in the 1960s and '70s, much of the technology that sent Neil Armstrong to the moon would trickle down to other industries and be utilized for the benefit of all.

The highly accelerated development of automotive electronic components in general, and ignition components in particular, is a direct by-product of American technology developed during the U.S. space program. Specifically, the development and subsequent sophistication of the transistor, diode and semiconductor, allowed the automobile to enter the space age.

ADVANTAGES

This electronic technology was applied to automobile ignition systems for several reasons. First, the solid-state electronic ignition system has enabled automobile manufacturers to standardize their ignition system components. Second, and more importantly, for the motorist, electronic ignitions significantly decrease maintenance requirements. Today's modern electronic ignition system virtually eliminates tuneups, greatly increases the interval at which sparkplugs must be changed, provides better fuel mileage and increases driveability and performance.

Moreover, electronic ignition systems are more efficient than the breaker point types they replace. They provide surer, quicker startups, and reduce exhaust emissions. Electronic ignitions provide a hotter, more uniform spark at a more precise interval than "points" type ignition systems. This promotes the more efficient burning of the air/fuel mixture in the combustion chambers, which is why these electronic systems are directly responsible for better mileage, fewer exhaust emissions, less maintenance, increased reliability, and more perky performance.

The term "solid state" in reference to ignition systems means that, although some moving parts are retained, the reciprocating mechanical motion in the system has been eliminated. The points, and the cam-like components that impart motion to them, have been eliminated. This simplification greatly minimizes required maintenance as well as malfunctions. In many systems, primary and secondary distributor-to-coil leads have also been eliminated, as in the General Motors HEI (High Energy Ignition) system.

Thus, in the engineering scheme of things, solid state components simultaneously maximize performance, efficiency, economy, and reliability, while minimizing harmful exhaust pollutants, thus saving the automobile manufacturers—and car buyers—money in the form of fewer required add-on exhaust emission controls.

Today, all domestic automobile manufacturers are supplying solid state electronic ignition systems as standard equipment on their cars and several imported automobile manufacturers are equipping their cars with electronic systems. The automotive electronic ignition system has arrived with a uniformly applied impact, as did hydraulic brakes, the one-piece windshield, power steering, rack and pinion steering, disc brakes, and radial tires before it.

A BRIEF HISTORY

Although Chrysler Corporation is generally credited with a domestic industry-wide first by introducing standardized electronic ignition systems on their 1971 cars, actual capacitor discharge ignition system research (using cold cathode tubes) goes back to 1936, and workable in-car electronic-type ignition systems were tested in 1948. The Delco-Remy division of General Motors began experimenting with semi-conductors in 1955 and Lucas Electric Company unveiled their transistorized ignition in 1958, utilizing a breakerless pulse pickup and a spark generator of the transformer type. This Lucas system was employed on 1962 Grand Prix engines of BRM and Coventry Climax manufacture.

Racing and racing research greatly advanced the introduction of standardized electronic ignition systems. The Ford Motor Company utilized the Lucas breakerless system on its 1963 Indianapolis 500 cars, ran a fleet test in 1964, and offered the Ford system as an option on particular Ford engines in 1965.

Naturally, parallel development was unavoidable and 1962 saw several aftermarket ignition systems that featured transistor-assisted, or boosted, spark timing. The Autolite Electric Transigniter 201 was such a system, and the first capacitor-discharge system, called the EI-4, was offered by Tung-Sol in 1962.

Although Fiat was the first worldwide manufacturer to utilize electronic ignition as standard equipment (1968), there were several earlier research, racing and "option" breakthroughs in this field, including engineer Earl W. Meyer's experiments at Chrysler Corporation in 1958-1961, and Chrysler's successful electronics application with their 1963-64 Hemi racing engines on the NASCAR circuit. Pontiac was the first to offer the Delcotronic transistorized system on its 1963 models, but only as an option.

The Delcotronic ignition system was of the breakerless, transistor-switched variety, and utilized a magnetic pulse generator. Chevrolet also later offered the Delcotronic system as an option on high performance engines.

Prestolite's CD-65 system was introduced in 1965 with an unprecedented 50,000-mile warranty and Delco followed suit in 1966 with a similar unit. Both operated on the CD, or Capacitor Discharge, theory. Both were breakerless. In 1967, the Delco CD system was offered as an option on Oldsmobiles, Pontiacs, and GMC pickup trucks. Nineteen-sixty-seven also saw the introduction of Motorola's breakerless CD system. GM introduced their HEI system on all models in 1975.

Chrysler's eventual 1972-73 standardization on electronic ignition systems was preceded by a 1971 offering of 1000 vehicles optionally equipped with this system, the 1972 "California V-8 Standardization," and optionally available elsewhere. Across-the-board standardization on all models, including six cylinder engines, came in 1973.

Ford followed a similar application program between May 1973 and final introduction. Eventually the entire line was standardized in 1975.

American Motors has used a 1972-introduced Prestolite system on its AMC cars since 1975. This Prestolite system, although breakerless, does not utilize the capacitor discharge theory.

Basic Definitions and Theory

Some basic definitions are in order, as well as a brief description of the fundamental theory of the electronic ignition system. The typical electronic ignition system does not utilize contact points of the type developed in 1922 for Delco by Kettering. The contact point system replaced the ancient (1880s) low-tension magnetos, and Ford's Model T flywheel magneto. With the electronic systems, contact points are replaced by a magnetic pickup device in the distributor that works in conjunction with an electronic amplifying device. The electronic amplifier utilizes transistors.

Actually, virtually all ignition systems are electrically operated, by the direct current provided by the car's battery. Electricity is similar to magnetism, *electric current* being the physical flow of electric charges through wires. *Electronics,* however, involves the actual movement of *electrons,* electrons being the negatively-charged particles of an atom. Because of their negative charge, electrons repel each other, but are attracted to positively charged matter. So *free electrons* (i.e., electrons that are not orbiting around an atom's nucleus) can be drawn along a conductor to a positive ion area, which, being "positive," is lacking in negatively charged electrons. Thus, in electronic ignition systems, the electronic signals utilize flow as a sort of "current," via a quick exchange of electrons between ions in the system's conductors.

Electronic science also, of course, utilizes other components like semiconductors, diodes and transistors. *Semiconductors* are solid components that function neither as conductors nor insulators, but perform the electronic function in the area between the two. A *diode* is a semiconductor with two connections. A diode functions as a check valve for the system's current. A *transistor* is basically two diodes wired in series. But the two diodes "work" in opposite directions.

The transistor has three leads emanating from its center, which is called its "base circuit." The three leads are called the base, collector and emitter. When a small current is passed from the emitter to the base, a larger current will flow from the emitter to the collector. This is why the transistor is used in some electronic ignition systems. It serves to ease the load on the contact breaker, thus functioning in the electronic system as a *high-speed relay.*

Transistorized ignition systems, however, were merely the first phase of the virtual electronic revolution in automobile ignition systems. The transistorized systems still utilize the pre-electronics type distributor with its attendant coil and contact breaker point system, the revolutionary aspect being composed of the transistor unit that functions between the coil and the breaker points. The transistor acts so only a minute amount of the low-tension current passes through the breaker points, while the collector circuit of the transistor switches the main load of the primary current around the contact breaker points, to the coil (in phase with the trigger current). Thus, transistorized ignition systems are really transistor-assisted spark timers, which vastly lengthen "point" life and improve high speed spark intensity. The transistorized systems did, nevertheless, permit the elimination of the condenser from the ignition system. And the advent of transistorized systems hastened the development of high energy coils. Moreover, transistorized (inductive) discharge systems feature a long spark-burn time, and thus lessen the emission of harmful pollutants.

An additional advantage of the transistorized (and CD) systems is that they provide more than enough voltage reserve because their output stays constant and to higher speeds. This prevents spark plug misfire due to inadequate voltage reserve.

There are two other broad types of electronic ignition systems—*capacitor discharge,* and *breakerless ignition.* Capacitor discharge (electronic) ignition systems use a capacitor to replace the condenser in the old-style breaker point systems. The *capacitor* is an energy storage unit of solid state construction. The capacitor is charged with up to 350 volts by an energy converter, and discharges into the coil. In this setup, the electrical current for spark timing is transistorized. The signal from the timing circuit goes to a thyristor—an electronic switch between the capacitor and the coil—which prevents the discharge until the precisely timed "right moment." Note that capacitor discharge systems can be employed either with or without breakerless ignition systems.

Usually the ignition system's induction coil is utilized as a spark generator on the conventional point system. But CD systems enable a *pulse transformer* to function as the spark generator, which means that CD systems do not exhibit current leakage inductance, nor stray capacitance. Furthermore, capacitor discharge ignition systems issue forth about 30% higher voltage than points systems and do so 80% quicker than a conventional ignition system. This higher voltage and rapid fire feature prevents current "bleedoff," which is the cause of fouled spark plug misfire or no-fire.

Breakerless ignition is the third basic type of electronic ignition system. Breakerless ignition, as the name implies, eliminates the conventional breaker points, as well as the breaker mounting plate, rubbing block, and the cam lobes on the distributor shaft. Additionally, the old style condenser is also eliminated. Whereas the breaker point setup is replaced by various devices, the distributor base, distributor cap, rotor, vacuum and/or centrifugal spark advance mechanisms are retained in the breakerless ignition system.

The devices that replace the breaker points can also work in conjunction with capacitor discharge (CD). But none of the American car makers utilize CD with their latest breakerless ignition systems. It is also important to note that breakerless systems do not eliminate the primary current which is transistor-switched. In other words, breakerless systems do not generate the spark, but rather control the spark timing, functioning as *spark timers.*

There are several distinct devices, or methods, of arriving at breakerless ignition: light-actuated pickup, metal-proximity detector, and magnetic pulse generator.

Magnetic Pulse Generator—In this type of spark timing arrangement, the cam and breaker points are replaced by a reluctor (or armature) and a pickup unit. The armature is a gear-like rotor, replacing the cam on the distributor shaft. The number of teeth on the armature corresponds to the number of spark plugs. The pickup unit is stationary, mounted in place of the breaker arm. It consists of a permanent magnet and a small coil wound around a tiny pole piece. When a tooth on the armature approaches the pole piece, the pole senses the velocity of change in the narrowing air gap, and the magnetic field in the pickup unit is strengthened. This induces a positive voltage in one terminal of the coil around the pole piece. As soon as the armature tooth passes the pole piece and moves away, the air gap widens and a negative voltage is induced in the same coil terminal.

Without some mechanical contact, a low strength pulse is generated in the magnetic field of the pickup. This signal is very accurately timed, and strong enough to trigger the electronic circuitry in the transistor-switch control unit and break the supply of primary current to the coil, which sends a high-voltage current to the right plug via the distributor rotor and cap terminals in the normal manner. The timing circuit in the control unit senses when the coil has fired and redirects the current to the primary winding in the coil, in readiness for the next firing.

Metal proximity detector—This is also a solid-state spark timing system, but does not rely on magnets to activate the circuitry. The cam and breaker-arm combination is replaced by a revolving trigger wheel and stationary metal detector. The trigger wheel has a number of tapered spokes, with the narrow end anchored at the hub, so that the wheel for an 8-cylinder distributor resembles two Maltese crosses superimposed on the same center and phased at 45 degrees. The metal detector is anchored in the periphery of the distributor housing. It is not speed-sensitive and relays the signals to a transistorized power switch which breaks off current delivery to the primary

winding in the coil. A metal detector will stimulate the circuitry down to about 12 rpm, while a magnetic speed detector may not develop optimum electrical output at low cranking speeds.

Light-actuated pickup—This type of breakerless ignition is solid-state, but relies on infrared light beams rather than voltage pulsations. The breaker arm is replaced by a lens unit containing a light-emitting diode. The beam is aimed at an infrared sensor (a silicon phototransistor unit). A slotted disc mounted on the distributor shaft cam interrupts the beam to produce an on/off signal for spark timing. The number of blades on the disc corresponds to the number of spark plugs. The light source is a gallium-arsenide lamp with a hemispherical lens which focuses the beam at the chopping point to about 0.05-in. diameter. The signal from the phototransistor is amplified by a bi-stable inverse-switching amplifier, and delivered to a transistor which switches the primary current and breaks the circuit to the primary winding in the coil. The ensuing events are the same as in conventional systems and other breakerless systems.

Transistorized Ignition

Review Questions for This Chapter on page 641

Ford Type

A transistor is a small electronic device that controls current through the peculiar conductive properties of an element called *germanium*. The transistor gets its name from the words "transfer resistance."

The transistor in an ignition system provides a means of furnishing greater current to the ignition coil with an accompanying greater secondary voltage for firing the spark plugs. At the same time it furnishes less current to the ignition contacts, thereby materially lengthening their life. This is accomplished by means of two inter-related primary ignition circuits whereas the conventional ignition system uses only one such circuit.

The system consists of a base circuit (low voltage trigger circuit), coil primary (collector) circuit (low voltage primary circuit), and a secondary (high voltage coil) circuit. The current will not flow in the primary circuit unless a current is

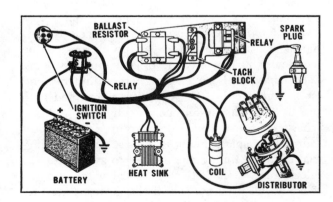

Fig. 1   Ford transistorized ignition system

flowing in the base circuit. A very small current in the base circuit allows a much larger current flow in the primary (collector) circuit.

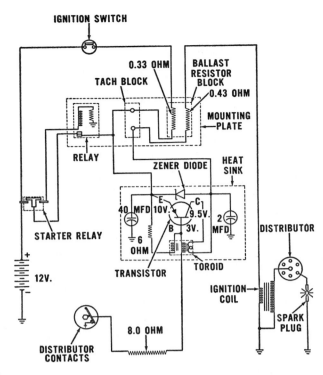

Fig. 2 Ford transistorized ignition circuit

Circuit Components, Figs. 1, 2, 3

Base Circuit:

1. Battery.
2. Ignition switch.
3. Ballast resistor (.33 ohm).
4. Tach block.
5. Heat sink (toroid, transistor, condenser [40 mfd], and resistor [6 ohm]).
6. Resistor.
7. Distributor.

Coil (Primary) Collector Circuit:

1. Battery.
2. Ignition switch.
3. Condenser (2 mfd).
4. Ballast resistor.
5. Tach block.
6. Heat sink (toroid and transistor).
7. Ballast resistor.
8. Coil primary.

Secondary Circuit:

1. Coil secondary.
2. Distributor cap and rotor.
3. Spark plugs.

Glossary

Transistor: A non-vacuum semi-conducting combination of chemical elements used in electrical circuits in place of vacuum tubes.

Zener Diode: A voltage regulating device and two element semi-conductor devices used to control the direction of electrical current flow.

Ballast: A current limiting element. A ballast resistor is a resistor to limit the current and keep it under control.

Tach Block: A block with exposed terminals for test equipment in or to an electrical circuit.

Heat Sink: A device for absorbing heat from electrical units and releasing the heat to the atmosphere to prevent overheating of the electrical units.

Toroid: A small doughnut shaped iron core around which coils of wire are wound similar to a transformer.

In actual operation, a ½ ampere current flows from the battery through the ignition contacts, the common .33 ohm emitter resistor, the ignition switch and returns to the battery.

Operation

The ignition coil primary is designed to draw a normal 12 ampere peak current or approximately 5.5 amperes average current in order to provide high spark plug voltage at the higher engine speeds.

This small control current permits the large 12 ampere current to flow from the battery, through the ignition coil, the .43 ohm collector resistor, the toroid, the transistor collector and emitter, the common .33 ohm emitter resistor, the ignition switch and back to the battery. Since this circuit supplies the power to the ignition coil, it is convenient to refer to it as the power circuit, or coil circuit.

During the dwell, when the ignition points are closed, current flows in both circuits. When the points begin to open, the collapsing magnetic field produced in the toroid sends a reverse pulse through the control circuit, stopping all electron flow in the control circuit. Because the control circuit triggers the power circuit, current stops in the power circuit also.

This causes the magnetic field, produced by the coil primary windings, to collapse, sending very high voltage into the secondary circuit to fire the spark plugs. In addition, high voltage is also in-

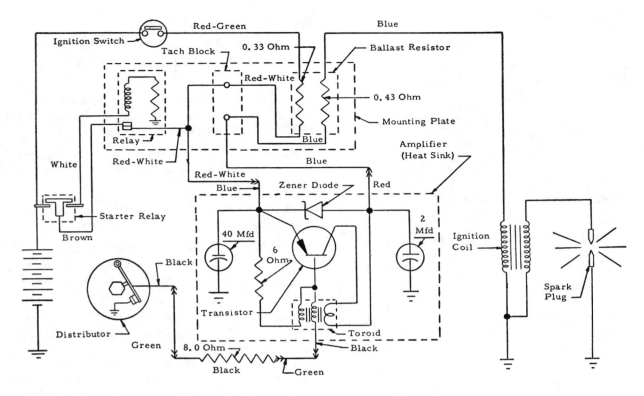

Fig. 3   Ford transistorized ignition wiring diagram

duced in the primary windings of the power circuit.

This would be sufficient to cause transistor destruction, inasmuch as the induced voltage instantly rises above the load line of the transistor until it reaches the breakdown level of the zener diode. For this reason the condenser is incorporated into the power circuit to absorb the high voltage current and keep it within load limits until the zener protection level is reached. Discharge of the condenser voltage from zener level to system voltage takes place by reverse flow through the power circuit, except that the transistor is by-passed by the diode which freely passes this reverse current.

Closing of the ignition points completes the cycle by closing the point circuit again and causing current flow in the power circuit.

Service Recommendations & Specifications

Because of the wide range and usage encountered by the types of vehicles using this ignition system, no set rules for service procedures can be established. However, general recommended procedure for diagnosing suspected ignition trouble in all but the 6-cylinder and dual point vehicles due to power loss is as follows:

1. If dwell is greater than 30 degrees, reset to 24 (plus or minus 1½) degrees.
2. If timing is as much as 4 degrees from specifications, reset to specifications.
3. Points normally need not be adjusted under 20,000 miles. Gap is .019–.021".
4. Spark plugs may be cleaned and regapped at 50,000 miles although experience indicates that a much greater mileage should be expected. Gap is .028–.032".
5. Use Chevron O.H.T. cam lubricant only.

Trouble Diagnosis

CAUTION: Do not use the conventional underhood short cuts such as the use of jumper wires and grounding devices except those described in the following instructions. The use of these "conventional" diagnostic methods can cause extensive

REFERENCE	CRANKING R.P.M.		IDLE (700 RPM)		NORMAL (1500 R.P.M.)	
	VOLT	AMP	VOLT	AMP	VOLT	AMP
A—BATTERY	10.4	——	13.2	——	14.7	——
B—EMITTER	9.5	4.6	13.0	5.6	14.0	12.5
C—EMITTER	8.0	4.6	11.2	5.6	10.0	12.5
D—COLLECTOR	4.0	4.2	4.6	5.0	9.5	12.0
E—COLLECTOR	2.0	4.2	2.2	5.0	4.5	12.0
F—BASE	4.0	0.4	5.8	0.6	3.5	0.5

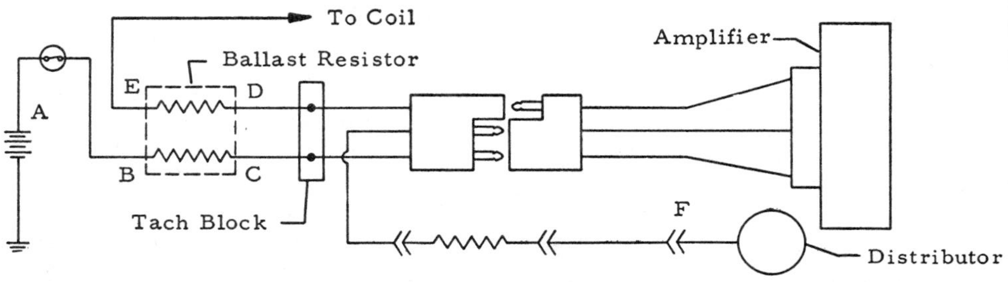

Fig. 4 Cranking amperages and voltages and simplified schematic diagram (Ford)

and unnecessary component damage to the transistor system.

1. Determine whether or not there is a spark by disconnecting coil-to-distributor high tension wire from the center terminal of the distributor cap. Turn on ignition and hold the loose end of the wire about ¼″ away from a good ground (clean engine block) while cranking engine.
   Note: Engine may be cranked by jumping the ignition switch with a jumper wire from battery terminal to "S" terminal of starting relay after disconnecting the red and blue lead (nearest battery terminal).

2. A good, strong spark, as measured by conventional standards, indicates satisfactory ignition operation. Reconnect coil high tension lead and test similarly for spark at spark plugs. A good spark at the plugs indicates the trouble to be other than ignition.

3. If the spark is weak or non-existent, remove cover from mounting plate and connect a dwell meter to the tach block (red to red and black to black). With ignition on, crank engine and observe meter reading. A dwell angle of less than 45 degrees indicates that the transistor is functioning properly and that the trouble is not in this component.

4. A dwell meter reading of zero indicates that the breaker points are contaminated or not closing and should be replaced or adjusted.

5. A dwell meter reading of 45 degrees indicates that: a) the ignition is not being supplied with power from ignition switch; b) breaker points are not opening; c) the transistor is malfunctioning. Disconnect bullet connector from distributor lead and again crank engine. A meter reading of zero indicates that the trouble is in the breaker points. A meter reading of 45 degrees indicates power source of transistor trouble. To determine which, connect a voltmeter or test lamp to the red/green lead terminal of the ballast resistor and crank engine. A meter reading of 45 degrees or steady light

indicates transistor malfunction. Replace the amplifier assembly (transistor). Absence of any indication on meter indicates an open circuit between battery and transistor.

6. A weak spark in Steps 1 or 2 indicates a weak coil. Turn ignition off, replace with a new coil and repeat Step 1. *Do not attempt to test coil because its low impedance contributes to inaccurate indications.*

7. To jump ignition switch from under hood, disconnect red with green chaser lead from ballast resistor and connect a jumper from positive battery terminal to vacated ballast resistor blade terminal. The transistor ignition system will now be supplied with battery power, providing

that the resistor is not open. Check this provision out before proceeding.

8. The only points from which to secure diagnostic information are the tach block terminals, ballast resistor terminals and the distributor primary bullet connector. *Do not pierce insulation elsewhere* with sharp probes or pointed objects for a more comprehensive analysis unless all other tests have failed to reveal a known ignition malfunction cause. By connecting a voltmeter and ammeter at these points, with a well charged battery and the brown lead disconnected from the cold start relay, cranking amperages and voltages may be obtained which should correspond to the data shown in Fig. 4.

Delco-Remy Type

Called the *Delcotronic Transistor Controlled Magnetic Pulse Ignition System,* it features a specially designed magnetic pulse distributor, an ignition pulse amplifier, and a special coil. The other units in the system are of standard design, consisting of the resistors or resistance wire, switch and battery. A typical distributor of this type is shown in Fig. 5.

Although the external appearance of the distributor resembles a standard distributor, the internal construction is quite different. As shown in Fig. 6, an inner timer core replaces the conventional breaker cam. The timer core has the same number of equally-spaced projections or vanes as engine cylinders.

The timer core rotates inside a magnetic pick-up assembly, which replaces the conventional breaker plate, contact point set, and condenser assembly. The magnetic pick-up assembly consists of a ceramic permanent magnet, a pole piece and a pick-up coil. The pole piece is a steel plate having equally-spaced internal teeth, one tooth for each cylinder of the engine.

The magnetic pick-up assembly is mounted over the main bearing of the distributor housing, and is made to rotate by the vacuum control unit, thus providing vacuum advance. The timer core is made to rotate about the shaft by conventional advance weights, thus providing centrifugal advance.

The ignition pulse amplifier shown in Fig. 7 consists primarily of transistors, resistors, diodes and capacitors mounted on a printed circuit panel board. Since there are no moving parts, the control unit is a completely static assembly.

Operating Principles

A wiring diagram showing the complete circuit for typical ignition system is illustrated in Fig. 8. Note that there are two separate resistors used in this type of circuit. The resistor connected directly to the switch is by-passed during cranking,

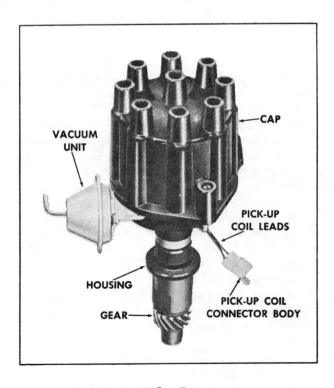

Fig. 5 Delco-Remy magnetic impulse distributor

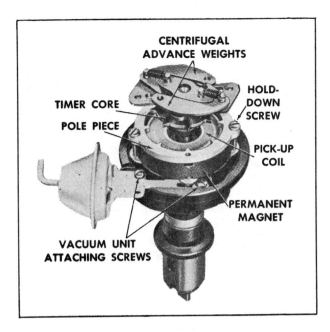

Fig. 6 *Partially exploded view of distributor with cap removed*

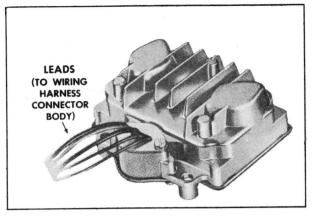

Fig. 7 *Ignition pulse amplifier*

whereas the other resistor is always in the circuit. The use of two resistors permits the required value of resistance to be by-passed during cranking.

In order to fire the spark plug, it is necessary to induce a high voltage in the ignition coil secondary winding by opening the circuit to the coil primary winding. In standard systems, this is accomplished by opening the distributor contact points. In this system, this is accomplished as follows:

When the switch is closed with the engine not running, the current flows through a part of the circuit, Fig. 9. The current flows from the battery through the switch and resistor R-6 to the amplifier. Current then flows through transistors TR-1 and TR-2, resistors R-1, R-2 and R-3, and the coil primary winding and resistor R-7 to ground, thus completing the circuit back to the battery. It is important to note that under this condition, full current flows through the coil primary winding, and capacitor C-1 is charged with the positive voltage towards transistor TR-2.

When the engine is running, the vanes on the rotating iron core in the distributor line up with the internal teeth on the pole piece. This establishes a magnetic path through the center of the pick-up coil, causing a voltage to be induced in the pick-up coil. This voltage causes transistor TR-3 to conduct, resulting in current flow in the circuit as shown in Fig. 10.

The current flow conditions shown in Fig. 10 exist until the charge on capacitor C-1 has been dissipated through resistor R-3. When this happens the system reverts back to the current flow conditions shown in Fig. 9. The system is then ready to fire the next spark plug.

Resistor R-5 is called a feedback resistor, and its purpose is to turn TR-3 off when TR-2 returns to the on-condition. Resistor R-1 is a biasing resistor which allows transistor TR-1 to operate. Zener diode D-1 protects transistor TR-1 from high voltages which may be induced in the coil primary winding. Capacitors C-2 and C-3 protect transistor TR-3 from high voltages which appear in the system.

TROUBLE SHOOTING

Faulty engine performance will usually be evidenced by engine miss, engine surge, or the engine will not run at all. When trouble shooting the system, it is recommended that the following checks be made in the order listed.

Engine Miss

If the trouble is not due to carburetion, check the timing and the condition of the spark plugs. All the wiring should be inspected for brittle or cracked insulation, broken strands and loose or corroded connections. The high tension leads in the coil and distributor cap should be checked to make sure they are pressed all the way down in their inserts. If rubber boots are used, they too should be tightly in place over the connections. Also, the outside of the distributor cap and the coil cover should be inspected for carbonized paths which would allow high tension leakage to

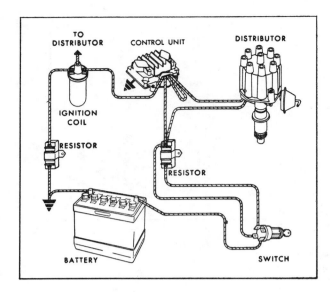

Fig. 8 *Wiring diagram of circuit*

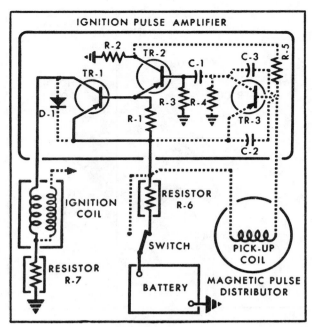

Fig. 9 *Solid lines show current flow with switch on and engine not running*

ground. Also, remove the distributor cap so the rotor and inside of the cap can be checked for cracks and carbonized paths.

Distributor—The pick-up coil in the distributor may be checked by separating the harness connector and connecting an ohmmeter across the coil. The resistance of the coil should be 300 to 400 ohms. If the reading is infinite (high off scale) the coil is open circuited; if the reading is low the coil is shorted. Remember that the resistance of the coil will increase slightly as the coil temperature rises.

Also, the pick-up coil may be checked for grounds by connecting the ohmmeter from either coil lead to the distributor housing. The reading should be infinite. If it is not the coil is grounded.

The distributor centrifugal and vacuum advance may be tested on a distributor testing machine. However, since this involves removing the distributor, delay this operation until after the remaining circuit checks are covered. Besides, it is not likely that the advance mechanism is the cause of the trouble.

Ignition Coil—The ignition coil primary can be checked for an open condition by connecting an ohmmeter across the two primary terminals. An infinite reading indicates the primary is open. For the engine to run but miss at times, the primary open must be of the intermittent type.

Also, the coil secondary can be checked for an open by connecting an ohmmeter from the high tension center tower to either primary terminal. To obtain a reliable reading, a scale on the ohm-

meter having the 20,000 ohm value within, or nearly within, the middle third of the scale must be used. If the reading is infinite, the coil secondary winding is open. *If a coil tester is available, make sure the tester is designed to test this SPECIAL coil.*

Ignition Pulse Amplifier—If all previous checks are satisfactory, and the amplifier is properly grounded, the engine miss is probably caused by a defective ignition pulse amplifier. Replacement of the amplifier will determine if the original amplifier is defective.

Engine Surge

An engine surge condition, of a nature much more severe than that characterized by a lean carburetor mixture, may be due to the two distributor leads being reversed in the connector body, or may be due to an intermittent open in the distributor pick-up coil.

The surge condition may result from the action of the vacuum unit causing a break in the distributor pick-up coil wiring to open and close intermittently. To check this, disconnect the vacuum line and observe engine behavior at idle speed.

To complete the checks on the pick-up coil, connect an ohmmeter to the two distributor pick-up coil lead terminals in the connector body. The

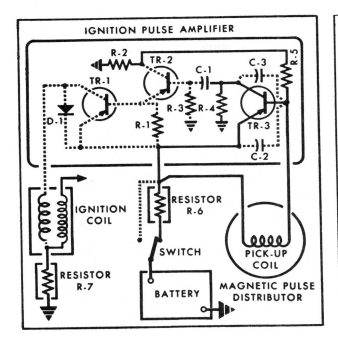

Fig. 10 *Solid lines show current flow when spark plug fires*

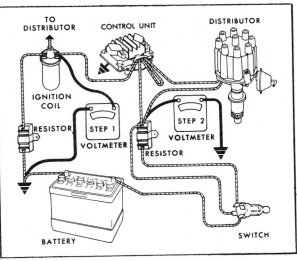

Fig. 11 *Voltmeter connections for circuit checks*

resistance should be 300 to 400 ohms. If the resistance is infinite the coil is open, and if the resistance is low the coil is shorted.

Also connect the ohmmeter from either terminal to the distributor housing. The reading should be infinite. If not, the winding is grounded.

Engine Will Not Run

If the engine will not run at all, remove the lead from one of the spark plugs and hold it about ¼″ away from the engine block while cranking the engine. If a spark occurs, the trouble most likely is not ignition. If a spark does not occur, check the distributor, wiring and ignition coil as previously described. If these check out correctly, further check the circuit continuity as follows:

1. Connect a voltmeter as shown in Step 1, Fig. 11. Observe reading with switch on.
   a. If reading is 8 to 9 volts, check the distributor as outlined above.
   b. If reading is battery voltage, there is an open in the circuit between this point and ground. This circuit consists of coil primary winding, resistor and wiring.

c. If reading is zero, there is an open in the circuit between this point and the battery. Proceed with Step 2 as follows:
2. Connect voltmeter as shown in Step 2, Fig. 11. Observe reading with switch on.
   a. If reading is zero there is an open between this point and the battery. This circuit consists of the resistor, ignition switch and battery.
   b. If the reading is battery voltage, there is an open in the circuit between this resistor and ignition coil. This circuit consists of ignition pulse amplifier and the wiring. If the wiring checks satisfactorily, replace the amplifier.

MAINTENANCE

Since the ignition pulse amplifier is completely static, and the distributor shaft and bushings have permanent-type lubrication, no periodic maintenance is required. The distributor lower bushing is lubricated by engine oil through a splash hole in the distributor housing, and a housing cavity next to the upper bushing contains a supply of lubricant which will last between engine overhaul periods. At time of engine overhaul, the upper bushing may be lubricated by removing the plastic seal and then adding SAE 20 oil to the packing in the cavity. A new plastic seal will be required since the old one will be damaged during removal.

Electronic Ignition

PRESTOLITE BID (AMERICAN MOTORS)

The American Motors Corporation system employs an electronic ignition system known as BID, or Breakerless Inductive Discharge, Fig. 1. The BID system is standard on all American Motors vehicles but not auxiliary engines.

The American Motors BID system is based on the Prestolite breakerless inductive ignition system. The Prestolite system has no breaker points and no cam or condenser. A sensor, a trigger wheel and an electronic control unit replace these parts, Fig. 2. The conventional distributor housing, drive assembly, advance mechanism, and distributor cap and rotor have been retained.

The BID system, thus, has five major parts—the electronic ignition control unit, ignition coil, distributor, sensor and trigger wheel. The electronic control unit is contained in a moisture-proof module that employs solid-state components, which are further protected by sealing them in an earthenware material to resist vibration and adverse environmental conditions. Moreover, all connections are waterproof and the module has built-in current regulation, reverse polarity protection, as well as transient-voltage protection.

There is no ballast resistor nor resistance wire used in the primary circuit, because the control unit features built-in current regulation.

The Prestolite system uses a metal proximity detector for spark timing and also uses a conven-

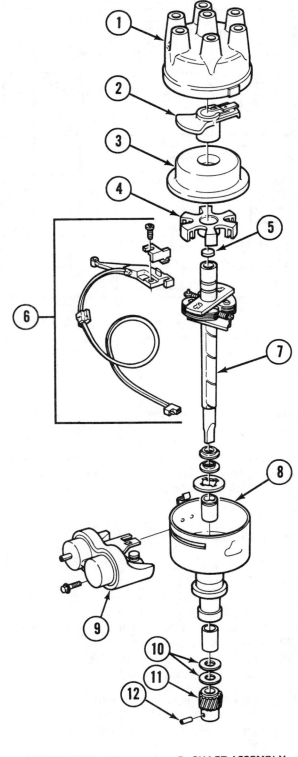

1. DISTRIBUTOR CAP	7. SHAFT ASSEMBLY
2. ROTOR	8. HOUSING
3. DUST SHIELD	9. VACUUM CONTROL
4. TRIGGER WHEEL	10. SHIM
5. FELT PAD	11. DRIVE GEAR
6. SENSOR ASSEMBLY	12. PIN

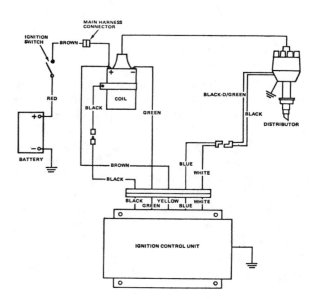

Fig. 1   Prestolite BID (American Motors) electronic ignition system

Fig. 2   Prestolite BID (American Motors) distributor

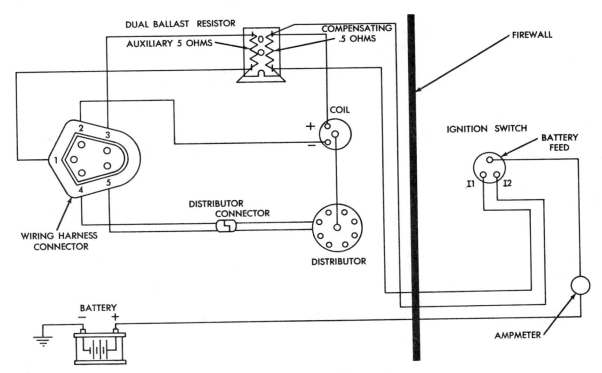

Fig. 3   Chrysler electronic ignition system

tional coil. In other words, magnets are not relied upon to activate the electronic circuitry, as on the Chrysler and GM systems. The BID system's electronic module contains an output transistor, an oscillator and a bi-stable demodulator.

The sensor detects the signal that causes the electronic control unit to operate the coil's primary circuit. The sensor is actually a small coil, wound with fine wire, which receives an alternating current signal from the electronic control unit.

A metal-sensitive electromagnetic field is then developed. The sensor detects the edges of the metal teeth of the trigger wheel and, when the leading edge of the trigger wheel tooth aligns with the center of the sensor coil, a signal is sent to the control unit to open the coil's primary circuit.

The BID system's coil delivers 25% higher voltage rise rate than older systems with condensers. This faster rise rate reduces emissions and enables the ignition system to successfully fire less than optimum condition spark plugs.

With this type of metal proximity detector system there are no contacting surfaces between the trigger wheel and sensor, which means there is virtually no wear. A lack of wear means that the dwell angle is always constant, thus requiring no adjustment at all. The dwell angle on this system is, in fact, determined by the control unit, and the angle between the trigger wheel teeth.

CHRYSLER CORPORATION

In the Chrysler Corporation electronic ignition system, Fig. 3, the breaker points have been replaced with a magnetic pickup device in the distributor, Fig. 4, and an electronic amplifying device that utilizes transistors. Essentially, Chrysler's Magnetic Impulse Ignition (MII) system is composed of a distributor, distributor cap, a sealed electronic control unit, a coil and a double ballast resistor.

The system's control unit is solid state, consisting of diodes, resistors and transistors. This electronic module serves the overall function of the old-type ignition points—deliver an electrical pulse to the coil. The points are, of course, eliminated. However, within the MII's distributor, the points themselves have been replaced by a pole piece and a pickup coil, which are mounted on a plate, as were the original points. Additionally, a reluctor has taken the place and function of the distributor cam, the reluctor's teeth functioning as the old cam's lobes. The reluctor is a pointed star wheel with a point directly corresponding to each engine cylinder. So a 6-cylinder engine has a 6-pointed reluctor, an 8-cylinder engine an 8-pointed reluctor, etc.

A stationary magnet replaces the points and condenser. As the reluctor rotates inside the distributor, a magnetic pulse is generated each time a point of the reluctor passes the stationary magnet.

That magnetic pulse is eventually timed and transformed into a spark at the spark plugs at precisely the right moment to ignite the air/fuel mixture.

The reluctor mounts on a centrifugal advance mechanism, whereas a vacuum-type advance mechanism operates the pickup plate. Each advance mechanism functions similarly in the MII system compared to the conventional ignition system.

A weak magnetic field, created in the pickup coil by the permanent magnet pole piece, is stationary when the reluctor is not turning because no voltage is induced in the coil. When the reluctor rotates, however, and approaches the pickup thereby decreasing the air gap between the two, the magnetic field in the pickup is increased. As the field strength increases, lines of magnetic force move out and over the pickup coil windings to induce a small positive voltage in the coil. This voltage continues to build until the reluctor tooth is directly opposite the pole piece. However, as the reluctor tooth begins to pass the pole piece, the air gap increases and magnetic field strength is decreased. It is then that the magnetic lines of force collapse, effectively reversing the polarity of the induced voltage, and the previously positive terminal now "appears" negative. It is this rapid increase and decrease of the magnetic field that induces a changing voltage.

Since the reluctor has the same number of teeth as the engine's cylinders, one voltage pulse will occur for each cylinder with each complete revolution of the reluctor. The magnetic pulse induced in the coil by the rotating reluctor is a precisely timed signal that triggers the control circuit in the module and cuts off the primary current. When the field in the ignition coil collapses, high voltage is delivered through the rotor and the high voltage wires, to fire the plugs.

The dual ballast in this system is a double resistor unit that protects the system from overload, simultaneously permitting maximum current to flow during engine cranking, assuring a strong spark for proper starting.

No condenser nor capacitors are used in the Chrysler MII system. Secondary current voltage varies between 20,000 and 30,000 volts and is dependent upon engine speed. And since the magnetic pickup, reluctor and control unit, or "black box," do not normally wear out nor require maintenance, engine timing and dwell does not require periodic adjustment. Ignition maintenance is thus simplified to just cleaning and/or replacing worn, fouled or misfiring spark plugs.

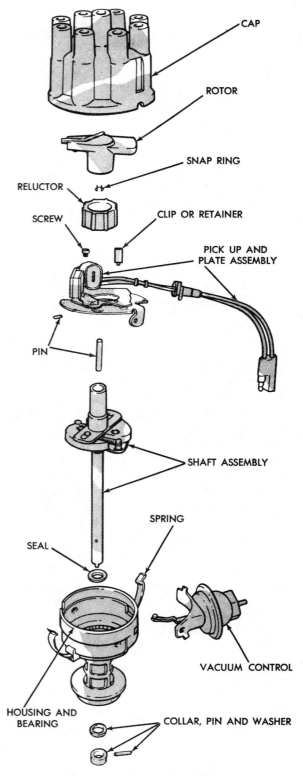

Fig. 4 *Chrysler electronic ignition system distributor*

With no points to "bounce," the MII system will operate at a higher rpm than a point-type system. The higher voltage in the electronic system provides faster startups and greater spark plug life—up to 20,000 miles. And since this pointless magnetic system has no parts rubbing together—only an air gap of .006-.012-inch between the stationary magnet and the reluctor—there is nothing to wear out as the miles pile onto the odometer. The point-type system, on the other hand, begins to wear as soon as it is adjusted, because its adjustment point coincides with its optimum efficiency point. This is why the Chrysler MII and similar pointless ignition systems have been labeled "maintenance free."

Maintenance has been further eliminated on the Chrysler MII system in that the electronic circuitry in the control unit actually determines timing dwell. With the MII system, dwell is not externally adjustable as on conventional ignition systems. Periodic ignition timing checks have also been eliminated because the dwell directly affects timing and the dwell is pre-set within the circuitry of the electronic brain or module.

FORD MOTOR COMPANY

The Ford Motor Company employs Motorcraft solid-state ignition systems on all production models from 1973, Fig. 5. The Motorcraft solid state system employs a magnetic pulse generator and a separately mounted coil and separate electronic module. As with most of these solid state electronic systems, the condenser and, in this case, the capacitor have been completely eliminated.

The coil on the Motorcraft system is a special unit in that it is oil-filled. The primary and secondary windings differ from the stock Motorcraft coil in that the oil-filled unit provides the necessary voltage rise rate, as well as secondary current up to 32,000 volts for starting. The current gradually drops to approximately 20,000 volts at 4000 rpm.

The Motorcraft system functions similarly to other magnetic pulse systems. An electronic module is connected between the ignition coil's primary circuit and the battery, via the ignition switch. The module enables battery current to flow into the coil's primary windings, and also acts to interrupt this current on a signal from the distributor. This interruption of current in the primary winding collapses the magnetic field and induces a high voltage in the coil's secondary winding. This high voltage is then transmitted to the spark plugs via the rotor, distributor cap, and the spark plug leads.

The Ford Motor Company recommends two

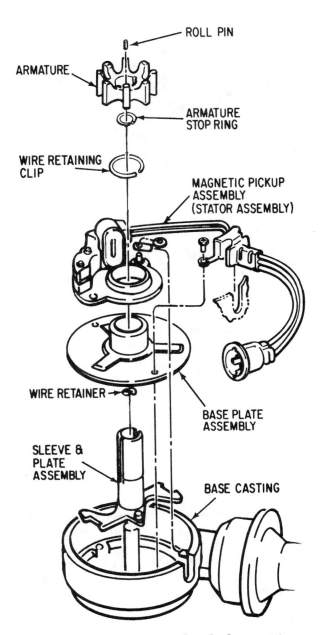

Fig. 5   Ford Motor Co. breakerless ignition distributor

varying service schedules for their cars because different emission control devices are used on various models. Therefore spark plug change is to be performed at 20,000-mile intervals on one schedule, and at 15,000 miles for another schedule. Moreover, some 1976 Continental Mark IVs carried a unique experimental modulated timing system (this system never went into production), which employed an additional electronic module that varied the spark timing according to inputs from such sources as coolant temperature signals, manifold vacuum and engine rpm signals that are generated from an electronic logic circuit. There-

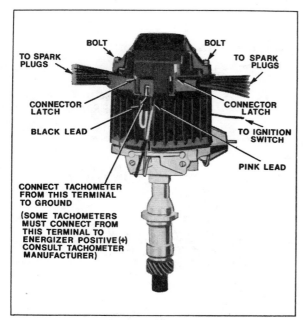

Fig. 6 *General Motors, Delco-Remy Unitized ignition distributor*

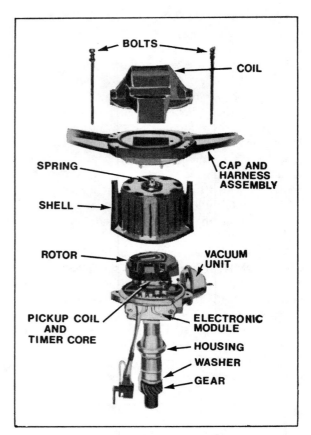

Fig. 7 *General Motors, Delco-Remy Unitized ignition distributor, disassembled*

fore, two separate timing pickups are fitted on the pulse generator—one for standard timing and one for advanced timing. The logic module can, in fact, switch between these two timing advance modes, thus providing super efficiency and preventing the engine from knocking due to insufficient spark retard.

In 1977 Ford also began utilizing two separate but identical appearing ignition systems on their 4-cylinder engines. All California passenger cars, except those with the 2.3-litre 4-cylinder engine, are equipped with the high energy Dura-Spark I ignitions. The California 2.3 L and all other 49-state engines (and Canada) use the Dura-Spark II system. Both systems use physically identical distributors. The Dura-Spark II system, introduced in 1976, is a solid-state system but has the ballast resistor value set to boost coil current and energy output. The II system uses the Dura-Spark I's rotor, distributor cap and adapter, wide gap spark plugs and ignition secondary wires to take advantage of the higher energy of the Dura-Spark I system.

GENERAL MOTORS

General Motors cars utilize Delco-Remy Division electrical components. GM uses two different electronic ignition systems—the High Energy Ignition (HEI) and, on some older models, the Unitized Ignition System. The HEI system, however, is the most modern, efficient system GM makes and has been standard on every GM car since 1975.

Unitized Ignition

The Unitized Ignition System, or Unit Distributor, utilizes an all-electronic module, pickup coil and timer core in place of the conventional ignition points and condenser, Figs. 6 and 7. Point pitting and rubbing block wear resulting in retarded ignition timing, is eliminated. Since the coil is part of the Unit Distributor there is no need for distributor-to-coil primary (breaker points to coil negative lead) or secondary lead (high voltage lead).

A magnetic pickup assembly located over the shaft contains a permanent magnet, a pole piece with internal teeth, and a pickup coil. When the teeth of the timer core rotating inside the pole piece line up with the teeth of the pole piece, the induced voltage in the pickup coil signals the all-electronic module to open the ignition coil primary circuit. When this occurs, the primary current decreases and a high voltage is induced in the ignition coil secondary winding. This high voltage is directed through the rotor and high voltage leads to fire the spark plugs.

The magnetic pickup assembly is mounted over the main bearing on the distributor housing, and

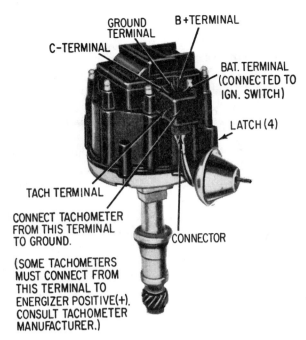

Fig. 8 General Motors, Delco-Remy High Energy Ignition distributor

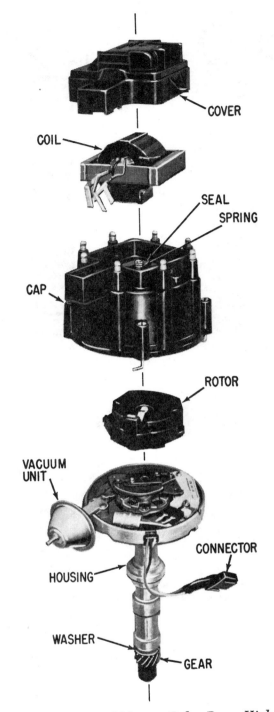

Fig. 9 General Motors, Delco-Remy High Energy Ignition distributor, disassembled

is rotated by the vacuum control unit, thus providing centrifugal advance.

No periodic lubrication is required. Engine oil lubricates the lower bushing, and an oil-filled reservoir provides lubrication for upper bushing.

High Energy Ignition System (HEI)

The HEI system uses an all electronic module, and a magnetic pulse distributor, Fig. 8, with a pickup coil and a permanent magnet sandwiched between a pole piece with internal teeth. The pickup coil generates a pulse signal and functions similarly to the pickup coil in the Chrysler system.

The major variation here is the unique feature of the HEI system, wherein the ignition coil is an integral part of the distributor cap assembly, Fig. 9. This completely eliminates distributor-to-coil primary and secondary leads.

The GM HEI system utilizes a full 12 volts and does not require a lead wire. A vacuum diaphragm is connected to the pole piece and advance weights rotate the timer core around the shaft to effectively provide the advancing mechanism. The HEI system efficiently provides high output secondary voltages of up to 37,000 volts for super spark plug firing capabilities. Like the Chrysler system, the HEI virtually is maintenance free. Spark plug life is, of course, greatly extended, as is the ability to fire worn or fouled plugs.

The maintenance-free HEI system calls for spark plug maintenance and/or change at 22,500-mile intervals.

With the HEI system, one wire connects to the ignition switch, and the spark plug leads complete the entire external electrical circuit. There is, however, a tachometer terminal provided. The HEI system also utilizes a new, closed-core coil that

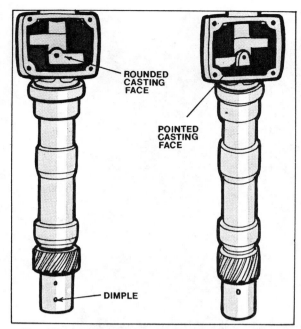

features a low-resistance primary winding, feeding on a 5.5-ampere (12 volt) primary current, and issuing forth a 37,000-volt secondary current. This closed-core coil enables further miniaturization of the entire package, but most particularly the control box and distributor, and the coil functions as a transformer.

Note that the HEI distributor when used on Cadillac vehicles equipped with the Electronic Fuel Injection System (EFI) incorporates the following modification.

A Speed Sensor, Fig. 10, is incorporated within the ignition distributor assembly (HEI) and consists of two components. The first has two reed switches mounted in a plastic housing. The housing is affixed to the distributor shaft housing. The second is a rotor with two magnets, attached to and rotating with the distributor shaft.

Fig. 10   General Motors, Delco-Remy High Energy Ignition distributor speed sensor location on vehicles equipped with Electronic Fuel Injection

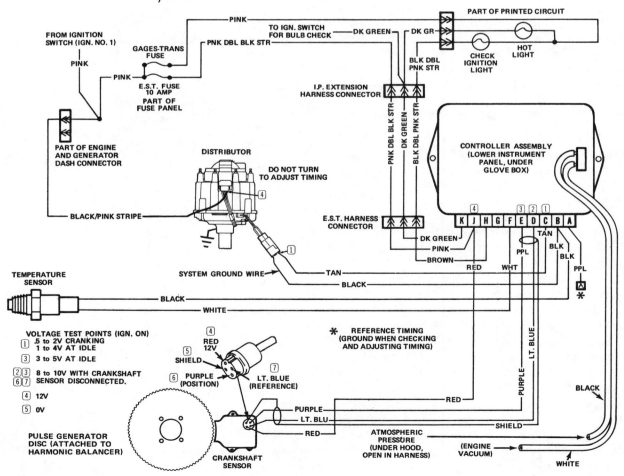

Fig. 11   General Motors, Delco-Remy Misar I Ignition System

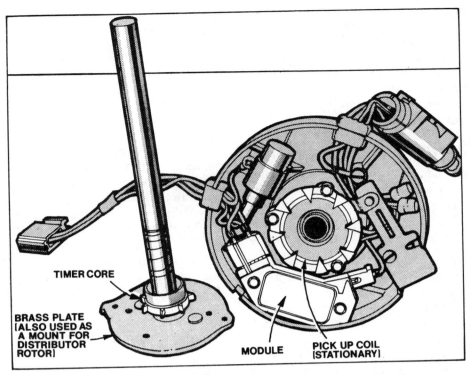

TIMER CORE

BRASS PLATE
(ALSO USED AS
A MOUNT FOR
DISTRIBUTOR
ROTOR)

MODULE

PICK UP COIL
(STATIONARY)

Fig. 12 General Motors, Delco-Remy Misar II ignition distributor

Micro-Processed Sensing & Automatic Regulation (MISAR) System

1977 Toronado MISAR I

This system uses the basic HEI distributor. However, the pick-up coil, pole piece and advance mechanisms have been deleted and the rotor has been redesigned, Fig. 11. This new system incorporates a crankshaft sensor, pulse generator disc, engine coolant temperature sensor and a controller unit.

The engine coolant temperature sensor resistance changes with changes in coolant temperature.

The electronic controller unit is mounted under the glove compartment and receives signals from the crankshaft sensor (crankshaft position and speed), engine coolant temperature sensor, engine vacuum and atmospheric pressure. The controller unit selects the most efficient advance as determined by the input signals and sends a signal to the distributor module to fire the spark plugs.

The electrical harness which connects these units together and to the vehicle harness, contains two vacuum hoses which are both connected to the controller unit. The white hose is connected to engine vacuum, while the black one is vented to atmospheric pressure outside of the vehicle.

The crankshaft sensor is mounted at the front of the engine with a disc located between the harmonic balancer and pulley.

A "Check Ignition" light, located in the instrument panel will go on under the following conditions:

1. Ignition switch is in start position. This mode provides a bulb check.
2. If system voltage is low and there is a heavy electrical load such as operation of power windows or other electrical accessories. Under this condition, the "check ignition" light will go off when the electrical load is removed, providing system voltage returns to normal. The alternator warning light may also go on under these conditions.
3. When checking reference timing and controller circuit is grounded.
4. If controller fails to advance spark timing.

1978 Toronado MISAR II

This system is similar to the MISAR I, except that the crankshaft sensor and pulse generator disc have been eliminated. The function of the crankshaft sensor and pulse generator disc is now performed by the pickup coil and timer core which is located in the distributor, and is similar to the HEI system, Fig. 12. The electronic circuitry in the controller assembly has also been redesigned. The rotor mounting plate is constructed of brass, which is a non-magnetic metal. The reason for this, is so that the brass mounting plate will prevent the signal produced by the pickup coil from being affected by magnetic attraction.

With these modifications, the engine timing can be set in the conventional manner by turning the distributor.

Imported Car Systems

Several imported car manufacturers are using electronic ignitions on their vehicles. Most are made by one of two companies—Lucas or Bosch.

Bosch Electric

The Bosch ignition system, known as TSZ, is of the breakerless variety, and is employed on V-8 powered Mercedes-Benz automobiles and on all models from Volvo. The Bosch TSZ system is based on a magnetic pulse generator and a transistorized switching unit.

The timing rotor is attached to the distributor shaft, directly below the rotor. The timing rotor is shaped like a starfish—with the tips of its long arms bent in a downward direction. The tips fill an open, annular space between a slotted outer stator, and the centrally located magnetic pickup unit. This pickup unit is different from American pickup units, in that the core and windings of the TSZ unit are symmetrically spaced in a coaxial arrangement with the shaft and rotor.

The pickup unit is connected at its base to the outer stator member. A pickup connection on the housing links it to the transistorized switching unit.

The remainder of the Bosch TSZ system is fairly conventional, employing standard coils and spark plugs. The recommended maintenance consists solely of lead wire inspections and 20,000-mile spark plug change intervals.

Lucas Electric Company

This employs a magnetic pulse generator system on Jaguar V-12 automobiles. With the Lucas magnetic pulse system, the timing rotor is shaped like a drum and contains ferrite rods spaced at 30-degree intervals around the drum. This is for the V-12 engine configuration. Fewer cylinders would require differently angled intervals. The ferrite rods are actually positioned so the drum's rotation will make the rods bridge the gap between the two poles in the pickup head.

The pickup module contains a small transformer which is composed of an E-shaped ferrite core and two series-input coils on the end limbs of the "E." The input coils are wound in opposite directions. Therefore, no flux is present in the center limb of the "E." The output winding on the center limb is normally uncoupled from the input windings. However, when the ferrite rods in the rotor contact two poles on the core, a bridge is created, thus connecting the input and output windings, producing an oscillating current in the output winding. This current is then delivered to an amplifier unit containing a fixed-frequency oscillator. The inductance of the output winding is tuned to the oscillator frequency by a capacitor. A well defined signal is thus created. The signal is amplified and switches the transistor in the primary circuit. The transistor then breaks off current supply to the coil, and this break triggers a high voltage current in the coil. The current is delivered to the spark plugs via the rotor, distributor cap, and spark plug leads.

Aftermarket Conversion Kits

There are a host of aftermarket electronic ignition systems available to those who wish to convert their present convention ignition system to electronic operation.

The Future

Current electronic ignition systems have solved a host of problems that were annoying the consumer—maintenance, repair and parts replacement. Fuel waste through misfiring has been minimized. What remains to be done is to optimize the spark for emission control. General Motors and Texaco are exploring new forms of spark control, going in different directions, looking for the same thing. That thing is longer spark duration. The idea is to keep the plug gap full of fire long enough to ignite all combustible elements in the charge and thereby assure more complete burning of the fuel.

Texaco Ignition System—Texaco's design is intended specifically for burning lean mixtures and will aid combustion in stratified-charge engines. The Texaco idea is to switch the spark back and forth as long as necessary, from the center electrode to the ground electrode, back again, forward again, in rapid alternation. The starting point is a current breakerless ignition system. With an energy oscillator added to it, aided by a distributor sensor and a transistorized transformer-type of spark generator, it will provide the alternating spark current. The oscillator can be energized in 70 microseconds and will oscillate until turned off. That means there is no limit to spark duration.

The control unit can regulate spark duration to correspond with rpm, shortening the arc time at high speed to minimize plug wear.

GM Programmable Energy Ignition—Using the HEI system as a starting point, GM Research Laboratories is experimenting with an arc-duration control unit as an addition, retaining the basic mode of operation. The second mode consists of transferring energy to the spark plug gap from the battery at about 1000 volts (through the coil, but without the breaks that produce the 35,000-volt-current).

The basic mode starts the spark and the second mode sustains the spark for an extended but limited duration. The real trick is to alternate rapidly between the two modes—that will produce unlimited spark duration. Spark duration can be controlled in three ways: 1) It can be time-controlled by the pulsing of a monostable multivibrator; 2) It can be sped-controlled by signals from the magnetic pulse generator, with or without the extra refinement of spark cutoff at a given crank position, such as top dead center; 3) It can be distributor-controlled by the limits on duration imposed by distributor geometry (about 45 degrees of crank angle in a V-8). Crossfire to the wrong plug will occur if the limit is exceeded.

Conclusion

Electronic ignition systems are just another step in the on-going electronic revolution in automobiles. We have long had transistorized auto radios and electronic fuel injection has been in production since 1968. In future cars, electronics will be used to control a multitude of functions such as instruments, signals and automatic control of wheelspin, locked-wheel braking and other driving phenomena. There will be electronic malfunction monitoring on a full-time basis—not just when plugged into a computer analysis station. One day there will be an on-board computer that coordinates almost everything.

Review Questions

1. What is a transistor? 625

2. What is a Zener Diode? 625

3. What is a ballast as used in transistorized ignition? 625

4. What is a heat sink? 625

5. What is a tach block? 625

6. What is a toroid? 625

7. Explain why distributor points used with transistorized ignition would outlast points used with conventional system. 624

8. How much current do the distributor points handle in a transistorized system? 625

9. How does the function of the condenser in a Ford transistorized system differ from its function in a conventional system? 626

10. Can the ignition coil used with transistorized ignition be tested on a conventional coil tester? 628

11. On the Ford type transistorized ignition, what are the points from which diagnostic information may be secured? 628

12. What is the basic difference between the Ford and Delco-Remy types of transistorized ignition? 628

13. What three conditions would indicate faulty performance of the Delco-Remy transistorized ignition? 629

14. What is the procedure for checking the ignition coil on the Delco-Remy system? 630

15. What is the first step in checking out a transistorized ignition system if the engine will not run? 631

D.C. Generators and Regulators

Review Questions for This Chapter on Page 668

GENERATOR SECTION INDEX

Brushes, replace 647
Construction, generator 642
Generator construction 642
Generator, motoring 653
Generator not charging 646
Generator, polarizing 653

Generator, removing 646
Generator, replacing with new. 652
Generator service 646
Motoring generator 653
Not charging, generator 646
Polarizing generator 653

Removing generator 646
Replacing brushes 647
Replacing with new generator . 652
Service generator 646

REGULATOR SECTION INDEX

Adjustments, Bosch electrical. 658
Adjustments, Bosch
    mechanical 658
Adjustments, Delco-Remy elec-
    trical 662
Adjustments, Delco-Remy
    mechanical 665
Adjustments, Ford electrical .. 658
Adjustments, Ford mechanical 668
Adjustments, Prestolite
    electrical 658
Adjustments, Prestolite
    mechanical 665
Bosch, electrical tests &
    adjustments 658
Bosch, mechanical
    adjustments 668

Delco-Remy, electrical tests &
    adjustments 662
Delco-Remy, mechanical
    adjustments 665
Electrical adjustments, Bosch. 658
Electrical adjustments, Delco-
    Remy 662
Electrical adjustments, Ford .. 658
Electrical adjustments,
    Prestolite 658
Electrical tests & adjustments. 658
Electrical tests, Bosch 658
Electrical tests, Delco-Remy .. 662
Electrical tests, Ford 658
Electrical tests, Prestolite 658
Ford, electrical tests &
    adjustments 658

Ford, mechanical adjustments. 668
Operation of regulator 654
Prestolite, electrical tests &
    adjustments 658
Prestolite, mechanical
    adjustments 665
Purpose of regulator 654
Regulator, operation of 654
Regulator, purpose of 654
Regulator, replace 658
Tests, Bosch electrical 658
Tests, Delco-Remy electrical .. 662
Tests, electrical 658
Tests, Ford electrical 658
Tests, Prestolite electrical ... 658

GENERATOR SECTION

In the early days of the gasoline automobile there were not many electrical devices and the generator had only a small capacity. As additional electrical equipment was added it was necessary to increase the capacity of the generator to take care of the increased load. While the battery capacity of the 6-volt electrical system has increased only slightly to take care of the starting and parking light load, the 6-volt generator capacity has increased from approximately 15 amperes on early engines to upwards of 50 amperes at which period 12-volt generators came into use.

Although of approximately the same physical size as the 6-volt units, 12-volt generators provide about one-fourth more electrical output to carry the entire electrical system load. This is possible because finer wires can be used to carry 12-volt currents; more armature conductors can be put into the same space; more powerful field coils, wound with more turns of finer wire, can be placed in the same size frame as used for the 6-volt generator.

How a generator works is explained in the chapter on Electricity.

GENERATOR CONSTRUCTION

The generator is mounted on the front of the engine, Fig. 1, and is driven by the belt which also rotates the fan and water pump.

The principal parts of a generator, Fig. 2, are its circular field frame, field poles, armature and end plates or frames.

Automobile generators invariably have two field poles and two brushes. Larger generators for trucks and busses may have four or six field poles and an equal number of brushes.

Fig. 3 shows the generator circuit in its simplest form. The current produced in the armature is

*Fig. 1   Typical automobile generator
driven by fan belt*

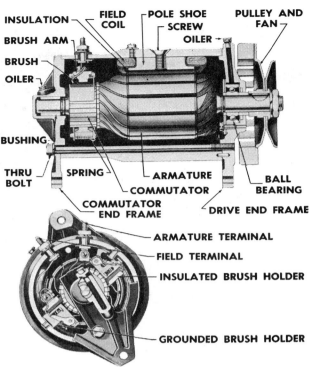

Fig. 2   Sectional view of generator

delivered to the brushes. Most of this current is sent to the external circuit (battery, lights, etc.) by way of the armature terminal. A small percentage is diverted (shunted) to the field coils to produce the magnetism in the field poles. Arrows show the paths through which the current flows.

The generator in Fig. 3 makes no provision for regulation of current and voltage. In order to install a regulator, an insulated field terminal must be added, Figs. 4 and 5, whereas in Fig. 3 the field winding is grounded. The generator regulator may be connected between the armature terminal and field terminal, Fig. 4, or it may be connected between the field terminal and ground, Fig. 5. In Figs. 4 and 5, the regulator is repre-

sented by a resistance in parallel with the points. Actually there may be two of these units in the regulator.

The regulators indicated in Figs. 4 and 5 control both voltage and current. They are called three-unit regulators because there is a unit for (1) voltage, (2) current, (3) a device for breaking the current to the battery when the engine is stopped,

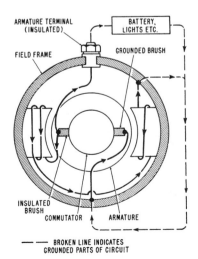

*Fig. 3   Simple diagram of current
flow in a shunt generator*

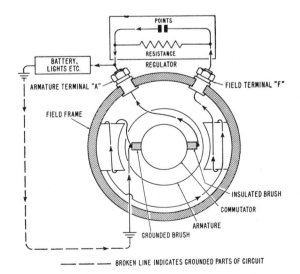

*Fig. 4   Regulator connected between
armature and field terminals*

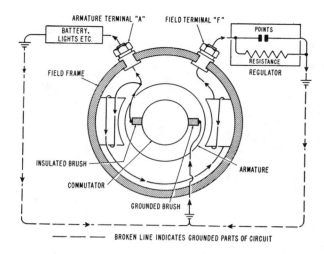

Fig. 5 *Regulator connected between field terminal and ground*

Fig. 6 *The centrifugal fan pulls air through the generator as shown by the arrows, thus cooling the armature*

called a reverse current cutout relay, or circuit breaker.

A sectional view of a typical generator is shown in Fig. 2. The field frame is a cylindrical steel tube. The commutator end frame and the drive end frame are attached to the two ends of the tubular field frame. Two long through bolts hold the two end frames in place. The bolts are threaded into holes in the drive end frame and are kept tight by lock washers under the heads of the bolts at the commutator end. The two end frames are "located" by dowels. The field poles, or pole shoes as they are often called, are each held in place by a husky pole shoe screw.

The armature shaft is mounted in a ball bearing at the drive end and in a bushing at the commutator end. A ball bearing is always used at the drive end because of the load imposed by the pull of the fan belt.

Automobile generators are equipped with a cooling fan which is attached to the drive pulley. This fan, Fig. 6, is a centrifugal design whose blades throw air outward. The consequent suction

at the fan hub draws air through the generator as shown and cools the armature and therefore increases the amperage the generator can be made to produce. The explanation is simple. The greater the armature amperage the more the copper armature windings are heated by the flow of current. If armature amperage exceeds a certain value, the increase in temperature will destroy the insulation. The cooling effect of the air helps keep the temperature down. Therefore, an air-cooled armature can carry more amperes than an uncooled one without danger of damaging the insulation.

The drive pulley is keyed to the armature shaft and is held in place by a nut and lock washer.

The armature windings are contained in slots in the armature core, Fig. 7. The core is not a solid

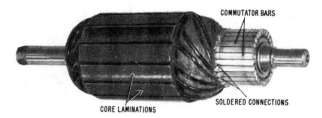

Fig. 7 *The ends of each armature coil are located to diametrically opposite armature slots*

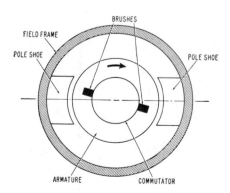

Fig. 8 *Because of distortion of the magnetic field the brushes are tilted with respect to the center line of the field poles*

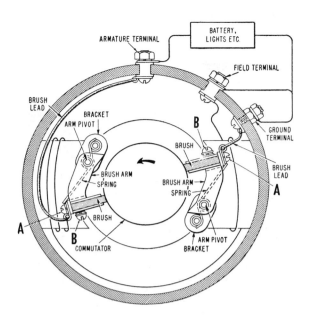

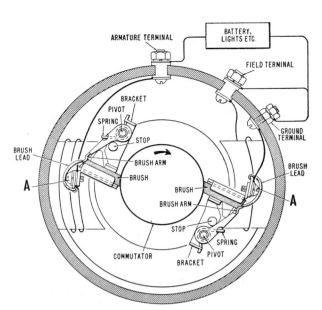

Fig. 9   Trailing swivel type brush holder.
The brush arm is pivoted to the bracket. The
brush is attached to the bracket by screw A.
The spring holds the brush against the
commutator. The brush lead is attached
by screw B

Fig. 10   Reaction swivel type brush
holder. The brush arm is inserted in a
holder in end of brush arm. The
spring causes the arm to press the
brush against the commutator. The
brush lead is attached by screw A

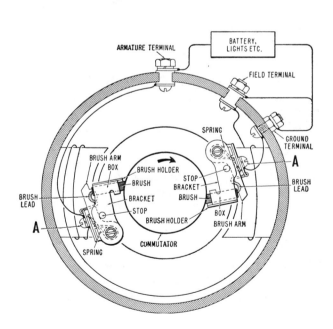

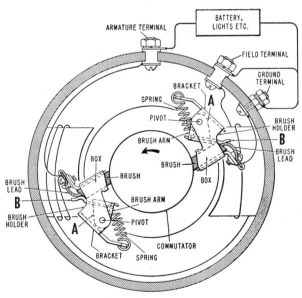

Fig. 11   Trailing box type brush holder.
The right end of the bracket is formed
into a box in which the brush can slide
freely. The spring presses the brush arm
against the brush and holds the brush
against the commutator. The brush lead
is attached to screw A

Fig. 12   Reaction box type brush holder.
The brush is mounted in a box at right
of bracket. The brush can slide freely
in the box. The brush is forced against the
commutator by part B which is firmly
attached to part A which is mounted
on a pivot. The spring pulls on part A,
causing part B to press against the brush

piece of iron mounted on the armature shaft. Instead it consists of numerous slotted discs of iron or soft steel assembled on the armature shaft. The discs are insulated from each other by special varnish or thin paper. This "laminated" armature core is necessary to avoid the production of stray currents in the armature core. Bear in mind that the rotating core acts like a conductor. Stray voltages and currents are produced in it. If the core were one piece, the currents produced in the core would seriously reduce the amount of current produced in the armature windings. The problem is solved by breaking the core up into many thin laminations which are insulated from each other—the net result being that the stray currents are negligible in value.

Each armature coil, Fig. 7 consists of several turns of wire which fit into two diametrically opposite slots in the armature core. All the coils, of course, are connected in series as shown in the chapter on *Electricity*. The ends of each armature coil are soldered in place, Fig. 7. The copper commutator bars are insulated from each other by sheet mica.

Brushes

In Fig. 3 the brushes are located in line with the centers of the pole pieces on the assumption that the armature coil being commutated is not cutting any magnetic lines of force. The coil being commutated is the one whose commutator bars are in contact with the brushes. This brush position is okay when the current in the armature is small with the result that the magnetic field created by the armature is too weak to distort the magnetic field of the pole shoes.

But actual armature currents of 20 to 40 amperes or more distort the magnetic field of the pole pieces, moving the field a few degrees in the direction of armature rotation. Therefore, since the armature coils which are being commutated must not cut any lines of force, it is necessary to relocate the brushes the same number of degrees in the direction of rotation, Fig. 8, rather than have them in the central position indicated in Fig. 3.

There are two methods of mounting brushes: (1) The brush is attached to a pivoted (swiveled) arm, Figs. 9, 10. (2) The brush is mounted in a box-like structure with open ends, Figs. 11, 12. If the commutator rotates away from the brush it is called a trailing type, Figs. 9, 11, whereas if the commutator rotates toward the brush, Figs. 10, 12, it is called a reaction type.

GENERATOR SERVICE

Generator Not Charging

1. To check out the trouble, first make sure all connections at the generator and the regulator are clean and tight. Then run the engine at a fast idle speed. If the generator still fails to show a charge, proceed with Step 2. But if a charge is indicated on the ammeter or if the indicator light goes out, there was a poor connection in the circuit.
2. If the generator field is internally grounded, connect a jumper wire from the regulator armature terminal to the field terminal. If the generator field is grounded externally through the regulator, ground the field terminal to the regulator base. In both cases, the regulator has been taken out of the circuit. Again run the engine, and if the generator now shows a charge the regulator is at fault.

Caution

The foregoing procedure should not be used with double contact voltage regulators. With external ground systems (Delco-Remy), disconnect the field lead and ground it to the regulator base. With internal ground systems, disconnect the field lead and hold it against the armature terminal of the regulator. If this is not done, the lower set of contacts will burn, thereby making the regulator inoperative.

3. If there is still no charge after the foregoing test, short out the circuit breaker and current regulator by connecting a jumper wire from the regulator armature terminal to the battery terminal. Again run the engine, and if the generator now shows a charge, the regulator is at fault. However, if there is still no charge, the trouble is probably in the generator itself, although it may be elsewhere in the charging circuit.

Removing Generator

1. Remove the leads from the generator terminals, Fig. 13. Identify each lead in order that it can be replaced to the proper terminal. Note that the condenser (for radio suppression) is always attached to the "A" terminal and never to "F" terminal.
2. The fan belt must be removed from the generator pulley. On some cars it is necessary to remove the nut holding the belt tension idler pulley in place. With tension loosened, the belt can be removed.

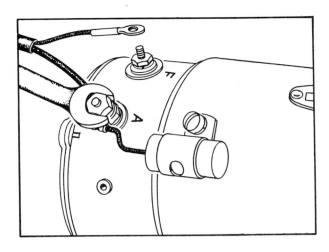

Fig. 13 *Note that condenser is always attached to the "A" terminal. If connected to the "F" terminal it would make the voltage regulator ineffective and the generator would overcharge*

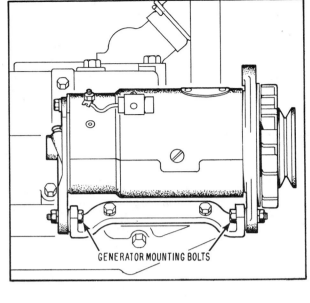

GENERATOR MOUNTING BOLTS

Fig. 14 *Typical generator mounting*

3. Remove the generator by removing the bolts holding the generator lugs to the engine mounting bracket, Fig. 14. On some cars, it is necessary to remove the stud holding the adjusting bracket to generator end frame, Fig. 15. Then loosen the nuts holding the generator to the engine mounting bracket. Move generator toward the engine to release the belt from the pulley. Remove mounting bolts and lift generator from bracket.

4. Some applications have a power steering pump driven by an extended generator shaft, Fig. 16. To remove this type generator, the pump must first be removed from the generator by loosening the two mounting screws. The pump can then be pulled off and the generator removed from the engine as previously described.

Replacing Brushes

1. Remove commutator end frame by first removing the through bolts. Some types of through bolts have hex heads whereas others have slotted screw heads. After bolts are removed, it may be necessary to lightly tap the end frame to remove it from the field frame, Fig. 17.

2. Remove drive end frame and armature assembly from field frame, Fig. 18.

3. With a cloth, clean inside of field frame assembly, Fig. 19. *Do not dip field coils in any cleaning solvent or puncture field coil insulation during the cleaning process.*

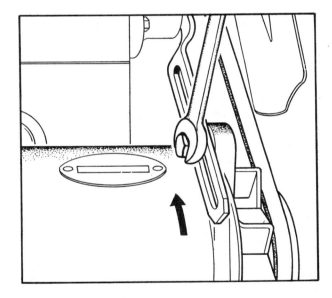

Fig. 15 *Removing stud holding adjustment bracket to generator end frame*

4. Clean commutator end frame, using a cloth dampened in solvent. Place a drop or two of oil in the bushing type bearing to facilitate reassembly.

5. Clean the armature and drive end assembly with an air hose. A clean dry rag may be used to wipe off dirt if an air hose is not available. *Do not use solvent of any kind on the armature.*

6. Examine commutator for high bars, high mica, pitted bars, or excessive wear. If any of these

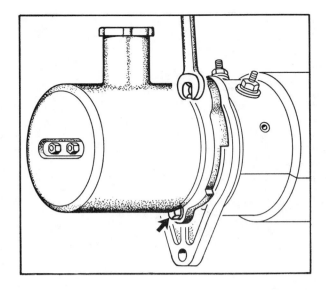

Fig. 16   *Removing power steering pump from generator*

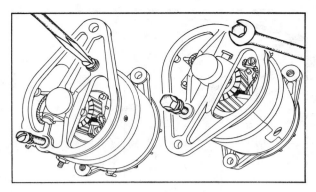

Fig. 17   *Removing generator end frame from field frame*

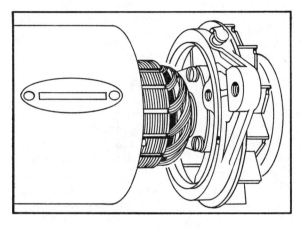

Fig. 18   *Removing drive end frame and armature assembly*

conditions are found, the commutator should be turned down on a lathe and the mica undercut, Figs. 20, 21, 22. Burned bars may indicate a defective armature which should be tested and replaced if necessary, Figs. 23, 24, 25.

7. If commutator appears to be in good condition except for dirt or minor corrosion, it may be cleaned with a fine grade of sandpaper. After cleaning, blow out abrasive particles left between commutator bars. *Never use*

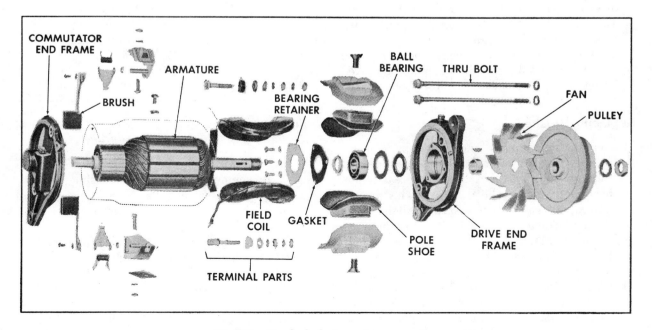

Fig. 19   *Exploded view of a generator*

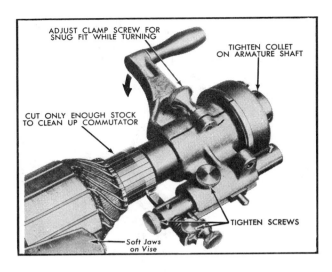

Fig. 20 *Turning down commutator with a lathe*

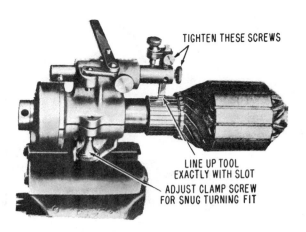

Fig. 21 *Undercutting mica insulation. Depth of cut should be 1/32"*

emery cloth for this purpose as particles of emery may become embedded in the commutator bars and cause a short.

8. Remove brushes from field frame. Wipe off brush holders with a rag dampened in solvent. Any corrosion on the brush bearing surface of the brush holders should be removed to permit freedom of brush movement.

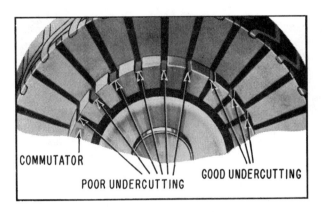

Fig. 22 *Example of proper and improper undercutting*

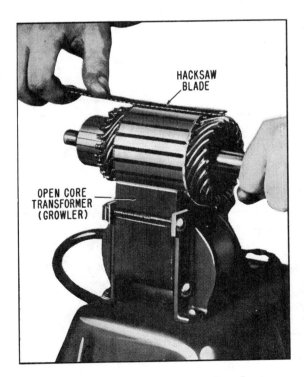

Fig. 23 *Testing armature for short circuit. As armature is rotated by hand, steel strip (hacksaw blade) will vibrate if short circuit exists*

9. If brush holders are mounted on the commutator end frame, remove necessary screws allowing complete removal of brushes.

10. Generators with cover bands need not be disassembled, unless so desired, for brush replacement. First remove the cover band and then remove the screws that attach the flexible brush leads to the brush holders. Then lift up on the brush spring arm and pull out the old brush. A new brush is then pushed down into position in the brush holder. The flexible leads of the brush are then re-fastened to the brush holder. If a field coil lead is present, it also must be fastened to the proper brush holder at this same time.

11. To insert new brushes into the brush holders attached to the field frame, push brush into holder from the bottom and shove it up all the way. This will allow the brush arm to lock the brush up into a position for installation over the commutator. Refasten flexible leads to commutator.

Fig. 24 *Armature test for ground. Using test lamp, place test prod lead on armature core and the other prod on each commutator bar. If lamp lights armature is grounded and must be replaced*

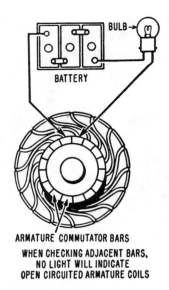

ARMATURE COMMUTATOR BARS
WHEN CHECKING ADJACENT BARS,
NO LIGHT WILL INDICATE
OPEN CIRCUITED ARMATURE COILS

Fig. 25 *Armature test for open circuit*

12. When brushes are installed, make sure the pre-formed angle on the brush matches the commutator contour.

13. Generators with brush holders mounted to the commutator end frame of the type shown in Fig. 26, should have their brushes inserted and flexible leads securely fastened in place before reassembling the end frame onto the field frame.

14. On some generators the brushes may be seated as shown in Fig. 27, using No. 00 sandpaper cut as wide as the commutator finished surface. Excessive use of sandpaper should be avoided since it will shorten the brush and decrease its life. Blow off abrasive dust and carbon after completing the seating process.

15. Another method of seating brushes after the generator is reassembled is to spread some brush seating compound on the commutator and then turn the armature by hand for 20 or 30 revolutions. Then blow out the carbon and dust residue left from the brush seating operation.

16. Before reassembling the generator, it may be desirable to check out the field coils and terminals. If so, refer to Figs. 28 to 31.

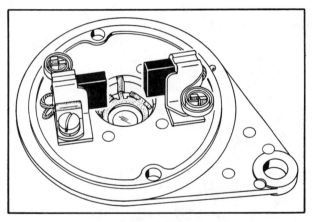

Fig. 26 *Brushes mounted to commutator end frame*

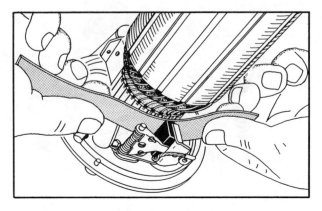

Fig. 27 *Seating brushes with sandpaper*

650

Fig. 28 *Field coil test for open circuit. Using test lamp, place one test lead on field terminal and the other on field coil lead to armature terminal. If test lamp does not light, field coils are open and must be replaced (unless a loose soldered connection is found at field terminal)*

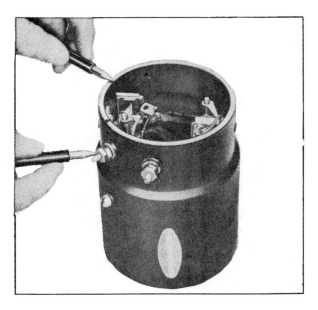

Fig. 30 *Armature terminal test for ground. Using a test lamp, place one test lead on armature terminal and the other on generator frame. Be sure loose end of terminal lead is not touching ground. If lamp lights, armature terminal insulation through generator frame is broken down and must be replaced*

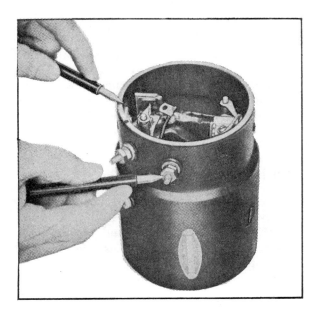

Fig. 29 *Field coil test for ground. Using a test lamp, place one test lead on generator frame (ground) and the other on field terminal. Be sure end of field wire is not touching ground and field terminal insulation is not broken. If test lamp lights, field coils are grounded. If ground cannot be located or repaired, field coils must be replaced*

Fig. 31 *Insulated brush holder test for ground. Using a test lamp, place one test lead on insulated brush holder and the other on ground. If lamp lights, brush holder is grounded due to defective insulation at the frame*

651

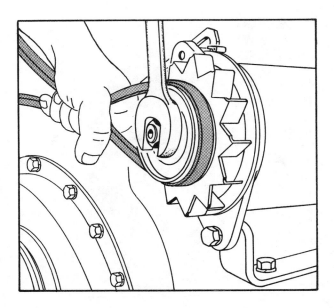

Fig. 32 *Holding pulley tight with fan belt while loosening or tightening pulley nut*

Fig. 33 *Removing generator pulley with special puller*

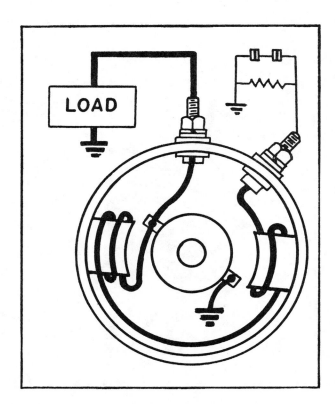

Fig. 34 *Wiring diagram of a generator having the field grounded externally through the voltage regulator. This can readily be identified by the fact that the grounded brush has only its own lead connected to it*

17. When reassembling generators, make sure brushes are out of the way when brush assembly is slipped over commutator. In the event brushes are held up by the brush arm, make sure they are seated down on commutator before installing generator on engine.

Replacing with New Generator

1. When exchanging the existing generator with a new or replacement generator, it is necessary to remove the pulley and fan assembly from the old generator for use on the new generator. After the generator has been removed from the engine, it may be possible to remove the pulley nut by holding the pulley armature shaft by a belt, Fig. 32.

2. In the event that the pulley nut cannot be loosened by the belt method, it is necessary to disassemble old generator to remove the pulley, armature and drive end frame assembly. This assembly can then be placed in a vise and the pulley nut removed.

3. A pulley puller is then used to remove the pulley from the armature shaft, Fig. 33. Care should be exercised not to bend or distort

either the pulley or fan blades during the removal process.

4. The fan and pulley assembly is then placed over the new generator shaft with the keyway in position over the key. Then lightly tap the pulley and fan assembly down on the shaft until the pulley nut can be started on threaded end of shaft.

5. After the generator is assembled and brushes properly seated the generator should be run as a motor, the procedure for which is given below. Mount the generator on the engine and tighten the pulley nut to force the pulley into position against the armature shaft shoulder. Use the vehicle belt as a holding device on the pulley.

6. Apply 8 to 10 drops of light engine oil to the oil cups if present on the generator. Sealed bearings do not require lube.

Polarizing Generator

After the generator is installed on the engine and all leads are connected, the generator must be polarized before starting the engine. The polarizing procedure depends upon whether the field is grounded externally or internally. Having determined which system is used, proceed as follows:

If the generator field is externally grounded, Fig. 34, momentarily connect a jumper wire from the "BAT" to the "GEN" or "ARM" terminals of the voltage regulator. Just a touch of the jumper to both terminals is all that is required.

If the generator field is internally grounded, Fig. 35, disconnect the field wire from the regulator and momentarily touch this wire to the regulator "BAT" terminal.

Motoring Generator

Run the generator as a motor by connecting the ground side of a battery to the generator housing. Connect the ungrounded side of the battery to the generator armature terminal.

On generators with externally grounded fields,

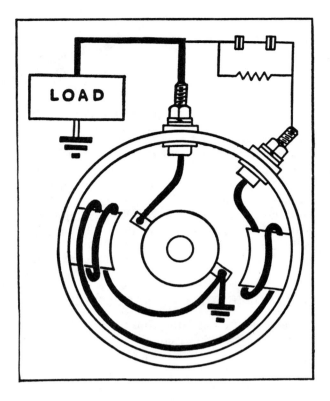

Fig. 35 *Wiring diagram of a generator having the field grounded internally through the grounded brush. This can readily be identified by the fact that the grounded brush has both its own lead and the field lead connected to it*

connect a jumper wire from the generator field terminal to the generator frame. For internally grounded units connect a jumper wire from the armature terminal to the field terminal.

While motoring, the armature should rotate slowly. If it does not, it may be due to improper bearing fit or alignment, mechanical interference between armature and field coil pole shoes or improper end play. If end play appears excessive, check the tightness of the pulley nut. Make sure that the pole shoe screws are securely tightened. Two or three sharp raps on the generator frame with a rawhide or plastic hammer will often help to free the armature.

Regulator Section

PURPOSE OF REGULATOR

A generator regulator is designed for one purpose only and that is to regulate or control the charging rate in the generator-battery circuit. When a good battery is low, the regulator will automatically increase the charging rate until the battery becomes fully charged. As soon as the battery is charged up, the regulator will automatically cut down the charging rate. That's all it does.

OPERATION OF REGULATOR

The modern regulator consists of three units shown in Fig. 36. The circuit breaker (or cutout relay) is the same type as and functions exactly like the conventional circuit breaker that has always been used on cars; it automatically closes the circuit between the generator and battery when the engine is running a little above idling speed, and it opens the circuit when the engine is idling or stopped. In other words, it is simply a magnetic switch.

The voltage and current regulators, regardless of type, are automatic magnetic switches that weaken or strengthen the generator field circuit according to the requirements of the battery. This is accomplished by means of resistances that automatically cut into or out of the field circuit.

In other words, when the battery is not fully charged *resistance is cut out of the circuit,* thereby allowing the generator to recharge the battery. When the battery becomes fully charged, *resistance is automatically cut into the circuit,* reducing the charging rate of the generator so that the battery voltage is held within safe limits.

Before we describe the operation of a typical regulator, you will recall if you have read the chapter on magnetism, that if a wire is wound around a soft iron rod, and a current is passed through the wire, the rod becomes a magnet. Its magnetism depends upon the number of turns of wire and the amperage. To put it briefly, therefore, the regulator consists of three electromagnetic switches which interrupt the circuits automatically and at the proper time. In an electromagnetic circuit interrupter, the magnet acts on a hinged spring-loaded piece called an armature, thereby closing or opening the circuit contacts at predetermined voltages and amperages.

How a Regulator Works

Although the following describes the operation of a modern Delco-Remy regulator, it has been

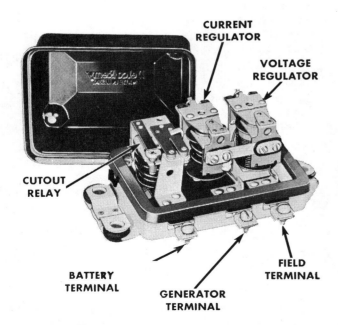

Fig. 36 *Delco-Remy generator regulator with cover removed*

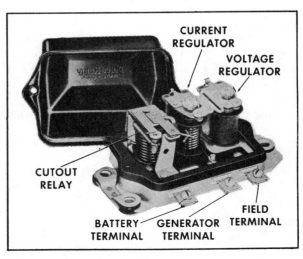

Fig. 37 *Delco-Remy generator regulator for cars without air conditioning*

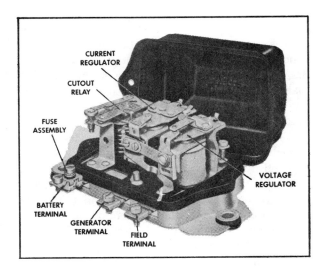

Fig. 38   *Delco-Remy generator regulator for air conditioned car*

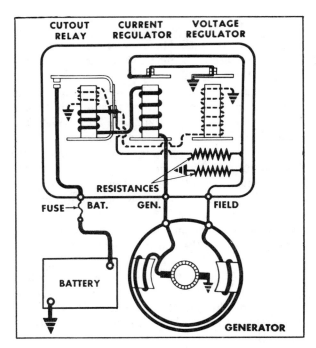

Fig. 39   *Generator system circuit for car without air conditioning*

selected because two types are in general use. The so-called single contact type, Fig. 37, is used on standard cars and the double-contact type is used on air conditioned cars, Fig. 38. Bear in mind, however, that the fundamental principles of operation also apply to the Prestolite, Ford and Bosch regulators even though their construction differs somewhat from Delco-Remy regulators. These differences will become apparent when we describe the service requirements of each regulator.

Usually the regulator is grounded through its attaching bolts, and to insure positive ground the base of the regulator is often connected by a wire to the generator frame or housing.

Cutout Relay

The cutout relay opens the circuit to prevent the battery from discharging to ground through the generator whenever the engine is stopped or generator is operating at such low speed that its voltage is less than voltage of battery. When the voltage of generator is slightly greater than battery voltage the relay closes the circuit so that generator can furnish current to the electrical system.

The cutout relay has a series of current winding of a few turns of heavy wire, and a shunt or voltage winding of many turns of fine wire, both assembled on the same core. The shunt winding is connected between generator armature and ground so that generator voltage is impressed upon it at all times. The series winding is connected so that all generator output current must pass through it. It is connected to a flat steel armature which has a pair of contact points

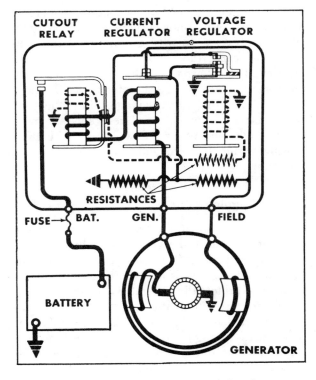

Fig. 40   *Generator system circuit for air conditioned car*

through which current passes to the battery and other electrical units. The contact points are held open by armature spring tension when the unit is not operating, Figs. 39, 40.

When the generator begins to operate, voltage builds up and forces current through the shunt winding, thereby magnetizing the core. When the voltage reaches the value for which the relay is set, the magnetism is strong enough to overcome the armature spring tension and pull the armature toward the core, thereby closing the contact points. Generator current now flows through the series winding of relay in the right direction to add to the magnetism holding the points closed, and passes on to the battery and other electrical units in operation.

When the generator slows to engine idling speed, or stops, current begins to flow from the battery back through the generator, reversing the current flow through the series winding. This reduces the magnetism of the relay core to the extent that it can no longer hold the contact points closed against armature spring tension. The points are separated and the circuit broken between the generator and battery.

Both standard and air condition regulators have a fuse in the generator charging circuit. This fuse connects to the battery terminal of the regulator and the battery lead connects to it in turn. The purpose of the fuse is to protect the generator and wiring should a stuck or welded cutout relay occur. Shorts or grounds occurring in the charging circuit or reverse polarity conditions of the generator can cause the cutout relay points to weld together. This allows the battery to discharge thru the generator when the generator is not developing greater than battery voltage. Since the generator has such low internal resistance, tremendously high current will flow from the battery causing wiring to burn and deterioration of the generator.

Current Regulator

The current regulator automatically controls the maximum output of the generator. When the current requirements of the electrical system are large and the battery is low, the current regulator operates to protect the generator from overload by limiting its output to a safe value.

The current regulator has one series winding of heavy wire through which the entire generator output flows at all times. This winding connects to the series winding in the cutout relay, described above. Above the winding core is an armature, with a pair of contact points which are held together by spring tension when the current regulator is not operating. When the current regulator is not operating and the contact points are closed, the generator field circuit is directly grounded so that the generator may produce maximum output unless further controlled by the voltage regulator described further on.

When the generator output increases to the value for which the current regulator is set, the magnetism of the current winding is sufficient to overcome the armature spring tension. The armature is pulled toward the winding core so that the points are separated. The generator field circuit must then pass through a resistance which reduces the flow through the field coils and thereby reduces the output of the generator. This reduces the magnetic strength of the current winding so that spring tension again closes the contact points, directly grounding the generator field circuit and increasing generator output. This cycle is repeated many times a second, and this action limits the generator output to the value for which the generator is set.

The current regulator has a bi-metal hinge on the armature for thermostatic temperature control. This automatically permits a somewhat higher generator output when the unit is cold, and causes the output to drop off as the temperature increases.

The current regulator operates only when the condition of battery and the load of current-consuming units in operation require maximum output of the generator. When current requirements are small, the voltage regulator controls generator output. Either the current regulator or voltage regulator operates at any one time; both regulators never operate at the same time.

Voltage Regulator

The voltage regulator limits the voltage in the charging circuits to a safe value, thereby controlling the charging rate of the generator in accordance with the requirements of the battery and the current-consuming electrical units in operation. When the battery is low, the generator output is near maximum but as the battery comes up to charge, and other requirements are small, the voltage regulator operates to limit the voltage, thereby reducing the generator output. This protects the battery from overcharge and the electrical system from high voltage.

The voltage regulator unit has a shunt winding consisting of many turns of fine wire which is connected across the generator. The winding and core are assembled into a frame. A flat steel armature is attached to the frame by a flexible hinge so that it is just above the end of the core. When the voltage regulator unit is not operating, the tension of a spiral spring holds the armature away from the core so that a point set is in contact which allows the generator field circuit to complete the ground through them, Figs. 39 and 40.

When the generator voltage reaches the value for which voltage regulator is set, the magnetic pull of the voltage winding is sufficient to overcome the armature spring tension so that the armature is pulled toward the core and the contact points are separated. The instant the points separate, the field current flows only through the resistance to ground. This reduces the current flow through the field coils and decreases generator voltage and output.

The reduced voltage in the circuit causes a weakening of the magnetic field of the voltage winding in the regulator. The resulting loss of the magnetism permits the spring to pull the armature away from the core and close the contact points again, thereby directly rounding the generator field so that generator voltage and output increases.

This cycle is repeated many times a second, causing a vibrating action of the armature, and holds the generator voltage to a constant value. By maintaining a constant voltage, the voltage regulator continues to reduce the generator output as the battery comes up to charge. When the battery reaches a fully charged condition, the voltage regulator will have reduced the generator output to a relatively few amperes.

The voltage regulator has a bi-metal armature hinge for thermostatic temperature control. This automatically permits regulation to a higher voltage when the unit is cold, and a lower voltage when hot, because a high voltage is required to charge a cold battery.

As previously stated, the current and voltage regulators do not operate at the same time. When current requirements are large, the generator voltage is too low to cause voltage regulator to operate, therefore the current regulator operates to limit maximum output of generator. When current requirements are small, the generator voltage is increased to the value which causes voltage regulator to operate. The generator output is then reduced below the value required to operate the current regulator, consequently all control is then dependent on the operation of voltage regulator.

Double Contact Voltage Regulator

Two sets of points are required in the voltage control to handle the high field current used in the heavy duty generator Fig. 40.

The voltage regulator armature has two contact points which are just over and under stationary contact points. When the voltage regulator unit is not operating, the tension of a spiral spring holds the armature away from the core so that the lower set of contacts is closed and the generator field current is completed directly to ground through them Fig. 40.

When the voltage regulator unit is controlling generator output, there are two operating conditions which result in entirely different action of the voltage regulator:

1. When the engine speed is low and there is a great demand for current by the accessories and/or battery, generator field current flow must be high. Under this operating condition, the voltage regulator vibrates on the lower set of contacts. When these contacts are closed, field current flows directly to ground; when they are open, current flows through a resistor to ground. Field current will therefore be somewhere between that allowed by the resistor and a direct ground.

2. When engine speed is high and there is little demand for current by the accessories or battery, generator field current flow must be regulated to a very low value; the resistance inserted in the field circuit when the lower contacts open is not sufficient to control the generator voltage. Under this operating condition, the voltage increases slightly (.1 to .3 volt), the armature is pulled farther down, and the voltage regulator operates on the upper set of contacts.

When these contacts are open, field current flows through the resistor to ground, when they are closed, field current is *stopped* due to current from the charging circuit bucking against the field flow, Fig. 40. Field current will therefore be somewhere between that allowed by the resistor and zero.

Resistances

The current or voltage regulator circuit both use the same resistance which is inserted in the field circuit when either regulator operates.

The sudden reduction in field current occurring when either the current or voltage regulator contacts open, is accomplished by a surge of induced voltage in the field coils as the strength of the magnetic fields change. These surges are partially dissipated by the two resistances, thus preventing excessive arcing at the contact points.

VOLTAGE REGULATORS
REPLACE

1. When working on voltage regulators, it is good practice first to remove the battery ground strap or cable from the battery post. This prevents any short circuits or accidental grounds from occurring.
2. To aid in correctly re-wiring to the replacement regulator, identify the wires in some manner that will aid in proper installation.
3. Remove lead or leads connected to the battery terminal of the regulator.
4. Remove lead or leads connected to the armature terminal of the regulator. This terminal is marked "GEN" on Delco-Remy units, and "A" or "ARM" on Autolite, Ford and Bosch units. *If a condenser is present, note that it must be connected to the armature terminal only.*
5. Some regulators have a fuse connected to the regulator battery terminal. This fuse should be removed for use with the replacement regulator. Before installing the fuse, however, it should be tested for continuity with a test lamp. This is to make sure the fuse is not defective or "blown" which would result in an open circuit.
6. After the new regulator has been installed in position, scrape all lead connections or terminals clean to provide a good metal-to-metal contact when re-connected to the regulator terminals.
7. After all leads are connected and before the engine is started, the generator must be polarized as outlined in the *Generator Section.*

ELECTRICAL TESTS ON REGULATORS

Service Notes

1. Do not attempt to adjust a regulator unless its operation is thoroughly understood and accurate meters are available. Even a slight error in the setting of the unit may cause improper functioning, resulting in a run-down or over-charged battery, or damage to the generator or regulator.
2. When electrical tests are made at the regulator terminals, there is great danger of short circuiting the regulator. Whenever possible it is better to connect test apparatus where there is no danger of damaging the regulator. For example:
3. When checking generator output on single contact voltage regulators (with external field ground), ground the field terminal at the generator rather than at the regulator to avoid

the possibility of grounding the regulator battery terminal with a screwdriver. *Caution: When making this test on double contact voltage regulators, disconnect the field lead from the generator and ground this lead to the generator frame.*
4. Timing or trouble lights should be connected at the starter solenoid terminal rather than at the regulator battery terminal. This will avoid danger of the clip connector slipping off and simultaneously touching the battery and field terminal of the regulator.
5. Before testing or adjusting the regulator, be sure all connections in the charging circuit are clean and tight, and that the battery is fully charged. Check the generator output and be sure the regulator is the correct unit for use with the particular generator. Be sure regulator is properly grounded.
6. Always make the voltage regulator test before making the current regulator test. When removing or installing the regulator cover, do not allow the cover to touch regulator parts, as this might cause a short circuit and damage the unit.
7. Before starting the tests make sure the contacts are clean and not rough or pitted. If this is not done, subsequent tests will only produce erratic meter indications. When cleaning Prestolite, Ford and Bosch contacts, they should be filed parallel to the armature as shown in Fig. 41, using a No. 6 American Swiss cut equaling file. Do not file crosswise as grooves may form which would tend to cause sticking and erratic operation. Delco-Remy recommends the use of a spoon riffler file, Fig. 42. After filing, use a strip of linen tape dampened with lighter fluid to clean the contacts; then run a strip of dry tape across the points to remove any residue left from filing.

PRESTOLITE, FORD & BOSCH
ELECTRICAL TESTS

Circuit Breaker

1. With an ammeter connected as shown in Fig. 43, and a voltmeter connected from regulator armature terminal to regulator base, disconnect field lead from regulator field terminal and insert a variable resistance between field lead and its terminal.
2. Run engine at about 800 rpm. Turn variable resistance to the "all in" position. Then slowly reduce the resistance, noting the voltage reading just as the circuit breaker closes. The voltmeter will give a sharp fluctuation at that point and

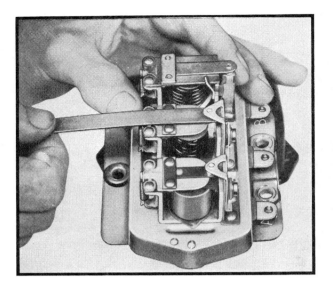

Fig. 41 *Prestolite contacts should be filed parallel with the length of the armature since they produce a wiping movement when in action*

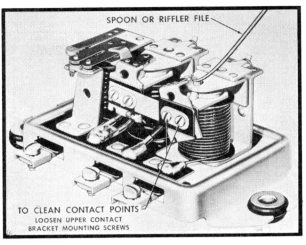

Fig. 42 *Cleaning Delco-Remy regulator contacts. To reach the contacts, loosen the contact bracket mounting screws and tilt the bracket to one side*

usually a slight click can be heard as the contacts close.

3. On Prestolite units, if the closing voltage is not within specifications, remove the regulator cover and change the armature spring tension by bending the lower spring hanger. Bending the spring hanger down increases spring tension and raises the voltage, and vice versa.

4. If closing voltage is not within specifications on Ford and Bosch units, bend the adjusting arm upward to increase voltage, and vice versa, Fig. 44.

Single Contact Voltage Regulator

1. Leave the ammeter connected as shown and move the voltmeter clip from the regulator armature terminal to the regulator battery terminal, Fig. 45.

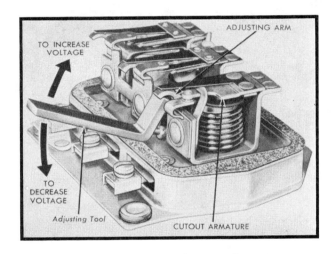

Fig. 44 *Adjusting closing voltage on Ford regulators*

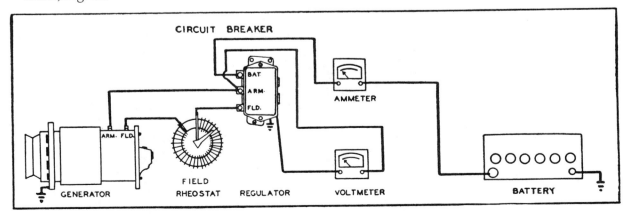

Fig. 43 *Circuit breaker test for Prestolite, Ford and Bosch regulators*

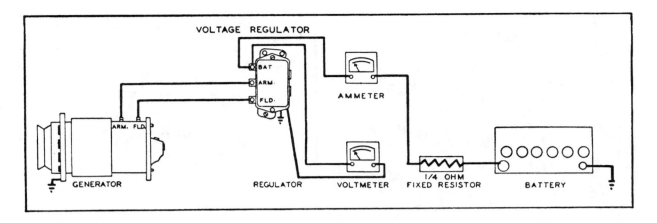

Fig. 45 *Voltage regulator test on Prestolite, Ford and Bosch single contact regulators*

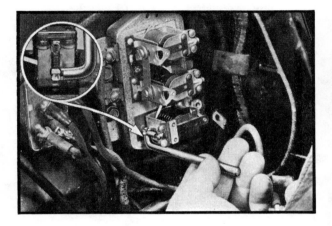

Fig. 46 *To adjust closing voltage on Prestolite regulators, change the spring tension by bending the lower spring hanger. Increasing tension raises closing voltage*

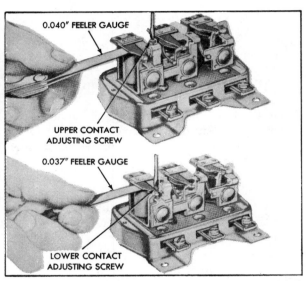

Fig. 48 *Adjusting Ford and Bosch double contact voltage regulator*

Fig. 47
Adjusting voltage regulator (left) and current regulator (right) on Ford and Bosch regulators

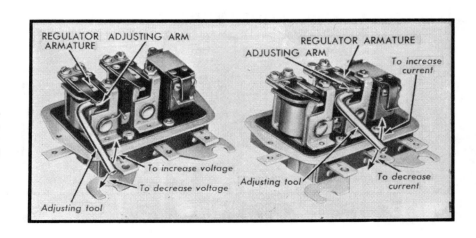

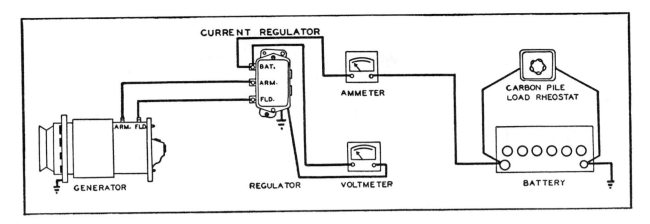

Fig. 49 *Current regulator test on Prestolite, Ford and Bosch regulators*

2. Disconnect the variable resistance from the regulator field terminal and reconnect the field lead. Then insert a ¼ ohm fixed resistor in series with the ammeter.

3. Increase engine speed to obtain about a 7 ampere charge and operate for about 15 minutes with regulator cover in place.

4. Stop and restart the engine to cycle the generator; then note voltage reading. If not within specifications, adjust as follows:

5. On Prestolite units, adjust by bending lower spring hanger, Fig. 46. Bending the hanger down increases spring tension and raises voltage and vice versa.

6. On Ford and Bosch units, increase spring tension by bending adjusting arm upward to raise voltage and vice versa, Fig. 47.

7. Be sure to stop and restart engine and replace cover after each setting before taking voltage readings.

Ford & Bosch Double Contact Voltage Regulator

If proper voltage cannot be obtained by bending the spring hanger as directed for single con-

tact voltage regulators, further adjustments are required as follows:

1. Disconnect ground terminal from battery.

2. Remove regulator from vehicle and take off cover.

3. Turn upper contact screw in or out to obtain a .040″ gap between armature core and armature, Fig. 48.

4. Turn lower contact screw in or out to obtain a .037″ gap between armature core and armature. While making adjustments, hold contact points together by pressing downward on screwdriver.

Current Regulator

1. Meters are connected as for the voltage regulator test except that the fixed resistor is removed and a carbon pile rheostat is connected across the battery as shown in Fig. 49 to allow full output. An alternate method for allowing the current regulator to operate is to operate the starting motor for 15 to 20 seconds cranking the engine and turning on lights and accessories.

2. Operate engine at about 1800 rpm. Using the

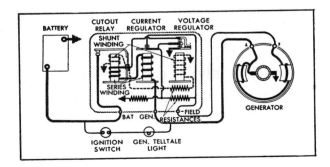

Fig. 50 *Delco-Remy charging circuit with single contact voltage regulator*

Fig. 51 *Delco-Remy charging circuit with double contact voltage regulator*

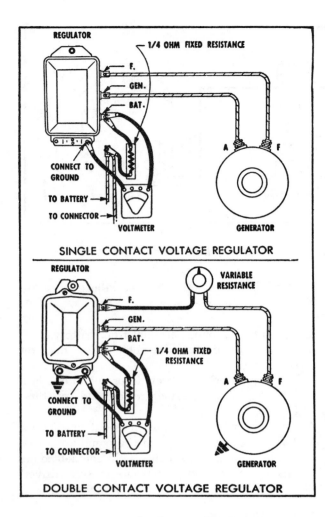

SINGLE CONTACT VOLTAGE REGULATOR

DOUBLE CONTACT VOLTAGE REGULATOR

Fig. 52   Checking Delco-Remy
voltage regulator setting

REGULATOR AMBIENT TEMPERATURE	VOLTAGE		
	LOW		HIGH
165° F	13.1	—	13.9
145° F	13.5	—	14.3
125° F	13.8	—	14.7
105° F	14.0	—	14.9
85° F	14.2	—	15.2
65° F	14.4	—	15.4
45° F	14.5	—	15.6
NORMAL SPECIFICATION RANGE			
■ INDICATES PUBLISHED SPECIFICATIONS			

Fig. 53   Delco-Remy single contact regulator
temperature and voltage factors

Fig. 54   Delco-Remy regulator
adjusting screws

carbon pile, increase the load to lower the regulated system voltage approximately one volt and allow the current regulator to operate.

3. When a steady reading is reached which cannot be increased by a slight increase in engine speed, this will be the current regulator setting.

4. After operating at full output for 15 minutes, the current regulator should be within specified limits at operating temperature.

5. Adjustment is made in the same manner as for the voltage regulator.

DELCO-REMY ELECTRICAL TESTS

Figs. 50 and 51 show the generator circuit using single contact voltage regulator and double contact voltage regulator, respectively. In making the following tests, the regulator must be at operating temperature before taking meter readings. Operating temperature may be assumed to exist after not less than 15 minutes continuous operation with a charging rate of about 10 amperes and regulator cover in place.

CAUTION: With charging circuits having the double contact voltage regulator, Fig. 51, it is extremely important never to ground the field terminal of the generator or regulator when these units are connected or operating together. To do so will burn up the upper set of voltage regulator contacts.

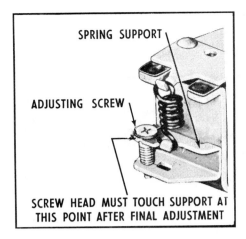

Fig. 55 *Proper contact between regulator spring support and adjusting screw on Delco-Remy regulators*

REGULATOR AMBIENT TEMPERATURE	LOW	VOLTAGE	HIGH
205° F	13.3	———	14.1
185° F	13.4	———	14.2
165° F	13.5	———	14.4
145° F	13.7	———	14.5
125° F	13.8	———	14.6
105° F	14.0	———	14.8
85° F	14.1	———	14.9
NORMAL SPECIFICATION RANGE			

Fig. 56 *Delco-Remy double contact regulator temperature and voltage factors*

For best results, the electrical tests must be made in the order given below.

Single Contact Voltage Regulator

1. Make meter connections as shown in Fig. 52 (upper view), and operate the engine at about 1600 rpm for 15 minutes with the ¼ ohm resistor in the circuit and cover in place to bring the regulator to operating temperature.

2. Cycle the generator by stopping the engine, then restarting and bringing generator speed back to 1600 engine rpm.

3. Note voltmeter reading and ambient temperature (temperature of air surrounding regulator ¼″ from cover). The voltage reading represents setting at ambient temperature. As shown in Fig. 53, setting will be different at other ambient temperatures. *If method of measuring ambient temperature is not available it may be assumed to be 40 degrees above room temperature.*

4. To adjust voltage setting, remove regulator cover and turn voltage regulator adjusting screw, Fig. 54. Turn screw clockwise to increase spring tension and raise voltage, and vice versa. *Final adjustment should always be made by turning screw clockwise to assure contact between screw head and spring support, Fig. 55.*

5. After each adjustment and before taking meter reading, replace cover and recycle generator, as in Step 2.

Double Contact Voltage Regulator

1. Make meter connections as shown in Fig. 52 (lower view).

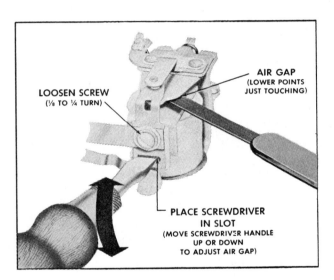

Fig. 57 *Adjusting lower set of contacts on Delco-Remy double contact regulator*

2. With variable resistance turned out (minimum resistance), operate generator at a speed so that the voltage regulator is operating on the upper set of contacts. Continue to operate for 15 minutes to establish operating temperature. Regulator cover must be in place.

3. Cycle generator by turning variable resistance to the "open" position momentarily, then slowly decrease (turn out all) resistance. Regulator should again be operating on the upper set of contacts.

4. Note voltmeter reading and ambient temperature, and see Fig. 56 for correction factors.

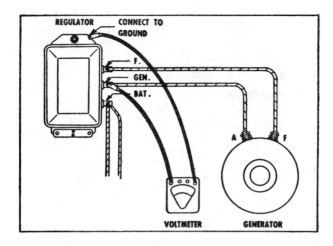

Fig. 58   Delco-Remy circuit breaker test

5. Increase (turn in) resistance slowly until voltage regulator begins to operate on the lower set of contacts. The lower set should operate at a lower voltage than the upper set of contacts.

6. To adjust voltage setting on upper set of contacts, do so in the same manner as directed for single contact voltage regulators above.

7. For the lower set of contacts, the difference in voltage between the upper set and lower set is increased by *slightly* increasing the air gap between the armature and center of core and decreased by *slightly* decreasing the air gap, Fig. 57. This adjustment is made while the regulator is operating. If necessary to make this adjustment, recheck the voltage setting of both sets of contacts.

Circuit Breaker Closing Voltage

1. Make connections as shown in Fig. 58.
2. Check closing voltage by slowly increasing generator speed and noting voltage at which points close. Decrease generator speed and make sure points open.
3. If not as specified, adjust voltage by turning adjusting screw clockwise to increase voltage and vice versa (see Fig. 54).

Current Regulator

1. Before making connections shown in Fig. 59, disconnect battery ground lead. After completing the hook-up, reconnect battery lead.
2. Turn on all lights and accessories and connect an additional load across the battery, such as a carbon pile or bank of lights, so as to drop the system voltage to 12.5–13.0 volts.
3. Operate generator at 1600 engine rpm for at

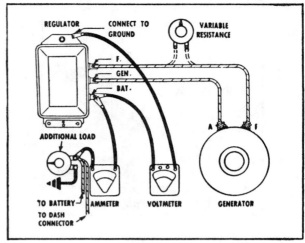

Fig. 59   Delco-Remy current regulator test

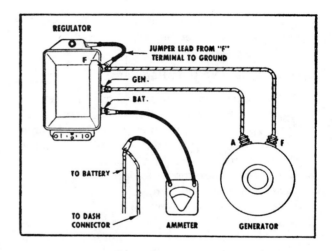

Fig. 60   Checking for oxidized regulator contact points. On double contact regulators do not use jumper wire as shown; instead remove lead from "F" terminal and ground this lead

least 15 minutes with cover in place to establish operating temperature.
4. Cycle the generator as previously directed and note current regulator setting.
5. Adjustment is made in the same manner as outlined for single contact voltage regulators (see Fig. 54).

Check for Oxidized Contacts

1. Oxidized contacts may be the cause of low generator output or a discharged battery. To check for this condition, connect an ammeter into the circuit as shown in Fig. 60 and turn on headlights.

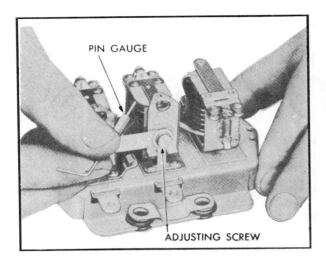

Fig. 61   *Checking air gaps on Prestolite regulators*

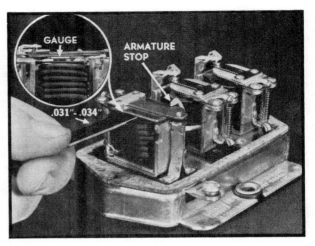

Fig. 62   *Checking Prestolite circuit breaker armature air gap with special gauge*

2. Operate generator at a speed that will produce a charge rate of 5 amperes.
3. Ground the "F" terminal of regulator as shown. *Caution: On double contact regulators, remove lead from "F" terminal and ground this lead.*
4. If generator output increases more than 2 amperes, oxidized contact points are indicated. Remove regulator from vehicle and clean points.

PRESTOLITE MECHANICAL ADJUSTMENTS

Current & Voltage Regulator Air Gaps

1. Use a pin type gauge which measures .048–.052". Insert gauge on point side of air gap and next to armature stop pin with contacts just separating. Fig. 61.
2. If an adjustment is necessary, loosen bracket screws and raise or lower contact point brackets until the foregoing clearance is obtained. Tighten screws securely after adjustment.
3. With armature held down so that stop rivet rests on magnet core, the point gap should be .015" when checked with a feeler gauge.

Circuit Breaker Air Gap

1. As shown in Fig. 62, use a flat gauge which measures from .031–.034". Insert gauge between armature and magnet core. Place gauge as near to hinge as possible.
2. To adjust, bend armature stop, Fig. 63, so that space between core and armature is within the foregoing limits. The stop must not interfere with armature movement.
3. Adjust the contact gap to .015" by expanding or contracting the stationary contact bridge,

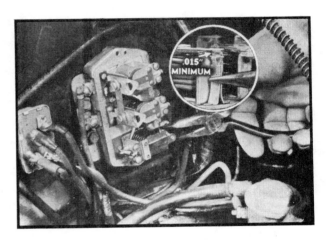

Fig. 63   *The minimum gap between the circuit breaker contacts on Auto-Lite regulators should be .015" and is adjusted by bending the bridge attached to the lower contact*

Fig. 63. When making this adjustment, keep the contact points in alignment.

DELCO-REMY MECHANICAL ADJUSTMENTS

Circuit Breaker Air Gap & Point Opening

1. Place fingers on armature directly above core and move armature down until points *just* close. Then measure air gap between armature and center core, Fig. 64. Air gap should be .020".
2. Check to see that both points close simultaneously. If not, bend spring finger so that they do.
3. To adjust air gap, loosen two screws at back of circuit breaker and raise or lower armature as

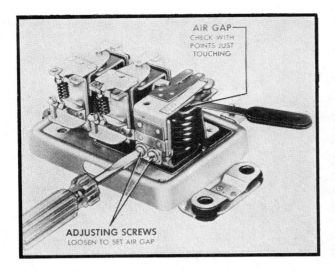

Fig. 64 *Checking and adjusting Delco-Remy circuit breaker armature air gap*

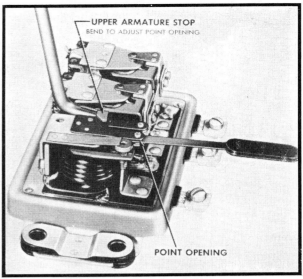

Fig. 65 *Checking and adjusting Delco-Remy circuit breaker point gap. If necessary, bend upper armature stop with tool shown*

required. Tighten screws securely after adjustment.

4. Check point opening with feeler gauge as shown in Fig. 65 and adjust to .020″ by bending upper armature stop.

5. After both adjustments have been made, recheck closing voltage and make any necessary adjustments.

Voltage Regulator Air Gap on Single Contact Units

1. Referring to Fig. 66, push armature down to core and release it until contact points *just* touch. Then measure air gap with pin gauge between armature and center of core. Air gap should be .075″.

2. On late units, adjust gap by turning nylon nut on top of regulator as shown. On earlier units, adjust as shown in Fig. 67.

3. After making adjustment, recheck voltage setting and make necessary adjustments.

Voltage Regulator Air Gap & Point Opening on Double Contact Units

Point Opening

1. With lower contacts touching measure point opening between upper set of contacts. The opening should be .016″.

2. On early units, adjust as shown in Fig. 68. On late units, adjust as shown in Fig. 69.

Air Gap

1. On late units, Fig. 70, with lower contacts touching, measure air gap and adjust as shown.

Fig. 66 *Adjusting air gap on late type Delco-Remy single contact regulator*

2. On early units, first make sure adjusting screw on top of armature is turned all the way in a clockwise direction. Then check and adjust the gap as shown in Fig. 71.

Current Regulator Air Gap

Check and adjust current regulator air gap in exactly the same manner as the single contact

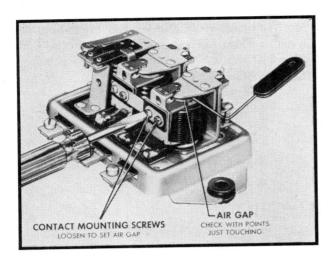

Fig. 67 *Checking and adjusting Delco-Remy voltage regulator armature air gap. Check current regulator in exactly the same way*

Fig. 68 *Adjusting Delco-Remy double contact voltage regulator point opening*

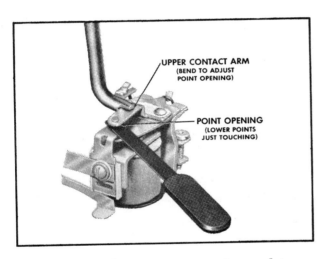

Fig. 69 *Adjusting point opening on late type Delco-Remy double contact regulator*

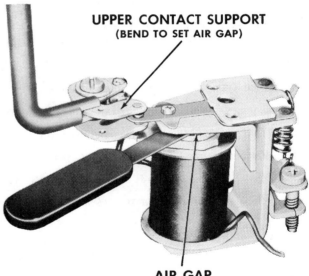

Fig. 71 *Checking and adjusting Delco-Remy double contact voltage regulator armature air gap*

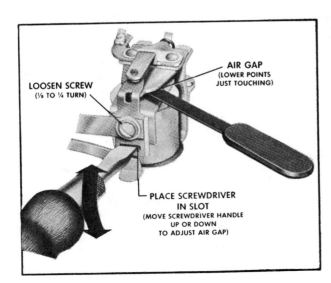

Fig. 70 *Adjusting air gap on late type Delco-Remy double contact voltage regulator*

voltage regulator. Air gap should be .075". After making the adjustment, recheck current setting and adjust as required.

FORD & BOSCH MECHANICAL ADJUSTMENTS

No mechanical adjustments are provided on these regulators as they are of riveted construction. However, on double contact regulators, the upper and lower voltage regulator contact sets are provided with an adjustment. See text in connection with Fig. 48.

Review Questions

GENERATORS

		Page
1.	On what part of the engine is the generator usually mounted?	642
2.	How many field poles are there usually in a D.C. auto generator?	642
3.	What are the principal parts of a D.C. auto generator?	642
4.	How is the generator driven?	642
5.	What keeps generator output at the desired level?	643
6.	What does each unit do in a three unit voltage regulator?	643
7.	Why is a ball bearing always used on the drive end of a generator while a bushing serves the purpose in most cases on the commutator end?	644
8.	What cools a generator?	644
9.	What type of cooling fan is most commonly used on generators?	644
10.	In what direction does air circulate through the generator?	644
11.	What is the name of the particular type of armature core used on auto generators?	646
12.	What insulates commutator segments from each other?	646
13.	Why is it necessary to tilt the brushes with respect to the center line of the field poles?	646
14.	When the commutator rotates away from the brush, it is then called what type of brush arrangement?	646
15.	When the commutator rotates towards the brush, it is then called what type of brush arrangement?	646
16.	What is the first step in checking for a "no charge" condition?	646
17.	What precaution should be observed while testing generator on systems using double contact voltage regulator?	646
18.	To which generator terminal should a radio suppressor condenser be attached?	646
19.	Why should emery cloth never be used to clean the commutator?	649
20.	How can it be established by inspection that there is a defective coil in an armature?	648
21.	What must never be done while cleaning generator components?	647
22.	How deep should mica be undercut between commutator segments?	649
23.	What device is most effective in testing for shorted coils in an armature?	649

24. What simple devices are used to test for defective windings in a generator armature? 650

25. What is the difference between a "grounded" and an "open" field winding? 651

26. When does it become necessary to polarize a generator? 653

27. How is the field grounded in a system with external field ground? 652

28. How is the field grounded in a system with internal field ground? 653

29. What is the procedure for polarizing a generator with an externally grounded field? 653

30. What is the procedure for polarizing a generator with an internally grounded field? 653

Review Questions

REGULATORS

Page

1. What is the function of a generator regulator? 654

2. What is the function of each of the three units in a modern generator regulator? 654

3. Voltage and current regulators, regardless of type, are said to be what kind of switches? 654

4. What material is used in the core of any voltage regulator? 654

5. What part of a voltage regulator is known as the armature? 654

6. What are the names of the two windings on the circuit breaker of a voltage regulator? 655

7. Which winding on the circuit breaker is made up of the heavier wire? 655

8. Which winding on the circuit breaker closes the circuit to the generator? 655

9. Why does field strength in the generator diminish when the voltage regulator points open? 657

10. Is voltage regulator operation influenced greatly by temperature? 657

11. What particular feature of the current regulator distinguishes it from the voltage regulator? 657

12. What particular unit in the charging system limits the amount of current that can flow through the generator fields? 657

13. What component of the regulator makes its operation sensitive to temperature? 657

14. Do the current and voltage regulating units in a regulator ever operate at the same time? 657

15. Besides reducing field current, what other function do the resistances in a regulator fulfill? 657

16. On the double contact voltage regulator, which contacts grounds the field through a resistor? 657

17. Describe a simple method of testing generator output on a system with externally grounded field. 658

18. How does the output test on systems with externally grounded field using double contact regulator differ from single contact regulator? 658

19. What measurement is used to determine whether or not the circuit breaker is operating correctly? 658

20. What is ambient temperature? 663

21. What is meant by the term "Recycle generator"? 663

669

22. As the ambient temperature increases, does voltage for a given regulator setting increase or decrease?   663

23. On the Delco-Remy double contact regulator, which set of contacts provides higher voltage?   664

24. At what point is circuit breaker air gap adjusted on a Prestolite regulator?   665

25. At what point is circuit breaker contact gap adjusted on a Prestolite regulator?   665

Alternator Systems

Review Questions for This Chapter on Page 729

INDEX

INTRODUCTION

Alternator construction   671
How system functions   672
Single phase alternating
   voltage   674
Three phase alternating
   voltage   675

Three phase connections   676
Functions of a diode   676
Changing A.C. to D.C.   676
Service precautions   676

ALTERNATOR SERVICE

Chrysler system   677
Delco-Remy system   690

Ford system   700
Leece-Neville System   712
Motorola system   717
Prestolite system   723

Introduction

Prior to their use on passenger cars, alternators, otherwise known as A.C. generators, were used mainly in buses, refrigerated trucks, trucks used in stop-and-go city delivery service, and other vehicles such as police cars, ambulances and taxicabs equipped with telephones. In all such vehicles, the conventional D.C. generator is unable to supply sufficient current for the heavy electrical loads required.

To understand how the A.C. system overcomes these limitations, it may be helpful to review briefly the operation and construction of the conventional D.C. system.

It may come as a surprise to some that the D.C. generator actually produces alternating current. As the armature rotates through the magnetic field produced by the field coils, alternating current is induced in the armature coil, then flows through the commutator and brushes, where it is converted to direct current. With this design, cooling can be a serious problem under high-output, low-speed conditions. Low speed output can be in-

creased and satisfactory cooling secured by changing the drive ratio (using smaller pulley) to turn the generator faster at low engine rpm. This scheme works fine if engine speed can be limited, but the generator will turn at excessive rpm if the vehicle is operated at cruising speed on the highway. The result is likely to be thrown armature windings and rapid wear of commutator and brushes.

The electrical system of a motor vehicle cannot use alternating current because the battery, ignition system and other electrical devices are designed for direct current use. Although the A.C. generator produces alternating current, direct current is provided by means of the rectifier as we shall soon see.

Alternators under discussion here are of the modern type with built-in silicon diode rectifiers. Under normal operating conditions, these alternators have a rating of 30 to 60 amperes, depending upon the requirements of the vehicle, and usually deliver 5 to 10 amperes at curb idle speed.

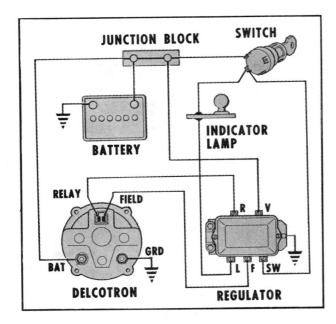

Fig. 1 Wiring circuit of a typical alternator charging system

tem, regulates the output of the generator-rectifier system.

Although the operating principles of all alternators are basically the same, we have adapted the following material furnished by Delco-Remy in describing its alternator known as the "Delcotron."

The charging circuit consists of the units shown in Fig. 1. These units work together as a team to supply electrical energy to the accessories in the vehicle electrical system and to charge the battery even when the engine is operating at idle or at slow speeds. This is an important factor in modern vehicles where the electrical requirements have increased due to added electrical accessories, and where considerable time is spent at engine idle and at slow speed driving due to traffic conditions. As a result, the battery is maintained in a higher state of charge which helps to insure easier starting and improved performance from the electrical system. Excessive battery discharge is also minimized which tends to increase battery life.

Construction

An alternator, Figs. 2 and 3, consists primarily of two end frame assemblies, a rotor assembly, and a stator assembly. The rotor assembly is usually supported in the drive end frame by a ball bearing and in the slip ring end frame by a roller bearing. These bearings are usually pre-lubricated, thereby eliminating the need for periodic lubrication.

The alternator is composed of the same functional parts as a D.C. generator; however, they operate differently. The field is called a rotor and is the turning portion of the unit. A generating part, called a stator, is the stationary member (comparable to the armature in a D.C. generator). The rectifier, which changes alternating current to direct current, can be compared to the commutator and brushes in a D.C. generator. The regulator, similar to those used in a D.C. sys-

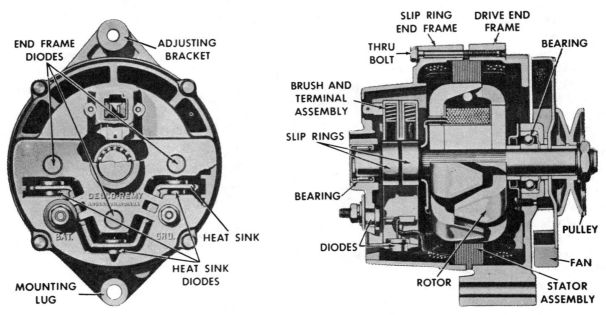

Fig. 2 Delco-Remy alternator known as the "Delcotron"

671

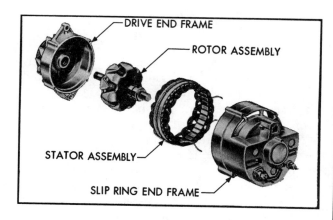

STATOR ASSEMBLY

DRIVE END FRAME

ROTOR ASSEMBLY

SLIP RING END FRAME

Fig. 3   Delcotron disassembled

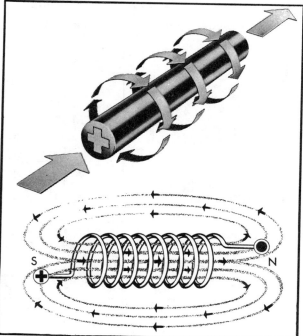

Fig. 4   Electromagnetic principles

The stator assembly is mounted between the two end frames, and consists of loops of wire wound into the slots of the laminated stator frame.

The rotor assembly contains a doughnut shaped field coil wound onto an iron spool. The coil and spool are mounted between two iron segments with several interlacing fingers which are called "poles". These parts are held together by a press fit on the shaft.

Two slip rings, upon which the brushes ride, are mounted on one end of the rotor shaft and are attached to the leads from the field coil.

Six electronic check valves called *diodes* are located in the end frame assembly nearest the slip rings. Three of these diodes are negative and are mounted directly to the end frame. Three positive diodes are mounted into a strip called a "heat sink," which is insulated from the end frame. These diodes change the alternating or A.C. voltages developed in the stator windings to a D.C. voltage which appears at the output (battery) terminal on the alternator. Note: The alternator just described is for use on a vehicle having a negative ground system. When used on a positive ground system, the negative diodes would be mounted in the heat sink.

How System Functions

To understand how an alternator works, let's review some electrical fundamentals. It is a fact that electricity and magnetism are closely related because when an electric current passes through a conductor, such as a copper wire, a magnetic field is created around the wire, Fig. 4. The magnetic field is illustrated as concentric circles around the straight wire. As the current in the wire increases, the strength or intensity of the magnetic field increases. However, the strength of the magnetic

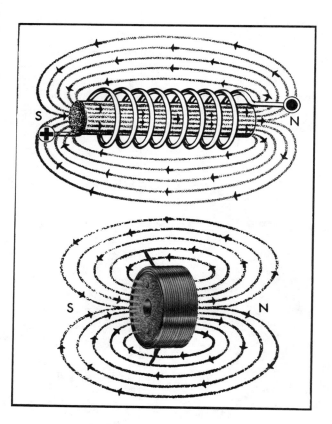

Fig. 5   Simple electromagnet shown in top view. Bottom view is an electromagnet in the form of an alternator rotor

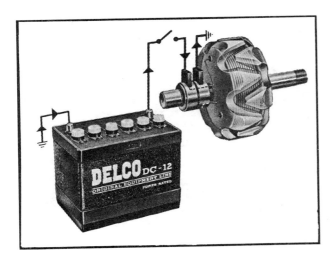

Fig. 6 *Battery connected to a rotor through brushes and slip rings*

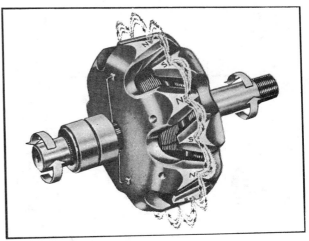

Fig. 7 *Magnetism created by field current causes poles on rotor to be alternately North and South poles*

field around the straight conductor is too weak to be of value for most applications. Therefore, a means for obtaining a stronger magnetic field must be found.

This is accomplished by winding the straight conductor in a series of loops to form a coil. When this is done, and the same current that was passed through the straight conductor is passed through the coil, a stronger magnetic field is produced. The magnetic field then takes the form shown in Fig. 4. A North ("N") pole is produced at one end of the coil, and a South ("S") pole at the other end. The magnetic lines leave the North pole and then re-enter the coil at the South pole, as shown by the arrows.

The strength of the magnetic field around the coil can be increased even further by placing an iron core inside the coil and passing current through the conductor. Since iron offers a much easier path for magnetism to pass through the air, the magnetic lines become more concentrated and consequently the magnetic field becomes stronger. An assembly of this type is called an electromagnet as shown in Fig. 5.

The same principle is used in the design of the rotor assembly of an alternator. As shown in the lower view, Fig. 5, a coil of wire is wound around an iron spool. When current is passed through the coil, an electromagnet is formed with its magnetic field surrounding the assembly. The strength of the magnetic field of an electromagnet can be altered by changing either the amount of current passed through the winding or by changing the number of turns in the coil. In an alternator, the coil in the rotor assembly is called the "field" coil and the current that passes through it is called the field current.

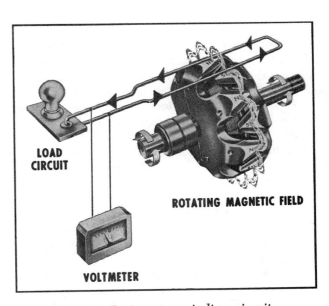

Fig. 8 *Basic rotor winding circuit, using only one loop of wire for illustration purposes*

In the rotor assembly, the coil and iron spool are placed over the rotor shaft, and two iron end pieces with interlacing fingers are also placed over the shaft. Two slip rings, which are connected to the coil, complete the rotor assembly, Fig. 6. Two brushes ride on the slip rings, and are connected through a switch to the battery.

When the ignition switch is closed, current from the battery passes through one brush, through the slip ring upon which the brush rides, and then through the field coil. After leaving the field coil, current flow continues through the other slip ring

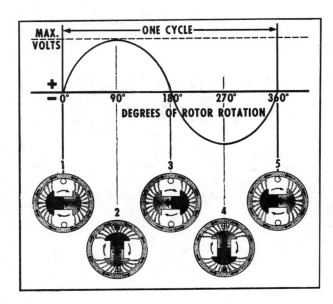

Fig. 9   *Single phase alternating voltage*

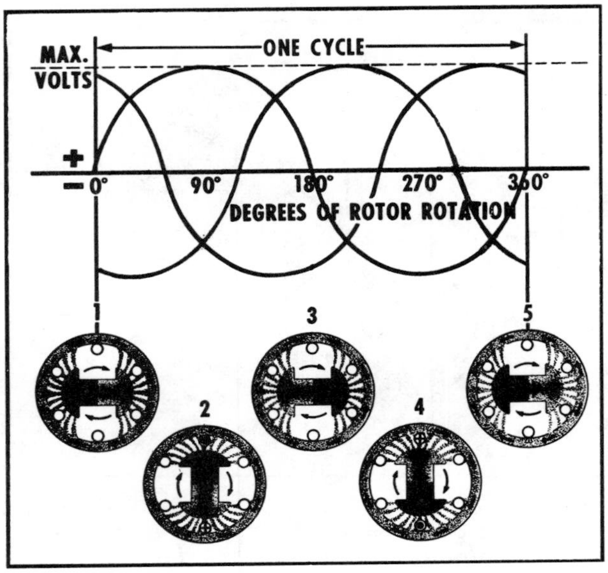

Fig. 10   *Three phase alternating voltage*

and brush before returning to the battery through the ground return path. This flow of electrical energy through the field winding is called *field current* and creates a magnetic field.

The magnetism created by the field current causes the poles on the rotor to be alternately North and South poles, Fig. 7. It will be shown later that this magnetic field is used to produce alternating voltages in the stator windings.

For purposes of illustration, the most basic stator windings can be represented by a single loop of wire placed over a rotor, Fig. 8. Connecting the ends of the loop to a load such as a light bulb completes the circuit. As the rotor turns inside the loop, the magnetic field from the North and South poles on the rotor cuts across each wire, causing a voltage to be induced in the loop. Since the wire is influenced alternately by a North and then a South pole, the voltage produced is called an alternating or A.C. voltage.

When a load, such as a light bulb, is connected to the ends of the loop, a current will flow first in one direction and then in the other, and the bulb will light each time. This is called alternating current or A.C. current. If a meter is placed in the circuit it will show that current flows first in one direction and then in the other by the fluctuation of the needle.

Single Phase Alternating Voltage

To illustrate further how an A.C. voltage is produced, consider a simple alternator made up of a two-pole rotor and a stator made of a single loop

of wire. Only the ends of the loop of wire are shown, Fig. 9.

Different positions of the rotor as it turns are shown in the circles below the horizontal line. The height of the curve above and below the horizontal line shows the magnitude of the voltage which is generated in the loop of wire as the magnetic lines cut across each side of the loop as the rotor turns. The entire curve shows the voltage output generated or the electrical pressure which can be measured across the ends of the wire, just as line voltage can be measured across the terminal posts of a battery.

With the rotor in the first position the voltage is zero. No voltage is generated in the loop of wire because there are no magnetic lines of force cutting across the wire. As the rotor turns and approaches position two, the rather weak magnetic field at the tip of the rotor starts to cut across the conductor, and the voltage increases. As the rotor turns, the voltage reaches its maximum value as shown above the horizontal line, Fig. 9, when the rotor reaches position two. The maximum voltage occurs when the rotor is directly under each wire in the loop. It is in this position that the loop of wire is being cut by the heaviest concentration of magnetic lines of force.

It should be noted in particular that the height of the voltage curve changes because the concentration of magnetic lines of force cutting across the loop of wire varies. This occurs because the magnetic field is rather weak at the tips of the poles, and the heaviest at the center of the poles.

As the rotor turns from position two to position

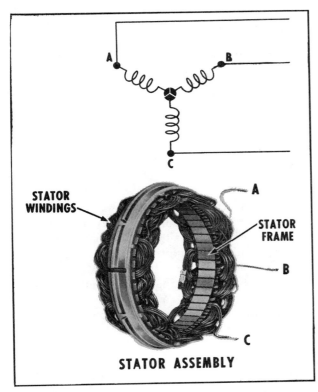

STATOR WINDINGS

STATOR FRAME

STATOR ASSEMBLY

Fig. 11   Three phase ("Y") connections

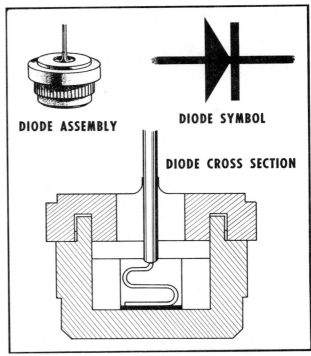

DIODE ASSEMBLY

DIODE SYMBOL

DIODE CROSS SECTION

Fig. 12   A diode changes alternating current to direct current

three, the voltage decreases until at position three it again becomes zero. It should be noted that between positions one through position two to position three, the *South* pole is on top, and the voltage curve *above* the horizontal line is called positive voltage. This means that the voltage will cause current to come out of the top part of the loop, and reenter the lower part when a load is connected to the loop of wire.

As the rotor turns from position three, through position four to position five, the *North* pole is on top, and the voltage curve is *below* the horizontal line. The voltage curve is negative, and current will *leave* the *lower* part of the loop and *re-enter* the *top* when the load is connected. Thus, as the top and bottom parts of the loop of wire are influenced alternately by North and South poles, the current flow through the loop of wire flows first in one direction and then in the other.

Three Phase Alternating Voltage

As shown in Fig. 10, when three separate loops of wire are spaced evenly around the rotor, the voltage developed in each loop can be represented by the three voltage curves shown. Note that the peak voltage for each loop of wire occurs at equal intervals as the rotor turns. This represents the most basic type of a three-phase stator winding.

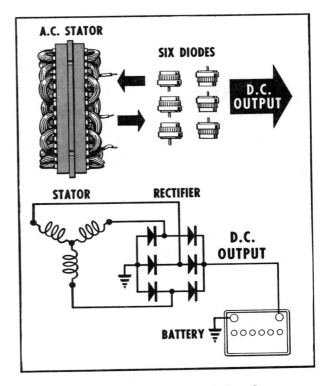

A.C. STATOR

SIX DIODES

D.C. OUTPUT

STATOR   **RECTIFIER**

D.C. OUTPUT

BATTERY

Fig. 13   Changing A.C. to D.C. voltage

Three Phase Connections

In the actual stator of an alternator, there are three separate windings, each consisting of many loops of wire. A schematic drawing of the way the three stator windings are usually connected at one end to form a three-phase unit is shown in Fig. 11. This is known as a "Y" type connection, although other methods are used. The other end of each winding is connected to a pair of diodes—one positive and one negative. The winding is then assembled onto a laminated iron frame to complete the stator assembly.

Function of a Diode

It has been shown that A.C. or alternating voltages are produced in the stator windings, and that alternating current flows through a load when connected to the stator windings. Also, we have seen that alternating current flows first in one direction and then in the other. Since the battery and all electrical accessories operate on direct or D.C. current, which flows in one direction only, it is necessary to change alternating current to direct current. This is the function of the diode.

The chemical composition of a diode is such that it will allow current to flow through itself in only one direction. A cross-sectional view of a typical diode is illustrated in Fig. 12, along with the diode symbol used in wiring diagrams. This symbol indicates by the arrow that current will flow only in the direction of the arrow. The diode case is sealed to prevent entry of moisture. Connections are made to the diode at the lead and the case.

Changing A.C. to D.C.

There are six diodes mounted in the slip ring end frame of the alternator. Three negative diodes are mounted in the end frame and three positive diodes are mounted in the heat sink which is insulated in the end frame. (This arrangement is used when the vehicle is negative grounded; for positive grounded systems the three negative diodes would be mounted in the heat sink.) These diodes act together to change A.C. voltages developed in the stator windings to a single D.C. voltage. Therefore, D.C. voltage appears at the output terminal of the alternator, consequently, the alternator supplies D.C. or direct current to charge the battery and to operate electrical accessories.

The method by which the diodes are connected to the stator are shown in Fig. 13. This type of circuit arrangement provides a smooth flow of direct current to the battery and other accessories connected to the alternator. Also, the blocking or one-way action of the diodes prevents battery discharge through the alternator, and thus eliminates the need for a cutout relay or circuit breaker.

SECOND GENERATION ALTERNATORS

Starting several years ago a new type alternator began appearing on American cars. Called "second generation" alternators, these new units take advantage of recent advances in electronic technology to provide voltage regulators that are integral with the alternator. Presently used on some General Motors Corp., Chrysler Corp. and Ford Motor Company products, the integral, solid state regulators are not serviceable or adjustable and are replaced as a unit if defective. Using fundamentally the same alternators as previously used with external voltage regulators—therefore, not greatly changing service procedures on the alternator itself—the new systems simplify overall maintenance procedures.

ALTERNATOR SERVICE

Precautions

1. Be certain that the battery polarity of the system is known so that the battery is connected to the proper ground. *Reversed battery polarity will damage rectifiers and regulator.*
2. If booster batteries are used for starting they must be connected to the vehicle battery properly to prevent damage to rectifiers and regulators. Negative cable from booster battery to negative terminal on vehicle battery and positive booster cable to positive terminal.
3. When a fast charger is used to charge a vehicle battery the vehicle battery cables should be disconnected *unless the fast charger is equipped with a special Alternator Protector,* in which case the vehicle battery cables need not be disconnected. Also the fast charger should never be used to start a vehicle as damage to rectifiers will result.
4. Unless the system includes a load relay or field relay, grounding the alternator output terminal will damage the alternator and/or circuits. This is true even when the system is not in operation since no circuit breaker is used and the battery is applied to the alternator output ter-

minal at all times. The field or load relay acts as a circuit breaker in that it is controlled by the ignition switch.

5. When adjusting the voltage regulator, do not short the adjusting tool to the regulator base as the regulator may be damaged. The tool should be insulated by taping or by installing a plastic sleeve.

6. Before making any "on vehicle" tests of the alternator or regulator, the battery should be checked and the circuit inspected for faulty wiring or insulation, loose or corroded connections and poor ground circuits.

7. Check alternator belt tension to be sure the belt is tight enough to prevent slipping under load.

8. The ignition switch should be off and the battery ground cable disconnected before making any test connections to prevent damage to the system.

9. The vehicle battery must be fully charged or a fully charged battery may be installed for test purposes.

Chrysler System

The Chrysler Corp. alternator, Figs. C1 and C2, is an A.C. generator with six built-in diodes that convert alternating current into direct current. Direct current is available at the "output" "BAT" terminal. A voltage regulator is used in the field circuit to limit the output voltage. The main components of the alternator are the rotor, the stator, the diode rectifiers, the two end shields and the drive pulley.

The only function of the voltage regulator is to limit the output voltage. The regulator accomplishes this by controlling the flow of current in the rotor field coil, and in effect controls the strength of the rotor magnetic field.

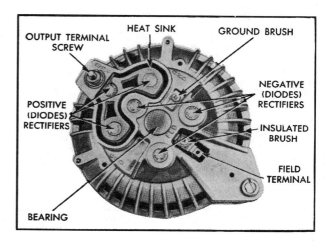

Fig. C1   End view of Chrysler alternator

TESTING SYSTEM ON VEHICLE

Field Circuit Resistance Test

1. Referring to Fig. C3, disconnect ignition wire at coil side of ballast resistor and connect a test ammeter and voltmeter in the circuit as shown. All lights and accessories should be turned off.

2. Turn ignition switch on and turn voltmeter selector switch to the low voltage scale and read the meter. The voltage should not exceed .55 volt. A reading in excess of .55 volt indicates high resistance in field circuit between battery and voltage regulator field terminal.

3. If high resistance is indicated, move negative voltmeter lead to each connection along the circuit to the battery. A sudden drop in voltage indicates a loose or corroded connection between that point and the last point tested. To test the terminals for tightness, attempt to move the terminal while observing the voltmeter. Any movement of the meter pointer indicates looseness. *Note: Excessive resistance in the regulator wiring circuit will cause fluctuation in the ammeter.*

4. Turn ignition switch off, disconnect test instrument and reconnect ignition primary wire at the coil side of the ballast resistor.

Charging Circuit Resistance Test

With battery in good condition and fully charged, first disconnect the battery ground cable to avoid accidental shorting of the charging or field circuit when making the test connections shown in Fig. C4.

1. With the test instruments connected as shown and with battery ground cable re-connected, start and operate engine at a speed to obtain 10 amperes flowing in the circuit.

2. The voltmeter should not exceed .3 volt. If a higher voltage drop is indicated, inspect, clean and tighten all connections in the charging circuit. A voltage drop test may be performed at

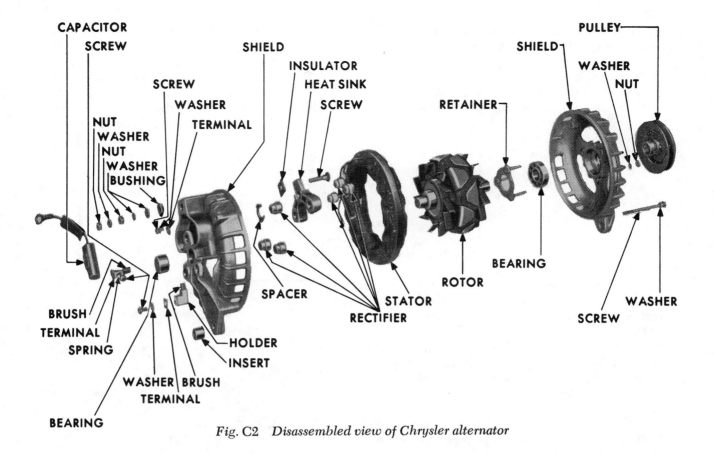

CAPACITOR
SCREW
NUT
WASHER
NUT
WASHER
BUSHING
SCREW
WASHER
TERMINAL
SHIELD
INSULATOR
HEAT SINK
SCREW
RETAINER
PULLEY
SHIELD
WASHER
NUT

BRUSH
TERMINAL
SPRING
BEARING
WASHER
TERMINAL
BRUSH
TERMINAL
HOLDER
INSERT
SPACER
STATOR
RECTIFIER
ROTOR
BEARING
SCREW
WASHER

Fig. C2   Disassembled view of Chrysler alternator

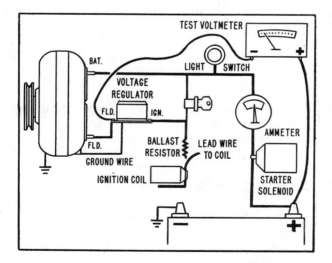
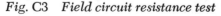
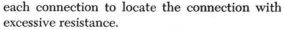

Fig. C3   Field circuit resistance test

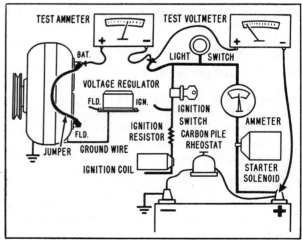

Fig. C4   Charging circuit resistance test

each connection to locate the connection with excessive resistance.

3. Turn ignition switch off. Disconnect ground cable at battery to avoid accidental shortening of the charging or field circuit when disconnecting the test instruments. Connect battery lead to alternator "BAT" terminal and tighten securely. Connect ignition lead to regulator

ignition terminal and re-connect ground cable at battery.

Current Output Test

1. With test instruments connected in circuit as shown in Fig. C5, connect an engine tachometer.

678

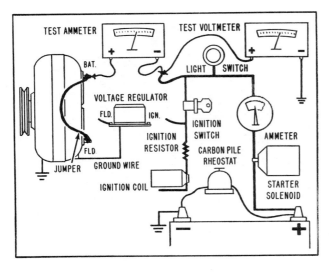

Fig. C5   *Current output test*

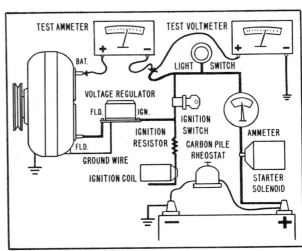

Fig. C6   *Voltage regulator test*

2. Start and operate engine at 1250 rpm.
3. Adjust carbon pile rheostat to obtain a reading of 15 volts on the test voltmeter.
4. Observe reading on test ammeter.
5. If the output is slightly less (5 to 7 amperes) than the rated output of the alternator, it may be an indication of an open-circuited diode or other internal alternator problem.
6. If the output is considerably lower than the rated output of the alternator, it may be an indication of a short-circuited diode or other internal alternator problem. In either case the alternator should be removed and tested. *Note: Turn off the carbon pile rheostat immediately after observing reading on test ammeter.*
7. If the alternator current output tested satisfactorily, turn off the ignition switch and remove the jumper lead from the alternator field terminal and output terminal.

Voltage Regulator Test

Upper Contact Test
1. With engine at normal operating temperature and test instruments connected as shown in Fig. C6, start and operate the engine at 1250 rpm. Adjust carbon pile to obtain a 15 ampere output as indicated on the test ammeter. *Note: No current reading on the ammeter would indicate either a low regulator setting or a blown fuse wire inside the voltage regulator between upper stationary contact and "IGN" terminal. Correct the cause and replace the fusible wire.*
2. Operate engine at 1250 rpm and a 15 ampere load for 15 minutes to make sure entire regulator system is stabilized.

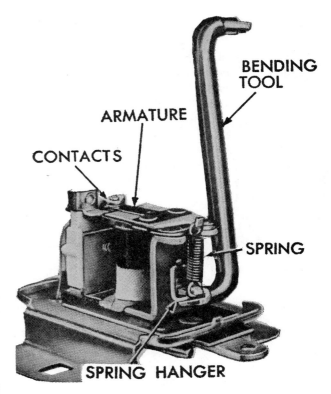

Fig. C7   *Adjusting voltage regulator spring tension*

3. Measure temperature at regulator by holding a reliable thermometer ¼ inch from regulator cover.
4. Read test ammeter. With fully charged battery and 15 amperes flowing in circuit, voltmeter readings should be within specifications.
5. If regulator operates within specifications, pro-

679

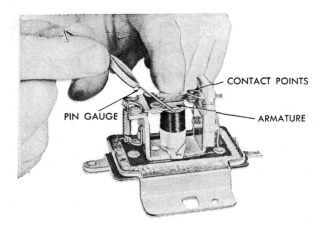

Fig. C8 *Measuring armature air gap*

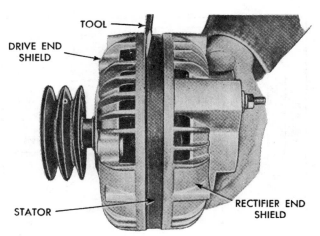

Fig. C9 *Separating drive end shield from stator*

ceed to the lower contact voltage test. If not, remove cover and adjust voltage setting as outlined under "Regulator Adjustments."

Lower Contact Voltage Test

1. Increase engine speed to 2200 rpm. Vary carbon pile to decrease current load to 7 amperes output as registered on test ammeter. The voltage should *increase* and amperage should *decrease*. *Note: There will be a slightly higher voltage at higher engine speeds above 2200 rpm. However, this increased voltage must not exceed the voltage specified by more than .7 volt at any temperature range.*

2. If the regulator setting is outside the specified limits, the regulator must be removed to remove the cover.

3. To adjust the voltage setting, bend the regulator lower spring hanger *down to increase voltage,* or *up to decrease* voltage setting, Fig. C7. The regulator must be installed, correctly connected, and retested after each adjustment of the lower spring hanger. *Note: If repeated readjustment is required, it is permissible to use a jumper wire to ground the regulator base to the fender splash shield for testing instead of reinstalling the regulator each time. However, it is important that the cover be reinstalled, the regulator connections correctly connected, and the regulator insulated to prevent grounding the regulator terminals or resistances. When testing, the regulator must be at the same attitude (or angle) as when installed on the vehicle.*

4. If the alternator and regulator tested satisfactorily, turn the ignition switch off. Disconnect battery ground cable, then the test instruments. Connect the leads to alternator and regulator. Finally re-connect battery ground cable.

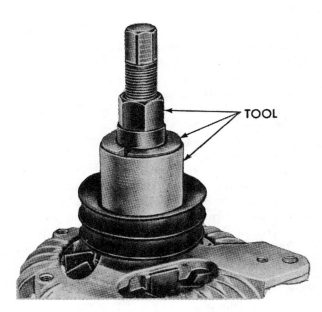

Fig. C10 *Removing pulley*

Regulator Adjustments

If the regulator cannot be adjusted for voltage control, or if the regulator performance is erratic or malfunctions, it may be necessary to adjust the air gap and contact point gap.

1. Remove regulator from vehicle and take off cover.

2. Insert a .048″ wire gauge between regulator armature and core, next to stop pin on spring hanger side, Fig. C8.

3. Press down on armature (not contact spring) until it contacts wire gauge. Upper contacts should just open. *Note: A battery and test light connected in series to the "IGN" and "FLD"*

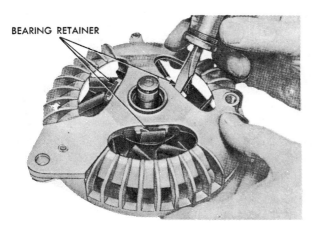

Fig. C11   *Disengaging bearing retainer from end shield*

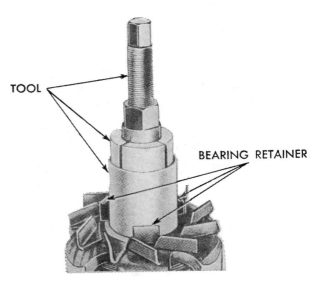

Fig. C12   *Removing bearing from rotor shaft*

terminals may be used to determine accurately the contact opening. When the contacts open, the test light will go dim.

4. Insert a .052″ wire gauge between armature and core, next to stop pin on spring hanger side.

5. Press down on armature until it contacts wire gauge. The contacts should remain closed and test lights should remain bright.

6. If adjustment is required, adjust air gap by loosening the screw and moving the stationary contact bracket. Make sure air gap is measured with attaching screw fully tightened. Re-measure the gap as directed above.

7. Remove wire gauge. Measure lower contact gap with feeler gauge, which should be .012 to .016″. Adjust lower contact gap by bending lower stationary contact bracket.

8. Install regulator cover and then the regulator.

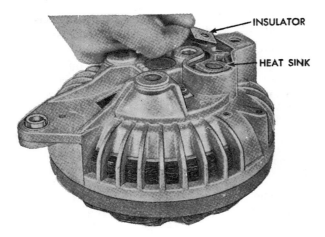

Fig. C13   *Removing heat sink insulator*

Finally, make electrical adjustments as outlined above.

BENCH TESTS

If the alternator performance does not meet current output specification limits, it will have to be removed and disassembled for further tests and servicing.

To remove the alternator, disconnect the battery ground cable and the leads at the alternator. Then unfasten and remove the alternator from the vehicle.

Field Coil Draw

1. Connect a test ammeter positive lead to the battery positive terminal of a fully charged battery.

2. Connect ammeter negative lead to the field terminal of the alternator.

3. Connect a jumper wire to negative terminal of battery, and ground it to the alternator end shield.

4. Slowly rotate alternator rotor by hand. Observe ammeter reading. The field coil draw should be 2.3 to 2.7 amperes at 12 volts. *Note: A low rotor coil draw is an indication of a high resistance in the field coil circuit (brushes, slip rings or rotor coil). A higher rotor coil draw indicates a possible shorted rotor coil or a grounded rotor.*

Testing Alternator Internal Field Circuit

1. To test the internal field circuit for a ground, remove the ground brush. Touch one test prod

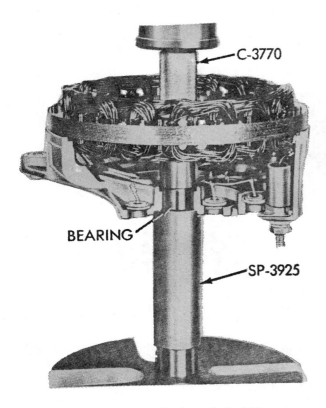

Fig. C14 *Removing diode end shield bearing*

"Y" CONNECTION (OPENED)

Fig. C15 *Separating the three stator leads*

end shields by removing the three through bolts.

3. Again test by placing one of the test prods to the slip ring and the remaining test prod to the end shield. If the lamp lights, the rotor is grounded and requires replacement. If the lamp does not light after removing the insulated brush and separating the end shields, the insulated brush is grounded.

4. Examine the plastic insulator and screw. *The screw is a special size and must not be substituted by another size.*

5. Install insulated brush holder, terminal, insulated washer, shake proof washer and screw. *If the parts were not assembled in this order or if the wrong screw was used this could be the cause of the ground condition.*

from a 110 volt test lamp to the alternator insulated brush terminal and the remaining test prod to the end shield. If the rotor or insulated brush is not grounded, the lamp will not light.

2. If the lamp lights, remove insulated brush (noting how parts are assembled) and separate

ALTERNATOR REPAIRS

Disassembly

To prevent possible damage to the brush assemblies, they should be removed before proceeding with the disassembly of the alternator. The insulated brush is mounted in a plastic holder that positions the brush vertically against one of the slip rings. A disassembled view of the alternator is shown in Fig. C2.

1. Remove retaining screw lockwasher, insulated washer and field terminal. Carefully lift plastic holder containing the spring and brush from the end housing.

2. The ground brush is positioned horizontally against the remaining slip ring and is retained in a holder that is integral with the end shield. Remove retaining screw and lift the clip, spring and brush from end shield. *Note: The*

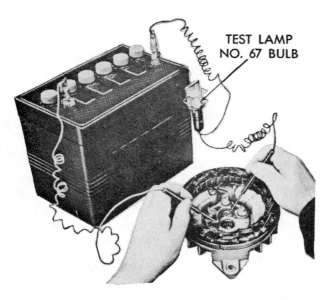

TEST LAMP NO. 67 BULB

Fig. C16 *Testing diodes with a test lamp*

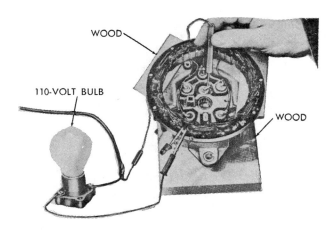

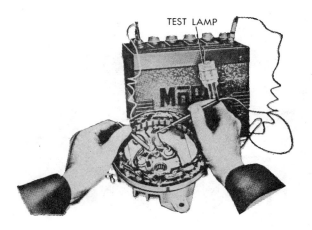

Fig. C17   Testing stator for grounds

Fig. C18   Testing stator windings
for continuity

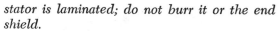

stator is laminated; do not burr it or the end shield.

3. Remove through bolts and pry between stator and drive end shield with a screwdriver. Carefully separate drive end shield, pulley and rotor from stator and diode rectifier shield, Fig. C9.

4. The pulley is an interference fit on the rotor shaft; therefore, a suitable puller must be used to remove it, Fig. C10.

5. Pry drive end bearing spring retainer from end shield with a screwdriver, Fig. C11.

6. Support end shield and tap rotor shaft with a plastic hammer to separate rotor from end shield.

7. The drive end ball bearing is an interference fit with the rotor shaft; therefore, a suitable puller must be used to remove it, Fig. C12.

8. Remove D.C. output terminal nuts and washers and remove terminal screw and inside capacitor (if equipped). *Note: The heat sink is also held in place by the terminal screw.*

9. Remove the insulator, Fig. C13.

10. The needle roller bearing in the rectifier end shield is a press fit. If it is necessary to remove the rectifier end frame needle bearing, protect the end shield by supporting the shield when pressing out the bearing as shown in Fig. C14.

Testing Diode Rectifiers

A special Rectifier Tester Tool C-3829 provides a quick, simple and accurate method to test the rectifiers without the necessity of disconnecting the soldered rectifier leads. This instrument is commercially available and full instructions for its use is provided. Lacking this tool, the rectifiers may be tested with a 12 volt battery and a test

lamp having a No. 67 bulb. The procedure is as follows:

1. Separate the three stator leads at the "Y" connection, Fig. C15. *Cut the stator connections as close to the connector as possible because they will have to be soldered together again. If they are cut too short it may be difficult to get them together again for soldering.*

2. Connect one side of test lamp to positive battery post and the other side of the test lamp to a test probe. Connect another test probe to the negative battery post, Fig. C16.

3. Contact the outer case of the rectifier with one

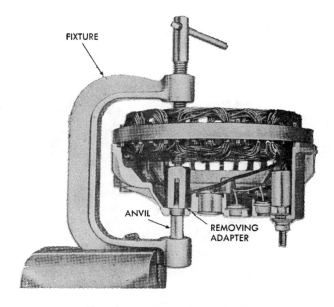

Fig. C19   Removing a diode

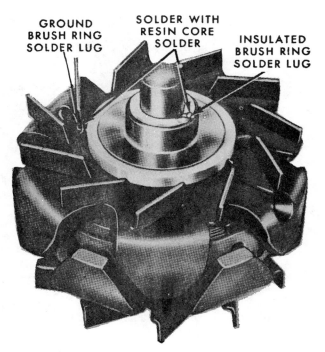

GROUND BRUSH RING SOLDER LUG • SOLDER WITH RESIN CORE SOLDER • INSULATED BRUSH RING SOLDER LUG

Fig. C20   Soldering points with slip ring installed

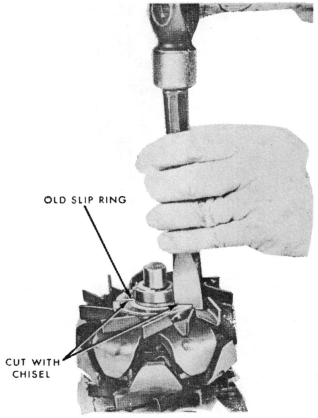

OLD SLIP RING

CUT WITH CHISEL

Fig. C21   Cutting old slip rings for removal

probe and the other probe to the wire in the center of the rectifier.

4. Reverse the probes, moving the probe from the rectifier outer case to the rectifier wire, and the probe from the wire to the case.

5. If the lamp lights in one direction but not in the other, the rectifier is satisfactory. If lamp lights in both directions, the rectifier is shorted.

SLIP RING

GUIDE WIRE

GUIDE WIRE SOLDERED TO LEAD

Fig. C22   Aligning slip ring with field wire and guide wire

If the lamp does not light in either direction, the rectifier is open. *Note: Possible cause of an open or a blown rectifier is a faulty capacitor (condenser) or a battery that has been installed on reverse polarity. If the battery is installed properly and the rectifiers are open, test the capacitor capacity, which should be .50 microfarad plus or minus 20%.*

Testing Stator

1. Unsolder rectifiers from stator leads.

2. Test stator for grounds using a 110 volt test lamp, Fig. C17. Use wood slats to insulate the stator from the rectifier shield.

3. Contact one prod of test lamp to the stator pole frame, and contact the other prod to each of the three stator leads. The lamp should not light. If the lamp lights, the stator windings are grounded.

4. To test the stator winding for continuity, connect one prod of the test lamp to all three stator leads at the "Y" connection. Contact each of the three stator leads (disconnected from diodes). The lamp should light when the prod contacts

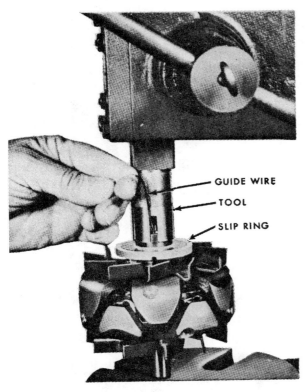

Fig. C23   *Installing slip ring*

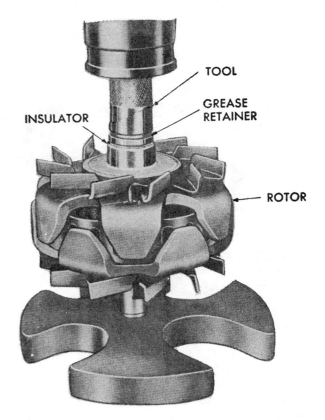

Fig. C24   *Installing bearing grease retainer*

each of the three leads. If the lamp does **not** light, the stator winding is open, Fig. C18.

5. Install new stator if one tested is defective.

Removing Rectifiers

1. Three diodes are pressed into the heat sink and three in the end shield. When removing the diodes, it is necessary to support the end shield and/or heat sink to prevent damage to these castings.
2. Install the tools shown in Fig. C19, making sure bore of tool completely surrounds diode.
3. Carefully apply pressure to remove diode from end shield.

Replacing Slip Rings

1. Cut through rotor grease retainer with a chisel and remove retainer and insulator.
2. Unsolder field coil leads at solder lugs, Fig. C20.
3. Cut through copper of both slip rings at opposite points with a chisel, Fig. C21.
4. Break insulator and remove old ring.
5. Clean away dirt and particles of old slip ring from rotor.

6. Scrape ends of field coil lead wires clean for good electrical contact.
7. Scrape one end (about 3/16″) of a piece of bare wire (about 18 gauge) three inches long to be used as a guide wire.
8. Tin the scraped area of the guide wire with resin core solder. Lap the tinned end of the wire over the field coil lead to the insulated ring and solder the two together.
9. Position new slip ring carefully over guide wire and rotor shaft so wire will lay in slip ring groove, Fig. C22. Groove in slip ring must be in line with insulated brush field lead to provide room for lead without damaging it.
10. Place installing tool over rotor shaft with guide wire protruding from slot in tool, Fig. C23.
11. Position rotor, slip ring and tool in arbor press, Fig. C23. Pull on guide wire, being careful to guide insulated field lead into slip ring groove. While guiding insulated field lead through groove, press slip ring on shaft. When slip ring is bottomed on rotor fan, end of field lead should be visible at solder lug, Fig. C20.
12. Unsolder guide wire from insulated brush slip ring lead. Press field lead into solder lug and solder to lug. *Caution: Be sure solder bead*

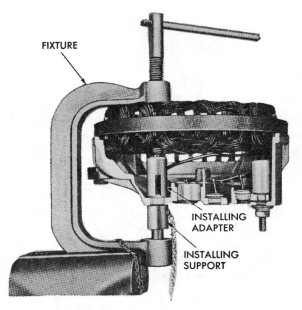

Fig. C25   Installing a diode

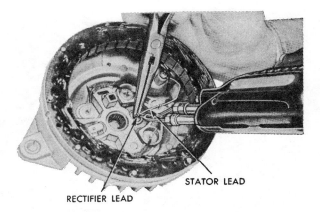

Fig. C26   Soldering diode and stator leads

does not protrude beyond surface of plastic material. Do not use acid core solder as a short may result and corrosion will definitely occur.

13. Coil ground brush field lead around solder lug and solder with resin core solder.

14. Test slip rings for ground with a 110 volt test lamp by touching one test lead prod to rotor pole shoe and remaining prod to slip rings. The lamp should not light. If lamp lights, slip rings are shorted to ground, possibly due to a grounded insulated field lead when installing slip ring.

15. If rotor is not grounded, lightly clean slip ring surfaces with No. 00 sandpaper and assemble to alternator.

16. Position grease retainer gasket and retainer on rotor shaft and press retainer on shaft, Fig. C24. Retainer is properly positioned when inner bore of installer tool bottoms on rotor shaft.

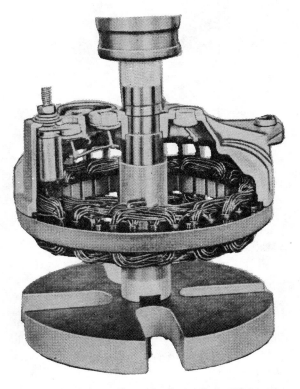

Fig. C27   Installing diode end shield bearing

Alternator Assemble

1. Install diodes as shown in Fig. C25. *Do not use a hammer to start diode in its bore in end shield. Do not hammer or shock diode in any manner as this will fracture the thin silicon wafer in the diode, causing complete diode failure.*

2. Clean leads and mate stator lead with diode wire loop and bend loop snugly around stator lead to provide a good electrical and mechanical connection. Solder wires with resin core solder. Hold diode lead wire with pliers just

below joint while soldering, Fig. C26. Pliers will absorb heat from soldering and protect diode. *Note: After soldering, quickly cool soldered connection by touching a damp cloth against it. This will aid in forming a solid joint.*

3. After soldering, stator leads must be pushed down into slots that are cast into end shield and cemented to protect leads against possible interference with rotor fans. Test each replacement diode to make certain it was not

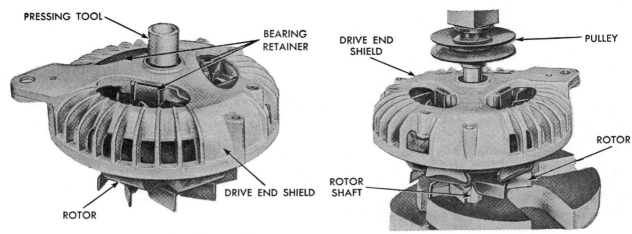

Fig. C28   *Installing drive end shield and bearing*

Fig. C29   *Installing pulley*

damaged by soldering or pressing operation. (Cement to use is Mopar No. 2299314.)

4. Install diode end shield bearing, Fig. C27.
5. Press bearing on rotor shaft until bearing contacts shoulder on shaft, Fig. C28.
6. Install pulley, Fig. C29. Do not exceed 6800 lbs. pressure. Press pulley on rotor shaft until pulley contacts inner race of drive end bearing.
7. Make sure heat sink insulator is in place. Then install capacitor stud through heat sink and end shield.
8. Install insulating washers, lockwashers and lock nuts.
9. Make sure heat sink and insulator are in position and tighten lock nut.
10. Position stator on diode end shield.
11. Position rotor end shield on stator and diode end shield.
12. Align through bolt holes in stator, diode end shield and drive end shield.
13. Compress stator and both end shields by hand and install through bolts, washers and nuts.

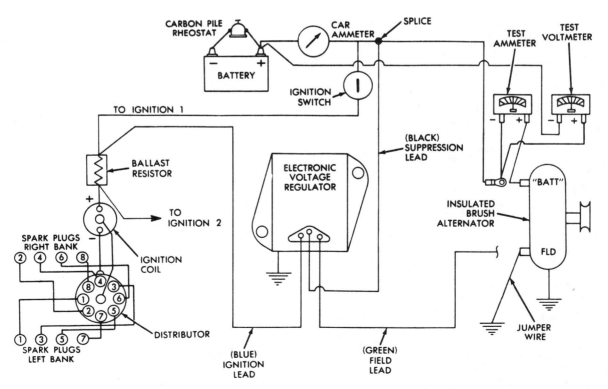

Fig. C30   *Charging circuit resistance test, insulated brush units*

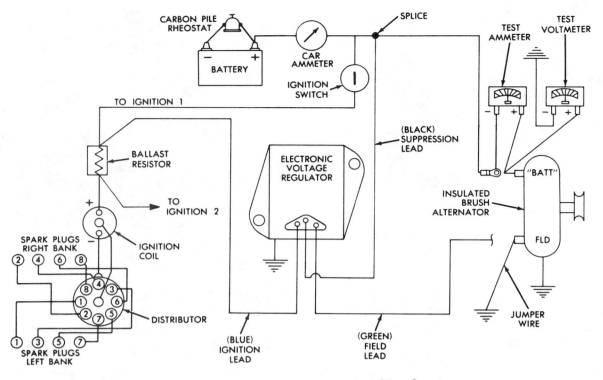

Fig. C31   *Current output test, insulated brush unit*

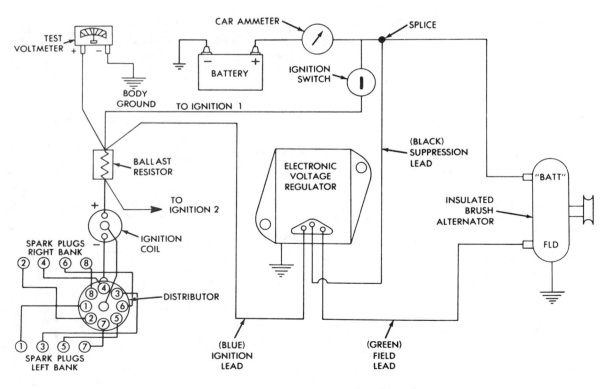

Fig. C32   *Voltage regulator test, insulated brush unit*

14. Install insulated brush in diode end. Place bronze terminal on plastic holder with tab of terminal in recess in plastic holder.
15. Place nylon washer on bronze terminal and install lockwasher and attaching screws.
16. Install ground brush and attaching screw.
17. Rotate pulley slowly by hand to be sure rotor fans do not touch diodes, capacitor lead and stator connections.
18. Install alternator and adjust drive belt.
19. Connect leads to alternator.
20. Connect battery ground cable.
21. Start and operate engine and observe alternator operation.
22. If necessary, test current output and regulator voltage setting.

INSULATED BRUSH ALTERNATOR

Used in conjunction with the electronic voltage regulator, the insulated brush alternator is serviced as follows:

Charging Circuit Resistance Test

1. Disconnect battery ground cable. Disconnect "Batt" lead at the alternator.
2. Complete test connections as per Fig. C30.
3. Connect battery ground cable, start engine and operate at idle.
4. Adjust engine speed and carbon pile to obtain 20 amps in the circuit and check voltmeter reading. Reading should not exceed .7 volts. If a voltage drop is indicated, inspect, clean and tighten all connections in the circuit. A voltage drop test at each connection can be performed to isolate the trouble.

Current Output Test

1. Disconnect battery ground cable, complete test connections as per Fig. C31 and start engine and operate at idle. *Immediately after starting, reduce engine speed to idle.*
2. Adjust the carbon pile and engine speed in increments until a speed of 1250 rpm and 15 volts are obtained.

Caution: While increasing speed, do not allow voltage to exceed 16 volts.

3. Check ammeter reading. Output current should be within specifications.

ELECTRONIC VOLTAGE REGULATOR

Of the silicon transistor type, this voltage regulator functions by varying the cycle of a series of voltage pulses to the alternator field. Frequency of the voltage pulses to the field is controlled by the ignition frequency of the engine because the regulator senses these through feedback from the electrical system. Since the frequency of operation is established by the ignition system—increasing with increasing engine RPM—the voltage regulator actually controls alternator output voltage by varying the on and off time between ignition firings. While the voltage across the field and the current through the output transistor is switching completely on and off, the field current through the alternator is only cycling through fractional changes.

Voltage Regulator Test

NOTE: Battery must be fully charged for tests to be accurate.

1. Make test connections as shown in Fig. C32.
2. Start and operate engine at 1250 rpm with all lights and accessories turned off.
3. As the engine starts, the instrument panel ammeter will deflect to the right. If it deflects less than ¼ scale, turn on high beams and heater blower. If deflection is greater than ¼ scale, turn on heater blower on high.
4. Voltage should be 13.8–14.4 if temperature at the regulator is 80 degrees F. and 13.3–14.0 at 140 degrees F.
5. If voltage is not correct, and the alternator is satisfactory, turn off ignition and disconnect regulator connector. Check for battery voltage at black and green leads. Turn on ignition without starting engine and check for battery voltage at blue lead. If voltage is not present at these connections, wiring is at fault.
6. Check regulator for good ground.
7. If regulator voltage is .5 volt away from specification, it must be replaced.

NOTE: The field circuit is grounded through the regulator. A good ground is established through the use of cup shaped washers on the regulator mounting screws which cut through body paint. These washers must be reinstalled whenever the regulator is removed.

Delco-Remy System

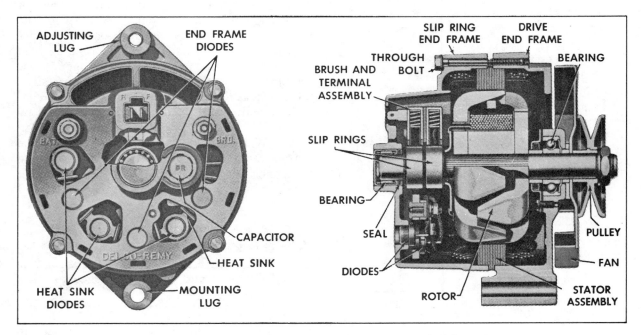

Fig. D1 *Sectional end and side views of "Delcotron" alternator*

DESCRIPTION

Alternator with External Regulator

The "Delcotron" (the trademark of the Delco-Remy alternator) is a continuous output, diode rectified alternating current generator, Fig. D1. The rotor is mounted on a ball bearing at the drive end, and a roller bearing at the slip ring end. Each bearing has a grease supply which eliminates the need for periodic lubrication.

Two brushes are used to carry current through the two slip rings to the field coil which is mounted on the rotor. The brushes are extra long and under normal operating conditions, will provide long periods of service.

The stator windings are assembled on the inside of a laminated core that forms part of the alternator frame. Fig. D2 illustrates the internal lead connections.

Regulator

The regulator, Fig. D3, similar to those used with D.C. generators, is made up of a double contact voltage regulator. Unlike the regulators used with D.C. generators, however, there is no need for a current regulator. And since the diodes permit current to flow through the circuit in only

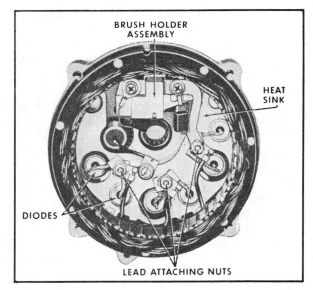

Fig. D2 *Alternator internal lead connections*

one direction, the need for a cutout relay is also eliminated.

When the vehicle has a charge indicating light, a field relay is incorporated, as shown in Fig. D3. When the vehicle is equipped with an ammeter, the field relay is not required. Fig. D4 shows a wiring diagram of the system.

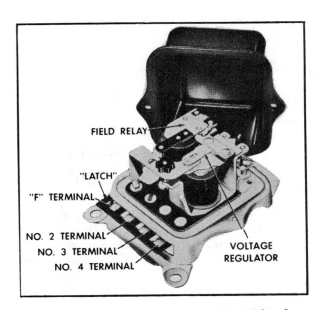

Fig. D3 *Voltage regulator. The field relay is used only on vehicles having a charge indicator light instead of an ammeter*

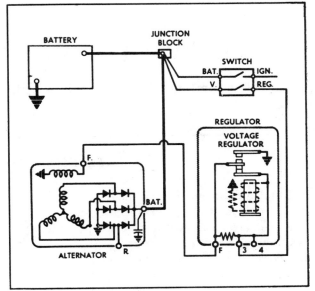

Fig. D4 *Wiring diagram of the alternator charging circuit*

I H VAN RELAY.   F = FIELD   3 = AMP MTR
2 = IGN. SW   4 = CAPIC. TO G ROUND

On some vehicles, for example the 1963 Pontiac with high performance engine equipped with Delco-Remy Transistorized Ignition, a Transistor regulator is furnished; more about this regulator later on.

TESTING SYSTEM IN VEHICLE

Current Output Test

1. Check and adjust belt tension if necessary.
2. Disconnect ground cable from battery.
3. Connect test ammeter between alternator "BAT" terminal and disconnected lead as shown in Fig. D5.
4. Connect tachometer from distributor terminal of coil to ground.
5. Reconnect battery ground cable and connect a voltmeter across battery.
6. Turn on all possible accessory load.
7. Apply parking brake firmly.
8. Start engine and adjust engine idle speed to the recommended setting (usually 500 rpm in Drive).
9. At this engine speed alternator output should be 5 amperes or more.
10. Shift transmission to Neutral. Then increase engine speed to 1500 rpm. Output should be 25 amperes or more.
11. Shut off engine and turn off all accessories.
12. If output is low in either all of the above tests, try supply the field directly to cause full alternator output. Unplug the connector from the alternator. Then connect a jumper wire

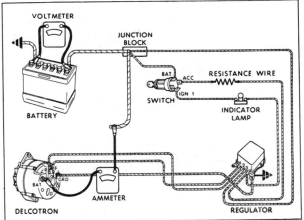

Fig. D5 *Alternator test connections*

from the alternator "F" terminal to the "Bat" terminal. Retest as described above. If the output is still low, the alternator is faulty and must be removed for bench tests and repairs.

13. If the output (using field jumper) is now satisfactory, the trouble is in the voltage regulator or wiring harness. Clean and test regulator, and check all wiring connections.
14. Remove field jumper and reinstall vehicle field connector.

Test & Adjust Regulator

1. Leave all test instruments in place, Fig. D5, but make sure field jumper is removed if one was used (see above).

Air Temperature at Regulator	85°	105°	125°	145°	165°
Voltage Setting	13.8–14.6	13.7–14.5	13.5–14.3	13.4–14.2	13.2–14.0

Fig. D6   *Voltage regulator settings*

2. Install a thermometer near the regulator.
3. Run engine at about 1500 rpm for 15 minutes. Make sure all electrical load except ignition is turned off.
4. Check ammeter reading. For an accurate voltage setting check, ammeter must read between 3 and 10 amperes. If ammeter reading is still high after 15 minutes, it may be necessary to substitute a fully charged battery.
5. Momentarily increase engine speed to 2000 rpm and read voltmeter and thermometer. See Fig. D6 to determine if upper voltage regulator setting is within limits for the existing temperature. If setting is within limits and battery condition has been satisfactory, *voltage setting should not be disturbed.*
6. If voltage regulator setting is not within correct limits, make a note of the change required to place voltage in the middle of the specified range. Remove regulator cover, carefully lifting it straight up. *Caution: If cover touches regulator unit, the resulting arc may ruin the regulator assembly.*

7. With cover off, voltage reading will change considerably. Starting with the changed voltage reading, increase or decrease voltage as required as shown in Fig. D7. *Caution: Always make final adjustment by increasing spring tension to assure contact between screw head and spring support.*
8. After making an adjustment, replace cover carefully. Cycle the regulator by unplugging connector from alternator. Reinstall connector in alternator and recheck voltage setting of regulator.

Tailoring Voltage Setting

It is important to remember that the voltage setting for one type of operating condition may not be satisfactory for a different type of operating condition. Vehicle underhood temperatures, operating speeds, and nighttime service all are factors which help determine the proper voltage setting. The proper setting is attained when the battery

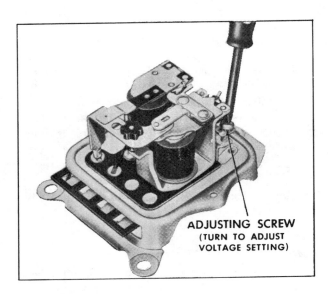

Fig. D7   *Adjusting voltage regulator setting*

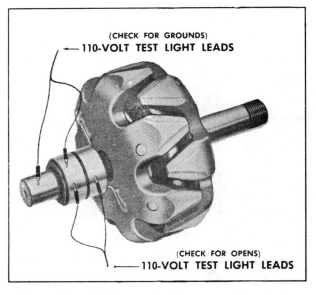

Fig. D8   *Checking rotor for opens or grounds*

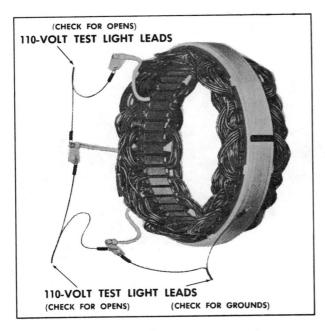

Fig. D9 *Checking stator for opens or grounds*

remains fully charged with a minimum use of water.

If no circuit defects are found, yet the battery remains undercharged, raise the setting by .3 volt, and then check for an improved battery condition over a service period of reasonable length. If the battery remains overcharged, lower the setting by .3 volt, and then check for an improved battery condition. However, never adjust the voltage setting out of the limits specified in Fig. D6.

ALTERNATOR SERVICE

If the system is not charging properly and the "In Vehicle Tests" indicated that the trouble is in

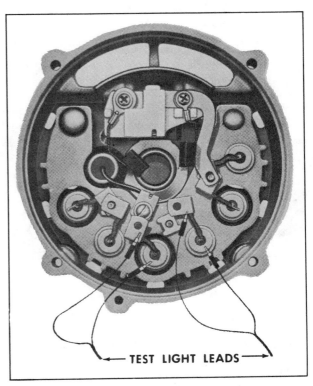

Fig. D10 *Checking diodes for opens or shorts*

the alternator, it need not be removed from the vehicle unless the drive end frame needs servicing because the slip ring end frame separates from the drive end frame by simply loosening the drive belt and removing the four through bolts. If the drive end frame must be serviced, remove and disassemble the alternator as follows:

Alternator Removal

1. Disconnect battery positive cable to avoid an injury from the hot battery lead at alternator.
2. Remove two leads at alternator.
3. Loosen adjusting bolts and remove drive belt.
4. Remove alternator retaining bolts and take off alternator.

Alternator Disassembly

1. If rotor, drive end frame bearings, or pulley and fan need replacement, remove and replace the shaft nut, using a strap wrench around the fan assembly. *Note: If the nut should happen to be cross-threaded or rusted and unusually difficult to remove, an alternate procedure is to use the strap wrench around the rotor.*

Fig. D11 *Removing a diode*

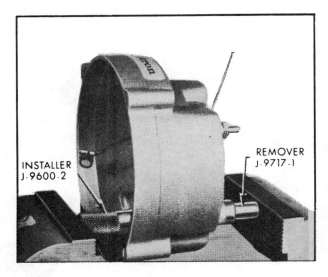

Fig. D12   *Installing a diode*

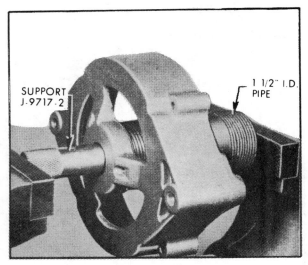

Fig. D13   *Removing drive end frame bearing*

2. Scribe a mark between the two halves of the alternator to help locate the parts in the same position during reassembly.
3. Remove four through bolts.
4. Separate the drive end frame and rotor assembly from the stator assembly by prying apart with a screwdriver at the stator slot. The fit between the two is not tight and the two can be separated easily. The separation is to be made between the stator and drive end frame. *Caution: As the rotor and drive end frame is separated from the slip ring frame, the brushes will fall down onto the shaft and come in contact with the lubricant. Brushes which come in contact with the shaft should be cleaned immediately to avoid contamination by oil, or they will have to be replaced.*

Rotor Checks

1. To check for grounds, connect a 110 volt test lamp from either slip ring to the rotor shaft, Fig. D8. If the lamp lights the field winding is grounded.
2. To check for opens, connect the test lamp to each slip ring. If the lamp fails to light the winding is open.
3. The winding is checked for short circuits by connecting a battery and ammeter in series with the two slip rings. The field current at 12 volts and 80° F. should be between 1.9 and 2.3 amperes. *Note: For vehicles with Delco-Remy transistorized ignition and transistor voltage regulator, the field current should be between 3.2 and 2.8 amperes.*
4. An ammeter reading above the values given

indicates shorted windings, and the rotor assembly should be replaced.

Stator Checks

1. To check the stator windings, remove all three stator lead attaching nuts and separate the stator from the end frame.
2. The stator winding may be checked with a 110 volt test lamp. If the lamp lights when connected from any stator lead to the frame, the windings are grounded. If the lamp fails to light when successively connected between each pair of stator leads, the windings are open, Fig. D9.
3. A short circuit in the stator windings is difficult to locate without laboratory test equipment due to the low resistance of the windings. However, if all other electrical checks are normal and the alternator fails to supply rated output, shorted stator windings are indicated.

Diode Checks

1. Each diode should be checked electrically for a shorted or open condition using a *test lamp of not more than 12 volts*, Fig. D10.
2. With the stator disconnected, connect the test lamp leads across each diode, first in one direction and then in the other.
3. If the lamp lights in both checks, or fails to light in both checks, the diode is defective.
4. When checking a good diode, the lamp will light in only one of the two directions.

Diode Replacement

1. To remove a diode, place slip ring end frame in a vise with the remover equipment mounted

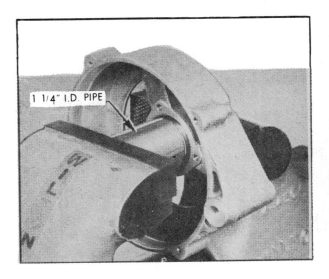

Fig. D14   *Installing drive end frame bearing*

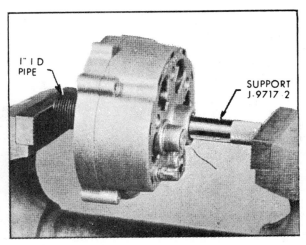

Fig. D15   *Removing slip ring end frame bearing*

as shown in Fig. D11. Tighten the vise to remove the defective diode.

2. To install a diode, place the new diode in the installer, Fig. D12. With the tools installed in the vise as shown, tighten the vise to install the new diode. *Caution: Never attempt to remove or install a diode by striking it as the shock may damage the other diodes.*

Slip Ring Service

If the slip rings are dirty they may be cleaned with No. 400 silicon carbide paper and finished polished with crocus cloth. Spin the rotor in a lathe, or otherwise spin the rotor, and hold the polishing cloth against the slip rings until they are clean.

Caution: The rotor must be rotated in order that the slip rings will be cleaned evenly. Cleaning the slip rings by hand without spinning the rotor may result in flat spots on the slip rings, causing brush noise.

Slip rings that are rough or out-of-round should be trued in a lathe to .002″ maximum runout as indicated on a dial gauge. Remove only enough material to make the rings smooth and round. Finish polish with crocus cloth and blow away all dust.

Bearing Replacement

1. The bearing in the drive end frame can be removed by detaching the retainer plate screws and then pressing the bearing from the end frame as shown in Fig. D13.

2. To install a new bearing, press in with a tube or collar that just fits over the outer race, Fig. D14. It is recommended that a new retainer plate be installed if the felt seal in the retainer plate is hardened or excessively worn.

3. The bearing in the slip ring end frame can be removed by pressing with a tube or collar that just fits inside the end frame housing. Press from the outside of the housing towards the inside as shown in Fig. D15.

4. To install the new bearing, place a flat plate over the bearing and press in from the outside towards the inside of the frame until the bearing is flush with the outside of the end frame. Support the inside of the frame with the pipe shown to prevent breakage of the end frame, Fig. D16.

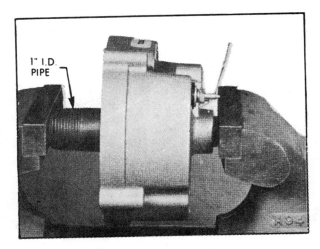

Fig. D16   *Installing slip ring end frame bearing*

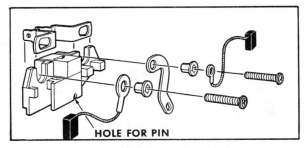

Fig. D17 *Assembling brush holder and related parts*

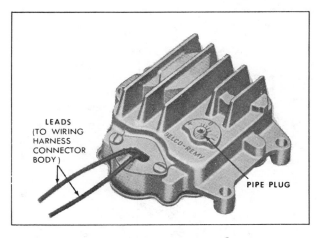

Fig. D19 *Transistor regulator*

5. Saturate the felt seal with S.A.E. 20 oil and reassemble the felt seal and steel retainer.

Brush Replacement

1. When the slip ring end frame assembly is separated from the rotor and drive end frame, the brushes will fall down onto the shaft and come in contact with the lubricant. If the brushes are to be re-used, they must be thoroughly cleaned with a soft dry cloth immediately. Also, the shaft must be thoroughly cleaned before reassembly.
2. The brush springs should be inspected for any evidence of damage or corrosion. If there is any doubt as to the condition of the brush springs, they should be replaced.
3. To install the springs and brushes into the brush holder assembly from the end frame by detaching the two screws.
4. Install the springs and brushes into the brush holder, and insert a straight wire or pin into the holes at the bottom of the holder to retain the brushes, Fig. D17.
5. Attach the brush holder assembly to the end frame, noting carefully the proper stack-up of

the parts as shown. Allow the straight wire to protrude through the hole in the end frame.

Heat Sink Replacement

1. The heat sink may be replaced by removing the "BAT" and "GRD" terminals from the end frame, and the screw attaching the condenser lead to the heat sink.
2. During reassembly, note carefully the proper stack-up of parts as shown in Fig. D18.

Alternator Reassembly

1. Reassembly is the reverse of disassembly. Refer to Fig. D2 for connection of internal leads.
2. When installing the pulley, secure the rotor in a vise only tight enough to permit tightening the shaft nut to a torque of 50–60 ft. lbs. If excessive pressure is applied to the rotor, the assembly may become distorted.

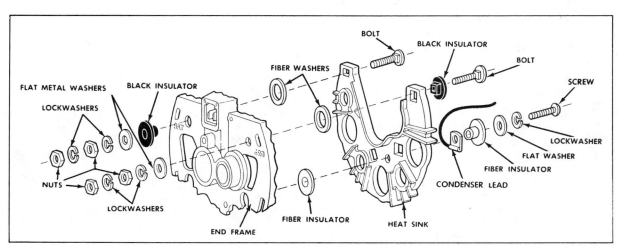

Fig. D18 *Disassembled view of heat sink and related parts*

696

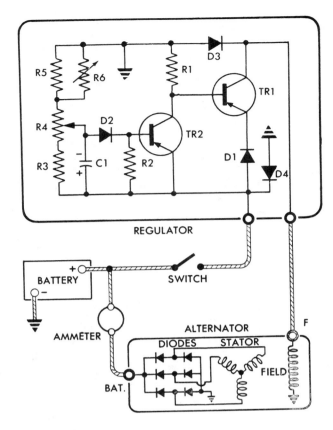

Fig. D20 *Wiring diagram of transistor regulator in charging circuit*

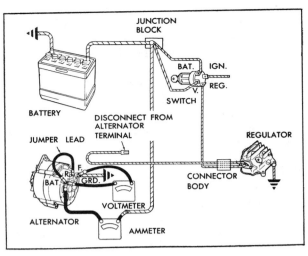

Fig. D21 *Checking charging circuit for undercharged battery condition*

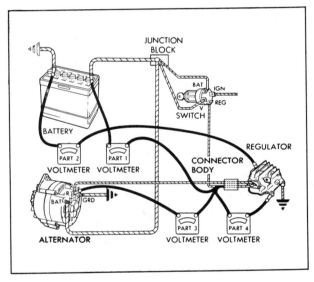

Fig. D22 *Checking charging circuit for overcharged battery condition*

3. To install the slip ring end frame to the rotor and drive end frame, remove the tape over the bearing and shaft (if used for protection upon disassembly) and make sure the shaft is perfectly clean.

4. Insert a straight wire as previously mentioned through the holes in the brush holder and end frame to retain the brushes in the holder. Then withdraw the wire after the alternator has been completely assembled. The brushes will then drop onto the slip rings.

TRANSISTOR REGULATOR

The transistor regulator, Fig. D19, is an assembly composed principally of transistors, diodes, resistors, a capacitor, and a thermistor to form a completely static unit containing no moving parts.

The transistor is an electrical device which limits the alternator voltage to a preset value by controlling the alternator field current. The diodes, capacitor and resistors act together to aid the transistor in controlling the voltage, which is the only function that the regulator performs in

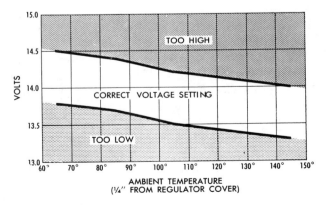

Fig. D23 *Temperature correction chart*

the charging circuit. The thermistor provides a temperature-compensated voltage setting.

The voltage at which the alternator operates is determined by the regulator adjustment. The regulator voltage setting can be adjusted externally by removing a pipe plug in the cover, Fig. D19, and turning the adjusting arm inside the regulator. This procedure is explained later on, and permits regulator adjustments without removing the cover.

Operating Principles

A typical wiring diagram showing internal circuits is shown in Fig. D20. When the switch is closed, current flows through diode D1 and transistor TR1 in the regulator to the alternator "F" terminal, and then through the alternator field winding to ground.

When alternator voltage reaches a preset value, the other components of the regulator cause transistor TR1 alternately to "turn-off" and "turn-on" the alternator field current. The regulator thus operates automatically to limit the alternator voltage to a preset value.

Checking Circuit

1. Connect test ammeter and voltmeter in circuit as shown in Fig. D21, and connect a jumper wire from alternator "F" terminal to alternator "BAT" terminal.
2. Operate alternator at specified speed, turn on accessories as required to obtain specified voltage and observe output. For example, the alternator used with this regulator on 1963 Pontiac has a rated output of 52 amperes at 14 volts at a speed of 5000 rpm.
3. If current output is low, remove and check the alternator as outlined previously.
4. If the alternator failure was caused by a defective stator or diodes, the repaired alternator may be installed back on the vehicle and no further checks are needed.
5. If the alternator failure was caused by a defective field winding, the repaired alternator may be installed back on the vehicle and the following checks must be made to locate possible damage to regulator.
6. Referring to Fig. D21, remove jumper lead and reconnect wiring harness connector to alternator "F" terminal.
7. Turn on ignition switch but do not start engine.
8. Connect voltmeter positive lead to battery positive terminal and negative lead into regulator (black lead) connector body to make connection to regulator (Part 1, Fig. D22). Record voltage drop.
9. Connect voltmeter to negative terminal of battery and ground on regulator (Part 2, Fig. D22). Record voltage drop.
10. If addition of voltage readings is greater than .3 volt, check ignition switch for poor contacts and system wiring for high resistance. If voltage difference is less than .3 volt, proceed as follows, referring to Part 3, Fig. D22.
11. Connect voltmeter positive lead to regulator positive terminal and voltmeter negative lead to alternator "F" terminal. Slide voltmeter positive lead into regulator connector body (black lead terminal) to make connection. Record voltage.
12. If the voltage is .9 volt or less, replace regulator, as transistor is shorted. If voltage is 2.0 volts or greater, replace regulator as transistor is open. If voltage is between .9 and 2.0 volts, proceed as follows:
13. Operate engine at approximately 1500 rpm for 10 minutes with low beam headlights on. Referring to Part 4, Fig. D22, with engine running at 1500 rpm, record voltage reading from regulator positive terminal to ground by sliding voltmeter lead into regulator connector body black lead terminal to make connection.
14. Compare with Fig. D23. Ambient temperature is temperature of air measured ¼" from regulator cover.
15. If voltage reading is within specifications, charging system is satisfactory but voltage setting may need to be changed to a different value to meet the requirements of driving conditions.
16. To do this, remove the pipe plug on regulator and insert a small screwdriver in adjustment slot. Turn counterclockwise for an undercharged battery one or two notches to increase setting.
17. For an overcharged battery, as evidenced by excessive water usage, turn clockwise one or two notches to decrease setting. For each notch moved, voltage setting will change by approximately .3 volt. Then check for an improved battery condition over a service period of reasonable length.
18. If voltage is not within specifications, check to see if the adjustment arm is in the center position. If the voltage reads out of specifications in the center position, replace the regulator.

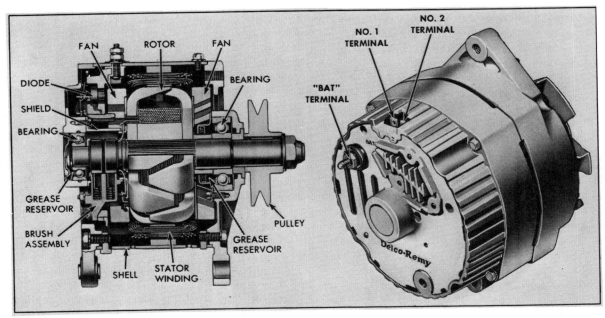

Fig. D24   *Delcotron Type S1 Integral Charging System*

ALTERNATOR WITH INTEGRAL REGULATOR

DESCRIPTION

This unit, Fig. D24, features a solid state regulator mounted inside the alternator slip ring end frame, Fig. D25, along with the brush holder assembly. All regulator components are enclosed in a solid mold with no need or provision for adjustment of the regulator. A rectifier bridge, containing six diodes and connected to the stator windings, changes A.C. voltage to D.C. voltage which is available at the output terminal. Generator field current is supplied through a diode trio which is also connected to the stator windings. The diodes and rectifiers are protected by a capacitor which is also mounted in the end frame.

No maintenance or adjustments of any kind are required on this unit.

SYSTEM TESTS

Alternator

The procedures for testing the alternator proper are similar to the standard Delcotron.

Diode Trio

1. With diode unit removed, connect an ohmmeter to the single connector and to one of the three connectors.

2. Observe the reading. Reverse ohmmeter leads.
3. Reading should be high with one connection and low with the other. If both readings are the same, unit must be replaced.
4. Repeat between the single connector and each of the three connectors.

NOTE: There are two diode units differing in appearance. These are completely interchangeable.

The diode unit can be checked for a grounded brush lead while still installed in the end frame by

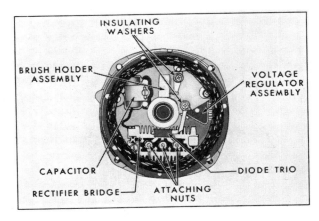

Fig. D25   *Slip ring end frame. S1 Delcotron*

connecting an ohmmeter from the brush lead clip to the end frame as in Steps 1 and 2 above. If both readings are zero, check for a grounded brush or brush lead.

Rectifier Bridge Test

1. Connect ohmmeter to the grounded heat sink and one of the three terminals.
2. Observe the reading then reverse leads.
3. Reading should be high with one connection

and low with the other. If both readings are the same, unit must be replaced.
4. Repeat test for each of the other terminals.

Voltage Regulator/Brush Lead Test

Connect an ohmmeter from the brush lead clip to the end frame, note reading, then reverse connections. If both readings are zero, either the brush lead clip is grounded or the regulator is defective.

Ford System

ALTERNATOR WITH EXTERNAL REGULATOR

TESTING SYSTEM IN VEHICLE

Alternator Output Test

1. Make connections as shown in Fig. F2. Be sure that the Generator Field Control is in the "open" position at the start of this test.

2. Close the battery adapter switch. Start the engine, then open the battery adapter switch. *All electrical accessories, including door operated interior lights must be turned off.*

3. Carefully increase engine speed to a tachometer reading of 2900 rpm. *Do not exceed this speed.*

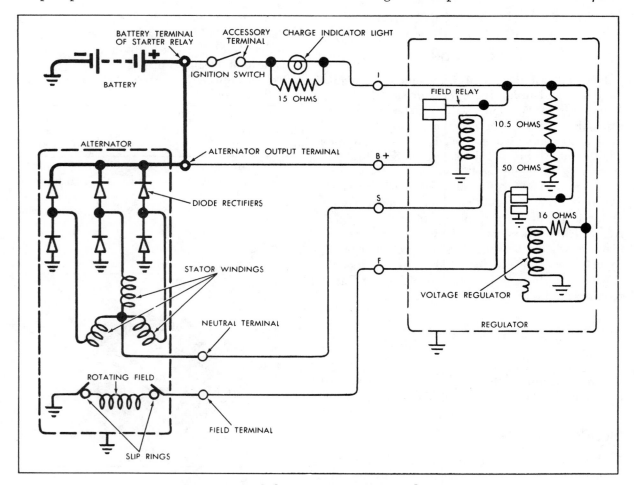

Fig. F1 Ford alternator system wiring diagram

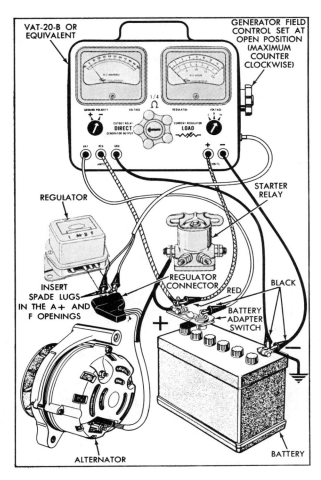

Fig. F2 *Alternator output test*

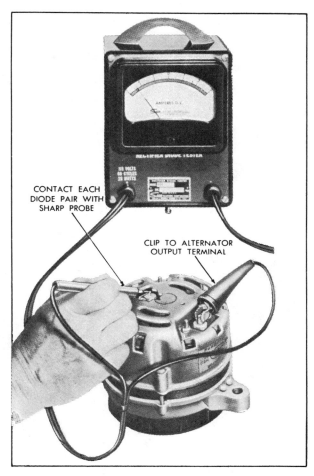

Fig. F3 *Positive diode test*

4. Adjust the Generator Field Control until the voltmeter reads exactly 15 volts. Observe the ammeter reading. Add 5 amperes to this reading to obtain total alternator output. The 5-ampere factor represents the field current and the ignition system current, and must be added to the ammeter reading as these currents are not indicated on the ammeter. *Make this test in the shortest possible time and do not exceed 2900 rpm. If the battery was fully charged, it might not be possible to obtain maximum current output. If specified current is not obtained, make the following test before condemning the alternator.*

5. Turn the Generator Field Control to the "open" position. Rotate the tester control knob to the "load" position. Maintain 2900 rpm engine speed.

6. Adjust the Generator Field Control and the "load" control, maintaining a voltmeter reading of 15 volts maximum, until the Generator Field Control is at its maximum clockwise position.

7. Readjust the "load" control until the voltmeter reads exactly 15 volts. Observe the ammeter reading. Add 5 amperes to this reading to obtain total alternator output.

8. Stop the engine, return the Generator Field Control to the "open" position and disconnect the test equipment.

Test Analysis

1. An output of 2 to 5 amperes below specifications indicates an open diode rectifier. An output of approximately 10 amperes below specifications indicates a shorted diode rectifier. *An alternator with a shorted diode will usually whine, which will be noticeable at idling speed.*

2. A shorted *positive* diode may sometimes be accompanied by alternate flashing of the oil pressure and charge indicator lights when the ignition switch is off. The field relay contacts will also be closed and the battery will be discharging through the field to ground.

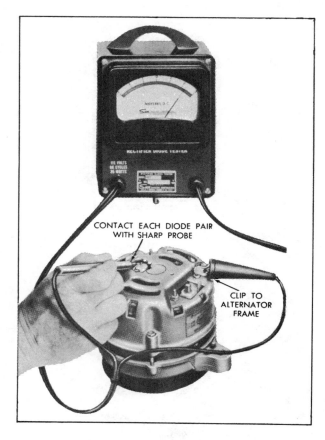

Fig. F4   Negative diode test

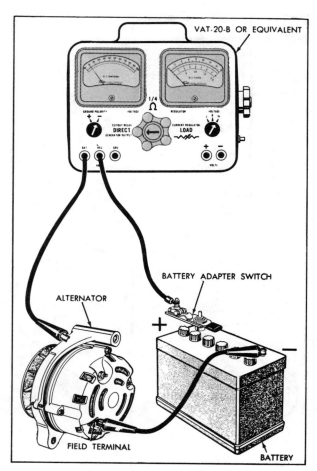

Fig. F5   Field open or short circuit test

3. Under this condition, the instrument constant voltage regulator will receive power through the charge indicator light. The operating of the constant voltage through the charge indicator light causes the alternate flashing of the lights. When the regulator contacts close, the oil pressure light becomes dim and the charge indicator light becomes bright. When the contacts open the oil pressure light becomes bright and the charge indicator light becomes dim.

Diode Test

1. To test the positive diodes, make connections shown in Fig. F3. Connect the probe to each diode lead. Make sure that the tip of the probe is sharp and that it penetrates the varnish at the diode terminal.
2. To test the negative diodes, make the connections shown in Fig. F4. Follow the same procedure as for positive diodes.
3. Good diodes will be indicated as on the meter in Figs. F3 and F4, that is, 2 amperes or more and readings alike within 2 scale divisions.

Field Open or Short Circuit Test

1. Make connections as shown in Fig. F5. The normal current draw, as indicated by the ammeter, should be about 3 amperes.
2. If there is little or no current flow, the field has a high resistance or is open, or the brushes are not making proper contact with the slip rings.
3. A current flow considerably higher than that specified (usually 2.9 to 3.1 at 12 volts) indicates shorted or grounded turns.
4. If the test shows that the field is shorted, *and the field brush assembly is not at fault,* the entire rotor must be replaced.

Field Relay Supply Voltage Test

The regulator field relay will close only if the voltage supplied by the neutral terminal of the alternator is sufficient to operate the relay. The wiring from the alternator neutral terminal to the relay "S" terminal also must be intact. The follow-

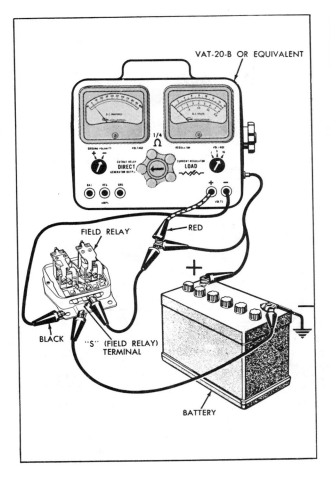

Fig. F6 Field relay test

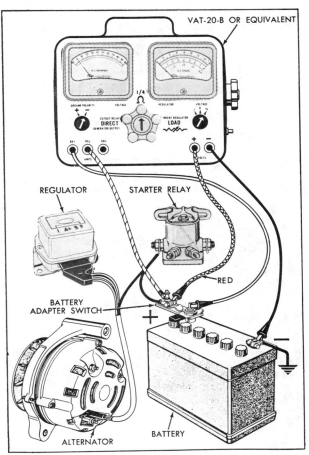

Fig. F7 Voltage limiter test

ing test will show that both sufficient voltage is available and that the wiring is in good condition.

1. Remove the connector plug from the regulator, remove the regulator, then reinstall the connector plug.
2. Connect the negative voltmeter lead to ground. Start the engine and operate it at 400 to 500 rpm.
3. Connect the positive voltmeter lead to a small screwdriver. Touch the screwdriver to the center rivet at the front of the regulator. *Use care to touch only the rivet or the rivet terminal so as not to short this point to ground or to the other nearby terminals.* The voltmeter should indicate at least 6 volts, and the relay contacts should be closed.
4. Low voltage at this point can be caused by a defective alternator or defective wiring.

REGULATOR TESTS

The following tests are to be made with the regulator in the vehicle. Be sure that the regulator

Voltage Regulation Setting (Volts)	Ambient Air Temperature °F
14.3-15.1	50
14.1-14.9	75
13.9-14.7	100
13.8-14.6	125
13.6-14.4	150
13.5-14.3	175

Fig. F8 Voltage regulation versus ambient air temperature

is at "normal" operating temperature. This is equivalent to the temperature after 20 minutes of operation with a 10-ampere load.

Field Relay Test

1. Disconnect regulator terminal plug and remove the regulator cover.

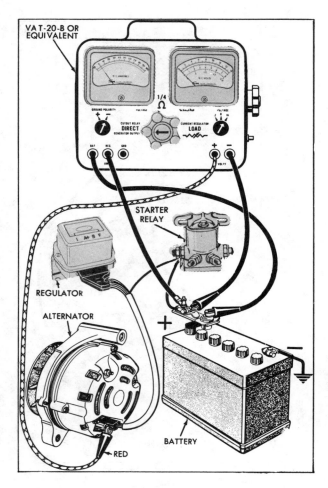

Fig. F9 *Voltage drop test from alternator to battery positive terminal*

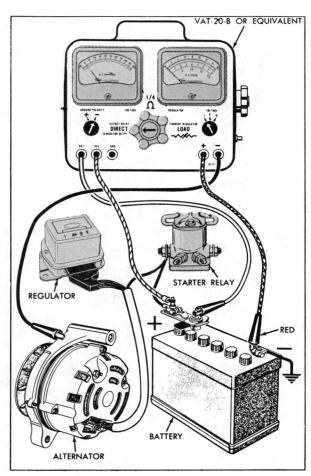

Fig. F10 *Voltage drop test from alternator to negative battery terminal*

2. Make connections shown in Fig. F6.
3. Slowly rotate field resistance control clockwise from the off position until the field relay contacts close.
4. Observe the voltmeter reading at the moment that the relay closes. This is the relay closing voltage.
5. If the relay closes immediately, even with the field resistance close to the "off" position, use a 6 volt battery for this test.
6. If the closing voltage is not within specifications, adjust the relay.

Voltage Limiter Test

For test purposes, the lower stage regulation is used (armature vibrating of the lower contact). Voltage limiter calibration test must be made with the regulator cover in place and the regulator at "normal" operating temperature (equivalent to temperature after 20 minutes of operation with a 10-ampere load).

1. Make test connections shown in Fig. F7.
2. Turn off all accessories, including door-operated dome lights.
3. Close battery adapter switch, start engine, then open adapter switch.
4. Attach the voltage regulation thermometer to regulator cover.
5. Operate engine at 2000 rpm for 5 minutes. Turn master control to "direct" position.
6. If the ammeter indicates more than 10 amperes, remove battery cables and charge the battery.
7. When battery is fully charged, and the voltage regulator has been temperature stabilized, rotate the master control to the "Voltage Reg." position, the ammeter should indicate less than 2 amperes.
8. Cycle the regulator as follows: Stop engine, close adapter switch, start engine, and open adapter switch.
9. Allow battery to normalize for a short time, then read the voltmeter.

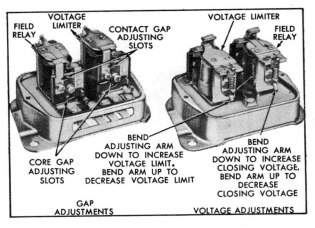

Fig. F11   *Voltage regulator adjustments*

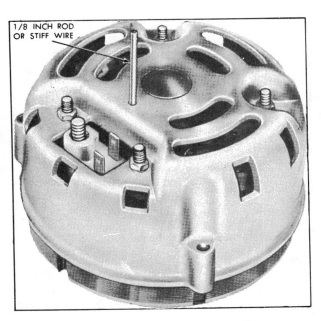

Fig. F12   *Retracting alternator brushes*

10. Read the thermometer, and compare the voltmeter reading with the voltage given in Fig. F8.
11. If the regulated voltage is not wihin specifications, make a voltage limiter adjustment. *After each adjustment, be sure to cycle the regulator before each reading. Readings must be made with cover in place.*

Circuit Resistance Test

For the purpose of this test, the resistance values of the circuits have been converted to voltage drop readings for a current flow of 20 amperes.

Alternator to Battery Positive Terminal
1. Make connections shown in Fig. F9.
2. Turn off all lights and electrical accessories.
3. Close battery adapter switch, start engine, then open battery adapter switch.
4. Slowly increase engine speed until ammeter reads 20 amperes.
5. Voltage should be no greater than 0.3 volts.

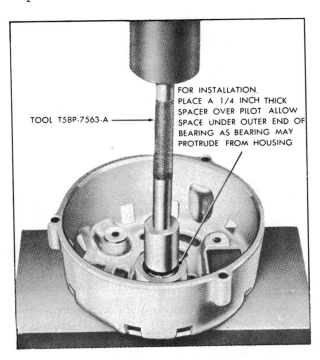

Fig. F13   *Rear bearing removal*

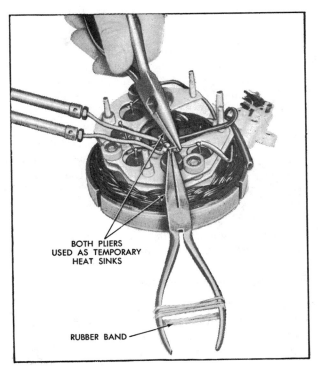

Fig. F14   *Soldering diode leads*

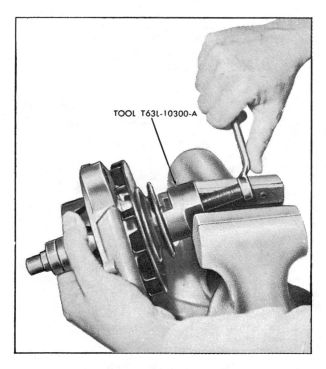

Fig. F15   Removing pulley

Alternator to Battery Ground Terminal

1. Make connections shown in Fig. F10.
2. Close battery adapter switch, start engine and open battery adapter switch.
3. Slowly increase engine speed until ammeter reads 20 amperes.
4. Voltage indicated should be less than 0.1 volt.

REGULATOR ADJUSTMENTS

Erratic operation of the regulator, indicated by erratic movement of the voltmeter during a voltage limiter test, may be caused by dirty or pitted regulator contacts.

Use a very fine abrasive paper such as silicone carbide, 400 grade, to clean the contacts. Wear off the sharp edges of the abrasive by rubbing against another piece of abrasive paper. Fold the abrasive paper over and pull it through the contacts to clean them. Keep all oil or grease from contacting the points. *Do not use compressed air to clean the regulator. When adjusting the gap spacing, use only hospital clean feeler gauges.*

Regulator Bench Adjustments

The difference between the upper stage and lower stage regulation (0.3 volt), is determined by voltage limiter point and core gaps.

Adjust point gap first. Referring to Fig. F11, loosen the left side lock screw ¼ turn. Use a screwdriver blade in the adjustment slot above the

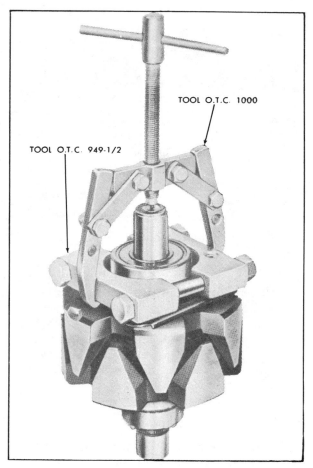

Fig. F16   Removing front bearing

lock screw. Adjust the upper contact until there is .010 to .015" gap between lower contacts. Tighten lock screw and recheck the gap.

To adjust the core gap, loosen the center lock screw ¼ turn. Use a screwdriver blade in the slot under the lock screw. Adjust the core gap to .045 to .052" clearance between armature and core at edge of core closest to contact points. Tighten lock screw and recheck core gap.

Regulator Voltage Adjustments

Final adjustment of the regulator must be made with the regulator at operating temperature.

The field relay closing voltage is adjusted by bending the spring arm, Fig. F11. To increase the closing voltage, bend the spring arm down. To decrease the closing voltage, bend the spring arm up.

The voltage limit is adjusted by bending the voltage limiter spring arm, Fig. F11. To increase the voltage, bend the adjusting arm downward. To decrease the setting, bend the adjusting arm upward.

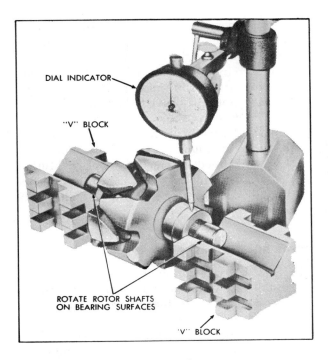

Fig. F17   *Checking slip ring runout*

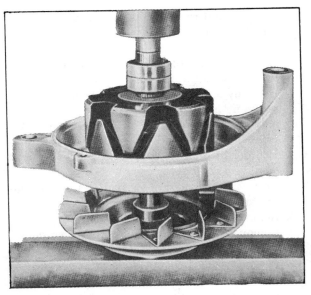

Fig. F18   *Installing pulley*

Before setting the voltage and before making a final voltage test, the alternator speed must be reduced to zero and the ignition switch opened momentarily to cycle the regulator.

ALTERNATOR REPAIRS

Disassembly

1. Mark both end housings with a scribe mark for reassembly. Reach through a ventilation slot, raise both brushes off slip rings and install a short length of ⅛″ rod or stiff wire through hole in rear end housing, Fig. F12, to hold brushes off slip rings.
2. Remove 3 housing through bolts and separate front housing and rotor from rear housing

and stator. *Make certain that brushes do not contact the greasy rotor shaft.*

3. Remove nuts from rectifier-to-rear housing studs and remove rear housing. Remove two spacer sleeves from rectifier plate studs.
4. Press bearing from rear end housing, Fig. F13.
5. Remove terminal spacer block from studs and unsolder neutral wire from spacer block neutral terminal.
6. If brushes are being replaced, straighten field brush, terminal blade locking tabs with a pair

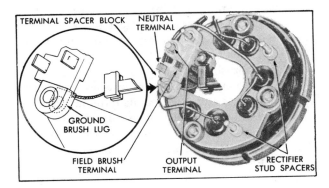

Fig. F19   *Stator, heat sink and terminal spacer block assembly*

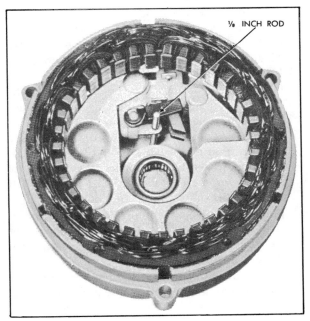

Fig. F20   *Brushes retracted for assembly*

of pliers and remove terminal blade from terminal spacer block assembly. Remove brushes and holders.

7. If either diode plate is being replaced, carefully unsolder leads from diodes, Fig. F14. Use only a 100 watt soldering iron. Leave the soldering iron in contact with the diode terminals only long enough to remove the wires. Both pliers are used as temporary heat sinks in order to protect diodes. *Excessive heat can damage a good diode.*

8. Remove 3 insulated diode plate screws and insulators, and separate diode plates.

9. Remove drive pulley, Fig. F15.

10. Remove 3 screws that hold front bearing retainer and remove front housing.

11. If the bearing is being replaced, remove it as shown in Fig. F16. Remove bearing retainer and spacer. It will not be necessary to remove the stop ring unless it has been damaged.

Inspection

1. The rotor, stator, diodes and bearings are not to be cleaned with solvent. These parts are to be wiped off with a clean cloth. Cleaning solvent may cause damage to electrical parts or contaminate the bearing internal lubricant. Wash all other parts with solvent and dry them.

2. Rotate front bearing on drive shaft. Check for any scraping noise, looseness or roughness that would indicate that the bearing is excessively worn. As the bearing is being rotated, look for any lubricant leakage. If any of these conditions exist, replace the bearing.

3. Place the rear end bearing on the slip ring end of the shaft and rotate the bearing on the shaft. Make the same check for wear or damage as for the front bearing.

4. Check the housings for cracks. Check the front housing for stripped threads in the mounting holes. Replace defective housings.

5. Pulleys that have been removed and installed several times may have to be replaced because of the increased bore diameter. A pulley is not suitable for reuse if more than ¼ of the shaft length will enter the pulley bore with light pressure. Replace any pulley that is bent out of shape. After installing the pulley, check for clearance between the fins and the alternator drive end housing.

6. Check all wire leads on both stator and rotor for loose soldered connections and for burned insulation. Resolder poor connections and replace parts that show burned insulation.

7. Check slip rings for damaged insulation. Check the slip rings for runout as shown in Fig. F17. If the slip rings are more than .0005″ out of round, take a light cut (minimum diameter limit ½″) from the face of the rings to true them up. If the slip rings are badly damaged, the entire rotor will have to be replaced as they are serviced only as a complete assembly.

8. Replace the terminal spacer block assembly if the neutral terminal is loose. Replace any parts that are burned or cracked. Replace brushes that are worn to less than .350″ in length. Replace the brush spring if it has less than 7 to 12 ounces tension.

Assembly

1. If the stop ring on the drive shaft was broken, install a new stop ring. Push the new ring on the shaft and into the groove. *Do not open the ring with snap ring pliers.*

2. Position the front bearing spacer on the drive shaft against the stop ring, and position the bearing retainer on the shaft with the flat surface of the retainer outward.

3. Putting pressure on the inner race only, press the new bearing on the shaft until it contacts the spacer.

4. Place the front housing over the shaft with the bearing positioned in the front housing cavity. Install the bearing retainer mounting screws.

5. Press the pulley onto the shaft until the hub just touches the inner race of the front bearing, Fig. F18. *A new pulley must be installed if more than ¼ of the shaft length will enter the old pulley bore with light pressure.*

6. If a new diode plate is being installed, mount the two plates together so that they are insulated from each other, Fig. F19. Solder the wire leads to the diodes as shown in Fig. 14, using only a 100 watt iron. *Avoid excessive heat as this can result to damage to the diode.*

7. Insert the new field brush terminal blade into the slot in the terminal spacer block with the brush pigtail extending toward the brush holder pivots, Fig. F19.

8. Install brush holders and brush spring to terminal block, then position brushes in holders.

9. Solder the neutral wire to its terminal. Position terminal spacer block on rectifier plate mounting studs, with the ground brush lug over the mounting stud farthest from the output terminal, Fig. F19.

10. Place spacers on rectifier mounting studs farthest from terminal block.

11. Install rear bearing so that its open end is

flush with the inner surface of the housing boss, Fig. F13. Allow for space under the outer end of the bearing during installation.

12. Place rear end housing over rectifier plate and stator assembly and mount rectifier plates to housing.

13. Retract brushes and insert a short piece of ⅛″ rod or stiff wire through hole in rear end housing to hold brushes in retracted position, Fig. F20.

14. Wipe clean the rear bearing surface of the rotor shaft.

15. Position rear housing and stator assembly over rotor and, after aligning marks made during disassembly, install housing through bolts. Remove brush retracting rod.

ALTERNATOR WITH INTEGRAL REGULATOR

DESCRIPTION

Alternator

The alternator used with this unit is basically the same as the standard Ford Autolite unit. Modifications have been made in the brush holder assembly, rear end frame and the stator assembly to accommodate the integral regulator. Except for tests outlined here, refer to preceding chapter for service procedures.

Integral Regulator

The integral regulator, Fig. F21, consists of an integrated, solid state circuit, made up of transistors, diodes and resistors, all connected by aluminum conductors and fabricated within a ⅛″ square silicon crystal. This is a one piece, non-adjustable unit which must be replaced if it malfunctions, or if it is not calibrated within specified limits of 13.5 to 15.3 volts between 50 and 125 degrees F.

SYSTEM TESTS

NOTE: Because the voltage sensing circuit is permanently connected across the charging system, resulting in a small but harmless current drain, a voltmeter cannot be connected in series with the battery for diagnosis. The integral regulator is not defective and should not be replaced because of leakage indicated by a voltmeter connected in series with the battery or regulator.

Voltage Regulator Test

1. Using a fully charged battery, turn off all lights and accessories. Be sure ignition switch is off and make test connections as shown in Fig. F22.

2. Open battery adapter switch. Ammeter should

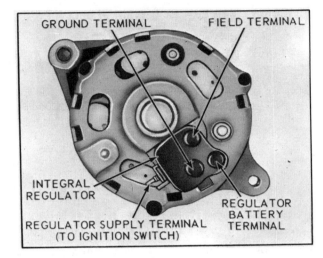

Fig. F21   Alternator with integral regulator

show zero amperes. A discharge (2 amperes) indicates a malfunction in the alternator field coil or the regulator. Refer to Field Circuit Tests. If ammeter shows zero, proceed as follows:

3. With transmission in neutral or park and parking brake applied, place tester master control at the ¼ ohm resistor position.

4. Close battery adapter switch and start engine. Be sure all lights and accessories are off and open battery adapter switch.

5. Operate engine at approximately 2000 rpm for 5 minutes, and check voltage. If voltage is between 13.3 and 15.3 volts, the regulator is functioning satisfactorily. If voltage does not rise above battery voltage, check regulator supply voltage. If voltage exceeds 15.3 volts, perform field circuit tests.

Supply Voltage Test

Check for voltage supply terminal of the alternator with a 12 volt test light or a voltmeter. If no

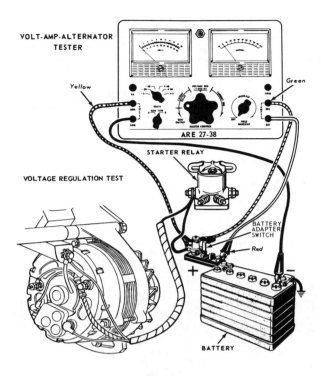

Fig. F22 *Voltage regulator test*

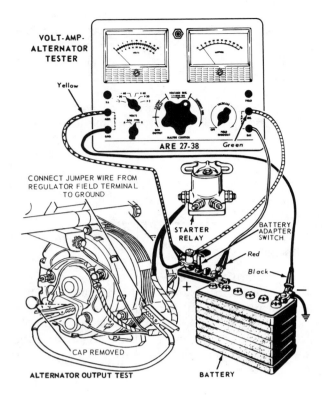

Fig. F23 *Alternator output test*

voltage is indicated, the supply circuit is disconnected or broken. If voltage is present, perform Alternator Output Test.

Alternator Output Test Off Vehicle

When using test bench, refer to manufacturer's procedures and be sure to disconnect battery cable as the alternator output connector is always connected to the battery.

Alternator Output Test on Vehicle

NOTE: Under no circumstances should the regulator battery terminal be connected to the regulator field terminal. To do so will damage regulator.

1. With transmission in neutral or park and parking brake applied, make test connections as shown in Fig. F23.
2. Close battery adapter switch. Start engine and reopen adapter switch. Voltage reading must be maintained between 10 and 15 volts.
3. Increase engine speed to 2000 rpm. Turn off all lights and accessories.
4. Turn master control clockwise until voltmeter

shows 15 volts. At 15 volts, ammeter should register 50 to 57 amperes. If alternator is working properly, regulator must be replaced.
5. Return engine speed to idle before releasing master control knob.
6. An alternator output of 2 to 8 amperes below minimum specification usually indicates an open diode rectifier. An alternator with a shorted diode will usually whine, most noticeably at idle speed.

Field Voltmeter Test

1. Turn off ignition switch and remove wire from regulator supply terminal.
2. Make test connections as shown in Fig. F24. Open battery adapter switch.

NOTE: If there was an ammeter drain that stopped when supply terminal was disconnected, an ignition switch or wiring problem is indicated. If discharge continues (2 or more amperes), proceed as follows:

3. Voltmeter should read 12 volts. If there is no voltage reading, the field circuit is open or grounded. Perform Alternator Field Ohmmeter Test. If ohmmeter tests show alternator field is

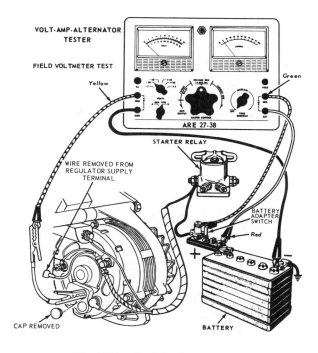

Fig. F24   *Field voltmeter test*

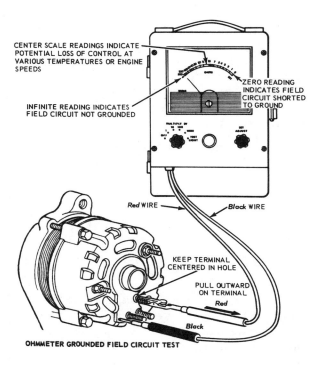

Fig. F25   *Ohmmeter field circuit test*

okay, the regulator is shorted and must be replaced.

4. If voltmeter reading in Step 3 is more than one volt but less than battery voltage, a partial ground in the alternator field circuit is indicated. Perform Field Ohmmeter Test to isolate trouble between alternator and regulator.

Field Ohmmeter Test

1. Disconnect battery ground cable and remove regulator from alternator.
2. Make ohmmeter connections as shown in Fig. F25.
3. If any of the conditions shown in Fig. F25 are found, remove and repair alternator. If alternator is okay, replace the regulator.

Diode Bench Test

1. Disassemble alternator and disconnect diode assembly from stator. Make test connections as shown in Fig. F26.
2. Touch one ohmmeter lead to diode plate and the other to each of the three stator lead termi-

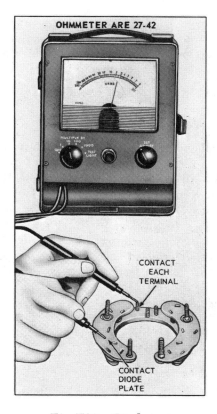

Fig. F26   *Diode test*

nals. Reverse probes and repeat test. Test the other set of diodes the same way.

3. All tests should show a low reading of approximately 60 ohms in one direction and an infinite reading with the probes reversed.

Open Stator Test

Connect ohmmeter probes between each pair of stator leads. If ohmmeter does not show equal readings between each pair of stator leads, the stator is open and must be replaced.

Grounded Stator Test

Connect ohmmeter probes between one of the stator leads and the stator core. If the ohmmeter shows any reading, stator is grounded and must be replaced.

Leece-Neville System

ALTERNATOR WITH DIODE RECTIFIERS

The alternator under discussion herewith is a 14-volt, 60-ampere heavy-duty unit with six diode rectifier cells mounted on the slip ring end housing and internally connected to the stator windings.

As shown in Fig. L1, three terminals (marked G−, F, G+) on the stator provide for connections to the electrical system of the vehicle or engine.

One slip ring brush is internally connected to the "F" terminal on the stator. The terminal of the other slip ring brush is grounded to the alternator frame.

A cover over the slip ring end protects the rectifiers from grease and dirt and also directs an air stream over the fins of the rectifier mounts. The cooling air is then drawn through the alternator.

Rectifier Tests

The following tests should be made when it is indicated that the rectifiers are not functioning properly. All leads should be disconnected from the G−, F and G+ terminals, Fig. L1, and the rectifier cover removed from the alternator. *No leads should be unsoldered from the rectifier elements or disconnected from the rectifier mounts until it is determined which of the rectifier elements must be replaced. Before making the test, any two of the three stator winding leads must be disconnected from the heat sinks.*

Fig. L2 shows the rectifier cell polarity markings, and Fig. L3 gives the details of the rectifier cell mounting and connections.

Positive Terminal Post Cells Tests, Fig. L4

1. With the negative prod of a polarized ohmmeter on the terminal post of cell No. 1 and the positive prod on the adjacent stator winding

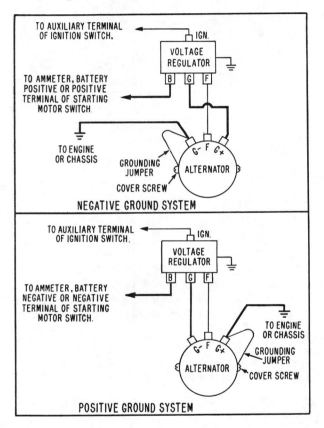

Fig. L1 Wiring hook-up for alternator with diode rectifiers

connection, the meter should indicate a low resistance (closed circuit).

2. With the positive prod of the ohmmeter on the terminal post of cell No. 1 and the negative prod on the adjacent stator winding connection, the meter should indicate a high resistance (open circuit).

3. Each of the positive terminal post cells (Nos. 1, 3, 5) should be checked in the above manner.

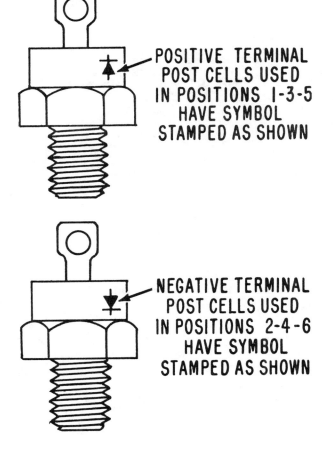

POSITIVE TERMINAL POST CELLS USED IN POSITIONS 1-3-5 HAVE SYMBOL STAMPED AS SHOWN

NEGATIVE TERMINAL POST CELLS USED IN POSITIONS 2-4-6 HAVE SYMBOL STAMPED AS SHOWN

Fig. L2 *Rectifier cell polarity markings*

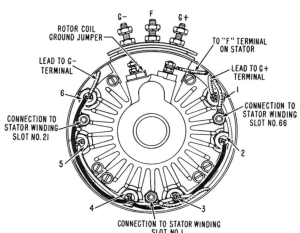

Fig. L3 *Rectifier cell mounting and connections*

Negative Terminal Post Cells Test, Fig. L5

All three cells (Nos. 2, 4, 6) should be tested in a similar manner. With the positive prod of the meter on the terminal post of the cell and the negative prod on the winding connection, the meter should indicate a low resistance (closed circuit). A high resistance or open circuit should be indicated when the meter prods are reversed.

Any cell or cells that the above tests indicate as not operating properly must be replaced. When replacing cells, care must be taken to disturb as few connections as possible. The metal-to-metal surfaces of the cell and terminal connections are given special treatment in assembly to insure good electrical contact and to prevent electrolysis and the formation of oxides. This same treatment must

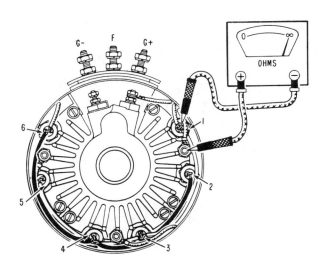

Fig. L4 *Positive terminal post cell test*

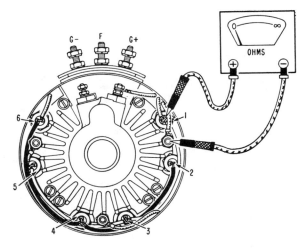

Fig. L5 *Negative terminal post cell test*

713

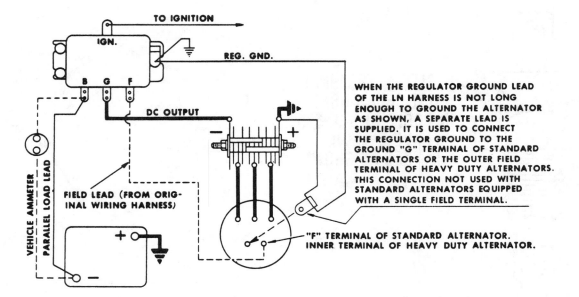

Fig. L6 Circuit diagram for positive ground system with vehicle ammeter

be applied when a new or replacement cell is installed.

Rectifier Mount Ground Test

With a 110-volt test lamp, check each of the three cell mounts for grounds to the alternator frame. If a ground or circuit is present between a rectifier mount and the alternator frame, check the insulation under the mount and the insulation of the attaching screws.

To check, remove the two screws holding the mount to the housing. Do not unsolder the leads from the cell posts. The flexible leads will permit

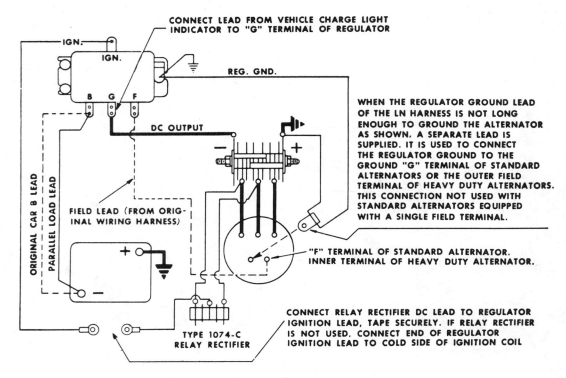

Fig. L7 Circuit diagram for positive ground system with charge indicator light

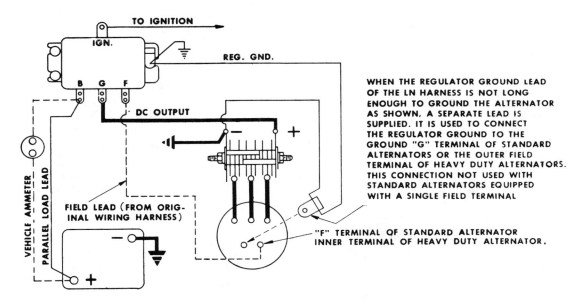

Fig. L8 *Circuit diagram for negative
ground system with vehicle ammeter*

the mount to be lifted slightly so that the insulator underneath may be pulled out. Check the insulating sleeve and washer on the screws, and replace broken or cracked insulators, sleeving or insulating washers.

When replacing the screws holding the mount to the housing, make sure the insulating washer is between the guard washer and cell mount.

Alternator Repairs

Before disassembling the alternator, test it for grounds, circuit continuity and phase as outlined previously.

Only when the unit is to be completely disassembled is it necessary to disconnect the stator winding leads from the rectifier mounts.

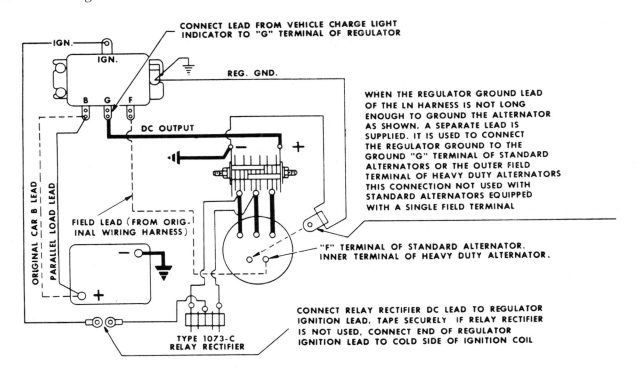

Fig. L9 *Circuit diagram for negative
ground system with charge indicator light*

TO IGNITION SWITCH

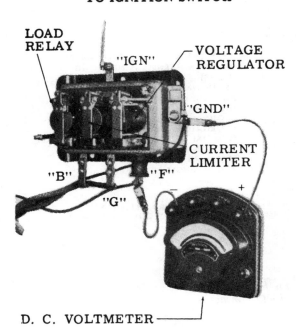

LOAD RELAY

"IGN"

VOLTAGE REGULATOR

"GND"

CURRENT LIMITER

"B"   "F"

"G"

D. C. VOLTMETER

Fig. L10 Hook-up for checking system voltage

TO IGNITION SWITCH

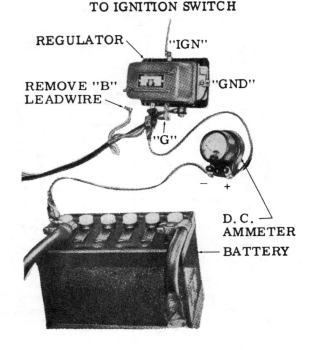

REGULATOR

"IGN"

"GND"

REMOVE "B" LEADWIRE

"G"

D. C. AMMETER

BATTERY

Fig. L11 Hook-up for checking current limiter setting

Disassembly—Disconnect stator leads from rectifier mounts and the "F" terminal lead from the brush holder. Unsolder the "G+" and "G−" leads from cells Nos. 1 and 6. The unit may then be completely disassembled.

Reassembly—When assembling the unit care must be taken to route all leads from the stator and terminals through the proper openings in the slip ring end housing.

Before connecting the stator leads to the rectifier mounts, the metal surface of the mounts immediately under the lead terminal must be cleaned with a wire brush and Special L-N Compound No. 56624. A light coating of the compound is applied to the clean surface immediately before securing the lead terminal to the mount.

Make sure the grounding jumper is installed on the inner slip ring brush holder, and the insulation band is in place inside the rectifier cover.

REGULATOR SETTINGS

Before we discuss this subject, it might be well to study the circuit diagrams shown in Figs. L6 to L9.

Voltage Setting

1. Start engine and allow it to run for about 15 minutes with battery load only. This permits regulator to be heated up to the desired temperature for adjustment purposes.

2. Referring to Fig. L10, connect a voltmeter across the "F" and "GND" terminals. This connection will show voltage across the alternator field.

3. Bring engine speed up to 2000 rpm. If battery is in fully charged condition, field voltage will be close to zero. This will be an indication that the regulator armature is operating on the bottom set of contacts. *If this test is used to determine bottom contact operation, do not forget to remove voltmeter lead from regulator "F" terminal and place it on the regulator "B" terminal for final regulator setting.*

4. When armature vibrates on bottom contact, operating voltage has been reached. The basic setting for the operating voltage is stamped on the name plate. Different climates and battery conditions will sometimes require this value to be changed slightly. The correct value is determined by finding the lowest possible voltage that will keep the battery charged. Ex-

perience has shown that 6.9 volts is an average setting for 6-volt systems, and 13.8 volts for 12-volt systems.

5. If regulator is to be re-set, maintain engine speed at approximately 2000 rpm. Operating voltage can be changed by adjusting the voltage regulator armature spring tension. *Restart engine and repeat this test after every adjustment.*

Current Limiter Setting

Referring to Fig. L11, connect an ammeter between regulator "B" terminal and the cable from the battery. Raise alternator speed above 2500 rpm. Turn on enough load (lights, heater, etc.) to start current limiter armature vibrating. Current output at this point should be within two amperes of the regulator rating. This can be adjusted by changing the armature spring pressure on the current limiter.

Rectifier Cell, Replace

1. Unsolder the leads from the cell to be replaced and unscrew the cell from its mounting hole.
2. With wire brush and a special compound (L-N 56624) clean area around cell mounting hole.
3. Immediately after cleaning, apply a thin coating of the above compound to the clean surface and install the new cell. Torque cell to 20–25 inch lbs.
4. Secure rectifier cell by staking the aluminum of the mount against the hex side of the cell.
5. Solder the leads to the new cell, applying only enough heat to insure a good connection to the cell post. Overheating may damage the cell.

Motorola System

DESCRIPTION

The electrical circuit of the alternator, Fig. M1, uses 6 silicon diodes in a full wave rectifier circuit. Since the diodes will pass current from the alternator to the battery or load but not in the reverse direction, the alternator does not use a circuit breaker. Fig. M2 shows the charging circuit.

The entire DC output of the system passes through the "Isolation Diode". This diode is mounted in a separate aluminum heat sink and is replaced as an assembly. The isolation diode is not essential for rectification. It is used to:

1. Provide an automatic solid state switch for illuminating the charge-discharge indicator light.

2. Automatically connect the voltage regulator to the alternator and battery when the alternator is operating.
3. Eliminate electrical leakage over the alternator insulators so that maximum leakage is less than one milliampere when the car is not in use.

Voltage Regulator

The voltage regulator is an electrical switching device sealed at the factory, requiring no adjustments. It senses the voltage appearing at the regulator terminal of the alternator and supplies the necessary field current for maintaining the system voltage at the output terminal.

TESTING SYSTEM IN VEHICLE

Alternator Output Test, Fig. M3

1. Close by-pass switch on battery post adapter, Fig. M4.
2. Start engine and adjust speed to 2000 rpm.
3. Open by-pass switch.
4. Rotate load control knob to the load position and adjust until voltmeter reads approximately 6 volts.
5. Rotate alternator field control to direct position.
6. Adjust load control to obtain exactly 15 volts.
7. Observe ammeter; it should indicate maximum output of alternator with 25 amperes mini-

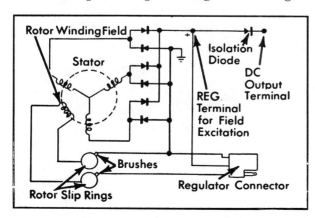

Fig. M1 Motorola alternator circuit diagram

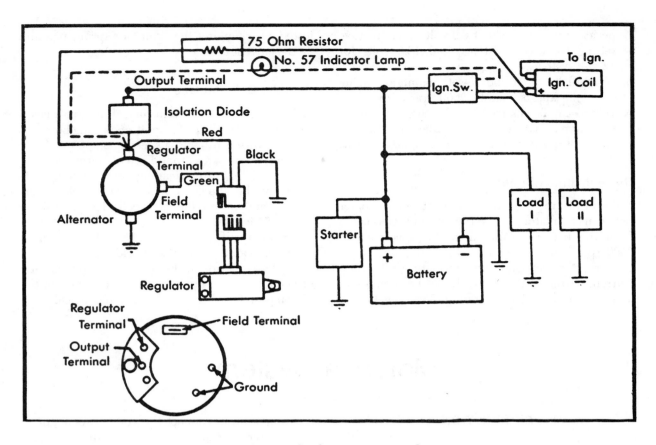

Fig. M2 Motorola alternator circuit diagram

mum. If no output is evident, observe voltmeter. *If there is over 12 volts at the regulator terminal and battery voltage is measured at the output terminal, the isolation diode is evidently open and should be replaced.*

8. Rotate alternator field control to open position.
9. Reduce engine speed to idle.
10. Rotate load control to direct position and stop engine.

Isolation Diode Test

If a commercial diode tester is used, follow the Test Equipment Manufacturer's instructions. If a commercial tester is not available, use a DC Test Lamp. *Caution: Do not use a 120 volt test lamp as diodes will be damaged.*

1. Connect test lamp to output terminal and regulator terminal of isolation diode.

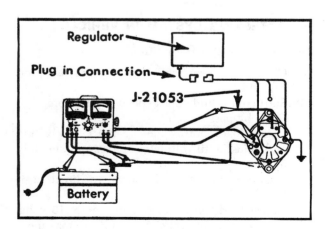

Fig. M3 Alternator output test connections

Fig. M4 Battery post adapter tool which provides a convenient method for connecting ammeter leads of volt-ammeter tester to the charging system

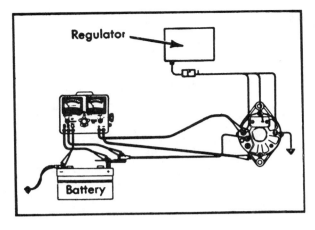

Fig. M5   Regulator terminal test connections

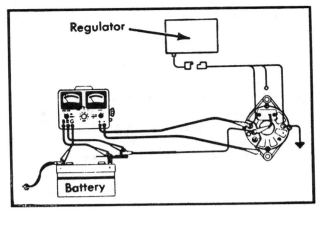

Fig. M6   Field current test connections

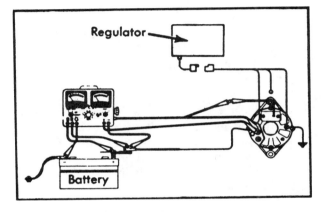

*Fig. M9   Insulated circuit resistance
test connections*

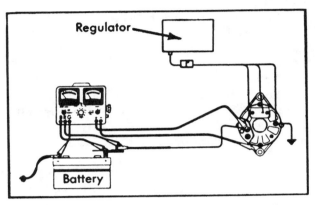

Fig. M7   Voltage regulator test connections

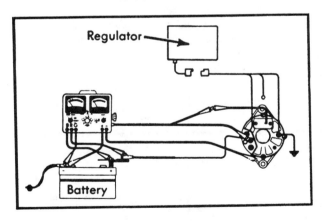

*Fig. M10   Ground circuit resistance
test connections*

0°-14.6-15.4	80°-14.0-14.8
20°-14.6-15.3	100°-13.8-14.6
40°-14.3-15.0	120°-13.7-14.5
60°-14.1-14.9	140°-13.6-14.4
160°-13.3-14.1	

*Fig. M8   Voltage at various ambient air
temperatures under a 10 ampere load*

Rectifier Diode Tests

Any commercial in-circuit diode tester will suffice to make the test. Follow Test Equipment Manufacturer's instructions.

Check diodes individually after the diodes have been disconnected from the stator. A shorted stator coil or shorted insulating washers or sleeves on positive diodes would make diodes appear to be shorted.

A test lamp will not indicate an open condition unless all three diodes of either assembly are open. However, a shorted diode can be detected.

2. Reverse test probes.
3. The test lamp should light in one direction but should not light in the other direction.
4. If the test lamp lights in both directions the isolation diode is shorted.
5. If the test lamp does not light in either direction, isolation diode is open.

This test is not 100% effective but can be used if so desired when an in-circuit diode tester is not available.

The test lamp should light in one direction but not in the other direction. If the test lamp lights in both directions, one or more of the diodes of the assembly being tested is shorted. If the test lamp does not light in either direction, *all three diodes in the assembly are open.* Check diodes individually after disassembly to ascertain findings.

Note: A shorted stator coil would appear as a shorted negative diode. Also check stator for shorts after disassembly.

Regulator Terminal Voltage Test, Fig. M5

1. Remove alternator field control leads.
2. Remove field jumper leads.
3. Connect voltage regulator plug-in connector.
4. Connect slip-on field connector.
5. Turn on ignition switch.
6. Voltmeter indicates regulator terminal voltage which should be ½ to 2 volts. *Note: If voltmeter indicates battery voltage of less than ½ volt, perform following steps to determine if voltage regulator is defective.*
7. Disconnect regulator plug-in connector.
8. Disconnect slip-on field connector and connect the field jumper lead to the alternator field terminal.
9. Connect field jumper clip to regulator terminal.
10. Voltmeter should now read 1 to 2 volts, which indicates voltage regulator is defective. *Note: If results are other than specified, continue with Field Current Test. If field current is within specifications, test voltage regulator circuit.*

Field Current Test, Fig. M6

1. Disconnect voltage regulator plug-in connector.
2. Turn off ignition switch.
3. Disconnect slip-on field connector and connect the field jumper to the alternator field terminal.
4. Connect the clip of the field jumper lead to the alternator output terminal.
5. Ammeter now indicates field current draw which should be 2 to 2½ amperes.
6. Disconnect field jumper lead from output terminal.

Voltage Regulator Test, Fig. M7

1. Remove field jumper lead.
2. Connect voltage regulator plug-in connector.
3. Connect slip-on field connector.
4. Connect positive voltmeter lead to alternator output terminal.
5. Start engine and adjust speed to 1500 rpm.
6. Rotate load control knob to the ¼ ohm position.
7. Voltmeter indicates voltage regulator setting which should be 14 to 14.8 volts at 75 degrees ambient temperature (see Fig. M8).
8. Return load control knob to direct position.
9. Reduce engine speed and shut it off.

Insulated Circuit Resistance Test, Fig. M9

1. Disconnect voltage regulator plug-in connector.
2. Connect field control leads, one to the output terminal and one to the field jumper lead.
3. Connect the negative voltmeter lead to the battery end of the positive battery cable.
4. Close by-pass switch and start engine, then open by-pass switch.

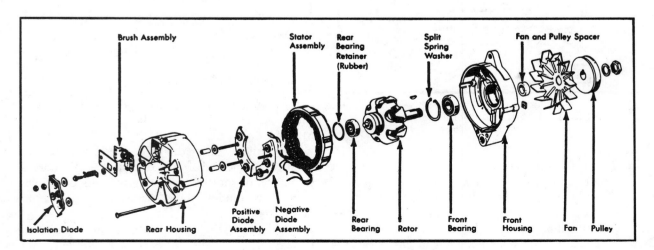

Fig. M11   Motorola alternator disassembled

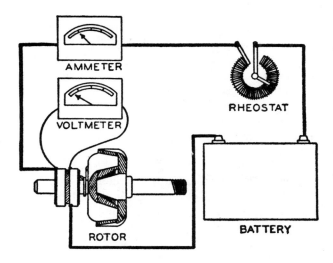

Fig. M12   *Field coil test*

5. Adjust engine speed to approximately 1500 rpm.
6. Rotate alternator field control slowly toward direct until ammeter indicates 10 amperes current flow.
7. Voltmeter now indicates the voltage drop in the alternator insulated circuit which should not exceed .3 volt.

Ground Circuit Resistance Test, Fig. M10

1. Connect negative voltmeter lead to alternator housing.
2. Connect positive voltmeter lead to negative battery post.
3. With ammeter indicating 10 amperes, voltmeter should indicate not more than .05 voltage drop in the alternator ground circuit.
4. Rotate field control to open position. Reduce engine speed to idle and stop engine.

ALTERNATOR REPAIRS

Disassembly, Fig. M11

Brush Assembly

The brush assembly can be removed in most cases with the alternator on the vehicle. The spring clip is bent back so that the field terminal plug can be removed. Remove the two self-tapping screws, field plug retainer spring and cover. Pull brush assembly straight out far enough to clear locating pins, then lift brush assembly out. The complete brush assembly is available for replacement.

Isolation Diode

Remove the 2 lock nuts securing the isolation diode to the rear housing and slide it off the studs. The diode is replaced as an assembly.

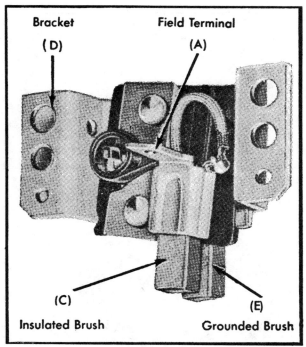

Fig. M13   *Brush assembly test*

Rear Housing

Remove the 4 through bolts and nuts. Carefully separate the rear housing and stator from the front housing by using 2 small screwdrivers and prying the stator from the front housing at 2 opposing slots where the "through bolts" are removed. Do not burr the stator core which would make assembly difficult. *Caution: Do not insert screwdriver blade deeper than 1/16″ to avoid damaging stator winding.*

Stator and Diode Assembly

Do not unsolder stator-to-diode wire junction. Remove stator and diode as an assembly. Avoid bending stator wire at junction holding positive and negative diode assembly from housing.

Remove 4 lock nuts and insulating washers. The insulating washers and nylon sleeves are used to insulate the positive plate studs from the housing. With the 4 nuts removed, the stator can be separated from the rear housing by hand.

Diode Replacement

In soldering and unsoldering leads from diodes, grasp the diode lead with pliers between the diode and stator lead to be removed. This will give better heat dissipation and protect the diode. Do not exert excessive stress on diode lead. *Make note of diode assembly to stator connections, and make sure replacement diode assembly connec-*

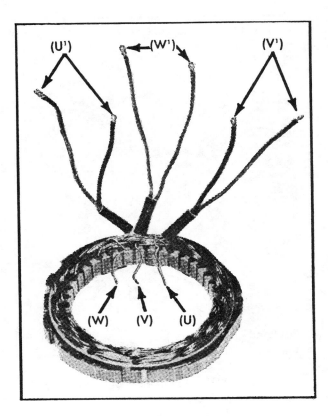

Fig. M14 *Stator coil shorts*
and continuity tests

ring and the other probe on the rotor core. If the bulb lights the rotor is grounded.

To test for shorted turns, check rotor field current draw as shown in Fig. M12. Slowly reduce resistance of rheostat to zero. With full battery voltage applied to field coil, the field current should be 1.4 to 1.9 amperes. Excessive current draw indicates shorted turn in field winding.

Brush Insulation Test
Connect an ohmmeter or a test lamp to the field terminal and bracket. Resistance should be high (infinite) or test lamp should not light. If resistance is low or if test lamp lights, brush assembly is shorted and must be replaced.

Continuity Test
Connect an ohmmeter to field terminal and brush. Use an alligator clip to assure good contact to brush, test points "A" and "C" in Fig. M13. *Caution: Do not chip brush.*

Resistance reading should be zero. Move brush and brush lead wire to make certain that brush lead wire connections are not intermittent. Resistance reading should not vary when brush and lead wire are being moved around. Connect ohmmeter to bracket and grounded brush, test points "E" and "D", Fig. M13. Resistance reading should be zero.

Stator In-Circuit Test
When making the in-circuit stator leakage test, some consideration must be given to the rectifier diodes that are connected to the stator winding. The negative diode assembly will conduct in one direction when properly polarized. A shorted diode in the negative diode assembly would make the stator appear to be shorted. For this reason, the rectifier diode plate assembly and stator must be checked individually after alternator has been disassembled if the problem is localized to the stator. *Caution: Use a special diode continuity light or a DC test lamp. Do not use a 120 volt test lamp as diodes will be damaged.*

1. Connect the test lamp to a diode terminal of the negative assembly and ground terminal.
2. Reverse test probes. The lamp should light in one direction but not in the other.
3. If the test lamp does not light in either direction, this indicates that all three rectifiers in the negative diode assembly are open.
4. If the test lamp lights in both directions, the stator winding is shorted to stator or one of the negative diodes is shorted.
5. Check stator again when it is disassembled from diode assemblies.

tions are the same. The positive diode assembly has red markings, the negative black markings.

Rotor
The rotor should only require removal from the front housing if there is a defect in the field coil itself or in the front bearing. Front and rear bearings are permanently sealed, self-lubricating type. If the front housing must be removed from the rotor, use a two-jaw puller to remove the pulley. The split spring washer must be loosened with snap ring pliers through the opening in the front housing. Remove the washer only after the housing is removed. The rotor and front bearing can be removed from the front housing by tapping the rotor shaft slightly. *Note: Make certain that the split spring washer has been removed from its groove before attempting to remove the front housing from the bearing.*

Alternator Bench Tests

Field Coil Test
The rotor should be tested for grounds and for shorted turns in the winding. The ground test is made with test probes connected in series with a 110 volt test lamp. Place one test probe on the slip

6. With alternator disassembled, connect an ohmmeter or test lamp probes to one of the diode terminals and to stator.

7. Resistance reading should be infinite or test lamp should not light.

8. If resistance reading is not infinite or test lamp lights, high leakage or a short exists between stator winding and stator. In either case, stator should be replaced.

Stator Coil Shorts Tests

1. This test checks for shorts between stator coil windings. The winding junctions must be separated as shown in Fig. M14. An ohmmeter or test lamp may be used.

2. Connect one of the test probes to test point "U" and the other to test point "V" and then to test point "W". Resistance should be infinite or test lamp should not light.

3. Connect test probes to test V and W. Resistance should be infinite or test lamp should not light. In either test, if the resistance reading is not infinite or the test lamp lights, high leakage or a short exists between stator windings. Stator should be replaced.

Continuity Test

1. Measure resistance of each winding in stator between test points U and U1, V and V1, W and W1, Fig. M14. Resistance should be a fraction of an ohm (approximately .1 Ohm). An extremely accurate instrument would be necessary to ascertain shorted turns. Only an open condition can be detected with a commercial type ohmmeter.

2. If the alternator has been disassembled because of an electrical malfunction, replace stator only after all other components have been checked and found to be satisfactory.

Assemble Alternator

1. Clean bearing and inside of bearing hub of front housing. Support front housing and, using a suitable driver, apply sufficient pressure to outside race of bearing to seat bearing.

2. Insert split spring washer hub of front housing, seating washer into groove of hub. *Note: Do not use a screwdriver or any small object to compress washer that can slip off and damage bearing seal. Make certain that split spring washer has been installed prior to assembling front housing and rotor.*

3. Use sufficient pressure to seat front bearing against shoulder on rotor shaft. The bearing drive tool must fit the inner race of bearing.

4. Install fan and pulley.

5. Use a $\frac{7}{16}$" socket to fit inside race of rear bearing and apply sufficient pressure to drive bearing against shoulder of rotor shaft.

6. Assemble front and rear housings.

7. Make certain that rear bearing is properly seated in rear housing hub and that diode wires are properly dressed so that rotor will not contact diode wires.

8. Align stator slots with rearing housing through bolt holes, then align front housing through bolt holes with respect to rear housing. *Note: The position of the brush assembly and belt adjusting screw boss must be in the same relative position to each other.*

9. Spin rotor to make certain that rotor is not contacting diode wires. Install through bolts and tighten evenly.

10. Before mounting isolation diode, make certain that positive rectifier diode plate has been properly insulated from housing.

11. Install brush assembly, cover and field plug retainer spring.

Prestolite System

TESTING SYSTEM IN VEHICLE

Charging Circuit Resistance Test

1. Make connections as shown in Fig. P1.

2. Adjust engine speed and electrical load to obtain 10 amperes in charging circuit. The voltage drop in this circuit should not exceed .2 volt. If more, locate and correct cause of high resistance before proceeding with tests.

3. With the same operating conditions as in Step 2, connect voltmeter leads from ground battery terminal to alternator frame, and from alternator to regulator base. In neither case should the voltage reading exceed .04 volt. If more, locate and correct cause of high resistance ground connections.

Alternator Output Test

1. Make test connections as shown in Fig. P2. *Be sure rheostat is "Off" before connecting leads.*

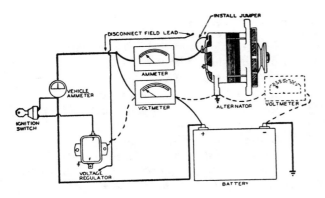

Fig. P1 *Meter connections for testing
voltage drop in charging circuit*

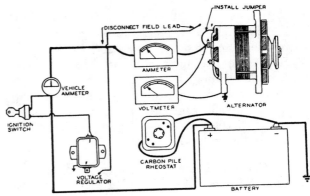

Fig. P2 *Meter connections for
testing alternator output*

2. Connect a tachometer and start engine. Adjust engine speed to 1750 rpm. Adjust rheostat to obtain 14.2 volts and observe ammeter reading. It should be as specified for the alternator being serviced.

3. If the alternator fails to reach rated output, it should be removed from the vehicle for repairs. A slightly low ammeter reading may indicate an open rectifier while a considerably lower reading may indicate a shorted rectifier.

4. If the output is only slightly low, a temperature check may be made to determine whether a rectifier is open. With the alternator operating as for the output test place a bulb-type thermometer (250° F scale) on the base of each rectifier heat sink for the bank being checked.

5. The rectifier temperature will normally be several degrees higher than the heat sink temperature. If a rectifier is open its temperature will be the same as that of the heat sink in which it is located.

6. If a rectifier is shorted the temperature of the entire heat sink in which it is located will be abnormally high.

Voltage Regulator Test

The first part of this test checks the operating voltage of the regulator when operating on the upper contact. Make test connections as shown in Fig. P3. *Be sure the ignition switch is off when connecting the field lead. Grounding of the field circuit while the ignition switch is on will damage the regulator.*

1. To test voltage regulator setting, start engine and adjust its speed to 750 rpm. Turn on lights and accessories to obtain a 10 ampere charge rate. Operate the system at this speed and load for 15 minutes to normalize the temperature. (This regulator is temperature compensated.)

2. Cycle the system by stopping and restarting the engine; then note the voltmeter reading. If seriously out of adjustment a rough setting may be made. The final setting is not made until the "spread" between operation on the upper contact and operation on the lower contact is established. This value is determined in the next part of the test which checks voltage when the regulator is operating on the lower contact.

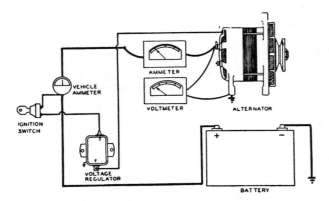

Fig. P3 *Meter connection for
testing voltage regulator*

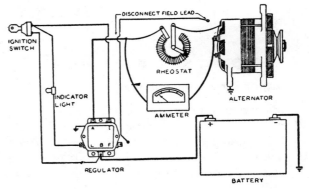

Fig. P4 *Meter connections for testing
indicator light relay operation*

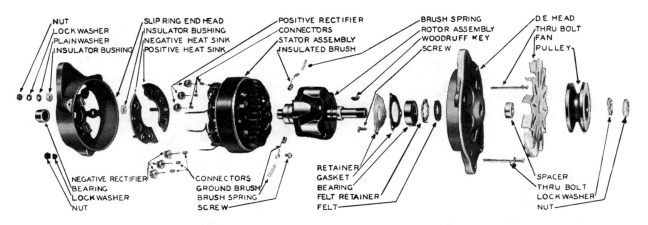

Fig. P5 Disassembled view of Prestolite alternator

NUT
LOCK WASHER
PLAIN WASHER
INSULATOR BUSHING

SLIP RING END HEAD
INSULATOR BUSHING
NEGATIVE HEAT SINK
POSITIVE HEAT SINK

POSITIVE RECTIFIER
CONNECTORS
STATOR ASSEMBLY
INSULATED BRUSH

BRUSH SPRING
ROTOR ASSEMBLY
WOODRUFF KEY
SCREW

D.E. HEAD
THRU BOLT
FAN
PULLEY

NEGATIVE RECTIFIER
BEARING
LOCK WASHER
NUT

CONNECTORS
GROUND BRUSH
BRUSH SPRING
SCREW

RETAINER
GASKET
BEARING
FELT RETAINER
FELT

SPACER
THRU BOLT
LOCK WASHER
NUT

3. Test connections remain the same as for the previous part of the test. Increase engine speed to 1500 rpm and turn off all lights and accessories. Voltage should increase and amperage should decrease. The "spread" in voltage between this reading and the reading noted in the first part of the test should be from .1 to .3 volt.

4. If the "spread" is greater or less than specified, remove the regulator cover and adjust by loosening the stationary contact support screw and moving the support up or down. Raising the stationary support will increase voltage "spread."

5. Replace the cover and reduce engine speed to 750 rpm and repeat the first part of the test with the regulator operating on the upper contact. If the voltage setting is not within spcifications, remove the cover and adjust by bending the lower spring hanger. Replace the cover and cycle after each trial adjustment to obtain accurate readings.

Indicator Light Relay Test

1. Make meter connections as shown in Fig. P4. *If the vehicle indicator light cannot be seen while making this test, connect a 12-volt test light, using a No. 57 bulb between the "L" and "B" terminals of the regulator.*

2. Turn the rheostat to the "resistance in" position and operate the engine at approximately 800 rpm. Slowly cut out resistance and observe the ammeter reading when the vehicle indicator (or test) light goes out. It should be 4 to 7 amperes.

3. To adjust relay contact opening amperage, remove the cover and bend the lower armature spring hanger up or down. Increasing the spring tension raises the setting. Replace the cover and recheck.

BENCH TESTS

If the regulator voltage is unstable or cannot be adjusted to specifications in the above test, remove the regulator from the vehicle for further tests and adjustments.

Voltage Regulator Contact Gap

Contact gap is checked by placing a .010″ gauge between the lower movable contact and the lower stationary contact. Adjust by bending the upper stationary contact arm up or down. Be sure that proper contact alignment is maintained.

Voltage Regulator Air Gap

To check the voltage regulator air gap, connect a No. 57 bulb in series with a 12-volt battery between the regulator field terminal and base.

Place a .034″ round wire gauge between the armature and core on the side of the brass stop rivet nearest the center of the core head.

The lower movable contact should barely touch the lower stationary contact when the armature is pressed down against the gauge and the lamp should light. With a .038″ gauge between the armature and core the light should go out.

To adjust to the above specifications, loosen the stationary bracket attaching screw and move the bracket up or down. *This is a preliminary adjustment only as the air gap may be changed when voltage "spread" is established as outlined previously.*

Servicing Regulator Contacts

If the contacts are rough or oxidized, they may be cleaned with an American Swiss No. 6 equal-

Fig. P6 *Removing drive end head and bearing*

Fig. P7 *Removing roller bearing from slip ring end head*

ling cut file. After filing, the contacts should be cleaned with a strip of linen tape saturated with a few drops of lighter fluid and drawn between the contacts. Then repeat with a dry strip of tape to remove fluid.

ALTERNATOR REPAIRS

When repairing the alternator, the complete disassembly may not be required. In some cases it will be necessary only to perform those operations which are required to repair or replace the defective part. However, the following material covers the complete overhaul procedure, Fig. P5.

Disassembly

1. Remove through bolts and tap lightly on end heads to separate them from stator.
2. Remove drive end head and rotor.
3. Remove nuts, etc. from rectifier bracket studs. Separate slip ring end from stator.
4. Both brushes are located in slip ring end housing and should be replaced if worn to $5/16''$ or less in length. Remove the brushes, being careful that the brush springs are not lost when brushes are removed from holders.

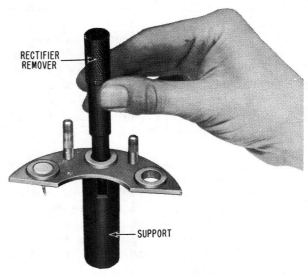

Fig. P8 *Removing diodes from heat sink*

5. Remove pulley with puller or arbor press. Remove fan.
6. Remove drive end head and bearing, Fig. P6.
7. Remove three screws from retainer plate and press bearing out of drive end head, using tool shown in Fig. P13.
8. Remove roller bearing from slip ring end head, Fig. P7.
9. If rectifiers must be removed, cut rectifier wire as near to crimped sleeve as possible. Then remove rectifiers, Fig. P8, using a press.

Inspection

1. When disassembled all parts should be wiped clean and visually inspected for wear, distortion

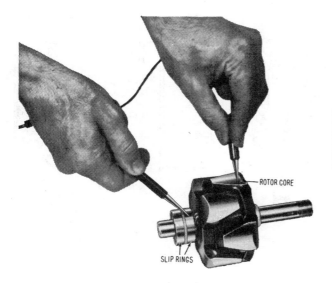

Fig. P9 *Testing rotor for grounds*

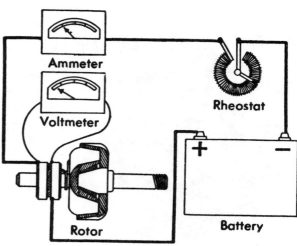

Fig. P10 *Meter connections for testing
rotor field current draw*

or signs of overheating or mechanical interference.

2. Stator windings and leads should be examined for insulation failures or defects. A shorted phase winding or rectifier will normally be evidenced by discoloration.

3. The stator can be checked for shorted windings with an internal-external growler. The test is made with the stator leads disconnected from the rectifiers.

4. The stator can be checked for grounded windings with a 110 volt test lamp and test probes.

Do not make this test with the rectifiers connected to the stator leads.

5. Test rectifiers as described below.

6. The rotor should be tested for grounds and for shorted turns in the winding. The ground test is made with the test probes connected in series with a 110 volt bulb as shown in Fig. P9. If bulb lights rotor winding is grounded. To test for shorted turns, check rotor field current draw as shown in Fig. P10. Using the rheostat, adjust voltage to 10 volts and read field current drawn on ammeter. Reading should be 2.34 to 2.42 at room temperature (70° F). Excessive current drawn indicates shorted turns in field winding.

Fig. P11 *Installing diodes in heat sink*

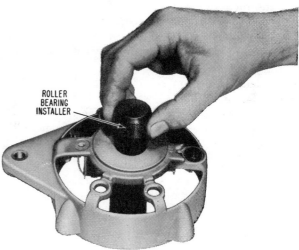

Fig. P12 *Installing roller bearing*

DRIVE END
BEARING
INSTALLER

Fig. P13 *Installing drive
end head and bearing*

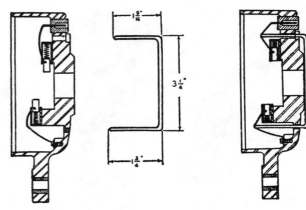

Fig. P14 *Brush spreader clip.
Right view shows clip installed*

Assembly

1. Press the rectifiers in the heat sink, Fig. P11. *Rectifiers are identified by red markings on the positive and black markings on the negative.* Strip the insulation back about ¼″ on stator leads and reconnect them to rectifiers. The connector sleeves should be crimped on rectifier leads. *Do not solder these connections as the excessive heat may damage rectifiers.*

2. Install roller bearing, Fig. P12. Enclosed end of bearing should be flush with outer surface of end frame when installed.

3. Install felt retainer and crinkle washer and press bearing into drive end frame. Use a flat block 2″ square so that pressure is exerted on outer race. Replace retainer plate and gasket on end frame. Make sure snap ring and retainer are in place on rotor shaft and press end frame on, Fig. P13. Press bearing down against snap ring retainer.

4. Replace spacer on shaft and install Woodruff key, fan and pulley.

5. Install brush springs and brushes. Eyelet on ground brush lead is fastened to end frame with a screw and the blade terminal on the insulated brush lead is pushed into slot in field terminal insulator bushing. A tab on blade terminal snaps into insulator bushing. If damaged, this bushing may be replaced.

6. To spread brushes so they will clear slip rings when end head is installed, a wire clip can be made and used as shown in Fig. P14. This clip can be fashioned from a coat hanger or ⅛″ welding rod cut and bent to the dimensions shown. File a "V" groove in ends of wire clip so that it will contact brush leads when inserted through holes in end head.

7. Install negative rectifier heat sink in slip ring end head and replace lockwashers and nuts. Install insulator bushings on positive heat sink studs and install heat sink in end head. Install outer insulator bushings, washers and nuts.

8. Manually press stator in position on slip ring end head and install assembled drive end head and rotor. Make sure that through bolt holes line up on the two end heads. Install and tighten through bolts.

9. Remove brush spreader clip and make sure brush leads do not drag on rotor and that rotor turns freely when rotated by hand. Test alternator when assembled.

TESTING DIODES

Follow the same procedure for testing diodes as outlined for Chrysler diodes.

Review Questions

Page

1. Why can't alternating current be used in the electrical system of an automobile? 670

2. What particular type of power requirements are best met by using an alternator? 670

3. When an alternator is used, how is alternating current changed to direct current? 671

4. What other name is commonly used for an alternator on motor vehicles? 671

5. Name three functional parts of an alternator. 671

6. Which part of an alternator corresponds to the field in a direct current generator? 671

7. Which part of an alternator corresponds to the armature in a direct current generator? .. 671

8. In an alternator system, what regulates power output? 671

9. What energizes the field windings on the alternator? 673

10. What is a distinguishing characteristic of A.C. current which its name implies? 674

11. What is the function of a diode? 676

12. What is the operating principle of a diode? 676

13. On a system that has a negative ground, which diodes are housed in the heat sink? 676

14. How does the use of diodes eliminate the need for a circuit breaker or cutout relay? 676

15. What precaution must be observed while using a battery charger on a car equipped with an alternator? 676

16. What is the first step in performing a charging circuit resistance test on a Chrysler alternator? 677

17. While performing a current output test on a Chrysler system, what would be indicated by a "no charge" reading on the test ammeter? 679

18. While performing a field coil draw test on a Chrysler alternator, what would a low current draw indicate? 681

19. Describe the procedure for testing rectifier diodes with a battery on Chrysler alternators. 683

20. While testing rectifier diodes with a battery, what indications of the test light would mean that a diode is serviceable? 683

21. What is a capacitor? 684

22. What type of solder should not be used for fastening leads in an alternator? 686

23. On a "Delcotron" system using charge indicating light, what extra element is used in the regulator? 690

24. At which terminal of the "Delcotron" is the current output test made? 691

25. Why is it necessary to measure the final voltage setting with the regulator cover on? 692

26. What is meant by "Tailoring Voltage setting"? 692

27. On a "Delcotron," what device is used to check for grounded or open rotor windings? 694

28. On a "Delcotron," what is the method of testing for short circuited coils? 694

29. Describe how it might be concluded that the stator windings of a "Delcotron" are shorted. 694

30. What precaution must be observed while replacing diodes? 695

31. Why is it advisable to rotate the rotor while cleaning slip rings by hand? 695

32. What is the maximum allowable runout on "Delcotron" slip rings? 695

33. What is the most singular feature of a transistor regulator? 697

34. What is the function of the "THERMISTOR" in a Delco transistor regulator? 698

35. Is it possible to alter the voltage setting on a Delco transistor regulator? 698

36. While performing alternator output test on a Ford system, what portion of current output is not indicated on the test ammeter? 701

37. What are some indications of a shorted diode in a Ford system? 701

38. What would be indicated by an excessive current reading for a field coil test on a Ford alternator? 702

39. After what operating conditions is the Ford regulator considered to be at "normal" operating temperature? 703

40. Is it advisable to clean Ford regulator points with compressed air? 706

41. On Ford regulators, what is the factor which controls the voltage difference between upper and lower stage regulation? 706

42. What is the procedure for "cycling" the Ford regulator? 707

43. What is the limit of allowable runout on the slip rings of a Ford alternator? 708

44. What instrument is used to test rectifiers on a Leece-Neville alternator? 712

45. How can negative and positive Leece-Neville terminal post cells be identified? 713

46. What device is used to perform rectifier mount ground test on a Leece-Neville alternator? 714

47. What is the function of the isolation diode in a Motorola alternator? 717

48. During an alternator output test on a Motorola system, what would be indicated by a higher than 12 volt reading at the battery terminal of the regulator while no output is shown on the ammeter? 718

49. While testing isolation diode on a Motorola alternator with a test lamp what condition would be indicated if the test lamp did not light in either direction? 719

50. Describe a simple test for an open rectifier that can be made on a Prestolite alternator. .. 724

51. What would be the result of grounding the field circuit in a Prestolite system while the ignition is turned on? 724

52. What are the means of identifying positive and negative rectifiers on Prestolite units? 728

Starting Motors

Review Questions for This Chapter on Page 758

INDEX

DESCRIPTION & OPERATION

Bendix drives 736
Construction of overrunning
   clutch 733
Construction of starting motor . 734
Description of starting motor .. 733
Fundamentals of starter
   operation 731
How engine is disconnected
   from starter 731
Overrunning clutch drive
   operation 733

CHECKING STARTER CIRCUIT ON CAR

Checking circuit with voltmeter . 738
If lights dim 737
If lights go out 737
Lights stay bright, no crank-
   ing action 738

CHRYSLER DIRECT DRIVE STARTER

Adjusting pinion clearance 744

Disassembly 742
Reassembly 743

CHRYSLER REDUCTION GEAR STARTER

Disassembly 744
Reassembly 745

DELCO-REMY STARTERS

Disassembling starter 751
Maintenance 750
Pinion clearance 751
Reassembling starter 752
Solenoid 749
Solenoid terminals 750

FORD STARTER WITH INTEGRAL POSITIVE ENGAGEMENT DRIVE

Description 752
Disassembly 753
Reassembly 753

FORD STARTER WITH FOLO-THRU DRIVE

Disassembly 754
Reassembly 755

PRESTOLITE STARTING MOTORS

Assembling starter 742
Disassembling Folo-Thru drive
   motor 741
Disassembling overrunning
   clutch motor 739

STARTER & DRIVE TROUBLES

Bendix Folo-Thru drive 757
Interpreting results of tests ... 758
Testing starters 757
Types of drive trouble 756
When Bendix drive fails 757
When clutch drive fails 757

FUNDAMENTALS OF STARTER OPERATION

As explained in the chapter on electricity, the starting motor circuit or starter circuit includes a small but powerful electric motor, a starter switch, a storage battery and the necessary conductors to carry the electric current. When the starter switch is closed, current flows from the battery through the starter, rotating the starter armature. On one end of the starter armature shaft is a pinion gear which is automatically meshed with a large gear on the rim of the flywheel. The starting motor drives the flywheel gear, which in turn rotates with the engine, Fig. 1.

The ratio between the flywheel gear and the pinion gear is about 15 to 1, which means that once the engine starts, the engine drives the starter armature 15 times as fast as the engine rotates. Even though the just-started engine ro-

tates at the moderate speed of say 1,000 revolutions per minute, the corresponding starter armature speed would be 15,000 rpm. This speed (or higher) might wreck the armature because of centrifugal force. Therefore, it is necessary to provide some means for disconnecting the armature from the engine the instant the engine starts.

How Engine is Disconnected from Starter

There are two methods for quick disconnection of the engine from the starter armature. One is known as the over-running clutch drive and the other is called the Bendix drive after the man who invented it. Both these drives will be described in detail later.

For the moment it is sufficient to explain that in the Bendix drive, the initial rotation of the armature forces the pinion into mesh with the flywheel gear and thus the starter drives the flywheel.

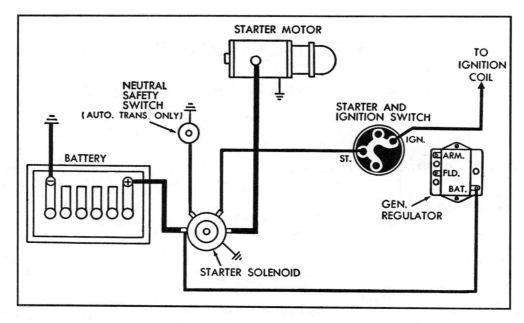

Fig. 1   Typical starting circuit diagram

When the engine starts, the flywheel tries to drive the armature pinion but this attempt results merely in the pinion being pushed out of mesh. With this system, all the car driver has to do is close the starter circuit. Then the pinion moves into mesh automatically and cranks the engine. When the engine starts, the pinion is automatically demeshed even though the driver still has the starter circuit closed.

The flywheel gear is often called a flywheel ring gear because any large gear is likely to be named a ring gear. A pinion gear or pinion, such as the starter pinion gear, is always a small gear, with about nine teeth or just a few more.

In the over-running clutch drive, depression of the starter pedal does two things almost simultaneously: (1) It shifts the pinion into mesh with the flywheel ring gear and (2) it closes the starter switch. The over-running clutch is a one-way automatic clutch. It permits the starter pinion to drive the flywheel gear but it won't allow the flywheel gear to drive the starter pinion (and the armature to which it is connected). In other words, the over-running clutch drives perfectly in the *right* direction and slips perfectly in the *wrong* direction. The over-running clutch is part of the starter pinion assembly. After the engine starts, the pinion remains in mesh with the flywheel ring

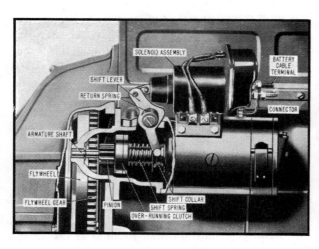

Fig. 2   Starter equipped with overrunning clutch drive

Fig. 3   When starter pedal is depressed the starter switch is closed and the pinion is moved into mesh with the flywheel gear

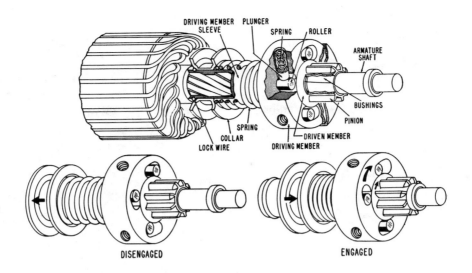

Labels on figure:

DRIVING MEMBER SLEEVE — PLUNGER — SPRING — ROLLER — ARMATURE SHAFT — BUSHINGS — PINION — DRIVEN MEMBER — DRIVING MEMBER — SPRING — COLLAR — LOCK WIRE

DISENGAGED   ENGAGED

Fig. 4   Details of overrunning clutch drive

gear until the driver releases the starter switch, then a spring moves the pinion out of mesh.

Overrunning Clutch Drive Operation

A starter with an over-running clutch drive is shown in Fig. 2 which shows the starter disengaged whereas in Fig. 3 the starter pedal has been depressed, the starter switch is closed and the pinion is in mesh with the flywheel gear.

When the starter pedal is depressed, the shift lever moves the collar to the left, compressing the "shift" spring which pushes on the over-running clutch to which the pinion is attached. The over-running clutch is splined to the armature shaft and slides freely on it.

The shift lever also depresses the starter button and closes the starter switch, Fig. 3. The purpose of the shift spring is to allow it to compress in the event that the teeth on the pinion butt against the teeth on the flywheel gear. In this case, the compression of the spring permits the starter switch to be closed even with teeth butting. Then the armature starts to rotate and enables the clutch spring to push the pinion into mesh.

Overrunning Clutch Drive Construction

The construction of the over-running clutch is shown in Fig. 4. The principal parts of the clutch are the driving and driven members (direction of rotation is shown by the arrows). A sleeve on the driving member is slidably splined on the armature shaft so that the drive pinion can be meshed and demeshed with the flywheel gear. The driven member is an integral part of the drive pinion.

There are four cylindrical rollers resting between the driving and driven members. The rollers are located in suitable slots. The outer surfaces of the slots are slightly sloped or tapered and each roller is forced toward the tapered end of its slot by a small coil spring so that when the driving member is rotated in a clockwise direction the rollers firmly wedge themselves between the driving and driven members, causing the driven member to be rotated positively by the driving member.

When the engine starts, it drives the pinion in a clockwise direction at many times the speed of the driving member. Thus the driven member runs away (over-runs) from the driving member, moving the rollers clockwise out of their jammed position and allowing the driven member to run free with respect to the driving member. Hence, the high speed of the pinion, which is in mesh with the flywheel gear, cannot be transmitted to the armature. The pinion is returned to its unmeshed position as soon as the driver releases the starting switch.

Description of Starting Motor

Fig. 5 is a simplified diagram showing the wiring within a starting motor. This particular motor has windings on just two field poles. Just beneath the starter terminal the current divides, half going through each field winding. Each winding is connected to an "ungrounded" brush on the commutator. From these brushes, current flows through the armature winding to the grounded brushes and from thence the current flows through the starter housing, through engine and

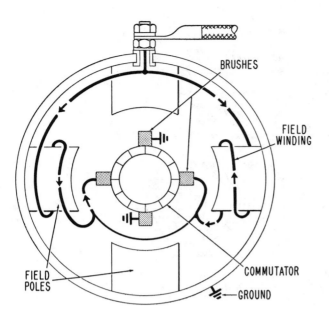

*Fig. 5 Wiring diagram of starter with
two field windings. Equalizer wire is
shown thinner*

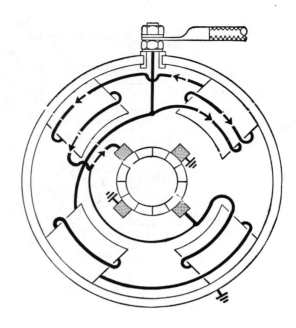

*Fig. 6 Wiring diagram of starter
with four field windings*

frame to battery ground strap to battery. The two ungrounded brushes are often connected by an equalizer wire which equalizes any slight variation in voltage at the two ungrounded brushes. The equalizer wire is invariably used on late model starters.

Many starters have windings on all four poles as shown in Fig. 6. The current divides as before. As shown by the arrows, the upper two windings are in series, current flowing from the starter terminal through the upper right winding, then through the upper left winding to the upper left ungrounded brush. Likewise, current flows from

the starter terminal through the lower left winding, through the lower right ungrounded brush.

The magnetic field strength (and cranking power) of a starter with only two field windings is almost as great as with four field windings because in the former case more turns per coil are used than in the latter.

Construction of Starting Motor

The starting motor, Fig. 7, has a round, tubular housing called the field frame. Attached to it at one end is the cast iron drive housing, while at the other end is the commutator end frame which carries the brushes. The commutator end frame may be made of cast iron, sheet steel or be die cast. "Through bolts," running from the commutator end frame to the drive housing, clamp the round field frame firmly between them to form a rigid three-piece housing.

The field poles are attached to the field frame by large screws and the field coils are a snug fit on the field poles. The armature shaft in Fig. 7 is supported in two bearings although some starters have three bearings. The bearings are usually thin bushings although a ball bearing is sometimes used at the drive end. If the commutator end frame is made of cast iron, the iron itself may be used as a bearing, as in Fig. 7. Bushings may be bronze (an alloy composed mostly of tin and copper) or they may be made of a porous copper material and the pores impregnated with oil. The

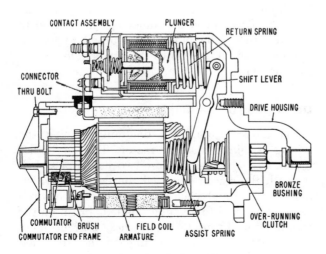

*Fig. 7 Sectional view of starter
with overrunning clutch drive*

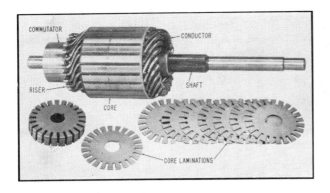

Fig. 8 *Starter armature and laminations*

latter is often called an oilless bearing or bushing because it does not require periodic oiling. Actually it is an "oilless" bushing just as long as the microscopic holes in its structure are filled with oil—as in the case of a sponge. The lubricant in an oilless bushing should last for the life of the car. On the other hand, when a starter is disassembled, the bushing should be soaked in engine oil.

The armature core, Fig. 8, is made of numerous laminations of soft iron about 1/32″ thick which are insulated by thin sheets of paper or by special varnish. If a solid iron core were used, this comparatively huge chunk of iron would generate local voltages in the various parts of its volume. This would result in rambling currents flowing here and there throughout the armature. These currents would represent wasted energy. They might heat up the armature to a point where the insulation on the windings might be charred, and they might distort the magnetic field of the armature sufficiently so that it would be unable to function at its designed efficiency. These difficulties are almost completely overcome with a

Fig. 9 *In this design brushes are mounted on pivoted arms*

core made of numerous insulated laminations, since the voltage and current generated in any individual lamination are negligible. Armature and shaft may be considered one piece because the laminated core and shaft are firmly pressed together.

There are two different methods of mounting the brushes as shown in Figs. 9, 10 and 11. In Fig. 9, it will be seen that the brushes are mounted on the ends of swiveled arms which are pivoted on pins attached to the commutator end frame. A coil spring surrounding the pin presses the brush against the commutator. In Figs. 10 and 11, the brush holder consists of a rectangular guide within which the brush slides. The brush is pressed against the commutator by a helical brush spring.

In both cases, the starter terminal connects with the field coils. The field coils in turn are connected to the "leads" running to the "live" brushes. The grounded brushes are grounded by attaching their leads to the field frame in the case of Figs. 10 and 11, and to the commutator end frame in the case of Fig. 9.

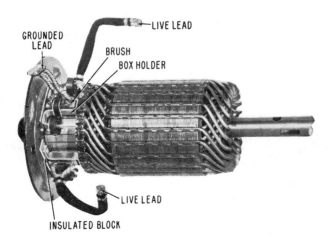

Fig. 10 *In this design brushes are slidably mounted in box-type metal guides*

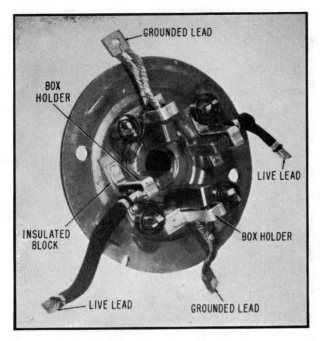

Fig. 11 Details of brushes and end frame shown in Fig. 10

BENDIX DRIVES

In the Bendix drive, the pinion is meshed and unmeshed by what may be described as the spinning nut principle. Everybody is familiar with the fact that if a bolt is held in one hand and a nut on the bolt is spun with the other hand that the nut will move along the bolt. It should be equally obvious that if the bolt is suddenly spun, the nut will move along the bolt. This is the principle of the Bendix drive. The pinion gear represents the outside of a large nut which is mounted on the threaded armature shaft. Sudden rotation of the starter armature moves the pinion into mesh with the flywheel gear. However, just as soon as the engine starts, it spins the pinion out of mesh because of the high speed at which the flywheel

gear rotates the pinion. (If at this instant the engine is running at 1,000 revolutions per minute, it spins the pinion at say 15,000 rpm.)

Fig. 12 illustrates what is known as the standard Bendix drive. [Bendix drives are also classified as to whether they are inboard or outboard.] The outboard type has bushings supporting both ends of the Bendix drive. In the inboard type the Bendix drive overhangs the starting motor. Another way of defining the inboard and outboard designs is to say that the outboard type has a drive housing whereas the inboard type does not. The drive pinion is threaded like a nut except the thread is a very coarse one. The pinion is mounted on a threaded sleeve which in turn is mounted on the armature shaft. The armature shaft drives the threaded sleeve. The pinion, Fig. 12, is weighted on the bottom so that when the armature suddenly starts rotating in the direction of the arrow, the weight holds the pinion against rotation and consequently the pinion is screwed to the right and into mesh with the flywheel gear.

As soon as the engine starts, it drives the pinion several times as fast as the armature speed and in so doing spins it back to the left out of mesh. When the armature is not in use, the light anti-drift spring holds the pinion to the left in the position shown, Fig. 12. If it were not for this spring, the pinion might accidentally be meshed on a rough road or a steep up-grade.

To prevent undue shock when the pinion starts driving the engine, the armature drives the threaded sleeve through a cushion drive spring, or drive spring for short.

CHECKING STARTER CIRCUIT ON CAR

When trouble develops in the starting motor circuit, and the starter cranks the engine slowly or not at all, several preliminary checks can be made

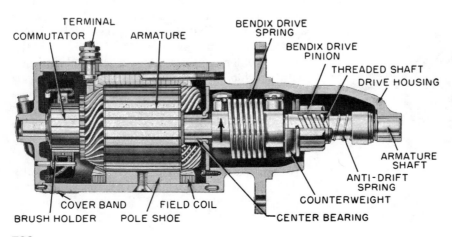

Fig. 12
Starter equipped
with standard
Bendix drive

to determine whether the trouble lies in the battery, in the starter, in the wiring between them, or elsewhere. Many conditions besides defects in the starter itself can result in poor cranking performance.

To make a quick check of the starter system, turn on the headlights. They should burn with normal brilliance. If they do not, the battery may be run down and it should be checked with a hydrometer (see Battery chapter).

If the battery is in a charged condition so the lights burn brightly, operate the starting motor. Any one of three things will happen to the lights: (1) They will go out, (2) dim considerably or (3) stay bright without any cranking action taking place.

If Lights Go Out

If the lights go out as the starter switch is closed, it indicates that there is a poor connection between the battery and starting motor. This poor connection will most often be found at the battery terminals. Correction is made by removing the cable clamps from the terminals, cleaning the terminals and clamps, replacing the clamps and tightening them securely. A coating of corrosion inhibitor (vaseline will do) may be applied to the clamps and terminals to retard the formation of corrosion.

If Lights Dim

If the lights dim considerably as the starter switch is closed and the starter operates slowly or not at all, the battery may be run down, or there may be some mechanical condition in the engine or starting motor that is throwing a heavy burden on the starting motor. This imposes a high discharge rate on the battery which causes noticeable dimming of the lights.

Check the battery with a hydrometer. If it is charged, the trouble probably lies in either the engine or starting motor itself. In the engine, tight bearings or pistons or heavy oil place an added burden on the starting motor. Low temperatures also hamper starting motor performance since it thickens engine oil and makes the engine considerably harder to crank and start. Also, a battery is less efficient at low temperatures.

In the starting motor, a bent armature shaft, loose pole shoe screws or worn bearings, any of which may allow the armature to drag, will reduce cranking performance and increase current draw.

In addition, more serious internal damage is sometimes found. Thrown armature windings or

Fig. 13 *Checking voltage drop between car frame and grounded battery terminal post*

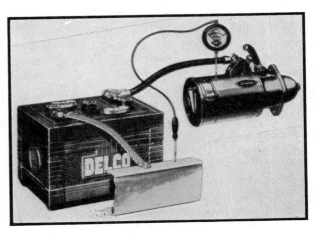

Fig. 14 *Checking voltage drop between car frame and starting motor field frame*

Fig. 15 *Checking voltage drop between insulated battery terminal post and starting motor terminal stud (or battery terminal stud on solenoid, if so equipped)*

737

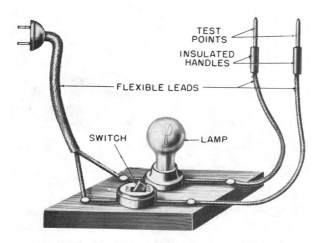

Fig. 16 *Illustrating a simple test lamp for use in making continuity and ground tests on armature and field windings*

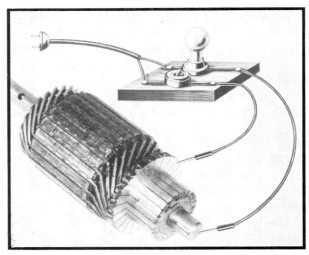

Fig. 17 *Checking armature for grounds. If lamp lights, armature is grounded and should be replaced*

commutator bars, which sometimes occur on over-running clutch drive starting motors, are usually caused by excessive overrunning after starting. This is the result of such conditions as the driver keeping the starting switch closed too long after the engine has started, the driver opening the throttle too wide in starting, or improper carburetor fast idle adjustment. Any of these subject the over-running clutch to extra strain so it tends to seize, spinning the armature at high speed with resulting armature damage.

Another cause may be engine backfire during cranking which may result, among other things, from ignition timing being too far advanced.

To avoid such failures, the driver should pause a few seconds after a false start to make sure the engine has come completely to rest before another start is attempted. In addition, the ignition timing should be reset if engine backfiring has caused the trouble.

Lights Stay Bright, No Cranking Action

This condition indicates an open circuit at some point, either in the starter itself, the starter switch or control circuit. If the car is equipped with a solenoid starter switch, the solenoid control circuit can be eliminated momentarily by placing a heavy jumper lead across the solenoid main terminals to see if the starter will operate. This connects the starter directly to the battery and, if it operates, it indicates that the control circuit is not functioning normally. The wiring and control units must be checked to locate the trouble. If the starter does not operate with the jumper attached, it will probably have to be removed from the engine so it can be examined in detail.

CHECKING CIRCUIT WITH VOLTMETER

Excessive resistance in the circuit between the battery and starter will reduce cranking performance. The resistance can be checked by using a voltmeter to measure voltage drop in the circuits while the starter is operated.

There are three checks to be made:

1. Voltage drop between car frame and grounded battery terminal post (not cable clamp), Fig. 13.
2. Voltage drop between car frame and starting motor field frame, Fig. 14.
3. Voltage drop between insulated battery terminal post and starting motor terminal stud (or the battery terminal stud of the solenoid), Fig. 15.

Each of these should show no more than one-tenth (0.1) volt drop when the starting motor is cranking the engine. Do not use the starting motor for more than 30 seconds at a time to avoid overheating it.

If excessive voltage drop is found in any of these circuits, make correction by disconnecting the cables, cleaning the connections carefully, and then reconnecting the cables firmly in place. A coating of vaseline on the battery cables and terminal clamps will retard corrosion.

NOTE—On some cars, extra long battery cables may be required due to the locations of the battery and starter. This may result in somewhat higher voltage drop than the above recommended 0.1 volt. The only means of determining the

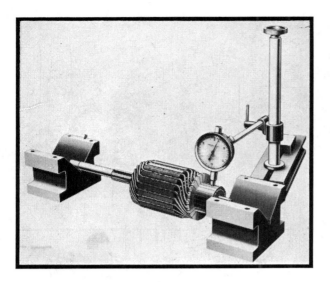

Fig. 18 *Measuring commutator runout with dial indicator. Mount shaft in "V" blocks and rotate commutator. If runout exceeds .003", commutator should be turned in a lathe to make it concentric*

Fig. 19 *Turning starting motor commutator in a lathe. Take light cuts until no worn or bad spots appear. Then remove burrs with 00 sandpaper*

normal voltage drop in such cases is to check several of these vehicles. Then, when the voltage drop is well above the normal figure for all cars checked, abnormal resistance will be indicated and correction can be made as already explained.

To obtain full performance data on a starting motor, or to determine the cause of abnormal operation, the starting motor should be submitted to a no-load and torque test. These tests are best performed on a starter bench tester with the starter mounted on it. There are several manufacturers of this equipment and each one furnishes complete instructions on how it should be used.

Of course, complete manufacturer's specifications for each starter tested are required.

From a practical standpoint, however, a simple torque test may be made quickly with the starter in the car. Make sure the battery is fully charged and that the starter circuit wires and terminals are in good condition. Then operate the starter to see if the engine turns over normally. If it does not, the torque developed is below standard and the starter should be removed for further checking.

Remove the starter from the engine, disassemble it as outlined further on and make the tests as suggested in Figs. 16 through 22.

PRESTOLITE STARTERS

Two types of starters are in general use. Heavy duty engines are equipped with the unit shown in Fig. 23 and light duty engines use the one shown in Fig. 24. In both motors the brush holders are riveted to a separate brush plate and are not serviced separately. Brush replacement is accomplished by removing the bearing end frame and armature.

Disassembling Overrunning Clutch Motor

1. Referring to Fig. 23, remove roll pin from shifting fork and solenoid coupling pin.
2. Remove solenoid.
3. Support solenoid coupling pin and drive out roll pin.
4. Remove solenoid rubber boot, plunger spring and coupling pin.
5. Remove thru bolts and take out commutator end cover, thrust washer and insulators.
6. Remove drive end housing, drive fork and armature from field frame.
7. Remove roll pin attaching shifting fork to pinion housing and remove retainer, dust cover and shifting fork.

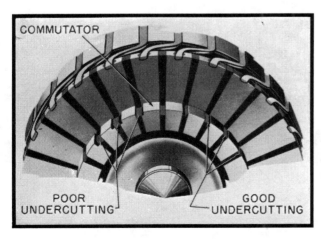

Fig. 20 *Good undercutting should be .002" wider than the mica, 1/64" deep and exactly centered so that there are no burrs on the mica*

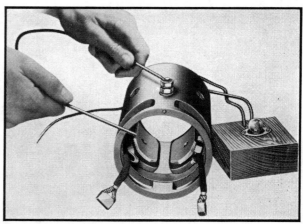

Fig. 22 *Testing field coils for grounds. A six-volt test lamp is shown but the test lamp illustrated in Fig. 29 can also be used. If a ground is present the lamp will light*

Fig. 21 *This illustrates the use of an open core transformer (growler) and a steel strip to test armature for shorts. Turn armature slowly and if a short is present the steel strip will vibrate rapidly*

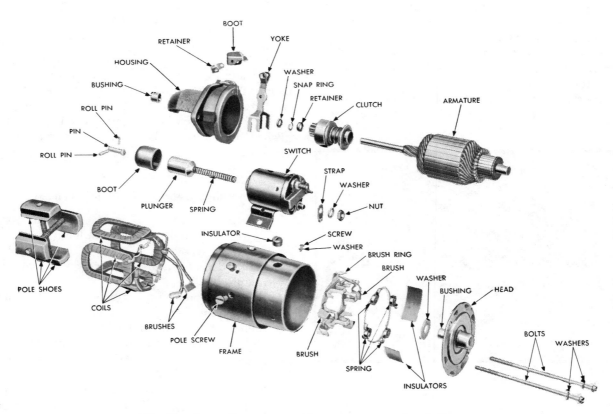

Fig. 23 *Prestolite starter with over-running clutch drive*

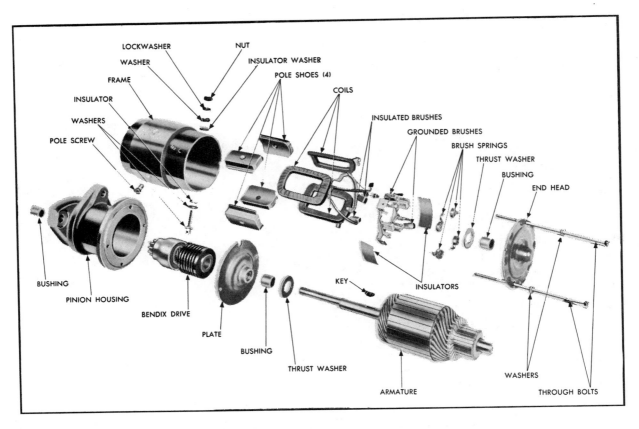

Fig. 24   *Prestolite starter with Folo-Thru drive*

8. Remove spacer and slide pinion gear toward commutator end of armature; then drive stop collar toward pinion and remove lock ring.

9. Slide starter drive from armature.

10. Remove brush holder ring from field frame (3 screws).

11. Disconnect field lead wire at brush holder ring, disengage the brushes from the holders and carefully slide brush holder ring from field frame.

Disassembling Folo-Thru Drive Motor

1. Referring to Fig. 24, remove thru bolts and tap commutator end frame from field frame.

2. Remove brush holder ring from field frame (3 screws).

3. Disconnect field lead wire at brush holder ring, disengage brushes from holders and carefully slide brush holder ring from field frame.

4. Remove set screw from drive and armature shaft by prying back drive spring to expose screw under spring.

5. Tap drive assembly from armature shaft and remove woodruff key.

6. Slide center bearing plate from armature shaft.

Overrunning Clutch

Place drive unit on shaft and, while holding armature, rotate pinion. The drive pinion should rotate smoothly in one direction (not necessarily easily), but should not rotate in the opposite direction. If drive unit does not function properly or pinion is worn or burred, replace drive unit.

Folo-Thru Drive

This type of drive, Fig. 25, is serviced only as an assembly. To clean and oil the drive with armature removed, use a screwdriver to pry back the drive spring to expose the attaching set screw. Back out the set screw far enough to clear arma-

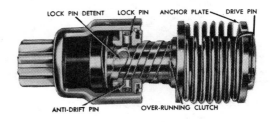

Fig. 25   *Folo-Thru starter drive with Bendix spring*

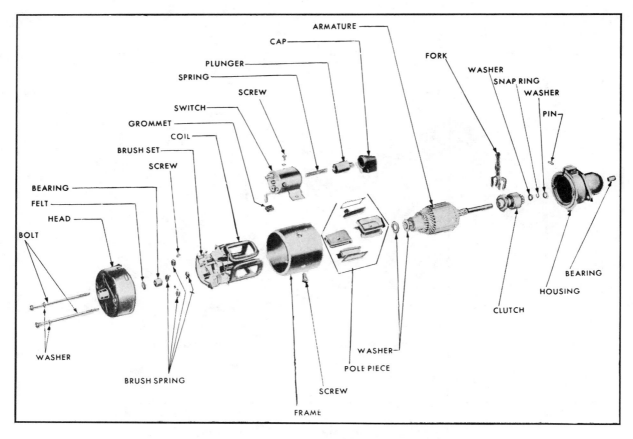

Fig. 26   Chrysler Corp. direct drive starter

ture shaft and slide drive off end of armature shaft. Wipe armature shaft clean and dry and re-oil with SAE 10W oil. Rotate pinion and barrel to the fully engaged position and apply clean kerosene with a brush to the screw shaft threads. Wipe dry and apply a light film of SAE 10W oil to the screw shaft threads and a few drops of oil under the spring on the thrust washer.

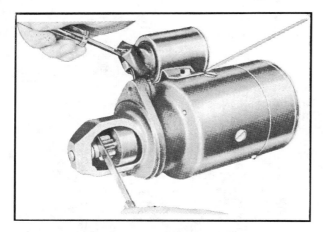

Fig. 27   Clearance between end of pinion and pin stop should be ⅛″ with plunger seated and pinion pushed toward commutator end

Assembling Starter

Reassemble procedure is the reverse of disassembly. However, on Folo-Thru drive motors, be sure the starter drive set screw is seated in the armature shaft; be sure the end frames are positioned on dowel pins; and be sure that the slots in the intermediate bearing and commutator end head line up with the indexing pins in the frame before installing and tightening thru bolts.

CHRYSLER DIRECT DRIVE STARTER

This Chrysler built starting motor, Fig. 26, is a four coil assembly with an over-running clutch type drive and a solenoid shift-type switch mounted on the motor. The brush holders are riveted to a separate brush plate and are not serviced individually. Brush replacement can be made by removing the commutator bearing end head.

Disassembly

1. Remove through bolts and tap commutator end head from field frame.

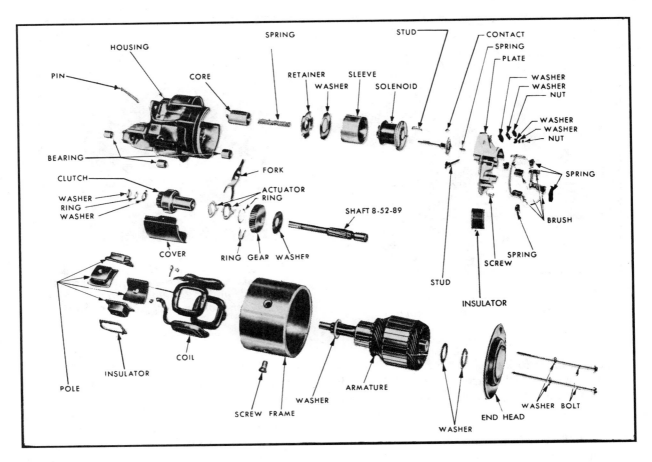

Fig. 28 *Chrysler Corp. reduction gear starter*

2. Remove thrust washers from armature shaft.
3. Lift brush holder springs and remove brushes from holders.
4. Remove brush plate.
5. Disconnect field leads at solenoid connector.

6. Unfasten and remove solenoid and boot assembly.
7. Drive out over-running clutch shift fork pivot pin.
8. Remove drive end pinion housing and spacer washer.
9. Note position of shifter fork on starter and remove fork.
10. Slide over-running clutch pinion gear toward commutator end of armature. Drive stop retainer toward clutch pinion gear to expose snap ring and remove ring.
11. Slide clutch drive from armature shaft.
12. If necessary to replace the field coils, remove screw that holds ground brushes and raise brushes with the terminal and shunt wire up and away from field frame. Remove pole shoe screws and take out field coils.

Reassembly

1. Lubricate armature shaft and splines with SAE 10W or 30W rust preventive oil.
2. Install starter drive, stop collar (retainer), lock ring and spacer washer.

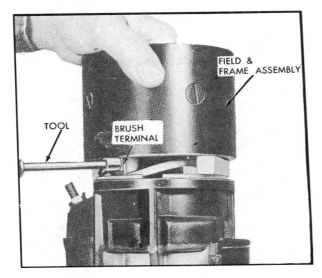

Fig. 29 *Removing brush terminal screw*

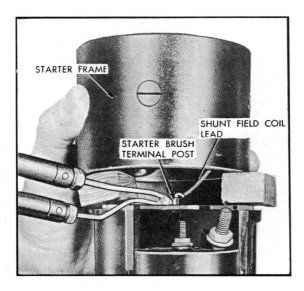

Fig. 30 *Unsoldering shunt field coil lead from starter brush terminal*

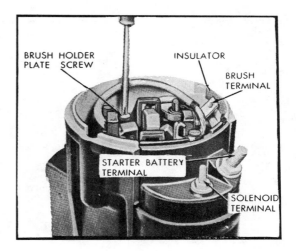

Fig. 31 *Remove brush insulator which prevents contact between brush terminal and gear housing. Remove screw attaching brush holder plate to gear housing*

3. Install shifter fork over starter drive spring retainer washer with narrow leg of fork toward commutator. *If fork is not positioned properly, starter gear travel will be restricted, causing a lockup in the clutch mechanism.*

4. Install drive end (pinion) housing on armature shaft, indexing the shift fork with slot in drive end of housing.

5. Install shift fork pivot pin.

6. Install armature with clutch drive, shifter fork and pinion housing. Slide armature into field frame until pinion housing indexes with slot in field frame.

7. Install solenoid and boot assembly and tighten bolts securely.

8. Install ground brushes.

9. Connect field coil leads at solenoid connector.

10. Install brush holder ring, indexing tang of ring in hole of field frame.

11. Position brushes in brush holders. *Be sure field coil lead wires are properly enclosed behind brush holder ring and that they do not interfere with brush operation.*

12. Install thrust washer on commutator end or armature shaft to obtain .010″ minimum end play.

13. Install commutator end head.

14. Install through bolts and tighten securely.

Adjusting Pinion Clearance

1. Place starter in vise with soft jaws and tighten vise enough to hold starter. *Place a wedge or screwdriver between bottom of solenoid and*

starter frame to eliminate all deflection in solenoid when making pinion clearance check.

2. Push in on solenoid plunger link, Fig. 27 (not fork lever) until plunger bottoms.

3. Measure clearance between end of pinion and pin stop with plunger seated and pinion pushed toward commutator end. Clearance should be ⅛″. Adjust by loosening solenoid attaching screws and move solenoid fore and aft as required.

4. Test starter operation for free running and install on engine.

CHRYSLER REDUCTION GEAR STARTER

This reduction gear starting motor, Fig. 28, has an armature-to-engine crankshaft ratio of 45 to 1, a 3½ to 1 reduction gear set is built into the motor assembly. The starter utilizes a solenoid shift. The housing of the solenoid is integral with the starter drive end housing.

Disassembly

1. Place gear housing of starter in a vise with soft jaws. *Use vise as a support fixture only; do not clamp.*

2. Remove through bolts and starter end head assembly.

3. Carefully pull armature up and out of gear housing, and starter frame and field assembly. Remove steel and fiber thrust washer. *The wire of the shunt field coil is soldered to the brush terminal. One pair of brushes are connected to this terminal. The other pair of brushes is at-*

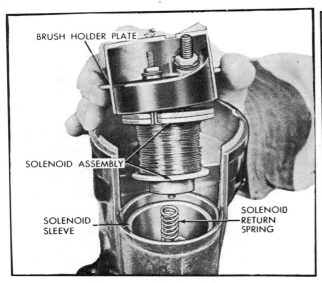

Fig. 32 *Remove brush holder plate with brushes and solenoid as a unit*

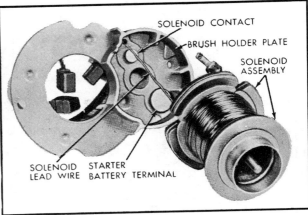

Fig. 34 *Remove nut, steel washer and nylon washer from solenoid terminal. Separate brush holder plate from solenoid*

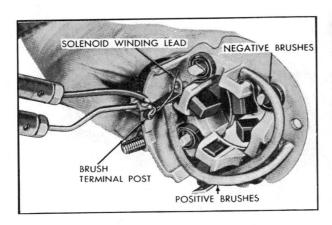

Fig. 33 *Unsolder solenoid winding from brush terminal*

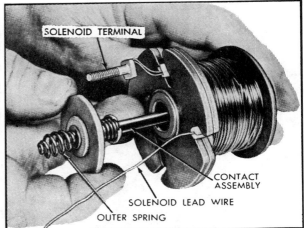

Fig. 35 *Remove nut, steel washer and nylon washer from starter battery terminal. Remove terminal from holder plate. Then remove solenoid contact assembly*

tached to the series field coils by means of a terminal screw. Carefully pull the frame and field assembly up just enough to expose the terminal screw and the solder connection of the shunt field at the brush terminal. Place two wood blocks between starter frame and gear housing, Fig. 30, to facilitate removal of terminal screw and unsoldering of shunt field wire at brush terminal.

4. Support brush terminal by placing a finger behind terminal and remove screw, Fig. 29.
5. Complete the disassembly procedure by referring to Figs. 30 through 44.

Reassembly

The shifter fork consists of two spring steel plates assembled with two rivets, Fig. 45. There

should be about $\frac{1}{16}''$ side movement to insure proper pinion gear engagement. Lubricate between plates sparingly with SAE 10 engine oil.

1. Position shift fork in drive housing and install fork retaining pin, Fig. 44. *One tip of pin should be straight, the other tip should be bent at a 15 degree angle away from housing. Fork and pin should operate freely after bending tip of pin.*
2. Install solenoid moving core and engage shifting fork, Fig. 43.
3. Enter pinion shaft in drive housing and install friction washer and driven gear.

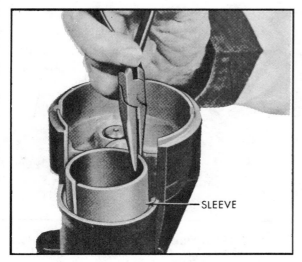

Fig. 36 *Remove solenoid coil sleeve*

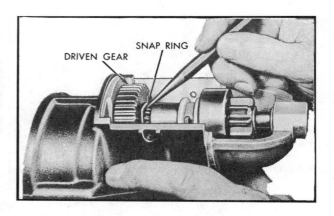

Fig. 39 *Release snap ring that positions driven gear on pinion shaft. This ring is under tension and a cloth should be placed over ring to prevent it from springing away after removal*

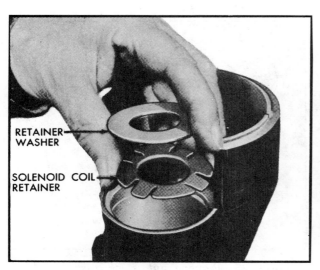

Fig. 37 *Remove solenoid return spring (Fig. 32). Then remove solenoid coil retainer washer and retainer from solenoid housing*

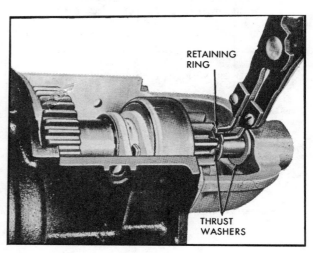

Fig. 40 *Release retainer ring at front of pinion shaft. Do not spread ring any greater than the outside diameter of pinion shaft, otherwise ring can be damaged*

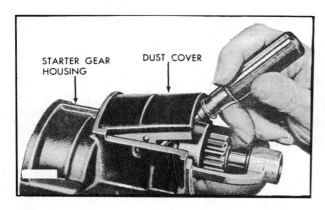

Fig. 38 *Remove dust cover from gear housing*

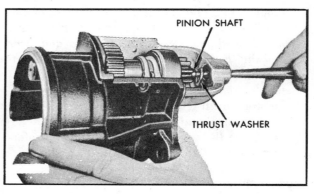

Fig. 41 *Push pinion shaft toward rear of housing and remove ring and thrust washers*

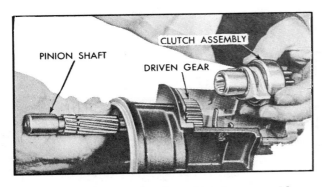

Fig. 42 *Lift out clutch and pinion assembly with the two shifter fork nylon actuators*

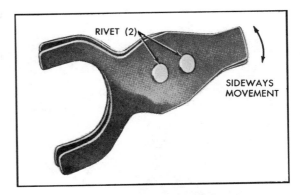

Fig. 45 *Shifter fork assembly*

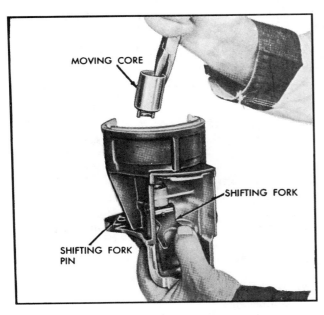

Fig. 43 *Remove driven gear and friction washer. Then pull shift fork forward and remove solenoid moving core*

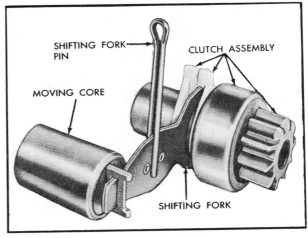

Fig. 46 *Shifter fork and clutch arrangement*

4. Install clutch and pinion assembly, Fig. 42, thrust washer, retaining ring, and thrust washer.

5. Complete installation of pinion shaft, engaging fork with clutch actuators, Fig. 46. *Friction washer must be positioned on shoulder of splines of pinion shaft before driven gear is positioned.*

6. Install driven gear snap ring, Fig. 40. Install pinion shaft retaining ring, making sure ring fits tightly in shaft groove.

7. Install solenoid coil retainer, Fig. 37, with tangs down. *Space retainer in housing bore so that the four tangs rest on ridge in housing bore and not in the recesses.*

8. Install solenoid retainer washer.

9. Install solenoid return spring, Fig. 32. *Inspect condition of solenoid switch contacting washer. If top of washer is burned from arcing, disassemble switch and reverse washer.*

10. Install solenoid contact into solenoid, Fig. 35. Make sure contact spring is positioned in

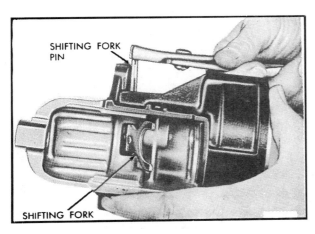

Fig. 44 *Remove shift fork retainer pin and take out shift fork*

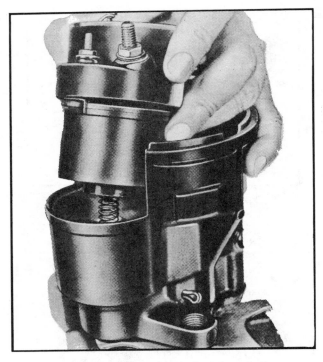

Fig. 47   *Installing solenoid coil and sleeve*

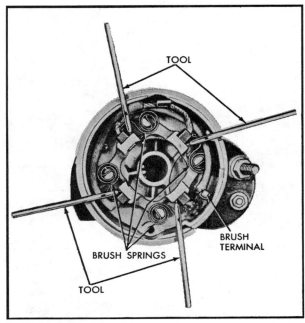

Fig. 48   *Positioning of brushes
with Tool Set C-3855*

solenoid contact. *Inspect condition in contacts in brush holder plate. If contacts are badly burned, replace brush holder with brushes and contacts as an assembly.*

11. Enter solenoid lead wire through hole in brush holder, Fig. 34, and solenoid stud, insulating washer, flat washer and nut.

12. Solder solenoid lead wire to contact terminal, Fig. 33. Wrap wire securely around terminal and solder with a high temperature solder and resin flux.

13. Carefully enter solenoid coil and coil sleeve into bore of gear housing and position brush plate assembly into gear housing, Fig. 47. Align tongue of ground terminal with notch in brush holder.

14. After brush holder is bottomed in housing, install attaching screw, Fig. 31. Tighten screw and install flat insulating washer and hold in place with friction tape.

15. Position brushes with tools shown in Fig. 48 or equivalent.

16. Position field frame to exact position and re-solder field coil lead, Fig. 30.

17. Install brush terminal screw, Fig. 29.

18. Install armature thrust washer on brush holder

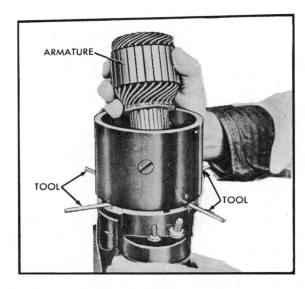

Fig. 49   *Installing armature*

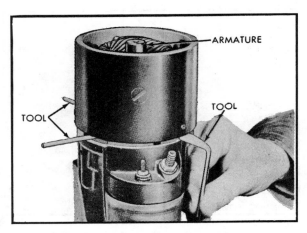

Fig. 50   *Removing brush positioning tools*

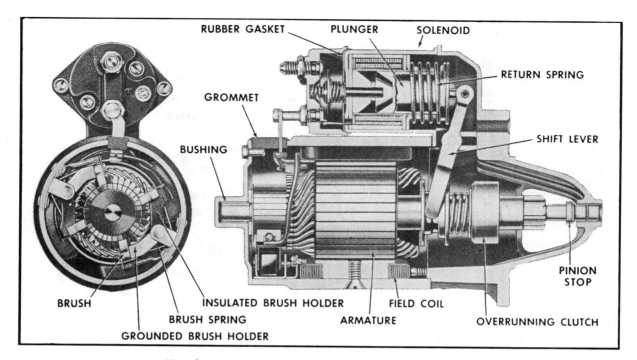

Fig. 51 Delco-Remy starter with enclosed shift lever

plate and enter armature into field frame and gear housing, Fig. 49. Carefully engage splines of shaft with reduction gear.

19. Remove brush positioning tools, Fig. 50. Install fiber and steel thrust washers on armature shaft.
20. Position starter end head, install screws and tighten securely.
21. Install gear housing dust cover. *Make sure dimples on cover are securely engaged in holes provided in gear housing.* Test starter for free running and install on engine.

DELCO-REMY STARTERS

This type starting motor, Fig. 51, has the solenoid shift lever mechanism and the solenoid plunger enclosed in the drive housing, thus protecting them from exposure to road dirt, icing conditions and splash. They have an extruded field frame and an overrunning clutch type of drive. The overrunning clutch is operated by a solenoid switch mounted to a flange on the drive housing.

Solenoid

The solenoid is attached to the drive end housing by two screws. The angle of the nose of the plunger provides a greater bearing area between the plunger and core tube. A molded push rod, Fig. 52, is assembled in the contact assembly. A shoulder molded on the push rod and a cup that

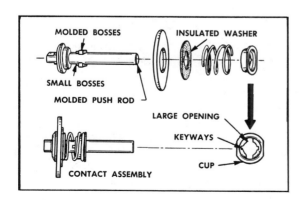

Fig. 52 Solenoid contact assembly

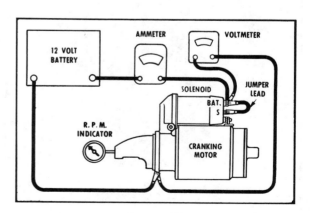

Fig. 53 Connections for checking free speed of motor

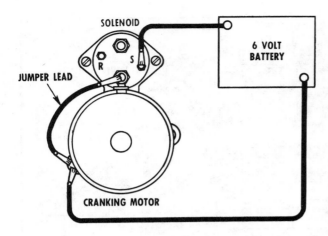

Fig. 54 *Connections for checking pinion clearance*

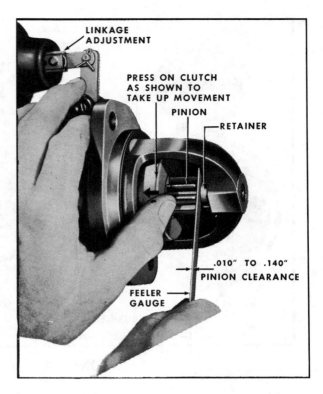

Fig. 55 *Checking pinion clearance*

can easily be assembled to the rod and locked into position over two molded bosses holds the contact assembly in place.

To disassemble the cup from the push rod, push in on the metal cup and rotate ¼ turn so the molded bosses on the rod are in line with openings in the cup; then slide the metal cup off the rod.

To assemble the metal cup on the rod, locate the parts on the rod as shown and align the large openings in the cup with the molded bosses on the rod; then push in on the cup and rotate it ¼ turn so the small bosses on the rod fall into the keyways of the cup.

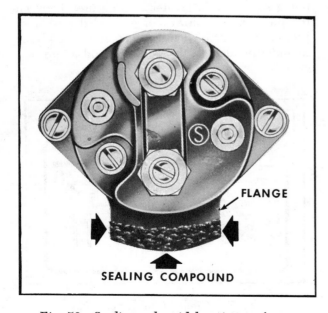

Fig. 56 *Sealing solenoid housing to frame*

Solenoid Terminals

The terminals of the solenoid are assembled in a molded cover. Some solenoids have an additional small terminal which is identified with the letter "R". To this terminal is attached a small metal finger which makes contact with a disc inside the solenoid when it is energized. On the vehicle, this terminal is connected to the battery side of the ignition coil. The purpose of this is to short out the ignition resistor during cranking and thereby provide high ignition coil output for starting the engine.

Maintenance

Most motors of this type have graphite and oil impregnated bronze bearings which ordinarily require no added lubrication except at times of overhaul when a few drops of light engine oil should be placed on each bearing before reassembly.

Motors provided with hinge cap oilers should have 8–10 drops of light engine oil every 5000 miles, or every 300 hours of operation. Since the motor and brushes cannot be inspected without disassembling the unit, there is no service that can be performed with the unit assembled on the vehicle.

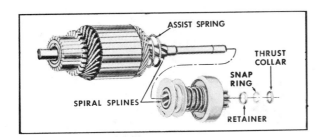

Fig. 57 Disassembled view of armature and over-running clutch

Free Speed Test

With the circuit connected as shown in Fig. 53, use a tachometer to measure armature revolutions per minute. Failure of the motor to perform to specifications may be due to tight or dry bearings, or high resistance connections.

Pinion Clearance

There is no provision for adjusting pinion clearance on this type motor. When the shift lever mechanism is correctly assembled, the pinion clearance should fall within the limits of .010 to .140″. When the clearance is not within these limits, it may indicate excessive wear of the solenoid linkage or shift lever yoke buttons.

Pinion clearance should be checked after the motor has been disassembled and reassembled. To check, make connections as shown in Fig. 54. *Caution: Do not connect the voltage source to the ignition coil terminal "R" of the solenoid. Do not*

use a 12-volt battery instead of the 6 volts specified as this will cause the motor to operate. As a further precaution to prevent motoring, connect a heavy jumper lead from the solenoid motor terminal to ground.

After energizing the solenoid with the clutch shifted toward the pinion stop retainer, push the pinion back toward the commutator end as far as possible to take up any slack movement; then check the clearance with a feeler gauge, Fig. 55.

Disassembling Motor

Normally the motor should be disassembled only so far as necessary to repair or replace defective parts.

1. Disconnect field coil connectors from solenoid "motor" terminal.
2. Remove thru bolts.
3. Remove commutator end frame and field frame assembly.
4. Remove armature assembly from drive housing. On some models it may be necessary to remove solenoid and shift lever assembly from the drive housing before removing the armature assembly. *Important: When solenoid is installed, apply sealing compound between field frame and solenoid flange, Fig. 56.*
5. Remove overrunning clutch from armature shaft as follows:
   a) Slide thrust collar off end of armature shaft, Fig. 57.

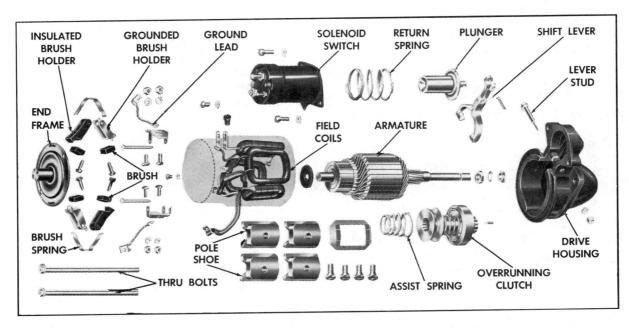

Fig. 58 Disassembled view of Delco-Remy starting motor

b) Slide a standard ½″ pipe coupling or other metal cylinder of suitable size onto shaft so end of coupling or cylinder butts against edge of retainer. Tap end of coupling with hammer, driving retainer toward armature and off snap ring.

c) Remove snap ring from groove in shaft. If snap ring is too badly distorted during removal, use a new one when reassembling the clutch.

d) Slide retainer, clutch and assist spring from armature shaft.

Reassembling Motor, Fig. 58

1. Lubricate drive end and splines of armature shaft with SAE 10 oil. *If heavier oil is used it may cause failure to mesh at low temperatures.*
2. Place "assist" spring on drive end of shaft next to armature, with small end against lamination stack.
3. Slide clutch assembly onto armature shaft with pinion outward.
4. Slide retainer onto shaft with cupped surface facing end of shaft.
5. Stand armature on end on wood surface with commutator down. Position snap ring on upper end of shaft and hold in place with a block of wood. Hit wood block with a hammer forcing snap ring over end of shaft. Slide snap ring into groove, squeezing it to ensure a good fit in groove.
6. Assemble thrust collar on shaft with shoulder next to snap ring.
7. Position retainer and thrust collar next to snap ring. With clutch pressed against assist spring, for clearance next to retainer, use two pairs of pliers at the same time (one pair on either side of shaft) to grip retainer and thrust collar. Then squeeze until snap ring is forced into retainer.
8. Place 4 or 5 drops of SAE 10 oil in drive housing bushing. Make sure thrust collar is in place against snap ring and retainer; then slide armature and clutch assembly into place in drive housing.
9. Attach solenoid and shift lever assembly to drive housing. Be sure lever buttons are located between sides of clutch collar.
10. Position field frame over armature, *applying sealing compound between frame and solenoid flange* (Fig. 56). Position frame against drive housing, using care to prevent damage to brushes.
11. Place 4 or 5 drops of SAE 10 oil in bushing in commutator end frame. Make sure leather

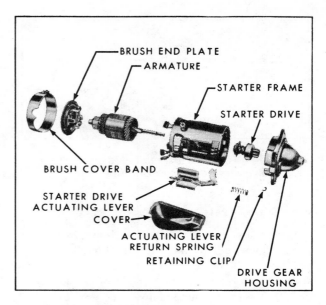

Fig. 59 Ford starter with integral positive engagement drive

brake washer is on armature shaft; then slide commutator end frame onto shaft.

12. Install thru bolts and tighten securely.
13. Reconnect field coil connectors to solenoid "motor" terminal.

FORD STARTER WITH INTEGRAL POSITIVE ENGAGEMENT DRIVE

This type starting motor, Fig. 59, is a four pole, series parallel unit with a positive engagement drive built into the starter. The drive mechanism is engaged with the flywheel by lever action before the motor is energized.

When the ignition switch is turned on to the start position, the starter relay is energized and supplies current to the motor. The current flows through one field coil and a set of contact points to ground. The magnetic field given off by the field coil pulls the movable pole, which is part of the lever, downward to its seat. When the pole is pulled down, the lever moves the drive assembly into the engine flywheel, Fig. 60.

When the movable pole is seated, it functions as a normal field pole and opens the contact points. With the points open, current flows through the starter field coils, energizing the starter. At the same time, current also flows through a holding coil to hold the movable pole in its seated position.

When the ignition switch is released from the start position, the starter relay opens the circuit to the starting motor. This allows the return spring to

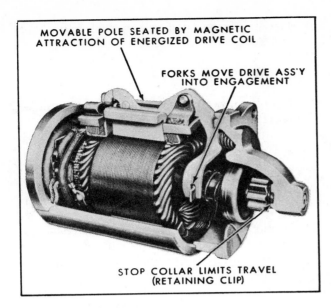

Fig. 60 *Starter drive engaged*

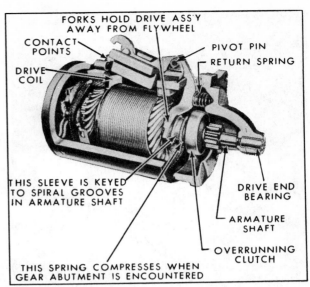

Fig. 61 *Starter drive disengaged*

force the lever back, disengaging the drive from the flywheel and returning the movable pole to its normal position, Fig. 61.

Disassembly

It may not be necessary to disassemble the starter completely to accomplish repair or replacement of certain parts. Thus, before disassembling the motor, remove the cover band and starter drive actuating lever cover. Examine brushes to make sure they are free in their holders. Replace brushes if defective or worn beyond their useful limit. Check the tension of each brush spring with a pull scale. Spring tension should not be less than 45 ounces. If disassembly is necessary, proceed as follows:

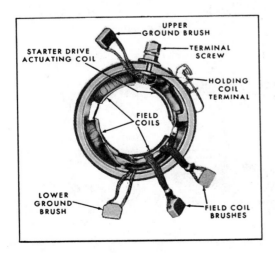

Fig. 62 *Field coil assembly*

1. Remove cover band and starter drive actuating lever cover.
2. Remove through bolts, starter drive gear housing, drive gear retaining clip cup and starter drive actuating lever return spring.
3. Remove pivot pin retaining starter gear actuating lever and remove lever and armature.
4. Remove and discard spring clip retaining starter drive gear to end of armature shaft, and remove starter drive gear.
5. Remove commutator brushes from brush holders and remove brush end plate.
6. Remove two screws retaining ground brushes to frame.
7. On the field coil that operates the drive gear actuating lever, bend tab up on field retainer and remove retainer.
8. Remove field coil retainer screws, Fig. 63. Unsolder field coil leads from terminal screw, and remove pole shoes and coils from frame.
9. Remove starter terminal nut and related parts. Remove any excess solder from terminal slot.

Reassembly

1. Install starter terminal, insulator, washers and retaining nut in frame, Fig. 62. Be sure to position slot in screw perpendicular to frame end surface.
2. Install field coils and pole pieces. As pole shoe screws are tightened, strike frame several sharp blows with a soft-faced hammer to seat and align pole shoes, then stake the screws.
3. Install solenoid coil retainer and bend tabs to retain tabs to frame.

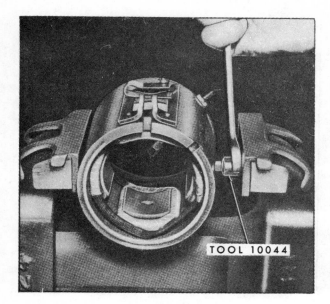

Fig. 63 Removing field coil pole shoe screws

4. Solder field coils and solenoid wire to starter terminal, using rosin core solder.
5. Check for continuity and grounds in the assembled coils.
6. Position solenoid coil ground terminal over ground screw hole nearest starter terminal.
7. Position ground brushes to starter frame and install retaining screws, Fig. 62.
8. Position starter brush end plate to frame with end plate boss in frame slot.

9. Install drive gear to armature shaft and install a new retaining spring clip.
10. Position fiber thrust washer on commutator end of armature shaft and install armature in frame.
11. Install starter drive actuating lever to frame and starter drive, and install pivot pin.
12. Position actuating lever return spring and drive gear housing to frame and install through bolts. Do not pinch brush leads between brush plate and frame.
13. Install brushes in holders, being sure to center brush springs on brushes.
14. Position drive gear actuating lever cover on starter and install brush cover band.
15. Check starter for free running and install on engine.

FORD STARTER WITH FOLO-THRU DRIVE

Disassembly

1. Remove starter drive, through bolts and rear end plate, Fig. 64. *Be sure to remove all burrs from shaft to prevent scoring rear end plate bushing.*
2. Remove armature, two thrust washers and cover band.
3. Remove brushes from their holders, and remove brush end plate.

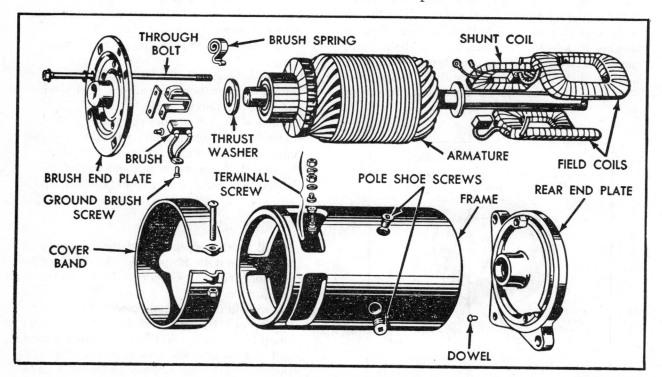

Fig. 64 Ford starter used with Folo-Thru drive

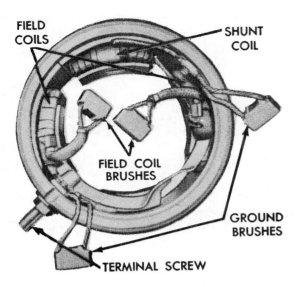

Fig. 65   *Field coil assembly*

4. Unscrew ground brush screws and remove brushes.
5. Remove pole shoe screws (see Fig. 63). Unsolder field coil leads from terminal screw and remove pole shoes and field coils.

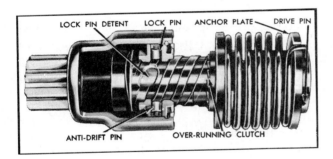

Fig. 67   *Folo-Thru starter drive with Bendix spring*

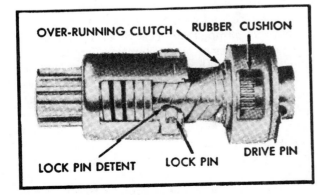

Fig. 68   *Folo-Thru starter drive without Bendix spring*

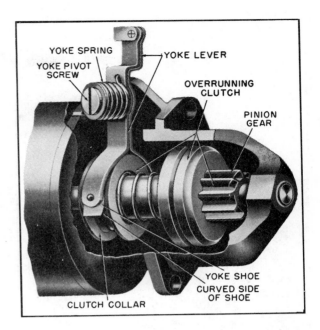

Fig. 66   *Over-running clutch drive. When disassembling, make sure curved sides of yoke shoes are toward gear end of clutch. Reversed yoke shoes can cause improper meshing of pinion*

6. Remove starter terminal. Remove any excess solder from terminal slot.

Reassembly

1. Install terminal screw with insulator washers and terminal nut. Be sure to position slot in screw parallel to frame end surface.
2. Install field coils. Use rosin core solder and be sure to position shunt coil ground lead under ground brush terminal farthest from starter terminal, Fig. 65. The other shunt coil lead is soldered to the series field coil lead farthest from starter terminal.
3. Install screws that connect ground brushes to starter frame.
4. Install brush end plate, making sure that brush plate boss is located in slot in starter frame. Do not pinch brush leads between frame and end plate.
5. Place thrust washer on each end of shaft, slide

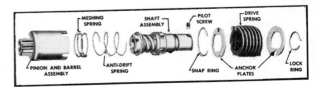

Fig. 69   *Barrel type Bendix drive*

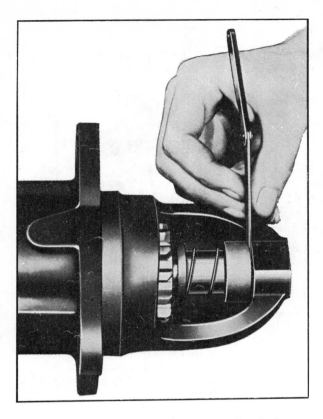

Fig. 70 *Measuring over-running clutch drive stop clearance. Do not compress anti-drift spring as this will give an incorrect clearance. If clearance is not present there is danger of the drive housing being broken as gear or collar slams back against it*

armature in place, and install rear end plate with dowel located in starter frame slot.

6. Install through bolts.

7. Install brushes in their holders, being sure to center brush springs on brushes.

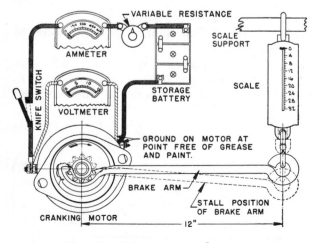

Fig. 72 *Diagrammatic layout of a starting motor torque tester*

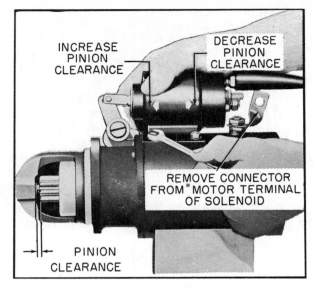

Fig. 71 *Adjusting pinion clearance on over-running clutch motor with an exposed solenoid having a non-adjustable plunger stud*

8. Install cover band and starter drive. Check starter for free running.

STARTER DRIVE TROUBLES

Starter drives fall into one or the other of two basic groups, the type that uses the principle of the over-running clutch, Figs. 66 and 68, and the Bendix, which uses the spinning nut principle, Figs. 67 and 69.

Starter drive troubles are easy to diagnose and they usually cannot be confused with ordinary starter difficulties. If the starter does not turn over at all or if it drags, look for trouble in the starter or electrical supply system. Concentrate on the starter drive or ring gear if the starter is noisy, if it turns but does not engage the engine, or if the starter won't disengage after the engine is started. After the starter is removed, the trouble can usually be located with a quick inspection.

Worn or chipped ring gear or starter pinion are the usual causes of noisy operation. Before replacing either or both of these parts try to find out what caused the damage. With the Bendix type drive, incomplete engagement of the pinion with the ring gear is a common cause of tooth damage. The wrong pinion clearance on starter drives of the overrunning clutch type leads to poor meshing of the pinion and ring gear and too rapid tooth wear.

A less common cause of noise with either type of drive is a bent starter armature shaft. When this shaft is bent, the pinion gear alternately binds and then only partly meshes with the ring gear. Most manufacturers specify a maximum of .003" radial run-out on the armature shaft.

When Clutch Drive Fails

The over-running clutch type drive seldom becomes so worn that it fails to engage since it is directly activated by a fork and lever, Fig. 66. The only thing that is likely to happen is that, once engaged, it will not turn the engine because the clutch itself is worn out. A much more frequent difficulty and one that rapidly wears ring gear and teeth is partial engagement. Proper meshing of the pinion is controlled by the end clearance between the pinion gear and the starter housing or the pinion stop, if one is used.

The clearance is set with the starter off the car and with the drive in the engaged position. To check the clearance, supply current to the starter solenoid with the electrical connection between starter and solenoid removed. Supplying current to the solenoid but not the starter will prevent the starter from rotating during the test. Take out all slack by pushing lightly on the starter drive clutch housing while inserting a feeler gauge between pinion and housing or pinion stop, Fig. 70.

On a number of cars, starting with 1957, the solenoids are completely enclosed in the starter housing and the pinion clearance is not adjustable. If the clearance is not correct, the starter must be disassembled and checked for excessive wear of solenoid linkage, shift lever mechanism, or improper assembly of parts.

On cars where the solenoid is exposed, the clearance can be adjusted either by loosening the screws holding the solenoid to the starter and moving the solenoid forward and backward, Fig. 71, or by screwing or unscrewing the link attached to the solenoid plunger.

Failure of the over-running clutch drive to disengage is usually caused by binding between the armature shaft and the drive. If the drive, particularly the clutch, shows signs of overheating it indicates that it is not disengaging immediately after the engine starts. If the clutch is forced to over-run too long, it overheats and turns a bluish color. For the cause of the binding, look for rust or gum between the armature shaft and the drive, or for burred splines. Excess oil on the drive will lead to gumming, and inadequate air circulation in the flywheel housing will cause rust.

Over-running clutch drives cannot be overhauled in the field so they must be replaced. In cleaning, never soak them in a solvent because the solvent may enter the clutch and dissolve the sealed-in lubricant. Wipe them off lightly with kerosene and lubricate them sparingly with SAE 10 or 10W oil.

When Bendix Drive Fails

When a Bendix type drive doesn't engage the cause usually is one of three things: either the drive spring is broken, one of the drive spring bolts has sheared off, or the screwshaft threads won't allow the pinion to travel toward the flywheel. In the first two cases, remove the drive by unscrewing the set screw under the last coil of the drive spring and replace the broken parts. Gummed or rusty screwshaft threads are fairly common causes of Bendix drive failure and are easily cleaned with a little kerosene or steel wool, depending on the trouble. Here again, as in the case of over-running clutch drives, use light oil sparingly, and be sure the flywheel housing has adequate ventilation. There is usually a breather hole in the bottom of the flywheel housing which should be kept open.

The failure of a Bendix drive to disengage or to mesh properly is most often caused by gummed or rusty screwshaft threads. When this is not true, look for mechanical failure within the drive itself.

Bendix Folo-Thru Drive

This type of drive, Figs. 67, 68, is in wide use on late model cars. It incorporates a device that keeps the pinion engaged to the flywheel until the engine reaches a specified rpm. When replacing one of these drives, be sure that you have the correct drive for the car. The drives are rated differently and the correct one must be used for the car being serviced. The Folo-Thru, incidentally, is not supposed to be repaired in the field because of the danger of incorrectly assembling the carefully calibrated springs in the pinion head.

TESTING THE STARTING MOTOR

On most motors the terminal stud is accessible and the motor can be connected for test without the switch in the circuit. However, on motors with manually-shifted clutch, the switch should be removed before making electrical tests.

Connect the motor terminal and frame to a battery of the correct voltage and operate on "no load" for two or three minutes to "seat" the brushes. Stop the motor and remove the burrs from the commutator with 00 sandpaper. Then blow out all dust with clean compressed air. If the motor will not operate, check the internal connections for shorts or binding.

NOTE—To obtain full performance data on a starting motor according to the manufacturer's specifications, certain equipment is required in addition to the specifications. There are a number of reputable manufacturers of this equipment and each one furnishes complete instructions for its use. Two tests are required to obtain full performance data, namely, no load test and torque test.

In the no load test, the starting motor is connected in series with a battery of the correct voltage and an ammeter capable of reading several hundred amperes. An RPM indicator (tachometer) should also be used to measure the armature revolutions per minute.

The torque test requires such equipment as illustrated in Fig. 72. The starting motor is mounted securely and the brake arm hooked to the drive pinion. Then, when the specified voltage is applied, the torque can be computed from the reading on the scale. If the brake arm is one foot long as shown, the torque will be indicated directly on the scale in pounds-feet. A high-current-carrying variable resistance should be used so that the specified voltage can be applied. Many torque testers indicate the developed pounds-feet of torque on a dial. The specifications are normally given at low voltages so that the torque and ammeter readings obtained will be within the range of the testing equipment available in the field.

Interpreting Results of Tests

Rated torque, current draw and no load speed indicates normal condition of starting motor.

Low free speed and high current draw with low developed torque may result from: (a) Tight, dirty or worn bearings, bent armature shaft or loose field pole screws which would allow the armature to drag. (b) Shorted armature. (c) A grounded armature or field.

Failure to operate with high current may result from: (a) A direct ground in the switch, terminal or fields. (b) Frozen shaft bearings which prevent the armature from turning.

Failure to operate with no current draw may be caused by: (a) Open field circuit. (b) Open armature coils. (c) Broken or weakened brush springs, worn brushes, high mica on commutator or other causes which would prevent good contact between brushes and commutator.

Low no-load speed with low torque and low current draw indicates: (a) An open field winding. (b) High internal resistance due to poor connections, defective leads, dirty commutator and causes listed under *Failure to operate with no current draw.*

High free speed with low developed torque and high current draw indicates shorted fields. There is no easy way to detect shorted fields,, since the field resistance is already low. If shorted fields are suspected, replace them and check for improvement in performance.

Review Questions

Page

1. What are the major components of a starting system? 731

2. What is the approximate ratio between the flywheel gear and the starter gear? 731

3. Why is it necessary to provide a means for disconnecting the starter drive from the engine? 731

4. What are the two methods used for quick disconnection of the starter drive from the engine? 731

5. What is the proper name for the drive gear on the starting motor? 731

6. What is the correct name for the large gear on the flywheel of an engine? 732

7. How does the over-running clutch type starter drive prevent the engine from driving the started armature? 732

8. What is the function of the shift spring on an over-running clutch starter drive? 733

9. Describe the construction of an over-running clutch starter drive. 733

10. After the engine starts, in what direction does it tend to drive the starter gear? 733

11. Explain the difference between an over-running clutch drive and a Bendix drive. 736

12. How many brushes does a starter have? 734

13. Where is the equalizer wire connected in a starter? 734

14. Why is the magnetic field in a starter with two field windings almost as strong as it would be with four windings? 734

15. What part of a starter constitutes the field frame? 734

16. What type of bushings are commonly used as starter bearings? 734

17. What is the name of the particular construction used in the starter armature core? 735

18. What two methods are used in mounting starter brushes? 735

19. The starter terminal usually connects to what part of the starter? 735

20. What is the name of the starter component which supports the brush holders? 735

21. Explain the difference between "inboard" and "outboard" starter drives. 736

22. The Bendix drive operates on what principle? 736

23. What actually disengages a Bendix drive from the flywheel? 736

24. What is the function of the anti-drift spring on a Bendix drive? 736

25. What is the function of the Bendix drive spring? 736

26. Why is there a counterweight on the Bendix drive pinion gear? 736

27. What two symptoms might indicate trouble in the starting system? 736

28. While investigating for troubles in a starting system, which component should be checked first? 737

29. Other than battery failure, what would be the probable cause for lights going out as the starter switch is closed? 737

30. Other than a weak battery, what could be the probable causes for lights becoming abnormally dim while the starter is being operated? 737

31. Give two reasons why a starter will crank an engine more slowly in cold weather than in warm weather. 737

32. A voltmeter is useful in testing for what particular defect in a starting system? 738

33. Is it necessary to operate the starter while testing for voltage drop in a circuit? 738

34. What three points in a starter circuit should be checked for voltage drop? 738

35. What steps can be taken to eliminate the causes of a voltage drop in a circuit? 738

36. What is the procedure for locating grounded windings on a starter armature? 738

37. What is the difference between a Chrysler direct drive starter and a reduction gear starter? 742, 744

38. What is unique about one of the field poles of a Ford starter with integral positive engagement drive? 752

39. Is it advisable to thoroughly clean a clutch type starter drive by soaking in solvent? 757

40. What is the name of the open core transformer used to test armatures? 740

41. What are three possible causes for mechanical failure in a Bendix drive? 757

42. What is the most prominent characteristic of the Bendix Folo-Thru starter drive? 757

43. Why is it not possible to service the Bendix Folo-Thru starter drive? 757

44. How many tests are required on a starter to obtain full performance data? 758

45. What are the names of the tests that are required on a starter to obtain full performance data? 758

46. What three measuring instruments are used to obtain full performance data on a starter? 758

47. If tests show that a starter has a low free speed and low torque with a high current draw, what defects would probably be found in the starter? 758

48. If tests show that a starter has high current draw and will not operate, what defects would probably be found in the starter? 758

49. If tests show that a starter has no current draw and will not operate, what defects would probably be found in the starter? 758

50. If tests show that a starter has low free speed and low torque with little current draw, what defects would probably be found in the starter? 758

Starting Switches

Review Questions for This Chapter on Page 769

INDEX

Carter vacuum switch 763
Carter vacuum switch service . 768
Magnetic switches 761
Solenoid switches 761
Starter switch service 766
Stromberg vacuum switch . . . 764

Stromberg vacuum switch
   service 768
Switches, magnetic 761
Switch operation, checking . . 766
Switch service, starter 766
Switches, solenoid 761

Vacuum switch, Carter 763
Vacuum switch service, Carter. 768
Vacuum switch, Stromberg . . . 764
Vacuum switch service,
   Stromberg 768

Magnetic and solenoid switches are designed to perform mechanical jobs electromagnetically such as closing a heavy circuit or shifting the starter drive pinion with the engine flywheel ring gear for cranking. Switches of this type consist basically of contacts and a winding (or windings) around a hollow cylinder containing a movable core or plunger. When the winding (or windings) is energized by the battery through an external control circuit the plunger is pulled inward, producing the necessary mechanical movement.

MAGNETIC SWITCHES

Figs. 1 and 2 illustrate two typical Delco-Remy switches. The switch shown in Fig. 1 is not designed for disassembly and must be replaced if defective.

In the switch shown in Fig. 2 the terminals are assembled into a molded terminal ring which is held in place on the switch case by the cover and screws. Gaskets on both sides of the ring seals the contact compartment as a protection against moisture and dirt. The winding assembly is not removable from the case on this unit although the contact disk, plunger and plunger return spring can be removed after the cover is taken off.

Fig. 3 is a heavy duty magnetic switch. It is completely serviceable and easy to disassemble and assemble. To disassemble, remove the four terminal plate nuts and washers and take off the terminal plate assembly. The contact disk may be removed by taking off the castellated nut. It is necessary to remove the spring and washers on the plunger rod only when the plunger rod needs to be disassembled. To remove the plunger, unscrew the large metal cover, take out the cotter pin in the plunger shaft, remove spring retainer washer and spring, and withdraw the plunger. The winding and switch case is an integral assembly. The only parts that can be removed are the switch terminals. Before removing the switch terminals, the winding leads must be unsoldered from the terminal studs. Whenever the switch is disassembled, upon reassembly, locate the contact disk properly by turning the castellated nut on the disk in or out as required to obtain the dimension shown in Fig. 3 between the contact disk and edge of housing.

SOLENOID SWITCHES

The solenoid switch on a cranking motor not only closes the circuit between the battery and the cranking motor but also shifts the drive pinion into mesh with the engine flywheel ring gear. This is done by means of a linkage between the solenoid switch plunger and the shift lever on the cranking motor. Some linkages are adjustable while others are not (see *Starting Motors* chapter). The linkage is not adjustable on the type shown in Fig. 4 but adjustment of the entire assembly is made by moving the switch on the motor frame.

Fig. 4 shows two views of a solenoid switch used on vehicles with 12-volt systems. Like other solenoid switches, this type is energized by the battery through a separate starting switch. Note, however, that the switch includes an additional small terminal and contact finger. This terminal has no functional duty in relation to the switch, but is used to complete a special ignition circuit during the cranking cycle only. When the solenoid is in the cranking position, the finger touches the

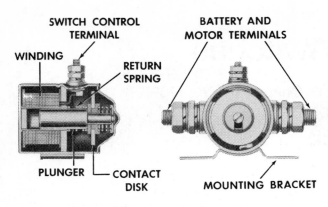

Fig. 1 End and sectional views of a typical magnetic switch

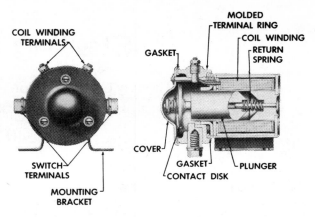

Fig. 2 End and sectional views of a sealed type magnetic switch which uses gaskets to seal the contact compartment

contact disk and provides a direct circuit between the battery and ignition coil.

Fig. 5 is an exploded view of the 12-volt solenoid switch shown in Fig. 4. When reassembling the switch the contact finger should be adjusted to touch the contact disk before the disk makes contact with the main switch terminals. There should be $\frac{1}{16}''$ to $\frac{3}{32}''$ clearance between the contact disk and the main terminals when the finger touches.

Fig. 6 is a wiring circuit of a typical solenoid switch. There are two windings in the solenoid; a pull-in winding (shown as dashes) and a hold-in winding (shown dotted). Both windings are energized when the external control switch is closed. They produce a magnetic field which pulls

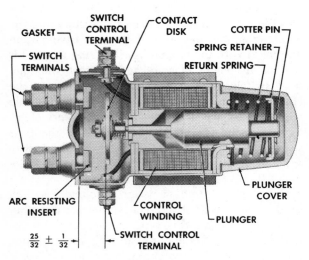

Fig. 3 Sectional view of a heavy duty sealed type magnetic switch. To adjust the location of the contact disk turn the nut on the disk in or out as required to obtain the dimension shown

the plunger in so that the drive pinion is shifted into mesh, and the main contacts in the solenoid switch are closed to connect the battery directly to the cranking motor. Closing the main switch contacts shorts out the pull-in winding since this winding is connected across the main contacts. The magnetism produced by the hold-in winding is sufficient to hold the plunger in, and shorting out the pull-in winding reduces drain on the battery. When the control switch is opened, it disconnects the hold-in winding from the battery. When the hold-in winding is disconnected from the battery, the shift lever spring withdraws the plunger from the solenoid, opening the solenoid switch contacts and at the same time withdrawing the drive pinion from mesh. Proper operation of the switch depends on maintaining a definite balance between the magnetic strength of the pull-in and hold-in windings.

This balance is established in the design by the size of the wire and the number of turns specified. *An open circuit in the hold-in winding or attempts to crank with a discharged battery will cause the switch to chatter.*

To disassemble the solenoid, remove nuts, washers and insulators from the switch terminal and battery terminal. Remove cover and take out the contact disk assembly.

When the solenoid has been removed from the starter motor for repair or replacement, the linkage must be adjusted to provide the correct pinion clearance or pinion travel when the solenoid is remounted on the motor. Some solenoids equipped with relays have an adjustable plunger stud, but others must be moved on the motor frame to adjust pinion travel.

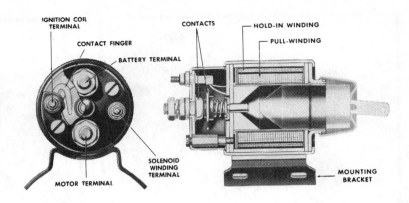

Fig. 4 *End and sectional views of a typical solenoid switch. The additional terminal and contact finger are used on 12-volt passenger car applications*

IGNITION COIL TERMINAL

CONTACT FINGER

BATTERY TERMINAL

CONTACTS

HOLD-IN WINDING

PULL-WINDING

SOLENOID WINDING TERMINAL

MOTOR TERMINAL

MOUNTING BRACKET

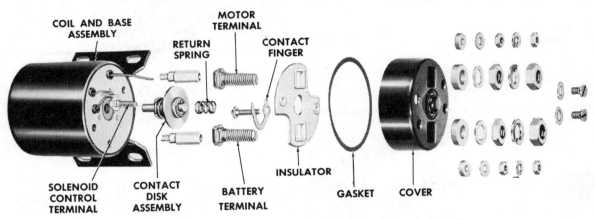

COIL AND BASE ASSEMBLY

MOTOR TERMINAL

RETURN SPRING

CONTACT FINGER

SOLENOID CONTROL TERMINAL

CONTACT DISK ASSEMBLY

BATTERY TERMINAL

INSULATOR

GASKET

COVER

Fig. 5 *Exploded view of solenoid switch shown in Fig. 4*

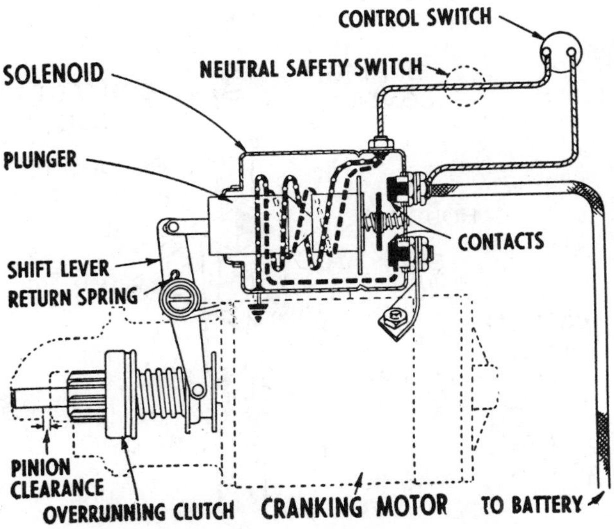

CONTROL SWITCH

SOLENOID

NEUTRAL SAFETY SWITCH

PLUNGER

CONTACTS

SHIFT LEVER
RETURN SPRING

PINION CLEARANCE

OVERRUNNING CLUTCH   CRANKING MOTOR   TO BATTERY

Fig. 6 *Wiring circuit of a typical solenoid switch*

CARTER VACUUM SWITCH

Referring to Fig. 7, it will be seen that with the throttle closed and engine stopped, the flat spot on the throttle shaft is in the position shown. A steel ball rests in the end of a bakelite plunger which is pressed against the ball by a coil spring.

When the throttle is partly opened, Fig. 8, the throttle shaft forces the ball and plunger to the right, causing the W-shaped contact spring to touch the two terminals, thus closing the solenoid relay circuit and causing the starter to crank the engine.

As soon as the engine starts, engine suction lifts the ball to the position shown in Fig. 9, enabling the coil spring to push the plunger to the left against the stop, thus breaking the solenoid relay circuit.

It should be noted that the stop prevents the plunger from closing the suction passage. Therefore, the ball is held in the position shown as long as the engine is running, thus preventing operation of the starter while the engine is in operation.

763

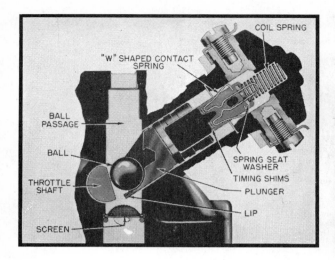

Fig. 7 Carter carburetor-mounted vacuum switch. This shows the position of the ball with the engine not running

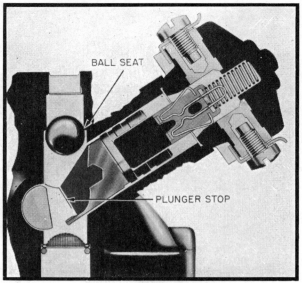

Fig. 9 Carter carburetor-mounted vacuum switch. This shows position of the ball when engine is operating

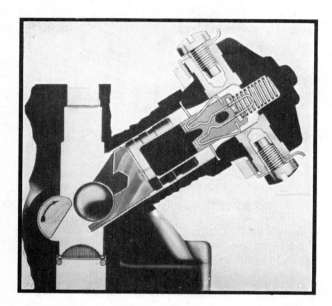

Fig. 8 Carter carburetor-mounted vacuum switch. This shows the position of the ball when throttle is partly opened and engine being cranked

STROMBERG VACUUM SWITCH

This switch, Fig. 10, consists of a housing which is flange mounted with a gasket to the carburetor throttle body and held in place by two screws. This housing is provided with a horizontal cylinder barrel to which vacuum is applied at one end by means of cored and drilled holes in the carburetor bodies. This end of the barrel is provided with a washer which forms a seal to prevent leaks when a piston opposed by a light spring is

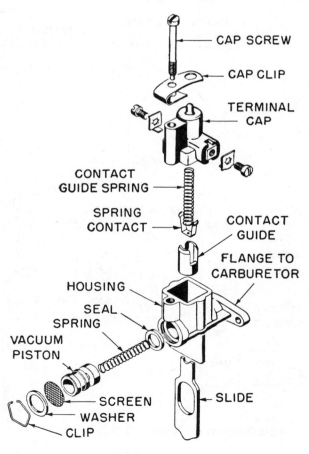

Fig. 10 Exploded view of Stromberg carburetor-mounted vacuum switch

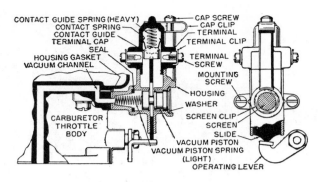

Fig. 11 *Details of Stromberg carburetor-mounted vacuum switch. The mechanism is shown with closed throttle, engine not running. Note that the slide is in contact with the operating lever*

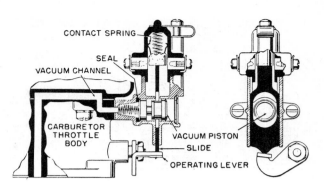

Fig. 13 *Stromberg carburetor-mounted vacuum switch in position it assumes when engine is running at idle speed*

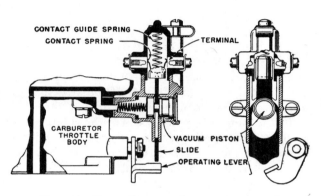

Fig. 12 *Stromberg carburetor-mounted vacuum switch. The mechanism is shown in the position it assumes with open throttle, cranking position*

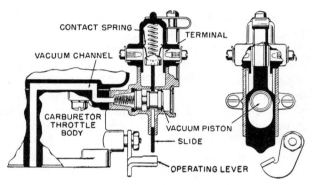

Fig. 14 *Stromberg carburetor-mounted vacuum switch in position it assumes when engine is running above idle speed*

drawn against it by vacuum. The opposite end of the barrel is vented to outside air through a fine mesh screen which is held in place by a screen clip. A flat slide, actuated by an operating lever on the throttle shaft, moves in a confined slot in the housing, and in a plane perpendicular to the axis of the cylinder barrel. This slide engages a cylindrical bakelite contact guide, the upward movement of which is opposed by a heavy contact guide spring. The contact guide carries a thin "U" shaped spring contact which moves up and down within a bakelite terminal cap to engage stationary contacts for opening and closing the starter control circuit. The terminal cap is held in place by a screw and clip.

Fig. 11 shows the throttle closed, the switch operating lever holding the slide in the upper position, thereby holding the U-shaped spring contact away from the stationary contacts in the terminal cap. This is the position the mechanism

assumes when engine is not running and throttle is closed.

Fig. 12 illustrates the mechanism in the position it assumes in cranking position with throttle open. Pressing down on the accelerator pedal causes the operating lever to move away from the slide. This allows the contact guide spring to move the slide and U-shaped spring contact down to a position to bridge the stationary contacts in the terminal cap, thus closing the circuit, the slide moving into the deeper of the two grooves in the vacuum piston which has been positioned against the screen by the vacuum piston spring.

Fig. 13 shows the position of the mechanism with engine running on closed throttle (idle range). When the engine starts and the throttle is allowed to close, the slide and U-shaped spring contact is moved upward by the switch operating lever, opening the circuit. With the slide in the up position, manifold vacuum pulls the vacuum piston inward until it seats against the seal. This

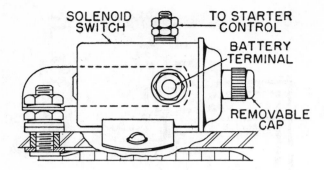

Fig. 15 *Auto-Lite one-coil solenoid switch with a removable cap so the switch can be operated manually by removing the cap and pushing the plunger in by hand*

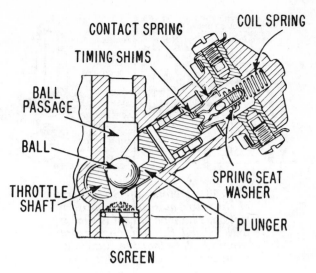

Fig. 16 *Details of Carter carburetor-mounted vacuum switch*

aligns the shallow groove in the piston with the slide.

Fig. 14 illustrates the position of the mechanism when the engine is running above the idle range (open throttle). When the throttle is opened beyond the idle range, the operating lever moves away from the slide which is then forced downward by the contact guide spring until it strikes the shallow groove in the vacuum piston. This acts as a stop and prevents the switch contacts from engaging while the engine is running. It also holds the piston in the inner position when engine load conditions cause the vacuum to become too low to hold the piston in.

STARTER SWITCH SERVICE

Checking Switch Operation

If the starting motor fails to crank the engine or if it cranks it too slowly, the switch, motor, battery and cables should be checked to find the cause. Consult other chapters in this book for the inspection procedures on the starting motor, battery and cables. But assuming that any of these are not causing the trouble, the operation of the switch may be checked as follows:

Use a heavy battery cable to eliminate the starting switch from the circuit by holding the jumper cable against the battery and starting motor switch terminals. Then if the motor cranks the engine, it indicates that the switch or its control circuit is at fault.

Checking Operation of One-Coil Switch Circuit

Should this type of switch fail to operate electrically by means of the instrument panel control

switch, it can usually be operated manually by removing the cap on one end of the switch, Fig. 15, and pushing the plunger in by hand.

If the starter operates when the plunger is pushed in by hand but will not operate when the control switch is closed, it is obvious that the trouble lies either in the control switch or the wire connecting this switch with the solenoid switch.

Check the control switch by placing a jumper wire across its terminals. Then if the starter oper-

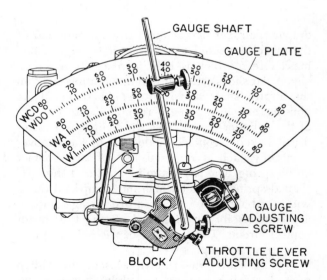

Fig. 17 *Carter starting switch gauge for determining throttle opening position in degrees at which switch contact is made*

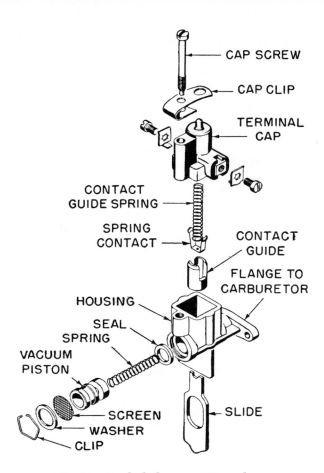

Fig. 18 *Exploded view of Stromberg carburetor-mounted vacuum switch*

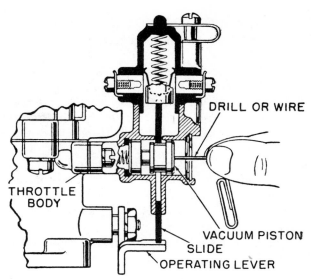

Fig. 19 *Using wire through screen to operate vacuum piston on Stromberg carburetor switch*

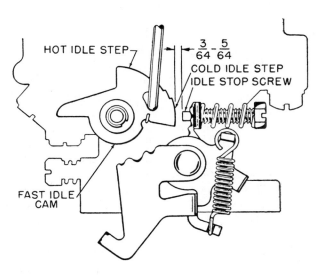

Fig. 20 *Checking Stromberg starting switch timing on car*

ates, the control switch is defective and should be replaced.

If by using the jumper wire the starter still does not operate, run a jumper wire from the control switch to its terminal on the solenoid switch. Then if the starter operates, the wire is defective and it should be repaired or replaced.

Checking Operation of Solenoid Switch with Relay Circuit

1. Be sure all connections are clean and tight.
2. Remove the relay cover and operate the instrument panel control switch. The relay points should close. If they don't, the trouble may be due to defective control switch, relay coil or faulty wire between control switch and relay.
3. Place a jumper across the terminals of the control switch. Then if the relay contacts close, the control switch is defective and should be replaced.
4. If by using the jumper wire the relay contacts do not close, place a jumper wire from the relay terminal to the battery terminal on the solenoid. Then if the contacts still fail to close, the trouble is in the relay coil and this assembly should be replaced.
5. If the relay contacts close and the starter fails to operate, clean the contacts with 00 sandpaper.
6. If the starter still fails to operate, all solenoid lead wire soldered connections should be

examined for looseness. If these connections are tight, the solenoid is defective and should be replaced.

7. If the starter drive pinion disengages from the flywheel after a start is made but the starter switch fails to break contact and the starter armature continues to revolve, the starter switch push rod may be stuck. If so, replace the solenoid.

8. If the starter tries to engage while the engine is running, look for trouble in the control switch or its terminals, as they may be vibrating into contact with each other. This trouble may also be attributed to a weak or broken return spring or hinge spring on the solenoid relay armature.

CARTER VACUUM SWITCH SERVICE

The W-shaped contact spring, Fig. 16, rests on two or more brass shims with square holes. These shims determine the point at which the switch contact is made. Contact should be made when the throttle valve is opened between 30 and 45 degrees. If not enough of these shims are in place, contact will not be made soon enough. Too many will cause the switch to function too soon (before 30 degrees) in which case there is danger that the switch may be in contact all the time.

In disassembling the switch, carefully remove these shims and put them aside in a safe place so they all will be returned to their proper position.

Between the W-shaped contact spring and the coil spring is a round washer with a square hole. This washer must not be confused with the timing shims. Neither the W-shaped spring nor the coil spring should be stretched or otherwise altered or the operation of the switch will be affected.

When reassembling the switch on the carburetor, be sure the plunger is placed in the position shown, Fig. 16. Never apply oil or grease to any of the switch parts as dust will collect and eventually cause the switch to stick.

When any new switch parts are installed, it is essential to use the special Carter starting switch gauge shown in Fig. 17. As can be seen, this gauge can be used on all model Carter carburetors on which this switch is installed. The gauge indicates the degrees of throttle opening at which the switch makes contact.

In using the gauge, we will use the WDO carburetor scale for an example.

1. Attach the gauge plate to the choke housing and tighten in position.
2. Connect the block to the throttle shaft lever by means of the screw as shown, making sure the block is tight.
3. Back out the throttle lever adjusting screw.
4. Hold the choke valve open to release the fast idle block, close the throttle valve tight and set the gauge shaft so that the pointer rests on line marked zero on WDO scale.
5. Tighten the gauge adjusting screw.
6. With the carburetor on the car and the switch connected, the switch should make contact when the throttle is opened so that the indicator has passed 30°, but the engine must start before the pointer has reached 45°. If it does not, the timing shims must be increased or decreased in number until the desired result is obtained.
7. When the carburetor is on the bench, it is necessary to attach a battery and a small bulb in series by wires to the two switch contacts. The point of contact of switch can then be determined by lighting of the bulb.

STROMBERG VACUUM SWITCH SERVICE

Before working on the switch, Fig. 18, always have the transmission gears in neutral and apply the parking brake. Then adjust the switch as follows:

1. Set the carburetor idle adjusting screw at 8 mph (hot idle).
2. With ignition off, insert a No. 65 drill (or a small size paper clip) through the center of the screen to operate the vacuum piston, Fig. 19. *Do not remove the screen.*
3. With throttle closed, first push the vacuum piston to its inner position and hold it there while opening the throttle. This will allow the slide to drop into the shallow groove in the piston and will lock it in the inner position and prevent the slide from dropping far enough to complete contact. Hold the throttle open to prevent release of the piston until completion of steps 4 and 5.
4. Remove drill or wire. Place a $5/64''$ spacer between the carburetor idle stop screw and the fast idle cam while holding the cam in the extreme cold idle position, Fig. 20. Close the throttle so that the spacer will hold the cam in this position. Turn ignition on, hold spacer and open the throttle—*the engine should not crank.*
5. While still holding the throttle open, place a $3/64''$ spacer between idle stop screw and fast idle cam while holding cam in extreme fast idle position. Close the throttle so the spacer holds the cam in this position, Fig. 20. Again open

the throttle with the ignition on—*engine should crank.*

6. If the ⁵⁄₆₄″ spacer causes the engine to crank, bend the tang on the operating lever downward. If the ³⁄₆₄″ spacer fails to cause the engine to crank, bend the tang on the lever upward. In making either adjustment, bend the tang a slight amount each time until by rechecking after each bend, the specified spacing is obtained.

Review Questions

Page

1. What is the basic function of a solenoid switch? 761

2. What is the purpose of the extra terminal and contact finger on some 12-volt solenoid switches? 761

3. How many windings are there usually on a solenoid switch? 762

4. What is each winding on a solenoid switch called? 762

5. Why is there more than one winding on most solenoid switches? 762

6. Proper functioning of the solenoid depends upon maintaining what kind of a balance? 762

7. What two factors in a starting system might cause the solenoid switch to chatter? 762

8. Correct starter drive pinion clearance is dependent on what part of an adjustable type solenoid switch? 762

9. How is a starter pinion clearance adjusted with a non-adjustable type solenoid switch? 762

10. Where is the vacuum starter switch usually mounted? 763

11. By what means does the driver operate the starter on a system with a vacuum switch? 763

12. There are two types of vacuum switches currently being used, what is each called? 763, 764

13. What prevents the vacuum switch from engaging the starter while the engine is running? 763

14. How can the starting switch be by-passed for test purposes? 766

15. What is controlled by adjustment on a Carter vacuum switch? 768

16. How is the Carter vacuum switch adjusted? 768

17. Is it good practice to lubricate a vacuum switch? 768

18. What is the name of the adjusting shims in the Carter vacuum switch? 768

19. The Carter vacuum switch should make contact before what throttle position is reached as the throttle is being opened? 768

20. What equipment would be needed to test the operation of a Carter vacuum switch on the bench? 768

21. What is the first step in adjusting the Stromberg vacuum switch? 768

22. What is the point in inserting a small rod or paper clip through the screen of the Stromberg vacuum switch? 768

23. Describe the method for checking starter switch timing on the Stromberg vacuum switch. 768

24. How is starting switch timing adjusted on the Stromberg vacuum switch? 768

25. Why is it of importance to have vacuum starter switches timed correctly? 768

769

The Storage Battery

Review Questions for This Chapter on Page 786

INDEX

Adding water 780	Battery troubles 776	Installing and removing
Batteries, care of new and	Capacity, battery 774	batteries 781
rental 785	Care of new and rental	Removing and installing
Batteries, removing and	batteries 785	batteries 781
installing 781	Charging, battery 782	Servicing batteries in the car.. 784
Batteries, servicing in the car. 784	Construction of a battery 771	Test, hydrometer 777
Battery capacity 774	Electrical tests 780	Tests, electrical 780
Battery charging 782	How a battery works 771	Troubles, battery 776
Battery, construction of a 771	Hydrometer test 777	Water, adding 780
Battery, how it works 771		

WHAT BATTERY DOES

The storage battery is a reservoir of electro-chemical energy. Fig. 1. Electrical current is drawn from the battery to turn over the starter and start the engine. When the engine is running at normal speed, the generator creates enough electricity to operate the ignition system, lights, and electrical accessories, and also supplies some additional electrical energy which is put back into the battery and stored up for future demands.

When the engine is running very slowly, the generator may not produce enough electricity to meet the demand and current must be drawn from the battery to make up the deficit. Any current used to operate lights or radio when the engine is turned off must be drawn entirely from the battery.

The relationship between the battery and the car's electrical system can be easily understood and remembered by comparing them with a city's water supply system.

The water reservoir corresponds to the storage battery—a full reservoir is like a well charged battery, and an empty reservoir like a discharged battery. The water pump, which operates only part of the time but must keep the reservoir full, corresponds to the generator, which works only while the engine is running but must keep the battery charged.

The water mains correspond to the car's wiring, the water outlets correspond to the lights, starter, ignition system and other users of electricity, and the faucets correspond to the switches.

Notice the similarities in these two systems:

If water is drawn from the tank while the pump is not operating, the pump must eventually start

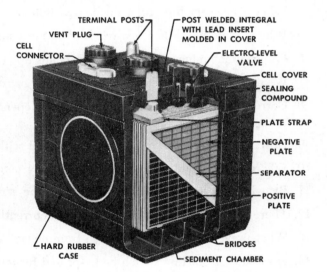

Fig. 1 *Typical 6-volt battery with end and side of one cell cut away to show details of cell construction. This battery has 3 cells; a 12-volt battery has 6 cells*

supplying water faster than it is being withdrawn in order to refill the tank. If current is drawn from the battery while the generator is not operating (as in starting), the generator must eventually supply more current than the electrical system uses, in order to recharge the battery.

If the outlets use more water than the pump can supply, the reservoir will eventually run dry regardless of its size. If the electrical system uses more current than the generator produces, the battery will eventually become discharged regardless of its size.

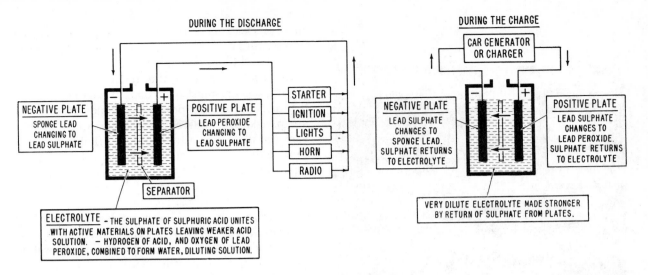

CAR GENERATOR
OR CHARGER

NEGATIVE PLATE
SPONGE LEAD
CHANGING TO
LEAD SULPHATE

POSITIVE PLATE
LEAD PEROXIDE
CHANGING TO
LEAD SULPHATE

STARTER
IGNITION
LIGHTS
HORN
RADIO

SEPARATOR

ELECTROLYTE - THE SULPHATE OF SULPHURIC ACID UNITES
WITH ACTIVE MATERIALS ON PLATES LEAVING WEAKER ACID
SOLUTION. - HYDROGEN OF ACID, AND OXYGEN OF LEAD
PEROXIDE, COMBINED TO FORM WATER, DILUTING SOLUTION.

NEGATIVE PLATE
LEAD SULPHATE
CHANGES TO
SPONGE LEAD.
SULPHATE RETURNS
TO ELECTROLYTE

POSITIVE PLATE
LEAD SULPHATE
CHANGES TO
LEAD PEROXIDE.
SULPHATE RETURNS
TO ELECTROLYTE

VERY DILUTE ELECTROLYTE MADE STRONGER
BY RETURN OF SULPHATE FROM PLATES.

Fig. 2   Electro-chemical action in a battery

If water is withdrawn steadily while the pump is not working, the tank will run dry. If the ignition switch or lights are left on while the generator is not running, the battery will become discharged.

If the water pump fails, the tank will run dry. If the generator fails, the battery will become discharged.

If there are leaks in the water mains, the tank will run dry. If there are leaks (short circuits) in the wiring, the battery will be discharged.

When the pump is not running, water from the tank would flow back through the pump if there were not some kind of valve to prevent it. When the generator is running slowly or not at all, current from the battery would flow back through the generator if there were not a "reverse current switch" to prevent it.

HOW A BATTERY WORKS

Strictly speaking, a battery does not store electricity—it stores chemical energy which is converted into electrical energy.

If two different metals are placed in an acid which can attack them, an electrical potential, or voltage, is created. If wires from the two metals are connected into a circuit, electrical current will flow through it. Eventually, however, the reaction of the acid will change the two metals into the same metal, and electrical action will cease.

This, Fig. 2, is the principle of the storage battery essentially, a charged battery consists of two different metals immersed in a solution of sulphuric acid. One of the metals—the positive plate—is lead peroxide. The other metal—the negative plate—is plain lead. The acid solution is called the electrolyte.

When the two plates are connected into a circuit, the sulphate splits away from the sulphuric acid and unites with the lead of both plates to form lead sulphate. The oxygen freed from the lead peroxide of the positive plate unites with the hydrogen which splits off from the sulphuric acid and forms water. This chemical reaction is what produces the electricity.

As this action continues, the acid solution gets weaker and weaker because the sulphuric acid is being taken away and water is being added. At the same time, both plates acquire a higher and higher proportion of lead sulphate. Eventually both plates are sufficiently coated with lead sulphate to stop the reaction, electrical output ceases, and the battery is then discharged.

To charge the battery, electricity from some outside source is driven through the battery to reverse the chemical reaction. The sulphate is broken away from the lead sulphate on both plates. It attaches itself to hydrogen from the water, thus forming sulphuric acid again. Meanwhile the oxygen freed from the water goes back to the positive plate to form lead peroxide again. This action continues until all the lead sulphate disappears; the battery is then back in its original condition and is again fully charged.

CONSTRUCTION OF A BATTERY

No matter how big the plates are made, a cell like the one described above will develop only

Fig. 3   A positive plate with part of active material removed to show grid structure

sheet of lead, consists of a spongy lead packed into a similar grid. The acid solution can penetrate the pores of the sponge lead.

In order to reduce the resistance which the electric current must overcome in passing through the electrolyte, the positive and negative plates are placed as close together as possible. To prevent the plates from touching and thus creating a short circuit that would discharge the battery, a separator, Fig. 4, is placed between them. The separator also helps hold the lead peroxide crystals in place on the positive plate.

Separators ordinarily are made of chemically purified wood, which is porous enough to permit free circulation of the electrolyte. Some separators are made of rubber with fine cotton threads running through them to make them porous. Occasionally a mat of spun glass or perforated rubber is placed between the separator and the positive plate to help retain the lead peroxide crystals and to protect the separator from the oxidizing effect of the positive plate.

The side of the separator next to the positive plate has vertical grooves in it to facilitate the circulation of the acid and to allow the escape of the gas which is generated around the positive plate while the battery is being charged.

A number of positive plates are welded to a post strap, Fig. 5, to produce a positive group, Fig. 6, and negative plates are welded into a similar negative group. The groups are put together as shown in Fig. 6, and separators are inserted between the plates. The completed assembly of plates, post straps, and separators is called an element, Fig. 7. Note that both outside plates are negative plates, which means that each element has one more negative plate than positive plates.

When an element is immersed in the electrolyte, the entire unit is called a cell. No matter how large the element is, each cell will have an open-circuit voltage of slightly more than 2 volts. An automobile battery consists of three cells con-

about 2 volts. However, the total amount of current a battery can deliver before becoming discharged is determined by the area of the plates in contact with the electrolyte. The modern battery, therefore, is designed to bring the largest possible plate area into contact with the acid in the most compact manner possible.

The positive plate, Fig. 3, is composed of very small crystals of lead peroxide wedged into a rectangular grid made of a lead-antimony alloy. Although the lead peroxide crystals are so small that they cannot be seen clearly even under a microscope, the acid solution is able to penetrate the spaces between them. The crystals are held together somewhat loosely and can be knocked off the plate rather easily.

The negative plate, instead of being a solid

Fig. 4   Showing separators of different materials

PORT ORFORD CEDAR

FIBROUS GLASS MAT

POROUS RUBBER

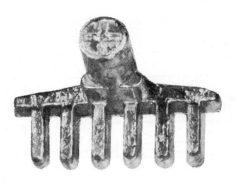

Fig. 5 *Battery post strap*

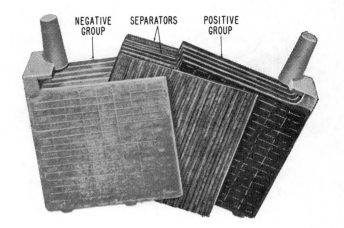

NEGATIVE GROUP   SEPARATORS   POSITIVE GROUP

Fig. 6 *Negative and positive group of plates with separators being nested together*

nected in series to provide a 6-volt current, or six cells, producing 12 volts.

The container or case, Fig. 8, is usually made of hard rubber or rubber composition, of a one-piece molded construction with partitions to separate the cells. At the bottom of each cell are ridges, or element rests, which raise the elements off the bottom and provide a space for the sediments which breaks off the positive plates. A short circuit is created if the accumulated sediment touches the plates. To minimize the chances of a short circuit, the plates have feet on the bottom, so arranged that the feet of the positive plates rest on one set of ridges while the feet of the negative plates rest on another set.

Over each cell is placed a cell cover, Fig. 9, usually made of molded hard rubber. The posts of the positive and negative plate groups project up through holes at each end of the cell cover to permit electrical connections. In the center is a larger hole or vent, with a screw cap which can be removed for testing or refilling the electrolyte. The cap has a small vent to allow gases to escape, and often has some sort of baffle plate to prevent acid from splashing out and a non-overfill device which automatically prevents the addition of too much water when the battery is being filled. An acid-proof sealing compound is used to secure the cover to the container and to prevent leaks around the posts.

The elements are placed in the container in such a way that the positive post of one cell is next

Fig. 7 *Assembled battery element*

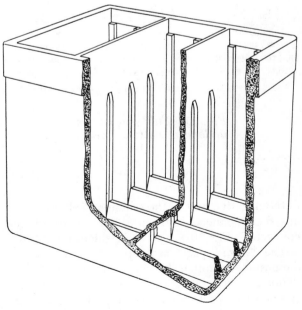

Fig. 8 *Battery case cut away to show partitions and element rests*

Fig. 9   Cell cover

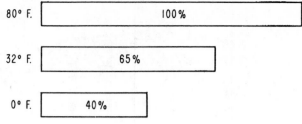

80° F.	100%
32° F.	65%
0° F.	40%

Fig. 12   *How a battery's cranking ability
decreases as its temperature drops*

Fig. 10   *A cell connector*

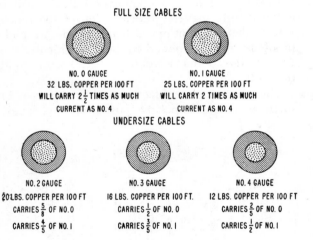

FULL SIZE CABLES

NO. 0 GAUGE
32 LBS. COPPER PER 100 FT
WILL CARRY $2\frac{1}{2}$ TIMES AS MUCH
CURRENT AS NO. 4

NO. 1 GAUGE
25 LBS. COPPER PER 100 FT
WILL CARRY 2 TIMES AS MUCH
CURRENT AS NO. 4

UNDERSIZE CABLES

NO. 2 GAUGE
20 LBS. COPPER PER 100 FT
CARRIES $\frac{5}{8}$ OF NO. 0
CARRIES $\frac{4}{5}$ OF NO. 1

NO. 3 GAUGE
16 LBS. COPPER PER 100 FT.
CARRIES $\frac{1}{2}$ OF NO. 0
CARRIES $\frac{3}{5}$ OF NO. 1

NO. 4 GAUGE
12 LBS. COPPER PER 100 FT
CARRIES $\frac{2}{5}$ OF NO. 0
CARRIES $\frac{1}{2}$ OF NO. 1

Fig. 11   *Full size versus
undersize battery cables*

to the negative post of the adjoining cell. The cells are then connected in series by means of cell connectors, Fig. 10, which are welded to the posts.

The battery posts are tapered to a standard dimension, the top of the positive terminal (11/16″ diameter) being slightly larger than the top of the negative post (⅝″ diameter). In addition, the positive terminal is marked with a POS, a plus sign, or red paint, while the negative terminal is marked with NEG, a minus sign, black or green paint.

If the markings are obscured, the polarity of the terminals can be determined by attaching a wire to each post and putting the free ends of the wires into a very weak solution of salt water or battery acid, without permitting the wires to touch. Bubbles of hydrogen gas will form around the negative terminal.

The cables which carry the current to the car's electrical system are provided with clamp-type terminals, which are slipped over the posts and bolted firmly into place. The cable terminals usually are made either of lead-coated brass, or of a corrosion-proof lead alloy. The cable which is grounded from the battery to the car's frame generally is uninsulated and usually is a flat strap of woven copper wires.

The cable must be large enough to carry the surge of 200–300 ampere current drawn by the starting motor with a minimum loss of voltage. Most cars require a No. 0 gauge or No. 1 gauge cable, Fig. 11. Undersize cables, like loose or corroded terminals, will cause a voltage drop that will prevent efficient starting.

BATTERY CAPACITY

Although all 3-cell auto batteries have a voltage of about 6, and 6-cell batteries have a voltage of about 12, their capacity varies widely.

The distinction between voltage and capacity can be clarified by comparing a storage battery and a water tank.

In a water tank, the pressure is determined by the depth of the water, and is measured in pounds per square foot. In a battery, the electrical "pressure," or potential, is measured in volts.

The amount of water flowing out of the tank through a pipe is measured in gallons per minute;

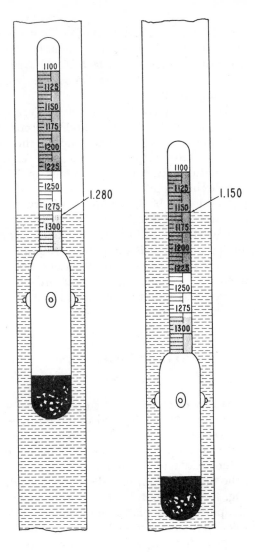

Fig. 13   *High float (left) means high specific gravity. Low float means low specific gravity*

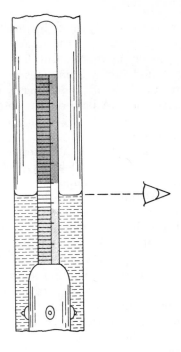

Fig. 14   *Correct method of reading a hydrometer. Eye must be on level with liquid surface*

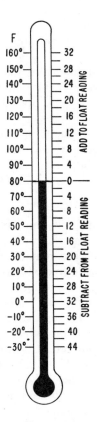

Fig. 15   *Temperature correction of specific gravity*

the amount of electrical current flowing out of a battery is measured in amperes.

If a large tank and a small tank both hold the same depth of water, the pressure is the same in both tanks. But if water is drawn from both tanks at the same rate—20 gallons per minute, for example—the little tank obviously will run dry first because its capacity is smaller.

In the same way, a large three-cell battery and a small three-cell battery both deliver 6 volts, but if electricity is drawn from both of them at the same rate—10 amperes, for example—the little battery will become discharged first because its capacity is less.

A battery's capacity is measured in ampere-hours. This figure is obtained by multiplying a given amperage output by the number of hours the battery can supply that output. For example, a battery that could supply 5 amperes for 20 hours, or 4 amperes for 25 hours, would have a capacity of 100 ampere-hours. A battery that could supply 4 amperes for 20 hours would have a capacity of 80 ampere-hours.

This does not mean that a battery capable of delivering 5 amperes for 20 hours (100 ampere-hours) could deliver 100 amperes for one hour, or 200 amperes for half an hour—as the amperage increases, the battery's capacity decreases. Cold weather also reduces a battery's capacity considerably, Fig. 12. At 32° F., a battery has only 65% of the cranking ability it has at 80° F., which is considered its normal operating temperature. At 0° F., the cranking ability drops to 40% of the 80° figure. To make matters worse, it takes about twice as much power to turn over a very cold engine because the oil congeals. Cold weather starting, therefore, is the toughest job a battery is called upon to perform.

The 12-volt batteries used in recent years have greater plate area, which makes more energy available for starting and also makes it possible to recharge the battery faster. These batteries offer low internal resistance to current flow to improve cold weather starting. In spite of being more powerful, 12-volt batteries are about the same size as 6-volt batteries.

Other factors influencing battery capacity are the number and size of the plates in each cell, the battery's design, its age, and the operating conditions.

Accurate measurements of battery capacity are provided by three tests developed by the Society of Automotive Engineers and accepted by the battery industry and the U.S. Government. They are:

1. Lighting ability (20-hour rating in ampere-hours): Each manufacturer rates his batteries on the basis of the maximum amperage they can produce continuously for 20 hours. If a 6-volt battery can produce 5 amperes for 20 hours, for example, its 20-hour rating is 100 ampere-hours. To test this rating, a fully charged battery is brought to a temperature of 80° and is discharged at a rate obtained by dividing the manufacturer's 20-hour capacity rating by 20. If the 20-hour rating is given as 120 ampere-hours, for example, the discharge rate would be 6 amperes. The battery is discharged continuously at this rate until its voltage drops to 5.25. The 20-hour rating in ampere

hours is then obtained by multiplying the discharge rate by the number of hours that elapse before the voltage drops to 5.25. The 20-hour rating of various auto batteries ranges from 80 to 220 ampere-hours.

2. Cranking ability (300 amperes at 0° F. rating): This is a test of the battery's ability to crank an engine in cold weather. The rating is the number of minutes required for a fully charged 6-volt battery to drop to a voltage of 3 volts when discharged at 300 amperes with a starting temperature of 0° F. Cranking ability ratings vary from about 3 to 9 minutes.

3. 5-second voltage rating: This is another test of cranking ability. It is the battery voltage taken 5 seconds after the start of the discharge at 300 amperes with a starting temperature of 0° F. On 6-volt batteries it varies from 4.2 to 4.8 volts with different makes of batteries.

BATTERY TROUBLES

Sulphation

As a battery discharges, the sulphuric acid splits up and combines with the lead in the positive and negative plates to form lead sulphate. If the battery is recharged promptly, this sulphate is driven out of the lead and again forms sulphuric acid. But if the plates remain in a discharged condition too long, the fine crystals of lead sulphate penetrate the pores in the plates and become hard and insoluble. This condition is called sulphation.

Prolonged charging at a very slow rate is required to restore a sulphated battery to serviceable condition—if a sulphated battery is charged at too high a rate, the plates will be buckled. If the sulphated condition has been neglected for too long, it is impossible to recharge the battery and it must be discarded. An idle battery must be charged about once a month to prevent sulphation.

Insufficient Electrolyte

If the level of the electrolyte drops below the top of the plates, the exposed portions of the plates will sulphate rapidly. The remaining electrolyte is more concentrated, which makes it harmful to the plates and separators. As the electrolyte becomes more concentrated, its electrical resistance increases. Current passing through this resistance creates heat, just as it does when passing through the resistance of a thin wire. The increased resistance therefore causes overheating of the battery.

Overheating

A battery can be ruined if it is allowed to operate when its temperature is 130° or above. The heat increases the chemical activity of the electrolyte, speeding up corrosion of the grids, damaging and possibly buckling the plates, and softening or charring the wood separators. The water in the electrolyte evaporates faster, and if it is not promptly replaced all the damages produced by insufficient electrolyte will appear.

Overcharging

If a battery is charged at too high a rate, or if charging continues after all the lead sulphate has been converted to acid, the excess current breaks up the water of the electrolyte into hydrogen and oxygen gas. The gas bubbles up violently, dislodging the active materials from the positive plate and shortening the life of the battery. The decomposition of the water leaves the electrolyte more concentrated, which damages plates and separators. The excessive gassing may force electrolyte out of the cells, causing corrosion of the terminals, carrier, and nearby parts; in addition, the current passing through the more concentrated electrolyte causes overheating.

Buckled Plates

Overheating, overcharging, extreme sulphation, or insufficient electrolyte may cause the positive plates to warp or buckle. The warped plates rub against the separators and frequently wear holes in them, short-circuiting the battery and ruining it.

Freezing

If the electrolyte is allowed to freeze, the ice dislodges the active material from the plates, buckles the plates, and cracks the container, ruining the battery. A fully charged battery will not freeze unless the temperature reaches about 85° below zero, but a battery which is 50% charged will freeze at 15° below, a battery 75% discharged will freeze at 5° above zero, and a completely discharged battery will freeze at 20° above. To avoid freezing, the battery should be kept fully charged.

If water is added to the battery in very cold weather, it should be added just before the car is run, so that the water will be immediately mixed with the electrolyte. Otherwise the water, being lighter than the electrolyte, will stay on top and may freeze, even if the battery is fully charged.

Corrosion

If electrolyte spills or splashes on the terminals, a green-colored deposit will form. To remove this corrosion, wash the parts with ammonia or a solution of bicarbonate of soda (ordinary baking soda), brush them with a stiff brush, then wash them with water and wipe them dry. Corrosion can be prevented by coating the terminals with vaseline or any of the corrosion resistant mineral oils prepared by the oil companies.

Vibration

If the battery is not securely fastened in its carrier, the excessive vibration and bouncing will shake the active material off the positive plate and shorten the life of the battery.

Battery "Dopes"

Concentrated acids and dopes of various kinds are peddled as battery "rejuvenators" or "rechargers." These dopes will injure and perhaps ruin a battery. Nothing but pure water, and sometimes sulphuric acid, should ever be added to the electrolyte.

HYDROMETER TEST

Sulphuric acid is heavier than water. Therefore the mixture of sulphuric acid and water used as the electrolyte is heavier than pure water.

When a battery is fully charged, the electrolyte has a specific gravity of about 1.250 to 1.290. That means that a given amount of the electrolyte will weigh 1.25 to 1.29 times as much as the same volume of pure water.

As the battery discharges, some of the heavier sulphuric acid is broken up and the sulphate unites with the plates. The remaining electrolyte, therefore, becomes lighter and lighter as the battery discharges. For this reason, it is possible to tell how much a battery has become discharged simply by measuring the specific gravity of the electrolyte.

The instrument used to measure the specific gravity is the *hydrometer*. It consists of a small float inside a glass barrel. A hydrometer reading is taken by squeezing the rubber bulb at the top, inserting the hard rubber nozzle into one cell of the battery, and releasing the nozzle to draw some of the electrolyte up into the barrel, Fig. 13. When the electrolyte has risen to a convenient level for accurate reading, it is prevented from rising farther by a small tube extending downward from the rubber bulb, which draws any excess electrolyte up into the bulb.

The lower the float sinks into the electrolyte, the lower the specific gravity. The exact figure is read off a small scale inside the float. To obtain an accurate reading, Fig. 14, hold the barrel vertical so the float does not touch the sides, and keep your eye level with the surface of the liquid. Disregard the slight upward curve of the liquid, caused by surface tension, where it touches the sides of the barrel and float. After taking the reading, return the electrolyte to the cell from which it was drawn by squeezing the rubber bulb.

When the temperature of the electrolyte is around 80°, this is what the hydrometer's specific gravity readings mean:

Specific gravity	Condition of battery
1.275–1.295	Fully charged.
1.250	75% charged. This is the normal driving condition.
1.225	50% discharged. Unless the battery is recharged, it may soon fail to start the car.
1.190	Only 25% charged—extreme sulphation commencing.
1.160	Very little useful capacity left.
1.130	Battery completely discharged.

Separate readings of each cell must be taken. If the specific gravity of one cell is more than 15 points lower than the others, that cell may be short circuited, or may have lost electrolyte through leakage or through overfilling and resulting overflow. Electrical tests, described below, may be used to check the condition of that cell, or an effort can be made to recharge the battery. If the cell still does not come up to the level of the others, it is defective. It must be opened up and repaired by a battery specialist, or a new battery must be substituted.

If a battery is very hot or very cold, the hydrometer readings must be corrected for temperature. This is because the electrolyte shrinks and becomes heavier (higher specific gravity) as it cools, and expands and becomes lighter (lower specific gravity) as it heats, regardless of the condition of charge.

Suppose a battery shows a specific gravity reading of 1.225 at 80° F. That battery is half discharged, and needs charging. But if it is cooled to 10° F., the specific gravity reading will go up to 1.253, although the battery is just as much discharged as ever. Hydrometer readings taken when the battery is very hot or cold, therefore, will be

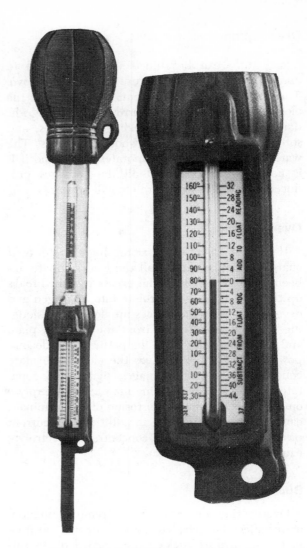

Fig. 16 *A hydrometer with a built-in thermometer and correction scale*

misleading unless they are corrected for temperature. If the battery is cold, a certain number of points must be subtracted from the hydrometer reading; if it is hot, points must be added.

Fig. 15 shows the number of points which must be added to or subtracted from the hydrometer reading for various temperatures of the electrolyte (not the air temperature).

The temperature of the electrolyte can be determined by inserting a thermometer into the middle cell. Some hydrometers, Fig. 16, have a built-in thermometer with a scale showing how many points must be added to or subtracted from the float reading.

An accurate hydrometer reading cannot be taken immediately after water has been added to the battery because the newly added water, being lighter than the electrolyte, will float on top of it until the battery has been charged or the car has

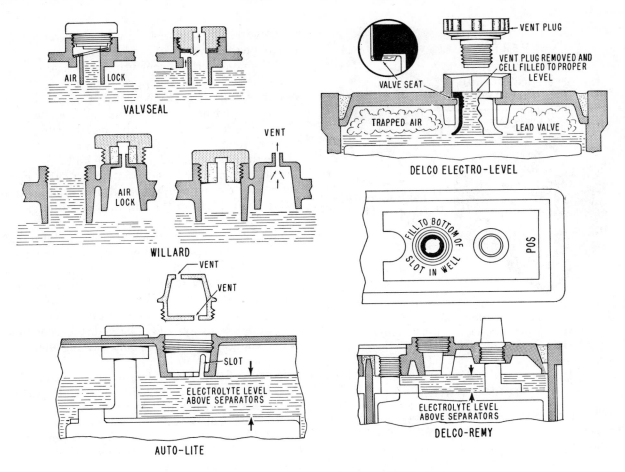

Fig. 17 Various types of non-overfill devices

been operated for a while. If the electrolyte is so low that it cannot be drawn up into the hydrometer, add enough water to bring the electrolyte to the proper level and, after the car has been driven a couple of hours, an accurate reading can be taken.

The hydrometer reading also will be misleading if it is taken immediately after the battery has been discharged at a high rate, as in prolonged cranking. Rapid discharging weakens the acid near the plates but temporarily leaves the rest of the acid stronger, so that acid drawn from the top of the cells will show a higher specific gravity than actually exists. The acid will slowly mix if the battery stands idle for several hours, or will be rapidly mixed if the battery is charged and gassing occurs.

The hydrometer barrel and float must be cleaned occasionally to prevent the float from sticking to the sides. Remove the rubber tip and float and wash all parts with soapy water. Inspect the float for cracks, which will permit electrolyte to seep inside the float and make its readings

inaccurate. If the paper scale inside the float is wet, the float leaks.

Accurate readings cannot be obtained with a small, cheap hydrometer. Use a laboratory or shop type.

The color of the electrolyte drawn into the hydrometer sometimes gives a clue to internal defects of the battery.

Brown or reddish electrolyte indicates that the positive plates are worn out—usually by lack of water, freezing, or a short circuit.

Milky white electrolyte, produced by small air bubbles, indicates that the generator is overcharging the battery.

Black specs in the electrolyte show that the wood separators have been charred by overcharging or overheating.

NOTE: In tropical climates, where freezing never occurs, an electrolyte with a lower specific gravity is used. A fully charged battery with this type of electrolyte will show a specific gravity reading of 1.200 to 1.225.

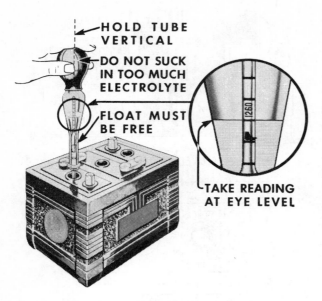

HOLD TUBE VERTICAL

DO NOT SUCK IN TOO MUCH ELECTROLYTE

FLOAT MUST BE FREE

1260

TAKE READING AT EYE LEVEL

Fig. 18 A battery filling syringe which automatically adjusts level of electrolyte

ADDING WATER

The electrolyte must be maintained at the proper level—it should never drop below the top of the separators, or be more than about ⅜″ to ½″ above them. If the level is too low, the exposed part of the plates will be badly sulphated and the lower parts will be corroded or overheated. If the electrolyte is too high, the expansion of the electrolyte when it becomes warmer will force some of it out through the vent plugs, corroding the terminals.

Water must be added to the electrolyte periodically to replace water which evaporates or is broken up into gasses by the electrical action. The sulphuric acid, however, does not evaporate, and for that reason it is never necessary to add anything but water, unless some of the electrolyte has been spilled.

Only distilled water, or drinking water which has been approved for batteries, should be used. Water which is perfectly safe to drink usually contain metals or chemicals which will ruin the battery. Many battery manufacturers will analyze samples of water without charge.

Battery water should be kept in a closed container of glass, earthenware, rubber, or lead.

The vent covers of many batteries have some sort of built-in, automatic non-overfill device, Fig. 17, which regulates the height of the electrolyte and makes it impossible to add too much water. After sufficient water has been added, an air vent is closed and no more water can enter. After refilling, be sure to screw the vent cover on tightly to reopen the air vent.

If the non-overfill mechanism is out of order, or if the battery is not equipped with one, a special battery syringe, Fig. 18, is useful. The syringe has a hard rubber nozzle with a hole ⅜″ from the tip. With the tip resting on top of the separators, the bulb is compressed and released and any water above the hole is sucked up into the bulb.

For use in low headroom spaces, the hard rubber tip can be gently heated over a flame at a point about 2″ from the tip, bent to an angle of about 60°, and chilled.

The frequency with which water must be added depends upon how hot the weather is and how much the car is driven and how fast. On the average, a battery should not need filling oftener than once a month. If the battery requires frequent filling, or takes an excessive amount of water, and is always fully charged, the generator is overcharging the battery and causing unnecessary wear on the plates.

If one cell takes considerably more water than the others, look for a loose-fitting vent plug, leaks in the sealing compound, or a crack in the container. The leakage should be immediately corrected by resealing or replacing the container. Otherwise the leaky cell will be totally ruined and the entire battery seriously damaged.

ELECTRICAL TESTS

Although the hydrometer test will determine how much a battery is charged or discharged, complete information about the condition and capacity of a battery can be obtained only by making an electrical test.

The open-circuit voltage (the voltage when the battery is not delivering any current to an outside circuit) of a fully charged three-cell battery is slightly over 6 volts. When current is being withdrawn from the battery, however, the voltage decreases. The voltage drop under load will be greater on a small-capacity battery than on a large-capacity battery, and it will be greater on a worn out or short-circuited battery than on a battery in good condition.

A battery's condition, therefore, can be determined from voltmeter readings taken when the battery is discharging. Its capacity can be determined by using an ammeter to measure the amperage a battery can discharge through a resistance.

There are two general types of electrical testers —the high discharge rate tester, and the low-reading voltmeter.

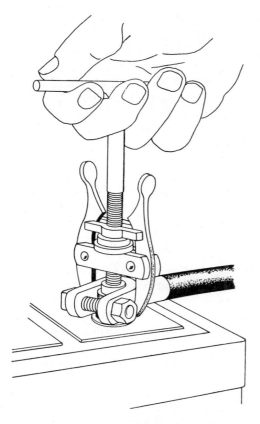

Fig. 19 *A screw-type terminal puller*

The high discharge rate tester is fairly complicated, but it provides the most accurate indication of the battery's condition. Usually it has both a voltmeter (often called a "Condition Meter" or some similar name), which is used to check cell conditions before charging the battery, and an ammeter (or "Capacity Meter"), which tests the capacity of each cell after the battery has been fully charged. Some testers cannot be used unless the battery is nearly at full charge; others require the battery to have a specific gravity reading of at least 1.220.

Battery testing equipment varies considerably in design, and it is important that the manufacturer's instructions be carefully followed to insure an accurate test and to avoid damaging the battery or the tester.

In general, the tester is connected across a single cell (usually a three-way switch is provided as a convenient means of checking each cell in succession) and current is drawn from the battery at about the amperage required by the starter. This amperage, which is adjusted by a rheostat, varies with the size of the battery—on some testers it is determined by the number of plates per cell, on others by tables giving the correct rate for all types of batteries. The rate usually is 26⅔ amperes for each positive plate in a single cell.

If, while current is being withdrawn at this rate, any two cells show a voltage variation of more than .15 volts, there is a short circuit in the low-voltage cell and the battery should be scrapped.

The capacity test is made not less than 12 nor more than 96 hours after the battery has been fully charged. The rheostat is adjusted to draw enough current to produce a specific voltage drop. If the ammeter shows that the current from any cell drops below a minimum point, which is determined by the size of the battery, the battery's capacity is low and it is worn out.

A less expensive but less accurate tester is the low-reading voltmeter, which has a scale reading from 0 to 3 volts. At the ends of the two wires from the voltmeter are long needlelike points which are pushed into the cell terminals to make the connection. The voltmeter must be connected to one cell at a time—if it is connected across all three, it may be ruined.

With the ignition switch off and the starter turning over, the voltage of each cell is read. A fully charged cell in good condition should show about 1.8 volts. A lower reading shows that the cell is either discharged or in bad condition. If the voltage drops slowly, the trouble is merely due to a discharged condition, but if it drops rapidly, the cell is defective.

If no instruments are available, a rough check of the battery's condition can be made by turning on all the car's lights and operating the starter with the ignition off. If the lights dim excessively and fluctuate, either the battery is weak or there is a poor connection between the battery and the cables. If the latter is the case, the terminals will feel hot to the fingers.

REMOVING AND INSTALLING BATTERIES

Careless installation of a new battery can ruin the battery. In removing the old battery, note the location of the positive battery post so the new or rental battery can be installed in the same position. Remove the ground cable first.

Use an open-end wrench to loosen the clamp. If the nut is very tight, use one wrench on the head of the bolt and the other on the nut, to avoid straining and possibly cracking the battery cover. A pair of battery pliers may be used to loosen the nut, but a wrench should always be used on the head of the bolt.

If a cable terminal is corroded to the post, don't

try to loosen it by hammering, or by resting a tool on the battery and prying—either method can break the battery container. Use a screw type terminal puller, Fig. 19, or spread the cable terminals slightly with a screwdriver.

Clean any corrosion from the cables, cradle, or hold-downs, and inspect them. Paint any corroded steel parts with acid-proof paint. Make sure the cable is of the correct size and that its insulation and clamp terminal are in good condition.

Be sure that the new or rental battery being installed has a specific gravity reading of at least 1.270—a capacity test is also advisable.

Put the new battery in the cradle, making sure it sits level, and tighten the hold-downs a little at a time, alternately, to avoid distorting and breaking the container. The hold-downs should be snug enough to prevent bouncing, but should not be too tight.

Before connecting the cables, check the battery terminals to be sure the battery is not reversed.

Clean the battery posts bright with sandpaper or a wire brush. Put a thin coat of vaseline or mineral oil on the posts, and also on the inside of the clamp terminals if they are of the lead-coated brass type. Non-corrosive cables of lead alloy should be scraped bright but not greased.

Don't hammer the terminals down on the posts —the cover may crack. Spread the terminals slightly if necessary. Connect the starter cable first and the ground cable last, tightening the terminal bolts after making sure that the cables don't interfere with the vent plugs or rub against the hold-downs.

As a final check, switch on the lights and see that the ammeter swings toward the "Discharge" or minus side to make sure the polarity of the battery hasn't been reversed. If the ammeter points toward charge, the battery is reversed and the car's electrical system will be damaged if the condition is not corrected.

Check the generator charging rate. Start the engine, leaving the lights and accessories off, and make sure the ammeter needle swings toward the "charge" side, indicating that the generator is working.

BATTERY CHARGING

Only direct current can be used to charge a storage battery. The positive cable of the direct current is attached to the positive terminal of the battery, the negative lead is attached to the negative terminal, and current is run through the battery.

This reverses the discharging action, driving the sulphate off the plates and causing it to unite with the water to form sulphuric acid. The ampere hours which have been withdrawn from the battery are replaced by means of the direct current and are stored in the form of chemical energy.

Direct current power is available in many cities, and can be used for battery charging if a rheostat and ammeter are connected into the line to regulate the current.

Instead of a rheostat, a "lamp bank" consisting of a number of ordinary light bulbs wired in parallel can be used. Since a known amperage flows through each lamp, the current can be increased by screwing in more lamps and decreased by screwing some of them out.

The current through each lamp is, roughly, the wattage of the lamp divided by the line voltage. For example, a 100 watt light in a 110-volt circuit will pass a current of about .9 amperes; four of them in parallel would provide a charging current of 3.6 amperes.

If only alternating current is available, it can be used to run a small A/C motor-generator which produces direct current, or the alternating current can be changed to direct current by means of a rectifier, either of the gas-filled bulb or dry disc type.

The room in which charging is done must be well ventilated, since the mixture of oxygen and hydrogen gasses released is very inflammable and explosive. Never smoke in the charging room, or permit any flame or spark in it. Turn off the current while connecting or disconnecting batteries to prevent sparking.

If there is any doubt about which lead is positive and which negative, dip both wires, without letting them touch, into a glass of water containing a teaspoon of ordinary salt or a few drops of battery acid. Bubbles will form around both wires, but the wire with the most bubbles will be the negative wire.

There are three general methods of charging batteries: (1) constant current charging, in which the amperage of the current passing through the battery is controlled; (2) constant potential charging, in which the voltage input remains constant but the battery takes a decreasing amount of amperage as it builds up its own counter-voltage; and (3) quick charging, or high-rate charging, in which the battery is rapidly charged without being removed from the automobile by running a high amperage through it for a short time.

If the battery is to be charged outside the car, by one of the first two methods, remove it from the automobile, put an identifying tag on it, and thoroughly clean the battery with a stiff whisk-broom and water to prevent dirt from dropping into it when the vent plugs are removed. Remove any corrosion with ammonia or a solution of baking soda and water. Fill the battery with distilled water ⅜″ to ½″ above the top of the separators.

If the battery has a non-overfill vent cap, be sure to screw the cap down tightly so the gasses generated during the charging can escape. If the top is a simple one without a baffle plate, it's a good idea to place it loosely over the opening without screwing it down; this allows gas to escape around the edges and prevents acid from spraying out through the vent hole.

Constant Current Charging

Turn off the charger. Connect the positive lead from the charger to the positive terminal of the battery, and negative to negative.

If several batteries are to be charged at once, the batteries are connected in series—that is, the negative post of one battery connected with the positive post of the next one, and so on.

Adjust the rheostat or lamp bank to give the correct charging rate. A good general rule is one ampere per positive plate in each cell. If a battery has 15 plates per cell, for example, each cell will have 7 positive plates and a safe charging rate would be 7 amperes.

Many batteries show the charging rate on the nameplate. Frequently two rates are given, a "starting" rate to be used when the charging begins and a lower "finish" rate to be used after the battery is nearly charged. Badly sulphated batteries can most safely be recharged at the "finish" rate only. Batteries can safely be left on charge overnight at about half the "finish" rate.

When several batteries are being charged in series, the rate must be adjusted to the smallest battery in the line. In general a 5-ampere rate is safe for a line of various sized batteries.

In constant current charging, the temperature of a battery should never be allowed to rise above 110° F. Check the temperature occasionally by putting a thermometer in the center cell of the smallest battery in the line. If no thermometer is available, feel a cell connector—it should never feel more than moderately warm. If any battery is overheating, reduce the charging rate. If a battery gets as hot as 120° F., stop charging until it has cooled.

After a battery has been on the line 8 to 10 hours, take its hydrometer reading and record the figure on the side of the battery with chalk. From then on, take a hydrometer reading every three hours—some manufacturers recommend hourly readings as the battery approaches the fully charged state.

If a vent plug has a non-overfill device, the gassing may force some of the electrolyte out of the cells while the hydrometer reading is being taken, or may prevent all the electrolyte from being returned to the cell. To prevent this, hold the non-overfill device in the open position with a small stick while taking the reading.

A battery is fully charged when all cells are gassing freely, and when there has been no change in the hydrometer reading for three hours. This is the only way of being sure a battery is fully charged, and it makes no difference what the hydrometer reading may be—if it remains the same for three hours, the battery is fully charged. Additional charging will only cause overheating and excessive gassing. A cell which does not gas either is not fully charged, or else is defective.

If the battery is fully charged, but the hydrometer reading is not between 1.270 and 1.295, the electrolyte contains too much or too little sulphuric acid. Before adjusting the acid content, however, be sure to correct the hydrometer reading for temperature—warm electrolyte will give a low reading.

If the gravity reading is too high, remove some of the electrolyte with the hydrometer and add water; if it is too low, remove some of the electrolyte and add battery acid of 1.400 specific gravity. Charge the battery for another hour to mix the electrolyte thoroughly and then take another hydrometer reading. If necessary, adjust the acid content again, continuing the process until a correct reading is obtained.

Unless some electrolyte has been spilled, it should not be necessary to add acid to a battery during its lifetime. Be sure not to add acid until the cell is gassing freely.

To mix 1.400 specific gravity battery acid, use seven parts of pure water to four parts chemically pure sulphuric acid of 1.835 specific gravity. Mix in a lead, glass, or earthenware vessel, never in a metal or composition container. *Add the acid to the water slowly,* because of the great amount of heat produced, stirring the solution with a glass rod. *NEVER ADD WATER TO THE ACID— THIS IS EXTREMELY DANGEROUS.* Always pour the acid into the water.

After a battery has been charged and removed from the line, examine it. If the sealing compound is rough and uneven, it's possible to seal and close it by going over it gently with a flame, meanwhile looking for any deep cracks. *Caution:* Before bringing any flame or spark near a battery, the explosive mixture of hydrogen and oxygen gasses in the cells should be gently blown out with a rubber bulb or syringe. If there are deep cracks, reseal the battery with new compound.

Leaks in the sealing compound or around the posts can easily be discovered by using a pressure pump made by running the tip of a battery syringe through a rubber stopper. Hold the stopper over the vent hole and gently squeeze the bulb to produce a moderate pressure.

Inspect the vent covers to make sure the non-overfill devices are in good condition, then screw the covers down tightly. Wipe off the battery top and terminals with ammonia or a baking soda solution, and put a little vaseline or mineral oil on the terminals.

Constant Potential Charging

In this system, batteries are connected in parallel across power lines which are kept at a constant voltage, the current usually being supplied by a motor-generator. Usually three power lines, or bus bars, are used—there is a potential of 15 volts between the two outside wires, for charging 12-volt batteries, and a 7-volt potential across one outside wire and the neutral wire, for charging 6-volt batteries.

Each battery draws a differing amount of current, depending on its condition and state of charge—the battery draws a large amperage at first, but this gradually decreases as the battery builds up its own counter-voltage.

Constant potential charging is faster than constant current charging because of the higher beginning rate, but it has a greater tendency to overheat batteries and their temperatures must be closely watched. Although constant potential charging is satisfactory for recharging batteries which are in good condition, constant current charging should be used if the battery is sulphated or in bad condition.

Each battery is connected individually to the bus bars, the positive terminal to the positive bus bar and negative to negative. Usually a resistance is connected in series with the battery to prevent an excessive surge of initial current.

When using constant potential charging, it is sometimes found that two cells of a battery may become fairly well charged but the third does not. If there is no internal defect in the battery, this condition results from insufficient current going through the cells because of the battery's counter voltage. The only solution is to put the battery on a constant current line until fully charged. Then,

if necessary, the acid can be adjusted. Do not add acid to batteries being charged on a constant potential line—there is no way of knowing for sure that a cell is fully charged.

High Rate Charging

Portable high-rate chargers, which can be wheeled up to the automobile to boost the battery in a half hour or so without removing it from the car, have become increasingly popular and profitable.

High-rate chargers usually are constant potential devices, producing current with a motor-generator or a rectifier of high enough capacity to provide an initial charging rate of 80 to 100 amperes.

This high rate will not harm a normal battery, unless its temperature rises above 125° or unless there is violent gassing. The chargers usually are equipped with a timing device which shuts them off after a predetermined time, or by a thermostatic control which cuts off the current when the battery's temperature reaches 125°.

Various methods are used for determining how much charging current to use, and how long to apply it. Some high-rate chargers contain a built-in high discharge rate tester; others test the battery voltage while the car's starter is turning over. Some rely entirely on the specific gravity reading; others take the battery's original capacity into consideration.

Although quick chargers cannot bring a battery to full charge in such a short time, they can speedily restore it to about 80% of capacity, and if the car's generator is functioning properly the battery will continue to give good service.

Caution: If the car is equipped with an alternator and a fast charger is used to charge the vehicle battery, the vehicle battery cables should be disconnected *unless the fast charger is equipped with a special Alternator Protector,* in which case the vehicle battery cables need not be disconnected. Also the fast charger should never be used to start a vehicle as damage to rectifiers will result.

SERVICING BATTERIES IN THE CAR

1. Use an apron to protect the car's upholstery or finish.
2. Clean the top of the battery with ammonia or a baking soda solution and a brush, and wash off with clear water.
3. Inspect the cables; suggest replacement if they are defective.

4. Inspect the terminal posts, and coat them with vaseline or mineral oil if they are of the lead-coated brass type.
5. Inspect the battery cradle and hold-downs; suggest replacement if they are unserviceable.
6. Make a hydrometer test. If the electrolyte is too low for a test, fill the battery and run the car for an hour or so to mix the electrolyte. Then make the test.
7. If the hydrometer readings show the battery is discharged, make further electrical tests to find the cause.
8. If the hydrometer reading is satisfactory, examine the vent plugs to make sure the non-overfill devices are in good condition. Wet them with water to insure a tight air seal and accurate leveling action.
9. Add pure water, avoiding overfilling.

CARE OF NEW AND RENTAL BATTERIES

Never stack batteries on top of one another, even if they are in cartons—place them on shelves or racks. If permanent racks are not available, temporary racks can be made from loose flat boards which need not be nailed together, since they are supported by the batteries themselves. Permanent racks with shelves 24″ apart, however, are handier when testing or charging stock batteries.

If the batteries already contain electrolyte, put the newer batteries behind the old ones, so the older ones will be sold first and can easily be reached for booster charging.

An idle battery containing electrolyte discharges very slowly. At temperatures of 70° to 80°, a fully charged battery will lose an average of about .001 in specific gravity reading per day. The self-discharge rate increases if the battery's temperature rises; "wet" stock batteries, therefore, should be kept in a cool place, away from steam or hot water radiators in the winter and shielded from the sun in the summer.

To counteract self-discharge, a booster charge must be given to "wet" stock batteries whenever the hydrometer reading drops below about 1.250, corrected for temperature. This will be about every 30 days in summer and less frequently in winter. Batteries in stored cars, or in displays, require the same attention.

Always be sure a new or rental battery is fully charged before installing it in a car.

Review Questions

		Page
1.	What does the storage battery do?	770
2.	The relationship between the battery and the cars electrical system can be easily understood by what comparison?	770
3.	Does a storage battery store electricity?	771
4.	What is the principle of operation of a storage battery?	771
5.	What is the main ingredient in positive battery plates?	771
6.	What is the main ingredient in negative battery plates?	771
7.	What is the name of the solution in a battery?	771
8.	The solution in a battery is a mixture of what two liquids?	771
9.	What change takes place in both negative and positive plates as the battery becomes discharged?	771
10.	What substance is being removed from the solution as the battery is discharging?	771
11.	How is the chemical reaction that has taken place in a completely discharged battery reversed?	771
12.	What determines the total amount of current a battery can deliver?	772
13.	How many volts does one cell develop?	771
14.	Where is the separator used?	772
15.	Name three different types of separators?	772
16.	What is the element of a battery composed of?	772
17.	What is the difference between an element and a cell?	772
18.	How many cells are there in a 12 volt battery?	773
19.	What is the purpose of the ridges at the bottom of each cell in a battery case?	773
20.	Why do battery caps have vent holes?	773
21.	How are the cells connected in a battery?	773
22.	If all means of identifying battery polarity should become obliterated, what simple test would establish correct polarity?	774
23.	What is battery capacity?	774
24.	How is battery capacity measured?	775
25.	How does a sharp increase in amphere draw affect battery capacity?	776
26.	How does weather affect battery capacity?	776
27.	What three tests are used to establish the capacity rating of a battery?	776
28.	What is battery sulphation?	776
29.	What condition causes most rapid battery sulphation?	776
30.	How can a sulphated battery be sometimes restored to service?	776

31. What might be the result of putting a sulphated battery on fast charge? 776

32. What are the harmful effects of solution level being too low in a battery? 776

33. What are the harmful effects of overcharging a battery? 777

34. Why does a fully charged battery resist freezing better than a discharged battery? 777

35. What is a hydrometer? 777

36. Explain how a hydrometer measures the charge level in a battery. 778

37. How does the temperature of the electrolyte affect a hydrometer's specific gravity reading? 778

38. Explain why a hydrometer reading would have no meaning if taken right after adding water to the battery? 778

39. What would be indicated by a brown or reddish electrolyte? 779

40. What is the proper level for electrolyte in a battery? 780

41. What makes it necessary to add water to a battery at regular intervals? 780

42. Why is distilled water better suited for use in a battery than ordinary drinking water? 780

43. What information is gained by measuring the voltage at each cell while the battery is under load? 780

44. Why is it advisable to remove corroded battery cable terminals with a puller? 781

45. What is the only type of current suitable for charging batteries? 782

46. A room where batteries are being charged should be well ventilated for what reason? 782

47. What precaution should be observed when acid and water are being mixed? 783

Dash Gauges

Review Questions for This Chapter on Page 807

INDEX

ELECTRIC GAUGE DESCRIPTION & OPERATION

AC gauges 788
  Fuel gauge 788
  Oil pressure gauge 790
  Temperature gauge 791
Auto-Lite gauges 792
  Thermal fuel gauge 792
  Magnetic fuel gauge 794
  Oil pressure gauge 794
  Temperature gauge 795
King-Seeley gauges 795
  Voltage regulator 796
  Fuel gauge 796
  Oil pressure gauge 797
  Temperature gauge 797
Stewart-Warner gauges 799

TESTING ELECTRIC GAUGES

Constant voltage type 799
  Voltage regulator 799
  Dash gauge 799
  Fuel tank gauge 799
  Oil gauge sending unit 799
  Temperature gauge sending
    Unit 799
Variable voltage type 800
  Grounded wire method ... 800
  Fuel tank gauge method ... 800

MISCELLANEOUS GAUGES, LIGHTS & ACCESSORIES

Ammeters 800
  Trouble shooting 800
  Fibre Optics 806

Generator indicator light 801
  Light circuit with D.C.
    generator 801
  Light circuit with alternator. 801
Oil pressure indicator light ... 802
  Trouble shooting 802
Temperature indicator light .. 802
  Trouble shooting 802
Pressure expansion type oil
  gauge 803
  Trouble shooting 803
Vapor pressure type tempera-
  ture gauge 803
Trouble shooting 804
Electroluminescent lighting .. 804
  Trouble shooting 805
Speedometers 805
Speedometer cable 805
Electric clocks 806
  Trouble shooting 806

Electric Gauge Description & Operation

All vehicles have at least four gauges or indicator lights to indicate the operating condition of certain units or systems that effect the operation of the engine. There are several makes of gauges in common use. AC, Auto-Lite and Stewart-Warner are of the variable voltage type, as was King-Seeley gauges prior to 1956. Starting with 1957, King-Seeley gauges are of the constant voltage design. Gauges and indicating lights are used in various combinations as follows:

Fuel: Dash panel and tank unit gauges to indicate the quantity of fuel in the tank.

Temperature: Dash panel and engine unit gauges to indicate whether the coolant is cold, at normal operating temperature, or overheated. When lights are used in conjunction with the engine unit, a green light will glow when temperature is normal. When an overheated condition exists a red light will glow.

Oil Pressure: Dash panel and engine unit gauges to indicate whether the oil pressure is low, normal or too high. When a light is used in conjunction

with the engine unit, a light will glow red when the pressure is zero or very low.

Charging Circuit: An ammeter to indicate whether the generator or alternator is charging or not charging. When a light is used it will glow red to indicate a "no charge" condition.

AC GAUGES

Fuel Gauge

Fig. 1 shows a wiring diagram of an AC fuel gauge. The dash unit consists of two coils spaced 90 degrees apart with an armature and integral pointer at the intersection of the coil axis. The dial has a scale in fractions between "Empty" and "Full".

The tank unit consists of a housing enclosing a rheostat (variable resistance) with a sliding brush which contacts the rheostat. The brush is actuated by the float arm, movement of which is controlled by the height of the fuel in the supply tank. Variations in resistance (height of fuel) change the

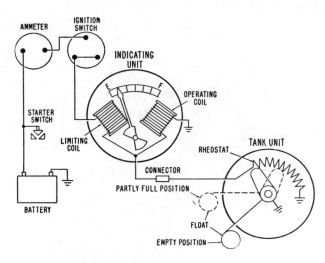

Fig. 1 *Wiring diagram of an AC fuel gauge*

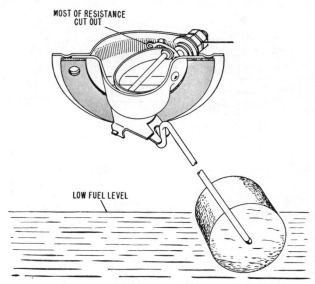

Fig. 2 *When the tank is empty or the fuel supply is low, the sliding brush has moved to eliminate all resistance in the tank unit*

value of the dash unit coils so that the pointer indicates the fuel available. A calibrated friction brake is included in the tank unit to prevent wave motion of fuel from oscillating the pointer on the dash unit.

Current from the battery passes through the limiting coil, Fig. 1, to the common connection between the two coils—which is the lower terminal on dash unit. At this point, the current is offered two paths, one through the operating coil of the dash unit and the other over the wire to the tank unit.

When the tank is empty or low on fuel, Fig. 2, the sliding brush cuts out all resistance in the tank unit. Most of the current will pass through the tank unit circuit because of the low resistance and only a very small portion through the operating coil of the dash unit, Fig. 3. As a result, this coil is not sufficiently magnetized to move the dash unit pointer, which is held at the "Empty" position by the limiting coil, Fig. 4.

If the tank is partly full, the float of the tank unit rises on the surface of the fuel and moves the

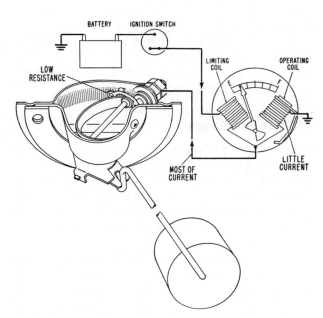

Fig. 3 *With low resistance in the tank unit, only a small part of the current flowing through the limiting coil goes through the operating coil*

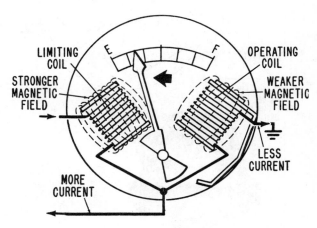

Fig. 4 *With more current flowing through the limiting coil, it becomes magnetically stronger than the operating coil. Thus the armature and pointer are pulled toward the limiting coil (to the left)*

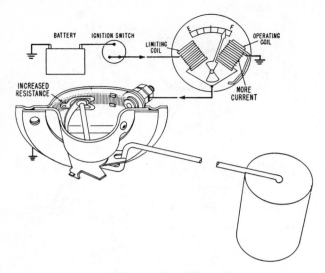

Fig. 5 *A higher fuel level moves the sliding brush along the rheostat, increasing the resistance at the tank unit. More current will then pass through the operating coil*

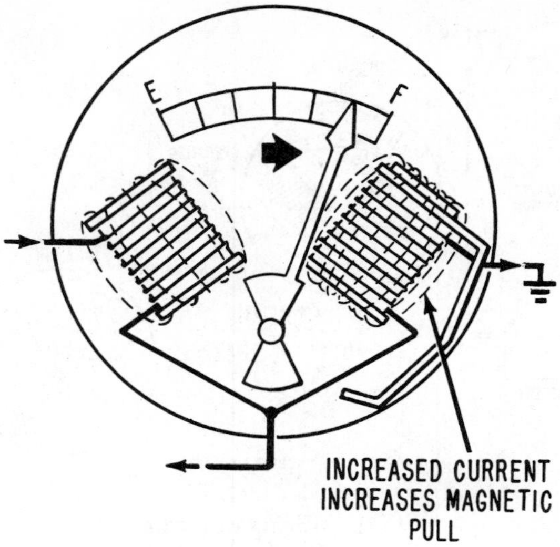

Fig. 6 *As the magnetic strength of the operating coil increases, the armature and pointer are pulled toward it (to the right) and a higher fuel level is indicated*

sliding brush over the rheostat, putting resistance in the tank unit circuit, Fig. 5. More current will then pass through the operating coil to give a magnetic pull on the pointer, which overcomes some of the pull of the limiting coil, Fig. 6.

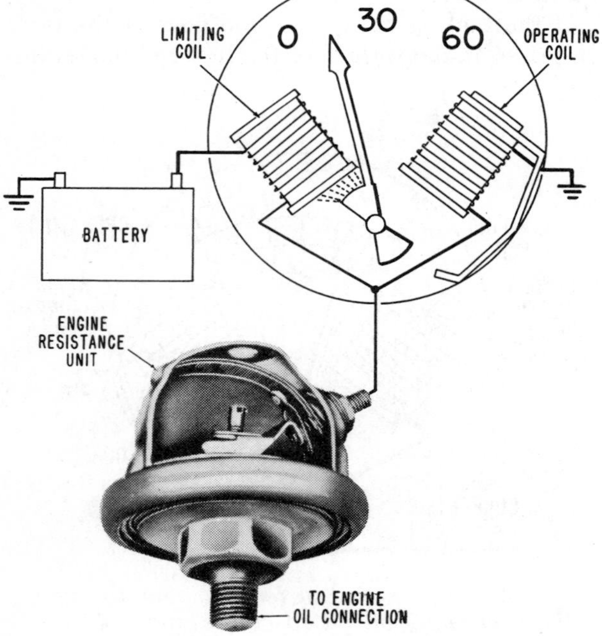

Fig. 7 *AC electric oil pressure gauge wiring diagram*

When the tank is full, the tank unit circuit contains the maximum resistance to the flow of current. The operating coil will then receive its maximum current and exert a maximum pull on the pointer to give a "Full" tank reading.

As the tank empties, the operating coil loses some of its magnetic pull while the limiting coil still has approximately the same pull so that the pointer is pulled toward the lower reading.

Since the operation of this gauge depends on the difference in the magnetic effect between two coils, variations in battery voltage will not cause an error in the gauge reading.

Oil Pressure Gauge

Figs. 7 and 8 illustrate this type gauge diagrammatically. Like the fuel gauge described previously, the dash unit consists of two coils spaced 90 degrees apart with an armature and integral pointer at the intersection of the coil axis. The dial has a scale to indicate the oil pressure in pounds per square inch.

The engine unit consists of a housing enclosing a rheostat (variable resistance) and a sliding brush which contacts the rheostat. The brush is actuated by the flexible diaphragm, movement of which is controlled by the oil pressure applied to it. Variations in resistance (amount of oil pressure) change the value of the dash unit coils so that the pointer indicates the oil pressure.

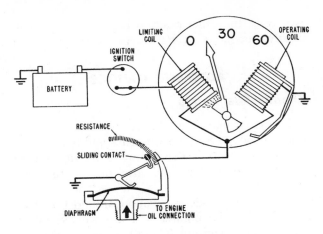

Fig. 8 *When the engine is running at idle speed, the engine unit diaphragm is flexed by oil pressure, causing the sliding contact to move along the resistance. More current will then pass through the operating coil to give a magnetic pull on the pointer to indicate the oil pressure*

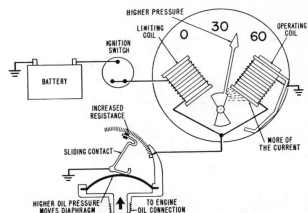

Fig. 9 *At intermediate and high speeds, higher oil pressure flexes the diaphragm further upward, causing the sliding contact to move farther along the resistance. Still more current passes through the operating coil, pulling the pointer across the dial*

Current from the battery passes through the limiting coil, Figs. 7 and 8, to a common connection between the two coils. At this point, the current is offered two paths, one through the operating coil of the dash unit and the other over the wire to the engine unit.

When the engine is idling, the diaphragm in the engine unit rises and moves the sliding brush over the rheostat, putting resistance in the engine unit circuit, Fig. 8. More current will then pass through the operating coil of the dash unit to give

a magnetic pull on the pointer, which overcomes some of the pull of the limiting coil.

When the engine is operating at intermediate and higher speeds, Fig. 9, the engine unit circuit contains the maximum resistance to the flow of current. The operating coil will then receive its maximum current and exert more magnetic pull on the pointer to indicate the increased oil pressure.

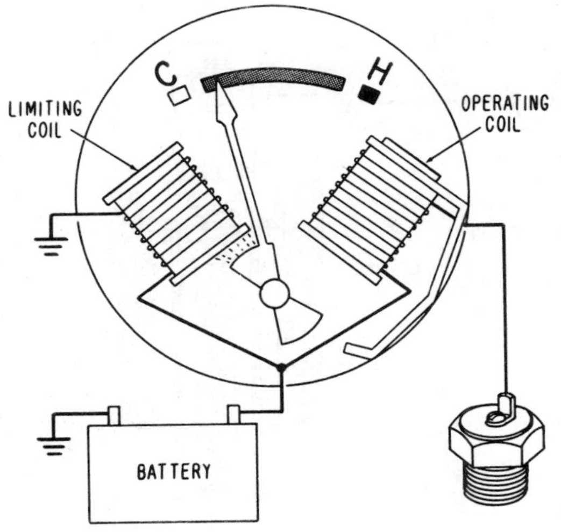

Temperature Gauge

Like the AC fuel gauge and electric oil gauge, the dash unit of this type gauge, Fig. 10 consists of two coils spaced 90 degrees apart with an armature and integral pointer at the intersection of the coil axis. The dial has a scale to indicate the coolant temperature directly in degrees F.

The engine unit has no moving parts and is essentially an electrical resistor which changes resistance with changes in temperature, Fig. 11. When cold the element contained within the unit has a high resistance value, and vice versa.

The limiting coil, Figs. 10 and 12, is connected directly across the battery (through the ignition switch) and the operating coil is connected to the battery and the resistance in the engine unit. The resistance allows more current to flow through the operating coil as the engine becomes warm, causing the operating coil to balance the constant magnetism of the limiting coil. Thus, the armature and pointer comes to rest somewhere between the

Fig. 10 *Diagram of AC electric temperature gauge*

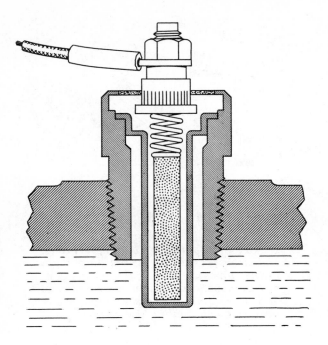

Fig. 11 AC electric temperature gauge engine unit. The element contained in this unit has a high resistance value when cold and a low resistance value when hot

two coils, the exact position depending upon the relative magnetic strength between the two coils.

AUTO-LITE GAUGES

Thermal Fuel Gauge

The dash unit, Fig. 13, has two bimetal strips that are wound with heating coils and two bimetal strips without coils that compensate the unit for external temperature and also protect the coils from overheating.

The heating coils are wound around the strips that actuate the pointer and are welded to the strip at the lower end. The other ends of the coils are connected to the terminals marked "1" and "2" respectively.

The left bimetal strip has a set of contact points at its lower end which are held together by a coil spring just below the hinge pivot. The right hand bimetal strip is held stationary at its lower end by an insulating stop or button.

The internal and external connections of the thermal type fuel gauge with its corresponding tank unit are shown in Fig. 14. The tank unit has two terminals that are marked "1" and "2" and are connected to the terminals on the dash unit which have the same markings. Both ends of the resistance are insulated while the ground connection is made through the sliding contact arm.

Movement of the float arm varies the resistance in the tank unit and therefore the current in the two heating coil circuits. The amount of bending

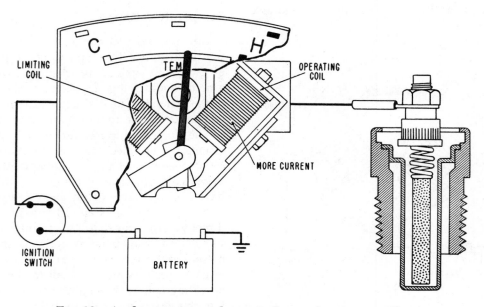

Fig. 12 As the engine unit heats, its lowered resistance allows more current to flow through the operating coil thus pulling the pointer toward the high or "HOT" side of the scale

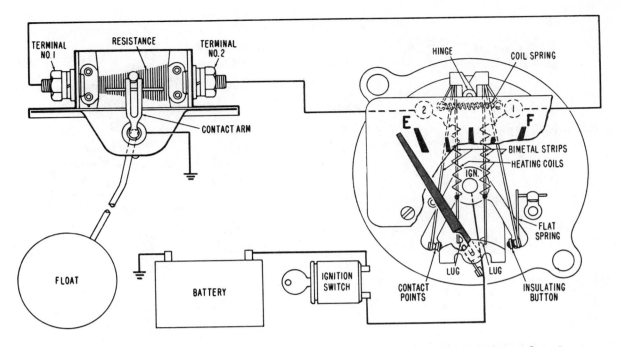

Fig. 13 *Auto-Lite thermal type fuel gauge. Note that both ends of the tank unit rheostat are connected to terminals while the sliding contact arm grounds the resistor. The dash unit requires a short period of time to heat up when the ignition is turned on or when a change occurs in the amount of fuel. This feature explains the absence of pointer vibration whenever the fuel sloshes from side to side in the tank*

or warping of the two bimetal strips depends upon the current in the heating coils. This bending causes the lower end of the strips to grip the pointer and move it into position.

The two outer bimetal strips are assembled so that any external temperature change causes the hinges to rotate slightly and move the strips that actuate the pointer so that the pointer is held stationary.

As the coils are heated the bending of the bimetal strips creates a pressure which rotates the

left-hand hinge slightly and opens the contact points. Cooling begins immediately and the coil spring closes the contacts, causing only a very slight movement of the pointer.

When the tank unit float is all the way down, all of the rheostat resistance is in terminal No. 2 circuit and none is in terminal No. 1 circuit. Most of the current will flow through terminal No. 1 circuit and the left-hand bimetal strip heating coil, Fig. 13. This current through the coil will heat the bimetal strip, making it deflect in such a manner

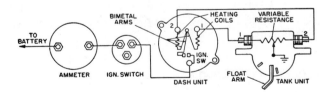

Fig. 14 *Wiring circuit of the Auto-Lite thermal fuel gauge. In this schematic diagram the circuit can be followed from the battery to the dash unit, through the contacts where it divides, with each half passing through a heating coil and part of the tank unit rheostat then to ground through the sliding contact arm shown realistically in Fig. 13*

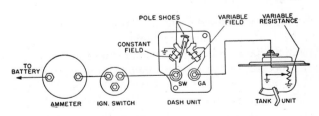

Fig. 15 *Wiring circuit of the Auto-Lite magnetic fuel gauge. The variable resistance (rheostat) in the tank unit controls the current flowing through the variable field coil in the dash unit and controls the relative magnetism between the two coils which act on the pointer*

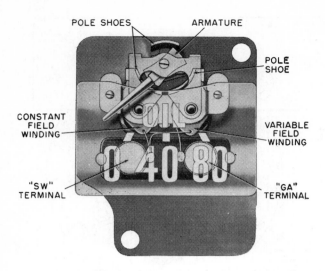

Fig. 16 *Auto-Lite electric oil pressure gauge dash unit. This unit has three magnetic poles, two of which have windings. The pointer assembly has a soft iron armature which is attracted by these poles and assumes a position between them, depending upon their relative strength. The two windings are wound in the same direction so that the two lower magnetic poles have the same polarity while the upper pole has the opposite polarity and completes the magnetic field*

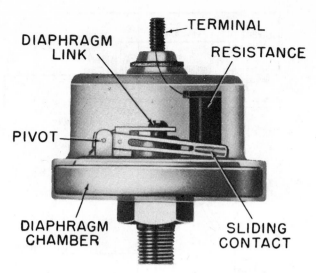

Fig. 17 *Auto-Lite electric oil pressure gauge engine unit. Oil pressure acts on the diaphragm to push upward on the link and raise the sliding contact. This reduces the resistance in the variable field circuit and increases its strength*

as to push the indicator pointer to "Empty" position.

When the tank unit float is raised all the way up, the rheostat resistance is zero in terminal No. 2 circuit and is all in terminal No. 1 circuit. The heating coil around the right-hand bimetal strip will then receive most of the current and become heated. The heated strip will then bend in such a manner as to cause the pointer to move toward the "Full" mark.

Magnetic Fuel Gauge

Fig. 15 shows the wiring circuit for this gauge. The dash unit has three magnetic poles, two of which have windings. One of these windings is connected to the ignition switch and ground and creates a steady magnetic pull towards the "Empty" position whenever the ignition switch is turned on. The other winding is also connected to the ignition switch but it is grounded by the tank unit. It creates a magnetic pull toward the full position, the strength of which is dependent upon the amount of resistance inserted in the circuit by the tank unit.

The tank unit contains a variable resistance

(rheostat) and a sliding contact arm which moves as the float moves. The tank unit case is grounded to complete the fuel gauge circuit.

The magnetic field around the variable windings in the dash unit changes with a change in the amount of fuel in the tank. As the float in the tank moves from "Full" to "Empty," the strength of the magnetic field is gradually reduced. When the float moves from "Empty" to "Full" the strength of the magnetic field is increased.

The pointer is mounted on a magnetic vane which is attracted by the two lower magnetic poles and assumes a position between them depending upon the combined magnetic field. A counterweight is mounted on the pointer to bring the reading back to "Empty" when the ignition is turned off.

Oil Pressure Gauge

The dash unit, Fig. 16, has three magnetic poles, two of which have windings. One of these windings is connected to the ignition switch and to ground and creates a steady magnetic pull toward the "Zero" position whenever the ignition switch is turned on. The other winding is also connected to the ignition switch but it is grounded by the sending (engine) unit. It creates a magnetic pull toward the maximum pressure position,

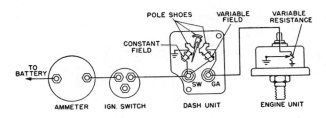

Fig. 18 *Auto-Lite electric oil pressure gauge. This diagram shows the relation and wiring hook-up between the various parts of the circuit. An increase in oil pressure reduces the resistance in the variable field circuit, thereby increasing its strength and moving the pointer toward the right*

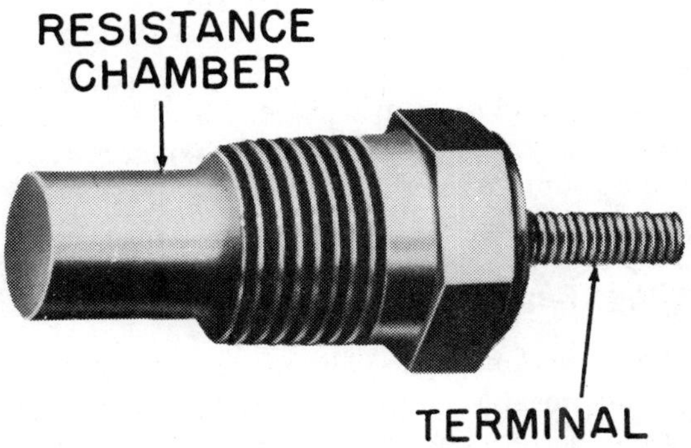

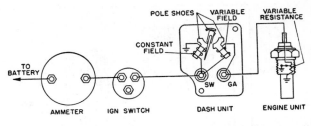

Fig. 20 *Wiring circuit of Auto-Lite electric temperature gauge. This circuit is similar to the oil gauge circuit shown in Fig. 18. The only difference is in the engine unit and in the resistance values of the various parts*

RESISTANCE CHAMBER

TERMINAL

Fig. 19 *Auto-Lite electric temperature gauge engine unit. It contains special metal oxides in the form of a flat disc that changes resistance as its temperature rises. It is connected from the terminal to ground and controls the current flowing in the variable field circuit of the dash unit*

the strength of which is dependent upon the amount of resistance inserted in the circuit by the sending unit.

The pointer is mounted on a magnetic vane which is attracted by the two lower magnetic poles and assumes a position between them depending upon the combined magnetic field. A counterweight is mounted on the pointer assembly to bring the reading back to zero whenever the ignition is turned off.

The engine unit, Fig. 17, has a resistance with a sliding contact which is actuated by the oil pressure. When oil pressure is applied to the diaphragm in the engine unit, resistance is shorted out.

The internal connections and hook-up of these units are shown in Fig. 18.

Temperature Gauge

This type gauge is similar to the electric oil gauge. However, it is calibrated in degrees and the sending unit is actuated thermally without moving parts.

The dash unit has three magnetic poles, two of which have windings. One of these windings is connected to the ignition switch and to ground and creates a steady magnetic pull toward the "Cold" side of the dial whenever the ignition switch is turned on. The other winding is also connected to the ignition switch but it is grounded by the engine unit. It creates a magnetic pull toward the "Hot" side of the dial, the strength of the magnetic pull dependent upon the amount of resistance inserted in the circuit by the engine unit.

The pointer is mounted on a magnetic vane which is attracted by the two lower magnetic poles and assumes a position between them depending upon the combined magnetic field. A counterweight is mounted on the pointer assembly to bring the reading back to the extreme left side of the dial whenever the ignition is turned off.

The engine unit, Fig. 19, is actuated by heat without moving parts. The resistance unit used in the engine unit is made of special metal oxides in the form of a flat disc that changes resistance as its temperature varies. When it is hot the resistance inserted in the variable field circuit is reduced and the pointer is attracted to the "Hot" indication.

The internal connections and hook-up of these units are shown in Fig. 20.

KING-SEELEY GAUGES

The King-Seeley "CV" (constant voltage) Telegage is an electric metering system for the remote

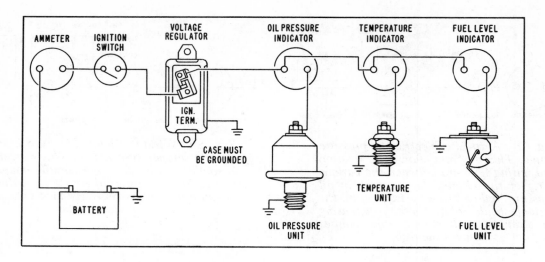

Fig. 21 Wiring diagram of King-Seeley constant voltage electric gauges

indication of fuel level, oil pressure or water temperature. It is a simple, single wire system between the dash panel indicator and the tank or motor unit, Fig. 21.

The voltage regulator is common to all three systems, that is, one regulator can be used to operate one, two or three gauge systems.

The tank and motor units are variable resistance units whose resistance is changed incidental to change in position or condition of the medium being measured (change in fuel level, change in oil pressure, change in water temperature). The dash panel indicators are milliammeters connected in series with the tank or motor units to indicate the current flowing in the resistance of that associated unit. The dash panel indicators are calibrated to indicate this current in terms of

pressure, level or temperature. All of these gauges require only an insulated wire to connect the tank or motor unit to the dash indicator.

Voltage Regulator

The function of the voltage regulator is to regulate the variable (input) voltage from the vehicle storage battery, or the charging system, to produce a constant 5.0 volts output to the gauges. The regulator, Fig. 22, is a simple device operating with a heater bimetal in conjunction with a pair of contacts. It is temperature compensated to produce correct constant voltage for the gauge systems at all anticipated ambient (surrounding air) temperatures, and for this reason it is mounted with or near the dash gauges at approximately the same ambient temperatures.

The voltage regulator does not produce a steady voltage output, but rather a pulsating voltage at an effective 5 volts D.C. The input source can, therefore, be D.C., intermittent or interrupted D.C. or A.C. just so long as the effective input voltage does not drop below 5.0 volts. Input voltage lower than 5.0 volts will result in proportionately low gauge indication.

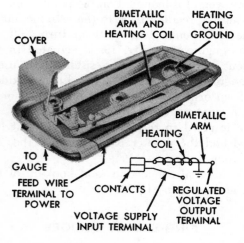

Fig. 22 King-Seeley constant voltage regulator

Fuel Gauge

The fuel gauge consists of a tank unit, Fig. 23, and a dash unit. When the tank is empty, Fig. 24, the float holds the slide rheostat (variable resistance) at maximum resistance, causing the gauge to read zero (with ignition on).

When the tank is full, Fig. 25, the slide rheostat is moved to the minimum resistance point causing

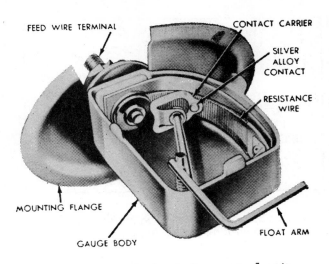

Fig. 23   *King-Seeley fuel gauge tank unit*

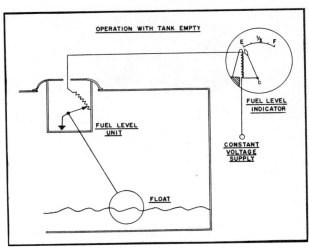

Fig. 24   *King-Seeley fuel gauge operation with tank empty*

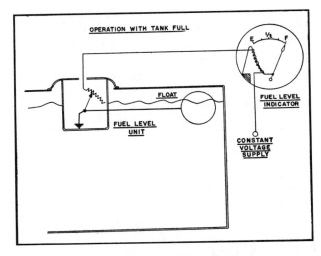

Fig. 25   *King-Seeley fuel gauge operation with tank full*

the gauge to read full (with ignition on). The use of the bimetal in the fuel indicator provides stability of reading and eliminates pointer fluctuation caused by fuel surging in the tank and the float bobbing on the surface of the fuel.

Oil Pressure Gauge

The construction of this instrument, Fig. 26, is similar to the fuel gauge except that the slide thermostat movement is caused by the diaphragm in the engine unit flexing due to varying oil pressure delivered by the engine oil pump.

Temperature Gauge

The operation of the temperature gauge system is identical to that of the fuel gauge system except

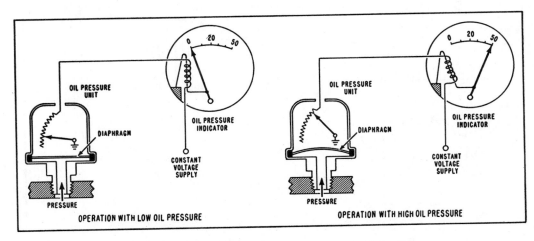

Fig. 26   *King-Seeley oil pressure gauge operation*

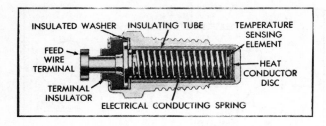

Fig. 27 *King-Seeley temperature gauge sending unit*

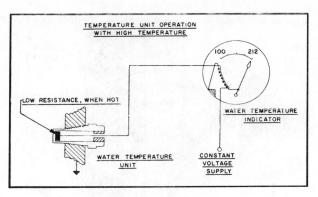

Fig. 29 *King-Seeley temperature gauge operation at high temperature*

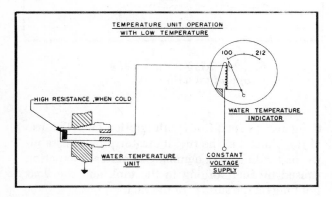

Fig. 28 *King-Seeley temperature gauge operation at low temperature*

for the method of varying the resistance of the engine (sending) unit, Fig. 27. Any change in the coolant temperature causes a like change in the resistor incorporated in the sending unit.

When the engine is cold the resistance of the disc in the sending unit is high and a low temperature will be indicated, Fig. 28. As the engine temperature increases, the resistance of the sending unit disc starts to decrease. A resultant increase in the current flow will occur, causing the

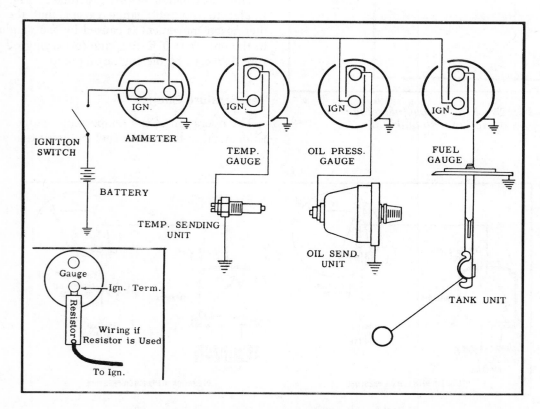

Fig. 30 *Stewart-Warner electric gauge wiring*

gauge pointer to indicate the increase in engine temperature, as shown in Fig. 29.

STEWART-WARNER GAUGES

STEWART-WARNER GAUGES

The Stewart-Warner gauge system, Fig. 30, operates in much the same manner as AC and Auto-Lite units.

Testing Electric Gauges

Caution

Gauge failures are often caused by defective wiring or grounds. Therefore, the first step in locating trouble should be a thorough inspection of all wiring and terminals. If wiring is secured by clamps, check to see whether the insulation has been severed thereby grounding the wire. In the case of a fuel gauge installation, rust may cause failure by corrosion at the ground connection of the tank unit.

CONSTANT VOLTAGE TYPE

Voltage Regulator Test

1. Turn on ignition switch.
2. Check voltage at gauge feed wire at one of the gauges.
3. Voltage should oscillate between zero and about 10 volts.
4. If it does not, voltage regulator is defective, or there is a short or ground between voltage regulator and gauges.

Dash Gauge Tests

1. Turn off ignition switch.
2. Connect the terminals of two series-connected flashlight batteries to the gauge terminals in question (fuel, oil or temperature).
3. The three volts of the batteries should cause the gauge to read approximately full scale.
4. If the gauge unit is inaccurate or does not indicate, replace it with a new unit.
5. If the gauge unit still is erratic in its operation, the sender unit or wire to the sender unit is defective.

Fuel Tank Gauge Test

1. Test the dash gauge as outlined above.
2. If dash gauge is satisfactory, remove flashlight batteries.
3. Then disconnect wire at tank unit and ground

it momentarily to a clean, unpainted portion of the vehicle frame or body *with ignition switch on.*
4. If the dash gauge still does not indicate, the wire is defective. Repair or replace the wire.
5. If grounding the new or repaired wire causes the gauge to indicate, the tank unit is faulty and should be replaced.

Oil & Temperature Sending Unit Tests

1. Test dash gauge as outlined above.
2. If dash gauge is satisfactory, remove flashlight batteries.
3. Then start engine and allow it to run to warm up to normal temperature.
4. If no reading is indicated on the gauge, check the sending unit-to-gauge wire by removing the wire from the sending unit and momentarily ground this wire to a clean, unpainted portion of the engine.
5. If the gauge still does not indicate, the wire is defective. Repair or replace the wire.

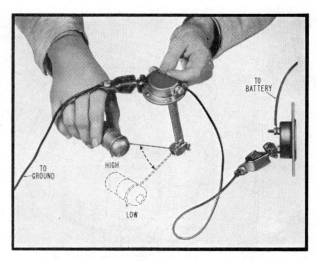

Fig. 31 Hook-up for testing the dash unit with an AC fuel gauge tester

6. If grounding the new or repaired wire causes the dash gauge to indicate, the sending is faulty.

VARIABLE VOLTAGE TYPE

The procedure given herewith applies to AC, Auto-Lite and Stewart-Warner systems. The following is the method of quickly checking the guage system to determine which component (sender or receiver) of a given system is defective.

Fuel Gauge Tank Unit Method

1. Use a spare fuel gauge tank unit known to be correct.

2. To test whether the dash gauge in question (fuel, oil or temperature) is functioning, disconnect the wire at the gauge which leads to the sending unit.

3. Attach a wire lead from the dash gauge terminal to the terminal of the "test" tank gauge, Fig. 31.

4. Ground the test tank unit to an unpainted portion of the dash panel and move the float arm.

5. If the gauge operates correctly, the sending unit is defective and should be replaced.

6. If the gauge does not operate during this test, the dash gauge is defective and should be replaced.

Miscellaneous Gauges, Lights & Accessories

AMMETERS

This instrument shows whether the battery is being charged by the generator or alternator or is being discharged by lights, radio, engine, etc. If a constant discharge is indicated on the ammeter, it is a signal that the battery is being run down. It is often a signal that the generator is out of order. Since both a charged battery and a working generator are very necessary—especially with vehicles equipped with many electricity-consuming devices such as heater, defroster, fog lights, radio, etc.—an inoperative ammeter should be given prompt attention.

The typical ammeter, Fig. 32, consists of a frame to which a permanent magnet is attached. The frame also supports an armature and pointer assembly.

When no current flows through the ammeter, the magnet holds the pointer armature so that the pointer stands at the center of the dial. When current passes in either direction through the ammeter, the resulting magnetic field attracts the armature away from the effect of the permanent magnet, thus giving a reading proportional to the strength of the current flowing.

Trouble Shooting

When the ammeter apparently fails to register correctly, there may be trouble in the wiring

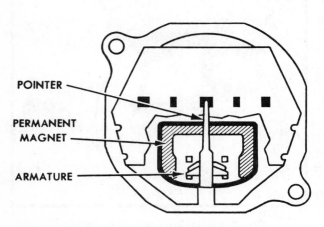

POINTER

PERMANENT MAGNET

ARMATURE

Fig. 32   Drawing of a typical ammeter

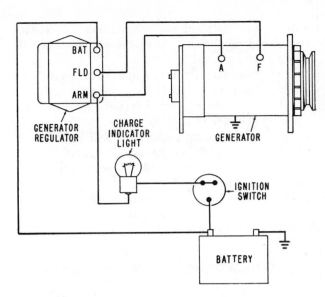

the generator or alternator is not supplying current.

The light should glow when the ignition is turned on and before the engine is started. If the bulb does not light, either the bulb is burned out or the indicator light wiring has an open circuit. After the engine is started, the light should be out at all times with the alternator system. With the D.C. generator system, the light should also be out at all times with the engine running except in cases where the engine idling speed is set too low; however, when the engine is speeded up the light should go out.

If the light fails to go out when the engine is running, the drive belt may be loose or missing on the generator or alternator or voltage regulator may be defective.

Light Circuit with D.C. Generator

The light is usually connected between the armature terminal of the generator regulator and the necessary terminal on the ignition switch, Fig. 33. If the ignition switch is on and the cutout relay contacts are open, the light will glow, indicating that the generator is not electrically connected to the battery. As soon as the generator is speeded up, the cutout relay contacts close. This by-passes the indicator light and thus indicates that the battery is electrically connected to the generator.

Light Circuit with Alternator

A double contact voltage regulator together with a field relay is used on Delco-Remy and Ford alternators when used with the indicator light. The circuit is as follows:

With the ignition switch turned on (engine not running), current flow is through the ignition switch through the indicator light on the dash panel. From there it goes to a terminal of the regulator (marked "4" or "L" on Delco-Remy or "I" on Ford). The circuit continues through the lower contacts of the voltage regulator (held closed by a spring), out the "F" terminal of the regulator, in the "F" terminal of the alternator, through a brush and slip ring, through another brush and slip ring to ground.

After the engine is started, the voltage output of the alternator immediately closes the field relay. This causes battery voltage from the battery terminal of the regulator (marked "3" or "V" on Delco-Remy, "B" on Ford) to be present at the "4", "L" or "I" terminal. Since battery voltage is present on both sides of the indicator light, the light goes out.

Fig. 33 Wiring diagram of a typical telltale light circuit. When the ignition is on the battery voltage lights the bulb. When the generator begins to charge the cut-out relay contacts close and the generator current by-passes the light and the light goes out

which connects the ammeter to the generator and battery or in the generator or battery themselves.

To check the connections, first tighten the two terminal posts on the back of the ammeter. Then, following each wire from the ammeter, tighten all connections on the ignition switch, battery and generator. Chafed, burned or broken insulation can be found by following each ammeter wire from end to end.

All wires with chafed, burned or broken insulation should be repaired or replaced. After this is done, and all connections are tightened, connect the battery cable and turn on the ignition switch. The needle should point slightly to the discharge (−) side.

Start the engine and speed it up a little above idling speed. The needle should then move to the charge side (+), and its movement should be smooth.

If the pointer does not behave correctly, the ammeter itself is out of order and a new one should be installed.

GENERATOR INDICATOR LIGHT

A red generator or alternator "no charge" light is used on many cars in lieu of an ammeter. This light flashes on if the battery is discharging and

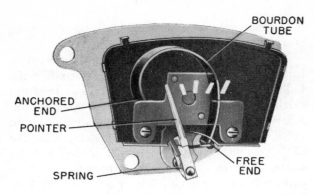

BOURDON TUBE

ANCHORED END

POINTER

SPRING

FREE END

Fig. 34 Auto-Lite pressure expansion type oil gauge. This type gauge is connected by a tube to the oil line and the oil pressure causes the Bourdon tube to straighten. The straightening of the tube moves the pointer

If the generator light comes on with the engine running, the charging circuit should be tested as soon as possible to determine the cause of the trouble.

OIL PRESSURE INDICATOR LIGHT

Many cars utilize a warning light on the instrument panel in place of the conventional dash indicating gauge to warn the driver when the oil pressure is dangerously low. The warning light is wired in series with the ignition switch and the engine unit—which is an oil pressure switch.

The oil pressure switch contains a diaphragm and a set of contacts. When the ignition switch is turned on, the warning light circuit is energized and the circuit is completed through the closed contacts in the pressure switch. When the engine is started, build-up of oil pressure compresses the diaphragm, opening the contacts, thereby breaking the circuit and putting out the light.

Trouble Shooting

The oil pressure warning light should go on when the ignition is turned on. If it does not light, disconnect the wire from the engine unit and ground the wire to the frame or cylinder block. Then if the warning light still does not go on with the ignition switch on, replace the bulb.

If the warning light goes on when the wire is grounded to the frame or cylinder block, the engine unit should be checked for being loose or poorly grounded. If the unit is found to be tight and properly grounded, it should be removed and

a new one installed. (The presence of sealing compound on the threads of the engine unit will cause a poor ground.)

If the warning light remains lit when it normally should be out, replace the engine unit before proceeding further to determine the cause for a low oil pressure indication.

The warning light sometimes will light up or will flicker when the engine is idling, even though the oil pressure is adequate. However, the light should go out when the engine is speeded up. There is no cause for alarm in such cases; it simply means that the pressure switch is not calibrated precisely correct.

TEMPERATURE INDICATOR LIGHTS

A temperature (bimetal) switch, located in cylinder head, controls the operation of a "Cold" temperature indicator light with a green lens and a "Hot" temperature indicator light with a red lens. When the cooling system water temperature is below approximately 110 degrees F., the temperature switch grounds the "Cold" indicator circuit and the green light goes on. When the green light goes out, the water temperature is high enough so that the heater can be turned on and be effective. *Note: The car should never be subjected to full throttle accelerations or high speeds until after the green light has gone out.*

If the engine cooling system is not functioning properly and the water temperature should reach a point where the engine approaches an overheated condition, the red light will be turned on by the temperature switch.

Note: As a test circuit to check whether the red bulb is functioning properly, a wire which is connected to the ground terminal of the ignition switch is tapped into its circuit. When the ignition is in the "Start" (engine cranking) position, the ground terminal is grounded inside the switch and the red bulb will be lit. When the engine is started and the ignition switch is in the "On" position, the test circuit is opened and the bulb is then controlled by the temperature switch.

Trouble Shooting

If the red light is not lit when the engine is being cranked, check for a burned out bulb, and open in the light circuit, or a defective ignition switch.

If the red light is lit when the engine is running, check the wiring between light and switch for a ground, temperature switch defective, or overheated cooling system.

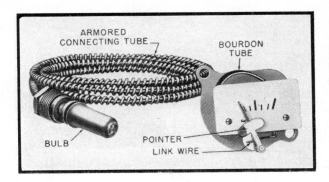

ARMORED
CONNECTING TUBE

BOURDON
TUBE

BULB

POINTER

LINK WIRE

*Fig. 35 Auto-Lite vapor pressure tempera-
ture gauge. This gauge is similar in construction to the pressure expansion type oil gauge
shown in Fig. 34. However, on the heat indicator, the capillary tube is included with
the unit and connects to a bulb on the lower
end. The bulb and tube are filled with a
fluid, such as ether, which expands with
heat and creates a pressure which acts on
the Bourdon tube to cause it to straighten
slightly and move the pointer across the
dial*

If the green light is not lit when ignition is on
and engine cold, check for a burned out bulb, an
open in the light circuit, or a defective temperature switch.

If the green light stays on after normal engine
warm-up period, check for a ground between light
and switch, defective temperature switch, or a
defective cooling system thermostat.

PRESSURE EXPANSION TYPE OIL
GAUGES

Made by both AC and Auto-Lite, the pressure
expansion type oil gauge, Fig. 34, consists of a
metal case enclosing a dial, frame and mechanism.
The dash unit is connected by a tube to the oil
line leading from the oil pump. Oil pressure from
the pump is carried by the tube to the dash unit
where it causes a Bourdon tube to straighten out
slightly. This straightening of the Bourdon tube
causes the pointer movement.

The Bourdon tube is an oval tube, bent in a
circular shape and closed except for a connection
to the oil pump. When oil pressure is applied to
the Bourdon tube, it tends to expand to make a
circular cross section, and in so doing straightens
out slightly.

The amount of expansion and straightening depends upon the amount of pressure applied to the
tube. One end of the tube is fixed while the other
is linked to the pointer. While no spring is necessary to return the pointer to zero, a spring is often
used to keep a slight tension on the pointer and to
take up any slack in the linkage, thus reducing
vibration or fluctuation of the pointer.

Trouble Shooting

Pressure expansion type gauges are subject to
two kinds of trouble: (1) The gauge shows a very
low pressure at normal engine speed and temperature; (2) the pointer is jumpy, sticky or uneven in
its movement.

If the pointer on the gauge is sticky, jumpy or
uneven in its movement, it cannot be fixed. Install
a new gauge of the same make as the one removed.

If the gauge is inoperative, disconnect the tube
from the back of the gauge. Check for possible
plugging of the tube by holding the disconnected
end over a pan or other receptacle and starting the
engine. The oil should flow from the tube at a
steady rate. If it does not, check the tube for
kinks, leaks and plugging. If none of these conditions is evident, the oil pump should be removed
and checked for damage or wear.

If the foregoing inspection shows that oil pressure is reaching the gauge, remove the gauge from
the instrument panel. Inspect to see if the small
hole leading into the Bourdon tube is open and
clean it out with a pin. Make sure there is no
binding on the pointer or other parts. Connect the
unit to the oil line and again check its operation.
If it still will not operate, replace it with a new
gauge of the same make as the one removed.

VAPOR PRESSURE TYPE
TEMPERATURE GAUGES

Made by both AC and Auto-Lite, this type
gauge, Fig. 35, is a remotely controlled thermometer which tells the temperature of the engine
coolant. The dash unit is the Bourdon tube type
having a capillary tube and bulb attached to the
unit. This tube and bulb are filled with a liquid
and sealed. When the bulb is heated by the
engine coolant, the confined liquid generates
vapor pressure which is transmitted through the
capillary tube to the Bourdon tube in the gauge.
This Bourdon tube is similar to that described for
the pressure expansion type oil pressure gauge
and is linked to a pointer in the same manner. The
complete gauge is calibrated to read directly in
degrees.

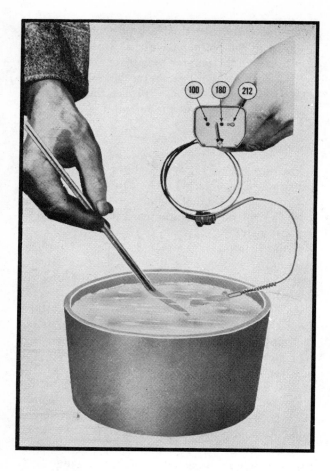

Fig. 36 *Testing vapor pressure temperature gauge in hot water and thermometer. This method may also be used to test electric temperature gauge engine units but the water must not be allowed to get above the threads as the unit may be ruined*

Trouble Shooting

If the gauge does not give a reasonably accurate indication of the coolant temperature and it is certain that no other part of the cooling system is at fault, the complete unit must be removed and tested as follows:

1. Drain the water from the cooling system.
2. Loosen the plug which holds the vapor pressure bulb in the engine block.
3. Pry loose the adapter which seats on the bulb. Do not pull up on the tube until the adapter and bulb are free as this may damage the capillary tube and bulb. A light tap on the adapter with a screw driver blade will often free the bulb for easy removal.
4. Remove the vapor pressure bulb from the engine block.

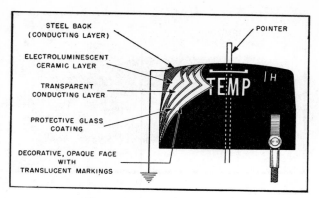

Fig. 37 *Electroluminescent instrument panel lighting*

5. Place the vapor pressure bulb in a pail of hot water, together with a thermometer which reads up to 200 degrees F. or higher, Fig. 36. The thermometer must be reasonably accurate in the hot water.
6. Leave the bulb and thermometer in the hot water long enough to allow the gauge to come to its indication. If the gauge is OK the pointer should register the same temperature as the thermometer.
7. If this test shows that the trouble is in the temperature gauge itself, install a new gauge of the correct make, as there is nothing to fix.

ELECTROLUMINESCENT LIGHTING

Used on late model Chrysler cars, the electroluminescent panel lighting achieves a soft uniform glow that illuminates the panel instruments without objectionable intensity or glare. Light level is adjusted by means of a manually controlled knob.

Electroluminescent lighting has no filaments or gases but instead is composed of laminated layers of material which glow when an alternating current is applied. A typical lamp is composed of the materials shown in Fig. 37.

The lamp is electrically a condenser. When A.C. voltage is applied between the steel plate and the transparent electrically conducting layer, the electric field excites the dielectric (non-conducting material) causing a solid state, which results in visible light. (The phosphorescent surface acts as a dielectric between the two conducting surfaces and it also has the property of glowing when excited by a high frequency high voltage current.) The layer principle is also applied to the instrument pointers. As a result the pointers are a light source in themselves, as are the instrument dials.

Electroluminescent is powered from a transistor oscillator which converts the 12 volt D.C. to 200

Fig. 38 *Electroluminescent powerpack unit*

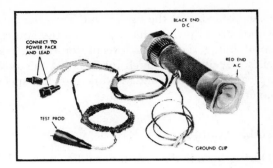

Fig. 39 *Electroluminescent testing tool C-3764*

volts A.C. at 250 cycles per second. This power pack unit, Fig. 38, is mounted on the cowl panel underneath the instrument panel.

Trouble Shooting

1. If one instrument fails to glow, look for a disconnected or broken lead to the instrument dial. If wiring is satisfactory, replace the instrument.
2. If an instrument glows but the pointer fails to glow, look for a broken lead to the pointer within the instrument. If the wiring is satisfactory, replace the instrument.
3. If all instruments fail to glow, look for a faulty panel light switch by testing the operation of the switch with a test lamp. Install a new switch if necessary. If the switch tests out satisfactorily, the power unit is probably defective.

Caution

Chrysler has available a testing tool C-3764, Fig. 39, for checking out the system when the entire panel fails to light and the cause cannot be found as in Step 3 above. However, before connecting the tool, always turn the instrument panel switch "Off". Failure to do this might result in an electric shock from the high voltage in the instrument power unit. The terminals are the protective type and the shock should not occur unless the terminal of a disconnected socket is inadvertently touched. Due to the low wattage, however, the shock should cause no harm. Instructions for the use of the tester is furnished with the tool.

SPEEDOMETERS

The following material covers only that service on speedometers which is feasible to perform by the average service man. Repairs on the units themselves are not included as they require special tools and extreme care when making repairs and adjustments and only an experienced speedometer mechanic should attempt such servicing.

The speedometer has two main parts—the indicating head and the speedometer drive cable. When the speedometer fails to indicate speed or mileage, the cable or cable housing is probably broken.

Speedometer Cable

Most cables are broken due to lack of lubrication, or a sharp bend or kink in the housing.

A cable might break because the speedometer head mechanism binds. If such is the case, the speedometer head should be repaired or replaced before a new cable or housing is installed.

A "jumpy" pointer condition, together with a sort of scraping noise, is due, in most instances, to a dry or kinked speedometer cable. The kinked cable rubs on the housing and winds up, slowing down the pointer. The cable then unwinds and the pointer "jumps".

To check for kinks, remove the cable, lay it on a flat surface and twist one end with the fingers. If it turns over smoothly the cable is not kinked. But if part of the cable flops over as it is twisted, the cable is kinked and should be replaced.

Lubrication

The speedometer cable should be lubricated with special cable lubricant every 10,000 miles. At the same time, put a few drops of the lubricant on the wick in the speedometer head.

Fill the ferrule on the upper end of the housing with the cable lubricant. Insert the cable in the housing, starting at the upper end. Turn the cable around carefully while feeding it into the housing. Repeat filling the ferrule except for the last six inches of cable. Too much lubricant at this point may cause the lubricant to work into the speedometer.

Installing Cable

During installation, if the cable sticks when inserted in the housing and will not go through, the housing is damaged inside or kinked. Be sure to check the housing from one end to the other. Straighten any sharp bends by relocating clamps or elbows. Replace housing if it is badly kinked or broken. Position the cable and housing so that they lead into the head as straight as possible.

Check the new cable for kinks before installing it. Use wide, sweeping, gradual curves where the cable comes out of the transmission and connects to the head so the cable will not be damaged during its installation.

Arrange the housing so it does not lean against the cylinder head because heat from the engine may dry out the lubricant.

If inspection indicates that the cable and housing are in good condition, yet pointer action is erratic, check the speedometer head for possible binding.

The speedometer drive pinion should also be checked. If the pinion is dry or its teeth are stripped, the speedometer may not register properly.

The transmission mainshaft nut must be tight or the speedometer drive gear may slip on the mainshaft and cause slow speed readings.

ELECTRIC CLOCKS

Since about 1957, regulation of electric clocks used on automobiles is accomplished automatically by merely resetting the time. If the clock is running fast, the action of turning the hands back to correct the time will automatically cause the clock to run slightly slower. If the clock is running slow, the action of turning the hands forward to correct the time will automatically cause the clock to run slightly faster (10 to 15 seconds a day).

Winding Clock When Connecting Battery or Clock Wiring

The clock requires special attention when reconnecting a battery that has been disconnected,
or when replacing a blown clock fuse. *It is very important that the initial wind be fully made.* The procedure is as follows:

1. Make sure that all other instruments and lights are turned off.
2. Connect positive cable to battery.
3. Before connecting the negative cable, press the terminal to its post on the battery. Immediately afterward strike the terminal against the battery post to see if there is a spark. If there is a spark, allow the clock to run down until it stops ticking and repeat as above until there is no spark. Then immediately make the permanent connection before the clock can again run down. The clock will run down in approximately two minutes.
4. Reset clock after all connections have been made. *The foregoing procedure should also be followed when reconnecting the clock after it has been disconnected, or if it has stopped because of a blown fuse. Be sure to disconnect battery before installing a new fuse.*

Trouble Shooting

If clock does not run, check for blown "clock" fuse. If fuse is blown check for short in wiring. If fuse is not blown check for open circuit.

With an electric clock, the most frequent cause of clock fuse blowing is low voltage at the clock which will prevent a complete wind and allow clock contacts to remain closed. This may be caused by any of the following: discharged battery, corrosion on contact surface of battery terminals, loose connections at battery terminals, at junction block, at fuse clips, or at terminal connection of clock. Therefore, if in reconnecting battery or clock it is noted that the clock is not ticking, always check for blown fuse, or examine the circuits at the points indicated above to determine and correct the cause.

FIBER OPTIC MONITORING SYSTEM

Fiber optics are non-electric light conductors made up of coated strands which, when exposed to a light source at one end, will reflect the light through their entire length, thereby illuminating a monitoring lens on the instrument panel without the use of a bulb when the exterior lights are turned on.

Review Questions

Page

1. In most cases, dash gauges inform the driver concerning what four items on an automobile?   788

2. What four makes of gauges are currently being used on automobiles?   788

3. How many wires connect the dash unit and the tank unit in an AC fuel gauge system?   789

4. Name the main parts of an AC fuel dash gauge.   788

5. What is in the housing of an AC fuel tank unit?   788

6. What are the names of the two coils in the dash unit of an AC fuel gauge system?   789

7. Which of the two coils in the dash unit of an AC fuel gauge does not get current when the gas tank is empty?   789

8. Does current go through the wiring to the tank unit of an AC fuel gauge system when the dash unit is on the "full" mark?   790

9. Variation in battery voltage does not affect the accuracy of the AC fuel gauge. Explain why.   790

10. What part of an AC electric oil gauge corresponds to the tank unit of an AC fuel gauge system?   790

11. In an AC electric oil gauge system, does the current flow in the wire between the dash unit and engine unit increase or decrease as the oil pressure increases?   791

12. In what way is the AC electric temperature gauge similar to the AC fuel gauge and oil pressure gauge?   791

13. In what way is the AC electric temperature gauge not similar to the AC fuel gauge and temperature gauge?   791

14. How many wires connect the dash unit and tank unit in a Prestolite thermal fuel gauge system?   792

15. How many bi-metal strips are there in the dash unit of a Prestolite thermal fuel gauge system?   792

16. How many bi-metal strips have windings in the dash unit of a Prestolite thermal fuel gauge system?   792

17. How are the terminals marked on the tank unit of a Prestolite thermal fuel gauge system?   792

18. What causes movement of the pointer within the dash unit of a Prestolite thermal fuel gauge system?   792

19. What is similar between an AC fuel gauge system and a Prestolite magnetic type fuel gauge system?   794

20. What is not similar between an AC fuel gauge system and a Prestolite magnetic type fuel system?   794

21. What keeps the voltage constant in constant voltage gauge system?   796

22. What makes the Prestolite electric temperature gauge sensitive to heat?   795

23. As engine temperature increases, current flow in the wire which connects the dash unit and engine unit of a Prestolite electric temperature gauge increases or decreases?   795

24. What does the tank unit contain in a King-Seeley fuel gauge system?   796

25. Is the flow of current at the maximum or minimum point in King-Seeley fuel gauge system when the fuel tank is empty? 796

26. What is the first step in locating electric dash gauge troubles? 799

27. As engine temperature increases, current flow increases or decreases in a King-Seeley temperature gauge system? 798

28. What information does the dash ammeter give to the driver? 800

29. What two methods are used to indicate a "charge" or "no charge" condition in a generating system? 800, 801

30. Explain how a charge indicator bulb is connected into the charging circuit. 801

31. What type of oil pressure indicating system does not have a dash gauge? 802

32. What is the name of the engine unit in an oil pressure indicating system that does not have a dash gauge? 802

33. What takes the place of the dash gauge in an oil pressure indicating system that does not have this component? 802

34. Describe the operation of an oil pressure indicator light system. 802

35. Why is it not good practice to use sealing compounds on the threads of oil pressure and temperature indicating light engine units? 802

36. Describe the operation of the pressure expansion type oil gauge. 803

37. Is the operation of the pressure expansion type oil gauge dependent on the electrical system? 803

38. What is a Bourdon tube? 803

39. Pressure expansion type oil gauges are subject to what two types of trouble? 803

40. Describe the difference between a vapor pressure type temperature gauge and an electric type temperature gauge. 803

41. Describe a simple method of testing a vapor pressure type temperature gauge. 804

42. Is it possible to service a vapor pressure type temperature gauge? 804

43. What is a dielectric material? 804

44. What is the source of alternating current for the Electroluminescent lighting system? 804, 805

45. What would be the probable cause of trouble if all instruments in a Electroluminescent system did not light? 805

46. What are three conditions that are apt to result in a broken speedometer cable? 805

47. Describe a simple test for kinked speedometer cable. 805

48. What precaution should be observed while lubricating a speedometer cable? 805

49. What precautions should be observed while routing a speedometer cable between speedometer and transmission? 806

50. What is the most probable cause of "blown" clock fuses? 806